VIERTER INTERNATIONALER KONGRESS FÜR ELEKTRONENMIKROSKOPIE

FOURTH INTERNATIONAL CONFERENCE ON ELECTRON MICROSCOPY

QUATRIÈME CONGRÈS INTERNATIONAL DE MICROSCOPIE ÉLECTRONIQUE

BERLIN 10.—17. SEPTEMBER 1958

VERHANDLUNGEN

HERAUSGEGEBEN VON

W. BARGMANN · G. MÖLLENSTEDT · H. NIEHRS
D. PETERS · E. RUSKA · C. WOLPERS

SPRINGER-VERLAG BERLIN HEIDELBERG GMBH

VERHANDLUNGEN BAND II

BIOLOGISCH-MEDIZINISCHER TEIL

HERAUSGEGEBEN VON

W. BARGMANN · D. PETERS · C. WOLPERS

MIT 650 ABBILDUNGEN

SPRINGER-VERLAG BERLIN HEIDELBERG GMBH

ISBN 978-3-642-49480-2   ISBN 978-3-642-49764-3 (eBook)
DOI 10.1007/978-3-642-49764-3

Inhaltsverzeichnis

---

* Übrige Präparationstechnik siehe Bd. I.

F. Ergebnisse der Elektronenmikroskopie in der Anatomie

1. Epithelgewebe

2. Muskelgewebe

3. Kollagen

4. Hartgewebe

5. Exokrine Drüsen

Mitarbeiterverzeichnis

Inhaltsübersicht

Band I · Physikalisch-technischer Teil

Eröffnungs-Ansprache. Von Ernst Ruska

Opening remarks. By V. E. Cosslett

Festvortrag. Geschichte des Elektrons. Von M. von Laue

A. Elektronen- und ionenoptische Elemente, Geräte und Verfahren:
1. Kathoden — 2. Linsen und Ablenksysteme — 3. Objekteinrichtungen — 4. Bildaufzeichnungsverfahren — 5. Photographische Emulsionen (und Elektronenwirkung auf Silbersalze) — 6. Stereoaufnahme — 7. Vakuum, Strahlspannung, Linsendurchflutung — 8. Durchstrahlungsmikroskope — 9. Reflexions- und Emissionsmikroskopie — 10. Interferenzmikroskopie und Interferometrie — 11. Röntgen-Projektionsmikroskopie — 12. Elektronen- und Röntgen-Rastermikroskopie — 13. Materialbearbeitung mit Elektronenstrahlen

B. Einwirkung des Objekts auf Strahl und Bild:
1. Streuung am Objekt und Bildkontrast — 2. Abbildung von Kristallgitter-Perioden — 3. Mehrfachbeugung am Objekt und Entstehung von Moirés

C. Elektronenmikroskopische Präparationstechnik*:
1. Trägerfolien — 2. Dünne Objektschichten — 3. Oberflächen — 4. Aufdampf- und Abdruck-Verfahren

D. Ergebnisse der Elektronenmikroskopie in der Technologie (Kristallographie, Metallographie, Chemie):
1. Kristallgitter-Strukturen — 2. Kristallwachstum — 3. Kristalloberflächen — 4. Kondensierte Schichten — 5. Kristallbau-Fehler und Versetzungen — 6. Umwandlungs- und Ausscheidungsvorgänge in Metallen — 7. Natürliche und künstliche technologische Fasern — 8. Verschiedene Produkte der chemischen Technik — 9. Staube und Rauche — 10. Spuren-Nachweis

E. Feldemissionsmikroskopie:**
1. Feldelektronen-Mikroskopie von Metalloberflächen — 2. Adsorptionsuntersuchungen an Feldkathoden — 3. Feldionen-Mikroskopie

Anhang

---

* Die speziell für biologische Präparation bestimmte Technik, insbesondere Mikrotomie, siehe in Band II.
** Feldemissionsmessungen siehe unter A./1. Kathoden.

BIOLOGISCH-MEDIZINISCHER TEIL

Festvortrag

Electron microscopy in morphology and molecular biology

Francis O. Schmitt

Biology Department, Massachusetts Institute of Technology, Cambridge, Mass. (USA)

Electron microscopy, in scarcely more than two decades, has led to revolutionary new concepts of cell structure and the mechanism of basic life processes. In many instances biological systems have been observed at or near the molecular level. These advances become the more significant because of profound parallel discoveries in biochemistry, biophysics, and biophysical chemistry in about the same period. The fusion of these sciences into a single unified effort has already been begun, leading toward what may appropriately be called molecular biology, a term early employed by one of the pioneers in the field, W. T. Astbury.

It will be our task to bring these recent advances into perspective with the fundamental cytological discoveries of the preceding century and to indicate the lines along which further profitable advances may be expected. In this brief paper we can provide only a perspective and a point of view, sketched in roughest outline. However, because it is urgent that contemporary morphologists and molecular biologists have a clear understanding of the limitations and potentialities of the various approaches to the study of life processes, I am grateful to Professor Ernst Ruska and the Program Committee of this International Congress on Electron Microscopy for giving me the opportunity to express my point of view on this timely subject.

Exactly a century ago, here at the University of Berlin, Rudolf Virchow (85) gave a lecture, „Die Cellularpathologie," in which he brought together the evidence that the unit of life, with which pathology must deal, is the cell. He had previously (84) announced his dictum omnis cellula e cellula. This point of view, which has been orthodox biological dogma for so long, was by no means universally accepted at the time, especially by those who sought the ultimate living units in subcellular, possibly molecular, particulates. Even to eminent cytologists, such as Martin Heidenhain, the cell was but one level of complexity in an organizational hierarchy in the realm of living things. Virchow, however, regarded the organism as an integrated federation of its constituent cells. For the effective development of pathology and of medical research generally he emphasized the necessity of integrating histology, pathology, and physiology, rather than pursuing each of these disciplines as sciences self-sufficient in its own right. For a provocative essay on Virchow's contribution in relation to the present-day situation see Bargmann (4).

We stand in need of a similar message today. If we are to achieve a "molecular biology", there must be a joining of forces among physicists, chemists, and biologists in a common assault on the fundamental problems of life science, of which medical science is but a branch. Pauling's (58) development of the concept of the "molecular disease" is a landmark on the road leading toward such integration. The amazing developments in molecular genetics, virus research, and the proof that a purified molecular entity (DNA) can be infective, focus attention upon molecular codes in transferring genetic — and also pathological — information. Despite these promising successes in the analytical approach to what may be termed biophysical and biochemical communications theory at the molecular level, there is need constantly to bear in mind the importance of the organization of the whole — not just the cell or tissue, but the organism as a whole — probably even the organization of the individuals within the biological community.

The electron microscope permits high resolution (10 Å) examination of sections of entire cells. Within the limits of the biochemical indeterminacies involved, the electron microscope is capable of dealing with subcellular particulates at the molecular level and, at the same time, with the entire cell as a whole (seen in section, of course). It should be possible, therefore, with the electron microscope to employ not only the analytical, but also the "systems" approach, so vital in dealing with biological problems.

The electron microscope has also revealed invaluable information concerning the interaction properties of highly organized macromolecular systems. Such data, together with other biophysical and biochemical evidence, should lead to an understanding of some of the intrinsic properties of systems which are at the basis of life processes. We shall try, in this paper, to bring some of these problems into focus.

I. A century of cytology — Historical perspectives

Cell and tissue structure has been studied primarily with three basic motivations:

1. As morphology for its own sake, independent of applications in physiology, pathology, medicine, developmental biology, biophysics, or biochemistry. As such, the work should be judged only as to its excellence as morphology, not by the extent to which it may have solved a basic problem, say, for example, the nature of gene action, muscle contraction, or nerve conduction.

2. To provide a basis for understanding function, either normal, as in physiology, or abnormal, as in pathology. This is the classical, pragmatic approach; the many basic contributions that have arisen from this motivating force explain why morphological studies have occupied a place of prominence in medical research and teaching for more than a century.

3. To search for fundamental life principles, such as those which may be involved in the replication and ordering processes by which physical and chemical information constituting the molecular message of life for the individual and for the continuity of living organisms is transferred within the microcosm of protoplasm. The motivation of this approach may be essentially independent of immediate or eventual applications in the biomedical sciences. It is found frequently in the work of physicists who have recently entered the field of biophysics and who, like physicists generally, search for ultimate causes.

The historical perspective will be better understood if interpreted in terms of the basic aims — such as those mentioned above — that motivated the investigators as well as by the actual results achieved. Clarity in this regard is important also for present-day investigators, not only for more effective orientation of their own efforts, but in better assessing the significance of the work of others in the broad area of the biomedical sciences.

With the advent, in mid-nineteenth century, of methods of fixation, staining, sectioning, and other histological techniques, there was a turning-away from the study of the living cell which characterized the biology of the early decades of the nineteenth century. There followed, in the Golden Age of Cytology (1870—1890), many noteworthy discoveries concerning the major organelles of the cell, such as the nucleus, nucleoli, chromosomes, mitochondria, Golgi apparatus, ergastoplasm, cell membranes, and so forth. Rapid strides were made also in the histological characterization of the specialized structures of particular tissues, such as muscle, nerve, and glands. By and large, these discoveries involved structures sufficiently interbonded chemically to permit reasonably satisfactory fixation and preservation. These historical discoveries were of salient importance in providing a structural basis for understanding physiological and pathological function. However, they also ushered in an intense search for the physical basis of life itself in terms of subcellular particulates and pseudo-crystalline aggregates [called micelles by Naegeli (53)]. There sprang up what might be characterized as the fibrillar, membranous, and granular schools of thought concerning that which is actually "alive" in the cell. Many new terms were coined to describe these hypothesized vital units, most of which have long since been forgotten. Replication and biochemical determination (coding) were considered primary life criteria. The followers of the granular theory, particularly Altmann (1, 2), proposed "omne granulum e granulo," paraphrasing Virchow's earlier dictum "omnis cellula e cellula."

Unfortunately in this rush to discover the physical basis of life in terms of microscopically resolvable objects, the investigators — competent morphologists though they were — neglected to evaluate the effect of the fixatives and microtechnical processes upon highly metastable protoplasmic systems. The artifactitious nature of many of the structures became glaringly obvious to the physiologists and biochemists and eventually even to the morphologists themselves. There followed an unfortunate eclipse of morphology during the early decades of this century in which morphological descriptions of fixed and stained preparations were given scant attention by physiologists.

Since the limit of light-microscope resolution is approximately 0.2 μ (2,000 Å), it was assumed that henceforth any information concerning smaller objects would have to derive from indirect evidence, as by the application to biological problems of polarization optics. This method, which could be applied to fresh, unfixed material, provided information about molecular orientation and the regularity of the internal organization of the micelles. For a useful introduction to this literature, the reader is referred to the books of W. J. SCHMIDT (68, 69) and FREY-WYSSLING (26, 27). It is an interesting commentary on this early work that conclusions about macromolecular organization of cells have been confirmed in all cases in which electron microscopy has subsequently been applied.

Another powerful indirect method, that of X-ray diffraction, developed through the theoretical work of the first speaker of this session, Professor VON LAUE (28). Beginning with the early applications of this technique to the study of tissue structure by investigators such as HERZOG (35) and SPONSLER (80, 81), X-ray diffraction has culminated more recently in extraordinarily fruitful investigations of the internal structure of proteins, nucleic acids, and other biologically significant macromolecules; for a valuable non-technical description of recent discoveries see PERUTZ (59).

The diffraction principle has now been adapted to the electron microscope, making it possible to obtain electron diffraction patterns of selected delimited areas observed directly in the electron microscope. This ability to identify objects by their diffraction pattern offers promise of reducing one of the main limitations of electron microscopy, namely that of identifying, crystallographically or chemically, extremely small objects observed in specimens with high-resolution electron microscopy.

The development of the electron microscope as a practical tool for the study of biological structures far smaller than could be resolved by the light microscope ushered in a new era in cytology or "ultrastructure" research. No longer was it necessary to rely on indirect methods to deduce the structure of "submicroscopic" objects. By the late 1930's it became clear that resolutions of the order of 10 Å would be feasible. Though World War II retarded development in this field, electron microscopes practical for use by non-physicists were manufactured in fairly large quantities by the mid-forties. The degree to which these technical developments have now been pushed in many countries, with the productions of a gratifying array of types of instruments, is graphically illustrated in the exhibits at this Congress. Until methods of obtaining ultrathin sections were developed in the 1940's, most of the relatively high resolution electron microscopy of biological materials was performed upon fragmented specimens, such as biological fibers and other macromolecular preparations, deposited as dispersed particulates on the grid film. Since biological materials are composed largely of atoms of low atomic number, it became necessary to develop "electron stains", i.e., compounds which combine specifically with the biological materials and which contain elements of high scattering power. The heteropolyacids, such as phosphotungstic acid, as well as uranyl salts, and osmium tetroxide, were found valuable for such staining. Heavy metal evaporation, to produce shadows, was another significant milestone in these early developments. The use of plastics for embedding fixed tissues and the mechanical improvements of microtomes, which permit reliable sectioning to as thin as 100 Å, opened the door for biological application generally. Within the span of one decade the application of electron microscopes, capable of reliable and reproducible resolutions of 10 to 15 Å and sufficiently simplified for use by biologists, revolutionized cytology.

II. Electron microscopy and the new golden age of protoplasmic ultrastructure research

As progress in cytology was accelerated a century ago when methods of fixation and staining were first invented, so in the mid-twentieth century, particularly after the importance of maintaining appropriate p_H of the fixative became appreciated, startling new discoveries revealed the structure of cellular organelles, especially those whose constituent molecules are so interbonded as to render them insoluble and stable after fixation, embedding and sectioning. A significant difference exists, however, between these investigations of cellular ultrastructure and those made in the last century by the light-microscope cytologists. The chemical nature and probable function of the structures observed with the light microscope had to be inferred from the rudimentary histochemical criteria then available and by attempting to arrange the structures in a meaningful, functional sequence from comparative studies and from their apparent behavior in different phases of physiological processes, such as secretion and contraction. Some twenty years ago it was discovered that cells could be macerated and, by appropriate methods of fractionation and differential centrifugation, relatively pure fractions of individual organelles, such as mitochondria and microsomes, could be obtained. The isolated organelles were therefore available for direct chemical investigation. By these methods it was shown that mitochondria are the site of oxidative phosphorylation and electron transport, leading to the formation of "energy rich" compounds, such as adenosine triphosphate (ATP), which are the biochemical fuel of the cell. Mitochondria thus became recognized as the "power plants" of the cell. By detailed biochemical investigation of mitochondria and their fragmented membranes *in vitro*, some deductions may be made about the way in which the enzyme assemblies are arranged within the planes of the lipoprotein membranes, so as to carry out the coordinated reactions of the citric acid (Krebs) cycle [see (*49*)]. From comparative studies and other indirect evidence the cytologists of a generation ago had concluded that the function of the mitochondria must be respiratory; this is now verified by the biochemical studies mentioned.

In the microsome fraction, containing fragmented bits of the intracytoplasmic membrane systems (endoplasmic reticulum, ergastoplasm), biochemists have demonstrated a system which, when associated with particles rich in ribonucleic acid ("ribosomes"), constitutes the mechanism by which important biomolecular compounds, such as proteins and steroids, are biosynthesized. The early light-microscopists realized that these regions in the cell, being basophilic, must contain an organic acid. This acid was proved by Caspersson (*10*) and by subsequent investigators to be nucleic acid. Moreover, since the basophilic material, as in the Nissl substance of nerve cells and in the basal area of secreting gland cells, occurs in the regions of active secretion or metabolism, it was concluded by the light microscopists that the basophilic material must be involved in biosynthesis.

This new knowledge from fractionated and purified organelles, or particulates derived therefrom, has made the current developments in the electron microscopy of cells interesting to biochemists; their active collaboration will doubtless help descriptive morphologists avoid misinterpretation based on fixation and other microtechnical artifacts.

Similar considerations apply also to cytological and histological studies of tissues, the major macromolecular constituents of which have been isolated and subjected to detailed biochemical and physicochemical studies. Thus mechanical properties as manifested by connective tissue are coming to be much better understood by virtue of investigations of isolated collagen macromolecules; interpretation of muscle has been enormously facilitated by physicochemical studies on isolated fibrous proteins of muscle; vast strides have been made in an understanding of the genetic mechanisms through a study of purified DNA.

The new discoveries in the biochemistry and physiology of cellular organelles place in new perspective the early cytologists' speculations about the physical nature of life and subcellular "living" entities. Though the membrane-limited structures (endoplasmic reticulum) of cytoplasm are not the vital units postulated by the nineteenth-century proponents of the membrane theories, they are, nevertheless, essential constituents of the biosynthetic mechanism without which life in the cell would be impossible. Similarly, the vital granule ("bioplast") theory of Altmann is now superseded; though the granules (mitochondria) are essential as generators of the biochemical fuel of the cell, they are not themselves "living" structures. Life requires the complex functional and structural organization of the types of these subcellular constituents, each of which subserves a specialized role vital to the cell function.

Recent investigations of molecular genetics, notably by the group of younger biophysicists, have focused attention upon the third category of structural units thought by some nineteenth-century cytologists to be living, namely fibrous structures, specifically nucleic acid and nucleo-protein structures. Thanks to the fruitful studies of plant, animal, and bacterial viruses, and to the crystallographic and physicochemical studies of purified DNA, the mechanism by which genetic segregation and recombination occurs, through alterations of the double helix of the DNA macromolecules, is rapidly becoming clearer. DNA is considered the primary code that carries the genetic information in the chromosomes, the unit of the code being the individual nucleotide residue types and their specific sequences in the DNA chains. This biochemical information transfer is not merely a speculative construct, but can be directly deduced from modern chemical genetic studies of microorganisms and can be directly demonstrated *in vitro* by experiments involving RNA (as a secondary code) in the biosynthesis of proteins and other biological compounds.

Nineteenth-century proponents of the theory that the living structures in the cell are fibrous may perhaps consider recent biophysical studies of DNA as the primary code to constitute proof of their postulate. DNA is undoubtedly regarded by some contemporary enthusiasts of chemical information theory as the master molecules, if not the actual vital units of the cell. Speculations have been offered in connection with problems of the origin of life in which DNA or its primeval counterpart was capable, as the *Urstruktur des Lebens*, to organize its environment to form systems of increasing complexities until cells as we know them were finally evolved. Such considerations are reminiscent of the ancient conflict between the concepts of preformation and epigenesis. Present-day advocates of the primacy of DNA conceive of this substances as the ultimate preformed code in the egg and sperm which directs all biochemical transformations occurring in the developing embryo as well as in the mature and aging organism.

Strictly interpreted, positivistic preformationists [see (*60*)] believe that the preformed DNA code, containing all the hereditary information in a real or accessible form, behaves like a superb machine capable of storing and transforming the information needed to bring about the adaptive and "purposive" biochemical reactions that occur in the development, growth, and aging of an organism. Others, such as ELSASSER (*14*), hold that, however effective the inherited code as a storer and transformer of information, the information is orders of magnitude less than the information content of an adult higher organism. Such adherents of modernized epigenetic theory maintain that this additional information is in principle unobservable and is not stored in the gamete. They attempt to justify their view on the basis of the methods and concepts of modern theoretical physics.

Biologists, having been misled in the past by enthusiastic, though sometimes biologically inexperienced advocates of particular molecules or macromolecules as sufficient codes for processes of growth and development, and having had experience with the impressive potentialities of the reacting system as a whole (including the codes and their organized products) are likely to take a wait-and-see attitude toward this controversy. Without invoking vitalism, there would seem to be room for the discovery of new principles concerning the physical basis for the field-like organizing tendencies that have been clearly demonstrated by investigators of developmental biology. Whether these systems properties are evolved by a truly epigenetic process or whether the faith of the mechanist is justified that all the informational bits needed are present in the gamete — if not exclusively in the DNA — and need only to interact as a complex automaton, employing biochemical and biophysical processes understood in principle today, remains to be seen.

Electron microscopy, after further enlargement of its instrumentational scope and adaptation to biological problems, will unquestionably play an important role in the physical attack on such vital problems because it is capable of dealing effectively with detailed structure at highest resolution — eventually observing intramolecular structure — and, at the same time, with intermolecular relationships and portions of the entire cell, i.e., with the exquisitely regulated reacting system as a whole.

III. Preparative artifacts — The new indeterminacies

The criteria by which to judge the quality of fixation in the electron microscopy of thin sections are similar to those used in light microscopy: degree of swelling, shrinkage, distortion, loss of material, staining characteristics, and so forth. In dealing with unfamiliar material, judgment must be guided by experience, but in some cases it becomes as much artistic as scientific; the investigator compares what he sees with what he thinks the structure really is — or ought to be!

Osmium tetroxide in a buffered saline is currently the fixative of choice. Structures that are stabilized, insolubilized, and interbonded with adjacent structures will survive the physical and chemical ordeal of preparative techniques and will be preserved for observation in ultrathin sections. Why osmium tetroxide is so superior as a fixative is not understood, because the chemistry of its interactions with typical biomolecules is only partially known. However, it is clear that the p_H and ionic strength of the medium at the time of fixation, as well as the duration of fixation, are important (*56*). Direct X-ray diffraction control, discussed in more detail below, has shown that infiltration of plastic and its polymerization in the substance of the fixed material cause substantial alterations in interlamellar separation presumably due to a mechanical effect of the polymerization process. In spite of these and many other difficulties and uncertainties of preparative technique, the experienced investigator can rather effectively judge the relative quality of preparations, because he is primarily concerned with the structure and interrelationship of components that are comparatively large compared with the resolution obtained with the preparation. Gestalt and the interrelationship of parts are by long odds the primary factors in judgment of quality.

With such criteria and with admittedly limited knowledge of the mechanism of fixation, new and unquestionably fundamental protoplasmic structure has been discovered, such as the lamellar structure of mitochondria, Golgi apparatus, and the cytoplasmic reticulum which apparently makes of the cytoplasm a diphasic system: one phase is continuous with the nucleoprotein of the nucleus and mitochondrial apparatus of the cytoplasm, the other phase containing products of biosynthesis for utilization either locally within the cell itself or for transport to extracellular regions. Still other fundamental structures may be discovered by the application of essentially similar methods.

However, cellular processes such as active ion transport, contractility, and biosynthesis involve localized reactions occurring inter- or intramolecularly as well as interactions between two or more immediately adjacent highly organized macromolecular systems. An excellent example of the fibrous type is muscle. According to the "two-filament" type of muscle theory (*33, 34, 41*), shortening involves a rapid, reversible interaction of actin and myosin macromolecular threads separated by considerable distance. Characteristic of the lamellar type is the ribosome-cytomembrane system. In secretory cells the ribosomes presumably interact in intimate relationship with the cytoplasmic membranes to synthesize and transport the products into the appropriate aqueous phase of cytoplasm ready for membranous packaging and removal from the cell.

In any system involving adjacent membranes, as in the intracytoplasmic membrane-limited system, in the space between adjacent cells, or in a stack of membranes, as in nerve myelin, chloroplasts, and the outer limbs of retinal rods, considered from a physiochemical standpoint, the distance between adjacent membranes is determined by the number of charges and their pattern of distribution on the apposed membranes and upon the ionic strength, p_H, and type of ions present in the intermembrane fluid. The distance is also dependent upon the biochemical and physiological integrity of the membrane material. When metastable colloidal systems of this sort are subjected to fixation, there is an immediate change in the composition of the aqueous environment of the membrane because the ionic contents of fixatives and the extracellular and intercellular fluids are never identical. The ionic alteration alone must influence the extent of the Helmholtz-Gouy layer around the membranes merely in a passive physicochemical sense. Some of the factors that determine equilibrium interlayer separations have been discussed by Verwey and Overbeek (*83*) and in the review by Booth (*7*). These ionic adjustments and change of distribution of water probably take place before definitive fixation of the membranes occurs. The latter process then proceeds to insolubilize and "preserve" the system as it has been altered by the abnormal chemical environment.

Insolubilization, interbonding, and ultimate fixation quickly and irreversibly alter the molecular organization of membranes responsible for physiological properties, such as specific differential permeability, active ion transport, and the bioelectrical processes that depend on them. As a result of this chemical alteration of the membrane a rapid adjustment of water and solute distribution must occur, probably even before fixation stabilization of the membranes is completed. Such processes would also change the normal membrane relationships which, after further alterations due to subsequent preparative procedures, would appear in the final structure observed in the high-resolution electron micrograph. Not only does fixation destroy the specific built-in physiological properties of the structures, but the chemical alteration of the membrane may be expected to change the number, type, and distribution of its charges, and this in turn will affect the diffuse ion atmosphere and interaction with adjacent structures. Destruction of enzyme action by the fixative must also lead to changes which are eventually observable electron microscopically. The time lapse after fixation involved in the processes discussed above may vary from a fraction of a second to many minutes.

The chemical changes with fixation in the structural matrices of the cell and the shifts in intracellular and extracellular solutes are inescapable and, at the molecular level, constitute fundamental indeterminacies which become the more serious the higher the resolution, i.e., the more clearly we visualize the molecules or parts of molecules that are being subjected to the abnormal environment (fixation).

Changes in the state, distribution, and binding of water are at the basis of many of the changes brought about by fixation. Since in the preparative techniques, water must be quantitatively removed, this indeterminacy could be minimized only by use of a non-chemical fixation such as rapid freezing, preferably in liquid helium, along lines being explored by FERNÁNDEZ-MORÁN [(21) and unpublished]. Even so it is necessary to infiltrate plastic or other substance into the tissue before sectioning.

IV. Fruitful alternatives offered by molecular biology

Though the indeterminacy described above would seem inescapable in fixed material, it may be possible by indirect methods to secure information valuable in interpreting fixed structures. These methods fall into two classes: 1. study of unfixed tissue or cells by optical means that have no perceptible influence on the system; and 2. the detailed study of models which permit simultaneous investigation by electron microscopy and by biophysical and biochemical methods applicable to unfixed material.

Optical studies of fresh material. To indicate the nature of this approach, mention need be made of but three methods: polarization optics, light scattering, and optical rotation.

Application of polarization optics to light microscopy has already been developed into an extraordinarily sensitive method for the detection of molecular orientation in cells and for chemical identification of molecular types giving rise to birefringence. For the latter purpose, dichroism, natural and artificial, is frequently informative, particularly in the ultraviolet region. However, further increase in contributions of polarization optics may be expected from improvements in instrumentation such as those recently made by SWANN and MITCHISON (82), INOUÉ (42, 43), and INOUÉ and HYDE (44). Particularly significant is the dynamic aspect made possible by microcinematography with a polarizing microscope of very high extinction coefficient, hence sensitivity to small retardations (ca. 0.1 Å).

Since the early exploratory application of darkfield, ultramicroscope principles to cytology, meager use has been made of this method for studying living cells and cell constituents, particularly systems of elongate macromolecules. Under favorable conditions visual evidence of fibrils (such as collagen or paramyosin) of the order of 500 Å in width may be obtained. Removal of dust, gas bubbles, and other scattering sources, and substantial increase in light intensity may increase considerably the scope of this application of light scattering. Preliminary investigations of such methods, in conjunction with electron microscopy, have been made in this laboratory by W. Nyborg (unpublished); the extent to which they may prove applicable to living systems, where unwanted scattering from various sources can be reduced only to a limited extent, remains to be seen.

Recent X-ray studies have shown that, in the native state, many macromolecules manifest a helical configuration of their constituent covalent chains. These may be one strand (α state of KMEF class), two strand (DNA), or three strand (collagen class) helices. Optical rotatory dispersion is a measure of the degree of helicity (*51*) and of changes thereof, as in denaturation, or in reversible helix-random coil transitions observed with certain polyamides (*50, 92*). Within restrictions imposed by the wave length of light, measurement of optical rotation of regions within cells and tissues may lead to invaluable information, especially regarding the problem of whether contractility involves a reversible change in polypeptide chain configuration.

Multilateral studies of model systems. Certain natural and artificial systems lend themselves well to study, under a variety of conditions, unfixed and fixed, by optical, X-ray diffraction, physicochemical, and biochemical methods. By application of these methods simultaneously with those of electron microscopy much may be learned about the behavior of such molecular systems in response to change in chemical environment and about interaction characteristics of macromolecular systems of varying degrees of complexity.

Two such models will be discussed: 1. membranous, layered structures, as illustrated by nerve myelin; and 2. fibrous systems composed of elongate macromolecules, as illustrated by collagen. These particular types were chosen because they represent two of the most significant types of ordering and because they are ones which the author has had most experience.

V. Membranous, layered systems

Studies of the properties of thin films or membranes and of mesomorphic multilayered systems of the smectic type, have been made by polarization-optical, X-ray diffraction, and physicochemical means (*16*). Detailed consideration has also been given to the forces and factors which determine the interaction and the distance between adjacent layers (*7*). With biological materials this distance may be of the order of several hundred Ångstrom units. With certain inorganic constituents it may reach several thousand Ångstrom units (Schiller layers).

Nerve myelin provides a natural model for studies of the properties of layered systems. Myelin is a lipo-protein material, contained in the myelin sheath and produced by the satellite cells which invest nerve axons of certain type. Interruptions in this insulating myelin sheath occur at regular intervals (order of a millimeter) at the so-called nodes of Ranvier. The myelin in each internode is produced by a single satellite (Schwann) cell. Current flows across the axon surface membrane at the nodes, permitting saltatory propagation of the electric action wave and accompanying high conduction velocity characteristic of myelinated fibers. Thus structural differentiation of the cells enclosing the axon has a significant biological function. Study of the molecular organization and mode of myelin formation has long engaged biologists; for more detailed recent accounts of these studies see the papers of Fernández-Morán (*20*), Schmitt and Geschwind (*76*), Schmitt (*73*), Robertson (*64, 65, 66*) and Engström and Finean (*15*).

Polarized-light and X-ray-diffraction data showed that myelin is a regularly layered structure consisting of smectic lipo-protein layers which, with associated water, have a thickness of 171 Å in amphibian nerves and about 186 Å in mammalian nerves. X-ray studies of purified individual lipids and mixtures of lipids in the dry and in the moist state revealed that, with certain lipids, thick water layers may separate bimolecular lipid leaflets (*57*) and that the radial repeating distance in myelin contains, besides water and a thin, possibly monomolecular layer of protein, *two* bimolecular layers of mixed lipids (*75*). Electron microscopic studies of myelin in thin sections (*19, 78*) confirmed the layered structure deduced from X-ray data.

Geren (*29*) discovered that, in the peripheral nerves studied, the lipid-protein layered structure represents the surface membrane of the satellite cell which, after surrounding the axon, infolds and wraps itself as a double membrane many times around the axon. After cytoplasmic material of the satellite cells is squeezed out, the layers become condensed, and the regularly layered myelin, giving five or six orders of small-angle X-ray diffraction, is formed.

This process explains why SCHMITT, BEAR, and PALMER (*75*) found *two* bimolecular layers of lipid per period: the infolded surface membrane of the satellite cell has an outer and an inner surface and the radial repeat distance includes the two membranes which, after infolding, are wrapped as pairs around the axon. The repeating double membrane thus has an inner, two outer, and another inner surface (four surfaces of two membranes). Illustration of these structure are given by ROBERTSON in his symposium lecture at this Conference (see ROBERTSON, p. 159 Vol. II).

The myelin system provides an excellent model to test the effect of fixation and other procedures in the electron microscopic method, because it is possible to obtain X-ray-diffraction data from the same nerve that is subsequently fixed, sectioned, and examined with the electron microscope. Such studies were first published by FINEAN, SJÖSTRAND, and STEINMANN (*25*). Detailed investigations were made by FERNÁNDEZ-MORÁN and FINEAN (*22*) and FINEAN (*23*). These investigators determined the X-ray small-angle radial repeating distance, first from the fresh rat sciatic nerve (178 Å), then from the same nerve after fixation in OsO_4 (160 Å), after dehydration in alcohol (130 Å), after treatment with plastic monomer (150 Å), and after polymerization and embedding (154 Å); finally the layer thickness, as seen in the electron microscope, was determined (123 Å). The studies showed that fixation substantially reduces the layer thickness characteristic of the fresh myelin. Dehydration further reduces it, but embedding in plastic increases it somewhat, presumably due to the polymerization of plastic in the interlayer region previously occupied by water.

The osmotic pressure of the fluid to which the nerve is exposed substantially influences the layer thickness. FINEAN and MILLINGTON (*24*) found increases of as much as 150 Å be produced by exposure to hypotonic media, and ROBERTSON (*66*) observed increases in the electron microscope under similar conditions. The water apparently enters at the interface represented by the outer surface of the satellite cell. The myelin system offers excellent opportunity for further detailed studies, jointly by X-ray and electron microscope means, of the effect of various factors on the structure of the membrane and the aqueous channels between membranes. If such information can be used to interpret the situation in invertebrate and "unmyelinated" nerve, where the satellite cell encloses the axon without forming many wrappings of membrane (myelin) around it, valuable conclusions may be drawn regarding the nature of the aqueous channels surrounding the axon surface membrane in the living fiber (*72*, *73*). It is of course through such aqueous channels that the bioelectric current is presumed to pass.

Investigation of purified lipids and lipo-protein complexes in the dry state and in emulsions or suspensions simultaneously by X-ray diffraction and electron microscope methods should aid greatly in interpreting electron micrographs of cellular systems and in determining the effect upon such systems, of ionic strength, p_H, and the presence of biologically active substances (specific ions, surface-active compounds, ATP, and other compounds). A beginning in this direction was made by GEREN and SCHMITT (*30*), and continuing programs are planued by FINEAN, ROBERTSON, and others. POLICARD, COLLET, and PRÉGERMAIN (*61*), REVEL, ITO, and FAWCETT (*62*) have studied fixed and sectioned myelin forms and have compared their appearance in the electron microscope with that of the endoplasmic reticulum and other membranous structures in cells. KAUTZ, DEMARSH and THORNBURG (*45*) compared the appearance of cell structure in the electron microscope with that of the same cells in the fresh state, viewed in phase contrast and with the polarizing microscope. The results lend confidence in the fidelity of the electron microscope to reveal structure closely resembling that in the living state.

Other multilayered structures, such as those of chloroplasts and the outer limbs of retinal rods and cones, present similar opportunities for study by X-ray and electron microscope methods. The results of these studies will not only contribute to an interpretation of fixation artifacts, but should also provide data important for interpreting function. To illustrate, we may mention only briefly the case of chloroplasts. The detailed layered structure observed in the electron microscope by Sjöstrand (*79*), Hodge, McLean, and Mercer (*38*), and Hodge (*36*) has been interpreted in the light of chemical analytical studies by Calvin (*8*) to indicate a polarized structure and function of the layers, consistent with his exciton migration theory of energy transfer across an asymmetric membrane. The pi electron interaction facilitates this energy transfer. The high degree of order in the membrane structure leads to efficient trapping of photons; facilitation of exciton migration is provided by stabilization of the membrane structure which contains the hypothesized conduction bands.

Basic studies, such as those described above, will assume added significance as the role of membranous structures becomes better known. There is a current tendency to regard all membranous apparatus in the cell (cell membrane, cytoplasmic reticulum, mitochondria, Golgi apparatus, and nuclear membrane) to have arisen from a common precursor; the biochemical and physiological differentiations in the various types due to enzymatic and other biochemical systems presumably associated with the membrane subsequently. High resolution electron microscopy reveals most such membranous structures to be composed of a typical "unit membrane" 60—70 Å thick which, in OsO_4-fixed material, appears as two dense regions separated by a middle less dense zone (*64, 67*). This presumably reflects the presence of a bimolecular layer of mixed lipids (with low density hydrocarbon chains), flanked on either side by thin layers of protein or other polar substances containing groups which combine with and reduce OsO_4.

From such a picture one might characterize the membranous structures as the "floor space" of the cellular factory. Upon this flooring are mounted the various enzymatic machines, each in specific relationship to other machines, to produce the assembly line of the biosynthetic departments.

The lipid types seemingly are similar in the various kinds of membranes, though their relative concentrations and proportions may vary. The question of whether a specific protein or class of proteins is common to intracellular membranes generally, though having certain attractive features, remains unanswered. In particular instances it is clear that proteins having properties related to the special function of the organelle occur in the membrane. Thus, according to Wald (*86*), opsin represents the major constituent of the protein moiety in the double layers of the outer limbs of the retinal rods. The structural protein of the mammalian erythrocyte envelope is an elongate macromolecule called *elinin* by Dandliker, Moskowitz, Zimm and Calvin (*13*), who, with their collaborators, have made important physicochemical studies of this and other substances in the red-cell envelope. The extent to which this protein is involved in the production of specific permeability, the anchorage of Rh factor, blood groups, and other immunologically active substances is unknown [see also (*52*)].

As a working hypothesis it would seem attractive to suppose that the unit membrane with its associated specifically positioned biocatalyzers may form macromolecular complexes, or mixed micelles, in the planes of the membranes, capable of manifesting rapid, reversible, and specific interaction with each other and with smaller molecules, ions, and water, thus providing the molecular basis of the dynamic properties known to characterize the cell surface.

VI. Fibrous systems of elongate macromolecules — the collagen class

Collagen has properties that make it exceptionally appropriate for a discussion of fundamental principles of fibrous ordering. Because of its chemical composition, collagen manifests a regular, banded structure as seen in the electron microscope, and this permits detailed investigation of its molecular structure. There is also considerable knowledge about the physicochemical properties of the collagen macromolecule with which to supplement the structural data. Indeed, research in the last two decades has made collagen one of the best understood of proteins. Finally, collagen represents the most abundant (*ca.* 30%) protein of the mammalian organism. Because it plays a vital role in growth and development and in abnormal, disease states, it has become a subject of major interest in biomedical research.

Collagen fibrils, when treated with an appropriate electron stain, such as phosphotungstic acid (PTA), manifest a banded appearance with an axial repeat of about 700 Å and an intraperiod fine structure consisting of a number of bands and interbands having characteristic position and density. This axial periodicity corresponds with that observed by small-angle X-ray diffraction (640 Å for dry collagen and about 700 Å for moist material). Efforts have been made to correlate the X-ray and electron microscope data by means of optical transforms of the band structure (*5*)].

Electron micrographs and text figures concerning the macromolecular basis of collagen structure and principles of ordering illustrated by the collagen system are given by HODGE and SCHMITT (*39*) and by HODGE (*37*) in this volume and may be referred to there (see page 119).

It was originally supposed that the 700 Å axial repeat represents the length of the collagen molecule. However, by dissolving the fibrils in dilute acid it was possible to cause the collagen macromolecules reversibly to reassemble in different states of aggregation, the so-called fibrous long-spacing (FLS) and segment long-spacing (SLS) forms, each of which has a characteristic and different intraperiod banded fine structure, but the same long period of 2800 to 3000 Å (*74, 77*). It was suggested that this represents the true length of the macromolecules; the period one quarter this long, characteristic of native fibrils, is produced by packing the macromolecules in parallel array, i.e., all "pointing" in the same direction, but staggered by one quarter of a molecular length with respect to molecular ends. It had become customary to associate the 640—700 Å repeat with "collagen" structure. Therefore, to distinguish between the collagen macromolecule itself and a particular (native) macromolecular aggregation pattern, the macromolecules were termed "tropocollagen" (i.e., capable of "turning into" collagen). This terminology avoided the question of the nature of the biochemical precursor of collagen, though the substance termed "procollagen" by OREKHOVITCH and his colleagues (*55*), and thought by them to be a precursor, is probably identical with tropocollagen.

The tropocollagen hypothesis has been confirmed by two independent lines of evidence: 1. BOEDTKER and DOTY (*6*) deduced from physicochemical data that the macromolecule has dimensions of 14 × 2800 Å; 2. by application of a new technique involving deposition of the protein upon freshly cleaved mica, HALL (*32*) was able to visualize the macromolecules directly in the electron microscope. The observed dimensions agreed well with those predicted by BOEDT-KER and DOTY and with the lengths predicted from electron microscopy, i.e., the "long-spacing" of FLS and SLS i.e., 2800 — 3000 Å.

The large-angle X-ray-diffraction pattern of collagen, which defines the collagens as a class and which is perhaps the most crucial single criterion for identification purposes, is best interpreted on the assumption that the tropocollagen macromolecule contains three chains coiled helically about each other to form a coiled coil (*11, 12*). This helical structure has no features inconsistent with known facts about collagen. Although the triple helix model has not been definitely established as correct, it is the best thus far proposed and may be considered correct until proven otherwise.

The high content of proline, hydroxyproline, and glycine is important in determining the structure of the collagen triple helix (intramolecular and intermolecular hydrogen bond stabilization). However, the acidic and basic side chains confer on this fibrous protein the ability to show regular and detailed band patterns in the electron microscope. These long side-chains interact with each other and with appropriate electron stains to produce the dense bands. The shorter amino and imino acid residues, that are abundant in collagen and that do not react with the electron stains, form the interbands. Because of the particular sequence of these two types of residues in each of the three chains in the triple helix, and because of the positioning of the three chains with respect to each other, their distribution can be visualized by the electron microscope. It is also possible to distinguish the acidic from the basic residues. The former combine with chrome salts, while arginine, the most abundant of the basic residues, combines stably with phosphotungstic acid, with which lysine also combines, though not stably in aqueous milieu (*48, 47*).

The segment long-spacing (SLS) type of macromolecular aggregation is the key to the analysis of collagen structure because, in this form, the macromolecules are all in register with respect to their ends; hence the band pattern of a segment is also that of the constituent macromolecules.

This band pattern is markedly asymmetric, due to the characteristic linear distribution of residues. The macromolecules are thus polarized and in SLS they are in parallel array, i.e., with their A ends all at one edge of the SLS and their B ends at the opposite edge (*39, 40, 74*). By contrast, in FLS the macromolecules are in antiparallel array, conferring on the fibril an unpolarized, symmetrical band-pattern. Native collagen (with an axial repeat of 640—700 Å), on the other hand, is not decisive for structural analysis because the axial stagger gives rise to an intraperiod band pattern which, though it results from the specific linear array in the tropocollagen macromolecules, cannot be used as a basis for deducing the full pattern in the native macromolecule.

Bands are produced when the aggregated macromolecules are in their native helical configuration. Denatured collagen (gelatin) does not produce banded aggregates. The measurement of optical rotation has proven a valuable means of determining the helical status of fibrous proteins, particularly collagen. From optical rotation and viscosity measurements, Nishihara and Doty (*54*) concluded that sonic irradiation fragments collagen into halves or quarters without substantial alteration of the native triple-helix structure of the constituent chains. Electron microscope studies (*39, 40*) confirmed this conclusion by the following experiment: SLS types were produced by addition of ATP to irradiated samples; from the length of the SLS and the band pattern it was possible to determine the length of each fragment type and the region along the length of the macromolecule from which the fragment (usually halves or quarters) was derived.

In irradiated samples, dimers and polymers of the macromolecules were commonly produced after addition of ATP. From the band pattern on either side of macromolecular junctions one could distinguish whether the junction was of the type A-A, B-B, or A-B, which is the normal one for collagen. From the fact that polymers are formed, and from the appearance of the junctional regions after treatment with electron stains that react either with basic or acidic side chains, it was deduced that a terminal chain protrudes from the helix at either end, that the lengths and amino acid composition of the two (A and B) ends are different, and that polymerization of the native tropocollagen macromolecules to produce protofibrils occurs by a coiling of A and B chains about each other in a specific configuration. Damage to end chains, as by sonic irradiation or by chemical action, by preventing polymerization into protofibrils, also prevents the formation of the collagen-type structure, because the latter involves a staggering by one quarter of a macromolecular length in adjacent polymer strands.

VII. Principles of ordering — Hierarchical ordering and life processes

From studies of model system, such as have been illustrated with a few examples in the preceding sections, on membranous and fibrous systems — a third important type dealing with globular molecules or macromolecules capable of crystallization, such as plant and animal viruses, might also have been included — much may be learned about the properties of these types, which determine ordering characteristics at each level of organization. The structural, physical, and chemical specificity which gives rise to the biological properties resides in the organized units, each at its particular level of complexity. Interaction of units leads to the production of still higher levels of organization, to macromolecules, organelles, cells, tissues, and finally to the organism.

Many of the ordering and interaction properties of biomolecular systems are understandable from the action of well-known forces, e.g., covalent, ionic, and hydrogen bonding, together with London-Van der Waals forces. Other, still poorly understood forces, which depend upon the structure and composition of the complex macromolecules, may play a significant role. Among these are proton migration, discussed by Kirkwood (*46*), and the tendency, discussed by Wiener (*90*), for the atoms within molecules, considered as non-linear oscillators, through random interaction to tend to group in a narrow band. This may be a significant factor in determining the specificity of macromolecular interaction, which is at the basis of life processes.

From the collagen studies there emerge certain ordering principles which probably apply in many systems of elongate macromolecules, including those such as nucleic acids, nucleoprotein, and muscle proteins, which may show little if any banding in the electron microscope, hence would be difficult to analyze by this means.

The specificity of macromolecular interaction is built into the macromolecule by virtue of its composition, the steric array of its constituent covalent strands and, if there are more than one strand per macromolecule, the relationship of these chains to each other. The manner in which the macromolecules will behave under particular conditions, i.e., whether they will orient themselves in parallel or in antiparallel array, in register or staggered with respect to macromolecular ends, depends upon the chemical environment at the moment.

It has been shown by GLIMCHER, HODGE, and SCHMITT (31) that calcification in bone, which results from the deposition of calcium and phosphate ions upon the collagen matrix in the crystalline form of hydroxyapatite, requires the macromolecular configuration characteristic of native collagen; none of the other forms including the long-spacing variety, is effective. Apparently the particular juxtaposition of side chains, in the staggered neighboring chains characteristic of the native collagen structure, facilitates the formation of nuclei of apatite; calcification then takes place spontaneously. The theory is capable of generalization to explain the manner in which so important a process as mineralization is under the control of the organism by regulation of the pattern of interacting macromolecules, which must eventually form the nuclei of mineralization.

In a system of pure collagen, where only a single fibrous macromolecular species is concerned, at least half a dozen different modes of aggregation may occur under different conditions. Where two or more macromolecular species are involved, e.g., nucleic acid and protein, as in chromosomes, there may be correspondingly more variation in the ways in which the built-in specificity of each type may manifest itself. It is possible that, though the genetic code depends primarily on the nucleic acid, i.e., upon the sequence of the nucleotide residues in the two DNA chains, the behavior at any particular time, e.g., whether a particular gene will be activated or suppressed, will depend on whether or not that locus within the DNA is combined with protein or other shielding material. This type of modulation may be of importance not merely ontogenetically, but also in "normal" aging and in pathological states. It also illustrates a process by which the linear DNA code may initiate chemical reactions, the products of which may in turn inhibit the same or other regions of the DNA from reacting with other substrates. From this type of feedback upon the preformed code may arise phenomena which may appear to be epigenetic in nature.

Cyclic changes in the chemical environment, such as those involving the formation and hydrolysis of ATP and other metabolites in muscle contraction, may rapidly and reversibly alter the mode of aggregation of one fibrous protein (myosin) with respect to another protein (actin), producing shortening of the fiber or tension. The degree to which such changes in macromolecular interaction, can account for contraction without change in intramolecular polypeptide chain configuration, as in the "two-filament" type of hypothesis (34, 41), remains to be seen. Essential also in the interaction is a cyclic change in the activity of an enzyme ATP-ase, which is situated in or on one of the macromolecular partners (heavy meromyosin); this change in its activity may result from steric changes brought about by the cyclic process itself.

Finally, the generation of large super-macromolecular repeat patterns, such as muscle sarcomeres, that may have lengths in the range of 2 to 15 μ, may involve a specific linear and lateral aggregation of a number of macromolecular species, none of which individually may have lengths more than a few thousand Ångström units. The aperiodic band pattern of the giant chromosomes must arise from a similar specific interaction of DNA, RNA, protein, and histones (71).

Formation of the microlattices of tissues in the appropriate place in cells and at the appropriate time requires first the synthesis of the constituent species of elongate macromolecules, then the production of the small-molecule environment that will assure the automatic and spontaneous aggregation of the macromolecules into the appropriate arrays, as required in each successive stage of growth and development. The interaction need not be exclusively inside cells, but may also occur at their surfaces, thus, through changes in the "molecular ecology" (87), causing the cells to assume geometric configurations characteristic of each tissue type, e.g., layers of flat cuboidal cells as in epithelia; wedged packing of tall columnar cells as in the neural tube and the acinus of exocrine cells; linear aggregations as in cardiac muscle, and so on (70, 88, 89).

14   Francis O. Schmitt:

The discussion above concerns primarily highly ordered linear or layered structure. However, it must be remembered that the fact that a particular intracellular region may appear completely amorphous or unstructured even at the highest resolutions of the electron microscope, does not mean that the region does not contain the macromolecular precursors of ordered structure. Thus it is possible *in vitro*, by relatively minor chemical alteration, such as addition of Mg ions, astonishingly rapidly to convert random coils of polynucleotides into DNA-type two-stranded macromolecular helices (*18*). Conceivably a similar process might be involved generally where two or more covalent strands form coiled coils to produce native macromolecules. If so, morphogenesis may depend not only upon the biosynthesis of the major macromolecular species, but also upon the availability of the appropriate inorganic ions and small-molecule cofactors — all part of the homeostatic feed back, "black box" mechanisms of biogenesis, growth, and development.

The examples cited above may suffice to illustrate the fundamental nature of the role which electron microscopy may play when used as one among many biophysical and biochemical tools in the armamentarium of molecular biology for the study of partial systems and purified constituents as models, from which to discover the reaction characteristics that underlie function, normal and abnormal. Without such studies, descriptive morphology — so attractive now that electron microscopes have been made relatively easy to use effectively — may not only fail to disclose the nature of the processes subserved by the observed structures, but may even grossly mislead because of the unavoidable indeterminacies of the method. Conversely, study of partial systems and molecular models alone can also be misleading and can never reveal the unique and full story of a biological process; for this one must study the living system as a whole. Within the limits of its inescapable indeterminacies, electron microscopy is ideal for the study of molecular biology, for it can visualize the entire cell as a whole, yet at sufficiently high resolution to deal effectively with the macromolecular lattices of the protoplasmic structure as a whole.

References

1. Altmann, R.: Arch. Anat. u. Physiol. 55 (1893).
2. — Die Elementarorganismen und ihre Beziehung zu den Zellen. Leipzig: Veit and Co.
3. Astbury, W. T.: J. Intern. Soc. Leather Trades Chem. **24**, 69 (1940).
4. Bargmann, W.: Dtsch. med. Wschr. **83**, 361 (1958).
5. Bear, R. S., and R. S. Morgan: In Connective Tissue (R. E. Tunbridge, Ed.). p. 321. Oxford: Blackwell Sci. Publ. (1957).
6. Boedtker, H., and P. Doty: J. Amer. chem. Soc. **78**, 4267 (1956).
7. Booth, F.: Progr. Biophys. **3**, 131 (1953).
8. Calvin, M.: From microstructure to macrostructure and function in photosynthesis. Symp. on Photochemical Apparatus: its Structure and Function. Brookhaven Nat. Lab. 1958.
9. — Biophysical Science. A Study Program (J. L. Oncley, Ed.). New York: John Wiley and Sons, in press. Also in Rev. modern Physics (in press). 1959.
10. Casperson, T.: Skand. Arch. Physiol. **73**, Suppl. 8, 1 (1936).
11. Crick, F. H. C., and A. Rich: Nature (Lond.) **176**, 780 (1955).
12. — In Recent advances in gelatin and glue research (G. Stainsby, Ed.). p. 20. New York.: Pergamon Press Inc. (1958).
13. Dandliker, W. B., M. Moskowitz, B. M. Zimm and M. Calvin: J. Amer. chem. Soc. **72**, 5587 (1950).
14. Elsasser, W. M.: The physical foundation of biology. New York: Pergamon Press 1958.
15. Engström, A., and J. B. Finean: Biological ultrastructure. New York: Academic Press Inc. (1958).
16. Faraday Society: Trans. Faraday Soc. **29**, 881 (1933).
17. — Disc. Faraday Soc. **6**, 1 (1949).
18. Felsenfeld, F., and A. Rich: Biochim. biophys. Acta **26**, 457 (1957).
19. Fernández-Morán, H.: Exp. Cell Res. **1**, 309 (1950).
20. — Progr. Biophys. **4**, 112 (1954).
21. — Metabolism of the nervous system (D. Richter, Ed.). New York: Pergamon Press. p. 1. (1957).
22. — and J. B. Finean: J. biophys. biochem. Cytol. **3**, 725 (1957).
23. Finean, J. B.: Exp. Cell Res. Suppl. **5**, 18 (1958).
24. — and P. F. Millington: J. biophys. biochem. Cytol. **3**, 89 (1957).

25. — F. S. Sjöstrand and E. Steinmann: Exp. Cell Res. **5,** 557 (1953).

26. Frey-Wyssling, A.: Submicroscopic morphology of protoplasm. New York: Elsevier Publ. Co. 1953.

27. — Die submikroskopische Struktur des Cytoplasmas. Protoplasmatologia. Handbuch der Protoplasmaforschung 2 A 2, 1 (1955).

28. Friedrich, W., P. Knipping and M. von Laue: Ann. Physik **41,** 971 (1913).

29. Geren, B. B.: Exp. Cell Res. **7,** 558 (1954).

30. — and F. O. Schmitt: J. appl. Physics **24,** 1421 (1953).

31. Glimcher, M. J., A. J. Hodge and F. O. Schmitt: Proc. nat. Acad. Sci. (Wash.) **43,** 860 (1957).

32. Hall, C. E.: J. biophys. biochem. Cytol. **2,** 625 (1956).

33. Hanson, J., and H. E. Huxley: Nature (Lond.) **172,** 530 (1953).

34. — Symp. Soc. exp. Biol. **9,** 228 (1955).

35. Herzog, R. O., and W. Jancke: Festschr. K. W. Ges. Förd. Wiss. S. 118. Berlin 1921.

36. Hodge, A. J.: J. biophys. biochem. Cytol. **2,** 221 (1956).

37. — This volume p. 119.

38. — J. D. McLean and F. V. Mercer: J. biophys. biochem. Cytol. **1,** 605 (1955).

39. — and F. O. Schmitt: Proc. nat. Acad. Sci. (Wash.) **44,** 418 (1958).

40. — This volume p. 343.

41. Huxley, H. E.: J. biophys. biochem. Cytol. **3,** 631 (1957).

42. Inoué, S.: Exp. Cell Res. **2,** 513 (1951).

43. — Exp. Cell Res. **3,** 199 (1952).

44. — and W. L. Hyde: J. biophys. biochem. Cytol. **3,** 831 (1957).

45. Kautz, J., Q. B. DeMarsh and W. Thornburg: Exp. Cell Res. **13,** 596 (1957).

46. Kirkwood, J. G.: J. cell. comp. Physiol. **49,** 59 (1957).

47. Kühn, K.: Leder **9,** 217 (1958).

48. — W. Grassmann and U. Hofmann: Z. Naturforsch. **13b,** 154 (1958)

49. Lehninger, A., C. L. Wadkins, C. Cooper, T. M. Delvin and J. L. Gamble: Science **128,** 450 (1958).

50. Moffitt, W., and J. T. Yang: Proc. nat. Acad. Sci. (Wash.) **42,** 596 (1956).

51. — Proc. nat. Acad. Sci. (Wash.) **42,** 736 (1956).

52. Moskowitz, M., W. B. Dandliker, M. Calvin and R. S. Evans: J. Immunol. **65,** 383 (1950).

53. Naegeli, C.: Micellartheorie. Ostwald's Klassiker. No. 227 (A. Frey, Ed.), Leipzig 1928.

54. Nishihara, T., and P. Doty: Proc. nat. Acad. Sci. (Wash.) **44,** 411 (1958).

55. Orekhovitch, V. N., and V. O. Shpikiter: Science **127,** 1371 (1958).

56. Palade, G. E.: J. exp. Med. **95,** 285 (1952).

57. Palmer, K. J., and F. O. Schmitt: J. cell. comp. Physiol. **17,** 385 (1941).

58. Pauling, L., H. A. Itano, S. J. Singer and I. C. Wells: Science **110,** 543 (1949).

59. Perutz, M. F.: Endeavour **17,** 190 (1958).

60. Platt, J. R.: Perspectives in Biol. and Med. **2,** 243 (1959).

61. Policard, A., A. Collet et S. Prégermain: Bull. Microsc. appl. **7,** 49 (1957).

62. Revel, J. P., S. Ito and D. W. Fawcett: J. biophys. biochem. Cytol. **4,** 495 (1958).

63. Robertson, J. D.: J. biophys. biochem. Cytol. **1,** 271 (1955).

64. — In Ultrastructure and cellular chemistry of neural tissue (H. Waelsch, Ed.). New York: Hoeber-Harper Publ. Co., 1 (1957).

65. — J. biophys. biochem. Cytol. **4,** 39 (1958a).

66. — J. biophys. biochem. Cytol. **4,** 349 (1958b).

67. — This volume p. 159.

68. Schmidt, W. J.: Die Bausteine des Tierkörpers in polarisiertem Lichte. Bonn: F. Cohen. 1924.

69. — Doppelbrechung von Karyoplasma, Zytoplasma und Metaplasma. Protoplasma Monographien. Berlin: Gebr. Borntraeger 1937.

70. Schmitt, F. O.: Growth **5,** 1—20 (1941).

71. — Proc. Amer. phil. Soc. **100,** 476 (1956).

72. — In Metabolism of the nervous system (D. Richter, Ed.). New York: Pergamon Press 1957.

73. — Exp. Cell Suppl. **5,** 33 (1958).

74. — In Biophysical science. A study Program (J. L. Oncley, Ed.). New York: John Wiley and Sons, Inc. Also Rev. Modern Physics **31,** 5 (1959).

75. — R. S. Bear and K. J. Palmer: J. cell. comp. Physiol. **18,** 31 (1941).

76. — and N. Geschwind: Progr. in Biophys. **8,** 166 (1957).

77. — J. Gross and J. H. Highberger: Symp. Soc. exp. Biol. **9,** 148 (1955).

78. Sjöstrand, F. S.: Experientia (Basel) **9,** 68 (1953).

79. — In physical techniques in biological Res. **3,** 241. (G. Oster and A. W. Pollister, Ed.). New York: Academic Press, Inc., 1956.

80. Sponsler, O. L.: Amer. J. Bot. **9,** 471 (1922).

81. — Quart. Rev. Biol. **8,** 1 (1933).

82. Swann, M. M., and J. M. Mitchinson: J. Biol. **27,** 226 (1950).

83. Verwey, E. J. W., and J. T. G. Overbeck: The theory of the stability of lyophobic colloid. New York: Elsevier Pub. Co. 1948.

84. Virchow, R.: Arch. path. Anat. Phys. **8,** 1 (1855).

85. — Die Cellularpathologie in ihrer Begründung auf physiologische und pathologische Gewebelehre. Berlin: August Hirschwald 1858.

86. Wald, G.: Exp. Cell Res. Suppl. **5,** 389 (1958).

87. Weiss, P.: The chemistry and physiology of growth. p. 135. Princeton Univ. Press 1949.

88. — Proc. nat. Acad. Sci. (Wash.) **42,** 819 (1956).

89. — Intern. Rev. Cytol. **7,** 391 (1958).

90. Wiener, N.: Non-linear problems in random theory. Cambridge, Mass.: Technology Press 1958.

91. Whittingham, C. P.: Progr. in Biophys. **7,** 320 (1957).

92. Yang, J. T., and P. M. Doty: J. Amer. chem. Soc. **79,** 761 (1957).

A. Elektronenmikroskopische Präparationstechnik in der Biologie*

1. Fixieren und Einbetten

Probleme der Fixation in Licht- und Elektronenmikroskopie

K. Zeiger †

Anatomisches Institut, Universitätskrankenhaus Eppendorf, Hamburg

Einige Begriffsbestimmungen

Fixieren heißt „festigen". Die Fixationsmethoden der Lichtmikroskopie sind vornehmlich aus dem Wunsch entstanden, das lebende Substrat durch besondere Kunstgriffe beim Abtöten zu stabilisieren, es wenn möglich in einen Zustand überführen zu können, der als naturgetreues, erstarrtes Zustandsbild des Lebens die unterschiedlichen Phasen der Herstellung eines sog. „Präparates" unverändert über sich ergehen ließe. Daß solch ideale Forderung auch nicht in einem bescheidenen Umfang verwirklicht werden kann, liegt auf der Hand. Man müßte denn schon in einem beliebigen Augenblick jedes Atom an den ihm zukommenden Ort bannen können. Dem widerspricht jedoch der Ordnungscharakter lebender Systeme. Die Zellstrukturen sind zum Mindesten teilweise keine freiwilligen Strukturen, wie sie sich auf Grund stabiler physikalisch-chemischer Gleichgewichte einstellen, sondern unfreiwillige. Ihre wichtigsten Strukturträger, die Proteine, besitzen einen bestimmten Grad von Unwahrscheinlichkeit in ihrem Aufbau, und sie werden nur mit Zwang in ihrer spezifischen Gestalt aufrechterhalten. Zu ihrer Stabilisierung bedürfen die Zellstrukturen ständig zugeführter Energie, und dieser Umstand bedingt ihre außerordentliche Labilität. Die submikroskopischen Strukturveränderungen, die sich im Cytoplasma bei Störung der Zellatmung einstellen, dokumentieren überaus sinnfällig die enorme *Strukturempfindlichkeit* solcher Stoffsysteme. Das labile Gleichgewicht des lebenden Zustandes muß also unter den massiven chemischen und physikalischen Einwirkungen von Fixierungsmitteln, selbst der besten, zusammenbrechen, und dabei antwortet das submikroskopische Gefüge zwangsläufig mit tiefgreifenden und irreversiblen Veränderungen. Alle Fixierer erzeugen also Artefakte.

Für die morphologische *Naturtreue lichtmikroskopischer Bilder* ist allein entscheidend, in welchem Ausmaß sich das reaktive Geschehen im submikroskopischen und amikroskopischen Raum an der sichtbaren Zell- und Gewebsorganisation bemerkbar macht. Am wenigsten dürften noch die Formteile aus steifen, irreversiblen Gelen verändert werden. Alles übrige, die Formteile aus weniger zähen oder aus dünnflüssigen Gelen, erst recht aber Gel-Lösungen und echte Sole werden ohne Schrumpfung oder Quellung, ohne Verflüssigungs-, Entmischungs- oder Coagulationsvorgänge, vor allem aber ohne Substanzverlagerung und ansehnliche Substanzverluste, in manchen Fällen aber auch ohne Substanzaufnahme niemals stabilisiert werden können.

Von diesem Gesichtspunkt aus kann man *fixationsstabile und fixationslabile Strukturen* unterscheiden [Zeiger (*1*, *2*, *3*)]. Viele „Strukturen" entstehen jedoch beim Ablauf des Fixationsvorganges mit chemischen Mitteln, an Orten, die im Leben lichtmikroskopisch homogen erscheinen, und hinter diesem Schleier verbergen sich Phänomene von verschiedener Wertigkeit. Wir stehen hier vor dem *Problem der vitalen Scheinhomogenität* [Zeiger (*2*)], und es ist notwendig, bei solchen Erscheinungen zu entscheiden, ob maskierte oder latente Strukturen vorliegen, und

---

* Übrige Präparationstechnik siehe Band I, 375.

was allein als reines Artefakt zu definieren ist. *Maskierte Strukturen* sind so, wie sie im Fixationsbild erscheinen, im Leben präformiert, wegen ihres Hydratationszustandes jedoch im gewöhnlichen Licht nicht sichtbar. Sie können im Phasenkontrast oder durch Entquellung sichtbar gemacht und formgetreu fixiert werden. *Latente Strukturen* hingegen sind die submikroskopischen Substrate von Äquivalentbildern im Sinne von Nissl. Das heißt: Sie bestimmen durch ihre lichtmikroskopisch *nicht* erkennbare Struktur das Erscheinungsbild des Äquivalents, das erst durch eine standardisierte Fixierung und Färbung sozusagen aus dem optischen Nichts hervorgezaubert wird.

Neurofibrillen, eine Äquivalentstruktur

Erlauben Sie mir, an einem Beispiel den Begriff der latenten Struktur bzw. des Äquivalentbildes näher zu kennzeichnen. Ich wähle die *Neurofibrillen* im Axoplasma der Nervenfaser der Wirbeltiere, denn hier läßt sich das Problem anhand polarisationsoptischer, elektronenoptischer und biochemischer Daten bis in die Dimension der Makromoleküle verfolgen. Kritische Beobachtungen haben erwiesen, daß diese Neurofibrillen keine präexistenten, also maskierten Strukturen sind. Unter optimalen Beobachtungsbedingungen ist nämlich der lebende Achsenzylinder im Hell- und im Dunkelfeld, im kurzwelligen UV und im Phasenkontrast, aber auch bei vitaler Fluorochromierung in situ optisch leer. Seine fibrillären Strukturen haben sich zumindesten als von submikroskopischer Größenordnung erwiesen. Die Neurofibrillen der klassischen Fixationshistologie sind also lichtmikroskopisch sichtbare Konglomerate aus submikroskopischen Elementarbestandteilen. Aber bislang kann nicht entschieden werden, ob die Bilder der nach Osmium-Fixierung elektronenmikroskopisch nachgewiesenen *Protofibrillen,* mehrere tausend in einer einzigen markhaltigen Faser, lebensgetreu sind oder auch nur durch die Präparationstechnik aus einem im lebenden Zustand noch homogener verteilten Material niedergeschlagen werden [Stoekkenius u. Zeiger (*4*)].

Was diesem Fall nun sein besonderes Gepräge gibt, sind noch einige andere Daten. Maxfield (*5*) hat im Axoplasma der Riesennervenfaser von *Loligo* mit Ultrazentrifuge und Elektrophorese verschiedene Proteine nachgewiesen und eine dieser Komponenten isoliert. Elektronenmikroskopische Bilder des Axoplasmas zeigen ähnliche *Filamente* wie hochgereinigte Proteinfraktionen. Die Fraktion der Filamente ließ sich, und zwar reversibel in Teilchen von niedrigerem Molekulargewicht und geringerer Asymmetrie, dissoziieren. Unter den Bedingungen des pH und der Ionenkonzentration des lebenden Axoplasmas könnten sie dort nicht völlig in Fadenform vorliegen, und deshalb muß angenommen werden, daß andere Komponenten des Axoplasmas ihre Fadengestalt im Leben stabilisieren. *Hier ahnen* wir, in wie hohem Grad die submikroskopische *Vitalstruktur von dem Zusammenwirken von Komponenten abhängig* ist. Sie stellt einen äußerst labilen Gleichgewichtszustand dar. Das klassische Fixierungsbild des Achsencylinders hat mit der Wirklichkeit des Lebens außer dem stark verzerrten und ungemein vergröberten Linearcharakter seiner Tektonik aber auch gar nichts gemeinsam. In diesem Zusammenhang ist nun weiter von Bedeutung, daß Maxfield (*6*) im Axoplasma von *Homarus* mit gleicher Methodik eine Fraktion von Axonfilamenten nicht darstellen konnte. Wenn dieser Befund sich bestätigt, würde das bedeuten, daß bei einer anderen Species die submikroskopische Organisation des Axoplasmas anderen Regeln folgt.

Bedeutung der Diffusionsfähigkeit der Fixierungsmittel

Fixierende Substanzen können ihre Wirkung nur auf dem Diffusionsweg entfalten. Ihre Diffusionsgeschwindigkeit ist von ihrer chemischen Konstitution, ihrer molekularen Konzentration und ihrer Temperatur abhängig. Osmiumtetroxyd dringt unter den üblichen Bedingungen nur 40 μ in 5 min vor [Zetterqvist (*7*)]. Andere Fixierer sind erheblich flinker. Eine wichtige Rolle spielt dabei das, was Tellyesniczky (*8*) „*membranogene Wirkung*" genannt hat. Sie beruht darauf, daß unmittelbar beim Eindringen des Mittels in der Oberfläche des Organstückes durch Coagulationsvorgänge oder durch Bildung von Hauptvalenzbrücken eine Verdichtungszone entsteht, welche das weitere Eindringen des Mittels ins Innere verzögert, und dies ist in hohem Grad von der Größe des Objektstückes abhängig. So kommt es zu einer schichtweise verschiedenartigen Wirkung mit oft erheblichen Differenzen der entstehenden Fixationsstruktur, denn gleiche Zell-

und Gewebsbestandteile werden wegen des Zeitunterschiedes in einem ganz verschiedenen physiologischen Zustand vom Fixationsmittel erreicht.

Post mortem-Veränderungen

Ältere lichtmikroskopische Studien haben bereits gezeigt, daß die ersten Anzeichen einer Zellschädigung sich durch Veränderungen an den Mitochondrien manifestieren [COWDRY (9), DUTHIE (10), ZOLLINGER (11) u. a.]. Sie sind Folgen des im isolierten Organstück einsetzenden Sauerstoffmangels. Elektronenmikroskopische Analysen solcher Vorgänge haben RHODIN (12) an der Niere sowie SJÖSTRAND und HANZON (13) am Pankreas angestellt. RHODIN fand eine Mitochondrienschwellung schon 5 min post mortem. Nach 15 min war dieses Phänomen verstärkt und von einer Vacuolisierung des Cytoplasmas begleitet. ZETTERQVIST (7) hat an der Darmepithelzelle die post mortem-Veränderungen in Intervallen bis zu 40 min geprüft und bestätigt. Eine unterschiedliche Strukturempfindlichkeit verschiedener Zellorganelle und verschiedener Zelltypen war evident. Die Zeitspanne zwischen Tötung des Versuchstieres und Einbringen des Objektes in die Fixierungsflüssigkeit bei 1—2° C Raumtemperatur sollte also bei cytologischen Untersuchungen, wenn möglich auf 2 min, beschränkt werden.

Bedeutung der osmotischen Konzentration der Fixierungsmittel

Schon in der Ära der Lichtmikroskopie wurde eingehend die Frage geprüft, ob die molekulare Konzentration einer Fixierungsflüssigkeit osmotische Wirkungen und damit Form- und Strukturveränderungen hervorrufen könne. Die Forderung einer Isotonie der Fixierungsflüssigkeiten ist erhoben, aber auch verworfen worden. Immerhin haben diese Untersuchungen schon gezeigt, daß Isotonie der Fixierungsflüssigkeit das Auftreten von Schrumpfungs- und Quellungserscheinungen nicht in jedem Fall verhindern kann [ZEIGER (2), RHODIN (12)]. PALADE (14), SJÖSTRAND (15) und RHODIN (12) haben das Problem an verschiedenen Objekten mit dem Elektronenmikroskop studiert, und ZETTERQVIST (7) hat ihre Ergebnisse bestätigt. Abweichungen des osmotischen Drucks des Fixativs beeinflussen die Struktur der Grundsubstanz des Cytoplasmas und der Cytoplasmavacuolen, aber nicht die der Mitochondrien. Hier zeigen sich also Unterschiede in der Fixationsstabilität verschiedener Zellbestandteile.

Wirkung der aktuellen Reaktion der Fixierungsmittel

Die Lichtmikroskopie kannte auch schon den Unterschied zwischen „sauren" und „alkalischen" Fixationsbildern. Der Übergang von einem zum anderen Bild erfolgt in einem bestimmten p_H-Bereich des Fixierungsmittels. Auch war bekannt, daß saure Lösungen niemals befriedigende Resultate geben. Diese Tatsache wurde dann in den Anfängen der Elektronenmikroskopie vielfach bestätigt [PORTER, CLAUDE and FULLAM (16), DALTON u. Mitarb. (17)]. Bei Versuchen mit Formalin ergaben neutrale Lösungen günstigere Ergebnisse als saure [PORTER (18), ROSZA and WYCKOFF (19)]. PALADE (14) hat zuerst eine 1% wäßrige Lösung von Osmiumtetroxyd mit Veronalacetat-Puffer auf ein p_H von 7,3—7,5 eingestellt und damit eine erhebliche Verbesserung an Ultradünnschnitten erzielt. RHODIN (12) hat an Nierenepithelzellen, ZETTERQVIST (7) an der Darmepithelzelle die außerordentliche Säureempfindlichkeit der Grundsubstanz des Cytoplasmas bestätigt, während die Mitochondrien sich als sehr viel fixationsstabiler erwiesen. Hier hat sich außerdem gezeigt, daß wir mit geringen physiologischen Schwankungen des Neutralwertes bei verschiedenen Zelltypen rechnen müssen.

Wirkung der Fixierungsmittel an einem Plasmamodell

Wir gehen nun dazu über, *die amikroskopischen Vorgänge bei der Fixation* ins Auge zu fassen. Dabei ist zu bedenken, daß zwischen den morphologischen Befunden der Elektronenmikroskopie und den biochemischen oder biophysikalischen Beschreibungen elementarer Lebensvorgänge, wie sie uns die Physiologie anbietet, und ihren morphologischen Substraten immer noch eine erhebliche Lücke klafft. Wir können sie zur Zeit nur mit Modellvorstellungen ausfüllen. Hier hat sich mir bei früheren Bemühungen die *Haftpunkttheorie* von FREY-WYSSLING (20) bewährt, ein Modell, das ganz allgemein die morphologische Organisation der Zelle an Plasmagele und damit an eine spezifische räumliche Ordnung bindet. In einer anhand neuerer Erkenntnisse modifizierten Form erlaubt diese Theorie, fibrilläre, laminare und räumliche Strukturen submikroskopischer Dimension aus in den Bereich des Elektronenmikroskopes hineinragenden Sphäroproteinen herzuleiten (Frey-Wyssling (21). Für unsere Betrachtungen sind aber nicht allein die Proteine von Bedeutung, ebenso wichtig erscheinen auch diejenigen Anteile der Lipoide und Fette, die als integrierende Baustoffe von Plasmastrukturen auftreten.

An unserem Plasmamodell (Abb. 1) müssen *Träger und Fassaden* unterschieden werden [Zeiger (3)]. Als Träger, also als relativ statische Elemente des Gefüges, wirken die submikroskopischen fibrillären, laminaren oder räumlichen Komplexe aus Proteinmakromolekülen, die durch polare Haftpunkte von Hauptvalenzcharakter miteinander aggregieren. In einer Außenzone um sie herum findet sich mehr oder weniger fest mit den Trägern verbunden eine deckende Fassade von Lipoiden und mit ihnen wegen des homöopolaren Charakters ihrer Kohäsionsbindung nur locker verbundene Neutralfette.

Es gibt eine Reihe von Argumenten, die dafür sprechen, daß wir mit diesem Spiel nicht ganz an der Wirklichkeit vorübergehen. Nur ein einziges möchte ich hier vorbringen. Die vitale Plasmafärbung der Pflanzenzelle ebenso wie die diffuse Färbung der tierischen Zelle beruhen auf einer Farbstoffspeicherung in den lipoiden Phasen des Plasmas. Dieser initiale Speichertyp spricht für eine Abschirmung der meisten freien Carboxyl- und Aminogruppen der Plasmaproteine und der Plasmaproteide durch Lipoide.

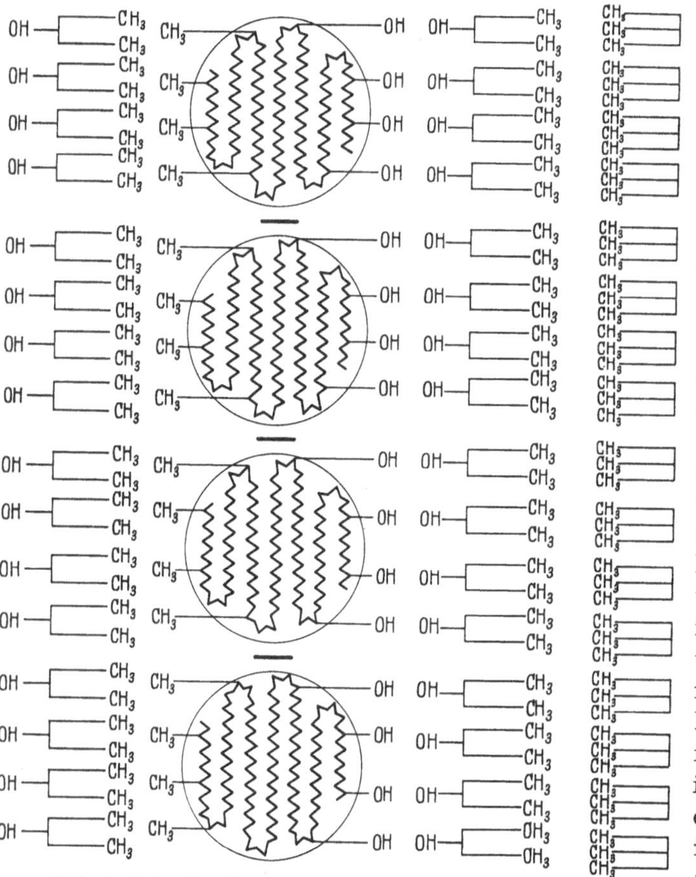

Abb. 1. Submikroskopische Proteinfibrille mit Lipoidbelag

Besonders instruktiv sind Ergebnisse mit Neutralrot. Hier zeigt das Cytoplasma bei inturbanter Färbung die Fluorescenzfarbe der in neutralen hydrophoben Medien, also in Lipoiden, gelösten Farbbase und nicht die der wäßrigen Farbbasenlösung [Toth (22)].

Viele unserer Probleme lassen sich an dem hier entworfenen Modell anschaulich entwickeln. So sind für das Ergebnis jeder Art von histologischer Fixation die Vorgänge in den Lipoidfassaden von ausschlaggebender Bedeutung. Je besser diese Fassaden erhalten und in sich stabilisiert werden, um so weniger wird die Proteinträgerstruktur destruiert. Fixierungsmittel, welche die Fassaden wenig verändern und durch chemische Brückenbildung gar noch verfestigen, sind also die idealen.

Über die Veränderungen bestimmter Haftpunktkategorien unter dem Einfluß von Fixierungsmitteln können auf Grund der Theorie Voraussagen gemacht und experimentell verifiziert werden [Zeiger (3, 23)]. Dazu habe ich folgende quantitativen Methoden benutzt:

1. die Ermittlung des *Quellungs- oder Schrumpfungsgrades* von Gewebselementen unter dem unmittelbaren Einfluß des Fixierungsmittels,

2. die Ermittlung der Verschiebung ihrer *Ladungsäquivalente* gegenüber einem neutralen Bezugswert,

3. die Ermittlung ihrer *Fixierungsstabilität*, also ihres Verhaltens gegenüber den Einwirkungen der Nachbehandlung, besonders der Dehydratisierung.

Als Objekte für die Feststellung von Volumenveränderungen dienen lebende Einzelzellen. Sie können unter dem Einfluß des Fixierungsmittels gemessen, die Ergebnisse variationsstatistisch ausgewertet werden. Die Ladungsäquivalente werden durch die Bestimmung des pH-Bereiches des Adsorptionsminimums für je einen nicht umladbaren basischen und sauren Farbstoff am Schnittpräparat bestimmt [PISCHINGER (*24*)]. Die Ermittlung der Fixierungsstabilität endlich beruht auf dem Größenvergleich meßbarer Gewebselemente im Verlauf der Nachbehandlung. Alle diese Verfahren ergeben *Bauschanalysen*, denn sie gestatten nur die Feststellung einer Totalreaktion der Molekularstruktur und nicht die Prüfung des Verhaltens bestimmter Haftpunkte.

Einteilung der Fixierungsmittel

Ich kann hier nicht die Ergebnisse meiner Versuche, die sich auf alle typischen Fixierer beziehen, im einzelnen vorführen. Ich will nur auf das grundsätzlich Wichtige eingehen und im Zusammenhang damit einige aktuelle Probleme streifen. Alle chemischen Fixierungsmittel, welche die Lipoide nicht zu stabilisieren vermögen, lassen das submikroskopische Gefüge des Plasmas zusammenbrechen. Sie wirken als Eiweißfäller, als *Proteincoagulatoren*. Dazu gehören Alkohol und Aceton, Säuren wie Essigsäure und Pikrinsäure, Schwermetallsalze wie Sublimat. Diese Substanzen können jedoch durch allzu starke Hydratation maskierte Strukturen sichtbar machen oder infolge ihrer submikroskopischen Abmessungen unsichtbare Strukturen durch Konglomeration von Elementarbestandteilen in ein charakteristisches lichtmikroskopisches Äquivalent verwandeln. Sie wirken also als *Revelatoren* [FREY-WYSSLING (*21*)], und insofern leisten sie in der Lichtmikroskopie wertvolle Dienste. Ihnen stehen die *Lipoidstabilisatoren* gegenüber [ZEIGER (*2, 3*)]: Osmiumtetroxyd, Formaldehyd und Kaliumbichromat. Wie sie wissen, haben sich in der Elektronenmikroskopie nur Vertreter dieser Gruppe bewährt. So ist es für uns wenig interessant, die Vorgänge bei der Fixierung mit Eiweißfällern hier näher zu analysieren. Ganz allgemein kann man die Wirkung von Proteincoagulatoren dahin charakterisieren, daß man sagt, ihnen fehle jede Fähigkeit, den Gelzustand nativer Substrate zu verstärken und damit auch die Potenz, ein Sol oder eine Gel-Lösung zu gelatinieren. Dies aber ist das entscheidende Kennzeichen der Lipoidstabilisatoren, von denen ich jetzt Formaldehyd und Osmiumtetroxyd etwas näher betrachten will.

Formaldehyd

Neutrales Formalin wirkt nicht wie die Schwermetall- oder Säurewasserstoffionen durch Entladung und die damit verbundene Denaturierung und Ausfällung hydratisierter Proteine, sondern durch *Vernetzung*. Es vermag, flüssiges Cytoplasma zu gelatinieren und hochlabile Gallerten in steife, also verhältnismäßig stabile Gele zu verwandeln. Dies beruht darauf, daß Form-

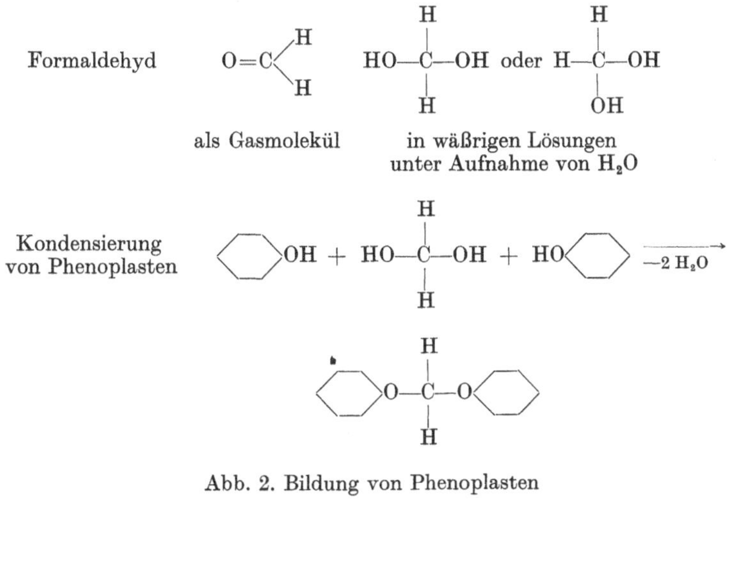

Abb. 2. Bildung von Phenoplasten

aldehyd imstande ist, zwischen benachbarten reaktionsfähigen Gruppen unter Wasseraustritt Brücken zu schlagen. Als solche aktiven Radikale wirken z. B. Phenylgruppen, die mit Formaldehyd zu Phenoplasten kondensieren (Abb. 2) [K. H. MEYER (*25*)]. Auch sekundäre Alkoholgruppen können vernetzt werden [DEUEL (*26*)]. Analog der Kondensierung von Phenoplasten ist auch zwischen benachbarten Phenylalanin-Seitengruppen eine Brückenbildung realisierbar (Abb. 3).

Aber nicht nur zwischen Seitenketten, auch zwischen den Polypeptidhauptketten können hauptvalenzmäßige Querverbindungen auftreten. Da man sich die Carbonylgruppe der Peptidbindungen auch in der Enolform vorstellen kann, ergeben sich hier reaktionsfähige tertiäre Hydroxylgruppen, die leicht mit Formaldehyd zusammentreten. Da die Imidogruppe der Peptidbindungen aber auch ähnlich wie ein Amid bei der Kondensation von Aminoplasten (z. B. Harnstoff und Formaldehyd) reagieren kann, wäre auch eine zweite Reaktionsart zu erwägen. Hier ist die übliche Wasserstoffbindung zwischen benachbarten Carboxylgruppen und Iminogruppen durch eine Oxymethylengruppe ersetzt, welche die beiden Ketten hauptvalenzmäßig untereinander verbindet (Abb. 3) [FREY-WYSSLING (*21*)].

Brückenbildung zwischen aktiven Phenylalanin-Seitenketten

Brückenbildung zwischen Polypeptid-Hauptketten

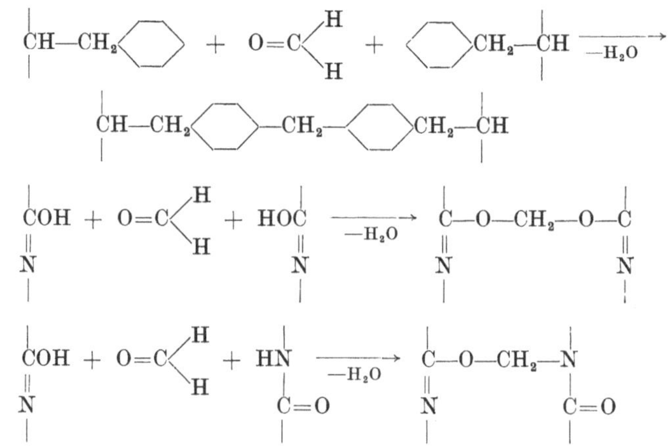

Carbonylgruppe — Imidogruppe — Oxy-Methylenbrücke

Abb. 3. Brückenbildung bei Formolfixierung

Eine Vernetzung findet aber auch in den *Lipoidzonen* der Substrate statt. Für die Realität solcher Vorgänge spricht die Tatsache, daß Fette und Lipoide durch Formalin in hohem Grad stabilisiert werden. Die Vernetzung erfolgt hier nicht allein nach dem Schema der Polyoxymethylen-Polymerisation (—O—CH_2—O—CH_2—). Es können auch sauerstofffreie Methylenbrücken (—CH_2—) entstehen. Mit fortschreitender Wirkung des Formaldehyds nimmt die Zahl der Hauptvalenzbrücken zu. Dabei erfahren die Gewebselemente keine Schrumpfung, sondern eher eine, wenn auch nur geringfügige Quellung [HERTWIG (*27*), ZEIGER (*3*)]. Wie die Gelatinierung flüssigen Cytoplasmas auf dem Einbau zahlreicher Brücken beruht, so wird die Quellung durch die damit verbundene Ausweitung der molekularen Struktur verständlich. Die Vernetzung ist zu einem nicht unbeträchtlichen Teil reversibel. Deshalb besitzen formolfixierte Objekte nur eine geringe Fixierungsstabilität. Mit sekundären Veränderungen formolfixierter Substrate nach allen Arten von Einbettung und mit einer Zunahme der Artefakte ist also hierbei zu rechnen.

Es ist lange bekannt, daß Formaldehyd ruhendes Chromatin und Mitosespindelfasern auflöst, während höhere Aldehyde, wie Acet-, Propion- und Butyr-Aldehyd, solche Strukturen erhalten, hingegen die in formol-fixierten Objekten sichtbaren Mitochondrien und die von den Botanikern Kernsaft genannten Bestandteile des Zellkerns zum Schwinden bringen [ZIRKLE (*28*)]. Noch bedenklicher ist die Feststellung, daß 5% Formollösung beträchtliche Stoffmengen aus Organstücken herauslöst. In 24 Std. gehen 10—17% der Trockensubstanz verloren [SYLVEN (*29*)]. Hingegen werden die meisten der nicht gelösten Stoffe unlöslicher, so vor allem die Lipoide. In der

Literatur kann man lesen, neutrales Formol liefere besonders feine Strukturen. Solche elektronenmikroskopischen Bilder zeigen jedoch sehr wenig Kontraste. Chromosomen erscheinen homogen, Cytoplasma und Mitochondrien hingegen besonders feinkörnig, Kernmembran und Spindel werden unsichtbar [Rozsa and Wyckoff (19)]. Man darf also vermuten, daß infolge der Vernetzung oder des Substanzverlustes bestimmte Einzelheiten der Zellstruktur nach Formolfixierung nicht zur Darstellung kommen, Das sind unabänderliche Nachteile, denn sie werden durch die spezifische Wirkungsweise des Formaldehyds verursacht.

Osmiumtetroxyd

In der Lichtmikroskopie gilt Osmiumtetroxyd seit langem als bestes Mittel für cytologische Untersuchungen, und z. Z. beherrscht es die elektronenmikroskopische Präparationstechnik. Auf Grund polarisations- und elektronenoptischer Erfahrungen wird heute allgemein die Auffassung vertreten, Osmiumtetroxyd erhalte die submikroskopische Organisation der Zelle ausgezeichnet. Solche Superlative entfalten leicht eine suggestive Wirkung, je länger sie wiederholt werden, um so nachhaltiger. Das sollte Veranlassung geben, die Methode sehr viel eingehender als bisher zu analysieren.

Schon vor 60 Jahren haben Mönckeberg und Bethe (30) festgestellt, daß Osmiumtetroxyd gelöstes Hühnereiweiß nicht coaguliert, sondern in eine Gallerte verwandelt, die weder durch chemische Mittel noch durch Hitze zur Gerinnung gebracht werden kann. Osmiumtetroxyd koaguliert also nicht wie ein typischer Eiweißfäller, es gelatiniert. Schon deshalb muß man annehmen, daß hierbei ein *Hauptvalenzgel* entsteht.

Osmiumtetroxyd löst sich in beschränktem Grad in Wasser, aber auch in Alkohol und Äther. Es ist kein Elektrolyt und verhält sich gegen Lackmus neutral. Es wirkt stark oxydierend und als Sauerstoffüberträger, und dies ist neben seiner evidenten Affinität zum Stickstoff entscheidend

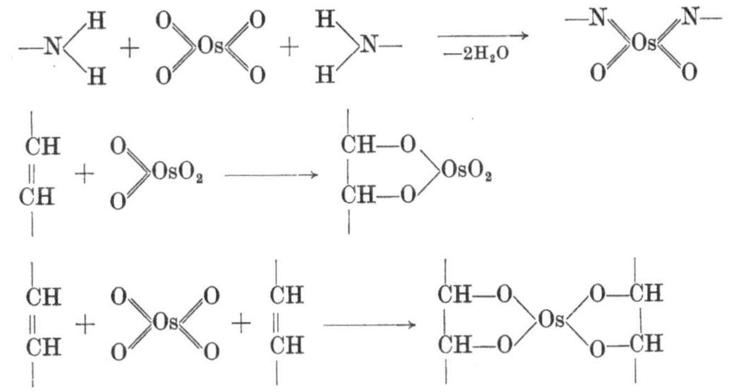

Abb. 4. Wirkungsweise des Osmiumtetroxyds

für seine Wirkung auf das molekulare Gefüge des Plasmas. Mit Berg (31) können wir eine primäre und eine sekundäre Wirkung unterscheiden. In einer ersten Stufe werden die Zellstrukturen in kürzester Zeit fixiert. Ihre Färbbarkeit bleibt dabei erhalten. Nach den Erfahrungen des histologischen Laboratoriums kann dieser Effekt schon durch eine kurz dauernde Räucherung oder durch kurzes Einbringen in eine wäßrige Lösung herbeigeführt werden. Man kann heute nur Vermutungen über diese primäre Wirkung äußern. Möglicherweise beruht sie auf der Bindung von Osmiumtetroxyd an N-haltige Gruppen. Dabei entstehen, wie ich vermute, *Ammine* (Abb. 4) [Zeiger (3)]. Es genügt hier auf die Neigung des Tetroxyds zur Reaktion mit Ammoniak hinzuweisen. Solche Komplexbildungen könnten zur Verfestigung von NH_2-Gruppen an benachbarten Seitenketten oder innerhalb einer Polypeptidkette beitragen. Dabei würden diese Radikale durch Anlagerung von Osmiumtetroxyd in dessen Kraftfeld einbezogen und eine Art von Verklammerung geschaffen.

Eine solche Abschirmung von NH_2-Gruppen macht auch die Tatsache verständlich, daß aus den gleichen Gründen wie bei der Formaldehydwirkung die Ladungsäquivalente der Plasmaproteine in extremer Weise nach niedrigeren pH-Werten verschoben werden [Zeiger (23)].

In einer zweiten Wirkungsstufe entfaltet Osmiumtetroxyd seine *oxydativen Eigenschaften*. Die Abspaltung von Sauerstoff erfolgt langsam, wie die nur allmählich eintretende Ablagerung der dunkel gefärbten Reduktionsstufen beweist. Der frei werdende Sauerstoff könnte im Bereich von Doppelbindungen zur Bildung von Epoxyd- und Peroxyd-Bindungen, unter Umständen auch zur Bildung von Sauerstoffbrücken führen (Abb. 5) [Zeiger (*3*), Frey-Wyssling (*21*)]. Doch gibt es hierfür noch keine Beweise.

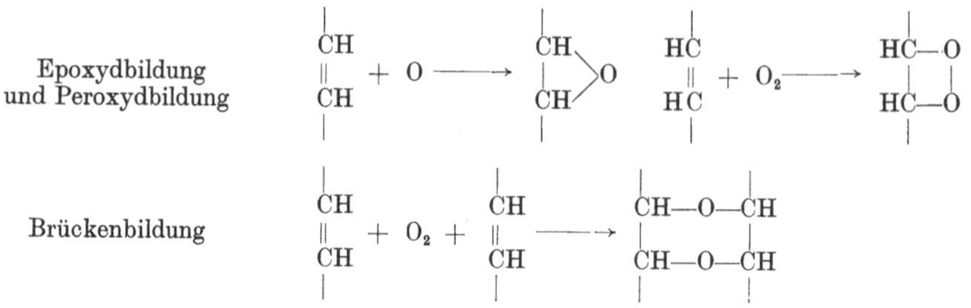

$$\text{Epoxydbildung und Peroxydbildung:}\qquad \begin{array}{c}|\\ \mathrm{CH}\\ \parallel\\ \mathrm{CH}\\ |\end{array} + \mathrm{O} \longrightarrow \begin{array}{c}|\\ \mathrm{CH}\\ \diagdown\\ \mathrm{CH}\diagup\mathrm{O}\\ |\end{array}\qquad \begin{array}{c}|\\ \mathrm{HC}\\ \parallel\\ \mathrm{HC}\\ |\end{array} + \mathrm{O_2} \longrightarrow \begin{array}{c}|\\ \mathrm{HC{-}O}\\ |\quad\ |\\ \mathrm{HC{-}O}\\ |\end{array}$$

$$\text{Brückenbildung:}\qquad \begin{array}{c}|\\ \mathrm{CH}\\ \parallel\\ \mathrm{CH}\\ |\end{array} + \mathrm{O_2} + \begin{array}{c}|\\ \mathrm{CH}\\ \parallel\\ \mathrm{CH}\\ |\end{array} \longrightarrow \begin{array}{c}|\qquad\quad|\\ \mathrm{CH{-}O{-}CH}\\ |\qquad\quad|\\ \mathrm{CH{-}O{-}CH}\\ |\qquad\quad|\end{array}$$

Abb. 5

Gewöhnlich wird im Anschluß an Arbeiten von Criegee (*32, 33*) eine einfache Anlagerung des Osmiumtetroxyds an die Doppelbindungen der ungesättigten Fettsäurereste von Lipoiden angenommen. Diese für verschiedene Olefine nachgewiesene Verbindung bildet sich beim Arbeiten in organischen Lösungsmitteln. Sie ist wenig stabil. In wäßrigen Medien entsteht hingegen, wie Criegee gezeigt hat, eine andere unlösliche, sehr stabile Verbindung mit zwei benachbarten Fettsäureketten. Sie wird als Diester der hypothetischen Osmiumsäure, H_4OsO_5, aufgefaßt (Abb. 6). So

$$\begin{array}{c}|\\ \mathrm{HC}\\ \parallel\\ \mathrm{HC}\\ |\end{array} + \mathrm{OsO_4} \rightarrow \begin{array}{c}|\\ \mathrm{HC{-}O}\\ |\qquad\diagdown\\ \qquad\mathrm{OsO_2}\\ \mathrm{HC{-}O}\diagup\\ |\end{array}\ \text{(I)}\qquad\qquad \begin{array}{c}|\qquad\ \mathrm{O}\qquad\ |\\ \mathrm{HC{-}O}\diagdown\ \parallel\ \diagup\mathrm{O{-}CH}\\ \qquad\quad\mathrm{Os}\\ \mathrm{HC{-}O}\diagup\ \quad\diagdown\mathrm{O{-}CH}\\ |\qquad\qquad\qquad |\end{array}\ \text{(II)}$$

Abb. 6

wäre eine Vernetzung benachbarter Fettsäureketten in Lipoidstrukturen möglich. Da auch aromatische Doppelbindungen mit Osmiumtetroxyd reagieren, ist auch eine Verknüpfung über die Doppelbindung des Cholesterins denkbar. Die bei der Bildung des Diesters (II) aus dem Monoester notwendige Oxydation der Doppelbindungen zu Diolen würde an diesen Orten zu einer Ablagerung niederer Oxydationsstufen des Osmiums und damit zum Auftreten einer Schwärzung und einer entsprechenden Steigerung des elektronenoptischen Kontrastes führen.

Diese Begleitreaktion erscheint deshalb von besonderer Bedeutung, weil nach den Angaben über den Sättigungsgrad der Fettsäuren in Phosphatiden bei ausschließlicher Bindung des Osmiums in Form des Diesters nur ein einziges Molekül Osmium pro Molekül Phosphatid unterzubringen wäre [Stoeckenius (*34*)].

Endlich ist zu beachten, daß infolge der hohen Sauerstoffspannung die reduzierten Stufen aller Redoxsysteme in die oxydative Form übergehen. Die *Verlagerung des Redoxpotentials* sollte labilere homöopolare Hauptvalenzbindungen des Molekulargefüges verfestigen [Zeiger (*3*)]. Falls Cystein-Schwefelbrücken in der Struktur des Cytoplasmas eine Rolle spielen, würde das Osmium auch in diesem Fall durch Wegoxydation von Wasserstoff zur Bildung neuer Brücken beitragen [Frey-Wyssling (*21*)].

Naturtreue und Osmiumfixierung

Bahr (*35*) hat unter relativ gut bekannten Bedingungen in vitro die Reaktionsfähigkeit von Osmiumtetroxyd gegenüber einer großen Reihe von biologisch wichtigen Substanzen geprüft. Dabei ergab sich, daß einige besonders intensive Reaktionen von den sterischen Bedingungen der betreffenden organischen Substanzen abhängig sind. Andererseits bleiben viele Eiweiße und Lipoide auch nach Osmiumbehandlung löslich. Natürlich lassen diese bemerkenswerten Ergebnisse keine Schlüsse in bezug auf die Lokalisation von Substanzen in Zellstrukturen zu. Doch sind solche

Erfahrungen von Bedeutung für den systematischen Aufbau von Strukturmodellen und die Interpretation der chemischen Vorgänge bei der Osmierung. BAHR (*36*) hat weiter gezeigt, daß die Aufnahme von Osmium aus Fixierungsflüssigkeiten bei verschiedenen Geweben nach einem charakteristischen Muster verläuft und nach verhältnismäßig kurzer Zeit abgeschlossen ist. Nach Versuchen von BAHR, BLOOM und FRIBERG (*37*) setzt in 1% Osmiumtetroxyd in Tyrode-Lösung bei einem p_H von 7,2 sofort eine rapide Volumenzunahme der Objekte ein, deren Halbwert schon nach 15 min erreicht wird. Entwässerung und Einbettung der Objekte führen anschließend zu unterschiedlicher Schrumpfung, die in Methacrylat am geringsten ist. Diese Quellung ist nicht durch das Fixationsmittel, sondern durch das wäßrige Lösungsmittel bedingt. Sie kann durch Zusatz von angemessenen Mengen kolloidosmotisch aktiver Substanzen vollständig aufgehoben werden.

Es erübrigt sich also, daß bei Fixierung mit Osmiumtetroxyd im Ablauf des üblichen Präparationsverfahrens erhebliche Schwankungen im Volumen, Gewicht und spezifischem Gewicht der Objekte auftreten, welche die ursprüngliche Form, Größe und auch Struktur der Zell- und Gewebsbestandteile beeinflussen können. Bei der unterschiedlichen Fixationsstabilität aller dieser Elemente, besonders aber der typischen Zellstrukturen, muß das auch Folgen in bezug auf die Naturtreue der elektronenoptischen Abbildung haben können, und diese Möglichkeiten sind noch nicht unter unserer Kontrolle. So werden Versuche verständlich, andere Fixierungsmittel in der elektronenmikroskopischen Technik zu verwenden, und es ist bemerkenswert, daß dabei typische Lipoidstabilisatoren, wie Chromsäure und ihre Abkömmlinge mit und ohne Kombination mit Osmiumtetroxyd [DALTON (*38*), LOW and FREEMAN (*39*)] sowie Permanganat [LUFT (*40*)], eine Rolle spielen.

Ausblick

Damit stehen wir wiederum vor dem entscheidenden methodischen Problem der Präparationstechnik. Es wird uns im elektronenmikroskopischen Bereich in seiner ganzen Bedeutung, aber auch in seiner ganzen Schwere ad oculos demonstriert, und zwar mit erschreckender Deutlichkeit. Im lichtmikroskopischen Bereich war es durchaus angebracht, eine «fixation morphologique» und eine «fixation des substances» zu unterscheiden; denn je mehr wir uns der molekularen Struktur annähern, um so mehr ist die morphologische Naturtreue untrennbar an die chemische gebunden, um so mehr fällt die Ortsrichtigkeit der Substanzen mit der Formrichtigkeit der Struktur zusammen.

Man kann diesen Tatbestand nicht genug ins Bewußtsein heben. Wer die histochemische und elektronenmikroskopische Literatur mit kritischem Blick durchmustert, muß sich wundern über die Selbstsicherheit mancher der uns angebotenen Deutungen. Schon in der klassischen Kolloidchemie sind die Gele viel weniger intensiv studiert worden als die Sole, eben deshalb, weil die meisten Gele, und erst recht die biologischen Gele, sehr veränderliche Systeme darstellen, und es keine Methoden gab, ihre Veränderungen quantitativ zu verfolgen. In der Elektronenmikroskopie stehen wir vor den gleichen Schwierigkeiten. Es liegt im Wesen dieser Methode, daß die naturgetreue Darstellung von labilen Gelstrukturen, etwa vom Prototyp des Cytoplasmas oder der Membranstrukturen, hier großen Schwierigkeiten begegnen muß. Die größte scheint mir eine schonende Entwässerung zu bieten, die im Hochvakuum bis zur völligen Dehydratation vorgetrieben werden muß.

Wie kann man diesen außerordentlichen Schwierigkeiten begegnen? Nun wohl dadurch, daß man sich zunächst einmal die aktuelle Situation verdeutlicht. Schon die Lichtmikroskopie hat die methodischen Engpässe, die uns hier Sorgen bereiten, klar erkannt und innerhalb ihrer bescheidenen Grenzen zu einem experimentell angreifbaren Forschungsgebiet gemacht. Eine ihrer wichtigsten Erkenntnisse lautet: Jedes Gewebselement besitzt dem Fixationsmittel gegenüber seine eigene Reaktionsnorm. Das dokumentiert sich auch im Bereich der Zelle, und zwar in der unterschiedlichen Strukturlabilität ihrer verschiedenen Bestandteile, und dies wiederum ist bedingt durch deren unterschiedliche Größe und Beständigkeit sowie die unterschiedliche Verwickelung ihrer morphologischen und chemischen Organisation [ZEIGER (*41*)].

Die ideale Fixierungsmethode, die jedes Atom an seinen Ort zwingen könnte, gibt es noch nicht, und es wird sie wohl auch nicht geben, obwohl noch wesentliche Verbesserungen möglich.

erscheinen, ebensowenig wie in der lichtmikroskopischen Präparationstechnik, die heute auf eine Erfahrung von hundert Jahren zurückschauen kann. Eine ideale Fixation gibt es offensichtlich nicht, weil sie dem eingangs geschilderten Ordnungscharakter lebender Stoffsysteme ganz und gar widerspricht.

Trotzdem brauchen wir nicht zu sagen: Lasciat' ogni speranza! Wir sollten vielmehr jetzt die Möglichkeit der experimentellen Erzeugung und Umbildung von physikalisch-chemisch gut definierten Strukturen ins Auge fassen. Es gibt ausgezeichnete Ansätze für diese Art von Untersuchungen. Man sollte also in planmäßigen Versuchen die Reaktionsnorm solcher Strukturmodelle studieren und die optimalen Bedingungen ermitteln. Freilich wird man nur in einer mühsamen Kleinarbeit den Anschluß an die biologischen Substrate gewinnen können. Auch hier gilt, was für die gesamte physikalisch-chemisch orientierte Cytologie stets Geltung besitzen wird: Eine Analyse im molekularen Bereich sollte immer versuchen, bei der Darstellung ihrer Tatbestände bis zur äußersten Grenze der Wirklichkeit und Genauigkeit vorzudringen, auch wenn die biologischen Aspekte dabei verlorengehen.

Literatur

1. Zeiger, K.: Die Methoden der histologischen Technik vom Standpunkt der Kolloidlehre. In: Med. Kolloidlehre. Herg. von L. Lichtwitz, R. E. Liesegang u. K. Spiro. S. 973—1016. Dresden u. Leipzig: Theodor Steinkopff 1935.
2. — Physikochemische Grundlagen der histologischen Methodik. Dresden u. Leipzig: Theodor Steinkopff 1938.
3. — Z. Zellforsch. **34,** 230 (1949).
4. Stoeckenius, W., u. K. Zeiger: Erg. Anat. **35,** 420 (1956).
5. Maxfield, M.: J. gen. Physiol. **37,** 201 (1954).
6. — J. gen. Physiol. **34,** 853 (1951).
7. Zetterqvist, H.: The ultrastructural organization of the columnar absorbing cells of the mouse jejunum. Stockholm 1956.
8. Tellyesniczky, K.: Fixation. In Krauses Enzyklop. d. mikrosk. Techn., Bd. 1, 3. Aufl. S. 750—785. Berlin u. Wien: Urban & Schwarzenberg 1926.
9. Cowdry, E. V.: Special cytology. 2nd Ed. New York: Paul B. Hoeber Inc. 1932.
10. Duthie, E. S.: J. Path. Bact. **41,** 311 (1935).
11. Zollinger, H. U.: Schweiz. Z. Path. **11,** 617 (1948).
12. Rhodin, J.: Correlation of ultrastructural organization and function in normal and experimentally changed proximal convoluted tubule cells of the mouse kidney. Stockholm Diss. 1954.
13. Sjöstrand, F. S., and V. Hanzon: Exp. Cell Res. **7,** 393 (1954).
14. Palade, G. E.: J. exp. Med. **95,** 285 (1952).
15. Sjöstrand, F. S.: J. cell. comp. Physiol. **42,** 15 (1953).
16. Porter, K. R., A. Claude and E. F. Fullam: J. exp. Med. **81,** 233 (1945).
17. Dalton, A. J., H. Kahler, M. J. Striebich and B. Lloyd: J. nat. Cancer Inst. **11,** 439 (1950).
18. Porter, K. R.: Anat. Rec. **106,** 311 (1950).
19. Rozsa, G., and R. W. G. Wyckoff: Biochem. biophys. Acta **6,** 334 (1950).
20. Frey-Wyssling, A.: Submikroskopische Morphologie des Protoplasmas und seiner Derivate. Protoplasma-Monographien Bd. 15. Berlin: Gebrüder Borntraeger 1938.
21. — Die submikroskopische Struktur des Cytoplasmas. In: Protoplasmatologia. Handbuch der Protoplasmaforschung. Bd. 2, A 2, S. 1. Wien: Springer-Verlag 1955.
22. Toth, A.: Protoplasma **41,** 103 (1952).
23. Zeiger, K.: Z. Zellforsch. **10,** 481 (1930).
24. Pischinger, A.: Z. Zellforsch. **3,** 169 (1926).
25. Meyer, K. H.: Hochpolymere Chemie. Bd. 2. Leipzig: Akad. Verlagsges. 1940.
26. Deuel, H.: Helv. chim. Acta **30,** 1269 (1947).
27. Hertwig, G.: Z. mikrosk. anat. Forsch. **23,** 484 (1931).
28. Zirkle, C.: Protoplasma **20,** 169 (1934).
29. Sylven, B.: Acta Un. int. Cancr. **7,** 708 (1951).
30. Mönckeberg, G., und A. Bethe: Arch. mikrosk. Anat. **54,** 135 (1899).
31. Berg, W.: Osmiumsäure. In Krauses Enzyklop. d. mikrosk. Technik. 3. Aufl., Bd. 3, S. 1742. Berlin und Wien: Urban & Schwarzenberg 1927.
32. Criegee, R.: Liebigs Ann. Chem. **522,** 75 (1936).
33. — B. Marchand u. H. Wannowius: Liebigs Ann. Chem. **550,** 99 (1942).
34. Stoeckenius, W.: Exp. Cell Res. **13,** 410 (1957).
35. Bahr, G. F.: Exp. Cell Res. **7,** 457 (1954).
36. — Exp. Cell Res. **9,** 277 (1955).

37. Bahr, G. F., G. Bloom and U. Friberg: Exp. Cell Res. **12**, 342 (1957).
38. Dalton, A. J.: Anat. Rec. **121**, 281 (1955).
39. Low, F. N., and J. A. Freeman: J. biophys. biochem. Cytol. **2**, 629 (1956).
40. Luft, J. H.: J. biophys. biochem. Cytol. **2**, 799 (1956).
41. Zeiger, K.: Verh. Anat. Ges. Marburg. Erg.-H. zum Anat. Anz. **99**, 9 (1952).

Fixation of plant tissue

J. D. McLean

Division of Chemical Physics, C. S. I. R. O. Chemical Research Laboratories, Melbourne (Australia)

Electron micrographs of mature plant cells fixed in neutral osmium tetroxide solutions frequently show disrupted cell membranes and an apparently dispersed cytoplasm. In the case of young non-vacuolated plant cells their fixation has been improved by the addition of various

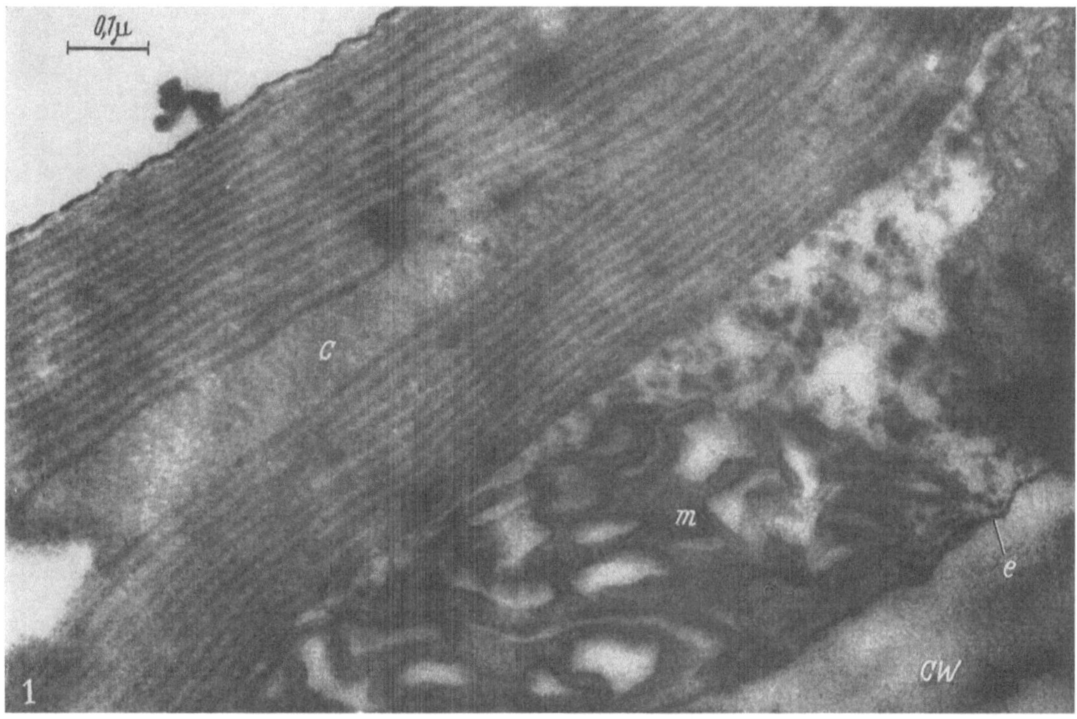

Fig. 1. Section of a parenchyma sheath cell in a maize leaf showing portion of a chloroplast (*c*), mitochondria (*m*), ectoplast membrane (*e*) and cell wall (*cw*)

agents such as Locke's medium and dextrin (*1*) or sucrose (*2*) to a buffered osmium tetroxide fixation medium. Presumably these additives tend to prevent the swelling of cell components. Another source of damage in turgid vacuolated cells may be the destruction of the differential permeability of the tonoplast during the early stages of fixation. This occurrence would be accompanied by a loss of turgor resulting in a disturbance of the partly fixed cytoplasm of the initially distended cell. Such damage may be expected to be apparent particularly at the cytoplasm-cell wall boundary. Further disruption of the cytoplasm could result from the sudden release of vacuolar fluid into the cytoplasm.

It has been found possible to improve the preservation of the cytoplasm and especially the limiting membranes of the plant cell by reducing the turgor of the cells prior to their fixation. The living plant cell is able to accommodate small changes in the osmotic pressure of its environment without apparent damage, as judged by the continuation of protoplasmic streaming. Plasmolysis of the cells was avoided.

Small pieces of tissue from maize leaves and bean roots were treated with glucose solutions of increasing concentration finally reaching 0.2 M. After about five minutes the tissue was transferred to an osmium tetroxide (2%) solution, buffered at p_H 7.2 with acetate-veronal (3) and made up to 0.2 M with glucose. Three hours in this medium at room temperature was found to give satisfactory fixation. The tissue was then washed briefly in a 0.2 M glucose solution, dehydrated in an ethanol series and embedded in butyl methacrylate. Sections were cut using a microtome

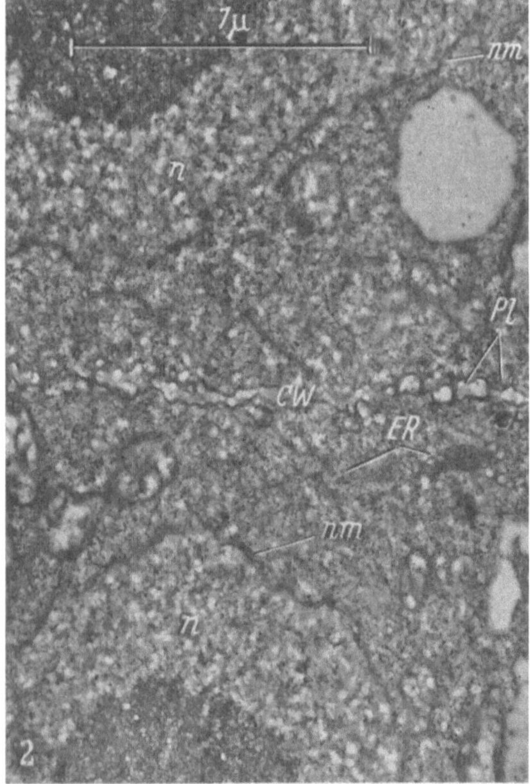

Fig. 2. Portion of two cells in a young bean root. Plasmodesmata (*Pl*) traverse the cell wall (*cw*) and the ectoplast of the adjacent cells is continuous through these pores. Nuclei (*n*) are surrounded by a pair of membranes (*nm*) and a number of elements of the endoplasmic reticulum (*ER*) appear in the cytoplasm

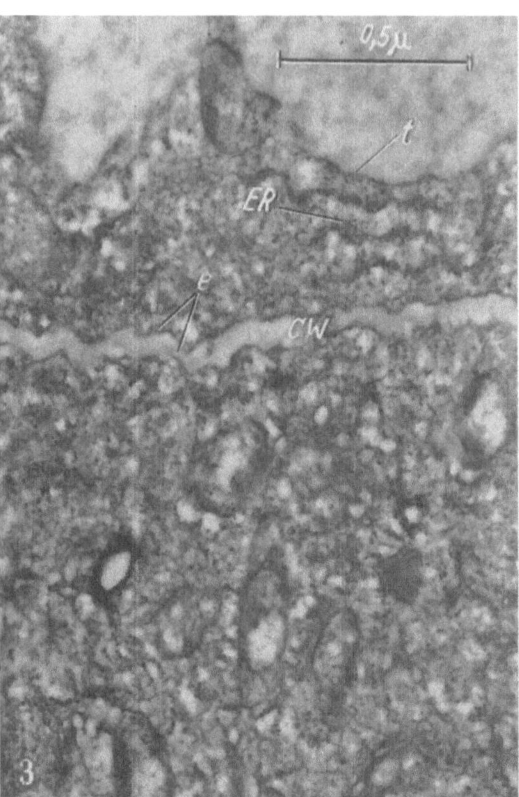

Fig. 3. Section of two cells in a vacuolated region of the bean root. The tonoplast (*t*) and ectoplast (*e*) membranes are similar in appearance

designed by Farrant and Powell (4) and examined in an RCA EMU-1 electron microscope.

Fig. 1 shows part of a mature vacuolated cell in a maize leaf fixed by the above procedure. The membrane bordering the cytoplasm next to the cell wall is intact and in close contact with the wall. The mitochondria appear satisfactorily preserved and show continuity of internal membranes. The sheath chloroplast possesses an ordered lamellar system and does not exhibit evidence of swelling.

In the cells of young bean root (Fig. 2) the cytoplasm and nuclear material similarly lack fixation damage. The external membrane or *ectoplast* of the cells is clearly defined and in places can be seen to be continuous with the ectoplast of the adjacent cell through pores or plasmodesmata. In vacuolated cells of the bean root (Fig. 3) the tonoplast appears similar to the ectoplast. The cytoplasm contains a number of membranous profiles of the endoplasmic reticulum.

Although in the tissues described the procedure of fixation outlined above gave satisfactory preservation of cell structure, difficulties have been encountered using this and other methods in old tissue with relatively thick cell walls. Here factors such as the rate of penetration through the cell wall by the components of the fixation medium may be contributing to their unsatisfactory preservation. Where such factors are not operating, fixation of the plant cell appears more satisfactory after reduction of the pressure. Such a procedure ensures that changes in cell volume are kept at a minimum during fixation.

References

1. Hodge, A. J., E. M. Martin and R. K. Morton: J. biophys. biochem. Cytol. **3,** 61 (1957).

2. Caulfield, J. B.: J. biophys. biochem. Cytol. **3,** 827 (1957).

3. Palade, G. E.: J. exp. Med. **95,** 285 (1952).

4. Farrant, J. L., and S. E. Powell: Proc. First Regional Conference in Asia and Oceania, Tokyo, p. 147 (1956).

KMnO₄-Fixierung von Blutelementen

Manfred Bayer

Bernhard Nocht-Institut für Schiffs- und Tropenkrankheiten, Abteilung für Virusforschung, Hamburg.

Bislang kam in der Präparation biologischen Materials für elektronenoptische Schnittuntersuchungen am häufigsten die Osmiumtetroxyd-Fixierung zur Anwendung (*1*). Als ein deutlicher Vorteil gegenüber anderen Fixierungen wird die geringe Artefaktbildung angesehen (*2*). Die Proteine werden dabei unter Bildung von Reduktionsprodukten des Osmiumtetroxyds in einen gelartigen Zustand übergeführt.

Auch in der Blutmorphologie hat sich das OsO₄ bewährt (*3*). Man konnte Strukturen, die von der Lichtoptik her bekannt waren, elektronenoptisch bestätigen und Vorstellungen vom Zellaufbau und von den Zelltypen erweitern.

Die Kaliumpermanganatfixierung, für die Lichtoptik von W. J. Schmidt (*4*) angewandt, in die Präparation für die Elektronenmikroskopie von Luft (*5*) eingeführt, bewirkt eine Reduktion des Permanganats an bestimmten Zellorten zu Manganoxyden und -hydroxyden.

Im vorliegenden Fall wurde die von Luft angegebene Permanganatfixierung in etwas abgeänderter Weise an Blutelementen von Mensch, Ratte und Huhn untersucht. Eingebettet wurde in „Plexigum" (*6*) und Gelatine.

Bei der elektronenoptischen Betrachtung KMnO₄-fixierter *Leukocyten* fällt in Übereinstimmung mit Luft die starke Zeichnung der Mitochondrien auf (Abb. 1). Die zwischen den Cristae liegenden Zonen sind häufig sehr hell. Das Cytoplasma erscheint sehr feingranulär. RNS-Granula, die für OsO₄-behandelte Zellen charakteristisch sind (*7*), fehlen. Das endoplasmatische Reticulum hebt sich scharf und kontrastreich von der Umgebung ab. Die Zellwand läßt sich oft als zarte Doppellinie erkennen. Im Leukocyten treten die von der Osmiumpräparation her bekannten, als „spezifische Granulation" bezeichneten Körper auch nach der Permanganatfixierung deutlich in Erscheinung. Auch ihre Feinstruktur, wie etwa der von Bargmann und Knoop (*8*) am OsO₄-fixierten eosinophilen Granulocyten gezeigte lamelläre Aufbau eines Granulums läßt sich mit dieser Methode nachweisen. In der spezifischen Granulation der basophilen Granulocyten der Ratte treten häufig dunkle zentrale Zonen deutlicher auf, als nach der OsO₄-Technik (*9*). Eine dünne doppelte Membran umhüllt meist die einzelnen Granula. Auch der Zellkern wird von einer oft sehr kontrastreichen doppelten Membran umgeben. Das Karyoplasma ist auffallend hell und feingranulär. Dunkle Zonen treten vorwiegend im Kerninneren auf.

Am *Thrombocyten* sind die Granula, die Elemente des endoplasmatischen Reticulums und die Mitochondrien kontrastreich dargestellt.

Im Schnittbild des *Erythrocyten* kommen bei unreifen Zellen die Mitochondrien, das endoplasmatische Reticulum und die Wand der Vacuolen deutlich zur Abbildung.

Das Zellinnere ist feingranulär. Die Zelle wird umhüllt von einer doppelt konturierten Membran (Abb. 2a), die eine Gesamtdicke von etwa 70 Å aufweist. Sie läßt sich in dieser Form bei den Erythrocyten von Huhn, Ratte und Mensch nachweisen.

Um mögliche Herauslösungen bei der Alkoholentwässerung und Einbettung zu umgehen, wurden Ratten-Erythrocyten nach der KMnO$_4$-Fixierung in ausschließlich wäßrigem Milieu weiterbehandelt, nach einer abgeänderten Verarbeitungsweise gegenüber GILEV (10) und FERNÁNDEZ-MORÁN und FINEAN (11) in Gelatine eingebettet und geschnitten. Auch dabei zeigte sich

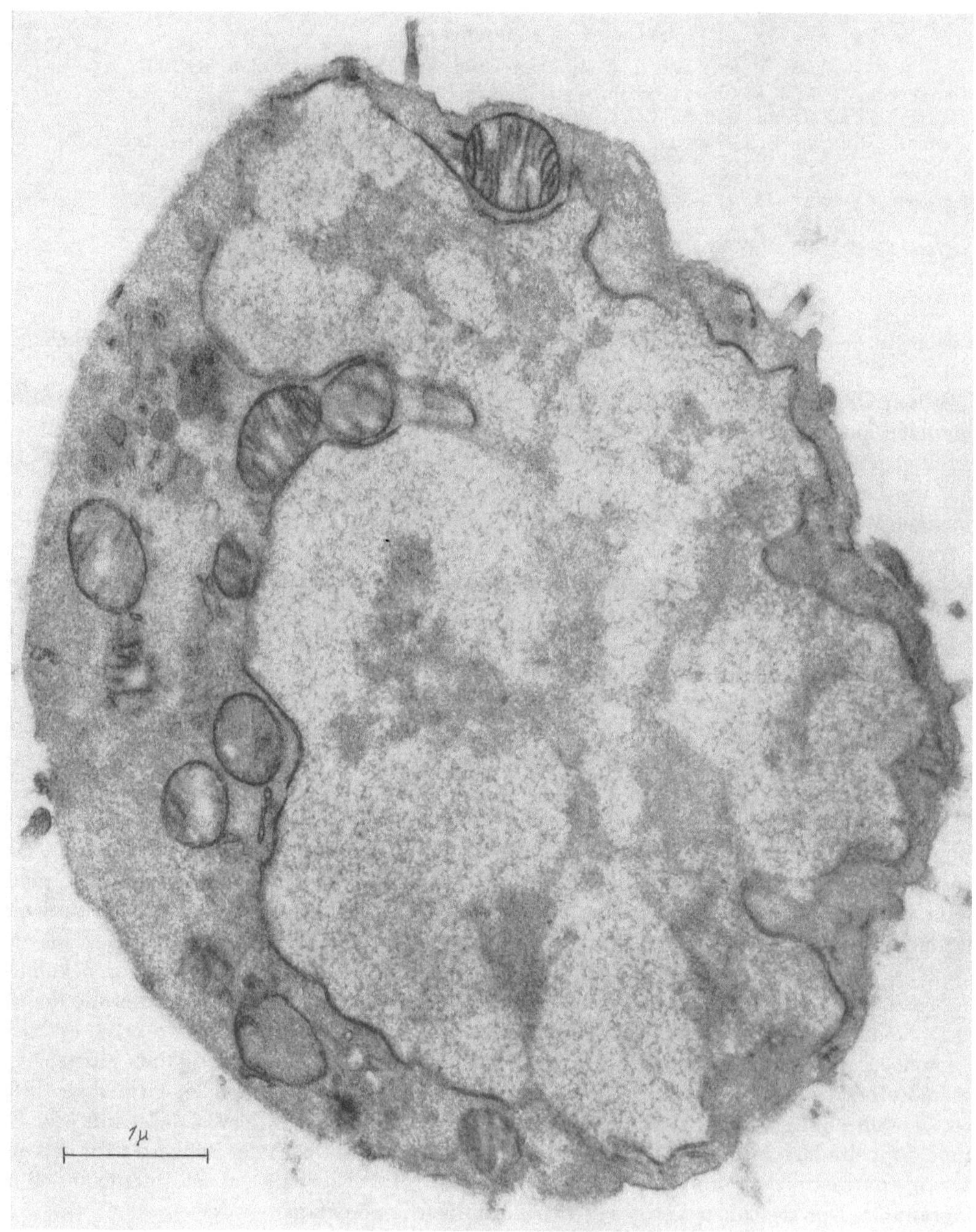

Abb. 1. Lymphocyt der Ratte, KMnO$_4$-fixiert

deutlich die doppelte Membranstruktur (Abb. 2b). Das Zellinnere hat hierbei eine körnige Beschaffenheit. Die Abmessung von etwa 70 Å für die Zellmembrandicke deckt sich mit den an verschiedenen anderen Objekten gefundenen Ergebnissen von ROBERTSON (12) und in der Dimension mit den von PARPART und BALLENTINE (13) nach anderen Methoden berechneten Werten.

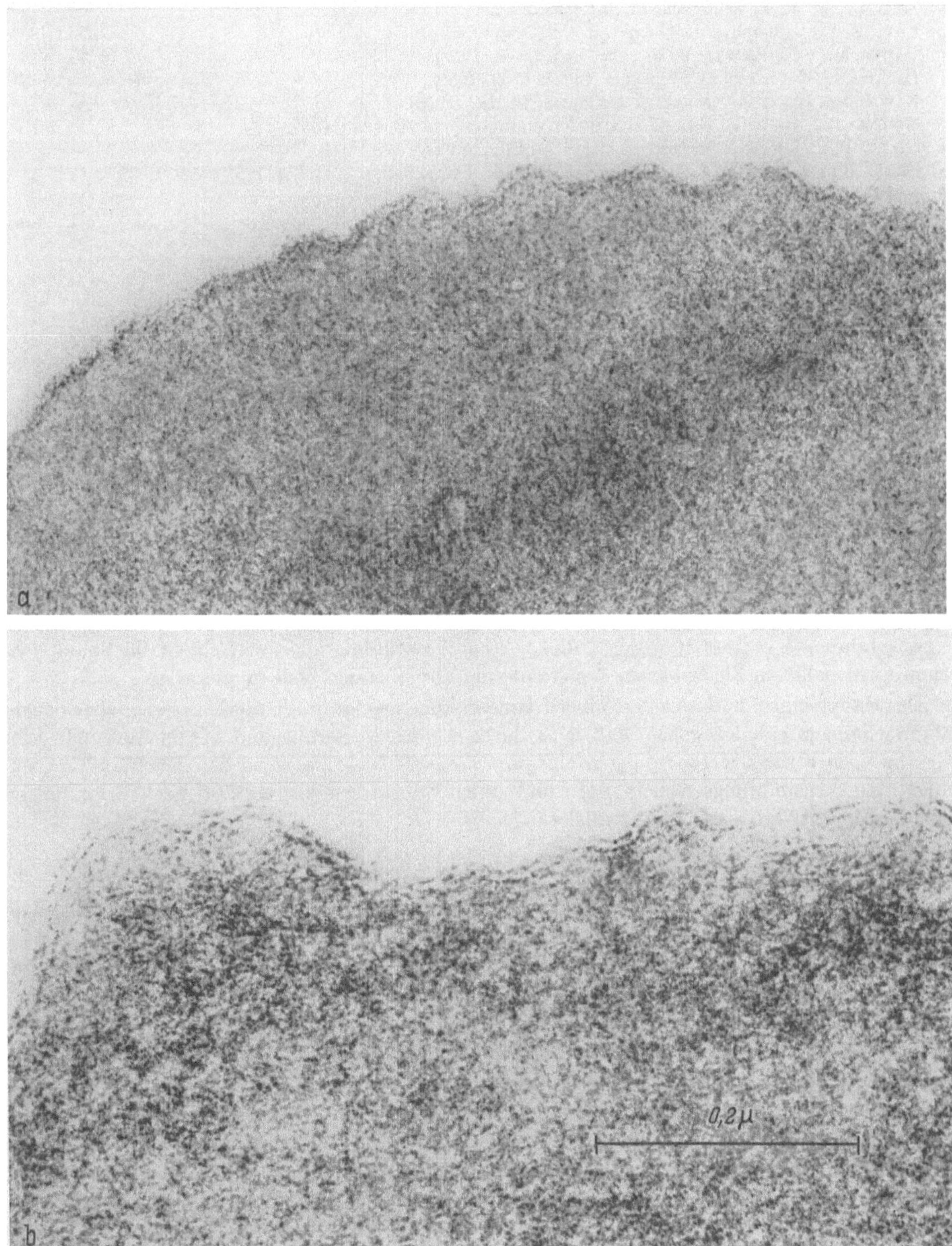

Abb. 2a u. b. a Doppelstruktur der Erythrocytenmembran nach KMnO₄-Fixierung und Einbettung in „Plexigum". b Doppelstruktur der Membran eines KMnO₄-fixierten Erythrocyten. Gelatine-Einbettung

Ich danke Herrn Dr. D. PETERS für Förderung und Unterstützung, der Deutschen Forschungsgemeinschaft für die Schaffung der materiellen Voraussetzungen.

Literatur

1. PALADE, G.: J. exp. Med. **95,** 285 (1952).
2. PORTER, K. R., and F. KALLMAN: Exp. Cell Res. **4,** 127 (1953).
3. BESSIS, M.: Cytology of the blood and blood-forming organs. New York, London: Grune & Stratton p. 107 1956.

4. Schmidt, W. J.: Z. Zellforsch. **23**, 261 (1936).
5. Luft, J. H.: J. biophys. biochem. Cytol. **2**, 799 (1956).
6. Bayer, M., u. D. Peters: J. Ultrastr. Res. **2**, 444 (1959).
7. Palade, G. E.: J. biophys. biochem. Cytol. **1**, 59 (1955).
8. Bargmann, W., u. A. Knoop: Z. Zellforsch. **44**, 282 (1956).
9. Goodman, J. R., E. B. Reilly and R. E. Moore: Blood **12**, 428 (1957).
10. Gilëv, V. P.: Electron Microscopy, Proc. Stockholm Conf. 1956, 113.
11. Fernández-Morán, H., and J. B. Finean: J. biophys. biochem. Cytol. **3**, 725 (1957).
12. Robertson, J. D.: Proc. physiol. Soc. **13**—**14**, P 58 (1957).
13. Parpart, K. A., and R. Ballentine: Trends in physiology and biochemistry, p. 135. New York: Acad. Press Inc. 1952.

Die Dehydratisierung

K. Mühlethaler

Eidgenössische Technische Hochschule Zürich

Im Gegensatz zur lichtmikroskopischen Technik, wo man heute in zunehmendem Maße lebendes Material im Phasenkontrastmikroskop untersucht, sind wir bei der elektronenmikroskopischen Präparation gezwungen, alle Objekte abzutöten und vollständig zu entwässern. Darauf beruht die Tatsache, daß auch heute noch viele Cytologen den Ergebnissen der Elektronenmikroskopie kritisch gegenüberstehen. Es ist nicht zu bestreiten, daß durch die Fixierung, Dehydratisierung und Einbettung Artefakte verursacht werden können. Die Präparationstechnik ist aber in den letzten Jahren so verbessert worden, daß wir bei Anwendung aller Vorsichtsmaßnahmen wohl schon recht nahe am Ziel sind, ein Äquivalentbild der lebenden Zelle herzustellen.

Der Haupteingriff in das Gefüge einer lebenden Zelle erfolgt durch die Fixierung, aber neuere Untersuchungen haben ergeben, daß auch durch die Entwässerung und die nachfolgende Einbettung noch Strukturveränderungen eintreten können. Diese bestehen hauptsächlich in Quellungs- und Schrumpfungserscheinungen und einem Substanzverlust an löslichen Verbindungen beim Durchgang durch das Dehydratisierungsmittel.

Grundsätzlich kann die Entwässerung entweder mit einer physikalischen oder chemischen Methode durchgeführt werden, wobei beide verschiedene Vor- und Nachteile aufweisen. Allgemein gibt man heute den chemischen Methoden den Vorzug, da sie in jedem Laboratorium ohne kostspielige Einrichtungen durchgeführt werden können und die Strukturen der Zellelemente besser erhalten bleiben als nach der Gefriertrocknung. Letztere weist aber den großen Vorteil auf, daß die chemische Zusammensetzung der Zelle erhalten bleibt. Für histochemische Untersuchungen ist sie daher unerläßlich.

1. Die chemische Entwässerung

Allgemein verwendet man als Dehydratisierungsmittel Alkohol, doch können auch andere mit Wasser mischbare Medien, wie Methanol, Aceton, Pyridin oder Dioxan dazu verwendet werden. Die Entwässerung kann entweder nach der Methode von Bernhard (*1*) kontinuierlich, oder stufenweise, in aufeinanderfolgenden Mischungsverhältnissen von 25—50—70—95 und 100% erfolgen. Wie eingangs erwähnt, verursacht der Wasserentzug eine starke Schrumpfung der Objekte. Messungen über die Veränderungen sind von Bahr, Bloom und Friberg (*2*) veröffentlicht worden. Wie aus Abb. 1 ersichtlich ist, erfolgt durch die Osmium-Fixierung eine Quellung von etwa 30%. Sie erfolgt sehr rasch und erreicht nach 15 min bereits die Hälfte des Endwertes. Bei der nachfolgenden Dehydratisierung wird diese Quellung rückgängig gemacht, so daß die Präparate vor dem Einbetten die ursprüngliche Größe nahezu wieder erreicht haben. Diese Schrumpfung ist, wie aus Abb. 2 hervorgeht, abhängig von der chemischen Zusammensetzung des behandelten Gewebes. Im Muskelgewebe ist sie z. B. doppelt so groß wie in der Leber. Daraus müssen wir den Schluß ziehen, daß auch innerhalb der Zelle die verschiedenen Strukturelemente je nach ihrer chemischen Beschaffenheit verschieden stark schrumpfen, wodurch sich die ursprünglichen Größenverhältnisse verändern. Es stellt sich nun die Frage, ob durch langsame oder

schnelle Entwässerung der Schrumpfungsprozeß verkleinert werden kann. Die Messungen von
BAHR, BLOOM und FRIBERG (2) — Abb. 3 — zeigen, daß bei direktem Eintauchen des Objektes in
abs. Alkohol die Entquellung plötzlich erfolgt, während bei Anwendung steigender Konzentra-
tionsreihen eine Verlangsamung zu beobachten ist. Der kritische Übergang erfolgt im Alkohol
zwischen 70% und 90%. Wie aus Abb. 3 ersichtlich ist, ergeben sich jedoch nahezu die gleichen

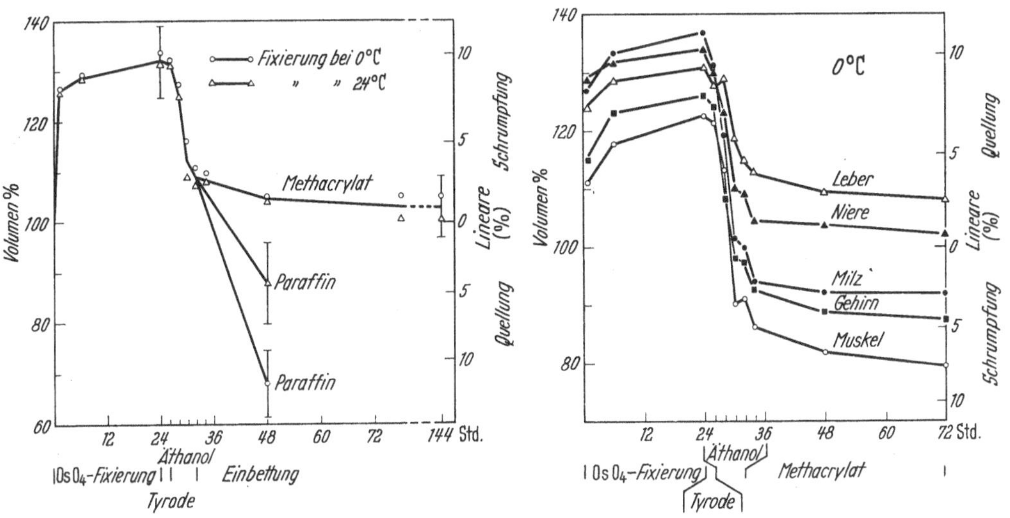

Abb. 1. Volumenveränderung von Gewebestücken während der Osmium-Fixation und Einbettung [nach BAHR, BLOOM und FRIBERG (2)]

Abb. 2. Volumenveränderung verschiedener Gewebe während der Fixation und Einbettung [nach BAHR, BLOOM und FRIBERG (2)]

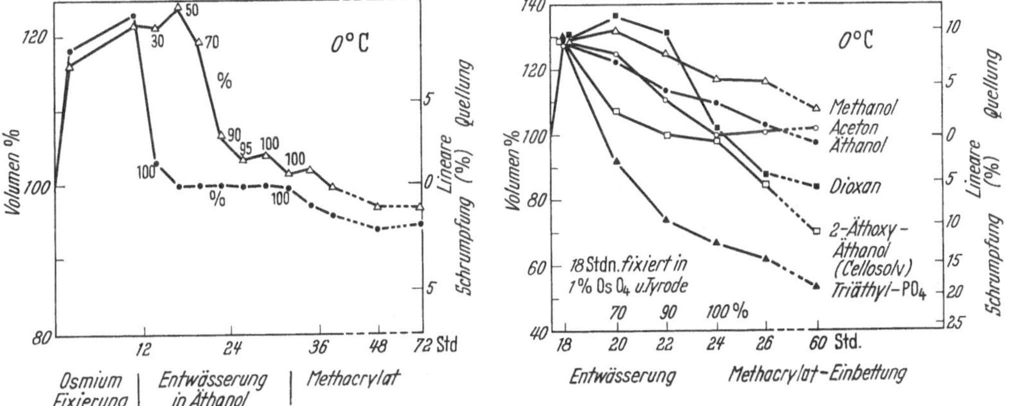

Abb. 3. Volumenveränderung bei schneller und lang-samer Entwässerung in Alkohol [nach BAHR, BLOOM und FRIBERG (2)]

Abb. 4. Entquellungskurven bei Verwendung verschiedener Entwässerungsmedien [nach BAHR, BLOOM und FRIBERG (2)]

Endwerte. Versuche mit verschiedenen Entwässerungsmedien (Abb. 4) zeigen, daß nur Methanol
bessere Resultate ergibt als Äthylalkohol. Morphologisch ist in Objekten, die mit verschiedenen
Medien dehydratisiert wurden, nur vereinzelt ein Unterschied festzustellen. So zeigt die Aceton-
Entwässerung zur Darstellung der Lamellensysteme in Chloroplasten (nach vorheriger $KMnO_4$-
Fixierung) deutlichere Bilder als eine Alkoholbehandlung. Das könnte darauf zurückzuführen
sein, daß mehr Lipoidsubstanzen herausgelöst wurden, wodurch das übrigbleibende Eiweißgerüst
dichter und damit kontrastreicher erscheint. Es empfiehlt sich daher, vor der Untersuchung eines
bestimmten Zellorganells abzuklären, welches Dehydratisierungsmittel am vorteilhaftesten ist.

Eine bessere Konservierung durch Beeinflussung des Quellungszustandes und der Löslichkeit
auswaschbarer Substanzen läßt sich durch Zugabe von Salzlösungen in die Entwässerungsreihe

erreichen. Man kann dazu die gleiche Pufferlösung verwenden, wie zur Fixierung, oder eine einfache Salzlösung mit spezifisch fällenden Eigenschaften. Nach Untersuchungen von Vatter (*3*) eignet sich dazu vor allem $MgCl_2$. Nur durch Zugabe einer 1% $MgCl_2 \cdot 6\,H_2O$-Lösung konnte er die innere Struktur von photosynthetischen Bakterien in einwandfreier Weise konservieren.

Über den Substanzverlust der Präparate während der Fixierung und Entwässerung sind nur wenige Untersuchungen veröffentlicht worden. Sie können aber, wie Sylvén (*4*) nachwies, recht beträchtlich sein. Seine Massenbestimmungen an Rattenleber ergaben einen Verlust von 40—60% der ursprünglichen Trockensubstanz. Neben Kohlenhydraten und anorganischen Ionen werden besonders Lipoide, Proteine und Nucleotide herausgelöst. In pflanzlichen Geweben gehen auch die Pigmente leicht in Lösung. Wolken (*5*) stellte fest, daß in *Euglena gracilis* nach der OsO_4-Fixierung durch den Alkohol 2—10% der Pigmente verlorengehen. Dauert diese Behandlung mehrere Tage, verschwinden sogar 15—40% der Farbstoffe. Um Substanzverluste zu vermeiden, muß daher die Entwässerungsdauer möglichst kurz bemessen werden.

Die praktischen Schlußfolgerungen, die wir aus den verschiedenen Resultaten ziehen können, sind folgende: In allen Entwässerungsmedien tritt eine Schrumpfung ein, welche die vorausgegangene Quellung während der Fixierung rückgängig macht. Der Gradient der Konzentrationsreihe und die Dauer der Überführung in den einzelnen Stufen ist nicht entscheidend. Im Hinblick auf den Lösungseffekt empfiehlt es sich, in den unteren Stufen anstelle von Brunnenwasser Salzlösungen (z. B. $MgCl_2$) zu verwenden und die Präparate nur kurze Zeit, d. h. etwa 10 min, in den verschiedenen Mischungen zu belassen. Damit das Wasser restlos entfernt wird, ist vor dem Einbetten nur frisches, über $CaCl_2$ destilliertes Dehydratisierungsmedium zu gebrauchen.

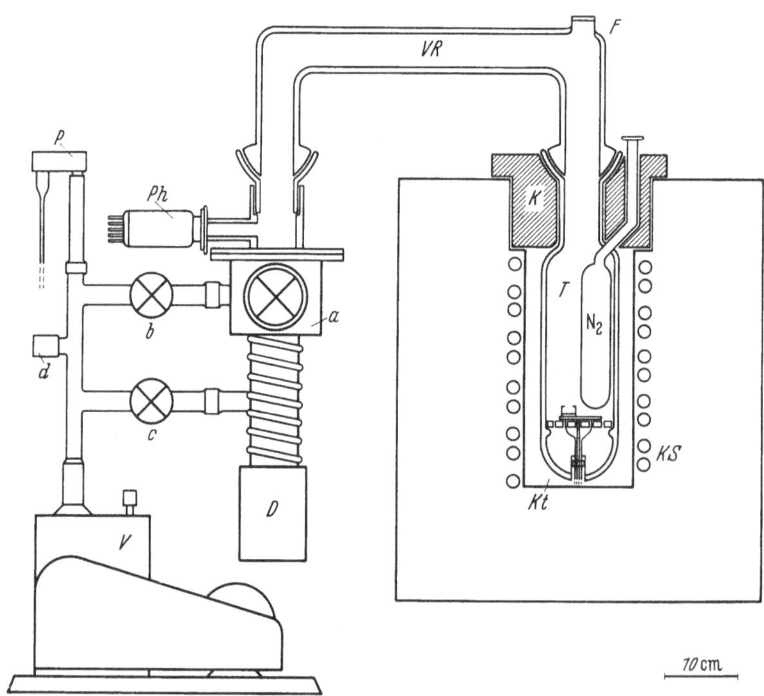

Abb. 5. Gefriertrocknungsapparatur [nach Müller (*6*)]:

V Vorvakuumpumpe	*VR* Verbindungsrohr	*T* Trocknungstubus	*KS* Kühlschlangen
D Diffusionspumpe	*Kt* Kühltruhe	*K* Korkdeckel	*a* Baffle-Hahn
P Pirani-Meßrohr	N_2 Stickstoffinger	*F* Beobachtungsfenster	*b, c* Vakuumhahnen
Ph Ionisationsröhre			*d* Flutventil

2. Gefriertrocknung

Die bei der chemischen Fixierung und Entwässerung auftretenden Fehlermöglichkeiten, wie Quellung, Lösung und Denaturierung, können bei Anwendung der Gefriertrocknung weitgehend vermieden werden. Die Methode besteht darin, die Objekte rasch in flüssigem Propan (—185° C) abzukühlen und anschließend das Eis im Hochvakuum bei Temperaturen von —50° C bis —80° C wegzusublimieren. Die von uns zu diesem Zweck gebaute Apparatur ist in Abb. 5 wiedergegeben. Das gefrorene Präparat wird in einem Metallbehälter in den Trocknungszylinder gebracht, der von einer Diffusions- und Rotationspumpe dauernd auf einem Endvakuum von 10^{-4} bis 10^{-5} mm Hg gehalten wird. Um die Sublimation des Eises zu beschleunigen, ist in der Nähe des Objektes ein Kühlfinger eingesetzt, der mit flüssigem Stickstoff gefüllt ist. Für die lichtmikroskopische Untersuchung können die Präparate nach der Trocknung durch Erwärmen eines darunter liegenden Paraffinblöckchens direkt eingebettet werden. Für die elektronenmikroskopische Technik konstruierten wir ein spezielles Gefäß, das unter Vakuum verschlossen und später mit dem Einbettungsmittel aufgefüllt wird.

Im Gegensatz zur chemischen Entwässerung geht der Trocknungsprozeß im Hochvakuum langsam vor sich und die Trocknungszeit steigt mit sinkender Temperatur stark an. MÜLLER (6)

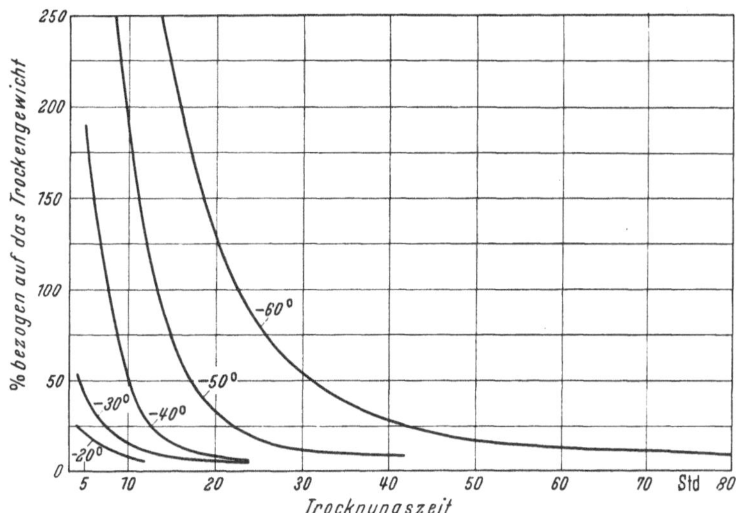

Abb. 6. Restwassergehalt in pflanzlichen Geweben nach der Trocknung bei tiefen Temperaturen
[nach MÜLLER (6)]

hat den Wassergehalt an Blattstücken von *Eucharis grandiflora* nach verschieden langen Trocknungszeiten und Temperaturen untersucht und dabei die in Abb. 6 wiedergegebenen Kurven gefunden. Man kann daraus entnehmen, daß nach einer Trocknungszeit von 80 Std. bei —60° C immer noch 10% Wasser, bezogen auf das Trockengewicht, zurückbleibt. Dieses Restwasser kann erst durch Erhöhen der Temperatur bis gegen 0° C entfernt werden. Die schnellste Trocknung erreicht man, wie die Kurve zeigt, bei Temperaturen wenig unter dem Gefrierpunkt, also bei möglichst hohen Sublimationsdrucken.

Die morphologische Erhaltung der Zellstrukturen wird durch den Rekristallisationseffekt gestört. Das äußert sich darin, daß während der Trocknung das Wachstum der großen Eiskristalle bevorzugt wird, so daß sie sich auf Kosten der kleineren vergrößern. Bei sinkender Temperatur nimmt die Rekristallisationsgeschwindigkeit ab, dafür verlängert sich die Trocknungszeit, so daß im Endeffekt die gleiche Umschichtung eintritt. Das Objekt weist dann, wie aus Abb. 7 ersichtlich, immer eine schwammähnliche Struktur auf. Die einzige Möglichkeit, die Trocknungszeit und damit die Kristallbildung einzuschränken, besteht in der Verkleinerung des Objektes. Dies ist wohl die Ursache, weshalb Virusteilchen, die in sehr kurzer Zeit getrocknet sind, eine gute Strukturerhaltung zeigen. Bei der Untersuchung von Geweben kommt noch eine weitere Schwierigkeit hinzu, nämlich die nachträgliche Einbettung der Objekte in Methacrylat oder Araldit. Die labile Struktur ist noch ungenügend verfestigt, so daß als Folge des Polymerisationsprozesses lipoidhaltige Zellorganellen in kurzer Zeit zu einer amorphen Masse verquellen. Diese Erscheinung ist, z. B. in Abb. 7, in den Chloroplasten am völligen Zerfall der Lamellensysteme deutlich zu erkennen.

Aus diesem Grunde hat auch die von Hancox (*7*) verwendete „Freeze-Substitution" keine Aussicht auf Erfolg. Seine Methode besteht darin, die gefrorenen Präparate bei —38° C in Butanol, statt im Vakuum zu trocknen. Der Endzustand des Objektes entspricht dann einer Butanol-Fixierung, die sehr schlecht aussieht.

Wohl die aussichtsreichste Anwendung der Gefriertrocknung ist die von Steere (*8*) vorgeschlagene Eis-Ätz-Methode. Hier wird unmittelbar nach dem Gefrieren das Objekt angeschnitten, eine dünne Oberflächenschicht durch Sublimation des Eises freigelegt und davon ein Abdruck hergestellt. Damit können die beiden wichtigsten Fehlerquellen, nämlich die Eiskristallbildung

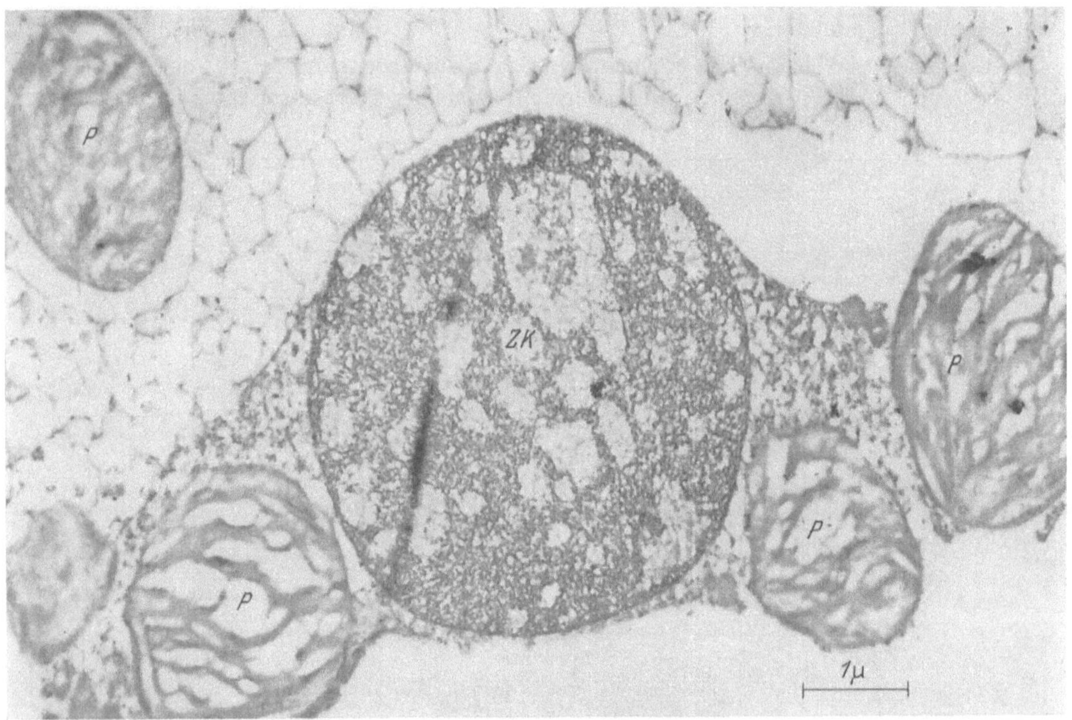

Abb. 7. Dünnschnitt aus dem Mesophyll von *Aspidistra elatior* mit Eiskristall- und Trocknungsartefakten, nach Gefriertrocknung und Methacrylateinbettung. *ZK* Zellkern; *P* Plastiden [nach Müller (*6*)]. Vergr.: 14000mal

durch langsame Trocknung und der Zerfall der Strukturen beim Einbetten, umgangen werden. Wie weit diese Methode für cytologische Untersuchungen zu gebrauchen ist, bedarf noch der Abklärung.

3. Die "Critical-Point"-Methode

Zum Schluß soll noch die von Anderson (*9*) ausgearbeitete "Critical-Point"-Methode besprochen werden. Sie dient dazu, labile räumliche Strukturen beim Auftrocknen auf die Folie vor dem Zusammenfallen zu bewahren. Das Kollabieren wird durch die Veränderung der Oberflächenspannung beim Durchgang einer Phasengrenze (z. B. Flüssigkeit-Luft) verursacht. Das Problem besteht also darin, die Bildung solcher Phasengrenzen während der Fixierung, Entwässerung und Trocknung zu vermeiden. Dies kann, wie Anderson (*9*), zeigte, durch ein Medium erreicht werden, das sich nahe seinem kritischen Punkt befindet. Als günstig erweist sich CO_2, dessen kritischer Punkt bei 31° C und 73 atm liegt. Zur Entwässerung dürfen nur Flüssigkeiten verwendet werden, die unter sich mischbar sind, wie z. B. die Reihe: Wasser—Alkohol—Amylacetat—CO_2 flüssig—CO_2 gasförmig. Der letzte Präparationsschritt wird in einer Druckkammer ausgeführt. Bei Erhöhung der Temperatur der Kohlensäure auf 45° C wird der kritische Punkt überschritten und die Flüssigkeit geht allmählich in die Dampfphase über. Das unter Druck stehende Gas kann anschließend durch ein Ventil entweichen, worauf das Präparat zur Beobachtung bereit ist. Die Methode ist hauptsächlich bei stereoskopischen Untersuchungen komplizierter räumlicher Gebilde von großem Nutzen. In der normalen Schnittmethode wird heute auf ein Herauslösen des Einbettungsmittels verzichtet, um die beim Verdampfen des Lösungsmittels auftretenden Phasengrenzen zu vermeiden.

Die hier gegebene Zusammenstellung der verschiedenen Entwässerungsmethoden zeigt, daß diese noch in keiner Weise vollkommen sind. Soll das zu Beginn erwähnte Ziel, ein Äquivalentbild der lebenden Zelle zu erhalten, erreicht werden, so muß auch den Vorgängen, die bei der Entwässerung die Struktur beeinflussen, vermehrte Beachtung geschenkt werden.

Literatur

1. BERNHARD, W.: Exp. Cell Res. **8**, 248 (1955).
2. BAHR, G. F., G. BLOOM and U. FRIBERG: Exp. Cell Res. **12**, 342 (1957).
3. VATTER, A. E.: J. appl. Physics **28**, 1381 (1957).
4. SYLVÉN, B.: "Freezing and drying", p. 169. Symposium London, 1951.
5. WOLKEN, J. J.: J. Protozool. **3**, 211 (1956).
6. MÜLLER, H. R.: J. Ultrastr. Res. **1**, 109 (1957).
7. HANCOX, N. M.: Exp. Cell Res. **13**, 263 (1957).
8. STEERE, R. L.: J. biophys. biochem. Cytol. **3**, 45 (1957).
9. ANDERSON, T. F.: C. R. 1, Congrès int. Micr. électr. p. 567. Paris 1953.

Problems in methacrylate embedding

DAN H. MOORE

The Rockefeller Institute New York 21, N. Y.

Although certain epoxy resins have been demonstrated to be excellent embedding media for many materials, electron microscopists will continue, no doupt, to use methacrylate, for many purposes. If properly handled methacrylate is an excellent embedding medium and has the advantage of rapid penetration and consequently a much shorter delay before specimens can be examined in the electron microscope. A study of acrylic polymerization, however, indicates that the embedding procedure employed in most laboratories would not be expected to yield satisfactory results. Much of the tissue derangement, vacuolization, and distortion seen in the electron microscope result from incomplete penetration of the methacrylate, non-uniform polymerization or even chemical interaction of the methacrylate with tissue components. Non-uniform polymerization will cause uneven contraction and therefore shifting and tearing of fine structure. The final ratio of solid matter to plastic in cells or cell structures is directly related to the original water content. The density of tissue thus affects the cutting properties and preservation of structure. Local variations in hardness due to the tissue density or incomplete polymerization may lead to distortion, compression, or chatter during microtomy.

Polymerization

The molecular weight of monomeric butyl methacrylate is 142. Addition-polymerization may be caused by heat, light, short wave irradiation, or chemical initiators. Linear polymer chains are formed varying in length all the way from a few monomeric elements to several millions depending upon conditions; however, most of them fall within much narrower limits.

A stretched out molecule of polymethacrylate would have a length of 1500 to 3000 Å. These molecules are, however, usually arranged in random intermeshing coils. If the tissue structure alone is considered, the plastic molecules should not cause physical destruction during controlled polymerization. Brittle or charcoal-like areas are sometimes encountered which indicate that chemical properties of the tissue network may either prevent permeation or interfere with polymerization. Wetting properties or capillarity is probably an important factor in penetration, and the amount of bound osmium may also be involved.

Initiators, though often referred to as catalysts or accelerators, are really neither. Unlike catalysts, these agents are consumed in the reaction and do not affect the rate of chain growth beyond the initial stage. Once a chain is initiated it goes to completion in approximately 10^{-3} seconds. Although the synthesis of an individual polymer molecule from unreacted monomer to polymer occurs within such a short time, the overall conversion of monomer to polymer may require hours. At any stage the reaction mixture consists almost entirely of unreacted monomer and high polymer. The proportion of actively growing chains is ordinarily very small.

The commonly used peroxide initiators decompose slowly at temperatures of 40 to 100 ° C, with release of free radicals:

$$\text{C}_6\text{H}_5\text{—C(=O)—O—O—C(=O)—C}_6\text{H}_5 \longrightarrow \text{C}_6\text{H}_5\text{—C(=O)—O} \cdot = \text{I} \cdot = \text{R} \cdot.$$

Some initiators may be split by chemical action as well as by heat or irradiation; e. g., dimethyl aniline may be used to split benzoyl peroxide into its two free radicals. The ideal initiator and conditions for its optimum use probably should be determined for each type of tissue. Luperco CDB (2,4-dichlorobenzoyl peroxide with dibutylphthalate) has given good results in many cases.

The steps of polymerization may be described by the following equations, in which I = initiator, $I \cdot = R \cdot$ = initiator free radical, M = monomer, and $M_1 \cdot = R M \cdot$ (i.e., a radical formed by the combination of initiator radical with monomer):

$$(1)\ \text{I} \xrightarrow{\ \text{U. V., heat, chemicals}\ } 2\,\text{R}$$
$$(2)\quad \text{R} \cdot \ + \text{M} \to \text{M}_1 \cdot$$
$$(3)\quad \text{M}_1 \cdot \ + \text{M} \to \text{M}_2 \cdot$$
$$(4)\quad \text{M}_n \cdot \ + \text{M} \to \text{M}_{n+1} \cdot$$

The rate of initiation is directly proportional to the concentration of initiator and the velocity of its dissociation as influenced by temperature, radiation, etc. The rate of conversion, however, is dependent upon the square root of the concentration of initiator. Thus degree of polymerization (chain length) may vary within wide limits since it is determined essentially by the ratio of the rate of conversion to the rate of initiation. Termination of chains may occur either by coupling (where the active ends of two growing chains join), or disproportionation (where a hydrogen atom from one growing chain is transferred to another growing chain, leaving the donor with a terminal double bond, while the receiver is terminated by the acceptance of the hydrogen atom). The former is dominant. As the viscosity increases, the migration of chains is reduced, and coupling cannot readily occur. Chain growth continues, however, since the movement of monomer is relatively less affected. Chain length diminishes with higher temperature, and the percentage conversion to polymer increases, making for a more brittle plastic. Usually the active center remains on the same molecular chain. Chain transfer agents may remove this active center, shorten the polymer length, and at the same time start a new polymer. This reduces the degree of polymerization and a higher percentage of final conversion occurs. Chemicals with sulfhydryl groups such as the mercaptans may act as chain transfer agents, because the SH reacts with the growing radical as follows: $M \cdot + SH \to MH + S$. Transfer agents provide a means of controlling chain length.

Because polymerization depends upon the action of a small number of active molecules during a long period of time, it is extremely sensitive to traces of substances active as initiators or inhibitors. Agents yielding concentrations of active centers in order of 10^{-12} moles may alter the course of the reaction (1). It is, therefore, imperative to maintain controlled standard conditions in the preparation of biological embedments. Much of the difficulty now prevailing probably could be eliminated through a more careful procedure. This includes the removal of dissolved oxygen and water. Oxygen may combine with low molecular weight polymer to form polymeric peroxide.

$$\text{M}_n \cdot + \text{O}_2 \ \to \ \text{M}_n\text{—O—O} \cdot. \tag{1}$$

This in turn reacts with more monomer.

$$\text{M}_n\text{—O—O} \cdot + \text{M} \ \to \ \text{M}_n\text{—O—O—M} \cdot. \tag{2}$$

The speed of reaction 1) is greater than the usual reaction $M_n \cdot + M \to M_{n+1}$, but the speed of reaction 2) is very slow and in this way O_2 acts as an inhibitor. As polymerization proceeds, however, chains of polymeric peroxide may split into free radicals.

$$R—M—O—O—M—R \to 2 [R\ M\ O \cdot].$$

These additional very reactive free radicals may cause a violent polymerization and consequent explosion of embedding. Thus the presence of molecular oxygen acting both as an inhibitor and accelerator, provides a very unstable and uncontrolled condition for polymerization.

Purification and handling of the methacrylate

Methacrylate is shipped commercially with hydroquinone inhibitor added, and since oxygen improves the inhibition of the hydroquinone the methacrylate is aerated to insure storage without polymerization. Before use, however, the inhibitor should be removed. Hydroquinone is removed by washing the methacrylate several times with equal parts of 2% NaOH, followed by at least three washings with equal parts of distilled water to remove the NaOH which may act as an initiator. The solubility of water in methacrylate is about 1% by weight. The water can be removed by filtration through anhydrous Na_2SO_4, and subsequent storage over molecular sieve, which is a strongly selective dehydrating agent, and will insure complete dehydration. Water tends to give the methacrylate monomer and polymer a clouded appearance. In order to remove all dissolved oxygen dried methacrylate should be evacuated to 10 mm Hg and flushed with nitrogen in a desiccator jar. It may then be stored under dry nitrogen (not more than 20 parts per million O_2) in a brown jar at 5° C where it may be kept pure for long periods. Refractive index measurements with an interferometer indicate only slight changes after several months. The refractive index is a most sensitive indicator of polymerization. An increment of less than 3×10^{-4} is considered insignificant.

Embedding procedure

The fixed and dehydrated tissue is soaked in two changes of pure monomer for 1 hr. The tissue is then placed in dry gelatin capsules which have been filled with a mixture of monomer and Luperco. This monomer-luperco solution should previously be thoroughly dried and deaerated. The gelatin capsules are covered and a small hole is punched in the cap. These are then set in small test tubes and placed upright in a desiccating jar which is slowly evacuated over a period of at least 1 hr. This removes free oxygen from the tissue, the methacrylate, and the gelatin capsule. It is necessary to remove the air very slowly, otherwise the tissue may be damaged. The vacuum is then slowly released by introducing nitrogen, and each glass tube is immediately stoppered. This leaves the gelatin capsules in a nitrogen atmosphere. The castings are polymerized at 40 to 45° C or under ultraviolet light for 18 hr.

Simplified procedure

If the use of nitrogen or other inert gas is inconvenient, castings are greatly improved simply by evacuating the methacrylate after the inhibitor and water have been removed. Upon evacuation the methacrylate undergoes vigorous bubbling which indicates removal of dissolved air. Evacuation should continue until the bubbling ceases. If not stired or poured excessively the methacrylate may be handled in the presence of air without undue reabsorption of oxygen.

The unpredictable castings, containing bubbled, distorted areas in the tissue, and plastic of varying hardness, previously obtained were found to be due to dissolved oxygen, water, and other impurities which readily enter the methacrylate unless precautions are taken at all stages of the process.

References: *1.* WALLING, C., and E. BRIGGS: J. Amer. chem. Soc. **68**, 1141 (1946).

Open face flat embedding technique

E. Borysko[*, **]

New York University College of Dentistry. The Murry and Leonie Guggenheim Foundation. Institute for Dental Research, New York

Preliminary studies of bacterial structure revealed, that these organisms were susceptible to damage during the polymerization of the acrylic embedding matrix, when a cure temperature of 60° C was used (*1*). Utilizing the principle that the internal stresses responsible for the damage could be relieved by maintaining the matrix in as fluid a condition as possible to the termination of the polymerization reaction (*2, 3*), the temperature of the curing oven was raised until, at a temperature of 95° C, polymerization damage of the bacteria could not be detected with an optical microscope.

This work was done in a completely enclosed flat chamber (*4, 5*) (Fig. 1 A). When polymerization temperatures of 95° C were used, however, this method proved to be highly impractical because of the large number of bubbles that developed during the curing of the matrix. Efforts to eliminate the bubbles led to the open face embedding method described here.

The following specimens were selected:

Bacteria: Escherichia coli. Metazoa-tissue culture: Fibroblast D-189, Fibroblast MaF, HeLa, Rat tooth germ. *Protozoa: Chaos chaos, Amoeba proteus, Tetrahymena sp., Paramecium aurelia, Blepharisma undulans.*

A total of 47 embeddings were made. Each flat embedding was equivalent to more than twenty gelatin capsule embeddings in terms of the quantity of specimen material available for sectioning.

The bacteria were grown as scattered clones on a thin film of agar on coverslips. The amoebae (*C. chaos* and *A. proteus*) were allowed to fasten to a microscope slide or to a thin film of agar on a microscope slide and were carefully examined with an optical microscope prior to fixation (*6*). *Tetrahymena* and *Paramecia* were included in the amoebal preparations as food organisms. Many of them were trapped alive in large food vacuoles in the amoebae or were enmeshed in the surface of the agar. The HeLa cells and the fibroblasts were grown on coverslips in Porter flasks, the rat tooth germ cells in Carrel flasks. They were fixed and embedded in situ. *B. undulans* cells were concentrated in a drop of water, fixed and suspended in warm (fluid) agar. The agar suspension was then poured on to a microscope slide and allowed to set before dehydrating and embedding.

All specimens were *fixed* for three to ten minutes with 1% OsO_4 in distilled water at room temperature. No attempt was made to adjust or control p_H, tonicity, osmolarity or temperature of the fixative. It has long been the experience of the author that a simple solution of osmium tetroxide in distilled water preserved the cells perfectly as judged by the optical microscope. *Dehydration* was accomplished by acetone in tap water (50, 70 and 100%) with three changes in pure acetone, allowing a minimum of 5 min in each change. Acetone was used instead of ethanol because it is soluble in both the monomer and the polymer. The possible adverse effect of trace amounts of acetone on the "cuttability" of the embedded tissues was avoided by subjecting the embeddings to high vacuum for several hours, so that the acetone diffuses to the surface and evaporates. The polymer-insoluble ethanol cannot be removed by this method (*7, 8*).

The cells were *impregnated* in three changes (minimum 5 min) of the following inhibitor-free monomer mixture[1] 35—40% methyl methacrylate, 60—65% n-butyl methacrylate, 0.50% (wt/vol.) Luperco CDB paste, 0.01—0.05% (wt/vol.) Aerosol OT.

The proportions were empirically adjusted to produce a polymer sufficiently hard so that a sheet about 2 mm thick will break with a brittle fracture, and yet soft enough so that relatively thin unshattered sections can be cut with a razor blade. It is important to note that the properties of this polymer partly depend on the thermal conditions of polymerization and the method used for preparing the partially polymerized embedding syrup. The Aerosol OT in the mixture acts as a parting agent so that adhesion between the embedding matrix and the mold is minimized (*7*). Concentrations less than 0.01% were not effective and concentrations above 0.05% would inhibit the polymerization reaction.

An *embedding* syrup was prepared by partially polymerizing a quantity of the monomer mixture in a drawn out test tube at 70° C with frequent violent agitation as previously described (*4*). Embedding molds were made by folding a rectangle of aluminium foil $1^1/_4'' \times 3^1/_4''$ (3.2 × 8.3 cm) around a standard $1'' \times 3''$ (2.5 × 7.5 cm)

---

* Present address: Ethicon, Inc., Somerville, New Jersey.

** This work was supported by Research Grants D-368 and D-448 from the US Public Health Service, National Institutes of Health.

[1] The inhibitor was removed from the monomer by washing once with 20—30% NaOH and once with tap water. The monomer was then dried over $CaCl_2$ and filtered before adding catalyst and parting agent.

Suppliers: Monomers — Röhm & Haas Company, Washington Square, Philadelphia 5, Pa., Luperco CDB — Lucidol Division, Novadel-Agene Corp., 1740 Military Road, Buffalo, 5, New York. Aerosol OT — Product of American Cyanamide Company, Sold by Fisher Scientific Company, 635 Greenwich Street, New York 14, N. Y.

microscope slide to form a shallow tray, as shown in Fig. 1. In embedding the protozoa, the microscope slides with the specimens already fastened were used as bottom of the mold (Fig. 1 *B* and *C*). In the cases of the tissue culture cells and bacteria the coverslips were placed on a clean microscope slide, specimen side up (Fig. 1 *D*). The molds were then filled with the viscous embedding syrup, bubbles were teased out with a dissection needle and the trays were placed in a cold oven without any attempt to shield the surface from air. Then the temperature was allowed to rise to 95—105° C. Fig. 2 shows rise in temperature used in this work. The embeddings were allowed to "soak" at high temperature for 18—24 hr to drive the reaction to completion. When the embeddings

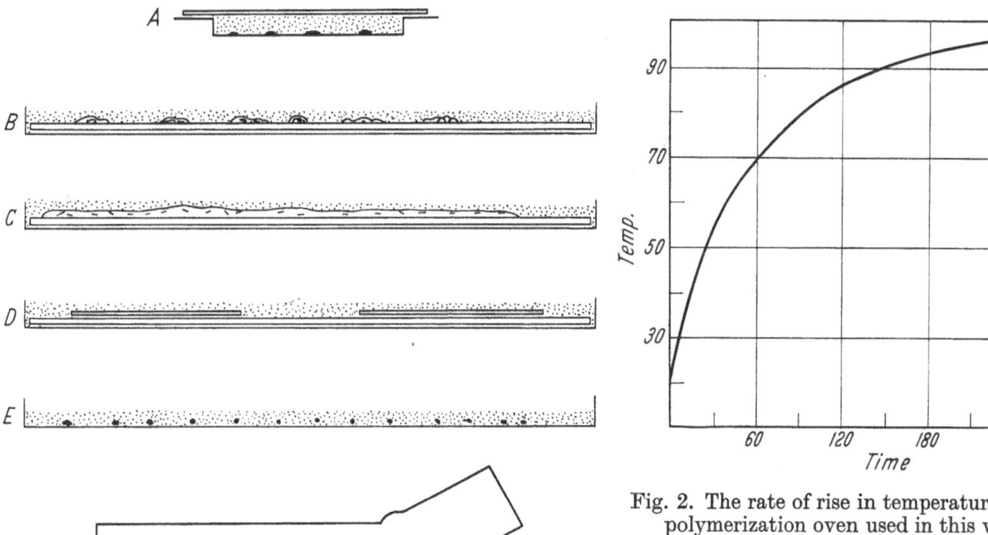

Fig. 2. The rate of rise in temperature of the polymerization oven used in this work

Fig. 1. Sectional views of various types of flat embeddings. The stippling represents the methacrylate embedding matrix. In *A* through *E*, the embedding mold consists of a shallow aluminium foil tray. *A*. Old method in which the tray was filled with embedding syrup and covered with a coverslip to shield the acrylic against the supposed inhibiting effect of oxygen. Bits of tissue were placed in the tray or, in other instances, tissue culture cells or bacteria were grown on the coverslip. Fixed, dehydrated and impregnated in situ and then placed on the embedding syrup, specimen side down. *B*. Open face embedding of amoebae fastened to a microscope slide while alive and fixed, dehydrated and embedded without any mechanical handling. One should use 50 to 100% excess syrup to compensate for evaporation and syrup flowing under the slide. Embedding thickness 1—2 mm. *C*. Open face embedding of ciliates that had been dispersed in warm (45—50° C) fluid agar after fixation. The agar was cast on to a slide in the form of a thin film which served to anchor the ciliates firmly during dehydration and embedding. *D*. Open face embedding of tissue culture cells grown on coverslips. After fixation, dehydration and impregnation, the coverslips were placed, specimen side up, on a slide. *E*. Open face embedding of bits of tissue in a simple aluminium foil tray. This method is probably the best for specimens that do not lend themselves to direct optical microscopic examination. *F*. Open face embedding of tissue culture cells grown in Carrel flasks. The cells were processed in situ and the flask was broken away from the methacrylate after polymerization of the matrix

had cooled to room temperature, the acrylic was separated from the glass by cooling the glass with a fragment of dry ice.

Then the embedded specimens were examined with an optical microscope and checked for hardness by cutting thin slices from their edges and by indenting with a thumb nail. All specimens appeared to be well preserved except for the shrinkage that invariably occurs during dehydration (*2, 3*). The polymer was of a good cutting consistency and there was no evidence of layering or rubberiness in the clear plastic. After subjecting the embeddings to high vacuum (see above) they were trimmed and sectioned as described earlier (*9*). A special clamp[1] was used for holding the flat embeddings. Sectioning was done on a Porter-Blum microtome (*10*) with a diamond knife (*11*)[2, 3].

For many years, the author was of the belief that oxygen was detrimental to the proper polymerization of acrylic embedding materials. This belief was supported by a number of studies (*5, 12, 13, 14*) and the general observation that a rubbery layer always appeared on the top of gelatin capsule embeddings when polymerization temperatures of 45—50° C. were used. This

---

[1] The flat embedding clamp is now commercially available as an optional attachment for the Servall microtome manufactured by Ivan Sorvall, Inc., Norwalk, Conn.

[2] The author is indebted to Dr. H. Fernández-Morán for the gift of a diamond knife.

[3] The sections were mounted serially on collodion films on slot-type grids and examined in a modified RCA EMU 3A electron microscope, using a one mil platinum objective aperture.

rubbery layer was presumed to be due to the formation of a polyperoxyde by a reaction between the monomer and dissolved oxygen or to the chain termination effect of molecular oxygen (*5, 7, 12, 13, 14*). However, when high polymerization temperatures were used (60—80° C), it was

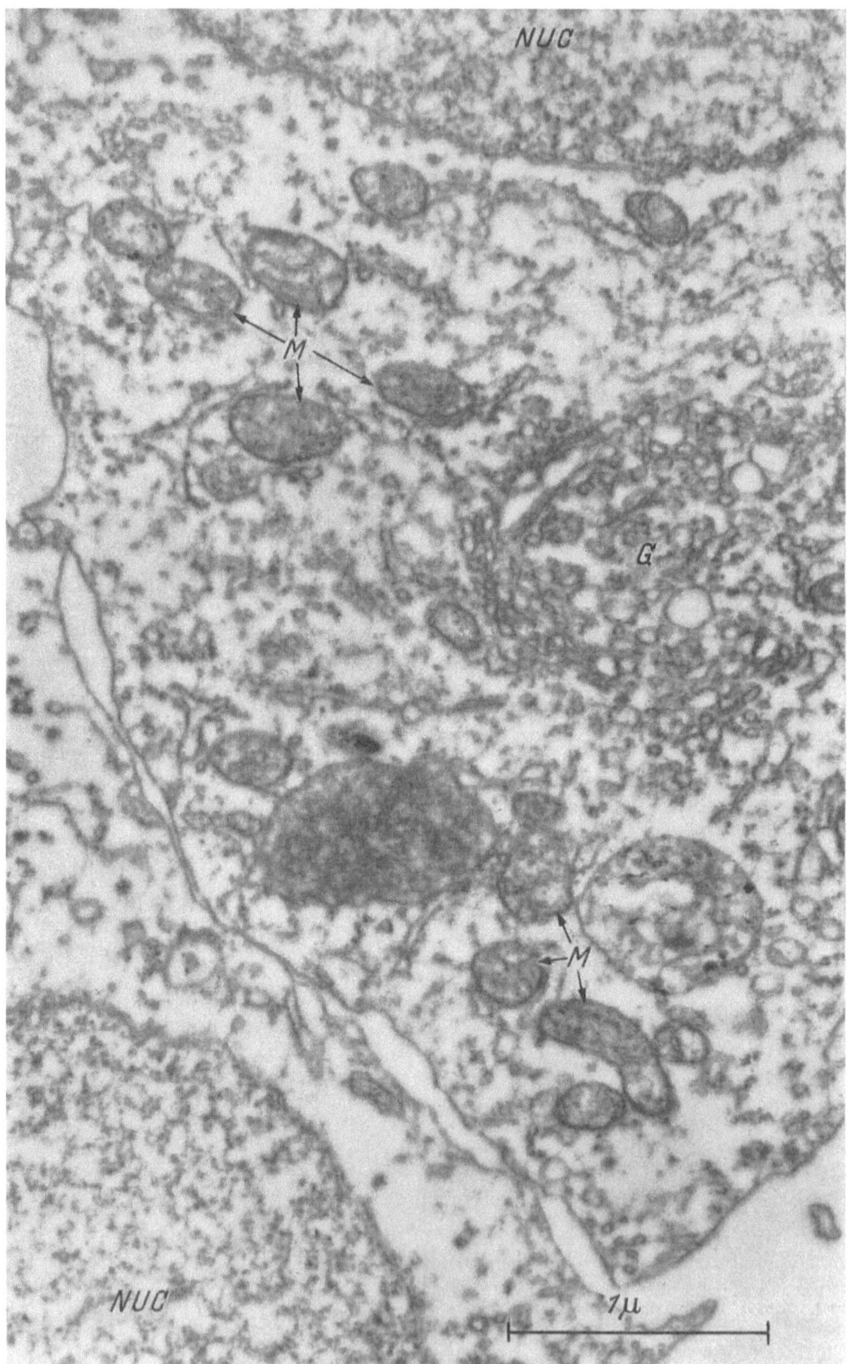

Fig. 3. Section of HeLa cells grown on coverslips in Porter flasks, showing portions of the nuclei (*NUC*) of two cells, numerous mitochondria (*M*) and the Golgi zone (*G*) of one of the cells. 35000 ×

noticed that the rubbery layer became very thin or was absent. Furthermore, it was also observed that rubbery layers formed at low temperatures could be hardened by subjecting the embeddings to temperatures above 60° C for several hours. It is quite probable that, at elevated temperatures, the inhibiting polyperoxide decomposes and acts as a catalyst for the completion of the polymerization reaction (*14, 15*). According to MAYO (*16, 17*), a polyperoxide molecule may be de-

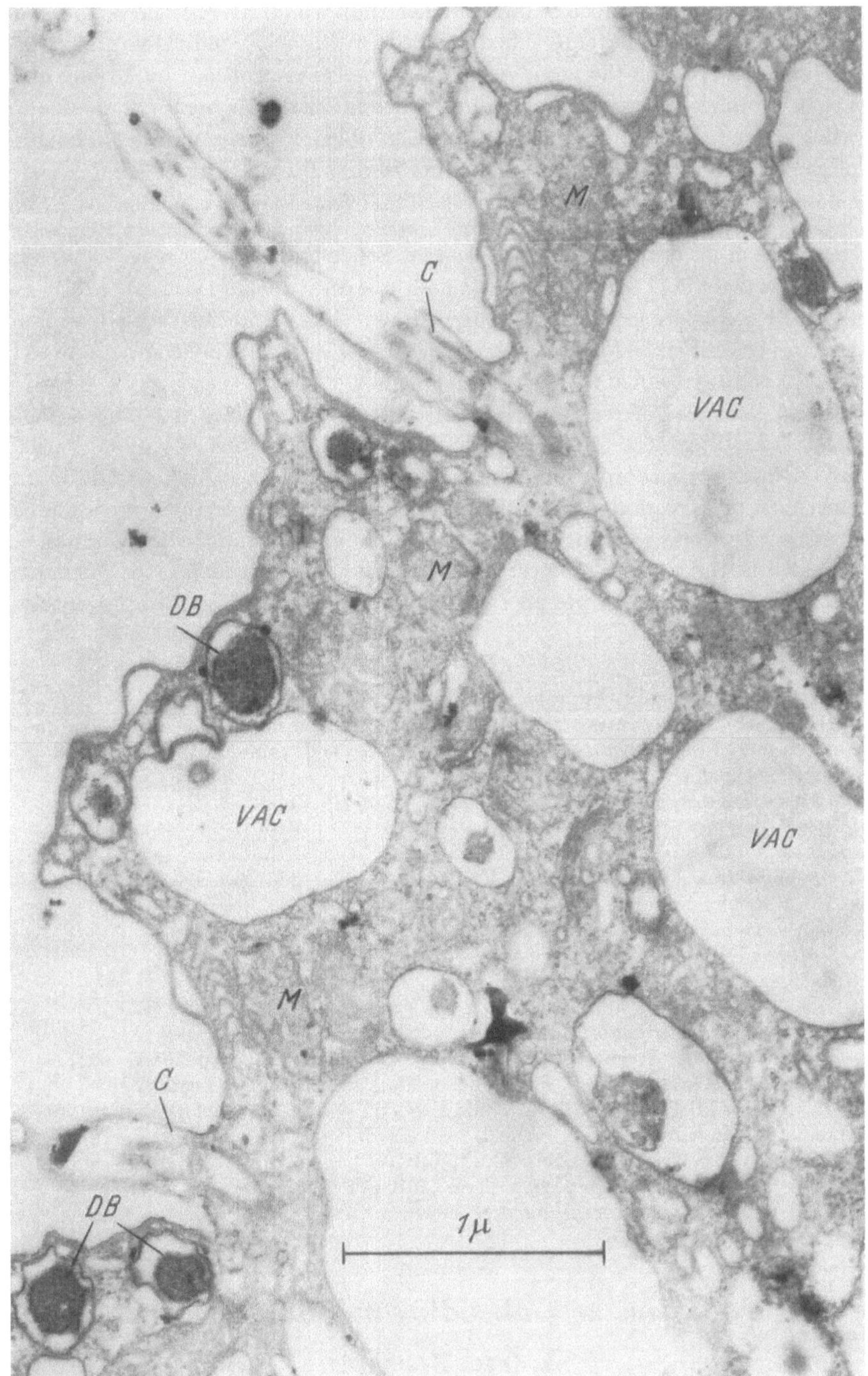

Fig. 4. Cross section of a pigmented ciliate, *Blepharisma undulans*, showing details in the cortical region, including the basal portions of two cilia (*C*), dense cortical bodies (*DB*), numerous vacuoles (*VAC*) and mitochondria (*M*). The low density of the mitochondria is a characteristic of this species. 35000 ×

composed into two free radicals by the action of heat or light. These radicals are capable of initiating two polymerization chains. Thus, although it is quite true that oxygen may act as a strong inhibitor in the polymerization of methacrylates in one set of conditions, it may also act indirectly to accelerate the reaction under other conditions. Indeed, several workers have considered a certain amount of oxygen to be essential for the thermal polymerization of acrylates (18, 19). It is therefore obvious that if the monomer fails to polymerize because of oxygen inhibition, it is only necessary to alter the conditions in a manner that would permit the oxygen to act as an initiator. The realiability of the cure schedule used in this work indicates that it is not at all necessary to replace oxygen in the monomer with nitrogen, as proposed by Moore and Grimley (20). If oxygen inhibition were an embedding problem, then frequent polymerization failures would be encountered. The author has not been troubled by polymerization failure since the adoption of the prepolymerization technique more than eight years ago (4).

The vigorous stirring used in this technique counteracts the effect of oxygen inhibition and produces a homogeneous embedding mass (2, 7). When highly viscous embedding syrup is used, polymerization will always go to completion, regardless of the final embedding temperature used in the range from room temperature to 110° C.

Contrary to a widely held superstition, the high temperatures in this method did not produce any optically visible changes in the appearance of the cells. In addition, the electron microscopic images of sections of cells embedded at high temperatures were entirely comparable to the images of well preserved cells embedded at lower temperatures.

The following examples of results obtainable with the open face flat embedding technique (Fig. 3 and 4) are representative of the type of pictures that one can rely upon. Selection of embeddings for sectioning was completely unnecessary because of their highly uniform quality with respect to specimen preservation and "cuttability". The flat shape enabled precise predetermination of the plane of sectioning through the specimen and even of the particular small portion of a cell to be included in the sections, as described elsewhere (9). In general, the pictures shown are from the first set of sections cut from an embedding. No attempt was made to obtain high resolution.

References

1. Borysko, E., and H. Hoffman: In preparation.
2. — An evaluation, by phase contrast microscopy, of the methacrylate embedding technique used in the electron microscopy of cells. Doctoral Dissertation, The Johns Hopkins University 1955.
3. — J. biophys. biochem. Cytol. Suppl. 2, 3 (1956).
4. — and P. Sapranauskas: Bull. Johns Hopk. Hosp. 95, 68 (1953).
5. Price, C. C.: J. Polymer Sci. 1, 83 (1946).
6. Borysko, E., and J. Roslansky: Ann. N. Y. Acad. Sci. Biology of Amoeba, 1958.
7. Lang, R.: Rohm & Haas Company, Bristol, Pa. Personal communication, 1955.
8. Karrer, H.: J. biophys. biochem. Cytol. 2, 38 (1956).
9. Borysko, E.: J. biophys. biochem. Cytol. Suppl. 2, 15 (1956).
10. Porter, K. R., and J. Blum: Anat. Rec. 117, 685 (1953).
11. Fernández-Morán, H.: J. biophys. biochem. Cytol. Suppl. 2, 29 (1956).
12. Barnes, C. E., R. M. Elofson and G. D. Jones: J. Amer. chem. Soc. 72, 210 (1950).
13. Anonymous: Rohm & Haas Comp. Bull. 169, Philadelphia, Pa. 1953.
14. Riddle, E. H.: Monomeric Acrylic Esters. New York: Reinhold Publ. Corp. 1954.
15. Dahlstrom, B.: Rohm & Haas Comp. Philadelphia, Pa. Personal communication, 1958.
16. Mayo, F. R.: Stanford Research Institute, Menlo Park, California. Personal communication, 1958.
17. — and A. A. Miller: J. Amer. chem. Soc. 80, 2493 (1958).
18. Dupont de Nemours Co.: Ind. Eng. Chem 28, 1160 (1936).
19. Schulz, G., V. and F. Blaschke:Z. physik. Chem. 50B, 305 (1941).
20. Moore, D. H., and P. M. Grimley: J. biophys. biochem. Cytol. 3, 255 (1957).

The use of epoxy resins as embedding media for electron microscopy

A. Birch-Andersen

Statens Seruminstitut, Copenhagen

During a study of fixation of bacteria for sectioning and electron microscopy using the conventional methacrylate embedding technique (1) a considerable number of blocks showing almost a 100% burst and swollen cells were noted. By carefully controlling each step in the procedure by means of light microscopy it could be demonstrated that the swelling and bursting occurred during the actual hardening of the methacrylate (2).

This observation is not new, however, BORYSKO and SAPRANAUSKAS (3) had reported that certain tissue preparations are completely disintegrated if embedding is started in liquid rather than prepolymerized and highly viscous methacrylate, and WEINREB (4) recommended ultraviolet irradiation of the capsules for initiating the polymerization in order to reduce polymerization artefacts. BORYSKO (5) later studied these artefacts of polymerization damage, and obtained better results when using prepolymerized monomer, a high concentration of initiator (benzoyl peroxide or a similar compound) and a high temperature of polymerization.

We tried several of these modifications, but never achieved completely satisfactory results. We felt that our difficulties arose from uneven polymerization throughout the block during the hardening process; as if the very cells we wanted to embed could act locally both as initiators and inhibitors of the curing process. Also the shrinkage which occurs during the setting seemed to be of importance. These concepts now seem to be widely accepted, and the general background for the attempts to find another and more suitable embedding substance has recently been adequately given by BIRBECK and MERCER (6). Simultaneously with our work KELLENBERGER et al. (7) have introduced the use of polyesters with similar intentions, and the results obtained with this type of embedding medium will be reviewed later in this session.

When looking for another plastic for our purpose it was obvious that ideally it should fulfill the following requirements: It should have in the monomer state the low viscosity of the monomers of methyl- and butyl-methacrylate; it should be fully compatible with alcohol so that the conventional fixation and dehydration procedures could be used; it should in the cured state have the same sectioning properties as polymerized methacrylate and the same low electron-scattering capacity; and at the same time it should harden in a homogeneous manner without changing volume. The material which gave us the most promising results completely satisfied only one of these requirements, however. It was a highly viscous epoxy compound (Hysol XL 2010-911) which, on addition of diethylene triamine (both ingredients obtained from Houghton Laboratories, Olean, N. Y.) hardened at about 60° C in 24—48 hr without changing volume and formed a light yellow resin from which we were able to cut sufficiently thin sections for electron microscopy.

The picture we obtained when using this resin for the embedding showed an improved preservation of structures (2). The cell wall remained evenly attached to the cells indicating that very little or no shrinkage had occurred. The cytoplasm appeared very smooth and completely without fissures or cracks indicating that no swelling had taken place. In the nuclear regions some dense tubular structures were observed. These structures had not been observed in the methacrylate embedded specimens, but they were constantly found in this material when we used the type of osmium fixation, which we at that time for various reasons considered the optimal condition of fixation. Finally it was observed that the highly viscous monomer apparently failed to penetrate these dense structures.

From these and many similar preparations the use of epoxy resins for embedding appeared rather promising, but on the other hand our material satisfied too few of the requirements needed for more general use. It was for instance not easily compatible with alcohol. We could, however, by intensive stirring prepare a suspension containing about 30% of alcohol, and this was actually used as the first soaking bath. The final mixture with hardener was highly viscous in the monomer state. The blocks obtained were rather hard and brittle and thus difficult to cut, and the polymer showed a somewhat higher electron-scattering capacity than methacrylate. In addition the resin did not sublime to the same extent as the methacrylate in the electron beam so extremely thin sections were needed. We did, however, succeed in obtaining blocks with improved cutting properties by adding some dibutyl phthalate to the resin as a plasticizer. At this point our preliminary results were published (2).

A few months later at the Conference on Electron Microscopy in Stockholm, September 1956, GLAUERT and BRIEGER were able to present some very promising results (8). The epoxy resin they used was based on the Ciba epoxy resin, araldite in combination with an aliphatic anhydride as hardener. This mixture had been worked out by GLAUERT et al. and has been described in a series of papers (*8, 9, 10*). It was found to be freely soluble in alcohol, and it has a much lower viscosity than ours. This is because the hardener, which is of high molecular weight, has to be added to the resin in almost equal amounts to ensure the right proportions for hardening. In our mixture, using diethylene triamine, we only needed 10% of the amine as a hardener. The high temperature, which is needed for curing when using an anhydride, could be reduced to a suitable value for the embedding process i. e. some 50—60° C, by adding an amine accelerator. The hardness of the final blocks is controlled by adding suitable amounts of a plasticizer (dibutyl phthalate).

Glauert and Glauert (*10*) have stated the following general formula of epoxy resins:

$$CH_2\!\!-\!\!CH\!\!-\!\!CH_2\!\!-\!\!O\!\!-\!\!\left[ROCH_2\!\!-\!\!\underset{\displaystyle |}{\overset{\displaystyle OH}{CH}}\!\!-\!\!CH_2\!\!-\!\!O\right]_n\!\!-\!\!ROCH_2\!\!-\!\!CH\!\!-\!\!CH_2$$

where R is commonly diphenylpropane,

$$\bigcirc\!\!-\!\!\underset{\displaystyle CH_3}{\overset{\displaystyle CH_3}{C}}\!\!-\!\!\bigcirc$$

and *n* an integer.

The resins are polyarylethers of glycerol with terminal epoxy groups. They can be obtained ranging from viscous liquids to fusible solids and can be cured by a number of setting agents which add across the epoxy groups to give three dimensional structures. The hardening can be made to occur with a volume shrinkage as low as 2% where the comparable figure for methacrylate is of the order of 15—20%. This small change in volume is due to the fact that the hardening is an addition process, and that the resin itself is highly associated.

On the other hand the presence of the many diphenyl propane groups may be responsible for the relatively high electron-scattering capacity of the epoxy resins. For this reason in our laboratory we have tried epoxy resins of the Shell epikote series which are said to be purely aliphatic. Our attempts have failed so far as these resins seem to set extremely rapidly at room temperature even when using anhydrides or citric and oxalic acids as curing agents.

The epoxy resins are generally used in industry as casting materials, surface coating agents and as adhesives. They also have good insulating properties.

Epoxy resins seem to have come into more general use as embedding media since the results on the araldite mixtures were presented. Dr. Glauert uses araldite mainly for sectioning bacteria, and so do we. Peters et al. (*11*) have used it for embedding He-La cells infected with vaccinia virus and Huxley (*12*), Robertson (*13*) and Birbeck and Mercer (*6*) for the embedding of various tissues. In particular Birbeck and Mercer (*6*) have carefully stated the advantages and disadvantages of the technique for this purpose.

Our own experience is mainly grained from our work on sectioning of bacteria, but a few examples from our work on sectioning of tissue from the small intestine of the mouse will also be presented. In general our embedding procedure and the mixture we use are similar to those described by Glauert and Glauert (*10*).

Araldite resin M	10.0 ml
Hardener 964 B	10.0 ml
Dibutyl phthalate	1.0 ml
Accelerator	0.5 ml

The resin, hardener and accelerator are obtained from the Aero Research Ltd., Duxford, Cambridge, England through our local Ciba company. We can broadly confirm what has been stated by Glauert and Glauert (*10*) and Birbeck and Mercer (*6*) in regard to soaking of the specimens, the time required for hardening of the blocks, and the actual cutting of the sections. The few examples to be shown serve as illustrations only, and in addition a more detailed description of the technique we have used will be given.

After fixation the bacteria are taken up in agar from which extremely small blocks (0.5—1 mm³) have been cut. For tissues also very minute blocks seem to be essential for good penetration of the embedding medium. Following normal dehydration through a series of alcohols we soak our specimens for one hour at 40—50° C in a 50% alcohol araldite mixture, the araldite prepared as above. Then we soak them for an hour in a freshly prepared araldite mixture at the same temperature, and repeat this step for an additional hour in fresh mixture. Clean, dry gelatine capsules are then filled with a freshly prepared araldite mixture, the specimens are introduced and allowed to soak in the capsules for at least 24 hr at room temperature before the curing of the blocks, which is carried out at 60° C. The blocks usually harden within 2—3 days. When changing from one soaking bath to another we have found it advantageous to remove superfluous monomer from the specimen blocks by filtration through a piece of filter paper on a small Büchner funnel to which a vacuum from an ordinary water jet pump is applied. After filtration the specimen blocks are scraped off from the filter paper by means of a steel spatula and put into the next bath or into the capsules. The usual care not to introducing air bubbles must be taken when filling the capsules. All our sections have been cut on the Sjöstrand microtome using glass knives, and the sections have been floated off on the surface of 20% alcohol in distilled water.

Generally speaking we find that the araldite mixture, although far superior to the original epoxy resin we used, is still not quite as easy to cut as methacrylate. Attempts to make sections which show the ripples mentioned by BIRBECK and MERCER (6), stretch better on the surface of the water by using the vapour of chloroform, xylene, dibutyl phthalate or other solvents have failed completely. We have the impression, however, that the sections cut better if we remove the static electricity produced during cutting simply by earthing the trough by means of a thin copper or platinum wire. We have not systematically tried other cutting angles on our knives or other cutting speeds as did BIRBECK and MERCER (6). Up till now we have always succeeded sooner or later in obtaining sufficiently good sections from our blocks after changing to a new knife.

The illustrations will demonstrate the type of results we obtain. For comparison an example of a methacrylate embedding will also be given.

Fig. 1 shows methacrylate embedded cells of Salmonella typhimurium fixed after the method of RYTER and KELLENBERGER (14). A rather pronounced shrinkage of the cells is seen, and the thin strands of the nucleoplasm seem to be aggregated to some extent. Small fissures or holes are also present in the cytoplasm. In contrast to this the same organism fixed and dehydrated in an identical manner but embedded in araldite (Fig. 2) shows a much better picture. Due to the prolonged fixation used the cells appear to show some shrinkage, but the cytoplasm is very smooth, free of fissures, holes or cracks, and the contents of the nuclear regions appear as delicate strands of finely dispersed material. This confirms RYTER's and KELLENBERGER's observations in polyester embedding (14).

In our study on bacteriophage infected E. coli cells (15) we used methacrylate embedding, and found it extremely difficult to get pictures of well preserved cells in the later stages of infection. Even reinfection which delays the ordinary time of lysis failed to improve this. We generally obtained pictures where the cells in addition to the phage infection were suffering from polymerization damage. In araldite embeddings E. coli cells at the same late stage of infection show no sign of polymerization damage. In the interior of the cells phage particles with well preserved hexagonal structures are present. This is in accordance with the results obtained by KELLENBERGER and RYTER in polyester embedded E. coli cells (16).

In view of these facts we find embedding of bacteria in araldite of advantage. The last two examples (Fig. 3 and 4) will demonstrate that it shows promise as an embedding material for sectioning tissues. Epithelial cells from the villi of the small intestine from the mouse have been used. The animals are fasted for 24 hr prior to decapitation and the method of fixation is the conventional 1% osmiumtetroxide (2 hr at room temperature). Sections with areas from 500 to 1 000 μ^2 have been cut. Such sections appear perfectly smooth without any indication of polymerization damage. No swelling of the nuclei has been seen. Adjacent cells always show good adhesion in contrast with methacrylate embeddings in which cracks between cells are frequently observed. The mitochondria show a very smooth general outline, and none of the other constituents of the cells show any obvious sign of polymerization artefacts. This can be seen on the two micrographs (Fig. 3 and 4), which show two different regions of epithelial cells in such a section.

Fig. 3 shows a part of a nucleus and its immediate viscinity. The nuclear membrane can only be faintly distinguished. The plasma membranes of the two adjacent cells are very distinct. A very dense mitochondrion in which the internal membranes are just resolved can also be seen. Fig. 4 shows a part of a cell particularly rich in mitochondria. The mitochondria here present an extremely smooth general outline. The mitochondrial matrix appears to be rather dense, and for this reason both internal and external membranes do not show up as distinctly as they do in methacrylate embeddings. The inner membranes are very tightly packed but their double con-

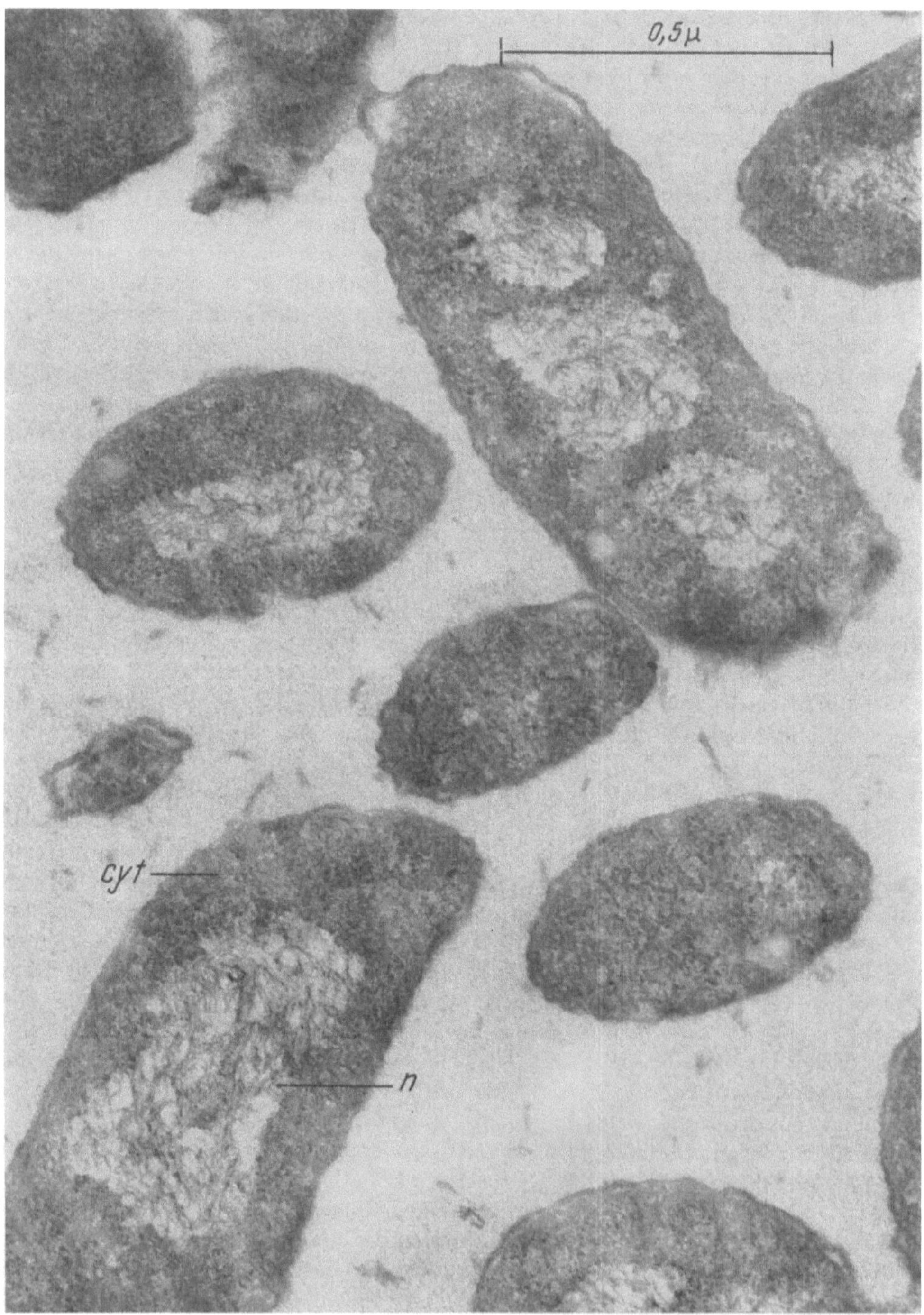

Fig. 1. Cells of Salmonella typhimurium. Methacrylate embedding. Osmium fixation according to RYTER and KELLENBERGER (14). The thin strands of the nucleoplasm (n) are aggregated to some extent. Fissures and holes present in the cytoplasm (cyt). Magnification 94,500 ×

tours as well as the double contours of the outer membranes can be seen. A part of a nucleus with its double nuclear membrane is also found in this field. In addition the endoplasmic reticulum is seen in between the mitochondria. It shows an abundance of very distinct PALADE granules.

In this paper an attempt has been made to draw attention to the use of araldite mixtures as an attractive alternative to the conventional methacrylate embedding. At present many laboratories

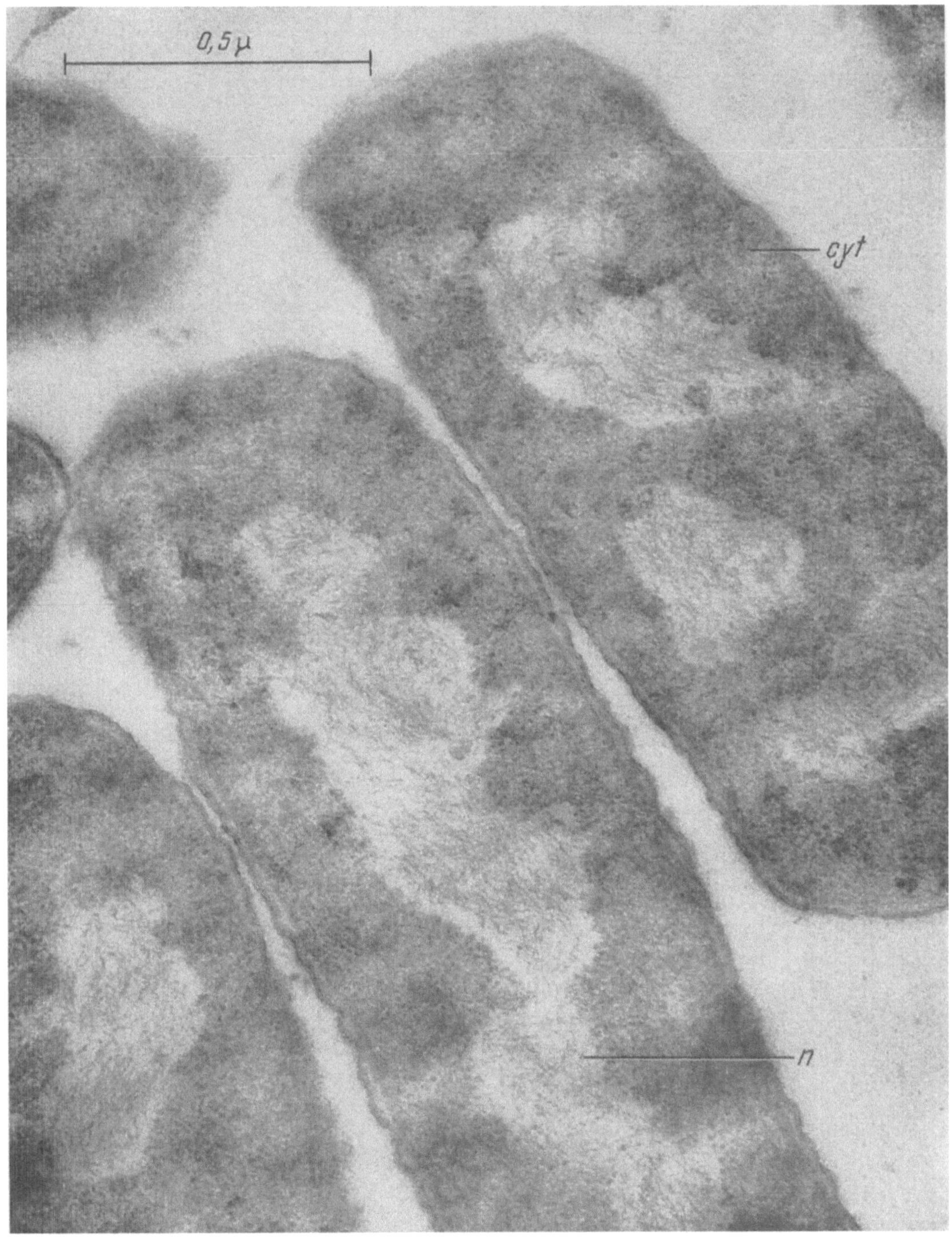

Fig. 2. Cells of Salmonella typhimurium. Araldite embedding. Osmium fixation according to RYTER and KELLENBERGER (14). The nucleoplasm (n) consists of finely dispersed, delicate strands. The cytoplasm (cyt) appears smooth. Magnification 94,500 ×

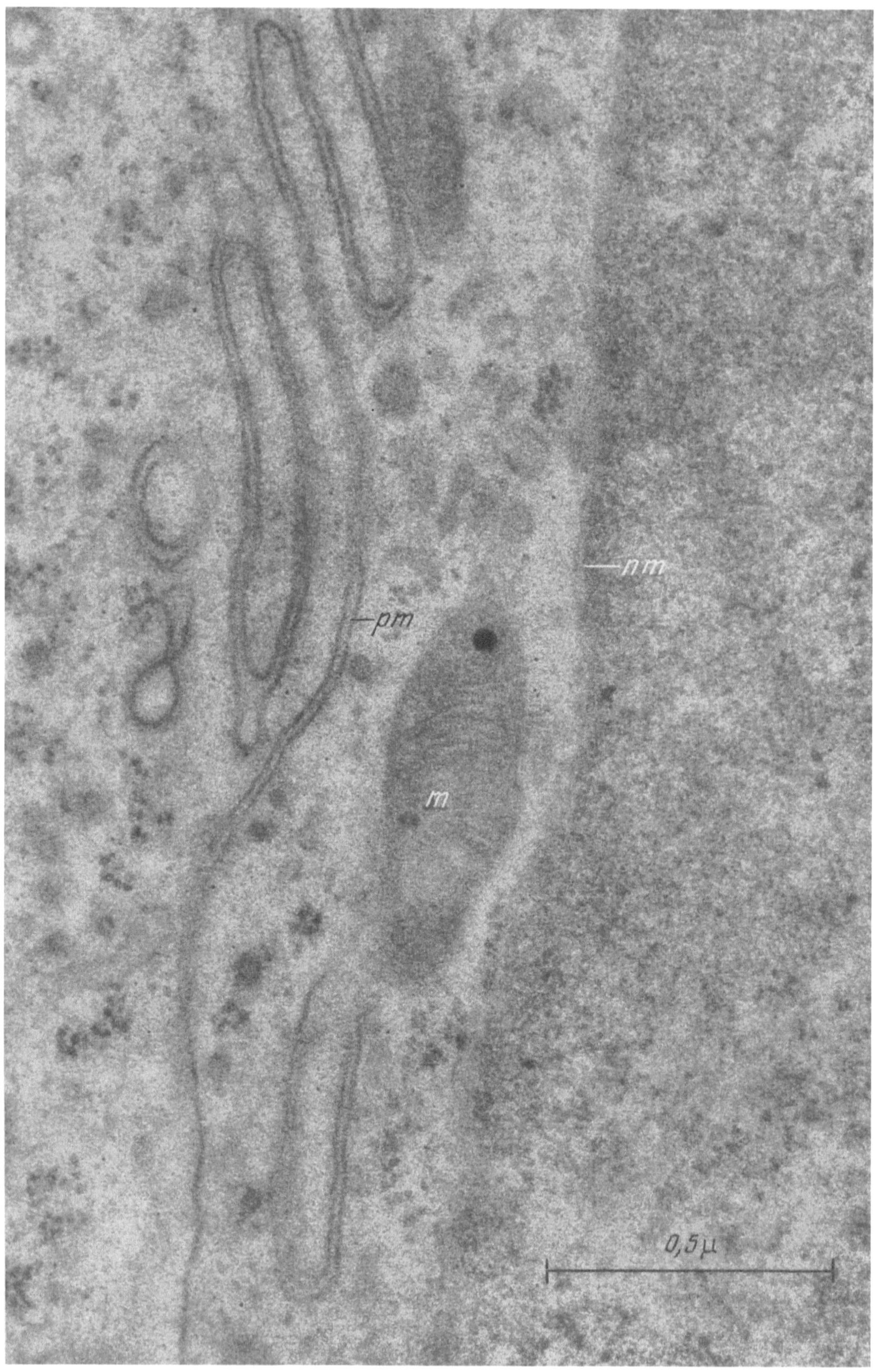

Fig. 3. Part of two epithelial cells of the central region of the villi of the small intestine from the mouse. Araldite embedding. Osmium fixation. Nuclear membrane (*nm*) only faintly distinguishable. Plasma membranes (*pm*) of the two adjacent cells are very distinct. The mitochondrion (m) appears very dense. Magnification 94,500 ×

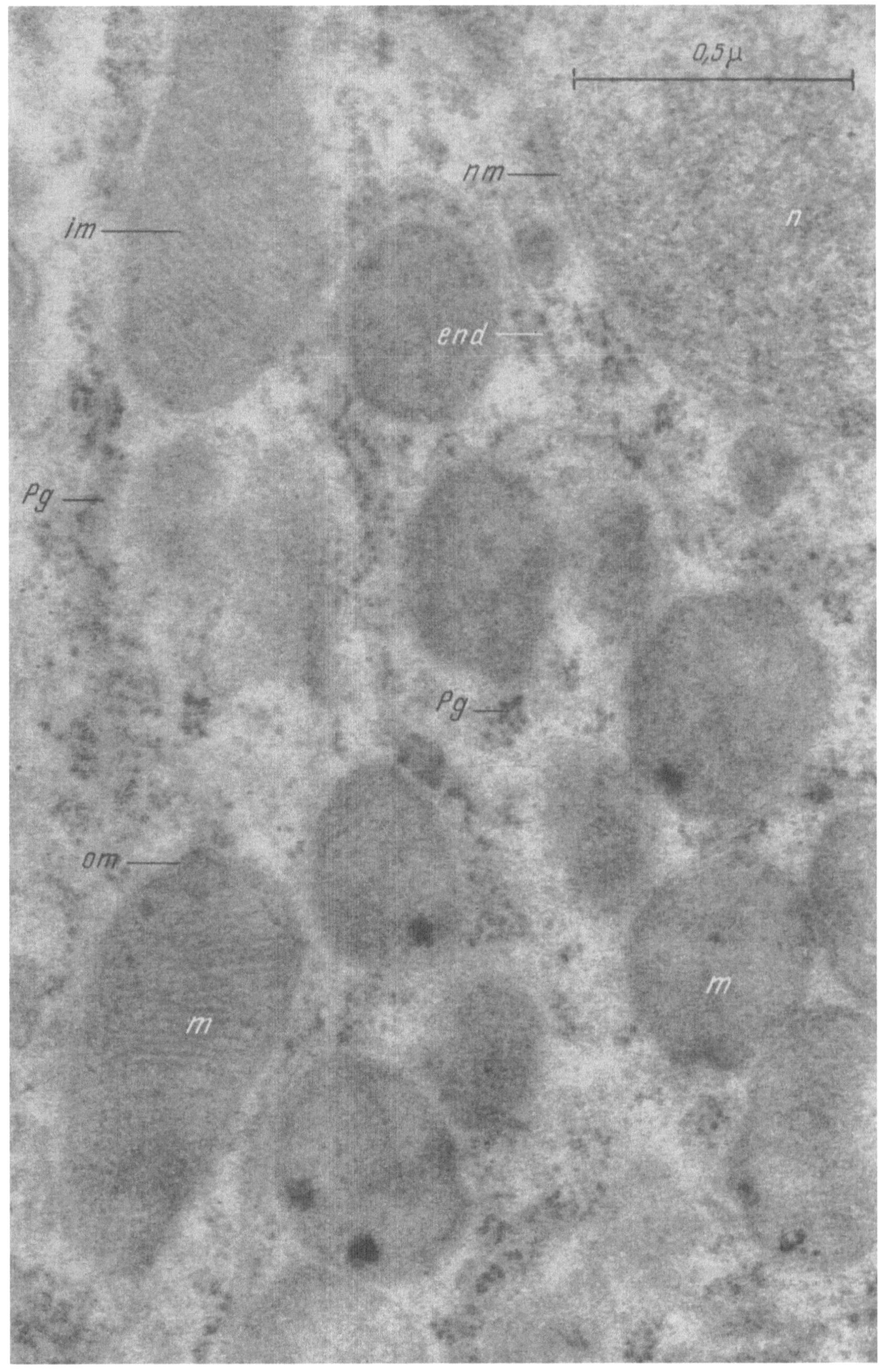

Fig. 4. Part of another epithelial cell from the same section as Fig. 3. The many mitochondria (*m*) present in this region all show an extremely smooth general outline. Mitochondrial inner membranes (*im*) and outer membranes (*om*) can be seen. The nucleus (*n*) shows a double contoured nuclear membrane (*nm*). The endoplasmic reticulum (*end*) with PALADE granules (*Pg*) is also seen. Magnification 94,500 ×

may benefit from using these resins for certain types of specimen which for some reason or other have proved to be difficult to deal with after methacrylate embedding.

Thanks are due to Miss ANNE-GRETE OVERGÅRD for expert handling of the photographic material, and to Mrs. HELENE RAVN for unlimited help in sectioning and electron microscopy.

The work on bacteria and bacteriophage has been carried out in collaboration with Dr. O. MAALØE, Statens Seruminstitut, and the work on tissue from the small intestine of the mouse in collaboration with Dr. H. MOE, Department of Anatomy, University of Copenhagen.

References

1. NEWMAN, S. B., E. BORYSKO and M. SWERDLOW: J. Res. nat. Bur. Standards **43**, 183 (1949).
2. MAALØE, O., and A. BIRCH-ANDERSEN: Bacterial Anatomy, 6th Symposium of the Society for General Microbiology, p. 261. Cambridge: University Press. 1956.
3. BORYSKO, E., and P. SAPRANAUSKAS: Johns Hopk. Hosp. Bull. **95**, 68 (1954).
4. WEINREB, S.: Science **121**, 774 (1955).
5. BORYSKO, E.: J. biophys. biochem. Cytol. Suppl. **2**, 3 (1956).
6. BIRBECK, M. S. C., and E. H. MERCER: J. roy. Microscopical Soc., Ser. III **76**, 159 (1956), published April 1958.
7. KELLENBERGER, E., W. SCHWAB et A. RYTER: Experientia (Basel) **12**, 421 (1956).
8. GLAUERT, A. M., and E. M. BRIEGER: Proceedings of the Stockholm Conference on Electron Microscopy 111 (1956).
9. — G. E. ROGERS and R. H. GLAUERT: Nature (Lond.) **178**, 803 (1956).
10. — and R. H. GLAUERT: J. biophys. biochem. Cytol. **4**, 191 (1958).
11. PETERS, D., G. NIELSEN and K. H. ANDRES: 7th International Congress for Microbiology. Abstracts of communications 245 (1958).
12. HUXLEY, H. E.: Paper at 15th meeting of electron microscopy society of America, Boston 1957.
13. ROBERTSON, J. D.: J. biophys. biochem. Cytol. **4**, 39 (1958).
14. RYTER, A., et E. KELLENBERGER: Z. f. Naturforsch. **13b**, 597 (1958).
15. MAALØE, O., A. BIRCH-ANDERSEN and F. S. SJÖSTRAND: Biochim. biophys. Acta **15**, 12 (1954).
16. KELLENBERGER, E., and A. RYTER: J. biophys. biochem. Cytol. **4**, 323 (1958).

Inclusion au polyester

ANTOINETTE RYTER et EDOUARD KELLENBERGER

Laboratoire de Biophysique, Université de Genève

Les gonflements rencontrés si fréquemment avec le méthacrylate, surtout dans les tissus végétaux et les bactéries, nous ont aussi poussés à chercher un nouveau matériel d'inclusion. Les polyesters (*1*) révélèrent des propriétés de polymérisation extrêmement régulières et leur application comme matériel d'inclusion fit l'objet d'une mise au point dans notre laboratoire durant ces trois dernières années (*2*). Après de nombreux essais avec des polyesters de différentes provenances, nous nous sommes arrêtés au Vestopal W qui possède les propriétés de pénétration et de coupe les plus adéquates. Sa polymérisation est amorcée par un initiateur et un activateur et l'élévation de la température aux environs de 60°. Le choix de l'initiateur et de l'activateur qui fit l'objet de nombreux essais se fixa sur le tert. butyl perbenzoate comme initiateur et le naphténate de cobalt comme activateur. Il fut déterminé d'une part par leur activité et leur stabilité favorables et d'autre part par la facilité de manipulation et d'incorporation dans le Vestopal que leur confère leur état liquide.

L'inclusion du tissu, précédée d'une déshydratation à l'acétone, se fait en laissant séjourner les fragments de tissu dans trois bains acétone — Vestopal dont la teneur en Vestopal croît de l'un à l'autre, puis dans du Vestopal dépourvu d'acétone mais additionné d'activateur et d'initiateur. Les morceaux de tissu sont ensuite déposés au fond de capsules de gélatine remplies de ce dernier mélange. Toutes ces manipulations se font à la température de la chambre. La polymérisation, elle, se fait à 60° en 12 à 24 hr.

La régularité de polymérisation du Vestopal se remarque déjà macroscopiquement dans la dureté identique de chaque bloc d'une série d'inclusion à l'autre. On ne voit jamais de bulles autour du tissu, quels que soient la nature du tissu et le fixateur employé. La régularité de polymérisation se manifeste également microscopiquement dans la conservation du tissu qui ne présente jamais aucun gonflement quelle que soit la nature du tissu (Fig. 1).

Les coupes s'obtiennent facilement avec des couteaux de verre; elles s'étalent immédiatement et parfaitement.

Au microscope électronique, elles offrent certaines différences avec celles de méthacrylate:

1) les détails sont en général très nets et la résolution reste bonne même sur les coupes relativement épaisses (or à la surface de l'eau) (Fig. 2);

2) le contraste est un peu plus faible;

3) les stries dues aux défauts du couteau sont plus visibles.

L'étude de la surface des coupes par ombrage avant et après exposition aux électrons nous a permis de voir que ces caractéristiques dépendent des comportements différents du polyester et du méthacrylate sous le faisceau électronique. Les coupes de méthacrylate et de Vestopal avant exposition aux électrons sont tout à fait semblables: leur surface est plane, le matériel inclus n'est pas en relief et les stries sont nombreuses et bien visibles dans les deux cas.

Les coupes de méthacrylate ayant passé sous le faisceau électronique avant d'être ombrées montrent une diminution d'épaisseur du matériel d'inclusion, provoquant la mise en relief du matériel inclus (bactéries) comme l'ont déjà montré plusieurs auteurs (3, 4). Les contours des bac-

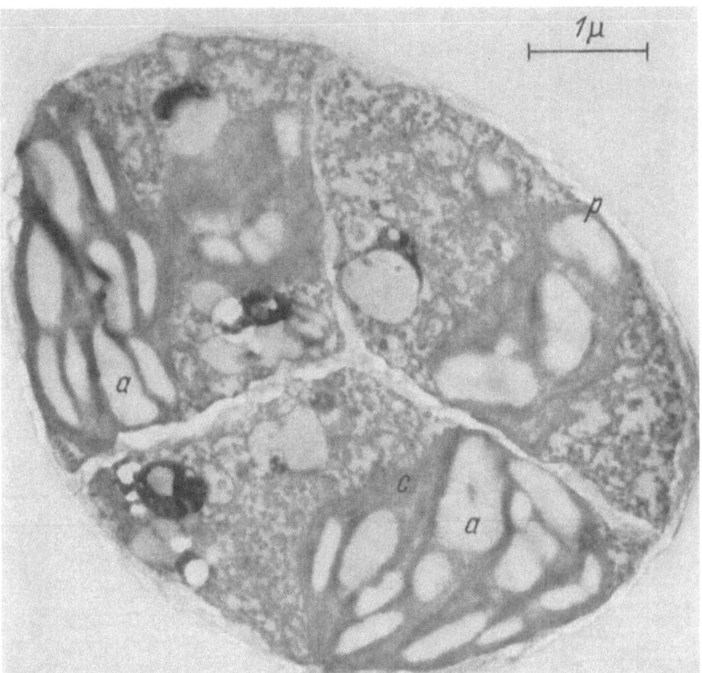

Fig. 1. Algue «Stichococcus diplosphaera». Les chloroplastes (c) contiennent plusieurs grains d'amidon (a) qui sont plus clairs que le fond de polyester. La paroi cellulaire (p) de nature cellulosique est également très claire

téries sont arrondis et un peu flous. Cet aspect s'accentue avec l'épaisseur de la coupe. Le fond de méthacrylate ne présente plus aucune strie, mais on les devine encore dans le cytoplasme bactérien.

Les coupes de Vestopal ayant passé dans le faisceau électronique puis ombrées révèlent un aspect tout à fait identique à celui des coupes ombrées avant exposition aux électrons c'est-à-dire que la surface est plane, que le matériel inclus ne présente aucun relief et que les stries sont bien visibles.

Nous avons déduit de ces résultats que le méthacrylate subit une fusion suivie d'évaporation sous le faisceau, tandis que le polyester subit une dénaturation conduisant à la «sublimation» d'une partie de ses constituants. Cette différence de comportement nous permet d'expliquer les différences d'aspect qui existent entre les coupes de polyester et celles de méthacrylate et que nous avons signalées plus haut.

1) La fusion du méthacrylate entraîne pour des raisons de tension superficielle des déformations dans les structures incluses et l'empâtement des parties saillantes, qui concourent tous deux à l'imprécision de l'image et produisent une diminution de la résolution très rapide en fonction de l'épaisseur de la coupe. Le polyester qui ne subit ni fusion ni changement d'épaisseur, conserve l'intégrité des structures incluses et permet d'obtenir une bonne résolution malgré une épaisseur relativement grande.

2) La constance d'épaisseur du Vestopal autour et à l'intérieur des structures incluses permet de comprendre en tous cas partiellement pourquoi le contraste des images électroniques est un peu plus faible.

3) La présence des stries et leur abondance sur les coupes de méthacrylate révélées par l'ombrage avant exposition aux électrons montre que les tissus inclus dans ce polymère ne sont pas

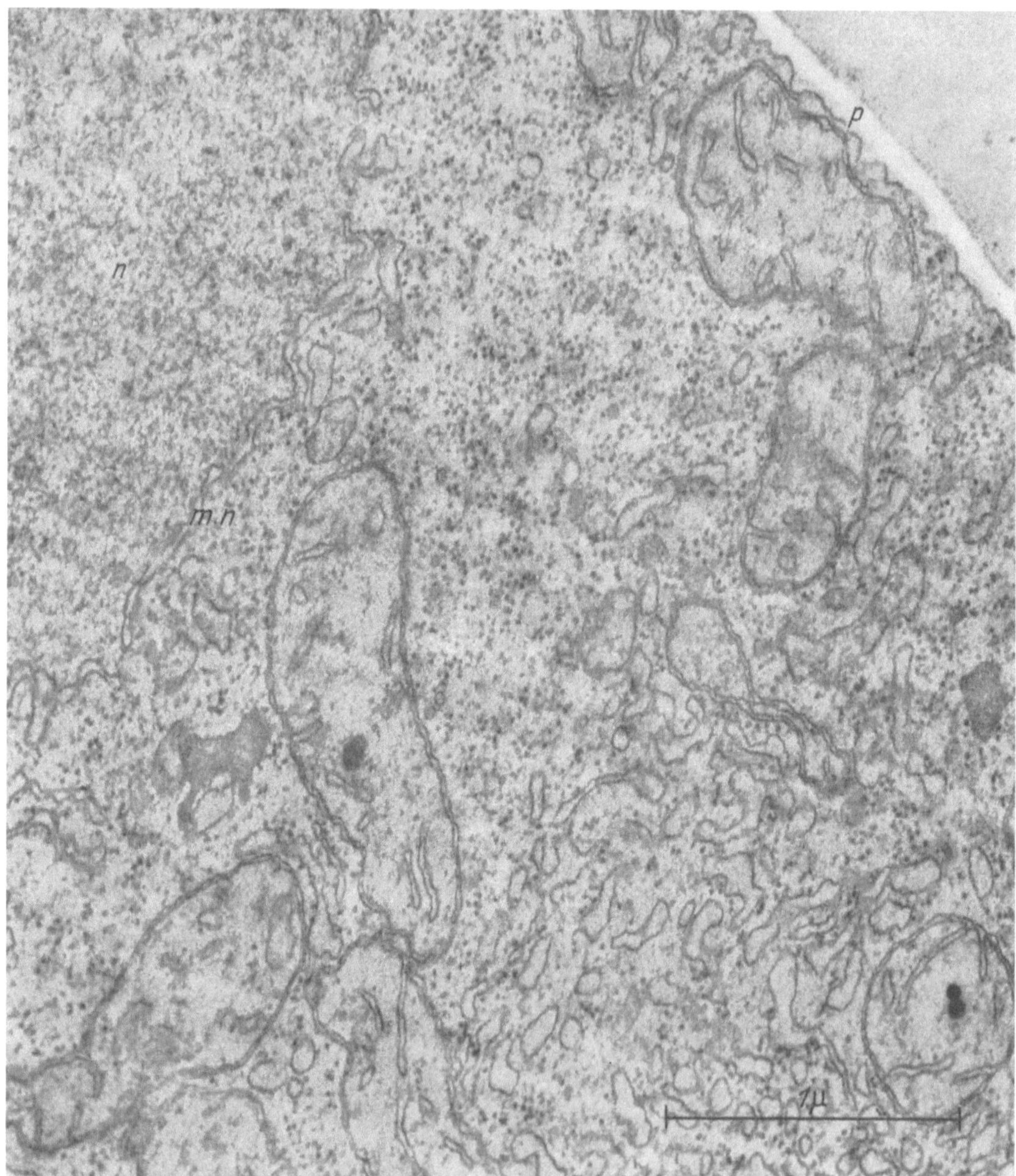

Fig. 2. Champignon «Allomyces macrogynn. Em.», cytoplasme de la plantule, avec une partie du noyau (n), limité par une membrane nucléaire double (m. n.). Les mitochondries sont nombreuses et leur structure bien visible. La paroi cellulaire (p) de nature cellulosique est plus claire que le found de polyester. (Préparation et micrographie B. Blondel)

moins endommagés par le couteau que ceux inclus dans le Vestopal, mais que la fusion du méthacrylate sous les électrons entraîne leur disparition sans toutefois pouvoir supprimer les dégâts qu'elles ont causés dans le tissu lui-même. On voit donc que la présence des stries sur les coupes de

Vestopal qui, au premier abord semblait être un inconvénient, permet au contraire de distinguer avec sécurité les coupes le moins endommagées par le couteau et qui sont exemptes de stries de celles qui en ont davantage souffert et qui risquent de présenter des arrachements et des déformations des structures fines souvent difficilement discernables sur l'écran du microscope électronique.

Le Vestopal[1] offre donc la possibilité d'entreprendre des études systématiques et comparatives avec n'importe quel tissu grâce à ses propriétés de polymérisation régulière et permet d'obtenir facilement d'excellentes images électroniques grâce à son comportement sous les électrons.

Bibliographie

1. Kellenberger, E., W. Schwab et A. Ryter: Experientia (Basel) **12**, 421 (1956).
2. Ryter, A., et E. Kellenberger: J. Ultrastr. Res. **2**, 200 (1959).
3. Williams, R. C., and F. Kallman: J. biophys. biochem. Cytol. **1**, 301 (1955).
4. Watson, J. H. L.: Lab. Invest. **5**, 451 (1956).

A water-miscible embedding resin for electron microscopy

I. R. Gibbons

Biological Laboratories, Harvard University, Cambridge 38, Mass. (USA)

Embedding media that are miscible with water have played a significant role in light microscopy, particularly in the study of lipids. The embedding media used for this purpose have mostly been based upon polyethylene glycols (*1*); they do not have suitable properties for cutting the specially thin sections needed for electron microscopy. On the other hand, the embedding media that have been developed for thin sectioning are not miscible with water and have necessitated treatment of the specimens with a separate dehydrating agent, such as acetone or ethyl alcohol, before embedding (*2, 3, 4*). Gelatin has been proposed as a water-miscible embedding medium for electron microscopy (*5, 6, 7*), but it seems likely that the great shrinkage that occurs during hardening will preclude its general use.

The present paper describes some preliminary work on a new embedding medium based upon a liquid epoxy resin, "Aquon", that is completely miscible with water. Cured with a suitable hardener the resin sets, without appreciably shrinking, to yield a solid with good properties for thin sectioning.

Preparation and Properties: Aquon resin is a water-miscible fraction of the commercial epoxy resin Epon 812 (Shell Chemical Co.). It has been prepared by extracting Epon 812 with water to obtain a solution of Aquon, from which the resin is crudely separated by salting it out with sodium sulphate. Residual water is removed by drying in a vacuum dessicator.

When prepared in this way, Aquon resin is a colorless hygroscopic liquid of fairly low viscosity (its coefficient of viscosity is approximately 100 centipoise at 25° C). It is completely miscible with water at temperatures below about 15° C; at slightly higher temperatures it is only partially miscible.

Like other liquid epoxy resins (*8*), Aquon can be cured to a solid resin by a variety of hardeners. Working with a non-water-miscible epoxy resin Glauert and Glauert (*3*) have reported that a system of resin plus acid anhydride hardener and tertiary amine accelerator is most suitable for an embedding mixture. This conclusion has been confirmed by the present study. The hardness of the cured resin can be controlled by the choice of acid anhydride hardener. The amount of tertiary amine accelerator determines the rate at which the resin sets. The cured resin can be sectioned with a conventional ultra-microtome using a glass knife and trough of distilled water. The cut sections are not dissolved by the water in the trough, though they do become somewhat softened. This softening permits much of the compression inevitably caused by thin sectioning to re-expand within a few seconds; it also facilitates ribboning of successive sections. Thin sections can be stained for electron microscopy by floating them on the surface of the staining solution (*9*); even thick sections (10 μ) cut for examination by light microscopy can readily be stained without removing the embedding material.

The following procedure has been used for embedding. The fixed and washed specimens were slowly dehydrated by passing them through increasingly concentrated solutions of *plain Aquon resin in water*. When they

---

[1] Le Vestopal avec initiateur et activateur testés dans notre laboratoire peut être obtenu commercialement par la Maison Jaeger, Vésenaz, Genève.

had been completely dehydrated by soaking in dry Aquon resin, the specimens were transferred to the complete embedding mixture (resin plus hardener and accelerator). After a short time for soaking they were transferred to fresh embedding mixture in gelatin capsules, and then placed in an oven at 50° C for four days. It should be emphasized that dehydration must be in the resin alone and not the complete embedding mixture, for the acid anhydride hardener is not compatible with water.

Results. Aquon resin has been used to dehydrate and embed osmium tetroxide fixed specimens of pancreas, retina, and testis, of bacteria (*E. coli*), and of plant root tips. Comparison specimens have been prepared firstly by dehydration in alcohol and embedding in Aquon embedding mixture,

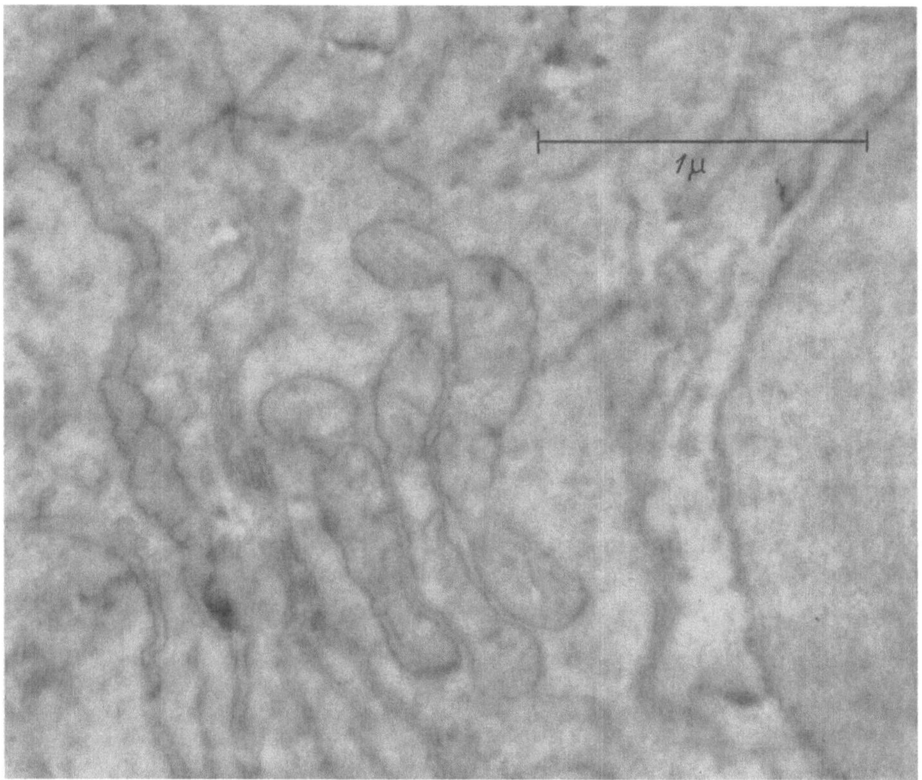

Fig. 1. Primary spermatocyte of a grasshopper, showing part of a nucleus and an area of cytoplasm with profiles of cytoplasmic membranes, mitochondria, and cell membranes

and secondly by dehydration in alcohol and embedding in Araldite epoxy resin. All these preparations were sectioned on a Porter-Blum microtome fitted with a glass knife and trough of distilled water. Sample micrographs are shown in Fig. 1—3[1].

In most cases the general quality of preservation in the test specimens dehydrated in Aquon resin appeared the equal of that in the control preparations. The characteristic organisation of the centrioles, granular and agranular cytoplasmic membranes, mitochondria, nuclei, and retinal rods appeared substantially the same in all cases; further work is required to decide whether any significant differences occur. Dehydration in Aquon resin must be carried out very slowly, for too speedy dehydration causes shrinkage of the nuclei and clumping of the ground substance of the cytoplasm.

These results indicate that dehydrating and embedding in Aquon resin is a practicable method of preparing specimens for thin sectioning. The method has the theoretical advantage of avoiding the use of conventional dehydrating agents with their strong solvent power. The range of applications in which this theoretical advantage will result in practical benefits still remains to be

---

[1] All figures show preparations fixed with 1% buffered osmium tetroxide; dehydrated and embedded with Aquon resin.

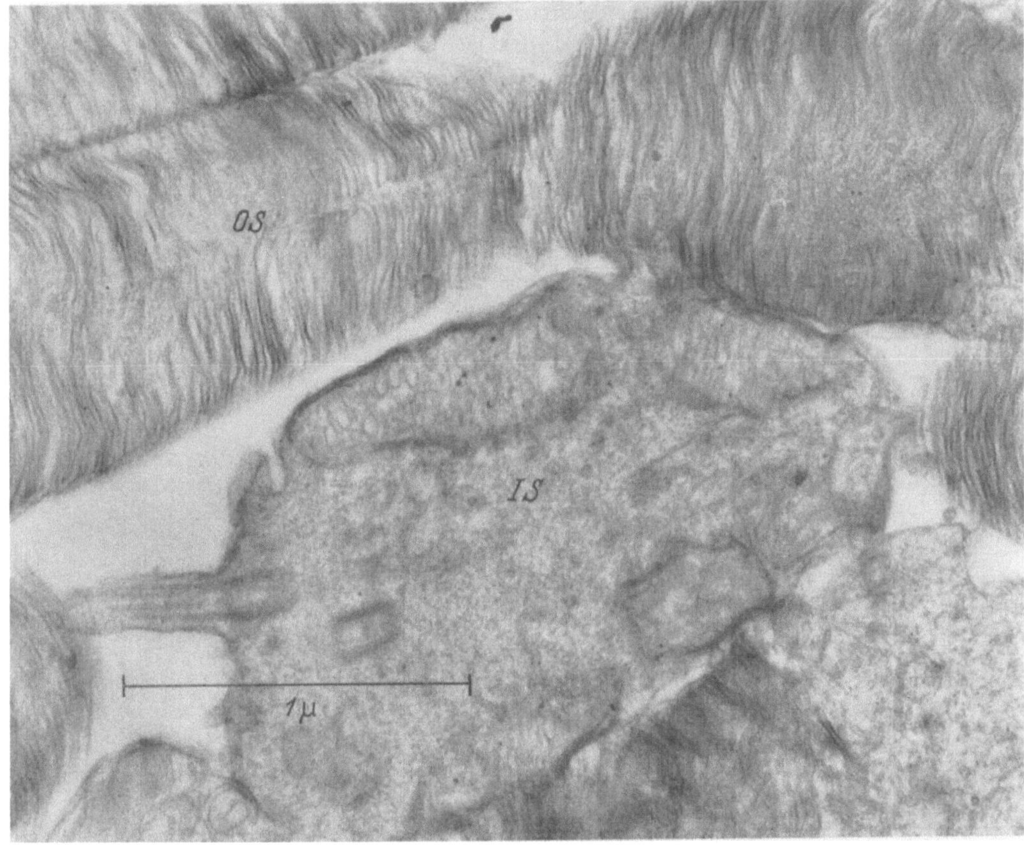

Fig. 2. Retina of a rat, showing the junction between the inner and outer rod segments. The section is not well oriented, but the two centrioles; the cilium connecting inner and outer segments, and the piled discs of the outer segments can be seen in different rods

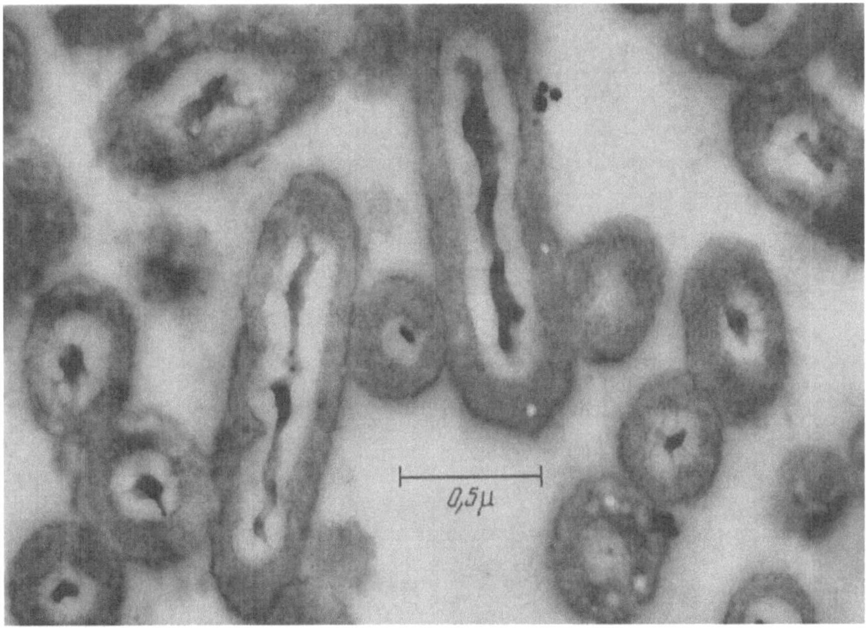

Fig. 3. Bacteria (*E. coli*) in logarithmic growth phase. The cell outline appears fairly smooth. The contracted appearance of the central nuclear apparatus is typical of that caused by plain osmium tetroxide fixation

determined. Two potential uses are on specimens fixed with formaldehyde and specimens stained
with organic compounds.

The greater part of this work was carried out at the Johnson Research Foundation of the University of
Pennsylvania. It was supported by a grant from the National Science Foundation to Dr. T. F. Anderson, to
whom I am grateful for his interest and encouragement.

References

1. Miles, A. E. W., and J. E. Lindner: J. roy. Micr. Soc. **72**, 199 (1952).
2. Newman, S. B., E. Borysko and M. Swerdlow: J. Res. nat. Bur. Stand. **43**, 183 (1949).
3. Glauert, A. M., and R. H. Glauert: J. biophys. biochem. Cytol. **4**, 191 (1958).
4. Kellenberger, E., W. Schwab and A. Ryter: Experientia (Basel) **12**, 421 (1956).
5. Gilëv, V. P.: Electron microscopy: Proceedings of the Stockholm Congress. p. 113. Stockholm: Almqvist and Wiksell 1957.
6. — J. Ultrastructural Res. **1**, 349 (1958).
7. Fernández-Morán, H., and J. B. Finean: J. biophys. biochem. Cytol. **3**, 725 (1957).
8. Lee, H., and K. Neville: Epoxy resins. New York: McGraw Hill 1957.
9. Gibbons, I. R., and J. R. G. Bradfield: Electron Microscopy: Proceedings of the Stockholm Congress. p. 121. Stockholm: Almqvist and Wiksell 1957.

Fixierungs- und Einbettungsstudien für die Ultrahistologie

Jörg Klima

Zentrallaboratorium für Elektronenmikroskopie, Medizinische Klinik der Tierärztlichen Hochschule, Wien

Einbettungsmaterialien. Zwei Gruppen von Plasten wurden verwendet und einige ihrer technologischen Eigenschaften untersucht: Methacrylat und ein Polyester („Vestopal", Chem. Werk Hüls), mit und ohne Styrolzugabe.

Polymerisationsgeschwindigkeit. In Abb. 1 ist vergleichend die Polymerisationsgeschwindigkeit von Vestopalpräparationen und Methacrylat aufgetragen. Als Katalysator wurden jeweils 0,4 ml einer 50% Cyclohexanonperoxydlösung auf 10 ml Monomeres zugesetzt; zu Vestopal zusätzlich

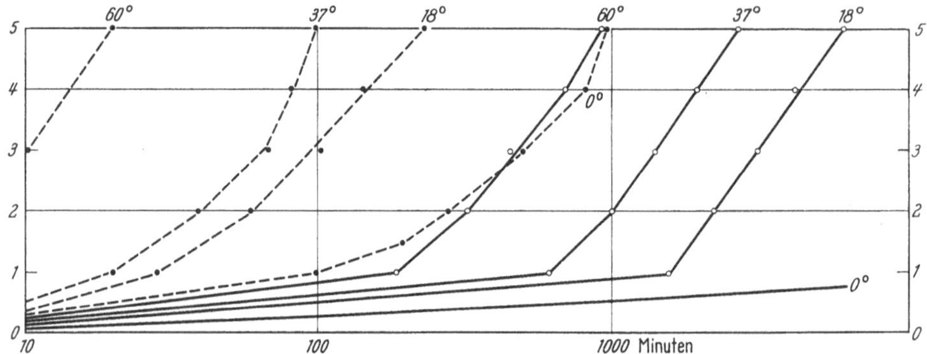

Abb. 1. Polymerisationsgeschwindigkeit. Abszisse = Zeit in Minuten, logarithmisch aufgetragen. Ordinate = Stufen zunehmender Polymerisation. ●·······● = Vestopalpräparationen (s. Text). ○————○ = Butyl:Methyl methacrylat 9 :1 (Katalysator s. Text)

noch 0,1 ml einer 10% Co-Naphthenatlösung in Styrol als Beschleuniger. In Methacrylat zersetzt sich der Beschleuniger bei Zugabe von Peroxyden. Vestopal polymerisiert rascher und noch bei tieferen Temperaturen. Ein Einfluß des Mischungsverhältnisses auf die Polymerisationsgeschwindigkeit ist bei Methacrylat sehr deutlich; bei Vestopal ist er nicht erkenntlich, solange man den Styrolzusatz nicht über 50% bei Vestopal H bzw. 15% bei Vestopal W steigert.

Mikrohärte. Plastikstoffe sind keine eigentlichen festen Körper, zeigen aber sehr unterschiedliche Härte. Mit Hilfe eines Mikrohärteprüfers gelingt es, die Härte eines Blockes genau zu definieren. Es besteht eine starke Abhängigkeit zwischen Dauer der Belastung und der Größe des

Eindruckes des Prüfkörpers bei derselben Last. Man kann diese Zeitabhängigkeit eliminieren, indem man die Belastungsdauer gerade solange wählt, daß die Mikrohärte (ihre Dimension $= \mathrm{kpmm}^{-2}$) dem Kickschen Ähnlichkeitssatz $P = a\,d^2$ folgt und belastungsunabhängig ist. Abb. 2 zeigt die so gemessene Härte von Methacrylatgemischen und einzelnen Vestopalpräparationen. Die Belastungsdauer betrug bei Butylmethacrylat 20 sec, bei Methylmethacrylat 70 sec und bei den Vestopalpräparationen 60 sec.

Schnittdeformation. Die Stauchung der Schnitte während des Schneidens ist streng mit der Mikrohärte korreliert, wie eigene und von H. SITTE (1958) vorgenommene Untersuchungen zeigen. Weiche Methacrylatgemische, die eine starke elastische Rückbildung des Eindruckes aufweisen, zeigen starke primäre Stauchungen, können aber anschließend wieder auf die volle oder nahezu die volle frühere Länge gestreckt werden. Harte Gemische, bei denen nur eine geringe oder überhaupt keine Rückbildung des Eindruckes auftritt, zeigen eine geringere primäre Stauchung, die aber unveränderlich bleibt. Vestopal zeigt dasselbe Verhalten; bei sehr harten Gemischen ist die primäre Stauchung nicht mehr nachzuweisen. Eine Rückbildung war bei Vestopal nicht zu erzielen.

Weitere Vorteile des Vestopal sind seine geringere und allseitige Schrumpfung in Gelatinekapseln sowie Unempfindlichkeit gegenüber Sauerstoff und vielen Chemikalien.

Entwässerungsmedien. Verwendet und z. T. neu erprobt wurden Alkohol (Al), Aceton (Ac), Isopropanol (Ip), Tetrahydrofuran (THF), Dimethylformamid (DMF), Dioxan (Di), und Pyridin (Pn). Isopropanol ist nur für Methacrylat geeignet; Vestopal W ist mit Alkohol nur ungenügend mischbar. Ein Zusatz bis zu 10% Entwässerungsmedium stört die Polymerisation nicht. Bei stark flüchtigen Verbindungen tritt vor der Polymerisation weitgehender Schwund des Mediums ein. Schwerer flüchtige Substanzen bleiben im Block gelöst und verleihen dem Plastikstoff eine größere Hydrophilie. Dies kann für biochemische Untersuchungen von Bedeutung sein und soll demnächst geprüft werden.

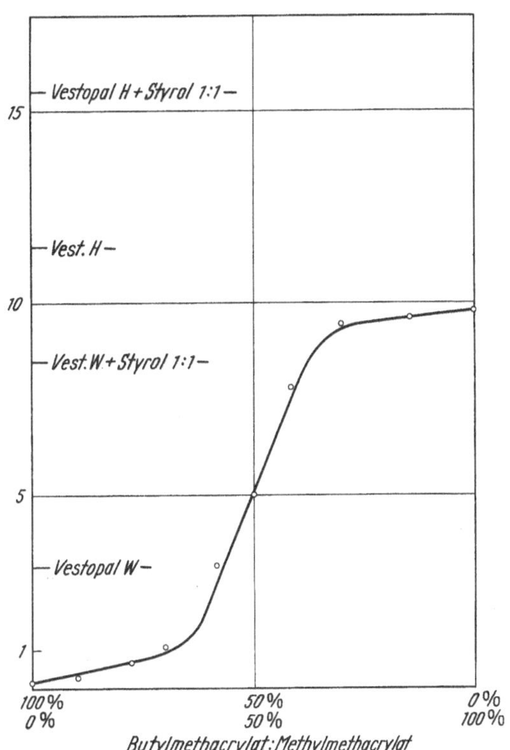

Abb. 2. Härte verschiedener Einbettungsmittel. Abszisse = Mischungsverhältnis Butyl:Methylmethacrylat. Ordinate = Mikrohärte (kpmm^{-2}) (Absoluter Betrag) bei Belastungsunabhängigkeit (Belastungsdauer bei den einzelnen Meßpunkten verschieden)

Die Entwässerungsmedien reagieren mit den Fixierungsmitteln, wie in vitro geprüft wurde. In der Tab. 1 stehen in den Spalten die Entwässerungsmedien, in den Zeilen die drei derzeit gebräuchlichsten Fixierungsmittel. Die Möglichkeit, auf die Fixantien verschieden angreifende Entwässerungsmedien anzuwenden, schien die Aussicht zu eröffnen, Strukturen, die auf dasselbe Fixierungsmittel ansprechen, feiner zu differenzieren. Jedem Histologen ist die sekundäre Osmiumtetroxydreduktion

Tabelle 1

Fixierungsmittel	Entwässerungsmedien						
	Al.	Ip.	Ac.	THF.	DMF.	Di.	Pn.
OsO_4	r. 1 h	r. 3 h	r. 12 h	l. —	l. —	l. —	K —
$KMNO_4$	r.	r.	r. L. u.	r.	r.	r.	r.
$K_2Cr_2O_7$	ul.	ul.	ul.	ul.	l.	ul.	ul.

Erklärung: — = Keine Reaktion in 50% Medium, x h = Anzahl der Stunden bis zu tiefer Schwärzung in 50% Medium, K = bildet Komplexe mit reduzierter Form (Os aus dem Block gelöst), L = langsam, l. = löslich. in abs. Medium, r. = wird reduziert, u. = Reaktion unvollständig, ul. = unlöslich.

in Alkohol bekannt. Gut sichtbar ist ihre Wirkung am Cutisgewebe der Haut. Mit Tetra-
hydrofuran kann diese sekundäre Reduktion verhindert werden. Bei Untersuchungen im
Elektronenmikroskop zeigte es sich, daß sämtliche Doppelmembranen der Zelle nach Entwässerung in Tetrahydrofuran keinen oder nur einen ganz geringen Kontrast zeigen. Eine nachfolgende
Kontrastierung mit 1% alkoholischer Phosphorwolframsäure im benachbarten Schnitt schließt
Boryskoschäden aus. Diese Membranen sind durch OsO_4 allein schon stabil gegen anorganische
Lösungsmittel geworden, in fixiertem Zustand aber erst durch die sekundäre Alkoholreduktion
kontrastiert worden. Eine Entscheidung zwischen den verschiedenen vorgeschlagenen Membran-
modellen läßt dieser Befund vorerst noch nicht zu. Doch müssen künstliche Modelle, die den
Anspruch erheben, den Zellmembranen ähnlich zu sein, dieses Verhalten zeigen.

Literatur

1. Die Mikrohärte. Herausgegeben von den Optischen Werken C. Reichert, Wien. 2. Aufl. Optische Werke
Reichert, Wien 1953.
2. FAKIR, P. G., and L. D. PEACHY: J. biophys. biochem. Cytol. **4,** 345 (1958).
3. PEACHY, L. D.: J. biophys. biochem. Cytol. **4,** 233 (1958).
4. RAMSTHALER, P.: Mikroskopie (Wien) **2,** 345 (1947).
5. SITTE, H.: Dieser Band S. 63.

Lichtmikroskopische, kontinuierliche Kontrolle
von Präparatveränderungen während der Fixierung, Kontrastierung,
Entwässerung und Einbettung bei Gewebekulturen

NORBERT WEISSENFELS

Zoologisches Institut der Universität Bonn

Die Methoden zur Präparation biologischer Objekte für die elektronenmikroskopische Untersuchung sind in den letzten Jahren immer mehr verfeinert worden. Bei bester Fixierung, Kontrastierung, Entwässerung und Einbettung treten aber immer noch Präparatveränderungen ein,
die z. T. schon lichtmikroskopisch sichtbar sind. Mit einer einfachen Methode können die Strukturveränderungen in der Zelle während der Präparation kontinuierlich kontrolliert werden.

Hierzu wurden gut ausgewachsene Kulturen von Hühnerherzfibroblasten verwendet, die auf
44 × 44 mm großen Immersionsdeckgläsern in Maximow-Glaskammern gezogen worden waren.
Für den Versuch wurde eine solche Deckglaskultur auf eine Durchströmungskammer (Abb. 1) um-
gesetzt. Diese besteht aus Leichtmetall, ist 76 mm lang, 48 mm breit und 1 mm dick. Sie besitzt in
der Mitte eine große, runde Öffnung, die auf der Kammerunterseite durch eine Rinne nach rechts
und nach links mit zwei Zapfenbohrungen in Verbindung steht (Abb. 1b). Die beiden Zapfen dienen
als Schlauchansätze. Die Metallkammer wird von oben durch das quadratische Deckglas, das zum
Kammerinnern hin die Kultur trägt, und von unten durch ein rechteckiges Deckglas so ver-

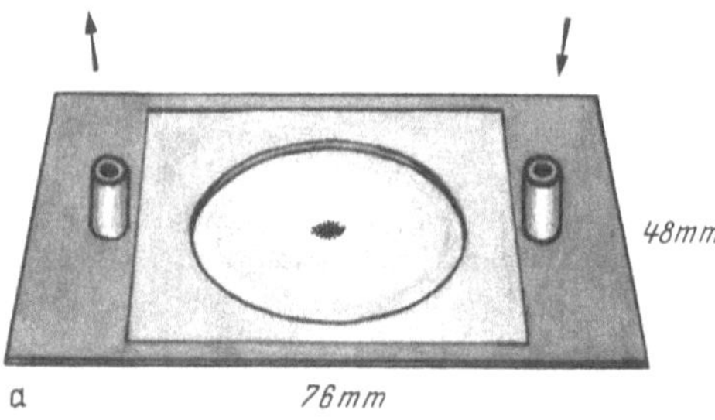

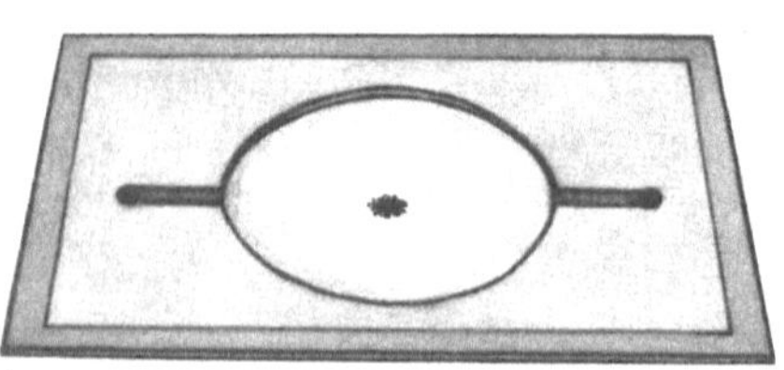

Abb. 1 a u. b. Ober- und Unteransicht
einer Durchströmungskammer

schlossen, daß von Zapfenbohrung zu Zapfenbohrung ein Flüssigkeitsdurchstrom möglich
ist. Die Deckgläser werden mit dem Glaszement „Stabilit" (Henkel-Werke, Düsseldorf)
angeklebt, der im harten Zustand physiologisch unwirksam und beständig gegen die üblichen

Präparationsmedien ist. Wo diese die Metallflächen der Durchströmungskammer berühren, wird außerdem eine Stabilit-Schutzschicht aufgetragen. Die Durchströmungskammer kann aus einem Tropftrichter über einen Verbindungsschlauch mit Nähr- und Versuchslösung in gewünschter Menge beschickt werden. Mit dieser Apparatur lassen sich Gewebekulturzellen unter sterilen Bedingungen bei regelmäßiger Erneuerung der Nährlösung und .physiologischer Temperatur ungewöhnlich lange ohne pathologische Veränderungen beobachten.

Für den eigentlichen Versuch läßt man die flüssige Nahrung aus der Durchströmungskammer ablaufen und ersetzt sie durch das zu erprobende Fixierungsmittel, das dann, ähnlich wie bei dem Entwässerungsapparat von BERNHARD, durch Nachströmen des jeweils nächsten Mediums und schließlich des Einbettungsmittels langsam verdrängt wird. Nachdem letzteres polymerisiert und die lichtoptische Untersuchung abgeschlossen ist, kann die Durchströmungskammer geöffnet, das Präparat auf einen Leerblock aufgeklebt und zur elektronenmikroskopischen Untersuchung weiter verarbeitet werden. Auf diese Weise lassen sich alle mikroskopisch sichtbaren Präparationsschäden und der Zeiltpunkt ihrer Entstehung erkennen und genau festlegen.

BAHR, BLOOM und FRIBERG (1) berichteten bereits 1956 über Volumenänderungen von Geweben während verschiedener Präparationen. Diese Ergebnisse kann ich im wesentlichen bestätigen. Darüber hinaus bietet die neue Methode nun aber auch die Möglichkeit, die durch die Präparation bedingten Veränderungen einzelner Zellstrukturen genau zu beobachten.

In mehreren Versuchen wurde jeweils ein Hühnerherzfibroblast

<table>
<tr><td>1. im Leben,</td><td>5. nach vollständiger Entwässerung im abs. Alkohol</td></tr>
<tr><td>2. nach einstündiger Fixierung,</td><td></td></tr>
<tr><td>3. im 70%igen Alkohol,</td><td>6. nach einstündiger Einwirkung von flüssigem Methacrylat</td></tr>
<tr><td>4. nach einstündiger Kontrastierung,</td><td></td></tr>
<tr><td></td><td>7. im polymerisierten Methacrylat</td></tr>
</table>

photographiert und dann auf seine Volumen- und Strukturveränderungen hin untersucht.

Schon nach den ersten Versuchen zeigte sich, daß die Güte der Fixierung ausschlaggebend ist für das Gelingen der Präparation. Schlechte Fixierer machen sich lichtmikroskopisch sofort durch die Entstehung einer Wabenstruktur im Grundcytoplasma und eine starke Mitochondrienquellung bemerkbar. Mit einem Gemisch 1%iger Lösungen von Osmiumtetroxyd und Kaliumbichromat (2), das mit 3% Trauben- und 15% Rohrzucker (1) versetzt und auf p_H 7,3 eingestellt worden war, konnten die Fibroblasten bisher am besten fixiert werden. Durch die Zugabe von Kaliumbichromat scheinen die Eindringungsgeschwindigkeit und -tiefe des Osmiumtetroxyds günstig beeinflußt zu werden. Trauben- und Rohrzucker verhindern in der angegebenen Konzentration weitgehend die Fixierungsquellung, die bei Verwendung der bisher üblichen Fixierungsmittel häufig eintrat.

Durch den Wasserentzug in der Alkoholreihe und die Einbettung in Methacrylat schrumpfen die Zellen, und zwar der Kern durchschnittlich um 10%, das Protoplasma dagegen um nicht mehr als 3%. (Hier werden lineare Werte angegeben, wie das auch bei den Größenangaben der Ultrastrukturen üblich ist.)

Das Kontrastierungsmittel, bestehend aus einem Gemisch von 1%iger Phosphorwolframsäure und 0,5%igem Uranylacetat in 70%igem Alkohol, verursacht keine lichtmikroskopischen Veränderungen in der Zelle. Durch seine Anwendung treten lediglich alle nach der Fixierung bereits sichtbaren Zellstrukturen, aber auch die dabei entstandenen Strukturvergröberungen in Kern und Cytoplasma deutlicher hervor.

In einem Beispiel für die Kontrolle der Präparatveränderungen kann hier nur der Zustand eines Fibroblasten im Leben (Abb. 2) mit dem nach erfolgter Einbettung (Abb. 3) verglichen werden. Die größten Veränderungen treten beim Zellkern und bei den Fetttropfen ein. Das Karyoplasma wird kontrastreich und ebenso wie der Nucleolus leicht granuliert. Die in der lebenden Zelle hell aufleuchtenden Fetttropfen werden dunkel und klein. Man erkennt beim Vergleich der beiden Abb. 2 und 3 geringe Form- und Lageveränderungen der ganzen Zelle und ihres strukturierten Inhalts, z. B. der Mitochondrien. Diese Veränderungen sind hier aber nicht durch die Präparation bedingt, sondern durch die Eigenbewegung der Zelle, die ja nach der Lebendaufnahme

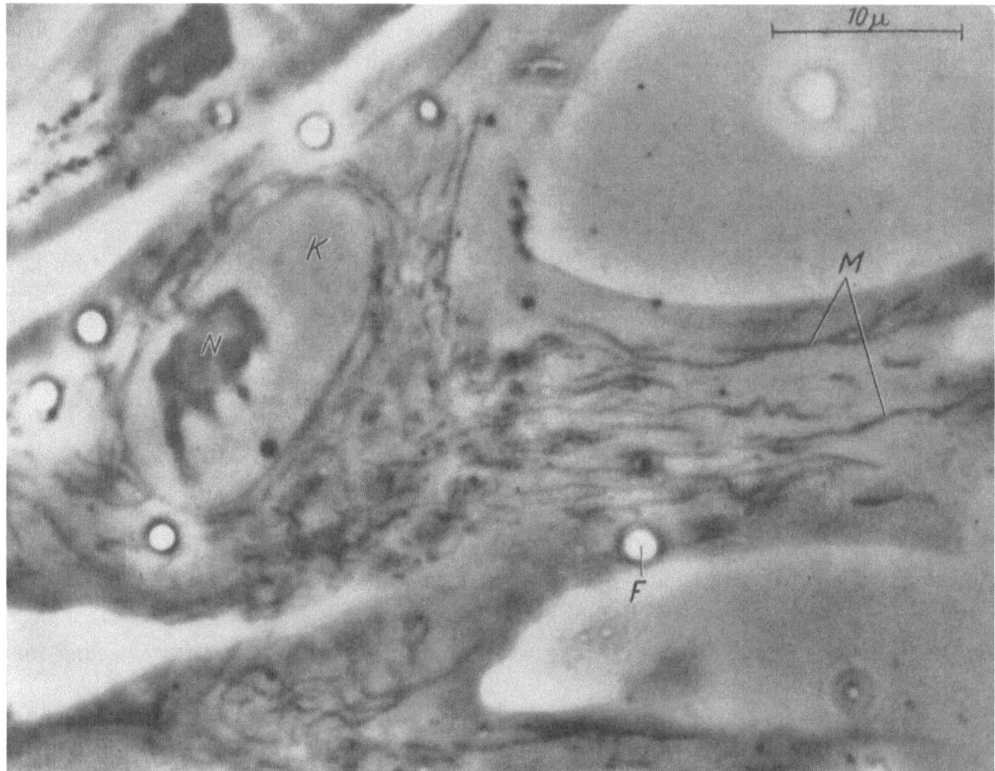

Abb. 2. Hühnerherzfibroblast im Leben. *K* Kern; *N* Nucleolus, *M* Fadenmitochondrium, *F* Fetttropfen

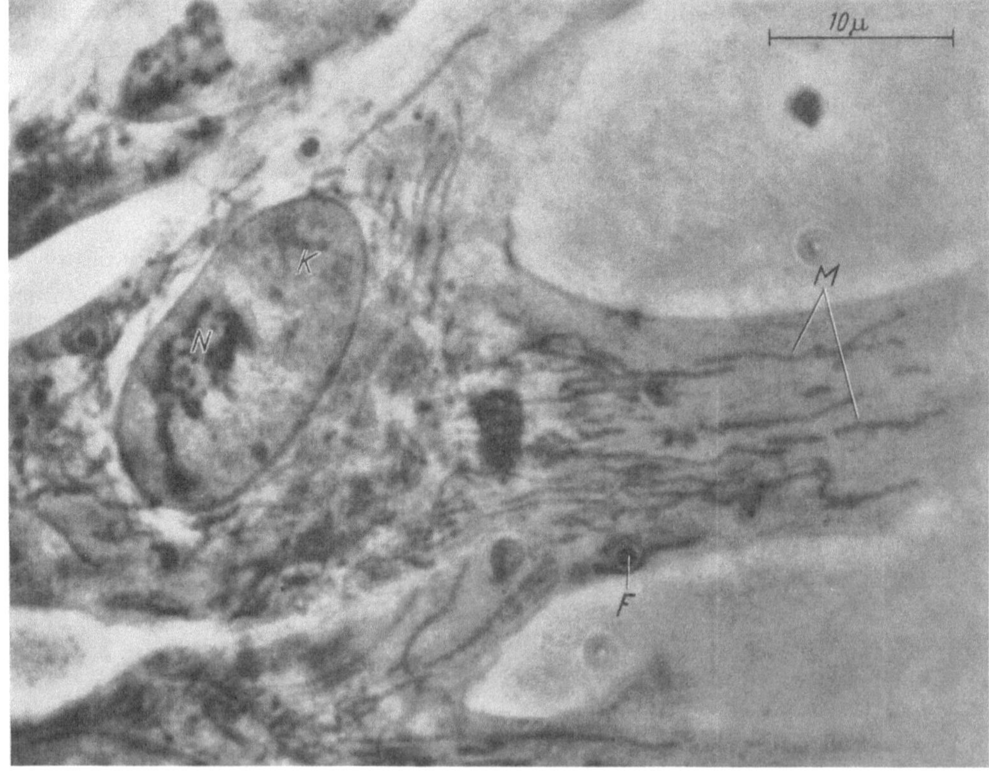

Abb. 3. Hühnerherzfibroblast nach erfolgter Einbettung (Zeichenerklärung s. Abb. 2.)

anhielt, bis in die unterdessen entleerte Durchströmungskammer das Fixierungsmittel eingefüllt wurde. Im übrigen kann die präparierte Zelle kaum von der lebenden unterschieden werden.

Das in der letzten Zeit für elektronenmikroskopische Zwecke sehr bewährte Einbettungsmittel Vestopal kann wegen seiner ungünstigen Lichtbrechung leider nicht zu lichtmikroskopischen Vergleichsversuchen verwendet werden.

Literatur

1. Bahr, G. F., G. Bloom and U. Friberg: Proc. First Europ. Conf. Electr. Microscopy. Stockholm 1956. S. 106. Stockholm: Almquist & Wiksell 1957.
2. Wohlfarth-Bottermann, K. E.: Naturwissenschaften **44**, 287 (1957).

2. Schneiden und Mikrotome

Physikalische Probleme bei der Herstellung von Dünnschnitten[1]

H. Sitte

Elektronenmikroskopische Abteilung der Medizinischen Fakultät, Pathologisches Institut der Universität Heidelberg

Auch ein ideales Ultramikrotom (reproduzierbarer inerter Vorschub, einwandfreie Präparatführung, vollkommene Unempfindlichkeit gegen Einfluß äußerer sowie schneidebedingter Kräfte usw.) sowie ein ideales Messer[2] und ein optimales Präparat[2] schützen nicht vor Störungen, die in der Natur des Schneidevorganges selbst liegen. Dies sind vornehmlich: Unregelmäßigkeiten der Schnittdicke, „Chatter" (*1, 2*) und Deformationen der Schnitte (*3*).

Unregelmäßigkeiten der Schnittdicke werden durch die Kräfte hervorgerufen, die zum Abtrennen des Schnittes notwendig sind; sie erreichen nach Bennet et al. (*4*) Werte bis zu 10 g und sind durchaus in der Lage, Messer und Präparat in einem Ausmaß zu verformen, das die Größenordnung der Schnittdicke erreicht. Das Ausmaß ist vor allem abhängig von: Präparatzuschnitt und -halterung, Anschnittfläche F, Bahngeschwindigkeit des Präparates an der Schnittstelle v_P, Klingenwinkel α, Messerneigung ε und Blockhärte; vgl. Abb. 1a. Am wichtigsten sind

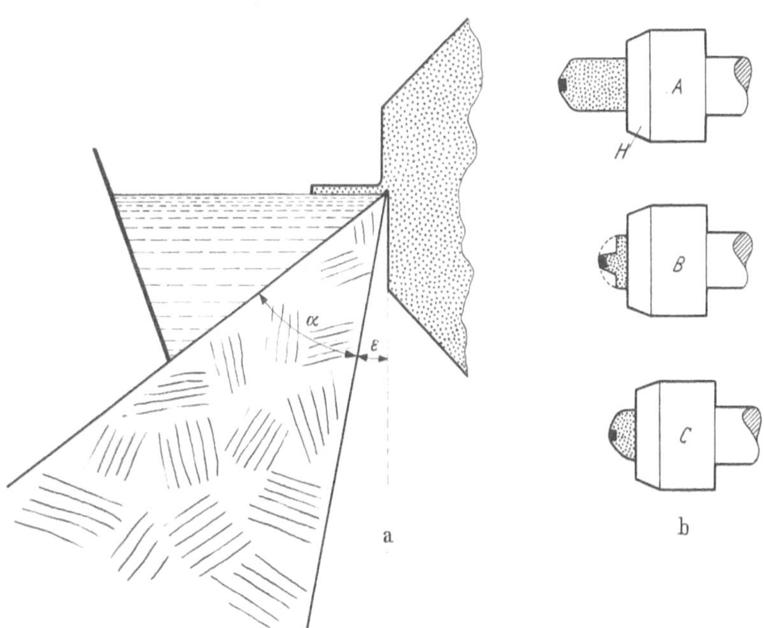

Abb. 1a und b. Erklärungen im Text

---

[1] Eigene Untersuchungen, die im Rahmen dieses Referates veröffentlicht werden, wurden durch die Deutsche Forschungsgemeinschaft unterstützt, der an dieser Stelle herzlich gedankt sei.

[2] Die eigenen sowie die zit. methodischen Untersuchungen schließen lediglich biologische Objekte bei Einbettung in Methacrylat ein, die mit Glasmessern (*5*) geschnitten wurden. Diamantklingen nach Fernàndez-Moràn (*6, 7, 8*) und neuartige Einbettungsmedien [vgl. u. a. (*9—14*)] sind in diese Betrachtungen nicht einbezogen. Diamantklingen sind wahrscheinlich stabiler; die beim Schneiden auftretenden Kräfte scheinen reduziert, da nach persönl. Mitteilungen (u. a. von G. Bergold und H. Ruska) Chatter und Deformationen gegenüber Glasmessern reduziert auftreten. Letzteres könnte unter Umständen als ein Zeichen verringerter Reibung (Kunststoff/Messer) aufgefaßt werden.

Blockzuschnitt und -halterung: Zu weit aus dem Halter (H; Abb. 1 b) herausragende Blöcke (A) und zu spitze Pyramiden (B) steigern die Störung; bei richtigem Zuschnitt (C) ragt der Block

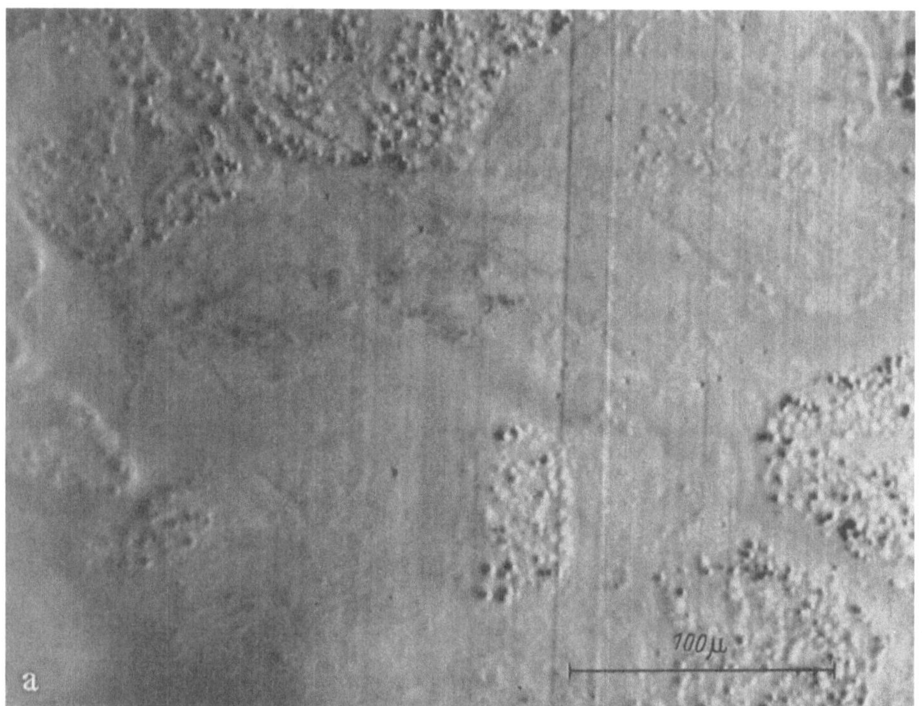

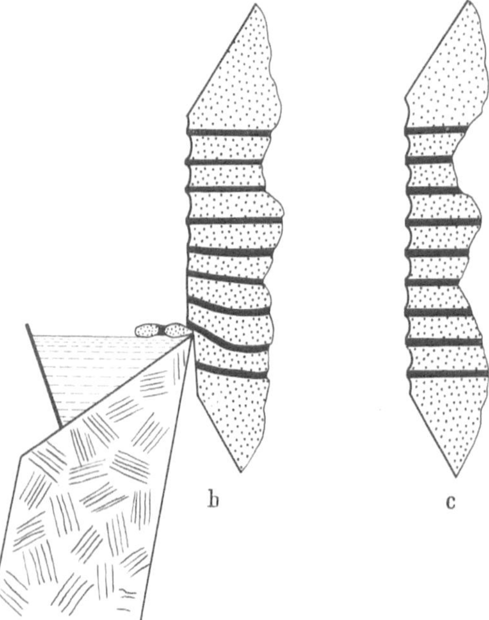

Abb. 2a—c. a Anschnittfläche am Block im schrägen Auflicht. Reichert-Epilum; 11:1; Objekt: Planaria alp. Dana, Plexiglaseinbettung, OsO$_4$-Fixierung nach Palade. b und c. Deformation und Entstehung der Anschnittflächenskulptur beim Schneidevorgang

nicht weiter als unbedingt nötig aus dem Halter heraus und die Zuschnittpyramide weicht kaum von der ursprünglichen Form des Blockes ab. Senkt man v_P, so nehmen die Unregelmäßigkeiten im allgemeinen zu, weil die zum elastischen Ausbiegen von Klinge und Präparat notwendige Zeit dadurch gegeben ist. F sollte nicht das notwendige Maß überschreiten, da bei zunehmendem F die schneidebedingten Kräfte steigen; die Schnittfläche wird am besten quadratisch zugeschnitten: Liegende Rechtecke senken die Stabilität, hochkant gestellte führen leicht zu Faltenbildung und Abreißen von Schnittbändern. α wird am besten in einem mittleren Maß gehalten: Besonders stumpfe Winkel erhöhen zwar die Stabilität der Klinge, mit ihr aber auch die Kräfte und das Ausbiegen des Blockes; spitze dagegen verringern die Stabilität der Klinge in größerem Ausmaß als die schneidebedingten Kräfte. ε sollte nicht über 5° gesteigert werden, da hierbei ($\alpha + \varepsilon$) zu große Werte erreicht und die Stabilität der Klinge spürbar gemindert wird.

Zusammenfassend wird als Faustregel für normale Objekte empfohlen: Trimmen und Einspannen des Blockes nach (C) (Abb. 1 b); v_P 0,5—2 cm/sec, F quadratisch 0,25—1 mm²; α 45—50°; ε 2—5°.

Die Ursachen der Schnittdickenschwankungen können nicht restlos ausgeschaltet werden; bestenfalls erreicht man ein von Schnitt zu Schnitt annähernd reproduzierbares Ausbiegen von

Klinge und Objekt und damit hinreichend gleichmäßige Schnitte. Das Ausmaß der elastischen Deformation zeigt Abb. 2a (Anschnittfläche in schrägem Auflicht): Präparatbedeckte und -freie Teile weisen Höhendifferenzen auf, die meines Wissens erstmals MILLER (persönliche Mitteilung, 1956) zum kontrollierten Zuschneiden der Blöcke benutzt hat. Abb. 2b erklärt dies anhand eines fiktiven Präparates, das Lamellen größerer Festigkeit (dunkel) aufweist, die in Plexiglas (punktiert) eingelagert sind. Die Lamellen leisten dem Messer einen größeren Widerstand und führen zu einer Deformation der jeweils über diesen Lamellen liegenden Plexiglasbereiche; nach dem Schnitt weicht das Plexiglas zurück (Abb. 2c) und bildet damit die in Abb. 2a sichtbaren Skulpturen.

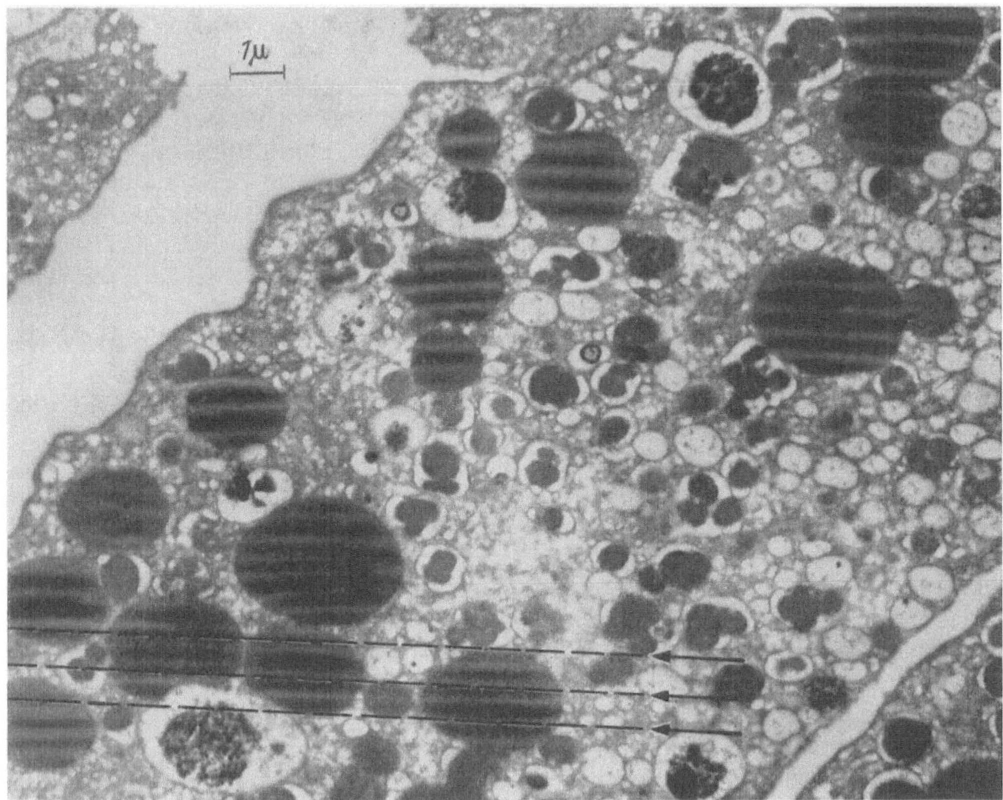

Abb. 3. Chatter in Fetttropfen (Planaria alpina DANA). El.-opt. 1500:1

Chatter treten entweder spezifisch begrenzt an Tropfen (Abb. 3), Erythrocyten, Nucleolen usw. oder bei ungünstiger Methodik im gesamten Präparat auf. Ihr Habitus läßt einen sicheren Rückschluß auf Schwingungen zu; die Frequenz liegt meist zwischen 10^4 und $3 \cdot 10^4$ Hz. Sie ist (2) zu hoch, um gerätebedingt zu sein. Der Chattervorgang läßt sich am besten bei strenger lokaler Begrenzung auf Tropfen oder dergleichen (Abb. 3) beurteilen. Die gechatterten Tropfen liegen zwischen Gewebspartien, deren Habitus den üblichen Forderungen nach Glätte und Regelmäßigkeit der Schnitte voll entspricht. Man kann feststellen, daß die Chattermarken in benachbarten Tropfen jeweils genau auf einer Linie liegen; dieser Befund gilt — soweit nicht lokale Deformationen vorliegen — für die gesamte Präparatfläche. Der Schwingungsvorgang ist demnach nicht örtlich begrenzt — er schließt die ganze Schnittfläche ein. Ungeklärt ist noch, ob die Schwingung primär das Messer[1] oder das Präparat betrifft oder ob beide Elemente in gleicher Weise beteiligt sind. Dagegen kann die Richtung der Schwingung festgelegt werden: Schwingungen in Vorschubrichtung scheiden aus; der Chatter müßte sich hierbei gleichmäßig über alle Strukturen erstrecken. Schwingungen in Richtung der Messerschneide kommen nicht in Betracht, weil keine Kräfte in

---

[1] REIF (1958, pers. Mitt.) hält den Vorgang für eine Messerschwingung, da Kunststoffe Schwingungen zu stark dämpfen.

dieser Richtung wirken und kein sinusförmiger Verlauf der Messerriefen beobachtet werden kann. Die Schwingungen verlaufen daher in Richtung des Präparatweges — sie verändern demnach periodisch die relative Präparatgeschwindigkeit. Wahrscheinlich erreicht unter bestimmten Voraussetzungen der v_P überlagerte Vektor im Scheitelwert die Größenordnung von v_P oder übertrifft diese sogar. Das Präparat wird also z. B. — relativ betrachtet — nicht mit gleichmäßiger Geschwindigkeit über die Messerschneide geführt, sondern mit einer zwischen 0 und 2 v_P mit 10^4 Hz wechselnden Geschwindigkeit. Dies erklärt den lokalisiert auftretenden Chattereffekt: Objekte großer Dichte in Plexiglas werden bei hohem v_P irreversibel plastisch deformiert (vgl. unten), während freies Plexiglas oder Präparatpartien, deren physikalisches Verhalten vor allem vom Methacrylat bestimmt wird, auch bei relativ hohem v_P keine signifikante plastische Deformation erleiden. Tropfen, Erythrocyten und Nucleolen sind dichte Gebilde, die relativ wenig Plexiglas enthalten und in ihrem physikalischen Verhalten wesentlich von den anderen Präparatpartien abweichen. Periodisch wechselnde Präparatgeschwindigkeiten führen daher bei ihnen zu periodisch auftretenden plastischen Deformationen, während die zwischen ihnen liegenden Plasmapartien dieses Phänomen im Elektronenbild üblicherweise nicht mehr zeigen. Es ist anzunehmen, daß geringfügige Spuren dieses Schwingungsvorganges auch im Plasma und dergleichen auftreten — zumindest diejenigen Skulpturen, die eine derartige schwingende Bewegung in jedem Fall erzeugt. Wahrscheinlich können die von MORGAN et al. (*15*; "periodic banding") beobachteten periodischen Skulpturen als morphologische Manifestation dieser Schwingungen betrachtet werden, da sie die gleiche Größenordnung aufweisen. Die hochfrequenten Schwingungen treten wohl bei jedem Schneidevorgang auf, führen aber nur im Falle zu hoher Scheitelwerte zu Störungen.

Der Chatter nimmt zu, je mehr die physikalischen Daten des eingebetteten Objektes von denen des objektfreien Plexiglases abweichen, d. h. je höher das Trockengewicht bzw. die „Trockendichte" ansteigt. Das Phänomen nimmt ferner zu beim Steigern von v_P, F, α und ε. Entgegen Vermutungen (*16*) sind Messerwinkel und -justierung von weit größerer Bedeutung als restlose Schartenfreiheit und Glätte der Klinge. Wahrscheinlich sind zudem Blockzuschnitt und -halterung (vgl. oben) sowie Härte des Einbettungsmediums entscheidend; scheinbar wird der Effekt durch Reduktion der Schnittdicke (D) gesteigert.

Die Abhängigkeit des Phänomens vom Schneidegut läßt kaum quantitative Aussagen zu. Im allgemeinen ist nach eigenen Messungen bei Methacrylatblöcken (0—15% Methylester) bei gutem Zuschnitt (quadratisch 0,25—1 mm²), $\alpha \leqq 45°$, ε 2—5°, v_P etwa 0,5—1 cm/sec und $D \geqq 400$ Å kein Chatter zu erwarten; tritt der Effekt dennoch objektbedingt auf, so kann er durch Reduktion von α, v_P oder F bzw. durch Erhöhen von D ausgeschaltet werden.

Deformation der Schnitte. Übt man auf einen Kunststoffblock einen hohen Druck aus, so wird er bei konstantem Volumen deformiert. Nach Aussetzen des Druckes erreicht er nicht — wie eine ideal elastische Feder — sofort seine ursprünglichen Maße, sondern dehnt sich langsam aus; wurde seine ursprüngliche Höhe h auf h' verringert, so ergibt sich für die Rückbildung das in Abb. 4a wiedergegebene Zeit/Höhe-Diagramm. Die Höhe nähert sich asymptotisch einem Wert $h'' < h$, den sie zur Zeit t annähernd erreicht. Die Zeitspanne 0 bis t sei definitionsgemäß die „Nachwirkung" (N). Die „Gesamtdeformation" (GD) gliedert sich in eine reversible elastische (ED) und eine irreversible plastische Deformation (PD). PD ist durch druckbedingte Veränderungen im molekularen Gefüge des Kunststoffes zu erklären.

Diese allgemein geltenden Gegebenheiten bestimmen das Verhalten der Objekte am Ultramikrotom: Die beim Schneiden auftretenden Kräfte führen notgedrungen mindestens zu einer ED, meist aber auch zu einer PD der Schnitte[1]. Definiert man die Abmessung der Anschnittfläche parallel zur Messerschneide als Breite, senkrecht dazu als Höhe des Schnittes, so ist zunächst festzustellen, daß die Breite unverändert bleibt, während sich die Höhe h zu h' verringert. Bei konstantem Volumen folgert hieraus eine Dickenzunahme von D zu $D' = (h/h') \cdot D$, wobei D der Schnittdicke bei $GD = 0$ entspräche. Durch die Deformation werden runde Strukturen (Kerne, Tubuli usw.) oval, isotrop im Objekt verteilte längliche Strukturelemente erhalten eine

---

[1] PEACHEY (*3*) deutet die Deformation als eine Folge periodischer „Wellen" im Schnitt — also nicht als Deformation im oben angegebenen engeren Sinn. Eigene Untersuchungen erbrachten bisher keine Hinweise dafür, daß derartige „Wellen" besondere Bedeutung besitzen.

Vorzugsrichtung usf. Dieser allgemein bekannte Sachverhalt wurde in den Arbeiten von PEACHEY (3, 17) sowie SATIR und PEACHEY (18) erstmals quantitativ betrachtet. Demnach stört die Deformation der Schnitte in zweierlei Hinsicht: erstens ergeben Messungen, welche nicht in Richtung der unveränderten Breite ausgeführt werden, falsche Resultate; zweitens mindert die erhöhte Schnittdicke die Auflösung (vgl. 19—21). Beide Fakten sind von so großer Bedeutung, daß alles darangesetzt werden muß, die ED vollständig rückgängig zu machen und die PD durch entsprechende Methodik möglichst gering zu halten.

Die Dauer der Nachwirkung (vgl. oben, N) ist weitgehend von physiko-chemischen Bedingungen abhängig. Die Kurve (I; Abb. 4b) gibt die Schnittspreitung auf H_2O dest. bei Raumtemperatur wieder; binnen 48 Std. ist h'' noch nicht erreicht ($N > 48$ Std.). Auf 10% Aceton (II) und 20% Alkohol (III; identisch mit II) beträgt N weniger als 6 Std. Eine weitere Verkürzung von N auf ≤ 30 min bewirkt eine Temperaturerhöhung (IV); dieser Kunstgriff ist bekannt (22, 23) und wurde von PEACHEY (3, 17) erstmals quantitativ erfaßt. Schließlich gelingt ein noch rascheres Spreiten (V; $N \leq 5$ sec) nach dem Prinzip von SOTELO (24) durch organische Solventien in Gasphase. SATIR und PEACHEY (18) haben hierfür besonders geeignete

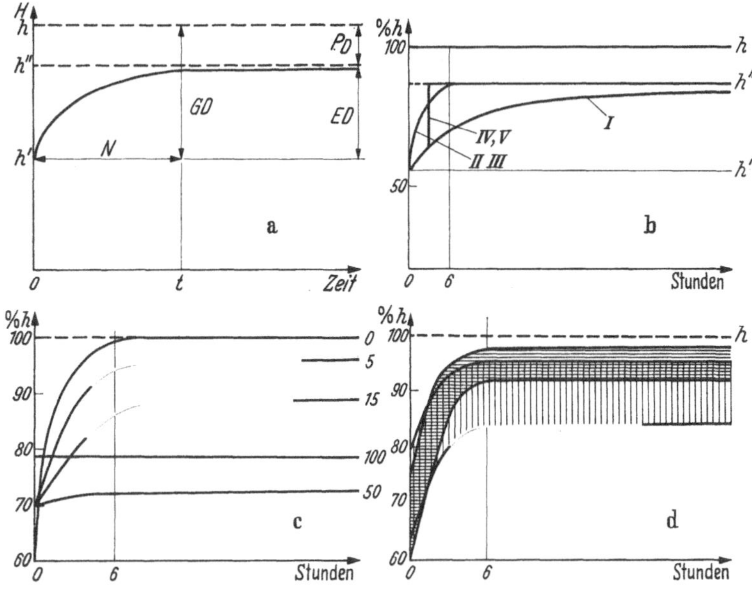

Abb. 4. Deformation und Spreitung von Schnitten, vgl. Text

Lösungsmittel (Xylol, Toluol) ausgewählt und quantitativ untersucht. Verbindliche Aussagen über den physiko-chemischen Mechanismus der Spreitung wurden bisher meines Wissens nicht vorgelegt. Eine zusätzliche Artefizierung des Schneidegutes tritt bei den Verfahren II bis V nach bisher vorliegenden Erfahrungen nicht ein.

Während demnach die ED routinemäßig ausgeschaltet werden kann, wurde die irreversible PD bislang kaum beachtet. Die PD ist abhängig von: Blockhärte (Mischung Butyl-Methylester), physikalisches Verhalten des eingebetteten Objektes, F, v_P, $\alpha(\varepsilon)$ sowie D. Unter vollkommen gleichen methodischen Bedingungen[1] zeigen Blöcke verschiedener Härte (Abb. 4c; angegeben Methylester-%: 0, 5, 15, 50 und 100) ein sehr unterschiedliches Verhalten. Während GD für $t = 0$ bei zunehmender Blockhärte abnimmt, steigt PD hierbei zunächst von 0 (reiner Butylester) auf fast 30% (50% Methylester); bei weiter gesteigerter Härte sinkt PD wieder geringfügig auf etwa 22% (100% Methylester). Hierbei streuen die Einzelwerte bei niederen Methylester-Zusätzen bis zu $\pm 7\%$ infolge der unregelmäßigen Polymerisation des Methacrylates und der damit sehr unterschiedlichen Härte der Blöcke. Diese Ergebnisse stimmen gut mit den nach Mikrohärteprüfung und Deformationsbestimmung von KLIMA (25) gewonnenen Ansichten überein. — Der Einfluß der Methodik wurde zunächst an leeren Blöcken (5% Methylester) untersucht; PD änderte hierbei seine Größe nicht signifikant. Variiert wurde innerhalb folgender Grenzen: F 0,25—0,5 mm²; v_P 0,5—4 cm/sec; α 37—75°; ε 1—5°; D 500—3000 Å; hierbei wurde abwechselnd jeweils eine Größe auf den angegebenen Minimal- oder Maximalwert gebracht, alle anderen auf einem mittleren Wert belassen. Sämtliche erhaltenen Spreitungskurven liegen im waagerecht

---

[1] $F = 0,25$ mm²; $v_P = 1$ cm/sec; $\alpha = 45°$; $\varepsilon = 5°$; $D = 750$ Å; Spreiten auf 10% Aceton. Das Spreiten durch Dämpfe organ. Solventien oder Temperaturerhöhung führt wahrscheinlich zu ähnlichen Resultaten, wurde aber bei der vorliegenden Untersuchung nicht verwendet.

schraffierten Bereich der Abb. 4d; die Streuungen sind vollkommen unspezifisch und im wesentlichen auf die unterschiedlichen Blockhärten zurückzuführen. — Wesentlich anders verhalten sich Methacrylatblöcke mit Objekt. Blöcke (5% Methylester) wurden so getrimmt, daß die Anschnittflächen (F 0,25—0,4 mm², quadratisch) zur Gänze von der Randpartie der Objektblöckchen eingenommen wurden; geschnitten wurden unter gleichen Bedingungen ($v_P = 1$ cm/sec; $\alpha = 45°$; $\varepsilon = 5°$; $D = 750$ Å) Nierenrinde, Nierenmark, Lunge und Pankreas von Vertebraten sowie Planariengewebe. Die Ergebnisse sind in Abb. 4d (senkrecht schraffierter Bereich) dargestellt. Die Streuung ist in diesem Fall etwas breiter (rund $\pm 6\%$), wiederum vollkommen unspezifisch; eine signifikante Abhängigkeit vom eingebetteten Gewebe als Ganzem konnte nicht festgestellt werden. Die PD erreicht im Mittel einen höheren Wert (rd. 11 %) als bei leeren Blöcken. — Weitere Versuche wurden — diesmal ausschließlich mit Nierenrinde — unter wechselnden methodischen Bedingungen durchgeführt. Wird im Gegensatz zu den unter Anm. 1, S. 67 angegebenen Bedingungen jeweils eine Größe — v_P ($\geqq 4$ cm/sec), α ($\geqq 70°$), F ($\geqq 1$ mm²) oder D ($\geqq 1500$ Å) — auf den in der Klammer angegebenen Grenzwert erhöht, so verursacht dies eine signifikante Zunahme der PD — teilweise bis über 40%. Es zeigt sich demnach eine Zunahme der PD bei Zunahme der Blockhärte sowie von v_P, $\alpha(\varepsilon)$, F und D. Wie beim Chatterphänomen ist auch hierbei im Gegensatz zu Vermutungen (16) nicht die Messerqualität innerhalb gesetzter Grenzen, sondern α (exakter $\alpha + \varepsilon$) der entscheidende Faktor. Dies konnte durch wiederholtes Schneiden eines Blockes (Eliminieren der variablen Blockhärte) mit Glasklingen verschiedener Güte bewiesen werden. Praktisch unabhängig von der Zahl der Messerriefen ergaben sich bei gleichem α und ε fast identische Werte der PD.

Zur Deformation der Schnitte ist demnach zusammenzufassen: ED sollte restlos rückgängig gemacht werden; PD wird durch das eingebettete Gewebe gegenüber dem freien Plexiglas deutlich erhöht. Daher werden Strukturelemente hoher Trockendichte (vgl. Chatter) stärker plastisch deformiert. PD kann bei folgenden Bedingungen auf ein Minimum eingeschränkt werden: Blockhärte so gering wie möglich (oder: falls hart, so hart wie möglich); $v_P \leqq 1$ cm/sec; $\alpha \leqq 45°$; $F \leqq 0{,}5$ mm²; $\varepsilon \leqq 5°$. Bei Messungen ist darauf zu achten, daß verschiedene Strukturen mit verschiedener Trockendichte verschieden deformiert werden und daß lediglich die Richtung parallel zur Messerschneide verbindliche Werte liefert.

Wie gezeigt wurde, sind auch bei idealen Voraussetzungen Unregelmäßigkeiten und Deformationen der Schnitte nicht zu vermeiden. Neben dem behandelten Fragenkomplex fällt in den Rahmen dieses Referates noch ein zweiter von ebenso großer Bedeutung. Er betrifft die **Reproduzierbarkeit und Art des Schneidevorganges** selbst. Zunächst müssen wir uns hierzu kurz der Schnittdicke (D) zuwenden. Dieses Problem wurde vor kurzem unabhängig von Bachmann und P. Sitte (26, 27) und Peachey (3, 17) sowie von Sechaud et al. (28) behandelt; im Vergleich mit der bislang gebräuchlichen Relation der Interferenzfarben im Reflex zu D

Tabelle

Interferenzfarbe im Reflex	Bachmann u. Sitte (26)	Peachey (3)	Elektr.-Mikrosk.	Licht-Mikrosk.
Grau	$\leqq 450$	$\leqq 600$	Hochauflösende Arbeiten	Anoptralkontrast nach Wilska
Silber	450— 800	600— 900	Routine-Arbeiten	
Gold	800—1300	$\geqq 900$		
Kupfer	1300—1600	$\leqq 1500$	Durchstrahlbar bei schlechter Auflösung	Phasenkontrast
Violett	1600—2000	1500—1900		
Blau	2000—2300	1900—2400		

(1) ergaben sich übereinstimmend höhere D-Werte (vgl. Tab.). Die neuen Werte stimmen gut mit den optischen Berechnungen überein; die Abweichungen liegen innerhalb der Streuung infolge subjektiver Farbabschätzung. D nimmt ab in Reihe Blau-Violett-Kupfer-Gold-Silber-Grau; dabei erreichen die Schnitte bei „Silber" eine maximale Helligkeit (ohne Farbton); „Grau" sind Schnitte (desgl. ohne Farbton), deren Helligkeit zwischen derjenigen von „Silber" und derjenigen der Flüssigkeitsoberfläche liegt. „Grau" erscheint nie dunkler als der Flüssigkeitsspiegel — eine Tatsache, die sowohl optisch-theoretisch wie experimentell leicht zu beweisen ist. Die Bezeich-

nung „Dunkelgrau" ist daher irreführend; die Annahme, daß die Schnitte bei Annäherung von D an 0 „Schwarz" würden, trifft nicht zu. Nach dem Habitus der veröffentlichten Schnittaufnahmen ist zu schließen, daß derzeit meist Schnitte mit D zwischen 400 und 1000 Å verwendet werden; nur in Ausnahmefällen liegt D unter 400 Å. Dünnste Schnitte (optimistisch: D rd. 250 Å) zeigen bei der üblichen Fixierung und Einbettung im allgemeinen nicht nur einen schwachen Kontrast; der Schnitt erscheint leer, die Feinstrukturen lösen sich in einzelne Grana auf und geben schwer deutbare, unansehnliche Bilder. Für die Praxis dürften daher derzeit die Richtwerte entsprechen,

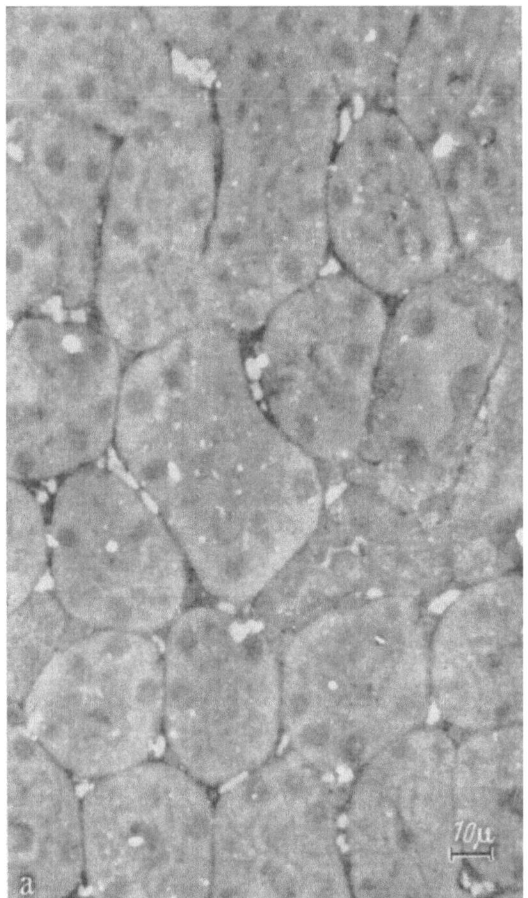

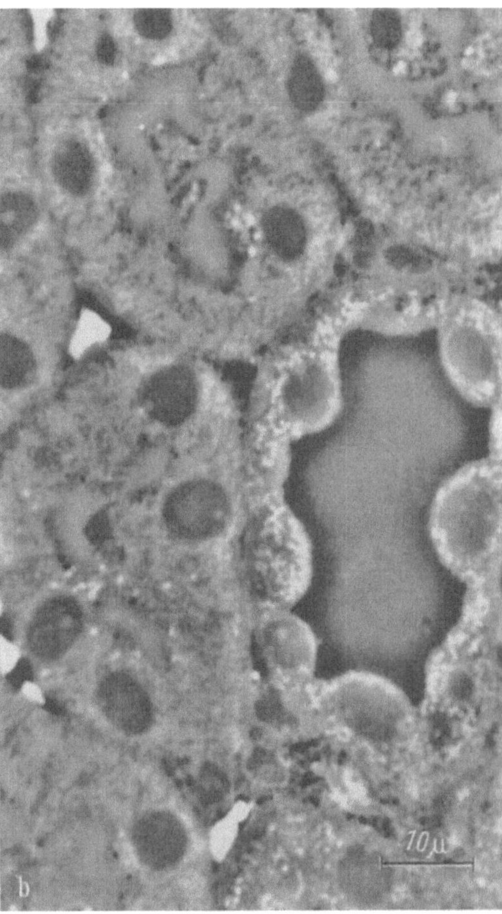

Abb. 5a u. b. Ultradünnschnitt durch Nierenrinde; Plexiglaseinbettung nach NEWMAN et al., OsO$_4$-Fixierung nach PALADE. Abb. mit Reichert-Anoptralkontrast; Obj. 20:1. Plexiglas nicht ausgelöst; Beobachtung ohne Einschlußmedium, jedoch mit aufgelegtem Deckglas. b) wie a), jedoch Obj. 45:1; Plexiglas mit Aceton ausgelöst

die aus der Tabelle hervorgehen: höchste Auflösung bei D zwischen rd. 400 und 600 Å (Grau bis Silber); hinreichend bei Routinearbeiten (mittlere Vergrößerung): D zwischen rd. 600 und 1200 Å (Silber bis Gold); Schnitte mit $D > 1200$ Å (Gold-Kupfer-Violett) sind zwar durchstrahlbar, aber wegen ihrer schlechten Auflösung nicht zu empfehlen. Soweit nicht interferometrische Kontrollen von D vor Beginn der elektronenoptischen Untersuchung [vgl. BACHMANN und P. SITTE (29)] vorliegen, sind am besten die Interferenzfarben anzugeben. — In einem kann die Frage beantwortet werden, welche Lichtoptik in diesem D-Bereich geeignet ist. Nach eigenen Erfahrungen eignet sich für $D < 1000$ Å im allgemeinen die Reichert-Anoptralkontrastoptik nach WILSKA (30); vgl. Abb. 5a und b. Bei $D > 1000$ Å liefert die Phasenkontrastoptik zufriedenstellende Abbildungen.

Zu der grundsätzlichen Frage tragen die neuen Erkenntnisse über D etwas bei. Mehrere Autoren (vgl. hierzu 31) stellten bei Serienschnitten von Schnitt zu Schnitt Veränderungen der Strukturen fest, die größer sind, als man zunächst annehmen mochte. Hierbei wird ein Teil — allerdings nicht der wesentliche — durch das falsche Abschätzen von D erklärt. Der wichtigste Faktor liegt im Fehlen einer räumlichen Betrachtung, auf deren Bedeutung in letzter Zeit in verschiedenem Zusammenhang hingewiesen wurde; man vgl. u. a. LENZ, „Tomatensalatproblem" (*32—34*), FUJIWARA, „Geometrischer Dickeneffekt" (*21*), oder die Hinweise von MORGAN et al. (*35*) oder PEACHEY (*3, 17*). Jede exakte Rekonstruktion auf Basis der Massendickeverteilung hätte, wie Abb. 6 aufzeigt, klar zum Ausdruck gebracht, daß 1. beim Anschneiden räumlicher Gebilde ($\varnothing$ z. B. 0,2—2 μ) selbst bei $D = 300$ Å (!) von Schnitt zu Schnitt sehr große Veränderungen der

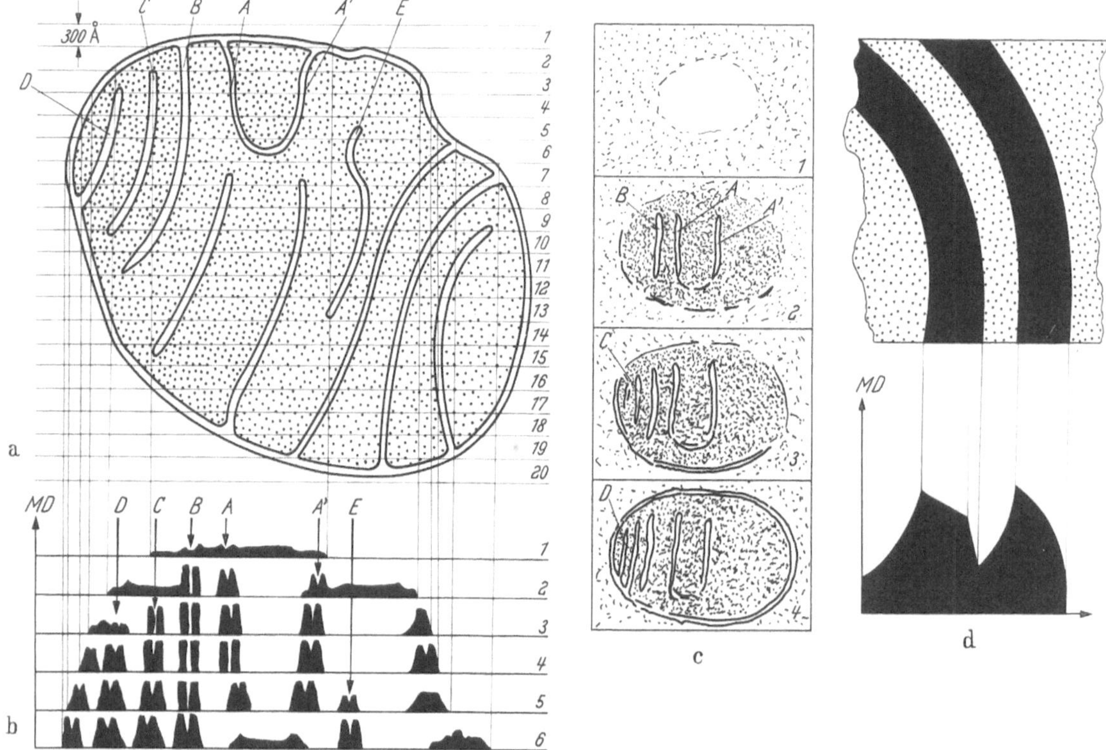

Abb. 6a—d. Schematische Darstellungen zur Deutung von Serienschnitten. a) Mitochondrium, zerlegt in Serienschnitte 1—20 (Dicke je 300 Å). b) Massendicke der Membranen für Schnitte 1—6. c) Rekonstruktion der ersten Schnitte durch das Mitochondrium (Schnitte 1—4). d) Massendicke (unten) von Doppellamellen, welche nicht durchgehend senkrecht zur Schnittfläche (waagerecht; oben) liegen

Struktur auftreten müssen; 2. daß feinere Elemente, z. B. Schläuche des endoplasmatischen Reticulums, oft lediglich in einem einzigen Schnitt auftreten können; 3. daß die klare elektronenoptische Abbildung einer einfachen oder mehrschichtigen Membran nicht den Rückschluß erlaubt, daß diese exakt senkrecht zur Schnittfläche liegt (Abb. 6d); sowie 4. daß demnach eine räumliche Rekonstruktion nach Serienschnitten [BANG und BANG (*36*)] bei Beachten dieser Tatsachen sehr wohl möglich ist.

Es kann daher zu Ansichten Stellung genommen werden, die bislang in diesem Zusammenhang unter anderen Voraussetzungen geäußert wurden. WILLIAMS und KALLMAN (*31*) versuchten, die sprunghaften Veränderungen von Schnitt zu Schnitt durch Materialverluste beim Schneiden und bei der elektronenoptischen Beobachtung zu erklären. Letztere sind heute sicher erwiesen [vgl. WATSON (*37*)]. Verluste beim Schneiden wurden durch Ungleichmäßigkeiten der Klinge erklärt. Demnach würde durch die Scharten stets Material entfernt, so daß stets an den gleichen Schnittstellen dünnere Partien beobachtet werden können; mit anderen Worten: Die Schnitte repräsentieren nicht das gesamte vom Block entfernte Material. Genaue Untersuchungen während des Schneidens sowie an Blöcken und Messern nach dem Schneiden gaben keine Anhaltspunkte für die Ablagerung verlorenen Materials. Die zitierte Ansicht trifft für den Großteil der Messerscharten nicht zu. Wird das Präparat jeweils gleichmäßig exakt über das Messer geführt, so erzeugt eine keilförmige Scharte (S; Abb. 7a; Schnitte 1 und 2) ein Profil, das bei hinreichender Stabilität der Schnitte zu keinen Massendickenunterschieden führt [vgl. FARRANT und POWELL (*38*)]; treten jedoch in der Präparatführung Unregelmäßigkeiten auf (Schnitt 3), so bilden sich in aufeinanderfolgenden Schnitten jeweils korrespondierend ohne Materialverlust Wülste und Rillen

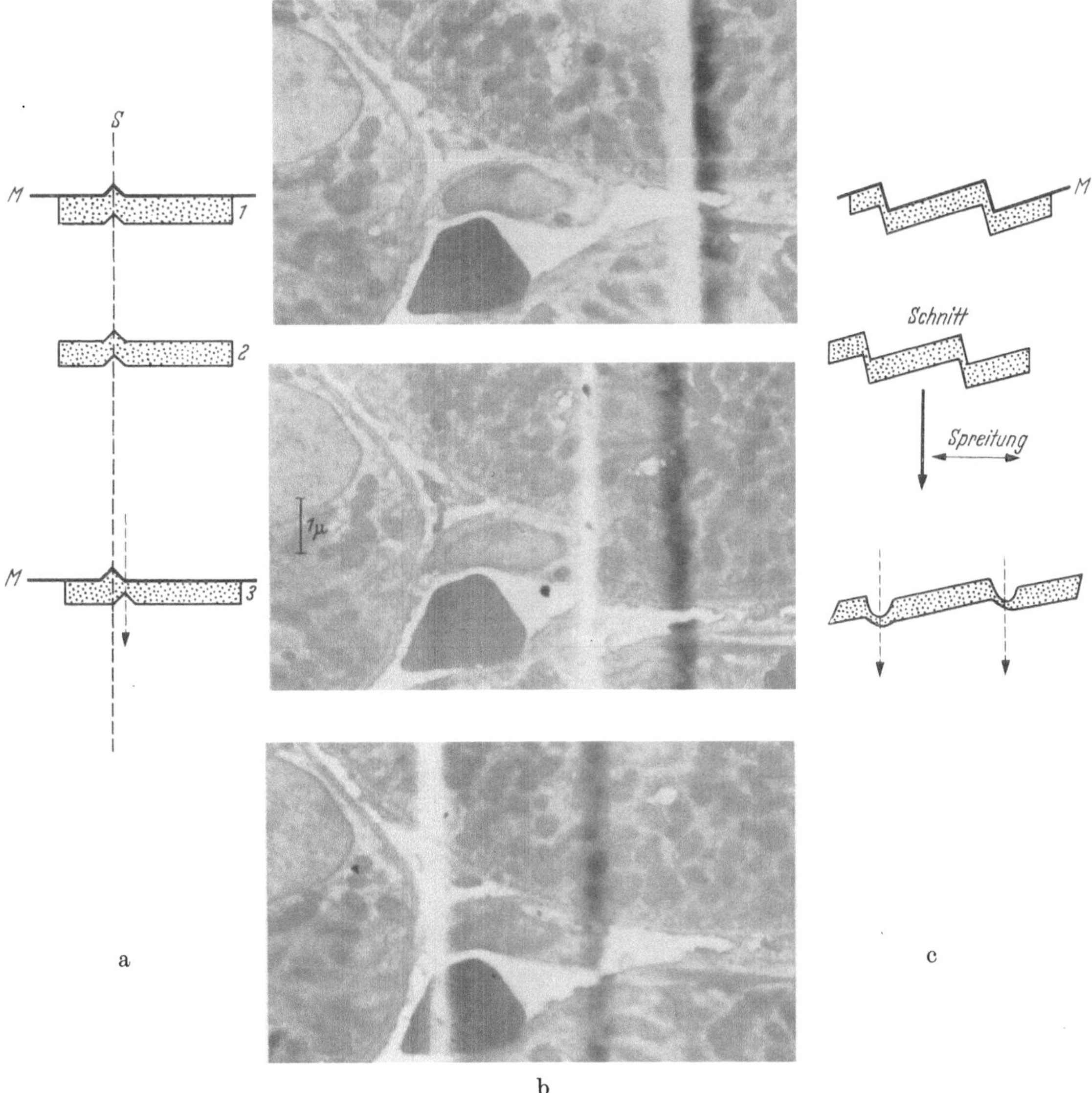

Abb. 7a—c. Erklärungen im Text

aus; vgl. Abb. 7b. Werden jeweils an entsprechenden Schnittstellen Streifen geringerer Massendicke beobachtet, so sind diese wohl nach Abb. 7c zu erklären; Substanzverluste lägen auch hierbei nicht vor.

Der zweite Teil der Betrachtungen kann wie folgt kurz zusammengefaßt werden: Neuere Arbeiten führten zu einer Revision der Schnittdicken/Interferenzfarben-Relation, die bislang zu optimistisch beurteilt wurde; Materialverluste treten beim Schneiden kaum auf — Serienschnitte zeigen also ein Kontinuum der Struktur; räumliche Rekonstruktionen sind daher bei Beachten der Gegebenheiten ohne weiteres möglich. Der Schneidevorgang kann nicht, wie oft geäußert, als ein „Kratzen", „Schaben" oder dergleichen ("ripping or gauging", "scraping") aufgefaßt werden; die beim Schneiden auftretenden Phänomene (Schwingungen, Deformationen usw.) liegen in der Natur des Vorganges und sind durch die hohen Vergrößerungen und die verfeinerte Präparationstechnik in vollem Ausmaß erkennbar geworden. Da man für die Ultramikrotomie die besten derzeit bekannten Klingen verwendet, wird man nicht fehlgehen, wenn man diesen Vorgang (soweit man das Wort „Schneiden" in seiner eigentlichen Bedeutung anwendet) als einen „Schneidevorgang par excellence" bezeichnet.

Literatur

1. Porter, K. R., and J. Blum: Anat. Rec. 117, 685 (1953).
2. Rhodin, J.: Correlation of structural organization and function in normal and experimentally changed proximal convoluted tubule cells of the mouse kidney. Stockholm, Karolinska Inst., Dept. of Anatomy, 1954.
3. Peachey, L. D.: J. biophys. biochem. Cytol. 4, 233 (1958a).
4. Bennet, H. St., N. J. Schaal, J. S. Quinton, J. S. Reddie and J. H. Luft: EMSA-Meeting, Santa Monica, USA; Abstr. B/23 (1958).
5. Latta, H., and J. F. Hartmann: Proc. Soc. exp. Biol. (N. Y.) 74, 436 (1950).
6. Fernández-Morán, H.: Exp. Cell Res. 5, 255 (1953).
7. — J. biophys. biochem. Cytol. 2, suppl. 29 (1956).
8. — Mikroskopie (Wien) 12, 81 (1957).
9. Birch-Andersen, A.: Dieser Band S. 44.
10. Glauert, A. M., G. E. Rogers and R. H. Glauert: Nature (Lond.) 178, 803 (1956).
11. — and R. H. Glauert: J. biophys. biochem. Cytol. 4, 191 (1958).
12. Grigg, G. W., and H. Hoffman: J. biophys. biochem. Cytol. 4, 331 (1958).
13. Kellenberger, E., W. Schwab et A. Ryter: Experientia (Basel) 12, 421 (1956).
14. Ryter, A., et E. Kellenberger: Dieser Band S. 52.
15. Morgan, C., D. H. Moore and H. M. Rose: J. biophys. biochem. Cytol. 2/4, suppl., 21 (1956).
16. Gelber, D.: J. biophys. biochem. Cytol. 3, 311 (1957).
17. Peachey, L. D.: Dieser Band S. 72.
18. Satir, P. G., and L. D. Peachey: J. biophys. biochem. Cytol. 4, 345 (1958).
19. Cosslett, V. E.: Brit. J. appl. Physics 7, 10 (1956).
20. — J. biophys. biochem. Cytol. 3, 815 (1957).
21. Fujiwara, T.: J. Electronmicr. (Japan) 6, 65 (1958).
22. Newman, S. B., S. Borysko and M. Swerdlow: J. Res. Nat. Bureau Standards 43, 183 (1949).
23. Nylen, M. U., and J. W. Holland: Exp. Cell Res. 13, 88 (1957).
24. Sotelo, J. R.: Exp. Cell. Res. 13, 599 (1957).
25. Klima, J.: persönl. Mitt., sowie: Dieser Band S. 58.
26. Bachmann, L., u. P. Sitte: Comparison between colors in reflected light and thickness of sections according Tolansky-measurements. (Ausführliche Darstellung von Methoden und Ergebnissen). Scientific exhibit and travelling exhibit, XV th EMSA-Meeting, MIT., Cambridge, Mass., USA. September 1957.
27. — u. P. Sitte: Mikroskopie (Wien) 13, 289 (1959).
28. Kellenberger, E., J. Sechaud et A. Ryter: Dieser Band S. 212.
29. Bachmann, L., u. P. Sitte: Dieser Band S. 75.
30. Wilska, A.: Mikroskopie (Wien) 9, 1 (1954).
31. Williams, R. C., and F. Kallman: J. biophys. biochem. Cytol. 1, 301 (1955).
32. Lenz, F.: Optik 11, 524 (1954).
33. — Physik. Verh. 6/2, 26 (1955).
34. — Z. wiss. Mikr. Techn. 63, 50 (1956).
35. Morgan, C., S. A. Ellison, H. M. Rose and D. H. Moore: Exp. Cell Res. 9, 572 (1955).
36. Bang, B. G., and F. B. Bang: J. Ultrastr. Res. 1, 138 (1957).
37. Watson, M. L.: J. biophys. biochem. Cytol. 3, 1017 (1957).
38. Farrant, J. L., and S. E. Powell: Proc. I st Reg. EM.-Conf. Asia and Oceania, p. 147. Tokyo 1956.

Section thickness and compression

Lee D. Peachey

The Rockefeller Institute for Medical Research, New York

The need for information about the thickness of sections used for electron microscopy is the direct result of electron optics, and arises from the fact that the depth of field is such that the whole thickness of the section is focused in the final image. The electron image is, then, a superposition of many profiles from various levels in the section. Analysis of such images requires knowledge of the depth of the specimen in addition to the lateral dimensions displayed in the image. The effect of specimen depth can be considerable in electron microscope images, where structures with dimensions of a few tens of mμ are often studied in sections with thickness in the range of 100 mμ. Therefore, consideration of section thickness is indeed essential for proper interpretation of electron images of sections.

The colors seen in light reflected from floating sections have provided a convenient measure of relative thickness, and the series: colorless, gold, purple, and blue, has become well known to microtomists as the colors of sections in order of increasing thickness. This is the same series as the colors seen in ordinary interference phenomena in thin films, and it seemed apparent that the colors of sections were interference colors of the usual sort due to interference between light rays reflected from the top and bottom surfaces of the sections.

Thus, the correlation between section thickness and color should agree with optical interference data. The estimates of section thickness that were available, however, in every case placed the thickness for a given color at a somewhat lower value than that predicted from the optical data. Since section thickness information is of value, it was decided to clarify this point by making direct thickness measurements on actual sections (1).

Sections were cut from methacrylate blocks using a Porter-Blum microtome and glass knives. The sections were mounted on glass slides that had previously been coated with chromium and a series of steps of barium stearate double layers. Section thickness was measured with an ellipsometer, using the double layers, the thickness of which is very accurately known, for calibration purposes.

In the ellipsometer, plane-polarized light incident on the slides becomes elliptically-polarized after reflection. The state of polarization of the reflected light is extremely sensitive to the thickness of the section, and measurements are theoretically possible on sections as thin as 0.1 mμ. An accuracy of ± 5 mμ is easily obtained throughout the range of thickness used here. This is greater than the accuracy with which the colors may be judged, because of the subjective nature of color judgement.

Table 1. *Color-Thickness Correlation*

Color	*Thickness range* Index of refraction = 1.5 *mμ*
Gray	less than 60
Silver	60 to 90
Gold	90 to 150
Purple	150 to 190
Blue	190 to 240
Green	240 to 280
Yellow	280 to 320

The results of these measurements are shown in Fig. 1*. The abscissa is a thickness scale and the color scale on the ordinate is based on optical theory and published optical data. The scales of the ordinate and abscissa are the same, so the 45° line represents the expected correlation between color and thickness

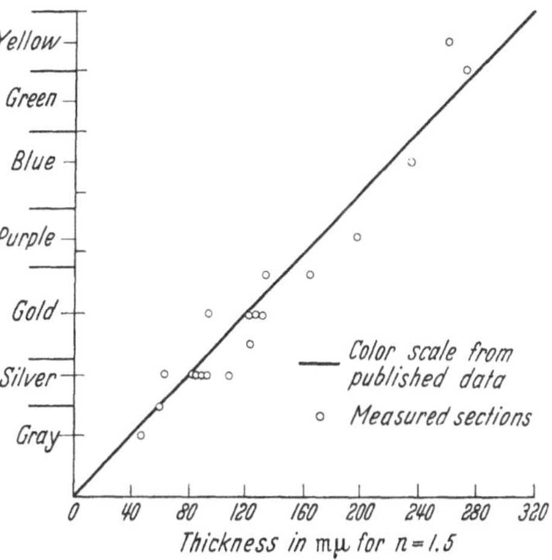

Fig. 1. Results of section thickness measurements and correlation with optical data

for interference phenomena. The results of the measurements on sections are indicated by the circles, which are plotted *opposite* the observed colors of the sections, and *above* their measured thicknesses. The correlation with the theoretical line is good and it can be concluded that the interference colors shown by sections do give a true measure of thickness and that the earlier estimates were in error.

Having these results, it was then possible to construct the color-thickness chart shown in Table 1. It should be emphasized that the color scale is continuous, and that the division of the chart into discrete colors is merely for convenience. Actually, various shades of these colors may be discerned and finer thickness distinctions made. For instance, within the gold range, from 90 to 150 mμ, shades of light gold near the silver end, pure gold near the center, and reddish

* All figures are reproduced with the permission of the Journal of Biophysical and Biochemical Cytology.

gold near the purple end may be observed, and the section thickness may be placed at the thin, median, or thick portion of the gold range, respectively. Thus the thickness of the section may be determined to within 10 to 20 mμ by comparison with this chart.

During the course of this study, measurements were made on the physical distortion of the section which resulted from the cutting process. It was already known that sections were shortened or "compressed" in the direction of cutting without a corresponding change in width, and our measurements indicated that this compression was often as high as 30 to 50%, or even higher. This shortening is undesirable for two reasons. First, dimensions and spatial relationships in the section are altered, and second, the specimen becomes unduly thick, which is undesirable for optical reasons. Since it appeared as if compression was an unavoidable concomitant of the sectioning process, we investigated methods of removing it from the sections before they were mounted on the specimen grids (2).

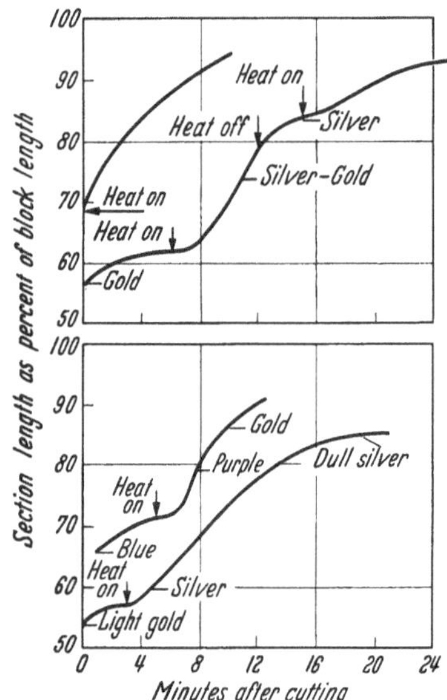

Fig. 2. Increase of section length with heat treatment

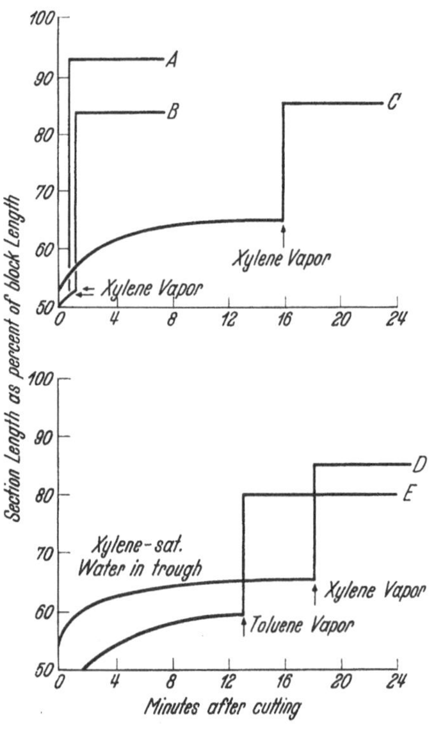

Fig. 3. Increase of section length with solvent vapor treatment

High concentrations of solvent in the collecting fluid were found to be effective in removing compression, but were not satisfactory because of excessive extraction of embedding material. Heating of the fluid (which could be water) about 10° C, however, removed most of the compression artifact in 5 to 10 min, as shown in Fig. 2.

A far superior method of removing the compression artifact was found to be the use of solvent in a *vapor* phase, thus avoiding extraction of embedding material and also avoiding the production of thermal effects in the microtome, which was a serious defect of the heating method. The solvent vapor is easily applied to the sections as they float on the collecting fluid by dipping a glass rod into a solvent such as xylene, and bringing the attached drop of solvent close to, *but not in contact with* the sections. The expansion is very rapid when the solvent is brought close enough, and should be observed through the viewing microscope to make sure that the expansion is complete. Solvents other than xylene are effective, but the use of *vapor* is essential. It is suggested, that if other solvents are to be used, they should at least be compared to xylene to make sure that they are giving complete expansion. The sections usually expand to 85—95% of the block length, leaving only a small residual compression (see Fig. 3).

The obvious effect of expansion on the appearance of sections in the electron microscope is the restoration of dimensions in one direction to more realistic values. This is especially obvious in the case of spherical structures such as some nuclei, which appear round instead of oval. It should not be forgotten that the specimen is also improved by being thinner without sacrifice of material.

In summary, we have measured the thicknesses of sections of various colors and have found that the thicknesses correspond to those expected from interference theory. A clor scale has been presented which can be used to estimate the thickness of a section to within 10 or 20 mμ. A simple method employing solvent vapor has been found for reducing the compression of sections which occurs during cutting, thus improving the spatial relationships of tissue structures included in sections. Knowledge of section thickness permits a more accurate interpretation of electron images of structures that have different profiles at different levels in the section.

References

1. Peachey, L. D.: J. biophys. biochem. Cytol. **4,** 233 (1958).
2. Satir, P. G., and L. D. Peachey: J. biophys. biochem. Cytol. **4,** 345 (1958).

Über Schnittdickenbestimmung nach dem Tolansky-Verfahren

L. Bachmann und P. Sitte

Aus dem Physikalisch-chemischen Institut und dem Labor für Elektronenmikroskopie der Universität Innsbruck

Seit die Ultramikrotomie breitere Anwendung findet, besteht das Problem der Schnittdickenbestimmung; es hat heute nichts an Aktualität verloren, weil sich mehr und mehr zeigt, daß die genaue Kenntnis der Schnittdicke an einer bestimmten, mikroskopisch kleinen Präparatstelle für die richtige Bilddeutung entscheidend sein kann — es sei hier nur an die Plasmastruktur (*1, 2*) erinnert. Eine exakte Schnittdickenmessung stößt jedoch auf unerwartet große Schwierigkeiten; wohl vornehmlich aus diesem Grunde liegen auch die Angaben der verschiedenen Autoren stark auseinander und vielfach zu tief [vgl. (*3*)].

Wenn man von Spezialfällen absieht, in denen durch günstig liegende Präparatelemente bekannter Abmessungen eine Bestimmung der Schnittdicke möglich ist (Cilien, Lipoidkugeln und dgl.), bleiben folgende allgemeiner verwendete Verfahren [vgl. etwa (*4*)]:

1. Abschätzen der Schnittdicke nach den am flottierenden Schnitt auftretenden Interferenz-
2. Messung der Schnittdicke am Schnittrand nach Beschattung;

erscheinungen;

3. Messung der Elektronenabsorption;
4. Stereoskopie von Schnitten;
5. Interferometrie.

Für die genaue Dickenbestimmung an einer konkreten, eng umgrenzten Präparatstelle lassen sich die meisten dieser Meßmöglichkeiten nicht anwenden. Die *Beschattungsmethode* liefert lediglich Angaben über die Dicke am Schnittrand, nicht aber an einer gegebenen Stelle in der Schnittfläche und scheidet daher für die Lösung des gestellten Problems aus; ebenso die 2. der genannten Methoden: Die *Interferenzerscheinungen* gestatten zwar eine rasche Abschätzung der mittleren Schnittdicke und somit eine erste Beurteilung der Schnittgüte. Sie eignen sich aber nicht für eine genaue Dickenbestimmung dünnster Schnitte, da sich gerade die Interferenzerscheinungen „grau" und „silber" über einen verhältnismäßig großen Dickenbereich erstrecken (*3*).

Auch die Dickenbestimmung aus der Messung der *Elektronenabsorption* (*5*) bringt große Schwierigkeiten mit sich: Zum einen ist genaue Eichung notwendig, so daß sich dieses Verfahren immer auf ein anderes stützen muß (z. B. Interferometrie). Weiter ist zu bedenken, daß diese Eichung dann nur für Objektbereiche ganz bestimmter Dichte gilt, in der Praxis meist für „leeres" (nicht präparathaltiges) Plexiglas. Nun kommt es aber gerade auf die präparathaltigen Stellen an, die sich in ihrer Dicke von den leeren erheblich unterscheiden können und in denen vor allem die Dichte starken Schwankungen im allerkleinsten Bereich unterworfen ist. Somit schwinden die Aussichten, anhand der Elektronenabsorption zu genauen Dickenangaben für bestimmte Präparatstellen zu kommen. Es bleibt noch die *Stereometrie* von Schnitten (*4, 6*), die aber nur während der Bestrahlung im Elektronenmikroskop möglich ist, wodurch die aktuelle Dicke gegenüber der ursprünglichen „Schnitt"-Dicke in nicht genau abschätzbarem Ausmaße verringert ist (*6*).

Es verbleibt noch die *Interferometrie* von Schnitten. Dabei ist von vornherein zu verlangen, daß die Messung unter dem Lichtmikroskop erfolgt, damit allenfalls die Dicke mikroskopisch kleiner Objekteinzelheiten angegeben werden kann; denn Ultradünnschnitte sind keineswegs über die ganze Fläche gleich dick (*3, 6*). Die Möglichkeiten der Interferenzmikroskopie zur Präparatdickenbestimmung sind vornehmlich in zwei Richtungen ausgenützt worden: einerseits unter Verwendung eines Sondenverfahrens (*7*), andererseits mit Hilfe des Tolansky-Verfahrens (*3, 8, 9*).

Das Tolansky-Verfahren ist für seine Genauigkeit bekannt und läßt sich in geeigneter Modifikation auch für die Vermessung von Schnittdicken leicht anwenden (*3, 8*). Es hat jedoch den Nachteil, daß dabei die Präparate etwa 1000 Å dick mit Silber bedampft werden müssen und daher nicht mehr im Elektronenmikroskop untersucht werden können.

Ziel der vorliegenden Arbeit war es, zu prüfen, ob die aufgedampfte Silberschicht nach der Tolansky-Messung wieder spurlos entfernt werden kann, ohne daß das Präparat dabei geschädigt wird.

Methodik. Für unsere Untersuchungen verwendeten wir als Objekte Ultradünnschnitte durch in Plexiglas eingebettete tierische und pflanzliche Gewebe, die mit dem Reichert-Ultramikrotom nach H. SITTE (*10*) hergestellt worden waren. Das hier verwendete Verfahren eignet sich aber prinzipiell auch für andere Präparate. — Die Schnitte wurden zunächst auf Glasobjektträger aufgebracht, die mit einer Formvarfolie überzogen waren, anschließend bei 10^{-5} Torr aus 145 cm Entfernung in einer Apparatur der Gerätebau-Anstalt Balzers mit Silber bedampft bis zu einer Lichtdurchlässigkeit von 2%. Nach der Bedampfung wurden die Schnitte nach der früher (*8*) angegebenen Weise vermessen, anschließend die Silberschicht wieder abgelöst und die Schnitte nach dem Zielpräparationsverfahren von H. SITTE (*11*) auf Athene-Netze aufgebracht und im Elektronenmikroskop beobachtet.

Ergebnisse. Die Untersuchungen ergaben, daß durch Quecksilber die Silberschicht unter Amalgamierung augenblicklich und ohne Rückstand aufgelöst wird. Dieses Verfahren ist schnell und einfach und läßt das Präparat, das vom Quecksilber nicht benetzt wird, völlig unverändert.

Das Silber läßt sich auch mit verdünnter Salpetersäure innerhalb weniger Sekunden entfernen. In orientierenden Untersuchungen konnten wir auch hierbei keine Objektveränderungen feststellen. Das Quecksilber-Verfahren ist aber sicherer und jedem nassen Verfahren vorzuziehen. Ein Versuch, das Silber von den Präparaten durch intensive Bestrahlung im Elektronenmikroskop wegzudampfen, blieb erfolglos.

Weiterhin wurde genau geprüft, ob beim Aufdampfen des Silbers unter den angegebenen Bedingungen infolge der Erwärmung Präparatschäden auftreten, wie es für zarte Objekte gelegentlich beschrieben wurde (*12*). Eine überschlagmäßige Berechnung der Präparaterwärmung aus dem Wärmeinhalt des aufgedampften Silbers zeigte, daß sie vernachlässigbar gering ist; auch die Strahlungswärme führt bei dem großen Abstand des Präparates von der Strahlenquelle zu keiner spürbaren Erwärmung. Es wurde versuchsweise Paraffin (Schmelzpunkt 42°) den Präparaten adhäriert und mitbedampft, wobei jedoch der Schmelzpunkt nicht erreicht wurde.

Übrigens hat sich zwischen unbedampften Kontrollen und den vermessenen Schnitten jeweils der gleichen Schnittserien nie irgendein Unterschied feststellen lassen; somit scheint erwiesen, daß die zur Schnittdickenbestimmung notwendigen zusätzlichen Präparationsgänge (Ver- und Entsilberung) ohne Objektänderung vorgenommen werden können. Es ist also — in Verbindung mit zweckmäßiger Zielpräparation — möglich, die Präparatdicke einer bestimmten, lichtoptisch identifizierten Objektstelle, selbst wenn sie nur wenige μ^2 groß ist, genau zu vermessen und *dieselbe* Stelle dann im Elektronenmikroskop zu untersuchen.

Die Abb. 1—3 geben ein Beispiel. Abb. 1a zeigt einen Ultradünnschnitt durch Gewebe der Erbsenwurzelspitze (aus einer Fixierungsstudie, P. SITTE); der Schnitt wurde nach TOLANSKY vermessen (Abb. 2), entsilbert und dann im Elektronenmikroskop untersucht (Abb. 3), wobei

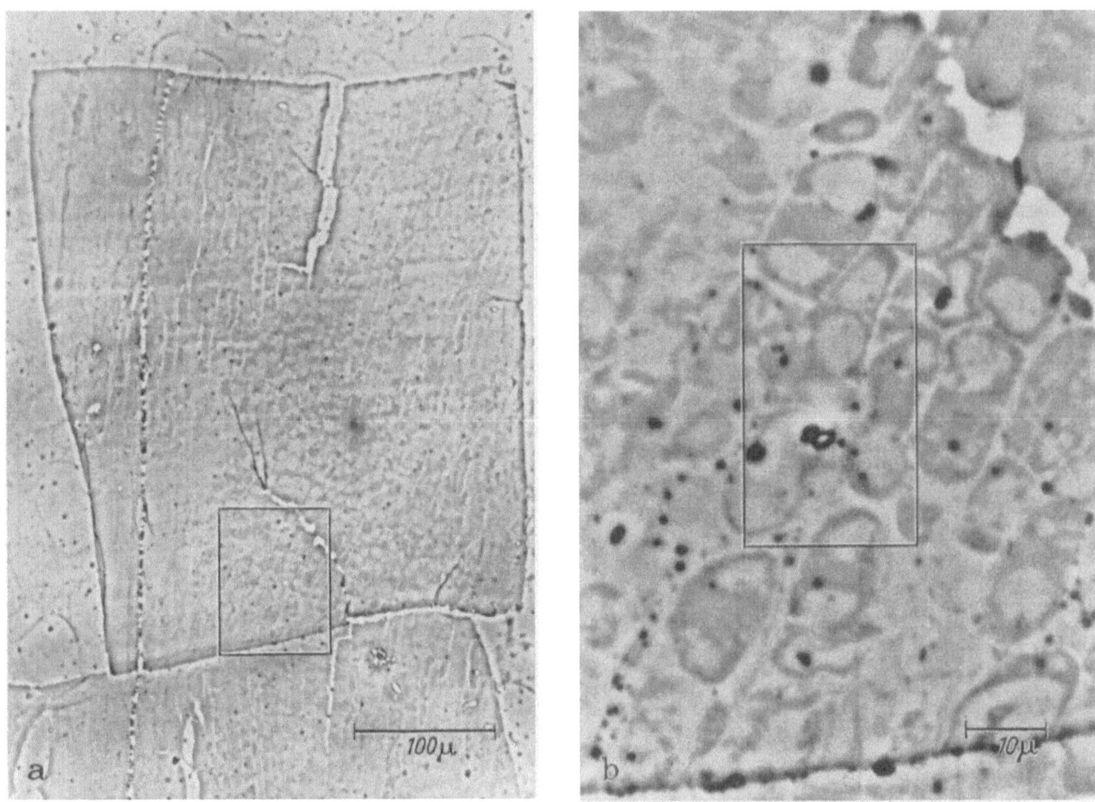

Abb. 1a u. b. Ultradünnschnitt durch Erbsenwurzelgewebe im Phasenkontrast, 195:1.
Der eingezeichnete Bereich in Abb. 1b in 1100facher Vergrößerung

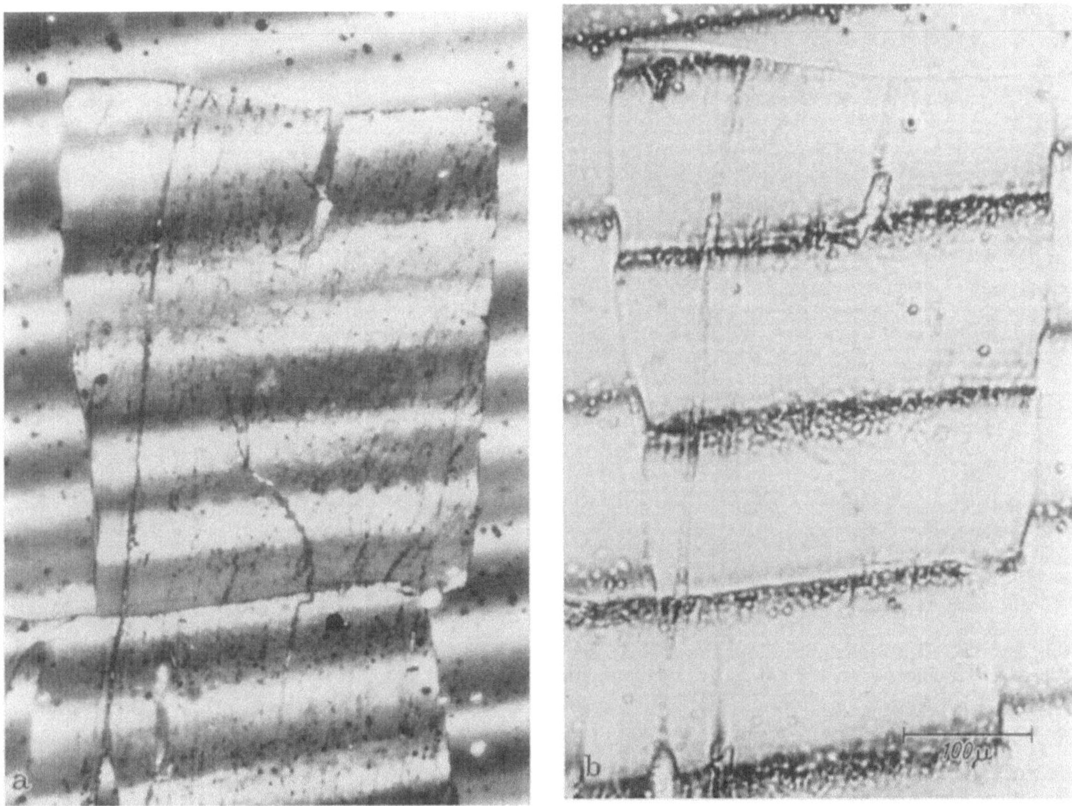

Abb. 2a u. b. Tolansky-Vermessung des gleichen Schnittes. 174:1. a) Durchlicht, ungefiltert
(Quecksilber-Höchstdrucklampe); b) Auflicht, monochromatisch ($\lambda = 2795$ Å)

leicht der in Abb. 1b eingezeichnete Ausschnitt (der auch am versilberten Schnitt während der Messung deutlich zu erkennen war) wiedergefunden wurde. An der Stelle der Abb. 3b ist der Schnitt 650 Å dick.

Der wesentlichste Mangel, der bisher dem Tolansky-Verfahren bei der Dickenbestimmung elektronenoptischer Präparate anhing — daß die vermessenen Präparate selbst durch die notwendige Versilberung für die weitere Beobachtung unbrauchbar waren —, ist somit behoben, und der breiten Anwendung dieses genauen und relativ einfachen Verfahrens steht nichts im Wege.

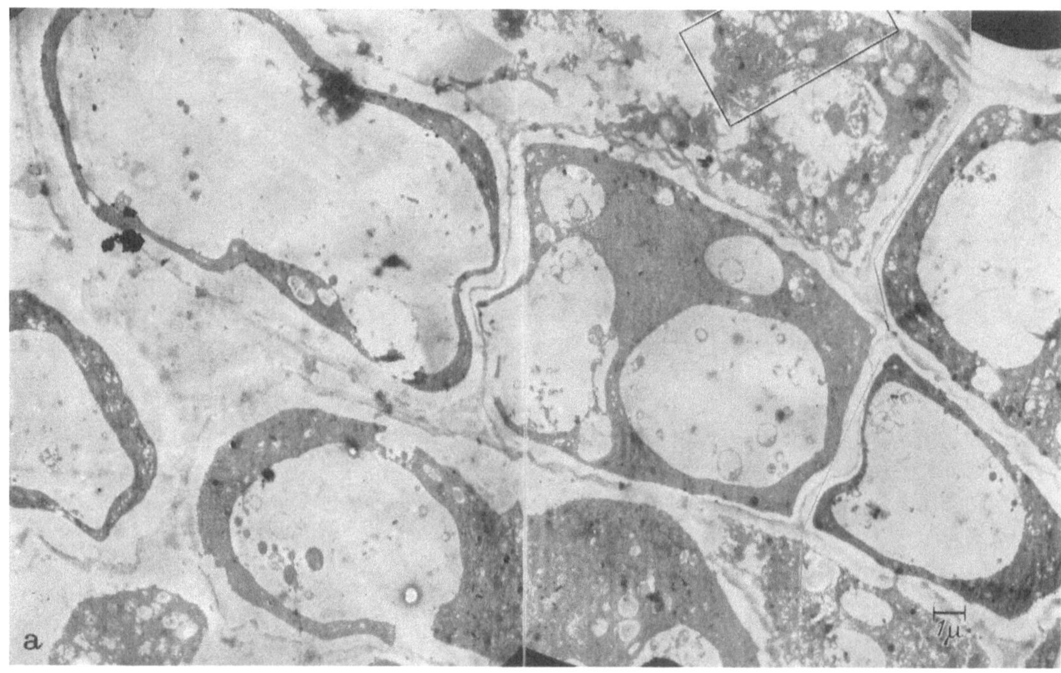
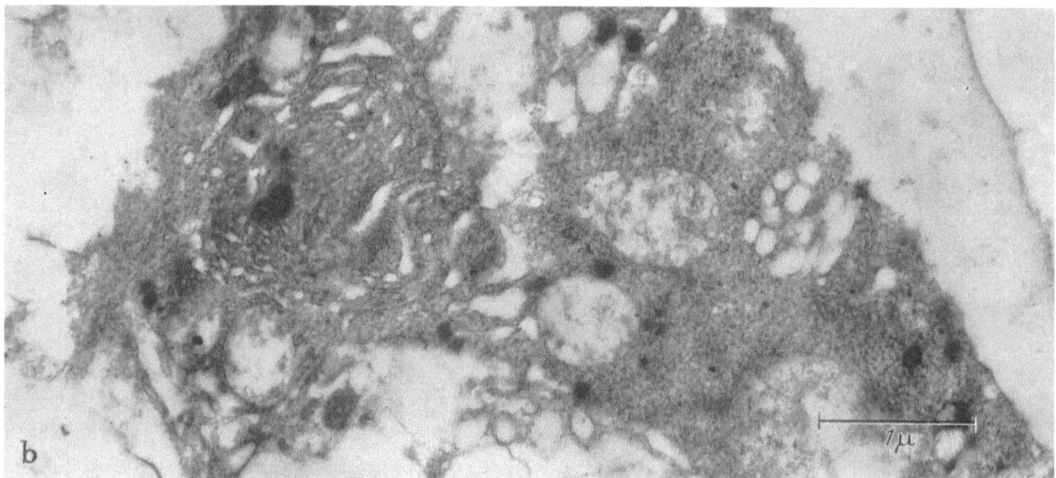

Abb. 3a u. b. Elektronenmikroskopische Aufnahmen des gleichen Schnittes. a) Übersicht, der Bereich ist in Abb. 1b eingezeichnet. 4400:1. b) Detailaufnahme des in a) eingezeichneten Ausschnittes. 23 000:1

Literatur

1. Strugger, S.: Ber. dtsch. bot. Ges. 70, 91 (1957).
2. Sitte, P.: Protoplasma 49, 447 (1958).
3. Bachmann, L., u. P. Sitte: Mikroskopie (Wien) 13, 289 (1959).

4. Gettner, M. E., and L. Ornstein: In G. Oster, u. A. W. Pollister: Physical techniques in biological research, New York **3**, 627 (1956).
5. Reimer, L.: Naturwissenschaften **44**, 335 (1957).
6. Williams, R. C., and F. Kallman: J. biophys. biochem. Cytol. **1**, 301 (1955).
7. Menzel, E.: Optik **14**, 151 (1957).
8. Bachmann, L.: Über die Untersuchung von Oberflächen mit Hilfe eines modifizierten Tolansky-Interferenzverfahrens. Mikroskopie (Wien) **13**, 258 (1958).
9. Geren, B. B., and D. McCulloch: Exp. Cell Res. **2**, 97 (1951).
10. Sitte, H.: Mikroskopie (Wien) **10**, 365 (1956).
11. — Experientia (Basel) **13**, 419 (1957).
12. Frey-Wyssling, A., K. Mühlethaler u. H. Moor: Mikroskopie (Wien) **11**, 219 (1956).

Gezielte Ultradünnschnitte durch beliebige Zellen aus jedem Gewebe*

Friedrich-Wilhelm Schlote und Gisela Föge

Anatomisches und Zoologisches Institut, Göttingen

Viele tierische Zellen sind in bezug auf den jeweiligen funktionellen Zustand ihrer Zellen inhomogen. Nicht einmal alle Bereiche ein und derselben Zelle sind einander stets gleichwertig. So ist die Kontraktionswelle, welche die Muskelfaser eines *quergestreiften* Muskels entlangläuft, nur einige Sarkomere lang. Die einzelnen Zellen eines *glatten* Muskels enthalten, wie das Polarisationsmikroskop (Abb. 2a) zeigt, scharf begrenzte, $0{,}5-1\,\mu$ breite Querstreifen (*1*), deren Doppelbrechung hier plötzlich verschwinden oder negativ in bezug auf die Längsrichtung der Zelle werden kann. In den *Gonaden* der Weinbergschnecke liegen die untereinander gleichalten Abkömmlinge einer Spermatogonienmutterzelle zu sovielen beisammen, wie der Anzahl der abgelaufenen Spermatogonienmitosen entspricht. Selten sind es mehr als 16 Zellen, die um eine polyploide Nährzelle herum versammelt sind. Dicht daneben befinden sich andere Nährzellen mit Spermatocyten anderen Alters. Die bisher üblichen Methoden der Dünnschnitt-Technik versagen, wenn Zellen oder Zellabschnitte aus derartigen Geweben untersucht werden sollen, um physiologischen Zustand und Feinbau streng einander zuzuordnen.

Wir haben deshalb in Anlehnung an Methoden von Borysko (*2*), dessen Verfahren sich nur für einzellige Schichten von Gewebekulturen eignet, eine Arbeitsweise entwickelt, die gestattet, jeden beliebigen Schnitt von 6—10 μ Dicke zunächst unter dem Lichtmikroskop bei hoher Apertur zu untersuchen und darauf Dünnschnitte aus besonders interessierenden Zonen herzustellen. Das Gewebe wird in der gewohnten Weise mit Osmium fixiert und in Methacrylat eingebettet. Nach dem Auspolymerisieren wird der Block auf ein Stück Pertinax aufgekittet, grob zugespitzt und auf einem Paraffinmikrotom Schnitte, die etwa 1 mm Seitenlänge besitzen und 6 μ dick sind, hergestellt. Die Schnitte werden der Reihenfolge nach auf einen Glasobjektträger gelegt und im Lichtmikroskop grob durchgemustert. Ihre Oberfläche ist jetzt noch rauh, das Bild daher schlecht. Soll geprüft werden, ob genügend Zellen der gewünschten Art vorhanden sind, so legt man auf einen der Schnitte ein Deckglas, welches zuvor mit etwas vorpolymerisiertem Methacrylat bestrichen wurde. Das Bild ist nun klar. Für Ultrastrukturuntersuchungen ist dieser Schnitt verdorben, denn er ist mit nicht auspolymerisiertem Methacrylat in Berührung gekommen. Die brauchbaren, noch frei auf dem Objektträger liegenden Schnitte werden nun jeder für sich auf ein etwa 2,5 mm dickes, 10×10 mm großes, planparalleles Stück Methacrylat desselben Härtegrades gebracht und mit absolutem Alkohol aufgeklebt, wobei sie sich zunächst strecken. Wird dieser Arbeitsgang unter dem Präpariermikroskop durchgeführt, so kann das Gewebe dabei orientiert werden. Der Schnitt sollte nach dem Aufkleben etwa 2 mm Abstand von den Seitenkanten des Plexiglasquaders haben. Sind alle Schnitte auf die entsprechende Anzahl von Plexiglasstücken geheftet worden, dann legt man diese „Objektträger" auf ebensoviele gläserne Objektträger, die beiderseits des Plexiglasstückes je einen Glasstreifen von 2 mm Dicke tragen.

---

* Die Arbeit wurde durchgeführt mit Unterstützung durch die Deutsche Forschungsgemeinschaft und die Göttinger Akademie der Wissenschaften.

Am Abend vorher wurden andere gläserne Objektträger mit einer etwa 50 μ starken Schicht aus stark vorpolymerisiertem Methacrylat derselben Mischung überzogen und dem Plexiglas gestattet, über Nacht bei $+80°$ C auszupolymerisieren. Diese gläsernen Objektträger tragen damit das „Deckglas" für den 6μ-Schnitt. Sie werden mit der Plastikschicht gegen den Schnitt über Glas-

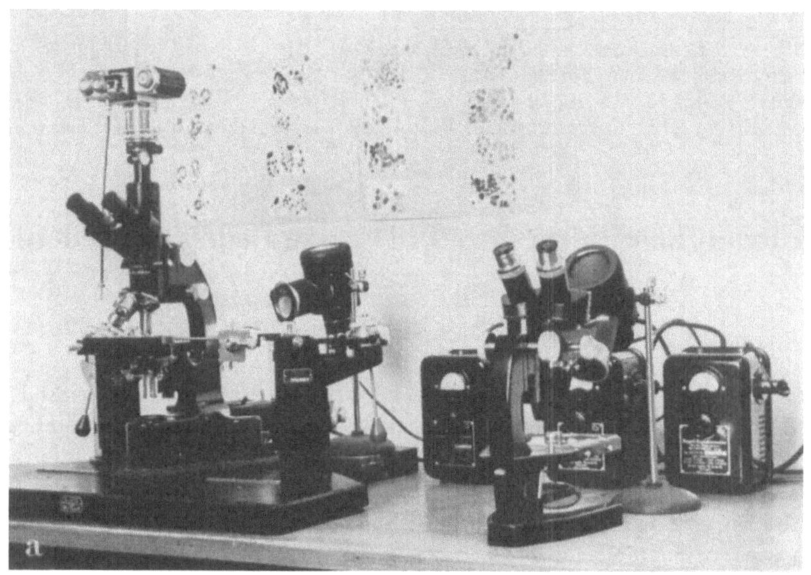

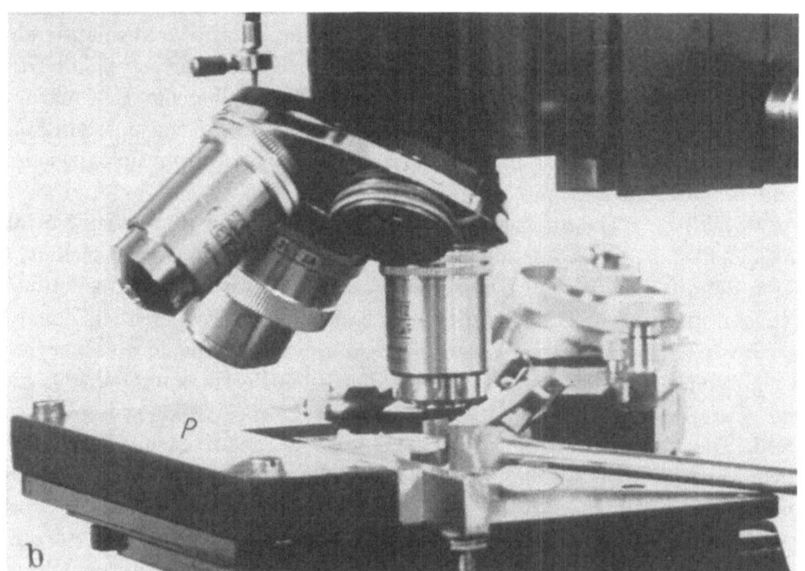

Abb. 1 a u. b. a) Die Anordnung zum gezielten Spitzen besteht aus einem Präpariermikroskop, einem Phasenkontrast- bzw. Polarisationsmikroskop mit Mikrophotoeinrichtung und arretierbarem Gleittisch und einem Mikromanipulator. b) Die Assistenten tragen je eine Messingstange mit einem Stück Rasierklinge an der Spitze. Auf dem Gleittisch ein Pertinaxwinkel, gegen den das Präparat geklemmt wird

streifen und Präparat gelegt und das Ganze mit einer Wäscheklammer zusammengedrückt. Im Exsiccator soll es bei 1 Torr und $+80°$ C über Nacht zusammenpolymerisieren. Die beiden 2 mm breiten Glasstreifen verhindern dabei, daß das thermoplastische Material in der Hitze zu stark oder schief zusammengepreßt wird. Am anderen Morgen wird das fertige planparallele Präparat von den gläsernen Objektträgern befreit und mit dem Schnitt nach oben auf einen Pertinaxquader von $8 \times 14 \times 25$ mm (für das Sjöstrand-Mikrotom passend) so aufgeklebt, daß der Schnitt noch gut 5 mm vom Rand des Halters entfernt bleibt.

Unter dem Phasenkontrastmikroskop (Abb. 1a) sucht man sich nun mit einem Objektiv hoher Apertur (z. B. Leitz Pv Apo 40/0.70 mit Deckglaskorrektur, dazu Großfeldokulare 16mal), notfalls mit der Glycerinimmersion (3) die gewünschte Zelle heraus und fertigt ein Lichtbild und eine Skizze an. Darauf spitzt man unter dem danebenstehenden Präpariermikroskop das Objekt

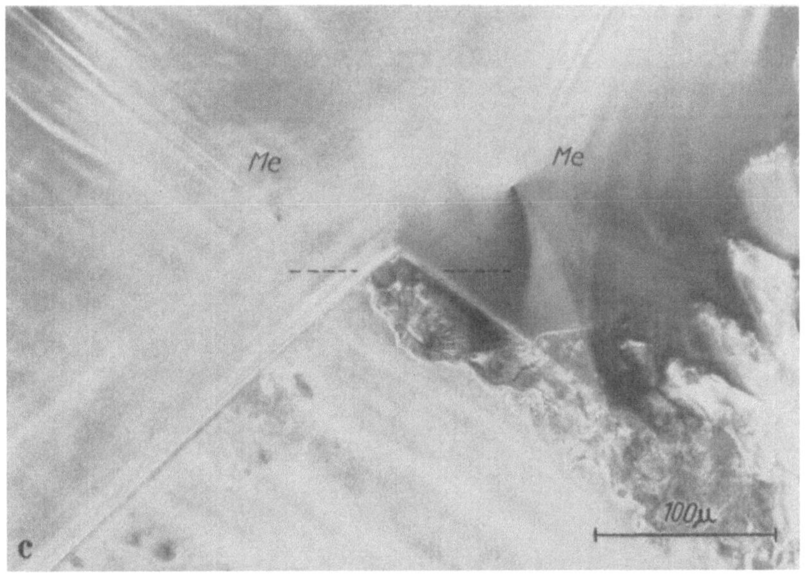

Abb. 1 c u. d. c) Das fertig zugespitzte Präparat, aufgenommen mit Objektiv Pv 10/0.28. Die gestrichelte Linie gibt die Zone an, von der Dünnschnitte hergestellt werden sollen. *Me* Messer. d) Vgl. Text

grob zu, bis die gewünschte Zelle von Spitze und Schenkeln des dabei entstehenden Winkels etwa 0,2 mm entfernt ist. Dann wird der Pertinaxhalter wieder auf den Gleittisch des Phasenkontrastmikroskopes (Abb. 1b) gelegt, dort befestigt, der Gleittisch in beiden Koordinaten arretiert und die Methacrylatplatte unter einem Objektiv mit großem Arbeitsabstand (z. B. Pv 10/0.28 oder Pv 25/0.50) so zugespitzt, daß der gewünschte Zellabschnitt, etwa der Kern, in einer Zone liegt, die bei 5 μ Abstand von der Spitze beginnt und bei 10 μ endet (Abb. 1c). Zum Spitzen dient der Mikromanipulator von Leitz, dessen Assistenten so verstellt werden, daß die Hauptvorschubrichtung unter 45° gegen die Sagittalebene des Mikroskopes zielt (Abb. 1a).

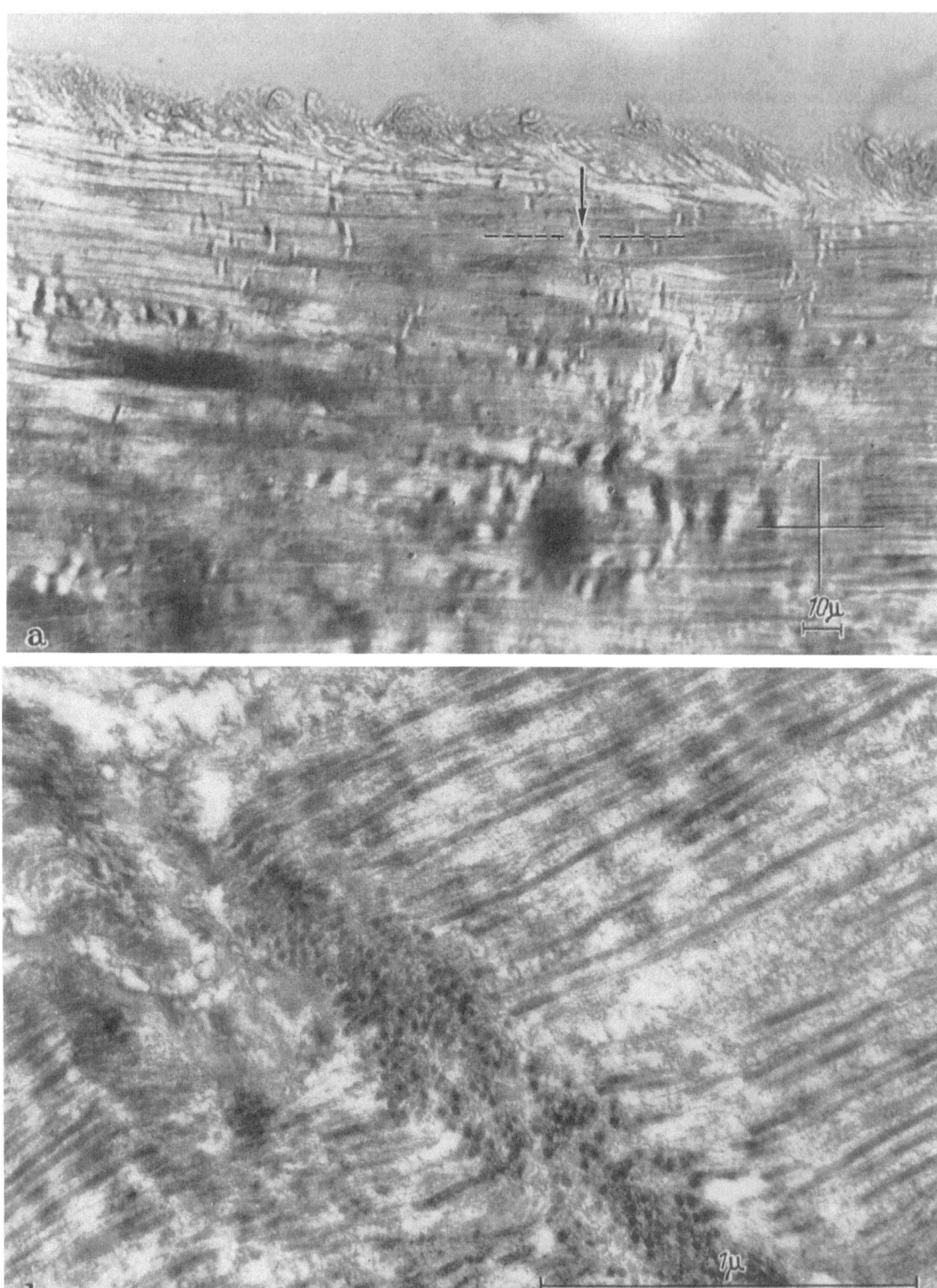

Abb. 2a u. b. a) Isometrisch gereizter glatter Muskel von Helix pomatia in Auslöschstellung zwischen gekreuzten Nikols, aufgehellt mit einem Kompensator. Die schwarzweißen Querstreifen sind Zonen, in denen die Myofilamente sich zopfartig umeinanderwinden und dabei zur Längsachse des Muskels senkrecht stellen. 6 μ-Schnitt auf Plexiglasobjektträger. b) Dünnschnitt durch die in Abb. 2a bezeichnete Zone

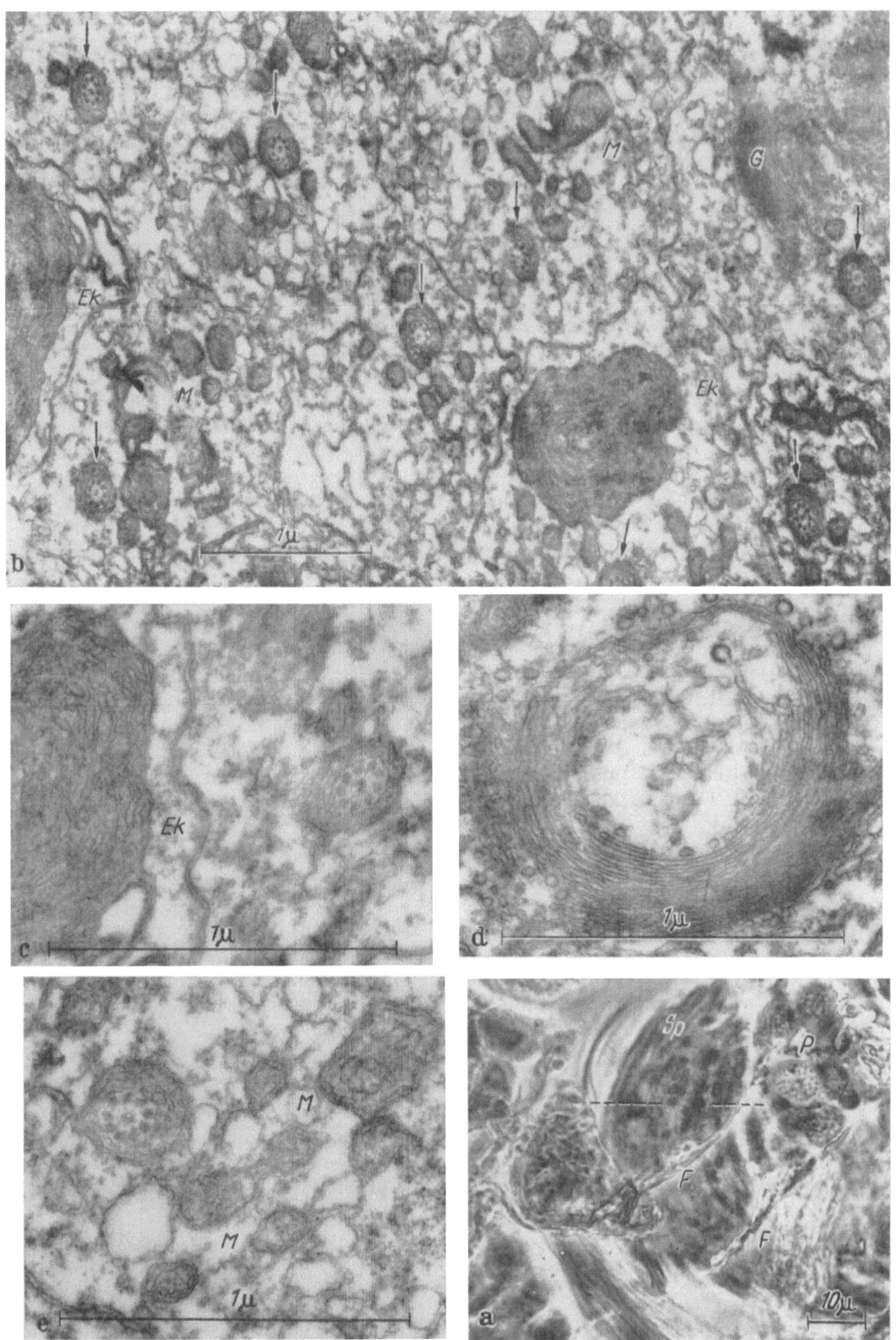

Abb. 3a—e. (Unterschrift siehe Seite 84 oben)   6*

Abb. 3a—e. a) Abschnitt von einem Bündel junger Spermatiden aus der Gonade von Helix pomatia. Es enthält eine Anzahl von dunklen Körpern unbekannten Baues, *Sp*. Die Dünnschnittebene (gestrichelte Linie) wird so gewählt, daß 2—3 dieser Körper getroffen werden müssen. 6 μ-Schnitt auf Plexiglasobjektträger. b) Dünnschnitt durch die in Abb. 3a bezeichnete Zone. Jeder Spermienschwanz (Pfeil) liegt in einer Zelle, deren Cytoplasma Mitochondrien mit konzentrischen Membranen (*M*), Golgi (*G*) und große Einschlußkörper (*Ek*) mit konzentrischen Lamellen enthalten kann. Die drei gesuchten dunklen Körper sind: zwei Einschlußkörper mit konzentrischen Lamellen und ein Golgiapparat. c) Spermienschwanz und Einschlußkörper. Ein Folgeschnitt. Die Zentralfilamente des Schwanzes sind durch dünnste Ausläufer mit den 9 peripheren Filamenten verbunden. d) Golgiapparat von Abb. 3b. Ein Folgeschnitt. e) Spermienschwanz und Mitochondrien mit konzentrischen Lamellen. Ein Folgeschnitt

---

Die Assistenten tragen an je einer Messingstange ein Stück Rasierklinge (Abb. 1b), welches mit Palavit dort in eine Kerbe hineingeklebt worden ist. Man stellt die Messer so ein, daß die Schneide nicht genau der optischen Achse des Mikroskopes parallel läuft, sondern gegen den Kondensor hin weiter vom Präparat fortzeigt. Das vom Kondensor kommende Lichtbüschel gelangt dann nahezu ungebrochen bis in die Zelle hinein, selbst wenn diese nur 10 μ unterhalb der Spitze liegt. Zum Schluß wird das Präparat hochkant gestellt und durch eine hinter dem Mikroskop stehende Lampe (Abb. 1a) von vorne angestrahlt. Dadurch tritt der 6μ-Schnitt schwarz gegen das umgebende Plexiglas hervor (Abb. 1d). Man spitzt nochmals, und zwar so, daß aus dem Dach eine Pyramide wird, deren Spitze in der Mitte des 6μ-Schnittes liegt. Nunmehr kann das Präparat, nachdem das vom Schneiden deformierte Plexiglas sich über Nacht hat setzen können, in den Halter des Ultramikrotoms gespannt werden. Sofern dies einen guten, fein arbeitenden Kreuztisch für die Messerverstellung besitzt, gelingt es nach einiger Übung leicht, mit dem ersten Schnitt nicht mehr als 5 μ abzuschneiden. Die Folgeschnitte sitzen dann mitten im gewünschten Zellbereich.

Das Verfahren arbeitet so genau, wie sich der cytologische Zustand einer Zelle im 6μ-Schnitt eines osmiumfixierten Präparates lichtmikroskopisch ermitteln läßt (vgl. Abb. 2a, b, Abb. 3a—e). Benutzt man statt der Phasenoptik einen Hellfeldkondensator, auf den von unten die Polarisationseinrichtung nach W. J. Schmidt (Leitz) aufgesteckt wird und ein Trägerstück mit Polarisationsobjektiven und Analysator, dann eignet sich das Verfahren auch gut für Muskeluntersuchungen und es gelingt ohne Schwierigkeiten, Dünnschnitte von Zonen mit besonders interessanter Doppelbrechung, wie z. B. von den obengenannten Querstreifen auf glatten Muskelzellen, herzustellen (Abb. 2a und b). Die Abbildungen zeigen, daß zusätzliche Artefakte durch die doppelte Hitzebehandlung nicht auftreten.

Literatur

1. Schlote, Fr. W.: Phys. Verh. **8**, 227 (1957).
2. Borysko, E.: J. biophys. biochem. Cytol. **2**, Suppl., 15 (1956).
3. Beermann, W.: Mündl. Mitteilung.

An ultramicrotome without bearings

T. Zelander and R. Ekholm

Department of Anatomy, University of Gothenburg (Sweden)

In recent years numerous ultramicrotomes have been designed. As electron microscopists, who have facilities to try every type of instrument, are rare, we can speak only for the Sjöstrand microtome (2) and an instrument of our own design (1) both based on the rotary principle. Both have given satisfactory results at this institute. However, rotary microtomes have three main drawbacks:

The cutting speed cannot be reduced below a minimum of about 10 cm/sec. The reason is that an oil film several times as thick as the sections must be interposed between the bearing surfaces. The thickness of the oil film will not remain constant unless the rotational speed exceeds a certain minimum. With such a relatively high cutting speed, the sections show a tendency to pleating and compression. This tendency becomes more marked the greater the section area.

2. Instruments based on the rotary principle are very susceptible to malfunction owing to dirt or air bubbles lodged between the bearing surfaces. Anything of this sort will cause the instrument to run out of true and wreck the results. Hence a rotary type of ultramicrotome almost constantly requires careful maintenance.

Fig. 1. The experimental model of the ultramicrotome

3. Rotary instruments are complicated and expensive to manufacture. The precision of the bearings must meet extreme demands and therefore special fabrication techniques are essential.

Because of these disadvantages of the rotary microtome we made it our aim to construct a microtome without bearings which should be simple to manufacture, simple to maintain, permit low cutting speed and yet should equal or surpass the rotary microtome in respect of section thickness and its constancy.

These aims are apparently realized in the instrument to be described. Essentially this microtome comprises an L-shaped arm supporting the specimen and a straight rod carrying the knife, the arm and the rod being mounted adjacently on a cast iron base (Fig. 1). The specimen arm is flexibly attached to a supporting pillar by means of a leaf spring 0.15 mm thick enabling the arm to pivot in the vertical plane, one section from the specimen on the free end of the arm being cut on each downstroke. The pivoting motion is accomplished by a hydraulic cylinder. The cutting speed is set by adjusting the oil flowrate. Means are provided for fine

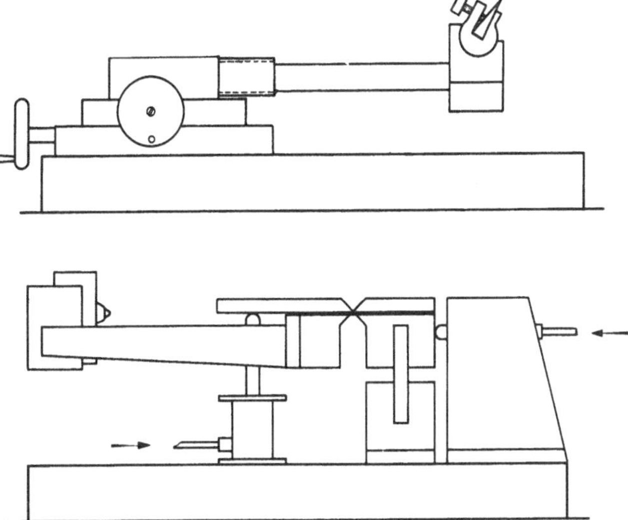

Fig. 2. A schematic drawing of the microtome. The upper part is a front view, the lower part is a back view. The arrows point to the inlet parts of the hydraulic cylinders

positioning of knife as well as specimen (Fig. 2.). To enable the specimen to clear the knife on the
return stroke, another hydraulic cylinder operates against the supporting pillar and causes it to
deflect slightly so that specimen arm moves horizontally at least 2—3 μ away from the knife. The
working cycles of the hydraulic cylinders can be suitably phased by means of adjustable cams
on the common drive shaft. The driving unit is mounted separately from the microtome and
the pipes are interconnected by flexible couplings in order to isolate the instrument from vibra-
tion (Fig. 3). The knife rod is affixed to a compound rest for initial positioning of the knife.
Feeding motion is imparted to the knife by passing a suitably adjusted current through a heating coil wound on the rod (Fig. 1).

Fig. 3. The motor-unit with sam discs and the drive cylinders

Up to now we have operated this microtome at a cutting speed of about 1 cm/sec and a feed
of approximately 200 Å per section. The instrument has been satisfactory in every way and
delivers sections as large als 0.2 mm squared without pleats or folds.

References

1. EKHOLM, R., and T. ZELANDER: Experientia (Basel) **12**, 195 (1956).
2. SJÖSTRAND, F.: Experientia (Basel) **9**, 114 (1953).

Eine magnetische Schneidekopflagerung
an einem Feinschnitt-Mikrotom

KURT ZAPF und GERHARD MAI

Deutsche Akademie der Wissenschaften zu Berlin, Institut für Medizin und Biologie, Berlin-Buch

Das in unserem Institut entwickelte Ultramikrotom nach JUNG und MAI mit thermischem
Vorschub hatte als Gleitlager zwischen Ausdehnungsstab und Schneidekopf planparallele Glasplatten, von denen eine rotierte. Die Koppelung der Platten erfolgte durch die Kohäsionskräfte
eines Fettfilmes, der von Zeit zu Zeit mechanisch erneuert werden mußte. Um eine gleichmäßige

Verteilung und Benetzung der Glasplatten und damit eine annähernd konstante Schichtdicke des Schmiermittels zu erreichen, war vor Beginn des Schneidens ein Einlaufen notwendig.

Durch die Konstruktion einer magnetischen Schneidekopflagerung (Abb. 1) wird eine gleichmäßige Übertragung der Stabausdehnung auf den rotierenden Schneidekopf garantiert.

Am freien Ende eines fest eingespannten Ausdehnungsstabes (Abb. 1/1) befindet sich ein mit einem geeigneten Polschuh (Abb. 1/2) versehener Ringmagnet (Abb. 1/3). Der Rotor (Abb. 1/4), an dem das zu schneidende Objekt (Abb. 1/5) an einem Konstruktionsteil (Abb. 1/6) exzentrisch befestigt ist, besteht aus einer Weicheisenscheibe, die mit einem Lagerring aus abriebfester Spezialbronze versehen, auf der freien Stirnseite des Magneten gleitende und drehende Bewegungen ausführen kann.

Durch eine am Sockel des Gerätes fest gelagerte Antriebscheibe wird dem Rotor über ein Dreipunktsystem die Drehbewegung aufgezwungen. Durch drei Stifte des Rotors, die in drei Mitnehmerflächen der Antriebsscheibe eingreifen, ist seine Drehachse festgelegt. Wirken Antrieb und Vorschub gleichzeitig auf den Rotor, so gleiten die Mitnehmerstifte in Vorschubrichtung auf den Mitnehmerflächen der Antriebscheibe. Um die Reibung bei dieser Vorschubbewegung auf ein Minimum zu beschränken, wurden in die Antriebscheibe Saphirsteine

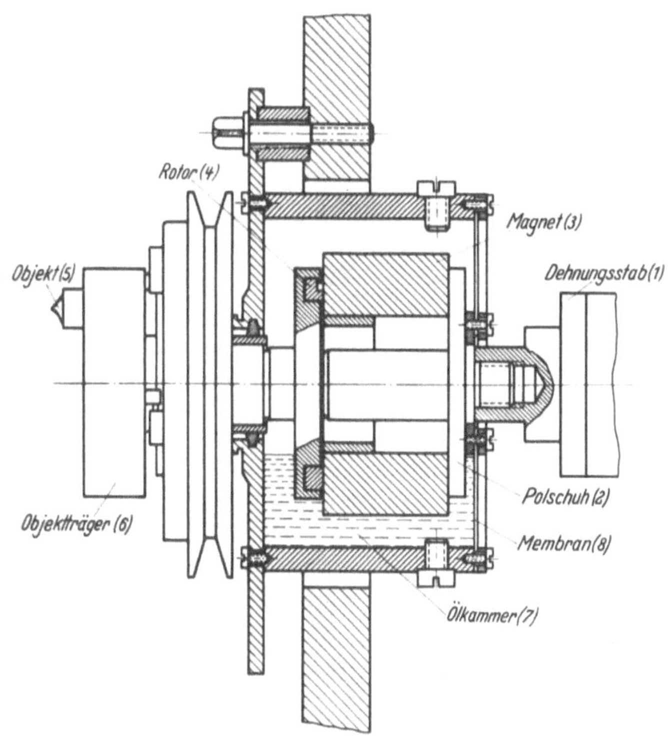

Abb. 1. Schnitt durch die magnetische Schneidekopflagerung

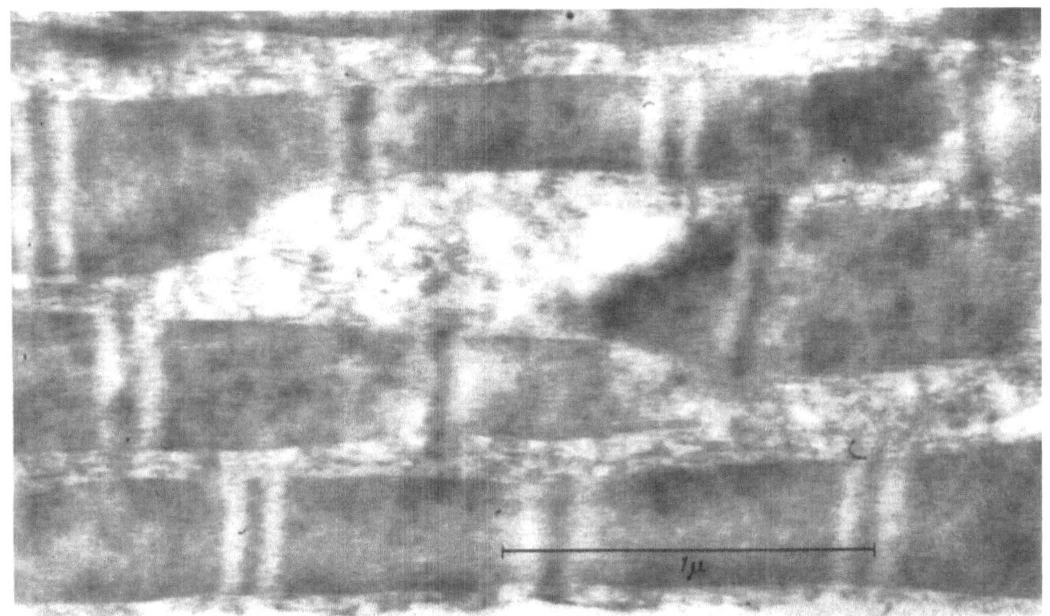

Abb. 2. Branchiostoma lanc. (Amphioxus) Quergestreifte Muskulatur. Elektronenopt. Vergr. 19600mal

eingesetzt, auf denen die gehärteten und polierten Stahlstifte des Rotors gleiten. Magnet und Weicheisenscheibe des Rotors sind in einer Ölkammer (Abb. 1/7) angeordnet, die mit einer geeignet ausgebildeten Membran (Abb. 1/8) den Vorschub nicht behindert. Durch geeignete Schmiernuten im Lagerring ist eine vollautomatische und gleichmäßige Benetzung der Gleitflächen mit Öl bei jeder Umdrehung gewährleistet. Damit entfallen sowohl mechanische Verschiebungen der Schneidekopflagerung durch nicht unmittelbar meßbare Schichtdicken, als auch der früher notwendige Einlauf des Gerätes vor Schneidebeginn. Betriebssicherheit und routinemäßiges Arbeiten sind dadurch ganz beträchtlich erhöht.

Das Dreipunktsystem bewirkt, daß die Objektzentrierung nur während der Rotierung wirksam wird. Um ein Abgleiten des Rotors bei Stillstand des Gerätes zu verhindern, wurde eine elastische Kuppelung zwischen Rotor und Antriebscheibe in Form von Gummipuffern angebracht.

Mit dieser Konstruktion ist ein geteiltes Lager entstanden. Die Zentrierung der Kreisbahn des Untersuchungsobjektes ist gewährleistet durch die Lagergenauigkeit der Antriebscheibe, ohne daß deren axiales Spiel den Vorschub beeinflußt. In Vorschubrichtung ist durch den gleichbleibenden magnetischen Kraftschluß die Spielfreiheit der Lagerung gewährleistet. Der auftretende Schnittdruck wird bei der beschriebenen Anordnung hauptsächlich von der Lagerung der Antriebscheibe aufgenommen. Der Ausdehnungsstab wird auch beim Schneiden harter Proben nicht auf Biegung oder Verdrehung beansprucht.

Wenn Einzelbilder auch keinen Beweis für die Güte und vor allen Dingen für ein rationelles Arbeiten mit einem Gerät darstellen, seien zur Demonstration einige Bilder von recht unterschiedlichen Objekten aus der Mikrobiologie und von tierischen und pflanzlichen Geweben gezeigt (Abb. 2).

Literatur

Jung, F., u. G. Mai: Dtsch. Gesundh.-Wes. 1955, 10, 1250.
— K. Zapf, I. Quasdorf u. G. Mai: Mikroskopie 1958, 13, 91.

B. Histochemie und Biochemie

Selective and cytochemical staining of frozen-dried preparations for study with the electron microscope

Isidore Gersh

Department of Anatomy, University of Chicago, Chicago, 37, Illinois

I should like first to show some examples from my laboratory of the use of organic and inorganic reagents for the selective and cytochemical staining of frozen-dried preparations.

The first group of examples includes certain instances where organic substances are used to facilitate identification. Chase (1) has used alcoholic acidified orcein to stain elastic fibers in mouse lung. The specificity of the elastic fiber stain at the level of the electron microscope was established by observation of thicker sections of the same specimen blocks with the light microscope.

Bondareff (2) has used leucofuchsin, another organic dye, to identify glycogen in liver cells when viewed with the electron microscope. Again, the appearance of stained sections viewed with the light microscope served as controls for sections viewed with the electron microscope. In addition, the use of salivary diastase to digest glycogen prior to staining served as an additional control.

The same dye was used also to stain some substrate in the pulmonary alveolar wall by Chase (1), as well as similarly reactive material in the matrix of epiphyseal cartilage by Durning (3). In thyroid gland cells, certain droplets are readily stained by this method [Gersh (4)]. Chase (5) has shown, however, that at the submicroscopic level there is marked alteration of the internal structure of the droplets. The Feulgen reaction for DNA, in which the same dye is used, is also accompanied by a redistribution of stainable material. A preparation of salivary gland chromosomes of Drosophila made by Isenberg (6) will be shown later to illustrate this point.

Malonate is an organic substrate which competes with succinate for succinic dehydrogenase by formation of an enzyme-inhibitor complex. Nelson (7) has shown that in rat sperm tail, the binding of the malonate is responsible for the enhanced contrast of the peripheral filaments. He also showed that tetrazolium derivatives were deposited as the reduced insoluble formazans in the same sites.

My last example of an organic reagent used in cytochemistry is the use by Nelson (8) of N-ethyl maleimide to detect sulfhydryl groups in rat sperm tail. Increased density appears in all structures of the sperm tail.

The second group of examples includes certain tests where inorganic substances are used to facilitate identification. The apatite crystals of cartilage and bone have a density sufficiently higher than that of the organic components of cartilage and bone so that no stains are necessary. Electron micrographs of sections of unstained rat cartilage and bone prepared by Durning (3) illustrate this clearly.

Calcium phosphate is similarly dense, and for this reason is a very satisfactory end product of the familiar Gomori method for alkaline phosphatase, and for ATPase. Nelson (9) applied this method to the study of rat sperm tail and found ATPase activity to be prominent in the peripheral filaments, while alkaline phosphatase activity was more prominent in the helical coil. In a similar way, reduction products of tellurite are sufficiently dense to serve as a good end product in the commonly employed method for succinic dehydrogenase. When the method was applied to rat sperm tail, NELSON (7) found the enhanced contrast due to the presence of tellurite reduction products most prominent in the peripheral filaments, the same sites found to be rich in succinic dehydrogenase activity with malonate and the tetrazolium dyes.

My last example of an inorganic reagent is the use of ferric ions to detect the highly soluble ferrocyanide ion after this is injected intravenously. CHASE (10) found that the Prussian blue droplets seen with the light microscope in the ground substance of the connective tissue of muscle is resolved with the electron microscope as clusters of small, sharply demarcated vacuoles.

The third group of examples includes certain tests where mixed organic and inorganic substances were used as reagents to facilitate identification. FINCK (11) has shown that gallocyanine-chromalum is useful for staining DNA and RNA for electron microscopy. The stain is self-terminating and requires no destaining or differentiation. Combined with the prior use of ribonuclease and deoxyribonuclease, the stain serves effectively to identify the substrates. The examples to be shown are from the work of FINCK (11) on liver, where RNA is identified as minute granules in the nucleolus and cytoplasm, and from unpublished work of ISENBERG (6) on salivary gland chromosomes of Drosophila where DNA occurs, also, as minute granules. By comparison, the Feulgen-positive granules are very much coarser.

My final example is the use of p-chlormercuribenzoate for the detection of sulfhydryl groups. When applied by NELSON (8) to rat sperm tail, the sites of increased density corresponded closely with those stained earlier with N-ethyl maleimide.

With these examples before you, I should like to make some general comments which seem important for the development of selective and cytochemical tests for use in electron microscopy. The first general comment is that fixation must be adequate. This means, at the least, that post-mortem changes should be minimal, that losses through chemical action of a fluid fixative should be minimal, and that diffusion of cellular and intercellular components should be minimal. These aims are frequently achieved for cytochemical studies with the light microscope by freezing and drying. The examples cited earlier show that the same advantages of fixation by freezing and drying are realized also in cytochemical work with the electron microscope. Of particular interest to electron microscopists is the fact that these aims are achieved with frozen-dried material while retaining a minimal contrast in the control preparations against which small increments in density are readily observed.

A second general comment concerns the requirements for achieving high contrast. As I have been at some pains to point out, contrast may result from the use of organic, inorganic, or mixed reagents. As ISENBERG (12) has shown, there seem to be two major factors: 1. a locally high concentration of staining reagent, organic or inorganic; and 2. a density of the reaction product greater than that of methacrylate or other embedding medium.

A third general comment is that all the difficulties that have faced histo-and cytochemists persist, usually in aggravated form, when using the electron microscope. Regardless of the method used to view the reaction product, the usual criteria for specificity, sensitivity, and precision of localization must be satisfied. In several of the examples cited earlier, a certain measure of specificity was attained by: a) the use of specific reagents (as, for example, the PAS method for glycogen, the Prussian blue reaction); b) the use of enzyme controls (as, for example, in the use of salivary amylase in the test for glycogen); c) comparison with light microscope images (as, for example, in the test for glycogen and for ferrocyanide); and d) the use of specific enzyme-inhibitor complexes (as, for example, in the use of malonate in the test for succinic dehydrogenase). Sensitivity in some of the examples cited earlier was improved through aiming at reaction products with high molecular weight and/or high concentration (as, for example, with the enzyme methods resulting in the deposition of calcium phosphate or the formazans), and low solubility product. Precision in localization was achieved in some of the examples cited earlier through use of brief incubation times (as with ATPase), and the formation of very insoluble reaction products (as, for example, the formazans, calcium phosphate, and ferric ferrocyanide). A very important factor in all methods is that they are progressive and/or self-terminating and do not require differentiation.

The aim of this presentation was to present simply some of the general procedures developed by students and colleagues working in my laboratory. Detailed descriptions of the method of freezing and drying are presented in other papers (Gersh (*13*), Gersh et al. (*14, 15*)] as well as in the reports cited. The latter also give details of post-fixation, staining, and embedding procedures.

The work reported here was aided throughout by grants from the Commonwealth Fund and the Clara A. and Wallace C. Abbott Memorial Research Fund of the University of Chicago.

References

1. Chase, W. H.: In preparation.
2. Bondareff, W.: Anat. Rec. **129**, 97 (1957).
3. Durning, W. C.: J. Ultrastructr. (In press).
4. Gersh, I.: J. Endocr. **6**, 282 (1950).
5. Chase, W. H.: In preparation.
6. Isenberg, I.: Unpublished.
7. Nelson, L.: Exp. Cell Res. (In press).
8. — In preparation.
9. — Biochim. biophys. Acta **27**, 634 (1958).
10. Chase, W. H.: Arch. of Path. (In press).
11. Finck, H.: J. biophys. biochem. Cytol. **4**, 291 (1958).
12. Isenberg, I.: Bull. Math. Biophys. **19**, 279 (1957).
13. Gersh, I.: J. biophys. biochem. Cytol. **2**, Suppl. 37 (1956).
14. — I. Isenberg, J. L. Stephenson and W. Bondareff: Anat. Rec. **128**, 91 (1957).
15. — — W. Bondareff and J. L. Stephenson: Anat. Rec. **128**, 149 (1957).

The combination of histochemistry and cytochemistry with electron microscopy for the demonstration of the sites of succinic dehydrogenase activity[1]

Russell J. Barrnett

Department of Anatomy, Harvard Medical School, Boston 15, Massachusetts[2]

This brief survey report is concerned with some of the experience gained during a series of experiments in which histochemistry and cytochemistry were combined with electron microscopy. In part, this paper will deal briefly with some of the experimental designs used to arrive at a compromise situation between the distinct and different requirements of the separate disciplines, histochemistry and electron microscopy, so that both can be used together. A second part is concerned with some of the results obtained during investigation on the sites of activity of the succinic dehydrogenase system of enzymes.

Although some of the experience gained in the applications of histochemistry to electron microscopy has recently been reviewed (*1*), it is pertinent to consider briefly some of the salient features. When these experiments were initiated, several working hypotheses were made on the basis of experience in the separate fields of histochemistry and electron microscopy as to what concessions could be made with the requisite conditions of each field. One of the more important of these hypotheses was that concerning the preparation of tissues. It was felt that small, fresh tissue blocks could be used for supravital incubation or the reactions could be carried on intravitally, provided no difficulties were encountered in the penetration of substrates and reagents into intact, fresh cells. In the case of the dehydrogenase system investigated, only fresh tissues could be used, since fixation destroys the enzymatic activity. However, for other enzymatic histochemical experiments, additional means of preparing tissue are available. Freeze-drying (*2*), freeze-substitution (*3*), or brief fixation in osmium tetroxide before incubation (*4*) may be used. We have had experience with only the latter two preparative procedures for demonstrating the activity of some of the more rugged hydrolases; and, as will be reported elsewhere (*5*), they are adequate. It should be pointed out, however, that, in the choice of preparative procedure, much will depend on the tissues and methods used and the aims of the investigation.

[1] This work was supported by a grant [A 452 (C5)] from the National Institute of Arthritis and Metabolic Diseases, National Institutes of Health, Department of Health, Education and Welfare, Bethesda, Maryland.

[2] Present address: Department of Anatomy, Yale University, School of Medicine, New Haven, Connecticut.

We also reasoned that the ingredients for an incubating medium and conditions of incubation can be obtained for the rapid and accurate deposition of final product with a minimum of destruction of the fine structure of cells. These factors were more necessities than hypotheses, and it was found that addition of sucrose to the incubating media added perceptibly in maintaining fine structural details.

Fortunately, in the case of dehydrogenase systems, the p_H of the medium used for the enzymatic activity, as well as rapid formation of final product, was close to the normal p_H environment of cells; whereas, wide variations in the p_H of the incubating medium, especially in the low acid ranges, are markedly destructive of tissue fine structure. In addition, it should be pointed out that the enzymatic histochemical methods that are applied to electron microscopy should involve a maximum of two reactions, the formation of the products by enzymatic activity and the reaction of a product with a reagent to form an insoluble final product. It is frequently the case that the conditions of incubation sometimes must favor the rapid formation of final product at the expense of the enzyme, substrate and incubation optima. Although many variables are implicit on the points, careful investigation of the histochemical reactions during incubation can facilitate the production of satisfactory results for the development of histochemical tests that can be used with electron microscopy.

In our initial experiments, we felt that the final product of the histochemical reactions should be metallic and hence opaque to electrons; however, recent experience has shown that this need not be the case. As will be presented in subsequent paragraphs, formazans, and in other work (5) azo dyes, may be visualized in the electron microscope. Needless to say, the final products must be stable to postincubation fixation, dehydration with organic solvents and embedding in plastic.

A final matter is that of postincubation fixation. Since the most exacting requirement of electron microscopy concerns the adequate preservation of the fine structural details of cells by fixation, and since osmium tetroxide (6) is still regarded as the most adequate means of fixation for electron microscopy, its use has been directly incorporated in the application of histochemistry to electron microscopy. Providing that the incubation conditions are satisfactory, postfixation in osmium tetroxide allows the adequate preservation of membranous material and the cytoplasmic matrix, especially its fibrillar and particulate differentiations, and the relationship of these structural entities to the final product of the histochemical reaction. It therefore provides for a frame of reference so that the histochemistry may be combined with electron microscopy as the latter is widely carried out and related to fine structure as it is now commonly accepted.

Examples of some of the combinations of histochemistry or cytochemistry with electron microscopy are provided by the following experiments concerned with the localization of the sites of activity of the succinic dehydrogenase system. As indicated, the choice of the experimental set for the histochemical portion of this work is restricted to carrying out the reactions supravitally on small blocks of frozen and thawed tissue or small blocks which have been washed in 0.44 M sucrose. Intravital or supravital reactions with fresh tissue blocks cannot be performed since endogenous dehydrogenase systems will also be demonstrated, whereas material which has been frozen and thawed or appropriately washed in sucrose shows no endogenous dehydrogenase activity. Since the tissues are subjected to strenuous manipulation by the freezing and thawing or washing, as well as by the subsequent incubation, the choice of tissue is somewhat restricted to a rugged one like muscle, especially cardiac muscle. This matter of the choice of tissues and variation of the conditions of incubation to meet the requirements of different tissues is one that is not commonly encountered in ordinary histochemistry.

According to the original working hypothesis that the final product of the histochemical reaction should contain an electron dense heavy metal, potassium tellurite was first investigated in a series of experiments in collaboration with Dr. G. E. PALADE (7). This reagent accepted electrons from respiratory enzyme systems as metabolites were oxidized by dehydrogenation, and the reduced tellurite (Te or TeO) thereby produced, fitted the requirements of being a final product with high electron scattering powers. Thus an incubating medium was prepared. Briefly, its contents were: potassium tellurite, sodium succinate, activators, SÖRENSEN's phosphate buffer (p_H 7.6) and enough sucrose to raise the overall osmolar concentration of the medium to 0.44 M.

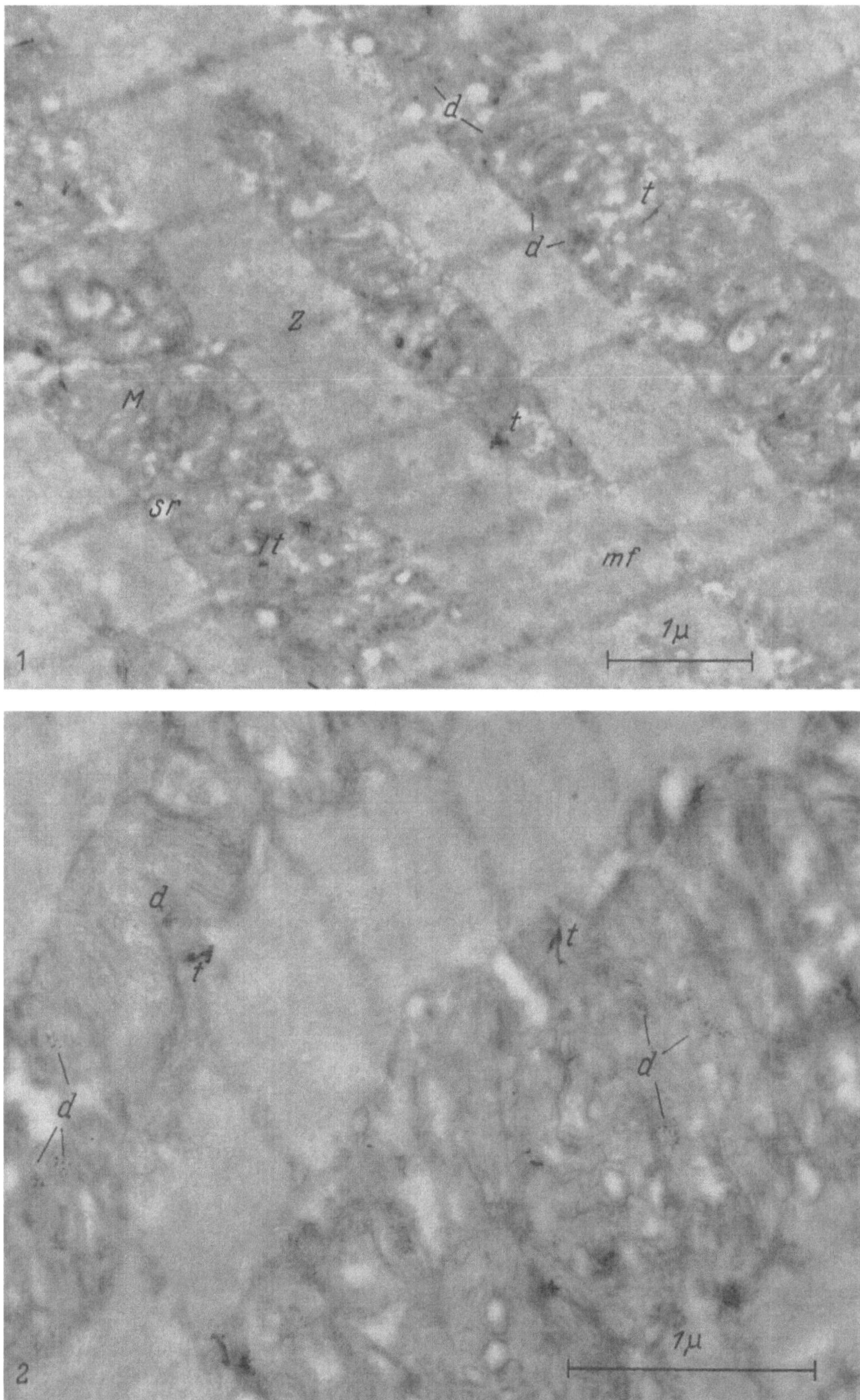

Fig. 1. Micrograph of heart muscle incubated for succinic dehydrogenase activity with potassium tellurite as reagent. The final product, reduced tellurite, occurs as fine deposits (*d*) or large needle-like crystals (*t*) within mitochondria. Contracted myofibrils (*mf*) have wide z bands (*Z*) and are free of final product. Although the mitochondria (*M*) show spotty extraction of the matrix and vacuolization and the sarcoplasmic reticulum (*sr*) is swollen, the membranous components remain intact and clearly defined 22,000 ×

Fig. 2. Micrograph of heart muscle incubated for succinic dehydrogenase activity with potassium tellurite as reagent, showing relationships of fine deposits (*d*) and crystals (*t*) of reduced tellurite to the fine structure of mitochondria. 38,000 ×

In some instances the incubations were carried out anaerobically, and in others potassium cyanide was added to the basic incubating medium.

Small blocks of frozen and thawed rat heart muscle incubated in the above medium turned black rapidly; and these were briefly fixed in osmium tetroxide, embedded in n butyl methacrylate, and sectioned on a Porter-Blum microtome. On observation of thin sections with the electron microscope, the electron-opaque deposits of the final product were localized almost exclusively to mitochondria and occurred as relatively large dense needles (100—200 Å at least 1000 Å) or extremely dense small particles 50—100 Å in size (Fig. 1, 2). The needle-like deposits occured singly or in sheaves, frequently oriented parallel to cristae, in apposition to their membranes, or on the outer limiting membrane of the mitochondria. Myofibrils, sarcoplasmic reticulum, sarcoplasm and nucleus were always free of crystals, but the final product was rarely deposited on the outer membrane of the nuclear envelope. The extramitochondrial location of final product could be due to a mitochondrion just out of the plane of section, since the topographical information in electron microscopy is largely restricted to two dimensions. The small dense particles occurred exclusively in the mitochondrial matrix in close relationship with the membrane outlining the cristae. They always occurred in groups or clusters, but not singly distributed throughout the mitochondria. Control experiments were performed to indicate that the histochemical reaction was enzymatic and that succinic dehydrogenase initiated the electron transfer. Similar results may be obtained in the localization of dehydrogenase systems that require the pyridine nucleotide coenzymes such as the lactic dehydrogenase system (8).

In order to cross check these results, another series of histochemical experiments in combination with electron microscopy was performed in which the final product was an organic molecule (formazan) that did not contain an electronopaque heavy metal. In these experiments tetrazolium salts were investigated as reagents as part of a study with Dr. A. M. Seligman and his co-workers. This aspect was made possible by the synthesis of a new tetrazole 2,2'-di-p-nitrophenyl-5,5'-diphenyl-3,3'-(3,3'-dimethoxy-4,4'-biphenylene) ditetrazolium chloride (Nitro BT) by Tsou, Cheng, Nachlas and Seligman (9). Since the respective formazan produced by reduction of Nitro BT (by acceptance of electrons from dehydrogenase systems) has substantive properties that bind it to protein, as well as poor solubility properties in organic solvents, allowing the cytological localization of end product (10), Nitro BT (0.5 mg/ml) was used as reagent instead of tellurite salts in incubating medium previously described for the demonstration of the activity of the succinic dehydrogenase system. After the incubation and brief fixation in osmium tetroxide, the formazan deposited in small blocks of heart muscle could be retained *in situ* during the process of dehydration and embedding in n-butyl methacrylate. Since it was possible that the catalyst used for polymerization of the plastic was a strong oxidizing agent that would oxidize the formazan to the tetrazole, α, α-Azo-di-iso-butyronitrile, a catalyst suggested to us by Dr. K. C. Tsou, was used in some experiments. These conditions and reagents are similar to those recently used by Sedar and Rosa (11). However, it should be pointed out that deposits of formazan are retained after embedding in n-butyl methacrylate with 2-4-dichlorobenzoyl peroxyde as catalyst.

In thin sections of heart muscle prepared as above, the formazan of Nitro BT was deposited nearly always in mitochondria, but the occurrence of formazan in other sites (e. g. myofibrils and nucleus) was more common than in the previous experiments with tellurite. The rate of deposition of the formazan was greater than that of the reduced tellurite, and the small blocks of heart muscle were adequately stained within 20 min, heavily stained within 40 min. The deposits of formazan varied in size, shape, and disposition in different parts of the block and in different experiments. Frequently they were roughly circular or oval in shape and were sometimes much larger than those obtained in the tellurite experiments (several hundred to over 1000 A in diameter). These deposits were sometimes so dense as to obscure the underlying mitochondrial fine structure, but at times they were smaller and less opaque and were seen to overlay the cristae (Fig. 3, 5). In the latter instances, portions of the cristae seemed unusually dense and appeared to be the foci about which the deposits grew. On occasion the deposits were small, no more than the diameter of a crista and usually related to these structures. Sometimes the deposits were linear and seemed to overlay or lie beside the cristae, making them appear thicker and partially obliterating the spaces

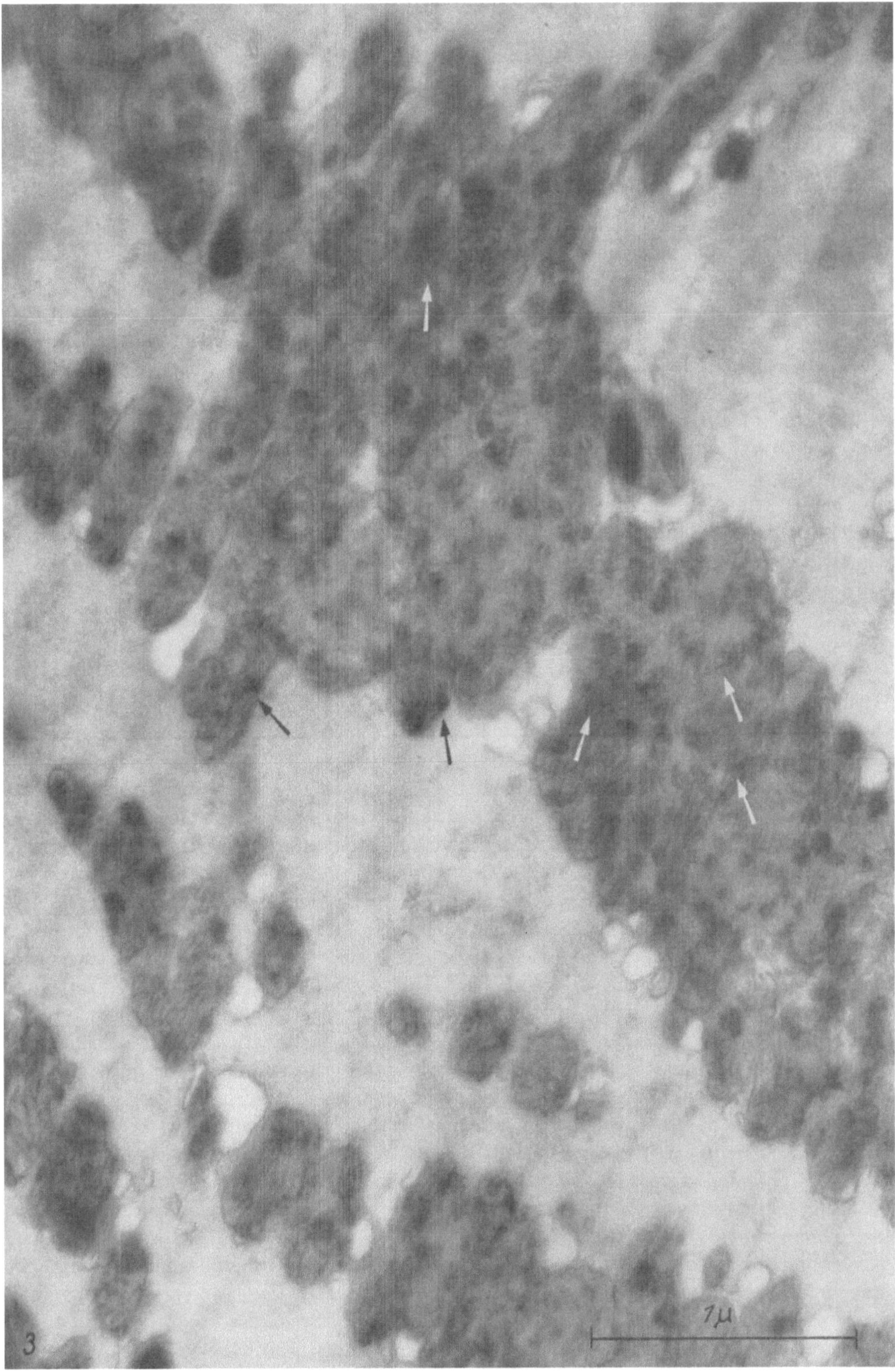

Fig. 3. Micrograph of hart muscle incubated for succinic dehydrogenase activity with Nitro BT as reagent. Numerous medium-sided circular deposits of formazan are deposited within the mitochondria. Although these deposits differ in their degree of density and size, the mitochondrial cristae seem to be foci about which they accumulate. In some instances where the deposits are light, the relationship to cristae can be made out (arrows).

41,000 ×

between the membranes. At some sites deposits were related to the surface membrane of a mitochondrion. These differences in the size and type of the deposits could occasionally be observed in adjacent groups of mitochondria. As in the tellurite experiments, numerous mitochondria in section showed no activity, whereas those that were active contained anywhere from one to many deposits. It is felt that the deposits represent accumulations of the formazan precipitating presumably at or near sites of high enzymatic activity. Their size is probably a function of the duration of incubation, composition of the incubating medium, availability and concentration of substrate and reagent at the site of the histochemical reaction and the rate of enzymatic activity. The fact that formazans containing only the molecular constituents of protein may be visualized and differentiated from protein is in agreement with theoretical considerations (*12, 13*).

Another tetrazole was tested as part of a series of experiments in collaboration with Dr. A. M. SELIGMAN on the histochemical use of organic reagents containing iodine in combination with electron microscopy. This tetrazole, 2,2'-di-p-nitrophenyl-5,5'-m-iodo-diphenyl-3,3'- (3,3'-dimethoxy-4,4'-biphenylene) ditetrazolium chloride (Iodo Nitro BT), one of a series of such compounds prepared by Drs. S. KARMARKAR, R. J. BARRNETT and A. M. SELIGMAN (*14*), was an adequate histochemical reagent for the demonstration of the activity of the succinic dehydrogenase system. When this compound was used in histochemical experiments in combination with electron microscopy, it was found that, although the distribution was similar to that obtained with the formazan of Nitro BT, there were fewer large, circular or irregular deposits of the iodo-formazan even though the duration of incubation was more than two times as long as with the latter reagent. In addition, the deposits appeared to be routinely somewahat larger but were not considered to be more electron opaque than those of the formazan of Nitro BT for similar thicknesses of section. Occasionally, as with some of the deposits of the formazan of Nitro BT, where the formation of the deposit was presumably early and not dense, the deposition of formazan on the membranes of mitochondria could be made out (Fig. 4).

It should also be pointed out that, although there was a moderately serious degree of disruption of the normal fine structure of tissues caused by the histochemical incubation in these experiments, enough of the fine structure was retained so as to be recognizable. Myofibrils of heart muscle were usually contracted; the myofilaments appeared finer, and their orderly disposition was lost to a varied extent. Granules in the nucleoplasm showed a curious clumping. Sarcoplasmic reticulum was always swollen; and the mitochondria showed a spotty and sometimes extensive extraction of the matrix, vacuolization and swelling. The sarcoplasmic matrix was to a large extent extracted.

A final series of experiments was conducted in collaboration with Dr. E. G. BALL. This consisted of an integrated morphological and biochemical study of a purified preparation obtained from heart muscle which contained the cytochrome system and displayed both succinate and reduced diphosphopyridinenucleotide (DPNH) oxidase activity (*15*). The procedure employed for the purification of this complex of oxidative enzymes [repeated blendings with water and centrifugation and differential removal of the enzyme particles by variation of the salt concentration of the suspending medium coupled with centrifugation (*16*)] was designed to yield a preparation of maximum activity per unit of dry weight; and no regard was paid in developing the procedure to the maintenance of the morphological integrity of any of the components of the muscle cell.

When this enzyme preparation was centrifuged and the pieces of the translucent, tan, gelatinous pellets obtained were fixed in osmium tetroxide, dehydrated and embedded in *n* butyl methacrylate. it was found on examination of thin sections to be completely composed of thin walled, membranous vesicular structures 400—10,000 A in diameter that were tightly packed (Fig. 6). These were mixed with short, straight or curved pieces of membranous material similar to that making up the walls of the vesicle-like structure. Rarely, intact and swollen, but recognizable, pieces of mitochondria were found in the pellet; and the membranes in these had an identical appearance with those in the walls of the vesicles. The walls of the vesicles of the straight pieces were characteristically composed of either one or two dense lines or membranes embedded in less dense, homogeneous ground substance. The dimensions of these lines or membranes in the walls of the vesicles were compatible with the dimensions of the membranous material found in intact isolated mitochondria. Although absolute purity cannot be claimed, it is felt on several

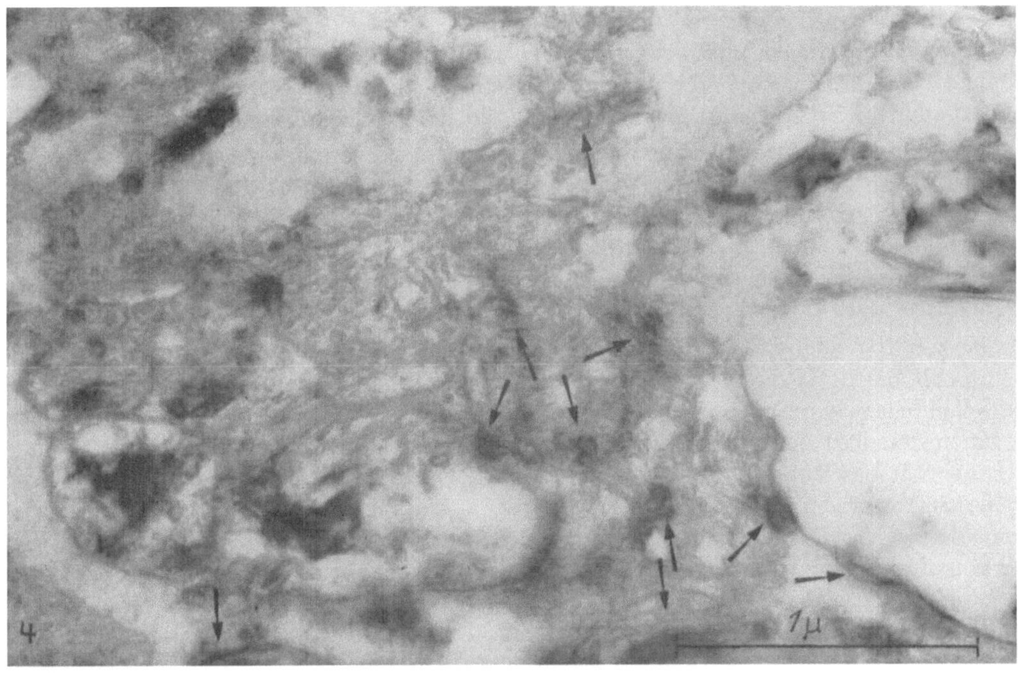

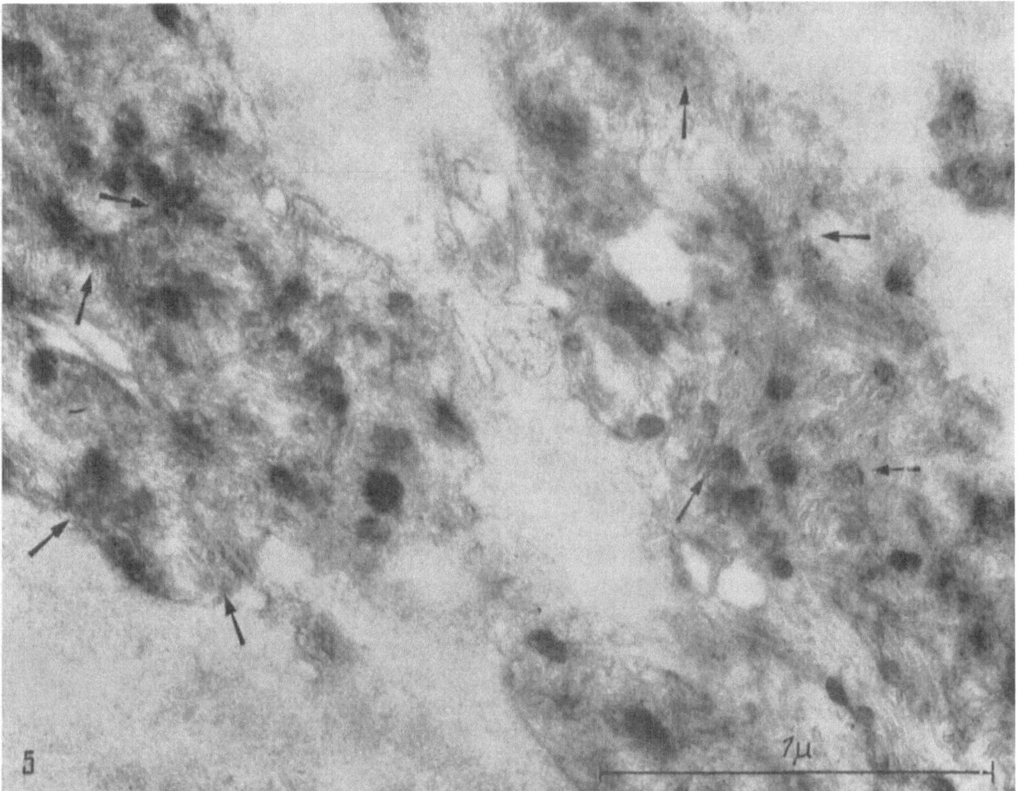

Fig. 4. Micrograph of heart muscle incubated for succinic dehydrogenase activity with Iodo Nitro BT as reagent. Although the mitochondria show more damage due to the longer incubation period, many deposits of the formazan appear to overlay cristae or surface membranes (arrows). 41,000 ×

Fig. 5. Micrograph of heart incubated for succinic dehydrogenase with Nitro BT as reagent. Many deposits of formazan are large and dense so as to obscure underlying fine structure, but in some favorable cases the deposits are clearly related to the membranes of the cristae (arrows). 54,000 ×

counts that this enzymatic preparation is primarily composed of swollen membranous mitochondrial fragments. The enzymatic activities displayed by the preparation were those known to be confined to mitochondria; and, taking into account the method of preparation, there is a remarkable similarity of the membranous material to that found in mitochondria.

It is of further interest that preliminary experiments have indicated that, in the presence of succinate, this membranous enzyme preparation will reduce Nitro BT directly to the corresponding formazan. It will not, however, reduce tellurite salts. Several cofactors, present in intact heart muscle but absent in the purified enzyme preparation, are required for the transfer of electrons from the enzyme system to tellurite. One of these is present in a boiled supernatant of blenderized heart muscle and can be substituted for by riboflavin.

It should be finally noted that approximately 44% of this enzyme preparation is lipide of which half, or 22% of the dry weight, is phospholipide (16). Since aqueous suspensions containing phospholipide may on fixation form membranelike material (17), the studies on the enzyme preparation described above after two extractions with 0.5% deoxycholate solution (15) may be pertinent. It was found as a result of the deoxycholate treatment that almost all the phospholipide was removed from the original enzyme preparation. Enzymatically, all the cytochrome c and some of the succinic dehydrogenase is extracted; but the preparation still contains a fair proportion of the cytochrome a, a_3, b, and c_1, and it can oxidize succinate if exogenous cytochrome c and the deoxycholate extract are added (18). When sections of the pellets of deoxycholate extracted material were prepared for electron microscopy, and examined, it was found that its composition was even more distinctly membranous than that of the unextracted enzyme pellet (Fig. 7); and the diameter of the membranes was similar or slightly smaller than these found in intact, unextracted mitochondria (15). Since all the material in the deoxycholate extracted pellets was obtained from the original vesicular enzyme preparation, it is suggested that partial extraction of the wall and collapse of the vesicles would account for the appearance of the membranes in the extracted preparation. It should be reiterated that the membranes of this preparation existed after most of the phospholipide was extracted.

In terms of the localization of the sites of the succinic dehydrogenase system as revealed by the histochemical and cytochemical experiments, it is felt that the latter information is more critical than the former. The histochemical experiments, while informative of the fact that histochemistry may be combined with electron microscopy, give little more indubitable information than that the oxydative enzyme systems are located in mitochondria, a fact already well known and substantiated. The size of the large dense deposits of the formazan of Nitro BT and the large crystals of reduced tellurite prohibit exact cytological localization in relation to the fine structure of the mitochondrion, even though it seems reasonable that a portion of the deposit or crystals may overlay a site of high enzymatic activity. The smaller dense deposits of reduced tellurite would be more adequate for a final product if it were not for the fact that the reduction of tellurite required several cofactors which could increase the hazard of false localization at the ultrastructure level.

However, this does not mean that experiments which combine histochemistry and electron microscopy are not worth while. It simply points out the faults observed by the early experimentation and the necessities of future experiments. Presumably a small, discretely localized endproduct having some of the virtues of the formazan of Nitro BT, or reduced tellurite, may regularly be obtained with another reagent and/or with more appropriate conditions of incubation favorable to the accurate localization of histochemical end product of the enzymatic reaction. Were this to occur, it is felt that localization of the sites of histochemical activity would be entirely restricted to membranes of the mitochondrion, cristae and possibly external limiting membrane, as it was in some favorable instances in the present experiments. These sites would be in aggreement with those of the cytochemical experiment. This implies that the results of the cytochemical experiments and significant portions of the histochemical experiments are taken to indicate that the succinic dehydrogenase system of enzymes including the cytochromes is located in or on the membranes of mitochondria. This is in agreement with similar suggestions by other workers (19—21).

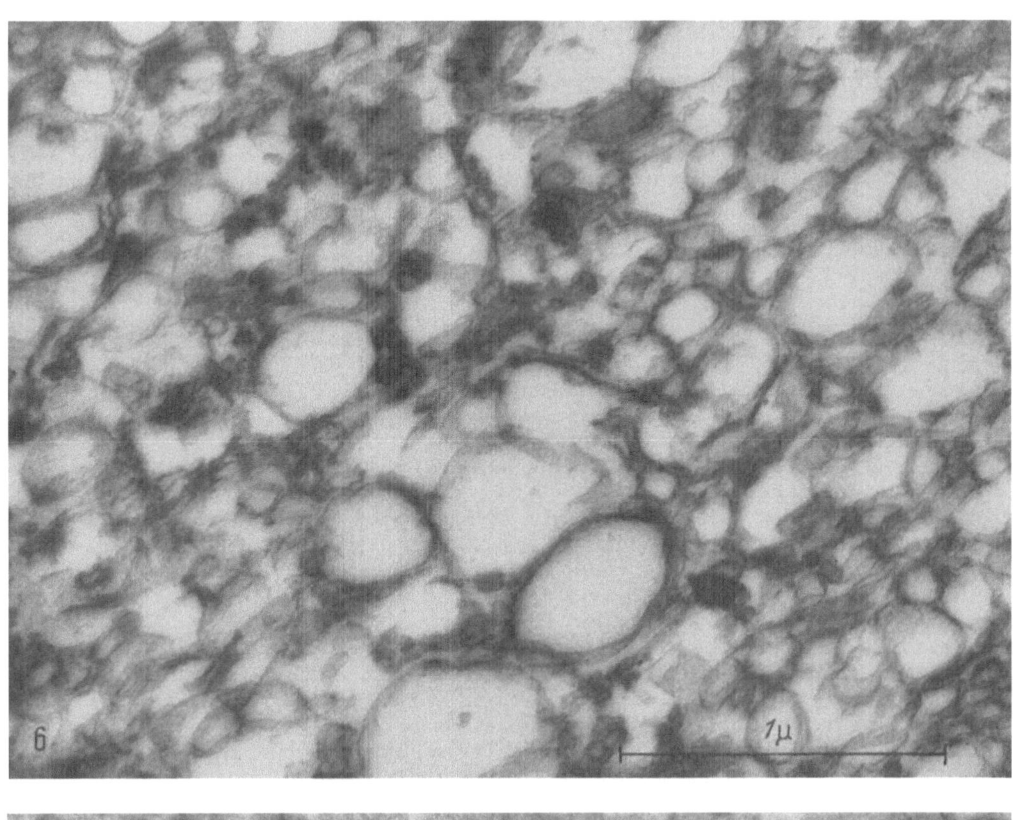

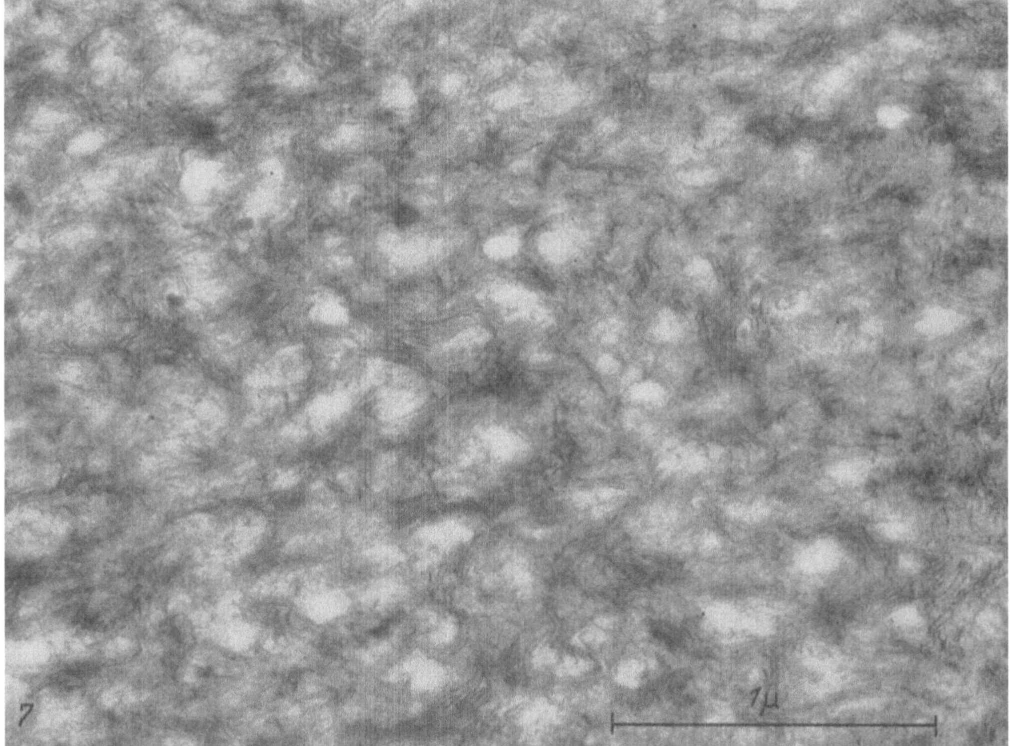

Fig. 6. Micrograph of thin section of enzyme preparation that contains the cytochromes and displays both succinate and DPNH oxidase activity. It is composed of membranous material arranged in either vesicles of varying sizes or straight pieces of membranous material. 45,000 ×

Fig. 7. Micrograph of thin section of enzyme preparation trecated with deoxycholate. Entire preparation consists of pairs of thin membranes. This preparation has lost most of its phospholipide but still retains some succinic dehydrogenase activity and the activity of the electron transmitter chain except cytochrome c. 45,000 ×

7*

In conclusion, it is hoped that the present experiments give some idea as to the current status of some of the work in which histochemistry was combined with electron microscopy. For the present, the overall view of the status is not an altogether happy one: but there is reason to believe that it is a possible integrating step in some cases that can lead toward a better understanding of the relationship of biochemical functions to cell components and to their fine structural elements.

References

1. Barrnett, R. J., and G. E. Palade: J. Histochem. Cytochem. **6,** 1 (1958).
2. Gersh, I., I. Isenberg, J. L. Stephenson and W. Bondareff: Anat. Rec. **128,** 91 (1957).
3. Feder, N., and R. L. Sidman: J. biophys. biochem. Cytol. **4,** 593 (1958).
4. Sheldon, H., H. Zetterquist and D. Brandes: Exp. Cell Res. **9,** 592 (1955).
5. Barrnett, R. J.: Exp. Cell Res. (1959). (In Press).
6. Palade, G. E.: J. exp. Med. **95,** 285 (1952).
7. Barrnett, R. J. and G. E. Palade: J. biophys. biochem. Cytol. **3,** 577 (1957).
8. — Anat. Rec. **127,** 395 (1957).
9. Tsou, K. C., C. S. Cheng, M. M. Nachlas and A. M. Seligman: J. Amer. chem. Soc. **78,** 6139 (1956).
10. Nachlas, M. M., K. C. Tsou, E. de Souza, C. S. Cheng and A. M. Seligman: J. Histochem. Cytochem. **5,** 420 (1957).
11. Sedar, A. W., and C. G. Rosa: Anat. Rec. **130,** 371 (1958).
12. Zeitler, E., and G. F. Bahr: Exp. Cell Res. **12,** 44 (1957).
13. Isenberg, I.: J. Histochem. Cytochem. **4,** 416 (1956).
14. Karmarkar, S. S., R. J. Barrnett and A. M. Seligman: J. Am. Chem. Soc. 1959 (In Press).
15. Ball, E. G., and R. J. Barrnett: J. biophys. biochem. Cytol. **3,** 1023 (1957).
16. — and O. Cooper: J. biol. Chem. **180,** 113 (1949).
17. Revel, J. P., I. Susumu and D. W. Fawcett: J. biophys. biochem. Cytol. **4,** 495 (1958).
18. Ball, E. G.: In Enzymes: Units of Biological Structure and Function (O. H. Gaebler, ed.) New York: Academic Press Inc. **1956, 433.**
19. Palade, G. E.: Anat. Rec. **114,** 427 (1952).
20. Cleland, K. W., and E. C. Slater: Biochem. J. **53,** 547 (1953).
21. Watson, M. L., and P. Siekevitz: J. biophys. biochem. Cytol. **2,** 639, 653 (1953).

Distribution of succinic dehydrogenase in the human spermatozoa as revealed in the electron microscope

N. J. Unakar and Satyavati M. Sirsat

Indian Cancer Research Centre, Parel, Bombay—12

Kothare and DeSouza (5) have reported the cytochemical distribution of succinic dehydrogenase in normal and abnormal spermatozoa using p-nitrophenyl substituted ditetrazole. They found the enzyme located in the middle piece and the head, but absent in the neck and tail. Thin sectioning studies have shown a complete absence of mitochondria in the head of the human spermatozoa (1). In view of this observation, and the close spatial association between this enzyme and mitochondria (2, 3, 7, 9) it seemed desirable to examine the exact localization of succinic dehydrogenase in the human sperm at the magnifications available in the electron microscope.

Ten samples of human semen were obtained from fertility clinics and processed within an hour of collection. The technique used for the demonstration of sites of dehydrogenase activity was a modification of the one described by Barrnett and Palade (2) using potassium tellurite as a hydrogen or electron acceptor. The sperm cells were washed in 3—4 changes of 0.44 M sucrose to halt endogenous activity. The samples were then incubated aerobically for two hours at 37° C in an incubating medium consisting of a) 20 cm³ Sörensen's phosphate buffer (pH 7.6), b) 0.1% potassium tellurite, c) 0.2 M sodium succinate as substrate, d) 0.5 cm³ of 0.6 M sodium bicarbonate as activator and e) 0.07 M sucrose.

Six control samples, taken from the same semen specimens were washed for 30 min in 0.44 M sucrose solution, kept for 5 min at 80° C in sucrose solution fixed in 80% ethanol and then incubated simultaneously with the test samples in an incubating medium from which potassium tellurite and the succinate were omitted.

After optimum incubation the experimental and control samples were fixed for 15 min in cold 1% osmium tetroxide (pH 7.6) washed in 0.44 M sucrose, dehydrated for small periods of time in graded ethanol, impregnated with monomer and embedded in a 1:4 mixture of n-butyl and methyl methacrylate monomers at 58° C with benzoyl peroxide as catalyst. Sections of 300—500 Å were cut on a Porter-Blum ultramicrotome, mounted on collodion coated copper grids and observed in an RCA EMU-2D electron microscope with an objective aperture of internal diameter .001″. For optical microscopy, smears were taken from experimental and control samples on clean glass slides, just after incubation, air dried and stained with dilute safranin O.

Results. Although potassium tellurite is very slow in reaction compared to the tetrazolium compounds, the experimental tubes showed gradually increasing darkening within half an hour of incubation. The control tubes showed no visible reaction.

Optical Microscopy. Dark granules were observed in the middle piece of almost all spermatozoa in the samples incubated with the tellurite. Similar granules or crystalline material were seen in the head region of some spermatozoa in each sample (Fig. 1). The deposit was never found intranuclear, but was always observed at the extreme edge of the sperm head apparently on the cytoplasmic membrane.

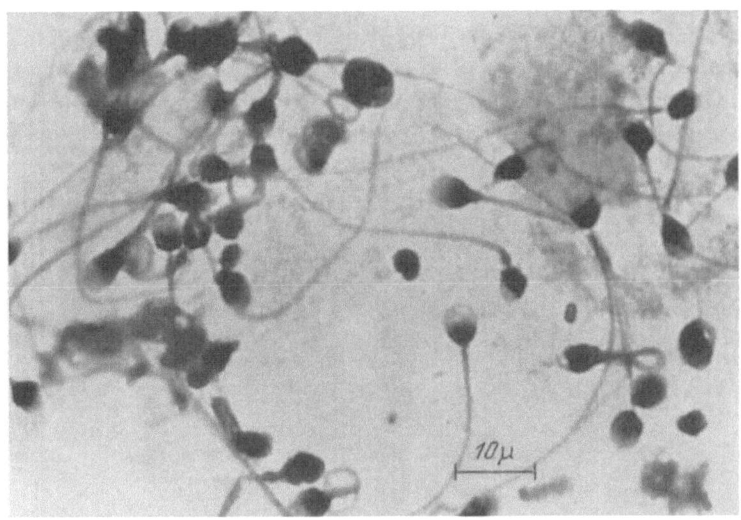

Fig. 1. Human Spermatozoa treated with potassium tellurite as seen under the optical microscope. The head and midpiece region shows dark granules due to the enzyme activity. 1120 ×

A few sperms in each test sample showed complete absence of dark granules in the middle piece and the head. Smears of sperms from the six control specimens showed a complete absence of the reduced tellurite in the body of the sperm.

Electron Microscopy. The specimens appeared well preserved. Dark granules were observed along the middle piece and tail in close association with the mitochondrial distribution (Fig. 2). The dark granules which were apparently situated in the head of the sperm in the smears seen in the light microscope, were localised only in mitochondria of the cytoplasmic remnants attached to some spermatozoa (Fig. 3). Normal mature sperms with no residual cytoplasm showed no sites of dehydrogenase activity in the head.

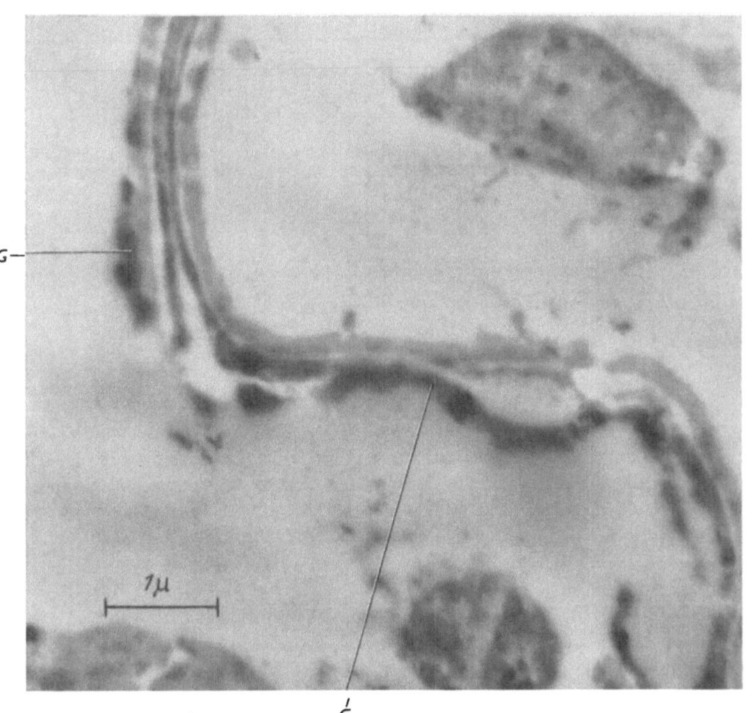

Fig. 2. Tail of the human spermatozoa. The black granular material (*G*) seems due to the enzyme activity. The activity is located in close association with the mitochondria. 15400 ×

Discussion. Reduction of potassium tellurite by living organisms was first described by KLETT (*4*). The correlation between this phenomenon and endogenous dehydrogenase was suggested by LAKON (*6*) and WACHSTEIN (*8*). The visualisation of the sites of dehydrogenase activity in the electron microscope is possible because the reduction product of tellurite is opaque to the electron beam. The modification of BARRNETT and PALADES (*2*) technique in this study were made to suit a single cell system, which would allow for rapid penetration of the reactive agents. The simultaneous use of all the activators enumerated in the original technique showed an increase both in the rate of tellurite reduction and intensity of reaction.

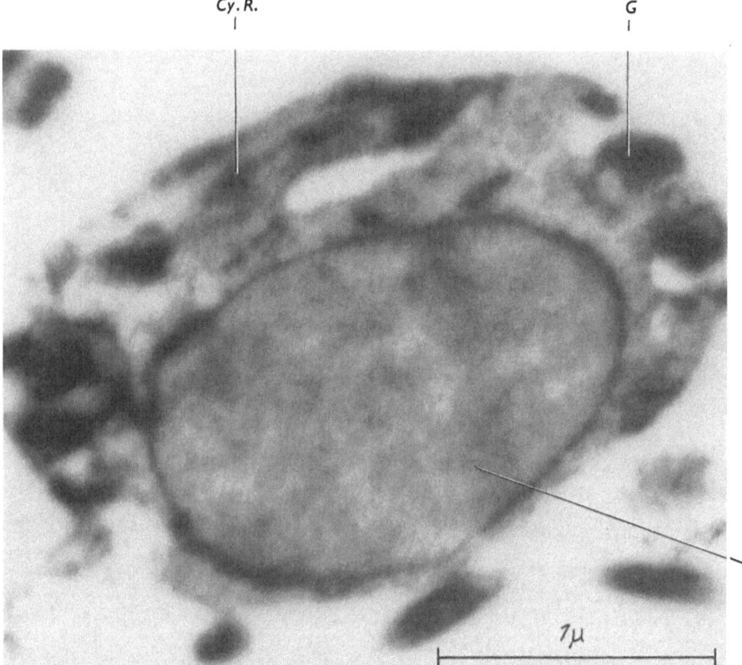

Fig. 3. Head of the immature spermatozoa with cytoplasmic remnants (*Cy.R.*) around the nucleus (*N*). Dark granular material (*G*) is visible in the cytoplasmic remnants denoting the activity of the enzyme in the mitochondria.
38 500 ×

KOTHARE and DESOUZA (*5*) have suggested the probable presence of mitochondria in the head of the sperm on the basis of their finding dehydrogenase activity in that location. Electron microscopic observations show that in confirmity with the absence of mitochondria, in the mature human sperm head, there is no succinic dehydrogenase activity there. Sites of tellurite reduction areas are found only in the mitochondrial inclusions present in the cytoplasmic remnants associated with the head region of some spermatozoa. Such immature germ cells are seen occasionally in all normal human semen, and increasingly so in clinically subnormal semen, also as that used in this investigation. The few spermatozoa which showed no enzyme activity at all, would be the non-viable ones found in all semen.

The consistent presence of this enzyme system in the middle piece indicates the independent respiratory mechanism of human spermatozoa.

This work was done under the direction of Dr. V. R. KHANOLKAR, Director, Indian Cancer Research Centre, Parel, Bombay. We wish to thank Dr. S. N. KOTHARE for some of the semen samples used in this study.

References

1. ANBERG, A.: Acta obstet. gynec. scand. **36**, Spl. 2 (1957).

2. BARRNETT, R. J., and G. E. PALADE: J. biophys. biochem. Cytol. **3**, 577 (1957).

3. HUGHES, A.: Quart. J. Micros. Sci. **97**, 165 (1956).

4. KLETT, A.: Z. Hyg. **33**, 137 (1900).

5. KOTHARE, S. N., and E. J. DE SOUZA: J. postgrad. Med. **4**, 70 (1958).

6. LAKON, G.: Ber. bot. Ges. **57**, 191 (1939); cit. in (*1*).

7. SIEKEVITZ, P., and M. L. WATSON: J. biophys. biochem. Cytol. **2**, 653 (1956).

8. WACHSTEIN, M.: Proc. Soc. exp. Biol. (N. Y.) **72**, 175 (1949).

9. — and E. MEISEL: J. biophys. biochem. Cytol. **1**, 483 (1958).

Coloration des coupes ultra-fines, au moyen de l'imprégnation à l'argent, pour la microscopie électronique

V. Marinozzi

Institut d'Anatomie et d'Histologie Pathologique de l'Université de Rome

L'imprégnation à l'argent a été largement employée en microscopie électronique pour colorer des tissus préparés par dilacération alors que très rarement elle a été appliquée à l'étude des coupes ultra-fines. Au cours de travaux précédents j'ai signalé qu'il était possible d'obtenir une bonne coloration des coupes ultra-fines par la méthode de Jones (acide périodique-argent-méthenamine) (*1*); l'observation de celles-ci pouvant se faire également au moyen du microscope optique (*2*) et du microscope électronique (*3*). Pour une bonne réussite de la méthode je recommandais d'éliminer le métacrylate des coupes par traitement à l'éther éthylique. Toutefois des recherches successives m'ont démontré que l'on pouvait obtenir une imprégnation complète, même sans éliminer le métacrylate, à condition d'augmenter la température d'incubation de la solution d'argent ammoniacal. En outre, le contrôle au microscope électronique des différentes phases du processus de coloration, m'a permis de constater qu'après fixation à l'acide osmique, l'imprégnation à l'argent se fait même sans traitement à l'acide périodique.

De petits fragments de tissus divers sont fixés à l'acide osmique au 2% (tamponné à p_H 7,2, selon Palade) et inclus en méthyle-butyle méthacrylate 1:20. Les coupes ultra-fines, d'épaisseur d'environ 300—400 Å, (obtenues au moyen d'un microtome Porter-Blum) sont montées sur des lames porte-objet, recouvertes d'une pellicule de formvar. Après les avoir passées à travers la série descendante des alcools et à l'eau distillée, les coupes sont soumises (pendant 2 ou 3 heures et à la température de 60° C) à l'action d'une solution d'argent ammoniacal, dont la composition est la suivante:

Nitrate d'argent au 0,25% . . . . 10 cm³
Méthenamine . . . . 0,3 g
Quand la méthenamine est complètement dissoute on ajoute:
Solution de borax au 5% . . . . 0,8 cm³
Eau bidistillée jusqu'à 30 cm³.

Les coupes sont ensuite soigneusement lavées à l'eau bidistillée, passées dans une solution de chlorure d'or au 0,2% et traitées pendant quelques minutes à l'hyposulphite de sodium au 0,5%. Après un nouveau lavage à l'eau bidistillée, les coupes et la pellicule de formvar sont alors détachées des lames et montées sur des grilles de cuivre. De celles-ci on découpe, à l'emporte pièce, les zones où sont distribuées les coupes et on passe à l'observation au microscope électronique.

Des coupes plus épaisses (500—1 000Å) montées sur des lamelles couvre-objet, ont été traitées par la même technique de coloration et employées pour l'observation au microscope optique.

Résultats. De nombreuses structures histologiques révèlent une affinité pour l'argent quand elles sont traitées par la technique ci-dessus. Je signalerai, en particulier, les membranes basales des glomérules et des tubules reinaux, les membranes basales (épithéliale et endothéliale) des alvéoles pulmonaires, les fibres élastiques et réticulaires, les globules rouges, les granules des mastzellen et des granulocytes éosinophiles, les granules de zymogène des cellules exocrines du pancréas et quelques inclusions cytoplasmique particulières, de signification douteuse.

Je me limiterai ici à présenter quelques-uns des résultats plus démonstratifs; des données plus complètes seront publiées prochainement.

La Fig. 1 montre un détail de la paroi d'un conduit alvéolaire de poumon de rat. La membrane basale épithéliale, rendue très opaque aux électrons par l'argent qui s'est déposé de façon homogène, est bien visible au dessous de l'expansion cytoplasmique des cellules. Dans l'interstice et en toute proximité d'une cellule de mésenchyme on aperçoit quelques fibres élastiques, imprégnées de façon homogène ainsi que de nombreuses fibres réticulaires. La photographie en haut et à gauche, montre des fibres élastiques d'aorte humaine, elles aussi fortement argentaffines, vues au microscope optique.

La Fig. 2 représente un détail de l'interstice d'un septum alvéolaire de poumon de rat. On reconnaît ici des fibres réticulaires, de l'épaisseur de 200 Å, dont la structure périodique est exaltée par l'imprégnation à l'argent . La grande période a une épaisseu moyenne de 270 à 300 Å.D'accord avec d'autres auteurs (*4*), je suis de l'opinion que les granules d'argent se déposent surtout au niveau de la bande A.

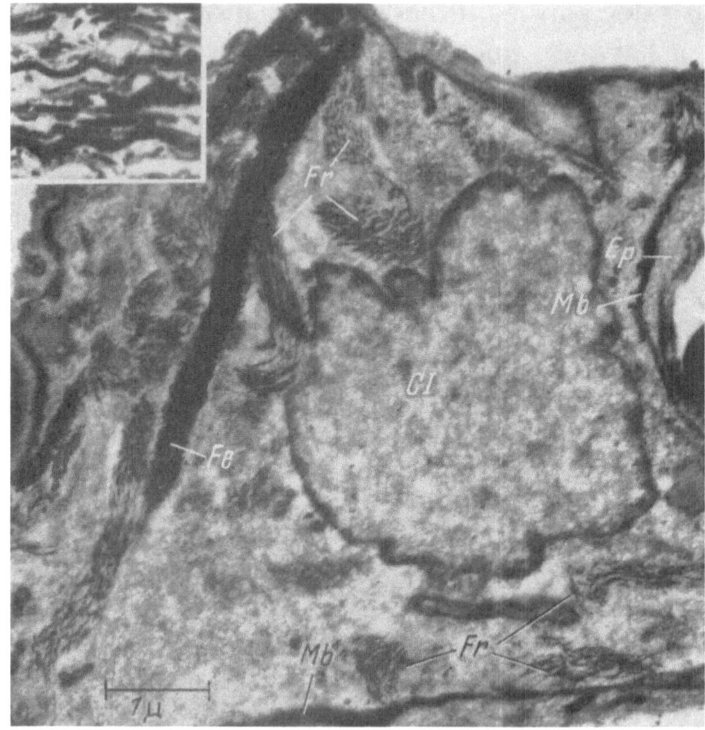

Fig. 1. *CI* = Cellule interstitielle; *Mb* = Membrane basale; *Ep* = Expansion cytoplasmique d'une cellule épithéliale; *Fe* = Fibres élastiques; *Fr* = Fibres réticulaires

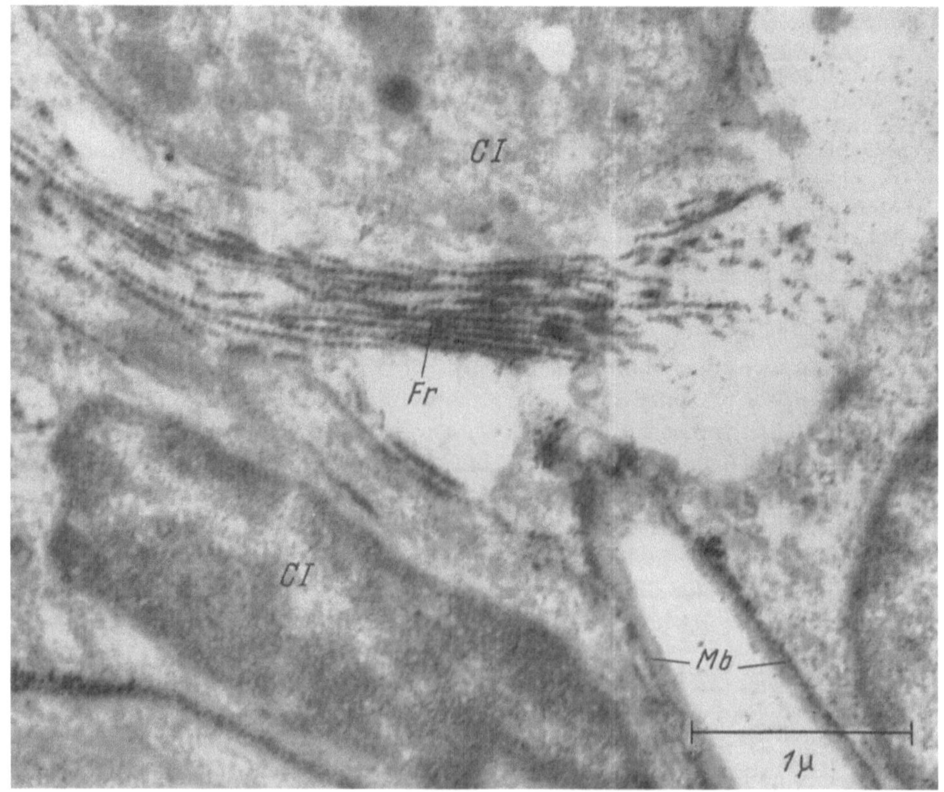

Fig. 2. *CI* = Cellules interstitielles; *Fr* = Fibres réticulaires; *Mb* = Membrane basale

La Fig. 3 A montre un macrophage endoalvéolaire dans un poumon de rat. Le cytoplasme est parsemé de corpuscules arrondis, très argentaffins, qui s'identifient avec les inclusions cytoplasmiques indiquées par SCHULZ (5) comme cytosomes. Ces inclusions sont surtout nombreuses et volumineuses au cours de la stase sanguine (Fig. 3 B), ce qui rend difficile, dans les sections colorées à l'argent, l'identification des granules férrugineux, notoirement abondants dans les cytosomes.

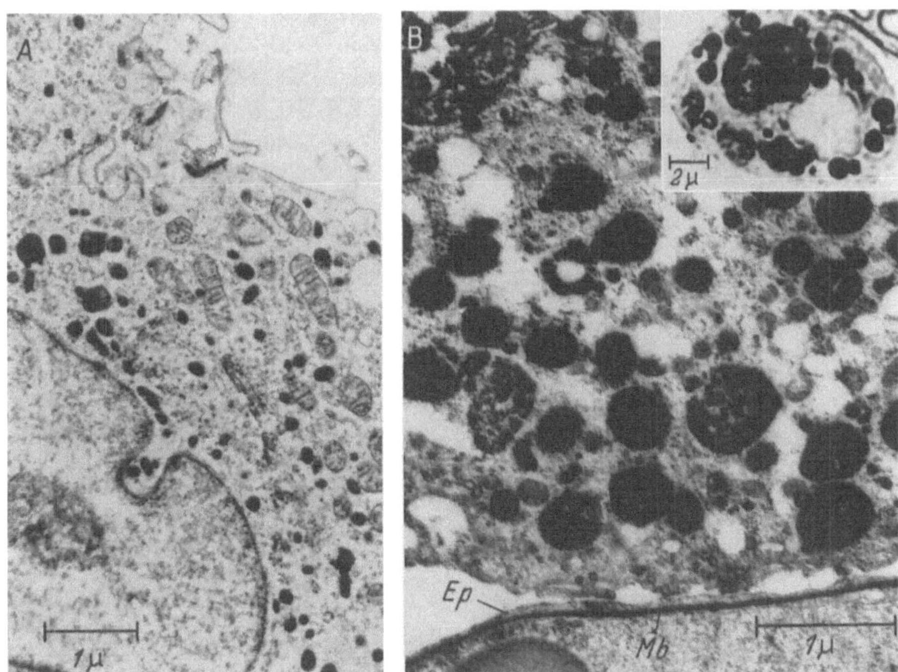

Fig. 3. A and B. *Mb* = Membrane basale; *Ep* = Expansion cytoplasmique d'une cellule épithéliale. En haut et à droite un macrophage endoalvéolaire vu au microscope optique

Occasionellement on observe aussi des structures argentaffines, semblables aux cytosomes, dans le cytoplasme des cellules épithéliales et endothéliales des poumons. Rarement les crêtes des mitochondries se révèlent fortement argentaffines.

Arrivé à ce point il est nécessaire de remarquer que la méthode de JONES se base sur la propriété qu'ont les groupes aldehydiques de réduire une solution d'argent ammoniacal lorsqu'ils sont libérés de leurs liaisons. On pense que le substrat responsable de cette réaction soit en grande partie représenté par les polysaccharides et les glyco-protéines.

La méthode d'imprégnation à l'argent que j'ai employée démontre, pour la première fois, que l'acide osmique est capable lui-même de révéler une argentaffinité (secondaire) de quelques structures tissulaires. Malheureusement nos connaissances sur le mécanisme de la fixation au moyen de l'acide osmique (6, 7) ne nous permettent pas d'établir la nature des groupes responsables de la réduction de l'argent, ainsi que les substrats organiques dont ceux-ci sont libérés par l'acide osmique. Toutefois des recherches préliminaires que je viens d'entreprendre à ce sujet, me consentent d'affirmer avec certitude que les groupes réducteurs sont, au moins pour la plus grande partie, différents des aldhéhydes et des cétones, par le fait qu'ils ne sont pas capables d'exercer une action réductrice sur le réactif de SCHIFF.

Bibliographie

1. JONES, B. D.: Amer. J. Path. **33,** 313 (1957).
2. MARINOZZI, V.: Atti Acad. naz. Lincei **24,** 600 (1958).
3. — Atti Acad. naz. Lincei **24,** 754 (1958).
4. BENEDETTI, E. L.: Sci. Med. Ital. **3,** 1 (1953).
5. SCHULZ, H.: Beitr. Path. Anat. **119,** 71 (1958).
6. PORTER, K. R., and F. KALLMAN: Exp. Cell Res. **4,** 127 (1953).
7. BAHR, G. F.: Exp. Cell Res. **7,** 457 (1954).

Die Wirkung der Elektronen auf natürliche organische Substanzen bei ihrer Untersuchung im Elektronenmikroskop

E. M. BELAWZEWA

Laboratorium für Elektronenmikroskopie, Akademie der Wissenschaften der USSR, Moskau

Die vorliegende Arbeit beschäftigt sich mit dem Studium der Wirkung mittelschneller Elektronen auf Kristalle des Gramicidins S, β-Carotins, Chlorophylls a + b und Jod-Gramicidins sowie auf amorphe Stoffe: völlig oxydiertes β-Carotin, kolloidales Chlorophyll a + b bei p_H 8,6. Die Untersuchung wurde mit dem Elektronenmikroskop UEM-100 und mit der Beugungskammer EG-100 ausgeführt. Die Stromdichte am Objekt wurde mit einem Mikroampèremeter gemessen, das mit einem Elektronen abfangenden Faraday-Käfig verbunden war.

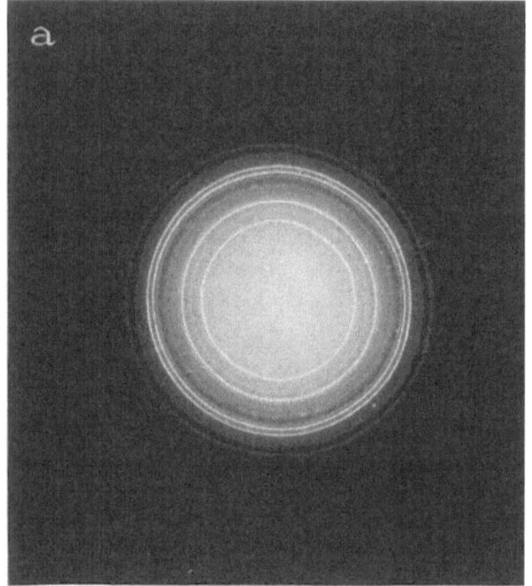

Abb. 1 a u. b. Elektronenbeugungsdiagramm (a) und elektronenmikroskopisches Bild (b) von β-Carotin

Bei einer Stromdichte am Objekt von $j = 1 \cdot 10^{-5}$ A/cm² zeigten die Elektronenbeugungsdiagramme der Kristalle von β-Carotin, Chlorophyll a + b, Gramicidin S und Jod-Gramicidin

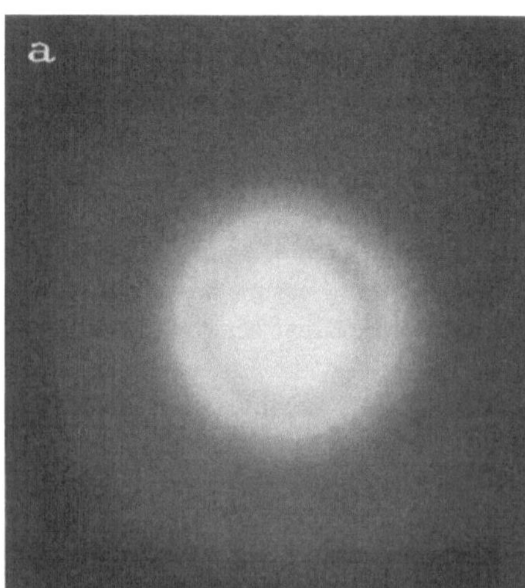
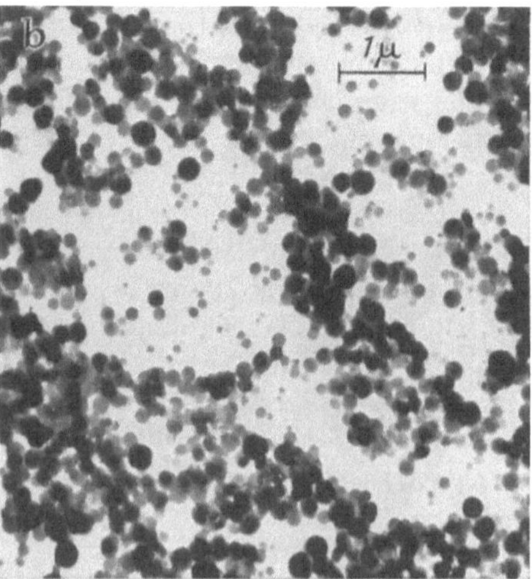

Abb. 2a u. b. Elektronenbeugungsdiagramm (a) und elektronenmikroskopisches Bild des Chlorophylls a + b bei p_H 8,6

scharfe Interferenzmaxima. Die Netzebenenabstände, welche auf Grund der Elektronenbeugungsdiagramme und entsprechender Röntgendiagramme berechnet wurden, hatten die gleichen Werte

(1). Abb. 1 zeigt ein Elektronenbeugungsdiagramm und ein elektronenmikroskopisches Bild von β-Carotin.

Die unter denselben Bedingungen erhaltenen Elektronenbeugungsdiagramme von oxydiertem β-Carotin und auch von kolloidalem Chlorophyll bei p_H 8,6 enthalten drei diffuse Ringe (Abb. 2).

Bei einer Stromdichte am Objekt von $j = 1 \cdot 10^{-3}$ A/cm² und höher sind an Stelle der scharfen Reflexe auf den Elektronenbeugungsdiagrammen der Kristalle von Carotin, Chlorophyll, Gramicidin und Jod-Gramicidin diffuse Ringe zu sehen.

Es wurde festgestellt, daß diese Substanzen eine unterschiedliche Empfindlichkeit gegenüber der Bestrahlung mit Elektronen besitzen. Unter diesen ist die am wenigsten resistente Substanz das Jod-Gramicidin, die resistenteste β-Carotin.

Auf Abb. 3 ist die Abhängigkeit der kritischen Arbeitsdichte (Minimalwerte) $(n_0 \tau)_k$, bei welcher der Verlust der Kristallinität des Carotins auftritt, von der auf das Objekt auffallenden Leistungsdichte n_0 bei drei verschiedenen Werten der Beschleunigungsspannung $U_B = 40\,\text{kW}$, $U_B = 60\,\text{kW}$, $U_B = 100\,\text{kW}$ gezeigt.

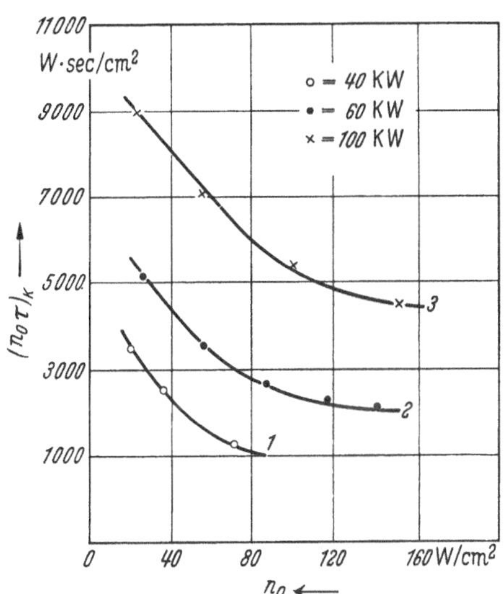

Abb. 3. Die kritische Arbeitsdichte (Minimalwerte) $(n_0\tau)_k$ als Funktion der auf das Objekt auffallenden Leistungsdichte n_0 bei $U_B = 40$ kW (Kurve 1), $U_B = 60$ kW (Kurve 2), $U_B = 100$ kW (Kurve 3)

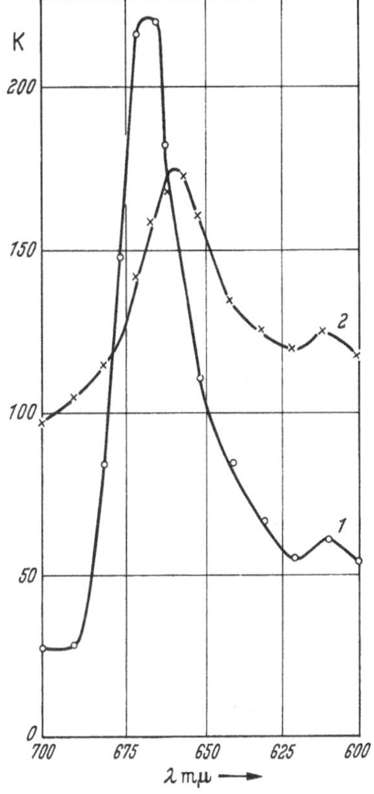

Abb. 4. Absorptionsspektrum einer Lösung von Chlorophyll a + b in Pyridin vor der Bestrahlung (Kurve 1) und nach der Bestrahlung im Elektronenmikroskop (Kurve 2)

Nach intensiver Bestrahlung mit Elektronen ($n_0 = 600$ W/cm²) gehen β-Carotin (oxydiertes und nicht oxydiertes), Gramicidin S und Jod-Gramicidin in den unlöslichen Zustand über. Bestrahltes Chlorophyll a + b löst sich nicht in Alkohol, Xylol und Petroläther, welche für das Chlorophyll gute Lösungsmittel darstellen. In Pyridin löst sich das bestrahlte Chlorophyll nur schwach. Vor der Bestrahlung liegt das Absorptions-Maximum des in Pyridin gelösten Chlorophylls bei $\lambda = 669$ mμ, nach der Bestrahlung bei $\lambda = 655$ mμ (Abb. 4).

Auf dem elektronenmikroskopischen Bild der bestrahlten und hierauf im Vakuum bis 300° erwärmten Objekte wurde kein Schmelzen der Kristalle beobachtet; es findet jedoch statt, wenn die Erwärmung ohne vorherige Bestrahlung vorgenommen wird. Es erfolgt also eine Veränderung der Schmelzbarkeit dieser Stoffe durch deren Bestrahlung mit Elektronen. Es wurde beobachtet, daß bei Bestrahlung mit Elektronen am Anfang ein Verlust der Kristallinität auftritt und hierauf eine Veränderung der Löslichkeit und der Schmelzbarkeit. Die Bestrahlung mit

Elektronen schwacher Intensität ($n_0 < 1$ W/cm²) rief keine merklichen Veränderungen dieser Eigenschaften der untersuchten Substanzen hervor.

Da das Elektronenbündel eine ionisierende und eine thermische Wirkung auf das Objekt ausübt, war es von Interesse, den Einfluß dieser beiden Faktoren im einzelnen auf Chlorophyll a + b und Gramicidin S zu studieren.

Zur Untersuchung der ionisierenden Wirkung wurden die Kristalle des Chlorophylls a + b und des Gramicidins S auf einem linearen Beschleuniger mit Elektronen einer Energie von 1 MeV bestrahlt. Die thermische Wirkung der Elektronen wurde auf Elektronenbeugungsdiagrammen und elektronenmikroskopischen Photographien der im Vakuum erwärmten Objekte untersucht. Es wurde festgestellt, daß Gramicidin S gegenüber der ionisierenden Strahlung stärker empfindlich ist, während Chlorophyll a + b eine größere Empfindlichkeit gegenüber der Erwärmung aufweist.

Bei einer Temperatur von 40° ist das Chlorophyll amorph. Die Veränderungen der Elektronenbeugungsdiagramme des Chlorophylls a + b sind nach Erwärmung im Ofen der Beugungskammer oder nach Bestrahlung im Elektronenmikroskop analog. Aus diesen Befunden folgt, daß der Verlust der Kristallinität des Chlorophylls bei der elektronenoptischen Untersuchung sowohl durch die thermische wie auch durch die ionisierende Wirkung der Elektronen verursacht sein kann.

Bei einer Temperatur von 250° befindet sich Gramicidin S noch im kristallinen Zustand, während es bei 280° amorph wird. Die Veränderung der Elektronenbeugungsdiagramme des Gramicidins im Temperaturintervall von 250—280° unterscheidet sich von ihrer Veränderung bei Bestrahlung mit Elektronen. Es soll bemerkt werden, daß während der Bestrahlung mit Elektronen bei einer Stromdichte am Objekt von $j_0 = 1 \cdot 10^{-3}$ A/cm², bei welcher ein schneller Verlust der Kristallinität des Gramicidins S eintritt, die Temperatur am Objekt (gemessen nach der Methode von Stojanowa und Belawzewa) nicht höher als 100° war. Die wahrscheinlichste Ursache des Übergangs des Gramicidins aus dem kristallinen in den amorphen Zustand ist daher die ionisierende Wirkung der Elektronen. Diese bedingt demnach eine starke Veränderung der Molekularstruktur des β-Carotins (des oxydierten und des nicht oxydierten), des Chlorophylls a + b und des Jod-Gramicidins bei der elektronenoptischen Untersuchung.

Es resultiert daraus, daß es bei genügend hoher Beschleunigungsspannung, bei geringer Leistungsdichte des Objekts und bei geringer Dauer der Bestrahlung möglich ist, natürliche organische Substanzen ohne merkliche Veränderung ihrer Struktur mit mittelschnellen Elektronen zu untersuchen.

Literatur

1. Belawzewa, E. M.: Biophysika (russ.) H. 2, 628 (1957).

An electron microscope study of the heat denaturation of human serum albumin in solutions

P. Bartl

Chemical Institute, Czechoslovak Academy of Science, Prague

The heat denaturation of human serum albumin has been studied by a number of authors (*1—5*). In most cases, indirect methods have been used to derive information about the mechanism of aggregation of the albumin particles by heat, and the products of this process. Since the system of particles formed by aggregation as a result of heat-induced denaturation is extremely complex it may be anticipated that the interpretation of the results obtained by indirect methods will be rather difficult. For this reason, we decided to investigate this problem by electron microscopy. We were, of course, quite aware that these results, too, would be subject to certain errors, mainly because of some of the conditions involved in the preparation of the samples. In the first stages of this work, it was our aim to get preliminary information about the basic features of this process.

The material used was an electrophoretically homogeneous commercial preparation of human serum albumin. Solutions of this protein in water, decinormal saline, and ammonium acetate were heated in a thermostated, closed glass vessel with efficient agitation. Samples were withdrawn at certain time intervals, dropped onto a collodion membrane, dried in a desiccator and shadowed with gold or a gold-palladium alloy at an angle of 30°. All the micrographs were recorded with a table-type electron microscope of Czechoslovak construction, with a resolving power down to 20 Å.

The resolution of details in the pictures was made rather more difficult by the fact that proteins, being highly surface active, tend to form a film at the interface between the solution and air. When the drop dries down this film forms a practically continuous cover on the collodion membrane and causes considerable electron scattering. In spite of this, the difference between films of native and denatured albumin is obvious at first sight. To avoid being misled by artefacts we considered only those particles in interpreting the pictures which were predominant in each particular sample, and were sufficiently raised over the level of the surrounding film to cast their own "shadows".

To get as much information about the process of aggregation as possible we investigated a number of samples prepared under various conditions. On the other hand, since we were only interested for the time being in the general features of the process we made no attempts to analyse the effect of indivual factors. For these reasons we chose conditions which seemed most favourable from the point of view of the technique for preparing the samples, even if this involved leaving out of account for instance such important conditions of reproducibility as constancy of the p_H and ionic strength.

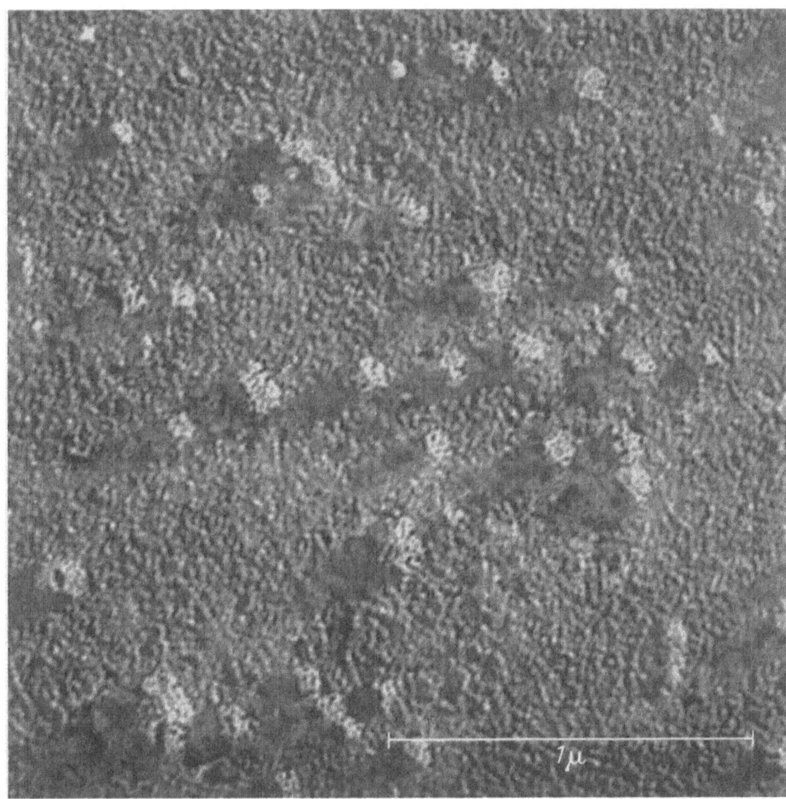

Fig. 1. Beaded chains and random coils formed by the aggregation of heat-denaturated human serum albumin molecules. Shadowed with gold-palladium

The pictures we have recorded permit the following description of the heat-induced aggregation of serum albumin particles: Initially, that is up to about 30 min from the beginning of denaturation at 60°, we find mainly aggregation of the particles of denatured protein into beaded chains which are visible mainly because of their shadows (Fig. 1). The segments of these chains are about 100 to 200 Å in size, that is close to the larger dimension of the native serum albumin molecule. The axial ratio is rather lower than in the native molecule. This can be explained by

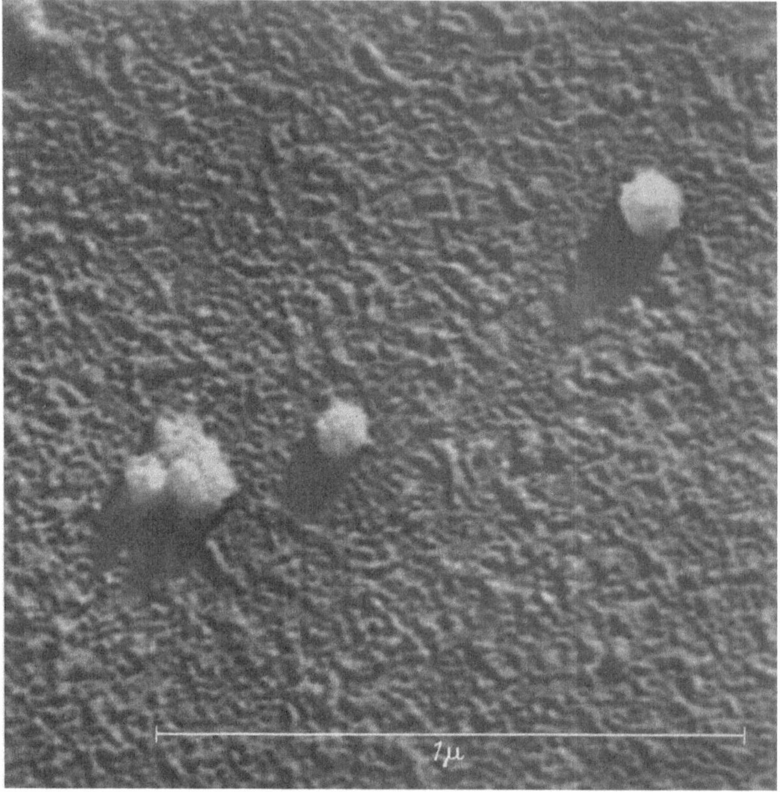

Fig. 2. Compact coils formed during further denaturation by heat. Shadowed with gold

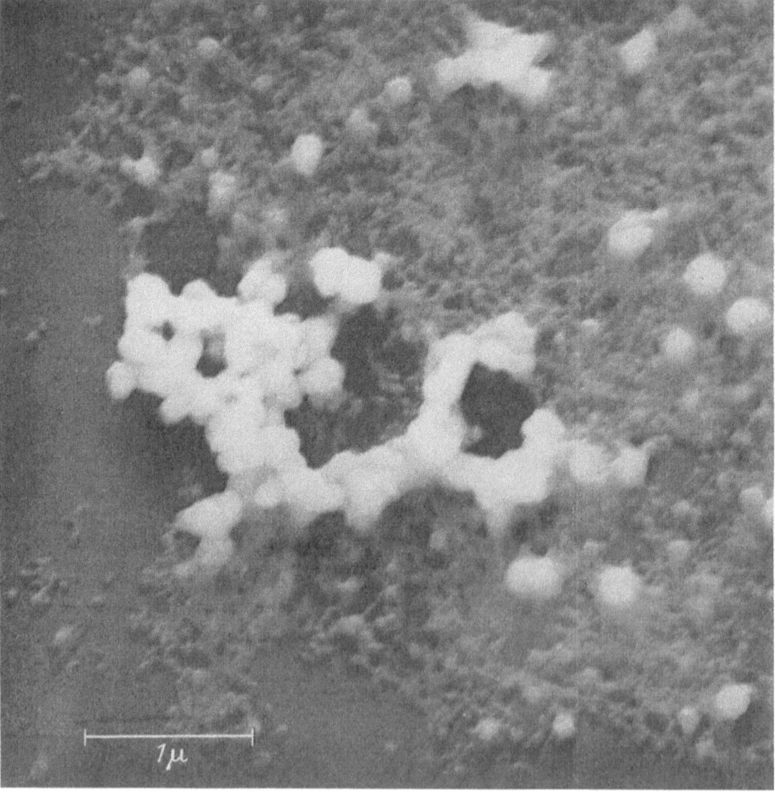

Fig. 3. Final aggregates of the denaturated human serum albumin molecules. Shadowed with gold-palladium

the fact that the layer of condensed metal is thicker across the width of the particle than along its length, or by a loosening of the coiled polypetide chain. These findings suggest that the linear aggregation of the denatured molecules is of the end-to-end type.

We also tried to determine the distribution of these chains by length (L). The distribution curve shows a maximum in the 400 to 800 Å region for both straight and branched chains (which form about 1/3 of the total). Calculating the various averages from these histograms we find $L_n = 785$, $L_w = 1020$, $L_z = 1270$, $L_{z+1} = 1420$ Å, where the different length averages are defined similarly as the averages used in calculations of molecular weight, i. e. $M_n = \Sigma c_i M_i / \Sigma c_i$, $M_w = \Sigma c_i M_i^2 / \Sigma c_i M_i$ etc. M_n here is called "number average molecular-weight", M_w "weight average", M_z "z-average" etc., and c_i is the concentration of particles of molecular-weight M_i. If we chose the value of L_w/L_n as a measure of the degree of polydispersity — in a similar way as M_w/M_n is often being used — we find a value of 1.3, which is rather low.

The formation of further particles, mainly loose coils, seems to be the second stage of aggregation (Fig. 1). Most of these coils are of practically the same size as the original chains and the distribution curve also has the same shape. It therefore seems that the coils arise mainly by the lateral joining of chains of a certain critical length, rather than by the coiling or intertwining of longer chains. Most of the coils are slightly ellipsoid, with an axial ratio of 3 : 4 to 2 : 3. On further heating, aggregation proceeds from the loose coils to compact coils of relatively high symmetry. Their diameters are in the region of 1600 to 2000 Å (Fig. 2). These in turn aggregate to chains, which of course have very much larger dimensions than the chains formed in the first stage of aggregation (Fig. 3). Finally, these chain-like structures aggregate into floccules which precipitate from solution. This is the final stage of heat denaturation which involves visible flocculation of the denatured protein.

In conclusion we may say that the method of electron microscopy has essentially confirmed the conclusions drawn from other independent methods, in particular from light scattering (6) and viscometry, for this system of human serum albumin undergoing denaturation by heat.

References

1. Anson, M. L.: Adv. Protein Chem. 2, 361 (1945).
2. Tongur, V. S.: Uspechi Sovremennoj Biol. 31, 391 (1951).
3. Putnam, F. W.: The Proteins. Vol. 1, Part B, p. 807. New York: Academic Press Inc. 1953.
4. Joly, M.: Progr. in Biophys. 5, 168 (1955).
5. Tsiperovich, A. S.: Uspechi Chim. 25, 1173 (1956).
6. Sedláček, B.: Chem. listy 50, 867 (1956).

The electron microscope as a quantitative instrument for dry mass determination

G. Bloom and E. Zeitler

The Institute for Cell Research and Genetics, Karolinska Institutet, Stockholm

The electron microscopical picture is a two dimensional presentation of a three dimensional object. In spite of the complex rules governing image formation it is worth a trial to derive some information about the third dimension and to a fuller extent exploit the possibilities of the electron microscope.

If we regard what happens in the electron microscope we can say in a rough manner that an object with a thickness or mass per unit area t, diminshes the original intensity of the electron beam I_0 to a value of I.

Hall and others (1, 2) have shown that with certain restrictions there is a simple relationship between these both intensities and the mass per unit area of the object.

If we define as contrast K the logarithm of the ratio I/I_0, the following simple formula is obtained:

$$K = \log \frac{I}{I_0} = \frac{t}{t_{10}} \tag{1}$$

From this we see that the interesting value t is directly proportional to the contrast K. The logarithmic definition of contrast has been chosen in order to be able to add the contrast of superimposed layers.

The factor t_{10} is the mass which reduces the original intensity I_0 to a value of 10% or in other words gives the contrast 1. The factor t_{10} is dependent on the chemical properties of the object as well as on the working data of the electron microscope (voltage and aperture). A detailed description of these influencing factors has been published by ZEITLER and BAHR (2).

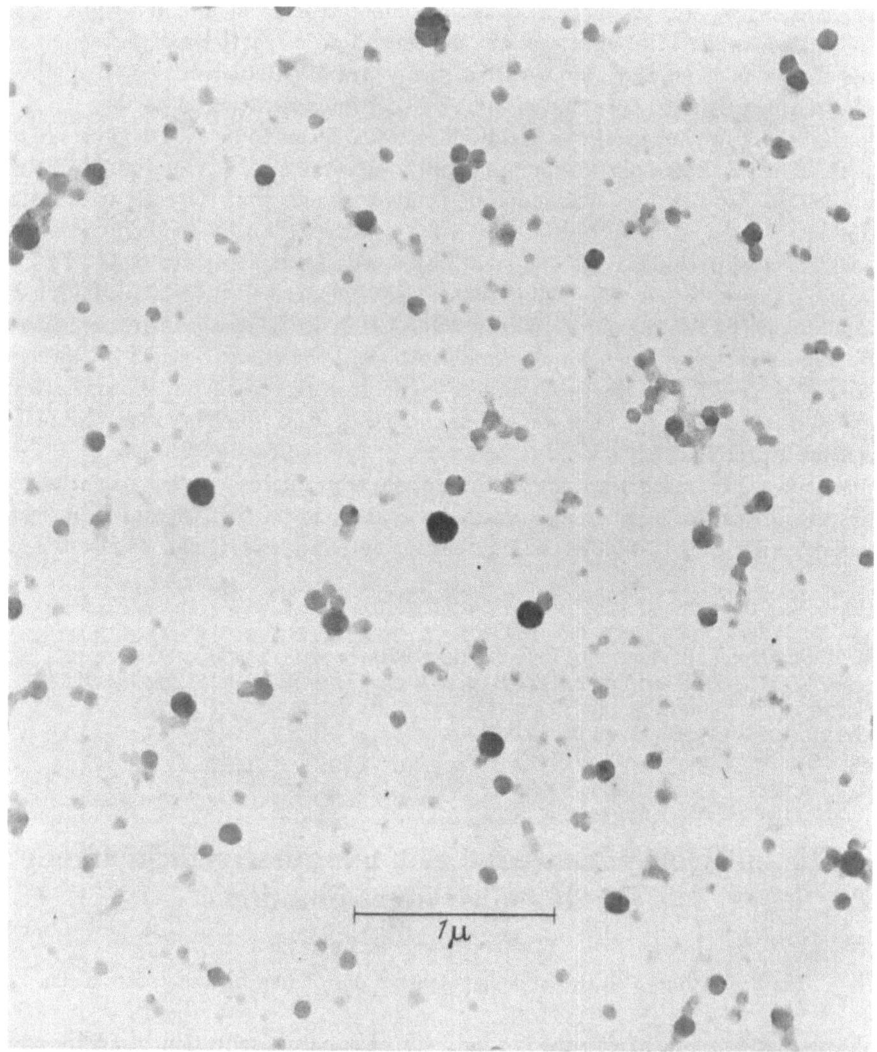

Fig. 1. Electron micrograph of casein particles from cow's milk. Magnification 27000 ×

Summarizing we can state that we can determine the mass per unit area of an object by aid of the measured contrast K and the known constant t_{10}.

In order to apply this method practically an experiment was performed in which the contrast distribution or as we can also say the mass distribution of biological objects was determined. Besides, the experiment was broadened to also include a comparison within a population of biological particles in order to find out if there exists a relationship between size and mass per unit area i. e. thickness.

In search for a suitable biological object on which to apply our method we chose the milk protein casein. The reasons for this choice are several: a) the material is easy to obtain and to prepare, (b) casein is often referred to as a protein model substance, c) the particle size lies within a suitable range for electron microscopy and lastly d) a large amount of electron microscopical work on this material has been carried out.

The casein was prepared according to the method given by NITSCHMANN and HOSTETTLER (3, 4). We exclude here the details of preparation but we wish to point out that in our experiment casein fixed in formalin was used.

Fig. 1 shows a typical electron micrograph of casein particles from cow's milk at a magnification of 27 000 ×.

In Fig. 2 the size distribution of 100 casein particles is depicted. The diameters D were determined with the aid of a measuring ocular on an enlarged paper copy. A coordinate system is chosen in which a normal Gauss distribution is represented by a straight line. The diameters vary from 40 to 240 mμ with a mean value of 146 mμ and a standard deviation of 40.8 mμ. From the resultant straight line we can verify that we here have a normal Gauss distribution.

In order to determine the contrast of our objects the photometrical method is chosen. With the aid of the density curve the density values are converted to intensity values. The physical behaviour of a photographical plate exposed to an electron beam facilitates this conversion as there is a linear relationship between both values, that is intensity and density.

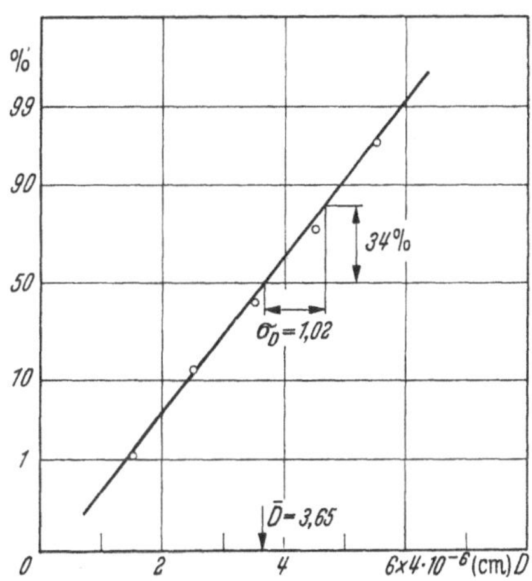

Fig. 2. Size distribution of casein particles (see text)

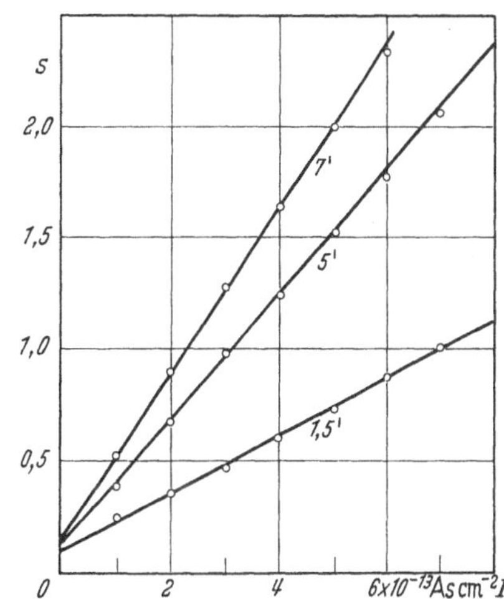

Fig. 3. Density curves of the photographic plates used exposed to an electron beam of 50 kV

Fig. 3 shows the density curves of Kodak Lantern Slides after different developing times. The linear relationship is valid up to densities of 1.7. Moreover this relationship is independent of the developing time and also of the temperature of the developer.

As the contrast is expressed as a function of a ratio of intensities, or as we have shown also of densities, the contrast becomes quite independent on the slope of the density curves, that is on the conditions of developing.

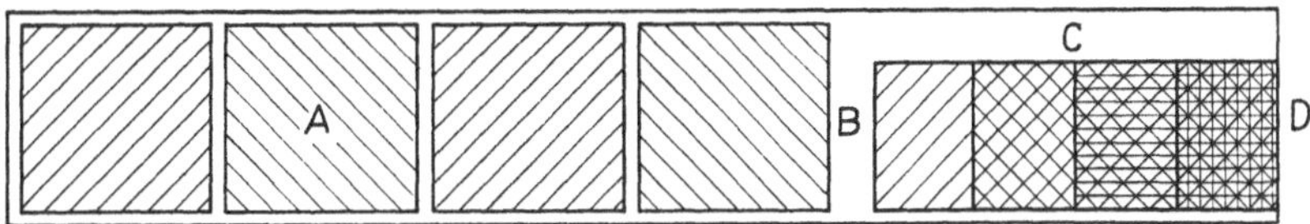

Fig. 4. Schematic figure showing the method for determining background density. A = micrographs of object; B = fog; C = area exposed only to scattered electrons; D = step wedge

A further detail of importance when converting our density values is to take into consideration the so called fog of the photographic plate. This fog is composed of two components: a) densities which appear without any exposure of the plate and b) density which is caused by electrons scattered uniformly over the whole image area. Fig. 4 shows a simple scheme of the evaluation of an exposed lantern slide. A represents micrographs of our objects, B is an entirely unexposed part of the plate, C shows an area exposed only to scattered electrons, due to screening by a grid bar which is naturally opaque to the electron beam. This is thus the fog value which we need in order to correct the obtained density values. D lastly, is a step-wedge made on each plate to control the mentioned linear relationship between density and intensity.

To measure all these densities we used a Zeiss Schnellphotometer II under constant conditions for the micrographs and the stepwedges.

Fig. 5 shows directly the contrast distribution of the investigated casein particle population. As in the size distribution we obtain here also a straight line representative of a normal Gauss distribution. The characteristical values are $\bar{K} = 0.215$ and $G_K = 0.061$.

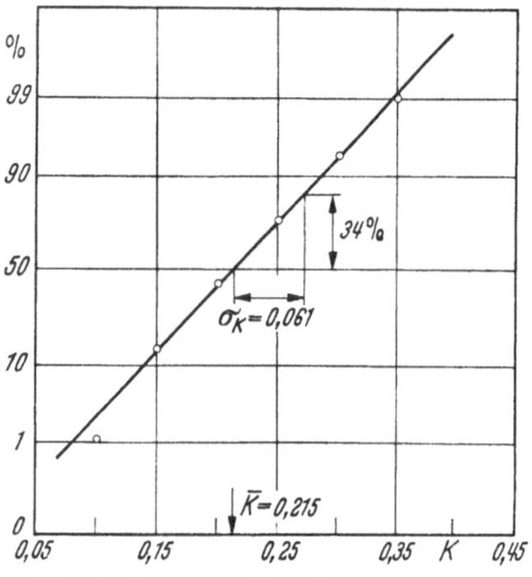

Now we come to the main point of our investigation, that is to convert contrast to mass per unit area. For this we need the constant t_{10}. Theoretically has been shown by Zeitler and Bahr (2) that t_{10} is the same for all organical objects with an error range of $\pm 5\%$. In order not only to rely on theoretical speculations we have calibrated our electron microscope at constant running data by means of a formvar reference system. We find in good agreement with the theoretical calculations that t_{10} for our microscope equals 70×10^{-6} g/cm².

From formula (1) can be derived that the mean mass per unit area is:

$$\bar{t} = \bar{K} \times t_{10} \tag{2}$$

Fig. 5. Contrast distribution of casein particles

Using the obtained values for both figures we get a mean mass per unit area of 15.1×10^{-6} g/cm². If we divide the mean mass value by the specific weight of casein as obtained from the literature we get the mean geometrical thickness

$$\bar{d} = \frac{\bar{t}}{\varrho} = 11.4 \times 10^{-6} \text{ cm} \tag{3}$$

Now knowing both the size distribution and the thickness distribution of casein particles it is of interest to ascertain if the shape of the particles is independent of their size. This being true, a constant relationship between thickness and diameter can be said to exist independent of the size of the particles. An arithmetical criterion for an existing linear relationship is the so called correlation coefficient r. We obtain for r the value of 0.95 and as the mutual dependence of both regarded values becomes stricter as r approaches 1, we can in our case speak of a pronounced linear correlation between thickness and size.

From this we gain the information that our casein particles have a similar shape with a thickness that is approximately 80% of the diameter. This deviation from unity may be caused by factors such as a) contrast determination gives the thickness averaged over a certain area surrounding the maximum value and b) a certain degree of flattening of the particles is bound to take place.

It appears as if this presented method makes possible not only determination of thickness and dry mass of biological specimens but naturally also the study of the influence of various preparatory techniques such as fixation, embedding and staining procedures. For thickness determination it is superior to for example the old shadowing method. We also believe that it will prove useful for quantitative histochemical purposes. The electron microscope appears thus to make possible a broadening of the field of quantitative microscopy into regions hitherto little investigated.

A detailed report of the investigation will appear shortly (5).

References

1. Hall, C. E.: J. biophys. biochem. Cytol. **1,** 1 (1955).
2. Zeitler, E., and G. F. Bahr: Exp. Cell Res. **12,** 44 (1957).
3. Nitschmann, H.: Helv. chim. Acta **33,** 1258 (1949).
4. Hostettler, H., and K. Imhoff: Landwirtschaftl. Jb. Schweiz **66,** 308 (1952).
5. Bloom, G., and E. Zeitler: To be published in Exp. Cell Res.

Detailed electron microscope studies on purified bacterial and viral nucleic acid (DNA and RNA) with some considerations on the relation of DNA to genetics

M. Terada

Department of Bacteriology, The Tokyo Jikei-kai School of Medicine, Tokyo

DNA and RNA have been extracted and purified from bacteria (acid-fast Kedrowsky strain) and bacterial virus (Salmonella sendai phage). After confirming them to be pure both physically and chemically, their elementary fibers of DNA and RNA have been studied with the electron microscope by applying chrome shadowing. The width of single DNA fibers is about 20—25 Å and that of RNA is between 14—17 Å which coincide with the theoretical values. These fibers have also been clearly observed by electron staining with $AgNO_3$. With the electron microscope the author demonstrated the following facts: DNA fibers are composed of various phases of network or branch-like fashion and end in an isolated elementary fiber, and the spiral-coiled figure is most specific. RNA also demonstrated similar electron microscope figures. However, RNA has a tendency of forming a more complicated net-work of entarngled elementary fibers as compared to DNA. Interesting figures of phages revealed long processes of DNA fibers protruding from the tails of phages. Genetic studies: A transformation of DNA which controls the color production of bacteria has been successfully carried out. The interesting data has been obtained that the recombinant phages were produced when mixed infection with phages DNA and another intact phage was carried out.

An extended version of this article will appear.

Morphologie gelöster Desoxyribonucleinsäure-Präparate und einige ihrer Eigenschaften in Oberflächen-Mischfilmen

A. Kleinschmidt und R. K. Zahn

Hygiene-Institut und Institut für vegetative Physiologie, Frankfurt a. M.

Es werden Ergebnisse über die Morphologie eines reinen Desoxyribonucleinsäure(DNS)-Präparates mitgeteilt, wobei das gelöste fädige Material in Mischfilmen von monomolekularer Struktur zusammen mit Proteinen, besonders mit Cytochrom c, untersucht und dargestellt wurde.

Ausgangsmaterial sind gereinigte Kerne von Seelachs-Spermatozoen. Aus diesen wird nach der Methode von Marko und Butler (*1*) die DNS mit Dodecylsulfat als viscöse Lösung dargestellt, in Äthanol mehrfach gefällt und ein Trockenpräparat gewonnen, das die in Tab. 1 dargestellten Reinheitskriterien besitzt.

Als filmbildende Proteine verwendeten wir reines Cytochrom c und Rinderserumalbumin (reinst) in einer 1%- bzw. 0,1%-Stammlösung, von der Trockensubstanz ausgehend.

Untersucht man ein DNS-haltiges Protein oder dieses allein als Oberflächenfilm auf wäßrigem Substrat, so ist für die quantitative Analyse der mechanischen Eigenschaften ein Langmuir-Trog geeignet (*2*). Die Flächenbesetzung des Proteinfilms bildet die Abszisse ($A = m^2/mg$), der Schub, mit dem der Film durch eine bewegliche Barriere komprimiert wird, die Ordinate

8*

(F = dyn/cm) der Kraft/Flächen-Kurven (Abb. 1). Sie sind in manchen Einzelheiten charakteristisch für das betreffende Protein. Durch graphische Konstruktion erhält man dabei die Dichtpunktfläche (3) [= limiting area nach GORTER (4)]; sie ist der bei verschiedenen Versuchen vergleichbare Wert eines Films beim Schub 0.

Zur Filmübertragung auf elektronenmikroskopische Blenden mit einer aufgedampften Kohlefolie wendet man Schübe an, die in der Regel größer als 0,1 dyn/cm sind. Dadurch erreicht man, daß die Wasseroberfläche mit dem Film voll besetzt bleibt und keine Fehlstellen aufweist. Dann kann man auch die durchschnittliche Oberflächenkonzentration DNS (in g) pro Flächeneinheit (in μ^2) angeben.

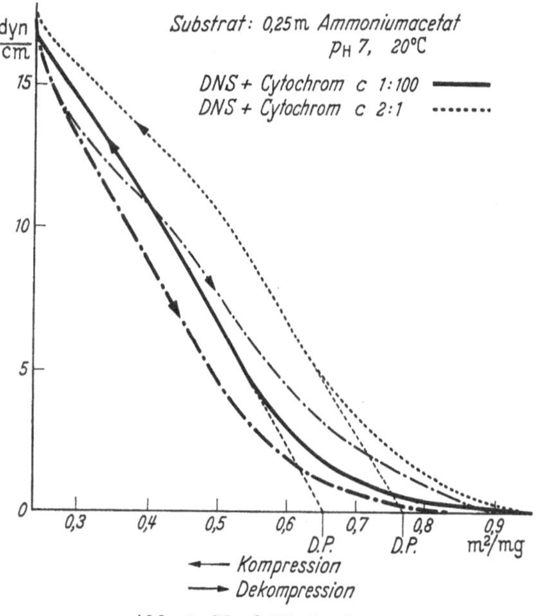

Abb. 1. Kraft/Flächenkurve

Tabelle 1. *DNS aus den Spermtozoenkernen von Gadus virens (Seelachs).* Gewichtsverhältnis $\dfrac{\text{Stickstoff}}{\text{Phosphor}} = 1,60 \pm 0,05$ Basenzusammensetzung, berechnet auf 100 Mol Phosphor:

Basen	Mol Basen
Adenin . . .	27,2
Thymin . . .	27,2
Guanin . . .	22,6
Cytosin . . .	22,9
Summe . . .	99,9

Das Verhalten der DNS-Moleküle im Film ist ferner davon abhängig, daß das Substrat eine genügende Salzkonzentration aufweist, die Komplexbildungen und Präcipitate verhindert. Der Salzgehalt von 0,25 m Ammoniumacetat war so gewählt, daß dabei sowohl für die DNS wie für Cytochrom c die untere Grenze für volle Löslichkeit etwa erreicht war. Zur Entfernung der Salze und für eine sichere Darstellung eines zu trocknenden Films benutzten wir die Isopentan-Methode [nach PETERS (5)]: Nachdem an der kohlebefilmten Platinblende ein Proteinfilm durch Abtupfen fixiert worden war, wurde die Schichtseite in die Oberfläche von Äthanol (30 sec) und dann in Isopentan (10 sec) gehalten. Dabei wird zunächst das salzhaltige Substrat durch Alkohol ersetzt und dann durch Isopentan verdrängt, dessen Oberflächenspannung um etwa das 6fache niedriger ist als die des Wassers. In Analogie zur Critical Point-Methode (6) vermeidet man unliebsame Kompressionserscheinungen beim Antrocknen. Luftgetrocknete Präparate werden mit Platin (4,5 cm Pt-Draht, 0,1 mm ⌀, 13 cm Abstand, Winkel 6—8°) schrägbedampft.

Ergebnisse

Mischungen von DNS und Cytochrom c wurden in der Regel frisch bereitet. Die Konzentrationen der Tab. 2 wurden in 2 m Ammoniumacetat hergestellt und 0,1 ml Lösung bei einem Schub von 0,1—0,5 (meist 0,3) dyn/cm auf 0,25 m Ammoniumacetat bei Zimmertemperatur nach der Spreitung (bis 15 min Dauer) übertragen.

Tabelle 2. *Spreitung verschiedener DNS-Cytochrom c-Konzentrationen*

Abb.	Konzentration		Verhältnis DNS : Cytochrom c	Teilchen/μ^2 u. Lagerung	Dichtpunktfläche m²/mg
	DNS	Cytochrom c			
2	0,0005%	0,05%	1:100	2,5 ± 0,5, einzeln	0,66
3	0,005 %	0,05%	1:10	über 20, Aggregate	0,76
4	0,05 %	0,05%	1:1	nicht bestimmbar	0,8
—	0 %	0,05%	0	keine Teilchen	0,67

Aus der Konzentration der Abb. 2 mit einzeln liegenden Teilchen leitet sich die *Molekulargewichtsbestimmung* ab. Man zählt die Teilchen pro $\mu^2 = a$, und berechnet das Gewicht der DNS

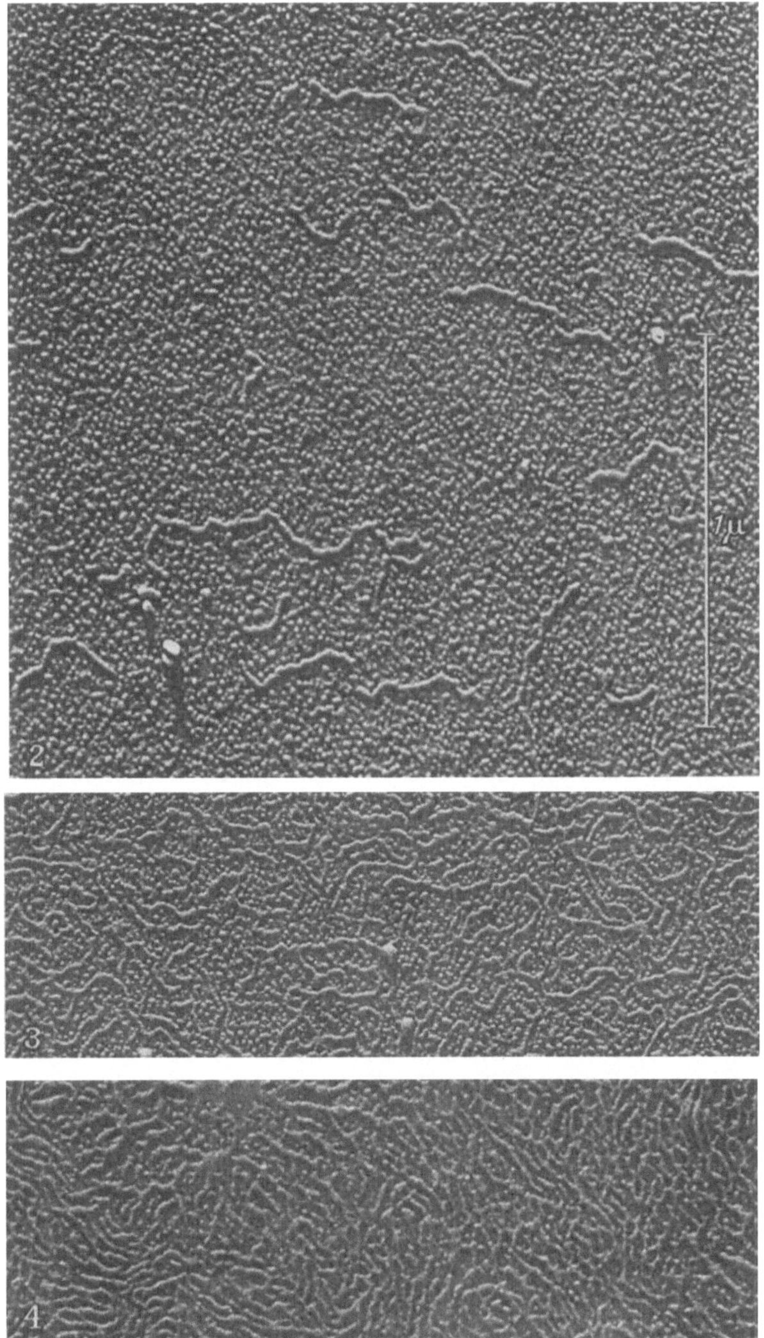

Abb. 2—4. DNS-Cytochrom c-Mischfilme. Konzentrationen 1:100 (Abb. 2), 1:10 (Abb. 3) und 1:1 (Abb. 4) bei gleichbleibender Cytochrom c-Konzentration entsprechend Tab. 2. Präparation s. Text. 52000mal

in g auf $1\ \mu^2 = b$. Dann ist das Molekulargewicht $\frac{b}{a} \cdot 6{,}02 \cdot 10^{23}$ (Loschmidtsche Zahl). Der Wert von a wird statistisch bestimmt durch Auszählen verschiedener Aufnahmen von mehreren Blenden. Das vorliegende Präparat ergab ein Molekulargewicht von 3,0 Mill., Einzelteilchen sind durchschnittlich $5 \cdot 10^{-18}$ g schwer.

Ferner ist es möglich, die Längenabmessungen mit den gefundenen Molekulargewichten zu vergleichen. Die Moleküle erscheinen kürzer, wenn man für das Zwillingsmolekül aus den Angaben über die Struktur der Nucleotideinheit das Molekulargewicht als das 200fache der gestreckten Länge (in Å) annimmt (7). Man kann also eine Verkürzung, am wahrscheinlichsten in Form eines überdrehten Seiles, berechnen. Die Ausmessung der queren Durchmesser ist nicht sicher, da bei der Schrägbedampfung unter flachem Winkel scheinbare Verbreiterungen auftreten.

Die Fähigkeit der Cytochrom c-DNS-Mischung, größere Flächen als mit Cytochrom c allein zu besetzen (Tab. 2 Dichtpunktflächen, Abb. 1), kann auf zwei verschiedenen Erscheinungen beruhen: Einmal ist bei höherem DNS-Gehalt ein tatsächlicher Flächenbedarf im Film dafür vorhanden, wie es etwa Abb. 3 und stärker noch Abb. 4 entsprechen kann; oder es wird das Cytochrom c durch eine (noch unbekannte) Wechselwirkung mit DNS veranlaßt, stärker auf dem Substrat zu spreiten. Für beide Erscheinungen können aus weiteren Versuchen Hinweise erhalten werden.

Der Vergleich mit Rinderserumalbumin als spreitendem Protein ergab bezüglich der Kraft/-Flächen-Kurven bei höheren DNS-Konzentrationen ähnliche Resultate: Nach dem Anstieg der Dichtpunktfläche nimmt bei hohem DNS-Albumin-Verhältnis (4:1) diese stark ab und wird (bei 8:1) nahezu 0. DNS-Lösungen ohne Protein sind auf wäßrigen Lösungen nicht filmbildend. Dagegen zeigt die morphologische Untersuchung der Mischfilme in gleichen Relationen wie bei Cytochrom c (Tab. 2) wesentliche Unterschiede. Bei Konzentrationen von 1:1 sind denaturierte Aggregate als einzeln liegende Strukturen von etwa gleicher Länge wie die in Abb. 2 dargestellten Moleküle vorhanden und bilden höchstens 1% der gesamten DNS-Menge. Über die Oberflächen-Mischfilme wird an anderer Stelle ausführlich berichtet.

Mit Unterstützung der Deutschen Forschungsgemeinschaft.

Literatur

1. MARKO, A. M., and G. C. BUTLER: J. biol. Chem. **190**, 165 (1951).
2. HARKINS, W. D.: The Physical Chemistry of Surface Films, p. 121. New York 1952.
3. MÜLLER, F. H.: Z. Elektrochem. **59**, 312 (1955).
4. GORTER, E., and F. GRENDEL: Proc. Akad. Wetensch. Amsterdam **29**, 1262 (1926).
   — Kolloid-Z. **136**, 102 (1954).
5. PETERS, D., u. R. GEISTER: Vortrag 7. Tagg. dtsch. Ges. f. Elektronenmikroskopie, Darmstadt 1957. Strukturaufklärung am Vaccine-Virus mit Hilfe organischer Lösungsmittel (persönl. Mitt.).
6. ANDERSON, T. F.: J. appl. Physics **21**, 724 (1950).
   — Trans. N. Y. Acad. Sci. II **13**, 130 (1951).
7. WATSON, J. D., and F. H. C. CRICK: Nature (Lond.) **171**, 737 (1953).

C. Ordnungsprinzipien in der Biologie

Principles of ordering in fibrous systems

Alan J. Hodge

Department of Biology, Massachusetts Institute of Technology, Cambridge 39, Massachusetts (USA)

The purpose of the present symposium is to identify and characterize those principles which determine order in biological systems in one, two, and three dimensions. However, such an analysis does not bear too close inspection, for, in reality, the fibrous systems possess considerable three-dimensional structure. Indeed, the presence of order in a system composed of macromolecules, usually of highly asymmetrical shape, implies a certain degree of three-dimensional crystallinity. The present paper will outline what is known of the principles involved in the aggregation of highly asymmetric macromolecules to form ordered fibrous structures, with attention being confined to the fibrous proteins. The "two-dimensional" membrane systems and the ways in which these structures are involved in the formation of higher order structures within the cell will be the province of the following articles.

The results of modern physical and chemical investigations indicate that the amino acid composition and sequence of a given protein determine its characteristic physico-chemical properties, including its ability under appropriate conditions to form crystalline or ordered quasi-crystalline structures. Thus, the amino acid sequence determines not only the degree of helicity present under specified conditions (e.g., the α-helical structure of many of the muscle proteins, and the triple helix configuration of the collagen macromolecule), but also the entire tertiary structure of the macromolecule, and consequently, its interaction properties under all environmetal conditions. For convenience, and without specific reference to forces involved in macromolecular interaction (such as electrostatic forces, fluctuation forces arising from proton or electron mobility, van der Waals forces, and hydrogen bonding) those characteristics of the macromolecule determining its various stable aggregation states may be referred to collectively as the "interaction profile" of the macromolecule. Since, for all but very small values of intermolecular distance, the relative orientation of a group of macromolecules will be determined mainly by the distribution of charged groups on their surfaces, the "interaction profile" is approximated well by a hypothetical closed surface surrounding the macromolecule (analogous to an equipotential surface), the shape of which will be determined at all points by the actual charge distribution on the surface of the molecule. Such an "interaction profile" will clearly be sensitive to variations in environmental factors such as p_H, ionic strength, temperature, and other physical parameters, as well as to changes in the charge distribution of the molecule resulting from reaction or adsorption of chemical agents with side-chain groups. It follows that all the molecules of a "pure" protein solution will exhibit identical "interaction profiles" under any given set of conditions, a concept which is supported by the marked change in properties which can accompany substitution of a single amino acid residue in a "sensitive region" of a protein, e.g., sickle cell hemoglobin (1). It can be seen in a general way, then, that the overall shape and the presence of elevations or other prominent features in the "interaction profiles" of the long thin macromolecules, to be considered here, will profoundly affect the probability of formation and the stability of any particular ordered aggregation state. Accordingly, it is the unique spatial distribution, particularly of the polar side-chains, and possibly other components such as carbohydrates, which determines the ability of a particular species of macromolecule to form various ordered aggregates under appropriate conditions.

The validity of these concepts will be illustrated by reference to the interaction properties of the macromolecules of soluble collagen and of certain muscle proteins.

Collagen

The native macromolecules extractable from a variety of collagenous tissues have been termed tropocollagen (TC) (2). In solution, they are long stiff rods having dimensions of about 14 × 2900 Å as determined by physical chemical methods (3). Evidence from X-ray diffraction data (4) and from studies on the denaturation of the native macromolecules (5) indicates that each macrmolecule contains three polypeptide strands of about equal length coiled about each other for the most part in a specific helical configuration, stabilized by hydrogen bonds between hydroxyproline and other residues.

One of the most striking properties of native collagen macromolecules is their ability to aggregate in a number of highly characteristic patterns, the type of pattern depending on the environment to which they are exposed. These ordered aggregation states are characterized by highly

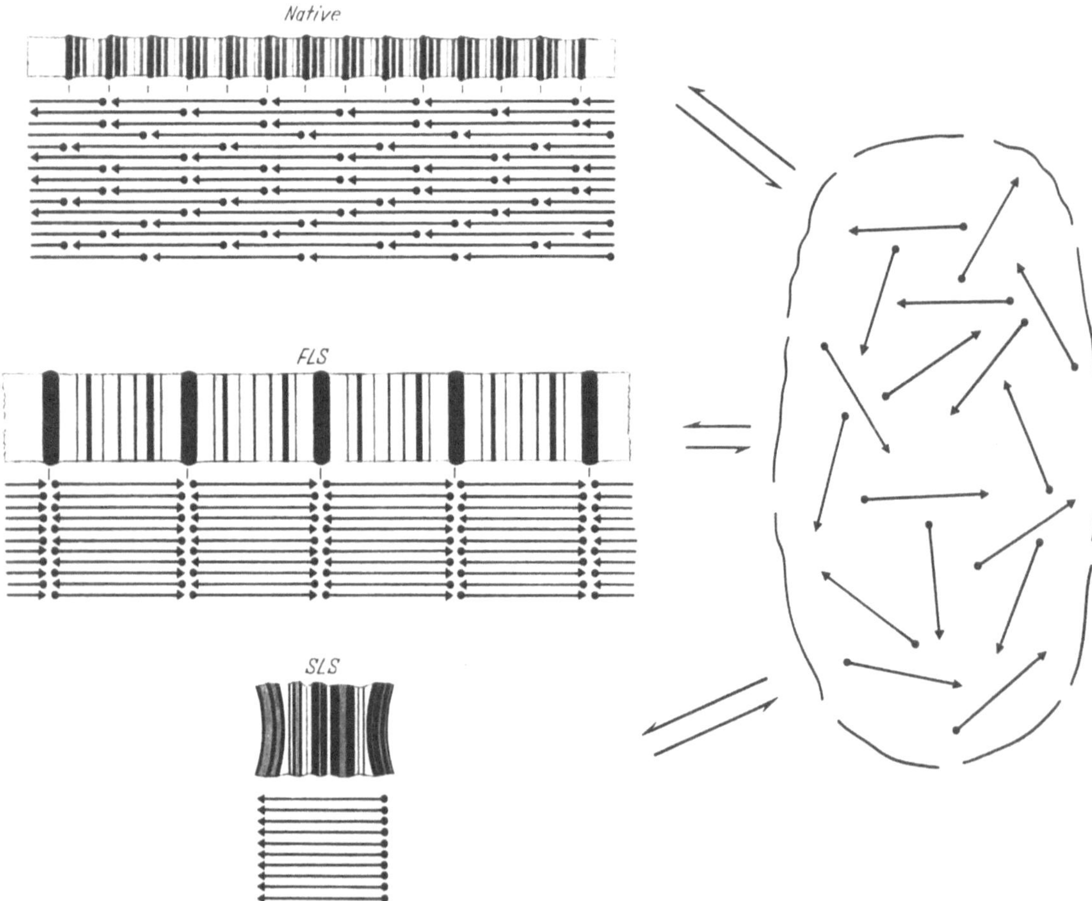

Fig. 1. Diagram to illustrate the reversible formation of native type, FLS and SLS ordered aggregates of collagen from the macromolecules in solution. The packing arrangements of the macromolecules or protofibrils (linear polymers of macromolecules) are indicated under the band patterns observed electron microscopically for each type of aggregate when stained with phosphotungstic acid

specific band patterns when viewed in the electron microscope, especially after treatment with an "electron stain" such as phosphotungstic acid (PTA). The capacity to form these ordered structures is lost when the native macromolecules are denatured by thermal or other means, a result which emphasizes the importance of the intact "interaction profile" in the formation of ordered aggregates. Since the work of KÜHN et al. (6) indicates that at least a part of the bound

PTA responsible for the band patterns seen in the electron microscope is specifically held by the guanidino-groups of arginine residues, it seems likely that PTA is bound only when the arginine (and possibly other basic groups such as lysine) are located in a specific steric array resulting from the orderly side-to-side packing of the TC macromolecules. To put it differently, this means that the mutual cooperation of several arginine and possibly other side-chains appears to be necessary for the binding of a single phosphotungstate ion. Since the specific steric configuration of these groups arises from the packing arrangement of the TC macromolecules in a particular aggregation state, it follows that bands will be exhibited only when the macromolecules retain their native helical configuration, and will not be observed in those regions where the polypeptide chains are uncoiled or otherwise disordered.

The various stable aggregation states of the TC macromolecule will now be considered individually. For reasons that will become clear later, the TC macromolecule is an "asymmetric" or "structurally polarized" unit, i.e., the distribution of density (or amino acid residues) along its length is such that one end is clearly distinguishable from the other. Hence, we may formally represent TC as an arrow, and label the head and tail A and B, respectively (Fig. 1). The segment long-spacing (SLS) type of ordered aggregate (7) is the simplest "crystalline" form of TC, and is

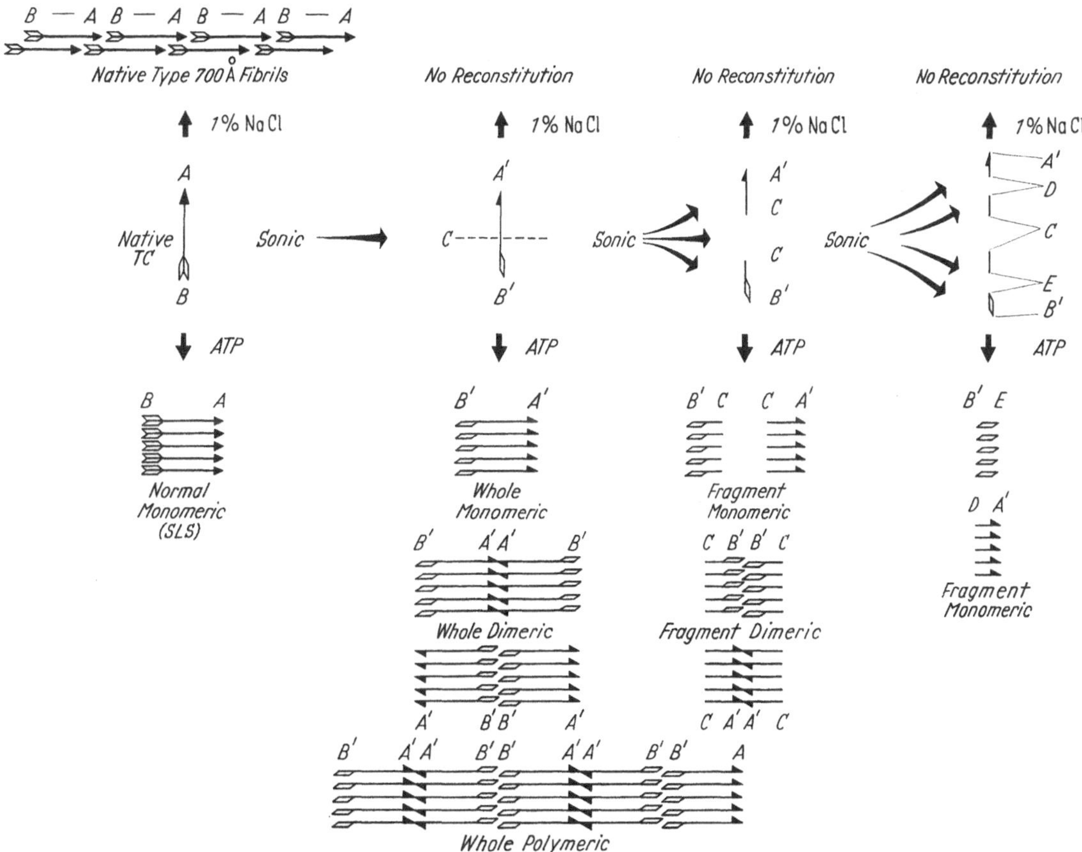

Fig. 2. Diagram illustrating the main effects of sonic irradiation on the tropocollagen macromolecule in solution. (9) Note that the terms monomeric, dimeric and polymeric refer only to the end-to-end linkages of the macromolecules or fragments

formed on addition of adenosine triphosphate (ATP) or certain other reagents to a solution of TC in dilute acid (Fig. 5). The ATP is thought to combine with the side-chains of lysine (2) or other residues, thus altering the "interaction profile" of the TC so that the favored packing arrangement is one in which the TC are in a parallel array with like ends in register (Fig. 1, 5, 6). Thus the asymmetric or "polarized" band pattern of SLS stained with PTA or other reagents is,

Fig. 3. Reconstituted native type fibril of calf skin collagen, stained with phosphotungstic acid, showing the polarized intraperiod band pattern. The axial period of ca. 700 Å is indicated by the row of dots, and the longitudinally staggered packing arrangement of the protofibrils (linear polymers of macromolecules) by the arrangement of arrows along the lower edge of the figure. Note that formation of protofibrils involves end-to-end linkage of the type A—B. 130,000 ×

Fig. 4. FLS type fibril (produced by dialysis vs. distilled water) from an acid solution of calf skin collagen with added serum glycoprotein and stained with PTA. The symmetrical axial period of ca. 2800 Å (indicated by dots) corresponds to the length of the collagen macromolecule. The protofibrils are packed in register in antiparallel array (indicated by the arrangement of arrows). 33,000 ×

Fig. 5. Segment type (SLS) ordered aggregates prepared by addition of ATP to an acid solution of calf skin collagen, and stained with PTA. The macromolecules are packed in parallel array with like ends in register (cf. Fig. 1), and their orientation is indicated by the arrows with ends labelled A and B. The short arrow, labelled (1) points to the F—G interband (7), one of the prominent features of the characteristically polarized band pattern of SLS. 65,000 ×

in effect, a kind of "molecular fingerprint" since it represents the distribution along the TC macro-molecule of polar groups responsible for PTA or other binding. Under other conditions, and in the presence of substances such as certain glycoproteins (7), the TC macromolecules or protofibrils aggregate in antiparallel array to form the so-called fibrous long-spacing (FLS) form (Fig. 1 and 4), in which the repeat period is again approximately equal to the length of the TC macromolecule but with a centrosymmetric distribution of bands in each repeat period. The other important aggregation state of TC is the *native* type found in tendon and other connective tissues. This form may be reconstituted from TC solutions under appropriate conditions (7), has an axial repeat period of about 700 Å (i.e., $^1/_4$ of the length of the TC macromolecule), and the intraperiod band pattern is characteristically asymmetric or "polarized" (Fig. 3)[1]. As indicated diagramatically in Fig. 1, the formation of native type fibrils most likely involves packing of protofibrils (each of which may be regarded as a linear polymer of TC arising from end-to-end linkage of the A—B type) so that adjacent protofibrils are displaced longitudinally with respect to one another by one-fourth of the molecular length.

Electron microscopic examination of SLS-type aggregates derived from solutions of TC after sonic irradiation has revealed a number of features of interest, particularly in relation to the mechanism by which linear polymers (protofibrils) of highly asymmetric macromolecules such as TC may be formed during fibrogenesis. The physical chemical measurements of NISHIHARA and DOTY (8) indicated that sonic irradiation of TC solutions results in fragmentation of the macromolecules into shorter pieces which, however, largely retain the orderly three-strand helical structure characteristic of the native macromolecules. NISHIHARA and DOTY concluded that the time-dependence of the molecular weight change was compatible with a preferential fragmentation of the macromolecules into halves and quarters. The EM observations confirm in general this conclusion (9) and, in addition, show that sonic irradiation produces profound effects other than scission of the macromolecules.

Fig. 2 summarizes the main effects of sonic irradiation on TC. Of these, the most important is a rapid alteration of "end-regions" without apparent change in length of the macromolecules and is manifested as a change in the end-to-end polymerization properties of the TC. Sonic irradiation gives rise to a heterogeneous population of macromolecules, whose diversity resides not only in particle length but also in their end-to-end polymerization properties. The most rapid process occurring during irradiation is this alteration of the "end-regions", indicated in Fig. 2 by labeling the ends A' and B' rather than A and B. One result of this is impairment of the capacity to form fibrils of native type (700 Å repeat), i.e., inhibition of end-to-end polymerization of the type A—B involved in the formation of "protofibrils". Another is a marked effect on the characteristics of the SLS-type precipitates formed on addition of ATP. Whereas the control unirradiated solutions usually yield single segments with only a few dimeric[2] or trimeric forms, the irradiated solutions yield progressively increasing amounts of dimeric and polymeric forms involving end-to-end interactions of the type A'—A' and B'—B' (Fig. 7, 9. 10, 11). The effects of sonication (other than fragmentation) are thus at least twofold: 1. inhibition of end-to-end interactions of A—B type under conditions favoring the formation of native type fibrils, and 2. enhancement of homologous end-to-end interactions of A'—A' and B'—B' type under conditions which usually favor the formation of single segments (whole monomeric type).

As irradiation proceeds, scission of the TC macromolecules becomes evident in the SLS type precipitates (Fig. 8, 10, 11, 13). Fragmentation is apparently not a random process and appears to occur initially in a well-defined region adjacent to the F—G interband (7), resulting in fragments A'C and B'C about 55 and 45% respectively of the original TC length, in approximate agreement with the physical chemical results of NISHIHARA and DOTY (8). These fragments are capable of forming monomeric (Fig. 13) and dimeric (Fig. 8) SLS-type aggregates, and of isomorphous growth on whole polymeric SLS aggregates (Fig. 10, 11). It is important to note that end-to-end

---

[1] Unless otherwise indicated, all figures are electron micrographs, and were taken with a Siemens Elmiskop I a electron microscope.

[2] The terms monomeric, dimeric, trimeric and polymeric are used in this context to denote the degree of end-to-end polymerization of the tropocollagen macromolecules in a particular SLS-type aggregate.

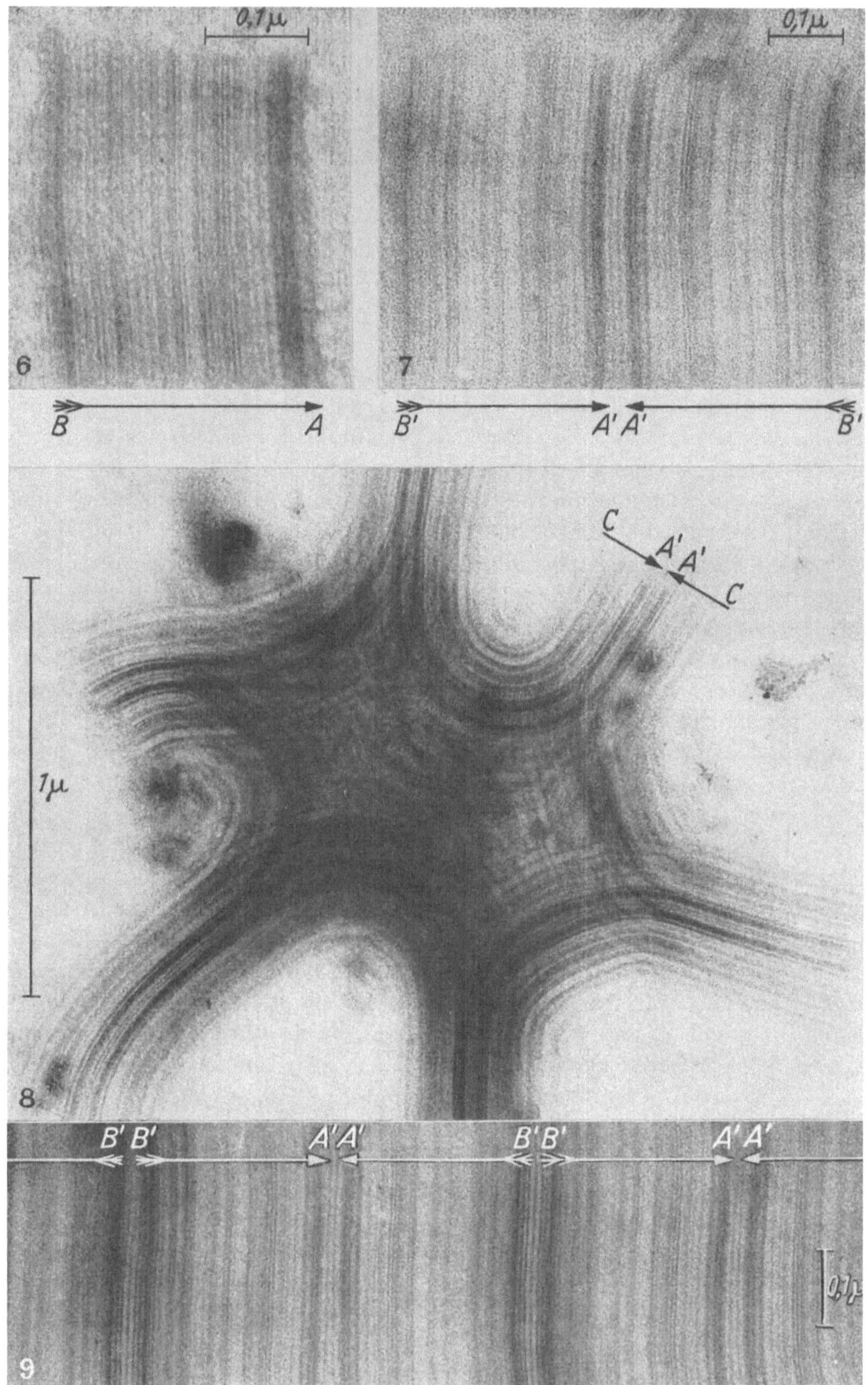

Fig. 6. Further enlargement of the segment in the upper left corner of Fig. 5, to show the detailed band structure. Orientation is indicated by the labelled arrow at bottom. 140,000 ×

Fig. 7. Segment type structure (whole dimeric form) from an acid solution of calf skin collagen irradiated sonically for 20 min prior to ATP addition. Orientation of the macromolecules is indicated along the lower edge of the figure. 100,000 ×

Fig. 8. SLS aggregates of the fragment dimeric type CA'—$A'C$ from a calf skin collagen solution irradiated sonically for 240 min, showing extensive lateral aggregation in the form of "ribbons". PTA stained (9). 55,000 ×

Fig. 9. SLS aggregate of whole polymeric type derived from a calf skin collagen solution irradiated sonically for 480 min. The type of end-to-end polymerization of the macromolecules is indicated by the arrows. The fundamental axial period of this structure is twice the length of the collagen macromolecule. Note the characteristic band structures in the A'—A' and B'—B' junctions. Stained with PTA. 100,000 ×

dimerization or polymerization of the scission products apparently never involves new ends produced by the fragmentation, but only the "original ends" of the intact TC. Thus, dimeric or polymeric forms involving linkages of the type C—C have not been observed. It seems, therefore, that the original ends of the TC macromolecules possess properties which can not be duplicated

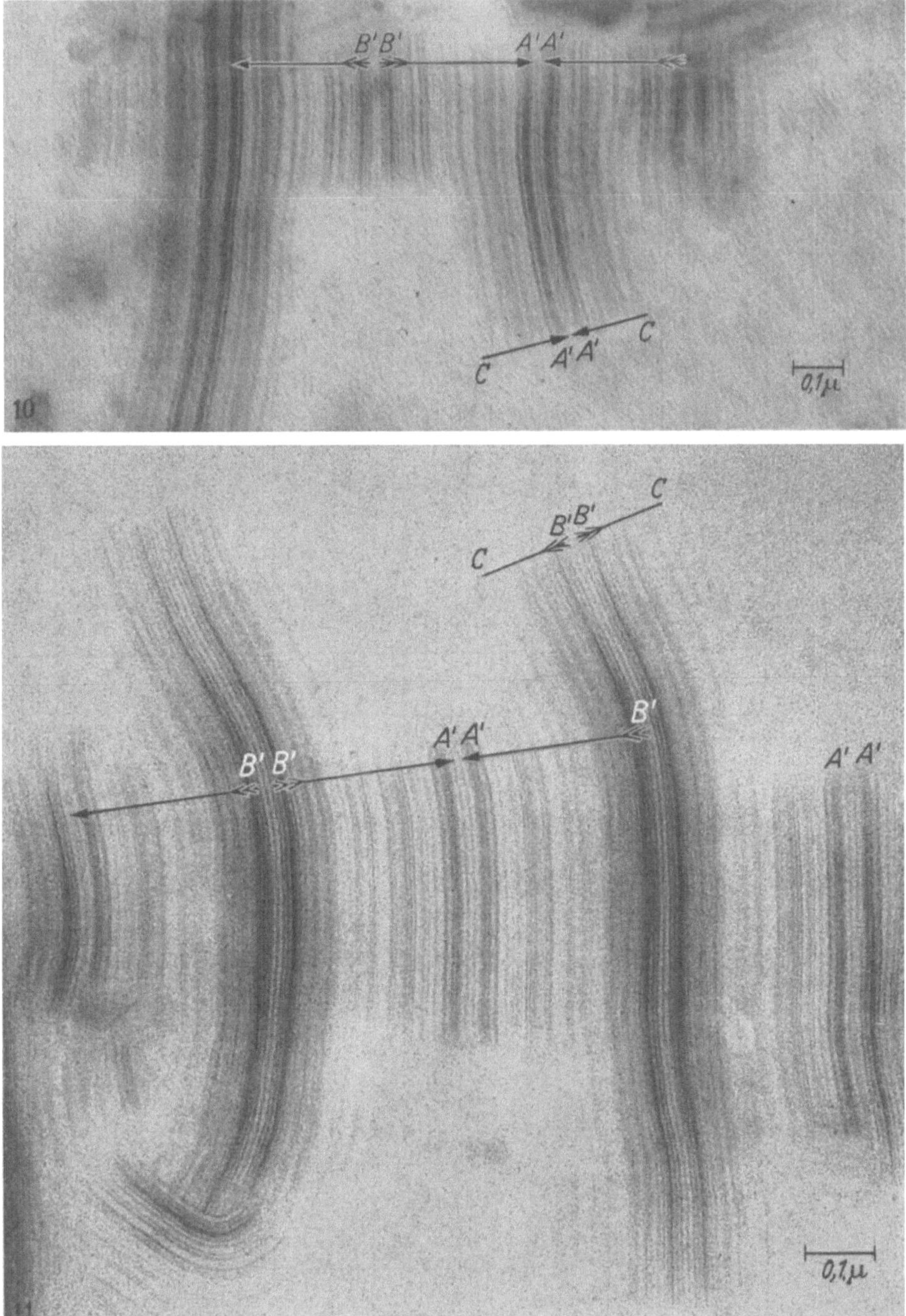

Fig. 10. Whole polymeric SLS form of calf skin collagen irradiated sonically for 240 min, showing isomorphous growth of fragment dimeric ribbons of the type CA'—$A'C$, (9) stained with PTA. 75,000 ×

Fig. 11. Whole polymeric SLS form of calf skin collagen irradiated sonically for 480 min, showing isomorphous growth of fragment dimeric ribbons of the type CB'—$B'C$, stained with PTA (cf. Fig. 10 and Fig. 2). 100,000 ×

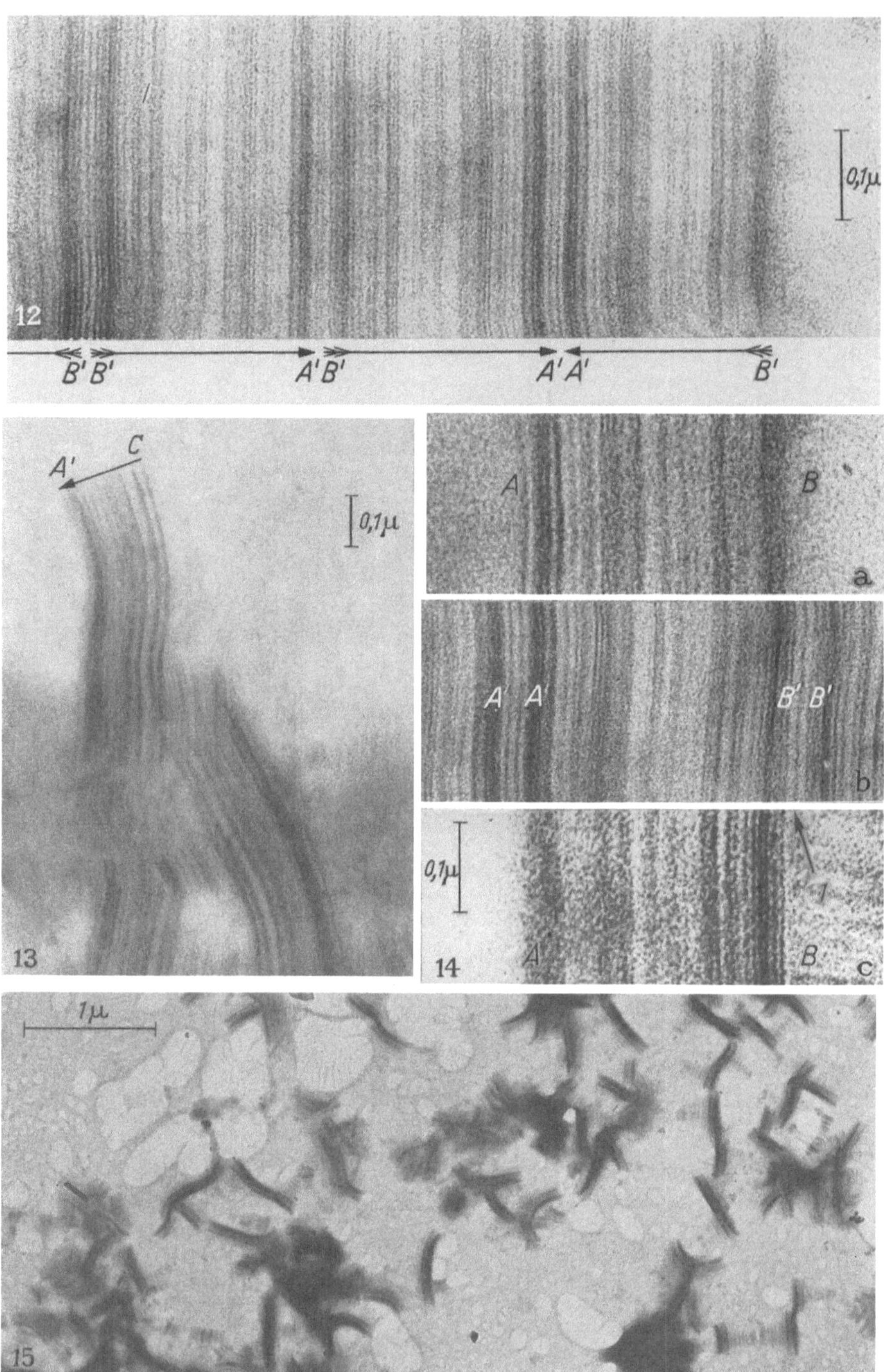

Fig. 12. Whole polymeric SLS form of calf skin collagen (480 min sonic irradiation) showing an $A'—B'$ junction as well as the more commonly observed $A'—A'$ and $B'—B'$ end-to-end linkages, stained with PTA. 120,000 ×

Fig. 13. SLS aggregates from calf skin collagen solution irradiated sonically 240 min, showing ribbon-like fragment monomeric forms of the type $A'C$ (9). 71,000 ×

Fig. 14a—c. Composite micrograph for comparison of the band patterns of whole monomeric forms (single segments a and c) with that of the whole polymeric form b. The use of two single segments is necessary since the band structures at one or both ends are often obscured by disordering during drying. Arrow 1 indicates the new band in the $B'—B'$ junction, not present in single segments and apparently resulting from end-to-end polymerization of the collagen macromolecules (9). Calf skin collagen, stained with PTA. 130,000 ×

Fig. 15. SLS aggregates from a calf skin solution irradiated for 335 min, stained with PTA. The ribbon-like structures are fragment monomeric forms of the types $A'D$ and $B'E$ (Fig. 2) (9). 18,000 ×

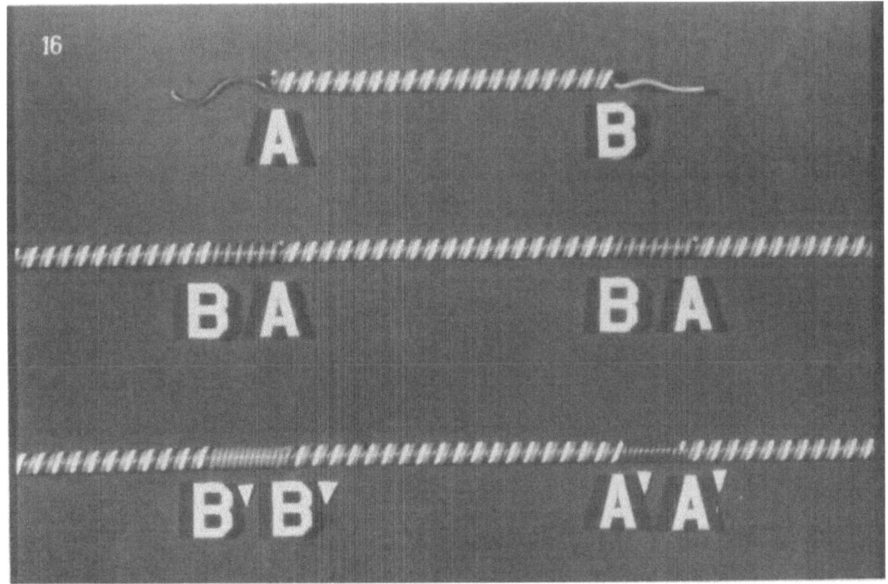

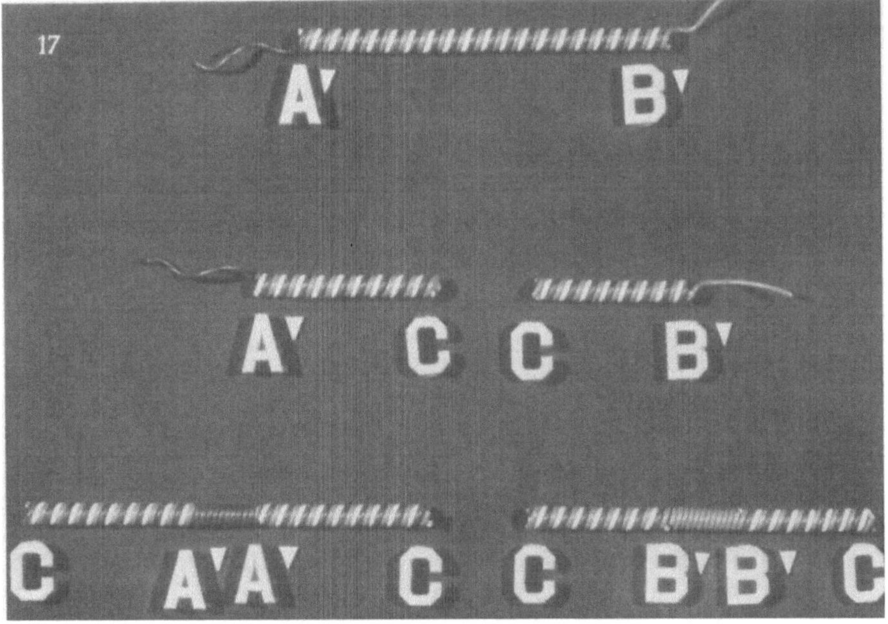

Fig. 16. At top is depicted a model of the collagen macromolecule in solution with its three polypeptide chains represented by wires (black, grey, and white respectively). In the body of the molecule, the three wires are coiled about one another in an orderly helical pattern, representing the type of structure deduced from wide angle X-ray diffraction studies of collagen. At the ends, the postulated end-chains are shown as randomly coiled, single polypeptide chains. Note that parameters such as the length and diameter of the molecule, the pitch of the helix and the lengths of the chains are not to scale. In the center is illustrated the heterologous type of end-to-end coiling of end-chains envisaged in the formation of protofibrils during "normal" fibrogenesis (leading to formation of native type collagen fibrils), while at the bottom is shown the homologous end-to-end coiling of end-chains presumably occurring during formation of whole polymeric segment type aggregates such as those shown in Fig. 9, 10, 11

Fig. 17. Photograph of a wire model similar to that in Fig. 16, illustrating the collagen macromolecule (top), its fragmentation by sonic irradiation (center) and the formation of fragment dimeric forms CA'—$A'C$ and CB'—$B'C$ (at bottom) on addition of ATP. Dimeric or polymeric forms involving C—C junctions have not been observed

by new ends formed by scission, a process which appears to involve simultaneous breakage across the entire triple helix structure of the native macromolecule. Extensive sonic irradiation produces smaller fragments of the type arising from scission at the points D and E (Fig. 2). Of the four scission products thus produced, only those which include A' or B' ends are capable of forming SLS type aggregates (Fig. 15), a result of some interest since inspection of the band structure of whole monomeric SLS indicates that the linear concentration of bands (and hence presumable of basic amino acid residues) is higher at the two ends than in the central regions. This suggests that the "interaction profiles" of fragments of the type CD and CE may be so "smooth" (i.e., lacking in prominent features) as not to favor the formation of ordered SLS structures. Again, this is of interest in view of the possibility that ATP modifies the interaction profile of the TC macromolecule by combining with basic amino acids such as lysine. In any event, it is clear that the basic amino acids are involved importantly in PTA binding (and possibly ATP binding), and are concentrated to a considerable extent toward the ends of the TC macromolecules.

A deduction of considerable importance has been provided by a detailed comparison of the band pattern of single segments (whole monomeric SLS form) with that of the whole polymeric form found after sonic irradiation, (see Fig. 14), which, it will be recalled, involves A'—A' and B'—B' linkages. In the A'—A' junction, the two bands in the polymeric structure corresponding to the A-terminal band of the monomeric SLS form appear as a doublet separated by an interband region about 100 Å wide. Similarly, in the B'—B' junction, the two corresponding terminal bands are separated by about 180 Å, and in addition a new band (arrow 1 in Fig. 14), not present in monomeric SLS forms, appears in the center of the junction. On the plausible assumption that well-defined bands (as already discussed) arise only in those regions where the polypeptide chains are in an ordered helical configuration (thus providing the specific steric array of basic side chains necessary for PTA binding), the sum total of these results points strongly to the following conclusions.

1. The main body of the TC macromolecule contains the polypeptide chains in the characteristic triple helical configuration deduced from X-ray diffraction studies, with short dangling chain appendages at both ends of the macromolecule having maximum lengths of about 100 and 200 Å for the A and B ends, respectively. In solution, these end-chains would be expected to be in the random coil configuration (Fig. 16).

2. If the correlation between intramolecular order and the presence of bands is valid, the occurrence of the new band in the center of the B'—B' junctions of whole polymeric SLS forms suggests that end-to-end polymerization involves an orderly coiling of these appendages about each other to form an ordered helical structure. The absence of such a band in the A'—A' junction would then be indicative of the paucity or absence of basic amino acids in the A end-chain. Indeed, it seems likely that the differences in specificity of the A and B end-chains lie in just such differences in amino acid composition, and that the alteration in end-to-end interaction properties induced by sonication is a result of a splitting off of a peptide or other fragment from the end-chains. The presence of such terminal structures is consistent with the evidence of BOEDTKER and DOTY (3), and of DOTY and NISHIHARA (5) that the three component polypeptide chains of the TC macromolecule are approximately equal in length. Hence the lengths of single- (or double-) chain appendages extending beyond the main triple helix "body" of the macromolecule would depend on the relative longitudinal displacement of the three chains with respect to one another. The various possible interactions of these end-chains in forming polymeric SLS type aggregates and the type of linkage presumed to occur in the formation of native type collagen fibrils are illustrated in Fig. 16 and 17. It is of interest to note that occasional junctions of the type A—B are observed in polymeric SLS precipitates derived from sonicated solutions of TC, and that the detailed band structure within these junctions (Fig. 12) correlates well with that already described for A'—A' and B'—B' junctions.

Mineralization of collagen

The mineralization of organic matrices in living organisms subserves vital physiological and adaptive functions, e.g., in the teeth and the exo- and endo-skeletal systems, and may have serious consequences if the inorganic material is laid down in normally unmineralized tissues. The process of mineralization (here restricted to the collagen-hydroxyapatite system) serves as a particularly fine example of stereochemical specificity (in this case resulting from a particular aggregation state of the TC macromolecules) influencing the order in a biological system. The experimental

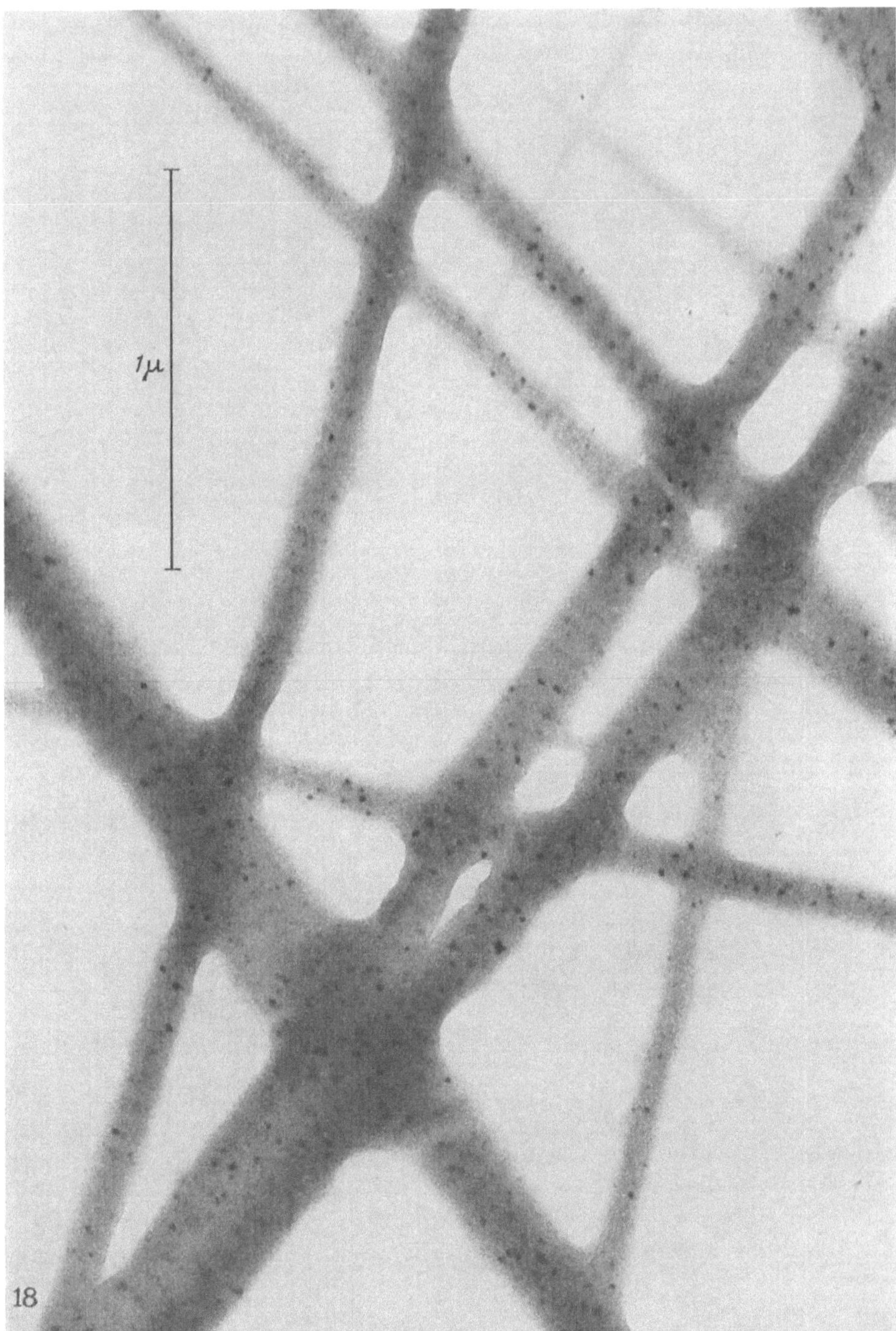

Fig. 18. Unstained reconstituted native type collagen fibrils after 20 min exposure to a metastable solution containing calcium and phosphate ions in low concentration, illustrating the steric specificity involved in the nucleation of hydroxyapatite crystals. Other ordered collagen aggregates (SLS, FLS, etc.) do not induce this crystallization of mineral. Note that the crystals appear in definite relation to the axial period of the collagen.
55,000 ×

evidence has already been given in some detail (*10, 12*), and attention will here be confined to the ability of various aggregation states of TC to induce crystallization of hydroxyapatite from appropriately metastable solutions containing Ca and PO_4 ions (i.e., solutions which, in the absence of a nucleating agent, do not give rise to a solid phase for indefinitely long periods of time).

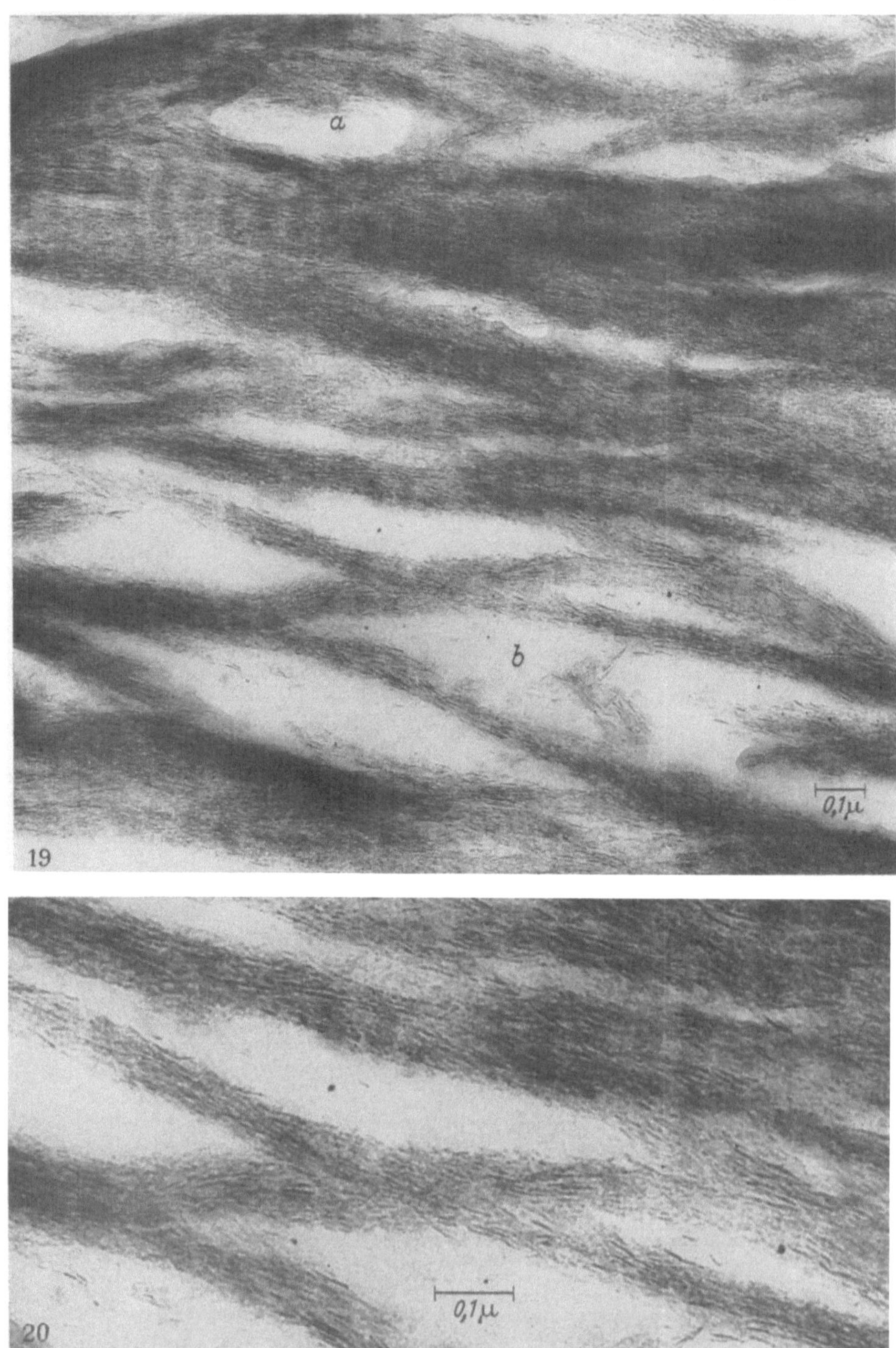

Fig. 19. Longitudinal section of fish bone (carp), fixed in buffered osmium tetroxide (pH 7.4) and embedded in n-butyl methacrylate showing (*a*) ordering of the needle-shaped hydroxyapatite crystals in relation to the 700 Å axial period of the collagen fibrils and (*b*) orientation of the crystals relative to individual collagen fibrils. 70,000 ×

Fig. 20. Enlargement of the lower part of Fig. 19, to show the crystals and their orientation more clearly. 110,000 ×

Various ordered and disordered aggregation forms of TC were prepared and exposed to the metastable Ca-PO$_4$ solutions. The forms investigated included structureless fibrils (no banding), native type (700 Å repeat), 220 Å repeat, FLS, and SLS. Of all these forms, the only one capable of initiating crystallization of apatite (or any other solid crystalline phase) as determined by chemical analysis electron microscopy and X-ray and electron diffraction, was the native type fibril (10, 12). Apparently, although the various specificities of the TC macromolecule are, so to speak, "built-in", i.e., are a direct result of the "interaction profile", which in turn is a consequence of the particular amino acid sequence of the three constituent polypeptide chains and of their relative positioning in relation one to another, these same specificities such as the ability to form banded structures and to initiate crystallization of apatite are not manifested unless the macromolecules are placed in a definite packing arrangement. Nucleation of apatite from such metastable solutions appears to require a precise juxtaposition of certain side-chains in a definite stereochemical configuration, a condition fulfilled only when the TC macromolecules are packed in the staggered parallel array characteristic of the native type aggregation state. The electron microscopic evidence supports this concept of specific "nucleation sites" arising from a particular packing arrangement of the TC macromolecules. In embryonic bone (11) and in studies of reconstituted native type fibrils (Fig. 18), nucleation appears to occur in a definite relationship to the characteristic intraperiod band structure of the fibrils. It should be added parenthetically that the presently available evidence (12) suggests that the failure of many collagenous tissues to calcify under normal physiological conditions is attributable to the presence of highly polymeric polysaccharide-protein complexes rather than to any difference in the stereochemical specificity of the collagen *per se*. Thus, mild extraction procedures and enzymatic depolymerization of ground substance components, which so far as is known have no effects other than depolymerization and extraction of certain polysaccharide components of the ground substance, render the collagen fibrils calcifiable in normally non-mineralized tissues such as skin and tendon (12).

Thin sections of highly mineralized skeletal tissues such as the fish "bone" shown in Fig. 19, 20, show that the hydroxyapatite occurs in the form of very thin needles or prisms, 15—30 Å wide and 200 Å or more in length. These needles are situated within and are oriented in relation to the collagen fibrils (lower part of Fig. 19, 20). Furthermore, the longitudinal distribution of the crystals is such that they reflect the 700 Å axial period of the collagen fibrils. These results are of considerable interest in relation to the process of mineralization since both *in vivo* (11) and *in vitro* (10, 12) studies show that the initial deposition of mineral within the fibrils is in the form of small rounded crystals which are not oriented relative to the fiber axis.

Muscle and the muscle proteins

It has already been shown in the case of collagen that the formation of ordered fibrous structures appears to require (a) a specific stereochemical distribution of side-chains, a property stemming directly from the amino acid sequences in, and the mode of coiling of, the polypeptide chains within the macromolecule, and giving rise to a characteristic interaction profile under any given set of conditions, and (b) a highly specific interaction between "end-chains" such that an orderly coiling of these about one another gives rise to linear polymers (protofibrils) in which particular groups, side-chains, or other features are spaced along the fiber axis at precise intervals. The particular environmental conditions prevailing, by virtue of their effect on the "interaction profile", then determine whether the fibrous aggregate is one involving a parallel, antiparallel, or staggered packing arrangement of the macromolecules (or protofibrils). We have also seen that the major clue in this analysis has been the discovery of the segment (SLS) structure, in which the macromolecules are packed in parallel array with their ends in register, giving rise to a characteristically asymmetrical band structure (or fingerprint) visible in the electron microscope when the aggregates are treated with a specific "electron stain" such as PTA. The situation is less well defined in the case of those muscle proteins which give rise to ordered fibrous structures, mainly because appropriate conditions have not so far been found for the production of SLS-type aggregates. However, since many of the macromolecular lengths are known fairly accurately, it is possible, by analogy with the better characterized collagen system, to make plausible guesses concerning the packing arrangements in a number of the ordered fibrous structures which can be formed from certain muscle proteins.

As shown by Bear (*13*), the X-ray diffraction patterns of muscles from various sources fall into two distinct groups as judged by the low angle diffractions present: 1. the Type I or paramyosin pattern, corresponding to an axial repeat of 725 Å, and so far found only in certain invertebrate muscles which possess the so-called "catch mechanism", and 2. the Type II diffractions corresponding to axial spacings of about 400 Å, which have been found in all muscles so far examined. The precision of this axial repeat is indicated by the fact that several orders of the fundamental reflexion can be obtained (*14*), and it has been observed frequently in electron microscopic investigations (e.g., *15*, *16*, *17*, *18*). At first, it was thought that the X-ray reflections, corresponding to a repeat of about 400 Å, were due to a single component, "myosin". However, it is now clear that most, if not all, of the fibrous components of striated muscle exhibit axial periodicities of this magnitude when isolated and purified [references are cited in (*19*)]. It would seem, therefore, that the total diffraction pattern from muscle results from an orderly interaction between the various elongate macromolecular components, which have lengths (or sub-units) of about 400 Å. Thus, it is possible to consider muscle as a multi-component, quasi-crystalline structure in which the various components interact in an orderly manner by virtue of having similar molecular lengths (and presumably compatible "interaction profiles") or specific reactive sites spaced about 400 Å apart, and in which contraction may be the result of a change in state or configuration of these axial repeating units, either synchronously, or in a sequential manner.

Myosin, the major component of most muscles, does not itself crystallize in a form showing ordered structure in the electron microscope. However, brief exposure to the action of certain proteolytic enzymes such as trypsin (*20*), chymotrypsin (*21*), and subtilisin (*22*) suffices to degrade the myosin molecule into two fairly well defined components, heavy meromyosin (HMM) and light meromyosin (LMM), of which the latter component "crystallizes" in the form of needles exhibiting a very striking transverse banding with an axial spacing of about 420 Å (*23*), corresponding to the molecular length as determined by physical chemical measurements. The high density of the transverse striations, even in the absence of "electron stains" such as PTA appears to be due to an occlusion of inorganic ions at sites located about 400 Å apart (possibly corresponding to the ends of the LMM molecules). This result is reminiscent of the distribution of bound mineral found in formalin-fixed myofibrils after electron-induced microincineration in the electron microscope (*17*). The intraperiod band structure of the LMM needles is symmetrically disposed, and it is therefore possible, by analogy with the situation in collagen already outlined, to desscribe these ordered aggregates as an FLS form (i.e., the molecules of LMM are packed together in antiparallel array with their ends in register). Alternatively, the LMM molecule may be centrosymmetric, with one half indistinguishable from the other.

The true crystals of tropomyosin, a protein first isolated and crystallized by Bailey (*24*), represent another instance in which a reasonable correlation can be made between physical chemical estimates of molecular length and the spacings observed in the crystals by electron microscopy. A study employing electron microscopy and small angle X-ray diffraction is currently in progress (*25*), and will be reported in due course. When a preparation of small tropomyosin crystals is placed on a grid and stained with buffered PTA, several types of ordered structures are observed in the electron microscope (*19*). The most common patterns observed are striations spaced about 200 Å apart, often with intraperiod lines (Fig. 22), or a crossed-grid network with similar spacings in two directions (Fig. 21, 23). The observed spacing of about 200 Å corresponds to one half the molecular length determined from physical chemical data (*24*). A less frequently observed pattern is a two-dimensional net structure with spacings of about 400 Å (Fig. 24). A variety of patterns is encountered when the tropomyosin crystals are fixed in buffered osmium tetroxide, stained with PTA and embedded in butyl methacrylate for thin sectioning (*19*), the structure observed depending on the relation of the plane of sectioning to the crystallographic planes of the crystals. However, patterns such as those shown in Fig. 25 and 26 demonstrate the three-dimensional net structure of the crystals (analogous to scaffolding). Such an open net-like packing of elongated macromolecules (essentially an ordered gel structure) serves to explain the

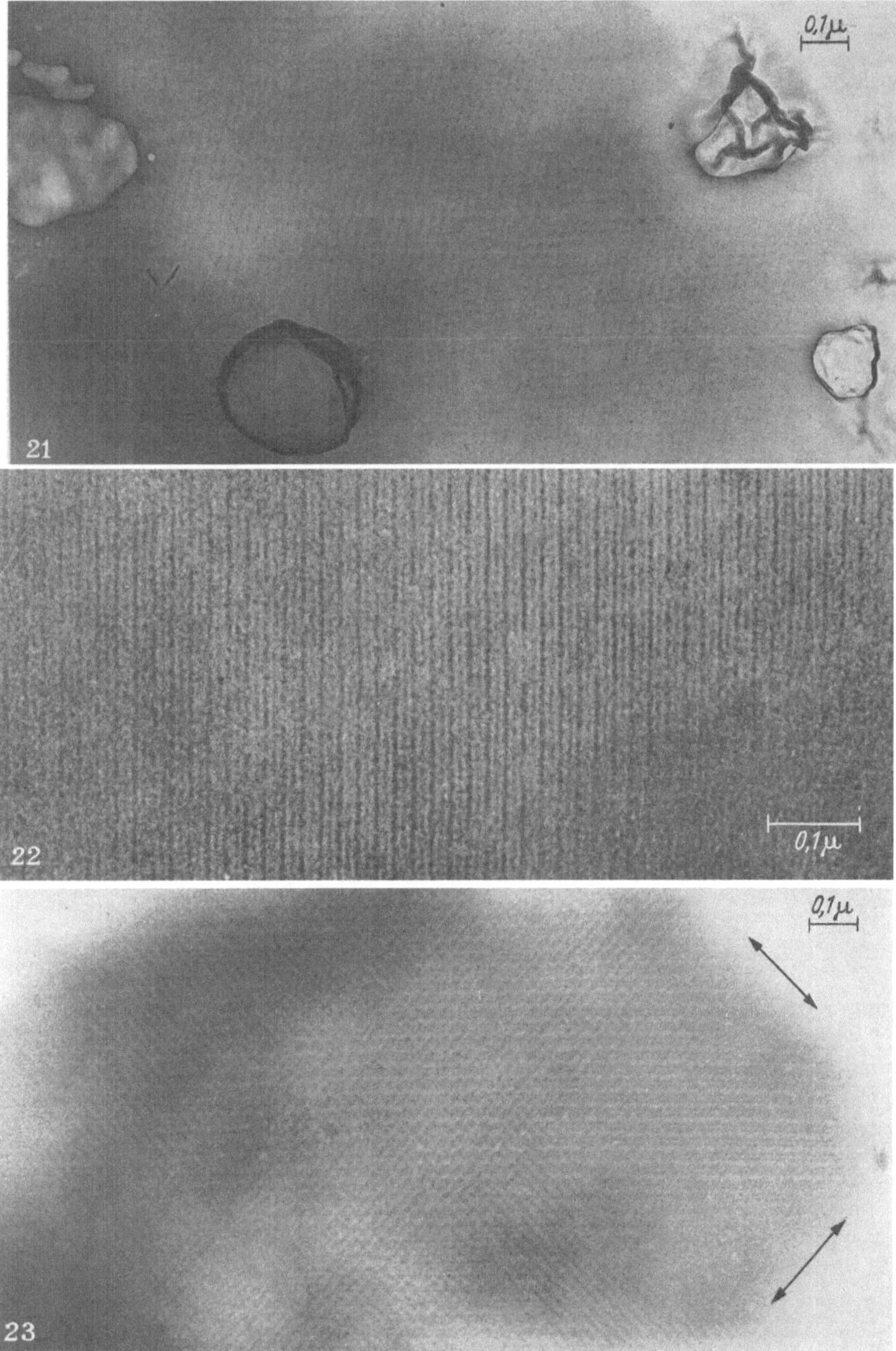

Fig. 21. Unstained rabbit tropomyosin crystal deposited on a carbon film from an ammonium sulphate suspension and allowed to dry without washing. The objects at right and lower center of the micrograph are "contamination replicas" of ammonium sulphate crystals evaporated by the electron beam. Note the highly ordered net-like structure of the tropomyosin crystal, the contrast presumably arising from occlusion of salt at specific points in the lattice (*19*). 70,000 ×

Fig. 22. Rabbit tropomyosin crystal after staining with PTA (*19*). The main spacing is about 200 Å and a well-developed intraperiod line is visible. 130,000 ×

Fig. 23. Crystal of rabbit tropomyosin stained with PTA (*19*). The spacings in the directions indicated by the two double-headed arrows are about 200 Å. 70,000 ×

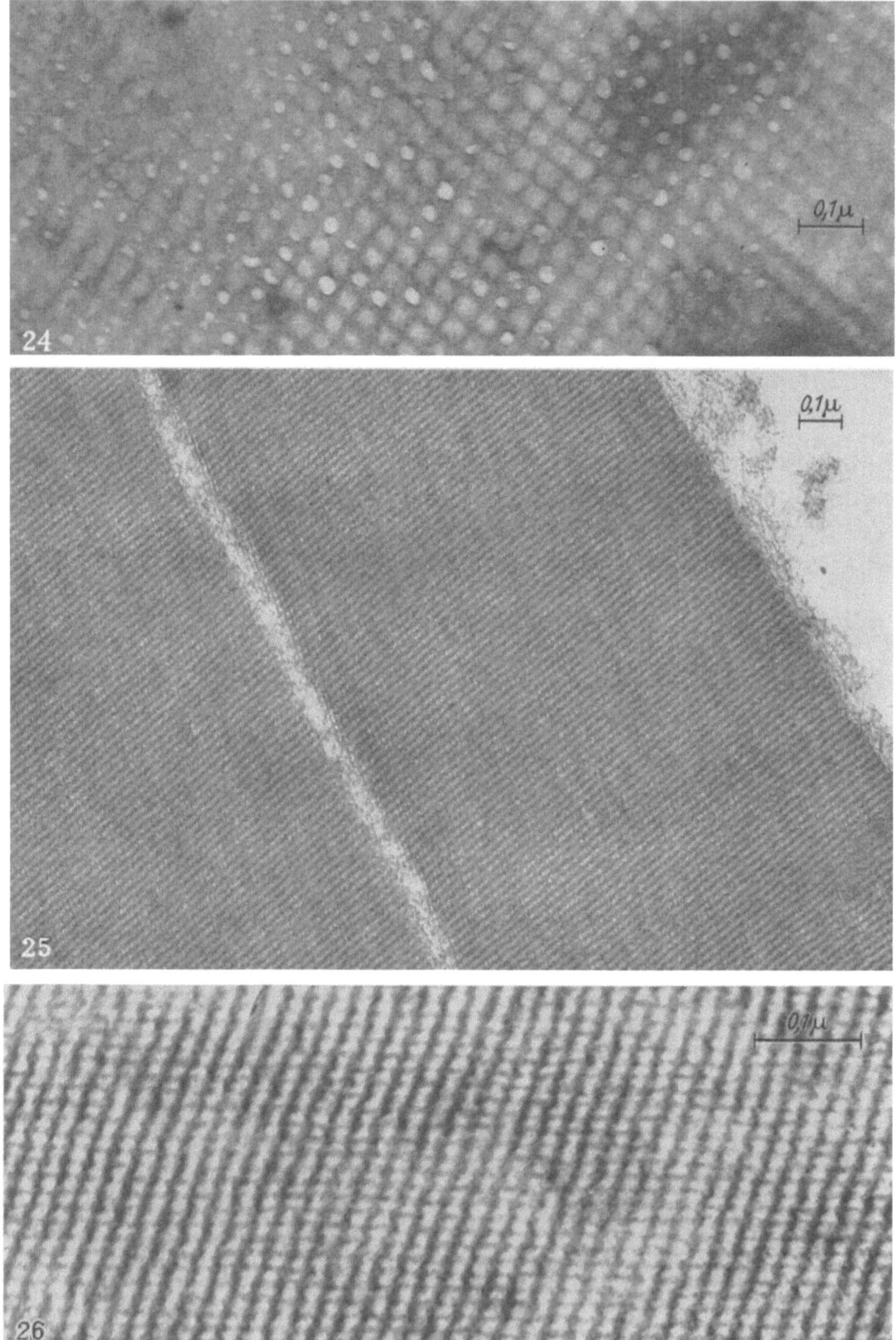

Fig. 24. Two- dimensional net-like structure observed in preparations of tropomyosin crystals (*19*). The spacing is about 400 Å, corresponding to the molecular length estimated from physical chemical data. 90,000 ×

Fig. 25. Thin section of a rabbit tropomyosin crystal preparation after fixation in buffered osmium tetroxide solution, staining with PTA in ethanol, and embedding in n-butyl methacrylate (*19*). The material at upper right is probably disordered tropomyosin adherent to the crystal face. 60,000 ×

Fig. 26. A small area of another thin section in which the plane of the section approximates one of the low index crystal planes, showing the net-like character of the tropomyosin crystal lattice (*19*). The difference in the spacings in the two directions is probably the result of compression of the section during cutting. Osmium fixation, PTA-stained, methacrylate embedding. Note that tropomyosin forms true three-dimensional crystals unlike the one-dimensional "crystals" of collagen and paramyosin. 150,000 ×

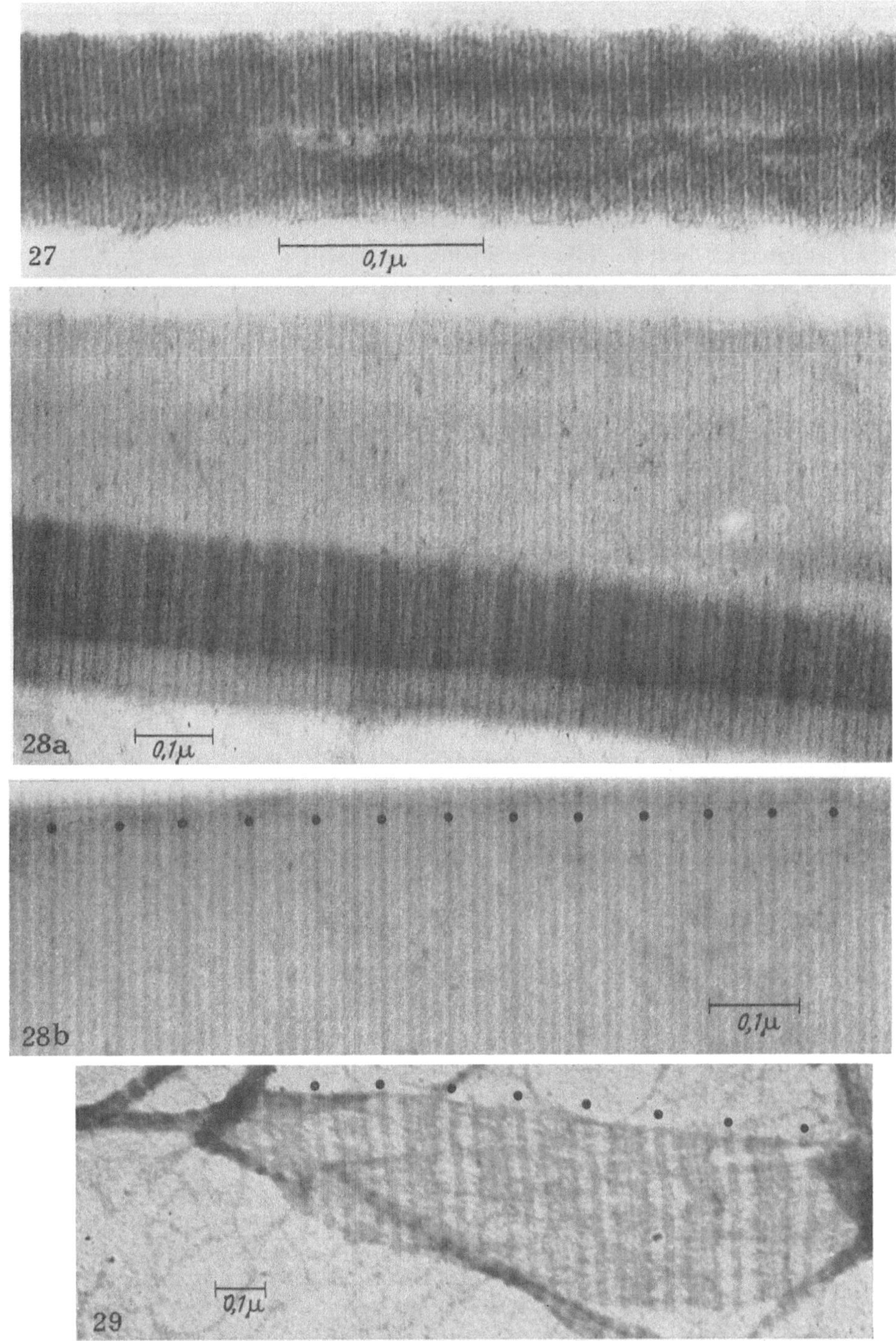

Fig. 27. Native paramyosin fibrils from the white portion of the adductor muscles of *Venus mercenaria*, isolated in 0.2 M KCl and stained with PTA, showing the apparent axial repeat of 145 Å. 190,000 ×

Fig. 28a and b. Reconstituted paramyosin fibrils obtained from a solution in 0.6 M KCl buffered with veronal-acetate (p_H 7.4) by reduction of ionic strength, stained with PTA (*19*). a) a fibril with a 145 Å period showing isomorphous growth of a second fibril having a fundamental repeat of 5 × 145 = 725 Å. 110,000 ×. b) another fibril with an axial period of 725 Å (indicated by dots). 130,000 ×

Fig. 29. Paramyosin fibril obtained by increasing the ionic strength of a paramyosin solution in dilute acetic acid (p_H 3.5) (*29*). The axial period is about 1400 Å, equal to the length of the macromolecules, and the intra-period band structure is symmetrical, indicating that this structure is the paramyosin equivalent of the FLS structure obtained with collagen (cf. Fig. 4). PTA stained. 70,000 ×

remarkably high degree of hydration observed for tropomyosin crystals. The results to date suggest that the interaction profile of the tropomyosin molecule includes prominences at the ends and in the middle, thus favoring packing arrangements with periods equal to, or one half of, the molecular length.

As a final example of a fibrous protein in which the molecular length as determined by physical chemical techniques can be correlated with the periodicities observed by X-ray diffraction and electron microscopy in ordered aggregates (both native and reconstituted), the case of paramyosin will be considered. This is a major component of certain specialized molluscan and annelid muscles which are distinguished by their ability to maintain a rigor-like contracture for long periods of time and were described in the older literature as possessing a "catch mechanism". Paramyosin is the component of these muscles predominantly responsible for the Type I low angle diffraction pattern obtained by BEAR (*13*), a pattern corresponding to a fundamental repeat period of 725 Å, with the fifth order strongly accentuated. Electron microscopic results (*26*) were in excellent agreement with the X-ray diffraction data, showing fibrils with transverse striations corresponding to a period of about 145 Å, with many of them exhibiting a characteristic system of nodes in a net-like structure with a fundamental repeat period of five times 145 Å = 725 Å. However, this spot pattern is frequently absent in native isolated fibrils (Fig. 27). Furthermore, the major component of these fibrils can be extracted and reconstituted into fibrils lacking spots but exhibiting band patterns with fundamental repeat periods of 725 Å (*19*) (Fig. 28). It seems probable therefore, that the nodes reflect the presence of components other than paramyosin within the fibrils. The recent observation of a laminated structure in thin transverse sections of molluscan adductor muscle fibrils (*27*) suggests that the fibrils are built up from alternating layers of paramyosin and other still unspecified components.

Native paramyosin fibrils from the clam, *Venus mercenaria*, were found to be soluble in dilute acid or alkali, and on raising the ionic strength of acid solutions of paramyosin, fibrous structures with an axial period of about 1400 Å (Fig. 29) and symmetrical intraperiod band structure were obtained (*28, 29*). By analogy with the type of analysis already described for the collagen macromolecule, it seemed clear that the molecules of paramyosin must be about 1400 Å in length and capable of packing in anti-parallel array to form a fibrous structure with symmetrical intraperiod band structure (Fig. 29). The existence of kinetic units of this order of magnitude was confirmed by the results of sedimentation, diffusion, viscosity, and light scattering investigations (*28, 29*) on acid solutions of paramyosin. These studies were necessarily rather crude because of the low ionic strength required to keep this protein in solution at low p_H values, and the electron microscopic data undoubtedly provided the more reliable estimate of molecular length. However, the physical chemical results indicated the presence in such acid solutions of particles of length between 1200 Å and 1500 Å with an axial ratio of about 70. The recent measurements of KAY (*30*), carried out on paramyosin in neutral solution at high ionic strength, appear to confirm the value of 1400 Å for the length of the paramyosin macromolecule.

BAILEY (*31*) isolated and purified paramyosin from molluscan muscles and, because its amino acid composition resembled that of tropomyosin, decided to use the term "insoluble tropomyosin" (*32*) to distinguish it from the original water-soluble tropomyosin already described. Similarly, KOMINZ *et al.* (*33*) referred to paramyosin as "tropomyosin A" to distinguish it from "tropomyosin B" (i.e., the original water-soluble tropomyosin of BAILEY, also found in molluscan muscles containing large amounts of paramyosin). It is at present difficult to justify the use of these newer terms rather than the original designation "paramyosin", especially in view of the fact that there exist well-defined differences in periodic structure as deduced from both X-ray diffraction and electron microscopic studies, as well as in crystalline form and solubility properties, which clearly differentiate this protein from the origin water-soluble tropomyosins of BAILEY which, unlike paramyosin, appear to be present in all types of muscle so far investigated.

As in the case of collagen, the macromolecules of paramyosin can be induced, under appropriate conditions, to form a variety of ordered aggregates. So far, fibrils exhibiting axial periods in the elctron microscope of *ca.* 145 Å, 725 Å, 1400 Å and 1800 Å (*19*) have been obtained in "reconstitution experiments", in which acid or approximately neutral solutions of paramyosin were

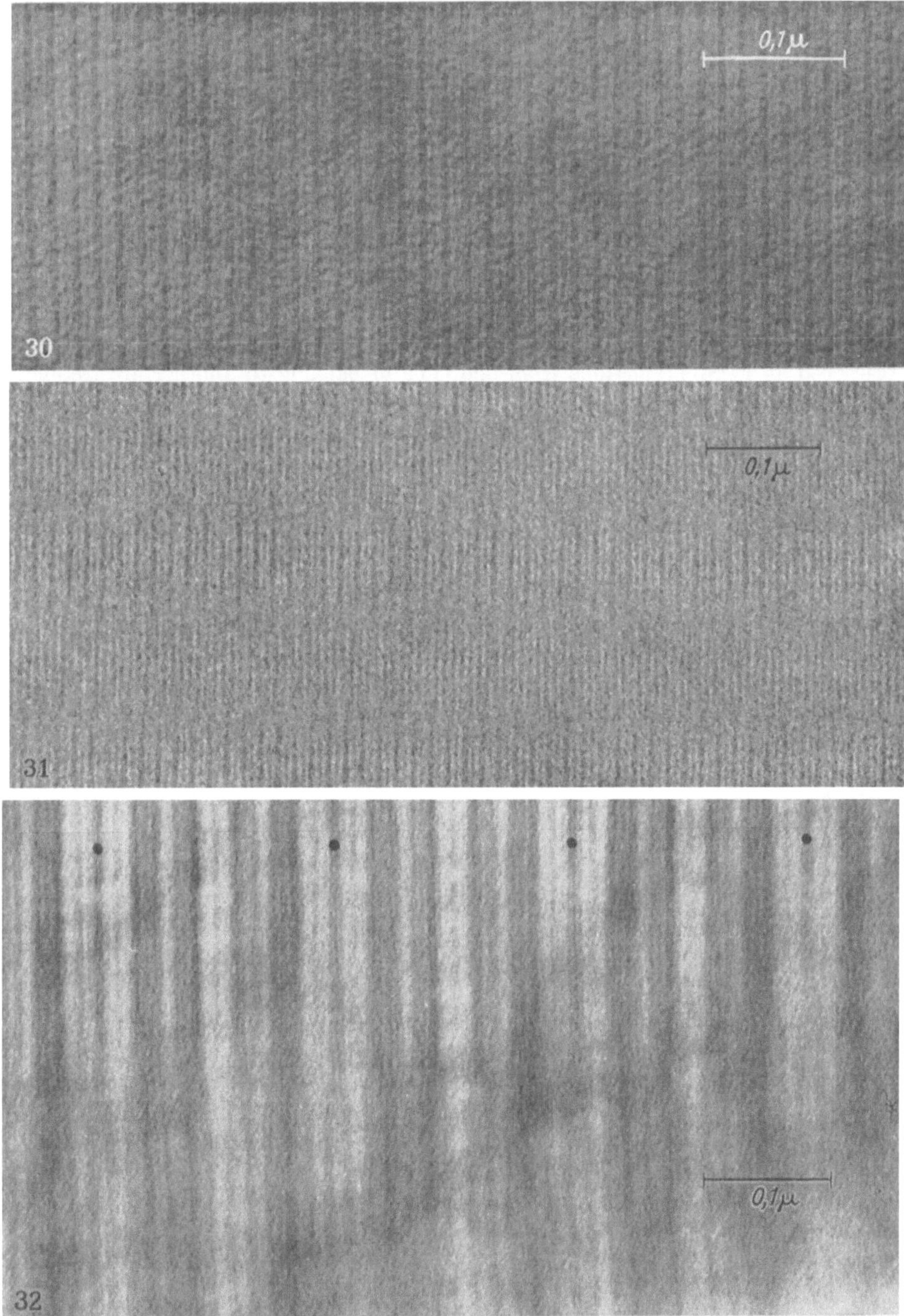

Fig. 30. A small area of a large flat paramyosin crystal from the same preparation as Fig. 28 showing a 145 Å repeat with detailed intraperiod band structure (*19*). Mounted directly on carbon supporting film. PTA stained. 200,000 ×

Fig. 31. Another crystal from the same preparation as Fig. 30, showing a rhythmic transition from a 145 Å period to an apparent 70 Å period. This pattern could be a MOIRE effect resulting from a small angular displacement of two overlapping thin crystals or of two adjacent layers within a single crystal (*19*). 160,000 ×

Fig. 32. A small region of a large reconstituted paramyosin "crystal" showing at top the 1 800 Å axial period first described by LOCKER and SCHMITT (*34*). Note the symmetrical band pattern within each period, and the transition in the lower part of the picture into a structure in which the 145 Å repeat is emphasized (*19*). 180,000 ×

variously treated to precipitate the protein (Fig. 28, 29, 30, 31, 32). LOCKER and SCHMITT (*34*) first described the 1 800 Å period in reconstituted fibrils, as well as a period of 360 Å. Reconstituted fibrils with periods of 145 Å and 725 Å have also been observed by others (*35*). Still missing from this list of ordered structures is the key one, namely the segment-type structure in which the macromolecules are packed in parallel array with like ends in register (cf. collagen, described earlier), thus revealing the distribution of stainable groups. Nevertheless, even in the absence of this "molecular fingerprint", it is possible to deduce, by analogy with the type of packing encountered in collagen, that these ordered structures must arise from (a) parallel staggered arrangements involving axial displacements of adjacent macromolecules by $^1/_{10}$ of their length, or integral multiples of this value, in the case of the 145 Å, 360 Å, and 725 Å periodicities, and (b) an antiparallel, in register, packing in the case of the symmetrically banded 1 400 Å spacing fibrils (Fig. 29). The latter type of ordered aggregate is the paramyosin equivalent of the collagen pattern described as FLS. Transitions from one type of packing to another are frequently observed (e.g., Fig. 32 shows such a transition from a spacing of 1 800 Å to one in which a 145 Å spacing is accentuated). Detailed examination of the band densities in the 1 800 Å repeat fibrils suggests the existence of a "super-lattice" with a true repeat period of $12 \times 1 800$ Å $= 21{,}600$ Å.

As has been shown, particularly in the case of the collagen macromolecule (tropocollagen), the interaction of elongated macromolecular monomeric units to form ordered fibrous structures exhibiting well-defined periodicities, determined both by X-ray diffraction and electron microscopic methods, involves primarily 1. an end-to-end linkage of the macromolecular units to form protofibrils (or linear polymers), a process probably involving an overlap and orderly coiling of end-chains, and 2. a side-to-side packing of those linear polymeric units with their relative axial displacements being determined by the charge distribution existing under any particular set of conditions, or in short by the "interaction profiles". While other factors such as intermolecular hydrogen bonding and other cross-links are involved in the further stabilization of ordered fibrous structures, it does not seem likely that they influence the patterns of organization determined by the forces already discussed.

References

1. INGRAM, V. M.: Nature (Lond.) **180**, 326 (1957).
   PAULING, L.: Amer. J. Psychiat. **113**, 492 (1956).
2. SCHMITT, F. O., J. GROSS and J. H. HIGHBERGER: Proc. nat. Acad. Sci. (Wash.) **39**, 459 (1953).
   GROSS, J., J. H. HIGHBERGER and F. O. SCHMITT: Proc. nat. Acad. Sci. (Wash.) **40**, 679 (1954).
3. BOEDTKER, H., and P. DOTY: J. Amer. chem. Soc. **78**, 4267 (1956).
4. CRICK, F. H. C., and A. RICH: in Recent advances in gelatin and glue research. New York: Pergamon Press 1958.
5. DOTY, P., and T. NISHIHARA: in Recent advances in gelatin and glue research. New York: Pergamon Press 1958.
6. KÜHN, K., W. GRASSMANN and V. HOFMANN: Naturwissenschaften **44**, 538 (1957).
   — Leder **9**, 217 (1958).
7. SCHMITT, F. O., J. GROSS and J. H. HIGHBERGER: Symposia Soc. exp. Biol. **9**, 148 (1955).
8. NISHIHARA, T., and P. DOTY: Proc. nat. Acad. Sci. (Wash.) **44**, 411 (1958).
9. HODGE. A, J., and F. O. SCHMITT: Proc. nat. Acad. Sci. (Wash.) **44**, 418 (1958).
10. GLIMCHER, M. J., A. J. HODGE and F. O. SCHMITT: Proc. nat. Acad. Sci. (Wash.) **43**, 860 (1957).
11. FITTON-JACKSON, S.: Proc. roy. Soc. (London) B, **146**, 270 (1957).
12. GLIMCHER, M. J.: The molecular biology of the mineralized tissues, with particular reference to bone. Rev. mod. Phys. **31**, 359 (1959).
   — A. J. HODGE and F. O. SCHMITT: The specificity of the macromolecular aggregation state of collagen in calcification. (In preparation).
   — — — Electron optical and X-ray diffraction studies of bone and of collagens mineralized *in vitro*. (In preparation.)
13. BEAR, R. S.: J. Amer. chem. Soc. **67**, 1625 (1945).
14. HUXLEY, H. E.: Proc. roy. Soc. (London) B **141**, 59 (1953).
15. HALL, C. E., M. A. JAKUS and F. O. SCHMITT: Biol. Bull. **90**, 32 (1946).
16. DRAPER, M. H., and A. J. HODGE: Aust. J. exp. Biol. med. Sci. **27**, 465 (1949).
17. — — Aust. J. exp. Biol. med. Sci. **28**, 549 (1950).
18. HODGE, A. J., H. E. HUXLEY and D. SPIRO: J. exp. Med. **99**, 201 (1954).
19. — The fibrous proteins of muscle. Rev. mod. Phys., **31**, 409 (1959).

20. Gergely, J.: J. biol. Chem. 200, 543 (1953).
   Mihalyi, E.: J. biol. Chem. 201, 197 (1953).
   — and A. G. Szent-Györgyi: J. biol. Chem. 201, 211 (1953).
21. Gergely, J., M. A. Gouvea and D. J. Karibian: J. biol. Chem. 212, 165 (1955).
22. Middlebrook, W. R.: The molecular architecture of the meromyosins. Abstr. of the meeting of the bio-physical Soc. p. 46, Boston 1958.
23. Philpott, D. E., and A. G. Szent-Györgyi: Biochim. biophys. Acta 15, 165 (1954).
24. Bailey, K.: Biochem. J. 43, 271 (1948).
   Tsao, T. C., K. Bailey and G. S. Adair: Biochem. J. 49, 27 (1951).
25. Hodge, A. J., A. G. Szent-Györgyi and Carolyn Cohen: Paper in preparation.
26. Hall, C. E., M. A. Jakus and F. O. Schmitt: J. appl. Physics 16, 459 (1945).
27. Elliott, G. F., J. Hanson and J. Lowy: Nature (Lond.) 180, 1291 (1957).
28. Hodge, A. J.: Ph. D. Thesis, Studies on paramyosin: The in vitro Reconstitution and transformation of periodic structure. Mass. Inst. Technol. 1952.
29. — Proc. nat. Acad. Sci. (Wash.) 38, 850 (1952).
30. Kay, C. M.: Biochim. biophys. Acta 27, 469 (1958).
31. Bailey, K.: Pubbl. Staz. Zool. Napoli 29, 96 (1957).
32. — Biochim. biophys. Acta 26, 612 (1957).
33. Kominz, D. R., F. Saad and K. Laki: Chemical characteristics of annelid, mollusc, and arthropod tropomyosins. Proc. Conf. on Chemistry of Muscular Contraction. Tokyo: Igakushoin, Ltd. 1957.
   — — — Nature (Lond.) 179, 206 (1957).
34. Locker, R. H., and F. O. Schmitt: J. biochem. biophys. Cytol. 3, 889 (1957).
35. Hanson, J., J. Lowy, H. Huxley, K. Bailey, C. M. Kay and J. C. Ruegg: Nature (Lond.) 180, 1134 (1957).

Ordering in lamellar systems

J. D. Robertson

Department of Anatomy University College, London

An analysis was given of the fine structure of the characteristic cellular membrane, as exemplified by the limiting envelope of typical tissue cells of vertebrates and invertebrates as well as by the cytoplasmic organelles of lamellar structure. An interpretation was given of the molecular organization of such membranes in terms of lipid and non-lipid constituents. The various types of membrane interaction and morphogenesis resulting from this specific molecular organization were illustrated by the case of myelination of nerve fibers (see page 159, this volume).

The basement membrane: substratum of histological order and complexity*

Daniel C. Pease**

Department of Anatomy, School of Medicine, University of California at Los Angeles, Los Angeles, and Veterans Administration Center, Los Angeles, California

Cytological Order

It has only been since the second great war that we finally are building a solid morphological bridge between the world of the biophysicists, populated in part by giant molecules, and the worlds of the cytologists, histologists, and embryologists where cells are the dominant populations. In previous meetings we have had ample demonstration of the important and often key role of electron microscopy in developing our knowledge of some families of macromolecules. People associated at one time or another with F. O. Schmitt at the Massachusetts Institute of Technology particularly have advanced macromolecular knowledge in studying collagen, muscle, myelin, and the cell surface.

---

* Aided in part by grants from the United States Public Health Service.
** The author wishes to acknowledge that he is indebted to several graduate students and visiting investigators who have worked with him in producing some of the results discussed in this article.

At the cytological level there have been extraordinary and detailed advances in our knowledge of mitochondria, the Golgi apparatus, and endoplasmic reticulum, to mention only those organelles with universal distribution. Chloroplasts, as a nearly universal organelle in plants, are comparable. All of these are "membrane systems" in which macromolecular organization is implied and about which we already comprehend many details. In Europe it has been the group associated with F. SJÖSTRAND at the Karolinska Institutet which has been most active in developing this field. In the United States, leadership has been with G. PALADE and K. PORTER at the Rockefeller Institute. This work has been reviewed several times including (36, 44), and is so generally well known that further comments are hardly necessary except, possibly, to call to the reader's attention a most thought-provoking effort by D. E. GREEN (14) attempting to relate quite specifically cytoplasmic membrane systems with enzymatic sterochemistry. However, it is important to realize that even within a single cell there are hierarchical stratifications of order that must ultimately be interpreted at different levels. Thus, cisterns of endoplasmic reticulum may be stacked in some cells in regular array to produce the "ergastoplasmic" and "Nissl" granules of conventional microscopy (2, 29, 30). GOLGI (7, 45) and chloroplast (16, 17, 21, 49) membranes often are organized similarly, and even such large units as mitochondria show amazingly complex patterns in the middle of spermatozoa (9).

Although it is likely that we ultimately can expect reasonably full understanding of some of these ordered systems at physico-chemical levels, others seem quite beyond the range of intermolecular forces. We seem to be without specific clues even though these patterns exist within single cells.

Histological order: Structure, functions and relations of basement membranes

There are still higher levels of organization where we must concern ourselves with the interaction of one cell with its neighbors, and with extracellular components of connective tissue. This is a world that the light microscopists have explored for a full century, and quite successfully up to the limits of their resolution. But now, quite suddenly, we are given an opportunity to invade this world with renewed vigor with vastly greater magnification. We can expect to discover new concepts and generalities, as well as to establish venerable hypotheses. This is the world I particularly would speak of here. More specifically I would like to develop the theme that many aspects of histological order and complexity depend upon basement membranes which act as durable microskeletons. This has far-reaching consequences that I would talk about. It is a subject partly considered by VOGEL (50).

It seems evident today that basement membranes are invariably present where connective tissue comes in contact with epithelial structures. The universal attachment of the one to the other is by way of the basement membrane. The basement membrane I speak of is not necessarily that of the light microscopist; for the term has not been used consistently. Indeed, most of the basement membranes that the electron microscopists see are altogether too thin to be visualized adequately, if at all, with the light microscope. Thus, in the past there have been serious doubts as to wether or not a basement membrane could be found underlying epidermis [Fig. 1, ref. (26, 42)]. Few light microscopists anticipated that thin walled capillaries invariably had a basement membrane [Fig. 3 and 10, ref. (27)]. On the other hand normal basement membranes always were easy to see with a light microscope in the kidney (Fig. 7—12), and they became particularly thickened in some pathological conditions (12).

Fig. 1. (Micron marks accompany this and subsequent figures.) Epidermo-dermal junction from the palmer skin of a distal phalanx of an adolescent Rhesus monkey. Arrows indicate where the basement membrane can be seen to advantage

Fig. 2. Corneal epithelium of a cat resting on Bowman's membrane which consists of fine collagenous fibers. The basement membrane, and the cement substance between it and the base of the epithelial cells is apparent. The cement substance can be followed into the greatly convoluted intercellular space

Fig. 3. Epithelium of the chorioid plexus (*choroid*) of a rat here lies close to a capillary (*cap.*). Thin basement membranes (*b. m.*) underlying both are apparent. Fenestrations in the endothelial sheet are indicated (*pores*). Part of a pial cell (*pia*) is interposed between the choroid epithelium and the capillary

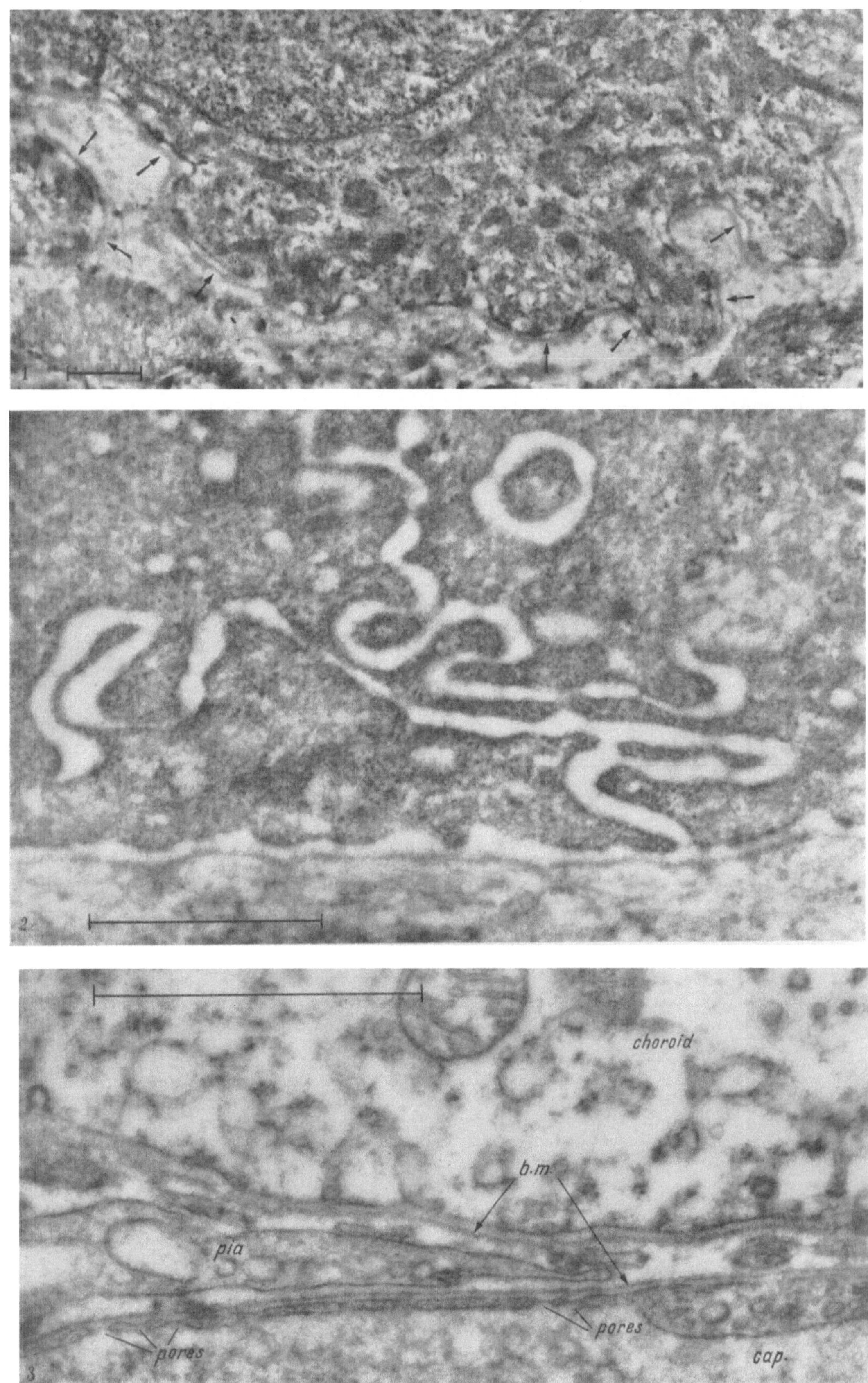

Fig. 1, 2 and 3 (see page 140)

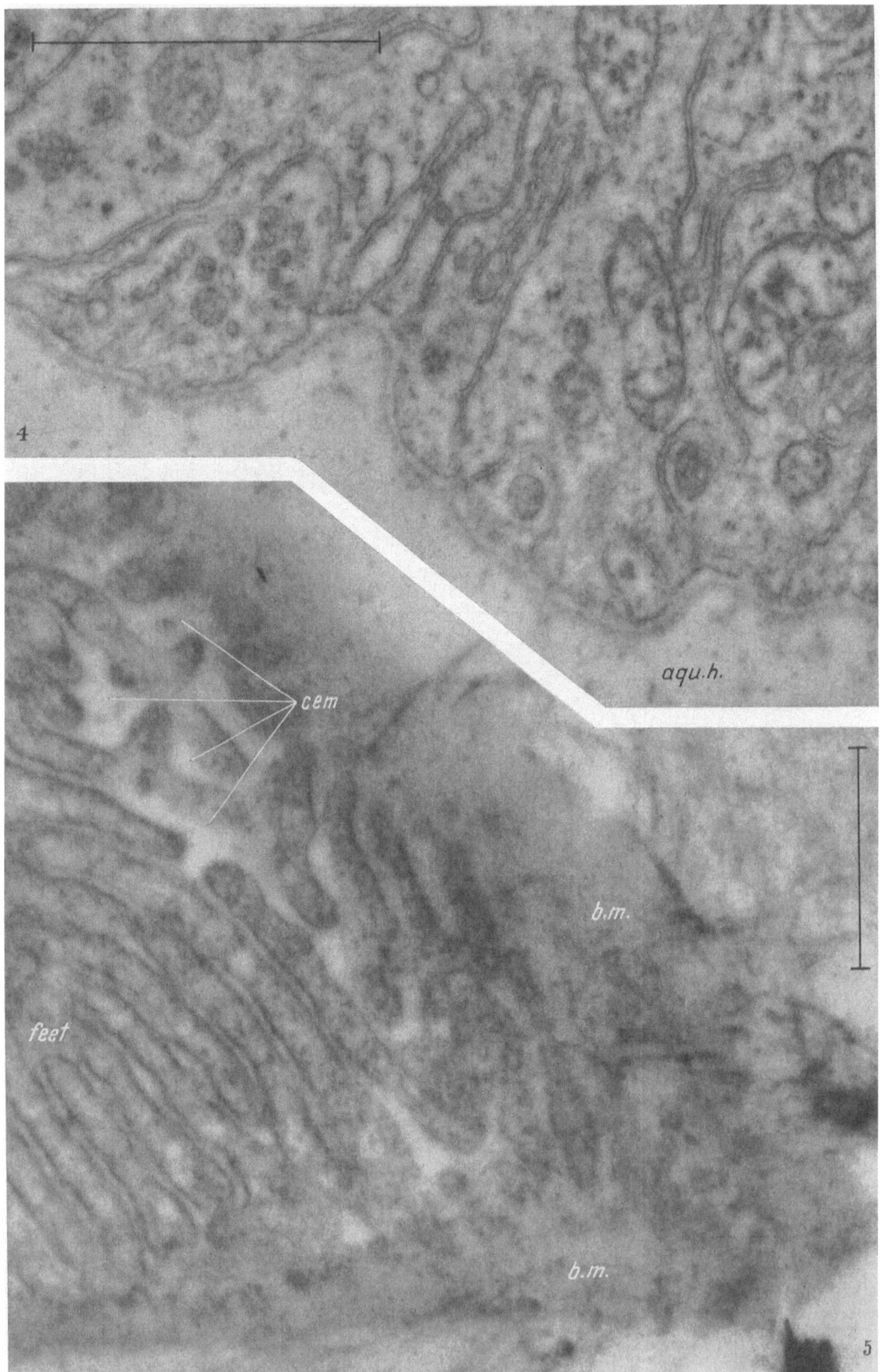

Fig. 4 and 5 (see page 144)

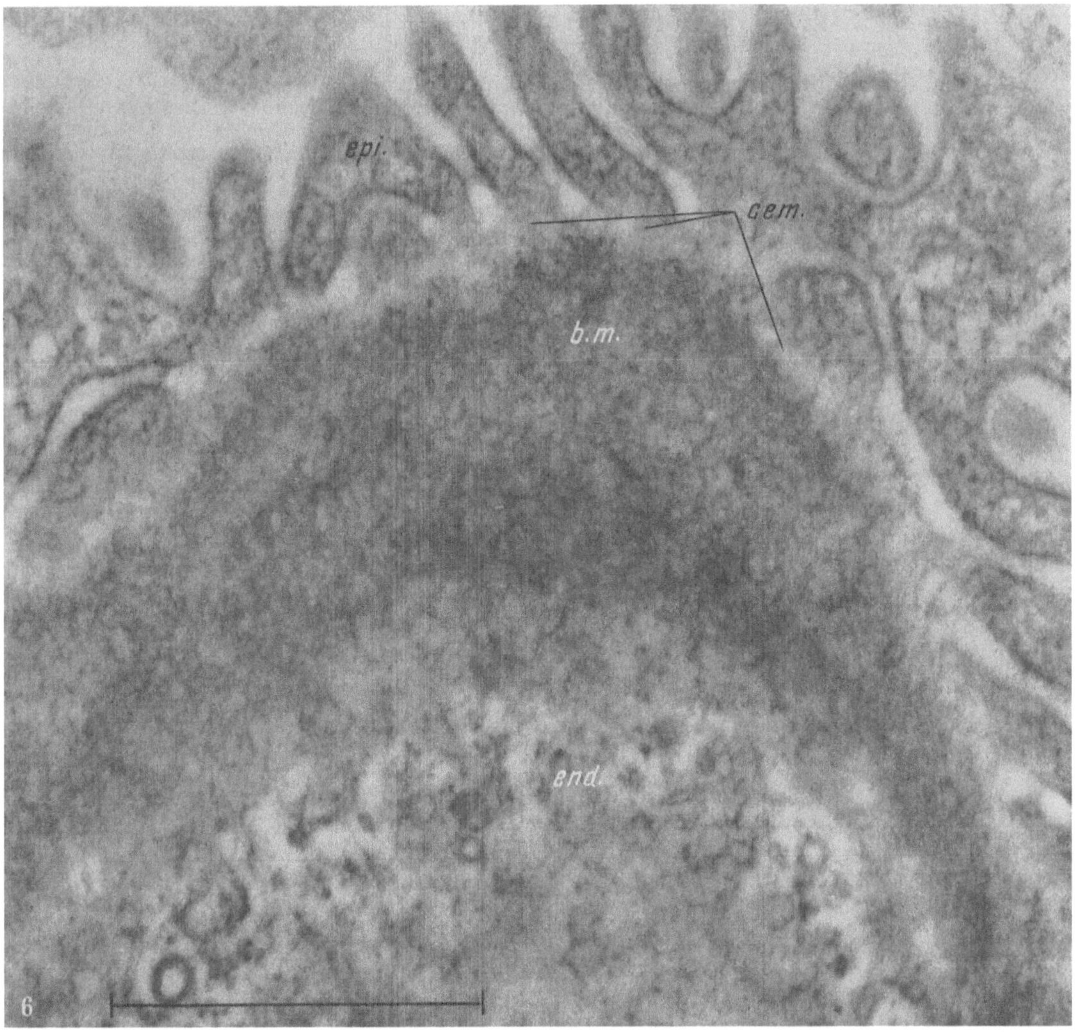

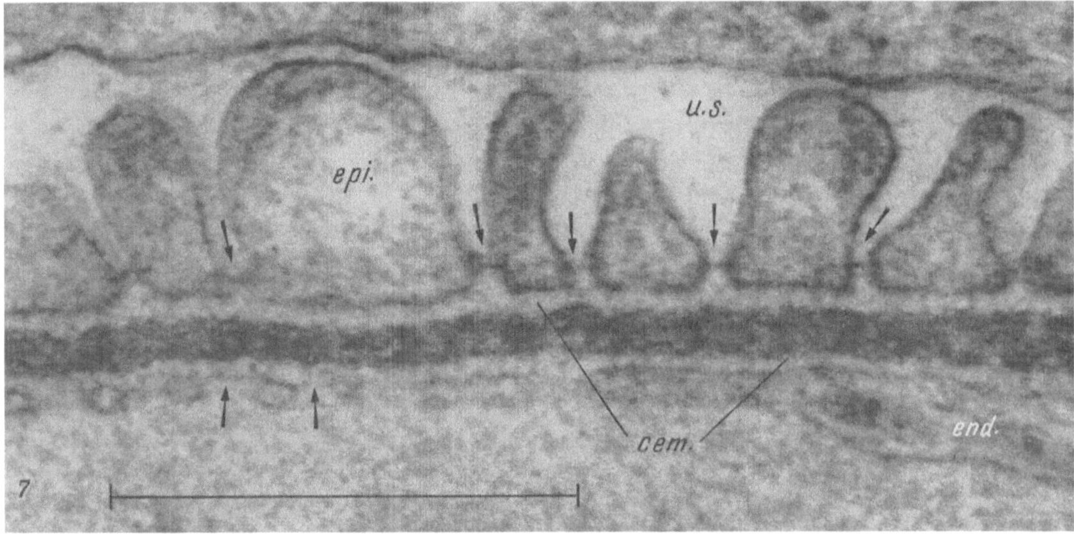

Fig. 6 and 7 (see page 144)

Fig. 4. The surface of the ciliary body of a rat eye is shown. An obvious basement membrane lies adjacent to the aqueous humor (*aqu. h.*). The surface of the epithelial cells is deeply infolded. The gap between these plasma membranes is continuous with the gap between basement membranes and the cell surface

Fig. 5. A tangential section through the base of proximal tubule cells and basement membrane in rat kidney. At the left are foot processes (*feet*) formed by the infolded basal cell surface. The homogeneous curving smudge is the basement membrane (*b. m.*). In some areas there are fairly wide expanses of cement substance (*cem.*) viewed in the horizontal plane. A few collagenous fibers are at the upper right

Fig. 6. A tangential section through a glomerular capillary of a rat. Epithelial feet (*epi.*) curve across the top of the figure. Endothelium (*end.*) fills the lower portion of the figure. Between lies the basement membrane (*b. m.*) and cement layers (*cem.*). This specimen and the next have been treated with 0.1% phosphotungstic acid after sectioning, and were mounted sandwiched between two evaporated carbon films somewhat in the manner proposed by Watson (*51*)

Fig. 7. A transverse section of a glomerular capillary of a rat stained and mounted in an identical manner with that of Fig. 6. In addition to the labels as above, note that the free margin of the cement layer is visible clearly between the epithelial feet (arrows), and its density is substantially different than that of the urinary space (*u. s.*). Similar features are barely visible on the endothelial side, within endothelial pores

At this point it is necessary to define terms. "Basement membrane" is used here to denote the lamella that is electron dense after osmium tetroxide fixation. The venerable term, "cement substance" is retained for the electron translucent layer between this and the dense plasma membrane of an adjacent cell surface (discussed below), and for the extension of this layer between neighboring cells. This corresponds to the "gap substance" of Robertson (*40, 41*). Histochemical tests, particularly the periodic acid-Schiff reaction, indicate that polysaccharides are the major components of basement membranes (*31*). As such, small, water soluble substances would be expected to diffuse through them easily. From the structural point of view, basement membranes have variously been reported as homogeneous or as consisting of a feltwork of extremely fine fibrils. Since one has to keep in mind the possibility of coagulation upon fixation, it probably is premature to decide which is the correct structural concept. After what seems to be "good" fixation with osmium tetroxide, basement membranes usually appear to be homogeneous, even under conditions of great resolution [Fig. 5, ref. (*54*)]. However, after post-osmic "staining" with phosphotungstic acid, they do not usually appear to be homogeneous (Fig. 6 and 7). Instead they might be described as vesiculated or even fibrillar. This may not be more meaningful than being the coagulum of what was in life a uniform gel. The thought of Bargmann, Knoop and Schiebler (*1*) that there is a vertically oriented repeating pattern seems unlikely.

In a physical sense it is likely that basement membranes can serve as barriers to particles of colloidal size. Indeed, in the kidney glomerulus (Fig. 7) there is nothing else interposed between blood and filtrate as has been indicated previously (*32*). That more than colloids may be retained by basement membranes is indicated by the important experiment of Ottoson, Sjöstrand, Stenström and Svaetichin (*28*) who demonstrated that the electrical potential of frog skin is maintained by this layer, which would prove it to be a barrier to ionic movement.

Traditionally basement membranes are regarded as connective tissue components. Yet these membranes are so intimately associated with the adjacent epithelia, and are so specialized in relation to the latter's function, that this is untenable. In fact, in certain situations, well developed basement membranes are to be found at remote distances from the nearest fibroblast, and must be manufactured by local cells. Examples include the basement membranes of the glomerular capillaries, shared by endothelium on one side and epithelium on the other (*32*). Alveolar walls of the lung are arranged similarly (*19, 22*). The basement membrane at the surface of the brain has astrocytic feet on one side, and pial cells on the other (*24*). Within the eye, the "inner limiting membrane" is a basement membrane (Fig. 4 ref. (*34*)]. There is no doubt but that the sarcolemma of muscle is a special case of basement membrane (*3, 38*). In the tunica media of muscular arteries there is a pure population of smooth muscle cells each embedded in such a membrane (Fig. 18 and 19). Schwann cells are intimately encased in basement membranes (Fig. 13 and 14) while endoneurial cells are haphazardedly and remotely scattered. There are, thus, many examples which attest the extremely close relationship between the basement membrane and the adjacent cell itself, and there often is no discernible association with connective tissue cells.

In both a phylogenetic and an ontogenetic sense, basement membranes appear early. They have been described in Invertebrates, a particularly good example being those of muscle in the wasp leg (*10*). They are associated with the embroyologically basal aspect of probably all epithelia irrespective of developmental distortions and contortions. Thus, the outside of the central nervous system derives from the basal aspect of the neural ectoderm. There exists here the basement membrane already referred to between nervous tissue and overlying pia (*24*). Similarly the lining of the retina comes from the basal aspect of the inner layer of the optic cup. The adjacent vitreous is originally mesenchyme. No doubt, basement membranes play a variety of functional roles, some of which are obvious, others, not so. Some are reasonably understood, others are obscure.

Perhaps the most general function of the basement membrane is that of attaching parenchyma to stroma. On the one hand it is intimately bound to an epithelium, and, on the other, is continuous with less specifically organized connective tissue components. Both of these relationships deserve particular attention. The attachment of basement membranes to adjacent cells is something that only the electron microscopists have visualized in some detail. It is evident that this attachment is essentially identical with that of one epithelial cell to its neighbor. When one cell is directly fastened to another, invariably one sees electron dense lines separated by a narrow gap that is electron translucent. The dense lines are approximately 75 Å. thick and the gap is approximately 150 Å wide to use the latest figures of ROBERTSON (*40, 41*). At the base of an epithelial cell exactly the same relations hold except that the basement membrane substitutes for the neighboring cell (cf. Fig. 2, of cornea first studied by JAKUS (*18*) and SHELDON (*43*)]. The translucent material of the gap is commonly referred to as "cement substance." It behooves us to know very exactly what this material is, for its character certainly influences intercellular permeability, to say nothing of having other more complex effects. SJÖSTRAND and RHODIN (*46*) were perhaps the first to suggest that lipids were its principal component. They felt that osmic acid did not necessarily render lipids insoluable, and they thought these subsequently might be extracted. It would seem that this is no longer a tenable hypothesis, however, as ROBERTSON (*40, 41*) has ably demonstrated. It is one of the triumphs of electron microscopy that now we can speak of and deal with 3-dimentional stacks of pure cell membrane. GEREN (*13*) demonstrated that myelin sheaths of peripheral nerves are formed by the successive spiralling of attenuated Schwann cell cytoplasm. Subsequently, ROBERTSON (*40, 41*) with great resolution and epoxy embedded material, fixed with permanganate solutions, demonstrated that each lamella consists of little more than a single layer of pure plasma membrane. The successive lamellae add the third dimension. This knowledge allows us to draw upon much biophysical evidence, and we now can say with considerable assurance exactly where the lipid is. It would seem to be contained within the electron dense line that one ordinarly observes at a cell's surface. With sufficient resolution and adequate preservation it is apparent that what we usually see as single line approximately 75 Å thick is, in fact, two dense lines with a narrow gap between, just wide enough to contain a bimolecular layer of lipid. The osmiophilic line at the surface of a cell is, then, a preserved derivative of the plasma membrane. The "cement substance" outside of the plasma membrane is something else again, and fundamentally extracellular.

In electron micrographs of conventionally preserved material the "cement substance" has such a low density that ones first impression is that there is nothing there. This is not so. One line of evidence that has been appreciated for some years is that intercellular gaps, as well as gaps between the bases of cells and basement membranes, ordinarily are of quite constant width, approximately 150 Å. These uniform gaps must be mainted by "something".

Apart from inferential evidence, however, it is quite possible to study and visualize the "cement material" directly in suitable situations. Glomerular capillaries of the kidney afford a particularly good opportunity; for this material is nakedly exposed to glomerular filtrate between the epithelial feet on the outer side of the basement membrane, and similary is exposed directly to blood plasma within the endothelial pores on the inner side of the basement membrane. At these exposed places a precise interface often can be seen and the density of the "cement substance" can be compared directly with the adjacent fluid systems (Fig. 7, arrows). It is an interesting and significant fact that gross fixation artifacts do not ordinarily rupture or otherwise involve the

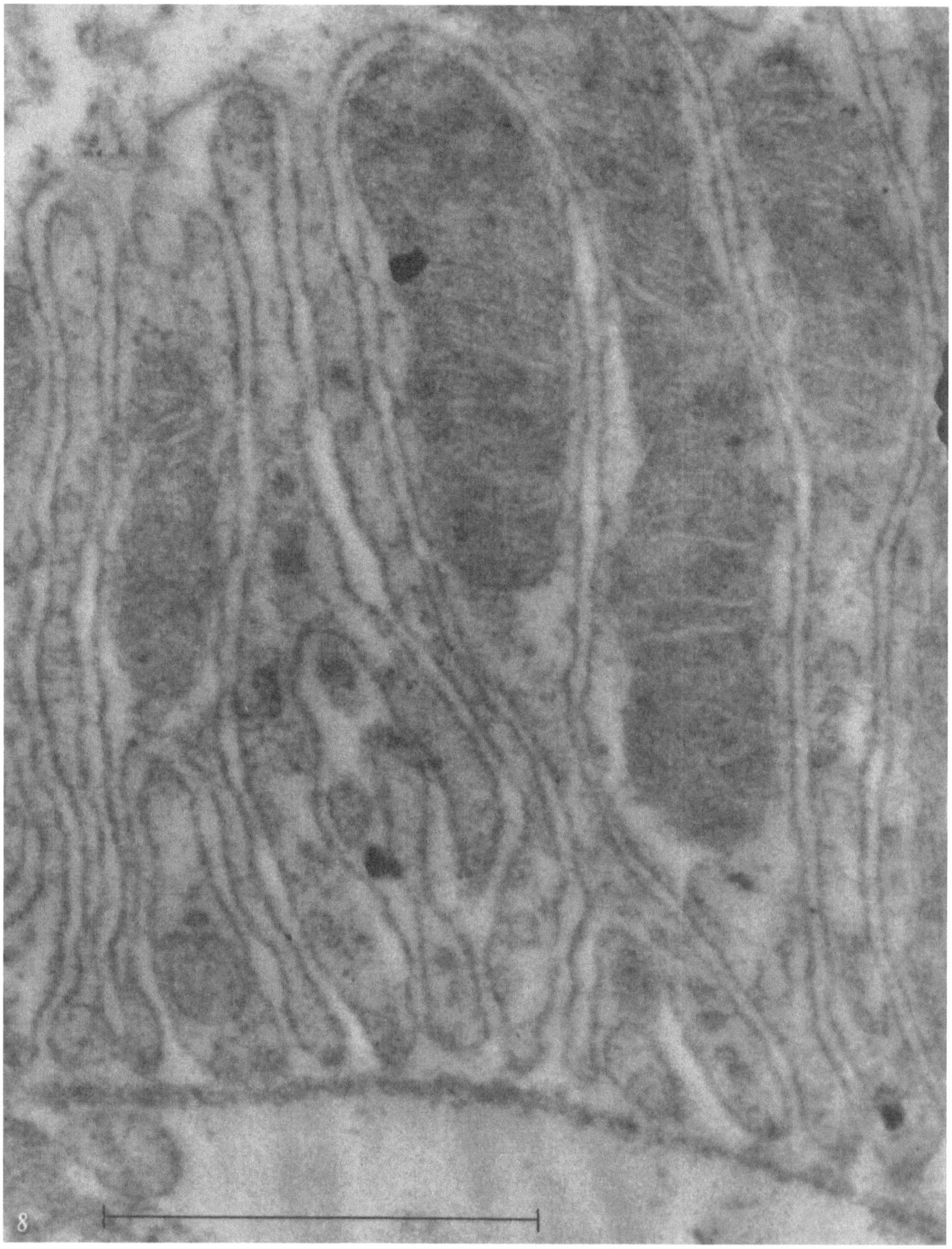

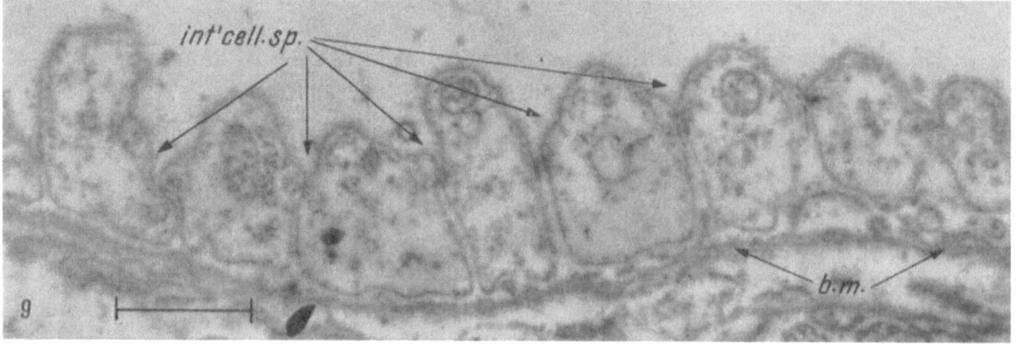

Fig. 8 and 9 (see page 148)

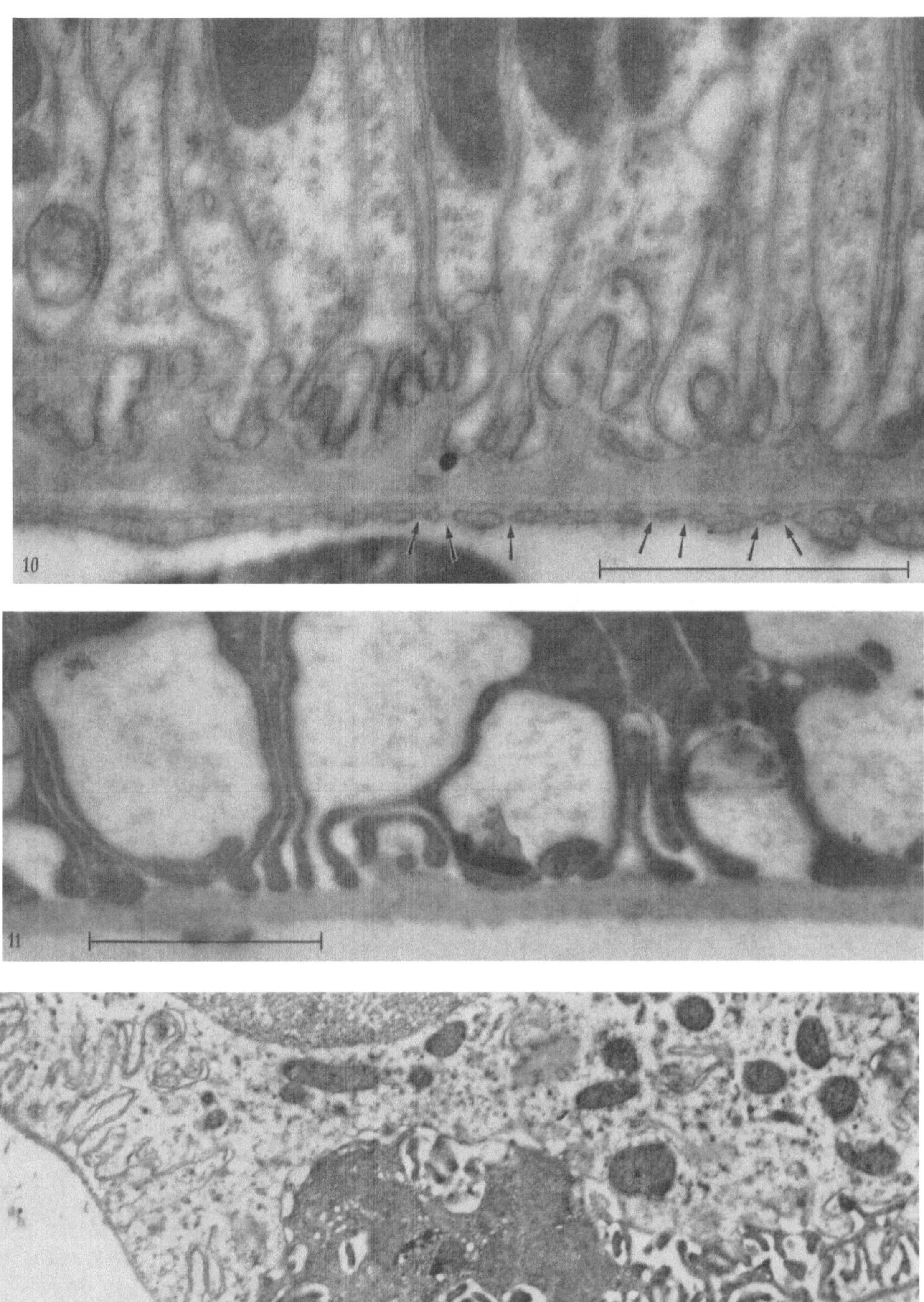

Fig. 10, 11 and 12 (see page 148)

10*

Fig. 8. The basal region of a cell of the distal tubule of rat kidney. Stained and mounted as the specimen of Fig. 6. The continuity of the cement substance adjacent to the basement membrane, and the material of the gap between infolded plasma membranes is apparent

Fig. 9. Transverse section of the thin segment of HENLE's loop from the medulla of a rat kidney. These flattened epithelial cells have deeply interdigitating margins so that the cross-sectional area of intercellular space (*int'cell. sp.*) is vastly increased

Fig. 10. Distal tubule of the rat kidney closely approximated to a capillary. Note that the basement membrane of the capillary is extremely thin, while that of the tubule is relatively massive. Furthermore, the inner surface of the tubular basement membrane is far from smooth in this instance. Substantial thickenings sometimes indent the base of the epithelial cells forming ridges that can be massive enough to be quite visible with a light microscope in suitable stained preparations. The deep infoldings of the basal surface of these cells shows here to advantage. The endothelial fenestrations are obvious, and arrows point to some of the pores

Fig. 11. Another transverse section of the basal surface of a distal tubule cell from a rat kidney. The potential spaces between the deep infoldings of the plasma membrane have been opened up in this instance, presumably by surface injury preceding fixation [see PEASE (*33*)]. The fact that they are extracellular compartments is obvious. Fluid, here, was being retained at least temporarily by the basement membrane

Fig. 12. A transverse section through the base of a collecting tubule in a rat kidney. A cell, or part of a cell, had been undergoing necrosis for unknown reasons. In spite of the local lesion, the basement membrane remained morphologically intact. The necrotic area was walled off by a plasma membrane belonging to the healthy neighboring cell

---

cement layers. Thus, when glomerular cells are poorly preserved with perhaps much shrinkage, the epithelial feet none the less remain firmly attached to the capillary basement membrane. Also it is an obvious fact that it is ordinarily difficult to tear even poorly preserved epithelial cells apart, one from the other, along intercellular lines. In the glomerular capillaries of the kidney, it is not difficult to section the cementing zones tangentially, so that portions of these layers are observed as flat sheets of material. Even under conditions of great resolution there are no consistent evidences of fibrous structures visible after preservation with osmium tetroxide (Fig. 5 and 6). Slight inhomogenities visible after treatment with phosphotungstic acid are attributable to a coagulation process. Thus, this is probably an amorphous material in life. Physiologists for many years have regarded intercellular spaces as a possible route for the migration of various materials through epithelial linings, particularly as a likely pathway for strong electrolytes through endothelium (*4, 25*). This implies the presence of an aqueous phase in these regions. Probably the most notable specialization of this sort is to be found in the thin segment of the mammalian nephron. The cross-sectional area of the intercellular substance is greatly increased by numerous, very deep, lateral interdigitations of adjacent squamous cells (Fig. 9). As an extension of this idea we know from electron microscopy that the basal plasma membranes of many cell types can be elaborately infolded to vastly increase the basal surface area. The original and most notable examples of this sort of specialization occur in the tubular epithelia of the kidney [Fig. 8 and 10, ref. (*33, 37, 46*)]. In increasing surface area the specialization is analogous to the development of brush borders at the apical ends of cells. Within the folds of plasma membranes we have an extension of the "cement substance" and a gap ordinarily about 150 Å wide. Presumably this specialization would be without physiological meaning if water soluble materials could not percolate through this material just as between cells. PEASE (*34*) proposed in fact, that this specialization is correlated with epithelia involved in water transport mechanisms, for he found this specialization in the epithelia of the choriod plexus, the ciliary body of the eye and the secretory ducts of salivary glands. It has since observed in a variety of places notably including the stria vascularis of the cochlear duct (*11, 47, 48*), the trophoblast of the placenta (*8*), and in sweat glands by this author. Thus, there is presumptive evidence that this is in fact an aqueous pathway.

There are conditions which allow epithelial cells to be separated one from the next. For many years it has been known that the removal of calcium and other bivalent ions reversibly softens the "cement substance". PEASE (*33*) demonstrated that water leaving the tubular cells of the kidney could dilate greatly the potential spaces between the basal folds of the plasma membrane (Fig. 11), although it is not known that this can occur within physiologically tolerable conditions. These various lines of evidence indicate that "cement substance" possesses a water phase and also contains a structural component. We can only guess as to the nature of the material. It is possibly a protein; although, if so, we would expect to see it precipitated after course fixation, and presumably it would possess substantial osmiophilia. Of the biological possibilities that we know of today, polysaccharides seem more plausible. This author has made an effort to visualize histochemically with the light microscope the basal folds of kidney tubular epithelium, utilizing material known to be well fixed by standards of electron microscopy, and employing the periodic acid-Schiff reaction. Staining was carried out in bulk before blocking, as well as in methacrylate sections of various appropriate thicknesses. The basement membrane proper showed a dramatically positive reaction. Intercellular zones and basal folds were weakly colored so that the reaction was uncertain, but actually the density of stainable material is so low that a vivid reaction hardly could be expected. It could be said that these zones stained as vividly as diffuse areas of connective tissue in the same section. Thus, it is the author's opinion that the reaction was positive, although confirmation from other sources is desirable.

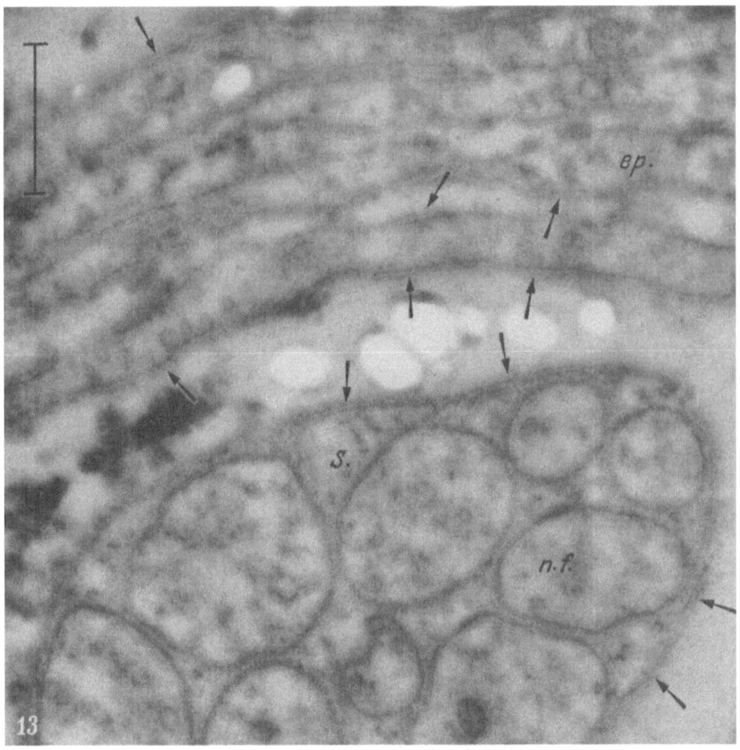

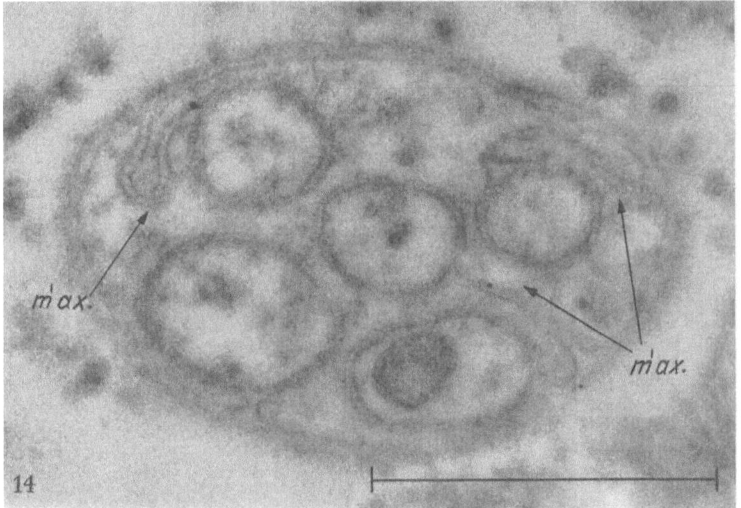

Fig. 13. Part of a cutaneous nerve and its sheath from ear skin of a rat. Unmyelinated fibers (*n. f.*) are enfolded into Schwann cell cytoplasm (*S.*). Basement membranes cover the Schwann cell, and also the attenuated cytoplasm of the epineurial cell (*ep.*). Arrows indicate where the basement membrane can be seen to advantage

Fig. 14. A small cluster of small unmyelinated axons embedded in Schwann cytoplasm from a Meissner's corpuscle of a monkey fingertip. The basement membrane at the surface of the Schwann cell is obvious, and the cement substance can be followed into the gap between plasma membranes in the mesaxons (*m'ax.*)

So far emphasis has been upon the relation of basement membranes and associated epithelia. We should now turn our attention to the relation between membrane and connective tissue. In a fundamental sense the phrase "connective tissue" can mean just what it says, and one can

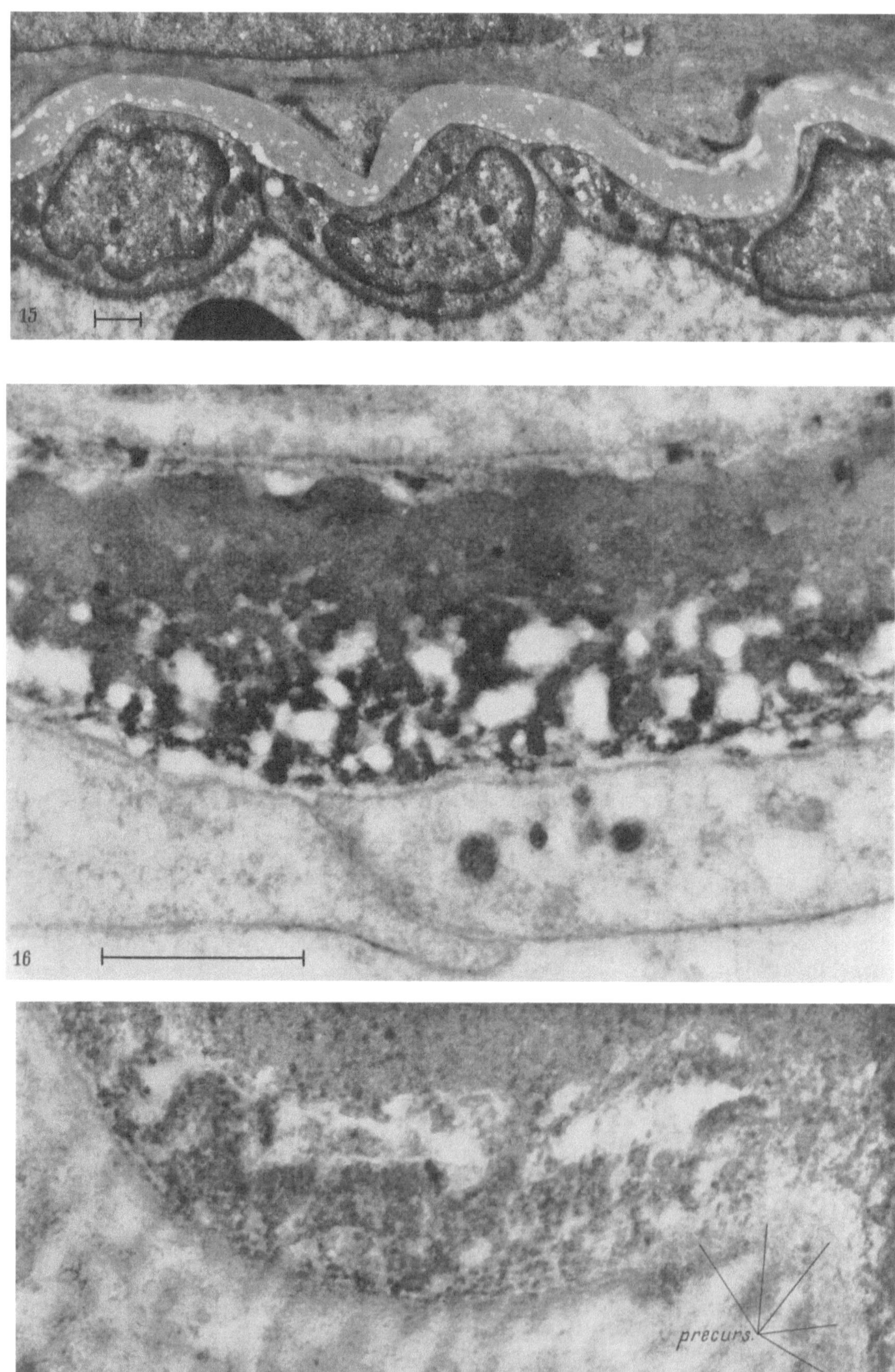

Fig. 15, 16 and 17 (see page 151)

think of the "cement substance" between two cells as being the minimum quantity of this. At the next level of complexity, basement membranes are added to the picture. A single basement membrane can be shared by two cells. The cells may be identical as, for example, adjacent smooth muscle cells in the tunica media of a small artery (Fig. 18) or they may be different types, as smooth muscle on one side and endothelium on the other, or, in the glomerular capillary, endothelium, and epithelium (Fig. 7), or, at the brain surface, astrocytes and pia (*24*). Arterial walls [and other aggregations of smooth muscle (*3*)] provide a particularly good place to observe the next order of complexity. A shared basement membrane sometimes can be seen to have central cavities that are in essence precursors of larger connective tissue spaces (Fig. 16). In other areas a single membrane frankly bifurcates to produce obvious connective tissues spaces between the two daughter membranes (Fig. 18). These spaces do not become large before other formed elements of connective tissue appear within them, most notably of course, collagen, elastin, and also some "fluff" (Fig. 18 and 19). The fluff is the same substance as the basement membrane, extending away from it in irregular networks. It is presumably poorly organized polysaccharide infiltrating the extracellular spaces. In our own laboratory we have been making a particular study of the relation of elastin to basement membranes. Definative elastin apparently is deposited within a precursor matrix of polysaccharide Phosphotungstic acid staining is important to distinguish the two (compare Fig. 15 with Fig. 16, 18 and 19). In arteries the heaviest deposits of elastin appear in the "fluff", rather than engulfing the basement membranes (Fig. 16 and 17), but here and there elastin becomes directly incorporated in basement membranes (Fig. 16 and 19). It is evident that there is a very close relationship between elastin and the simpler polysaccharides, of basement membranes and connective tissue generally, and that the "fluff" can become elasticized under some conditions by the addition of a second definative material. Although highly organized and oriented layers of collagenous fibrils are sometimes found just below basement membranes (*52*, *53*), more commonly a poorly organized layer of "reticular" fibers is present (Fig. 1 and 2). It is conventionally thought that these fibers blend with basement membranes to attach the membranes to the connective tissue, but this rarely can be demonstrated at electron microscopic levels. Rather, the attachment of basement membrane to connective tissue seems to be way of the essentially amorphous polysaccharide "fluff" mentioned previously, and under some circumstances by way of direct continuity with elastin as COOPER (*6*) has shown in skin, studied histochemically by light microscopy.

We can except to learn something of the lability and/or stability of basement membranes from pathology, and eventually experimental modification. At the present time this is an area of research just beginning. It is already evident that certain persistent disease processes result in greatly thickened basement membranes (*12*). It is also apparent that basement membranes can survive for a time quite apart from being associated with living cells. In studying the kidney we have several times observed minute spontaneous lesions which had denuded the basement membranes of glomerular capillaries or tubules (Fig. 12). We have no knowledge of how long such naked membranes might survive, but they may well provide a mould for regeneration. Thus the

---

Fig. 15. The tunica intima of a muscular artery from the pia of a cat. The internal elastic membrane fairly completely fills the connective tissue space between the muscularis and the endothelium

Fig. 16. The tunica intima of a muscular artery from the pia of an adolescent Rhesus monkey, "stained" with 0.1% phosphotungstic acid after sectioning. Elastic tissue regularly invades and breaks up the basement membrane of the endothelium, but tends to leave the basement membrane of the overlying smooth muscle intact. Only locally, as at the right of the figure, is the elastic layer directly a part of this basement membrane. Definitive elastin apparently is deposited in a polysaccharide matrix. Some of this precursor material, having only a moderate electron density, may be seen between and around the dense elastin masses. It is more apparent in Fig. 17

Fig. 17. The inner margin of the internal elastic membrane from the pia of an old Rhesus monkey, showing evidence of sclerosis and having much elastic "fragmentation". Section "stained" with 0.1% phosphotungstic acid and mounted sandwiched between two evaporated carbon fibers. The elastin is very electron dense, and is fundamentally granular in character. Polysaccharide precursor material of relatively low density, and also somewhat granular in character, is evident, particularly at the left of the figure (*precurs.*)

glomerular capillary might be reconstituted by the migration of healthy epithelial and endothelial cells over preexisting basement membranes. It is possible that the survival of basement membranes associated with Schwann cells might be the principal guide path in the regeneration of peripheral axons, but this is purely speculative at the present time. In our laboratory we have begun a study of intimal hyperplasia in arteriosclerosis. It is clear that basement membranes thicken, and that much polysaccharide "fluff" is produced. In turn, new elastin is deposited in this augmented matrix (Fig. 17). Thus we see that basement membranes respond to dynamic conditions, yet also provide a more less persistent framework upon which cells can be expected to differentiate and specialize. It is also pertinent to consider that basement membranes are adapted to specific functions. Thus the glomerular capillaries of the kidneys consist only of endothelium, epithelium and basement membrane. They are unusual capillaries in that there is no further support from surrounding connective tissue. These vessels have basement membranes unusually thick for capillaries (Fig. 7), and it is reasonable to think that they provide the main structural support of the vessel. The glomerular capillaries are subjected to relatively high arterial pressure. The peritubular capillaries, on the other hand must sustain only low venous pressures, and their basement membranes are notably thin (Fig. 10). In the cerebral cortex arterial capillaries can be identified by incomplete coverings of what appear to be primitive muscle cells. These blood vessels, subject to relatively high internal pressures, have notably thick basement membranes, while venous capillaries have extremely thin membranes (*24*), again suggesting a structural relationship to an applied force. Alveolar capillaries in general have no other support but the basement membrane which is relatively thick (*19, 22*). The sarcolemma of striated muscle has long been assigned a role in limiting the passive extension of muscle fibers. Its considerable residual strength can be demonstrated after destroying the living muscle fiber within the envelope, and no doubt is related to its thickness.

Role of basement membranes

I would like, in my concluding remarks, to return to the theme that basement membranes are an essential substratum for much histological order. They provide the mechanical bases upon which cells can develop complexities. The apical ends of epithelial cells can differentiate almost without limit, and produce brush borders, cilia, microvilli, secretory blebs, cuticles etc. But this is not so of the basal end of the cell. Here the primary consideration must be that of mechanical attachment to the underlying connective tissue. This necessity limits the possibilities of morphological specialization. But one of the most surprising findings of electron microscopy has been that there can be great basal specialization, expanding this surface by deep infolding, yet keeping the mechanical attachment secure by the great numbers of feet created by the folds. Examples mentioned above include kidney tubules (Fig. 8 and 10, ref. (*33, 37, 46*)], choriod plexus, secretory ducts, ciliary body (*34*), placental trophoblast (*8*), utricle (*47*), secretory parts of the cochlear duct (*11, 48*), etc. In somewhat the same category is the development of large subepithelial spaces and basal pseudopods, particularly in endocrine epithelia (*20*).

In the embryological development of kidney tubules at least, basal specialization does not, and presumably could not, begin until after basement membranes are established by simple cuboidal cells (*5*). That the basement membrane is an essential embryological precursor to basal

---

Fig. 18 and 19. Connective tissue zones between smooth muscle cells in the tunica media of muscular arteries from a cat's pia. Sections "stained" with 0.1% phosphotungstic acid, after which treatment the sarcolemnic basement membranes no longer appear homogeneous. Note to the right in Fig. 18 that two adjacent smooth muscle cells most simply can share a single basement membrane. Also note that basement membranes can bifurcate and so open up a patent connective tissue space. Collagen (*coll.*) and elastin (*el.*) may appear in such a space. Elastin may be directly incorporated in a basement membrane as is suggested in Fig. 18, and clearly evident in Fig. 19. In general, however, the basement membranes are not buried in elastin except as already noted on the endothelial side of the elastica interna of muscular arteries

Fig. 20. A developing arteriole from the most superficial, and only partly differentiated, cerebral cortex of an 8-day old rat. The endothelium and the overlying myoblasts lack basement membranes. It is evident that the surfaces of the cells have not yet been stabilized for they are rough contoured, and the muscle layer is still far from complete

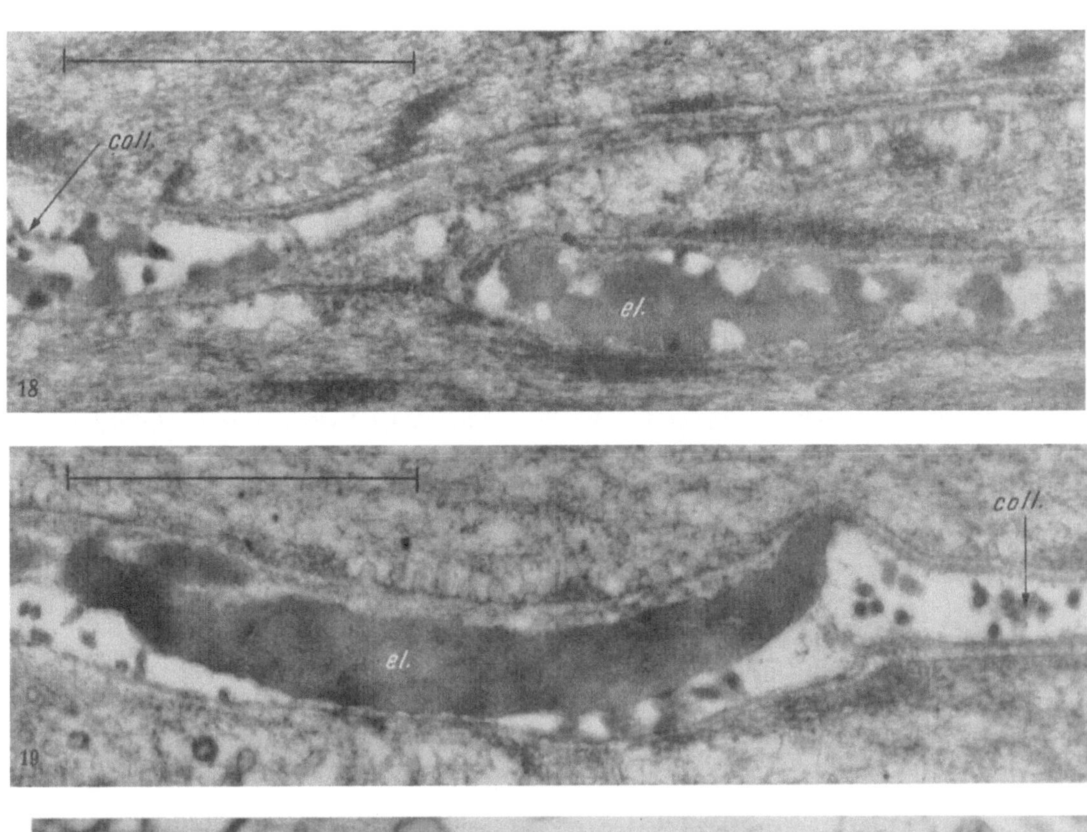

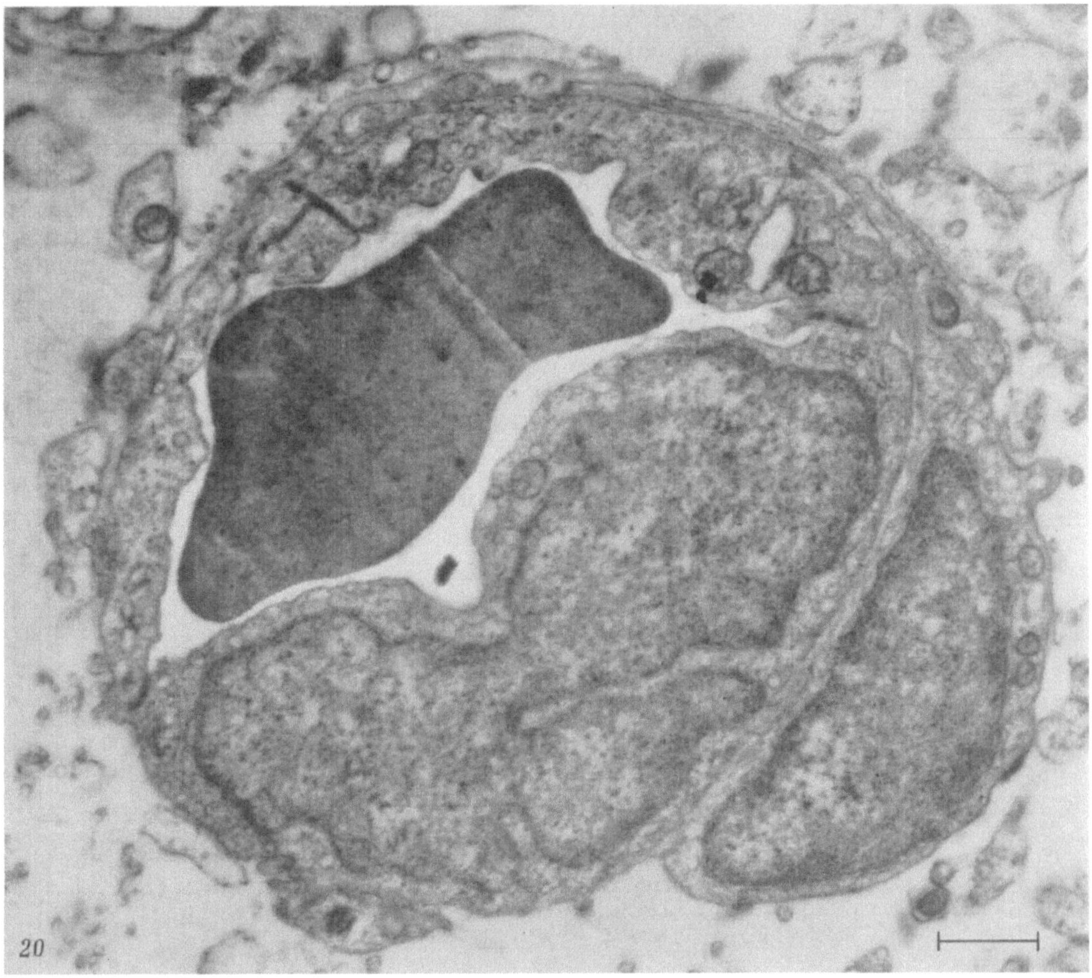

Fig. 18, 19 and 20 (see page 152)

specialization seems obvious. In the kidney glomerulus we have spoken of a complex arrangement of epithelial cells with their myriads of feet interdigitating with those of neighboring cells. It is impossible to imagine the complex precision of this system without the mechanical substrate of the basement membrane and the associated cement substance. In pathological conditions, these epithelial cells are seen to dedifferentiate to simple shapes (*12*). The healing process can be expected to recapitulate embryological development, and regenerative specialization probably is possible if the basement membrane survives. In some places in the body, capillary endothelium becomes extraordinarily attenuated, even to the point of becoming fenestrated [Fig. 3, 7 and 10, ref. (*15, 23, 32*)]. Who can believe that such thin walled capillaries possibly could exist without the structural support of a basement membrane. The complex lateral interdigitations of the thin segment of the mammalian nephron also surely require skeletal support from the basement membrane. In other situations where thin cellular lamellae are important components, we once again find basement membranes. Work in the writer's laboratory has demonstrated that the outer circumferential lamellae of Pacinian corpuscles have these (*35*), as does the capsule of the Meissner's corpuscles of the fingertips, and even the epineurial sheath of peripheral nerves (Fig. 13). Schwann sheaths and muscle represent greatly elongated strands of protoplasm. No doubt their shape in part is maintained by the basement membranes both possess (neurilemma, Fig. 13 and 14, sarcolemma, ref. (*38, 39*)].

Unfortunately we still know very little about the development and regeneration of basement membranes. Fig. 20 is of a partially differentiated arteriole, still without basement membranes. The external endothelial surface is somewhat irregular. The myoblasts have not as yet formed a continuous tunic, and have highly irregullar processess. The form of the vessel obviously is not as yet fully stabilized. That presumably would follow the formation of its basement membrane. Basement membranes are also known to precede final, complex differentiation in the development of epidermal (*26*) and corneal (*43*) epithelia. It is apparent from the work of WEISS and FERRIS (*53*) that basement membranes can regenerate quickly in tadpole skin. But we still have very little comparative information on this subject, which no doubt is of great importance in understanding regenerative processes. In short, it is the theme of this review to emphasize that basement membranes make possible cytological complexity within the basal parts of cells. They support directly or indirectly complex shapes of individual cells, and they play a major role in allowing populations and complex groupings of cells (epithelia) to be constructed upon them. Their appearance and their unexpectedly widespread presence as a microskeleton has far-reaching effects at several hierarchical levels of organization which we only now can begin to visualize.

References

1. BARGMANN, W., A. KNOOP u. T. H. SCHIEBLER: Z. Zellforsch. **42**, 386 (1955).

2. BERNHARD, W., and C. ROUILLER: J. biophys. biochem, Cytol. **2**, (suppl.) 73 (1956).

3. CAESAR, R., G. A. EDWARDS and H. RUSKA: J. biophys. biochem. Cytol. **3**, 867 (1957).

4. CHAMBERS, R., and B. W. ZWEIFACH: J. cell. comp. Physiol. **15**, 255 (1940).

5. CLARK, S. L.: J. biophys. biochem. Cytol. **3**, 349 (1957).

6. COOPER, J. H.: A.M.A. Arch. Derm. **77**, 18 (1958).

7. DALTON, A. J., and M. D. FELIX: J. biophys. biochem. Cytol. **2**, (suppl.) 79 (1956).

8. DEMPSEY, E. W., and G. B. WISLOCKI: J. biophys. biochem. Cytol. **2**, 743 (1956).

9. DE ROBERTIS, E., and H. F. RAFFO: Exp. Cell Res. **12**, 66 (1957).

10. EDWARDS, G. A., H. RUSKA and E. DE HARVEN: J. biophys. biochem. Cytol. **4**, 107 (1958).

11. ENGSTRÖM, H., F. S. SJÖSTRAND u. H. SPOENDLIN: Pract. Oto-rhino-laring. **17**, 69 (1955).

12. FARQUHAR, M. G., R. L. VERNIER and R. A. GOOD: J. exp. Med. **106**, 649 (1957).

13. GEREN, B. B.: Exp. Cell Res. **7**, 558 (1954).

14. GREEN, D. E.: Harvey Lect. 177; The Harvey Society of New York 1957.

15. HALL, B. V.: Proc. V. Ann. Conf. Nephrotic Syndrome, Sponsored by the Nat. Nephrosis Found. Inc. New York, 1, 1953.

16. HODGE, A. J., J. D. McLEAN and F. V. MERCER: J. biophys. biochem. Cytol. **1**, 605 (1955).

17. — — — J. biophys. biochem. Cytol. **2**, 597 (1956).

18. JAKUS, M. A.: Amer. J. ophthal., **38**, no. 1, pt. 2, 40 (1954).

19. KARRER, H. E.: J. biophys. biochem. Cytol. **2**, 241 (1956).

20. LEVER, J. D.: J. biophys. biochem. Cytol. **2**, (suppl.) 293 (1956).

21. Leyon, H.: Exp. Cell Res. **7**, 265 (1954).

22. Low, F. N.: Anat. Rec. **117**, 241 (1953).

23. Maxwell, D. S., and D. C. Pease: J. biophys. biochem. Cytol. **2**, 467 (1956).

24. Maynard, E. A., R. L. Schultz and D. C. Pease: Amer. J. Anat. **100**, 409 (1957).

25. Mende, T. J., and E. L. Chambers: J. biophys. biochem. Cytol. **4**, 319 (1958).

26. Menefee, M. G.: J. Ultrastructure Res. **1**, 49 (1957).

27. Moore, D. H., and H. Ruska: J. biophys. biochem. Cytol. **3**, 457 (1957).

28. Ottoson, D., F. S. Sjöstrand, S. Stenström and G. Svaetichin: Acta physiol. scand. **29**, 611 (1953).

29. Palade, G. E.: J. biophys. biochem. Cytol. **2**, (suppl.), 85 (1956).

30. Palay, S. L., and G. E. Palade: J. biophys. biochem. Cytol. **1**, 69 (1955).

31. Pearse, A. G. F.: Histochemistry, London, 1 (1953).

32. Pease, D. C.: Anat. Rec. **121**, 701 (1955).

33. — Anat. Rec. **121**, 723 (1955).

34. — J. biophys. biochem. Cytol. **2**, (suppl.) 203 (1956).

35. — and A. Quilliam: J. biophys. biochem. Cytol. **3**, 331 (1951).

36. Porter, K.: Harvey Lect. 175—228; Harvey Soc. New York 1956.

37. Rhodin, J.: Correlation of ultrastructural organization and function in normal and experimentally changed proximal convoluted cells of mouse kidney. Dept. of Anat., Karolinska Institutet, Stockholm. (1954).

38. Robertson, J. D.: J. biophys. biochem. Cytol. **2**, 369 (1956).

39. — J. biophys. biochem. Cytol. **2**, 381 (1956).

40. — J. biophys. biochem. Cytol. **3**, 1043 (1957).

41. — J. biophys. biochem. Cytol. **4**, 349 (1958).

42. Selby, C. C.: J. biophys. biochem. Cytol. **1**, 429 (1955).

43. Sheldon, H.: J. biophys. biochem. Cytol. **2**, 253 (1956).

44. Sjöstrand, F. S.: in Vol. 3, Cells and Tissues, ed. G. Oster and A. W. Pollister. New York: Academic Pres. Inc. 241 (1956).

45. — and V. Hanzon: Exp. Cell. Res. **7**, 415 (1954).

46. — and J. Rhodin: Exp. Cell Res. **4**, 426 (1953).

47. Smith, C. A.: Ann. Otol. Rhinol. Laryngol. **65**, 450 (1956).

48. — Ann. Otol. Rhin. Laryngol. **63**, 521 (1957).

49. Steinmann, E., and F. S. Sjöstrand: Exp. Cell Res., **8**, 15 (1955).

50. Vogel, A.: Verh. dtsch. Ges. Path. 41, 285 (1957) und in diesem Band S. 286.

51. Watson, M. L.: J. biophys. biochem. Cytol. **3**, 1017 (1957).

52. Weiss, P., and W. Ferris: Exp. Cell Res. **6**, 546 (1954).

53. — — J. biophys. biochem. Cytol. **2**, (suppl.) 275 (1956).

54. Zetterquist, H.: The ultrastructural organization of the columnar absorbing cells of the mouse jejunum. Dept. of Anat., Karolinska Institutet, Stockholm, 1956.

Diskussionsbemerkungen

zu den Vorträgen

Alan J. Hodge, J. D. Robertson und Daniel C. Pease

K. Kühn

Eduard-Zintl-Institut der Techn. Hochschule, Darmstadt

Zu den Ausführungen von Dr. Hodge möchte ich über den Einfluß der Aminosäurezusammensetzung und der Aminosäureanordnung auf das Verhalten und die Struktur der Faserproteine das Folgende bemerken: *Seide* verfügt über einen sehr hohen Gehalt an leichten und unpolaren Aminosäuren. Jede zweite Aminosäure ist Glycin, jede vierte Alanin und jede achte Serin. Der Gehalt an polaren Aminosäuren ist klein. Die Folge davon ist eine hochgeordnete Fibrille, bei der die einzelnen Peptidketten durch Wasserstoffbrückenbindungen von Peptidgruppe zu Peptidgruppe fest zusammengehalten werden. Die Fibrillen quellen nicht in Salzlösungen oder Säuren. *Myosin* oder *Fibrin* haben dagegen einen hohen Gehalt an polaren Aminosäuren (45,6 bzw. 38,2 Mol.-%). Die sperrigen Seitenketten der polaren Aminosäuren bedingen eine lockere Struktur. Die Peptidketten werden hauptsächlich durch salzartige Bindungen zwischen den positiven Karboxylationen und den negativen Amoniumionen der Seitenketten zusammengehalten. Bei Zugabe von Salzlösungen bildet sich eine diffuse Doppelschicht von Gegenionen zwischen den Karboxylat- und Amoniumionen aus, der Zusammenhalt zwischen den Peptidketten wird geschwächt, die Fibrillen beginnen zu quellen und gehen in Lösung. Eine Zwischenstellung nimmt

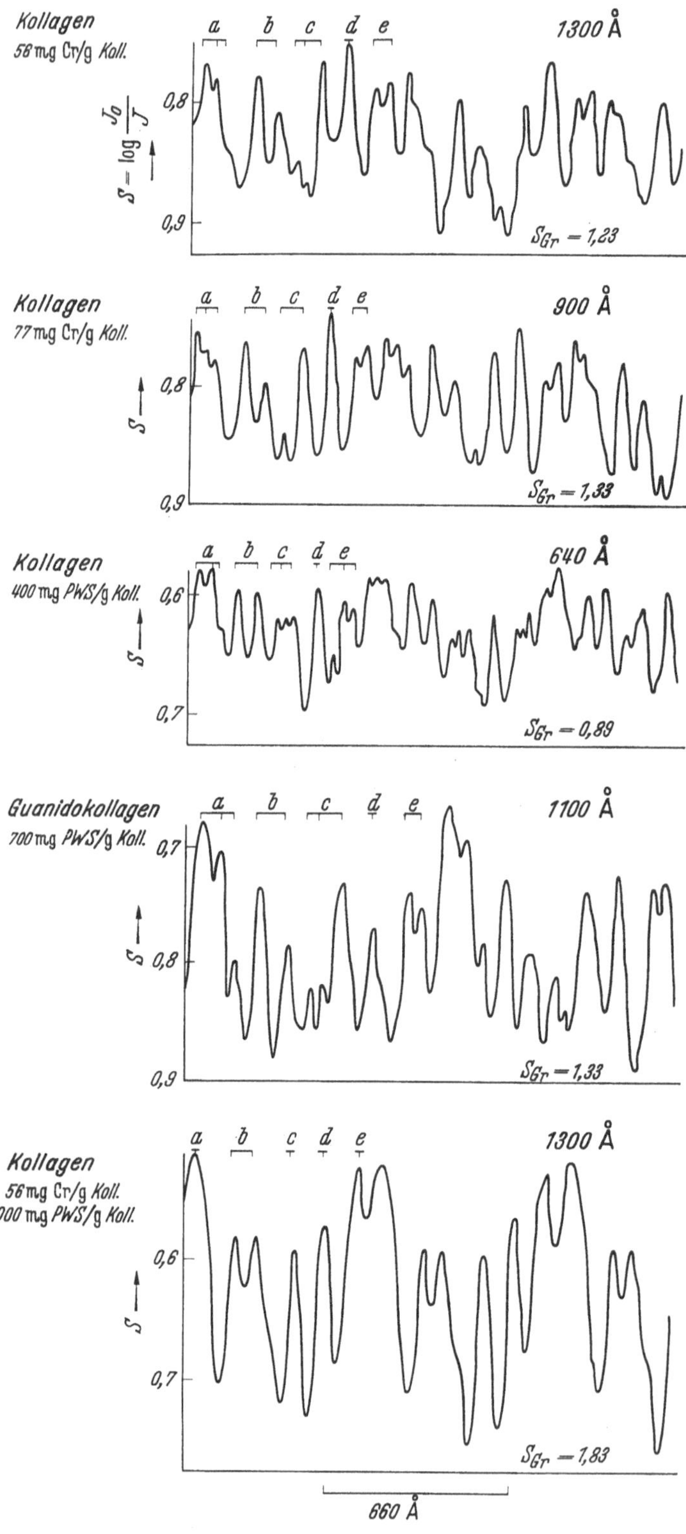

Abb. 1. Photometerkurven von Kollagenfibrillen aus Achillessehne (Rind) mit verschiedenen Chrom- und PWS-Gehalten. S = Schwärzung der Platte, linear aufgetragen, je kleiner S, um so größer die Massendicke der Fibrille an dieser Stelle. Bei jeder Fibrille oben rechts die Breite der Fibrille in Å, unten rechts S_{Gr} = Grundschwärzung neben der Fibrille

das lösliche *Kollagen* ein. Es enthält einen hohen Prozentsatz an leichten Aminosäuren, jede dritte Aminosäure ist Glycin, und einen verhältnismäßig hohen Gehalt an polaren Aminosäuren (20,9 Mol.-%). Daraus resultiert, neben der komplizierten Struktur, die gegenüber den Muskelproteinen geringere Löslichkeit. Die Ausbildung einer diffusen Doppelschicht reicht nicht mehr aus, die Kollagenfibrillen in Lösung zu bringen. Behandelt man dagegen mit verdünnten Säuren, so wird die Dissoziation der Karboxylgruppen zurückgedrängt, und die negativen Ladungen der Tropokollagenmoleküle (TC-Moleküle) gehen verloren. Durch die elektrostatische Abstoßung der nur noch positiv geladenen Teilchen fängt die Fibrille an zu quellen und geht in Lösung.

Am Beispiel des Kollagens soll weiter gezeigt werden, daß neben dem Gehalt an polaren Aminosäuren vor allem aber ihre Anordnung längs der Peptidketten für die Struktur der Fasern von besonderer Wichtigkeit ist. Nach BEAR (*1*) enthalten im Kollagen die dunklen Querstreifen (bands) vorwiegend schwere und polare Aminosäuren, die hellen Querstreifen (interbands) vorwiegend leichte Aminosäuren. Diese Annahme wird durch chemische Untersuchungen über die Aminosäureanordnung des Kollagens gestützt (*2*). Wir konnten diese Theorie von der elektronenmikroskopischen Seite her sehr wahrscheinlich machen. Bei der Untersuchung über die Bindung von Phosphorwolframsäure (PWS) (*3*) und von basischen Chrom III-Salzkomplexen (*4*) an Kollagen, beides Reagentien, mit denen man besonders gut eine hochunterteilte Querstreifung erzielen kann, stellte sich im Gegensatz zu den bisherigen Annahmen heraus, daß der Kontrasterhöhung durch die Masse dieser Anfärbemittel bei der Einlagerung in die Kollagenfibrillen keine entscheidende Rolle für die Ausbildung der hochunterteilten

Querstreifung zukommt. Wenn auch bei der PWS-Anfärbung mindestens 400 mg PWS/g Kollagen eingelagert werden müssen, um eine unterteilte Querstreifung zu erzielen, so genügt bei Behandlung mit basischen Chrom III-Salzlösungen schon eine Einlagerung von 5 mg Chrom/g Kollagen, um eine ebenso scharfe, wenn auch nicht so kontrastreiche Querstreifung zu verursachen. Der für die Ausbildung einer Querstreifung verantwortliche Teil der PWS wird durch die stark basischen Guanidogruppen des Arginins gebunden, wobei ein PWS-Ion stöchiometrisch mit drei Aminosäureresten reagiert. Bei den Chromkomplexen wird der wirksame Anteil durch die Karboxylgruppen der Asparaginsäure und der Glutaminsäure gebunden. Wir glauben, daß die entscheidende Wirkung der PWS sowie der Chrom III-Komplexe auf das Kollagen ein Ordnen der Peptidketten ist. Dies geschieht bei der PWS so, daß ein dreibasisches PWS-Ion drei Guanidogruppen aus drei benachbarten Dreikettenspiralen bindet und sie in einer Ebene fixiert. Dadurch kommen die Anhäufungen der polaren und schweren Aminosäuren genau nebeneinander zu liegen. Die Chrom III-Komplexe bewirken dasselbe über die Karboxylgruppen, die koordinativ in die Komplexe mit einbezogen werden.

Die Unterschiede in den durch PWS und Chrom hervorgerufenen Querstreifungen, verursacht durch die verschiedenen Masseeinlagerungen, konnten wir durch photometrische Ausmessungen von Fibrillen mit verschiedenen PWS- und Chromgehalten fassen (5,4) (Abb. 1). Bei den mit PWS behandelten Fibrillen zeigen die a-Streifen die größte Intensität. Steigert man die PWS-Aufnahme durch Guanidieren der ε-Aminogruppen des Lyins und Oxylysins oder indem man zuerst mit Chromsalzlösungen und anschließend mit PWS behandelt, so nimmt die Intensität der a-Streifen weiter zu. Bei den mit Chrom behandelten Fibrillen heben sich die a-Streifen nicht besonders über die benachbarten Streifen hinaus. Am dunkelsten ist hier der d-Streifen. Diese Untersuchungen deuten darauf hin, daß die PWS bevorzugt in den a-Streifen abgelagert wird und daß somit die basischen Aminosäuren in dieser Gegend besonders stark angereichert sind. Auf diese Weise wurde zum ersten Male experimentell der Nachweis einer Anreicherung bestimmter Aminosäuren an bestimmter Stelle der Periode geführt.

Für das in Lösung befindliche TC-Molekül bedeuten unsere Untersuchungen folgendes: Auf Grund der Aminosäureanordnung im Kollagen sind die negativen und positiven Ladungen nicht statistisch über die Oberfläche des TC-Moleküls verteilt, sondern an bestimmten Stellen nach einem besonderen Muster angereichert und angeordnet. Diese Anordnung zwingt die TC-Moleküle, mit Hilfe elektrostatischer Anziehungskräfte sich in ganz bestimmter Weise, wie Dr. Hodge zeigte, um ein Viertel ihrer Länge gegeneinander versetzt, so anzuordnen, daß die Anreicherungen der polaren Aminosäuren nebeneinander zu liegen kommen. Wie Dr. Hodge zeigte, kann durch bestimmte Kunstgriffe, wie Einbringen von negativen Ladungen in das in Säure gelöste TC-Molekül mit Hilfe der Polysäure ATP die ursprüngliche Ladungsverteilung so verändert werden, daß die Bildung anderer Kollagenformen möglich wird.

Allgemein, nicht nur für Kollagen, wird man sagen können: die Struktur der Faserproteine wird wesentlich von der Aminosäureanordnung in den Peptidketten bestimmt. Es ist nicht möglich, daß sich bei einer statistischen Aminosäureverteilung immer wieder mit absoluter Sicherheit dieselben Faserstrukturen ausbilden.

Literatur

1. Bear, R. S.: Advanc. Protein Chem. 7, 69 (1952).
2. Schroeder, W. A., u. a.: J. Amer. chem. Soc. 76, 3556 (1954).
   Kronner, Th. D., u. a.: J. Amer. chem. Soc. 74, 4084 (1953); 77, 3356 (1956).
   Grassmann, W., u. a.: Hoppe-Seylers Z. physiol. Chem. 306, 123 (1956).
3. Kühn, K., W. Grassmann u. U. Hofmann: Z. Naturforsch. 13b, 154 (1958).
4. — Leder 9, 217 (1958).
5. — U. Hofmann, W. Grassmann u. E. Gebhardt: Naturwissenschaften 45, 521 (1958).

W. Bernhard

Institut de Recherches sur le Cancer, Villejuif/Seine (France)

I admire very much the work you have just discussed and illustrated with magnificent electron micrographs. The method chosen for the study of "Principles of Order" in biology is, to-day, probably the only possible experimental approach. Crystal-like patterns are relatively easy to analyse morphologically. In addition, any periodicity, any geometrically arranged structure, discovered in biological material is aesthetically more satisfactory than "lawless" cell architecture. However, we should not forget that irregular biological structure and random distribution of cell organelles also follow principles of order which are, by far, more complex, more subtle, and therefore, more fragile and vulnerable. Too great a geometrical order in living systems is an expression of a low degree of organization. It is something like the memory of life for its inorganic precursors. Too great an order does not tolerate compromises either in human society or in molecular populations. A living cell is alive because it tolerates compromises around a median line of behaviour. But as function and structure are intimately linked, this means asymmetry, slight or even considerable disorder at the structural level. This disorder is only accepted to a certain degree. Beyond this, a regulatory principle pushes everything back to the normal average. Let us look at a normal liver cell. At first sight, there is nothing exciting about it, no periodicity, no crystal in its cytoplasm. Its mitochondria are distributed at random and have a variable size, oscillating around a median value. However, in the liver cancer cell, the ordered principle is lost: there may be too many mitochondria, or they may, on the contrary, be too few. They may reach a size ten times larger than normal, or be very tiny with abnormal density. Therefore, order and disorder are very relative in Biology, just as they are relative in human behaviour.

The charwoman, cleaning Pasteur's laboratory every morning, was probably horrified seeing the apparent disorder of her director's working room. But without any doubt, the beakers, bottles and Petri dishes were in the place where the brain and the reflexes of the genius wanted them. Something comparable to this exists in a living cell. Behind its apparent lack of highly ordered structure are hidden ordered principles which will be difficult to study, but which seem to be essential for highly organized living systems. Therefore, may I praise asymmetry and sing a hymn to (apparent) disorder.

D. Membranen und Membranmodelle

A molecular theory of cell membrane structure

J. David Robertson
University College, London

Electron microscopy of peripherical nerve fibres in recent years has resulted in the establishment of the general patterns of organisation of both unmyelinated and myelinated nerve fibres (Gasser 1952, 1954, 1956; Geren 1954, 1956; Robertson 1955, 1957, 1958, 1959). From the relationships between the satellite Schwann cells and axons in myelinated fibres, it is possible to make certain deductions about the molecular structure of Schwann cell membranes. These depend on a correlation of electron microscope findings with polarization optical and X-ray diffraction evidence. It is the purpose of this short communication to review the findings related to this topic and to present certain evidence suggesting that the pattern of structure characterizing Schwann cell membranes is similar for many, perhaps all cell membranes and membranous cell organelles. Certain studies of lipid model systems are presented since they aid in the interpretation of electron micrographs of membranes.

The appearance of Schwann cell surface membranes after $KMnO_4$ fixation may be made clear at the outset. These are consistently seen as in the through-focus series in Fig. 1a—c. In micrographs near to exact focus, (Fig. 1b), the membranes measure about 75 Å across and consist of a pair of dense lines $\sim$ 20 Å thick bounding a light core $\sim$ 35 Å thick. In underfocused images the membrane appears thicker (Fig. 1a) and in overfocused micrographs a spurious image is obtained (Fig. 1c). This $\sim$ 75 Å structure is called a "unit" membrane because it is the essential unit of more complicated paired membrane structures such as the mesaxons of unmyelinated nerve fibres.

The same type of unit membrane structure seen in Fig. 1 is characteristic of all cell membranes and membranous cell organelles so far examined after $KMnO_4$ fixation. OsO_4 fixation occasionally shows up the same structure and it appears highly probable that the $\sim$ 75 Å unit membrane structure is of general occurrence in animals of all phyla and in plant cells. This follows from a survey of a number of different tissues reported by the author (Robertson 1959a and b) and is supported by many observations of others. Only two examples will be given here because of space limitation. Fig. 2 shows the paired membrane structure between two smooth muscle cells in mouse intestine (note the $\sim$ 100 Å gap between the two $\sim$ 75 Å unit membranes). Fig. 3 shows the unit membranes of the endoplasmic reticulum of a Schwann cell. The same units are seen in mitochondria and other membranous cell organelles, including so-called "synaptic vesicles".

The two unit membranes of the mesaxons of adult nerve fibres run roughly parallel to one another and often are separated by a gap $\sim$ 150 Å across, which connects with the outside. The gap appears to contain a finely granular or delicately fibrillar material, which is continuous with basement membrane substance and with intercellular ground substance (see Fig. 5d). Basement membranes appear to represent a concentration of this intercellular material. The relationships of the unit membranes in nerve fibres are significant in that they give important clues about molecular structure.

Vertebrate peripheral nerve fibres are all composed of axons enveloped to a greater or lesser degree by a satellite Schwann cell. This generalisation applies to both unmyelinated and myelinated nerve fibres. Both these apparently greatly dissimilar types of nerve fibre appear to be derived developmentally from similar precursors. The stages in the evolution of both types of fibres may be seen during the first two weeks of development after birth in mouse sciatic nerves. The development of unmyelinated fibres will be considered first.

In developing mouse sciatic nerve two major groups of fibres may be distinguished. One type represents future unmyelinated nerves, and these can be roughly divided into three classes. The

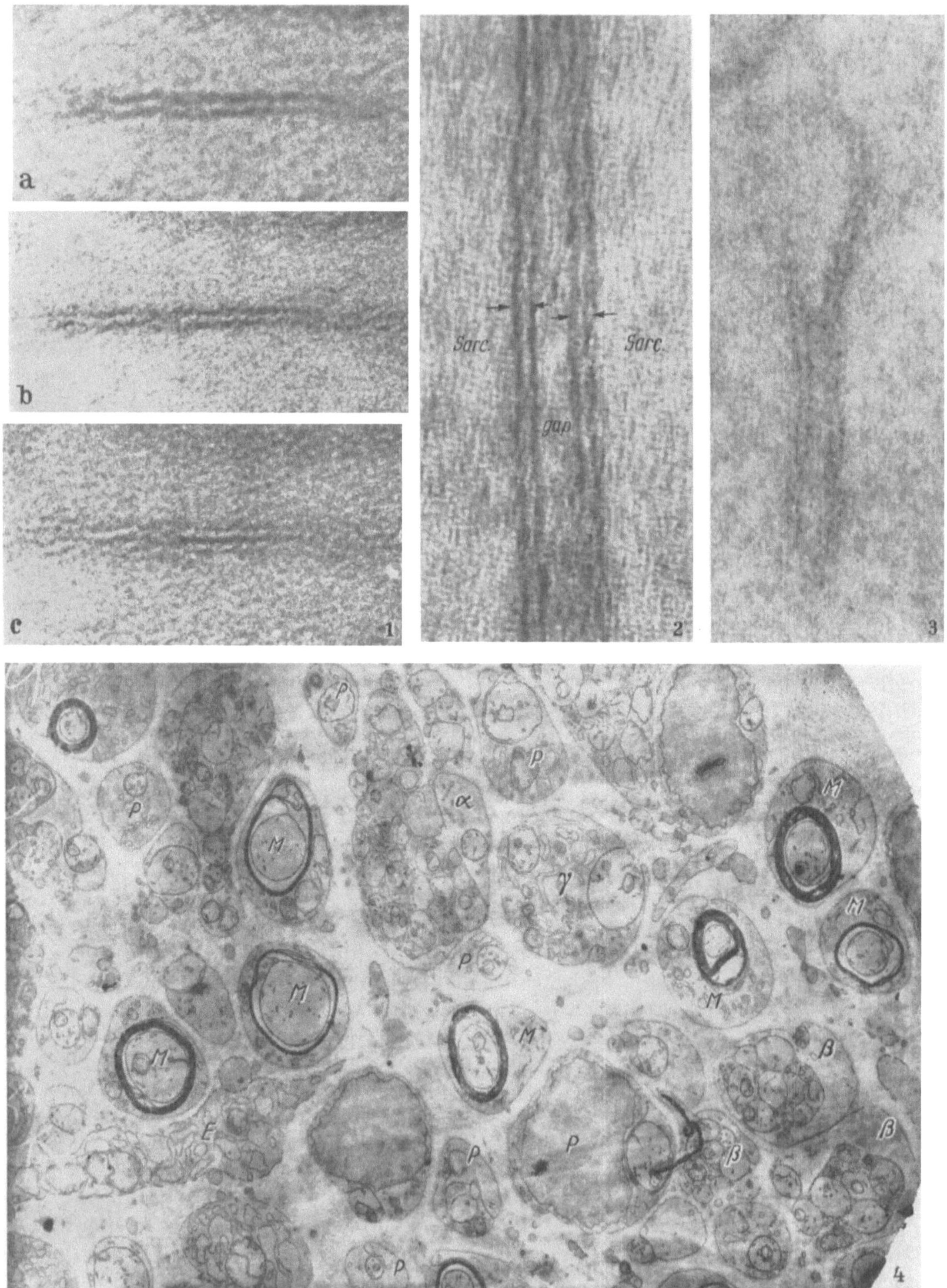

Fig. 1, 2, 3 and 4 (see page 161)

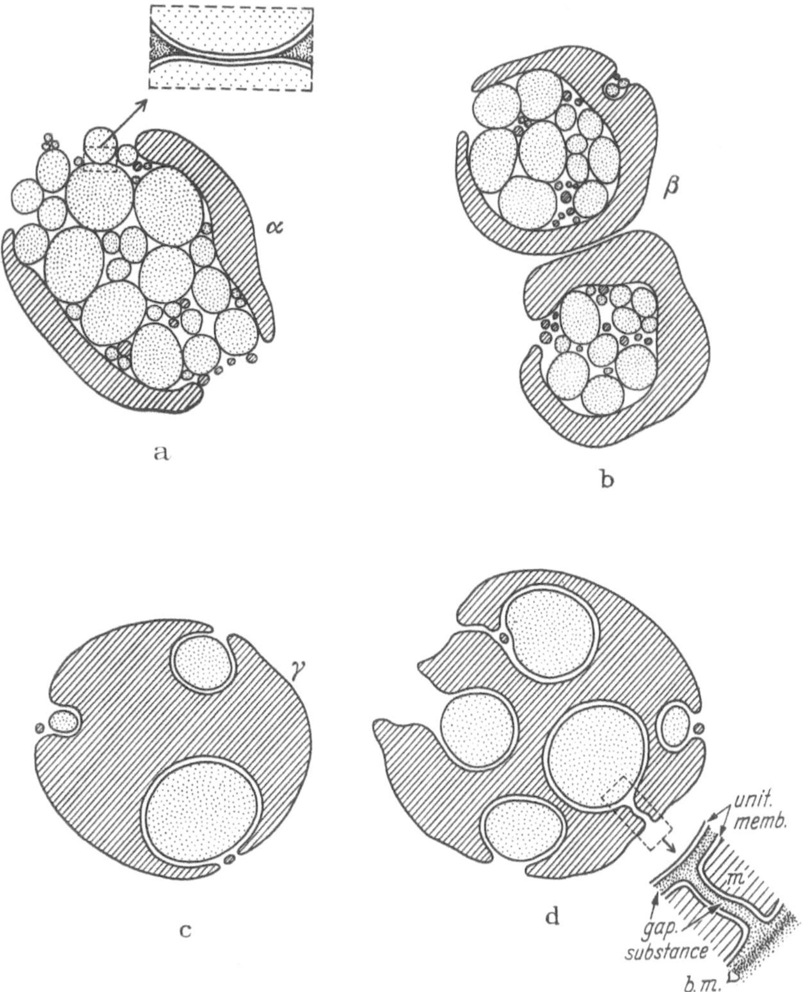

Fig. 5 a—d. Diagrams illustrating three types (α, β and γ in a, b and c resp.) of developing unmyelinated nerve fibres seen in young mouse sciatic nerves and a typical adult unmyelinated fibre (d). The unit membranes of the α and β fibres are in contact in many places making external compound membranes as shown in the enlargement in a. Axoplasm is lightly stippled and Schwann cytoplasm cross hatched. Gap substance is heavily stippled in the enlarged areas. α, β and γ correspond to the fibres so marked in Fig. 4. The rectangular area including the mesaxon m is enlarged to the lower right to show the gap substance between the ∼75 Å unit membranes and its continuity with basement membrane substance (b. m.)

Fig. 1 a—c. Segment of a Schwann cell surface membrane in frog peripheral nerve in three states of focus. a is a slightly under focus, b is nearly in focus, c is over-focused. The membrane measures ∼75 Å across. Its apparent thickness is increased in a. In c a completely spurious image is obtained. Mag. 450,000 ×

Fig. 2. Section of a portion of the boundary between two smooth muscle cells in mouse intestine. Sarcoplasm (Sarc.) lies to the right and left bounded by the membranes of each muscle cell indicated by the alined arrows. A gap of about 100 Å width is present between the two cell membranes. Both these cell membranes and the gap between them together would have been referred to as a "double membrane" until recently. This entity is now referred to as a "paired membrane structure". Mag. 450,000 ×

Fig. 3. A cysterna of the endoplasmic reticulum from a Schwann Cell showing the ∼75 Å unit membranes. Mag. 320,000 ×

Fig. 4. Section of developing mouse sciatic nerve showing particularly unmyelinated fibres but also several stages in the development of myelinated fibres. A few protofibres (P) and myelinating fibres (M) are seen. Three different kinds of developing unmyelinated fibres may be distinguished. One type (α) can be distinguished in which fairly large groups of closely aggregated axons and small round profiles are seen between areas of apparent Schwann cell sheets which are relatively thin and do not completely enclose the axons or round profiles. Another type (β) consists of Schwann cells more often closely associated with other Schwann cells. These contain groups of axons and small round profiles enclosed in one or more separate invaginations of the Schwann cell. A third type (γ) consists of a single isolated Schwann cell in which several axons are enclosed by individual invaginations of the surface. This type resembles mature myelinated fibres and may represent the latest stage of development seen here. Endoneurial cells (E) with abundant endoplasmic reticulum are also seen. Mag. 8,000 ×

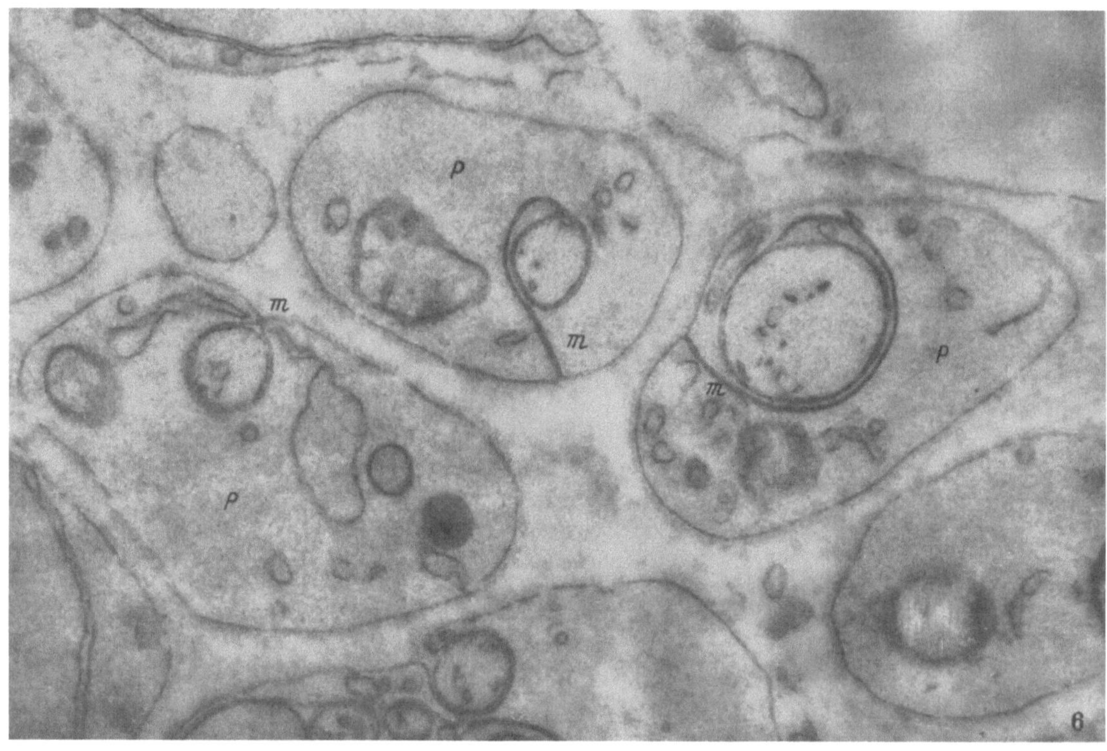

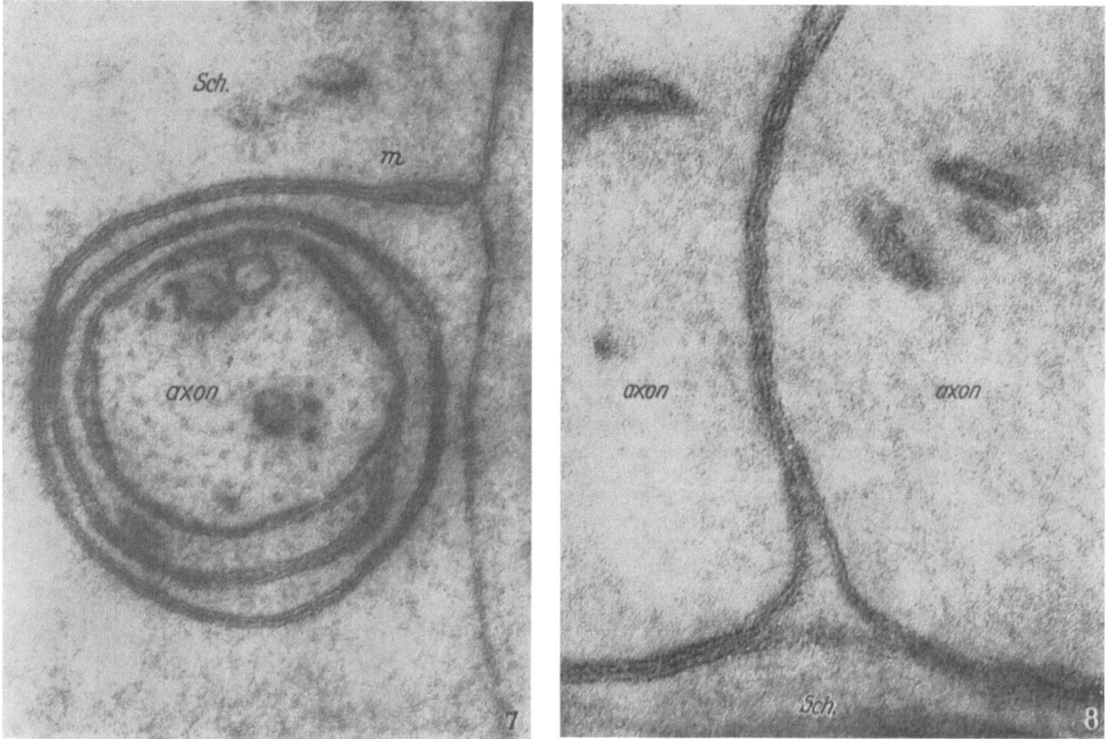

Fig. 6. Young mouse sciatic nerve showing three protofibres (*P*) with mesaxons (*m*) of varying length. Mag. 29,500 ×

Fig. 7. High magnification view of an intermediate myelinating fibre with a mesaxon (*m*) elongated into almost two complete loops about the axon. Note the strata of the surface membrane and mesaxon of the Schwann cell (*Sch.*). Mag. 125,000 ×

Fig. 8. Portion of a cross section of a developing unmyelinated fibre in mouse sciatic nerve showing two axons in contact with one another and with a Schwann cell (*Sch.*) below. Note the closure of the gaps between the unit membrane pairs each set of which here run perpendicular to one another in the section. Mag. 200,000 ×

first class (Fig. 5a and Fig. 4α) shows aggregated groups of axons associated with several Schwann cytoplasmic areas. The axons are only partially enveloped by thin sheets of one more satellite cells, which often do not meet to form mesaxons. Another type (Fig. 5b and Fig. 4β), appears in cross sections as groups of several axons in closely packed bundles. These whole aggregates of axons are completely surrounded by one satellite Schwann cell, whose enveloping lips may meet around the bundle to form a single mesaxon. Several such fibres are sometimes closely associated with their adjacent Schwann cells in apposition. These fibres resemble those in adult olfactory nerves (GASSER, 1956). Still another class may be distinguished (Fig. 5c and Fig. 4γ). Here individual axons lie in separate trough-like invaginations of the surface of a single Schwann cell.

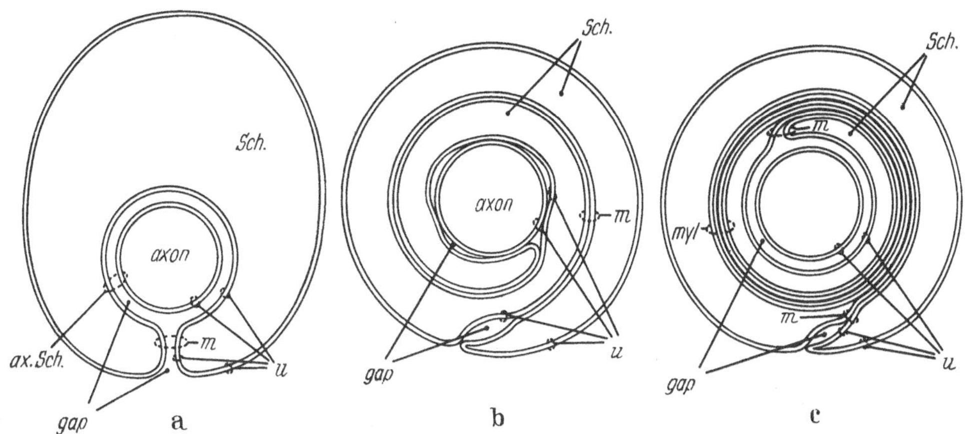

Fig. 9a—c. Diagrams illustrating the evolution of nerve myelin (myl.). a shows the earliest stage. Here a small axon is enveloped by a relatively large Schwann cell (Sch.) to make a protofibre. The combination of the unit membranes (u.) to make a mesaxon (m) and the axon-Schwann membrane (ax-Sch.) are shown. At the earliest stages the gap between the unit membranes is sometimes present. b shows an intermediate fibre. Here the unit membranes of the mesaxon and to some extent of the axon have come together with a closure of most of the gap. Their line of contact is the future "intermediate line" of myelin. c shows a later stage in which a few layers of compact myelin have formed by contact of the cytoplasmic surfaces of the mesaxon loops to make the major dense lines of myelin

In such cases each axon, if completely enveloped, has only one mesaxon, providing it with access to the exterior. This latter type resembles adult unmyelinated fibres and may represent a later stage of development. The general pattern of structure of adult unmyelinated fibres is shown in Fig. 5d.

Classes α and β of developing unmyelinated fibres characteristically show two peculiar features. Interspersed between the aggregated axons there are seen small (~ 500—1,000 Å) round or oval profiles, which often appear more dense in $KMnO_4$ fixed preparations. Some of these have been identified as finger-like projections of the Schwann cells but some might represent growing axon tips. The other peculiar feature is the frequent absence of gaps between the membranes of the axons and Schwann cells. The two ~ 75 Å unit membranes are often in contact, making external compound membranes as in Fig. 8. These peculiar features are not often seen in the γ class fibres or in adult fibres.

Whatever may determine the future course of development, myelinating fibres characteristically appear in newborn mouse sciatic nerves as single axons enveloped by a single Schwann cell (HESS, 1955). The Schwann cells are not aggregated in large groups such as may be seen in developing unmyelinated fibres. The mesaxon is progressively elongated as myelin develops. The stages in the formation of myelin are illustrated in Fig. 9. It is formed simply by a spiral elongation of the mesaxon around the axon many times and the aggregation of the membrane lopps into a compact membranous structure. Stages in this developmental process have been arbitrarily designated as follows: the earliest fibres in which the mesaxon forms less than one complete loop (Fig. 9a and Fig. 4p and Fig. 6) are called protofibres. The next stage in which more than one mesaxon loop appears but in which each loop is separated from its neighbour by a thin layer of

Schwann cytoplasm are called intermediate fibres (Fig. 9b and Fig. 7). The fibres are called myelinated after the mesaxon loops have come together with the obliteration of the thin layer of Schwann cytoplasm in between them. A young myelinated fibre showing the two ends of the mesaxon is shown in Fig. 10. As development proceeds the layer of Schwann cytoplasm next

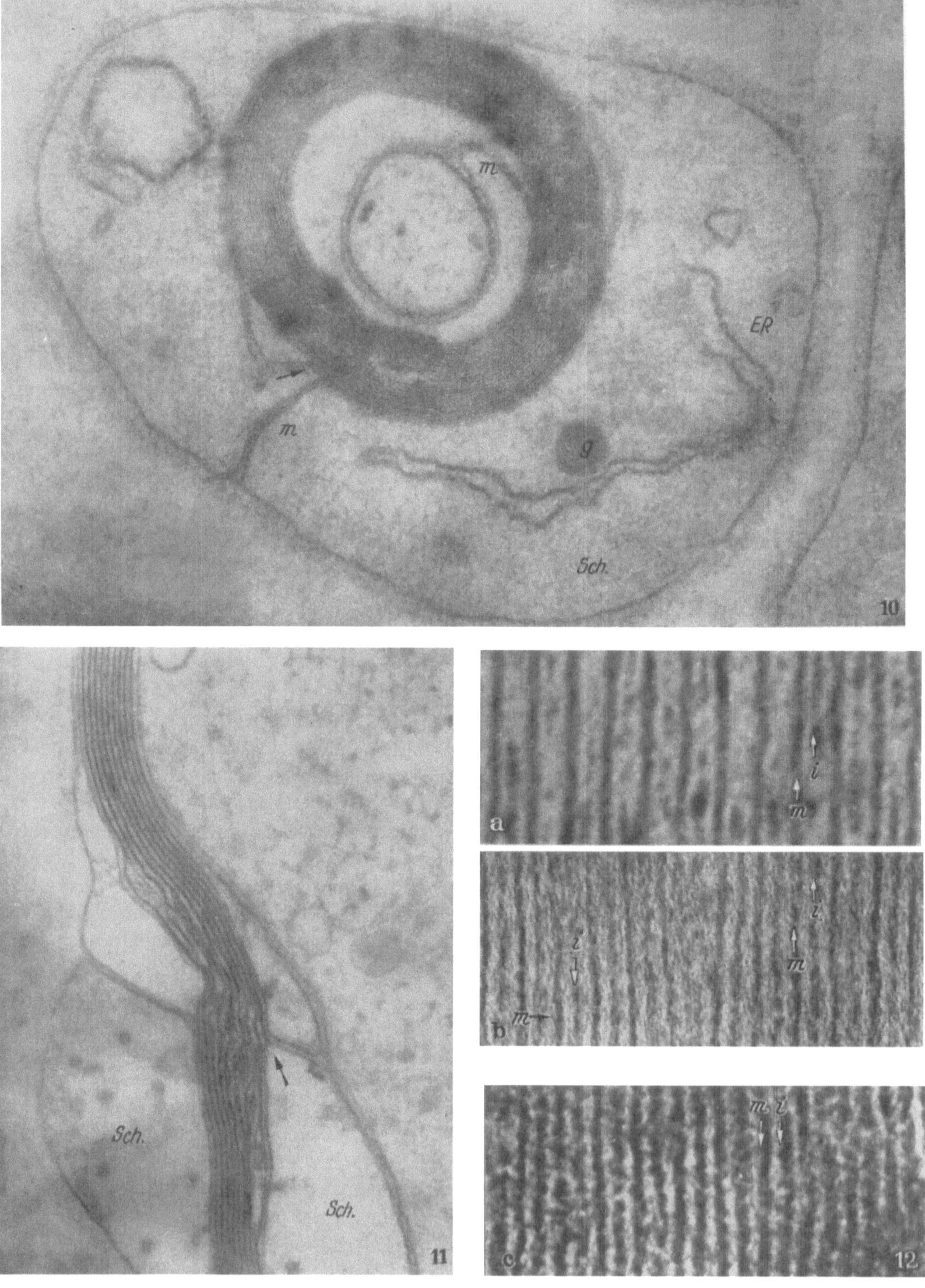

Fig. 10, 11 and 12a—c (see p. 165)

to the axon becomes thinner and the external cytoplasmic layer also is reduced in thickness except in the region of the Schwann cell nucleus, which incidentally always remains external to the compact myelin structure in vertebrates. No discontinuities in the paired membrane structure constituting the mesaxon have been observed during development. Adult nerve fibres consistently show both external and internal mesaxons appropriately oriented to suggest the maintenance of a simple spiral configuration throughout all stages of maturation. Both outer and inner mesaxons are shown in Fig. 11 in an adult frog myelinated fibre after OsO_4 fixation.

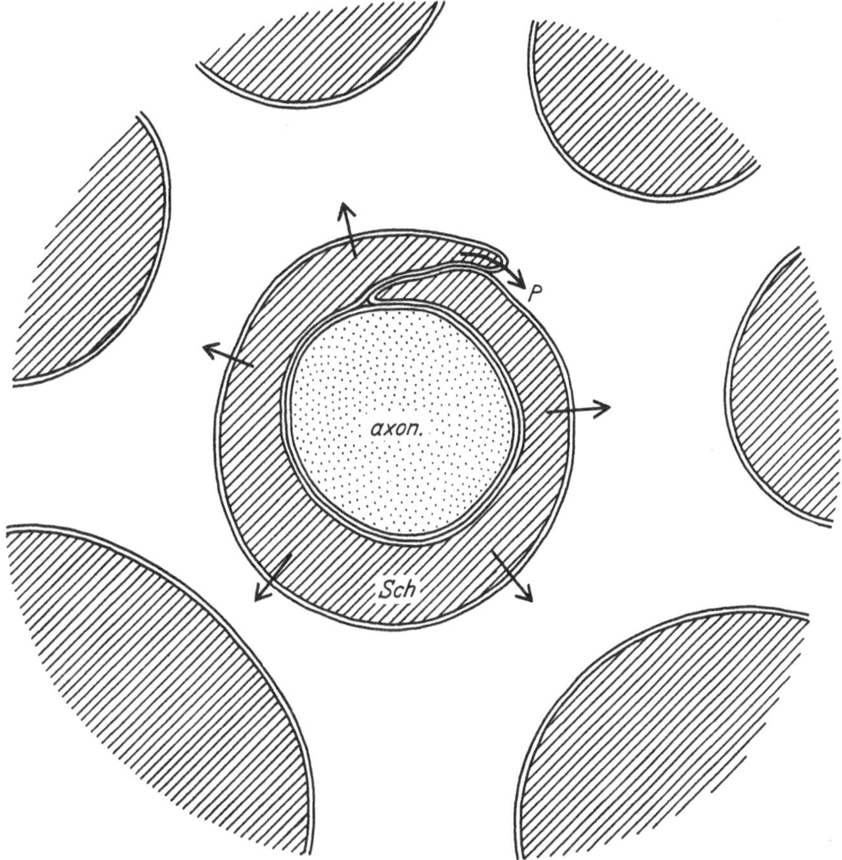

Fig. 13. Diagram illustrating a possible mechanism of myelination. The Schwann cell (*Sch.*) is presumed to be held to the axon by a chemotactic mechanism. Contact of the ~ 75 Å unit membranes of the Schwann cell either with itself or the axon is assumed to bring about "contact inhibition". Schwann cells are "migratory" cells moving by pseudipodial extension. Any pseudopodium extended in the direction of the unlabeled radial arrows would soon be immobilized by contact inhibition from the neighbouring cells. The only pseudopodium having a high probability of indefinite extension would be *P*. Its extension, drawing the nucleus behind it, would result in myelination

---

Fig. 10. Young myelinating fibre from a developing mouse sciatic nerve is which thirteen layers of myelin have formed. Note the outer and inner mesaxons (*m*). The junction of the mesaxon with myelin at the arrow shows the origin of the major dense line by apposition of the cytoplasmic surfaces of the mesaxon loops. The relatively abundant Schwann cytoplasm (*Sch.*) shows components of the endoplasmic reticulum (*ER*) and a granule (*g*) of unknown significance. Mag. 30,000 ×

Fig. 11. Segment of OsO_4 fixed myelin showing an outer and inner mesaxon (*m*) and fairly thick layers of outer and inner Schwann cytoplasm (*Sch.*). Compare the region of junction designated by the arrow with that in Fig. 10. Note that the same relations exist here after OsO_4 fixation. Mag. 86,000 ×

Fig. 12a—c. In a the appearance of myelin after fixation with OsO_4 is shown. Note the irregularity of the intra-period line (*i*) as compared with the major dense line (*m*). After fixation in potassium permanganate for about three hours a similar structure is seen but the intraperiod lines are more regular and prominent as indicated in b. If fixation in permanganate is prolonged to six or more hours the intraperiod lines and major dense lines become indistinguishable as indicated in c. Mag. 550,000 ×

It may be significant that the Schwann cell unit membranes of the mesaxon are in contact making an external compound membrane in developing myelinated fibres (Fig. 7) (ROBERTSON, 1958). Also the axon and Schwann cell membranes are in similar intimate contact. It is possible that such intimate membrane contacts signify the operation of the process of contact inhibition (ABERCROMBIE and HEAYSMAN 1954; WEISS, 1958). If this were so only the free edges of the Schwann cell would be capable of pseudopodial extension. If these were extended radially contact inhibition between adjacent Schwann cells would soon occur. The only pseudopodium having a chance of indefinite extension would be the one designated P in Fig. 13. The indefinite circumferential extension of this pseudopodial lip would result in myelination. It must be supposed that the Schwann cell nucleus, follows this lip as it elongates. The tissue culture findings of PETERSON, CRAIN and MURRAY (1958) are consistent with this suggestion.

A comparison of OsO_4 fixed and $KMnO_4$ fixed myelinated fibres is revealing (Fig. 12). In both cases a regular radial repeat period of -120 Å is seen in the myelin. The repeating unit in OsO_4 fixed material is a major dense line (m, Fig. 12a) ~ 30 Å thick together with a less dense zone ~ 90 Å thick bisected by a discontinuous thin dense line or row of granules referred to as the

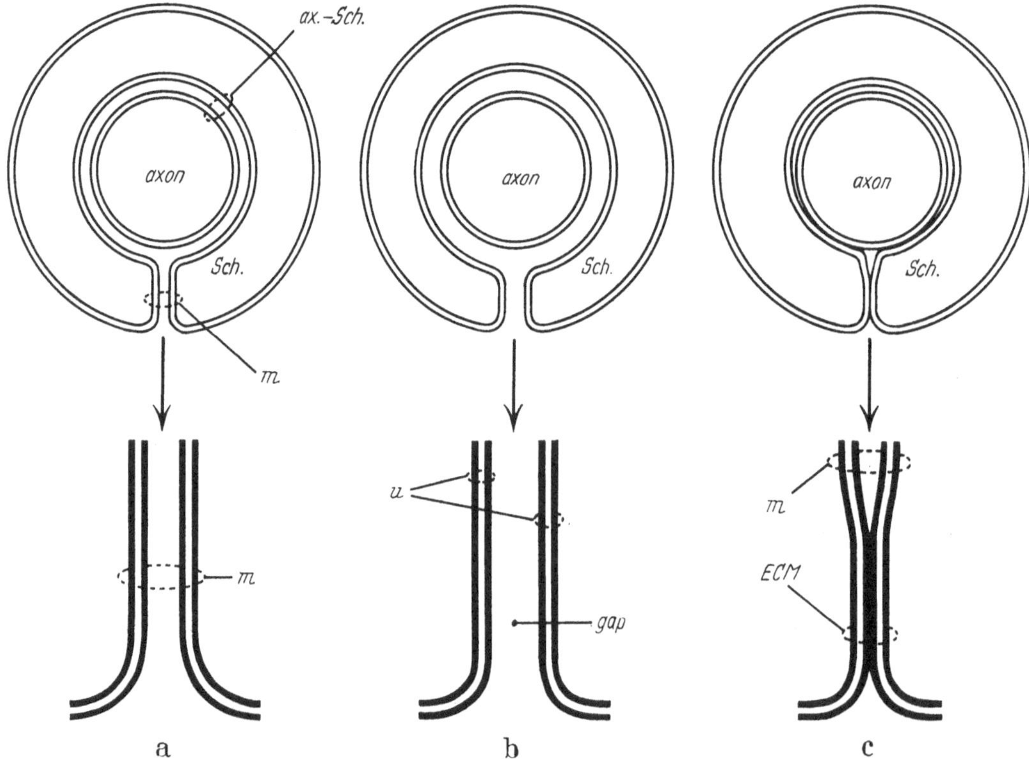

Fig. 14a—c. Diagrams indicating the effects on unmyelinated nerve fibres of immersion in hypotonic (b) or hypertonic media (c) before fixation. In (a) the appearance of control fibres is indicated. Here a single axon is shown surrounded by a single Schwann cell (*Sch.*). Note the gap between the unit membranes (*u*) of the axon-Schwann membrane (*ax.-Sch.*) and the mesaxon (*m*). The mesaxon is enlarged below. After soaking in hypotonic media the gap is widened as indicated in (b) and after soaking in hypertonic media it is closed as indicated in (c). After the gap is closed the paired membrane structure is converted into a combined structure referred to as an external compound membrane (*ECM*)

intraperiod line (i, Fig. 12a). Corresponding zones appear in $KMnO_4$ fixed preparations. The major dense lines are about the same but the less dense zone can be better described as two ~ 30 Å thick light zones separated by a dense line ~ 30Å thick resembling the major dense line except for a slightly lower density. After relatively long treatment with $KMnO_4$ (> 6 hours) the density of the intraperiod line becomes as great as that of the major dense line and the two are indistinguishable (Fig. 12c). The period is then halved in the micrographs (FERNANDEZ-MORAN and FINEAN, 1957; ROBERTSON, 1957a) It is clear from the developmental findings that the major dense line originates by contact of the cytoplasmic surfaces of the mesaxon loops as they come together to form compact myelin. The intraperiod line originates by contact of the apposed outside surfaces of the unit membranes of the mesaxon. Examination of the regions designated by the arrows in Fig. 10 and 11 will make clear the fact that these two repeating dense layers, despite their different characteristics in $KMnO_4$ and OsO_4 fixed material, represent identically the same underlying structures. They result respectively from contact of the inside and outside surfaces of ~ 75 Å unit membranes. In the case of $KMnO_4$ fixed material the outside strata of these unit membranes can be traced right out from compact myelin to the free surface of the Schwann cell along with the parallel inside ~ 20 Å stratum. In OsO_4 fixed material only the cytoplasmic half of the unit membrane structure can be seen at the free surfaces. The outside half is poorly preserved and usually not seen in micrographs. Its reality is evident nevertheless and it is clear that $KMnO_4$ gives a better preservation of the whole ~ 75 Å unit membrane structure. An important corollary of these findings is he obvious fact that despite their apparently symmetrical nature after prolonged $KMnO_4$ fixation, the inside and outside surfaces of the unit membranes must be chemicaly different. This point will be made more clear by certain experiments with hypotonic and hypertonic solutions.

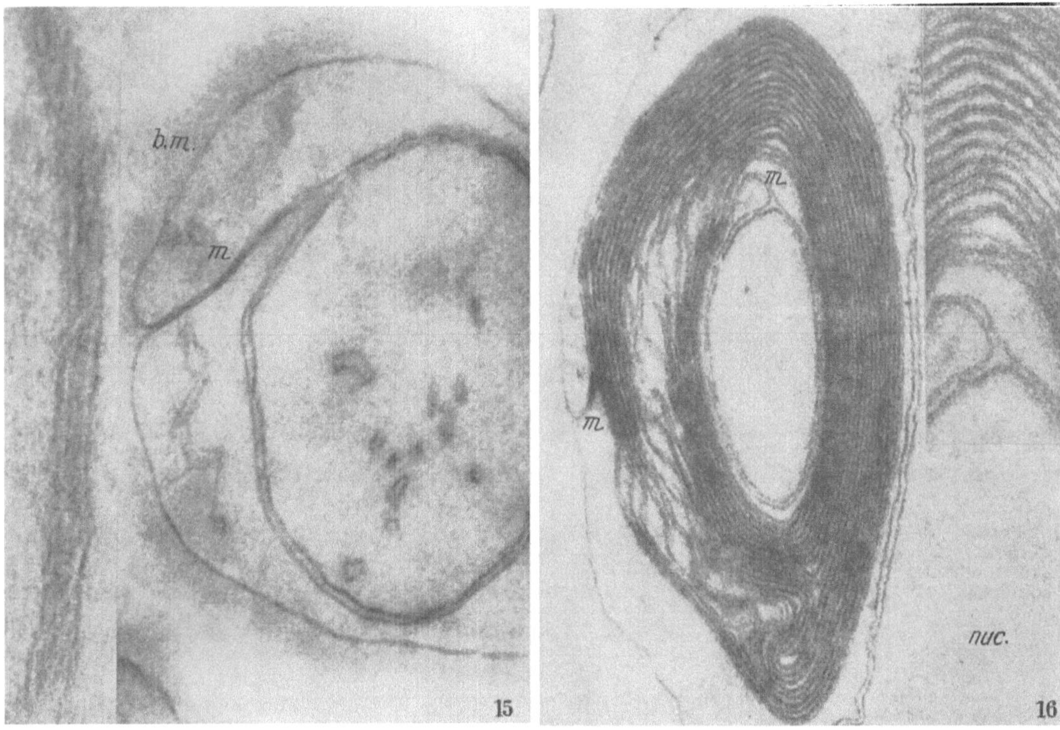

Fig. 15. Portion of an unmyelinated fibre in frog sciatic nerve soaked in hypertonic Ringer solution before fixation with permanganate. Note the closure of the gap between the unit membranes of the mesaxon (m) enlarged to the left. Note the hazy intercellular ground substance in the basement membrane region ($b.\ m.$). This fibre was soaked in Ringer solution made hypertonic by the addition of sucrose. Mag. 63,000×. Inset Mag. 425,000×

Fig. 16. Portion of myelinated fibre soaked in distilled water before fixation with permanganate. The myelin lamellae have separated by a splitting of the intraperiod line as shown in the inset to the upper right. The outer and inner mesaxons (m) are still visible. Part of the Schwann nucleus with the paired nuclear membrane is seen to the right ($nuc.$). Mag. 106,000×. Inset Mag. 450,000×

It is possible to bring about intimate contact of the unit membranes of the mesaxons of adult unmyelinated fibres by soaking them in hypertonic Ringer solutions before fixation (Fig. 14 and Fig. 15). The overall width of the paired membrane structure is reduced from ~ 250—300 Å to ~ 150 Å when this is done. The whole of the reduction seems to be taken up in the gap between the two apposed ~ 75 Å unit membranes which are then separated by no more than 20 Å. The gap is thus reduced in thickness by ~ 90%. The resulting ~ 150 Å thick external compound membrane is like those seen normally in developing nerve fibres (Fig. 8). The gap between such paired membrane structures as normally seen (Fig. 2) contains a substance which appears delicately granular after $KMnO_4$ fixation and is continuous with basement membrane substance and the intercellular ground substance. It is possible to interpret the collapse of this gap substance in hypertonic solutions as a syneresis phenomenon occurring in a highly hydrated polysaccharide gel. Alternatively, or perhaps in addition, the collapse might be due to an alteration of long range forces acting between the adjacent membranes. At any rate the collapse of the gap substance in hypertonic solutions along with the stability of the unit membranes is entirely consistant with the idea that the lipid components of paired membrane structures are all located within the unit membranes. This was originally suggested by the fact that all the strata of the unit membranes of mesaxons can be traced into compact myelin whereas the ~ 100—150 Å inter-membrane gap substance cannot (Robertson, 1957).

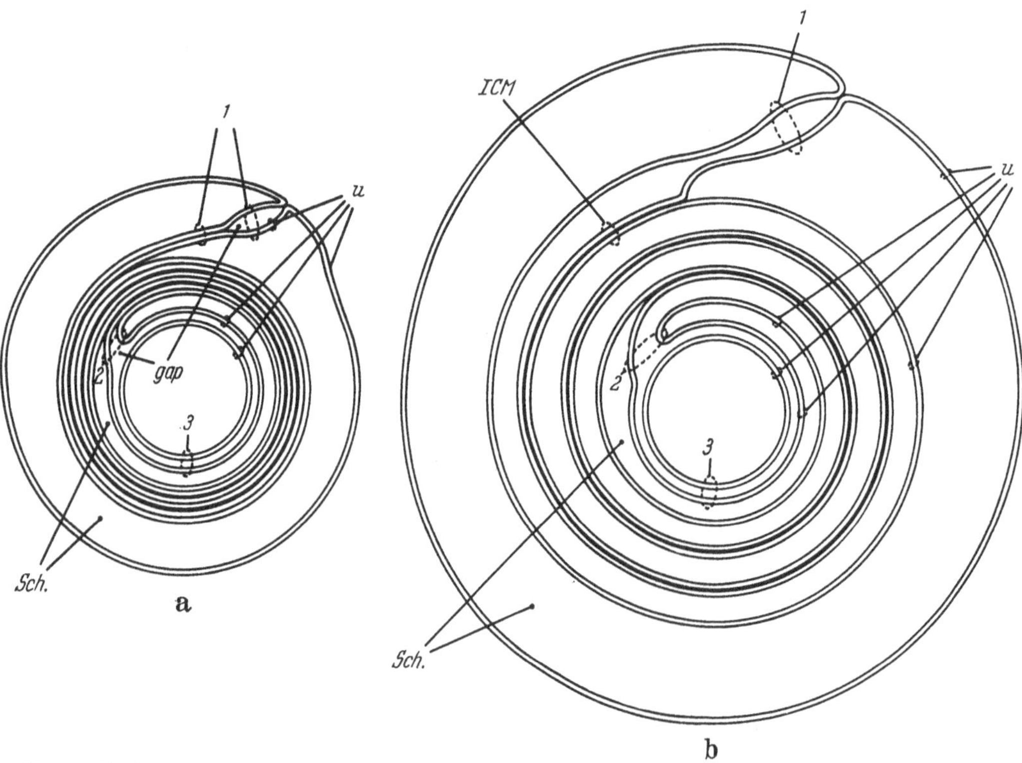

Fig. 17a and b. Diagram indicating the way in which the compact myelin structure indicated in (a) opens by a splitting apart from the outside surfaces of the Schwann cell membrane of the mesaxon after treatment with distilled water. The unit membranes (u) in compact myelin make up the outer (1) and inner (2) mesaxon and the axon-Schwann membrane (3). 100—150 Å gaps are sometimes present in these paired membrane structures and a thin layer of Schwann cytoplasm ($Sch.$) bounds the compact myelin structure. After soaking in distilled water the membranes are all split apart except where they were united along their cytoplasmic surfaces during development to make the major dense line of myelin. The compound membrane structure seen in myelin after this treatment is called an internal compound membrane (ICM)

After soaking nerve fibres in distilled water before fixation, other interesting changes are seen in myelin as well as paired membrane structures. The gaps between the paired membrane structures are widened by this treatment but the unit membranes remain unaltered. Finean and Millington (1956) showed that the myelin period increased after soaking in hypotonic Ringer solution and it has been found by electron microscopy that this increase is due to an opening up of the gap between the unit membranes of the mesaxon throughout compact myelin (Fig. 17 and Fig. 16). The unit membranes united developmentally along their cytoplasmic surfaces do not separate. This is further evidence that the outside surfaces of the unit membranes are different from the inside surfaces. Furthermore, since the unit membrane does not split apart in distilled water, one might suppose that its internal molecular components are bound together by strong bonds and perhaps that the central components are hydrophobic.

X-ray diffraction studies of myelin (SCHMITT, BEAR and CLARK, 1935; SCHMITT, BEAR and PALMER, 1941; FINEAN 1953) have resulted in the detection of a radial repeat unit of ~ 170—180 Å. Polarisation optical (SCHMIDT, 1936) and chemical analyses (see ROBERTSON 1959b) have suggested that this radially repeating unit consists of smetic bimolecular leaflets of mixed lipids interspersed between monomolecular layers of tangentially oriented protein molecules. Combined electron

microscope and X-ray diffraction studies of changes introduced in myelin during preparatory procedures such as those used in the preparation of thin sections have shown that these lead to a reduction in this periodic repeat unit to ~ 140—150 Å in the embedded material (FINEAN, 1953; FERNANDEZ-MORAN and FINEAN, 1957). A certain amount of shrinkage occurs when sections are examined in the electron microscope and it can be concluded that the ~ 120 Å repeat period seen in sections by electron microscopy corresponds to the ~ 170—180 Å period detected in fresh myelin (ROBERTSON, 1956). Furthermore, on symmetry grounds alone it is clear that the mesaxon, consisting of two ~ 75 Å unit membranes in contact along their outside surfaces, is the repeating unit in myelin. Therefore, it is reasonable to superimpose the diagrams of the molecular structure of the myelin repeating unit proposed by X-ray diffractionists on the elctron micrographs of mye-

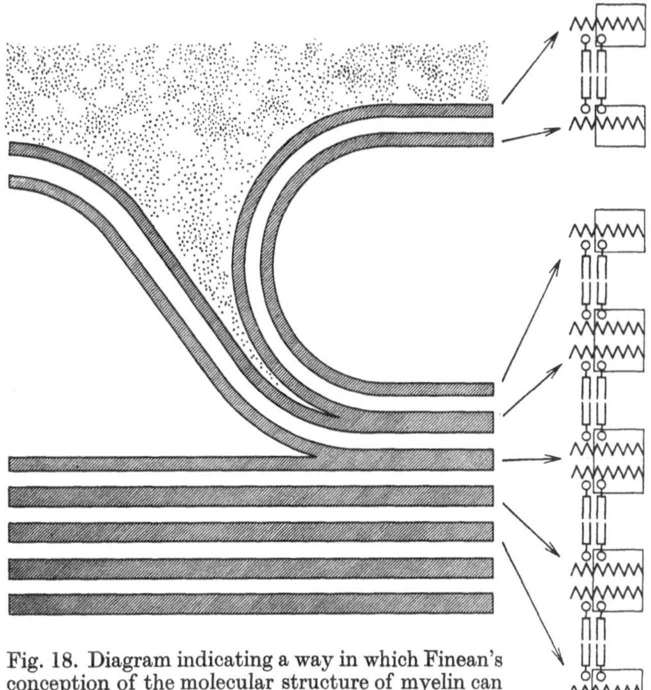

Fig. 18. Diagram indicating a way in which Finean's conception of the molecular structure of myelin can be superimposed on the electron micrographs showing the mesaxon leaving compact myelin and splitting to form the surface membrane of the Schwann cell. Since the cell membrane comprises half of the repeating unit its molecular structure may be inferred as indicated. The stippled area above represents intercellular ground substance (probably mucopolysaccharide)

linated fibres as in Fig. 18. From this it appears probable that ~ 75 Å unit membrane contains a single bimolecular leaflet of lipid with its two polar surfaces directed outwards and each covered by a monolayer of nonlipid material. This comparison of X-ray diffraction and electron microscope evidence leads to a model of cell membrane structure corresponding roughly to that proposed by DANIELLI and DAVSON from other lines of evidence (DANIELLI and DAVSON, 1935; DAVSON and DANIELLI, 1942). The conclusion about molecular structure is of course only valid insofar as the X-ray diffraction evidence is valid. The details of the arrangement of the molecular species present is not yet settled with as much certainty as the diagram might suggest. Therefore the molecular model is to this extent speculative.

The dimensions of the unit membrane structure are such that it is possible to include a single bimolecular leaflet of mixed lipids within the structure as well as monolayers of non-lipid material on the polar surfaces of the leaflet. SCHMITT, BEAR and PALMER (1941) found that mixed nerve lipids extracted with alcohol-ether gave X-ray diffraction patterns indicating a fundamental spacing of ~ 65 Å. They demonstrated that such lipid model systems could be hydrated with an increase in the spacing to about twice this value. Chemical studies of the reaction of OsO_4

with lipids and proteins (Edelhoch and Robertson 1952; Bahr, 1954) have indicated that proteins as well as lipids react vigorously with OsO_4 and both might therefore be expected to show up as high electron scattering components in fixed tissues. The double bonds of unsaturated fatty acids and the phosphatidyl moities (phospho-choline, phospho-serine and phospho-ethanolamine) common in nerve lipids all react with OsO_4 and might be expected to produce densities in electron micrographs. Since all these components react it is not immediately apparent why alternating

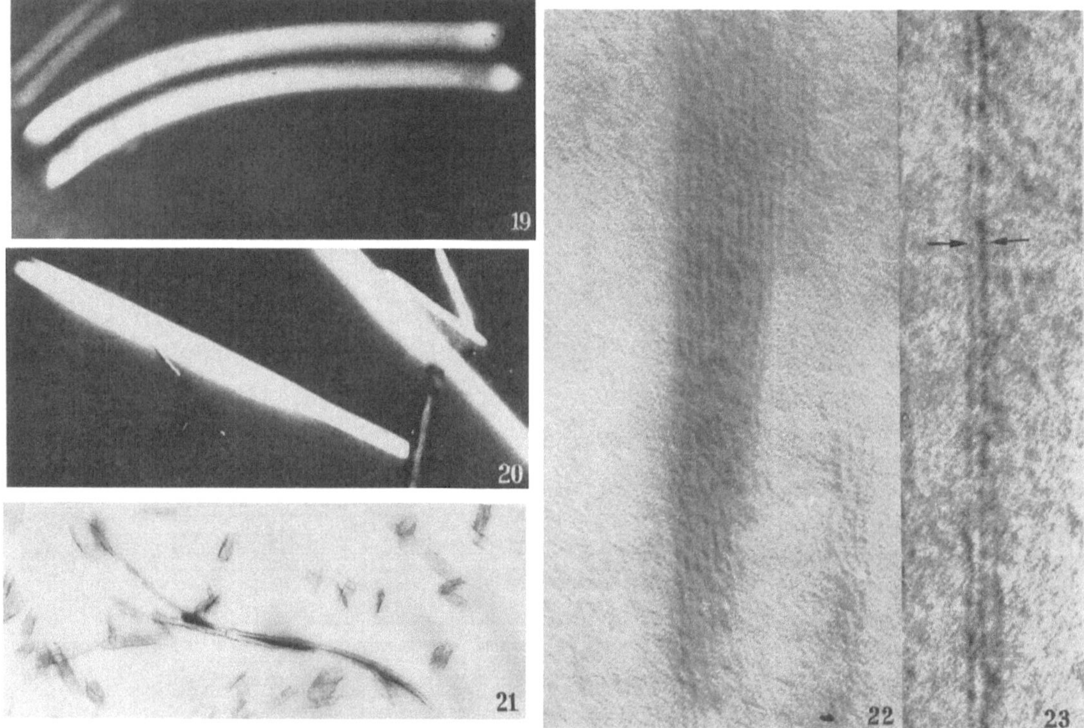

Fig. 19. A myelin form which grew out of the waxy residue of an alcohol-ether extract of cat brain. It is viewed in polarised light with a Köhler compensator arranged to display the negative birefringence of the form with respect to its long axis. Note the change in sign of the birefringence at each end of the form

Fig. 20. Flat plate-like crystals of lipid extracted from cat brain viewed in polarised light with a Köhler compensator arranged to display their negative birefringence with respect to their long axes

Fig. 21. Electron micrograph of a section of brain lipid fixed with $KMnO_4$ and embedded in araldite. Mag. 13,000×

Fig. 22. Higher power micrograph of an area in a section. Note the dense layers repeating at a period of about 40 Å. Mag. 324,000 ×

Fig. 23. A structure observed in a permanganate fixed brain lipid extract showing a plaque-like region tapering down to a single unit consisting of a pair of dense lines (arrows) together measuring about 50 Å across. Mag. 470,000×

dense and light strata are seen in myelin unless some reaction products have more electron scattering power than others. It is important then to fix nerve lipids with fixing agents and examine the reaction products by embedding and examining thin sections by electron microscopy. Accordingly, alcohol-ether extracts of cat nerves were prepared by soaking slices of brain and assorted peripheral nerve fibres in cold ($\sim 4°$ C) alcohol-ether. A waxy fraction was obtained on evaporating the supernatant. This waxy material was biuret negative and yielded myelin forms such as that in Fig. 19 when mixed with water. White crystaline material appeared as a precipitate in the supernatant material and this was also biuret negative. This material consisted of long plate-like negatively birefringent crystals (Fig. 20) and was water soluble. Each of these presumably lipid products was fixed with both OsO_4 and $KMnO_4$, dehydrated and embedded in araldite for sectioning. In all cases layered structures repeating at a period of ~ 40 Å were observed. Fig. 21 shows the appearance of the crystalline fraction after fixation in $KMnO_4$. Paired layers of material are seen spaced a few tenths of a micron apart and running randomly in the sections. These show thickenings which, when enlarged (Fig. 22), show parallel dense lines less than 20 Å thick repeating at a period of ~ 40 Å. These thickened regions taper down to single units as in Fig. 23 consisting of two parallel dense layers less than 20 Å thick separated by a light central zone of comparable thinness. Such units are ~ 50 Å in overall thickness and resemble the unit membrane structure seen in cells except for the reduced dimensions. No regions were found in which such unit structures could be seen to taper convincingly into a single dense line. The appearance of this unit structure and its dimensions are compatible with the assumption that it represents a single bimolecular leaflet of lipid and that the dense layers represent the aligned polar ends of the lipid molecules. It is interesting to speculate that the addition of monolayers of protein to each of these surfaces might be expected to alter this unit structure so that the appearance of the ~ 75 Å unit membrane of cells would result. The experiments of STOECKENIUS reported elsewhere in this volume could be interpreted as supporting this speculative point. However, STOECKENIUS interprets his findings differently and the matter cannot yet be considered settled. Further studies of lipid and protein model systems are required. Nevertheless, one important point can be made at this time. Lipid molecules seem to manifest themselves after fixation with either $KMnO_4$ or OsO_4 in electron micrographs as asymmetric densities. One end of the lipid molecule seems to produce greater density than the other. If this were not the case aggregated lipid components, whatever their precise molecular orientation, would not be expected to yield alternating light and dense layers of the dimensions and spacing found. Further studies of lipid and protein model systems may be expected to yield information that will allow definitive conclusions to be drawn regarding the precise molecular orientations which result in the observed densities.

The author wishes to acknowledge the constant support of Professor J. Z. YOUNG, and the technical assistance of Mrs. ROSE WHEELER. The work was supported by grants from the Nuffield Foundation, the Rockefeller Foundation and the Welcome Foundation.

Bibliography

ABERCROMBIE, M., and JOAN E. N. HAEYSMAN: Exp. Cell Res. **6**, 293 (1954).
BAHR, G. F.: Exp. Cell Res. **7**, 457 (1954).
DANIELLI, J. F., and H. A. DAVSON: J. cell comp. Physiol. **5**, 495 (1935).
DAVSON, H. A., and J. F. DANIELLI: *The permeability of natural membranes*. Cambridge: University Press 1943.
EDELHOCH, H., and J. D. ROBERTSON: Unpublished data.
FERNANDEZ-MORAN, H., and J. B. FINEAN: J. biophys. biochem. Cytol. **3**, 725 (1957).
FINEAN, J. B.: Exp. Cell Res. **5**, 202 (1953).
— Exp. Cell Res. **6**, 283 (1953).
— and MILLINGTON, P. F.: J. biophys. biochem. Cytol. **3**, 89 (1957).
GASSER, H. S.: Cold Spr. Harb. Symp. quant. Biol. **17**, 32 (1952).
— J. gen. Physiol. **38**, 709 (1955).
— J. gen. Physiol. **39**, 473 (1956).
GEREN, B. B.: Exp. Cell. Res. **7**, 558 (1954).
— in *Cellular mechanismus in differentiation and growth*. Ed. by D. Rudnick, Princeton Univ. Press, p. 213 (1956).
HESS, A.: Proc. roy. Soc. B **144**, 496 (1956).
PETERSON, E. R., S. M. CRAIN and M. R. MURRAY: Annat. Rec. **120**, 357 (1958).
ROBERTSON, J. D.: J. biophys. biochem. Cytol. **1**, 271 (1955).
— in *Ultrastructure and chemistry of neural tissue*, ed. by H. Waelsch, p. 1 (1956).
— J. biophys. biochem. Cytol. **3**, 1043 (1957).
— J. Physiol. (Paris) **137**, 6 (1957).
— J. biophys. biochem. Cytol. **4**, 349 (1958).
— The ultrastructure of cell membranes and their derivatives. Biochem. Soc. Symp. **16**, 3—43 (1959).
— The molecular structure and contact relationships of cell membrane. Progr. Biophys. X (in press) (1959).
SCHMIDT, W. J.: Z. Zellforsch. **23**, 657 (1936).
SCHMITT, F. O., R. S. BEAR and G. L. CLARK: Radiology **45**, 131 (1935).
— — and PALMER: J. cell. comp. Physiol. **18**, 31 (1941).
STOECKENIUS, W.: This volume p. 174.
WEISS, P.: Int. Rev. Cytol. **7**, 391 (1958).

Artificial models of biological membranes

E. H. Mercer

Chester Beatty Research Institute (Institute of Cancer Research: Royal Cancer Hospital) London

Although there is much to suggest that biological membranes have a lipoprotein character, their actual composition and structure has not been established beyond question. The electron microscopic image of these membranes is usually a thin, dense line splitting at higher resolution into two; but this image itself tells us little about the nature of the membrane except that it reacts strongly with various fixatives. A review of these matters has been given by Robertson (4).

Some progress follows from experiments, such as those of Bahr, on the reactivity *in vitro* of typical model substances. This sort of experiment can be made more realistic, as far as membranes go, by preparing the test material in membrane form and studying its actual morphology after various chemical treatments. In this way the author has examined films of protein (*1*) and Revel, Ito and Fawcett (*2*) and Stoeckenius (*5*) phospholipid membranes.

Films of proteins. Membranes of some proteins can be very simply prepared by spreading the protein on water in a small Langmuir-type trough. The film is then swept up to one end of the trough by means of a waxed barrier and the crumpled sheet lifted out on a wire support as a thin thread. Cross-sections of this thread for electron microscopy can be obtained by dehydrating, embedding and sectioning in a routine way. The effect of fixatives and stains on such films may be studied by treating the thread, mounted on its support, in appropriate solutions.

The thickness of protein films spread on water varies with p_H and other factors of the substrate. A film of polypeptide thickness is obtained when spreading is a maximum (*3*). However a certain amount of collapse occurs during the sweeping up of films leading to partial refolding and even, in the case of serum albumen, to re-solution. The thinnest films obtained with egg albumen were 50—60 Å thick but no relation between p_H and thickness was noted. Possibly uncontrolled factors in the rate of sweeping and collection determine the thickness. Fig. 1 shows an example of a section of a crumpled film of egg albumen spread at p_H 2 and fixed with osmium tetroxide.

Alcohol fixation yields the least dense films; osmium tetroxide and potassium permanganate stain to about the same degree (see Fig. 2). Films about 200 Å thick often appear doubled when fixed with permanganate. Intense staining was produced by phosphotungstic acid and uranyl acetate (Fig. 3). Thus, in respect of their reaction with permangante and osmium tetroxide, simple films resemble natural membranes; they differ however in being more strongly stained with PTA and UA which, of course, stain known protein components, such as myofilaments.

Compound films. By dissolving one component in the substrate and spreading another on the surface, it is possible to prepare films formed by the reaction of the two components. The most interesting combination biologically is lipid + protein.

A compound film between lecithin and egg albumen was obtained by dissolving the protein in the water, sweeping the surface clean from the film which rapidly forms, and immediately spreading a drop of lecithin on the fresh surface. With the particular sample of lecithin used, the drop spread immediately to give a film, silver grey in reflected light, i. e., probably less than 500 Å thick.

Evidence of combination with the protein was obtained when the surface was swept. Instead of seeing the usual succession of interference colours, which occurs when a liquid film is compressed and thickened on water substrate, a grey, crumpled, solid film forming a thread was obtained, which could be handled, fixed and stained, as were the films of protein, but proved more brittle. The compound film reacted vigorously with osmium tetroxide and permanganate. When sectioned, it proved thicker than expected from the area spread and the film colour (Fig. 4a). This increase in thickness is probably due to flow and collapse when the film is swept up and its occurence is a great defect in this otherwise simple method of preparing films. One side of the compound film was more smooth than the other and was assumed to be the protein layer. These films also showed a tendency to form branches (Fig. 4b), lobes (Fig. 4c) and tubes, due to a coalescence of folds. This behaviour may be analogous to that displayed by biological membranes when cells are broken up.

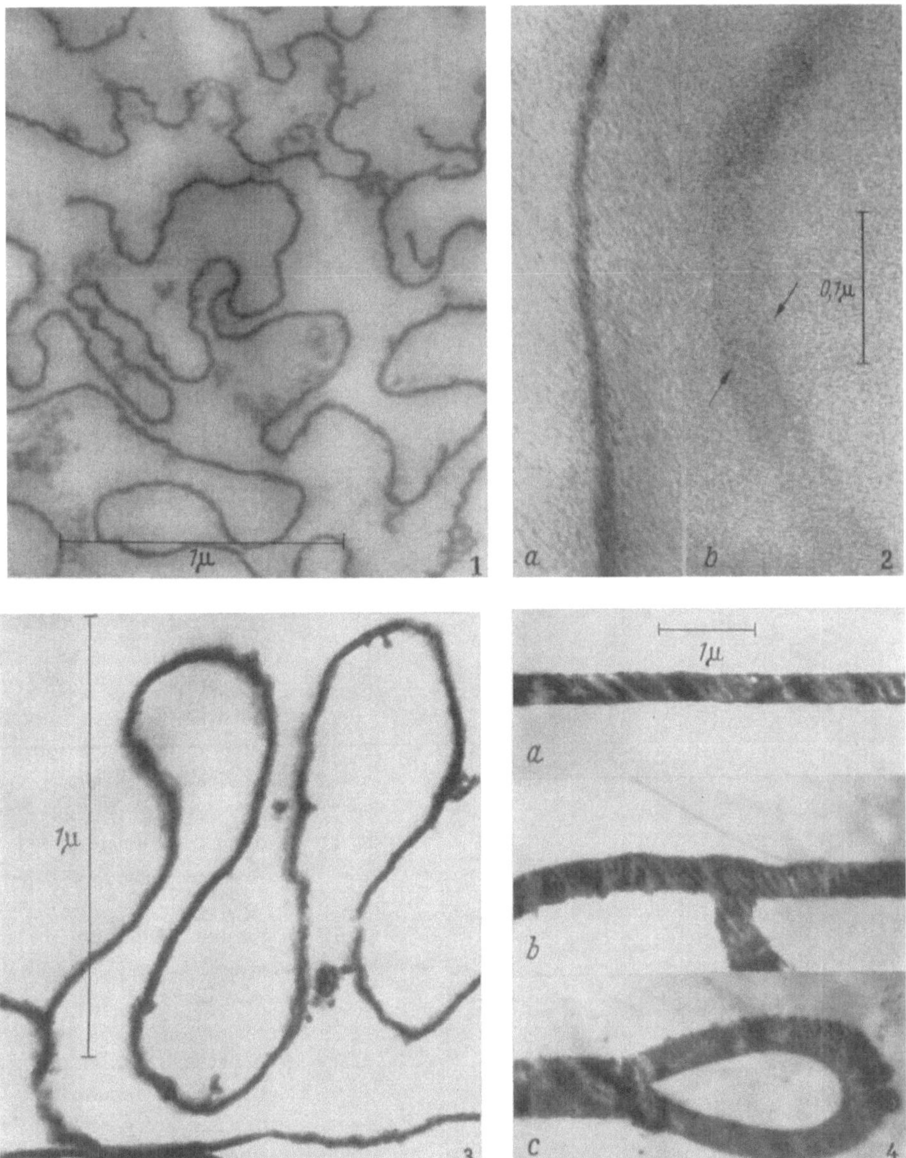

Fig. 1. Cross-section of a film of egg albumen spread on water at pH 2. Fixed in OsO₄

Fig. 2a and b. a). An albumen film spread at pH 8, fixed in osmium tetroxide to compare with. b) A thicker film spread under similar conditions and fixed with permanganate. Note the double appearance

Fig. 3. Film stained with phosphotungstic acid. Notice the increased density

Fig. 4. Examples of lipoprotein films (see Text) showing branching and looping

The work has been supported by grants to the Chester Beatty Research Institute (Institute of Cancer Research: Royal Cancer Hospital) from the British Empire Cancer Campaign, the Jane Coffin Childs Memorial Fund for Medical Research, the Anna Fuller Fund and the National Cancer Institute of the National Institutes of Health, US Public Health Service.

The author is very grateful for the photographic assistance of Mr. M. J. Docherty.

References

1. MERCER, E. H.: Nature (Lond.) **180,** 87 (1957).
2. REVEL, J. P., ITO, S. and D. W. FAWCETT: J. biochem. biophys. Cytol. **4,** 495 (1958).
3. ASTBURY, W. T., F. O. BELL, E. GORTER and J. VAN ORMONDT: Nature (Lond.) **142,** 33 (1938).
4. ROBERTSON, D.: This volume, p. 159.
5. STOECKENIUS, W.: This volume, p. 174.

Fixierung von Myelinfiguren aus Phosphatiden und Eiweiß mit OsO₄ und KMnO₄

WALTHER STOECKENIUS

Physiologisches Institut der Universität Hamburg

Eine Phosphatidfraktion aus menschlichem Gehirn, die etwa aus 52% Kephalin, 35% Lecithin und 13% Inositphosphatid bestand, wurde in Wasser zur Quellung gebracht und mit OsO_4-Dampf oder einer $KMnO_4$-Lösung (*6*) fixiert. Die Entwässerung erfolgte in Alkohol und Aceton, die Einbettung in Methacrylat. Dünnschnitte dieses Materials wurden mit dem Elmiskop I untersucht.

Bei dem OsO_4-fixierten Material zeigten sich dichte Bänder sehr unterschiedlicher Breite, die bei höherer Vergrößerung stellenweise ein sehr regelmäßiges Muster paralleler dunkler Linien von etwa 18 Å Breite aufwiesen. Der Abstand von der Mitte einer dunklen Linie bis zur Mitte der nächsten betrug etwa 40 Å (Abb. 1). Aus polarisationsoptischen Arbeiten (*8*) und den eingehenden Untersuchungen von SCHMITT und seinen Mitarbeitern (*1, 9, 10*) mit der Röntgenkleinwinkelbeugung sind wir über den molekularen Aufbau der bei der Quellung von Phosphatiden entstehenden Myelinfiguren gut orientiert. Sie bestehen aus bimolekularen Schichten, in denen die hydrophoben Fettsäurereste der Moleküle parallel liegen und einander zugekehrt sind, während die polaren Gruppen nach außen zeigen.

Abb. 1. Dünnschnitt durch die reine Phosphatidfraktion nach Quellung in Wasser und Fixation mit OsO_4 · 240 000 : 1

Zwischen die polaren Oberflächen benachbarter bimolekularer Lamellen können sich Wasserschichten einlagern.

Der von uns im Schnittbild gefundene Abstand von etwa 40 Å stimmt gut mit der im Röntgendiagramm beobachteten 43 Å-Periode der Phosphatide (*1*) überein. Eine ähnliche Übereinstimmung fanden schon GEREN und SCHMITT (*3*) bei Acetalphosphatiden. Aus diesem Grund dürfen wir annehmen, daß die 18 Å breite, dunkle Linie einem bestimmten Abschnitt jener bimolekularen

Lamelle entspricht, die sich auf Grund der Einlagerung von Osmium durch eine stärkere Elektronenstreuung gegenüber dem restlichen Molekül auszeichnet.

Es ist bekannt, daß OsO_4 mit den Doppelbindungen ungesättigter Fettsäuren reagiert. Wir haben daher in unserer Phosphatidfraktion durch Bromierung die Doppelbindungen blockiert. Danach war eine Fixierung mit OsO_4 nicht mehr möglich. Dieser Befund zeigt, daß die Fixierung der Lipoide auf einer Reaktion des Osmiums mit den Doppelbindungen beruht. Außerdem hat auch FINEAN (2) angegeben, daß synthetisches Kephalin mit gesättigten Fettsäuren nur schwach mit OsO_4 reagiert. Wir können also annehmen, daß in unseren Myelinfiguren die 18 Å breite dunkle Linie die Fettsäureketten der bimolekularen Lamelle markiert, und zwar wegen der bevorzugt am Kettenende liegenden Doppelbindungen deren mittleren Anteil (Abb. 2).

Wurde den gequollenen Phosphatiden eine wäßrige Lösung von Globin zugesetzt, das aus Rinderhämoglobin durch Fällung in mit HCl angesäuertem Aceton gewonnen war, oder ließ man die Quellung in einer etwa 4%igen Globinlösung vor sich gehen, so ergab sich im elektronen optischen Bild, im Gegensatz zu den Ergebnissen der Röntgenbeugung (10) keine Vergrößerung der Periode. Im Inneren der breiten dunklen Bänder zeigte sich vielmehr genau dieselbe Struktur wie bei der reinen Phosphatidfraktion. Auf ihrer Außenseite aber waren sie durch eine kräftige dunkle Linie begrenzt, deren Breite zwischen 25 und 50 Å schwankte. Dabei war die innere Begrenzung dieser Linie scharf, die äußere häufig unscharf und unregelmäßiger (Abb. 3) Augenscheinlich handelt es sich um angelagertes Eiweiß.

Auffällig war, daß der Abstand zwischen dem inneren Rand dieser „Eiweißlinie" und der benachbarten 18 Å breiten „Lipoidlinie" immer derselbe war, wie der der „Lipoidlinien" untereinander. Man sollte erwarten, daß ein auf der hydrophilen Außenseite der Lamelle angelagertes und durch seine Osmiumaufnahme kontrastreich dargestelltes Eiweiß nur etwa den halben Abstand von der nächsten „Lipoidlinie" aufweist. Eine Erklärung für das beobachtete Verhalten bieten Untersuchungen an monomolekularen Kephalinfilmen, in denen gezeigt werden konnte, daß die hydrophoben Seitenketten von Proteinen zwischen die Phosphatidmoleküle eindringen können, während die basischen Gruppen des Proteins mit den sauren des Lipoids reagieren (7). Außerdem

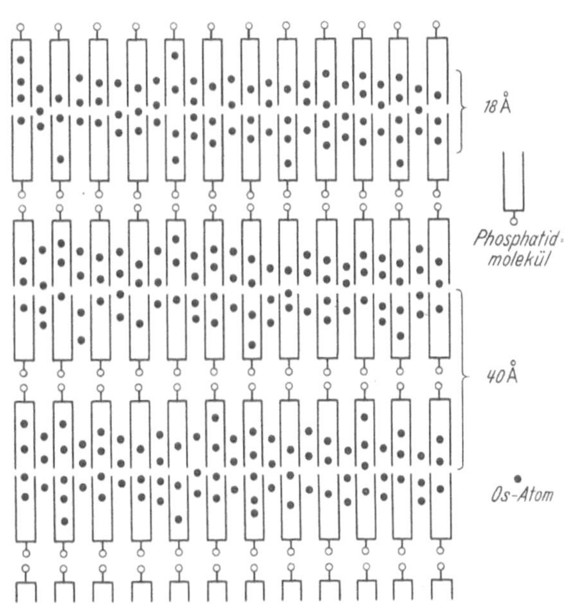

Abb. 2. Schematische Darstellung des molekularen Aufbaues und der Osmiumverteilung in den fixierten Phosphatiden

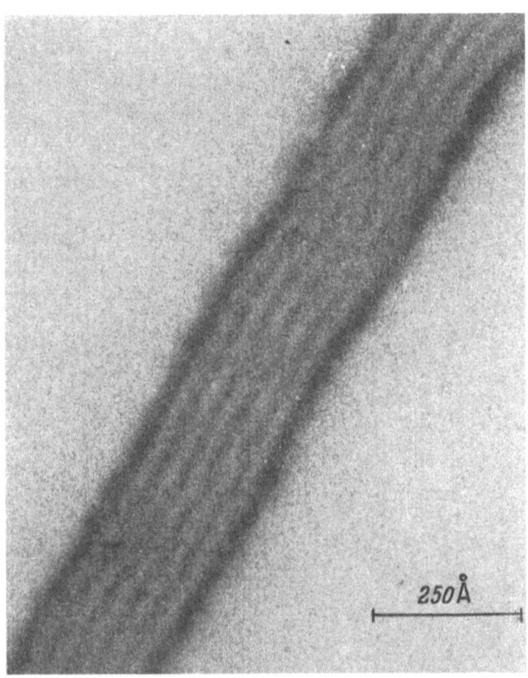

Abb. 3. Dünnschnitt durch eine, in einer wäßrigen Globinlösung entstandenen Myelinfigur. Das auf der Oberfläche des Phosphatids angelagerte Eiweiß erscheint als dunkle Linie. 820000 : 1

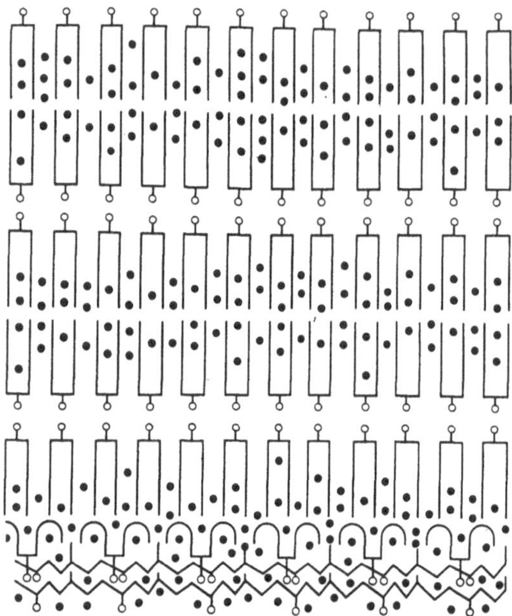

Abb. 4. Wie Abb. 2, aber am unteren Rand mit angelagertem Eiweiß. Die Polypeptidkette ist durch eine Zickzacklinie, eine polare Seitenkette durch einen Strich mit einem Kreis, eine apolare durch einen einfachen Strich dargestellt. Die Fettsäurereste der Phosphatidmoleküle in der dem Eiweiß benachbarten Lamelle sind umgebogen

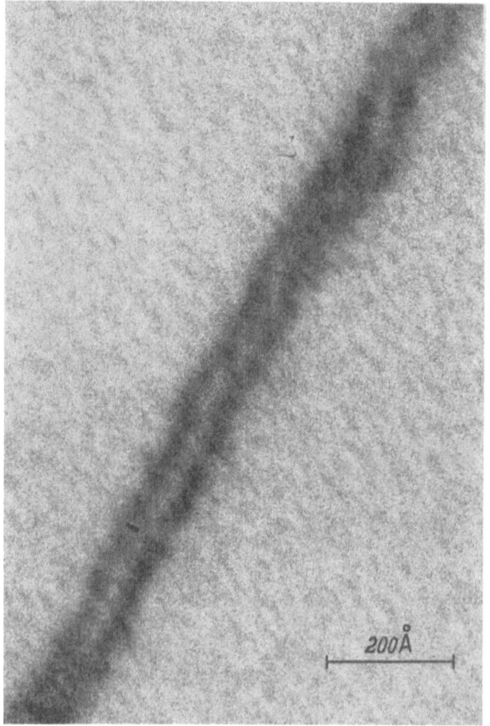

Abb. 5. Zwei bimolekulare Phosphatidlamellen mit angelagertem Globin. Der molekulare Aufbau ist in Abb. 6 schematisch dargestellt. 875000:1

muß man noch annehmen, daß sich die Fettsäureketten der auseinandergedrängten Lipoidmoleküle nach außen umbiegen und so den erweiterten Raum zwischen den Molekülen ausfüllen, da die hydrophoben Seitenketten des Proteins dafür zu kurz sein dürften. Dieses Umbiegen der Ketten ist aus der Penetration von Seifenmicellen durch Alkohole mit kürzerer Kohlenstoffkette bekannt (4, 5)[1]. Wenn wir diese Vorstellung auf unsere Myelinfiguren anwenden, so zeigt sich, daß die Ketten der Fettsäuren so dicht an das Eiweiß heranrücken, daß im elektronenoptischen Bild eine Verschmelzung der „Lipoid-" und der „Eiweißlinie" zu erwarten ist, und daß diese Linie von der nächsten „Lipoidlinie" denselben Abstand haben muß wie die „Lipoidlinien" untereinander (Abb. 4). Tatsächlich können wir beobachten, daß die äußerste „Lipoidlinie" und die „Eiweißlinie" auch an unvollständig mit Eiweiß besetzten Myelinoberflächen gelegentlich in gleicher Höhe liegen.

Einen weiteren Beweis für diese Auffassung liefert die Abb. 5. Sie zeigt die dünnste bisher in unseren Präparaten aufgefundene Lamelle. Diese besteht nur aus zwei kräftigen „Eiweißlinien", die denselben Abstand von etwas über 20 Å voneinander haben wie die reinen „Lipoidlinien". Handelte es sich hier um eine einzelne bimolekulare Lipoidlamelle, die auf beiden Seiten mit nur adsorbiertem nicht penetriertem Eiweiß besetzt wäre, so müßte man zwischen den beiden „Eiweißlinien" noch die durch die Doppelbindungen der Fettsäureketten hervorgerufene 18 Å breite „Lipoidlinie" sehen. Da dieses nicht der Fall ist, glaubten wir, ihr die in Abb. 6 angegebene Struktur zuschreiben zu müssen, d. h. einen Aufbau aus zwei bimolekularen Lipoidlamellen die auf den Außenseiten von Eiweiß penetriert sind.

Das Bild dieser beiden „Eiweißlinien" entspricht aber in auffallender Weise dem Bild der von ROBERTSON (11) in vielen Zellen gefundenen "unit membrane". Es wäre also zu diskutieren, ob dieser nicht eine entsprechende molekulare Struktur zuzuschreiben ist.

————————

[1] Herrn Dr. K. HECKMANN (Max-Planck-Institut für Physikalische Chemie, Göttingen) danke ich für eine eingehende Diskussion dieser Vorstellungen, für Literaturangaben und den Hinweis auf die Möglichkeit des Umbiegens der Kettenenden. Er hat auch an einem Kalottenmodell gezeigt, daß selbst für hochungesättigte Fettsäuren mit 5 Doppelbindungen ein Umbiegen der Kette um 180° möglich ist.

Im Vergleich zur OsO$_4$-Fixierung ergab eine Fixierung der Phosphatide mit KMnO$_4$ sehr viel schlechtere Resultate. Der Kontrast des Bildes war gering und die Lamellenstruktur war nicht so gut erhalten. Phosphatide und Eiweiß gaben dagegen kontrastreichere Bilder von annähernd parallelen Lamellen unterschiedlicher Dicke, die aber eine viel weniger regelmäßige Anordnung zeigten als nach OsO$_4$-Fixierung und durch Ablagerung von unregelmäßig geformtem Material gleichen Kontrastes überdeckt wurden.

Zu großem Dank verpflichtet bin ich Herrn Prof. H. RUSKA (Institut für Elektronenmikroskopie der Med. Akademie Düsseldorf), Herrn Dr. D. PETERS (Tropeninstitut Hamburg) und Herrn Dr. H. KÖLBEL (Tuberkuloseforschungsinstitut Borstel) für die Erlaubnis, ihre Mikroskope für diese Untersuchungen zu benutzen, Herrn Dr. E. LINDNER (Institut für topographische Anatomie der Med. Akademie Düsseldorf) und Herrn Dr. K. MANNWEILER (Institut zur Erforschung der Poliomyelitis und spinalen Kinderlähmung, Hamburg) für die Erlaubnis, ihre Mikrotome zu benutzen.

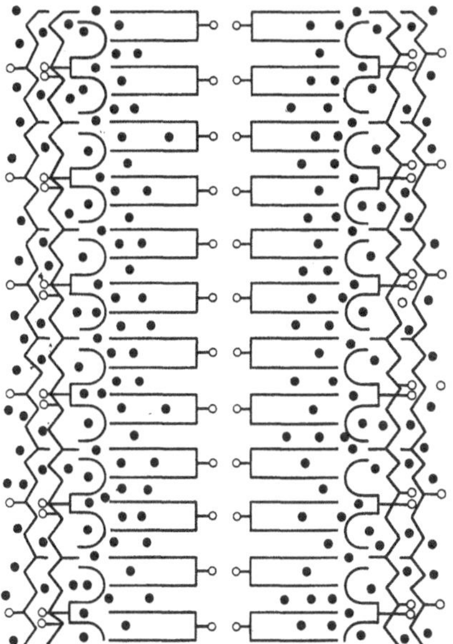

Abb. 6. Schematische Darstellung von zwei bimolekularen Phosphatidlamellen, denen sich auf beiden Außenseiten Eiweiß angelagert hat

Literatur

1. BEAR, S. R., K. J. PALMER and F. O. SCHMITT: J. cell. comp. Physiol. **17**, 355 (1941).
2. FINEAN, J. B.: Exp. Cell Res. **6**, 283 (1954).
3. GEREN, B. B., and F. O. SCHMITT: J. appl. Physics **24**, 1421 (1953).
4. HARKINS, W. D., and R. MITTELMANN: J. Colloid Sci. **4**, 367 (1949).
5. — and H. OPPENHEIMER: J. Amer. chem. Soc. **71**, 808 (1949).
6. LUFT, J. H.: J. biophys. biochem. Cytol. **2**, 799 (1956).
7. MALATON, R., and J. H. SCHULMAN: Disc. Faraday Soc. No. 6, 27 (1949).
8. NAGEOTTE, J.: Actual. Sci. industr. Nr. 431—434, Paris 1936.
9. PALMER, K. J., and F. O. SCHMITT: J. cell. comp. Physiol. **17**, 385 (1941).
10. — — and E. CHARGAFF: J. cell. comp. Physiol. **18**, 43 (1941).
11. ROBERTSON, J. D.: J. biophys. biochem. Cytol. **4**, 349 (1958).

Spectrophotometric and electronmicroscopic experiments with organic built-up films exposed to osmium tetroxide

H. J. TRURNIT, G. COLMANO and G. SHIDLOVSKY

Research Institute for Advanced Study (RIAS), Baltimore 12, Maryland

Electron micrographs of cellular structures, fixed and stained with OsO$_4$ or other agents, do not give any chemical information about the details of the image. Results from experiments in vitro on the reaction between lipides or proteins and OsO$_4$ may be misleading. We have, therefore, begun a study of the interaction between OsO$_4$ and artificially deposited multimolecular layers of known chemical substances. The approach is two-fold. Since OsO$_4$ has a well defined absorption band with a peak at 2460 Å it should be possible to estimate the amount of OsO$_4$ taken up by the multilayers. Furthermore, multilayers deposited on a suitable base may be prepared for examination in the electron microscope.

1. Spectrophotometric studies. It was found that a base consisting of a thin cellophane membrane (0.025 mm) coated with a thin paraffine wax layer, fulfilled the 3 essential requirements for the spectrophotometric studies. This base does not bind any OsO$_4$, it has a low absorption in the 2500 Å region, and it does not swell in water. The latter requirement is called for during the dipping process on the film balance (Langmuir-Blodgett technique) by which successive monolayers are deposited from a water surface onto the supporting membrane. For easier handling this base was attached to a 10 ×35 mm window in a thin metal strip. After preparing two slides by depositing a series of monomolecular films of various materials in a specific sequence, one was inserted into the sample compartment and the other into the blank compartment of a Cary (Model 14) Spectrophotometer and a zero line between 2000 Å and 3500 Å was taken. Then one of the slides was suspended for ten minutes in saturated OsO_4 vapor. Thereafter the treated slide was aerated under a fume hood, for a given period of time, to permit the loosely adsorbed OsO_4 to evaporate. Then the treated slide was put back into the spectrophotometer and the UV absorption spectrum was taken. The peak absorbance near 2500 Å after different intervals of aeration and for different film compositions is shown in Fig. 1.

The gelatine films were deposited by dipping the slides at 30° C into a 0.25% solution of calfskin gelatine (Kodak) and withdrwaing them slowly. Independent measurements with an ellipsometer, using Langmuir-Blodgett stearate test slides as a base, showed that under these conditions the gelatine film thickness is about 50 Å $\pm$ 10 Å[1]. The slides emerge dry from the gelatine solution. The object of this experiment was to compare the OsO_4 binding capacity of a saturated fatty acid (stearic), an acid of the same chain length containing one double bond (vaccinic), and a protein. This comparison (per unit thickness) is given in column III of table 1 in arbitrary units. The absorbance per unit thickness of OsO_4 treated gelatine is about 100 times higher than that of vaccinic or stearic acid. The unexpected phenomenon, that the absorbance of a gelatine layer on top of saturated or unsaturated fatty acid multilayers is so much smaller than the absorbance of gelatine alone may be caused by a diffusion of the fatty acids into the gelatine, thereby changing the interaction between gelatine and OsO_4. Fatty acids are known to penetrate and combine with certain protein molecules easily.

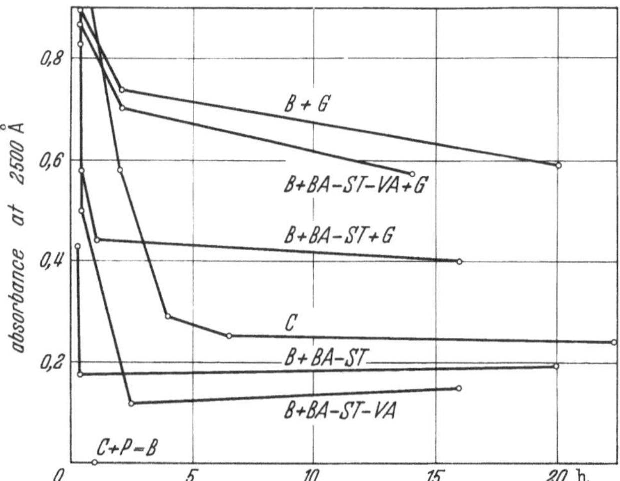

Fig. 1. The absorbance of film preparations, treated with OsO_4, as a function of time elapsed after treatment. The films are indentical with those described in Table 1. For explanation of symbols see Table 1

The initial rapid drop in absorbance shown in Fig. 1 is probably mostly due to desorption, and evaporation of OsO_4 from the sample. An additional cause could be the reduction of OsO_4 to lower oxides. The latter, however, have characteristic absorption bands in the visible region and we were unable to find any clear cut indications of these bands. A reduction of part of the OsO_4 to metallic osmium might be indicated by the following observation: the base itself (cellophane and paraffine) does not bind any OsO_4 (see Fig. 1) and remains transparent. All preparation where films were deposited and treated with OsO_4 look more or less gray like an exposed and developed photographic plate. The absorption in the visible showed no characteristic bands, but an increased absorption throughout the visible region, which gradually increases towards shorter wave lengths. Inspection of preparations with high magnification light microscopy failed to reveal any dark particles (possible metal deposits), which would be the most likely explanation of the gray appearance of the films. It may be that the metal deposits are too small for resolution in the light microscope.

The data of table 1, column I, show that the absorbance of a 100 Å thick gelatine film, 15 hr after a 10 min exposure to OsO_4 vapor, is 0.62 for the 2500 Å peak. The molecular extinction coefficient $\left(\alpha_m = \dfrac{\log \frac{1_0}{1}}{c\,d} \right.$, where c is given in moles per liter and d in cm) of OsO_4 (for the absorp-

---

[1] The thickness of the gelatine film obtained in this way is an inverse function of the solution temperature and depends on the concentration. These films, while on a water surface, consist of two layers: one monolayer of surface denatured protein, the other layer of adsorbed undenatured protein. When such a film of gelatine is deposited onto a hydrophobic surface, such as stearic acid, by dipping, the resulting structure is composed of two such films having their undenatured hydrophilic surfaces adjacent to each other.

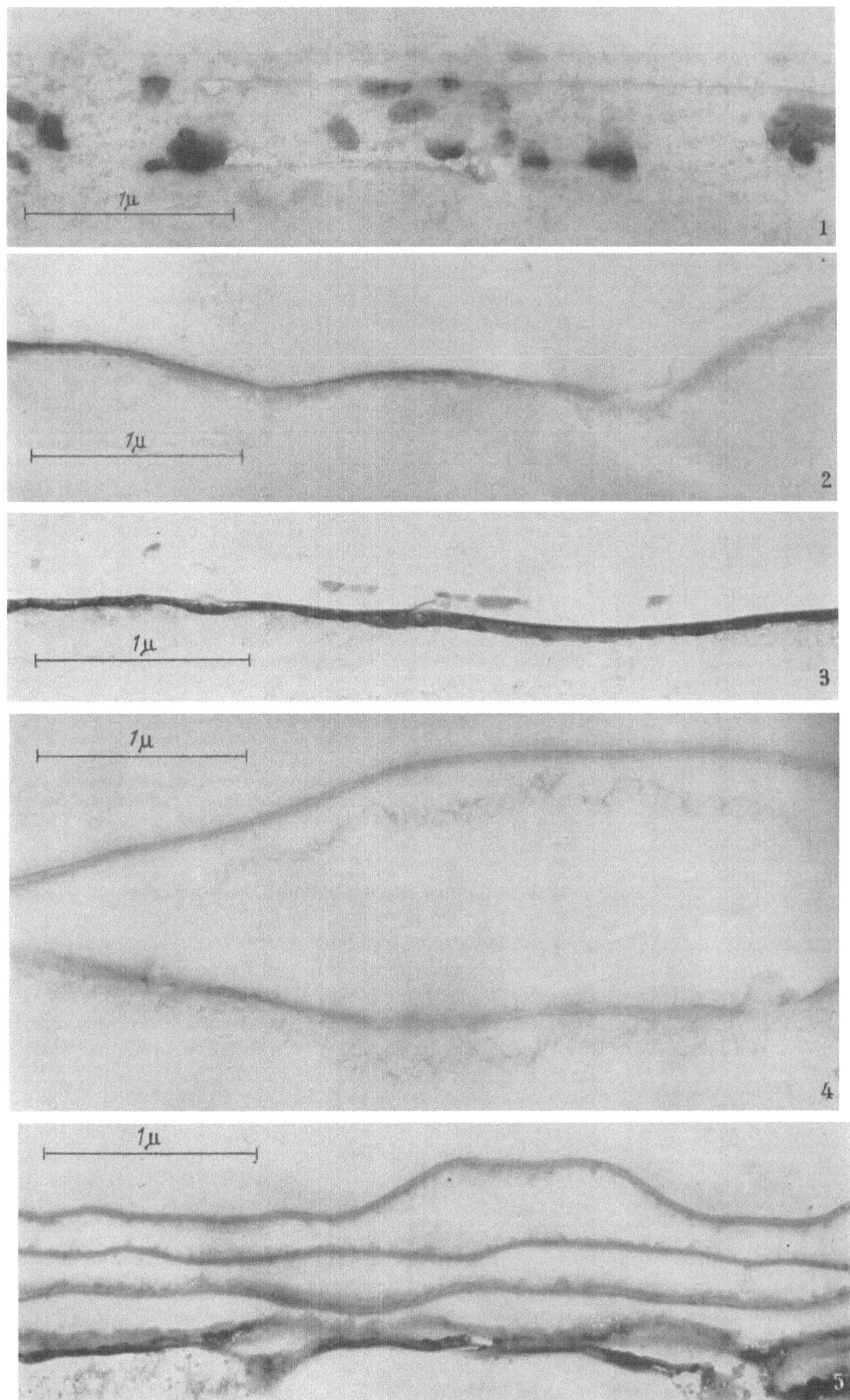

Fig. 2. Each electron micrograph represents the cross-section of the samples, the composition of which is shown in Fig. 3. The dense lines in pictures 2 to 5 are interpreted as the gelatine layers. The scattered densities along the gelatine layers are probably what remains of the barium stearate. The gelatine layers in pictures 4 and 5 are approximately 250 Å thick.

tion peak at 2460 Å)[1] as vapor in air or dissolved in H_2O was determined to be, in both cases, 3.3×10^3 at 25° C. From these data it would follow that the molar concentration of OsO_4 in the gelatine film was 180 moles/l. However, the molar concentration of OsO_4 crystals, as calculated from the crystal specific weight (4.9) and molecular weight (254.2) is only 19.3 moles/l. The high peak absorbance of 0.62 at 2500 Å must, therefore, be due predominately to some other phenomenon caused by the interaction between gelatine and the OsO_4 molecule. We are inclined to think that the gray appearance of the specimens, mentioned above, may be connected with this high background absorbance in the UV region. The presence, in the films of OsO_4 as such, is bornout by the absorption peak at 2460 Å.

Table 1. *Absorbance at 2500 Å in composite layer structures after 10 min exposure to saturated OsO_4 vapor at room temperature and 15 hr thereafter in open air*

System	I	II	III
1. C	0.25	—	0.001
2. C + P	0.0	—	—
3. C + P + G	0.62	100	6.2
4. C + P + Ba-St	0.18	4000	0.045
5. C + P + Ba-Va-St	0.15	2000	0.075
6. C + P + Ba-St + G	0.4	1000 + 100	0.364
7. C + P + Ba-Va-St- + G	0.56	2000 + 100	0.266

Column I   : absorbance of specimen after 15 hr in air.
Column II : combined thickness in Ångstrom of film deposited on both sides of base membrane.

Column III: $\dfrac{\text{Column I}}{\text{Column II}} \times 10^3$

C = cellophane: 0.025 mm, P = paraffine: ∼ 0.005 mm, G = gelatine: ∼ 50 Å. Ba-St = Barium stearate-stearic acid multilayers. Ba-Va-St = same, but formed from monolayers of equimolecular mixtures of stearic and vaccinic acid. Since films of pure unsaturated acid soaps do not have good transfer properties, we used equimolecular mixtures with stearic acid. These mixed films transfer well from water onto the supporting base.

2. Electronmicroscopic studies. Electron microscopic inspection of a number of different combinations of multilayer systems, prepared as described, so far has not given a clear correlation between the known composition of multilayers and the cross-sectional image in the electron microscope. One further requirement for the film base, additional to the three mentioned above, would be that the base should be suitable for embedding in methacrylate and for thin section cutting. We are working on developing a base of this type, so that it might be possible to correlate quantitatively the spectrophotometric and electronmicroscopic data on the same sample. So far we have deposited the films for electron microscopy on a thin base of methacrylate. However, this type of base binds OsO_4 and is therefore unsuitable for spectrophotometry.

As an example of the general applicability of the method we show in Fig. 2 electron micrographs of cross-sections of multilayer systems built according to the schemes in Fig. 3. Although it appears that the monolayers of barium stearate do not withstand the preparative treatment for electron microscopy (possibly because they do not bind much OsO_4 as shown by the spectrophotometric measurements), it seems that the correlation between the known composition of the samples and the evidence obtained in the electron microscope, as far as the gelatine is con-

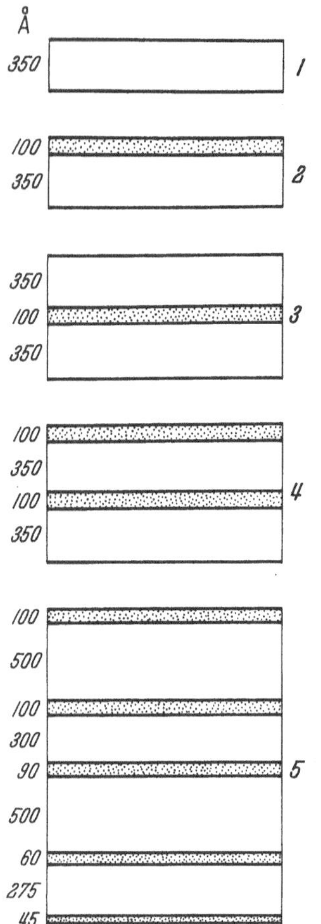

Fig. 3. The five blocks are diagrammatice representations of the layer sequences of gelatine and barium stearate (clear area: barium stearate, dotted area: gelatine) as prepared for the samples examined in the electron microscope and shown in Fig. 2. The base is always at the bottom of each block. The right hand figures correspond to the numbers on the electron micrographs. The left hand figures represent the thicknesses of the deposited layers as measured in the ellipsometer

[1] Peak of absorption band. This peak shifts close to 2500 Å in more condensed systems.

cerned, substantiates the spectrophotometric data. Much swelling is apparent and the dimensions do not agree with those obtained in the ellipsometer. The appearance of each gelatine layer as a double line may be due to the fact that the two outside surfaces of the gelatine layer consist of denatured protein, whereas the inner part is undenatured. The former may possibly bind more osmium.

Das gegenseitige Verhalten von „artifiziellen Zellmembranen" und synthetischen Tumorauslösersubstanzen tensionsaktiver Natur, elektronenoptisch untersucht

K. Setälä, O. Äyräpää, M. Nyholm, L. Stjernvall und Y. Aho

Pathologisches Institut und Institut für Elektronenmikroskopie der Universität Helsinki.
Chemische Abteilung der Technischen Hochschule, Helsinki

Beim Studium des Mechanismus der Tumorgenese in der Haut von Mäusen haben wir als Tumorauslöser sowohl technische [z. B. (1—6)] als auch von uns selbst synthetisierte tensionsaktive Substanzen vom Typ der Polyol-Fettsäureester [z. B. (7, 8)] sowie reine Fettsäuren [z. B. (9)] benutzt. In methodischem Sinne haben wir insbesondere mit nichtionisierbaren tensionsaktiven Substanzen vom Typ der Spane und der Tweene gearbeitet. Diese zeigten im Tierversuch jeweils besondere, gewissermaßen gegensätzliche Wirkungen, was uns veranlaßte, von einem „Span 60-", „Tween 20-" bzw. „Tween 60-Effekt" — im Falle von Tween 60 auch von einem durch Cholesterol-Zusatz hervorgerufenen „Schutzeffekt" — zu sprechen. Auf Grund unserer Untersuchungen nehmen wir an, daß die durch diese Substanzen von *dipoler* Natur zustandegebrachte Tumorauslösung als eine Folge von Veränderungen an den protoplasmatischen Phasengrenzen aufgefaßt werden kann.

Bei weiteren Untersuchungen über die Wirkungsweise dieser Substanzen *in vitro* haben wir uns der sog. Monomolekulartechnik [vgl. (10—12)] bedient. Mit Hilfe einer modifizierten Langmuir-Adams-Oberflächenwaage stellten wir verschiedenartige Lipo-Protein-Filme her (eingehendere Methodik vorläufig nicht veröffentlicht) und untersuchten dann das gegenseitige Verhalten dieser Zellmembran-Modelle und der Tumorauslösersubstanzen unter verschiedenen Bedingungen durch Messung der Veränderungen von Oberflächenspannung und Oberflächenpotential sowie durch elektronenoptische Analysierung der Strukturen.

An dieser Stelle werden nur die Wirkungen der folgenden von uns selbst synthetisierten Substanzen, allein und zusammen mit Cholesterin, auf den Cholesterin-Gliadin-Film besprochen: Tween 60 (Polyoxyäthylen-sorbitan-monostearat, im Tierversuch ein außerordentlich leistungskräftiger Tumorauslöser), Span 60 (Sorbitan-monostearat, als Tumorauslöser wirkungslos) und Tween 20 (Polyoxyäthylen-sorbitan-monolaurat, als Tumorauslöser nur schwach wirksam).

Um bei der Entnahme der Proben für die elektronenoptische Untersuchung den durch das Verdunsten des zwischen Film und Träger zurückbleibenden Wassers entstehenden Artefakten [vgl. (13)] möglichst auszuweichen, haben wir die Probe stets mit dem Pl-Ring so auf den Träger gebracht, daß die Wasserphase nach oben, der Film also direkt auf den Träger zu liegen kam. (Wir haben also den Film einfach umgekehrt.) Unmittelbar anschließend wurde 20 min in OsO_4-Dampf fixiert. Metallbedampfung: Au + Pd, 18°.

Elektronenmikroskop: Philips EM 100, 60 kV, Polschuhe ⌀ 1,8 mm, Objektivaperturblende 25 μ, Positivkopien.

Ergebnisse: Unter einschlägigen Versuchsbedingungen hergestellte Cholesterin-, Gliadin- und Cholesterin-Gliadin-Filme zeigen eine ziemlich regelmäßige, für jeden Film charakteristische Isotherme. Bei den Lipo-Protein-Filmen ist der Typ der Isotherme wesentlich vom Zustand des Proteins (ob denaturiert, teilweise denaturiert oder nativ) abhängig.

Spritzt man unter den Cholesterin-Gliadin-Film bei geringer Kompression (z. B. 5 dyn/cm) eine Wasserlösung von *Tween 60*, so sinkt der Oberflächendruck bis nahe 0. Die nach 60 min Wirkungsdauer ermittelte Isotherme gleicht am ehesten der des Gliadins. Das Cholesterin ist demnach aus dem ursprünglichen Film entweder völlig oder zum größten Teil veschwunden („Tween 60-Effekt"). Spritzt man aber eine Tween 60-Wasserlösung, die in einem bestimmten

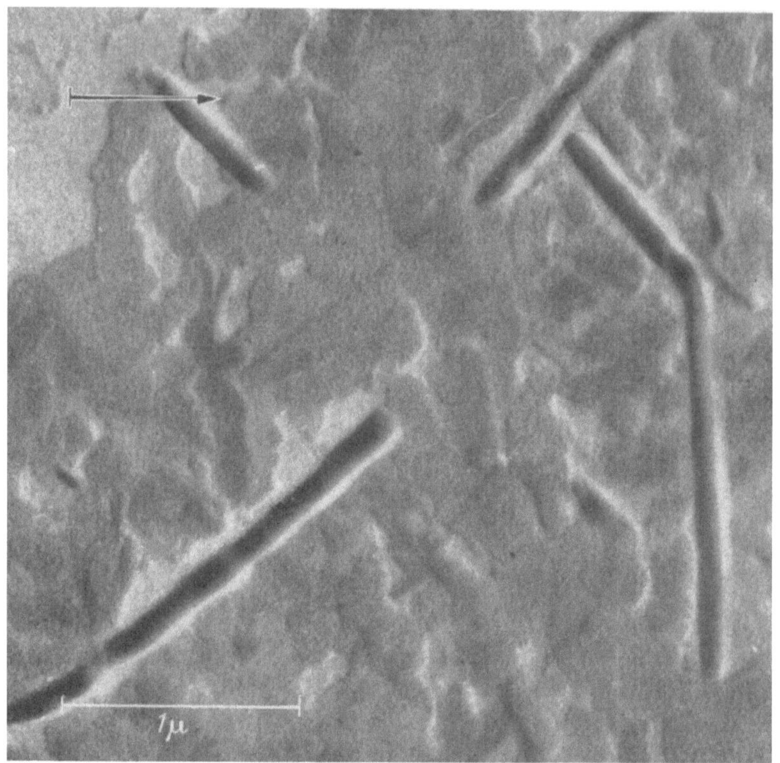

Abb. 1. Bei 34 dyn/cm Kompression kollabierter Cholesterin-Film. Das Cholesterin tritt in Form von verschieden großen Schollen auf. Typische Kollapsfalten. Das Erscheinungsbild steht im Einklang mit den entsprechenden physikochemischen Befunden. Pfeil: Richtung der Metallbedampfung

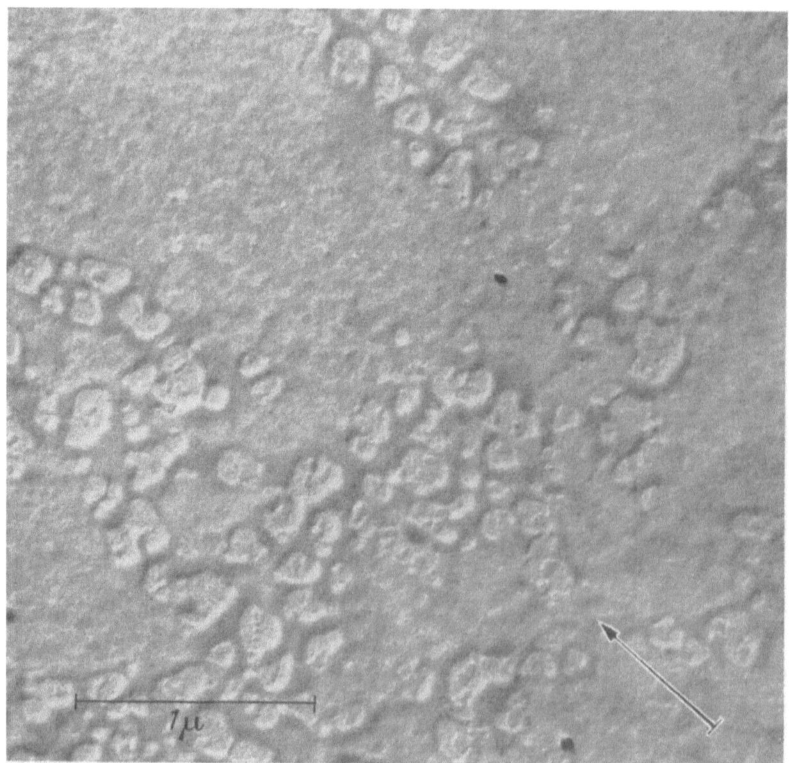

Abb. 2. Typischer Cholesterin-Gliadin-Film bei 29 dyn/cm Kompression unmittelbar vor dem Zusammenbruch. Helle Flächen: Lücken im Film, die Ränder wallartig erhöht (vielleicht Cholesterin). Kontinuierliche Fläche: intakter monomolekularer Film

— auf Grund von Hämolyseproben kalkulierten — Verhältnis Cholesterin solubilisiert enthält, so erhält man eine Isotherme vom Cholesterin-Gliadin-Typ, der sog. Präventivzusatz von Cholesterol hat demnach den ursprünglichen Film gegen die Einwirkung des Tween 60 geschützt. Die elektronenmikroskopischen Befunde stützen, wie es scheint, das im Obigen gewonnene physiko-chemische Bild. Bei geringer Kompression konstatiert man kontinuierliche und diskontinuierliche Gebiete. Die Strukturen der monomolekularen Filme treten als verschieden große Inseln oder Schollen von unregelmäßiger Form auf, deren gegenseitiger Abstand bei steigender Kompression abnimmt, bis er in der Nähe des Kollapsdruckes verschwindet (vgl. Abb. 1, 2 u. 3). Auf dem

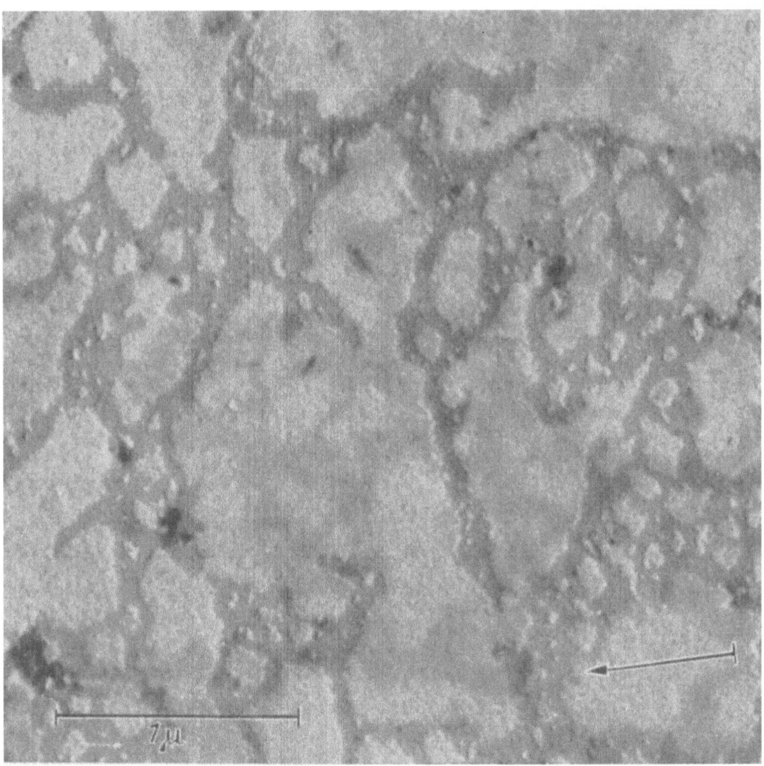

Abb. 3. Cholesterin-Gliadin-Film nach Unterspritzung von Tween 60-Wasserlösung bei Ausgangskompression von 5 dyn/cm und fortgesetzter Kompressionserhöhung bis 12 dyn/cm nach 60 min. Gliadin als großmaschiges Netzwerk übrig, Cholesterin verschwunden („Tween 60-Effekt")

elektronenoptischen Bild ist um so weniger zu sehen, je gelungener und kontinuierlicher der monomolekulare Film ist. Die Begriffe „monomolekular" und „kontinuierlich" sind unbedingt auseinanderzuhalten. Sie haben als Eigenschaften der in Rede stehenden Filme nichts miteinander zu tun.

Wird unter den Cholesterin-Gliadin-Film in Wasser dispergiertes *Span 60* bei 5 dyn/cm gespritzt, so nimmt die Oberflächenspannung zu, und der Film wird steif. Dies äußert sich u. a. in einem schroffen Anstieg der Isotherme. Dieser „Span 60-Effekt" beruht wahrscheinlich darauf, daß die Moleküle des Span 60 nach dem Oberflächenfilm hin wandern, so daß es unter demselben vielleicht zu der Bildung von Span 60-Schichten kommt. Wird die Unterspritzung mit Span 60 + Cholesterin vorgenommen, so verstärkt sich der obige versteifende Effekt noch mehr. Das äußert sich darin, daß der Film schon innerhalb der Versuchszeit von 60 min kollabiert. Die elektronenmikroskopische Untersuchung bestätigt diesen sog. Span 60-Effekt (Abb. 4).

Spritzt man unter den Cholesterin-Gliadin-Film bei 5 dyn/cm Kompression in Wasser gelöstes *Tween 20*, das auf rote Blutkörperchen stark hämolysierend wirkt, so wiederholt sich grundsätzlich dieselbe Erscheinung wie vorher im Falle von Tween 60 (Abb. 5). Die Wirkung ist jedoch deutlich stärker, was man u. a. daraus ersieht, daß sich das Cholesterin auch bei stärkerer Kom-

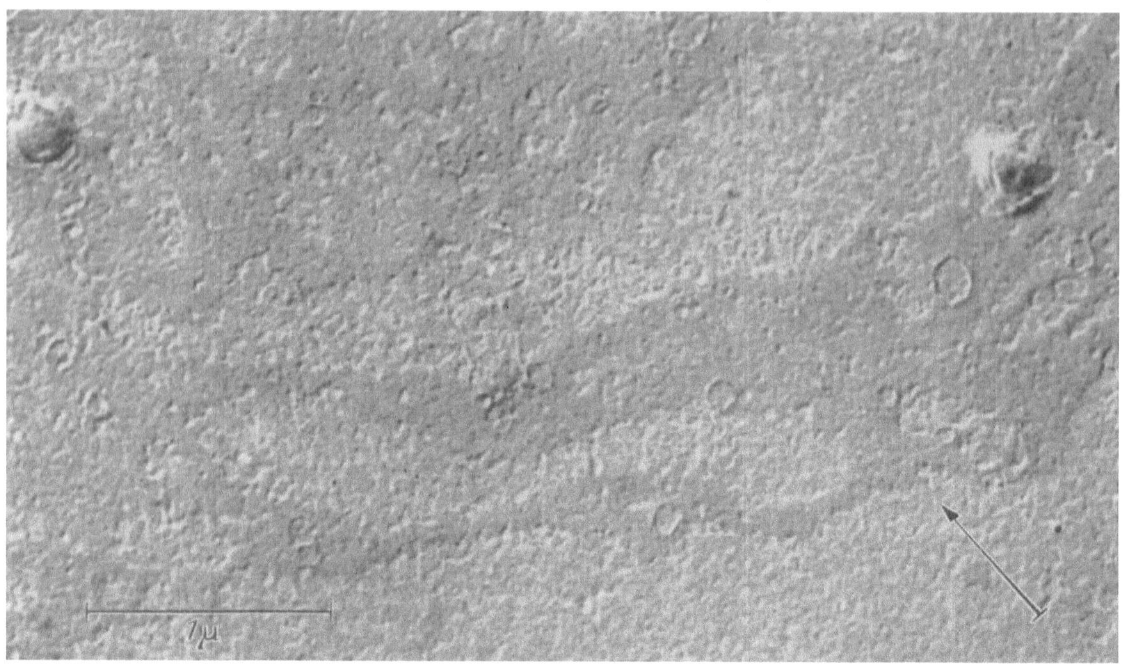

Abb. 4. Cholesterin-Gliadin-Film nach Unterspritzung von Span 60-Wasserdispersion bei Ausgangskompression von 5 dyn/cm und fortgesetzter Kompressionserhöhung bis 30 dyn/cm nach 60 min. Cholesterin durch umgekehrtes Auflegen des Films auf den Träger als runde Inselchen sichtbar geworden. Für die Rigidität des Films spricht dessen trotz starker Kompression immer noch netzartige Struktur („Span 60-Effekt")

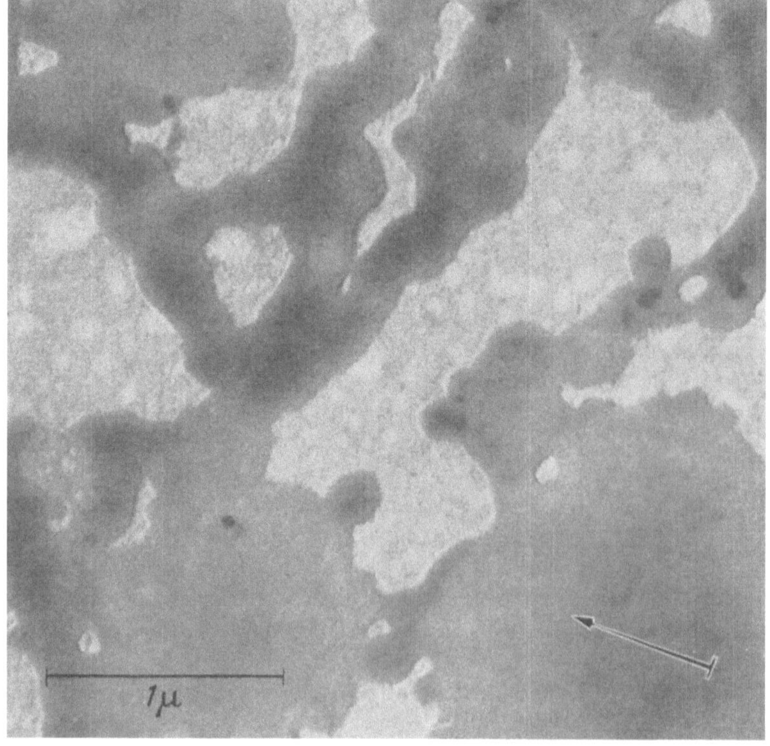

Abb. 5. Cholesterin-Gliadin-Film nach Unterspritzung von Tween 20-Wasserlösung bei Ausgangskompression von 5 dyn/cm und fortgesetzter Kompressionserhöhung bis 12 dyn/cm nach 60 min. Cholesterin offenbar verschwunden. Dieses Erscheinungsbild wiederholte sich ähnlich in allen Versuchen. Dies u. a. betrachten wir als stärksten Beweis dafür, daß kein Artefakt vorliegt („Tween 20-Effekt")

pression (22,5 dyn/cm) mehr oder minder vollständig vom Cholesterin-Gliadin-Film löst („Tween 20-Effekt"). Unterspritzt man Tween 20 + Cholesterin, so erhält man eine Isotherme vom Gliadin-Typ; der Präventivzusatz von Cholesterin hat demnach den Cholesterin-Gliadin-Film nicht gegen die Einwirkung des Tween 20 geschützt.

Die relativ groben Maßnahmen bei der Präparation der Lipo-Protein-Filme für die elektronenoptische Analyse könnten zur Entstehung von Artefakten führen. Den bisherigen Ergebnissen kann bei aller Vorsicht eine richtunggebende Bedeutung beigemessen werden, da die physikochemischen Messungen und die elektronenoptischen Befunde selbst unter verschiedenen Versuchsbedingungen einander adäquat sind.

Ein eingehender Bericht über die physikochemischen Befunde wird später erfolgen.

Mit Unterstützung der Stiftung *Sigrid Juselius' Stiftelse*, Helsinki, Finnland; sowie durch das Stipendium C-2930 M & G, *Institutes of Health, Public Health Service*, USA.

Literatur

1. SETÄLÄ, K., H. SETÄLÄ and P. HOLSTI: Science **120**, 1075 (1954).
2. — P. HOLSTI, S. LUNDBOM and H. SETÄLÄ: Co-carcinogenic lipids. Biochemical problems of lipids, p. 496. London 1955.
3. SETÄLÄ, H.: Acta path. microbiol. scand. Suppl. **115**, 1 (1956).
4. SETÄLÄ, K., H. SETÄLÄ, L. MERENMIES u. P. HOLSTI: Z. Krebsforsch. **61**, 543 (1957).
5. — P. HOLSTI, S. LUNDBOM, L. MERENMIES u. K. DAMMERT: Z. Krebsforsch. **61**, 569 (1957).
6. — u. L. STJERNVALL: Naturwissenschaften **45**, 203 (1958).
7. — — L. MERENMIES, P. HOLSTI, P. KAJANNE u. Y. AHO: Acta Un. int. Cancr. (im Druck) (1958).
8. — L. MERENMIES, L. STJERNVALL, P. KAJANNE u. Y. AHO: Acta Un. int. Cancr. **15**, 224 (1959).
9. HOLSTI, P.: Naturwissenschaften **45**, 394 (1958).
10. SCHULMAN, J. H.: The monolayer technique. Cytology and Cell Physiology. Ed. G. Bourne, p. 119, London 1952.
11. — and E. K. RIDEAL: Proc. roy. Soc. B **122**, 46 (1937).
12. ELEY, D. D., and D. G. HEDGE: Disc. Faraday Soc. **19—21**, 221 (1955).
13. RIES, H. R., and W. A. KIMBALL: Proc. 2nd int. Congress of Surface Activity **1**, 75 (1957).

E. Ergebnisse der Elektronenmikroskopie in der Zellmorphologie

1. Zellkern, Chromosom und Centriol

Problems in the study of nuclear fine structure

Keith R. Porter

The Rockefeller Institute, New York

The nucleus, both during interphase and mitosis, has come to be regarded as one of the most difficult of biological objects to study by methods of electron microscopy. Apart from the nuclear envelope there are few land marks and little evidence of order such as we find in collagen or muscle. What organization there may be is distributed in depth in the whole nucleus and so defies all but the most painstaking efforts in serial thin sectioning. It is not surprising, therefore, that the problem has proved discouraging except for the most determined and interested investigators. Of these we have a representative few on the program of this symposium and they will give you some measure of the progress that has been made to date. In view of the difficulties, their achievements must be regarded as impressive.

As a prelude to their presentation I have thought it would be valuable to review briefly some of the problems that are encountered and mention a few observations made on parts of the nucleus and its envelope that will not be discussed in the other papers in this symposium. These may help to fill out the total story and thus record the current state of our knowledge of these cell components.

The chromosomes. If we assume, as seems reasonable, that the physical arrangement of the nuclear elements is significant in the functioning of the nucleus, it is difficult to think of a more important object for investigation than the interphase nucleus. Here, in growing cells, DNA and its associated protein is reproduced before each division and here is synthesized the RNA and associated protein which serves, possibly, to transmit information in chromosomes to sites of active expression in the cytoplasm (*1, 2, 3, 4*). The processes taking place in this structure and the structural basis for their accomplishment, represent major problems of biology. It was naturally the hope of many biologists that the electron microscope would help to elucidate these problems.

The results, while perhaps not disappointing, have not been very helpful. Micrographs from a variety of materials prepared in several different ways have been consistent in providing evidence of the fibrillar nature of chromatin material. But, the reported dimensions of these have varied substantially. In nuclei of salivary gland cells (*Drosophila*) they measure 200 to 500 Å thick (*5*). And from the examination of nuclei from a number of different sources Ris concludes that a coiled microfibril about 200 Å thick is a universal structural unit of chromosomes (*6, 7*). Later evidence suggests that this is in turn constructed of two subunits (microfibrils) of nucleoprotein (see Ris — this symposium). The interchromosomal material of the interphase nucleus, where it can be identified in electron micrographs, is less dense than the chromatin and shows no definable structure except scattered dense particles which are similar to those identified as RNP particles in the cytoplasm (*8, 9, 10*). It is generally assumed that these may represent a product of nuclear synthesis and that from this point they find their way into the cytoplasm to influence there the synthesis of protein. Chromosomes in meiotic or mitotic prophase have been preferred to those in the interphase nucleus for studies of fine structure because they are much easier to identify with certainty. There is, furthermore, a much larger body of knowledge available on which to draw for instructive guidance and comparisons. If any order exists in chromosomes or even a pattern of structure, one might reasonably expect it to develop in this phase. Apart from the core structure which Moses has found in meiotic chromosomes of a number of forms (see paper in this symposium) the fine structure of the prophase chromosome also is described as fundamentally fibrillar. The size of the fibrils seems to vary from about 70 Å (*10, 11*) to a range between 200 and 500 Å (*7, 12, 13*). These measurements, it must be noted, were made on prophase chromosomes of different species which may account for the variation. It seems probable, however, that a part of the variation should be referred to preparation procedures and microscopy.

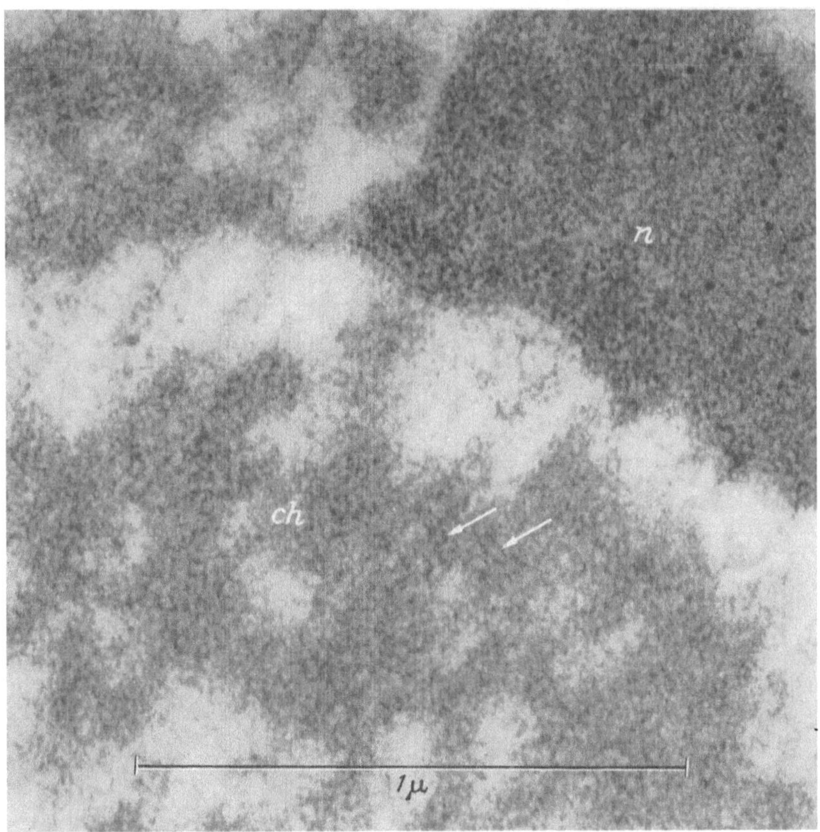

Fig. 1. Micrograph of small area of nucleus from cell of onion root tip, showing at upper right a portion of a nucleolus (*n*) and elsewhere parts of chromosomes (*ch*). The material of the chromosome appears fibrous in places (arrows). The preparation was stained with Pb(OH)$_2$ before microscopy. Mag. 75,000 ×

Some measure of the problems involved may be obtained from a brief discussion of a few illustrations. Fig. 1 shows a micrograph of a small part of a prophase chromosome in an onion root tip cell. The image is not worse than many published. A corner of the nucleolus is included for reference purposes. The cortical part of the nucleolus is obviously granular in nature (see below). The granules measure about 15mμ. The texture of the adjacent prophase chromosomes is obviously different and recognizably fibrillar. That none of the microfibrils can be followed more than a few hundred Ångstrom units suggests that, if long, they follow an irregular course. There is nothing to suggest that they are coiled or comparable to the helical units observed by Pappas in the nucleus of *Amoeba chaos* (*14*). Measurements can be made on selected fibrils, but if made completely at random the values would vary greatly. Obviously something better in the nature of an analytical approach is required for the study of this kind of material. The picture following osmium fixation, as shown here, doubtless has some meaning but it is influenced so markedly by operations unrelated to fixation that one is obliged to question the value of high resolution studies on such material unless all factors are carefully controlled.

Fig. 2. Portion of a metaphase chromosome from a cell of mouse sarcoma 180. The section, supported on a carbon film, was examined in an Elmiskop I *without* covering or staining, The "microfibrillar" elements of chromosome fine structure are obvious. Mag. 84,000 ×

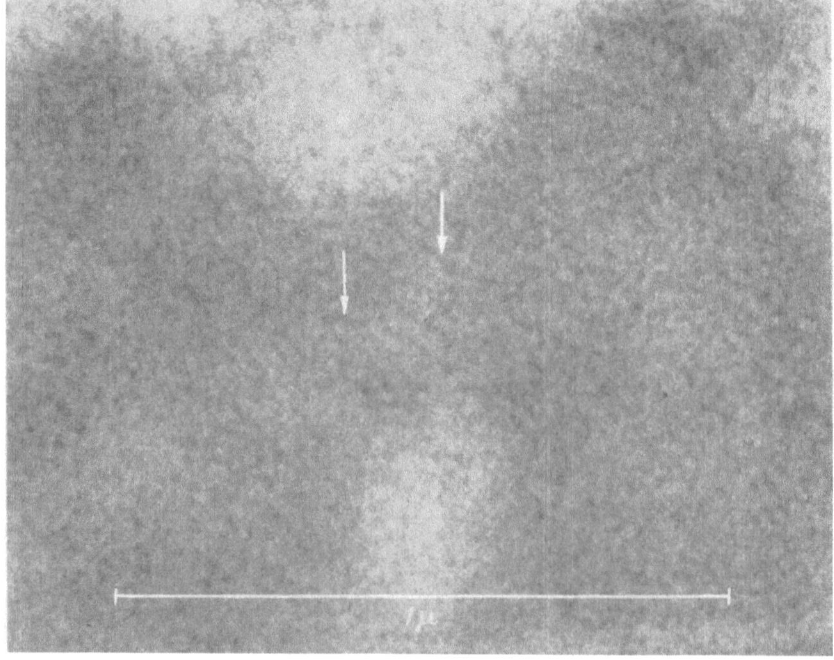

Fig. 3. Portion of same metaphase chromosome from a section adjacent to that used for Fig. 2. Here the section was "protected" against beam damage by a thin cover, or blanket, of formvar. It is now more difficult to discern the "microfibrillar" components of the chromosome. Two are designated by arrows. Mag. 84,000 ×

Brief attention to one aspect of electron microscopy will serve to illustrate this point. WATSON has recently reported that a specimen may change very appreciably in its fine structure under the electron beam (*15*). Apparently a part of the methacrylate may leave the specimen and smaller units of the embedded material may migrate to the larger. Some control over this effect of the beam is achieved by covering the section with a blanket of carbon or formvar. We have examined the effect of this operation on the appearance of metaphase chromosomes in cells of the same type. The effects are obviously striking (see Fig. 2 and 3) and one gains the impression that the blanketed preparation is the more faithful image of the fixed material even though it provides a less distinct (contrasty) image of the chromosome "fibrils". Exactly what techniques may eventually be

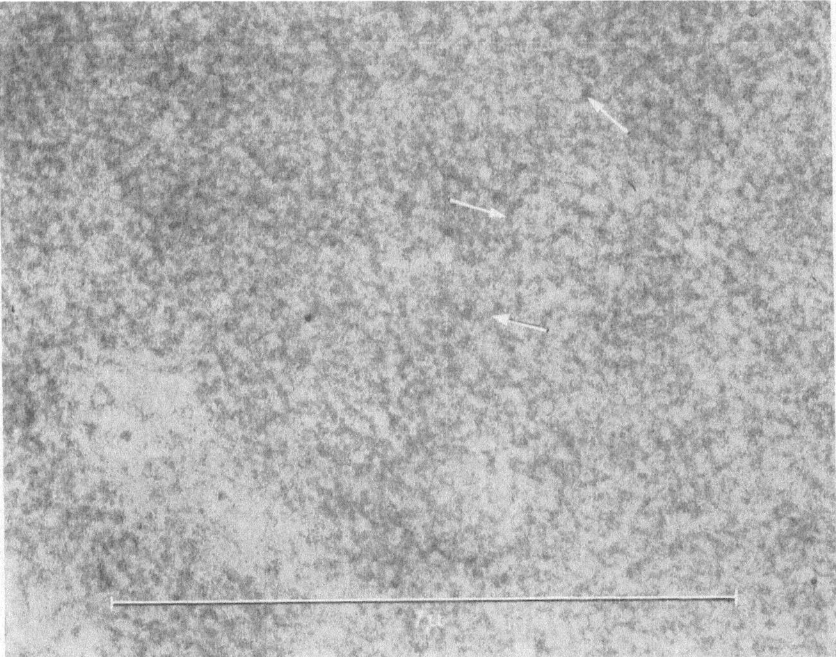

Fig. 4. Same material as shown in Fig. 2 and 3. Here the section was stained with Pb(OH)$_2$ before it was blanketed with formvar. Obviously the staining has served to increase the density of the microfibrils (arrows). Mag. 85,000 ×

devised to help electron microscopy unravel the details of chromosome fine structure are, at this point, a little hard to guess. One may suppose that selective extraction or enzyme digestion of certain components will provide useful information, or that this combined with "staining" intelligently employed may give interpretable images (*16*) (see Fig. 4). At the moment it is easy to recognize that there is much room for improvement in so far as the major part of chromosomal studies are concerned. Selected materials, known to experienced biologists, are at present yielding the most information (*17, 18*), and possibly from what is learned here, the truth about chromosomes in general will emerge.

The nucleolus. Studies on the fine structure of the nucleoli of interphase nuclei are perhaps more rewarding than investigations on chromosomes. Here, at least, a few definable components appear consistently and seem not to be the product of any steps in the usual procedures. One of these resolvable elements is similar in character to the RNP particles of the cytoplasm and has been interpreted as the product of RNA and protein synthesis taking place in the nucleus. It is reasonable, in line with generally accepted evidence on nuclear (and nucleolar) origin of cytoplasmic RNA (*19, 20*), to suggest that particles such as those found in the nucleolus and in the nuclear sap, may eventually find their way into the cytoplasm either at the time of cell division and nucleolar breakdown or by migration through pores in the nuclear envelope. Nucleoli are, of course, well known as prominent bodies of the interphase nucleus. They vary greatly in number, size and morphology. In some instances, as generally in oocytes, they are numerous and appear simply as spheres. More commonly they are present in small numbers and show some evidence of the nucleonemata described by ESTABLÉ and SOTELO (21, 22).

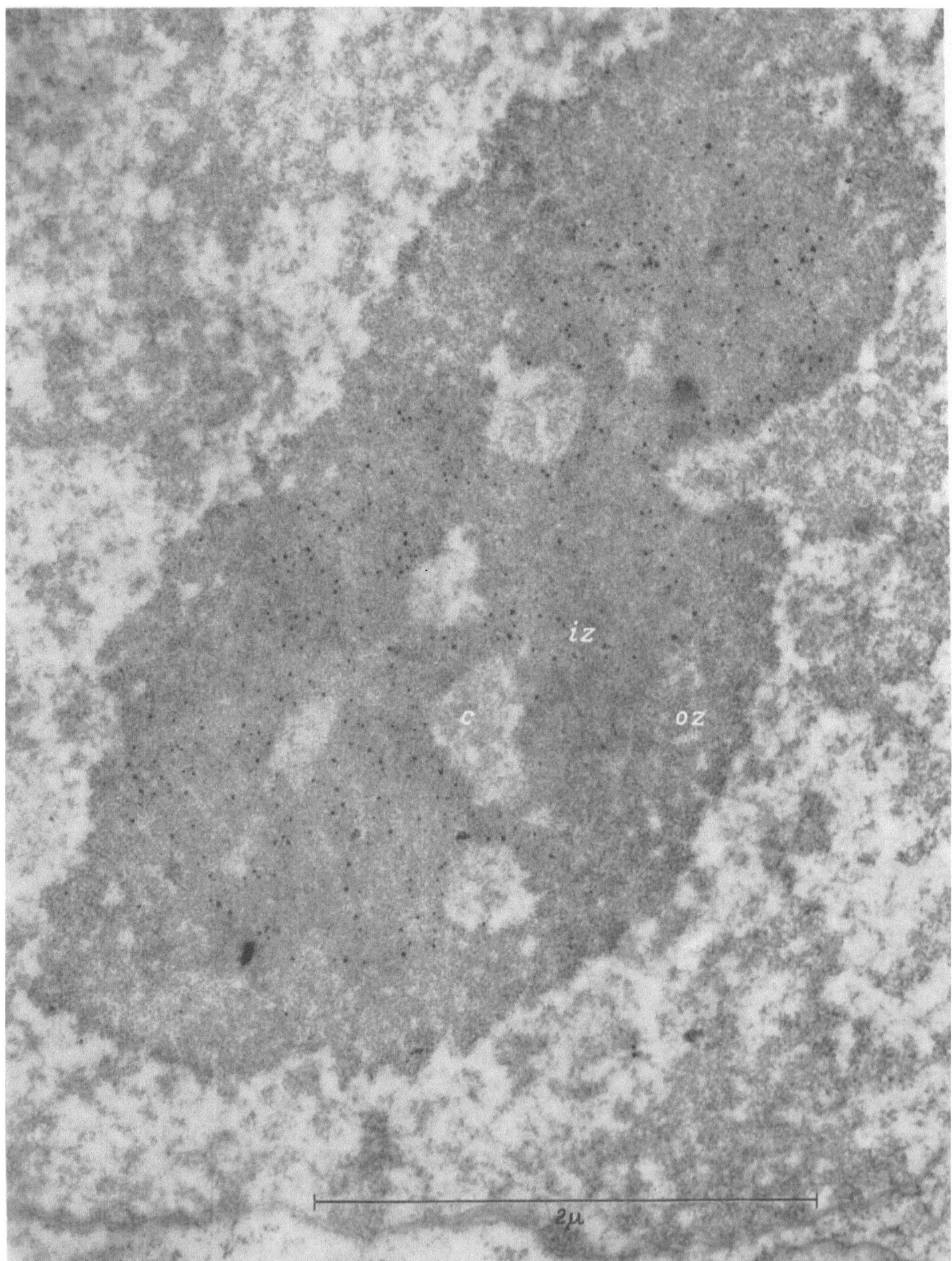

Fig. 5. A nucleolus in an interphase nucleus of a cell from onion root tip. Core material of the nucleolus is evident in at least 4 large spots (c). The balance of the structure identified as the cortex is divided into 2 zones the inner of which (iz) is the more dense and is characterized by the presence of extremely dense particles noted earlier by LAFONTAINE (27). The outer zone (oz) is more homogeneous and seems to be composed of particles of uniform density and size (see Fig. 6). Mag. 38,500 ×

Light microscopy has described nucleoli as consisting of two parts or zones of different density and staining properties: an outer cortical zone (*pars amorpha*, of Estable) and an inner zone or core (*23, 24, 25*) and there is evidence that the cortical zone is rich in RNA (*25, 26*). These same gross features of structure appearing as a cortex of high density and a central region or regions of

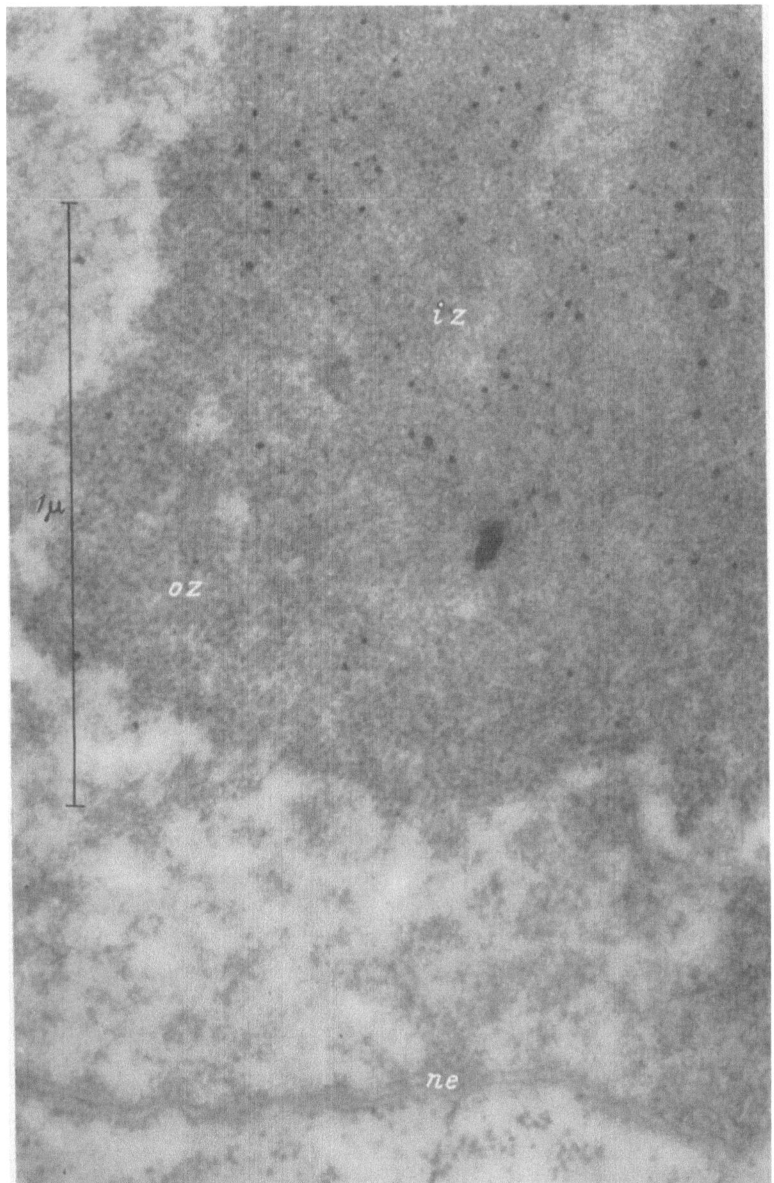

Fig. 6. Portion of nucleolus in Fig. 5 — enlarged to show fine structure of cortex. The outer zone (*oz*) is obviously made up largely of uniform particles ∼150 Å in diameter that resemble closely particles in cytoplasm at bottom of micrograph. The fine structure of the inner zone (*iz*) is not so readily analyzed. The nuclear envelope is at *ne*. Mag. 80,000 ×

low density can be observed in electron micrographs from several sources (*27, 28*). (Fig. 5). The core material frequently has the density and texture of the surrounding chromatin and may therefore be interpreted as the nucleolar organizer of chromosomal origin (*29*). The dense cortical portion of the nucleolus has attracted more attention from electron microscopists. Thus there are several reports that this zone is made up of granules (15 to 20 mμ in diameter) densely packed together (*8, 10, 27, 30*). As pointed out above, these bear a strong resemblance in size and density to granules prominent in the cytoplasm of cells actively engaged in protein synthesis. That both represent nucleoprotein seems reasonably well established. The general zones of the nucleolus are depicted in Fig. 5, which represents that body as it is commonly found in interphase nuclei of the onion root tip. The particulate composition of the outer zone of the nucleolar cortex is depicted in

Fig. 7. Nucleolus observed in muscle fiber of larval form of *Amblystoma*. One end of the nucleolus is contiguous with nuclear envelope (*ne*). Two masses of particulate material at opposite ends of nucleolus correspond to outer zone (*oz*) of nucleolus shown in Fig.5 and 6. Denser material constituting patterned portion of nucleolus resembles inner zone (*iz*) of nucleolus in Fig. 5. Mag. 26,000 ×

Fig. 6. A membrane, if present around the nucleolus as sometimes mentioned (*19*), has not been preserved in this preparation. The margin composed solely of particles has been noted in this material to vary in width from extremely narrow in telophase to broad in early prophase. This suggests that there is some synthesis of these particles during interphase. More centrally in the nucleolus, i. e. deeper in the central zone, these particles blend with other more finely divided material including the extremely dense particles noted first by Lafontaine (*27*). It is assumed that all this cortical portion of the nucleolus is rich in RNA, but whether it is distributed evenly throughout the cortex remains in doubt. It is important to note that essentially the same particulate materials can be recognized in nucleoli from very different sources. Fig. 7, 8 and 9, for example, depict the fine structure of the nucleolus commonly found in muscle fibers of larval forms of *Amblystoma* (*8*). It is evident in Fig. 7 that this nucleolus is also made up of two dense components. One of them, present in two masses at opposite ends of the nucleolus, consists of particles identical to those in the outer cortical zone of onion nucleoli (compare Fig. 6 and 8). The other dense component shows evidence of the same particles embedded in a dense matrix and resembles the material making up the inner cortical zone of the onion nucleolus. Here (Fig. 7) this part of the nucleolus is organized into a complex pattern, the full nature of which is evident from the additional view of it (cross section) in Fig. 9. This degree of complexity is unique and

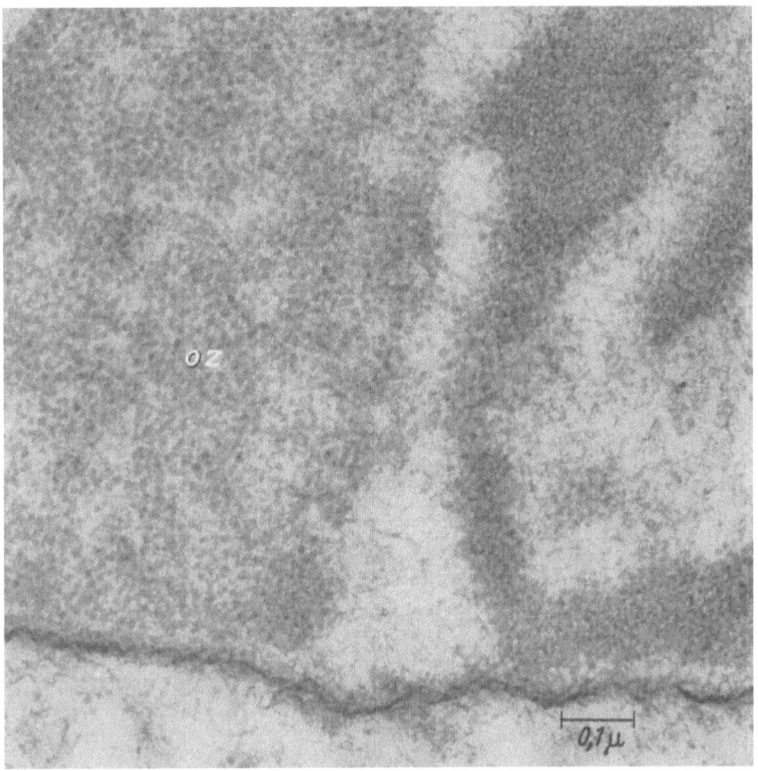

Fig. 8. Higher magnification of part of nucleolus shown in Fig. 7. Particles of outer zone (*oz*) appear to have centers of lower density. Mag. 100,000 ×

is to be regarded as a differentiated form characteristic of these muscle cells. The order achieved by this nucleolar material — presumably in response to the nucleolar organizer (*29*) — may reflect a similar degree of order present but not evident in other components of the nucleus.

The nuclear envelope. Finally, in an attempt to arrive at a fairly complete picture of our contemporary knowledge of the nucleus, I feel that something should be said about the nuclear envelope. Whether this structure should be regarded as part of the nucleus or the cytoplasm is debatable, but for present purposes we may, I think, include it with the nucleus. Further comment on this point is contained in what follows. As repeatedly described, the nuclear envelope (*9*) appears to consist always of two membranes enclosing (as in a vesicle) an intervening narrow space. The continuity of this double structure is interrupted at certain points by pores (30 to 70 mμ in diameter) which are patent and which establish a relatively free pathway for the possible movement of large particles between the nuclear sap and the matrix of the surrounding cytoplasm. The outer or cytoplasmic surface is usually covered with particles; the nuclear surface is contiguous with the chromosomes. These facts are now well established through the observations of WATSON and many other microscopists (*31, 32*). The number of pores seems to vary as does also their morphology, but the essential features of their existence, along with the three part structure of the envelope, are fundamental constants. This structure or a recognizable varient of it is found around the nucleus of all kinds of cells, including those of the fungi (*33*). The bacteria and blue-green algae are the only known exceptions (*34*). Were they known, the reasons for these exceptions

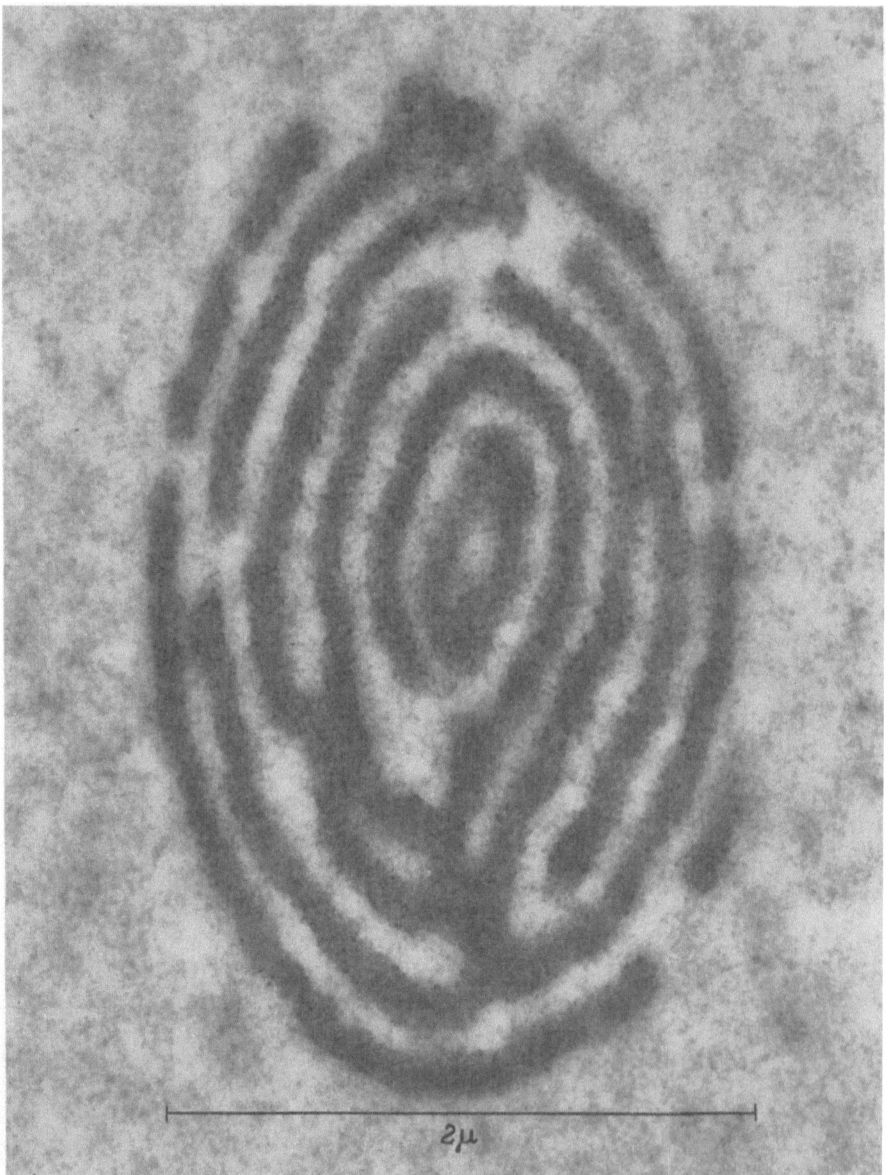

Fig. 9. Micrograph of transverse section through nucleolus of type shown in Fig. 8. It depicts the complex organization characteristic of these nucleoli. Mag. 40,200 ×

might provide some insight into the functions of this remarkable structure which limits the inter-phase nucleus.

The role of this envelope in the interchanges which must take place between nucleus and cytoplasm, while not known at the moment, seems within the scope of experimental investigation. In this connection certain recognizable structural relations between the envelope and cyto-plasmic elements provide some useful ideas. It is now clearly evident, e. g., that the outer mem-brane of the envelope is continuous with the membrane-limited vesicles and tubules found in the cytoplasm, which are referred to as elements of the endoplasmic reticulum (*er*) or ergastoplasm.

Thus these elements may be regarded as outpocketings of the nuclear envelope and obviously share in common the properties of the space within the envelope and the outer of the two membranes. It matters little, for purposes of discussion, whether the continuity with the envelope is

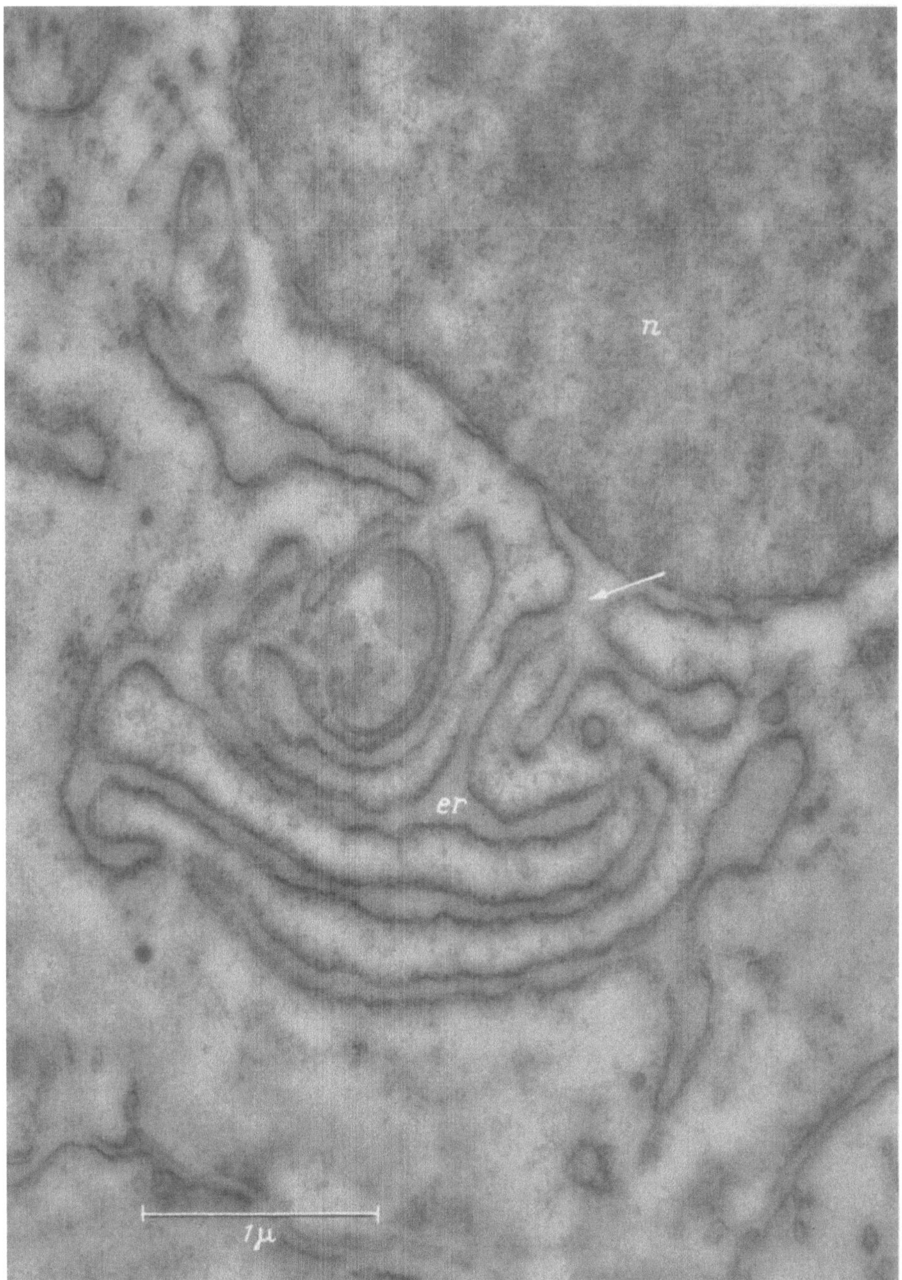

Fig. 10. Micrograph depicting connection between endoplasmic reticulum and the nuclear envelope in chorda cell of *Amblystoma* larva. It is obvious that the perinuclear space within the envelope is continuous at arrow with the space enclosed by membranes of *er*. Nucleus is indicated at *n*. Mag. 32,000 ×

maintained, the fact of its having existed makes these cytoplasmic elements essentially extensions of the envelope. In some instances the membrane-limited units of the cytoplasm repeat, even to patterns of pores, the structure of the envelope (*35*). Most commonly a connection between the two seems to be maintained (Fig. 10). This clearly provides for the channeling of metabolites between the space in the envelope and the extreme margins, if need be, of the cytoplast. It also provides for the flow of membranes (assuming they have this property) from intimate contact with chromosomes at the nuclear surface to the outer reaches of the cytoplasm. Thus, patterns

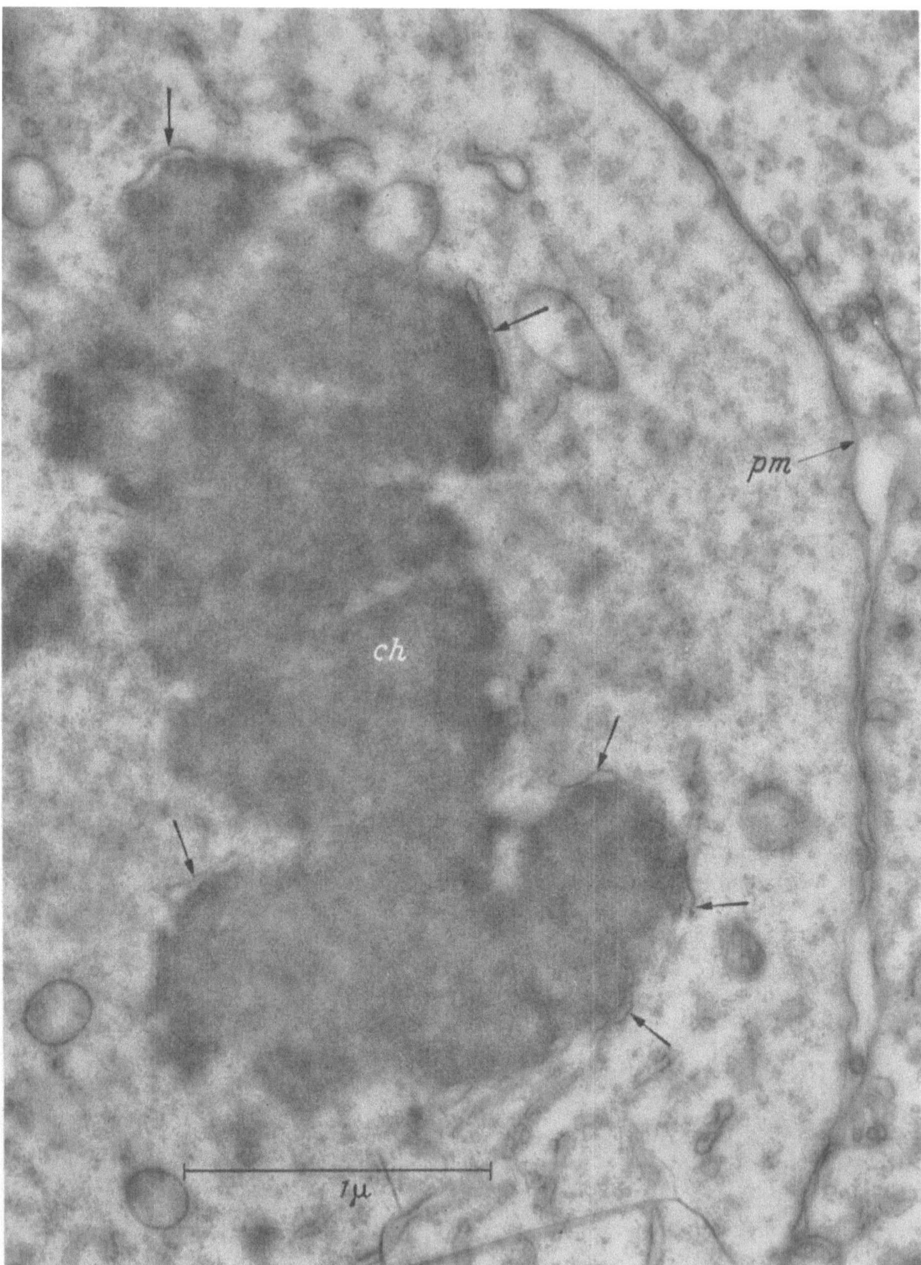

Fig. 11. Late anaphase chromosomes (*ch*) in cell from mouse sarcoma 180. The cell margin (*pm*) is at the right. Profiles of vesicular elements are evident (at arrows) at various points on chromosome surface. These represent an early stage in the reformation of the nuclear envelope. Mag. 42,000 ×

of structure and composition built into the inner membrane of the envelope could conceivably, at a later moment, be transmitting the information to a growing or differentiating part of the cytoplasm. Though this is highly speculative, it is to be noted that there is morphologic evidence of elaborate differentiations within parts of the *er*, some for association with RNP particles (*36*),

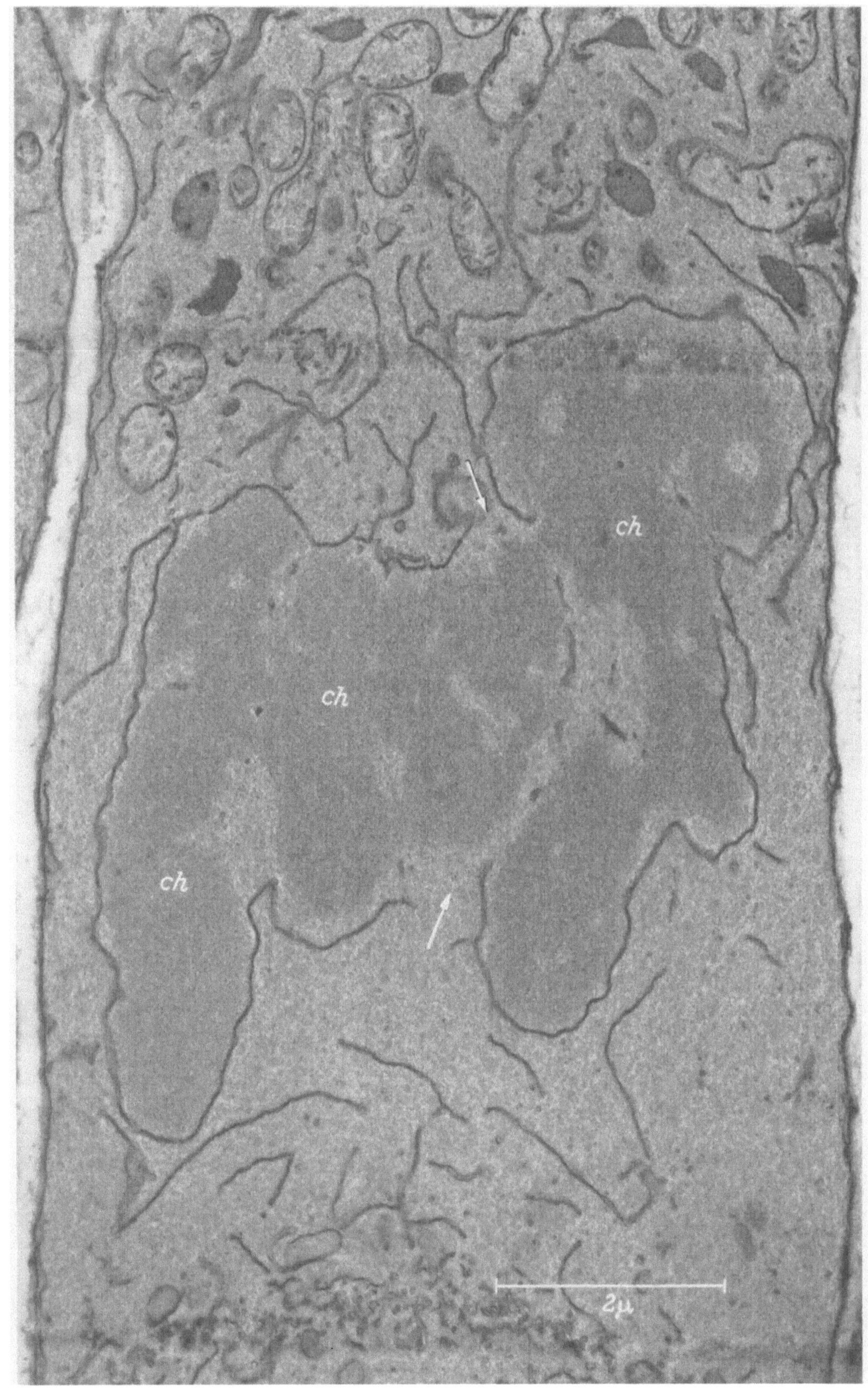

Fig. 12. Section through nucleus of onion root tip cell in early telophase. The nuclear envelope is in the final stages of formation. Only one or two gaps are still evident in the continuity (arrows) of the envelope. The elements going into its construction are not noticeably different from those remaining in the cytoplasm as parts of the endoplasmic reticulum and continuity is evident at a number of points. Telophase chromosomes are marked (*ch*). The new cell plate is included at the bottom of the figure. The material was fixed in $KMnO_4$. Mag. 18,000 ×

some for complex association with myofibrils (*37*) and some with characteristic structure but without obvious reason. In a sense, all of the forms of these *er* are based on the nuclear envelope.

It has occurred to many investigators, I am sure, that a better understanding of the significance of the envelope might be derived from an examination of its behavior during mitosis. Do parts of it retain a connection with individual chromosomes and, if not, do they persist anywhere in recognizable form to be divided between the two daughter cells, there to reform the new envelope? We made some attempts at an investigation of this problem a few years ago in an examination of thin sections of growing animal tumors (*38*). It was found, as has now been observed by others (Moses — this Congress), that in late prophase when the nuclear membrane (classical terminology) breaks down, fragments of the envelope persist as vesicles in the cytoplasm. In the tumor cells (fixed with OsO_4) these vesicles were small and widely scattered and appeared to lose their identity in the general population of *er* elements. The latter point is of course difficult to show with certainty. Later, as the chromosomes moved toward the poles of the spindle in late anaphase, profiles of vesicles reappeared at their surfaces (Fig. 10). Quite rapidly, it would seem, these either increased in surface area or coalesced with other vesicles to yield the continuous envelope.

In a similar study of onion root tip cells, recently made in collaboration with Raul Machado (with $KMnO_4$ as a fixative), it was found that fragments of the nuclear envelope are incorporated in the spindle and retain a position which brings them into intimate contact with the telophase chromosomes for envelope formation (Fig. 11). Thus the envelope seems to retain a degree of integrity through mitosis and to be divided according to some plan between two daughter nuclei.

These observations fall into a class of preliminary findings where, unfortunately, a large part of our information about the nucleus now resides. It will, however, be obvious from what follows in this symposium program that very substantial progress is being made and that before the next gathering of this International Society we should witness discoveries in this area paralleling in importance those that have characterized explorations of the cytoplasm during the period since the last meeting.

References

1. Allfrey, V. G., A. E. Mirsky and S. Osawa: J. gen. Physiol. **40**, 451 (1957).

2. — — Proc. nat. Acad. Sci. (Wash.) **43**, 821 (1957).

3. Pelc, S. R., and A. Howard: Exp. Cell. Res. **10**, 549 (1956).

4. Brachet, J.: Biochemical Cytology. New York: Academic Press 1957.

5. Gay, H.: J. biophys. biochem. Cytol. **2** (suppl.), 407 (1956).

6. Ris, H.: J. biophys. biochem. Cytol. **2** (suppl.), 385 (1955).

7. — Chromosoma structure. A symposium on the chemical basis of heredity. W. D. McElroy and B. Glass, editors. Baltimore: Johns Hopkins Press 1957.

8. Porter, K. R.: J. Histochem. Cytochem. **2**, 346 (1954).

9. Watson, M. G.: J. biophys. biochem. Cytol. **1**, 257 (1955).

10. DeRobertis, E.: J. biophys. biochem. Cytol. **2**, 785 (1956).

11. Moses, M. J.: J. biophys. biochem. Cytol. **4**, 633 (1958).

12. Gall, J. G.: Chromosomal differentiation. A symposium on the chemical basis of development. W. D. McElroy and B. Glass, editors. Baltimore: Johns Hopkins Press 1957.

13. Kaufmann, B. P., and D. N. De: J. biophys. biochem. Cytol. **2**, (suppl.), 419 (1956).

14. Pappas, G. D.: J. biophys. biochem. Cytol. **2**, 221 (1956).

15. Watson, M.: J. biophys. biochem. Cytol. **3**, 1017 (1957).

16. Watson, M. L.: J. biophys. biochem. Cytol. **4**, 727 (1958).

17. Grassé, P. P., N. Carasso et P. Favard: Ann. sci. nat. Zool. et Biol. anim. **18**, 339 (1956).

18. Nebel, B. R.: J. Hered. **48**, 51 (1957).

19. Caspersson, T. O.: Cell growth and cell function. p. 103. New York: W. W. Norton & Co. 1950.

20. Goldstein, L., and W. Plaut: Proc. nat. Acad. Sci. (Wash.) **41**, 874 (1955).

21. Establé, C., et J. Sotelo: Inst. Inv. Cien. Biol. Publ. **1**, 105 (1951).

22. — and J. R. Sotelo: In: Fine structure of cells. p. 170. Groningen: Noordhoff Ltd. 1954.

23. Hogben, L. T.: Proc. roy. Soc. B. **91**, 305 (1920).

24. Zirkle, C.: Cytologia **2**, 85 (1931).

25. Chayen, J., H. G. Davies and U. J. Miles: Proc. roy. Soc. B **141**, 190 (1952).

26. Thorell, B.: Acta med. scand. **117**, 334 (1944).

27. Lafontaine, J. G.: J. biophys. biochem. Cytol. **4**, 777 (1958).

28. Sotelo, J. R., and K. R. Porter: J. biophys. biochem. Cytol. **5**, 327 (1959).

29. McClintock, B.: Z. Zellforsch. **21**, 294 (1934).
30. Bernhard, W., A. Bauer, A. Gropp, F. Haguenau et Ch. Oberling: Exp. Cell Res. **9**, 88 (1955).
31. Bahr, G. F., and W. Beermann: Exp. Cell Res. **6**, 519 (1954).
32. Hartmann, J. F.: J. Comp. Neurol. **99**, 201 (1953).
33. Shatkin, A. J., and E. L. Tatum: J. biophys. biochem. Cytol. (in press).
34. Bradfield, J. R. G.: Symp. Soc. Gen. Microbiol. **6**, 296 (1956).
35. Rebhuhn, L. I.: J. biophys. biochem. Cytol. **2**, 93 (1956).
36. Palade, G. E.: J. biophys. biochem. Cytol. **1**, 59 (1955).
37. Porter, K. R., and G. E. Palade: J. biophys. biochem. Cytol. **3**, 269 (1957).
38. — In: Fine structure of cells. p. 236. Groningen: Noordhoff Ltd. 1954.

Patterns of organization in the fine structure of chromosomes*

Montrose J. Moses**

The Rockefeller Institute, New York

When observations on the chromosomes were limited to light microscopy, the failure of cytologists to reach a consistent formulation of chromosome organization much below the level of the chromatid could be laid largely to insufficient resolution in the microscope and to variations in procedures for preparing the biological material. Attempts to hypothesize a common structural pattern that would account for such divergent modifications as seen in lampbrush, giant polytene and mitotic chromosomes were at best difficult to test unequivocally with the means at hand. Even answers to such questions as whether the anaphase chromosome is split longitudinally once, twice or not at all were in disagreement (*1*). The lack of solutions to such problems lay directly in the path to an understanding of the structural basis for the genetic organization and function implicit in the gross structure that we know as the chromosome.

The advent of the electron microscope with its inherent hundredfold increase in resolution promised a fresh approach to these problems. But at first the methods for preparing biological material were crude and the results left much to be desired. Observations rested heavily on individual interpretation. As preparative methods improved, particularly with the introduction of osmium fixation, plastic imbedding and thin sectioning, the elegance and detail of information about the fine structure of the cell in all its parts, increased apace. But whereas the nucleus was the focus of many light microscopists who, with geneticists, had accumulated a disproportionately large body of information about its structure and function, it is now information about cytoplasmic organelles that has accrued under the electron microscopist's attention.

One reason may be that of all the cell structures in an electron image, chromatin is rarely arresting at first glance and its organization can be discovered only with considerable difficulty. Two major handicaps are contributory: there are seldom sharp boundaries separating grosser structural phases (with the possible exception of the nucleolus) within the nucleus, nor are delimiting membranes to be found, except at the nuclear periphery. Secondly, the finest structural unit of chromatin, so far amenable to observation with the electron microscope, is of the order of 100 Å, roughly two orders of magnitude below the chromosome itself. Bridging this gap involves ingenuity and considerable labor, and imposes an uncertainty in interpreting the underlying pattern of organization.

The one observation that emerges consistently, however, is that a microfibril of unknown length and varying in diameter (with different reports) from about 50 to 200 Å, appears to be the smallest visible subunit of the chromosome (*2, 3, 4, 5, 6, 7, 8*). Such a microfibril begins to approach the dimensions of a molecule of the deoxyribonucleic acid (DNA) that is practically a universal component of chromosomes and that presumably imparts genetic specificity.

Substantial efforts have been made to bridge the gap between microfibril and chromosome, making use of such technical variations as reconstruction of serial sections, modifications of

---
* Portions of the work discussed in this paper were supported in part by a Special Purposes Grant (INSP-85*a* and *b*) from the American Cancer Society, Inc.
** Present address: Department of Anatomy, *Duke* University School of Medicine, *Durham, North Carolina.*

fixation, extraction and shadowing of sections, careful selection of biological material, etc. Nevertheless, conclusions of contemporary workers are presently far from being in agreement (see discussion in *3* and *7*). While it is unlikely that the pattern by which the microfibrils are assembled into a chromosome will be unequivocally revealed with ease, recent observations on meiotic chromosomes, which will be discussed below, hold promise of providing valuable clues to the solution.

There are certain difficulties that must be taken into account in interpreting the electron images of chromosomes. In the first place, it is well known that variations in fixation produce

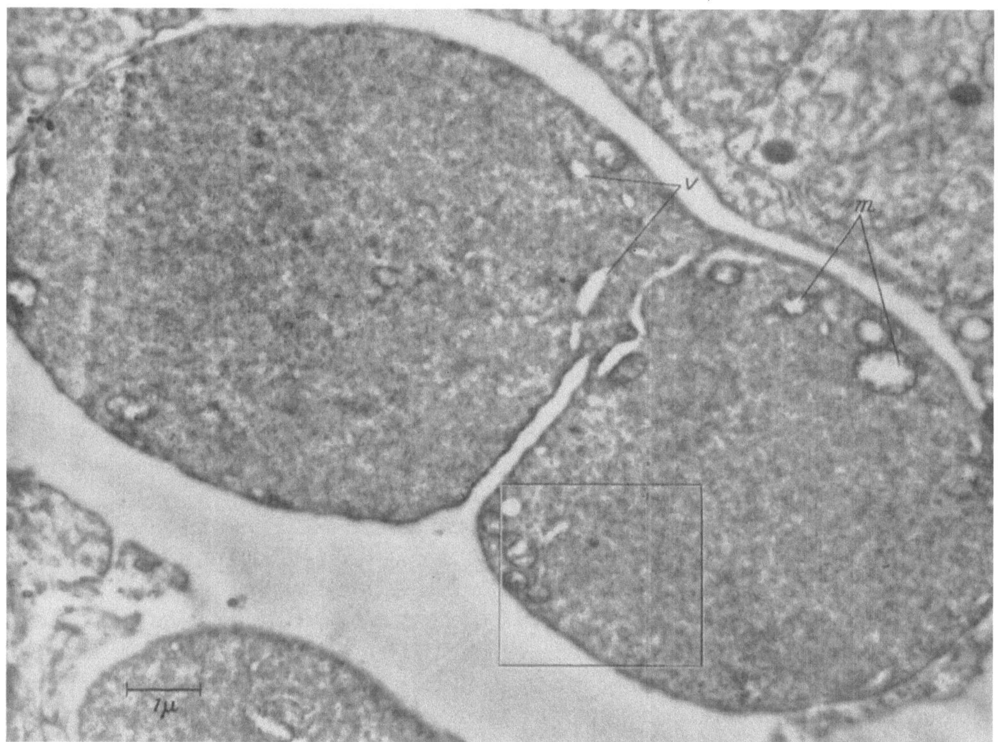

Fig. 1. Electron micrograph of two cells (presumably erythroblasts) from the liver of a newborn rat. Except for the cell wall, a few mitochondria (*m*) and some vesicular elements (*v*), structures larger than the granular material filling the cells are obscure. Mag. approx. 10,000 × (Note: The material in this and succeeding micrographs was fixed in buffered 1% OsO_4.)

profound changes in the appearance of chromosomes in the light microscope. While the nuclei in osmium fixed preparations resemble closely those in the living state, as seen in the light microscope, this mode of preservation has been one of the least satisfactory for seeing chromosome organization clearly in electron images (*9*). Secondly, although it has been shown that DNA and chromosomes at the light microscope level are preserved in osmium preparations (*9*) a cautious view should be taken about the extent of preservation (cf. *10*). Particularly since DNA appears to be non-reactive to osmium-fixation (*11*) there remains the unproven possibility that some DNA may be lost or translocated within the cell, and fine structure altered.

A third problem is imposed by the nature of thin sections of nuclear material in the electron microscope. This is best illustrated in Fig. 1 and 2. Fig. 1 is an electron micrograph of two erythroblasts from an osmium-fixed section of new-born rat liver. Very little detail is visible at the 1 to 0.1 μ level, except for a few mitochondria easily recognizable because they are membrane bound areas of low density. At higher magnification, there are abundant structures at the 100 to 10 mμ level, mostly of a fibrous and granular nature, and these are clustered and distributed differentially in a vague fashion that presumably reflects some pattern of organization (Fig. 3). But from such an image it is practically impossible to decide what the pattern may be. Fig. 2 is a 2 μ section of the same cells, cut immediately adjacent to those in Fig. 1 and stained by the Feulgen procedure (*9*). It is immediately apparent that both cells are in prometaphase of mitosis and contain perfectly recognizable chromosomes with at least a large part of their DNA intact. Thus, while there is actually present in these nuclei a structural pattern reflecting a functional state, it is visibly differentiated in the light microscope preparation and hidden in the electron image (compare comparable regions containing chromosomes, Fig. 2 and 3). The aforementioned lack of membranes and large size of the chromosome (1 μ) compared with the thickness of the section (0.1 μ), together with the obvious fact that the electron scattering power (hereafter referred to as density) of the chromosomal material is here similar to that of the interchromosomal material, all contribute to the camouflaged electron image.

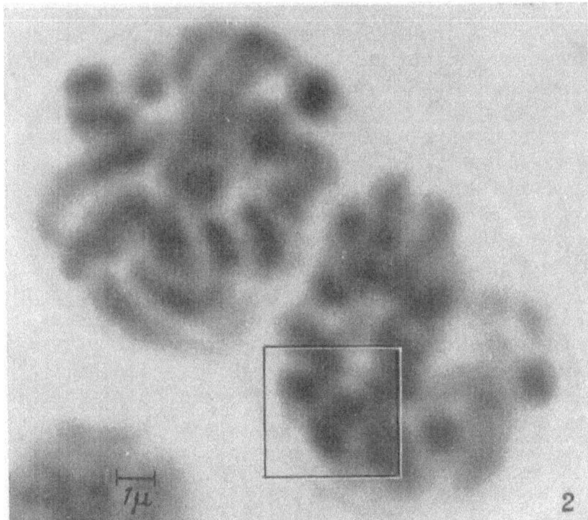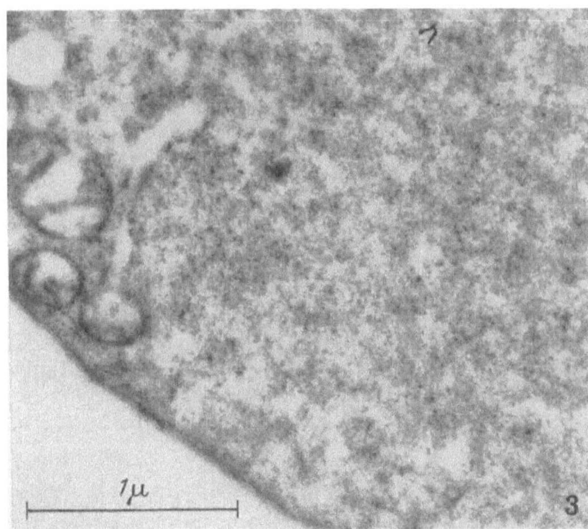

Fig. 2. Light micrograph of the same two cells in an immediately adjacent 2 μ section which has been stained by the Feulgen procedure. Typical chromosomes are clearly apparent. Their arrangement suggests that both cells are in prometaphase of mitosis. Mag. approx. 5,100$\times$

Fig. 3. Enlarged portion of the cell outlined in Fig. 1 and 2. Although oblique sections of at least two chromosomes are present in this area as judged from Fig. 2, it is difficult to discern their presence or location in the electron micrograph. Mag. approx. 26,000$\times$

Clearly, the foregoing is an extreme example that does not hold for all nuclei. But it emphasizes that the apparent lack of structural organization one or two orders of magnitude above the finest structures that can be discerned, may belie the very distinct pattern that is actually there. To reconstruct the true pattern from the electron image alone would be a most laborious and hazardous undertaking. An analogy may be extended to the chromosome at metaphase, where, for example, there is a characteristic absence of obvious structure intermediate between the chromosome itself and its fine structure (*12*). It is well known from light microscopy that such organization is there; that there are at least two strands per chromatid and that at least two hierarchies of coiling are largely responsible for its condensed appearance. But without special treatment resolution of such detail approaches the limits of light optics. Unfortunately, adjacent thick and thin sections are of little help in clarifying electron images of structure that cannot be clearly resolved in the light microscope. We are as yet without alternative means of accentuating chromosome structure in the electron image. The necessity for selecting material where the greatest amount of structural differentiation will be visible in the electron microscope, and as far as possible, to make careful correlations with the light microscope image, is obvious.

Since the pattern of linear organization in somatic chromosomes is elusive and appears to be obscured by coiling, it is logical to expect it to be more apparent in uncoiled chromosomes. In the prophase of the first (heterotypic) meiotic division, chromosomes are known to be in their most extended state (eg. *13*). From the precise point by point pairing of homologous chromosomes along their lengths at synapsis, one might predict a more highly refined degree of linear organization than at other stages.

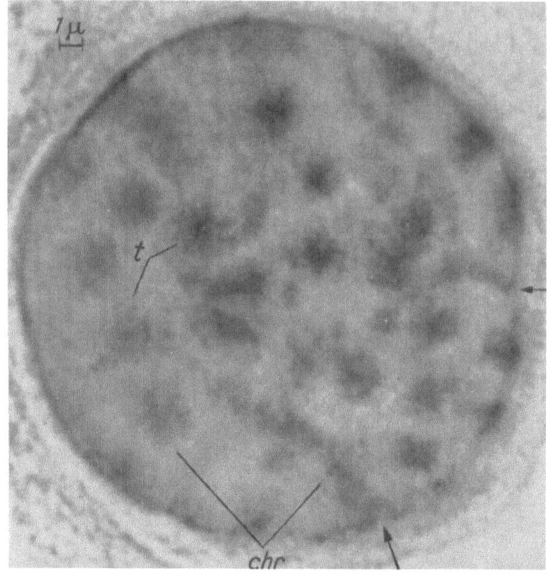

Fig. 4 (left). Light micrograph of a Feulgen stained 2.5 μ section of a primary spermatocyte prophase nucleus from the salamander, *Plethodon cinereus*. From several lines of evidence, it has been determined that these nuclei are at zygotene (*18*) and that the chromosomes seen here in cross and longitudinal sections (*chr*) are for the most part synapsed (bivalents). In most of the cross sections, the central region of the chromosome is more Feulgen positive than the periphery. Tufts of chromatin radiate from the axis (*t*). Two chromosomes, seen in longitudinal section, terminate characteristically at the nuclear envelope (arrows). The interchromosomal region is palely Feulgen positive. Mag approx. 2,600 ×

Fig. 5 (s. below). Electron micrograph of a nucleus similar to that in Fig. 4; this is a relatively thick section. Here, chromosomes are not easily distinguished because of camouflaging by dense granular interchromosomal material (*ic*). Two chromosomes, seen in longitudinal and cross section, (*chr, l* and *x*) have been marked out roughly along the boundary between the fine chromatin (*ch*) material and interchromosomal granules. In the longitudinal section, as elsewhere in the nucleus, a dense line passes along the chromosome axis. This is a single element of the synaptinemal complex; it cannot be decided from this micrograph alone whether the element is the axis of an unpaired chromosome, or whether it is part of the complex with the remaining elements lying above or below it out of the section. Cross sections of both single and complex structures are visible; that outlined [*chr (x)*] is complex. The two lateral elements of the complex can be seen in longitudinal and cross sections at the arrows. Mag. approx. 7,200 ×

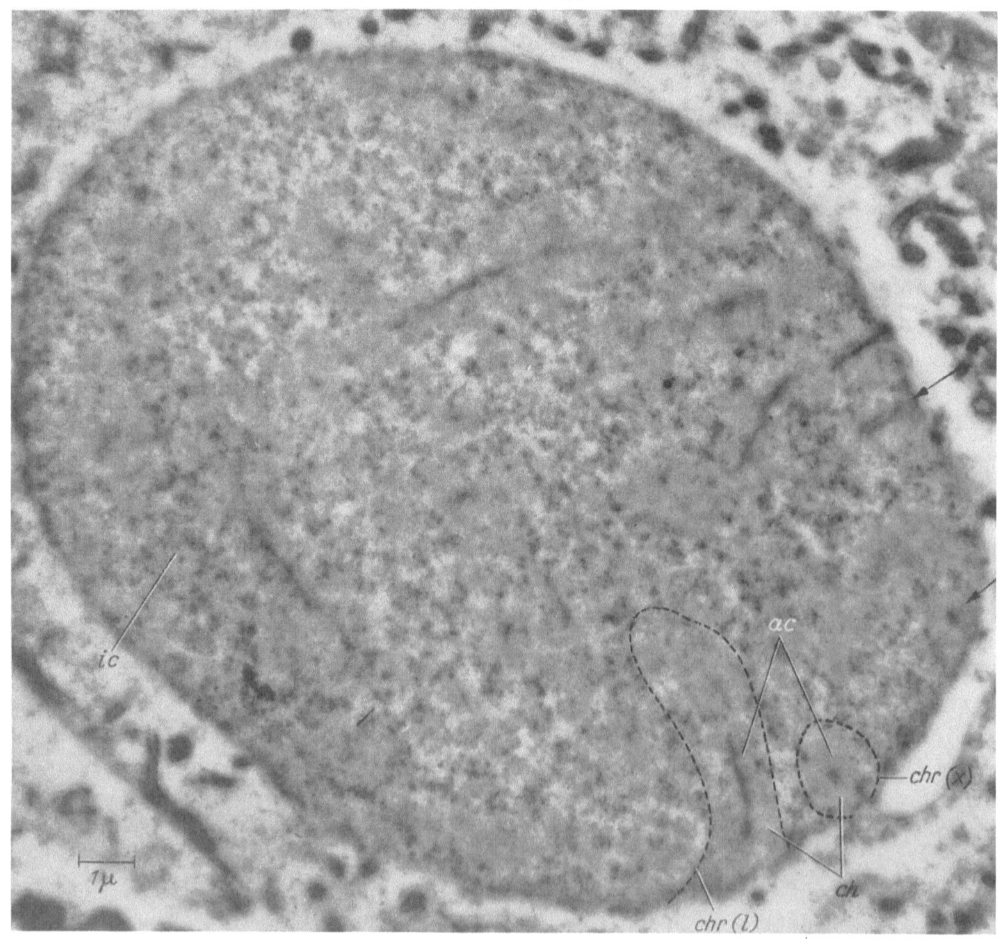

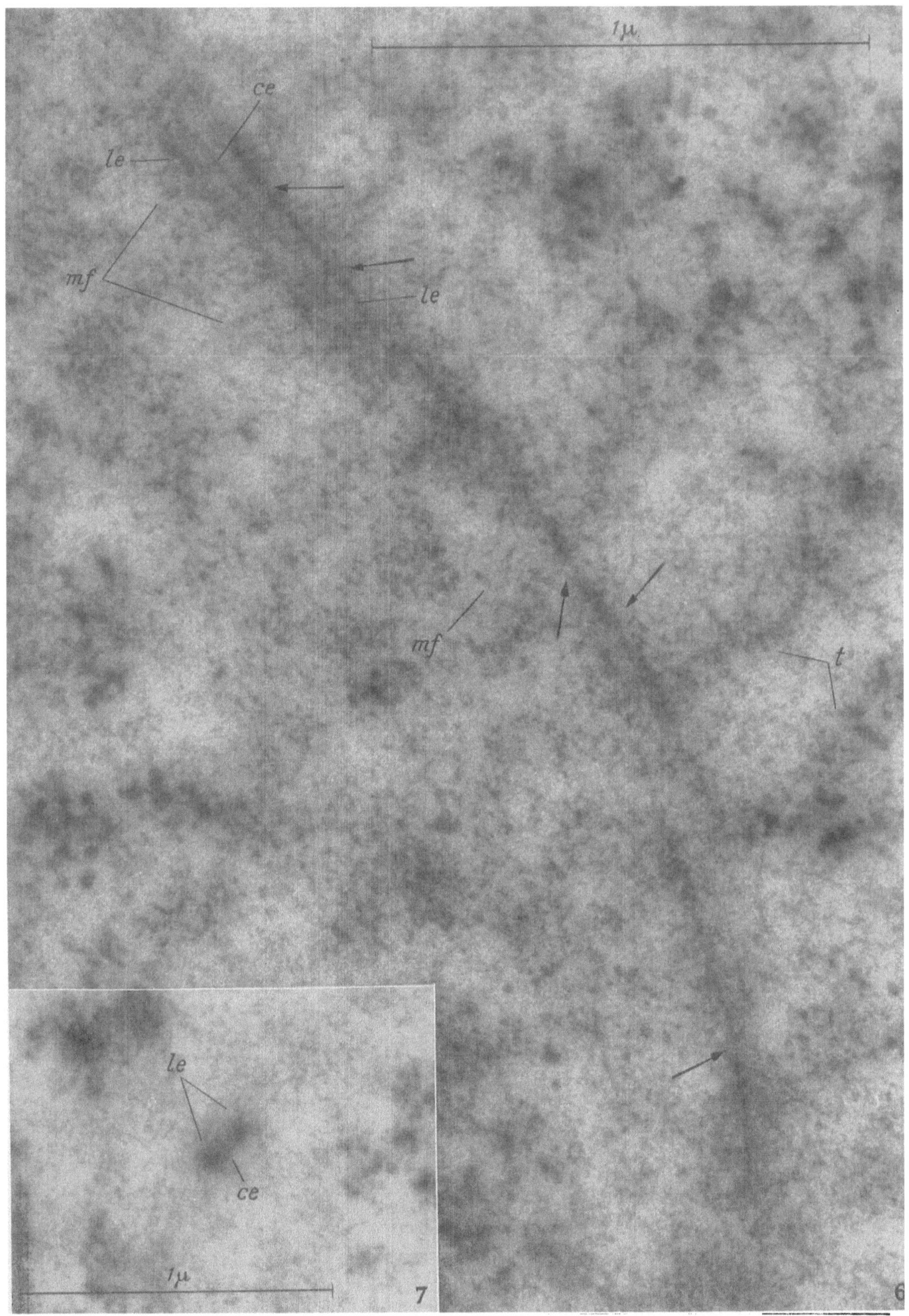

Fig. 6. Primary spermatocyte, prophase chromosome (bivalent) of *Plethodon*. The axial (synaptinemal) complex is seen in detail. Two parallel, dense, filamentous strands (*le*) flank a central element (*ce*). The complex apparently twists with the chromosome, and portions of it lie outside the section. Discontinuities in the lateral elements (*le*) can be seen at arrows. Twisted microfibrils 50—100 Å in diameter (*mf*) emerge from the lateral elements in a direction roughly normal to the axis along the complex. Frequently they are aggregated into tufts (*t*). The lateral elements appear to be formed of a tight aggregation of the microfibrils, especially at the discontinuities where the elements appear to fray into loose tangles of fine threads. Where the central element is present, occasional microfibrils connect it with the lateral elements; it could also be composed of fibrillar material. Mag. approx. 80,000×

Fig. 7. (insert). Cross section of the axial complex from another area of the same section shown in Fig. 6. The lateral elements (*le*) and central element (*ce*) are seen transversely as dense round spots, indicating the filamentous nature of the elements of the complex. Fine microfibrils of chromatin surround the complex. Mag. approx. 50,000×

In fact, it is precisely at these stages that long axial structures have been observed in chromosomes. First described in detail in primary spermatocytes of crayfish (*14*), they have been reported in such vertebrates as pigeon, cat and man (*15*), rat (*16, 9, 17*), *Xenopus* (*9*), a fish (*17*), two salamanders (*18*), mouse (*19*), and in invertebrates: locust (*20*), grasshopper (*9*), a spider (*17*), and two pulmonate snails (*21*). In addition, according to unpublished observations of a number of workers, the complex has appeared in a variety of other primary gametocyte (male as well as female) chromosomes of both animals and plants[1]. The appearance and dimensions of the complex

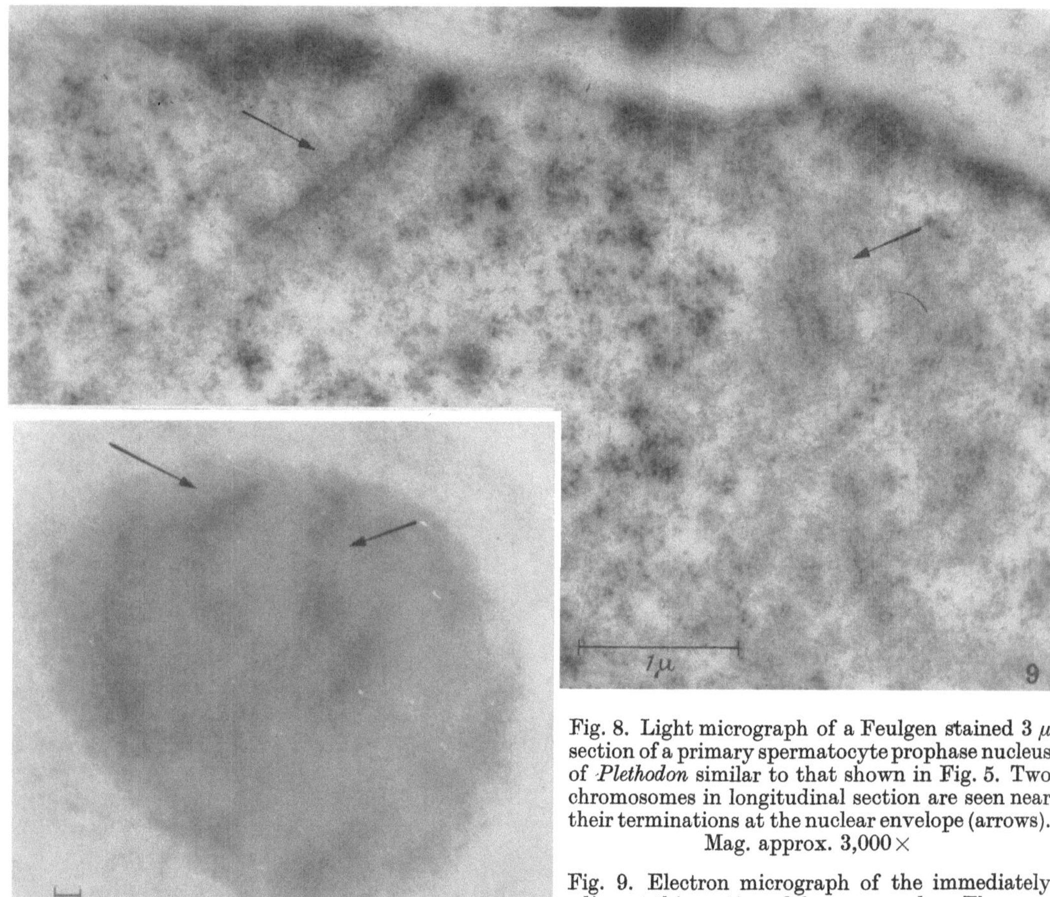

Fig. 8. Light micrograph of a Feulgen stained 3 μ section of a primary spermatocyte prophase nucleus of *Plethodon* similar to that shown in Fig. 5. Two chromosomes in longitudinal section are seen near their terminations at the nuclear envelope (arrows). Mag. approx. 3,000×

Fig. 9. Electron micrograph of the immediately adjacent thin section of the same nucleus. The same two bivalent chromosomes can be seen; each contains portions of the axial complex. The correspondence between the Feulgen positive material (DNA) and the region containing microfibrils (chromatin) is also evident. Mag. approx. 22,000×

are remarkably consistant (with the exception of crayfish, as noted below) in the species studied. It thus does not seem premature to propose that the structure is a regular concomitant of meiotic (heterotypic) prophase.

In the case of the salamander *Plethodon cinereus*, the stage at which the axial complex appears has been identified, through the use of adjacent thick and thin sections, as zygotene and early pachytene (*18*). Although in most other instances, the stages have unfortunately not been determined objectively, they are probably ones accompanying synapsis. For this reason, and because of its thread-like appearance, the structure has been termed the "synaptinemal complex" in preference to the less accurate and more casual term "core" (*18*).

---

[1] Since this paper was presented, the author has observed precisely the same structures in zygotene and early pachytene chromosomes of *Tradescantia* microsporocytes.

The complex is seen for greater [up to 8 μ has been observed (*18*)] or shorter lengths in thin sections, depending on the degree to which the chromosomes weave through the nucleus. In the salamander, for example, the synaptic chromosomes are polarized; their ends terminate at the nuclear envelope (Fig. 4). The chromosomes are relatively straight in the regions between the loop and the ends (Fig. 4). Thus, longitudinal sections of the complex are abundantly visible for considerable lengths in the electron microscope (Fig. 5). The complex usually appears as three parallel stands lying in a single plane (Fig. 6). In sections seen in sufficient length, the complex twists gently along the chromosome (*18*). From cross sections, the three elements appear to be filaments (Fig. 7). The two flanking, or lateral elements are dense and 250—300 Å in diameter. The central element varies in size and density, is generally of the order of 100—200 Å, and in some cases, may not be visible at all. The center to center separation of the lateral elements is remarkably consistent among various species, being 0.1—0.15 μ.

So far, the best documented light and electron microscope study of the complex has been reported for *Plethodon* (*18*). Here Feulgen positive chromosomes, seen clearly in cross and longitudinal sections in the light microscope (Fig. 4), are only distinguishable from the interchromosomal material ("sap") in the electron image by virtue of differences in texture (Fig. 5). The nuclear sap is palely Feulgen positive in thick sections (Fig. 4), indicating that not all the chromatin is condensed into the volume that is the chromosome. In some species, however, chromosomes in the electron image are clearly distinct from the sap, which has considerably less structure [e.g. crayfish spermatocytes (*9*,*14*)]. In all cases, the texture of the chromosome is finely fibrillar, in contrast to the sap which contains a lesser concentration of similar fibrils and varying numbers of dense granules several hundred Å in width. The dense synaptinemal complex is always seen to lie in the chromosome axis and in serial thick and thin sections has been shown in the crayfish (*14*) to be imbedded in Feulgen positive material (i.e. DNA). Fig. 8 and 9 are light and electron micrographs respectively of adjacent

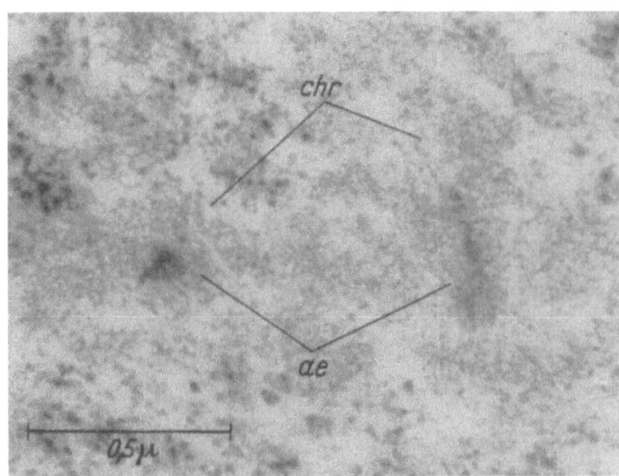

Fig. 10. Two unpaired chromosomes from material similar to that shown in preceding figures are seen in cross section (*chr*). Two single axial elements (*ae*) are cut transversely (left) and obliquely (right). The elements are continuous with the microfibrils of chromatin that surround them. Mag. approx. 48,000 ×

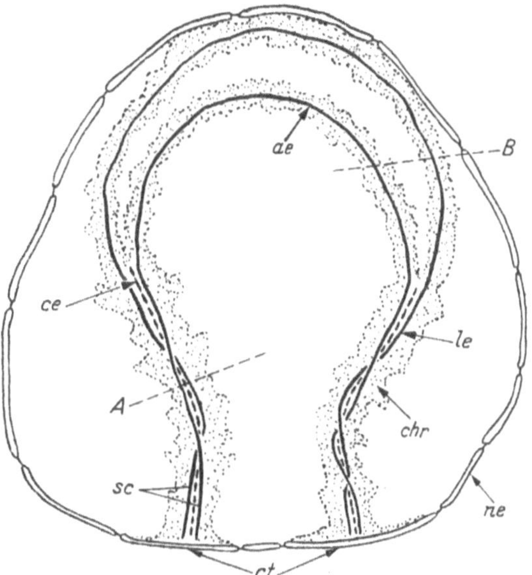

Fig. 11. Schematic diagram to illustrate the probable orientation of a pair of homologous chromosomes of the material illustrated in Fig. 4—10. In the "bouquet" stage, the chromosome ends are anchored (*ct*) at the innemembrane of the nuclear envelope (*ne*). In the region where chromosomes (*chr*) are synapsed, the tripartite synaptinemal complex (*sc*) is axial to the bivalent. A section through *A* would appear as shown in Fig. 7. At the top of the nucleus, the homologues are unpaired and a single axial element (*ae*) is seen in each homologue. The single element is continuous with the lateral element of the complex. The central element (*ce*) is not present except in the complex. A section through *B* would appear as shown in Fig. 10

sections of the same nucleus in the salamander. Two bivalent chromosomes (arrows) can be followed in both sections; the complex can be seen to be part of the Feulgen positive chromosome.

It has been demonstrated that the tripartite complex occurs where homologous chromosomes are paired and that single elements are found in regions where the chromosomes are unpaired (*18*). Fig. 10 shows cross and oblique sections of such elements which have dimensions similar to the lateral elements. It is probable that they belong to homologous chromosomes which have not yet completely paired (i. e. in a region such as that indicated in the diagram, Fig. 11, B). In such instances the single element is situated in the center of the chromosome, and is thus axial to the unpaired chromosome, as the complex is axial to the bivalent. The origin of the central element which appears only in the bivalent, is unknown, but it may be a condensation product of pairing, as has been suggested (*15*). It cannot be said whether either the complex or the individual elements form such a condensed structure all along the length of the chromosome, or whether they extend along the length of all chromosomes. Indeed it would be surprising if they did; events such as pairing, coiling, DNA synthesis (*22*), etc. are not timed simultaneously along or among chromosomes and it is not to be expected that a transient differentiation such as is represented by the synaptinemal complex would exhibit similar independence.

Microfibrils of the sort that have been suggested to be the fundamental structural unit of the chromosome (*7*) are abundantly evident in the chromosome around the synaptinemal complex (Fig. 6, 7 and 12). They, together with structures which may be either small granules or cross sections of the fibrils[1] are the only visible structural component of the chromosome, except for occasional masses of the interchromosomal material which invade and are associated with the microfibrils. The microfibrils are between 50 and 100 Å in diameter and can be followed for distances up to 0.2 μ (Fig. 6). They are highly twisted and kinked, and appear to branch and anastomose, though more critical microscopy is required to decide if the latter is an illusion of overlapping. In our experience, the main structural component of areas that can be shown to be Feulgen positive is the

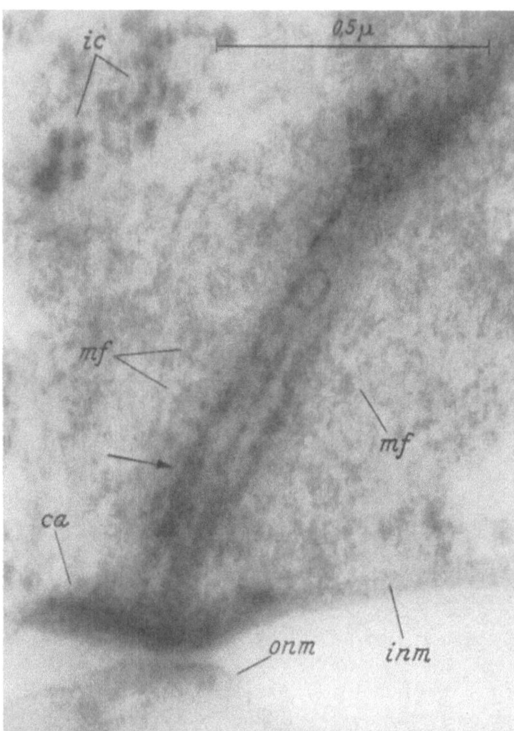

Fig. 12. A portion of a bivalent chromosome from material shown in previous figures, is seen longitudinally at its attachment (*ca*) to the inner nuclear membrane (*inm*) in a relatively thick section. Twisted microfibrils surround the complex and in places appear to emerge (*mf*) from the lateral elements. Dense interchromosomal granules (*ic*) lie outside the chromosome. At the arrow, the lateral element appears double. The dense material that spreads laterally from the chromosome foot corresponds to the Feulgen positive "boutons" seen in the light microscope in similar regions. The two membranes of the nuclear envelope are separated, possibly by artifact, except at the chromosome terminus. There is no evident structure at this point to hold the outer (*onm*) and inner (*inm*) membranes together. Mag. approx. 76,000×

microfibril. It is most reasonable to assume that it contains the DNA, though there is at present no direct evidence to confirm it. The circumstantial evidence, however, and particularly the observation that there is a rough parallel between concentration of Feulgen dye in a thick section and that of the fibrils (and not, for example of the less dense material between them) in a corresponding area in the electron microscope (e. g. interphase euchromatin vs. heterochromatin, vs. prophase, vs. metaphase), leaves very little room for an alternative localization of the DNA.

Careful examination of the complex reveals that the elements are not independent structures but are continuous with the microfibrils surrounding them. In both transverse and longitudinal sections microfibrils can be traced back to their origins in the lateral elements (Fig. 6, 7, 12); these latter can often be seen to be connected to the central elements by similar microfibrils

---

[1] In stereo micrographs, many of the "dots" can be seen to be end-on views of microfibrils.

(Fig. 6). The axial elements are actually discontinuous in places along their length (Fig. 6, arrows). In very thin sections, and in places where a lateral element is grazed as it passes out of the section (Fig. 6), it appears composed of a tight mat of twisted fibrils. Along the complex it appears that microfibrils emerge in various numbers from the lateral elements in a direction that is roughly perpendicular to the chromosome axis. Favorable cross sections in which the fibrils can be seen

clearly support the idea that they are radially oriented. The bristle-like arrangement is most obvious in stereomicrographs of sections. In places, groups of microfibrils emerge in tufts (Fig. 6) which give the chromosome its bushy appearance in the light microscope (Fig. 4). It is impossible to follow all microfibrils back to their origins and so it cannot be said whether all microfibrils insert to the axial elements. But if the elements constitute chromosome spines as is hypothesized, then they probably do.

The organization of such chromatin microfibrils into a structure axial to the chromosome represents a transition between two levels of organization that mark the gap between structures approaching the dimensions of the DNA molecule and those integral with the linear chromosome. It is yet to be shown, of course, how the molecule is related to the microfibril, but it is not difficult to envisage. It may be, for example, as RIS has suggested (7): that the microfibril has structural analogy to a tobacco mosaic virus particle. The tufts of fibrils (Fig. 6), visible also in the light micrograph (Fig. 4), which often give the impression of spiral arrangement about the complex would then reflect a secondary organization of the microfibrils, for which a function is obscure. They undoubtedly represent the projections often described as loops, or as chromomeres, depending on fixation, seen with the light microscope in squashes of similar chromosomes. They do not appear at this point to have an essential role in maintaining the chro-

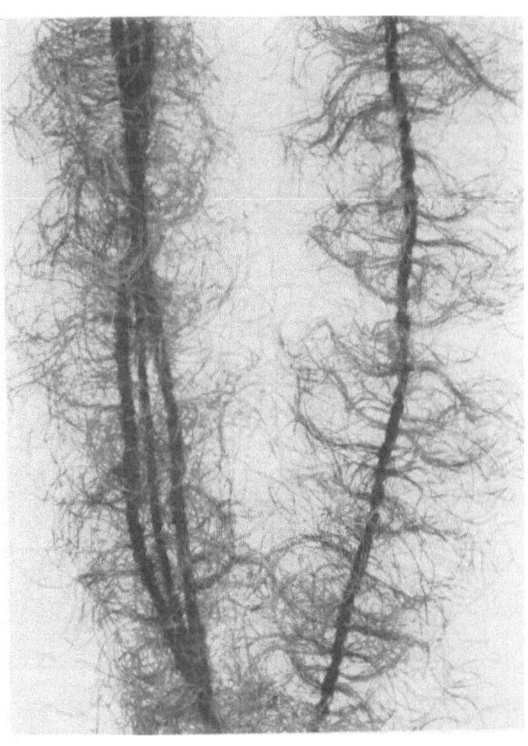

Fig. 13. Model constructed to interpret meiotic chromosome structure as seen in the electron microscope; compare with previous figures. On the left is a portion of a bivalent chromosome showing the tripartite complex and associated microfibrils. On the right is a segment of unpaired chromosome with single axial element. The figure on the left was made by apposing two lengths of that on the right, and adding a central filament. The microfibrils which emerge from the axial structures in tufts, are shown as free-ending, though this cannot be ascertained from the electron image; they may well be loops

mosome axis. A model has been constructed which represents a simplified three-dimensional interpretation of what can be seen in the electron image; paired and unpaired chromosomes are shown (Fig. 13).

It now becomes feasible to examine the possibility that the axial element of the single chromosome is the chromonema of light microscopy. The obvious corollary is that similar organization should also exist in mitotic chromosomes. The fact that to date searches have failed to find it outside of synapsis does not negate the hypothesis. Just as the structure is generally lost from view later at pachytene when coiling occurs, so it may also be obscured by the coiling of somatic chromosomes. In addition, it is likely that a further refinement of the basic organization may occur as a concomitant of synapsis, thus rendering the axial structure more obvious in meiosis than it would be in mitosis. The task, then, is the difficult one of discovering the pattern and of making it visible in somatic nuclei.

Of immediate concern is whether the observed structure can satisfy certain demands placed upon it by morphological and functional evidence presently at hand. A fundamental problem concerns the number of strands in the chromosome. Morphological evidence about the strandedness of leptotene and cytogene chromosomes is difficult to obtain at best and has provided no basis of agreement (e. g. discussion in Swanson (*23*), p. 120—121]. There are relevant data concerning the functional subdivisions, however.

Cytophotometric and autoradiographic determinations of the timing of DNA synthesis show that replication is complete by leptotene in grasshopper (*24*) and in *Lily* (*25*), and by cygotene

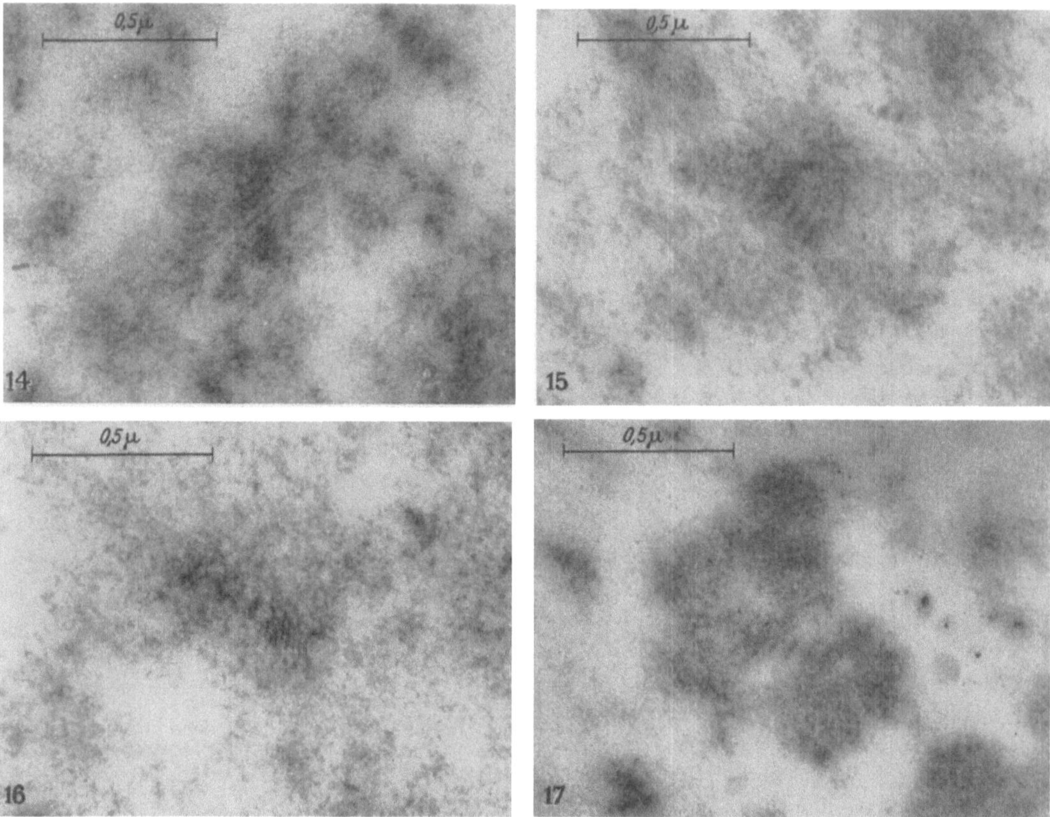

Fig. 14. Short longitudinal segment of a bivalent chromosome from a primary spermatocyte of the crayfish *Cambarus clarkii*. The axial complex here consists of a dense central element flanked by parallel paired structures integral with the fibrillar material of the chromatin. Mag. approx. 48,000×

Fig. 15. Oblique section of a chromosome similar to that in Fig. 14. The same pentapartite arrangement of the complex is apparent. Microfibrils are condensed and distributed in tufts. Mag. approx. 48,000×

Fig. 16. Transverse section of chromosome similar to those in Fig. 14 and 15. Here the central element, seen as a dense "dot" in the cross section, is flanked by two large dense masses of chromatin (microfibrils) within which the doublet organization can hardly be discerned. Mag. approx. 48,000×

Fig. 17. Oblique section of a late meiotic prophase chromosome (possibly late diplotene in a region where the homologues are still paired) of *Cambarus*. The two homologues are clearly defined and separated by the persistent central element. Only the faintest trace of the double lateral elements is visible. Mag. approx. 48,000×

or earlier *Tradescantia* (*26*). Where it has been possible to compare X-ray fragmentation evidence with the cytochemical observations (e. g. *26, 27*), pre-replication (interphase) chromosomes behave as single units while those that have completed replication (leptotone in *Lily*, prophase in *Tradescantia microspores*) react as double entities. The shift from chromosome to chromatid type breaks coincides with the synthesis period (*26,27,30*). Sub-chromatid damage does not appear in meiotic chromosomes until they are irradiated at pachytene (*27, 28, 29*).

Insofar as these observations can be extrapolated to meiotic chromosomes in general, they indicate that by cygotene, when the axial structures are clearly evident, chromosome replication (with respect to its DNA) is complete and the chromosome is functionally double. A morphological counterpart of this duality should be expected in any structure that comprises the linear backbone of the chromosome.

Contrary to expectation, axial elements generally appear single. There are, however, notable exceptions. Occasionally one observes regions where the lateral elements do appear double (Fig. 12, arrow). Indeed, Dr. YVES CLERMONT has permitted me to examine and cite an electron micrograph taken by him of a long segment (several microns) of the synaptinemal complex in a mouse spermatocyte, where the two lateral elements are loosely twisted and each element is clearly double along its entire length. The half-elements are also loosely twisted. Such evidence suggests that while the axial structures of other forms may actually be double functionally, this duality is only resolvable morphologically under particular circumstances which are presently unknown.

Strongest support for the doubleness of the lateral elements comes from the structure of the synaptinemal complex in crayfish spermatocytes (*9, 14*). Here also the structure follows the chromosome axis for considerable distances and has been shown to be integral with the DNA (*9*) and 100 Å microfibrils that make up the chromosome (Fig. 14—17). In longitudinal (Fig. 14) and oblique (Fig. 15) sections the complex usually appears as five parallel lines of approximately equal length. At some stages two dense masses, each about the width of the two outermost elements combined, flank a central element (Fig. 16). The central element is generally more prominent than in other forms, and is about 250 Å in width. In cross sections it appears as a dense dot and hence is filamentous. As in other forms, the central element is a synaptic structure that marks the axial juncture of two homologues. This is apparent from oblique and cross sections of late prophase (late diplotene, in places where the chromosomes have not yet separated) where, unlike the condition in many other forms, the complex persists. As shown in Fig. 17, the central element separates two homologues; only traces of the lateral elements are visible.

It is difficult to find cross sections of the lateral elements which show as much structure as oblique sections. The profiles are not those of filaments; they appear either as two dense masses (Fig. 16), each of which sometimes appears double, or as laminate structures surrounding the central element (*14*). Oblique and longitudinal sections normal to the plane in which central and lateral elements lie and sagittal to the lateral element show a broad band of dense material which in thin sections can be seen as aglomerations of microfibrils; these tend to be oriented parallel to each other and perpendicular to the plane joining the elements.

Thus, the structures flanking the central element are homologous to the lateral elements of the more common form, and like them are composed of microfibrils about 100 Å in width which often emerge in tufts (Fig. 14 and 15). But in the crayfish, the flanking structures may take three forms; as laminate masses encircling the central element (*14*), as masses on either side of the central element (Fig. 16 and 17), and in its most common form as double laminate structures lateral to the central element (Fig. 14, 15). The temporal or functional relationships of these variants are not clear. The separation of the lateral elements into two submits may well represent the structural duality visible only rarely in the more common form. Why it exists here is unknown. That it is a reality is clear, however, and it is possible that the doubleness may represent the chromatid split that should be present in a chromosome axis.

If we pursue the notion that the single axial (i. e. lateral) element maintains the chromosome's linear organization and is composed of chromosomal microfibrils possibly together with additional material, then some model for assembling the structure must be postulated which will conform to what is already known about chromosome behaviour and at the same time be consistent with our electron microscope observations. For example, it should be possible to make a transition, via the model from meiotic to somatic, to polytene, to lampbrush chromosomes, and to other chromosomes in various physiological states, and at the same time to satisfy the demands of chromosomal replication and segregation. It is not possible here to discuss the evolution of such a model in detail, but mention of its probable form is in order.

The simplest picture is of a strand (unit axial element) formed by the close association of radiating microfibrils at their origins, possibly with the aid of some other substance such as protein serving as a "glue". If the microfibrils are free-ending, like bristles (Fig. 13), the chromosome's morphological linearity as well as the genetic sequence along its length (assuming genetic specificity to reside in the DNA-containing microfibrils) is then maintained entirely by the organizing force or structure that anchors the attached ends. With respect to DNA replication, such a model resembles the first of two alternatives proposed by TAYLOR (*31*) to explain his chromosome replication results.

TAYLOR's second scheme, patterned after one by FREESE (*32*), is more satisfactory on a number of counts and could also fit with our observations. Here the double helix DNA molecules are attached end to end; the alignment of the junctions, again with the possible aid of accessory material, forms loops of the molecules and a backbone for the chromosome. An analogous situation could account for the axial elements of meiotic chromosomes. It is impossible to follow microfibrils for sufficient distances in electron micrographs to say whether they are actually free-ending or loops: they could perfectly well be the latter. In this scheme the genetic sequence is maintained independently of the arrangement of the joined ends into a linear structure, and would persist if the structures were to become disorganized, as might be the case, for example, in interphase. Thus the fundamental unit of the chromosome would be a single chain of DNA-protein molecules (100 Å microfibrils) of varying length, attached end-to-end. At least two such chains would comprise a single chromosome. Polyteny would result from a multiplicity of these chains. Structural and functional differentiations along the length of the chain could be produced from the degrees of coiling or kinking of the microfibrils and differences in amounts and kinds of associated protein, etc. This model bears a strong resemblance to the one that emerges from GALL's discussion of chromosome organization (*3*) in which he attempts to reconcile apparently disparate chromosome patterns.

At this point, it would be fruitless to speculate about the function of the synaptinemal complex in the process of pairing and in events such as crossing-over that have genetic consequences. Any such consideration will have to be based on a firmer knowledge of chromosome organization and on considerably more electron microscopic observations of the complex under various situations where the genetic factors are known. The most reasonable experimental approach would embrace the careful choice of biological materials, together with a study of chromosome modifications, both natural (e. g. asynaptic chromosomes, non-disjunction, triploids) and experimentally induced (e. g. uncoiling treatments, fragmentation by radiation, treatment with substances that effect crossing-over, etc.).

The attractiveness of the chromosome model discussed lies in its potentialities as a unifying concept of chromosome organization. But its actual value is in serving as a working hypothesis upon which to base further investigations of chromosome fine structure. It is encouraging at this stage of our ignorance of chromosome organization to find a structure within the chromosome that is well preserved by our cytological procedures and distinctly visible in the electron microscope, and one that at the same time provides a reasonable clue to an underlying pattern on which the chromosome may be built.

Bibliography

1. KAUFMANN, B. P.: Bot. Rev. **14**, 57 (1948).

2. DE ROBERTIS, E.: J. biophysic. biochem. Cytol. **2**, 785 (1956).

3. GALL, J.: In Chemical Basis of development. Ed. W. D. McElroy and B. Glass, p. 103. John Hopkins Press 1958.

4. GAY, H.: J. biophysic. biochem. Cytol. **2** (Suppl.), 407 (1956).

5. KAUFMANN, B. P., and M. R. MCDONALD: Cold Spr. Harb. Symp. quant. Biol. **21**, 233 (1956).

6. NEBEL, B. R.: J. Hered. **48**, 51 (1957).

7. RIS, H.: In Chemical basis of heredity. Ed. W. D. MCELROY and B. GLASS, p. 23. Baltimore: John Hopkins Press 1957.

8. YASUZUMI, G., and A. KONDO: J. Hered. **42**, 219 (1951).

9. MOSES, M. J.: J. biophysic. biochem. Cytol. **2** (Suppl.), 397 (1956).

10. LEUCHTENBERGER, C., P. F. DOOLIN and A. H. KUTSAKIS: J. biophysic. biochem. Cytol. **1**, 385 (1955).

11. BAHR, G. F.: Exp. Cell Res. **7**, 457 (1954).

12. PORTER, K. R.: These proceedings.

13. MANTON, I.: Biol. Rev. **25**, 486 (1950).

14. MOSES, M. J.: J. biophysic. biochem. Cytol. **2**, 215 (1956).

15. Fawcett, D. W.: J. biophysic. biochem. Cytol. **2**, 403 (1956).
16. Watson, M.: Spermatogenesis in the adult albino rat as revealed by tissue sections in the electron micro-scope. Univ. Rochester Atomic Energy Report UR-185, 1952.
17. Sotelo, J. R., and O. T. Trujillo-Cenòz: Exp. Cell Res. **15**, 1 (1958).
18. Moses, M. J.: J. biophysic. biochem. Cytol. **4**, 633 (1958).
19. Nebel, B. R.: These proceedings.
20. Gibbons, I. R.: Proc. X. int. Congress Genetics **2**, 96 (1958).
21. Roth, L. E.: These proceedings.
22. Taylor, J. H.: Exp. Cell Res. **15**, 350 (1958).
23. Swanson, C. P.: Cytology and cytogenetics. New York: Prentice Hall 1957.
24. Swift, H. H., and R. Kleinfeld: Physiol. Zool. **26**, 301 (1953).
25. Taylor, J. H., and R. D. McMaster: Chromosoma **6**, 489 (1954).
26. Moses, M. J., and J. H. Taylor: Exp. Cell Res. **9**, 474 (1955).
27. Mitra, S.: Genetics **43**, 771 (1958).
28. Crouse, H. V.: Science **119**, 485 (1954).
29. Wilson, G. B., A. H. Sparrow and V. Pond: Amer. J. Bot. **46**, 309 (1959).
30. Thoday, J. M.: New Phytologist **53**, 511 (1954).
31. Taylor, J. H.: Proc. X. int. Congress of Genetics **1**, 63 (1958).
32. Freese, E.: Cold Spr. Harb. Symp. quant. Biol. **23**, 13 (1958).

RNA and nuclear fine structure

H. Swift

Department of Zoology, University of Chicago, USA

Ribose nucleic acid (RNA) and deoxyribose nucleic acid (DNA) have been localized in micrographs of Triturus and Drosophila tissues, with the use of ferric ion binding. In Triturus liver and pancreas nuclei, RNA-containing material was distributed in the nucleoplasm, surrounding areas of compact DNA-containing chromatin. Where nucleoplasm was in contact with the nuclear membrane, discontinuities in the membrane (annuli) were apparent. RNA in the nucleoplasm was associated with a finely filamentous component. No structures of comparable morphology were evident in the RNA-containing structures of the cytoplasm. — RNA was also localized in specific bands of Drosophila salivary chromosomes. These areas were prepared for study by osmium fixation prior to smearing in acetic acid. Ferric binding demonstrated a finely filamentous or reticular structure in the RNA-containing regions. In the abscence of obvious morphological similarities between the RNA-containing components of nucleus and cytoplasm, we have concluded that these fractions are distinct, or that aggregation changes occur during nuclear — cytoplasmic exchange.

Fine structure of the nucleus during spermiogenesis

H. Ris

Department of Zoology, University of Wisconsin, Madison, Wisconsin, USA

Several investigators have reported microfibrils in chromosomes of various plants and animals. Fibrils about 100 Å thick have been found in all chromosomes studied. Often these are associated in pairs and form fibrils 200—250 Å thick. By dissolving calf thymus chromosomes isolated in saline-versene of p_H 6.5 in water we have obtained a nucleoprotein in the form of fibrils about 100 Å thick and identical in appearance to the fibrils found in sections through osmium fixed chromosomes. We assume therefore that the microfibrils making up the chromosomes represent the nucleoprotein. If calf thymus chromosomes are left in saline-versene of p_H 8 before dissolving in water the non-histone protein autolyzes and is lost from the fibrous nucleoprotein. In electron micrographs the 100 Å fibrils are now seen to consist of two 40 Å fibrils which seem to correspond to the nucleohistone. — During spermiogenesis the non-histone protein generally disappears from the nucleus. Electronmicroscopic studies of spermiogenesis in a large number of species show that at this time the 100 Å fibrils become visibly double, consisting of two 40 Å fibrils. In sperm where dichroism and X-ray diffraction indicate orientation of DNA, these fibrils are oriented in the same way. No other structures are visible in the sperm nucleus and in the mature sperm these fibrils fill the nucleus solidly. We conclude that the microfibrils correspond to the DNA with associated basic protein and that the DNA molecule is arranged in the long axis of the microfibrils. Late in spermiogenesis the 40 Å fibrils associate into large units. In some species progressively thicker bundles are formed which may be oriented in the long axis of the sperm or irregularly twisted through the nucleus. In other species the fibrils associate side by side into sheets 40 Å thick which are randomly folded and become progressively more closely packed until the nucleus is uniformly dense (s. a. H. RIS: „Die Feinstruktur des Kerns während der Spermiogenese", in „Chemie der Genetik", 9. Colloquium der Gesellschaft für Physiologische Chemie, 17.—19. 4. 1958 in Mosbach/Baden, Berlin, Springer 1959).

Das Nucleoplasma der Bakteriennucleoide verglichen mit der DNS von vegetativen und reifen Phagen

EDOUARD KELLENBERGER, JANINE SÉCHAUD und ANTOINETTE RYTER

Biophysikalisches Laboratorium der Universität Genf

Werden osmiumfixierte Bakterienzellen im Elektronenmikroskop beobachtet, so findet man den Inhalt der Kernvacuolen in den verschiedensten Zuständen. Je nachdem treffen wir auf stark elektronenstreuende unregelmäßig geformte Körper in einer elektronenoptisch leeren Vacuole oder mehr oder weniger feine fädige Zustände, die die Vacuole mehr oder weniger ausfüllen [Lit. vgl. (1)]. Diese große Unregelmäßigkeit der Darstellung hängt mit der Fixierung und zum geringeren Teil mit der Methode des Einbettens zusammen (2, 3). Wir haben eine Reihe von Fixationsmitteln (Formol, Bouin, Chromsäure) untersucht, die aber alle unbefriedigend waren, zum Teil weil das Plasma schlecht fixiert war. In einer systematischen Untersuchung (1) haben wir festgestellt, daß mit Osmium absolut regelmäßige Resultate erhalten werden können, sofern die Fixierung unter ganz bestimmten Bedingungen ausgeführt wird. Die neulich vorgeschlagene Permanganatfixierung (4) gibt ähnliche Resultate (5), wenn auch in unseren Händen die Regelmäßigkeit noch zu wünschen übrig läßt. Wir möchten hier einige Resultate besprechen, die mit der Osmiumfixierung erhalten worden sind. Die Einzelheiten dieser Experimente sind anderswo veröffentlicht (1, 6).

Sind in der Hauptfixierung der p_H-Wert bei 6, Ca- oder Mg-Ionen (0,01 m) und ungefähr 0,1% Aminosäuren (Casoaminosäuren Difco oder Bacto-Trypton) vorhanden, dann erhält man nach einer 16stündigen Fixierung ein sehr feinfädiges Nucleoplasma. Zu diesen Standardbedingungen gehört ferner noch, daß 2 Std. in Veronal-Acetatpuffer von p_H 6 gewaschen wird, der 0,5% Uranylacetat enthält, bevor mit Aceton entwässert und in Polyester eingebettet wird. Uranylacetat kann ersetzt werden durch 1% Lanthannitrat oder 0,01 m Ca. Unter diesen Bedingungen erhalten wir ein feinfädiges Nucleoplasma auf 5 verschiedenen untersuchten Stämmen und Arten. Abweichungen von diesen Bedingungen führten zu groben und gröbsten Einschlüssen. Wird insbesondere in einem Veronal-Acetatpuffer gewaschen, der 0,06 m Versen enthält, dann erhält man regelmäßig eine Coagulation. Da diese Versenbehandlung nur auf gewisse DNS-haltige Plasmen wirkt, haben wir sie zu einem Test gemacht, der uns gestattet, in Parallelexperimenten rasch spezifischen Aufschluß über die Natur dieses Plasmas zu erhalten. Wir haben dann diese Fixierung unter Standardbedingungen auf andere Plasmen, die DNS enthalten, angewendet, nämlich auf das Plasma der vegetativen Bakteriophagen und ihre reife intracelluläre Form (6). Es zeigte sich, daß unsere Fixation auch hier ein feinfädiges Plasma ergibt und daß sie allein fähig ist, die T2-Phagenköpfe in ihrer typischen polyedrischen Form darzustellen. Für das Plasma des vegetativen Phagen ist der Versentest ebenfalls positiv, d. h. es coaguliert.

Da es prinzipiell unmöglich ist, auf Grund der Morphologie allein zu entscheiden, welche der beiden Erscheinungsformen den biologisch signifikativen Zustand darstellt und welche den

Artefakt, müssen wir auch die verschiedenen Argumente für und gegen abwägen, die sich aus den biologischen Folgerungen ergeben. Da die Eigenschaften und die Vermehrung der DNS der Phagen verhältnismäßig besser bekannt sind als diejenigen des Bakterienkerns, möchten wir vorerst die diesbezüglichen Beobachtungen zeigen und besprechen.

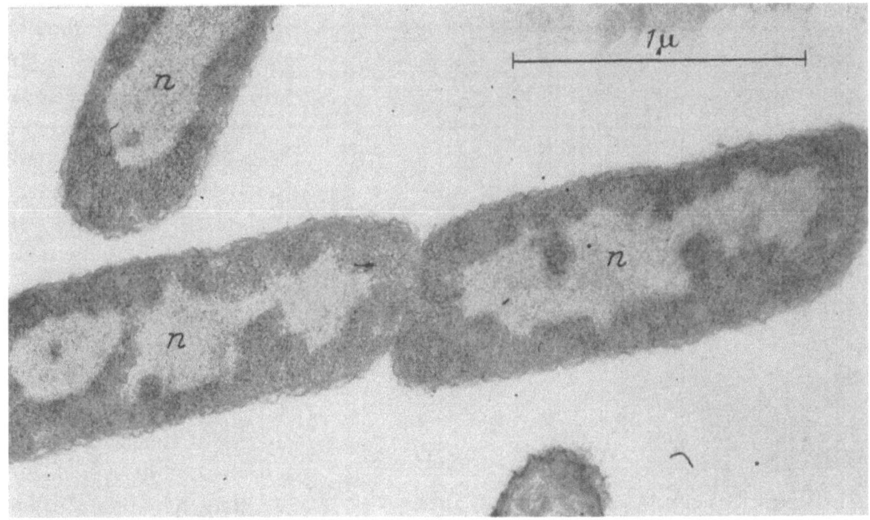

Abb. 1. Zellen von E. coli in aktivem Wachstum. Nucleoide mit „feinem" Nucleoplasma (n). 40000 mal

Die Darstellung
der vegetativen und reifen Phagen

Es ist schon längere Zeit bekannt, daß in T2-infizierten Zellen die Bakteriennucleoide sich rasch verändern, indem sie marginale Vacuolen bilden [Lit. vgl. (6)]. Da die DNS des Phagen T2 Hydroxymethylcytosin enthält anstelle des in der Bakterien-DNS vorhandenen Cytosins, läßt sich der Metabolismus der beiden Nucleinsäuren gesondert verfolgen (7, 8). Es zeigt sich, daß die Bakterien-DNS verschwindet und daß 8 min nach der Infektion die Phagen-DNS auftritt und stetig zunimmt. Gleichzeitig bildet sich, ausgehend von den marginalen Vacuolen, eine im Dünnschnitt gut sichtbare „Vacuole", die wiederum feinfädiges Material enthält (9). Diese Vacuole wächst ständig und dehnt sich nach innen aus. Man findet sie später in unregelmäßiger Verteilung und unscharf begrenzt überall im Inneren der Zelle (Abb. 3). Ungefähr 4 min später (d. h. 12 min nach der Infektion) finden wir in dieser „Vacuole" die ersten unreifen Phagen und später auch die reifen (Abb. 5). Wird 8 min nach der Infektion Chloramphenicol zugegeben (10), dann reifen die Phagen nicht mehr, weil die Proteinsynthese inhibiert

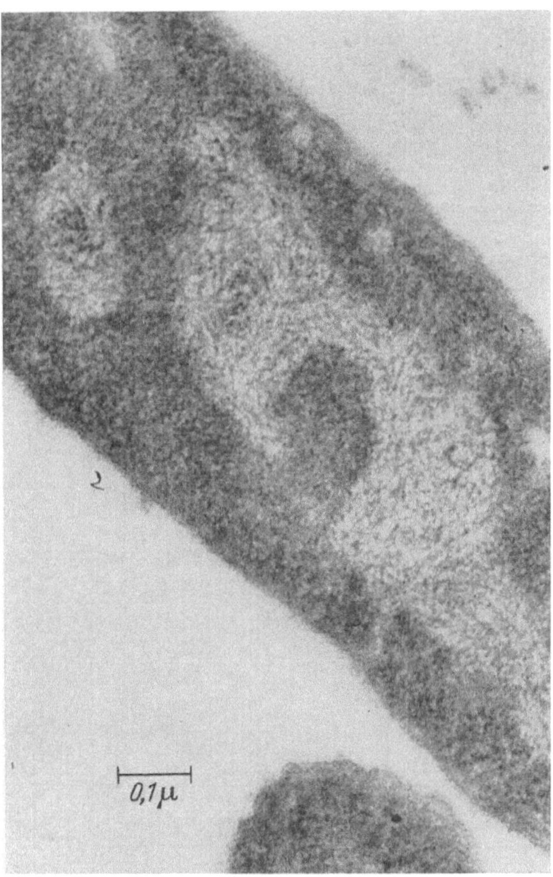

Abb. 2. Detail eines Nucleoids einer E. coli-Zelle, dessen Nucleoplasma orientierte Fibrillen zeigt. 100000 mal

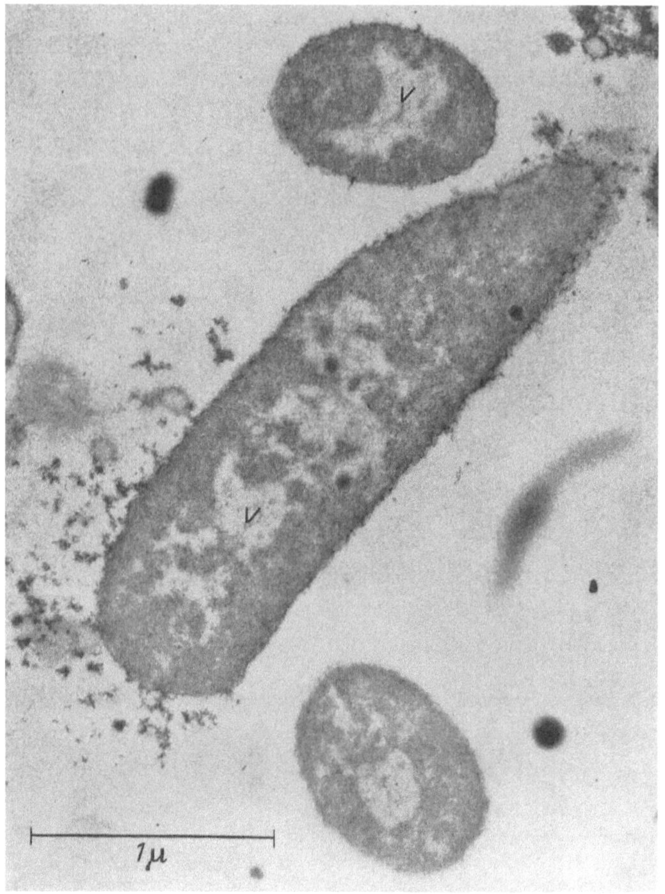

Abb. 3. Mit T2 infizierte Bakterien 12 min nach der Infektion. Deutlich sichtbarer Pool (*v*) des vegetativen Phagen. Zwei unreife Phagenpartikel sind sichtbar. 33000mal

ist, die Phagen-DNS jedoch vermehrt sich unverändert weiter. Morphologisch ist dann eine hypertrophierte Vacuole zu finden (*6*, Abb. 3). Es besteht demnach wenig Zweifel, daß diese Vacuole den „Pool" der Phagen-DNS darstellt. Die genetischen Experimente über die Rekombination der Phagen haben gezeigt, daß dieser Vorgang mit der Phagen-DNS ausgeführt wird, selbstverständlich bevor sie in einem fertigen, reifen Phagen organisiert ist. Die sich vermehrenden Phagengenome finden sich alle in einem genetischen Pool, wo sie sich alle paar Minuten kreuzen und wahrscheinlich gleichzeitig auch vermehren. Aus diesem Pool werden die Phagengenome entnommen, um zu fertigen Phagen ausgereift zu werden. Alles spricht dafür, daß der genetische Pool mit dem chemischen DNS-Pool gleichzusetzen ist und daß daher auch der morphologische Pool, d. h. unsere Vacuole, damit zu identifizieren ist. Dieser Pool (Abb. 4) enthält fädige Elemente von 30—60 Å kleinstem Durchmesser. Sie bilden ein ziemlich homogenes Gel.

Eine gewisse parallele Orientation der Fäden ist jedoch meistens sichtbar. Dieses Bild entspricht genau dem, was man sich unter dem genetischen Pool vorstellt: Man weiß, daß die vegetative

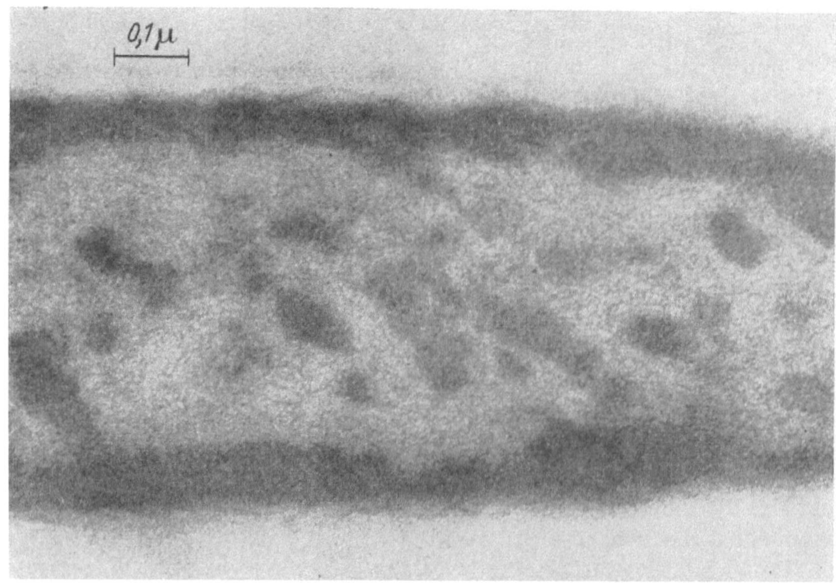

Abb. 4. Detail eines vegetativen Phagen-Pools. 100000mal

Phagen-DNS fädig ist (*11*) und daß die Energieverhältnisse es gestatten, eine Replikation in dieser Form anzunehmen (*12*). Wenn sich einige Fäden an homologen Stellen kreuzen, dann kann ein genetischer Austausch stattfinden.

Man wird nun mit Berechtigung einwenden, warum denn in diesem Falle die DNS in den Phagenköpfen soviel kontrastreicher ist als im Pool der vegetativen Phagen. Die Elektronenstreuung per Flächeneinheit in einem Dünnschnitt resultiert aus der Streuung im Einbettungsmaterial und der Streuung in den eingebetteten Gewebebestandteilen. Meistens ist die spezifische Streuung dieser letzteren Bestandteile größer als die des Einbettungsmaterials, wodurch sie sichtbar werden. Die Streuung per Flächeneinheit hängt also ab 1. von der spezifischen Stärke der Streuung der Gewebebestandteile, d.h ihrer chemischen Beschaffenheit und Zusammensetzung, 2. von der Konzentration dieser Gewebebestandteile relativ zum Einbettungsmaterial. Dieser zweite Faktor ist genau proportional dem ursprünglichen Wassergehalt des Gewebes. Da der Wassergehalt der verschiedenen Zellstrukturen stark variiert, ist es einleuchtend, daß dieser Faktor eine wesentliche Rolle spielt. Um auf unser Problem zurückzukommen, können wir schließen, daß die DNS in den Phagenköpfen viel konzentrierter ist als im Pool, wo sie sehr stark hydriert ist. Dazu kommt allerdings noch, daß im Phagenkopf die DNS mit einem Polypeptid oder Protein assoziiert ist (*13*, *14*), das möglicherweise mehr Osmium aufnimmt als die Elemente des Pools.

Das Nucleoplasma der Bakteriennucleoide

Mit unseren Fixierungsbedingungen erhält man Kernvacuolen, die gleichmäßig mit einem feinfädigen Nucleoplasma ausgefüllt sind (Abb. 1). Die feinsten Fibrillen sind wiederum 30—60 Å im Durchmesser

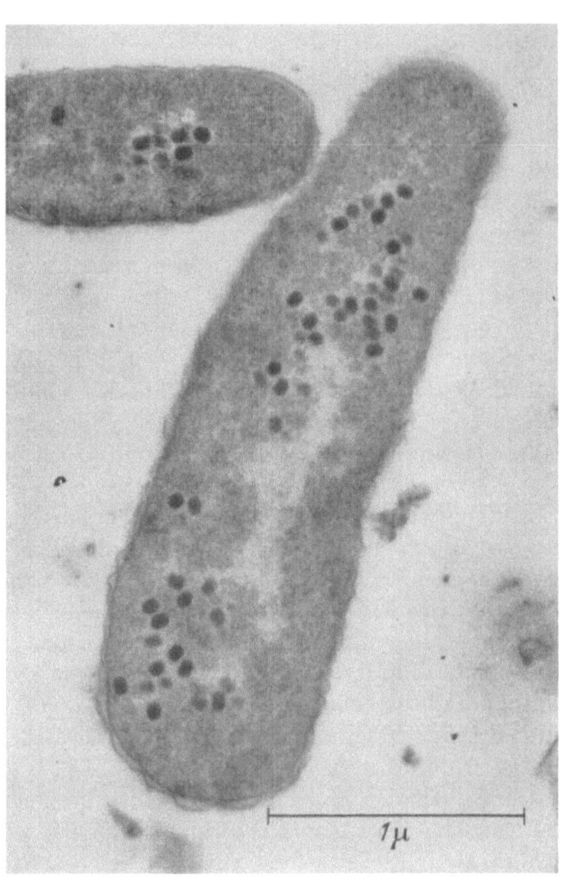

Abb. 5. Bakterien mit intracellulären reifen T2-Phagen. 35 000 mal

und meistens sehr ähnlich angeordnet wie im Phagen-Pool; man findet ebenfalls quer- und längsgeschnittene Regionen, in denen sie parallel laufen (Abb. 2). Sehr oft ist die Anordnung der Fibrillen aber auch mehr netzartig. Wir haben bis jetzt keine Möglichkeit, zu entscheiden, wie weit diese netzartige Struktur reell ist oder ein Artefakt. Abgesehen von der Anordnung der Fibrillen ist das Nucleoid vollkommen homogen ausgefüllt; keine sichtbaren, anders strukturierten Körper können darin gefunden werden. Verschiedene untersuchte physiologische Zellzustände (Karenzierung, Antibiotika, UV) haben keinen sichtbaren Einfluß auf die Kernfeinstruktur, jedoch auf die Form der Vacuole. Insbesondere ist der Einfluß der Salzkonzentration (*15*) in der Nährlösung zu erwähnen. Wird in Trypton ohne NaCl-Zusatz bebrütet, so erhält man Zellen mit sehr zerklüfteten, unregelmäßigen Nucleoiden. In 2% NaCl hingegen ist der Kern kompakt und von einfacher Form (*1*). Diese beiden, stark verschiedenen Formen haben jedoch keinen Einfluß auf die Teilungszeit der Zellen, die in beiden Fällen völlig gleich ist.

Diskussion

Als erste Frage stellt sich, inwieweit das feinfädige Plasma „richtiger" ist als die groben Einschlüsse. Folgende Argumente sprechen für die „Richtigkeit" der Fixation:

a) Nur unsere Bedingungen für die Osmiumfixation geben polyedrische Phagenköpfe, so wie sie schon von mindestens vier anderen guten Präparationsmethoden her bekannt sind [Lit. vgl. (6)].

b) Nur ein feinfädiger Pool kann mit den Vorstellungen des genetischen Phagen-Pools in Einklang gebracht werden.

c) Es ist unwahrscheinlich, daß die Fixation bestehende Strukturen zerstört, die dann durch die Versenbehandlung nachträglich wiederhergestellt werden können.

d) Die Permanganatfixierung erzeugt ebenfalls ein feinfädiges Nucleoplasma in den Bakteriennucleoiden und gestattet, den Kopf der Phagen polyedrisch darzustellen.

Dagegen sprechen die Schwierigkeiten, mit denen man sich einen Bakterienkern dieser Struktur funktionell vorstellen kann. Wir werden später darauf zurückkommen.

Als zweite Frage stellt sich: Falls wir annehmen, daß die feinfädige Struktur signifikativ ist, stellen die Fibrillen wirklich Desoxyribonucleinsäure mit einem neutralisierenden Polyamin, Protamin, Histon oder Protein dar? Es besteht kein Zweifel, wie schon in früheren Arbeiten dargelegt wurde, daß die Kernvacuole mit den Stellen des Feulgen-positiven Materials grob übereinstimmt (16). Wir haben aber keinen direkten Weg, um auszuschließen, daß das sichtbare Plasma ein Grundplasma darstellt, indem eine im Elektronenmikroskop nicht sichtbare Form der DNS eingelagert wäre. Da dieses Grundplasma elektronenoptisch sichtbar ist, wäre anzunehmen, daß es ein Protein oder Ribonucleoproteid ist. Dagegen spricht: a) Proteine und Ribonucleoproteide des Cytoplasmas werden leicht mit Osmium fixiert schon innerhalb einer halben Stunde, während das Nucleoplasma noch nicht genügend fixiert ist (1); b) die Ribonucleoproteide der Bakterien werden zum weitaus größten Teil als kugelige 150—200 Å große Teilchen gefunden (17); c) lysierte Spheroplasten geben eine Lösung sehr hoher Viscosität (18), die nach DNS-Behandlung verschwindet. Fädiges Protein oder Ribonucleoproteid kann daher nicht in großen Mengen vorhanden sein; Farbreaktionen, die einen chemisch heterogenen Aufbau des Nucleoids zeigen, können als Beweis kaum gebraucht werden, nachdem wir wissen, daß die bisher verwendeten Fixiermethoden in den weitaus meisten Fällen ein grobes Coagulat erzeugen. Sie können erst dann wieder berücksichtigt werden, wenn parallel dazu im Dünnschnitt die Feinheit des Nucleoplasmas beobachtet wird. Es scheint uns evident, daß eine im Elektronenmikroskop unsichtbare DNS-Komponente ebenfalls von fädiger Natur sein muß und entsprechend die Coagulation des dichten Grundplasmas gezwungenermaßen mitmacht. Nur eine globuläre Komponente könnte sich bei diesem Prozeß entmischen.

Ohne die zweite Frage definitiv beantworten zu können, möchten wir nun noch kurz zeigen, was man sich unter dem Bakteriennucleoid vorstellen kann, wenn die feinfädige Struktur tatsächlich als DNS betrachtet wird. Ein solches Nucleoid kann nicht als eigentlicher Zellkern betrachtet werden, weil es sich in wesentlichen Punkten unterscheidet: 1. besitzt es keine Kernmembran, 2. ist das Chromatin gleichmäßig verteilt und 3. ist es stark formveränderlich. Diese Eigenschaften passen aber gut auf das Chromosom selbst. Damit würde der unbekannte Prozeß der Teilung des Nucleoids auf den ebenso unbekannten und komplexen Prozeß der Teilung eines Chromosoms zurückgeführt! Der Vergleich hinkt aber insofern, als das Chromosom der höheren Zelle chemisch nicht homogen zusammengesetzt ist. Das Bakteriennucleoid würde entsprechend nur ein äußerst primitives Chromosom darstellen. Naiv betrachtet scheint das ziemlich einleuchtend zu sein: Die Natur hat wohl zuerst das Chromosom erfunden, bevor sie Kerne mit mehreren Chromosomen, Spindel und Centrosomen entwickelt hat. Der von ROBINOW (19, 20) in seinen ersten grundlegenden Arbeiten beschriebene Teilungsprozeß und selbst der von ihm verwendete Name „Bakterienchromosome“ würde damit wieder zum Ausgangspunkt, nachdem in den letzten Jahren in der Bakteriencytologie versucht wurde, einen komplizierten Kern zu finden!

Falls ein Nucleoid ein Chromosom darstellt, so muß man annehmen, damit die genetische Kontinuität gewährt bleibt, daß entweder nur eine einzige Bindungsgruppe existiert, oder andernfalls kann uns eine Polytänie die Lösung bringen. Die Beantwortung dieser Fragen ist jedoch verfrüht, und es wird sich vorerst darum handeln, zu untersuchen, ob in den Kernen höherer Organismen auch solche feinfädigen Nucleoplasmen zu finden sind, die auf den Versentest positiv reagieren. Indem progressiv von primitivsten zu den höher entwickelten Lebewesen geschritten wird, wird es vielleicht möglich werden, die sukzessive Komplizierung festzulegen. Wir denken dabei sowohl an die Art und Weise, wie die DNS strukturell in das eigentliche Chromosom eingebaut ist (*21*) als auch an die neutralisierende Komponente, die voraussichtlich im einfachsten Fall ein Polyamin (*22*) und erst in den höchst entwickelten Kernen ein Protein ist.

Unsere Arbeiten wurden unterstützt durch den Schweizerischen Nationalfonds für wissenschaftliche Forschung.

Literatur

1. Ryter, A., u. E. Kellenberger: Z. Naturforsch. **13b**, 597 (1958).
2. Maaloe, O., and A. Birch-Andersen: Bacterial Anatomy. Cambridge: Univ. Press. (1956). p. 261.
3. Kellenberger, E., u. A. Ryter: Experientia (Basel) **12**, 420 (1956).
4. Luft, J. H.: J. biophys. biochem. Cytol. **2**, 799 (1956).
5. Mercer, E. H.: Nature (Lond.) **181**, 1550 (1958).
6. Kellenberger, E., A. Ryter and J. Séchaud: J. biophys. biochem. Cytol. **4**, 671 (1958).
7. Cohen, S. S.: J. biol. Chem. **174**, 281 (1948).
8. Hershey, A. D., J. Dixon and M. Chase: J. gen. Physiol. **36**, 777 (1953).
9. Kellenberger E., J. Séchaud and A. Ryter: Electron microscopical studies of phage multiplication. IV. Virology (im Druck)
10. Hershey, A. D., and N. E. Melechen: Virology **3**, 207 (1957).
11. Watanabe, I., G. S. Stent and H. K. Schachman: Biochim. biophys. Acta **15**, 38 (1954).
12. Levinthal, C., and H. R. Crane: Proc. nat. Acad. Sci. (Wash.) **42**, 436 (1956).
13. Hershey, A. D.: Virology **4**, 237 (1957).
14. Levine, L., J. L. Barlow and H. van Vunakis: Virology **6**, 702 (1958).
15. Whitfield, J. F., and R. G. E. Murray: Canad. J. Microbiol. **2**, 245 (1956).
16. Kellenberger, E. u. A. Ryter: Schweiz. Z. allg. Path. Bakt. **18**, 1122 (1955).
17. Schachman, H. K., A. B. Pardee and R. Y. Stanier: Arch. Biochem. Biophys. **38**, 245 (1952).
18. Bolle, A., u. E. Kellenberger: Schweiz. Z. allg. Path. Bakt. **21**, 714 (1958).
19. Robinow, C. F.: Nuclear apparatus and cell structure of rod-shaped bacteria. In R. J. Dubos, The Bacterial Cell. Cambridge Massachusetts: Harvard Univ. Press 1949, p. 353.
20. — Bact. Rev. **20**, 207 (1956).
21. Taylor, J. H.: Sci. Amer. **198**, 37 (1958).
22. Ames, B. N., D. T. Dubin and S. M. Rosenthal: Science **127**, 814 (1958).

L'ultrastructure du centriole
et d'autres éléments de l'appareil achromatique

W. Bernhard et E. de Harven

Institut de Recherches sur le Cancer à Villejuif (Seine) et Laboratoire d'Anatomie Pathologique de l'Université Libre de Bruxelles

L'étude de la mitose reste au premier plan des préoccupations des morphologistes. Et pourtant, les progrès récents des techniques de la microscopie électronique n'ont pas encore permis d'étendre notre compréhension de ce phénomène fondamental de la cytologie. L'étude de la structure des chromosomes s'est révélée d'une difficulté particulière, et l'impression prévaut que nos techniques n'y sont pas encore parfaitement adaptées. En revanche, certains constituants de l'appareil achromatique ont fait l'objet d'observations précises. Parmi ces constituants, le centriole a été particulièrement étudié au cours des trois dernières années. En 1956, Bernhard et de Harven (*2, 6*) ont décrit cet organite tel qu'il apparaît dans différentes cellules de Vertébrés. Ces observations ont été rapidement confirmées par plusieurs auteurs (*3, 22*). Bientôt cependant, les travaux de Bessis et coll. (*4*) ainsi que ceux de Yamada (*21, 22*), devaient montrer que le centriole est d'une complexité plus grande qu'il n'était apparu tout d'abord. Le schéma proposé en 1956 (*6*) est resté valable mais il est indispensable de le compléter par ce que l'on peut appeler les *structures péricentriolaires*. Celles-ci correspondent à des corpuscules accessoires ou «satellites» du centriole, et sont très étudiées à l'heure actuelle. Il est toutefois prématuré d'en donner une description définitive. C'est pour tenter de faire le point que les aspects suivants seront envisagés dans le présent rapport:

1. Le centriole et les structures péricentriolaires.
2. Le nombre de centrioles par cellule et les rapports des centrioles entre eux.

3. Les images suggérant un mode de réduplication centriolaire.
4. Les rapports du centriole avec l'appareil de Golgi.
5. Les rapports du centriole avec le noyau au repos.
6. Les rapports du centriole avec la membrane cytoplasmique.
7. Le centriole dans les cellules en mitose.
8. La présence d'un centriole de structure identique dans toutes les espèces animales.
9. Essai de nomenclature du centriole et des structures péricentriolaires.

Les structures centriolaires des spermatides, ainsi que les corpuscules basaux ciliaires ne seront pas considérés ici.

1. Le centriole et les structures péricentriolaires

Rappelons que le centriole est une structure de forme cylindrique dont le diamètre est approximativement de 150 mμ et la longueur de 300 à 500 mμ (Fig. 1 et 2). Le centre de ce cylindre paraît dépourvu de toute substance dense aux électrons, ce qui contraste fortement avec la grande densité électronique de la paroi du cylindre centriolaire. Cette paroi comprend un certain nombre de tubules parallèles d'un diamètre moyen de 15 à 20 mμ, et dont la lumière est également dépourvue de substance dense.

Les travaux récents de Bessis et coll. (*4, 5*) ont fait apparaître la nécessité de compléter cette description. Il paraît certain que le nombre 9 soit la clé du système centriolaire. Il s'agit cependant de 9 groupes de tubules, et non de 9 tubules isolés. Chaque groupe peut contenir de 1 à 3 tubules ou davantage. Les unités à tubules doubles semblent les plus fréquentes. Exceptionnellement, tous les groupes sont constitués de trois éléments et leur ensemble évoque une *disposition en spirale* (Fig. 1). Cette image est cependant loin d'être constante. Le plus souvent il est très difficile de compter le nombre exact des tubules et de distinguer les 9 groupes d'éléments tubulaires. Les corpuscules basaux présentent d'ailleurs les mêmes particularités dans les cellules ciliées de Vertébrés (tumeurs rénales du hamster et muqueuse utérine normale de la femme, observations non publiées).

Le centriole est cependant plus complexe encore. Dans son voisinage immédiat, on voit des structures «péricentriolaires» ou «satellites» que Bessis et Breton-Gorius (*4, 5*) ont appelées «massules» et «ponts». L'existence de ces structures a été confirmée (Fig. 3 à 7). Les «satellites» sont caractérisés par leur forte densité électronique, leur centre paraissant généralement plus contrasté que la périphérie; leur forme est arrondie, leurs contours sont assez diffus, sans aucune membrane limitante, et leur diamètre approximatif est de 70 mμ. Les ponts relient un «satellite» au centriole voisin (Fig. 4 à 7). Il est cependant fréquent d'observer des centrioles typiques, coupés selon des incidences variées, et qui sont dépourvus de toute différenciation péricentriolaire. Cette observation paraît trop fréquente pour que la minceur des coupes puisse en rendre compte. *Les structures péricentriolaires correspondraient donc plutôt à des aspects transitoires,*

Fig. 1. Coupe transversale d'un centriole exceptionnellement bien conservé. A noter l'arrangement en spirale des tubules formant la périphérie du cylindre: 9 unités à 3 éléments chacune, enrobées dans une masse finement grenue. Gross. 174 000 ×

Fig. 2. Coupe longitudinale d'un centriole typique. Lumière des tubules périphériques nettement visible. Gross. 174 000 ×

Fig. 3. Coupe légèrement oblique d'un centriole, Corpuscules péricentriolaires très denses (1—6), d'apparence amorphe et sans contact visible avec le centriole. Gross. 87 000 ×

Fig. 4. Coupe transversale d'un centriole avec trois extensions rayonnant perpendiculairement par rapport à son centre. L'une d'entre elles se termine dans un corpuscule péricentriolaire (——→). Gross. 87 000 ×

Fig. 5. Centriole semblable à celui de la Fig. 4. (1—3) : corpuscules péricentriolaires reliés par des ponts (centrodesmes?) au cylindre centriolaire. Le n° 3 est entouré à son tour par quatre corpuscules denses (——→). Gross. 87 000 ×

Fig. 6. Centrosome double (diplosome), l'un coupé longitudinalement (C_1), l'autre transversalement (C_2). Un corpuscule péricentriolaire est relié par un pont à la partie médiane du centriole C_2. Gross. 87 000 ×

Fig. 7. Diplosome coupé longitudinalement, avec plusieurs corpuscules péricentriolaires en position perpendiculaire par rapport à C_1 et C_2. Image suggérant la transformation progressive des corpuscules péricentriolaires en centrioles filles (f_1 f_2). Gross. 87 000 ×

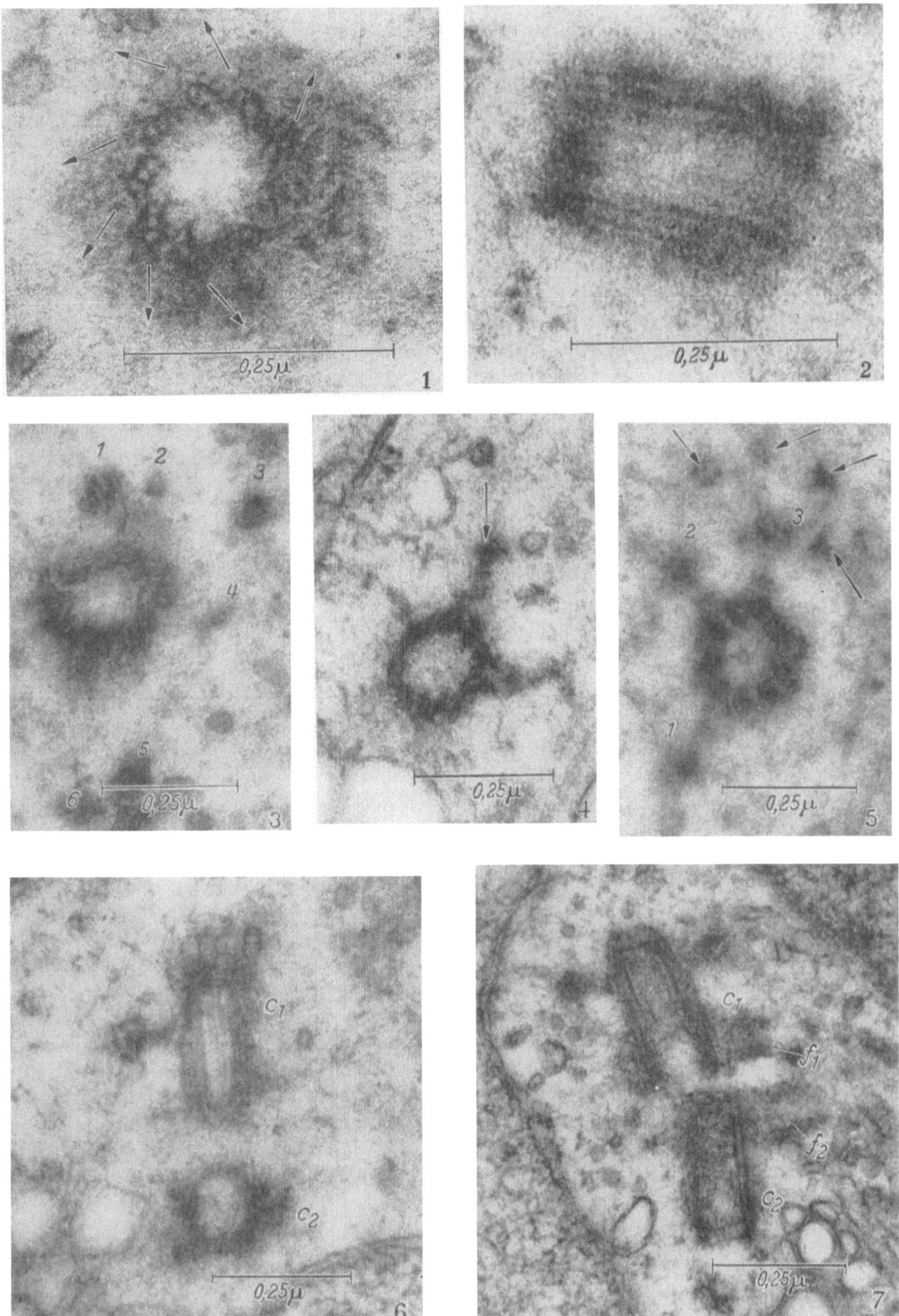

Fig. 1—7. Centrioles de cellules réticulaires de la rate de jeunes poussins (Fig. 1—6) et d'une rate de souris leucémique (Fig. 7). Coupes transversales et longitudinales. (Légende p. 218)

vraisemblablement liés à une des phases de l'activité centriolaire. D'autre part, lorsque les »satellites« sont présents, ils ne sont pas toujours reliés au centriole par un «pont», étant alors situés au voisinage immédiat ou à une distance plus éloignée du centriole, mais sans aucune attache avec celui-ci. Enfin, le nombre des structures péricentriolaires est très variable, et leur disposition n'est pas aussi systématique que Bessis l'a supposé. En effet, les deux couronnes de 9 «massules» chacune, que décrit cet auteur, n'ont jamais été vues dans l'ensemble du matériel que nous avons examiné. D'autre part, si les «ponts» sont attachés de préférence aux parties distales du centriole (Fig. 7), on en trouve également avec un point d'attache situé au milieu du cylindre (Fig. 6).

2. Le nombre de centrioles par cellule, et les rapports des centrioles entre eux

Dans le voisinage d'un centriole, il est extrêmement fréquent d'observer un autre centriole. Dans le cytoplasme de toute cellule diploïde au repos, on peut observer deux centrioles appariés. Cette paire de centrioles a reçu depuis bien longtemps le nom de *diplosome.* En 1894, Heidenhain avait décrit une structure reliant les deux centrioles d'un diplosome à laquelle il avait donné le nom de *centrodesme.* Dans un article récent, Tanaka et coll. (*20*) ont décrit 9 canalicules paraissant prolonger les tubules centriolaires et reliant les deux centrioles d'un diplosome. Notre matériel n'a pas permis l'observation de structures semblables, pouvant être interprétées comme centrodesmes. En revanche, il est certain que les deux centrioles formant un diplosome occupent, au sein de la centrosphère, des positions relativement bien définies. Deux aspects sont fréquemment rencontrés: *ou bien les deux centrioles sont placés dans le prolongement l'un de l'autre, ou bien ils sont disposés à angle droit*, image répondant sans doute aux observations anciennes de diplosomes en V. Cette deuxième éventualité paraît la plus fréquente. Elle est probablement liée au mode de réduplication du centriole.

3. Les images suggérant un mode de réduplication du centriole

On sait que le centriole est capable de se diviser, et cependant le processus intime de ce phénomène reste incompris. Cette division survient tout au début de la prophase et elle a même été considérée par Schreiner comme le premier signe morphologique du cycle mitotique (*18*). Théoriquement, il doit donc être possible de voir quatre centrioles dans les cellules en imminence de division mitotique, c'est-à-dire en pré-prophase. La technique des coupes ultrafines a permis de démontrer la présence de 3, voire même de 4 centrioles dans une même cellule (Fig. 7). La rareté de ces images correspond vraisemblablement à la rapidité de ce stade particulier de la pré-prophase. L'intérêt de ces observations est indiscutable, car elles permettent d'imaginer le mode de réduplication du centriole. Dans les coupes d'incidence particulièrement favorable, on voit nettement que cette réduplication se fait par un processus de bourgeonnement latéral du centriole-mère (Fig. 7). Le centriole-fille a un diamètre légèrement inférieur ou identique à celui du centriole-mère, mais il est cependant beaucoup plus court et semble accolé aux portions distales de la paroi du centriole-mère. Son orientation est typiquement perpendiculaire à celle du centriole-mère. La disposition si fréquente de deux centrioles à angle droit résulterait de ce mode de réduplication. Le point de départ du bourgeonnement latéral d'un centriole-fille reste à préciser, mais il est très tentant de voir dans les corpuscules péricentriolaires des précurseurs de centrioles-filles. La Fig. 7 paraît suggestive à cet égard. A l'heure actuelle, on voit encore mal comment un «satellite» pourrait se transformer en nouveau centriole, mais il n'est pas exceptionnel de constater que les ponts présentent une fine striation longitudinale qui pourrait faire penser à l'apparition des tubules centriolaires. Quoi qu'il en soit, cette possibilité d'une relation entre les structures péricentriolaires et le bourgeonnement des centrioles-filles est présentée ici à titre d'hypothèse et nécessitera des observations plus détaillées.

Fig. 8. Diplosome (C_1, C_2). Le centriole C_1 semble inactif. C_2 au contact de la membrane cellulaire a élaboré un cil qui se prolonge dans l'espace intercellulaire (*e.i.*) (——→). Gross. 87 000 ×

Fig. 9. Centriole avec ébauche d'un cil. Image montrant la continuité structurale entre tubules centriolaires et tubules ou fibres ciliaires (——→). Vacuole (*v*) caractéristique, formant une gaine autour de la partie basale du cil. Gross. 174 000 ×

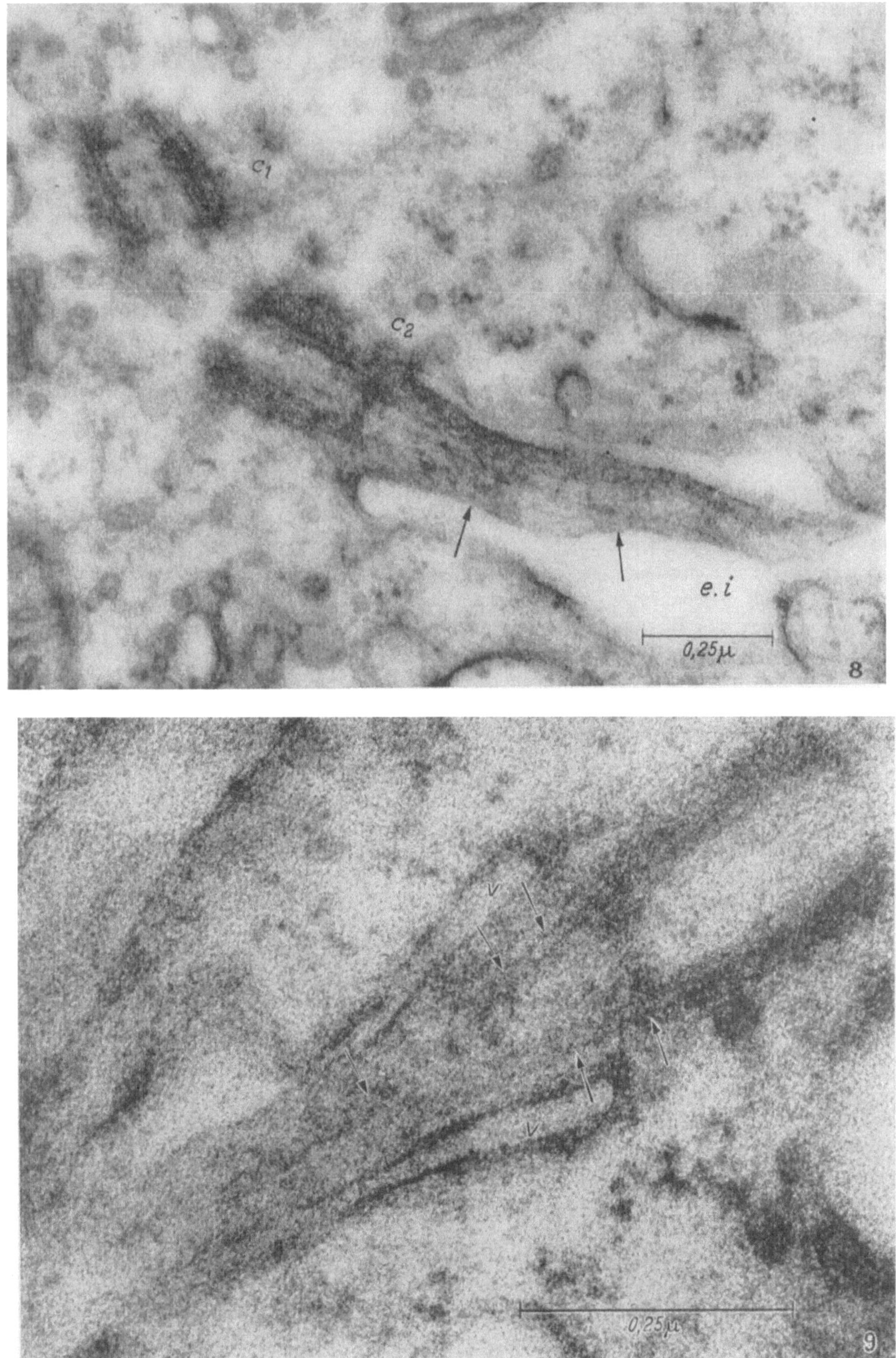

Fig. 8 et 9. Coupes longitudinales des centrioles de la rate de jeunes poussins révélant leur rapport avec la surface cellulaire. (Légende p. 220)

4. Rapports du centriole avec l'appareil de Golgi

Il est bien connu que pendant l'interphase le corpuscule central se trouve de préférence au milieu ou au contact de la zone de Golgi (centrosphère) dont les constituants ont été définis pour la première fois grâce à la microscopie électronique. Les rapports entre ces deux organites ont fait l'objet d'un travail récent de Policard et coll. (*15*). L'appareil de Golgi, de même que les chondriocontes, peut décrire une sorte de couronne autour du diplosome. Les structures golgiennes se présentent alors en amas relativement bien individualisés qui ménagent entre eux des espaces ne paraissant contenir aucun organite démontrable par les techniques actuelles. Notre matériel n'a montré aucune polarité des structures golgiennes par rapport au centriole. Aucune fibre continue unissant les centrioles au Golgi, et comparable aux «protofilaments golgiens» de Tanaka et coll. (*20*), n'a été observée. Exceptionnellement, on voit un faisceau de filaments sortant au contact du centriole et se prolongeant à travers la zone golgienne. Il s'agit probablement de l'ébauche d'un aster se formant au début de la prophase (Fig. 11). Aucun élément ne permet de voir dans ces filaments l'origine de l'appareil de Golgi dont le polymorphisme est incompatible avec l'hypothèse des auteurs japonais, selon laquelle le nombre 9 serait également la clé du système golgien.

5. Rapports du centriole avec le noyau au repos

Le centriole se trouve le plus souvent dans le voisinage immédiat de la membrane nucléaire. Une hypothèse récemment émise par Lettre et Lettre (*14*) suggère la possibilité de la persistance des fibres fusoriales pendant toute l'intercinèse. Ces fibres uniraient de façon permanente les centrioles aux chromosomes, en passant par les pores de la membrane nucléaire. Aucun document permettant de confirmer cette hypothèse n'a pu être recueilli dans l'ensemble de notre matériel. L'image que Bessis et coll. ont publiée (*5*) montre cependant une fibre unissant la membrane nucléaire à la région de la centrosphère. Cette image nous paraît correspondre à cette continuité fréquemment observée entre la membrane nucléaire externe et le réticulum endoplasmique, images sur lesquelles Porter (*16*) et Epstein (*8*) ont déjà insisté. Quoi qu'il en soit, le centriole occupe souvent une région du cytoplasme qui paraît déprimer la surface du noyau. Ce n'est pas une règle absolue, mais il est difficile de préciser des rapports constants pour définir la position d'un organite dont on sait qu'il est capable de se déplacer dans le cytoplasme.

6. Les rapports du centriole avec la membrane cellulaire

Des images suggérant la formation d'un cil unique et probablement fort court ont pu être observées récemment dans des cellules au repos de deux tissus différents: rate embryonnaire de poussin et tumeur ovarienne de la souris. La microscopie optique n'y avait pas, à notre connaissance, mis en évidence une telle différenciation ciliaire. On peut voir en effet, à l'une des extrémités d'un des centrioles d'un diplosome typique, un faisceau de canalicules parallèles paraissant en continuité directe avec les tubules centriolaires (Fig. 8 et 9). Au point où ces canalicules sortent du centriole, ils apparaissent flanqués d'une vacuole qui les entoure comme une gaine. Des cas plus avancés montrent que cet ensemble de canalicules franchit la membrane cytoplasmique pour se prolonger au dehors de la cellule, à la manière d'un petit cil (Fig. 8). On voit aussi très clairement que la vacuole qui paraît s'accoler au faisceau de canalicules est en rapport avec la surface cellulaire au contact de laquelle elle semble s'ouvrir, formant ainsi un anneau au centre duquel émerge le cil.

Dans les différents cas où ces images ont pu être observées, il est clair que ce n'est qu'un seul des deux centrioles du diplosome qui manifeste une telle activité ciliaire; l'autre centriole est le plus généralement dans le prolongement du premier, et paraît inactif. Ces observations donnent à penser que l'activité ciliaire des centrioles est probablement plus fréquente que la microscopie optique n'avait pu le démontrer. La signification de ce phénomène mérite certainement d'autres études; en particulier, il serait important de connaître le rapport de ce cil unique de cellules mésenchymateuses et épithéliales avec la «Centralgeissel» que Zimmermann (*24*) a décrite dans des cellules cylindriques de l'anse de Henlé du rein et dans les conduits excréteurs de glandes variées.

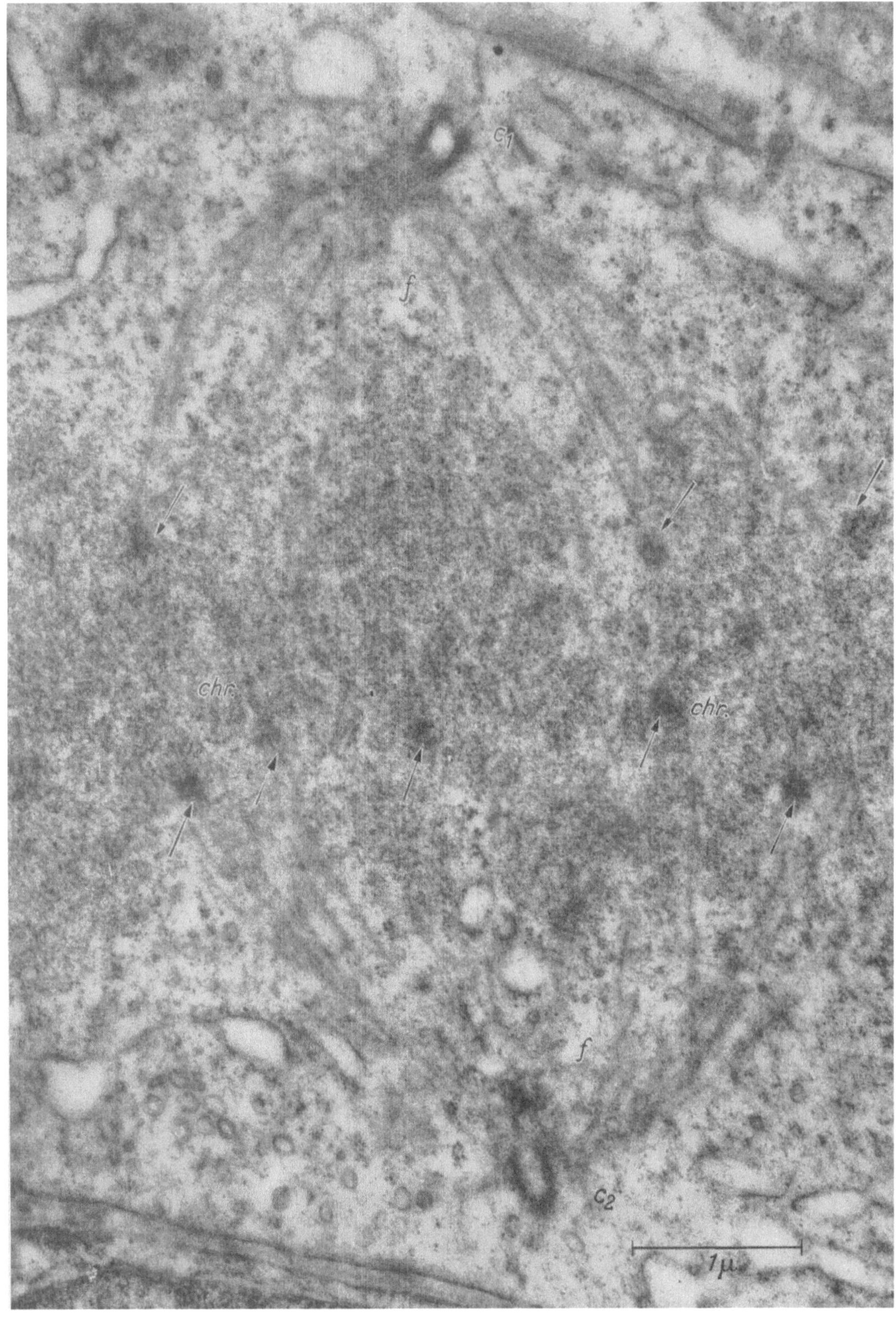

Fig. 10. Cellule réticulaire d'une rate de poussin en mitose. Métaphase permettant de voir tout l'appareil achromatique: les centrioles C_1C_2 aux deux pôles fusoriaux, les fibres fusoriales (f), et leurs points d'attache (cinétochores) (⸻⟶) aux chromosomes (*chr.*). Gross. 27 000 ×

7. Le centriole pendant la mitose

Les coupes qui permettent d'observer une figure mitotique comprenant le centriole ne sont pas fréquentes, mais leur étude permet déjà d'affirmer que, dans ses grandes lignes, la structure d'un centriole situé au pôle d'un fuseau mitotique est identique à celle qu'il présente dans le cytoplasme d'une cellule en intercinèse (Fig. 10). Cette observation coïncide avec celle publiée par Yamada (*23*) mais est opposée à celle de Amano d'après laquelle le centriole revêtirait, durant la mitose, un aspect sphérique (*1*).

Comme nous l'avons déjà montré (*6*) les *fibres fusoriales* ont un aspect très caractéristique. Elles se présentent comme de longs canalicules d'un diamètre assez constant de 15 à 20 mμ environ, et d'une longueur pouvant atteindre plusieurs microns (Fig. 10—12). On peut dans certains cas distinguer les fibres chromosomiques rattachées aux chromatides au niveau des centromères (cinétochores) (Fig. 10), des fibres inter-polaires qui traversent toute la masse des chromosomes et relient les deux pôles fusoriaux. Les diamètres et la structure fine de ces deux types de fibres semblent identiques. Les fibres astériennes visibles dans le cytoplasme avant la dissolution de la membrane nucléaire sont identiques (Fig. 11).

Le *point d'insertion* des fibres chromosomiques au niveau des *centromères* est d'un intérêt particulier. La structure des centromères est en effet loin d'être comprise (*19*). La seule chose que l'on puisse dire dès maintenant, c'est que le centromère représente une zone arrondie d'un matériel granulaire sensiblement plus dense que les chromosomes environnants, et contenant certains profils de structures tubulaires (Fig. 10). La situation paraît nettement plus confuse à l'autre extrémité des fibres fusoriales. En effet, le mode d'insertion des fibres sur le centriole n'est pas clairement démontrable par les techniques actuelles, et les structures satellites n'y semblent jouer aucun rôle. Il faut noter que la teminaison des fibres s'observe souvent à une certaine distance du centriole, à moins que ces fibres n'entrent en contact plus direct avec cette substance dense diffusément répartie autour du centriole. Les rapports des fibres fusoriales et des centrioles paraissent donc plus indirects qu'on aurait pu l'imaginer. Enfin, il est à remarquer que le grand axe du centriole ne coïncide pas toujours avec celui du fuseau, avec lequel il fait parfois un angle d'ouverture variable.

8. La présence d'un centriole de structure identique dans toutes les espèces du règne animal

La microscopie optique avait établi la notion de l'ubiquité du centriole dans toutes les espèces animales. Cette notion n'a reçu jusqu'ici qu'une confirmation très partielle en microscopie électronique. Il suffit pour s'en convaincre de comparer d'une part, la liste établie en 1907 par Heiden-hain (*12*) de tous les tissus et de tous les animaux chez lesquels le centriole avait été décrit à cette époque, et d'autre part, la liste que l'on peut dresser aujourd'hui des différents tissus qui ont fait l'objet d'une étude électronique du centriole. Dans la mesure où notre information est complète, cette liste comprend:

Rate de pleurodèle (*6*)	Moelle osseuse de rat (*6*)
Rate de triton (*6*)	Sarcome de Rous (*6*)
Rate d'embryon de poulet (*6*)	Tumeur de Murray-Begg du poussin (*6*)
Bourse de Fabricius d'embryon de poulet (*6*)	Tumeur de Brown-Pearce du lapin (*6*)
Rate de souris (*6*)	Ganglion lymphatique de souris (*20*)
Moelle osseuse de souris (*6*)	Leucocytes humains (*4*)
Rate du rat (*6*)	Ovaire de la souris (*22*)
	Cartilage trachéal d'embryon de souris (*22*).

Cette comparaison doit inciter à la prudence, et un matériel très abondant doit encore être examiné avant de présenter une conclusion générale sur la structure du centriole. En 1956, celle-ci a pu être démontrée chez le rat, la souris, le poussin et le triton (*6*). Les travaux de Bessis, de Amano et de Yamada ont porté essentiellement sur les organes hématopoïétiques des Mammifères. En 1957, de Robertis et coll. (*7*) publiaient un article consacré à l'étude des testicules de la sauterelle. Dans le cytoplasme d'une spermatide, ces auteurs observent un groupe de grains fort denses aux électrons et qu'ils identifient au centriole. D'autre part, dans des spermatides d'Helix pomatia (*9*) chez Spirotreptus castaneus (*10*), chez Pyrsonympha (*11*) et chez Chromulina (*17*), le

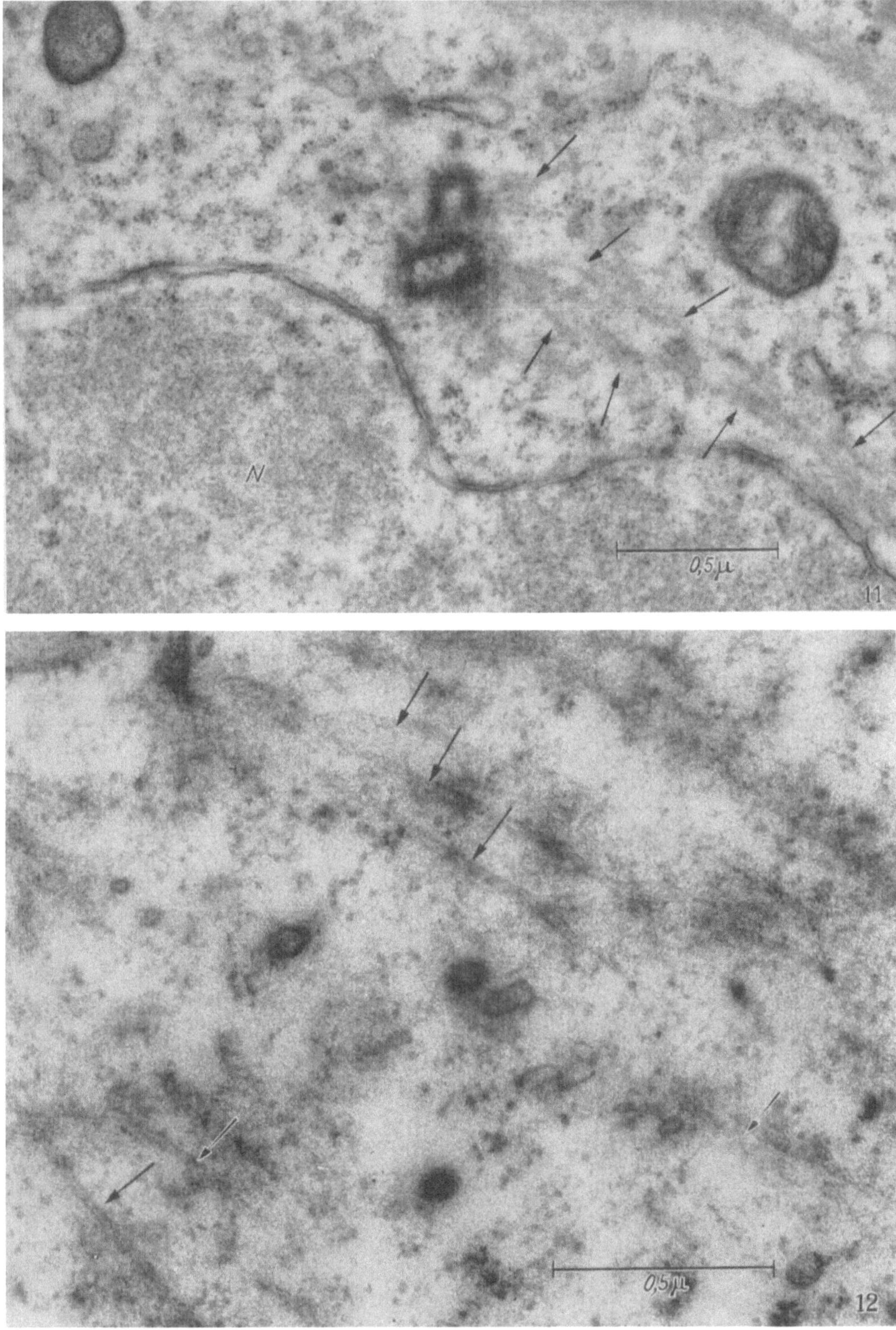

Fig. 11. Cellule réticulaire d'une rate de poussin. Diplosome formé de 2 centrioles rangés en angle droit (position très fréquente). A noter la chevelure de filaments très minces à droite des centrioles (———→) (ébauche d'aster ?). (N) = Noyau. Gross. 50 000 ×

Fig. 12. Cellule amniotique humaine. Quelques fibres fusoriales vues à un plus fort grossissement. Il s'agit en réalité de canalicules dont la lumière est nettement visible sur cette image (———→). Document du Dr. DOURMASHKIN. Gross. 68 000 ×

centrosome a été identifié à une masse osmiophile homogène. Aussi pensons-nous qu'en attendant de nouvelles études, il ne sera guère possible de généraliser nos conclusions pour tout le règne animal. La présence chez les Invertébrés de centrioles identiques à ceux décrits chez les Vertébrés est pourtant démontrée. Lehmann (*13*), étudiant la structure du Tubifex dit ne pas pouvoir mettre en évidence chez cet oligochète de structure centriolaire du type cylindrique. Cette opinion repose malheureusement sur un matériel fixé selon une technique très particulière. Cependant, si l'on fixe à l'osmium selon les procédés habituels, on peut observer chez Tubifex des centrioles typiques formant un diplosome entouré d'une couronne de structures golgiennes.

La question de l'ubiquité du centriole peut se poser sous un autre aspect. En effet, il n'est pas encore nettement établi que, chez un même animal, le centriole soit présent dans tous les tissus, indépendamment de leur activité mitotique. Il faut admettre que les premiers travaux ont tous porté sur des tissus à activité mitotique élevée: moelle osseuse, rate, thymus ou tissus cancéreux. Bessis et coll. ont cependant montré des images de centriole dans des leucocytes polynucléaires (*3*). Cette observation ne paraît pas apporter une réponse suffisante à la question que nous nous sommes posée. En effet, la durée de vie des leucocytes polynucléaires est fort courte. La chance qu'un leucocyte polynucléaire soit observé peu de jours après sa dernière division de maturation est donc élevée et l'on peut se demander si un centriole observé dans une telle cellule ne peut pas être considéré comme le vestige d'une activité mitotique récente. En l'absence d'autres documents il est donc prématuré de conclure à la permanence d'un centriole de structure constante dans les cellules à activité mitotique faible ou nulle. Un argument de grande valeur serait évidemment l'observation d'un centriole dans un neurone d'un animal adulte, mais, à notre connaissance, cette observation n'a pas encore été faite.

Un point mérite d'être souligné ici. Dans la cellule hépatique et dans les cellules tubulaires rénales de Mammifères, le centriole a été décrit en microscopie optique. Il est donc surprenant de constater que les spécialistes électroniciens de ces deux tissus ne semblent pas y avoir observé de centriole. On pourrait invoquer une difficulté particulière à reconnaître le centriole dans des cellules dont le cytoplasme est encombré d'organites variés. Ici encore, de nouvelles recherches s'imposent. Signalons toutefois que, récemment, étudiant un foie de souris leucémique, nous avons observé un centriole typique dans une cellule hépatique. Ce foie n'était pas un foie normal. Il était infiltré d'éléments leucémiques, et il n'est pas exclu que dans ce cas, une certaine activité de régénération du parenchyme hépatique restant soit intervenue, augmentant donc l'activité mitotique des cellules hépatiques.

Il est probable que les nouveaux matériaux d'inclusion, ainsi que le raffinement de techniques de fixation et de coupe donneront prochainement des documents analysables avec plus de finesse que ceux obtenus jusqu'ici. D'autre part, la comparaison de l'ultrastructure de centrioles typiques de cellules somatiques avec celle de corpuscules basaux et des centrioles de spermatides, sujets volontairement écartés de cette étude, permettra sans doute de définir très exactement l'identité et la dissemblance des détails structuraux d'une famille d'organites cellulaires extrêmement intéressante.

9. Essai de nomenclature du centriole et des structures péricentriolaires

Le problème de l'ultrastructure du centriole apparaît donc plein de promesse, à la condition cependant de ne pas surcharger une nomenclature qui existe déjà, du moins pour les structures principales, et d'assurer ainsi la continuité des observations. Ce dernier point mérite une attention toute spéciale. Le lecteur trouvera ici un bref essai de corrélation entre la nomenclature ancienne et celle d'aujourd'hui. Le livre de Wilson (*21*) contient un glossaire des principaux mots employés en cytologie. Parmi ceux-ci nous relevons:

Centriole: ce terme s'applique aujourd'hui à une structure cylindrique composée de neuf tubules parallèles ou de neuf groupes de tubules.

Centrodesme: ce terme ne s'applique à aucune structure démontrable par les techniques électroniques actuelles.

Centrosome: ce terme s'appliquait à l'ensemble du centriole et de la zone péricentriolaire, caractérisée par une densification diffuse de la structure fondamentale. Les *formations péricentriolaires* («satellites», «massules») en font partie.

Diplosome: Ce terme désigne une paire de centrioles.

Centrosphère: région cytoplasmique de forme sphérique et limitée à la périphérie par l'appareil de Golgi avec lequel elle peut se confondre. La centrosphère présente en son centre le diplosome et les corps péricentriolaires. Contrastant avec la forte densité électronique des centrioles, elle paraît beaucoup moins osmiophile que ceux-ci. L'aster se forme à son niveau.

Bibliographie

1. Amano, S.: Cytologia **22,** 193 (1957).
2. Bernhard, W., et E. de Harven: C. R. Acad. Sci. (Paris) **242,** 288 (1956).
3. Bessis, M., et J. Breton-Gorius: Bull. Micr. appl. **7,** 54 (1957).
4. — — C. R. Acad. Sci. (Paris) **246,** 1289 (1958).
5. — — et J.-P. Thiery: Rev. d'Hémat. **13,** 363 (1958).
6. Harven, E. de, et W. Bernhard: Z. Zellforsch. **45,** 378 (1956).
7. Robertis, E. de, and H. Franco Raffo: Exp. Cell Res. **12,** 66 (1957).
8. Epstein, M. A.: J. biophys. biochem. Cytol. **3,** 851 (1957).
9. Grasse, P. P., N. Carasso et P. Favard: Ann. sci. Nat. Zool. **18,** 339 (1956).
10. — O. Tuzet et N. Carasso: C. R. Acad. Sci. (Paris) **243,** 337 (1956).
11. — Arch. Biol. **67,** 595 (1956).
12. Heidenhain, M.: Plasma und Zelle (1. Abt.) Jena: Gustav Fischer 1907.
13. Lehmann, F. E., and V. Mancuso: Exp. Cell Res. **13,** 161 (1957).
14. Lettre, H., et R. Lettre: Rev. d'Hémat. **13,** 337 (1958).
15. Policard, A., M. Bessis, J. Breton et J.-P. Thiery: Exp. Cell Res. **14,** 221 (1958).
16. Porter, K. R.: Harvey Lect. **51,** 175 (1957).
17. Rouiller, Ch., et E. Faure-Fremiet: Exp. Cell Res. **14,** 47 (1958).
18. Schreiner, A., u. K. E. Schreiner: Arch. Biol. **21,** 183 (1905).
19. Tanaka, H., F. Uchino, S. Dohi and S. Amano: Acta haemat. Jap. **19,** 571 (1956).
20. — M. Hanaoka and S. Amano: Acta. haemat Jap. **20,** 85 (1957).
21. Wilson, E. B.: The cell in development and heredity. New York: MacMillan Cy. edit. 1925.
22. Yamada, E.: The fine structure of centriole in some animal cells. Proc. 1ʳᵗ regional Conference in Asia and Oceania. Tokyo 1956.
23. — Kurume med. J. **5,** 36 (1958).
24. Zimmermann, K. W., cité d'après W. v. Moellendorff: Handbuch der Mikroskopischen Anatomie des Menschen. Bd. VII., Teil I. Berlin: J. Springer 1930.

On the structure of mammalian chromosomes during spermatogenesis and after radiation with special reference to cores[1]

B. R. Nebel

Division of Biological and Medical Research, Argonne National Laboratory, Lemont, Illinois, USA

The material used in this study was obtained from testes of C_{57} black mice showing pathobiosis after 1000 r local acute X-rays, from control mice, and from biospy material of an oligospermic human male. Spermatocytes of the mouse survive 1000 r, and the timing and prophasic structure of meiotic chromosomes is not changed. In first meiotic metaphase the pathobiotic material shows kinetochores with more contrast than the control: the kinetochore regions are larger and the microfibrillar bundles of the chromosome body are more plainly defined (5). From such material the structure of the mouse chromosome can now be described with certainty. These meiotic bivalents of the mouse under low power electron microscopy correspond to the light microscope images published in the '30s for *Trillium* and *Tradescantia* (3).

The *Bauplan* or architectural scheme of a piece of a mitotic metaphase chromatid is illustrated in the model represented in Fig. 1. The clear plastic parts merely constitute the supporting structure for the model and should be disregarded. The model was constructed from two pieces of wire

---

[1] This work was performed under the auspices of the US-Atomic Energy Commission.

each originally 60 feet long. The Watson-Crick spiral is represented only figuratively by wool threads, which are well shown in the top center of the figure. The wool threads are surrounded by metal wire spirals. These smallest spirals, called the first-order spirals, represent the 30 Å-fibril seen in the electron microscope, probably the protein moiety of the ultimate units. Upon this first order of coiling, which is very stable as to its dimensions in animals and plants, a second order coil is superimposed: this coil varies with stage, material, and probably with the degree of microfibrillar bundling or packing. Its gyres may measure 200 — 500 Å. Upon this, a third-order coil is superimposed, which measures 1000 — 2000 Å; it is the largest and final coil of somatic chromosomes, often visible in the light microscope. In meiosis a fourth order of coiling measuring 10000 to 30000 Å in gyre width is superimposed on the others. This is not represented in Fig. 1; to show it, it would be necessary to extend the entire model by several feet and to throw the entire cylinder into a fourth-order coil.

It is thus obvious that the metaphasic chromosomes of mouse and man are not branched structures (5). Instead the genomic microfibrillae are represented by a roughly parallel array of 32 threads per chromatid (4) which show highly efficient paranemic bundling and packing.

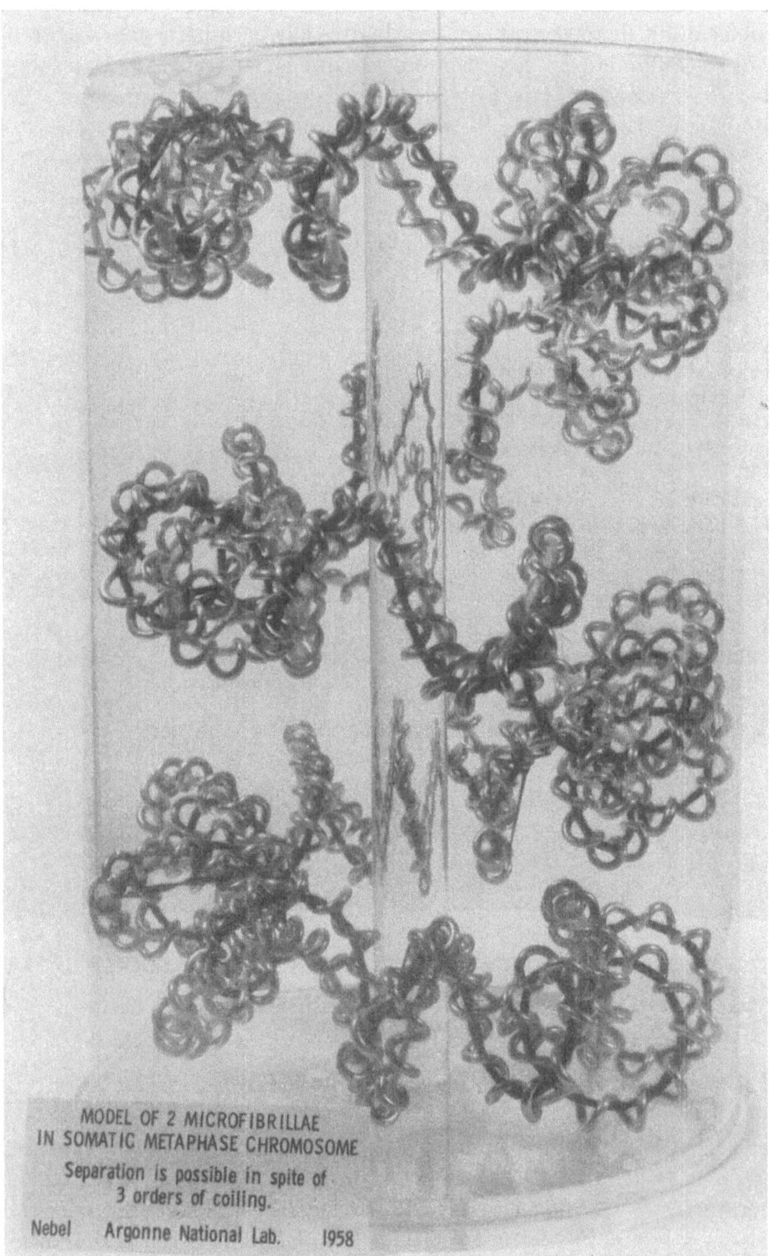

Fig. 1. A wire model illustrating the three successive orders of coiling which occur in somatic metaphase chromosomes. Only two microfibrillae are shown, a chromatid actually contains about 32

The model of Fig. 1 shows only two of the 32 microfibrillae contained in a chromatid. However, in spite of the presence of three or four orders of coiling superimposed upon one another the separation of subunits, e. g., chromatids and half chromatids is not only not difficult, but becomes self-regulatory, since it is geared to contraction and expansion of the bundles involved, as seen in many recent cinephotographs of other authors in light microscopy.

Much less is known concerning chromosome structure in interphase and early somatic and meiotic prophase. Perhaps these stages can be understood from the following. Impressions gained from light and electron microscopy have been synthesized into a working hypothesis which fits theory and facts as far as they are known.

The cores of meiosis (1) are considered to be not separate nuclear organelles, but rather specializing groups of the microfibrillar genonemata proper. The scheme is illustrated in the diagrams of Fig. 2[1].

During preleptotene we know from autoradiographic studies of many authors as well as my own that DNA synthesis and chromatid duplication occurs (3). The future bivalent then consists of four physiological units, the chromatids of the future first meiotic division. Such a single chromatid is illustrated in Fig. 2a. The lines running at right angles to the dark edge (the core component) are considered to be not the true microfibrillae but rather the cross links of the gene strings, emphasized by fixation. The chromatid unit of Fig. 2a is in sheet form; this is suggested to be the arrangement for every duplication, assuring the clean separation of daughter products, i. e., half chromatids (2) as described though not so named by TAYLOR et al. (6). The microfibrillae are considered to be despiralized at leptotene except for the first order of coiling represented in Fig. 2a, 2b and 2d by wavy lines (these wavy lines are intended to represent the regular 30 Å-spiral, not overall pathway, although the latter in nature is obviously never straight). In mammals this first order of coiling can apparently not be lost. The leptotene units may extend over 50 to 200 times the overall length of the metaphase chromosome. The genomic microfibrillae run parallel to the leading edge, which in electron photomicrographs is thickened and electron-

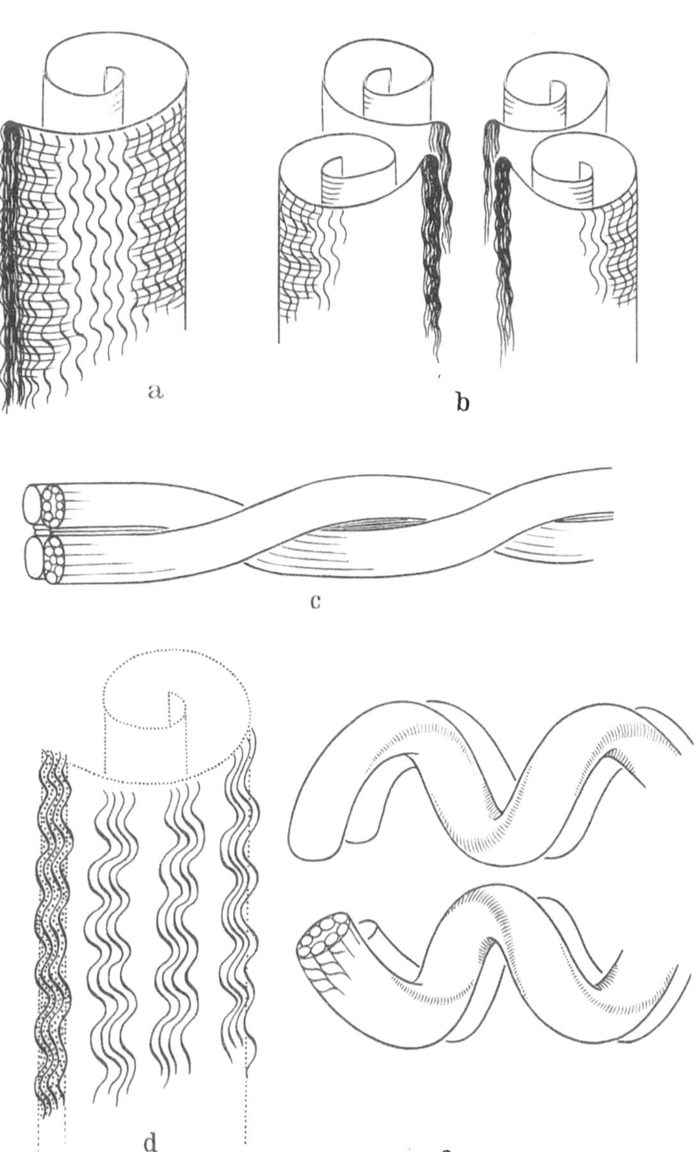

Fig. 2a—e. Interpretative drawings of meiotic chromosome cores in the mouse. a) A single chromatid unit in leptotene. The core component is a leading edge consisting of microfibrillae. b) Four of the units shown in 2a are placed in apposition for pairing as in early cygotene. The chromosomes are extremely long now, as at least two orders of coiling have been obliterated. c) Strepsitene. The microfibrillae are now in bundles, not in sheets. Cores are still visible but twists of the cores are implied where the chromatids twist. d) This suggests the manner by which the sheet arrangement may be changed to a bundle. e) As higher orders of coiling are imposed, the cores disappear (in mouse). With respect to the number of microfibrillae, the drawings suggest that 32 microfibrillae per chromatid are present during prophase of meiosis. At first these are grouped into eight bundles of four each, but with further packing and coiling resolution of the ultimate components is lost, and only the compound strands can be distinguished (Fig. b—e)

---

[1] For this drawing I am indebted to Assessor HILDE HEYDUCK, Academy of Art, Kassel.

dense, but is believed to consist of microfibrillae temporarily specialized. The sheet form is a sheet in principle and function, however the diagram is not proposed as a literal equivalent. Fig. 2 b shows the compound end of a future bivalent during pairing in late leptotene. Sister cores are now very close, with homologous cores opposed in parallel array. Fig. 2 b shows the compound end of a future bivalent during pairing in late leptotene. Sister cores are now very close, with homologous cores opposed in parallel array. Fig. 2 c shows the core array and its course in zygostrepsitene. This figure illustrates the continuously central position of the cores around which the participating chromatids may wind, twisting the cores proper as seen so often in the electron microscope. This model may also explain the diagonal clouds of microfibrillae passing over and under the cores as shown by MOSES (1). From Fig. 2 d it can be seen how easily the sheet array might be changed to a multiple-bundle array. This figure again is an illustration of principle, not a literal model. Fig. 2 e suggests that as the coil of the next higher order is formed, the cores disappear or have disappeared. This happens in late diplotene in the mouse, observations on which furnished the basis for the model presented here.

This study represents an extension of the light microscope image reaching macromolecular and approaching molecular dimensions. The structure of meiotic mammalian chromosomes at metaphase appears well established. However, observations on meiotic prophase and somatic interphase chromosomes still require an element of conjecture to make them understandable.

The present scheme is not in accord with the ex-cathedra model of TAYLOR (6), which has no electron microscope equivalence of any kind. I must differ with MOSES (1) in order to define more clearly my own model, which I consider more compatible with the classical findings of light and electron microscopy and with the basic tenets of genetics and cytogenetics.

References

1. MOSES, M.: This volume, p. 199.
2. NEBEL, B. R.: Botan. Rev. 5, 563 (1939).
3. — Cold Spring Harbor Symposia Quant. Biol. 9, 7 (1941).
4. — J. Heredity 48, 51 (1957).
5. — Proceedings of the International Congress of Radiation Research. Radiation Research Suppl. 1, (1958) (in press).
6. TAYLOR, J. H., P. S. WOODS and W. L. HUGHES: Proc. Nat. Acat. Sci. US 43, 122 (1957).
7. — Sci. American 198, 58, 36 (1958).

Breakdown and reformation of the nuclear envelope at cell division

MONTROSE J. MOSES

The Rockefeller Institute, New York/N. Y.

In all higher animal and plant forms so far studied, the nucleus is surrounded by an envelope consisting of two membranes separated by a variable distance of the order of 20 to 40 mμ. The continuity of the membranes is interrupted by circular differentiations which have been called "pores" (1) and "annuli" (2, 3). Four lines of evidence have been presented in support of the concept that the nuclear membranes are part of, and at times continuous with the vesicular component of the cytoplasm (endoplasmic reticulum): a) the cytoplasmic surface of the envelope is occasionally covered with dense particles similar to those that cover the flattened vesicles of the endoplasmic reticulum in actively synthesizing cells, as in the pancreas (1); b) actual continuity between the outer membrane and the endoplasmic reticulum has been observed in which the cisternal cavity of the reticulum is continuous with the space between inner and outer nuclear membranes (1); c) strong resemblance has been reported between the nuclear envelope and elements of the stacks of basophilic "annulate lamellae" particularly prominent in the cytoplasm of egg and sperm-forming cells of certain invertebrates (4); d) outpocketings of the nuclear envelope in Drosophila salivary gland cells (5) and in spermatocytes and spermatids of crayfish (6) have been observed which could conceivably give rise to membranous elements of the cytoplasm.

Additional evidence of a relationship comes from observations of the nuclear envelope at late prophase and telophase of cell division. The relationship between nuclear envelope, cytoplasmic membranes, and structures in the spindle region of dividing cells have been discussed by PORTER (7). Observations reported here bear on the first two points. In electron micrographs of primary spermatocyte prophases of the crayfish (*Cambarus clarkii*), fixed with buffered OsO_4 (PALADE),

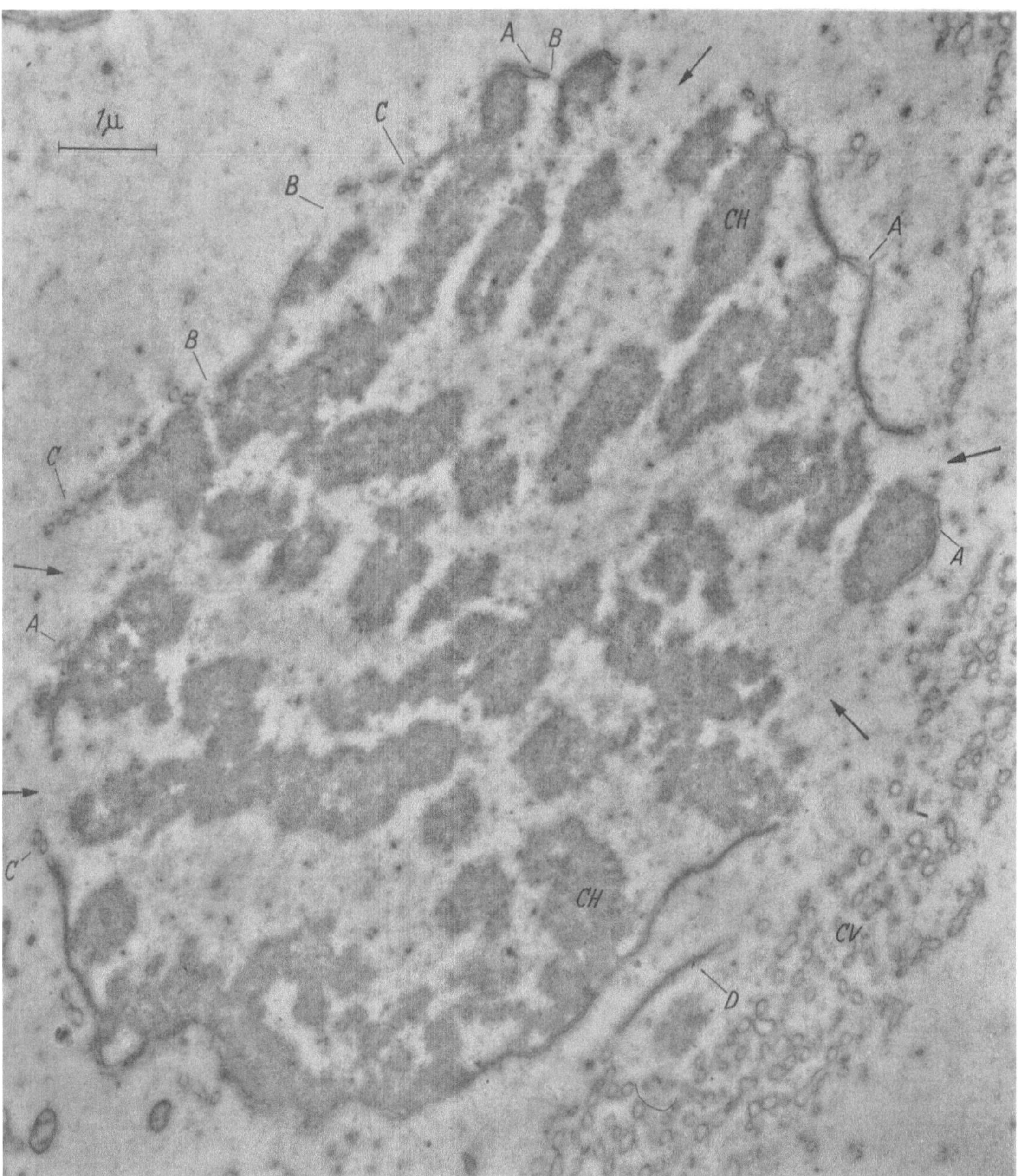

Fig. 1. Late prophase, primary spermatocyte of *Cambarus*. Condensed chromosomes (*CH*) are closely applied to the inner membrane of the nuclear envelope in several places. The envelope is interrupted by large gaps (arrows) and smaller spaces (*B*) where the continuity of the envelope is suggested by small vesicles like those in the cytoplasm (*CV*). Similar vesicles trail off the ends of broken (*C*) or detached (*D*) pieces of envelope. Where the envelope is intact, some portions show uneven separation of the two membranes (*A*). Neighboring cells in the same cyst represent earlier and later stages of prophase as determined from appearance of the chromosomes alone; in earlier stages, the envelope is more intact, in later ones it is gone. Thus, this figure depicts the normal process of envelope breakdown and is not a case of explosion or other damage in preparation. Mag. 14,200 ×

the intact nuclear envelope is characteristic in appearance (*1, 2*). Throughout prophase, chromosomes lie against and terminate on the inner membrane (*6*). At late prophase, the envelope begins to fragment, particularly in places where the inner membrane is free of adhering chromosomes (Fig. 1; arrows). The sequence of events appears to be: a) localized irregularities in the profiles

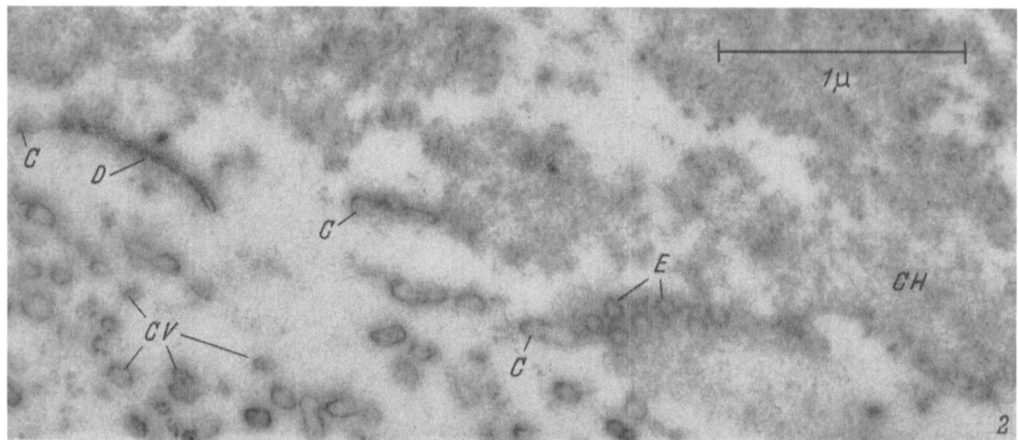

Fig. 2. Late prophase, primary spermatocyte, *Cambarus*. A portion of the nuclear envelope, still attached to the chromatin (*CH*) is cut obliquely (*E*) and reveals an array of annuli. A detached portion of envelope is shown at *D*. The envelope appears to be giving rise to vesicles (*C*) similar to those of cytoplasm (*CV*). Mag. 33,000 ×

of both membranes and uneven separation between them (Fig. 1, 3; A); b) interruptions in the nuclear envelope (Fig. 1: B) associated with c) the pinching off, at the edges, of small (100—150mμ) vesicles (Fig. 1, 2, 3; C) indistinguishable from those of the cytoplasm (Fig. 1, 2, 3; CV), or detachment of whole sheets of double membrane which often appear to break up and run off into small vesicles at their ends (Fig. 1, 2, 3; D). In oblique section, both intact nuclear envelope

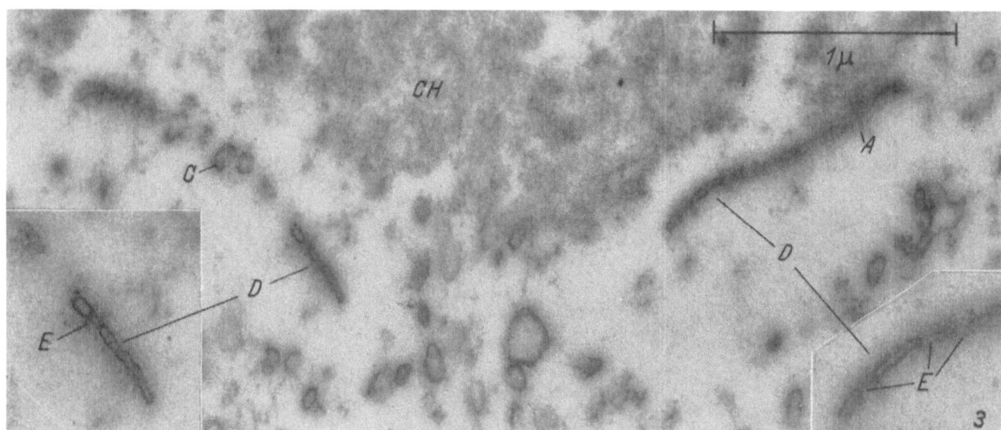

Fig. 3. Late prophase, primary spermatocyte, *Cambarus*. Two segments of nuclear envelope (*D*) are detached from the chromatin (*CH*). On the right, where a short region of the envelope is still attached, the profile and spacing of the two membranes are uneven (*A*). On the left, the contour of the envelope persists as a chain of vesicles (*C*) like those of the cytoplasm (*CV*). The intact portions of envelope (*D*), shown at higher magnification (50,000 ×) in the insets, are interrupted by pores (*E*) which are filled with dense material extending beyond the thickness of the envelope. The external surfaces of the membranes are also covered with a diffuse dense substance.
Mag. 33,000 ×

and detached segments contain annuli (pores) [Fig. 2, 3 (insets); E]. The segments (Fig. 1; D and Fig. 3; insets) resemble elements of the stacked "annulate lamellae" found in the cytoplasm (*4*). They do not appear to persist through division (cf. 2). It is not clear what role is played by the pores, or what happens to them in the process of breakdown. Intact portions of the disintegrating envelope may or may not contain pores; regions of uneven separation between inner and outer membranes generally lack them. The fairly uniform size of the associated vesicles, and their arrangement with respect to intact annulate envelope (Fig. 2 and 3; C) suggest that the pores may be "weak" areas from which small segments of the double membranes bounding them break away as vesicles (Fig. 3; C, D and E). Regions of the envelope associated with chromatin are apt to detach or break up into vesicles last (Fig. 1). The membrane profiles are always continuous and surround an intermembrane space; no single free edges are seen (Fig. 2 and 3). The process is apparently one of pinching off of portions of the envelope which eventually dedifferentiate to a basic vesicular element indistinguishable from that of the cytoplasm.

At metaphase, no recognizable portions of the envelope are left attached to the chromosome though small vesicles, individually or in chains, may align in the spindle axis and associate with the chromatin.

At telophase, groups of chromocomes form karyomeres. Agglomerations of vesicles, like those in the cytoplasm appear at the surface of the chromosomes. These vesicles fuse into flattened cisternae which eventually surround the chromatin. The process appears to be completed first at the leading, or poleward, edge of the chromatin masses. Karyomeres then fuse and frequently trap pieces of membrane. In late telophase nuclei, which are disc-shaped, various profiles of the trapped membranes are seen, depending on how many layers of fused vesicles are included. The trapped membranes persist well into interphase in spermatocytes and throughout a good part of spermatid development (6) (Fig. 10, 11).

With the exception of the membrane-trapping phenomenon, which appears to occur chiefly with the formation of karyomeres, the process of nuclear envelope breakdown and reformation is basically similar in a number of organisms observed. In spermatogonia and spermatocytes of salamander, grasshopper and rat, and in meristematic cells of root tips of *Allium* and *Vicia*, the disorganization of the nuclear envelope produces a vesicular component resembling the cytoplasmic membrane system (endoplasmic reticulum). Essentially, a reversal of this process occurs as the envelope reforms. The relation between chromatin and membrane appears somehow to be essential to these events.

Bibliography

1. Watson, M. L. J. biophys. biochem. Cytol. **1,** 257 (1955).
2. Afzelius, B.: Exp. Cell. Res. **8,** 147 (1955).
3. Swift, H.: J. biophys. biochem. Cytol. **2** (Suppl.), 415 (1956).
4. Rebhun, L. I.: J. biophys. biochem. Cytol. **2,** 93 (1956).
5. Gay, H.: Cold Spr. Harb. Symp. quant. Biol. **21,** 257 (1956).
6. Moses, M. J.: J. biophys. biochem. Cytol. **2** (Suppl.), 397 (1956).
7. Porter, K. R.: *In* Fine structure of cells. Symposium held at the 8th Congress of Cell Biology, Leyden 1954; Interscience, New York. p. 236, 1955.

Membrane interrelationships during meiosis

R. Barer, S. Joseph and G. A. Meek
Department of Human Anatomy, University of Oxford

It is well known that mitochondria are frequently to be found clustered around the nuclear membrane. The subject has been reviewed by Barer and Joseph (*1*) and Frederic (*2*). We have endeavoured to follow with the electron microscope the sequence of relationships between the mitochondria and the nuclear membrane from early prophase until metaphase, using the spermatocytes of the locust *Schistocerca migratoria*. Especial care was taken to identify the stages by means of alternate thick and thin sections and the use of the phase contrast microscope.

Dissolution of the nuclear membrane. Previous work on the living spermatocytes of the grasshopper *Chortippus parallelus* by phase contrast microscopy, observing cells immersed in a protein medium of variable refractive index (*1*), suggested that the mitochondria accumulate near the nuclear membrane at the end of prophase. At diakinesis, mitochondria frequently appeared to be attached to the nuclear membrane, and dissolution of the latter occurred soon afterwards. These observations were substantiated by the present work. Using thin sections of osmium-fixed testicular follicles of locust embedded in methacrylate, the mitochondria at late prophase were almost invariably seen to be distributed around the nucleus. Some were generally seen to be in close contact with the nuclear membrane, and very often at these contact points or closely adjacent to them, breaks were seen in the nuclear membrane. This supports the suggestion that the mitochondria are actively engaged in the process of dissolution (Fig. 1).

Reconstitution of the nuclear membrane. Hughes (3) and Schultz (4) suggested from light microscope studies that the nuclear membrane is re-formed after division at the interface between the cytoplasm and the vesicular chromosomes, and that the latter participate in some way in its formation. In our electron microscope study, the main constituent of the reformed nuclear membrane appears to be the endoplasmic reticulum (γ-cytomembranes).

In the living cell at telophase, the chromosomes became surrounded by a cloud of material which appears to be derived from the mitochondria. Under the electron microscope, however, the material was seen not to be mitochondrial in nature, but was composed of vesicular elements and paired membranes resembling agranular endoplasmic reticulum (γ-cytomembranes). These vesicles appeared to line up and fuse together, forming a continuous membrane around the individual chromosomes, which at this stage form a dense cluster. Later, the chromosomes become less densely stained by osmium and the whole daughter nucleus expands, the spaces between the chromosomes becoming filled with nuclear sap (Fig. 2). This process can be seen more clearly by fixing the material in 1% $KMnO_4$ solution, which leaves the chromosomes unstained and may even extract them, while the membranes of the ER are densely stained (Fig. 3).

Relationship between nuclear membrane and ER. Swift (5) and Rebhun (6) have pointed out certain morphological relationships between annulate "stacks" of lamellae found in oocytes and the nuclear membrane. Ruthmann (7) has shown similar structures in crayfish spermatocytes. We have frequently observed similar "stacks" at certain stages in the present material. In spermatogonia and early primary spermatocytes both the "stacks" and the nuclear membrane appeared to be made up of small vesicles as described by Swift, but in secondary spermatocytes these structures were composed of roughly parallel, agranular lamellae. Both types of stack when close to the nuclear membrane were almost invariably parallel to it, and the outer layer of the double nuclear membrane was elevated in the neighbourhood of the "stack" (Fig. 4). The spacing between the two membranes of the nucleus became the same as the spacing between the paired membranes of the "stack". In places, it seemed that the expanded nuclear membrane pairs were undoubtedly becoming part of the "stack". The impression is given of a definite shedding of nuclear membrane material into the cytoplasm. The nucleolus was frequently seen in close association with the point of formation of the "stack". The close similarity between the structure of the "stacks" and the structure of the nuclear membrane — vesicular in the spermatogonia, lamellar in later stages — strongly supports the suggestion that the "stacks" are formed directly from the nuclear membrane. Since also the "stacks" bear a close similarity to the ER, which sometimes appears to be continuous with their edges, it may be that the membranes of the ER are also formed directly from the nuclear membrane.

Fig. 1. Locust secondary spermatocyte; fixed in buffered osmium at p_H 7.2, embedded in butyl methacrylate. Mitochondria clustering around dissolving nuclear membrane (m) and "stack" (s). Late prophase

Fig. 2. Locust secondary spermatocyte; fixed in buffered osmium at p_H 7.2, embedded in Araldite. Early reforming nuclear membrane around chromosomal mass. Advanced telophase

Fig. 3. Locust secondary spermatocyte; fixed in 1% $KMnO_4$ at p_H 7.2, embedded in Araldite. Re-forming nuclear membrane and ER are now more clearly seen. Advanced telophase

Fig. 4. Locust secondary spermatocyte; fixed in buffered osmium at p_H 7.2, embedded in Araldite. Outer of the nuclear membrane pair appears to be lifting into the cytoplasm to form the "stack" in the neighbourhood of the nucleolus (nu). In places, the membranes of the "stack" appear to be continuous with the ER. Late prophase.

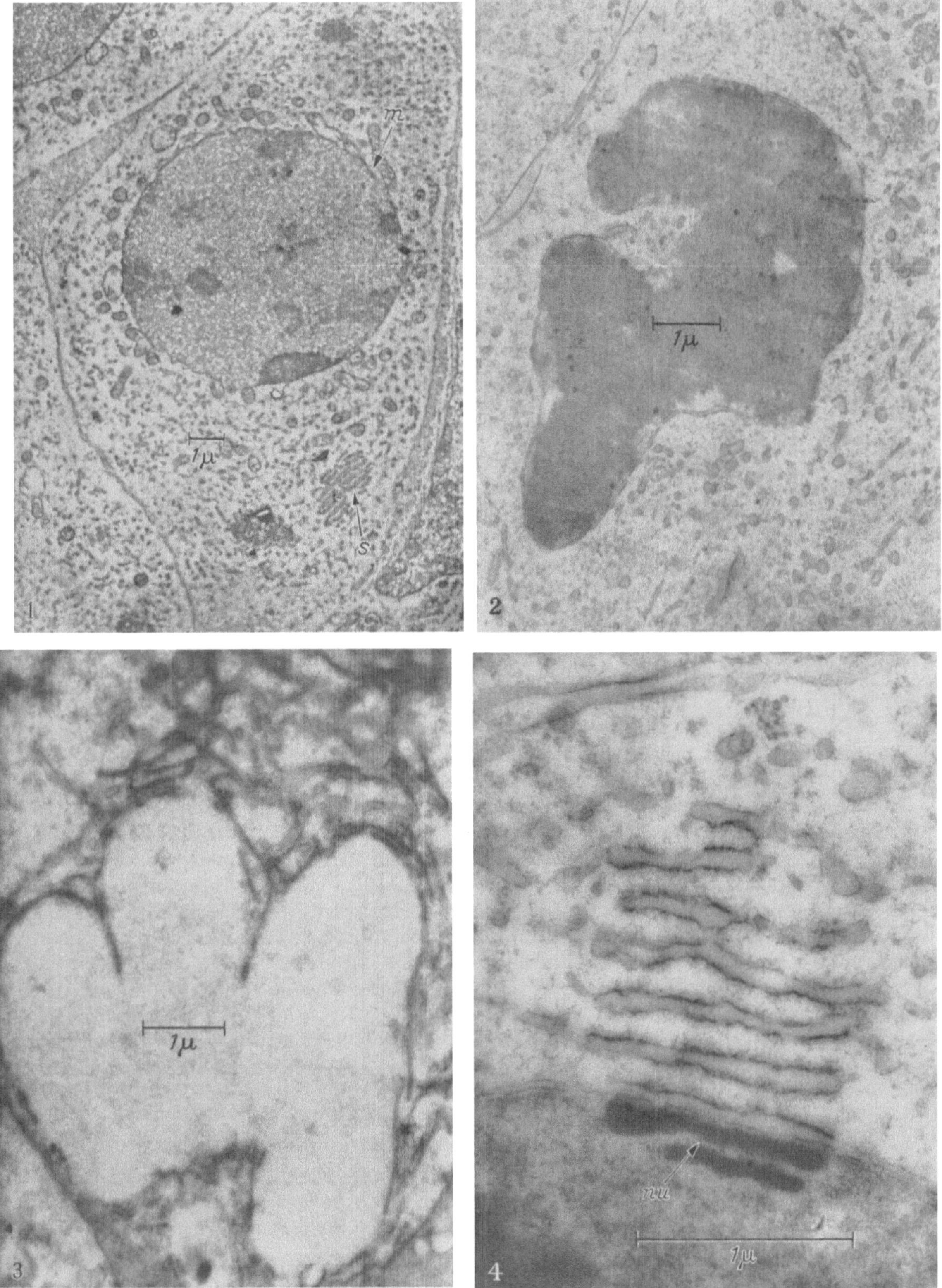

Fig. 1, 2, 3 and 4 (see page 234)

In view of the evidence summarised above, the hypothesis could be put forward that in prophase, nuclear membrane material is shed into the cytoplasm and stored there in the form of ER. In telophase, the ER is laid down over the chromosomal mass and becomes transformed once again into the nuclear membrane.

This work was carried out with a Siemens Elmiskop I kindly loaned to the Department of Human Anatomy by the Wellcome Foundation.

References

1. Barer, R., and S. Joseph: Symposia Soc. exp. Biol. **10**, 160 (1957).
2. Frederic, J.: Arch. Biol. **69**, 167 (1958).
3. Hughes, A.F.W.: The mitotic cycle. London: Butterworth 1952.
4. Schultz, J.: Exp. Cell Res. Suppl. **2**, 17 (1952).
5. Swift, H.: J. biophys. biochem. Cytol. **2**, No. 4 Suppl. 415 (1956).
6. Rebhun, L.: J. biophys. biochem. Cytol. **2**, 93 (1956).
7. Ruthmann, A.: J. biophys. biochem. Cytol. **4**, 267 (1958).

A comparative electron microscopic study on developing nuclei of various animals

G. Yasuzumi

The Electron Microscope Research Laboratory, Department of Anatomy, Nara Medical College, Kashihara, Nara Pref., (Japan)

In so far as the sperm nucleus carries genetic material, it would be of interest to know what changes in submicroscopic organization of chromosomes are associated with metamorphosis of the spermatid. Recent advances in electron microscopy have revealed a surprising variety of changing structural arrangements within the spermatid nuclei of various animals at different stages of maturation (*2, 3, 4, 6, 7, 8, 9, 11, 12, 13, 14,* and *15*). The present report deals with a comparative electron microscopic study on such developing spermatid nuclei as the grasshopper *Gelastorrhinus bicoler* de Haan, the pond snail *Cipangopaludina malleata* and the crab *Eriocheir japonicus*.

The development of the grasshopper spermatid

The nuclei of grasshopper spermatids are spherical in the early stages of metamorphosis, but later they become conical and then more and more elongated until they are long slender rods, rounded at the base and tapering at the tip to a sharp point. Concurrently with these changes in the nuclei, remarkable changes occur in the fine structure of the karyoplasm. In the early stages of maturation of the spermatid in which the nebenkern is divided into two parts separated by the developing tail filament, the content of the spherical nucleus is very dense, so that its internal fine structure is scarcely visible. In the elongated, slender, rod-shaped spermatid nulcei, it is possible to see a fairly high degree of order. It appears, in cross-sections, as a close packing of light areas enclosed by dense walls. The light areas measure 140 to 220 Å in diameter. The structure resembles most closely a hexagonal honeycomb. In the longitudinal sections of elongated spermatid nuclei it is possible to see the parallelly arranged, dense lines 70 Å in diameter. Such appearances provide clear evidence that both images (the honeycomb and parallel lines) belong to the same structure seen in different planes of section. As differentiation of the spermatid proceeds further, the spaces between the lines become narrower and narrower until it can no longer be resolved.

The development of the pond snail spermatid

We know that there are two types in spermatozoa of the pond snail, namely: a normal flagellate type having head and tail, and abnormal worm-shaped type having a small quantity of chromatic substance in its head and many tail flagella.

The development of the normal spermatid

An unusual structure has been found in sections through the mitochondria in the cytoplasm of the early normal spermatid. Observations of many electron micrographs, including some of serial sections, indicate that the mitochondria are doughnut shaped with a central space free of electron scattering material. In the pond snail the nebenkern forms in a way different from that of insects (*1, 5, 10* and *13*). In the latter case, there occurs an aggregation of all the mitochondria which retain their limiting membranes. They then coalesce and the cristae mitochondriales disappear. In the spermatid of the former the doughnut-shaped mitochondria first lose their limiting membranes. Following this, the cristae aggregate in two spherical masses: the nebenkern. The persisting cristae appear tightly packed, but randomly disposed. Later the cristae appear again disposed radially, leaving in the center of the nebenkern a light space which seems to be surrounded by a limiting membrane. As the differentiation of the spermatid proceeds further, the mitochondrial bodies increase in diameter and encircle completely the beginning of the tail filaments. Finally each nebenkern derivative takes the form of an elongated body wrapped helically around the bundle of tail filaments. Each of these mitochondrial bodies contains numerous tightly packed crista-like elements disposed parallel to one another and perpendicular to the long axis of the tail filaments. In spermatids in advanced stages of differentiation it is possible to see a fairly organized structure within the conical nuclei. In longitudinal sections the compact elements of this structure appear as dense lines varying in thickness from 100—160 Å, and generally disposed parallel to the long axis of the nucleus. In cross-sections through the same elements appear as small, dense round profiles. While the spermatid nuclei increase in length and decrease in width, the structural elements described become progressively twisted and give the impression of being disposed in a loose helix. At the same time it appears the aggregation of the originally thin fibers into progressively thicker strands. During further differentiation into spermatozoa, the nuclear strands increase in diameter and the spaces between them progressively reduced until finally the strands seem to fuse and interspaces are no longer resolvable with the microscope used in this study.

Development of the abnormal spermatid

The early abnormal spermatids appear larger in size than their normal counterparts, their nuclei are irregularly shaped and their cytoplasmic structure is highly characteristic. The nucleus contains a considerable number of small dense granules about 300 Å in diameter which in general appear to be disposed at random although in some places they seem to be arranged in short, coiled strands. At the periphery of the cell body, a characteristic and apparently new cytoplasmic component is encountered in relatively large number. It is vesicular structure 0.45 to 0.70 μ in diameter limited by a well defined membrane and containing one to four dense granules 80 to 230 mμ in diameter. It is particularly interesting that the granules give a Feulgen positive reaction. As the metamorphosis of the abnormal spermatid proceeds, the nucleus becomes smaller than that of a normal cell and its shape changes gradually into a crescent. During the elongation of the spermatid the bulk of the cytoplasm shifts to the posterior part of the cell while the nucleus moves towards the anterior pole of the cell body. At this stage the vacuoles containing dense granules, originally restricted to the periphery of the cell increase in number and appear districted to the periphery of the cell increase in number and appear distributed throughout the entire cytoplasm with the exception of zone immediately surrounding the developing tail filaments. Many of these vacuoles lie in close association with the cell membrane and appear to open through a pore in the intercellular spaces. In this way the dense material of the granules could reach the extracellular medium. As the spermatid develops, the diminishing crescentic nucleus migrate progressively to the periphery of the cell until it comes in contact with the cell membrane. It has been revealed that there are at least two types of head structure in abnormal spermatozoa: one is composed of tail flagella, having the centrioles at the anterior end of each tail flagellum, and surrounded by a thickened margin which originated from the nucleus which is devoid of DNA; the other consists of a bundle of 7 to 17 tail flagella which are enveloped by a double membrane.

Development of the crab spermatid

In the early spermatids the content of the spherical nuclei is very dense, so that the internal fine structure is scarcely visible. The first evident step in the transformation of the spermatid is the appearance of vesicles of low density in the cytoplasm next to the nucleus. These are small at first but by coalescing they soon form a large, clear vesicle on one side of the nucleus. While the vesicle (secondary nucleus) and the structures within it are assuming their mature form, the primary nucleus has become less dense and settles down upon the secondary nucleus like a cap. It must be noticed that in the peripheral portion of the mature spermatozoon head there is an evidence of lamellar structure parallel to the nuclear envelope, consisting of four to eight layers about 120 Å thick separated by a lighter space about 130 Å in width. In higher magnification, each linear profile is composed of three layers: two dense lines separated by a lighter space. Each layer measures about 40 Å on an average.

References

1. Beams, H. W., T. N. Tahmisian, R. L. Devine and L. E. Roth: Biol. Bull. **107,** 47 (1954).
2. Burgos, M. H., and D. W. Fawcett: J. biophys. biochem. Cytol. **1,** 287 (1955).
3. — — J. biophys. biochem. Cytol. **2,** 223 (1956).
4. Dass, C. M. S., and H. Ris: J. biophys. biochem. Cytol. **4,** 129 (1958).
5. Robertis, E. de, and H. Franco Raffo: Exp. Cell Res. **12,** 66 (1957).
6. Gibbons, I. R., and J. R. G. Bradfield: J. biophys. biochem. Cytol. **3,** 133 (1957).
7. Grassé, P. P., N. Carasso and F. Favard: Ann. sci. nat. zool. **28,** 339 (1956).
8. Rebhun, L. I.: J. biophys. biochem. Cytol. **3,** 509 (1957).
9. Sjöstrand, F. S., and B. A. Afzelius: Proc. Stockholm Conf. on Electron Microscopy p. 164, 1956.
10. Tahmisian, T. N., E. L. Powers and R. L. Devine: J. biophys. biochem. Cytol. **2,** No. 4, suppl., 325 (1956).
11. Yasuzumi, G.: Exp. Cell Res. **11,** 240 (1956).
12. — J. biophys. biochem. Cytol. **2,** 445 (1956).
13. — Proc. First Regional Conf. Asia and Oceania, p. 251, Tokyo 1956.
14. — W. Fujimura and H. Ishida: Exp. Cell Res. **14,** 268 (1958).
15. — and H. Ishida: J. biophys. biochem. Cytol. **3,** 663 (1957).

Chromosomal structure in primary spermatocytes of the locust

I. R. Gibbons

Johnson Foundation, University of Pennsylvania, Philadelphia, USA

During the early to middle diplotene stages of meiotic prophase, the chromosomes in locust spermatocytes contain a pair of dense fibrils, each about 250 Å thick, running parallel to the long axis of the chromosome. These fibrils usually appear to run parallel to each other, 700 Å apart (centre to centre); they have occasionally been observed twisted around each other in a very loose spiral. In some preparations there appears to be a third fibril running midway between the two fibrils just described above; this central fibril is thinner (about 100 Å thick), and less well-defined than the outer pair. All these fibrils, considered together, form only a small part (of the order of 0.5 per cent by volume) of the chromosomal substance.

An extended version of this article will appear elsewhere.

Observations on cells of the ovotestis of a pulmonate snail*

L. E. Roth

Argonne National Laboratory, Biological and Medical Research Division, Lemont, Illinois, USA

The numerous cytological problems in spermatogenesis and yolk formation have been studied intensively by light microscopists, but it is only recently that electron microscopy has been utilized in an effort to understand the profound nuclear and cytoplasmic changes which take place. The cells of the ovotestis of pulmonate snails offer excellent material for study in these areas. The following report gives an account of studies on both male germ cells and nurse cells of the snail *Otala vermiculata* including observations for both nuclear and cytoplasmic structures.

---

* This work was performed under the auspices of the US Atomic Energy Commission.

Ovotestes were fixed in 1% osmium tetroxide buffered with McIlvaine's citric acid-phosphate buffer at p_H 8.0, 0.05 M with 0.72% sodium chloride for 30 min at room temperature. After 25 min changes in 50%, 75%, 95%, and absolute ethyl alcohols and in methacrylate monomer (three changes), polymerization was carried out by ultra-violet irradiation. Four parts ethyl and six parts n-butyl methacrylate with 1% luperco CDB constituted the plastic mixture. Sections were cut on a Porter-Blum microtome, mounted on carbon membranes or coated with carbon after cutting, and observed in either an RCA EMU 3 A or Siemens Elmiskop I electron microscope.

Male germ cells. Spermatocytes and spermatids in several stages have been observed and a report will be given elsewhere of some cytoplasmic structures (1); only nuclear contents and structures closely associated with the nuclear membrane will be described here.

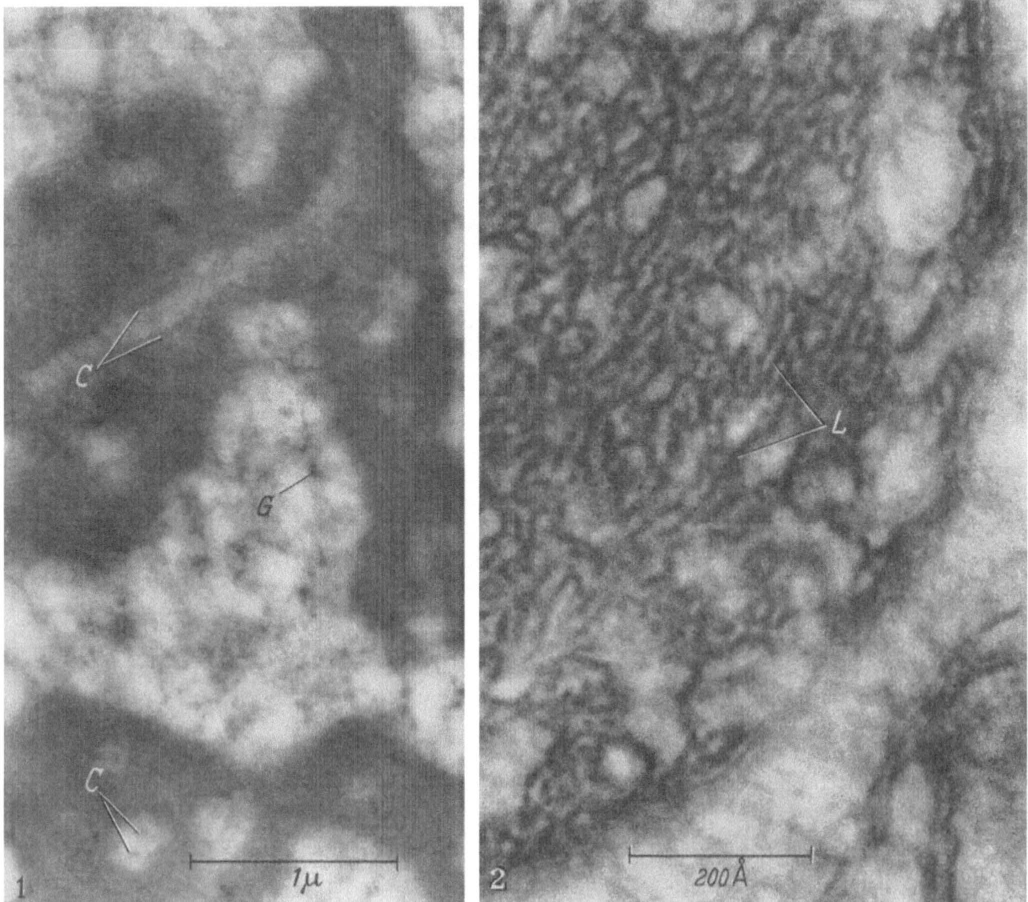

Fig. 1. Chromosomes with cores in first meiotic metaphase. 28,000 × — Fig. 2. Spermatid nucleus. 123,000 ×

Chromosome structures. The spermatocyte nucleus in meiotic prophase has chromosomal units which contain "cores" similar to those previously described by Moses (2) and Fawcett (3); attachment of the core fibrils at darkened areas of the nuclear envelope is seen in some cases. These cores persist from prophase into the first metaphase of meiosis (Fig. 1C). The two outer fibrils are 30 mμ in diameter while the smaller fibril which is located centrally equidistant from the outer ones is about 13 mμ in diameter; together the fibrils form a rubbon-like structure about 175 mμ wide.

In a very few cases, fibrils were observed oriented perpendicularly to the larger core fibrils and extending across the space which surrounds the core to the remainder of the chromosomal material. In this outer portion of the chromosome, numerous small filaments which have a diameter of 4 to 6 mμ have been observed; cross-sections have a dense outer layer with a less dense central portion.

Accumulations of dense granules are scattered in the nucleoplasm or arranged at the chromosome periphery in spermatocytes. In meiotic prophase, some of these granules are located at the periphery of the chromosomes; however, in metaphase the amount is reduced so that only a few are present scattered in the cytoplasm near the chromosomal group (Fig. 1 G).

In later stages of spermatogenesis, spermatids show the complex nuclear changes known from several studies [e. g. (4, 5)]. Filaments which have a diameter of about 5 mμ have been observed in early spermatids; later, at about the time when maximal chromatin orientation is first reached,

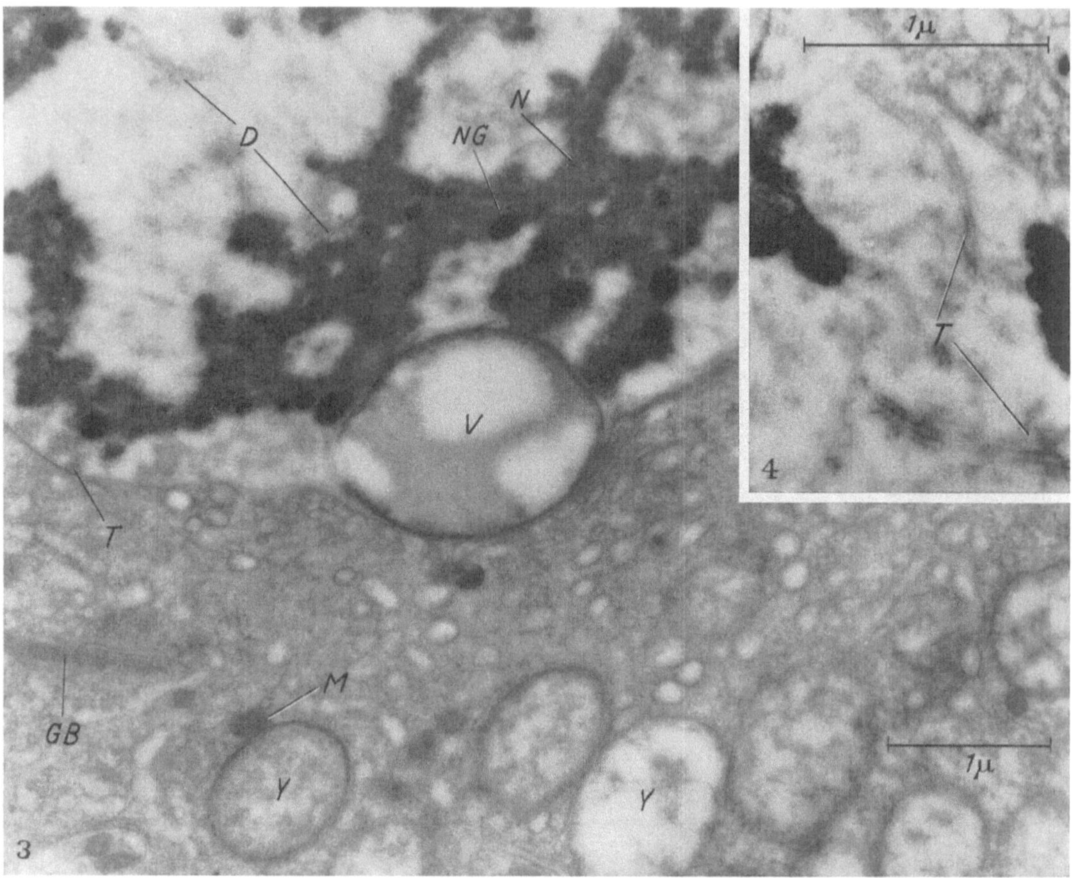

Fig. 3. Portion of yolk nucleus and cytoplasm of slightly older nurse cell. 22,000 ×

Fig. 4. Tubular elements in yolk nucleus. 33,000 ×

layers about 5 mμ thick have been formed which are covered by a less dense material to form a total thickness of about 8 mμ (Fig. 2 L).

Cytoplasmic filaments. Located immediately outside of the nuclear envelope, there are numerous filaments which measure 21 mμ in diameter. These elements are present in numerous spermatozoa; filaments similar in both size and structure are generally found near cilia and beneath the pellicle in most protozoa thus far studied (6).

Nurse cells. The other prominent cell type in the ovotestis is the nurse cell in which yolk is formed and around which the germ cells cluster to obtain nutrients. Shortly after differentiation and enlargement of this cell, a large yolk nucleus measuring about 15 μ in diameter appears; it contains numerous components: 1. large vesicles which measure up to 1.5 μ in diameter (Fig. 3, V); 2. a fine fibrillar component which forms a coarse, obvious network on which are attached (Fig. 3, N) 3. numerous, dense, homogeneous granules which measure up to 0.3 μ in diameter (Fig. 3, NG); 4. numerous low density filaments (Fig. 3, D) 16 mμ in diameter; and 5. a few larger filaments which measure 30 mμ in diameter and have a tubular cross-section and ladder-like appearance in longitudinal section (Fig. 3 and 4, T). A continuous single-layered membrane surrounds the yolk nucleus.

In the cytoplasm, the first appearance of young yolk granules is correlated with this morphology of the yolk nucleus. Each granule is represented by a nearly-empty vesicle with a single-layered membrane (Fig. 3, Y) which is not distinguishably different from the membranes of the endoplasmic reticulum. Frequently, mitochondria (Fig. 3, M) are closely applied to the surface of the yolk granules and rather large amounts of Golgi substance (Fig. 3, GB) are present. Yolk granules progress until they become filled with dense materials and finally provide nutrients for the germ cells by a progressive breakdown into smaller granules.

Numerous, dense granules which have a diameter about 6 mμ are scattered throughout the nurse cell nucleoplasm and cytoplasm even before yolk granules appear. They are conspicuously absent from the cisternae of the endoplasmic reticulum and the extra-cellular spaces. They are similar to the granules described by FAVARD and CARASSO (7) in oocytes of *Planorbis*, but no regular arrangements in yolk granules have been observed here.

Discussion and Conclusions. In several stages of male germ cell nuclei and in nurse cell nuclei, filaments which are about 5 mμ in diameter have been observed and are believed to represent a widely found chromosomal fibril; it possibly is the deoxyribose nucleoprotein complex per se.

These filaments are probably similar to those described by RIS (8). However, the arrangement of two, spiralized 4 mμ filaments described by RIS do not correlate well with this and previous observations of spermatid nuclei, e. g. GRASSÉ, FAVARD and CARASSO (4) who described 7 to 9 mμ lamellae in spermatids of *Helix pomatia*. A correlation may be possible if the 4 to 5 mμ filaments parallel each other in a flat sheet overcoated with a slightly less dense material causing the two components, thicker layer observed in this study. Thin sections which will allow not only good resolution but optimum contrast will be necessary to visualize carefully this fine detail.

Chromosome cores are present not only in meiotic prophase but remain into metaphase I of meiosis. Supposedly involved in meiotic pairing, they are probably lost (at least from their ribbon-shaped configuration) in the first meiotic division; however, the inability to definitely recognize secondary spermatocytes has not yet allowed conclusive observations.

Nurse cells in the ovotestis contain a large yolk nucleus which contains several morphological elements and which progresses through a series of morphological stages. These stages correlate closely with the appearance and maturation of yolk granules in the cytoplasm.

References

1. GATENBY, J. BRONTE and L. E. ROTH: In preparation.
2. MOSES, M. J.: J. biophys. biochem. Cytol. **2,** 215 (1956).
3. FAWCETT, D. W.: J. biophys. biochem. Cytol. **2,** 403 (1956).
4. GRASSE, P. P., N. CARASSO et P. FAVARD: Ann. Sci. Nat. Zool. Biol. Anim. 18, 339 (1956).
5. REBHUN, L. I.: J. biophys. biochem. Cytol. **3,** 509 (1957).
6. ROTH, L. E.: J. Ultrastruct. Res. **1,** 223 (1958).
7. CARASSO, N., and P. FAVARD: This volume, p. 431.
8. RIS, H.: This volume, p. 211.

Observations on division stages in the
protozoan hypotrich *Stylonychia**

L. E. ROTH

Argonne National Laboratory, Biological and Medical Research Division, Lemont, Illinois, USA

The study of mitotic events by electron microscopy has been greatly retarded by inability to select dividing cells among the relatively much greater proportion of cells that are not in division. The protozoa, even though their micronuclear division represents a specialized type of mitosis,

* This work was performed under the auspices of the US Atomic Energy Commission.

offer the possibility of isolation and observation of selected cytological events and therefore provide choice material for such observations. In addition, the ciliate protozoa exhibit special macronuclear morphology at the time of deoxyribose nucleic acid synthesis, a highly specialized division of the macronucleus and, further more, engage in the formation of a rather large number of cilia in a short time. Therefore, dividing ciliates offer the possibility for study of the four events, micronuclear mitosis, macronuclear nucleic acid synthesis, macronuclear division, and ciliary formation.

The organism *Stylonychia* which was chosen for this study possesses two macronuclei in interphase which are easily visible under light microscopic enlargements of 100 × or less (Fig. 1); in preparation for division into two daughter organisms, the macronuclei fuse, divide once, and then a second time so that marking the division period macronuclear number follows the sequence: two, one, two, four, and two. The resulting easy categorization of mitotic stages allows a mass culture of organisms to be processed by standard single-cell techniques and the selection of dividing organisms from capsules after complete processing and polymerization. A selected organism is trimmed free of all others, its macronuclear stage noted, and then sectioned for observation in the electron microscope.

Cultures in active growth were chosen and fixed in 1% osmium tetroxide buffered with veronal acetate to pH 8.5 with 0.9% sodium chloride for 90 min at room temperature. After 15 min changes in 50%, 75%, 95%, and absolute ethyl alcohols and in methacrylate monomer (three changes), polymerization was carried out in number two gelatin capsules at 60° C. Four parts ethyl and six parts n-butyl methacrylate with 1% luperco CDB constituted the plastic mixture. Sections were cut on a Porter-Blum microtome, mounted on carbon membranes, and observed in an RCA EMU 3 A electron microscope.

Micronuclear structure. — The interphase micronucleus is characterized by 40—50 mμ fibers which are irregularly connected into a rather coarse network; each fiber is composed of finer elements which are about one-tenth the fiber diameter but which show no regular arrangement in the fiber.

In mitotic stages, this fiber network is no longer present; finer elements measuring 5 to 7 mμ in diameter are again observed (Fig. 2, F) and on some occasions show an arrangement into circles which measure about 10 to 13 mμ in diameter. The structural units which are probably the chromosomes of light microscopists show little individual integrity in these preparations and are usually united or fused together along the greater part of their length. No spindle elements have been observed.

The interphase micronuclear envelope has a typical double-layered morphology with occasional layers closely apposed on the cytoplasmic side. In mitotic stages, however, the envelope is reduced to discontinuous remnants some of which still show a two-layered structure (Fig. 2, E).

Macronuclear structure. — The interphase macronucleus consists of a coarse, irregular network of chromatin with interspersed nucleolus-like spheres. The nuclear envelope has a typical double-layered appearance; it has not, in any of the several functional states observed, shown a period of breakdown into discontinuous fragments.

As the dividing macronucleus begins to pinch in two, the chromatin shows an alignment which is probably passive in the region of constriction; at the same time, a filamentous component about 30 mμ in diameter appears oriented parallel to the long axis of the greatly elongated macronucleus. As division is nearing completion, the chromatin alignment has become more obvious and the filamentous components have aggregated at the constriction point to form a rather large bundle which is located at the point which Summers (*1*) showed in his light microscope study to be Feulgen-negative (Fig. 3, MF).

The reorganization bands which traverse the interphase macronucleus and which Gall (*2*) has shown in *Euplotes* to coincide with the time of deoxyribose nucleic acid synthesis have been observed on several occasions. The separate solution and reconstruction zones are much less pronounced than in some organisms (*3*) but it is apparent that the obvious, large chromatin network is broken down into very fine components; filaments measuring about 6 mμ are observed at or near the level of break-down and in the solution plane. The nucleolus-like bodies pass through this zone with little or no alteration.

Cilia and Ciliary formation. — Numerous new membranelles appear at the time when the macronuclei have fused into a single structure. Organisms sectioned at this time show aggregations of structures which are very similar to the fibrillar pattern of the shafts of the cilia with the exception that some fibrils are frequently absent (Fig. 4). The number in one aggregation varies

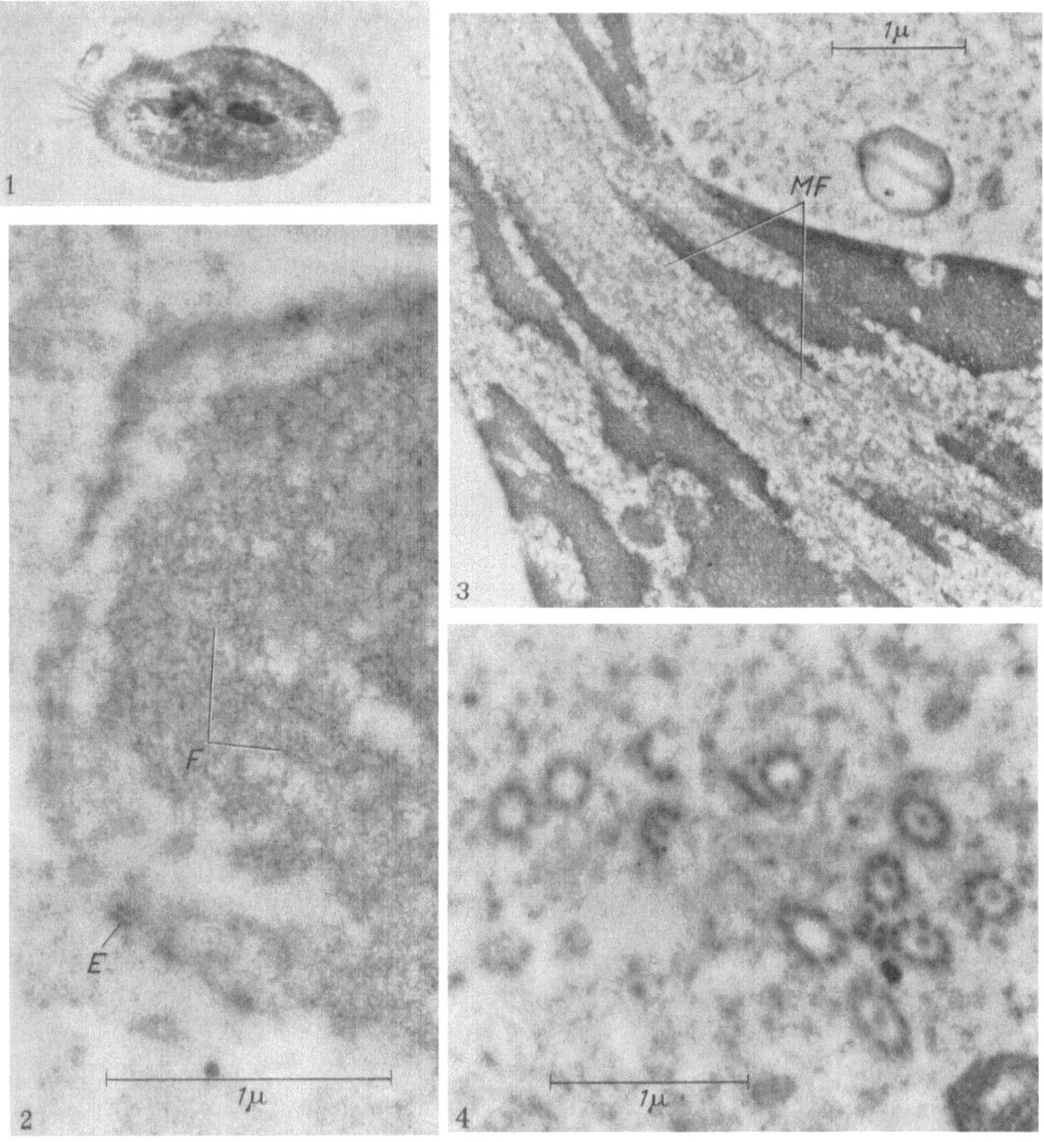

Fig. 1. Light micrograph of *Stylonychia*. 300 ×

Fig. 2. Dividing micronucleus with small portion of interphase macronucleus. 39,000 ×

Fig. 3. The constriction region of a macronucleus in a late division stage. 18,500 ×

Fig. 4. Structures in cytoplasm with ciliary fibril pattern in various states of completeness. 27,000 ×

from three to twelve; of the thirty-eight such structures observed in four of the twelve dividing organisms studied, eleven showed all central and peripheral fibrils, twenty-two showed all peripheral and no central fibrils, and five showed from two to four missing peripheral fibrils (four with central fibrils and one without). On most cases, there was a tendency to orient perpendicularly to the ventral surface of the organism (where almost all cilia are located), to approximately parallel each other, but not to be arranged in orderly rows as are the cilia of cirri and membranelles

16*

in this organism. In one case, serial sections allowed an estimate of length of about one micron for one aggregation of four fibril bundles.

In both non-dividing and dividing organisms, central fibrils are occasionally found in the bases of cilia although they are usually lacking in the bases and present only in the shafts. The shaft length is usually about 0.4 μ long.

Discussion. The several nuclear states seen here allow varied observations on nuclear contents. Filaments of a diameter about 5 mμ have been observed in interphase and dividing micronuclei and in interphase macronuclei including the reorganization bands. It seems possible that this represents the desoxyribose nucleoprotein complex per se of which the osmium fixative has bound to and darkened an outer protein shell. Additional chromatin arrangements have not been observed frequently enough to allow descriptions of any regular patterns.

The micronuclear envelope becomes highly discontinuous during division, while the macronuclear membrane always remains continuous.

The observation of cilium-like fibril bundles in the cytoplasm of organisms which are producing many cilia seems to argue against the long-established kinetosome duplication concept (especially if interpreted as a duplication of the basal bodies of existing cilia). In fact, since the structures observed here are longer than ciliary bases, are sometimes present rather deep in the cytoplasm as incomplete fibril bundles, are not ordered in rows as the existing cilia and are not intimately located close to existing cilia, it seems more likely that cilia in this organism are formed away from existing cilia.

References

1. Summers, F. M.: Arch. Protistenk. 85, 173 (1935).
2. Gall, J. G.: Biol. Bull. 113, 322 (1957).
3. Roth, L. E.: J. biophys. biochem. Cytol. 3, 985 (1957).

Helical structures in the nuclei of free-living amebas*

George D. Pappas and Philip W. Brandt

Departments of Anatomy and Ophthalmology, College of Physicians and Surgeons,
Columbia University, New York

Helical structures have been described in the Feulgen positive area of the nucleus of *Amoeba proteus* (1, 2) and more recently in the nuclei of *Pelomyxa carolinensis* (*Chaos chaos*). The helices are usually in clusters arranged radially from a central axis. Clusters of helices are found in the area of the nucleus subadjacent to the peripheral ring of "nucleoli" or dense granules characteristic of these two species.

P. carolinensis was fixed by pipetting individual cells directly into 1% OsO_4 buffered to p$_H$ 8.6 with diluted veronal acetate (1/1400 M). Fixation was carried out at 4° C for 2—3 min. The amebas were then rapidly dehydrated and embedded in n-butyl methacrylate. Large numbers of *A. proteus* were packed into a pellet by centrifugation. They were fixed, dehydrated and embedded as a pellet. Thin sections were cut on a Porter-Blum microtome, and collected on formvar-coated grids. An RCA-EMU-3C electron microscope was used for examining the sections.

The helices in Fig. 1 are arranged bristle-like around a central axis. The length of the cluster is about one micron. The length and pitch of the helices varies greatly depending on the thickness of the preparation and the plane of the section. The longest individual helix is about 3000 Å and the filament makes 8—10 complete turns. This corresponds to a period of 300—375 Å per turn. When seen in cross section (Fig. 2) the helix or coil has the profile of a torus or doughnut. The diameter of the coil is 300—350 Å. The thickness of the filaments making up the coil is 120—150 Å. These filaments are probably not simple unit structures, because they appear to have a more complex double structure. This can be seen particularly well when the end of the filament making

* This investigation was supported in part by research grant B 1202 (c) from the National Institute of Neurological Diseases and Blindness of the National Institutes of Health, Public Health Service.

up the coil appears split into subfilaments as in Fig. 3a. In Fig. 3b, the separated sub-filaments are coiled. The subfilaments illustrated in Fig. 3a and b are about 70 Å thick.

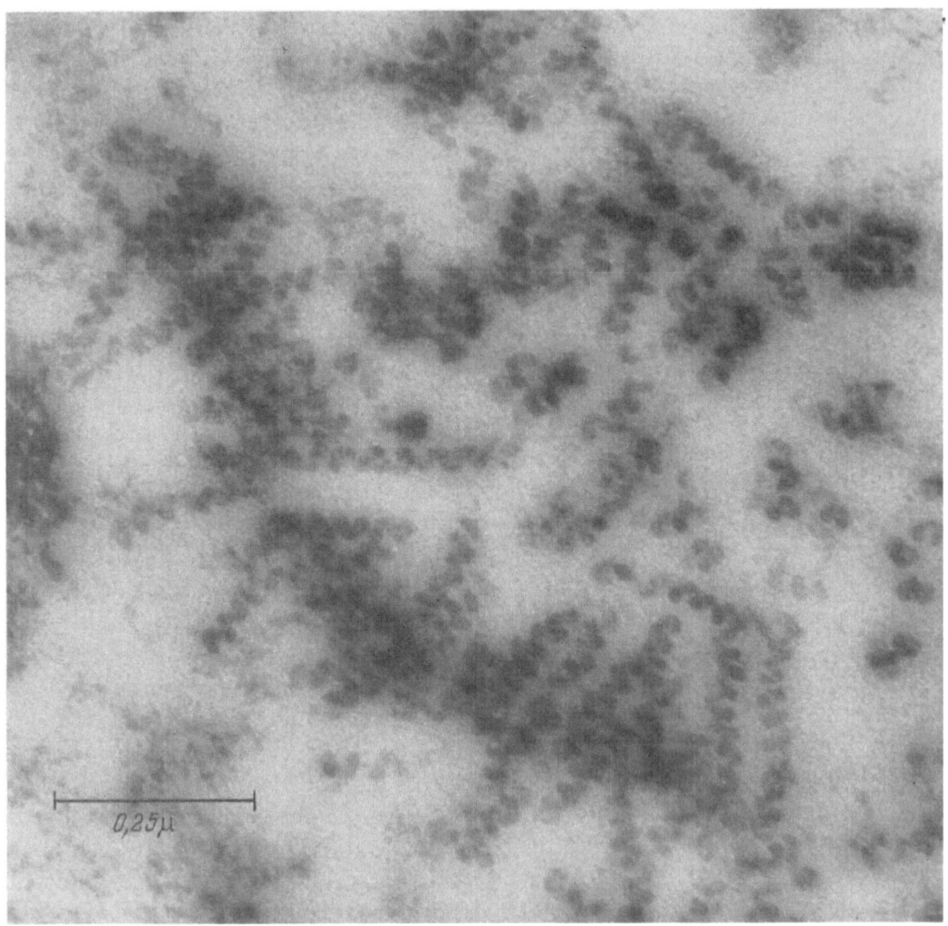

Fig. 1. Section from a portion of the Feulgen-positive region of the nucleus of *Amoeba proteus*. Helical structures are arranged bristle-like around a central axis. The thickness of the filaments making up the coils is 120—150 Å.
110,000 ×

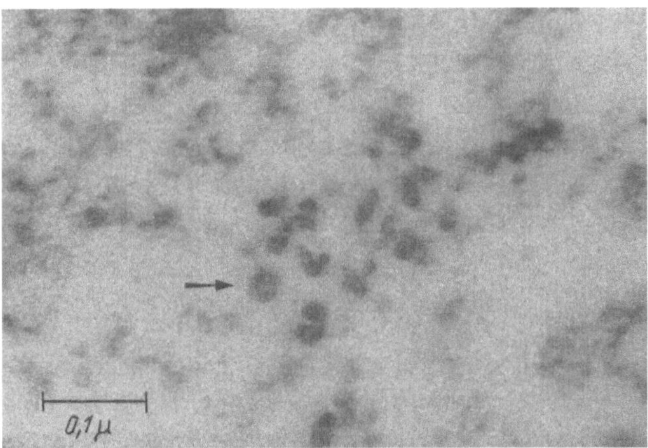

Fig. 2. Helical structures found in the nucleus of *A. proteus*. When seen in cross section (at pointer) the helix or coil has a profile of a torus or doughnut 300 Å in diameter. 140,000 ×

In the first report on the helices in *A. proteus* (*1*) the filaments were reported to be 70 Å, and no sub-filaments were observed. The helices shown in Fig. 1 may be made up of two sub-filaments paranemically coiled permitting separation without uncoiling.

Many questions remain to be resolved about the structure and functions of these helices. Helices are not observed in all sections of nuclei studied. It is difficult to estimate the helical content of a single nucleus. On the one hand, very thin sections of the order of 60—80 mμ are necessary to resolve the helices and their sub-units. Cosslett (*3*) has stated that the resolution can not be expected to exceed $^1/_{10}$ of the thickness of the section for biological material. On the other hand in a thin section, if the axis of the helix is not parallel to the plane of section the coil would not be identified. This limits the number of structures which can be positively identified as helical. Therefore even though considerable material of the same density and general dimensions as those described may be seen in the nucleus it can not be identified as helical. It is quite possible that well-formed helical structures may not be present in all stages of the nuclear cycle.

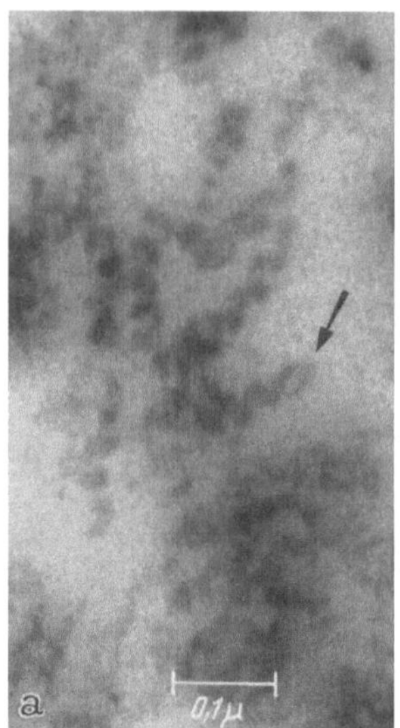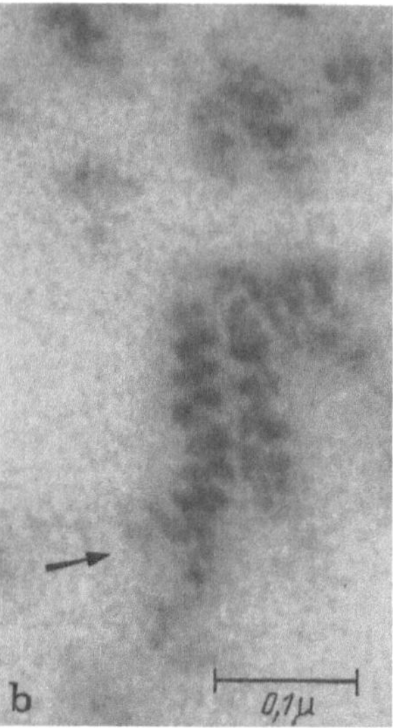

Fig. 3a and b. The filaments making up the coils appear to be divided into sub-filaments (at pointer) about 70 Å thick, as in Fig. a (*Pelomyxa carolinensis*). In Fig. b (*Amoeba proteus*) the separated sub-filaments appear coiled. a) = 100,000 × ; b) = 140,000 ×

Since helices lie in the Feulgen positive area of the nucleus, it is possible that they represent the conjugated deoxyribonucleoprotein components of these nuclei. Typical chromosomes have not been described in either *A. proteus* (*4*) or *P. carolinensis* (*5*). The chromatin material in *A. proteus* is organized, during division at least, into very small granules (*4*). The cluster of helices in Fig. 1 could very well represent one such chromatin "granule".

References

1. Pappas, G. D.: J. biophys. biochem. Cytol. **2**, 221 (1956).
2. — Ann. N. Y. Acad. Sci. 1958. (in press).
3. Cosslett, V. E.: J. biophys. biochem. Cytol. **3**, 815 (1957).
4. Chalkley, H. W.: J. Morphol. **60**, 13 (1936).
5. Hinchey, M. C.: Biol. Bull. **73**, 368 (1937).

Die submikroskopische Morphologie der Endothelzellen der Corneahinterfläche, mit besonderer Berücksichtigung der Centriolen

G. Millonig

Institut für normale Anatomie des Menschen und Histologie, Padua/Italien

Elektronenmikroskopische Untersuchungen ergaben, daß das Centriol einiger Zelltypen [Milzzellen (*1*), Blutelemente (*1, 2, 3, 4, 5*), Zellen experimenteller Tumoren (*1*)] einem ungefähr 0,5 μ langen Hohlzylinder gleicht, dessen Wand aus kleinen parallel gestellten Röhrchen (Tubuli) aufgebaut ist. Die Zahl dieser Tubuli wird meist mit neun angegeben; manchmal jedoch sieht man an Querschnitten neun aus 2, 3 oder 4 Tubuli zusammengesetzte Gruppen (*5*). Bei weiteren Untersuchungen an Plasmazellen (*2*) und normalen und leukämischen Leukocyten (*5*) konnten an den

Centriolen seitliche Querfortsätze (Appendices) nachgewiesen werden. Das oder die Centriolen sind von einem hellen Cytoplasmasaum (Centrosom) umgeben und befinden sich paranucleär im Zentrum der Golgi-Zone (6). Das Centrosom kann sich strahlenförmig in die umgebende Golgi-Zone ausbreiten und auf diese Weise den Aster bilden (5). Kürzlich konnten außerdem Verbindungen als wahrscheinlich erkannt werden, die sich zwischen der Centrosphäre und der Kernmembran ausbilden (5).

Die Untersuchung erstreckt sich auf das Endothel der Hornhauthinterfläche der neugeborenen Katze und des Schweines. Es wurde 5 min "in situ", noch vor Entnahme der Hornhaut, durch Einspritzen einer gepufferten 1%igen Lösung von Osmiumtetroxyd bei p_H 7,8 und einer osmotischen Konzentration von 1,1% in die vordere Augenkammer fixiert. Die weitere Behandlung erfolgte nach der in der Elektronenmikroskopie üblichen Methode.

Die bisherigen submikroskopischen Untersuchungen der Hornhaut haben keine Strukturen gezeigt, die Centriolen entsprechen könnten (8, 9, 10, 13). Bei der Durchsicht meiner Präparate gelang es mir, ungefähr 50 Centriolen zu beobachten; ich fand bei ihnen zahlreiche strukturelle Unterschiede. Trotz-

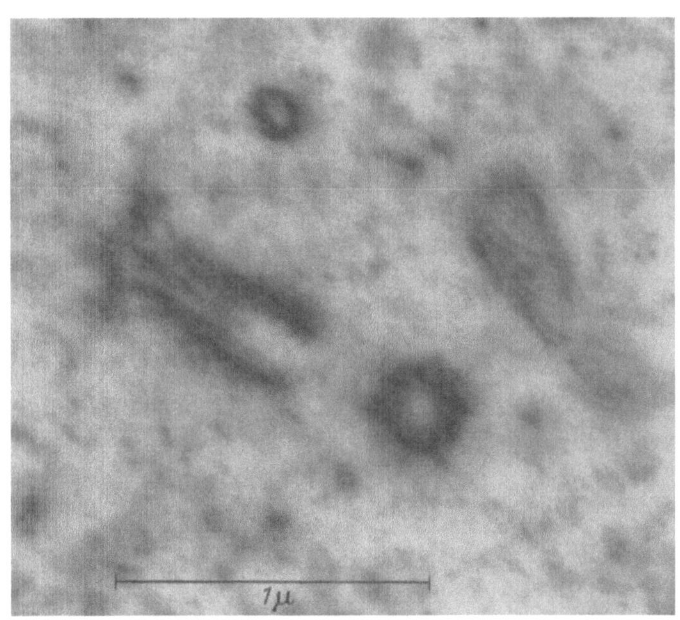

Abb. 1. Zwei Centriolen einer Hornhautendothelzelle des Katzenauges

dem kann ich die Untersuchung nicht als abgeschlossen betrachten, zumal ich die verschiedenen morphologischen Strukturen des Centriols bisher nicht mit einer Funktion in Zusammenhang bringen konnte.

Auf den Schnitten fand ich in den Zellen meist 2 Centriolen; nur relativ selten trifft man ein einziges angeschnitten. Man kann deshalb mit BALLOWITZ (11) annehmen, daß jede Zelle ein Diplosom (2 Zentralkörperchen) besitzt. Diese sind fast immer senkrecht zueinander angeordnet und das Ende des einen ist etwa 1000 Å vom Körper des anderen entfernt; nur selten sind sie zueinander parallel gestellt. Die Wand des Zylinders, die den Körper des Centriols bildet, ist aus neun Röhrchen aufge-

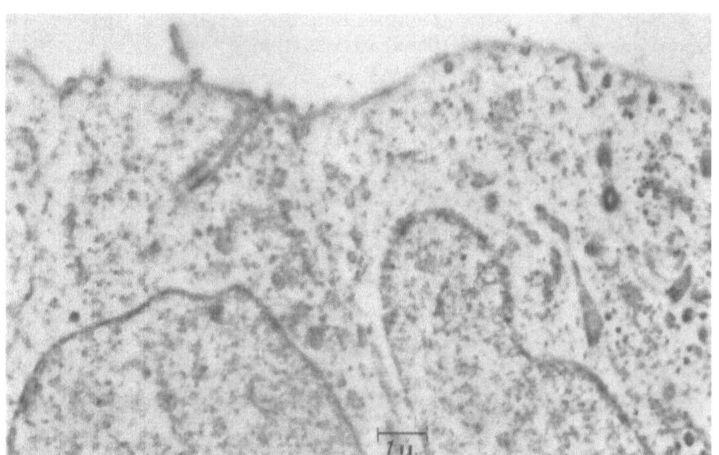

Abb. 2. Teilansicht zweier Endothelzellen der Cornea des Schweines. In der rechten Zelle ein Diplosom, in der linken ein Centriol, durch zwei streifenförmige Fortsätze mit der Zelloberfläche verbunden

baut. Die Appendices des Centriols sind in verschiedener Anzahl anzutreffen: in vielen Schnitten sind sie überhaupt nicht zu sehen, in manchen Querschnitten sieht man ein oder zwei Appendices und in Längsschnitten bis zu vier. Wenn das Centriol längs getroffen ist, hat man den Eindruck, als ob sich einige Tubuli der Centriolenwand nach außen biegen würden, um sich am

Aufbau der Appendices zu beteiligen (Abb. 1). Auf Querschnitten scheint diese Verbindung doppelt und das äußere Ende der Appendix in zwei kurze Fortsätze gespalten zu sein.

Das Diplosom ist häufig von einer hellen protoplasmatischen Substanz umschlossen, welche sich manchmal strahlenförmig in das umgebende an Körnern und ergastoplasmatischen Lamellen reiche Cytoplasma ausbreitet. Diese Anordnung im Aster ist jedoch keinesfalls als Regel anzusehen; die Centriolen können auch frei im Cytoplasma vorkommen. Oft fand ich die Centriole in einer cytoplasmatischen Zone sehr nahe der Zelloberfläche; ich möchte einige besondere Aspekte hervorheben, die auf Zusammenhänge zwischen Centriol und Zelloberfläche zurückzuführen sind.

In einem Falle befand sich eines der 2 Centriolen, das tiefer im Cytoplasma gelegene, parallel zur freien Zelloberfläche und das andere, darüber und vertikal zu diesem gelegen, ragte mit seinem distalen Ende aus der Zelle heraus. Eine ähnliche Beobachtung konnte ich in einer Endothelzelle machen, die keine direkte Beziehung mit der vorderen Augenkammer hatte. In dieser Zelle, in der das oberflächlichere Centriol mit seinem Ende in Berührung mit der Zelloberfläche gekommen war, schien dieses die Zellmembran auszustülpen und sie für eine kurze Strecke in die Nachbarzelle vorzuwölben. In einem anderen Schnitt kam das Centriol — nur eines war zu sehen — nicht mit seinem Ende, sondern mittels seiner lateralen Wand mit der freien Zelloberfläche in Berührung.

Einen anderen Fall schließlich betraf ein längs getroffenes Centriol, in schräger Position zwischen Kern und freier Zelloberfläche (Abb. 2). Vor seinem peripheren Ende zogen zwei parallele Streifen (ungefähr 1 μ lang) zur Zellmembran. Sie schienen eine Fortsetzung des Centriolenkörpers zu sein. Sie waren nicht so osmiophil wie das Centriol selbst und man erkannte bei starker Vergrößerung, daß jeder Streifen aus zwei dunklen Linien bestand, die von einem hellen Zwischenraum, 200 Å breit, getrennt waren.

Diese Ergebnisse zeigen, daß das Centriol in manchen Fällen sehr enge Beziehungen mit der Zelloberfläche einzugehen vermag, die zweifelsohne Ausdruck einer besonderen funktionellen Aktivität darstellen und deren Ursache und Finalität durch weitere Untersuchungen bei verschiedenen physiologischen Zuständen zu klären sind.

Literatur

1. HARVEN, E. DE, et W. BERNHARD: Z. Zellforsch. mikr. Anat. **45**, 378 (1956—57).
2. RONDANELLI, E. G., e G. MILLONIG: Monit. Zool. Ital. **66**, Suppl., 502 (1957).
3. BESSIS, M., et J. BRETON GORIUS: Bull. Micr. appl. **7**, 54 (1957).
4. YAMADA, E.: Acta anat. (Basel) **29**, 267 (1957).
5. BESSIS, M., J. BRETON GORIUS et J. P. THIERY: Rev. d'Hémat. **13**, 363 (1958).
6. POLICARD, A., M. BESSIS, J. BRETON GORIUS et J. P. THIERY: Exp. Cell. Res. **14**, 221 (1958).
7. LETTRÉ, H., et R. LETTRÉ: Rev. d'Hémat. **13**, 337 (1958).
8. GRIGNOLO, A.: Boll. Ocul. **33**, 513 (1954).
9. ROUILLER, CH., D. DANON et A. RYTER: Acta anat. **20**, 39 (1954).
10. JAKUS, M. A.: J. biophys. biochem. Cytol. **2**, 243 (1956).
11. BALLOWITZ, E.: Arch. mikrosk. Anat. **56**, 230 (1900).
12. BERNHARD, W., et E. DE HARVEN: Dieser Band, S. 217.
13. JAKUS, M. A.: Dieser Band, S. 344.

Licht- und elektronenoptische Untersuchungen am Pilzkern
(Polystictus versicolor)

MANFRED GIRBARDT

Institut für Mikrobiologie und experimentelle Therapie Jena der Deutschen Akademie der Wissenschaften zu Berlin

Morphologische Untersuchungen am Interphasekern der Pilze sind relativ selten. [Zusammenfassende Darstellungen: (1, 2, 3, 4, 5).] Im allgemeinen besteht die Vorstellung, daß der Pilzkern für lichtoptische Untersuchungen zu klein sei, um über seine Struktur Aussagen machen zu können. Die meisten Autoren begnügen sich damit, die Anzahl der Kerne im fixierten Präparat nachzuweisen.

Die phasenoptische Lebendbeobachtung der Pilzzelle hat gezeigt, daß die vegetativen Hyphen infolge ihrer geringen Dicke ($\varnothing$ etwa 5 μ) für das Phasenkontrastverfahren prädestiniert sind. Besonders bei Verwendung von Medien optimaler Brechungsindices (*6, 7, 8, 9*) können nicht nur Zellkern, Chondriosomen und andere Zellorganelle mit Sicherheit identifiziert, sondern auch ihr Lebendverhalten studiert und mikrokinematographisch analysiert werden (*10, 11, 12*). Es besteht daher die begründete Hoffnung, daß die Pilzzelle in den Kreis jener Objekte rückt, an denen nicht nur cytomorphologische, sondern auch experimentelle zellphysiologische Untersuchungen mit Aussicht auf Erfolg durchgeführt werden können. Hierfür ist weiter von besonderem Vorteil, daß die Pilze leicht auf chemisch definierbaren Nährböden jederzeit gezüchtet werden können.

Die vegetativen Zellkerne des Basidiomyceten *Polystictus versicolor* sind Zellorganelle, die zwar alle morphologischen Charakteristika eines echten Interphasekernes besitzen, bei deren Teilung jedoch wahrscheinlich einfachere, nichtmitotische Vorgänge ablaufen (*3, 5, 13*). Es ist daher von besonderem Interesse, das Lebendverhalten der Pilzkerne mit dem echter Kerne, etwa in Gewebekulturen, zu vergleichen.

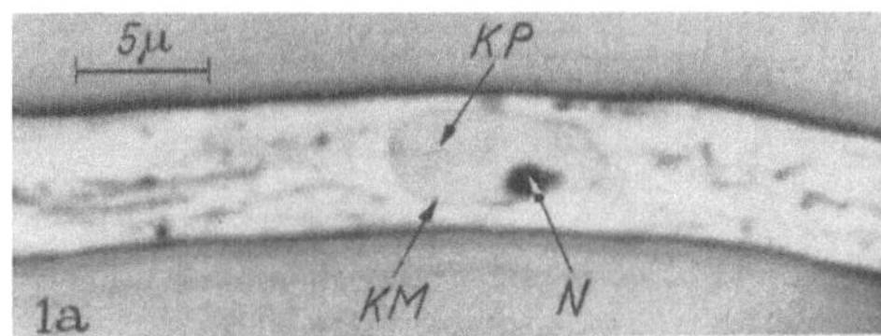

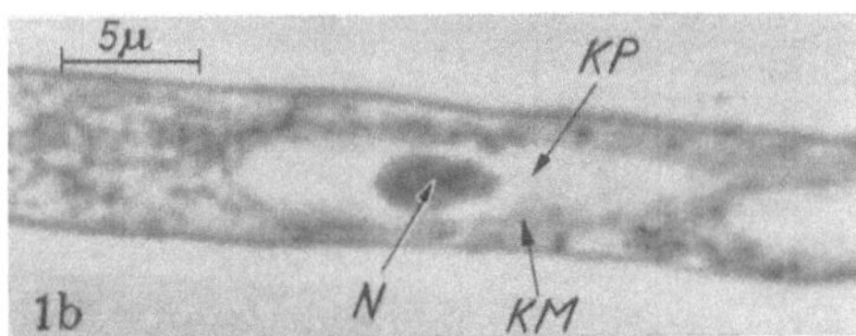

Abb. 1 a u. b *Polystictus versicolor*. a) Nicht wandernder Kern, Kernmembran (*KM*), Karyoplasma (*KP*) und Nucleolus (*N*) sind sichtbar. Das Karyoplasma erscheint dunkler als das stark hydratisierte Cytoplasma. b) Wanderkern. Bezeichnungen wie in 1a. Der Nucleolus ist phasenoptisch inhomogen. Das Karyoplasma erscheint hell gegenüber dem dunklen Cytoplasma. Phako lebend 2000:1

Material und Methodik. Für die Lebenduntersuchung wurde das vegetative Mycel auf Objektträgern, die mit 14%iger Malz-Gelatine (n_D 1.360) überzogen waren, in feuchten Kammern bei 20—22° C gezüchtet. Nach 3tägiger Inkubation erfolgten Beobachtung und Filmaufnahmen (Agfa-Superpan, 35 mm) mit Phako Zeiss, Jena (Obj. PH HI 90, Okul. 15mal). — Die Präparation für die elektronenoptischen Untersuchungen ist ausführlich bei GIRBARDT (*14*) angegeben. Die elektronenmikroskopischen Aufnahmen wurden mit dem elektrostatischen Gerät (Modell D) vom VEB Carl Zeiss, Jena bei 45 kV gemacht.

Am nicht wandernden, mehr oder weniger abgerundeten Interphasekern sind Kernmembran, Karyoplasma und Nucleolus lebend nachweisbar (Abb. 1a). Der Wanderkern ist spindelförmig gestreckt (Abb. 1b), sein Nucleolus erscheint phasenoptisch inhomogen. Die netzartige Nucleolarstruktur ist außerordentlich labil und dürfte der morphologische Ausdruck für reversible Sol-Gel-Umwandlungen sein. Die Struktur wird durch Fixation mit chromsäurehaltigen Fixierungsgemischen vergröbert, so daß sie lichtoptisch nach Färbung mit Hämatoxylin (HEIDENHAIN) gut nachweisbar ist. Das elektronenoptische Bild zeigt, daß der Nucleolus entweder aus globulären Einheiten (Abb. 2) oder nucleolonemataähnlichen (*15*) Strängen (Abb. 3) besteht. Der Vergleich mit dem Lebendverhalten der Nucleolarsubstanz macht es wahrscheinlich, daß die Stränge durch Aneinanderlagerung der globulären Einheiten entstehen. Die Nucleolusstrukturen bestehen im wesentlichen aus stark kontrastierten granulären Bestandteilen, deren Durchmesser etwa 100 Å beträgt.

In Abb. 1a liegt der Kern im stark hydratisierten (negativer Bildkontrast) Plasma. Da das Karyoplasma dunkler als das umgebende Cytoplasma erscheint, dürfte sein Gehalt an Trockengewichtssubstanz in dieser Phase, verglichen mit dem des Cytoplasma, höher sein. Umgekehrt liegen die Verhältnisse in Abb. 1b. Ein Vergleich der beiden Abbildungen lehrt, daß der Quellungsgrad des Karyoplasmas weitgehend unabhängig vom Quellungsgrad des umgebenden Cytoplasmas ist. Dies scheint anzuzeigen, daß entweder vorwiegend gebundenes Wasser vorliegt oder daß kein freier Austausch von Wasser durch die Kernmembran möglich ist.

Der Nucleolus des Interphasekernes zeigt lebend einen starken Formenwandel. Diese Erscheinung läßt zunächst eine amöboide Beweglichkeit vermuten. Die mikrokinematographische Analyse offenbart jedoch folgendes:

1. Bei Drehungen des gesamten Kernes folgt der Nucleolus diesen Drehungen und wird dabei häufig tropfenförmig ausgezogen.

2. Nach vollzogener Teilung entsteht der neue Nucleolus an einer bestimmten Stelle der Kernmembran und scheint mit dieser Stelle, der Nucleolus-Ansatzstelle (NA) in Verbindung zu bleiben.

3. Beim Übertritt des Kernes aus dem Schnallenauswuchs in die Haupthyphe eilt die NA allen Bewegungen des übrigen Kernes voraus. Dabei wird der Nucleolus fadenförmig ausgezogen.

4. Vor dem Einwandern des Kernes in den Schnallenauswuchs ist die NA starken bewegenden Kräften ausgesetzt. Der Nucleolus folgt verzögert diesen Bewegungen unter charakteristischen Formänderungen.

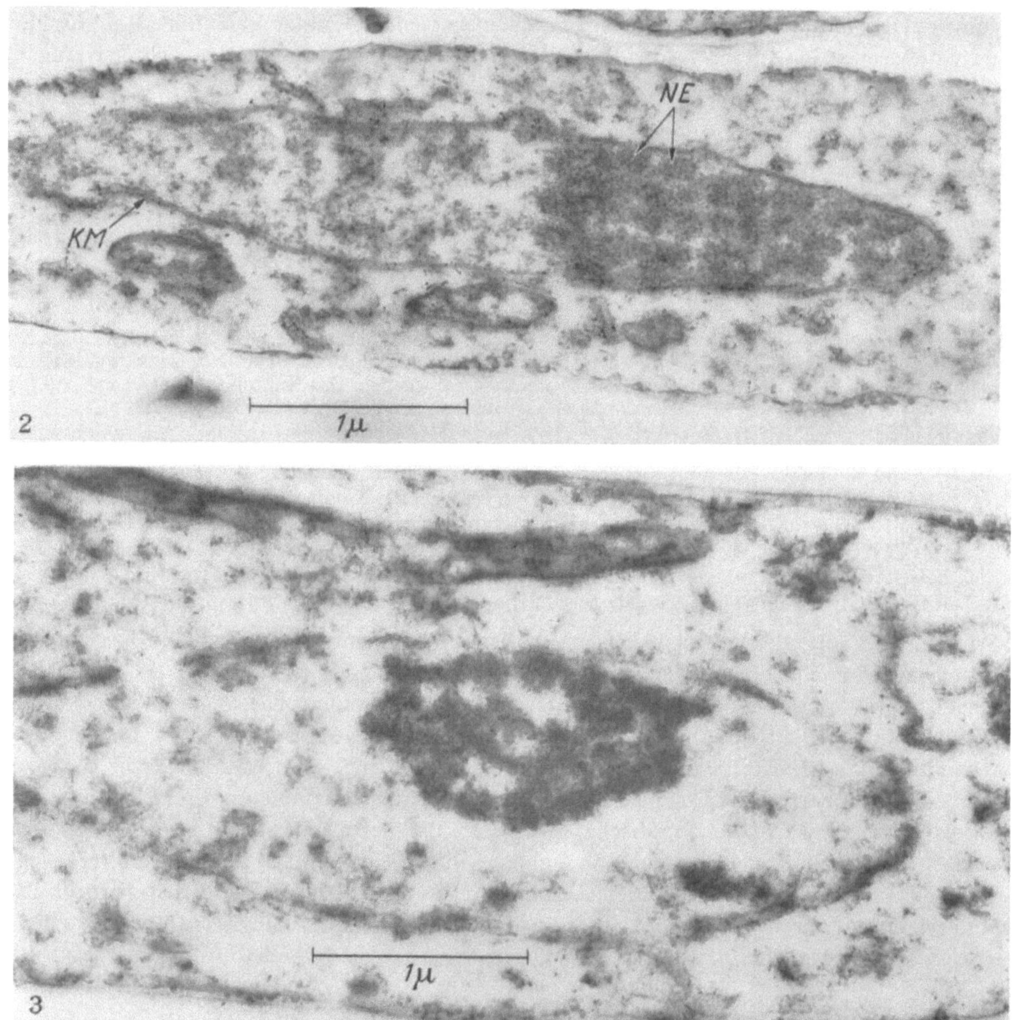

Abb. 2. *Polystictus versicolor*. Längsgeschnittener Zellkern. Der Nucleolus besteht aus globulären Einheiten (*NE*). Elektronenopt. 11000mal nachvergr. auf 30000mal

Abb. 3. *Polystictus versicolor*. Längsgeschnittener Zellkern. Der Nucleolus besteht aus nucleolonemataähnlichen Strängen. Elektronenopt. 11000mal, nachvergr. auf 30000mal

Als Ergebnis der Untersuchungen ergibt sich, daß die Formvarianz des Nucleolus kausal verständlich wird, wenn man postuliert, daß das Kernkörperchen ständig mit einer bestimmten Stelle der Kernmembran verbunden ist. Die den Kern bewegenden Kräfte scheinen an dieser Stelle anzugreifen. Der Formenwandel des Nucleolus wird daher als morphologische Manifestation der Reaktionen eines zähflüssigen Körpers auf die von außen am Kern ansetzenden Kräfte gedeutet.

Literatur

1. Cutter, V. M.: Ann. Rev. Microbiol. **5**, 17 (1951).
2. Olive, L. S.: Bot. Rev. **19**, 439 (1953).
3. Robinow, C. F.: Canad. J. Microbiol. **3**, 771 (1957).
4. — Canad. J. Microbiol. **3**, 791 (1957).
5. Bakerspigel, A.: Amer. J. Bot. **45**, 404 (1958).
6. Barer, R., and S. Joseph: Quart. J. Microsc. Sci. 3, Ser. **95**, 399 (1954).
7. — — Symp. Soc. exp. Biol. **10**, 160 (1957).
8. — — J. appl. Bact. **21**, 146 (1958).
9. Müller, R.: Mikroskopie **11**, 36 (1956).
10. Girbardt, M.: Flora **142**, 540 (1955).
11. — Planta **50**, 47 (1957).
12. — Film HF 201 (1957).
13. — Naturwissenschaften **43**, 429 (1956).
14. — Arch. Mikrobiol. **28**, 255 (1958).
15. Estable, C., and J. R. Sotelo: Fine Structure of Cells. Symposium Leiden 1954. p. 170.

Vergleichende Untersuchungen über einige Reaktionen der Chromosomen von Bacillus megaterium und Amphidinium elegans

P. Giesbrecht

Hygiene-Institut der Universität Bonn, Medizinisch-parasitologische Abteilung

Der Zellkern von Bacillus megaterium enthält offenbar nur ein einziges Chromosom, dessen zwei etwa 300 Å starke Chromatiden zwischen zwei Kernteilungen in fast vollständig spiralisiertem Zustand vorliegen („Großschrauben"; 4,5); diese Großschrauben lassen noch Feinstrukturen in makromolekularer Größenordnung erkennen („Kleinschrauben"). Ein solcher Feinbau

der Bakterienchromosomen ähnelt weitgehend dem Bau der Chromosomen des marinen Dinoflagellaten Amphidinium elegans (6). Nunmehr wurde versucht, weitere Aufschlüsse über den Bau dieser Strukturen zu gewinnen.

Als Standard-Fixierungsmittel fand 1% OsO$_4$ + 1% K$_2$Cr$_2$O$_7$ in 2,5% Saccharose (pH 6,7—7,2) Anwendung [vgl. (2, 6, 5)], als Kontrolle die übliche 1% OsO$_4$-Lösung in Veronalacetatpuffer (pH 6,7—7,2). Beide Lösungen wurden variiert, wie jeweils in Text und Bildlegende angegeben.

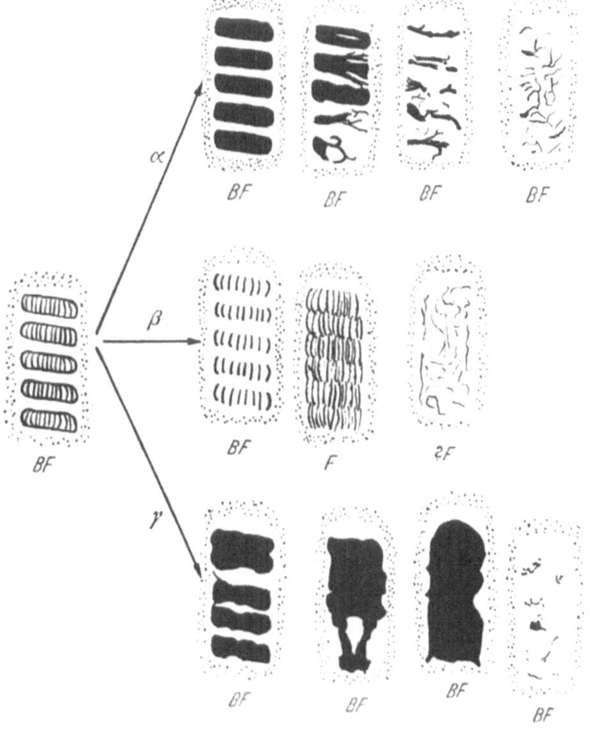

Abb. 1. Veränderungen der Chromosomen von Bac. megaterium und A. elegans (B = bei Bakterien, F = bei Flagellaten). α) Längsaufspaltung der Chromosomen (homogene Fäden — Längsaufspaltungen — feinere Fäden — ungeordnetes, feinfädiges Chromosomenmaterial). β) Auflockerung der Chromosomen (Auflockerung der Kleinschrauben — Vergrößerung des Windungsdurchmessers der Kleinschrauben — ungeordnetes, feinfädiges Material). γ) Coagulationen der Chromosomen (Coagulation benachbarter Schraubenwindungen — beginnende Strangbildung — Chromosom als undifferenzierte Masse — Chromosomenmaterial weitgehend herausgelöst)

Einbettung in Methacrylsäureester. Für Aufnahmen mit dem Siemens-Gerät (Typ 100 d) haben wir Herrn Doz. Dr. WOHLFARTH-BOTTERMANN zu danken.

Die Chromosomen-Großschraube der Bakterien und die der untersuchten Dinoflagellaten zeigten in den verschiedenen Präparationen so unterschiedliche Feinstrukturen, daß zumindest ein Teil davon als Struktur-veränderungen infolge unge-eigneter Präparationsbedin-gungen angesehen werden müs-sen. Auffälligerweise erwiesen sich jedoch diese Strukturver-änderungen in den Chromoso-men beider Organismen als prinzipiell identisch. Doch wäh-rend bei den Dinoflagellaten meist recht einheitliche Prä-parationen zu erhalten waren (einheitliches Erscheinungs-bild des Großteils der Chro-mosomen), reagierten bei Bac. megaterium oft nur ein Teil der Bakterienchromosomen in entsprechender Weise.

a) Längsaufspaltung der Chromosomen (vgl. Abb. 1, α). In einem Teil der mit Chromosmi-um fixierten Präparationen ließ sich die Kleinschraube der Chromosomenfäden nicht klar darstellen: Diese war entweder nur angedeutet erkennbar oder die Chromosomenfäden er-schienen völlig homogen (Abb. 2a u. b). Nach Anwendung der üblichen OsO_4-Fixierung, besonders aber dann, wenn man die OsO_4-Konzentration oder den osmotischen Wert der Fixierungslösungen verän-derte (2c—f), wiesen die homo-genen Chromosomenfäden fein-fädige, parallel zur Längsachse auftretende mit Aufspaltungen mit z.T. baumartigen Verzwei-gungen auf (Abb. 2c u. d; vgl. 11, 3, 9). Schließlich war von dem hohen Ordnungsgrad „Großschraube mit Klein-schraube" nichts mehr zu er-

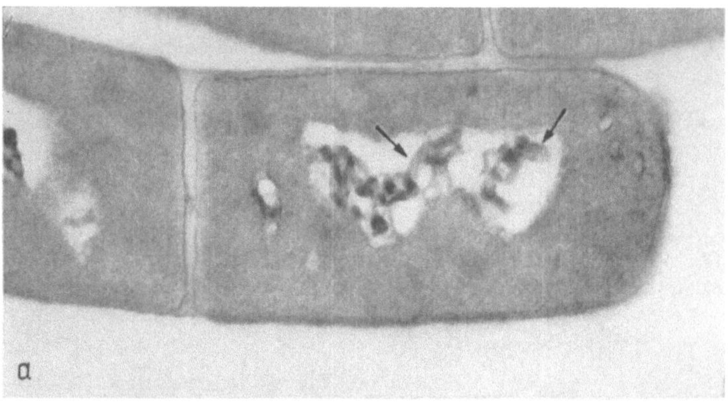

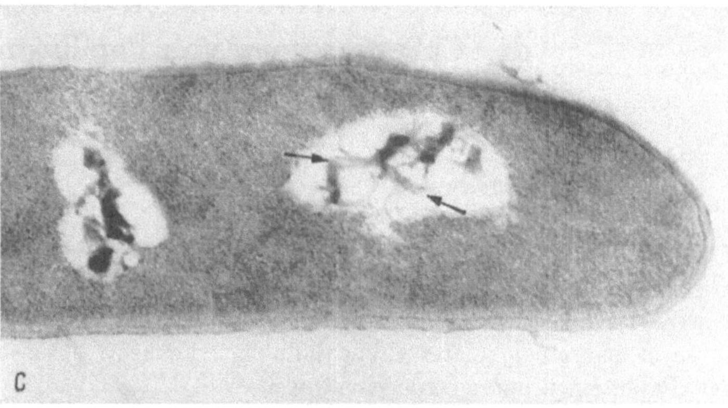

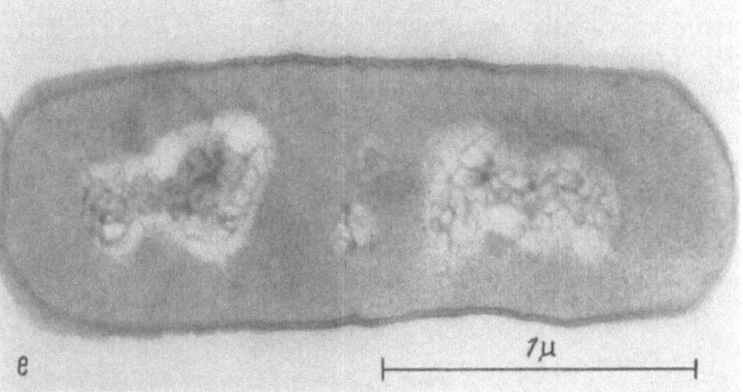

Abb. 2a—f. a) u. b) Bac. megaterium und Chromosom von A. elegans. Kleinschrauben der Chromosomen nur noch angedeutet erkennbar (Pfei-le!). *Fix.* a: 1% OsO_4 + 1% $K_2Cr_2O_7$ in 2,5% Saccharose, pH 6,7; 1 Std. bei 4° C. *Fix.* b: 1% OsO_4 + 1% $K_2Cr_2O_7$ in 3% Dextran + 7% Sac-charose, pH 7,2; 30 min bei Zimmertemperatur. c) u. d) Bac. megaterium und Chromosom von A. elegans. Längsaufspaltung der Chromosomen-fäden (Pfeile!). *Fix.* c: 1% OsO_4 + 1% $K_2Cr_2O_7$ in Aqua bidest., pH 7,0;

kennen, und im Bereich der Chromosomen sah man nur noch ungeordnete feinfädige Strukturen (Abb. 2e u. f). Solche Bilder erinnern an das «matériel nucléaire» von E.coli, wie es KELLEN-BERGER und RYTER (7) beschrieben haben. Bei Bac. megaterium wies jedoch auch nach Anwen-dung der Standard-Fixierung aus noch ungeklärten Gründen häufig ein Teil der Zellen gleichfalls solche Längsaufspaltungen oder feinfädiges Chromosomenmaterial auf.

b) Auflockerung der Chromosomen (vgl. Schema 1, β). Bei Anwendung der üblichen OsO_4-Fixierung konnten die Kleinschrauben der Flagellatenchromosomen wie in einen kompakten Chromosomenkörper eingelagert erscheinen (Abb. 3a). Ein ähnlicher kompakter Chromosomenkörper bzw. eine zwischen den Großschraubenwindungen der Chromosomen befindliche Zellstruktur war bei Bakterien bisher nicht aufzufinden, jedoch gleichen die Kleinschrauben weitgehend denen, die bei Bakterien dargestellt werden konnten. Bei Anwendung einer Chromosmiumfixierung, insbesondere in etwas höherer Konzentration ($1{,}5\%\ OsO_4 + 1{,}6\%\ K_2CrO_7$), fehlten jedoch meist diese zwischen den Großschraubenwindungen liegenden Zellanteile. Die Großschrauben waren dann quer zur Längsachse ihrer Chromosomenfäden mehr oder weniger stark aufgelockert, so daß die Baueinheiten der Kleinschrauben deutlicher erkennbar wurden (Abb. 3b). Solche Auflockerungen kommen (wenn auch relativ selten) bei Bakterien ebenfalls vor (vgl. Abb. 7 in 5). Bei den Flagellatenchromosomen war darüber hinaus noch eine weitergehende Auflockerung der Chromosomenfäden quer zu ihrer Längsachse zu beobachten, die häufig von einer so starken Vergrößerung des Windungsdurchmessers der Kleinschrauben begleitet war, daß sich die Kleinschrauben zweier benachbarter Windungen der Großschraube berührten und damit fast eine über den ganzen Chromosomenkörper laufende und die einzelnen Windungen der Großschraube verdeckende Längsstrukturierung vortäuschten (Abb. 3c). Solche Auflockerungen, in denen der Ordnungszustand „Großschraube mit

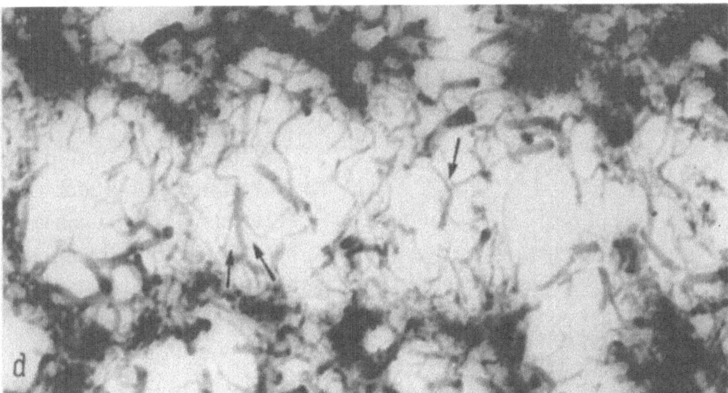

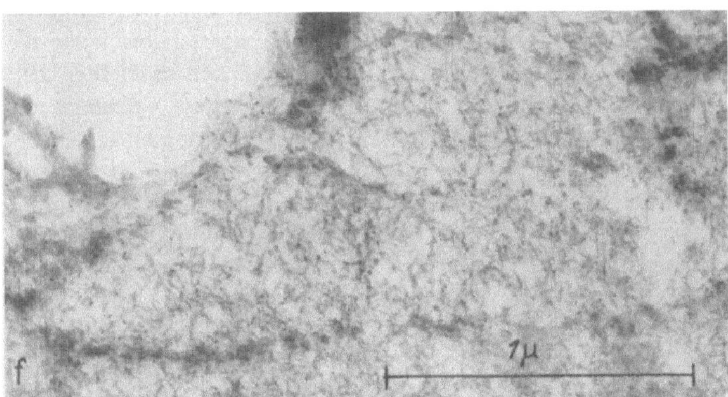

1 Std. bei 4° C. *Fix.* d: $0{,}5\%\ OsO_4$ in 50% Veronalacetatpuffer $+ 50\%$ Meerwasser-Kulturlösung, p_H 7,2; 1 Std. bei 0° C. e) u. f) Bac. megaterium und Chromosom von A. elegans. Ungeordnetes, feinfädiges Chromosomenmaterial. *Fix.* e: wie 1c. *Fix.* f: $1\%\ OsO_4$ in 50% Veronalacetatpuffer $+ 50\%$ Meerwasser-Kulturlösung, p_H 7,0; 1 Std. bei Zimmertemperatur. Die Präparationen 2a, c, e, f und 4a wurden mit Phosphorwolframsäure und Uranylacetat, 2d mit Phosphorwolframsäure nachkontrastiert

Kleinschraube" noch erkennbar ist, können offensichtlich solche Ausmaße annehmen, daß nur noch ungeordnete, feinfädige Strukturen erkennbar bleiben (Abb. 3c, rechte Seite), die z. T. sogar das Erscheinungsbild parallel liegender Fibrillen vortäuschen („Quer-Fibrillen").

c) Coagulationen der Chromosomen (vgl. Abb. 1, γ). Wurden zur Vermeidung von Volumenänderungen bei der Fixierung (*1*) anstelle von Salzlösungen größere Mengen unpolarer Stoffe

(10% Saccharose oder 3% Dextran + 17% Saccharose) dem Fixierungsmittel zugesetzt, so erschien ein Teil der Großschrauben ohne erkennbare Feinstrukturen und glich damit weitgehend denen in Abb. 2a u. b. Oft ließ sich jedoch eine Coagulation benachbarter Schraubenwindungen erkennen, wobei z. T. strangförmige Gebilde entstanden (Abb. 4a u. b). Diese Coagulationen konnten solche Ausmaße annehmen, daß das gesamte Chromosom nur noch als eine undifferenzierte Masse erschien (Chr in Abb. 4c u. d). Besonders solche Massen ließen sich relativ leicht durch längeres Waschen mit Phosphatpuffer nach der Fixierung herauslösen, ohne daß noch andere Zellveränderungen erkennbar wurden.

Bei unseren Befunden erscheint es uns vor allem von Wichtigkeit, daß all diese Veränderungen der Chromosomen zweier verschiedener Organismen zum Verwechseln ähnlich sind. Damit wird unsere Auffassung gestützt (4, 5), daß Bakterien echte Chromosomen besitzen, die bei manchen möglichen Unterschieden im einzelnen mit den Chromosomen einer vergleichsweise höher organisierten Zelle nicht nur in ihrem Bau prinzipiell übereinstimmen, sondern sogar vergleichbare Veränderungen aufweisen. Die Längsaufspaltung der Chromosomenfäden und die offenbare Querunterteilung derselben Fäden in Kleinschrauben lassen sich scheinbar kaum miteinander vereinbaren. Man könnte in diesem Zusammenhang die Hypothese diskutieren, daß die außerordentlich langen

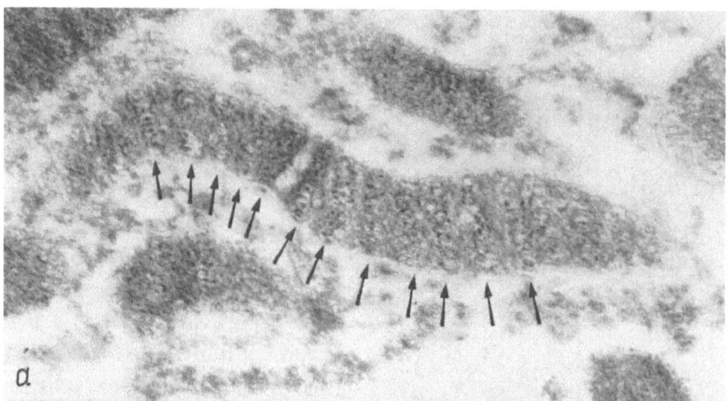

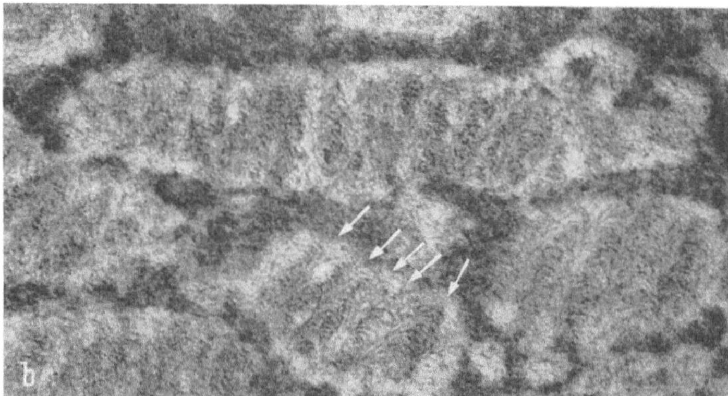

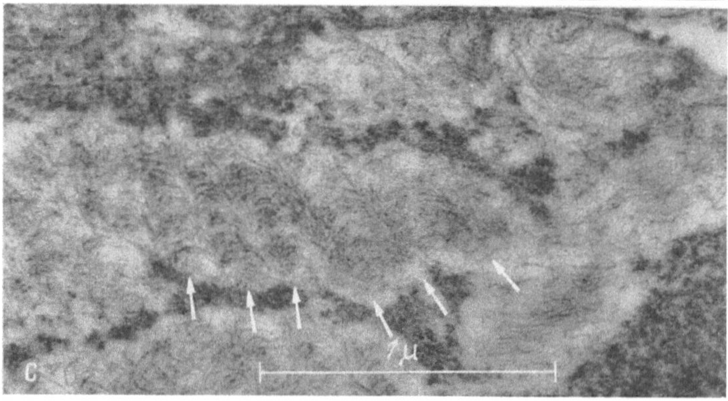

Abb. 3a—c. a) Chromosom von A. elegans. Windungen der Großschraube (Pfeile!) durch „Zwischenschraubensubstanz" voneinander getrennt. Kleinschrauben nur schwer erkennbar. *Fix.* a: 1% OsO₄ in Veronalacetatpuffer, pH 7,0; 1 Std. bei Zimmertemperatur. b) Chromosom von A. elegans. Windungen der Großschraube (Pfeile!) klar gegeneinander abgegrenzt. Kleinschrauben aufgelockert. *Fix.* b: 1,5% OsO₄ + 1,6% K₂Cr₂O₇ in 30% Meerwasser-Kulturlösung, pH 7,2; 17 Std. bei 4° C. c) Chromosom von A. elegans. Windungen der Großschraube (Pfeile!) unscharf gegeneinander abgegrenzt. Starke Vergrößerung des Windungsdurchmessers der Kleinschrauben. Im rechten Teil der Aufnahme ungeordnetes, feinfädiges Chromosomenmaterial. *Fix.* c: wie b

Fäden der Kleinschraube noch aus kleineren Bausteinen bestehen, die unter bestimmten Bedingungen frei werden und sekundär zu fädigen Gebilden parallel zur Längsachse der Chromosomenfäden aggregieren können. Homogen erscheinende Chromosomenfäden, wie sie auch bei Kernisolierungen (10) auftraten, wären damit als Übergangsstadium zwischen Längsaufspaltung und Querlockerungen anzusehen.

Literatur

1. BAHR, G. F., G. BLOOM, and G. FRIBERG: Electron Microscopy, Stockholm 1957, p. 106.

2. DALTON, A. J.: Anat. Rec. **121**, 281 (1955).

3. GIESBRECHT, P.: Diss. Math.-Naturwiss. Fakultät, Univ. Bonn 1957.

4. — Naturwissenschaften **45**, 473 (1958).

5. — u. G. PIEKARSKI: Arch. Mikrobiol. **31**, 68 (1958).

6. GRELL, K. G., u. K. E. WOHLFARTH-BOTTERMANN: Z. Zellforsch. **47**, 7 (1957).

7. KELLENBERGER, E., u. A. RYTER: Experientia (Basel) **12**, 420 (1956).

8. MAALØE, O., and A. BIRCH-ANDERSEN: Sympos. Soc, Gen. Microbiol. **6**, 261 (1956).

9. SCHLOTE, F. W.: Vortrag, 6. Tagung der Österreichischen Gesellschaft für Mikrobiologie und Hygiene, Semmering 1958.

10. SPIEGELMAN, S., A. I. ARONSON and P. C. FITZ-JAMES: J. Bact. **75**, 102 (1958).

11. TOMLIN, S. G., and J. W. MAY: Aust. J. exp. Biol. **33**, 249 (1955).

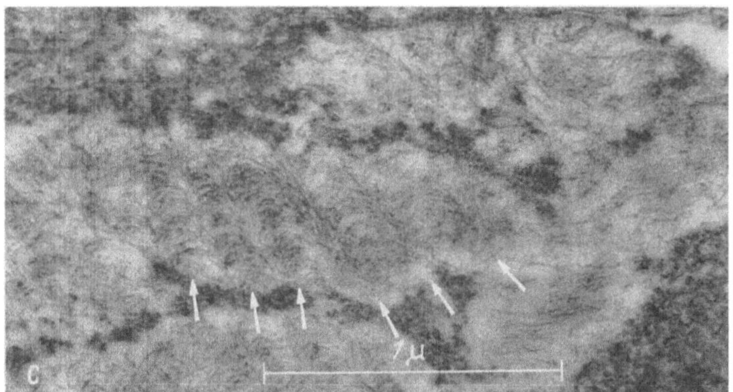

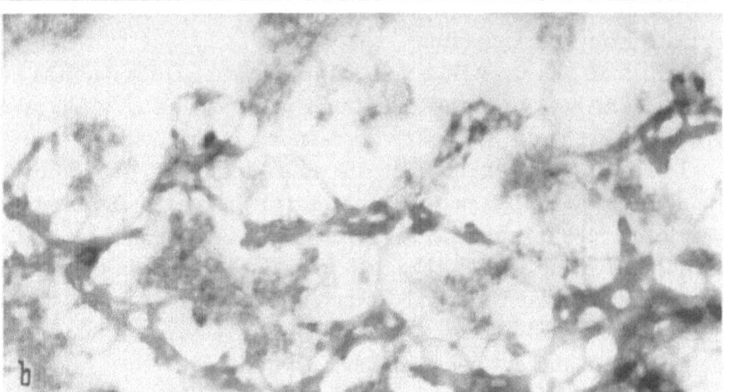

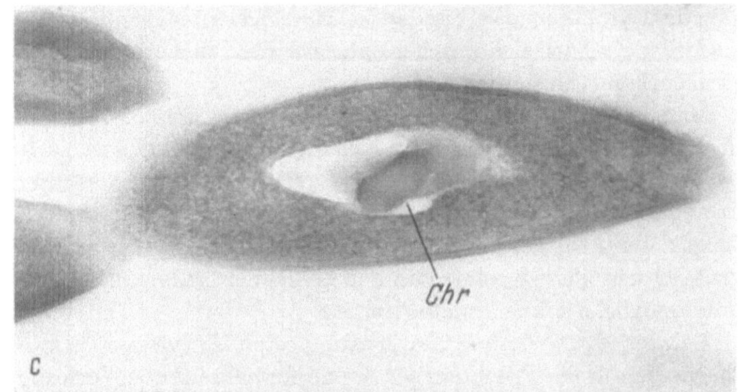

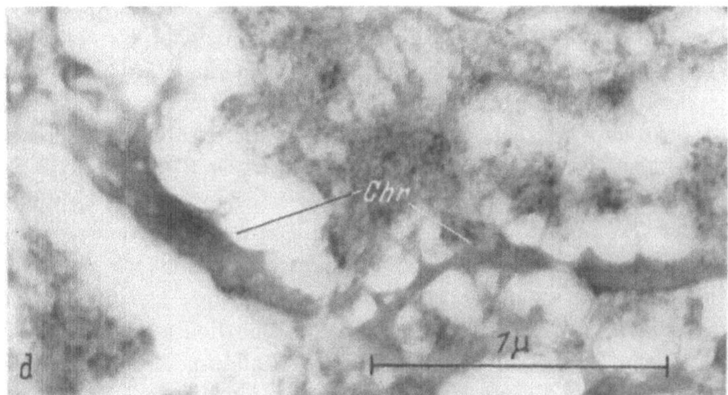

Abb. 4a—d. a) u. b) Bac. megaterium und Chromosom von A. elegans. Strangförmiges Chromosomenmaterial, dessen Entstehung aus den zwei Chromatiden noch erkennbar ist. *Fix.* a: 1% OsO$_4$ + 1% K$_2$Cr$_2$O$_7$ in 10% Saccharose, pH 6,7; 1 Std. bei 4° C. *Fix.*b: 1% OsO$_4$ + 1% K$_2$Cr$_2$O$_7$ in 17% Saccharose + 3% Dextran, pH 7,2; 30 min bei Zimmertemperatur. c) u. d) Bac. megaterium und Chromosom von A. elegans. Chromosomen (= Chr) nur noch als undifferenzierte Massen erkennbar. *Fix.* c: 1% OsO$_4$ + 1% K$_2$Cr$_2$O$_7$ in 10 % Saccharose, pH 7,2; 30 min bei Zimmertemperatur. *Fix.* d: wie b

2. Cytoplasma und Zellorganellen

Gestattet das elektronenmikroskopische Bild Aussagen zur Dynamik in der Zelle?

K. E. Wohlfarth-Bottermann

Zentral-Laboratorium für angewandte Übermikroskopie der Universität Bonn

Wir kennen heute allgemein verbreitete cytoplasmatische Strukturelemente wie z. B. das „Endoplasmatische Reticulum" (*1, 2*) und die sog. „Particuläre Komponente", jedoch wissen wir noch wenig über ihre biologische Bedeutung. Da im Elektronenmikroskop nur fixiertes Cytoplasma untersucht werden kann, fehlen uns vor allem Kenntnisse über den Zusammenhang von Feinstruktur und Dynamik.

Will man aus dem Bild von Dünnschnitten Rückschlüsse auf cytoplasmatische Bewegungsvorgänge ziehen, so lautet die Frage: Lassen sich Cytoplasmen mit verschiedener Bewegungsintensität im Strukturaspekt voneinander unterscheiden?

Das Cytoplasma kann sich wie eine Flüssigkeit aber auch wie ein fester Körper verhalten. Beide Zustände lassen sich wechselseitig ineinander überführen. Dieser Vorgang wird als Plasmasol- ⇋ Plasmagel-Transformation bezeichnet. Es ist sicher, daß hierbei intra vitam Strukturveränderungen auftreten. Fraglich ist jedoch, ob diese Strukturveränderungen elektronenmikroskopisch nach Fixierung und Einbettung sichtbar gemacht werden können.

Die Untersuchung einiger geeigneter Objekte erbrachte eindeutige Korrelationen zwischen Feinstruktur und cytoplasmatischer Dynamik:

Die *Wärmelähmung* einer Zelle bedeutet eine irreversible Dehydratation, die mit einer Gelierung des Cytoplasmas verbunden ist. Ein solches irreversibel geliertes Cytoplasma von *Paramecium caudatum* zeichnet sich durch Kontrastarmut, eine erhebliche Dichtezunahme und eine auffällige Strukturhomogenisierung aus.

Man kann *Paramecium*-Zellen durch einen *Elektroschock* vorübergehend bewegungsunfähig machen. Bei richtiger Dosierung wird die Schockwirkung in 5—10 min überstanden und die Zellen behalten ihre Teilungsfähigkeit. Das Cytoplasma der geschockten Zellen zeigt strukturelle Transformationen des „Endoplasmatischen Reticulums" in erstaunlichem Umfange. Man erkennt am Beispiel einer solchen reversiblen Zellschädigung, wie groß die Veränderungen der Cytoplasmastruktur sein können, ohne daß das Leben der Zelle erlischt. Wir erhalten damit einen Einblick in die mögliche Transformationsbreite.

Bei der *Sporenreifung* des Myxomyceten *Didymium nigripes*[1] strömt das Cytoplasma der Plasmodien in die Fruchtkörper. Seine fortschreitende Verfestigung zur reifen Spore prägt sich elektronenmikroskopisch in einer kontinuierlichen Strukturverdichtung aus. Prinzipiell gleiche Transformationen laufen wohl bei jeder pflanzlichen Samenbildung ab. Umgekehrt stellt sich die Verflüssigung des Cytoplasmas bei der *Sporenkeimung* als eine Verminderung der Strukturdichte bei der quellenden Spore und fortschreitend bis zu der aus der Spore schlüpfenden Amöbe dar (*3*).

Bei dem klassischen Objekt zum Studium cytoplasmatischer Bewegungsvorgänge, der Amöbe, interessiert besonders der Unterschied zwischen dem relativ festen Hyaloplasma (*Ektoplasma*) und dem flüssigeren „Körnchenplasma" (*Endoplasma*). Untersucht wurde eine Amöbenart vom Limax-Typ, bei der das hochviscose Ektoplasma phasenkontrastmikroskopisch von dem sehr beweglichen und niedrig viscosen Endoplasma deutlich getrennt war. Das Endoplasma (Abb. 1) ist charakterisiert durch eine „normale" lockere Cytoplasmastruktur, während das dynamisch wenig aktive Ektoplasma (Abb. 2) durch eine dichte Packung der sog. „Partikulären Komponente" imponiert. Kleine, in Bildung oder Rückbildung begriffene Ausläufer der Zelle (Pseudopodien) weisen ein Cytoplasma auf (Abb. 3), das deutlich lockerer ist als das normale Ektoplasma, sich jedoch wie dieses nur aus der sog. „Partikulären Komponente" zusammensetzt.

---

[1] Die Ergebnisse an *Didymium* verdanke ich meinem Mitarbeiter Dr. L. Schneider sowie der freundlichen Hilfe von Prof. A. L. Cohen, Oglethorpe-University, Georgia, USA.

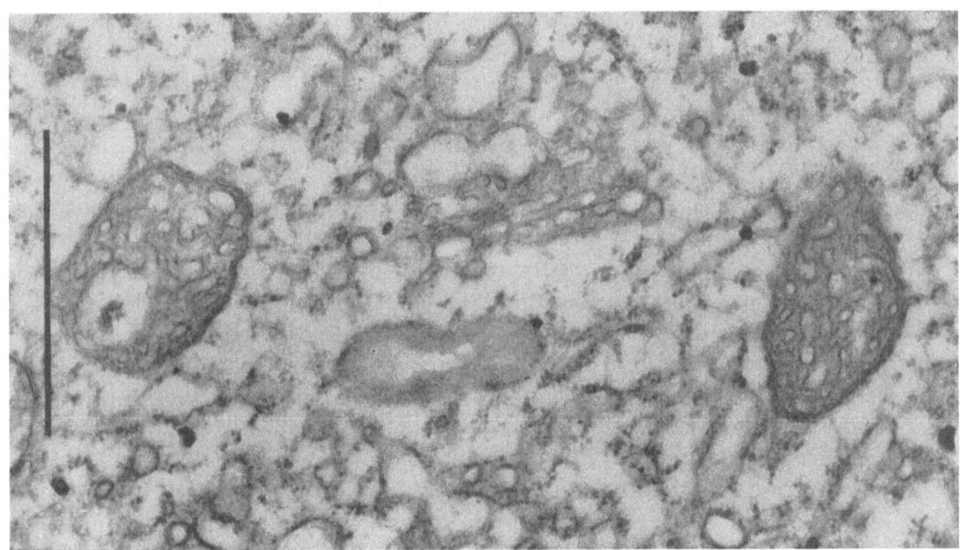

Abb. 1. Niedrig viscoses, relativ flüssiges Cytoplasma („Endoplasma", „Körnchenplasma") von Amoeba spez. (Limax-Typ). E. V. 40000:1

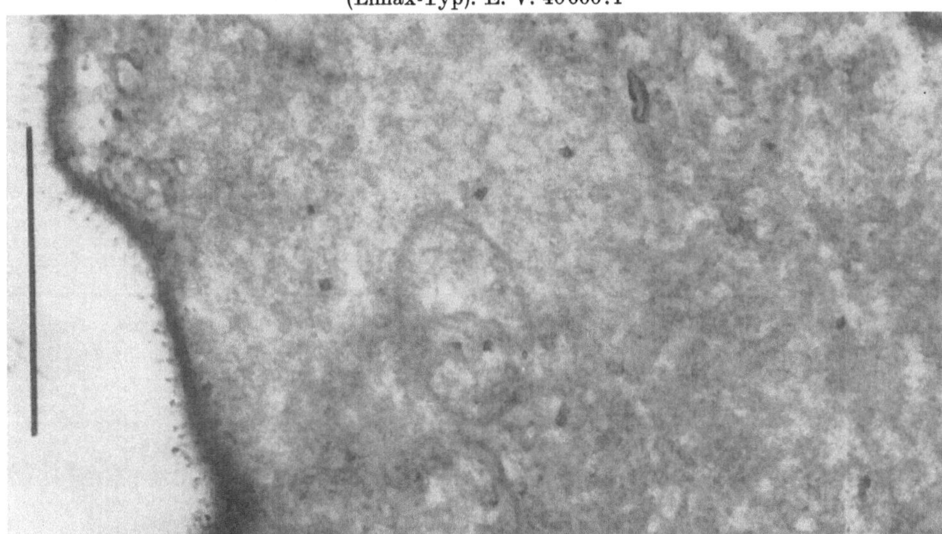

Abb. 2. Hoch viscoses, relativ festes Cytoplasma („Ektoplasma", „Hyaloplasma") der gleichen Zelle. E.V. 40000:1

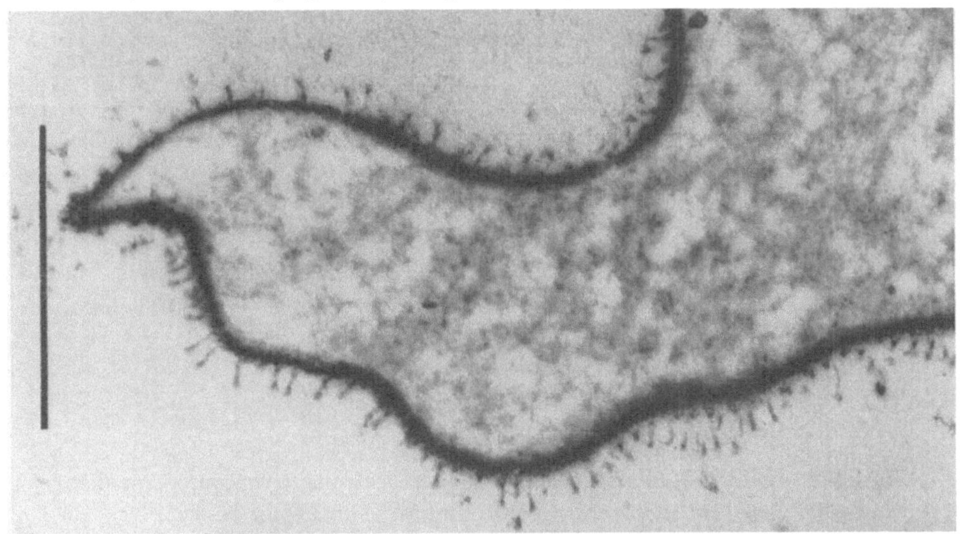

Abb. 3. Cytoplasma eines in Bildung oder Rückbildung begriffenen Ausläufers (Pseudopodium) der gleichen Zelle. E. V. 40000:1

An Heliozoen (*Actinophrys Sol*) kann man lichtmikroskopisch das schnellflüssige *Rheoplasma* an der Zellperipherie von dem sehr verfestigten und starren *Stereoplasma*, den Achsenfäden im Innern der Axopodien, unterscheiden. Beide Modifikationen des Cytoplasmas sind schnell ineinander überführbar. Das Rheoplasma besitzt als flüssiges Cytoplasma einen ausgesprochenen vacuolären Strukturaspekt, dagegen ist das viscosere Cytoplasma in Kernnähe besonders reich an (ungeordneten) Komponenten des „Endoplasmatischen Reticulums". Ebenfalls aus schlauchförmigen Strukturen, die jedoch ausgerichtet und gebündelt sind, besteht das verfestigte Stereoplasma, das temporär statische Funktionen in der Zelle erfüllt (*4*).

Auch an *Zellen höherer Organismen* lassen sich Korrelationen zwischen Feinstruktur und Dynamik auffinden: Bei Hühnerherzmyoblasten aus Gewebekulturen sieht man im Phasenkontrastmikroskop, daß die Mitochondrien und der flüssige Anteil des Cytoplasmas der Zellen in ganz bestimmten „Kanälen" strömen, die von verfestigtem Cytoplasma begrenzt werden. Dies wird besonders deutlich, wenn man die Cytoplasmaströmung durch hypertonische Medien verlangsamt bzw. durch hypotonische Medien beschleunigt, wodurch die Konsistenz des Cytoplasmas verfestigt bzw. verflüssigt wird. Die dynamisch inaktiven Cytoplasmaanteile mit statischer Funktion stellen Bündel von Fibrillen dar, deren Feinstruktur besonders gut nach Einbettung der Gewebekulturzellen in das von Kellenberger empfohlene Vestopal analysiert werden kann. Auch an Zellen höherer Organismen lassen sich also Unterschiede in der Dynamik der verschiedenen Zellbereiche elektronenmikroskopisch erkennen (*4*).

Die Analyse von Korrelationen zwischen Feinstruktur und cytoplasmatischer Dynamik bietet, wie wir gezeigt haben, Ansatzpunkte für morphogenetische Studien, die zu einer funktionellen Charakterisierung der verschiedenen cytoplasmatischen Strukturkomponenten beitragen werden.

Die Untersuchungen wurden mit Unterstützung der Deutschen Forschungsgemeinschaft durchgeführt.

Literatur

1. Palade, G. E.: J. biophys. biochem. Cytol. **2**, 85 (1956).
2. Porter, K. R.: Harvey Lect. Ser. L I, 175 (1957).
3. Schneider, L., u. K. E. Wohlfarth-Bottermann: Naturwissenschaften **45**, 140 (1958).
4. Wohlfarth-Bottermann, K. E.: Z. Zellforsch. **50**, 1 (1959) (Ausführliche Darstellung).

Étude au microscope électronique de la lipophanerose cytoplasmique

A. Policard, A. Collet et S. Pregermain avec la collaboration de Y. Pelanne, P. Pomes et C. Reuet

Centre d'Etudes et Recherches des Charbonnages de France et Service de Microscopie électronique du CNRS, Faculté des Sciences (Pr. P. P. Grasse, Directeur)

Sous l'influence d'agressions diverses, les lipides, normalement contenus dans le cytoplasme sont «démasqués» et deviennent apparents microscopiquement. C'est le processus bien connu de la «lipophanérose». Au microscope électronique ces corps lipidiques libérés offrent des aspects variables. Dans les images que nous avons pu recueillir, au cours d'observations de lésions pulmonaires et épiploïques provoquées par la silice ou le bacille tuberculeux sur des pièces fixées par OsO_4, nous avons rencontré cinq catégories morphologiques d'images:

1) Des formations lipidiques en gouttelettes ou «flaques», avec ou sans vacuoles;
2) Des structures denses feuilletées;
3) Des formations lamellaires concentriques fines;
4) Des figures en grille;
5) Des formations composites.

1) Gouttelettes lipidiques: homogènes, denses aux électrons, à contours irrégulières festonnés, elles ont été décrites par de très nombreux auteurs. Elles mesurent environ de 0,1 à 1 μ de diamètre. Une variété moins fréquente nous est apparue dans des lésions pulmonaires tuberculeuses expérimentales chez le Rat et le Cobaye. Une vacuole claire est nettement dessinée dans la gouttelette dense. Elle est plus ou moins excentrée, renferme parfois un petit corpuscule dense (Fig. 2) (*1*) et correspond aux vacuoles annexées aux gouttelettes lipidiques décrites en cytologie classique.

2) Structures denses feuilletées: Tres abondantes dans les lésions pulmonaires, elles apparaissent beaucoup plus rarement dans les granulomes réactionnels d'autres organes. Ces structures

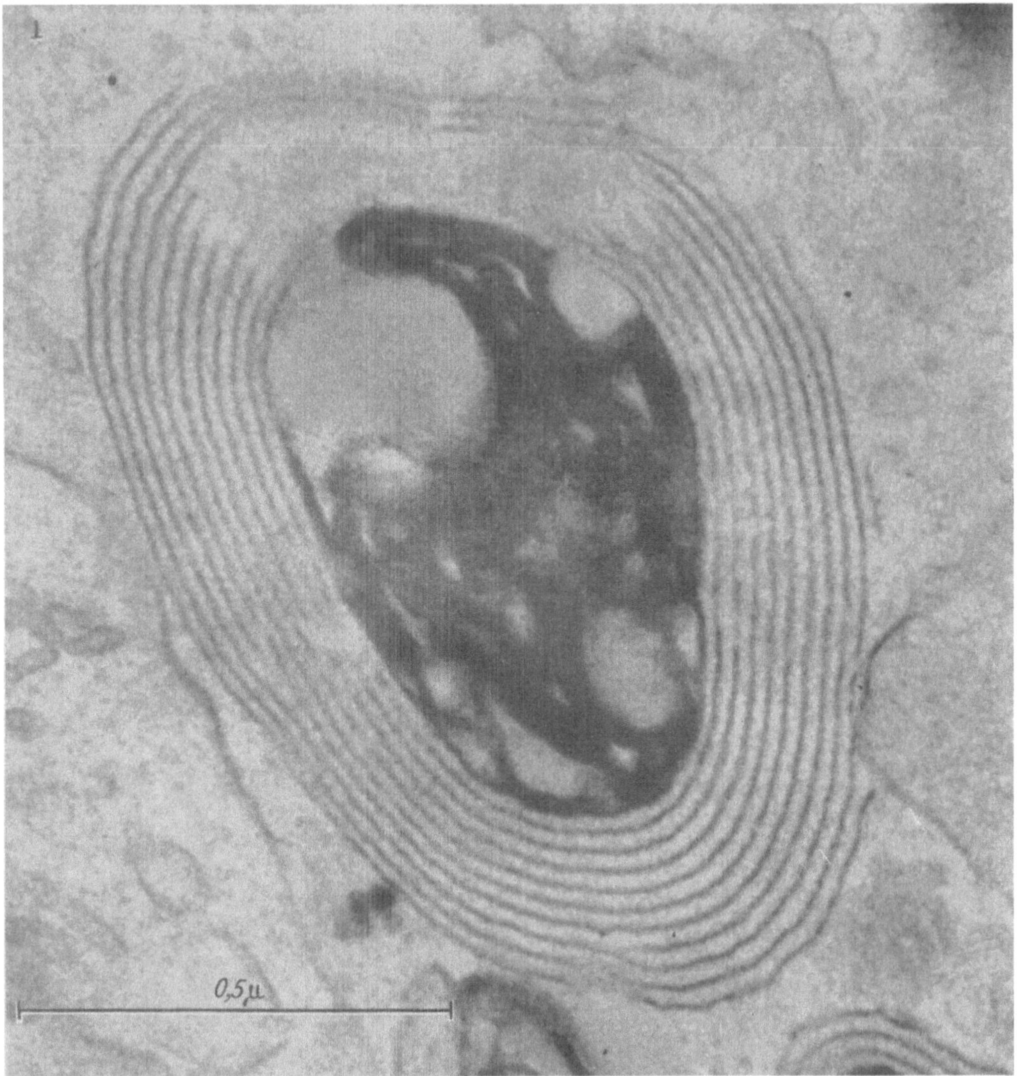

Fig. 1. Formation concentrique fine. Poumon (Rat) après injection de Silice Colloïdale, 48 h. Inclusion:Vestopal

mesurent entre 0,2 et 2 μ de diamètre, avec un contour irrégulier. Elles sont constituées par des feuillets très denses aux électrons, relativement épais (200 Å environ), légèrement festonnés, grossièrement concentriques, et plus ou moins au contact les uns avec les autres. A ce point de vue, on peut distinguer une variété à feuillets serrés à disposition régulière (Fig. 3) souvent à double contour, avec un espace intermédiaire plus clair, fréquemment moniliforme, et une variété à feuillets, largement et irrégulièrement espacés (Fig. 4). Entre ces deux variétés existent des intermédiaires. Nous avons précédemment discuté (*2*) leurs relations avec les figures myéliniques de VIRCHOW. Des formations semblables ont été étudiées au microscope électronique dans certaines cellules alvéolaires du poumon normal par SCHLIPKÖTER (*3*), KISCH (*4*), KARRER (*5*), SCHULZ (*6*)

17*

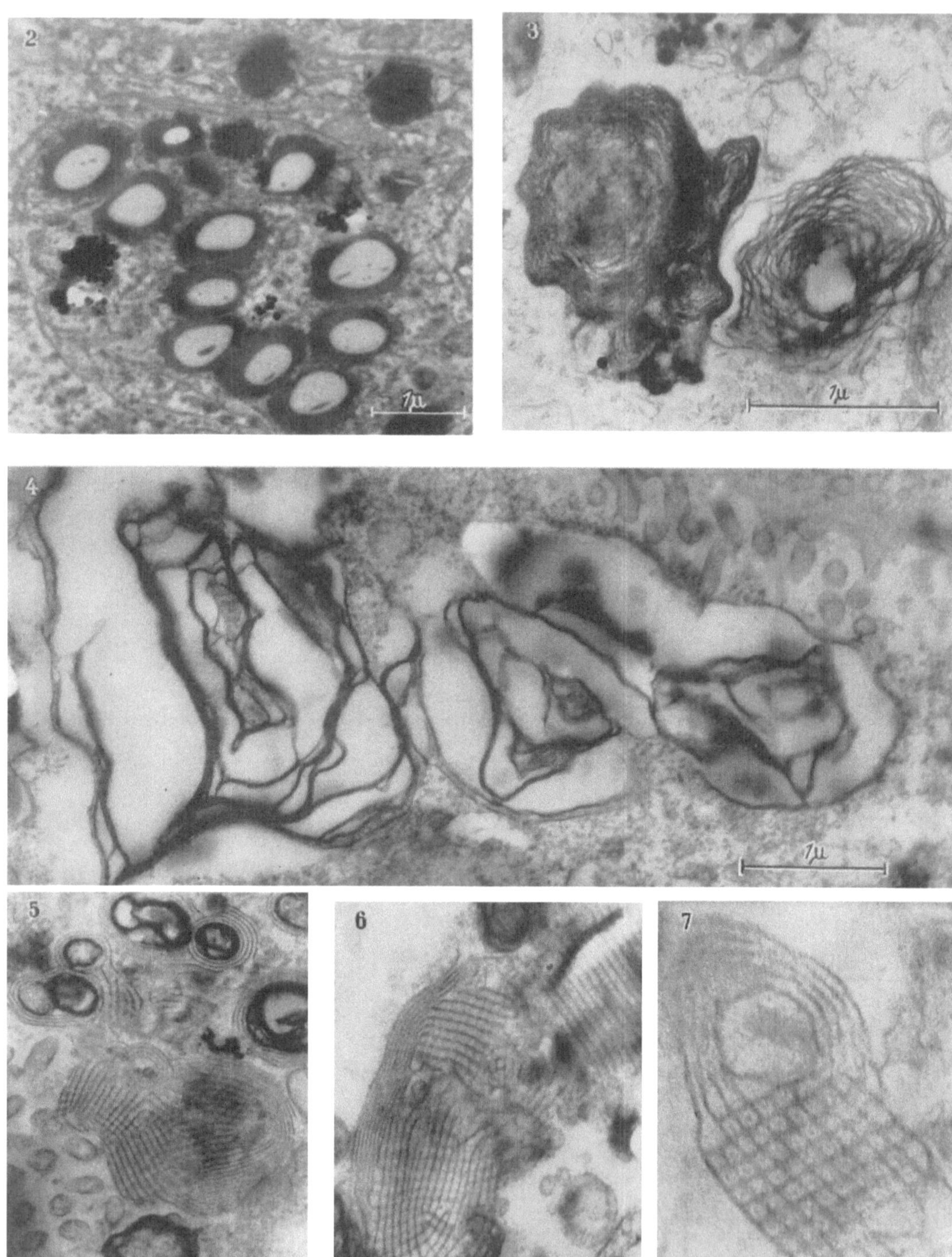

Fig. 2. Gouttelettes lipidiques avec vacuoles. Poumon (Rat) après injection de bacilles de Koch + silice colloïdale, 15 jours. Inclusion: méthacrylate de butyle

Fig. 3. Structures denses feuilletées. Poumon (Rat) après injection de silice colloïdale, 48 h. Inclusion: Vestopal

Fig. 4. Structures denses feuilletées avec lamelles espacées. Poumon (Rat) après injection de silice colloïdale, 7 jours. Inclusion: Vinox

Fig. 5. Figures en grille composite — Poumon (Rat) après injection de silice colloïdale, 48 h. Inclusion: Vestopal

Fig. 6. Figures en grille. Poumon (Rat) après injection de silice colloïdale, 7 jours. Inclusion: Vinox

Fig. 7. Figures en grille composite. Poumon (Rat) après injection de silice colloïdale, 21 jours. Inclusion: Vinox

et nous-mêmes (*7—8*). Le caractère physico-chimique principal de ces structures est leur capacité de gonfler.

3) Formations lamellaires concentriques fines: Ces formations feuilletées ont été précédemment décrites en détail (*2*) dans des lésions nécrotiques silicotiques expérimentales du poumon de Rat (Fig. 1). Elles sont constituées par de fines lamelles assez denses aux électrons de 50 à 60 Å environ d'épaisseur, continues, régulièrement disposées en couches parallèles, de 4 à 20 couches. Les espaces clairs mesurent approximativement 200 Å. Ces structures présentent quelques irrégularités de détail. L'ensemble mesure entre 0,2 et 1 μ, rarement plus. Ces structures ressemblent aux dispositifs de myéline des nerfs. Cependant, les feuillets de ces dernièrs sont plus serrés (100 Å) et plus fins. De telles figures ont été décrites dans des matériaux très différents: Stoeckenius (*9*) dans la Rate du Lapin, André et Rouiller (*10*) dans l'oocyte de l'araignée, Policard, Bessis et Breton-Gorius (*11*) au niveau de globules rouges phagocytés par des macrophages de la moelle osseuse, Charles (*12*) dans des faisceaux nerveux de la peau.

4) Figures en grille [ou en empreintes digitales («finger-print»]. Elles apparaissent soit isolées, soit associées à d'autres formations (Fig. 5, 6 et 7). Leur aspect est celui d'une grille ou d'un réseau, dessinés par des lignes denses aux électrons, d'épaisseur égale ou un peu supérieure à celle des structures concentriques, séparées par un espace de 500 Å en moyenne. Les mailles peuvent être carrées ou rectangulaires, avoir des angles droits ou quelconques. Parfois un petit corpuscule mal limité occupe les mailles. Certains réseaux s'assemblent avec des irrégularités. L'ensemble évoque des structures cristallines, avec des plans réticulaires.

5) Figures composites: Sauf les gouttelettes lipidiques, les formations ci-dessus sont le plus souvent composites. On observe très fréquemment l'association de lamelles fines concentriques, entourant des structures denses feuilletées ou une structure en grille (Fig. 5). D'autre fois une image en grille est entourée par des feuillets denses.

Cette description morphologique devra être complétée par une étude visant à la caractérisation chimique de ces structures. Il est toutefois difficile d'aller aujourd'hui plus loin en microscopie électronique, d'autant plus que nous connaissons encore mal les bases de l'osmiophilie (*13*). On rapprochera cependant ces images de figures obtenues expérimentalement avec des lécithines (*14*). Ces formations en tous cas apparaissent constamment en relation avec la dégénérescence cytoplasmique.

Bibliographie

1. Policard, A., A. Collet et S. Pregermain: C. R. Acad. Sci. (Paris) **246**, 3177 (1958).
2. — — — Bull. Micr. appl. **7**, 49 (1957).
3. Schlipköter, H. W.: Dtsch. med. Wschr. **79**, 1658 (1954).
4. Kisch, B.: Exp. Med. Surg. **13**, 101 (1955).
5. Karrer, H. E.: J. biophys. biochem. Cytol. **2**, 241 (1956).
6. Schulz, H.: Virchows Arch. path. Anat. **328**, 582 (1956).
7. Policard, A., A. Collet et S. Pregermain: Proc. Stockholm Conference on Electron Microscopy 1956, p. 244.
8. — — — Sem. Hôp. Paris **1957**, n° IV, 385.
9. Stoeckenius, W.: Exp. Cell Res. **13**, 410 (1957).
10. André, J., et Ch. Rouiller: J. biophys. biochem. Cytol. **3**, 977 (1957).
11. Policard, A., M. Bessis et J. Breton-Gorius: Exp. Cell. Res. **13**, 184 (1957).
12. Charles, A.: Exp. Cell. Res. **14**, 440 (1958).
13. Ornstein, L.: J. biophys. biochem. Cytol. **3**, 809 (1957).
14. Revel, J. P., S. Ito and Don W. Fawcett: J. biophys. biochem. Cytol. **4**, 495 (1958).

L'ultrastructure des thrombocytes du sang humain normal

R. Feissly, A. Gautier et I. Marcovici

Centre de Microscopie Electronique de l'Université de Lausanne (Suisse)

Les faibles dimensions des thrombocytes humains ont fait de ces éléments sanguins l'un des premiers sujets d'étude de la cytologie électronique. Depuis le premier travail de Wolpers et H. Ruska[1], les publications furent nombreuses: nous en avons récemment analysé plus de 80 (*1*).

---

[1] Voir les références bibliographiques complètes dans notre liste analytique (*1*).

Mais la fragilité des plaquettes rend difficile l'observation d'éléments non altérés. Par la technique des coupes minces, Bernhard et Leplus, Rinehart, Pease, Sueyasu et Tageshige, Feissly et coll., Schulz, Watanabe, Yamada et Goodman et coll. ont pu mettre en évidence les points suivants:

les thrombocytes, limités par une membrane semblable aux membranes cellulaires habituelles, possèdent un hyalomère non structuré et un granulomère complexe, constitué de mitochondries, de granulations denses et d'un ensemble d'éléments clairs polymorphes. Ces auteurs donnent des descriptions différentes de ce dernier groupe et l'assimilent parfois à divers organites comme l'ergastoplasme, le appareil de Golgi, etc.

La technique que nous avons mise au point précédemment (2) comporte une fixation immédiate et brève du sang complet dans un mélange acide osmique-Complexon III (Siegfried), suivie d'une surfixation du culot plaquettaire au mélange de Palade, puis d'une inclusion au polyester (actuellement: polyester Vestopal W.) Cette technique de routine nous permet de préciser maintenant la description de l'ultrastructure plaquettaire.

Nos préparations montrent un grand nombre de thrombocytes isolés ou accolés dont les dimensions moyennes des sections sont de 3.200 sur 2,100 mμ. Leurs pourtours sont parfois

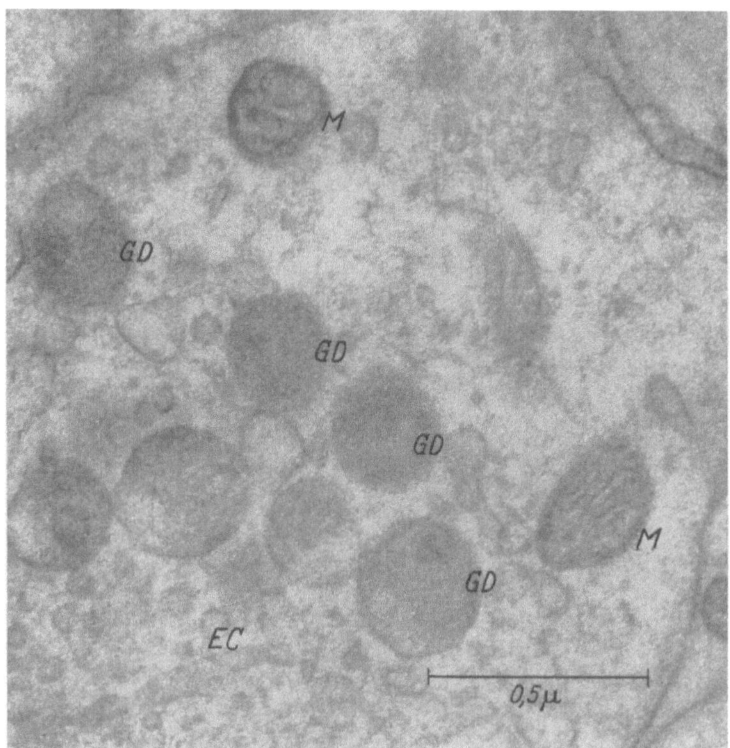

Fig. 1. Thrombocyte humain normal: mitochondries de petite taille (*M*), granulations denses (*GD*) et éléments clairs (*EC*). Observation au M. E. RCA. type EMU 3C, 1957. 50 kV

étrangement sinueux, mais on observe peu d'images de vrais pseudopodes. La finesse de la trame du hyalomère, ainsi que la concentration ou la dispersion du granulomère dans l'élément varient légèrement d'une préparation à l'autre. La membrane qui entoure le thrombocyte présente les caractères ultrastructurels habituels des membranes cellulaires: disposition en triple feuillet dense-clair-dense et images évoquant la présence éventuelle de pores. Son épaisseur varie de 5 à 50 mμ. — Nous avons prêté une attention particulière aux trois grands groupes de constituants du granulomère: mitochondries, granulations denses et éléments clairs.

Les mitochondries ont été décrites dans toutes les publications ci-dessus (sauf Goodman). En moyenne, nous en avons observé 1 à 2 par coupe d'élément (avec un maximum de 6). Les membranes et les cristae de ces organites ont la structure classique en triple feuillet. La très petite taille des mitochondries est frappante: les dimensions moyennes de leur section sont de 270 sur 220 mμ, alors que les dimensions moyennes des mitochondries du parenchyme hépatique, par exemple, sont de 1.100 sur 850 mμ (3). Cette si petite taille des mitochondries explique le faible nombre des cristae que chacune contient: généralement 2 à 3, comme l'indique Sueyasu. Cette petite taille explique aussi, sans doute, la difficulté de la mise en évidence du chondriome thrombocytaire au microscope optique et elle doit être mise en rapport avec le faible taux respiratoire des plaquettes (4).

Tous les auteurs soulignent la présence de *granulations denses*. Rinehart, Pease, Sueyasu et Watanabe les assimilent aux granulations azurophiles, ce que semble confirmer le travail récent de Hemmeler. Une membrane entoure souvent ces granulations qui contiennent, parfois, des masses internes plus denses. Elles représentent toujours le constituant le plus abondant du granulomère (avec un maximum de 27 par section de plaquette). Leurs profils circulaires ont un diamètre compris entre 80 et 320 mμ, valeurs très voisines de celles indiquées par Watanabe et Goodman.

Tous les autres constituants du granulomère — que nous groupons sous le nom d'*éléments clairs* — sont moins contrastés que les mitochondries et les granulations denses. Ils ont des dimensions et des aspects très divers (vacuoles, microvésicules, systèmes sacculaires, etc.), et montrent tant d'images intermédiaires entre ces différentes formes qu'une classification en sous-groupes distincts nous paraît absolument arbitraire. Leur périphérie est parfois couverte de granules noirs qui pourraient être des dépôts d'osmium.

L'attribution d'un élément précis à l'un des trois grands groupes de constituants du granulomère est parfois difficile: on observe, en effet, certains éléments qui présentent les caractéristiques de deux groupes différents. Les *images de transition* les plus fréquentes sont les formes intermédiaires entre mitochondries et granulations denses: Bernhard, Rinehart et Yamada y ont vu une origine mitochondriale possible de celles-ci. On remarque, mais moins fréquemment, des images de transition entre mitochondries et éléments clairs, comme entre ceux-ci et les granulations denses. Rinehart, Sueyasu, Feissly et Watanabe ont assimilé certains éléments clairs aux structures de *l'appareil de* Golgi: l'analogie morphologique, certes, peut être frappante, mais cette assimilation reste, pour l'instant, du domaine de l'hypothèse. Le reticulum endoplasmique *(ergastoplasme)* a été décrit dans les plaquettes par Pease et Yamada. Certains de nos clichés vont à l'appui de cette thèse, mais l'ergastoplasme n'est certes ni constant, ni important dans les thrombocytes.

Il est plus difficile encore d'affirmer à la suite de Pease et de Bernhard, l'existence de grains de Palade libres (granules de ribonucléoprotéines) dans le hyalomère, une identification morphologique étant extrêmement aléatoire en l'absence de certitude biochimique (4).

Le problème de l'origine mitochondriale des granulations denses ou des éléments clairs, malgré les aspects de transition nombreux entre ces trois groupes, ne nous semble pas actuellement pouvoir être tranché par l'étude submorphologique des thrombocytes du sang humain normal.

De même — si l'équivalence des granulations denses avec les granulations azurophiles de la cytologie classique s'impose — cette analyse ne nous permet pas actuellement *d'assimiler les granulations denses, ni les éléments clairs, à des groupes d'organites, tels qu'ils sont définis dans d'autres types cellulaires par la cytologie électronique.*

Bibliographie

1. Marcovici, I., et A. Gautier: Liste analytique des travaux de cytologie électronique traitant de l'ultrastructure des thrombocytes. Haematologica (Pavia) 1959.
2. Feissly, R., A. Gautier et I. Marcovici: Rev. Hématol. **12**, 397 (1957).
3. Gautier, A., J. Frei et H. Ryser: Ces C. R. p. 275.
4. Maupin, B.: Les plaquettes sanguines de l'homme. Paris: Masson et Cie 1954.

Elektronenmikroskopischer Nachweis von Strukturveränderungen des Thrombocyten während der Gerinnung

E. Kuhnke

Physiologisches Institut der Universität Bonn

Da die Zerfallsbereitschaft der Thrombocyten eng mit ihrer Funktion zusammenhängt, besteht großes Interesse an einer Kenntnis der Thrombocytenstruktur und an den morphologischen Veränderungen, die während der Gerinnung erfolgen (2).

Bei einer geeigneten präparativen Vorbehandlung wird der physiologische Thrombocytenzerfall erst mit dem Gerinnungsbeginn eingeleitet. Dieser Zeitpunkt ist daher leicht festzulegen. Die Unterbrechung dieses Vorganges kann jederzeit durch die Fixation erfolgen. Daher bestehen zur Erlangung eines für einen bestimmten Zeitpunkt nach dem Gerinnungsbeginn charakteristischen Materials günstige Voraussetzungen.

Als Ausgangsmaterial benutzten wir nur frisch gewonnenes Blut ohne jeden Zusatz von z. B. gerinnungshemmenden Stoffen. Die ersten Kubikzentimeter des nach der Gefäßpunktion ausströmenden Blutes wurden verworfen. Das in einem vorgekühlten Zentrifugenglas aufgefangene Blut wurde in einer Kühlzentrifuge sofort zentrifugiert. Nach Sedimentieren der Erythrocyten und Leukocyten wurde das überstehende thrombocytenreiche und gerinnungsfähige, aber noch *flüssige Plasma* auf mehrere Reagenzgläser verteilt. Während das Plasma in einem Glas sofort mit der Fixierlösung durchmischt wurde, verblieben die übrigen Gläser in einem Wasserbad von 37° C. Die Fixation jeweils eines der gebildeten Gerinnsel erfolgte zu verschiedenen Zeitpunkten der Gerinnung und Retraktion (physiologische Gerinnselverkleinerung). — Als Fixationsmittel benutzten wir 1%iges isotonisches Osmiumtetroxyd, 1%iges K-Bichromat und 1%iges Osmiumtetroxyd oder neutralisiertes Formalin. Bei der Entwässerung haben wir nach WOHLFARTH-BOTTERMANN (3) auf der Stufe des 70%igen Alkohols mit 1%iger Phosphorwolframsäure und 0,5%igem Uranylacetat kontrastiert. Einbettung: Methacrylat. Dünnschnitte: Porter-Blum-Mikrotom. Aufnahmen: Siemens Übermikroskop 100d

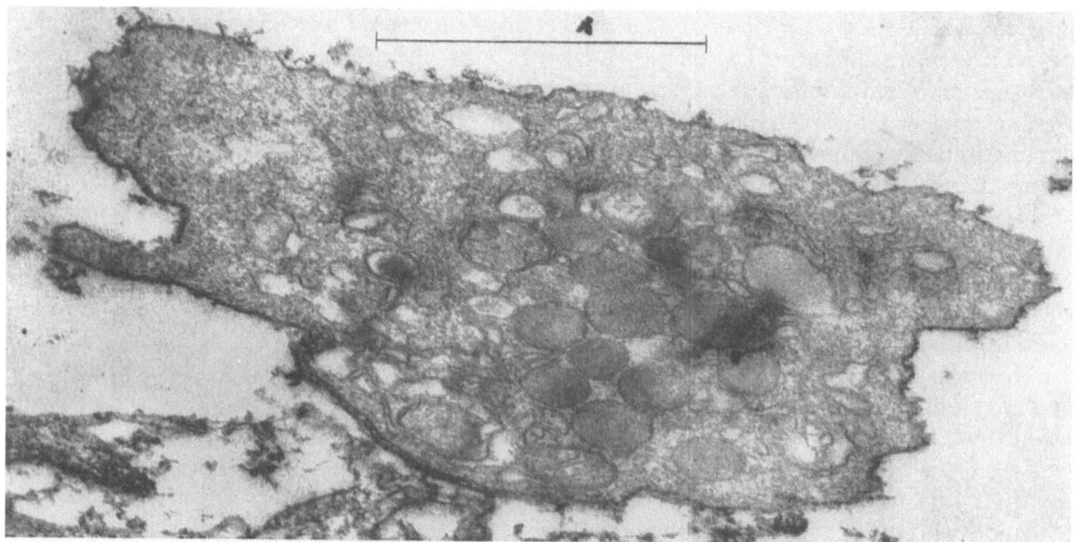

Abb. 1. Thrombocyt (Mensch) im Plasma vor der Gerinnung, Schnitt. E. V. 44000:1

Der noch *vor der Gerinnung*, also im flüssigen Plasma fixierte Thrombocyt (Abb. 1) hat eine relativ abgerundete, wenig gebuchtete Form. Nur an einigen Stellen sind Pseudopodienansätze erkennbar. In der Zelle liegen zahlreiche Granula, die in ihrer Gesamtheit das „Granulomer" bilden. Dem Granulomer wird, morphologisch und funktionell, das weiter in der Zellperipherie liegende „Hyalomer" (= Hyaloplasma) gegenübergestellt. Im Hyalomer liegen neben feinen membranartigen Strukturen kleine Granula. Außer den nicht sehr zahlreichen, aber fast in jedem Schnitt zu findenden Mitochondrien weist das Cytoplasma Vacuolen auf. Ihre Größe und Zahl ist sehr unterschiedlich und wahrscheinlich von dem augenblicklichen Funktionszustand der Zelle abhängig.

Kurz *nach der Gerinnung* zeigt der Thrombocyt (Abb. 2) zerklüftete Grenzen mit oft tief in den Zellkörper reichenden Buchten und zahlreiche Pseudopodien. Die Fibrinfibrillen lassen sich teilweise bis weit in den Thrombocyten hinein verfolgen oder liegen der Zelle bzw. den Pseudopodien an. Man gewinnt den Eindruck, daß diese Zellen in einem Stadium lebhafter Bewegung fixiert worden sind. Diese Ansicht wird durch die Untersuchungen von BORN (1) gestützt, nach denen der vor dem Beginn der Gerinnung außerordentlich hohe ATP-Gehalt, welcher als Ursache der „Zellruhe" und „Zellabrundung" angesehen wird, mit Gerinnungsbeginn sehr schnell abfällt.

Auf die Fibrillenbildung folgt bei Anwesenheit von überlebenden Thrombocyten im Gerinnsel die *Retraktion*. Ist sie abgelaufen, hat der Thrombocyt (Abb. 3) ein völlig verändertes Aussehen: Die Zellgrenzen fehlen weitgehend, die Granula des Granulomers vollständig. Vereinzelt sind noch

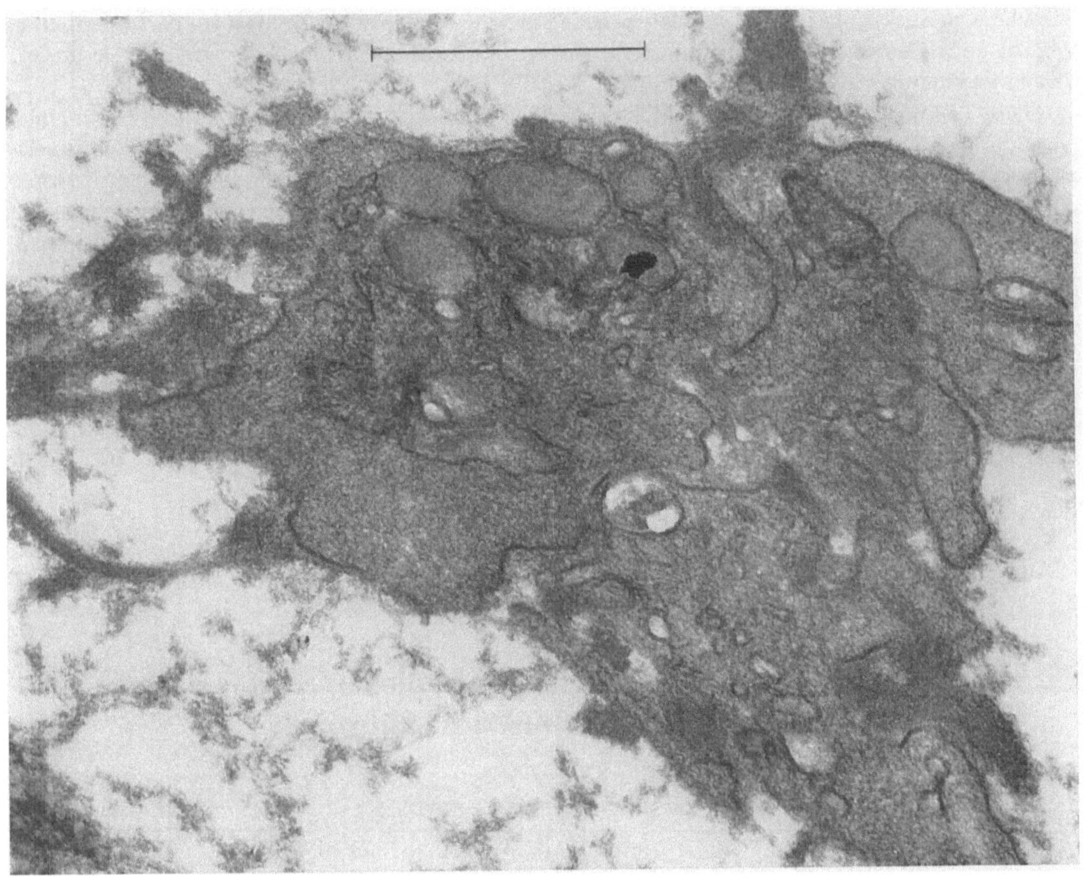

Abb. 2. Ein im Maschenwerk des Fibrins liegender Thrombocyt (Mensch) kurz nach der Gerinnung. Schnitt. E. V. 37000

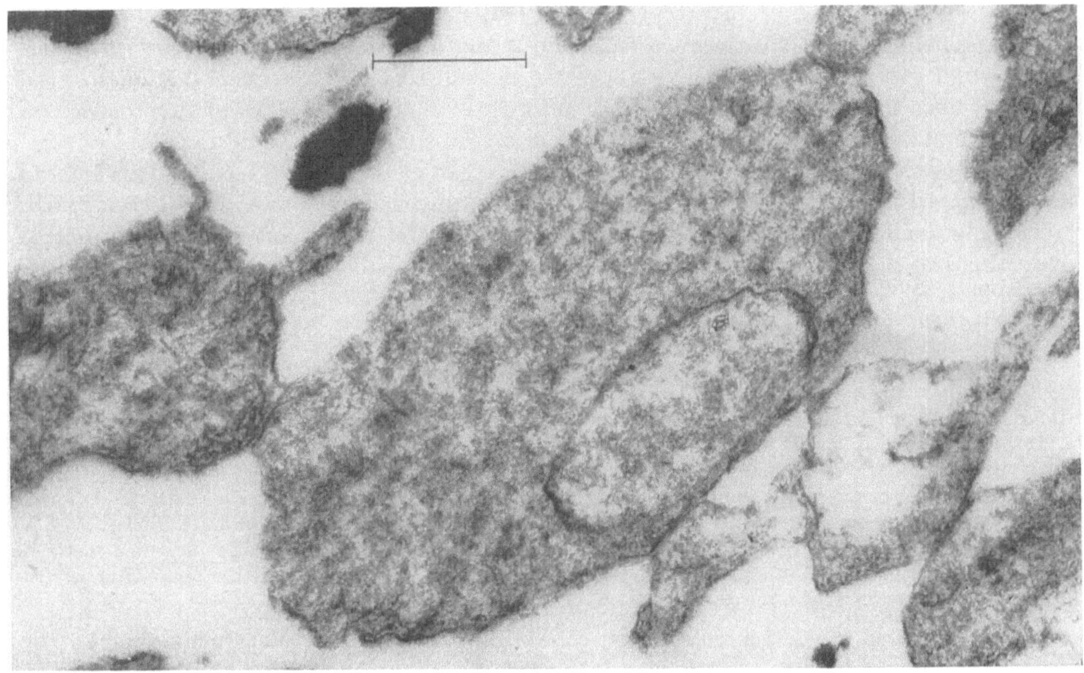

Abb. 3. Restthrombocyt nach beendeter Retraktion des Gerinnsels. Schnitt. E. V. 20000:1

Mitochondrien erkennbar. Auch das Cytoplasma weist nur noch Reststrukturen auf. — Im weiteren Verlauf wird die Restzelle immer massen- und strukturärmer. Schließlich sind im Gerinnsel Thrombocytenreste kaum noch auffindbar.

Diese Veränderungen der Form und Struktur der Thrombocyten bei ihrem auf die Gerinnung folgenden physiologischen Zerfall sind von den artefiziellen Zellzerstörungen, wie sie z. B. nach der Behandlung des Blutes mit isotonischen Zitrat-, Titriplex- oder ähnlichen Lösungen auftreten, scharf zu trennen. Bei den artefiziellen Zellzerstörungen kommt es neben der Bildung von großen Vacuolen zur Auswaschung des Cytoplasmas, während das Granulomer fester in der Zelle verankert zu sein scheint. Die beim physiologischen Zerfall schon zu einem frühen Zeitpunkt der Gerinnung nur noch an einzelnen Stellen nachweisbare Zellmembran erweist sich bei der Waschung als stabilste Struktur, die sich nach dem Verlust des Zellinhalts nicht selten zu im Schnitt spiralig aussehenden Bändern aufrollt.

Die Aufnahmen wurden im Zentrallaboratorium für angewandte Übermikroskopie (Leiter: Doz. Dr. Wohlfarth-Bottermann) gemacht. Der Deutschen Forschungsgemeinschaft danke ich für die Unterstützung.

Literatur

1. Born, G. V. R.: J. of Physiol. **133,** 61P (1956).
2. Kuhnke, E.: Pflügers Arch. ges. Physiol. **286,** 87 (1958).
3. Wohlfarth-Bottermann, K. E.: Naturwissenschaften **44,** 287 (1957).

Pleomorphic cytoplasmic inclusion bodies in tissue cultures of the otocyst exposed to dimycin

I. Friedmann

with the technical assistance of

E. S. Bird

Department of Pathology, The Institute of Laryngology and Otology, The British Postgraduate Medical Federation, University of London

The ototoxic action of streptomycin has been recognised ever since the introduction of this potent antibiotic into the treatment of tuberculosis and other infections. There have been numerous publications describing the results, often contradictory, of experimental studies of the site of the toxic action of streptomycin in the inner ear of animals. The effect of streptomycin on tissue cultures of fibroblasts and of the skin has also been studied.

The present observations, however, are based on organ cultures of the isolated fowl embryo otocyst exposed to the effect of Dimycin (Glaxo), a streptomycin-dihydrostreptomycin compound.

Previous studies of in vitro cultures of the otocyst (*1*) and subsequent electron microscopic observations of the ultrastructure of the in vitro developing otocyst (*2*) have shown that the isolated fowl embryo otocyst possesses great powers of self-differentiation both at light- and electron microscopical level.

The *method* of tissue culture here employed has already been described in the publications mentioned. Briefly the watch glass method developed by Fell and Robison (*3*) has been used, the isolated otocysts supported during cultivation on acetone treated Terylene voile strips, a modification of Shaffer's voile technique (*4*).

Dimycin was added usually 48 hr before the termination of the otocyst cultures on the eleventh or twelfth day in vitro. Previous investigations have suggested that the principal sensory areas of the otocyst have attained their completion or at any rate a high degree of differentiation between the eighth and twelfth day in vitro.

The cultures were fixed in 1% osmic acid at p_H 6 to 7,5 (buffered with veronal-acetate) containing 3% Dextrane and 0.25% sucrose (*5*) and embedded into methacrylate or araldite (*6*). The presented material consists of tissues embedded in araldite. The cytological changes were examined in thin sections under the electron microscope (Siemens Elmiscop I).

Cultures exposed to Dimycin in the final concentration of 1000 or 2000 microgrammes per ml for 24 to 48 hr showed various changes, some of which could be interpreted as retardation of differentiation, others as degeneration in character. Only the latter will be discussed here.

The degenerative changes have manifested themselves in the form of pleomorphic cystic bodies, "inclusion bodies", for want of a better term, of varied size and shape. The average size was similar to that of mitochondria and measured about 0.2 to 1 μ in diameter. These inclusion

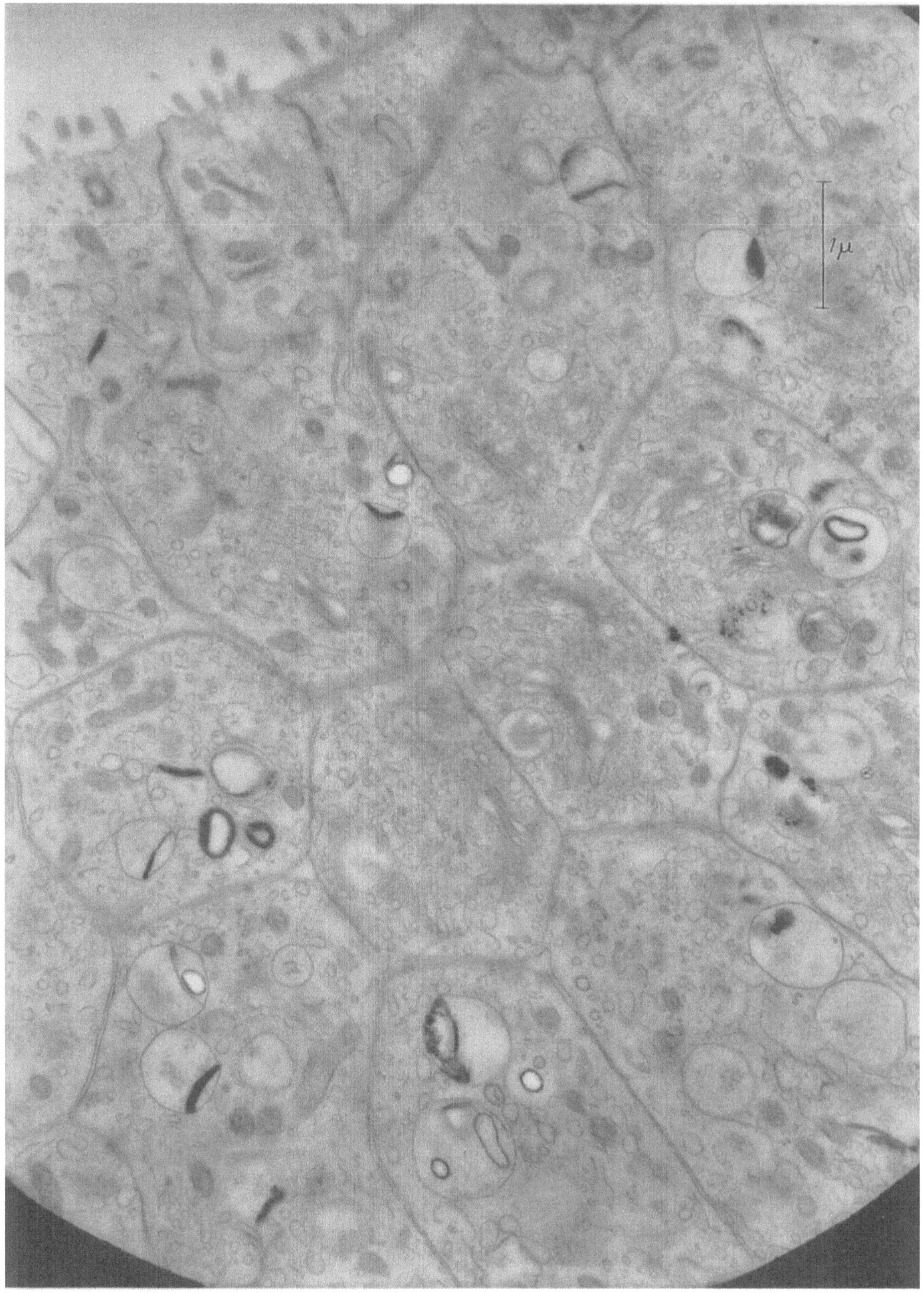

Fig. 1. Survey picture of sensory epithelium of in vitro culture of otocyst exposed to 1000 U/ml Dimycin (24 hr) showing large numbers of pleomorphic cytoplasmic inclusions. TM 19,500 $\times$

bodies were often lamellated and appeared to contain dense concentric lamellae or several layers of straight lamellae lining the inner wall of the cystic bodies. Many of these bodies were polycystic when they consisted of numerous small grapelike cysts. The dense lamellae would often form

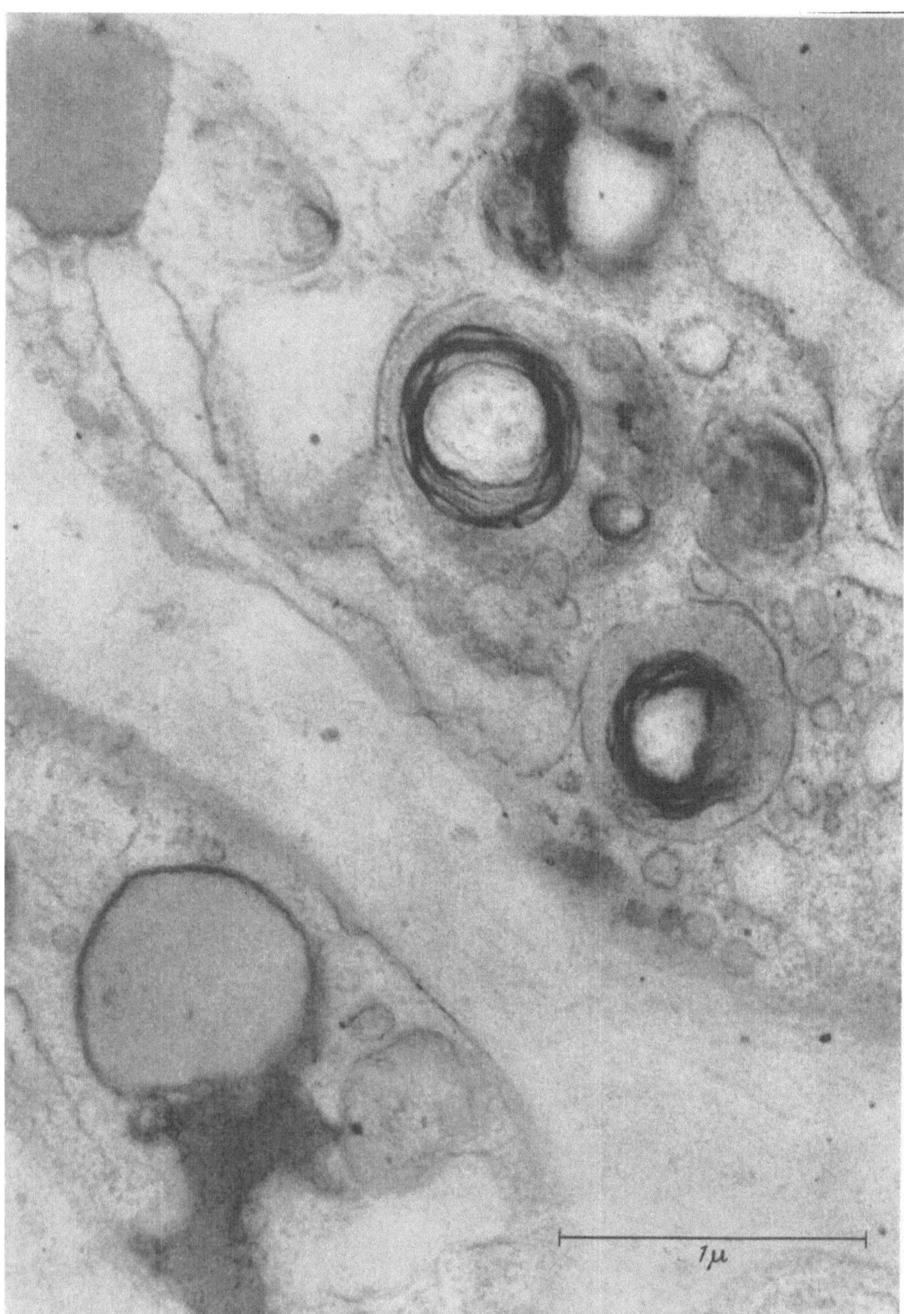

Fig. 2. Torcid lamellated inclusions in fibroblasts of in vitro culture of otocyst exposed to 1000 U/ml Dimycin for 24 hr. Note remnants of mitochondrial cristae. TM 42,000×

single ring-like annular structures inside the cystic body or skins producing serpiginous structures. Inclusion bodies were numerous in the sensory cells (Fig. 1) but also occurred, in fairly large numbers, in fibroblasts (Fig. 2) supporting the epithelial lining of the otocyst and also in the cartilage cells of the otic capsule. Degenerative changes have also been observed in nerve fibres of the richly innervated sensory areas of the otocyst but this aspect of the investigations requires further study. It might perhaps be mentioned in this place that other embryonic tissues such as embryonic liver exposed to the effect of Dimycin have also revealed similar inclusion bodies in thin sections.

Discussion

Pleomorphic cytoplasmic inclusion bodies have been described by several workers under various conditions which can be summarised in three groups: a) inclusions, microbodies, liposomes found in apparently normal tissues, b) in neoplastic cells: Hela, tumors, c) inclusions bodies formed in tissues exposed to the effect of various cytotoxic agents. The terminology of the structures is confusing and no attempt has been made to compile a complete list. Inclusion bodies, Microbodies, Dense Bodies have been described in the course of electronmicroscopical studies of lysoome-rich fractions of the rat liver by NOVIKOFF et al. (7). KARRER (8) described them in the normal mouse lung, ROUILLER and BERNHARD (9) studied microbodies in connection with mitochondrial regeneration in liver cells. Globoid or toroid inclusion bodies or microbodies have been noted in the cytoplasm of rapidly growing Hela cells by HARFORD and associates (10). Similar structures have been described by LAIRD (11) in a spontaneous transplantable mouse hepatoma and we have seen them occasionally in biopsy specimens of some human neoplasms (unpublished observations).

The exposure of cells to agents toxic to organelles may lead to similar changes (12). Thus DEMPSEY and WISLOCKI (13) observed the formation of silver aggregates in mitochondria in the cells of rats kept on drinking water containing silver nitrate, over a long period of time. WEISS (14) noted that intravitally injected neutral red and other dyes were segregated in the mitochondria of pancreatic cells causing swelling vacuolisation and disruption of the internal membranes. LEVER (15) induced mitochondrial changes with large doses of adrenocorticotrophic hormone. The development of polylaminated limiting membranes in the mitochondria of ACTH-treated hamsters (16) is of great interest. ANDERSON and VAN BREEMEN (17) described the disorganisation of the ultrastructure of spinal ganglion cells of the frog following the injection of malononitril. The increase in size of neurinal mitochondria coupled with a varying degree of pleomorphism was considered by these authors as evidence of functional change reflecting induced biochemical alterations.

In our own experience healthy tissue cultures of the otocyst or other organs of the fowl embryo do not show „inclusion bodies" except in very small numbers. When, however, exposed to the effect of Dimycin in the concentration described, large numbers of these structures make their appearance and can be found with no difficulty in thin sections. They can be more readily spotted in tissues embedded in araldite than in methacrylate sections. The action of streptomycin is not specific as the inclusion bodies have occurred in sensory cells, supporting cells, in fibroblasts and cartilage cells of the otocyst. As regards the cytoplasmic site of action the presented observations would suggest that both mitochondria and the endoplasmic reticulum may be affected. There was evidence of the change of mitochondria into "inclusion bodies" as well as of endoplasmic membranes undergoing swelling and morphological alterations. Polylamination of the contents was common and might indicate involvement of the Golgi membranes according to LEVER (16). On the whole the findings are in favour of LEVER's suggestion that liposomes are formed in degenerating mitochondria and not as BELT (18) suggested that mitochondria and liposomes may themselves be developed from some problematical microbodies.

References

1. FRIEDMANN, I.: Ann. Otol. (St. Louis) 65, 98 (1956).
2. — J. biophys. biochem. Cytol. 5. 263 (1959).
3. FELL, H. B., and R. ROBISON: Biochem. J. 23, 767 (1929).
4. SHAFFER, B. M.: Exp. Cell Res. 11, 244 (1956).
5. ENGSTRÖM, H., and J. WERSÄLL: Exp. Cell Res. 14, 414 (1958).
6. GLAUERT, A. M., and R. H. GLAUERT: J. biophys. biochem. Cytol. 4, 101 (1958).
7. NOVIKOFF, A. B., H. BEAUFAY and C. DE DUVE: J. biophys. biochem. Cytol. 2, Suppl., 179 (1956).
8. KARRER, H. E.: J. biophys. biochem. Cytol. 2, 241 (1956).
9. ROUILLER, C., and W. BERNHARD: J. biophys. biochem. Cytol. 2, Suppl., 355 (1956).
10. HARFORD, G. C., A. HAMLIN, E. PARKER and JAN. T. RAVENSWAY: J. biophys. biochem. Cytol. 2, Suppl., 347 (1956).

11. LAIRD, H. M.: Internat. Conf. of Cancer, London, July 1958, Abstracts, p. 133.
12. DEMPSEY, E. W.: J. biophys. biochem. Cytol. **2,** Suppl., 305 (1956).
13. — and G. B. WISLOCKI: J. biophys. biochem. Cytol. **1,** 99 (1955).
14. WEISS, J. M.: J. exp. Med. **101,** 203 (1955).
15. LEVER, J. D.: Amer. J. Anat. **97,** 409 (1955).
16. — J. biophys. biochem. Cytol. **2,** Suppl., 313 (1956).
      — Endocrinology **58,** 163 (1956).
17. ANDERSON, E. and V. L. VAN BREEMEN: J. biophys. biochem. Cytol. **4,** 83 (1958).
18. BELT, W. D.: J. biophys. biochem. Cytol. **4,** 337 (1958).

L'origine des mitochondries pendant le développement embryonnaire de Rana esculenta L

GIULIO LANZAVECCHIA

Institut de Zoologie et Anatomie comparée, Université de Milan, Italie

Les transformations que subissent les plaquettes vitellines ont été etudiées au microscope électronique, pendant les premiers stades du développement embryonnaire de la grenouille

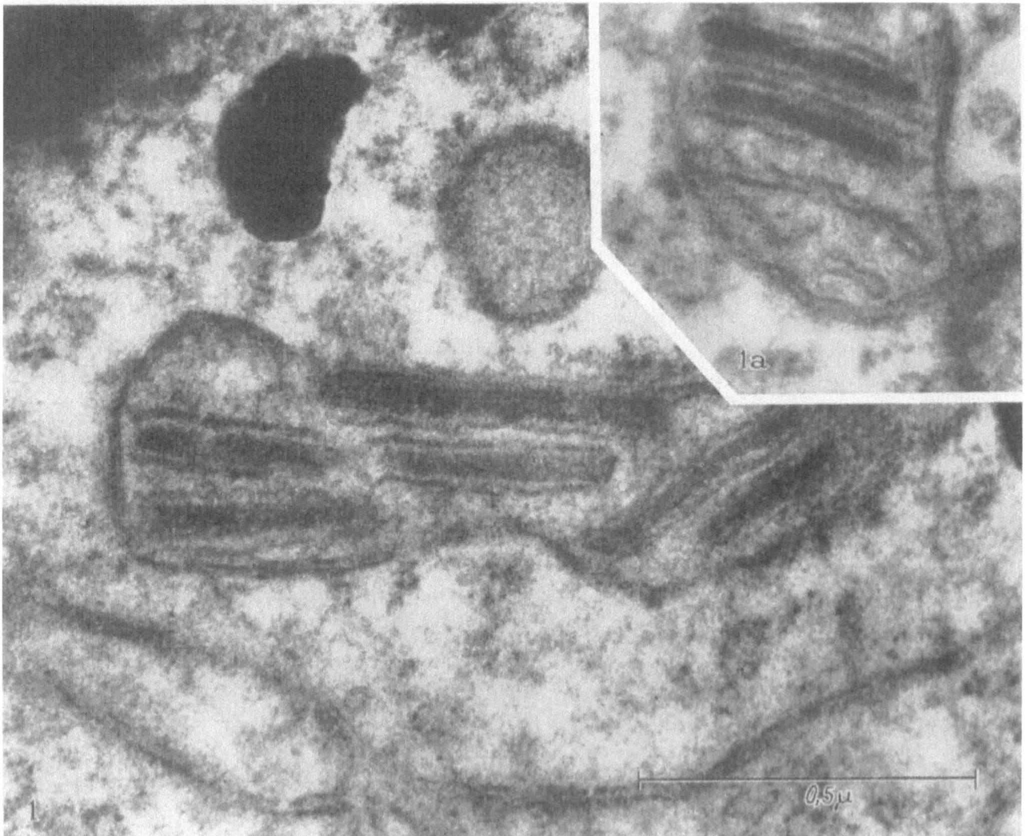

Fig. 1. Mitochondrie en voie de formation, avec des crêtes dilatées et remplies d'une masse de vitellus. 1 a—à l'interieur de deux crêtes, le vitellus est complètement réabsorbé. Magn.: 90 000:1

[du 14ème au 23ème de RUGH (*1*)], dans le but de déterminer leur sort. Le processus de démolition suit les mêmes modalités dans toute les cellules et les tissus observés, alors qu'il est nettement irrégulier dans le temps, d'un élément à l'autre. En effet, tandis que dans une même cellule il existe encore des plaquettes vitellines parfaictement intactes, d'autres apparaissent dans diverses étapes de leur processus de démolition.

Les fragments de tissus à étudier, fixés, déshydratés et enrobés selon les techniques ordinaires, ont été sectionnés avec l'ultramicrotome de HAANSTRA (2), et observés au microscope électronique Philips 75 kV.

Au microscope électronique, les plaquettes vitellines apparaissent comme des ellipses d'une forme assez régulière, entourées d'une unique membrane très mince, au-dessous de laquelle se trouve une couche à petits grains, d'un aspect plutôt irrégulier et d'une épaisseur variable. Le nucléus central de la plaquette, qui en constitue la plus grande partie, apparaît comme une masse opaque uniforme, qui dans des photographies à haute résolution est constituée par une

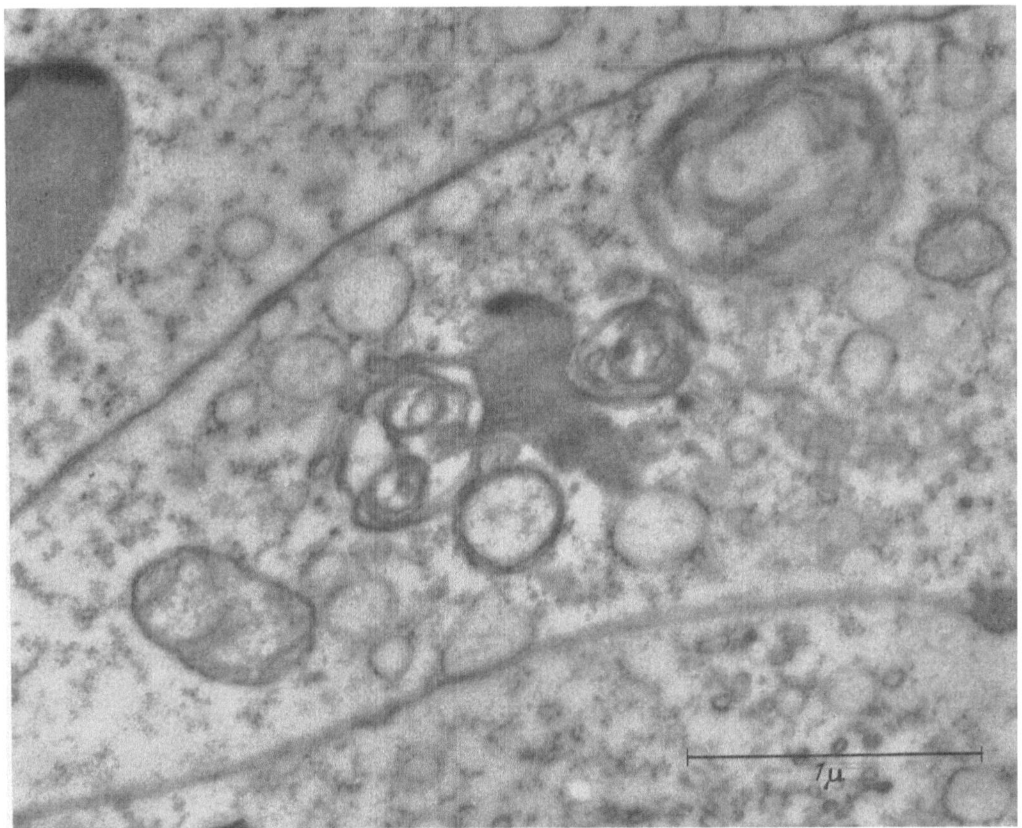

Fig. 2. Plaquette vitelline presque complètement exfoliée en lames enroulées d'une manière grossièrement concentrique. A gauche un mitochondrie presque complètement différencié. Magn.: 40,000:1

trame cristalline de lames minces d'environ 40 Å d'épaisseur, et distantes entre elles de 70 Å, d'un centre à l'autre (3, 4). Cette partie est constituée, du point de vue chimique, essentiellement par des lipoprotéines, tandis que la partie granulaire extérieure est très riche en acide ribonucléique.

La première partie des plaquettes vitellines qui subit un processus de démolition pendant le développement embryonnaire, est l'enveloppe extérieure de matériel granulaire qui se divise et va dans le cytoplasme sous forme de petits éléments presque ronds d'environ 150 Å de diamètre, et d'un aspect identique aux grains de PALADE.

Ensuite les processus commencent dans le matériel lipoprotéique, qui se brise en deux ou plusieurs parties d'une forme irrégulière, tandis que la membrane périphérique reste intacte. A partir de ce moment, la démolition des plaquettes vitellines suit deux chemins différents, qui aboutissent tous les deux au même résultat, c'est-à-dire la formation de mitochondries.

Dans le premier mécanisme de formation, c'est la plaquette vitelline elle-même qui se transforme directement en un mitochondrie, en perdant la plus grande partie de son propre matériel et en subissant une série de remaniements morphologiques. Dans la plaquette il se vérifie une perte graduelle de matériel lipoprotéique, de telle façon qu'il se forme un espace très transparent aux électrons au-dessous de la membrane périphérique. En outre celle-ci, qui dans les plaquettes vitellines est simple, devient double, en prenant ainsi un aspect identique à celui de la membrane des mitochondries.

Par la suite les dimensions de la plaquette vitelline continuent à diminuer à cause d'une perte ultérieure de matériel lipoprotéique, alors que à l'intérieur les crêtes mitochondriales se différencient, d'abord sous forme de lames à disposition irrégulière, ou grossièrement concentriques,

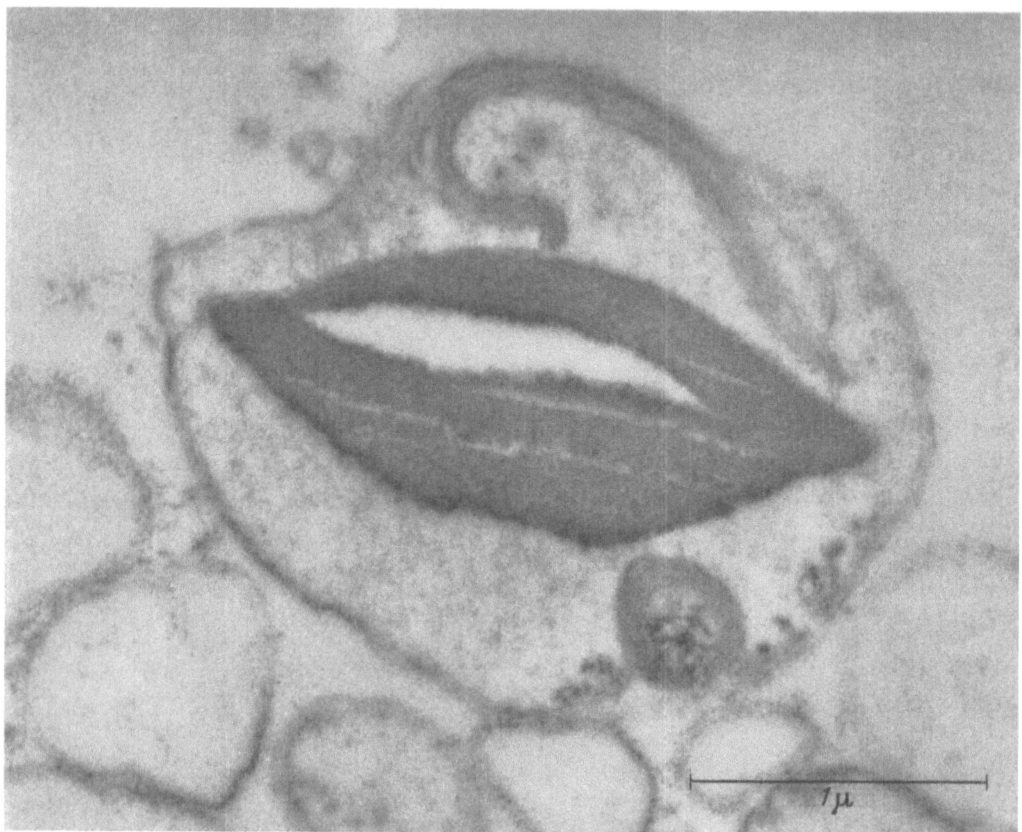

Fig. 3. Plaquette vitelline dont une grande partie de matériel lipoprotéique a déjà été perdu. En haut un faisceau de lames lipoprotéiques se sont séparées de la masse centrale de vitellus. En bas un corps formé par des lames enroulées concentriquement, passant à travers la membrane périphérique, va dans le cytoplasme. Magn.: 40000:1

puis selon la morphologie typique des mitochondries différenciés. Souvent quelques parties de vitellus, d'une forme plus ou moins discoïde, restent enfermées dans les lames qui tendent devenir des crêtes mitochondriales; c'est pourquoi il est facile d'observer des mitochondries en voie de formation, et déjà anvancés dans le processus de différenciation, avec des crêtes dilatées et remplies d'une masse de vitellus (Fig. 1, 1 a). Par la suite, ce matériel est complètement réabsorbé, et enfin on a des mitochondries d'un aspect normal, c'est-à-dire répondants aux descriptions communes.

Le second processus de formation des mitochondries commence par une espèce d'exfoliation laminaire du vitellus, avec une formation de corps constitués de lames très minces, enroulées d'une manière grossièrement concentrique, qui vont dans le cytoplasme en abandonant le résidu de la plaquette vitelline. D'ordinaire, dans ce cas la membrane de la plaquette se dissout complètement, et toute la plaquette vitelline se désagrège par ces formations laminaires (Fig. 2), lesquelles se transforment ensuite en mitochondries, en se regonflant dans le cytoplasme et en organisant leurs lames lipoprotéiques.

Les deux processus peuvent parfois avoir lieu en même temps, c'est-à-dire qu'une unique plaquette vitelline peut donner origine à plusieurs mitochondries, selon les deux chemins indiqués. Alors que la membrane périphérique de la plaquette est déjà devenue double, et une grande partie du vitellus à l'intérieur a été éliminée, on peut observer des formations laminaires de vitellus, enroulées concentriquement, qui en passant à travers la membrane périphérique, vont ensuite dans le cytoplasme en formant des mitochondries, en conformité de la deuxième méthode illustrée, tandis que la plaquette vitelline, évoluant ultérieurement, va former elle même un mitochondrie, selon la méthode directe (Fig. 3).

Bibliographie

1. Rugh, R.: Experimental Embryology. Burgess publishing company. Minneapolis 1948. p. 61.
2. Haanstra, H. B.: Rev. techn. Philips **17**, 195 (1955).
3. Wischnitzer, S.: J. biophys. biochem. Cytol. **3**, 1040 (1958).
4. Karasaki, S., and T. Komoda: Nature (Lond.) **181**, 407 (1958).

The mitochondria in human normal and cholestatic liver

R. Ekholm and Y. Edlund

Department of Anatomy and Department of Surgery II, University of Gothenburg (Sweden)

Extrahepatic biliary obstruction is accompanied by reduced glycogen content of the liver cells, probably due to impaired glycogenesis as well as enhanced glycogenolysis. The impaired glycogenesis is most likely caused by reduced ATP synthesis, in turn resulting from faulty

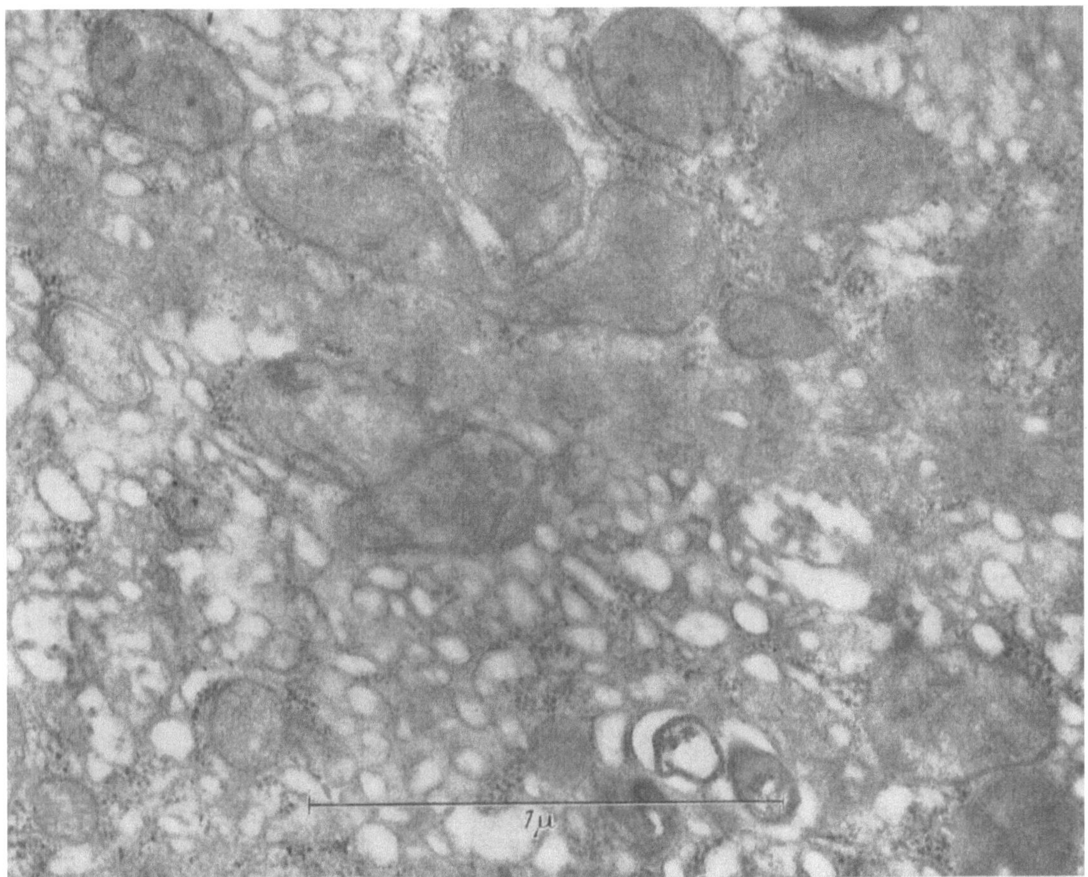

Fig. 1. Mitochondria from a normal human liver. The mitochondria are bordered by double edged outer membranes and contain a few double edged inner membranes. — Magnification 63,000 ×

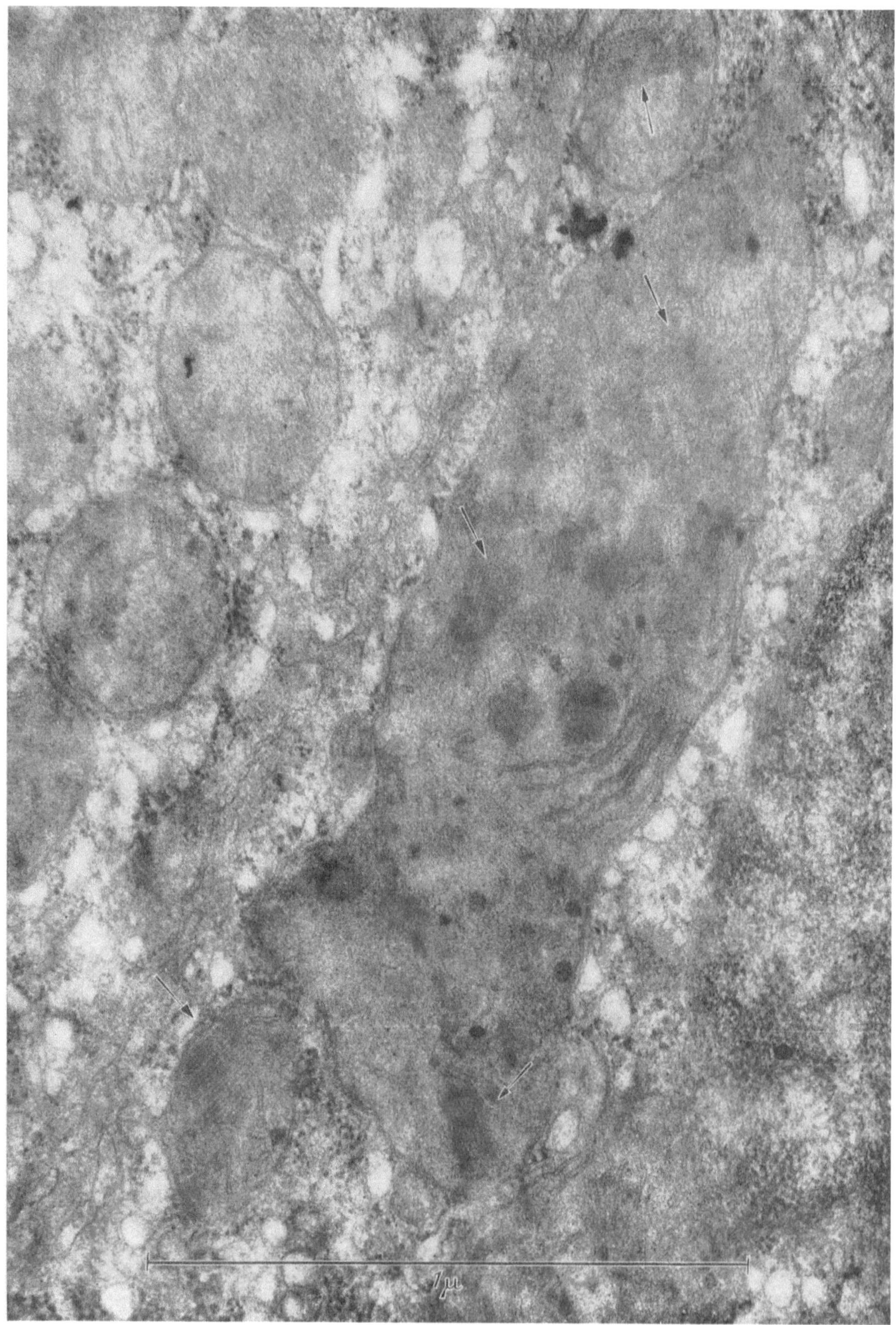

Fig. 2. Mitochondria from a human liver in biliary obstruction. One mitochondrion is very large and contains, in addition to double edged inner membranes, stripy regions (at the arrows). Two of the mitochondria of normal size also have well developed stripy regions. — Magnification 95,000 ×

carbohydrate dissimilation, probably owing to disturbance at some point in the isocitric acid cycle (*1*). Considering that the enzymes of the Krebs' cycle seem to be more or less completely located in the mitochondria it seemed justified to ascertain whether biliary obstruction gives rise to morphological changes in liver mitochondria.

Specimens from 5 persons with normal livers and 5 patients with biliary obstruction were studied. The specimens were taken from the left liver lobe immediately after the abdominal cavity had been opened and not more than about 5 min after induction of anesthesia. The specimens were fixed in 1% osmic acid, dehydrated in alcohol, embedded in methacrylate, sectioned on an ultramicrotome, and examined in an RCA EMU 3b microscope.

A survey picture of a normal human liver shows a large number of round, oval or somewhat irregular shaped mitochondria. They are of comparatively uniform size and only exceptionally their longest axis exceeds 1 micron. At a higher magnification the mitochondria in the normal liver cells (Fig. 1) are seen to be bordered by an outer double edged membrane composed of two dark lines separated by a less opaque space. In the interior of the mitochondria a system of inner membranes is observed. These membranes, double edged like the outer ones, are usually fairly few and of highly varying length and direction.

Survey pictures of mitochondria from cholestatic livers give a more polymorphous impression because mitochondria of normal size and shape are mingled with a number of very large mitochondria. The majority of these are extremely elongated and some are irregularly rounded. The major axis of these giant mitochondria is often 3 microns long. In biliary obstruction all the mitochondria in liver cells are, like those in normal liver cells, surrounded by a double edged surface membrane and contain short inner membranes. In addition to these normal structural details, it is found in a large percentage of the former mitochondria a stripy region never observed in normal livers (Fig. 2). This region has alternate dark and light, usually very straight parallel lines. The number of lines varies within wide limits but usually 10—20 dark lines can be counted. The thickness of the dark and light lines is about the same and varies between 50 and 100 Å. Met with in all types of mitochondria, this stripy region is most common and best developed in the elongated giant ones. Such mitochondria may contain two or even more stripy regions. In elongated mitochondria the stripy region is often oriented with its lines parallel to the long axis. The appearance of the stripy region brings to mind the submicroscopic structure of a crystalloid.

We presume that the observed morphological divergencies between the mitochondria from normal livers and those from cholestatic livers reflect biochemical differences.

References: *1.* Edlund, Y.: Acta chir. scand. **96,** suppl. 136 (1948).

Essais d'estimation quantitative des variations morphologiques des mitochondries hépatiques au cours d'une carence vitaminique

A. Gautier, J. Frei et H. Ryser

avec la collaboration technique de

V. Fryder

Centre de Microscopie Electronique de l'Université et
Laboratoire de Biochimie de la Clinique Médicale Universitaire Lausanne (Suisse)

Les variations morphologiques des mitochondries ont fait l'objet de plusieurs travaux récents de cytologie électronique. On a cherché à relier ces variations aux modifications du métabolisme, particulièrement dans le cas de perturbations biochimiquement localisées au niveau de la «fraction mitochondriale». Plusieurs auteurs ont cherché notamment à établir une relation entre la chute du taux d'oxydation phosphorylante et la présence d'un nombre important de mitochondries tuméfiées. Cette «tuméfaction» se caractérise par un gonflement des mitochondries, par la diminution en nombre et en longueur des cristae et par leur reflux vers la périphérie des mitochondries. Or, deux d'entre nous ont montré (*1, 2*) qu'une carence vitaminique B_1 expérimentale totale produit, au cours de la 4ème semaine, une diminution brutale de la résynthèse oxydative de l'ATP, de la respiration et du quotient de l'oxydation phosphorylante dans les homogénats hépatiques de rats. Le phénomène est réversible *in vivo*: après injection intrapéritonéale de vitamine B_1 ou de cocarboxylase, les animaux retrouvent leur pouvoir de phosphorylation. Par contre, *in vitro* l'adjonction de ces facteurs aux homogénats hépatiques de rats carencés est totalement inopérante. D'autre part, l'analyse néphélométrique des suspensions de mitochondries de rats carencés, selon la technique de RAAFLAUB, montre un *gonflement* caractéristique. Il semble donc que l'avitaminose B_1 soit la cause non seulement d'un trouble de la décarboxylation oxydative, mais d'une véritable lésion des mitochondries.

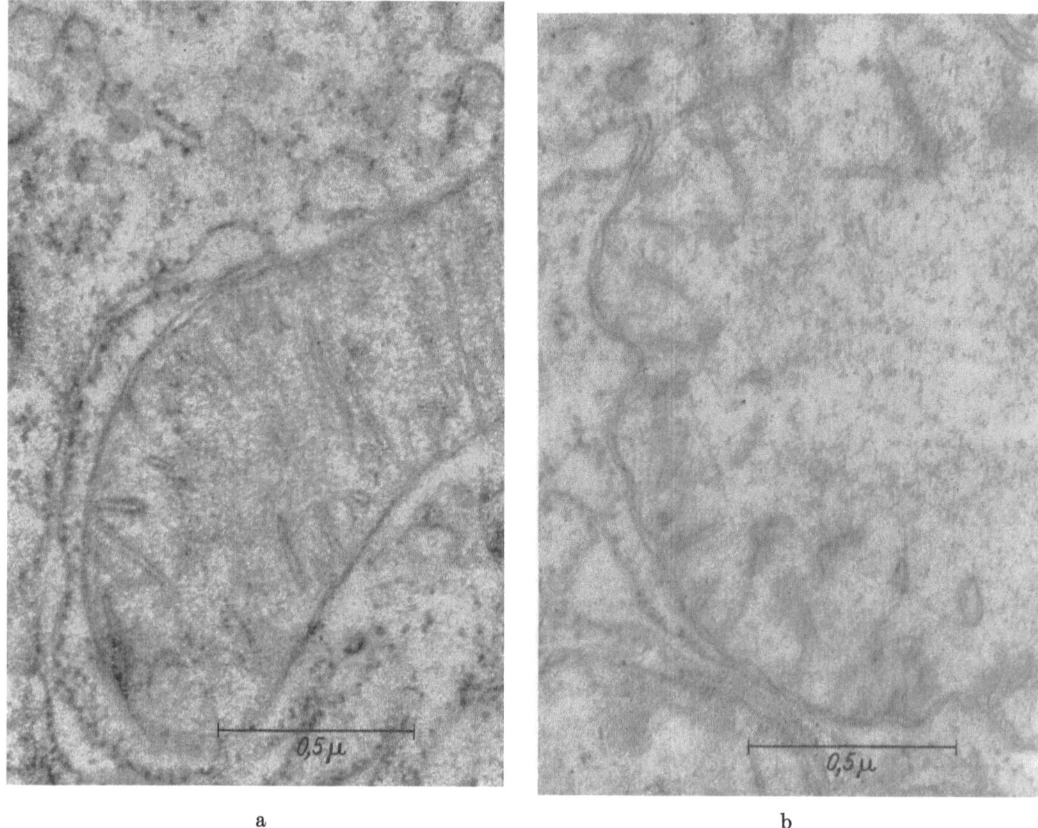

a   b

Fig. 1 a et b. *Mitochondries du parenchyme hépatique.* a) rat témoin, aspect habituel, b) rat en état de carence avancée (4 semaines): image de «tuméfaction»

Technique — 19 jeunes rats provenant d'un ensemble homogène, mis à notre disposition par F. Hoffmann-La Roche et Cie, ont été sacrifiés à divers stades (1, 2, 3 et 4 semaines) de carence vitaminique B_1 totale. 4 autres animaux ont servi de témoins. Mis à jeûn 16 à 24 h. avant le sacrifice; prélèvement immédiat; fixation au mélange de PALADE 2 h.; inclusion au polyester Vinox; observation au microscope électronique RCA, type EMU 3C.

Observations qualitatives. — Nous avons fréquemment trouvé, chez les témoins, des images caractéristiques de mitochondries normales (Fig. 1 a) et chez les animaux carencés, de mitochondries tuméfiées (Fig. 1 b). On observe donc bien une variation de la morphologie des mitochondries, mais aucune image d'altération aussi prononcée que dans d'autres cas pathologiques connus (*3, 4*) dont nous n'avons jamais trouvé l'équivalent chez ces animaux sub mortem. Mais, ces animaux carencés montrent également des mitochondries d'aspect normal: ROUILLER (*5*) cite des observations semblables dans divers cas de pathologie hépatique.

Observations dimensionnelles. — 11 animaux étudiés statistiquement ont été répartis en trois groupes:

groupe T: témoins   4 rats
groupe D: début de carence (1 à 3 semaines)   3 rats
groupe A: carence avancée (4 semaines)   4 rats

Dans chaque groupe, l'analyse porte sur près de 300 mitochondries. Pour estimer la variation de volume des mitochondries, leur tendance à la sphérisation et la variation de surface des cristae, nous avons choisi les critères déjà utilisés par SCHULZ et coll. (4): longueur et largeur des sections

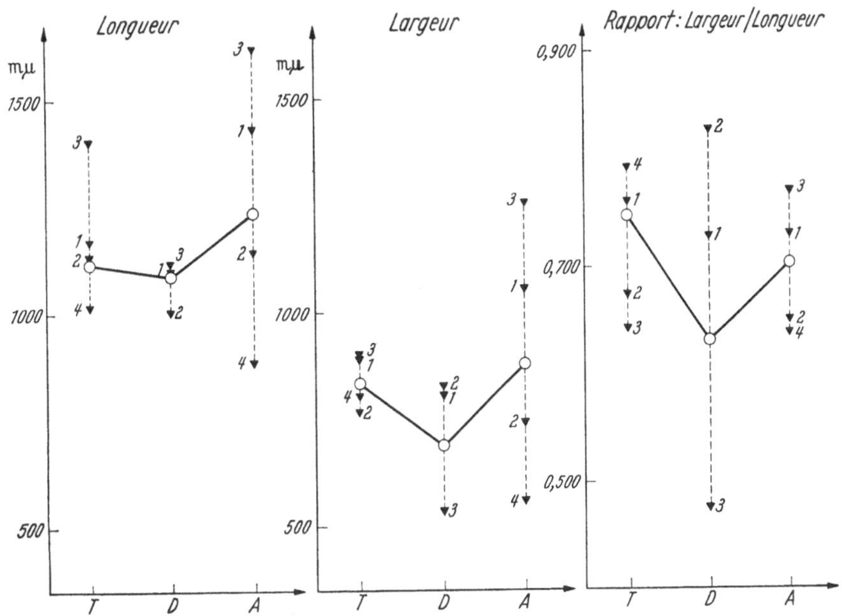

Fig. 2. *Schéma 1. Dimensions des mitochondries.* Dispersion des valeurs moyennes par animal (▼) et par groupe (O) d'animaux témoins (*T*), en début de carence (*D*) et en carence avancée (*A*)

de mitochondries et quotient de ces deux valeurs (Fig. 2, schéma 1), nombre de cristae par mitochondrie et longueur des cristae (Fig. 3, schéma 2). Ces deux schémas donnent pour chaque animal des trois groupes la moyenne arithmétique et pour chaque groupe T, D et A, la moyenne arithmétique (pondérée) des 5 critères choisis. L'ensemble des mesures porte sur 987 mitochondries comprenant 30.879 cristae. *La dispersion des résultats au sein de chaque groupe est frappante.*

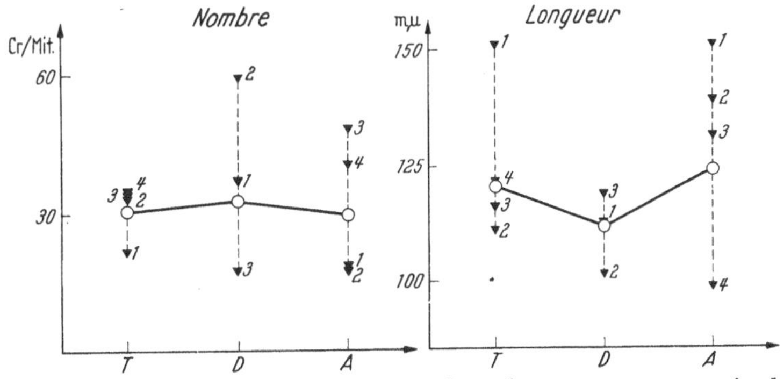

Fig. 3. *Schéma 2. Nombre et longueur des cristae.* Dispersion des valeurs moyennes par animal (▼) et par groupe d'animaux (O)

Les variations de groupe à groupe — inférieures d'ailleurs à celles indiquées par SCHULZ et coll. dans l'hyperthyréose expérimentale (4) — sont *négligeables par rapport à la dispersion à l'intérieur des groupes.* Cette dispersion des résultats nous a retenu d'étudier statistiquement l'épaisseur des cristae, ainsi que les dimensions des feuillets denses et clairs qui constituent l'ultrastructure des membranes et des cristae mitochondriales.

Discussion. L'avitaminose expérimentale a été menée avec les soins habituels dont l'analyse biochimique a montré la régularité des résultats (marge de sécurité de l'ordre de 99%). Le groupe A présentait tous les signes caractéristiques de la crise aigüe de béri-béri. On a voué, d'autre part, la plus grande attention à ce que chaque échantillon pour la cytologie électronique soit préparé «dans des conditions identiques». Les diverses erreurs de mesures — dues à la séparation plus ou moins fine des clichés, aux imprécisions de mesures des grossissements électroniques, à l'influence de l'épaisseur des coupes, etc. — nous semblent statistiquement négligeables. Les erreurs dans la détermination des cristae (longueur minimale observée: 8,5mμ) sont inévitables, mais corrigées par le nombre important des mesures (> 30.000), et la faible variation finale de groupe à groupe montre bien le peu d'influence de ces erreurs. Notre choix volontairement arbitraire des mitochondries étudiées néglige la localisation histologique du prélèvement, ainsi que leur localisations intralobulaire et intracellulaire. Mais, à notre connaissance, on n'a jamais indiqué une *atteinte localisée* du béri-béri dans le foie, et nos observations optiques et électroniques ne nous l'ont pas fait suspecter.

Conclusions. — Si l'on trouve *dans l'avitaminose B$_1$ expérimentale* (poussée à l'extrême limite de la résistance des animaux) des images de «tuméfaction» de mitochondries, on ne trouve pas cependant la confirmation de la lésion mitochondriale prévue par l'analyse biochimique. *Dans le cas étudié*, aucune variation morphologique statistiquement mesurable ne correspond à la chute particulièrement brutale du taux d'oxydation phosphorylante. Cette chute ne peut aucunement être mise en rapport avec une variation de volume des mitochondries, ni avec une variation de surface des cristae, ce que pouvait seule démontrer une analyse statistique portant sur un nombre suffisant de mesures. La dispersion des valeurs moyennes entre rats de même groupe laisse supposer que le caractère «tuméfaction» des mitochondries est la représentation de phénomènes chimiques bien plus complexes que l'on ne l'a admis jusqu'ici.

Nous tenons à remercier ici le Prof. A. Vannotti pour l'intérêt qu'il a porté à cette étude.

Bibliographie

1. Frei, J.: Helv. physiol. Acta **14**, 59 (1956).
2. Frei, J., et H. Ryser: Experientia (Basel) **3**, 105 (1956).
3. Gansler, H., et C. Rouiller: Schweiz. Z. Path. Bakt. **19**, 217 (1956).
4. Schulz, H., H. Löw, L. Ernster u. F. S. Sjöstrand: Proc. Stockholm Conf. E. M., 134 (1956).
5. Rouiller, C.: Ann. Anat. path. **2**, 548 (1957).

Untersuchungen an isolierten Mitochondrien

J. Kiendl und G. Schimmel

Battelle-Institut, Frankfurt am Main

Mitochondrien wurden nach verschiedenen Verfahren isoliert und elektronenmikroskopisch untersucht. Die besten Ergebnisse lieferte das modifizierte Verfahren von Birbeck und Reid, mit dem es gelang, isolierte Mitochondrien ohne Einbettung bei der Präparation intakt zu halten. Es wurden Versuche angestellt, ob es möglich wäre, isolierte Mitochondrien unfixiert zu trocknen und wieder aufzuschwemmen. Die Ergebnisse wurden diskutiert.

Siehe: Mikroskopie **14**, 1 (1959).

Vergleichende Untersuchungen isolierter Mitochondrien nach Kieselsäure-Inkubation

H. W. Schlipköter, Hj. Staudinger, K. Krisch und J. Lehmann

Institut für Hygiene und Mikrobiologie der Medizinischen Akademie Düsseldorf
und Hauptlaboratorium der Städt. Krankenanstalten Mannheim

1954 konnten wir berichten (*6,7*), daß die Mitochondrien in den Zellen von Quarzgranulomen geschwollen sind. Eine solche Mitochondrienschwellung ließ sich auch nach Injektion von amorpher Kieselsäure beobachten (*7,8*). Dieser Befund, von Policard u. Mitarb. (*3*) und Klosterkötter

u. Mitarb. (2) bestätigt, veranlaßte uns, Überlegungen darüber anzustellen, ob dieser Schwellungsvorgang unspezifisch oder ein Zeichen für die Schädigung enzymatischer Mitochondrienfunktionen durch die Kieselsäure ist. Zwischen der Struktur von Mitochondrien und ihrer biochemischen

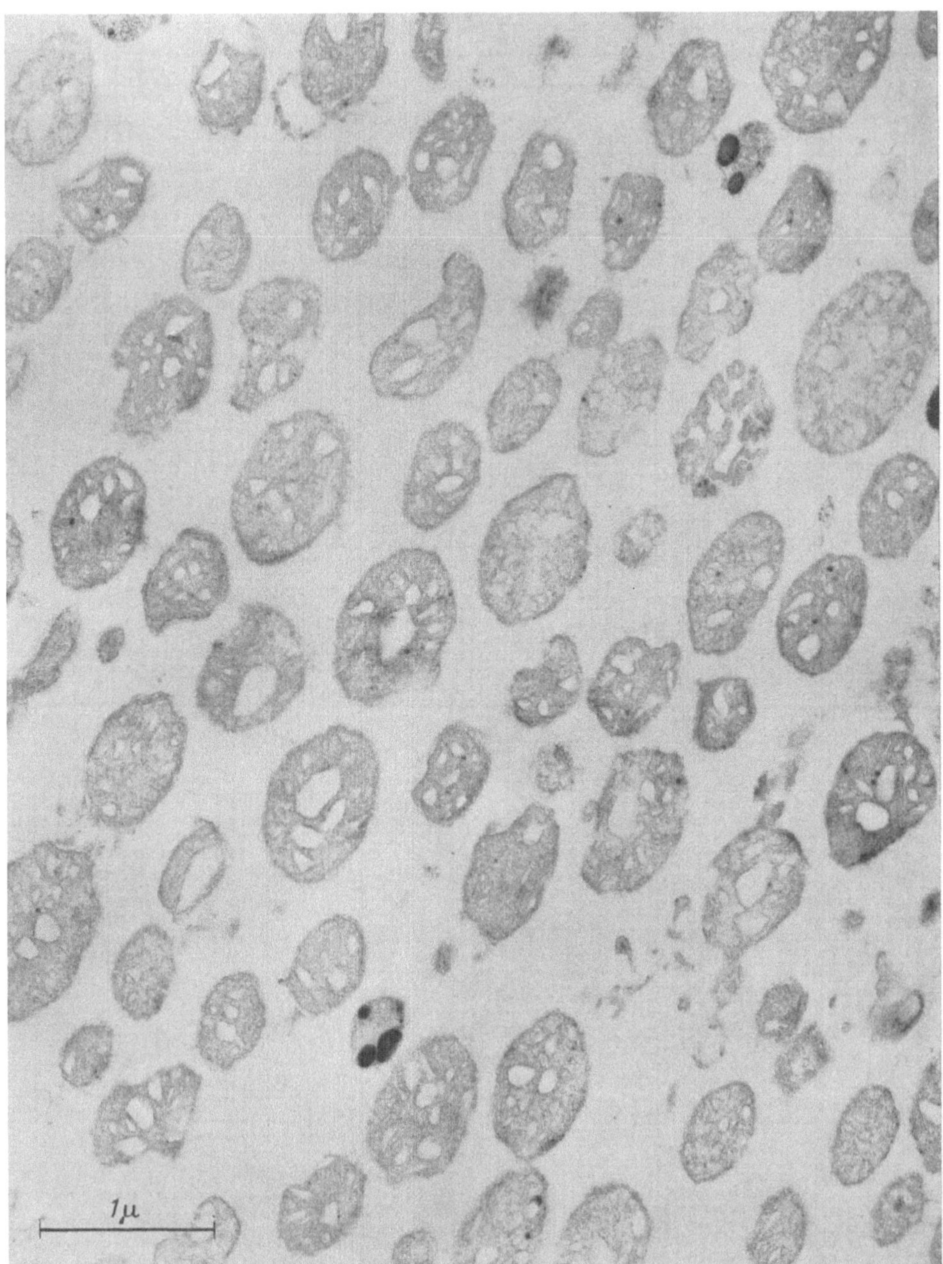

Abb. 1. Isolierte Mitochondrien aus Rattenleber. Arch.-Nr. 5029/58

Aktivität bestehen Wechselwirkungen. Es sind Faktoren bekannt, die eine Aufnahme von Wasser in den intramitochondrialen Raum fördern. Dazu gehören z. B. Thyroxin, Trijodthyronin, Calcium, p-Chloromercuribenzoat, aber auch längeres Stehenlassen der Mitochondrien bei Zimmertemperatur. Es ist eine Aufgabe der Mitochondrien, die bei der Oxydation frei werdende Energie in Form energiereicher Phosphatverbindungen zu speichern. Alle Faktoren, die zu einer Mitochondrienschwellung führen, schädigen die oxydative Phosphorylierung und verhindern damit die Energietransformation. Staudinger, Kersten und Krisch (5) haben festgestellt, daß in Gegenwart von 10 mg-% Kieselsäure sich eine verminderte Sauerstoffaufnahme und eine

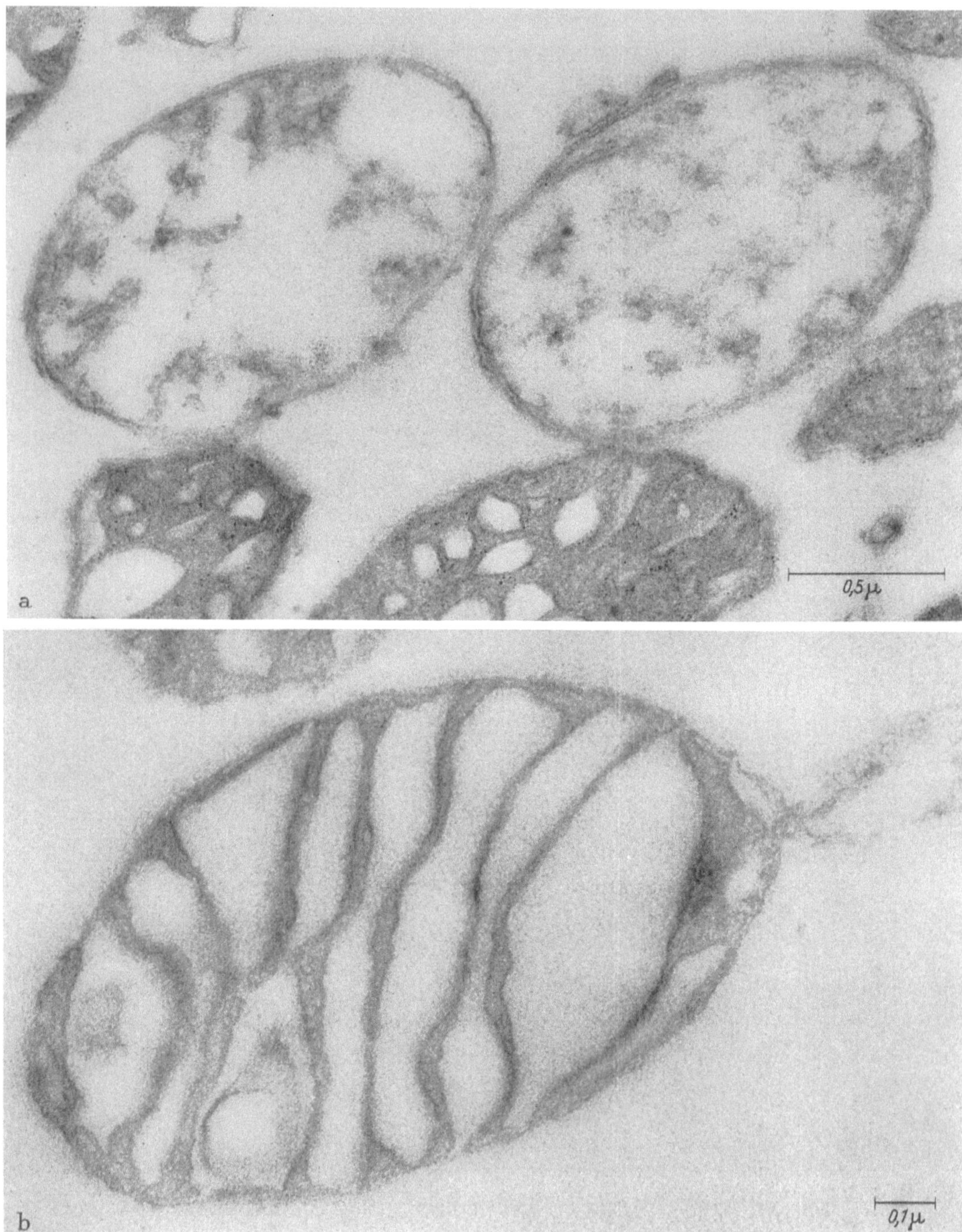

Abb. 2a u. b. Isolierte Mitochondrien aus Rattenleber. a) Matrixschwellung, Arch.-Nr. 5032/58, b) Membranschwellung, Arch.-Nr. 5192/58

Herabsetzung des $\frac{P}{O}$-Quotienten zeigt. Außer dieser Schädigung der oxydativen Phosphorylierung konnte die Schwellung von Mitochondrien durch Abnahme der optischen Dichte photometrisch verfolgt werden. Durch 3—10 mg-% Kieselsäure wird die Schwellung von Mitochondrien im Vergleich zu Kontrollansätzen beschleunigt, besonders bei ganz frischen Mitochondrien.

Wir versuchten, Unterschiede an isolierten Mitochondrien nach Kieselsäure-Inkubation elektronenoptisch darzustellen.

Hierzu wurden Rattenleber-Mitochondrien nach HOGEBOOM, SCHNEIDER und PALADE (1) in isotonischer Saccharoselösung isoliert und dabei 14 g frische Leber von zwei ausgewachsenen, gesunden Albinoratten verarbeitet. Die isolierten Lebermitochondrien wurden zu 50 % in isotonischer Saccharoselösung aufgenommen und zu weiteren 50 % in derselben Lösung mit einem zusätzlichen Gehalt von 10 mg-% Oligokieselsäure suspendiert. Die Kieselsäure wurde nach SCHWARZ u. Mitarb. (4) durch Hydrolyse des gereinigten Orthokieselsäure-Tetramethylesters hergestellt, wobei das freigesetzte Methanol durch anschließendes Erhitzen auf 90° C entfernt wurde. Die Molybdat-Reaktion diente zur colorimetrischen Kontrolle. Sofort nach Zusatz der Kieselsäure wurden beide Suspensionen im Höppler-Ultrathermostat bei 25° C inkubiert.

Zu verschiedenen Zeitpunkten und zwar sofort, nach 10, 20, 30, 60 und 120 min wurden 0,6 ml der beiden Ansätze in je 4,0 ml eiskalte Fixierlösung pipettiert. Nach Einbringen in die Fixierlösung zentrifugierten wir die Suspension niedertourig ab und konnten den Bodensatz wie ein Gewebsstückchen weiter behandeln. Am schonendsten erwies sich eine Weiterbearbeitung in den Zentrifugenröhrchen. Erst zum Einbetten und Polymerisieren wurde der Bodensatz zu Bröckchen aufgeteilt und in Gelatinekapseln gebracht. Fixierzeit: 2 Std. Dünnschnitte mit Porter-Blum-Mikrotom. Siemens-Elektronenmikroskop bei 80 kV.

Die Bilder zeigen vielgestaltige Mitochondrienformen. Neben kompakten Formen, die den normalen Mitochondrien ähnlich sind, waren Schwellungserscheinungen zu beobachten, die mit zunehmender Versuchsdauer deutlicher wurden. Hierbei zeigten sich vor allem zwei Formen: 1. sah man eine Membranschwellung; die Doppelmembranen waren auseinandergedrängt und bildeten Vacuolen (Abb. 1 u. 2b); 2. ließ sich eine Schwellung der Matrix mit einer entsprechenden Aufhellung erkennen (Abb. 2a). Diese Schwellung erschien uns bei dem mit Kieselsäure inkubierten Ansatz zu bestimmten Zeitpunkten ausgeprägter vorhanden zu sein. Als Endzustand beobachtet man weitgehend aufgelöste Mitochondrienstrukturen mit völligem Schwund der Matrix.

Um Angaben über eine möglicherweise vorhandene Veränderung der Mitochondriengröße nach Kieselsäurebehandlung zu erhalten, wurden die größte Länge und Breite von 16000 Mitochondrien durch jeweils 2 Personen doppelt ausgemessen. Der Mittelwert der beiden Resultate wurde zur Auswertung verwendet. Von jedem Objekt wurden Summenkurven über die Häufigkeitsverteilung in Wahrscheinlichkeitsnetze gezeichnet, wobei sich Geraden ergaben, deren steiler Verlauf im Wahrscheinlichkeitsnetz auf ein relativ enges Teilchengrößen-Spektrum hindeutet.

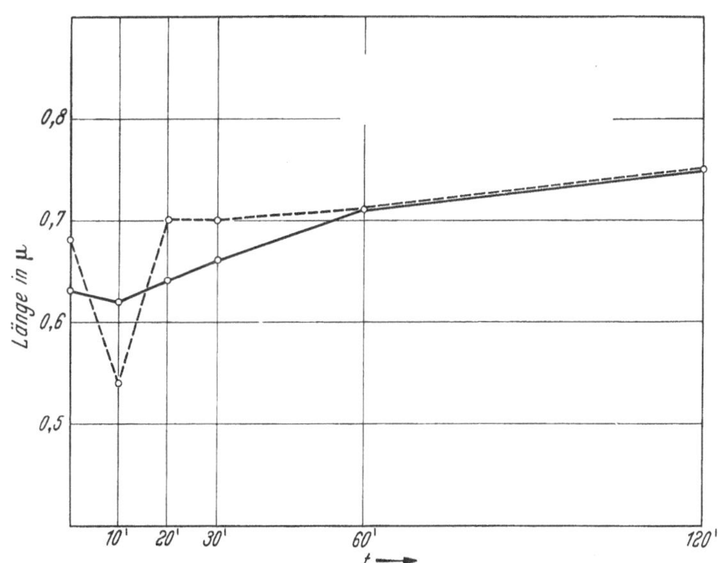

Abb. 3. Elektronenoptische Untersuchungen isol. Mitochondrien.
Versuch 1. Länge bei 50 % Summenhäufigkeit.
——— Kontrolle, ----- Inkubation mit oligomerer Kieselsäure

Die Lage der jeweiligen beiden Geraden zueinander zeigt die Unterschiede der Mitochondriengröße. Man sieht, daß zu allen Zeitpunkten der Unterschied durchaus nicht der gleiche ist. Auf der einen Seite erkennt man, daß die Mitochondrien nach Kieselsäureinkubation größer als die Vergleichsmitochondrien sind, auf der anderen Seite scheinen sie kleiner im Vergleich zu den Kontrollansätzen zu sein. Liest man die Größenverhältnisse bei 50% aus den Wahrscheinlichkeitsnetzen der Versuche ab und trägt sie zusammenfassend nach der Versuchsdauer in ein einfaches Koordinatensystem ein, so erhält man einen Überblick über das Verhalten der Mitochondriengrößen (Abb. 3). Man sieht, daß die Vergleichsmitochondrien, also die Leerkontrollen, allmählich schwellen. Gibt man jedoch den isolierten Mitochondrien 10 mg-% Kieselsäure zu, so erkennt man in den ersten 10 min eine plötzliche, eindrucksvolle Schrumpfung, die dann aber unmittelbar in eine entsprechende Schwellung umschlägt. Die Schwellung der Mitochondrien ist bei 20—30 min im Vergleich zu den Kontrollansätzen deutlich beschleunigt. Dieser Befund ist mit der verstärkten Mitochondrienschwellung vergleichbar, die in vitro photometrisch an Hand einer Abnahme der optischen Dichte verfolgt werden konnte. Nach einer Stunde erreicht die Kurve der Leerkontrollen den Wert nach Kieselsäureinkubation, um dann parallel mit ihr zu verlaufen. Der Versuch wurde in der gleichen Weise nochmals wiederholt. Es konnte hierbei genau die gleiche Schrumpfung in den ersten Minuten und die darauf folgende stärkere Schwellung der Mitochondrien sichtbar gemacht werden. Auch der Schnittpunkt bei 60 min ließ sich reproduzieren, lediglich war bei dem zweiten Versuch auch nach 2 Std. eine stärkere Mitochondrienschwellung im Vergleich zu den Kontrollansätzen deutlich.

Es ist besonders hervorzuheben, daß bei dieser Gemeinschaftsarbeit mit Biochemikern am gleichen Objekt mit verschiedenen Methoden dieselbe Aussage gewonnen werden konnte; nämlich, daß 10 mg-% Kieselsäure in vitro zu einer Größenveränderung und Schädigung der Mitochondrien führt.

Es wäre verfrüht, diesem Befund schon jetzt Spezifität, insbesondere im Hinblick auf die Pathogenese der Silikose, zuzumessen. Immerhin steht fest, daß 10 mg-% Kieselsäure keine für die Zelle indifferente Substanz darstellen und daß gerade Mitochondrien gegen Kieselsäure besonders empfindlich sind, während viele andere Enzyme und Zellfunktionen durch niederkonzentrierte Kieselsäure nicht beeinträchtigt werden.

Literatur

1. Hogeboom, G. H., W. C. Schneider and G. E. Palade: J. biol. Chem. **172,** 619 (1948).
2. Klosterkötter, W., u. H. Themann: Wissenschaftl. Forschungsber. B. 66 (Staublungenerkrankungen Bd. 3) 373 (1958).
3. Policard, A., A. Collet et S. Pregermain: Exp. Cell. Res. **12,** 400 (1957).
4. Schwarz, R., u. K. G. Knauff: Z. org. allg. Chem. **275,** 176 (1954).
5. Staudinger, Hj., W. Kersten u. K. Krisch: Hoppe-Seylers Z. physiol. Chem. **313,** 109 (1958).
6. Schlipköter, H. W.: Proc. Intern. Conf. Electron Microscopy, London, 527 (1954).
7. — Bekämpfung der Silikose Bd. II, Glückauf Essen, S. 20 (1956).
8. Kikuth, W., H. W. Schlipköter, P. Schroeteler: Proc. Stockholm Conf. Electron Microscopy, 246 (1956).

F. Ergebnisse der Elektronenmikroskopie in der Anatomie

1. Epithelgewebe

Die Oberfläche der verhornten Zelle der Epidermis im Reliefbild

Jan Wolf

Laboratorium für Elektronenmikroskopie der Tschechoslowakischen Akademie der Wissenschaft, Prag II

Um eine volle mechanische und chemische Widerstandsfähigkeit zu erreichen, muß das Hornblatt der Epidermis eine möglichst homogene Struktur besitzen. Der Organismus löst die gegebene Aufgabe, indem die ehemalig polyedrischen Zellen im Niveau der Hornschicht abflachen, sich an ihrer Oberfläche mit einem Relief von feinen Höckerchen versehen und mit Hilfe einer Kittsubstanz und Desmosomenresten verkleben. Dasselbe gilt grundsätzlich für die oberflächlichen Elemente der mehrschichtigen Pflasterepithelien. Die Oberfläche der Hornzelle der Epidermis, wie der mehrschichtigen Epithelien überhaupt, ist durch eine Menge von grampositiven Ernstschen Granula (*1, 2*) charakterisiert. In einer früheren Arbeit konnte ich den Beweis bringen, daß diese Körner auch durch Versilberung hervorgehoben werden können und somit keine Kunstprodukte sind, wie bisher angenommen wurde. Ihre Aufgabe ist unbekannt. Es ist notwendig, alle drei genannten Merkmale genetisch und funktionell in gegenseitigen Zusammenhang zu bringen. Zu diesem Zwecke wäre es vorteilhaft, unsere Kenntnisse über die Oberfläche der Hornzelle durch Flächenbilder ihrer Oberfläche im Elektronenmikroskop zu ergänzen.

Dieses Relief, welches ich vor Jahren bei der Klassifizierung der Hautreliefe „primäres Relief" nannte (*3*), tritt natürlich in den elektronenoptischen Replikabildern viel deutlicher hervor als in den bisherigen lichtoptischen. Wie aus Abb. 1 ersichtlich ist, handelt es sich hier um konvexe Höckerchen oder kurze Kämmchen, denen analoge konkave Gebilde entsprechen. Der Charakter dieses Reliefs ändert sich den einzelnen Facetten gemäß. Vor einigen Jahren habe ich versucht (*5*), die Bedeutung dieses Reliefs aufzuklären. Es handelt sich um eine Verzahnung der Höckerchen und Vertiefungen der Nachbarzellen ineinander, so daß jedes Auseinandergleiten der Zellen verhindert wird. Horstmann und Knoop (*6*) haben diese Vorstellung inzwischen mit Hilfe ultradünner Schnitte bestätigt. Nach diesen Autoren erfolgt die gegenseitige Verklebung der Hornzellen durch Desmosomenreste und durch intercelluläre Kittsubstanz.

Mein Interesse galt der Frage, ob und wie sich diese beiden Substanzen, die sich in ihrer Herkunft unterscheiden, in den elektronenoptischen Replikabildern darstellen. Schon Abb. 1 zeigt an der Oberfläche des primären Reliefs scheinbare flache Defekte, welche vermutlich den Desmosomenabdrücken entsprechen. Diese Defekte, die wahrscheinlich eingesunkene Felder sind, werden bei stärkerer Vergrößerung deutlicher sichtbar (Abb. 2). Die Felder erwecken den Eindruck eines abgefallenen Wandbezuges. Ihre Größe schwankt zwischen 0.1 μ und 0,6 μ. Es gibt lokale Größenunterschiede, da neben den großmaschigen auch kleinmaschige Strukturen vorkommen. Die Gestalt variiert zwischen sphärisch und halbmondförmig. Die Tiefe der eingesunkenen Felder entspricht der von Horstmann und Knoop angegebenen Dicke der Desmosomenplatten. Der Boden der Felder ist gewöhnlich grobkörniger als die Umgebung.

Bei dem dritten Element, welches die Oberfläche der verhornten wie der nicht verhornten Zellen der epidermoidalen Epithelien charakterisiert, d. h. bei den grampositiven und argyro-

JAN WOLF:

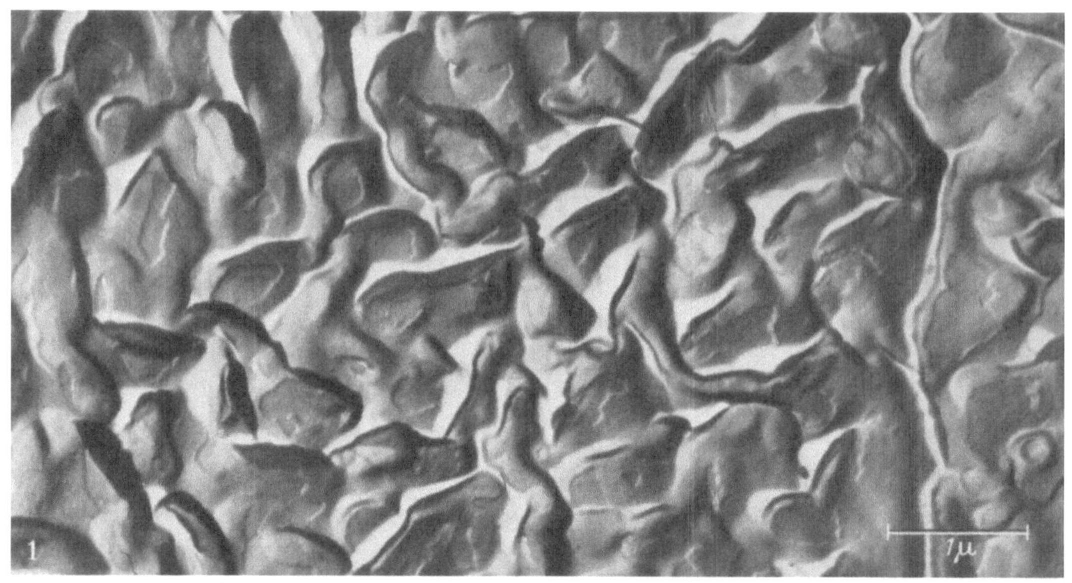

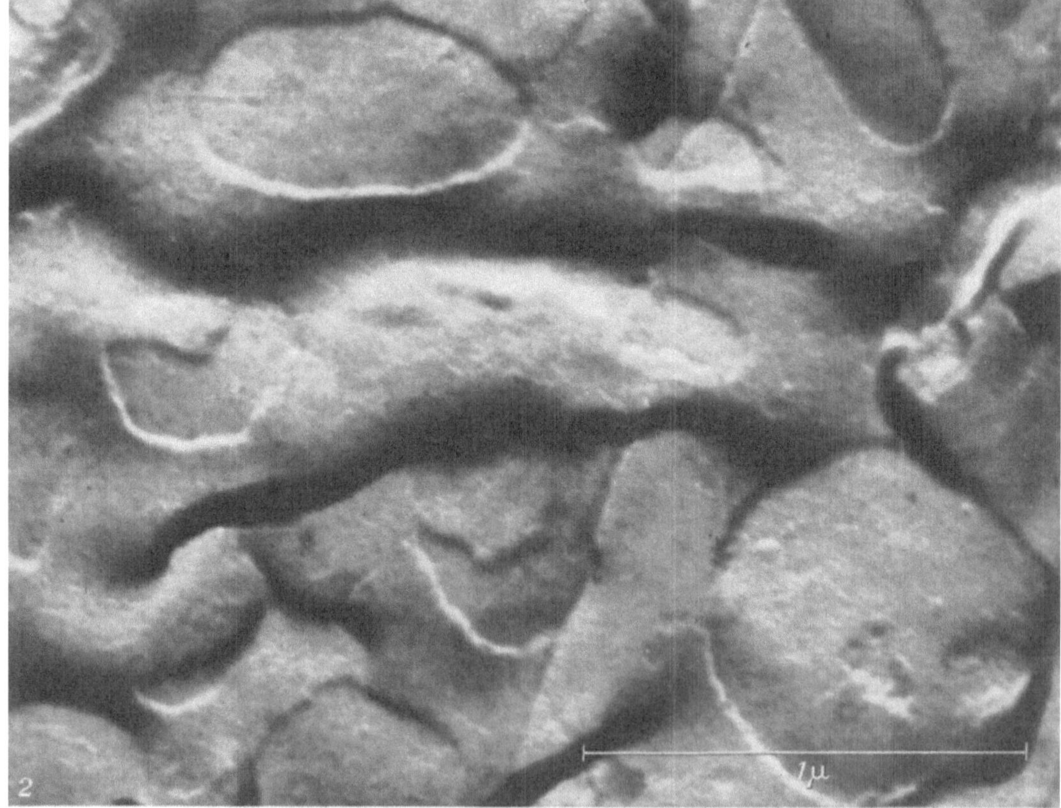

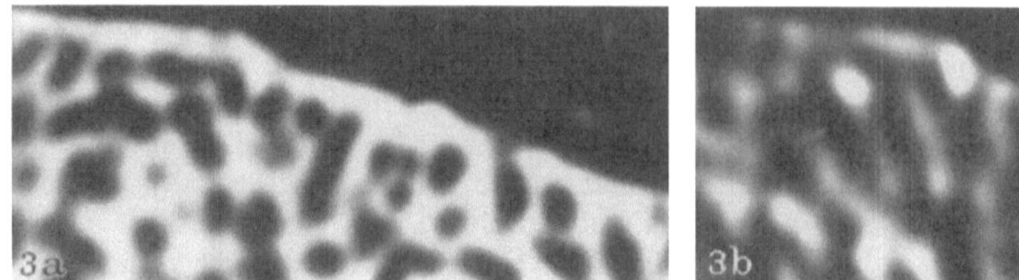

Abb. 1, 2, 3a und b. (Legenden siehe S. 285)

philen Ernstschen Granula (*2, 4*) ist es notwendig, ihre Beziehung zum primären Relief zu studieren. Über dieses prinzipielle wichtige Verhalten der bis heute bestrittenen Ernstschen Granula kann man sich allerdings nur anhand lichtoptischer Bilder überzeugen, denn die grampositiven

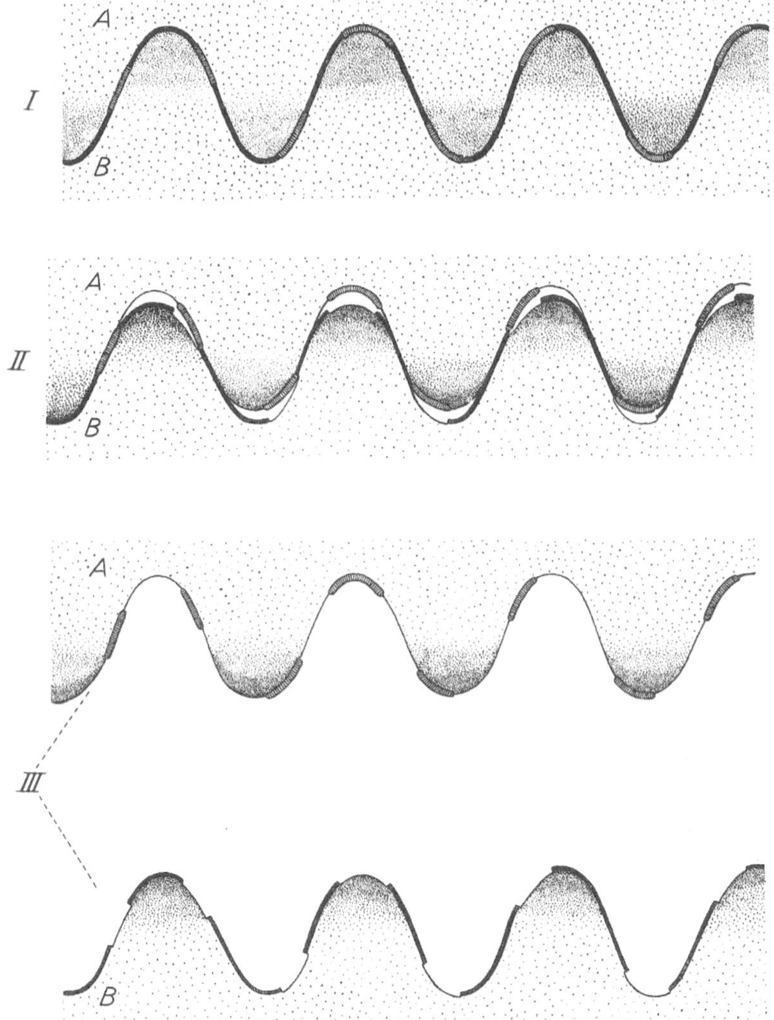

Abb. 4. Schematische Darstellung der Zellenverbindung (*I.*), der beginnenden Desquamation (*II.*) und der vollendeten Desquamation (*III.*). Schwarze dicke Linie = intercelluläre Kittsubstanz, quergestreifte Linien = Desmosomenreste, Punktierung = Hornsubstanz, dichte Punktierung = grampositive Granula. Nähere Erklärung im Text (Vergr. etwa 40000fach)

Granula sind in den elektronenmikroskopischen Schnittbildern als solche nicht zu erkennen. Die zum Vergleich herangezogenen lichtoptischen Abb. 3a und 3b zeigen bei stärkster Vergrößerung (8000mal), daß die grampositiven Granula (Abb. 3a) nur den Höckerchen und nicht den Vertiefungen des primären Reliefs (Abb. 3b) entsprechen.

Abb. 1. Primäres Relief der Hornzelle aus kleinen Höckerchen und Kämmchen bestehend. An der Oberfläche sind kleine rundliche Felder zu sehen. Rechts die interfazetare Rinne, welche dem Rand der Nachbarzelle entspricht

Abb. 2. Primäres Relief derselben Zelle. Die rundlichen Felder scheinen den abgerissenen Desmosomenresten zu entsprechen

Abb. 3a u. b. Lichtmikroskopischer Beweis, daß die grampositiven Ernstschen Granula der Hornzelle a den Höckerchen des primären Reliefs b entsprechen; 8000mal

Rekonstruiert man die gegenseitige Verbindung der abgeflachten Hornzellen sowohl anhand der elektronenoptischen Replikabilder wie anhand der lichtoptischen Ergebnisse nach Gram-Färbung, so ergibt sich folgendes Schema (s. Abb. 4). Die Mäntel der benachbarten Zellen A und B sind gegenseitig mit den Höckerchen und Vertiefungen des primären Reliefs ineinander gepreßt. Die Verklebung der Zellen geschieht mittels intercellulärer Kittsubstanz (schwarze Linien) und der Desmosomenkittplatten (quergestreifte Linien). Die grampositiven Ernstschen Granula sind örtlich nur an die Gipfel der Höckerchen und nicht an die Vertiefungen gebunden. Diese Tatsache halte ich für grundsätzlich wichtig für die Entstehung des primären Reliefs einerseits und der spontanen Desquamation andererseits. Meines Erachtens sind es nur die grampositiven, durchaus regelmäßig verstreuten und genau am Gipfel der Höckerchen lokalisierten Ernstschen Körner, welche einzig und allein die ihnen örtlich entsprechenden Unebenheiten des primären Reliefs mittels ihrer speziellen Substanz modellieren können. Es ist nicht ausgeschlossen, daß die Ein-schrumpfung, welche die Umwandlung von Cytoplasma in Hornsubstanz begleitet, im Bereiche der Ernstschen Körner wegen der Anwesenheit ihrer der Grundsubstanz beigemischten speziellen Substanz ein wenig verzögert wird. Für diese Modellierung können die Desmosomen nicht ver-antwortlich sein, wenigstens nicht allein, da sie manchmal ohne jeden örtlichen Zusammenhang, ohne Rücksicht auf die Höckerchen an der Zelloberfläche liegen. — Ähnlich läßt sich auch die spontane Desquamation erklären. Es ist anzunehmen, daß im Inneren der Körner Vorgänge ablaufen, welche zur Volumenänderung, d. h. zur Einschrumpfung im Bereiche der Körner führen, d. h. die zu Anfang im Bereiche der Ernstschen Körner verzögerte Einschrumpfung läuft erst jetzt ab. Durch diese Verzögerung würde demnach die Schnelligkeit der Desquamation bestimmt. Ein solcher Fall ist an unserem Schema in der zweiten Bilderreihe eingezeichnet. Dabei muß es natürlich zu Verschiebungen zwischen den Höckerchen der einen Zelle und den Vertiefungen der anderen Zelle und somit zur Beschädigung der gegenseitigen Verklebung kommen. — Die dritte Reihe der schematischen Abbildung veranschaulicht endlich den Zustand nach der vollendeten Desquamation, wo sich die obere Zelle A von der unteren B gelöst hat.

Literatur

1. ERNST, P.: Arch. mikrosk. Anat. **17**, 669 (1896).
2. WOLF, J.: Z. mikrosk.-anat. Forsch. **46**, 170 (1939).
3. — Z. mikrosk.-anat. Forsch. **47**, 351 (1940).
4. — Biol. Listy, Suppl. II, 191 (1951).
5. — Čs. morfol. **2**, 48 (1954).
6. HORSTMANN, E., u. A. KNOOP: Z. Zellforsch. **47**, 348 (1958).

Zum Feinbau der Intercellularbrücken nach Kontrastierung mit Phosphorwolframsäure

ALFRED VOGEL

Histopathologisches Institut der Universität Zürich

Intercellularbrücken sind Strukturen, welche die Zellen der Epidermis und der mehrschichti-gen Pflasterepithelien untereinander verbinden. In den früheren elektronenmikroskopischen Untersuchungen von OTTOSON et al. (1), PORTER (2) und SELBY (3) wurden Befunde erhoben, die wir zu den gesicherten Tatsachen über die Intercellularbrücken rechnen (Abb. 1):

Die Zellen sind voll individualisiert. Auflösbare Plasmodesmen sind nicht zu erkennen. Wie überall, wo Zellen zusammenstoßen, sind auch in der Intercellularbrücke ihre opaken Begren-zungsschichten durch eine weniger dichte Zone, die als *Intercellularfuge* (4) bezeichnet wurde, voneinander getrennt. Der Bereich der Zellperipherie, welcher innerhalb der Intercellularbrücke die zu beschreibende hoch spezialisierte Ausbildung aufweist, soll hier kurz als *Kontaktzone* be-zeichnet werden. Die lichtmikroskopischen Desmosomen setzen sich aus zwei intracellulären, den Zellmembranen längs der Kontaktzone anliegenden, Hälften zusammen. Die dichten Fibrillen des Cytoplasmas, die SELBY (3) zum Unterschied von den lichtmikroskopisch erkennbaren Tono-fibrillen als *Tonofilamente* (tonofilaments) bezeichnet hat, überqueren die Intercellularfuge nicht.

Die folgende Beschreibung bezieht sich auf Intercellularbrücken der mittleren Schichten des Portioepithels, welches während 1—2 Std. mit einer 1%-Lösung von Phosphorwolframsäure in 70% wäßrigem Aceton behandelt wurde.

Innerhalb der Kontaktzone fällt ein Paar, im Abstande von 250—300 Å parallel verlaufender dichter Linien (Abb. 2) von 100 Å Breite auf. Wir wollen sie als c-Schichten bezeichnen. An

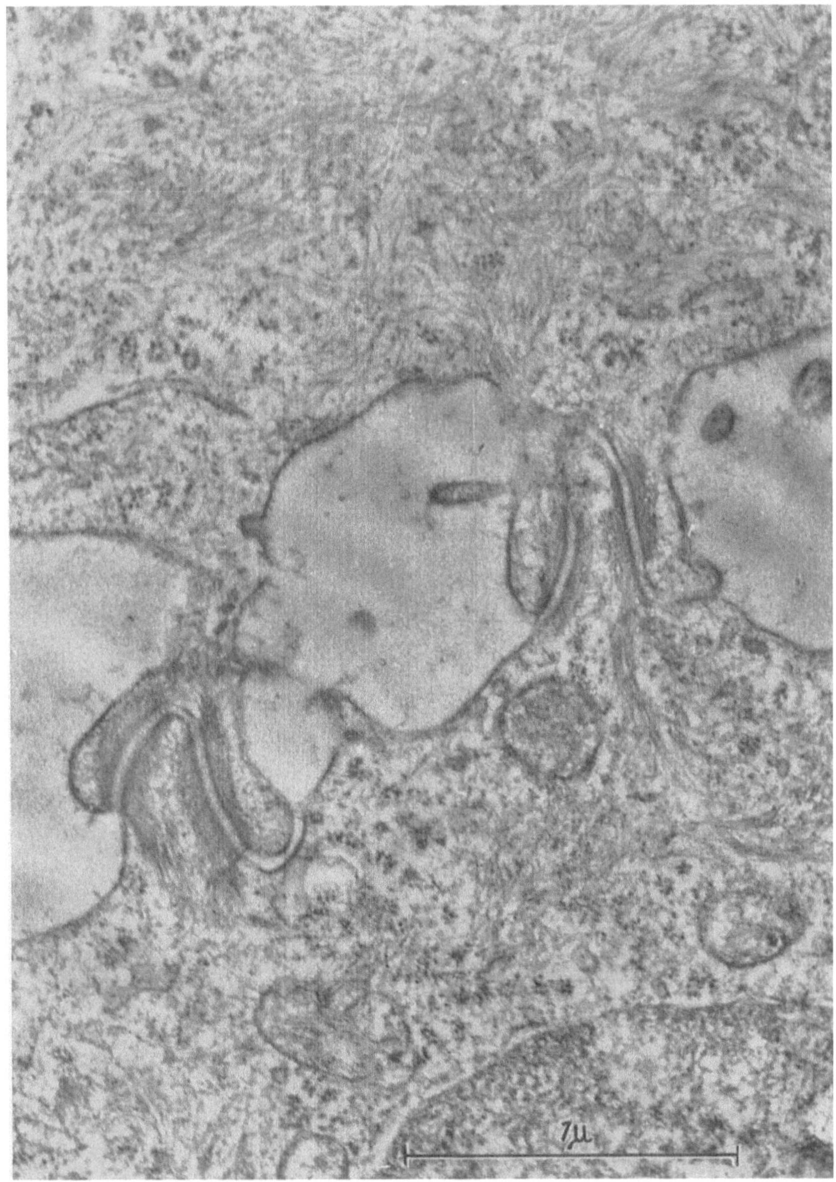

Abb. 1. Portioepithel, zwei Intercellularbrücken, dazwischen Intercellularraum. Einbettung in Vinox. 45000 mal

diese Formation schließt gegen das Zellinnere zu ein Raum geringerer, doch immer noch relativ hoher Dichte an (k). Darin ist die Substanz des lichtmikroskopisch nachweisbaren Brückenknötchens zu lokalisieren. In dieser Zone ist ein weiteres Dichtemaximum schwach zu erkennen, das — im Abstande von 200 Å — parallel zur c-Schicht verläuft und als d-Schicht bezeichnet werden soll. Die in der Kontaktzone der Intercellularbrücken ansetzenden Tonofilamente sind, nach Einbettung des Gewebes in Araldit, lediglich an der Streifung der angrenzenden Cytoplasmapartien nachzuweisen. Im Streifen zwischen den c-Schichten erkennt man leicht zwei dichte, parallele Linien, die als *a-Schichten* bezeichnet werden sollen. Endlich sind, in der Symmetrieachse der Kontaktzone, zwischen den a-Schichten, diskontinuierliche Verdichtungen (m) in einer Reihe geordnet. Sie erscheinen als Linie, wenn das Photo unter einem flachen Winkel in der Richtung des parallelen Liniensystems betrachtet wird. Nach stärkerer Bestrahlung eines Schnittes (Abb. 3) treten sowohl die m-Schicht als die a-Schichten deutlicher hervor. Nach den bisherigen Ausführungen ist somit der Raum zwischen den markanten c-Schichten in drei dichtere und vier weniger dichte Schichten von je etwa 40 Å gegliedert. Ein wichtiges Problem, besonders im Hinblick auf eine Deutung der Befunde, bilden die Beziehungen der Schichten entlang der Kontaktzone zu den Begrenzungsschichten der übrigen Zelloberfläche. Die a-Schicht und der äußerste Teil

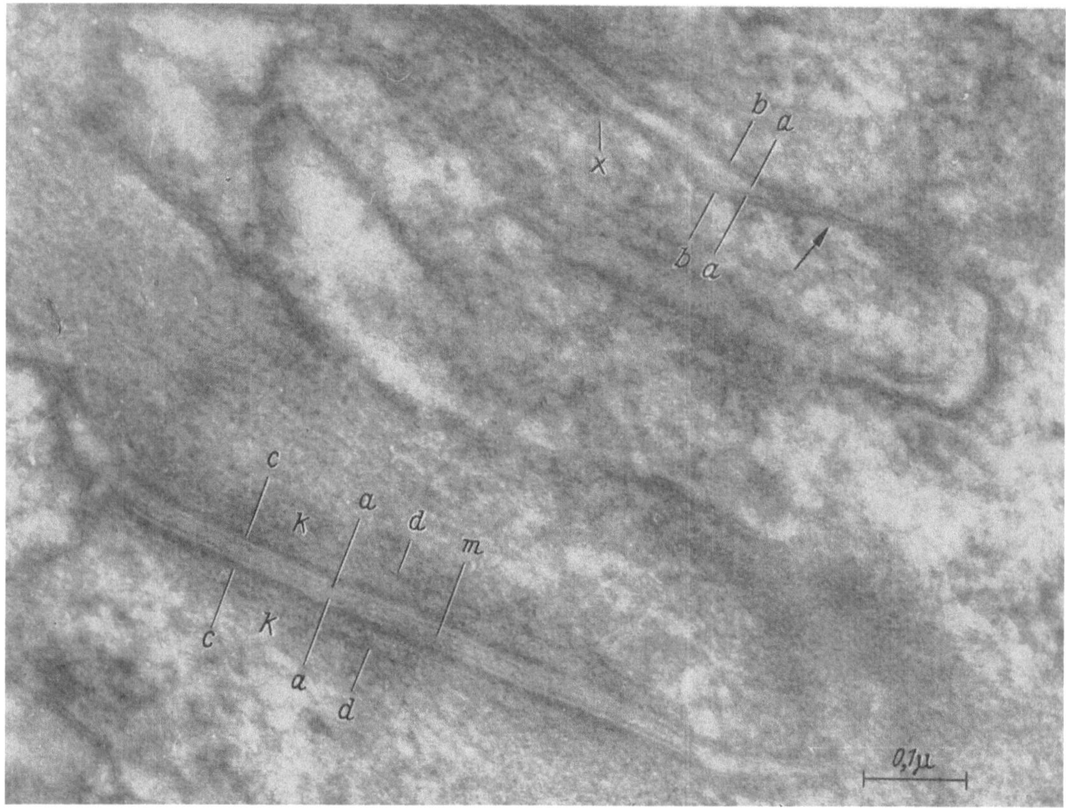

Abb. 2. Portioepithel, Kontaktzonen in Intercellularbrücken. Araldit-Einbettung. 140000 mal

der c-Schicht, der außerhalb der Kontaktzone als *b-Schicht* bezeichnet werden soll, setzen sich über die Kontaktzone hinaus fort (Abb. 2). Die Zellgrenzschicht gegen den intercellulären Raum ist überall aus den zwei dichten Schichten a und b aufgebaut, welche durch eine weniger dichte Zwischenschicht getrennt sind. Die m-Schicht und der größte Teil der c-Schicht enden dagegen auf ungefähr gleicher Höhe und begrenzen so die Kontaktzone. Die a-Schichten divergieren meistens außerhalb der Kontaktzone und begrenzen dann den intercellularen Raum. An einigen Stellen scheinen sie jedoch zu einer einzigen Schicht zu konvergieren (Abb. 2 und 3).

Die vorliegende Untersuchung wurde an den Intercellularbrücken der mittleren Schichten des Portioepithels durchgeführt. Der geschilderte Aufbau der Kontaktzone wurde jedoch darüber hinaus auch in der Epidermis der Fußsohle und des Rückens der Maus gefunden. Seit Intercellularbrücken, Schlußleisten einschichtiger Epithelien (5) und Glanzstreifen des Herzmuskels (6) elektronenmikroskopisch untersucht wurden, hat man auf Ähnlichkeiten im Aufbau dieser Zellverbindungs-Strukturen hingewiesen. Ein Vergleich unserer Befunde mit der ausführlichen Arbeit von SJÖSTRAND et al. (7) präzisiert, daß der Bau der Kontaktzone in den Intercellularbrücken den als „*S-Regionen*" bezeichneten Abschnitten des Glanzstreifens gut entspricht. Die

Übereinstimmung erstreckt sich auf die opaken Schichten a und c sowie auf das beobachtete Zusammenfallen der konvergierenden a-Schichten außerhalb der Kontaktzone. Neu hinzu kommt in den Intercellularbrücken die dichte, wenn anscheinend auch diskontinuierliche m-Schicht in der Symmetrieebene der Kontaktzone. Die d-Schicht, welche in der S-Region des Glanzstreifens mit gleichem Gewicht wie die c-Schicht in Erscheinung tritt, ist in der Intercellularbrücke nur angedeutet und häufig gar nicht zu sehen. Die Mittelwerte aus je 30 Messungen der Abstände zwischen den c- bzw. a-Schichten liegen etwas über den Abmessungen, die Sjöstrand et al. (7) im Glanzstreifen fanden. Die starke Streuung unserer Werte hängt möglicherweise mit einer Differenzierung der Intercellularbrücken von den unteren nach den oberen Schichten des Epithels zusammen. Eine Neuuntersuchung von Intercellularbrücken in nicht nachkontrastierten Geweben mit guter Auflösung hat ergeben, daß die a-Schicht auch hier, wenn auch sehr schwach, zu erkennen sein kann. Dagegen waren die m-Schicht sowie die beiden d-Schichten nur im phosphorwolframsäure-behandelten Gewebe zu beobachten. Ob es sich bei dieser „Färbung" um eine spezifische Reaktion von Gewebebestandteilen handelt oder lediglich um eine allgemeine Kontrastvermehrung durch eine Art proportionaler Verstärkung, entsprechend bestimmter Verstärkungsverfahren für photographische Negative, läßt sich noch nicht entscheiden. Das fakultative Zusammenfallen der a-Schichten zu einer einzigen Schicht außerhalb der Kontaktzone läßt vermuten, daß jene die äußerste Zellbegrenzung darstellen. Ihr konstanter Abstand innerhalb der Kontaktzone weist auf eine unpaare, beiden Zellen gemeinsame Zwischenschicht unaufgeklärter Zusammensetzung, doch von hoher molekularer Ordnung hin, die mit den übrigen Bauelementen der Kontaktzone in noch unbekannter Weise zusammenwirkt und so die mehrschichtigen Epithelien zusammenhält.

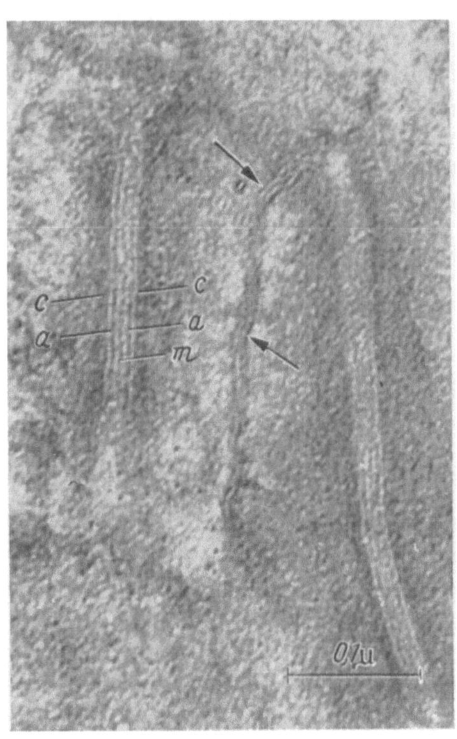

Abb. 3. Portioepithel. Kontaktzonen in Intercellularbrücken. Araldit-Einbettung. 185000mal

Literatur

1. Ottoson, D., F. S. Sjöstrand, S. Stenström and G. Svaetichin: Acta physiol. scand. **29,** 611 (1953).

2. Porter, K. R.: Proc. 3rd Int. Conf. on Electron Micr. London 1956, p. 539.

3. Selby, C. C.: J. biophys. biochem. Cytol. **1,** 429 (1955).

4. Vogel, A.: Verh. dtsch. Ges. Path. **41,** 285 (1958).

5. Zetterqvist, H.: The ultrastructural organization of the columnar absorbing cells of the mouse jejunum. Stockholm 1956.

6. Sjöstrand, F. S. and E. Andersson: Experientia (Basel) **10,** 369 (1954).

7. — Ebba Andersson-Cedergren and M. M. Dewey: J. Ultrastructure Res. **1,** 271 (1958).

Structural relationship between epithelial cells in Hydra

R. L. Wood

Department of Anatomy, University of Washington, Seattle, Washington, USA

The plasma membranes of epithelial cells in Pelmatohydra oligactis are characterized by two dense lines about 30 Å apart, corresponding to the unit membrane structure described by Robertson. At certain localized areas of apposing cell surfaces, continuities between the outer dense components of the respective plasma membranes form an oriented series of parallel plate-like lamellae perpendicular to the cell surfaces. Each lamella consists of two dense lines about 25 Å in thickness separated by about 25 Å. The lamellae occur at intervals of 100—150 Å. Fixation in buffered OsO_4 followed by postdehydration exposure to a solution of 0.1—0.5% phosphotungstic acid in absolute ethanol effectively demonstrates these relationships, as this treatment leads to sharp contrast between the dense membrane system and the pale interlamellar regions. Plate-like intercellular connections are seen after permanganate fixation but the interlamellar regions appear much more dense. Attempts have been made to obtain further information on the molecular structure of these lamellae by treating the tissues with crystalline enzyme preparations. Evidence at this time favors the view that these oriented lamellae function for intercellular attachment. An additional possibility is that they act as a permeability. barrier or seal, detering ambient water from passing between the cells into the intercellular fluid.

An extended version of this article will appear in J. biophpys. biochem. Cytol. 6.

2. Muskelgewebe

Einführende Bemerkungen zur Struktur und Funktion der Muskulatur

H. Ruska

Institut für Elektronenmikroskopie, Medizinische Akademie Düsseldorf

Der morphologische Vergleich von Muskeln mit verschiedenen funktionellen Eigenschaften, von krankhaft veränderten Muskeln und von Muskelfibrillen, die unter verschiedenen Bedingungen dargestellt sind, soll uns zeigen, wie weit gegenwärtig die Funktion aus dem Feinbau zu verstehen ist, oder wie wir den Feinbau im Hinblick auf die Funktion interpretieren können.

Es ist die Aufgabe der Muskeln, auf eine mit Änderungen des elektrischen Potentials an der Faser- oder Zelloberfläche einhergehenden Erregung, durch Spannung oder Verkürzung gegen einen äußeren Widerstand eine Kraft zu entfalten oder Arbeit zu leisten. Mit der Arbeit und der Erholung des Muskels ist die Produktion von Wärme verbunden. Bei der Erregung tritt eine Depolarisierung an den im einzelnen unterschiedlich ausgebildeten Zellmembranen ein. Außerdem gibt es Argumente für die Annahme, daß phasentrennende Membranen innerhalb der Zelle Potentiale besitzen und sich an Potentialänderungen und Ionenverschiebungen beteiligen. Kraftentfaltung und Längenänderung spielen sich am fibrillären Material ab, die oxydative Wärmeentwicklung in der Erholungsphase an den Mitochondrien. Aufschlußreich ist, daß örtliche Verteilung und Mengenverhältnisse dieser Komponenten offensichtlich mit unterschiedlichen funktionellen Eigenschaften der Muskelfasern oder Zellen zusammenhängen. Es ist für den Morphologen wichtig, zu wissen, ob die Hauptaufgabe einer bestimmten Muskelfaser die Entwicklung einer Spannung, einer kurzdauernden, möglichst großen Arbeit, einer kontinuierlichen Arbeit oder einer hochfrequenten Bewegung ist. Die verschiedenen Fasertypen zeigen außerdem sehr unterschiedlich gebaute neuro-muskuläre Verbindungen und können einfach oder mehrfach innerviert sein. Das spezifische Verhalten einer Muskelfaser resultiert aus den morphologischen und biochemischen Besonderheiten von Nerv, neuro-muskulärer Verbindung und Faser.

Um der Gefahr zu entgehen, daß sich die Interpretationen in der Morphologie zu weit von physiologischen Vorstellungen entfernen, habe ich Herrn Prof. Rothschuh gebeten, die Beziehungen zwischen Struktur und Funktion vom Standpunkt des Physiologen aus zu erörtern. Wir sind ihm besonders dankbar, daß er der Einladung gefolgt ist, als Gast zu uns zu sprechen.

Beziehungen zwischen Struktur und Funktion an der Muskelfaser

K. E. Rothschuh

Physiologisches Institut der Universität Münster

Seit etwa 15 Jahren beginnt die Morphologie für uns Physiologen erneut wieder interessant zu werden, denn die Elektronenmikroskopie dringt in Dimensionen ein, in der heute sehr aktuelle physiologische Forschungsaufgaben liegen. So ist uns alles äußerst wichtig, was die elektronen-mikroskopische Forschung heute über den Feinbau der Zelle und vor allem der Grenzflächen ermittelt. Und wir bemühen uns sehr, unsere physiologischen Befunde über die Lebensäußerungen mit den neuen morphologischen Befunden zu korrelieren. Am heutigen Tage ist es meine spezielle Aufgabe, diesen Beziehungen an der Muskelfaser nachzugehen. Natürlich ist dieses Thema un-erschöpflich. Ich greife daher einige Fragen heraus, welche für Morphologen und Physiologen gleich interessant sind, und werde sie aus der Sicht des Physiologen diskutieren. Im Anschluß daran sei über einige eigene Befunde berichtet und dazu ein Film gezeigt.

a) Die 2 Arten von Skeletmuskelfasern

Schon vor vielen Jahren haben SOMMERKAMP (1) und WACHHOLDER (2) nachgewiesen, daß es beim Frosch 2 funktionell verschiedene Arten von Muskelfasern gibt, nämlich einerseits die nicht-tonischen oder tetanischen oder phasischen Fasern und andererseits die tonischen Muskelfasern. Seitdem sind außerordentlich viele Untersuchungen, zumal von pharmakologischer Seite, über das Verhalten dieser beiden Arten von Muskelfasern gemacht worden. Neuerdings unterscheidet ST. W. KUFFLER (3, 4, 5) auf Grund seiner elektrophysiologischen Untersuchungen zwei funktio-nell verschiedene Muskelfaserarten, die er als „Twitch-System" bzw. als "Slow-Fiber-System" bezeichnete. Gewisse Muskeln, z. B. der Sartorius und der Adductor longus des Frosches bestehen vorwiegend aus *nichttonischen Twitchfasern* und werden durch dicke Nervenfasern innerviert, die eine Leitungsgeschwindigkeit von etwa 8—40 m/sec haben. Solche Muskelfasern haben mehrere motorische Endplatten. Sie bekommen also die Erregung an mehreren Stellen zugeleitet und ver-mögen sie über die ganze Faser fortzuleiten. Sie haben ein Ruhepotential von etwa 90 mV und eine Aktionsspannung von etwa 120 mV. Auf Gleichstromdurchströmung, ACh-Einwirkung oder KCl-Behandlung machen sie eine kurzdauernde Kontraktion und kehren dann trotz fortdauernder Einwirkung der Substanzen wieder in ihre Ruhelänge zurück. Das *tonische Slowfibersystem*, ver-treten etwa durch das Tonusbündel des Musculus iliofibularis oder den M. rectus, besteht aus relativ dünnen Muskelfasern. Sie werden versorgt von langsamleitenden (2—8 m/sec) dünnen Nervenfasern und besitzen sehr zahlreiche Nervenendigungen. Elektrisch haben diese Fasern ein relativ niedriges Membranruhepotential von etwa 50 mV. Sie entwickeln bei der Erregung nur ein niedriges Spitzenpotential von etwa 10 mV, aber ein langes positives Nachpotential. Im Bereich der Nervenendigungen erfolgt lediglich eine lokale elektrische Reaktion und eine nichtfort-gepflanzte, lokale Reaktion des contractilen Apparates. Eine mehrfache Reizung depolarisiert auch nicht über 30—35 mV. Unter Gleichstrom gehen die tonischen Fasern kathodisch in eine langdauernde Verkürzung über. ACh bewirkt lang dauernde Verkürzungszustände, Kalium unter-hält lang dauernde Kontrakturen. Das Bemerkenswerteste in dem Verhalten dieser tonischen Fasern ist aber der Befund, daß sie elektrisch überhaupt nicht reizbar, ja daß sie geradezu elektrisch unerregbar sind [BURKE u. GINSBORG (6)]. Es ist vielmehr so, daß an den zahlreichen Nervenendigungen der tonischen Muskelfasern eine Überträgersubstanz freigesetzt wird, die lokal außer der Kontraktion eine nicht spezifische Permeabilitätssteigerung für alle Ionen macht, ohne daß sich die Erregung oder die Aktivierung über die Faser fortpflanzt [BURKE u. GINS-BORG (7)].

Finden die Morphologen Strukturdifferenzen, welche die erheblichen physiologischen Unter-schiede beider Faserarten befriedigend erklären ?

KRÜGER hat (8, 9, 10) mit seinen Mitarbeitern auf zwei lichtmikroskopisch unterscheidbare Formen von Muskelfasern hingewiesen. Er unterscheidet Fasern mit „Fibrillenstruktur" und Fasern mit „Felderstruktur". Die Fasern mit Fibrillenstruktur zeigen im Querschnitt eine sehr gleichmäßige Ausbildung von Fibrillen, die rings von dem Reticulumsystem umgeben sind. Die Fasern mit Felderstruktur sind nicht gleicherweise fibrillär differenziert und die contractile Substanz ist mehr in Form zusammenhängender Massen in die Fasern eingelagert. Auch haben die tonischen Fasern ein weniger dichtes endoplasmatisches Netz. Offensichtlich decken sich KUFF-LERs Twitch-Fasern mit jenen von Fibrillenstruktur und die Slow-Fasern mit denen von Felder-struktur. Für den Warmblüter ist die Existenz dieser beiden Fasertypen noch umstritten [vgl. KUSCHINSKY u. Mitarb. (11)]. Zweifel äußerten BRECHT und FENEIS (12). Die unterschiedliche

Struktur des endoplasmatischen Reticulums wurde durch die elektronenmikroskopischen Beobachtungen von H. Ruska (13), Bennett (14), Porter und Palade (15) bestätigt und genauer differenziert. So berühren sich die physiologischen und morphologischen Befunde sehr eng und eröffnen ein neues Verständnis der großen funktionellen Unterschiede zwischen tonischen und nichttonischen Fasern.

b) Zur Verteilung der funktionellen Leistungen auf die morphologischen Elemente der Muskelfaser (Sarkolemm, Fibrillen)

Die äußerste Umhüllung der Muskelfaser, das *Sarkolemm*, ist augenscheinlich entgegen früheren Auffassungen [vgl. Fenn in Höber (16)] nicht das wesentliche Element für die Elastizität der Muskelfaser. Der elastische Widerstand, um es so auszudrücken, ist weniger eine Sache des Sarkolemms als der Myofibrillen, wie besonders Natori (17) nachgewiesen hat. Das Sarkolemm dürfte also nicht der zähelastische Strumpf für die ganze Muskelfaser sein, wie man meistens annimmt. Wesentlicher ist seine Schutzfunktion für den Fibrillenraum. Denn wo immer das Sarkolemm verletzt wird (Abb. 1a), kommt es zur fortschreitenden Strukturzerstörung der Fibrillen [„Phänomen der fortschreitenden Desorganisation" Rothschuh (18)]. Bleibt aber die

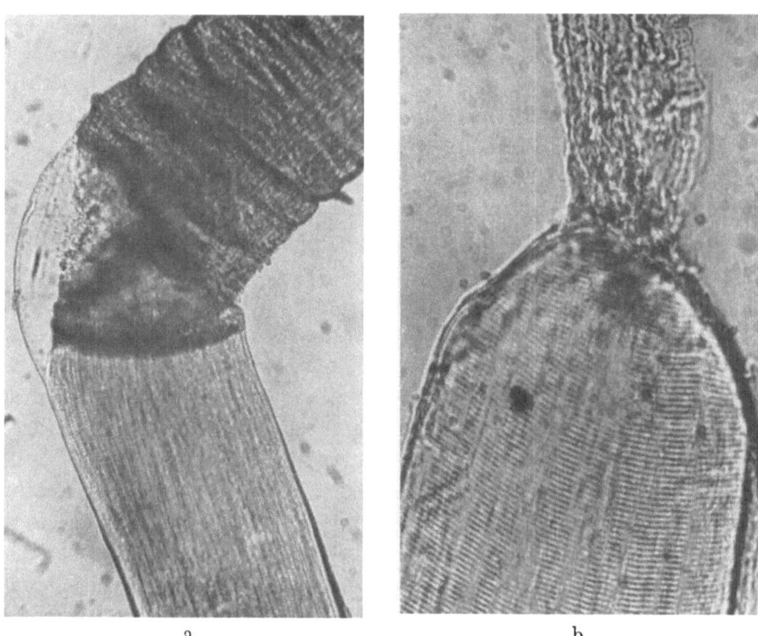

a　　　　　　　　　　　b

Abb. 1a u. b. Muskelfasern mit verletztem Sarkolemm und zerrissenen Myofibrillen zeigen eine fortschreitende strukturelle Desorganisation, die vom Ort der Verletzung ausgeht, b) die Fasern dagegen, deren Fibrillen zerrissen wurden, zeigen keinen Strukturzerfall, wenn sie ihren Sarkolemmüberzug behielten [„Sackphänomen" nach Rothschuh (40)]

Sarkolemmhaut um die zerrissenen Myofibrillen erhalten, so tritt die fortschreitende Desorganisation nicht auf („Sackphänomen" Abb. 1b). Es ist unklar, wem die entscheidenden Permeabilitätseigenschaften des Sarkolemms zukommen, dem äußeren Blatt oder dem inneren von Ruska (13) als „Plasmamembran" bezeichneten Sarkolemmblatt. Eine feste Verankerung der Z-Scheiben am Sarkolemm liegt bei vielen Muskelfaserarten z. B. beim Frosch sicher nicht vor, ebenso wenig bilden die Z-Scheiben eine die Fibrillen festverkoppelnde Struktur. Man sieht nämlich in lebenden Muskelfasern häufig, daß sich benachbarte Fibrillen erheblich gegeneinander verschieben können, dieses Phänomen des „Fibrillenfließens" wird später im Film gezeigt werden. Über das Sarkolemm als Sitz einer elektrischen Polarisation wird gleich ausführlicher zu sprechen sein. Zuvor sei noch etwas über physiologische Versuche mit isolierten, überlebenden Myofibrillen berichtet, die vor allem Natori (17, 19) ausgeführt hat. Nach Natori kontrahieren sich isolierte Myofibrillen nicht, wenn man sie bei der Präparation mit einer nicht leitenden Flüssigkeit (am besten Walöl) bedeckt.

Sobald aber irgendeine Flüssigkeit wie Leitungswasser, Salzlösung usw. an die Myofibrille heran-gelangt, kontrahiert sie sich an der betreffenden Stelle (*17*). Die Kontraktionsdauer oder der Kontraktionsvorgang einer solchen Myofibrille ist unter diesen Umständen sehr langsam und kann mehrere Sekunden dauern. Dabei ist das Plateau kurz, die Erschlaffung nimmt etwa 1—3 sec in Anspruch. Demnach scheint sich die Fibrille immer dann zu kontrahieren, wenn sie von einer leitenden Flüssigkeit getroffen wird. Da nun die herauspräparierten Fibrillen immer zerrissene Enden haben, möchte NATORI annehmen, daß es sich hier um eine Kontraktion der Myofibrillen durch die Schließung eines fibrilleneigenen Verletzungsstromes handelt. Zwischen einem kontra-hierten und einem erschlafften Myofibrillenabschnitt wurde von ihm (*19*) eine Potentialdifferenz bis zu 50 mV gemessen. Wenn sich diese Beobachtungen bestätigen, müßte man neben der Sarko-lemmpolarisation noch eine Fibrillenpolarisation annehmen. Ferner kann man mit der Anode örtlich an der Myofibrille eine Kontraktion hervorrufen [NATORI (*20*)]. Ähnliches sah ich selbst an isolierten Muskelfasern, an denen man anodisch eine lokale partielle Kontraktur erzeugen kann [ROTHSCHUH (*21*)].

c) Die Muskelgrenzflächen und die elektrischen Vorgänge bei der Muskelkontraktion

Unsere Vorstellungen über die elektrischen Vorgänge an der Muskelfaser sind in ungewöhn-lichem Maße von der Geschichte der Bioelektrizität beeinflußt [vgl. ROTHSCHUH (*22, 23*)], also durch die Entwicklung der Theorie von der Präexistenztheorie über die Alterationstheorie zur Membrantheorie und Ionentheorie. Die Bernsteinsche Membrantheorie (1902) wurde von HODGKIN u. Mitarb. (*24, 25*) zu einer Ionenpermeationstheorie erweitert, nach der am Nerven die Na-Ionen für den Erregungsanstieg und für den "overshoot" des Aktionspotentials eine besonders bedeut-same Rolle spielen, während die K-Ionen mehr für die Entstehung des Membranruhepotentials maßgeblich sind. Das ließ sich auch für die Skeletmuskelfaser beweisen, deren Ruhepotential in annähernd voraussagbarer Weise von der K-Außenkonzentration abhängig ist [NASTUK u. HODG-KIN (*26*), LING u. GERARD (*27*), ADRIAN (*28*)], während im Na-armen Medium die Höhe des Aktionspotentials stark absinkt. Nach dieser z. Z. besten, wenn auch sicher nicht abgeschlossenen Theorie der bioelektrischen Prozesse sehen wir in den Konzentrationsunterschieden von Ionen (elektrochemischen Gradienten) zwischen Muskelfaser und Faserumgebung eine der Vorbedingun-gen für die Ausbildung elektrischer Prozesse an der Muskelfaser (Abb. 2). Zu hinreichenden Potentialdifferenzen an biologischen ruhenden Grenzflächen kommt es nur dann, wenn erstens Ionengradienten hinreichender Größenordnung zwischen zwei durch eine Grenzfläche getrennten Medien vorliegen und wenn zweitens die Grenzfläche außerdem spezifische Permeabilitätsbedin-gungen für diese Ionen besitzt (Abb. 3). Damit aber eine solche elektrisch geladene Grenzfläche auch noch erregbar ist, muß sie zusätzlich erstens durch äußere Anstöße (Reize) zu einer plötzlichen Änderung ihrer Ionenpermeabilität fähig sein und zweitens über einen restituierenden Mechanis-mus mit Ionenpumpe verfügen. Aus diesen 4 Momenten ergibt sich, daß nicht ohne weiteres jede morphologische Membran polarisiert und erregbar im Sinne der Elektrophysiologie zu sein braucht. Die derzeitige Elektrophysiologie verlegt das Erregungsgeschehen ganz an die Oberfläche der Skeletmuskelfaser. Das geht nach KATZ (*36*) u. a. daraus hervor, daß elektrotonisches Poten-tial und Aktionspotential ohne scharfe Grenze ineinander übergehen. Auch ist die normale Alles-oder Nichts-Reaktion der Muskelfaser leichter verständlich, wenn die Aktivierung eine Sache der gemeinsamen Oberfläche ist. Ferner mißt man mit der Mikroelektrode, die in die Faser eingesto-chen wird, stets die gleiche Potentialdifferenz. Die Physiologie hat also gute Gründe, die wesent-lichen Prozesse bei der initialen Muskelerregung an die Oberfläche zu lokalisieren. Wahrscheinlich gibt es auch innerhalb der Faser elektrische Ladungsunterschiede, wie gleich zu zeigen sein wird, doch sind nur an der Faseroberfläche hinlänglich große und schnell reversible Ionenverschiebungen wahrscheinlich.

Es gibt übrigens gewisse *Schwierigkeiten* für die *Ionentheorie*. Nach SHAW u. Mitarb. (*29*) geht das Ruhe-potential dem Ionengradienten durchaus nicht immer in der erwarteten Weise parallel. Nach LÜLLMANN (*29a*) könnte das auf Verschiebungen der Permeabilitätskoeffizienten für K, Na und Cl beruhen. Salzinjektionen in die Muskelfaser hatten ebenfalls nicht den erwarteten Effekt [FALK u. GERARD (*30*)]. Auch sind die Beob-achtungen und Meinungen darüber, ob in der Faser das Muskel-K in gebundener oder freier Form vorliegt, noch sehr geteilt [USSING (*31*), SEGAL (*32*)]. Dann ist das Problem, wie das K gegen das Konzentrationsgefälle in die

Faser hinausgelangt, noch ungelöst. Jedoch ist ein „aktiver" Ionentransport thermodynamisch möglich [USSING (31)]. Es mag nicht unerwähnt bleiben, daß in neuester Zeit von SEGAL (32) wiederum eine Art von Alterationstheorie des Erregungsvorganges versucht wird, die sich auf die Untersuchungen von NASSONOW u. Mitarb. (33) und SEGALs eigene Versuche (34) stützt.

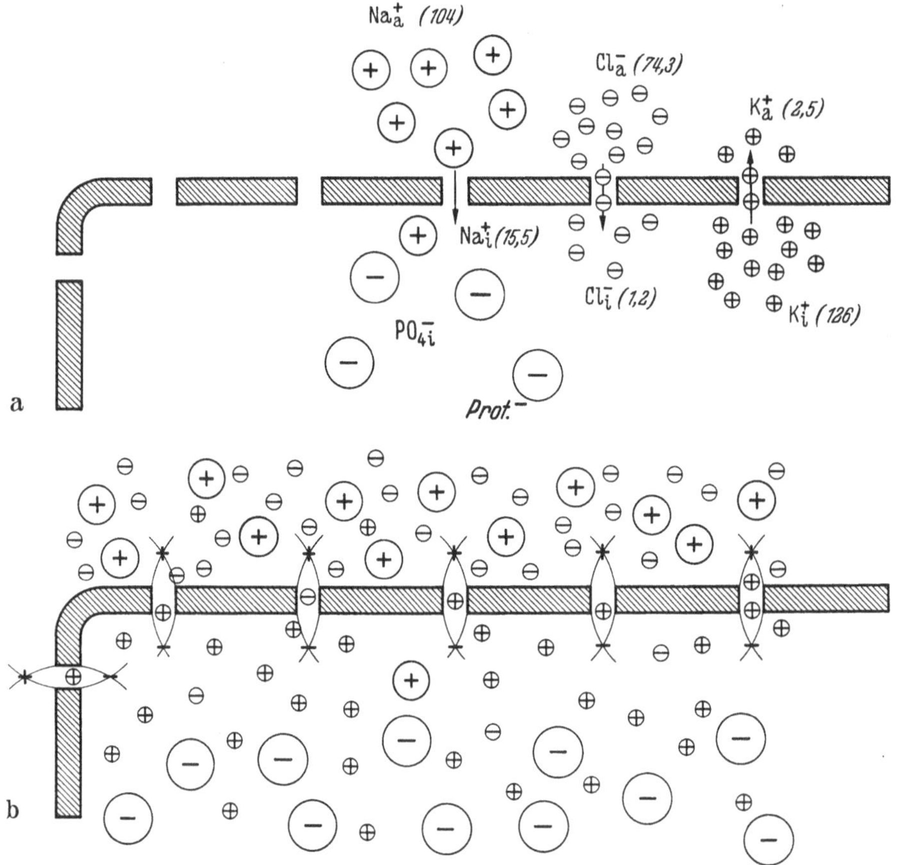

Abb. 2a. Ein Ruhepotential an Grenzflächen entsteht durch die elektrochemischen Ionengradienten zwischen dem Außenmedium und dem Innenmedium, die sich an der Grenzfläche gegenüberliegen. An der ruhenden Skeletmuskelfaser bestehen drei solcher Gradienten für $\dfrac{Na_a}{Na_i}$, für $\dfrac{K_i}{K_a}$ und $\dfrac{Cl_a}{Cl_i}$, welche die Richtung möglicher Ionenwanderungen bestimmen. Die Zahlenwerte geben die Konzentration in mM/l an

Abb. 2b. Diese Ionenverteilung bedingt eine Polarisation der Membran, welche in Ruhe außen relativ positiv und innen relativ negativ geladen ist. Die Durchtrittsorte für Ionen sind aus anschaulichen Gründen als Porenöffnungen dargestellt worden. Der resultierende Spannungsgradient an der ruhenden Membran beträgt an der Skeletmuskelfaser etwa 80 mV über eine Strecke von wenigen Å

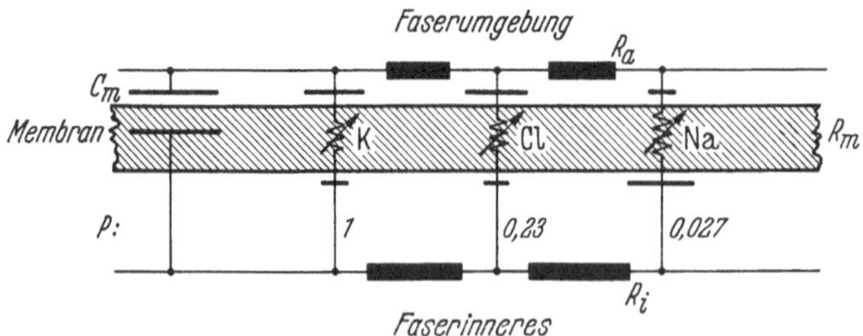

Abb. 3. Elektrisches Ersatzschema der Muskelmembran. Die 3 Ionengradienten an der ruhenden Muskelgrenzfläche sind 3 parallel geschaltete, teils gleich, teils entgegengesetzt gepolte Spannungsquellen. Die Permeabilitätsverhältnisse für K, Cl und Na sind sehr unterschiedlich. Der Membranwiderstand (R_m) für die Ionenpassage ist für K = 1, für Cl nur 0,23 und für Na 0,027. Die ruhende Membran ist also für Na vergleichsweise fast undurchlässig. Jede passive oder aktive elektrische Beeinflussung der Membran muß mit diesem R_m sowie dem äußeren (R_a) und dem inneren Widerstand der Faser (R_i) rechnen. An der Membran liegt parallel zur Spannungsquelle die Kapazität C_m

Alle Arten von Muskelfasern liefern bei ihrer Aktivität die bekannten Aktionspotentiale, welche bemerkenswerte Unterschiede aufweisen, für welche z. Z. noch keine rechte strukturelle oder funktionelle Erklärung zu geben ist. Die Erregungsdauer bewegt sich z. B. zwischen wenigen msec (Skeletmuskelfasern), vielen msec (Herz) und einigen Sekunden (bei manchen glatten Muskeln). Am Herzmuskel schwankt die Aktivitätsdauer je nach Art, Herzfrequenz und Temperatur außerordentlich, ohne daß man die Ursachen angeben könnte. Mitunter geht die Dauer der elektrischen Erscheinungen der Dauer der mechanischen Erscheinungen einigermaßen parallel, z. B. am Herzen. Hier möchte man meinen, daß die elektrischen Prozesse die mechanischen nicht nur einleiten, sondern auch unterhalten [ROTHSCHUH (35)]. Beim Skeletmuskel ist das Spitzenpotential viel kürzer als die Dauer der Kontraktion, doch schließen sich langdauernde negative Nachpotentiale an das Spitzenpotential an, so daß hier doch eine größere Parallelität von elektrischer und mechanischer Aktivitätsdauer existieren könnte, als gewöhnlich angenommen wird.

Der Elektrophysiologe wird versuchen, die Verschiedenheiten des Erregungsgeschehens mit einer Verschiedenheit der passiv elektrischen Eigenschaften der Muskelfasern in Zusammenhang zu bringen, also z. B. mit R_m, R_i, R_a, C_m usw. Diese Koeffizienten kann man z. B. ermitteln, indem man elektrische Stromstöße auf die Membranen einwirken läßt und dann prüft, in welcher Form eine solche örtliche Aufladung wieder abklingt, über welche Weglänge sie sich erstreckt und welche Widerstände die Fasermedien dem Stromdurchgang entgegensetzen. Solche Untersuchungen sind in neuerer Zeit viel gemacht worden [z. B. KATZ (36), vgl. H. SCHAEFER (37)], und es gibt gute Gründe, ein sog. ,,Ersatzschema" (Abb. 3) für das elektrische Verhalten der Muskelfaser zu entwerfen. Dabei wird eine Spannungsquelle überbrückt durch einen parallel geschalteten Membranwiderstand und durch eine Kapazität, die ebenfalls in die äußeren Grenzschichten zu lokalisieren ist. Sowohl die Spannungsquelle, also die Ruheladung der Membran, als auch die Widerstände der Membran und ihre Kapazität sind nun bei den einzelnen Faserarten verschieden und können bis zu einem gewissen Grade die Eigenarten des elektrischen Erregungsvorganges, insbesondere die *Geschwindigkeit* der Erregungsleitung erklären [vgl. MEVES (38)].

d) Erregungsleitung und Struktur

Es dürfte zweckmäßig sein, bei der Diskussion der Erregungsleitung an der Skeletmuskelfaser zweierlei auseinander zu halten, erstens die Fortpflanzung des Erregungsvorganges auf der Faseroberfläche und zweitens die Aktivierung des Fibrillenapparates. Da es am Nerven Erregung und Erregungsleitung ohne einen contractilen Fibrillenapparat gibt, liegt es nahe anzunehmen, daß der Erregungsvorgang auch am Muskel ohne Beteiligung der Fibrillen erfolgen kann. Dann wäre die Membranruheladung nur Sache der äußersten Fasergrenzflächen. Es spricht vieles dafür, daß es so ist, denn sicher liegt die größte Ionenkonzentrationsdifferenz zwischen dem Muskelfaserinhalt und zwischen dem Außenmilieu (Abb. 4a), und hier dürfte auch der größte Potentialsprung bestehen. Aber es ist durchaus nicht unwahrscheinlich, daß es daneben Inhomogenitäten in der Verteilung der Ionen innerhalb des Fibrillenaufbaus gibt. Selbst wenn die dahinweisenden Beobachtungen mit dem Verfahren der Mikroveraschung [DRAPER u. HODGE (39)] nicht ganz stimmen, so dürften immerhin innerhalb des morphologischen Fibrillengerüstes gewisse Ionenkonzentrationsdifferenzen und daher auch Potentialunterschiede bestehen, über deren Lagerung und Größe man allerdings nur Hypothesen aufstellen kann. Es ist beinahe wahrscheinlich, daß zwischen den Fibrillen und den intrafibrillären Sarkoplasmaräumen gewisse Potentialdifferenzen bestehen (Abb. 4b). Dafür sprechen die erwähnten Beobachtungen NATORIS (19) und die eigenen Feststellungen, daß die Muskelfasern vom Frosch, von der Wanderheuschrecke und anderen Insekten nicht selten partielle Fibrillenkontraktionen zeigen. Verletzte Fasern zeigen oft ungleichzeitige, zuckende Teilkontraktionen nebeneinander gelegener Fibrillenbündel, besonders unter NaCN-Einwirkung. Aber auch weit von verletzten Stellen können in isolierten Fasern Fibrillenkontraktionen auftreten, die ich als ,,Fibrillenzucken" beschrieben habe [ROTHSCHUH (40)]. Es gibt also sicher Partialkontraktionen an Skeletmuskelfasern, daher muß eine Aktivierung einzelner Fibrillenbündel durch irgendeinen Mechanismus erfolgen, der vielleicht elektrischer Natur ist. Neben der ,,Fibrillenpolarisation" wäre noch eine ,,Sarkomerenpolarisation" zu erwägen (Abb. 4c). Buchthal (*41*) hat einmal dahin weisende Beobachtungen gemacht. Auf die Möglichkeit einer Polarisation des endoplasmatischen Reticulums werde ich gleich zurückkommen.

Halten wir uns an die gebräuchliche Auffassung, daß die gesamten elektrischen Phänomene an der Muskelfaser Oberflächenphänomene sind, dann lassen sich die elektrischen Bedingungen für die Erregungsleitung nach Offner, Weinberg und Jung (*42*) in folgender Formel zusammenfassen:

$$V = \frac{\lambda}{\sqrt{2 \cdot R_m}} \cdot \frac{\sqrt{r}}{C_m \cdot \sqrt{R_i}}$$

Darin bedeutet λ einen Sicherheitsfaktor, R_m den spezifischen Membranwiderstand, C_m die Membrankapazität in $\mu\mathrm{F/cm^2}$ und r den Faserradius. Demnach würde die Leitungsgeschwindigkeit proportional der Wurzel aus dem Radius wachsen. Sie sollte ferner um so größer sein, je kleiner C_m, also die Speicherfähigkeit der Membran ist; denn dann wird die Membran von den Stromschleifen der Nachbarschaft schneller entladen und zur Erregung gebracht. Dazu paßt, daß die Membrankapazität beim schnelleitenden Nerven (bis 80 m/sec) außerordentlich klein (internodal $1{,}6 \cdot 10^{-5} \mu\mathrm{F}$), beim langsamer leitenden Skeletmuskel (bis 3 m/sec) um vieles größer ($6\text{—}8 \mu\mathrm{F/cm^2}$) ist. Hakansson (*43*) hat in mehreren neueren Arbeiten an einzelnen Skeletmuskelfasern vom Frosch die Leitungsgeschwindigkeit gemessen und gefunden, daß sie in Luft durchweg um die Hälfte niedriger liegt als in Ringerlösung. Das erklärt sich aus dem veränderten Verhältnis zwischen dem äußeren und inneren Leitungswiderstand. Ferner zeigte er, daß die Leitungsgeschwindigkeit nicht mit der Wurzel aus dem Faserradius, sondern linear mit dem Faserradius ansteigt, d. h. die dickeren Muskelfasern leiten beträchtlich schneller als die dünneren. Der Größenordnung nach liegen die Geschwindigkeiten etwa zwischen 1 und 3 m/sec. Ferner wächst die Aktionsamplitude bei Fasern mit größerem Radius stark an, d. h. die Stromabgabe pro Flächeneinheit der Faser ist an dicken Fasern höher als an dünnen Fasern. Hakansson (*44*) fand ferner, daß Dehnung bis zu 50% Verlängerung die Leitungsgeschwindigkeit

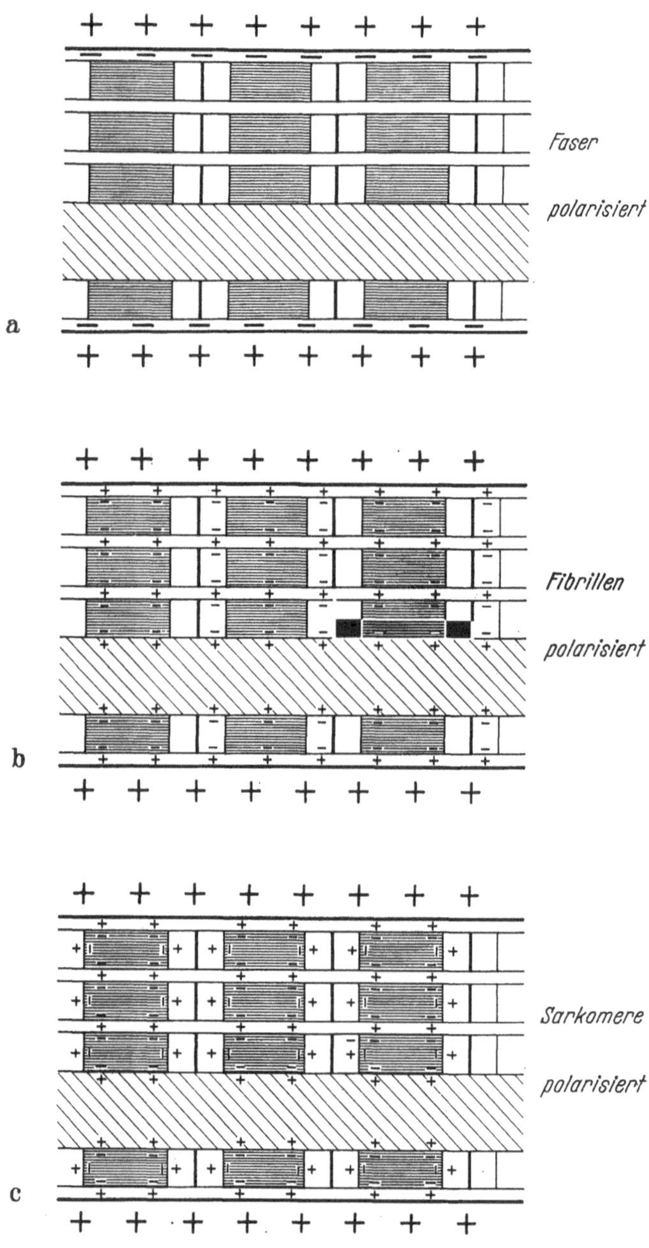

Abb. 4a—c. Die Verteilung elektrisch polarisierter Flächen bei der Skeletmuskelfaser. a) Die maßgebliche Spannungsdifferenz an der Skeletmuskelfaser liegt an der Grenze zwischen Fibrillen und Außenmedium (= Faserpolarisation); b) Gewisse Befunde, z. B. Partialkontraktionen einzelner Fibrillen und Fibrillenbündel sowie die gelegentliche Fortleitung des Kontraktionsvorgangs entlang einzelner Fibrillenbündel lassen auch an eine elektrische Spannungsdifferenz zwischen den Fibrillen und den interfibrillären Spalträumen denken (= Fibrillenpolarisation); c) Da die Ionenverteilung zwischen den verschiedenen Abschnitten der Sarkomeren kaum homogen ist, wäre auch hier an die Existenz gewisser Spannungsquellen z. B. zwischen I und A zu denken (= Sarkomerenpolarisation)

der Erregung in der Muskelfaser nicht ändert. Bei 100% Verlängerung aber sinkt das Potential auf etwa 40% der Ausgangsspannung, wider Erwarten wächst dabei die Leitungsgeschwindigkeit bis auf das 1,5fache des Ausgangswertes an, anstatt abzunehmen.

Wir werden diese Erfahrungen der Elektrophysiologen im Auge behalten müssen, wenn wir gleich zur Diskussion neuerer Auffassungen kommen, welche besonders von den Elektronenmikroskopikern [PORTER (*15*), BENNETT (*14*), KRÜGER u. GÜNTHER (*10*), MOORE u. RUSKA (*45*), RUSKA (*45a*)] über die Erregungsausbreitung über das endoplasmatische Reticulum und von anderen Autoren über eine helikoidale Erregungsleitung in gewissen Skeletmuskelfasern vertreten werden [ENGELHARDT (*46, 47*), MATTHAEI u. TIEGS (*48*)].

e) Die Aktivierung des Fibrillenapparates

Fraglos muß es irgendeinen Mechanismus geben, welcher die Aktivierung des contractilen Apparates bewirkt und vom Erregungsapparat ausgeht, denn die Erregung geht dem contractilen Geschehen um einige Zeitteilchen voran. Läßt man die Aktivierung der Fibrillen von der äußeren Fasermembran ausgehen, so ist die Geschwindigkeit, mit der die ganze Faser in Kontraktion übergeht, bemerkenswert groß. Noch ist die Frage umstritten, welche Seite des Erregungsgeschehens als aktivierender Faktor für den Fibrillenapparat maßgeblich ist (Abb. 5). Nach BAY und SZENT-GYÖRGYI (*49*) ist es die Längskomponente des elektrischen Faserfeldes ("windowfield"). Doch ist seit langem bekannt [W. BIEDERMANN (*50*),] daß eine Gleichstromdurchströmung

nur an den Ein- und Austrittsstellen der Faser auf den contractilen Apparat wirkt. Das haben auch neue Versuche von STEN-KNUDSEN (*51*) bestätigt. Wichtiger dürfte vielleicht die Querkomponente sein, die als Potentialgefälle (Feld) oder Aktionsströmung mit Ionenverschiebungen wirken könnte. SANDOW (*52*) erzielte mit einer simultanen Querdurchströmung des Muskels eine Erregung und Kontraktion. STEN-KNUDSEN (*51*) bewirkte an mit Procain unerregbar gemachten Fasern im Wechselstromfeld bei Querdurchströmung eine kräftige Kontraktion. Der erschlaffende Effekt der Anode am schlagenden Herzen oder der kathodische Kontraktureffekt am Skeletmuskel beschränken sich auch immer auf die Ein- und Austrittsstellen des Stromes. Nach FLECKENSTEIN

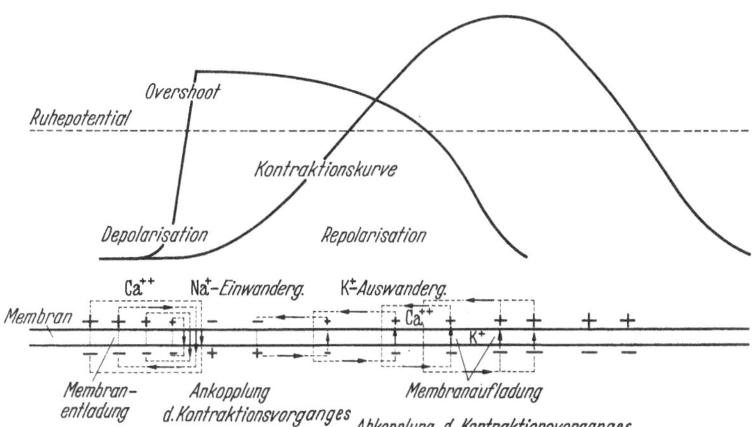

Abb. 5. Die zeitliche Zuordnung von elektrischen und mechanischen Vorgängen ist an der Herzmuskelfaser auffallend eng; das legt die Vermutung nahe, daß der Beginn der Kontraktion bei einem gewissen Grade der Membranentladung einsetzt, und zwar bei Na-Einwärtsbewegung (Ankopplung der Kontraktion). Die anfängliche Ladungsabgabe am Kopf der Erregung scheint mit Erschlaffung einherzugehen [Latenzzeit-Relaxation (*72, 73*)]. Die Dauer des Kontraktionsvorgangs ist am Herzen weitgehend von den elektrischen Prozessen an der Membran abhängig; der Gipfel der Kontraktionskurve fällt in der Regel mit dem Ende des monophasischen Erregungsvorgangs zusammen. Das Überwiegen der Wiederaufladung der Membran über die Entladung scheint den Kontraktionsvorgang abzukoppeln. Es gibt also quer zur Membran und längs der Membran gelagerte elektrische Spannungsdifferenzen und Aktionsströmungen

(*53*) ist die Erniedrigung des Membranpotentials der Anlaß für die Ankopplung des contractilen Apparates. Doch kann sich hinter gleichstarken Minderungen des Membranpotentials im Erregungsanstieg die umgekehrte Ionenverschiebung verbergen wie in der Repolarisationsphase. „Depolarisation" ist also zu unbestimmt. Sie ist mehr Folge als Ursache, nämlich Folge von Permeabilitätsänderungen, Widerstandsänderungen und Ionenbewegungen. Diese stehen dem contractilen Prozeß näher als die örtliche Membranspannung [ROTHSCHUH (*35*)]. Vor allem läßt sich die regelmäßige Beziehung zwischen K-Depolarisation und Kontraktur lösen. Wenn man einem Frosch EDTA injiziert und das Ca abbindet, tritt ähnlich wie unter Curare eine vollständige

Muskellähmung auf. Der Muskel hat noch volles Membranpotential, er läßt sich auch durch K stark depolarisieren, aber er geht trotz Depolarisation nicht in Kontraktur über. Das spricht sehr dafür, daß in dem Prozeß der Aktivierung des contractilen Apparates durch das Membrangeschehen dem Ca eine besondere Rolle zukommt (unveröffentlichte eigene Versuche). Für die besondere Bedeutung des Ca als eines Kopplungsgliedes zwischen Erregungsapparat und contractilem Apparat sprechen auch die Befunde von HEILBRUNN (54), DENTON (55), NIEDERGERKE (56—59), NASSONOW (33) und eigene Beobachtungen [ROTHSCHUH (40)]. Vielleicht ist das Ca an der Faseroberfläche oder Fibrillenoberfläche elektrostatisch gebunden und spielt hier für den Eintritt von Ionenbewegungen eine maßgebliche Rolle [vgl. FRANKENHAEUSER u. HODGKIN (60)].

Die elektronenmikroskopischen Befunde über das endoplasmatische tubulovesiculäre reticuläre System innerhalb der Muskelfasern haben zu der Annahme geführt, daß hier der Weg für die Ausbreitung der Erregung in das Innere der Muskelfaser zu suchen sei. Besonders RUSKA (45a) hat diese Theorie neuestens eingehend begründet. Es gibt in der Tat eine ganze Reihe einleuchtender Argumente dafür, z. B. das Kontinuum von Membranen, die Wahrscheinlichkeit einer verschiedenen Ionenkonzentration in den reticulären Räumen und dem angrenzenden Cytoplasma, die Existenz partialer Fibrillenkontraktionen und manches andere. Doch braucht nicht jede morphologisch sichtbare Membran nach dem oben (S. 293) Gesagten eine elektrisch erregbare Membran zu sein. Auch dürfte der Stoffaustausch über die engen endoplasmatischen Gänge zur äußeren Faseroberfläche zu langsam vor sich gehen, um die Ionengradienten aufrecht zu erhalten. Man wird aber annehmen dürfen, daß eine derart organisierte Struktur wie das endoplasmatische Reticulum keine unwesentliche Rolle in der Muskelfunktion haben wird. Man wird weitere Experimente abwarten müssen.

f) Der helikoidale Bau der Skeletmuskelfaser und das Erregungsgeschehen

In diese ganze Diskussion kommt von morphologischer Seite her der nicht zu unterschätzende neue Gesichtspunkt, daß viele Skeletmuskelfasern einen helikoidalen Bau besitzen. Das ist schon früher aus der sog. Noniusverschiebung erschlossen worden, neuerdings hat ENGELHARDT (46) an der Einzelfaser des M. cutaneus dorsi vom Frosch den Weg der Kontraktionswelle verfolgen können. Er erfolgt nach ihm auf schneckenförmigen Wegen (vielleicht zwischen zwei Z-Streifen) durch die Faser. Die benachbarten Sarkomeren wurden im Abstand von 2 msec aktiviert. Dann würde sich ein Kontraktionsvorgang, der geradlinig eine Scheingeschwindigkeit von 2 m/sec bei 50 μ Faserdicke besitzt, in Wirklichkeit im Wendelverlauf mit einer Geschwindigkeit von 60 m/sec über die Faser fortpflanzen. Diese helikoidale Fortpflanzung haben auch MATTHAEI und TIEGS (48) beobachtet. RUSKA und EDWARDS (61) entwarfen auf Grund elektronenmikroskopischer Befunde einen neuen Bauplan der helikoidalen Muskelfaser. Der helikoidale Weg der Erregungswelle durch die Muskelfaser ist z. Z. aus der Elektrophysiologie nicht erklärbar, da wir anzunehmen pflegen, daß der Erregungsvorgang auf der elektrisch homogen polarisierten Oberfläche des die Fibrillen einhüllenden Membransacks gleichmäßig und achsenparallel fortschreitet. Eine helikoidal gebänderte Membranpolarisation für die Muskelfaser anzunehmen, fällt uns schwer. Das endoplasmatische Reticulum hat auch keinen helikoidalen Verlauf. Ich bin auch noch nicht überzeugt, ob einem helikoidalen Bau auch stets eine helikoidale Erregungsausbreitung entspricht, denn bei eigenen unveröffentlichten Beobachtungen über das sog. „Wellenspiel" in isolierten Arthropodenmuskelfasern sah ich in helikoidal gebauten Fasern nicht selten achsenparallel fortschreitende Partialkontraktionen in den Fasern.

Zum Schluß möchte ich noch kurz über das neuerdings genauer analysierte Phänomen fortschreitender Desorganisation an lebenden verletzten Skeletmuskelfasern sprechen [ROTHSCHUH (18, 40, 62, 63)], welches durch den nachfolgenden Film (64) näher veranschaulicht werden soll.

g) Das Phänomen der fortschreitenden Desorganisation der Muskelfaser und das Relaxationsphänomen (mit Film)

Es ist seit langem [HERMANN (65), THOMA (66—68), STÜBEL (69), ZEIGER (70), SPEIDEL (71), NASSONOW (33)] bekannt, daß von verletzten Stellen der Muskelfaser ein Prozeß fortschreitender Strukturdesorganisation ausgeht, der sich langsam und dekrementiell über die Faser fortpflanzt.

Er beginnt damit, daß am Orte der Verletzung die zerrissenen Fibrillen in Kontraktion bzw. Kontraktur übergehen. Das zieht die Fibrillenbündel auf den Ort der Verletzung zu („Fibrillenfließen"). Kurze Zeit später verlieren sie ihre Struktur und zerfallen oder verklumpen zu unorganisierten Massen. Dieser Vorgang bedarf der Mitwirkung des Verletzungsstromes und wird durch Ca-Ionen wesentlich beschleunigt. Dabei scheint der elektrophoretische Eintransport von Ca entscheidend zu sein. Eine anodische faserparallele Gleichstromdurchströmung beschleunigt den Vorgang beträchtlich [ROTHSCHUH (40)]. In der Umgebung der Faserverletzung läßt sich unter bestimmten Bedingungen ein „Relaxationsphänomen" beobachten, welches der fortschreitenden Desorganisation vorauslaufend ebenfalls über die Faser langsam fortwandert (Abb. 6) [ROTHSCHUH (62)]. Das Phänomen entspricht vielleicht der Latenzzeit-Relaxation, welche nach RAUH (72) besonders durch A. SANDOW (73) ausführlich untersucht wurde. Eine befriedigende Erklärung des Phänomens steht noch aus.

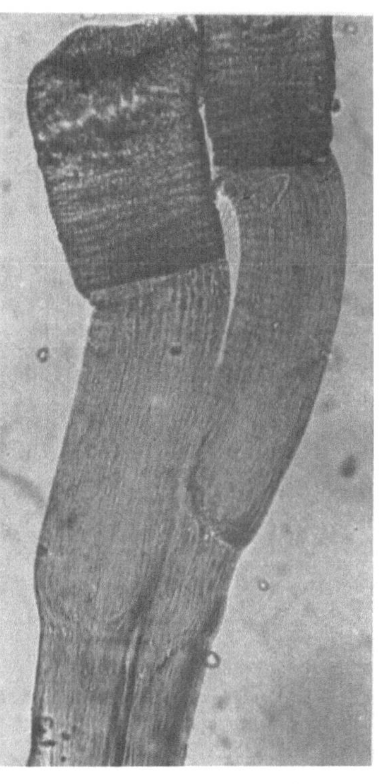

Am Herzmuskel fehlt das Phänomen der fortschreitenden Desorganisation [ROTHSCHUH (74)]. Das hängt wohl mit dem cellulären Aufbau des Herzmuskels zusammen. Ich habe schon 1950/51 (75, 76) zeigen können, daß der Herzmuskel aus potentiellen selbständigen elektrophysiologischen Elementen aufgebaut ist. Sie dürften den Abschnitten zwischen 2 Glanzstreifen entsprechen [den „Segmenten" von LINZBACH (77, 78)]. Das paßt zu den elektronenmikroskopischen Befunden von VAN BREEMEN (79), WEINSTEIN (80), SJÖSTRAND (81), POCHE und LINDNER (82) und MOORE und RUSKA (45), wonach die Fibrillen an den Glanzstreifen des Herzens endigen. Die Glanzstreifen selbst bilden morphologisch eine Diskontinuität und dürften als Zellgrenzen aufzufassen sein [MOORE u. RUSKA (45), HÄGGVIST (83)]. Dieser diskontinuierliche elektrophysiologische und anatomische Aufbau des Herzmuskels macht dem Physiologen erhebliche Schwierigkeiten bei der Erklärung der Erregungsleitung (Abb. 7) im Herzmuskel [H. SCHAEFER (84)], solange wir nicht wissen, wie die Ladungsverteilung an den Glanzstreifen ist (Abb. 8).

Abb. 6. Das Relaxationsphänomen [nach ROTHSCHUH (62)]. An frischen Faserverletzungen entwickeln sich, besonders deutlich in K-Ca-reichen Lösungen, vor dem Ort der fortschreitenden strukturellen Desorganisation (dunkle obere Zone) verdickte erschlaffte Faserasbchnitte, die ebenfalls über die Faser fortwandern. Sie scheinen mit dem Vorgang der beginnenden Membranentladung verbunden zu sein. Bei stärkerer Entladung geht die Faser in Kontraktur und Desorganisation über

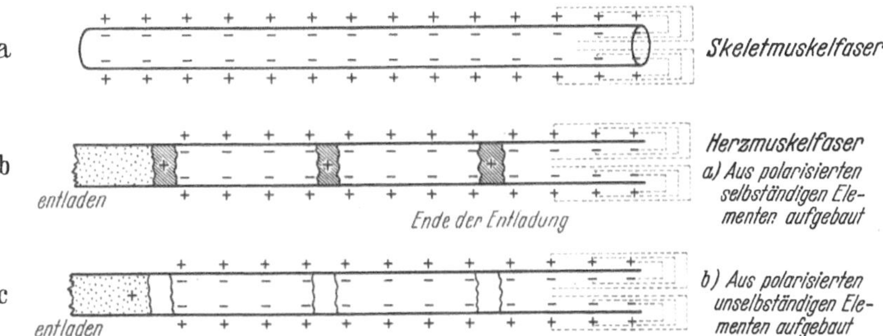

Abb. 7 a—c. Das elektrische Verhalten der verletzten Skeletmuskelfaser (oben) ist charakterisiert durch eine von verletzten Enden ausgehende Entladung, welche die ganze verletzte Faser ergreift. Von der verletzten Herzmuskelfaser entlädt sich nur ein ganz kurzes Stückchen. Das läßt entweder die Deutung zu, daß der Herzmuskel aus polarisierten selbständigen Elementen aufgebaut ist, deren Grenzen durch die Glanzstreifen dargestellt werden (a), oder daß er aus potentiell selbständigen Elementen aufgebaut ist (b)

Physiologische Funktionsanalyse und elektronenmikroskopische Strukturforschung begegnen sich also auf dem Muskelgebiet an vielen Punkten. Daß noch viele Probleme ungelöst blieben, möge als Anreiz zur Beschäftigung mit diesem reizvollen Fragenkomplex wirken.

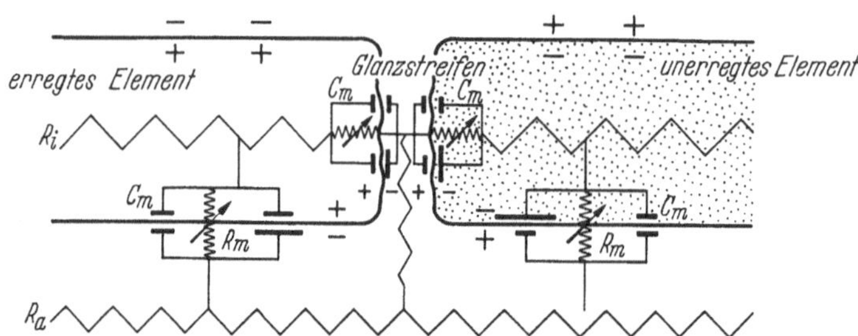

Abb. 8. Die Untergliederung des Herzmuskels in zelluläre Elemente durch die Glanzstreifen (oben) läßt die Frage entstehen, wie die Erregung von einem zum anderen Element hinübergelangt. Das kann humoral oder elektrisch gedacht werden. In letzterem Falle ist sie vielleicht kapazitiv (unten). (Symbole wie in Abb. 3)

Literatur

1. Sommerkamp, H.: Naunyn-Schmiedebergs Arch. exp. Path. Pharmak. **128**, 99 (1928).
2. Wachholder, K.: Pflügers Arch. ges. Physiol. **226**, 255, 274 (1931).
3. Kuffler, St. W.: Naunyn-Schmiedebergs Arch. exp. Path. Pharmak. **220**, 116 (1953).
4. — and E. M. Vaughan Williams: J. Physiol. **121**, 289 (1953).
5. — — J. Physiol. **121**, 318 (1953).
6. Burke, W., and B. L. Ginsborg: J. Physiol. **132**, 586 (1956).
7. — — J. Physiol. **132**, 599 (1956).
8. Krüger, P.: Tetanus und Tonus der quergestreiften Skeletmuskeln der Wirbeltiere und des Menschen. Leipzig 1952.
9. — u. P. G. Günther: Z. Biol. **109**, 41 (1956).
10. — — Acta anat. (Basel) **28**, 135 (1956).
11. Kuschinsky, G., H. Lüllmann u. Mitarb.: Anat. Anz. **103**, 116 (1956).
12. Brecht, K., u. H. Feneis: Z. Biol. **103**, 355 (1950).
13. Ruska, H.: Z. Naturforsch. **96**, 358 (1954).
14. Bennett, H. St.: Amer. J. phys. Med. **34**, 46 (1955).
15. Porter, K. R., and G. E. Palade: J. biophys. biochem. Cytol. **3**, 269 (1957).
16. Höber, R.: Physikalische Chemie der Zellen und Gewebe. 2. Aufl. Bern 1947.
17. Natori, R.: Jikeikai med. J. **1**, 18 (1954).
18. Rothschuh, K. E.: Pflügers Arch. ges. Physiol. **260**, 437 (1955).
19. Natori, R., and Ch. Isojima: Jikeikai med. J. **4**, 27 (1957).
20. — Jikeikai med. J. **2**, 1 (1955).
21. Rothschuh, K. E.: Pflügers. Arch. ges. Physiol. **263**, 589 (1956).
22. — Geschichte der Physiologie. Berlin-Göttingen-Heidelberg 1953.
23. — Entwicklungsgeschichte physiologischer Probleme in Tabellenform. München 1953.
24. Hodgkin, A. L., and B. Katz: J. Physiol. **108**, 37 (1949).
25. — Biol. Rev. **26**, 339 (1951).
26. Nastuk, W. L., and A. L. Hodgkin: J. cell. comp. Physiol. **35**, 39 (1950).
27. Ling, C., and R. W. Gerard: Nature (Lond.) **165**, 113 (1950).
28. Adrian, R. H.: J. Physiol. **133**, 631 (1956).
29. Shaw, F. H., Sh. E. Simon and B. M. Johnstone: J. gen. Physiol. **40**, 1 (1956).
29a. Lüllmann, H.: Pflügers Arch. ges. Physiol. **267**, 188 (1958).
30. Falk, G., and R. W. Gerard: J. cell. comp. Physiol. **43**, 393 (1954).
31. Ussing, H. H.: In Ion transport across membranes. Ed. H. T. Clarke and D. Nachmansohn. New York: Acad. Press 1954.
32. Segal, J.: Die Erregbarkeit der lebenden Materie. Jena 1958.
33. Nassonow, D. W., u. D. L. Rosenthal: J. gen. Biol. (Russisch) **8**, 281 (1947).
34. Segal, J.: Wiss. Z. Humboldt-Univ., Math.-Naturwiss. Reihe **3**, 243 u. 257 (1953/54); **4**, 19 u. 235 (1954/55).
35. Rothschuh, K. E.: Klin. Wschr. **34**, 1049 (1956).
36. Katz, B.: Proc. roy. Soc. B. **135**, 506 (1948).
37. Schaefer, H.: Elektrophysiologie. Bd. 1, 2. Wien 1941/42.
38. Meves, F.: Elektromedizin **2**, 225 (1957).
39. Draper, M. H., and A. I. Hodge: Nature (Lond.) **163**, 576 (1949).
40. Rothschuh, K. E.: Pflügers Arch. ges. Physiol. **261**, 557 (1955).
41. Buchthal, Fr., u. T. Péterfi: Pflügers Arch. ges. Physiol. **234**, 527 (1934).

42. Offner, F., A. Weinberg and G. Young: Bull. Math. Biophys. 2, 89 (1940).
43. Hakansson, C. H.: Acta physiol. scand. 39, 291 (1957).
44. — Acta physiol. scand. 41, 1 (1957).
45. Moore, D. H., and H. Ruska: J. biophys. biochem. Cytol. 3, 261 (1957).
45a. Ruska, H.: Experientia (Basel) 14, 117 (1958).
46. Engelhardt, A.: Z. Biol. 105, 313 (1953).
47. — Anat. Anz. 101, 233 (1954/55).
48. Matthaei, E., and O. W. Tiegs: Phil. Trans. roy. Soc., Ser. B. 238, 349 (1954/55).
49. Bay, Z., M. C. Goddall and A. Szent-Györgyi: Bull. Math. Biophys. 15, Nr. 1 (1953).
50. Biedermann, W.: Elektrophysiologie. Bd. 1. Jena 1895.
51. Sten-Knudsen, O.: J. Physiol. 125, 396 (1954).
52. Sandow, A.: Fed. Proc. 7, 107 (1948).
53. Fleckenstein, A.: Der Na-K-Austausch als Energieprinzip in Muskel und Nerv. Berlin-Göttingen-Heidelberg 1955.
54. Heilbrunn, V. L.: The Dynamics of Living Protoplasm. New York 1956.
55. Denton, E. J.: J. Physiol. 107, 32 P (1948).
56. Niedergerke, R.: J. Physiol. 128, 12 P (1955).
57. — J. Physiol. 134, 584 (1956).
58. — J. Physiol. 134, 569 (1956).
59. — Nature (Lond.) 179, 1066 (1957).
60. Frankenhaeuser, B., and A. L. Hodgkin: J. Physiol. 137, 217 (1957).
61. Ruska, H., and G. A. Edwards: Growth 21, 73 (1957).
62. Rothschuh, K. E.: Pflügers. Arch. ges. Physiol. 262, 354 (1956).
63. — Pflügers Arch. ges. Physiol. 263, 411 (1956).
64. — (Film): Fortschreitende Strukturveränderungen an verletzten Skeletmuskelfasern. Inst. wiss. Film. Göttingen (1956).
65. Hermann, L.: Allgemeine Muskelphysik. Handbuch der Physiologie, Bd. I, S. 235 (1879).
66. Thoma, R.: I. Virchows Arch. path. Anat. 186, 64 (1906).
67. — II. Virchows Arch. path. Anat. 195, 93 (1909).
68. — III. Virchows Arch. path. Anat. 200, 22 (1910).
69. Stübel, H.: Pflügers Arch. ges. Physiol. 180, 209 (1920).
70. Zeiger, K., u. H. Schreiber: Z. Zellforsch. 4, 617 (1927).
71. Speidel, C. C.: Amer. J. Anat. 65, 471 (1939).
72. Rauh, F.: Z. Biol. 76, 25 (1922).
73. Sandow, A.: J. cell. comp. Physiol. 24, 221 (1944).
74. Rothschuh, K. E.: Pflügers Arch. ges. Physiol. 266, H. 1 (1957).
75. — Pflügers Arch. ges. Physiol. 252, 445 (1950).
76. — Pflügers Arch. ges. Physiol. 253, 238 (1951).
77. Linzbach, J.: Virchows Arch. path. Anat. 318, 575 (1950).
78. — Verh. dtsch. Ges. Path. 32, 143 (1951).
79. Breemen, V. L. van: Anat. Rec. 117, 355 (1953).
80. Weinstein, H. J.: Exp. Cell. Res. 7, 130 (1954).
81. Sjöstrand, F. J., u. E. Andersson: Experientia (Basel) 10, 369 (1954).
82. Poche, R., u. E. Lindner: Z. Zellforsch. 43, 104 (1955).
83. Hägqvist, G.: Gewebe und Systeme der Muskulatur. Handbuch der mikroskopischen Anatomie des Menschen. Bd. 2, Teil 3 u. 4. Berlin 1931 u. 1956.
84. Schaefer, H.: Pflügers Arch. ges. Physiol. 255, 251 (1952).

Comparative studies on the fine structure of motor units

George A. Edwards

From the Division of Laboratories and Research, New York State Department of Health, Albany (USA)

The problem in the consideration of the comparative aspects of motor units is that of determining whether muscle response is due to one given mechanism involving components of like structure, or to several mechanisms and unlike components in both nerve and muscle. Not long ago this problem was considered essentially physiologic, i. e., that variation in muscle response depended upon differences in mode of innervation of fixed muscle types and also upon the inherent properties of contractile substances.It is now evident that modifications occur in the fundamental intracellular components in relation to specific functional differences, and that differences exist among axons and neuromuscular junctions, as well as among muscle fibers.

1. Contractile material

The contractile material of the muscle cell apparently consists of interacting protein chains, morphologically represented as either one or two types of myofilaments [cf. (*1, 2, 3, 4*) for discussion] which may vary in thickness, axial period, and arrangement. Striated muscle falls into two major categories on the basis of the arrangement of the contractile material within the cell, viz., afibrillar and fibrillar.

The afibrillar (tonic) muscle fibers, e. g., anterior latissimus dorsi of the pigeon (Fig. 1), soleus of the cat, and some abdominal and visceral muscles of insects, are characterized by the arrangement of the contractile material into "fields" (*5*), or irregular masses, of contractile filaments separated by a scarce, irregular endoplasmic reticulum and few mitochondria. The cytolemma of the afibrillar muscle fiber is variable in width, but usually thicker than that of fibrillar fibers. The cell border is often highly scalloped or deeply rugose, particularly near nerves, and may show considerable pinocytosis. The cytoplasmic area surrounding the nucleus contains a Golgi apparatus, few mitochondria and numerous granules (Fig. 1), but very few myofilaments, thus resembling the residual cytoplasmic area of smooth muscle [cf. (*6*)]. The mitochondria are scarce, long, irregular and possess few cristae. The fields of myofilaments are widely separated by membranes of the endoplasmic reticulum and by dense granules. The contractile material consists of irregular bundles of thick and thin filaments interconnected by short bridges.

A given muscle may be purely afibrillar, e. g., the anterior latissimus dorsi, or mixed afibrillar and fibrillar, e. g., rat diaphragm, or may be purely fibrillar as for example the gastrocnemius. Fibrillar (phasic, tetanic) muscle fibers, which comprise the majority of muscles of the body, are characterized by the arrangement of the contractile material into distinct myofibrils which are usually cylindrical in cross sectional outline. The fibrillar muscle fiber type may be further divided into several categories. High frequency fibers are found in hummingbird flight muscle (*7*), cicada tymbal muscle (*8*) and all flight muscle of higher insect orders. The myofibrils of the high frequency fibers are smoothly cylindrical, large diametered, and are surrounded by large mitochondria and few elements of the endoplasmic reticulum. The myofibrils do not unite or split.

The more general low frequency fibrillar fibers are characterized by distinct myofibrils, which may unite or split. The fibrils may be closely packed, so as to give an appearance of an almost continuous "Grundmembran", as in the iliofibularis of the frog (Fig. 3), or may be more widely separated, as in the semitendinosus of the mouse (Fig. 4). The fibrils are surrounded by an abundant endoplasmic reticulum. The mitochondria are usually small, may be non-uniformly distributed among the fibrils (Fig. 3), but more often surround the I region of the sarcomere in pairs (Fig. 4).

A variant of this type contains band-shaped myofibrils, i. e., in the form of flattened cylinders. In cross section such muscle fibers may show, as in wasp femoral muscle (*9*), a radial array of fibrils around a central nucleus. The endoplasmic reticulum may be well-developed, and the mitochondria surround the fibrils in pairs at the I region. An unusual type of banded fiber is seen in dragonfly flight muscle (Fig. 2), in which the fibrils are separated by large, cristae-filled mitochondria, each of which may have a length equal to one or more sarcomeres.

It is evident from the above brief survey that variations in contractile material, endoplasmic reticulum, and mitochondria occur within striated muscle. The particular arrangement of these components in a given muscle fiber is directly related to the specific function of that fiber. The variations in contractile material are related to the mechanical events within the cell. The size, form and distribution of mitochondria are associated with the energy requirements for the reconstitution of the cell potentials and structure. The endomembrane variations among the several cell types can most logically be linked to differences in intracellular potentials and to the transmission of impulses into and throughout the cell (*10*).

2. Peripheral nerve

It has long been proposed, on the basis of electro-physiologic measurements, that differences in muscle function could be explained by type of innervation, i. e., fast, slow, or inhibitory axons to a group of mixed muscle fiber types, or to a given muscle fiber (multiterminal innervation).

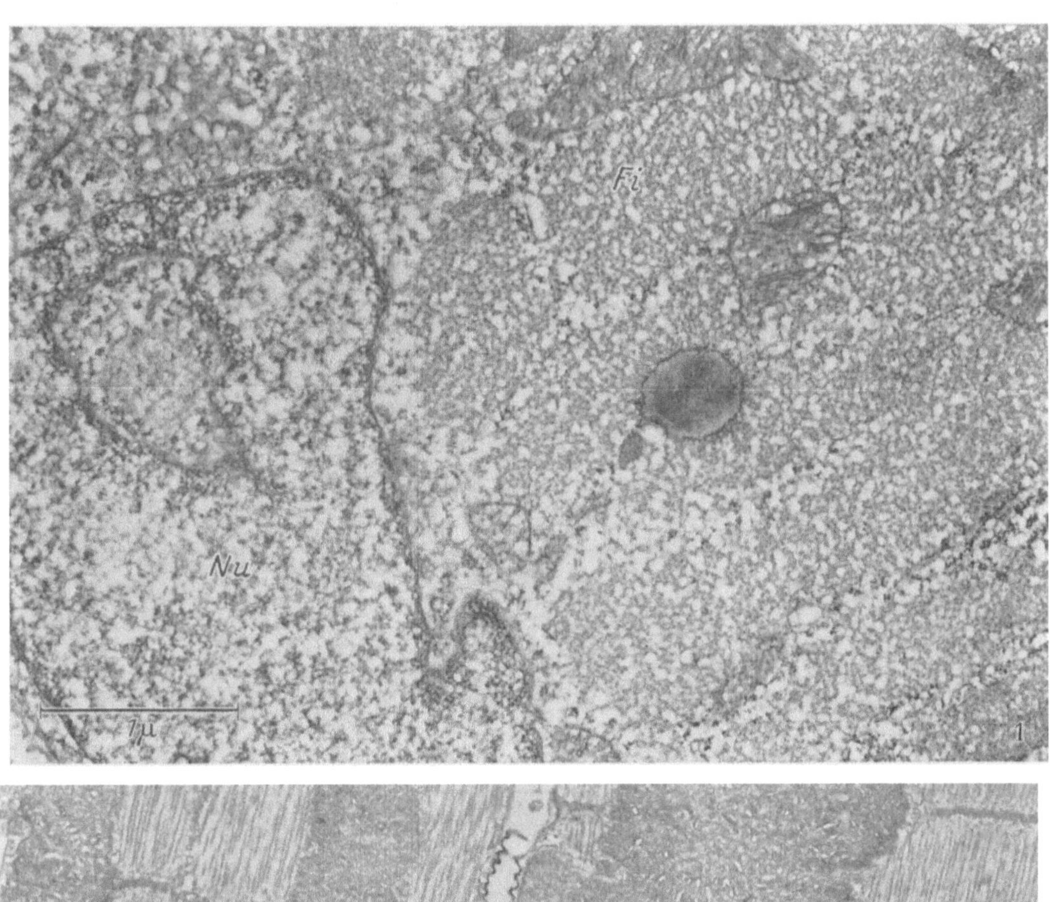

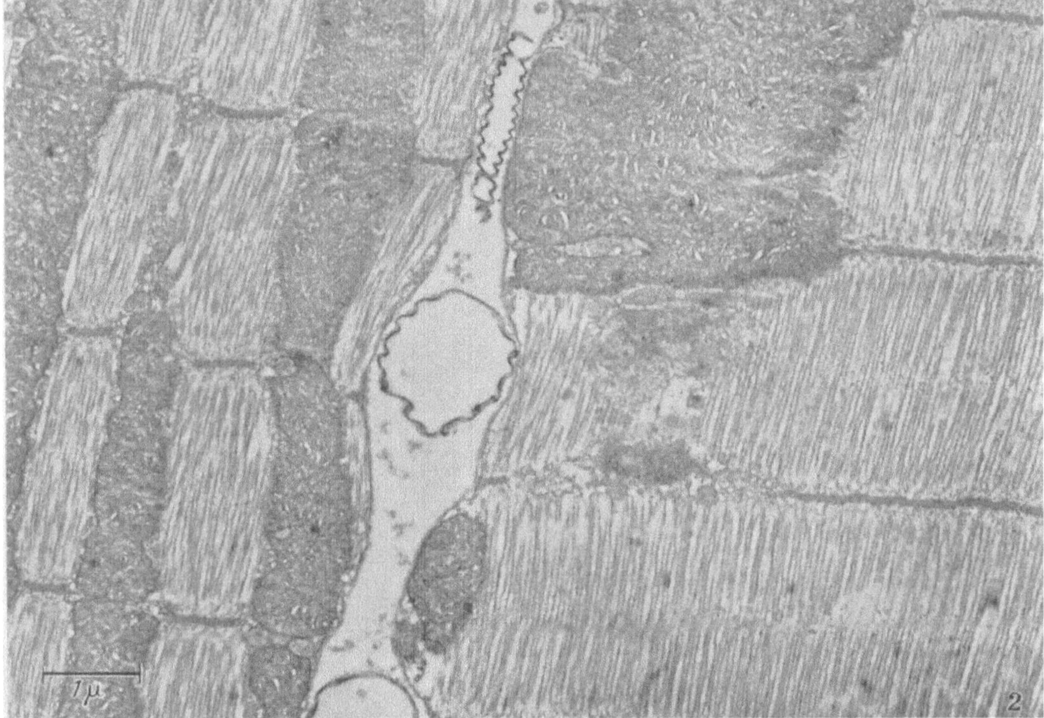

Fig. 1. Cross section of anterior latissimus dorsi of pigeon, showing nucleus (*Nu*), relatively empty perinuclear aera, and "fields" of contractile material (*Fi*), separated by granules and scarce mitochondria. 27500 ×

Fig. 2. Longitudinal section through two adjacent banded fibers of dragonfly flight muscle. Plane of section passes through short axis of fibrils in fiber at left, and through long axis in fiber at right, demonstrating that fibrils are flattened cylinders. 13500 ×

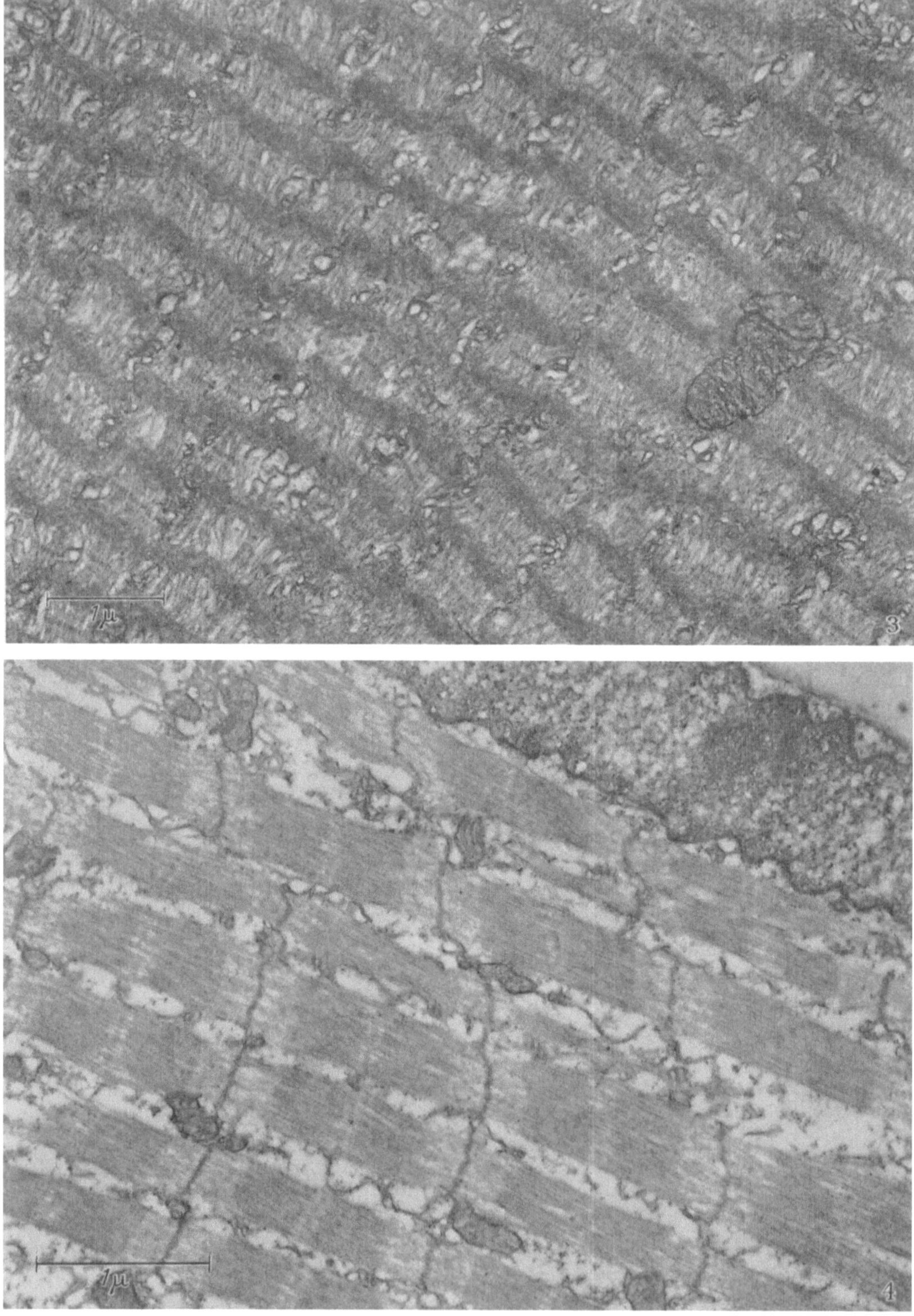

Fig. 3. Longitudinal section through iliofibularis of frog. Although a fibrillar fiber, the fibrils are so closely packed as to give the impression of a tonic fiber. 18 000 ×

Fig. 4. Longitudinal section of typical fibrillar fiber of the semitendinosus of mouse. This muscle being mixed, other fibers may show the "field" structure illustrated in Fig. 1. 27 000 ×

Electron microscopic studies of peripheral nerves indeed show differences in numbers of axons to a given muscle and also differences in size, tunication or myelination, and cytoplasmic content of

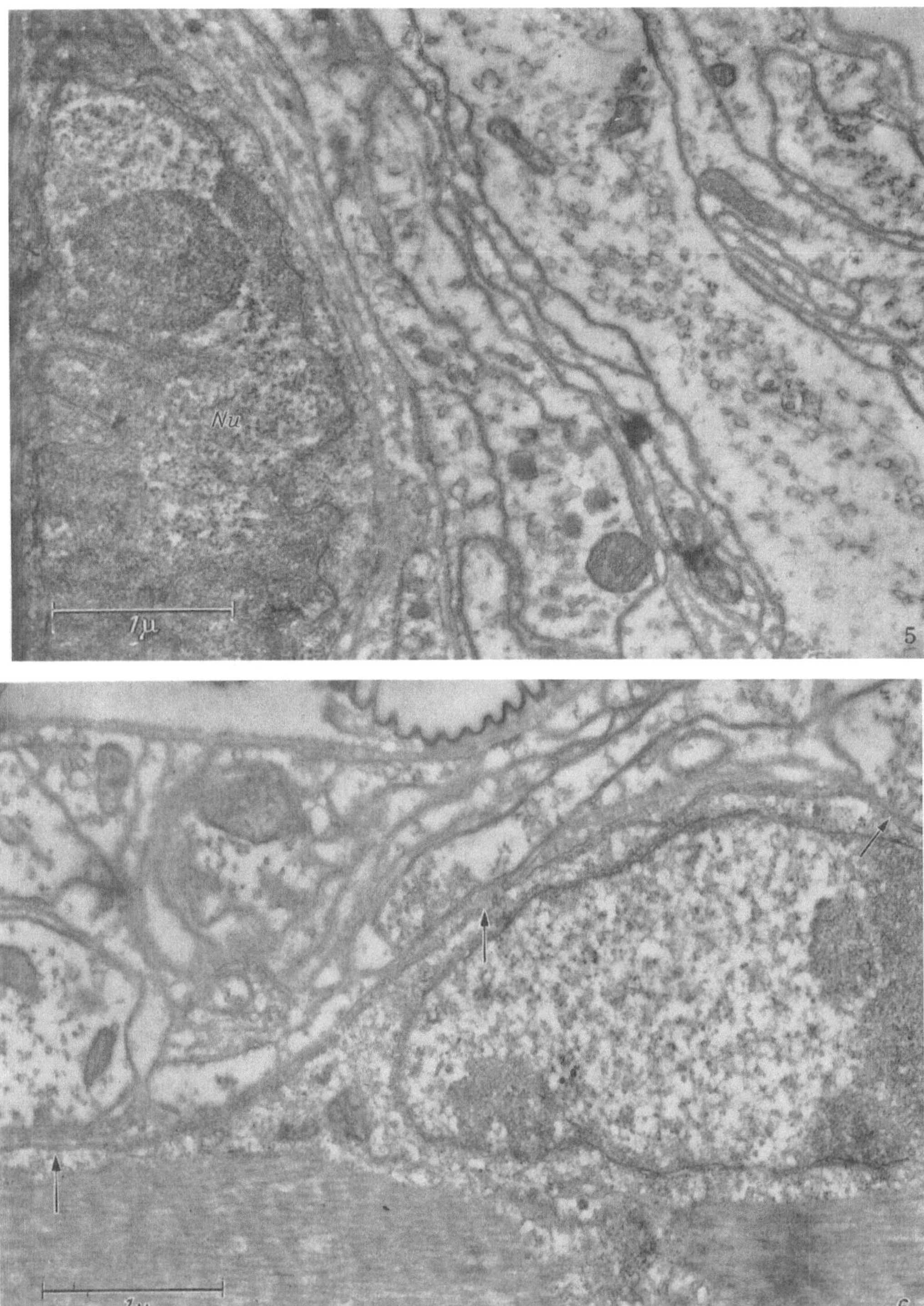

Fig. 5. Longitudinal section through peripheral nerve contacting dragonfly muscle fiber whose nucleus (*Nu*) is seen at left. The contacting nerve is multiaxonal and the axons contain both filaments and vesicles. 27000 ×

Fig. 6. Multiterminal innervation of a dragonfly thoracic longitudinal muscle. Three synapses are visible along the fiber (arrows). 27000 ×

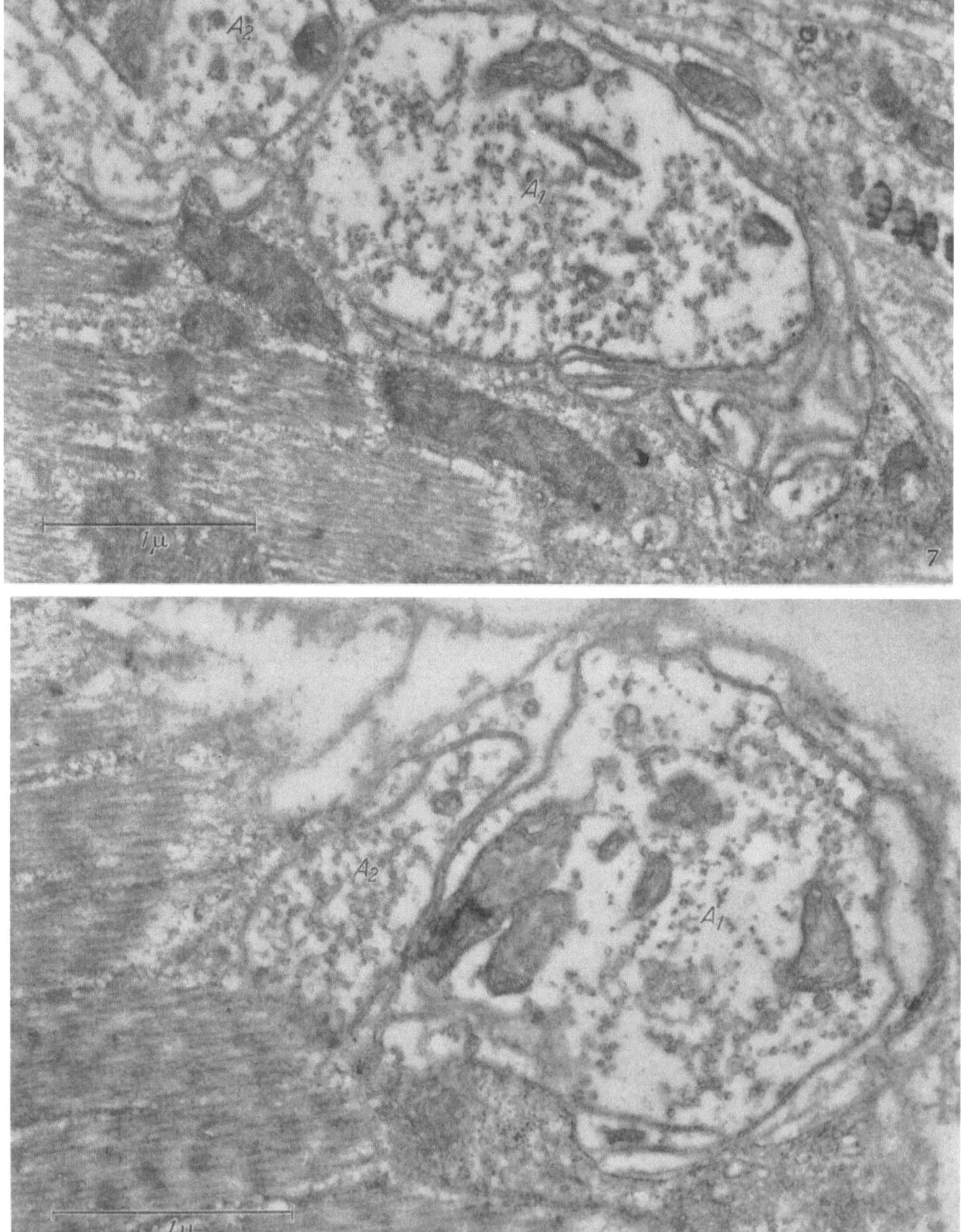

Fig. 7. Synapse of a pair of axons (A_1 and A_2) with the muscle fiber in the dragonfly. Noteworthy is the system of membranes in the muscle cytoplasm in the synaptic region. 31500 ×

Fig. 8. Synapses of nerve and a pair of axons (A_1 and A_2) in the thoracic longitudinal muscle of the dragonfly. Such a pair could represent the "slow" and "fast" axons demonstrated physiologically. 36000 ×

the axons of a given nerve as it approaches, contacts, and synapses with the muscle cell. The axons are always ensheathed by the membranes and a variable amount of cytoplasm of the lemnoblast. A given lemnoblast may ensheath one or a number of axons, either myelinated or unmyelinated (*11*). The mesaxon may be a simple pocket-like invagination of the plasma membrane, it may make several loose turns around the axon as in insect tunicated nerve, or it may form a number of tightly wound layers as in the myelinated axons of vertebrates. While still distant from the muscle fiber the axons contain mitochondria and neurofilaments but no vesicles. As the nerve contacts the muscle fiber (Fig. 5) the axons contain vesicles as well as filaments. In the terminal, synapsing portions of the axon, mitochondria and vesicles are present in large numbers, but no neurofilaments are observed (Fig. 6, 7 and 8).

3. Neuromuscular junctions

At least three types of neuromuscular junctions are observed in striated muscle. Uniaxonal, unijunctional synapses occur in vertebrate fibrillar muscle fibers (*12*). Multiaxonal, unijunctional synapses have been observed in wasp femoral muscle (*9*) and dragonfly flight muscle. Evidence of multiaxonal, multijunctional synapses has now been obtained from the longitudinal abdominal muscle of the cockroach and the longitudinal thoracic muscles of the dragonfly (Figs. 5—8). In the cockroach the peripheral nerve innervating several muscle fibers generally contains one large central and several small peripheral axons, each more or less tunicated. The number of synapses with a given muscle fiber varies from one to four, the synapses being spaced about 15 to 25 mμ apart. A given synapse involves one axon only.

The contact and synapses of the peripheral nerve and muscle fiber in the dragonfly are more similar to those expected on the basis of physiologic and light microscopic studies (*13*). The multiaxonal nerve contacts the muscle fiber at right angles, then spreads around and along the fiber (Fig. 5) as in a Doyere's hillock [cf. (*14*)]. As in the nerve of the cockroach, the components are generally one large central axon and several smaller peripheral axons, all tunicated, within a given lemnoblast. Contact of the surfaces of nerve branches and muscle is usually mediated by a tracheoblast. Different from the situation in the cockroach, the spreading nerve branches along the dragonfly muscle fiber usually contain several axons (Fig. 5 and 6). Multiple synapses occur between axons and a single muscle fiber (Fig. 6). Perhaps most significant is the fact that the final synapses are between pairs of axons and the synaptic regions of the muscle fiber, usually one member of the pair being larger than the other (Fig. 7 and 8). Such a pair could represent "fast" and "slow" nerve fibers. The synapsing axons are characterized by large numbers of mitochondria and myriads of synaptic vesicles. The synapse is by apposition of axonal and muscle fiber plasma membranes only. No basement membrane material intervenes. The muscle cytoplasm in the synaptic region contains aposynaptic granules, numerous tubules and vesicles of the endoplasmic reticulum, and many mitochondria. In some areas of the synapse (Fig. 8) the axonal vesicles appear disrupted, the plasma membranes incomplete, and the aposynaptic granules aggregated close to the synapsing membranes. These structural modifications must be related to the release of transmitter substance, its transmission from nerve to muscle and the reaction of the muscle cytoplasm.

This study was aided by grants from the National Heart Institute of the Department of Health, Education and Welfare, Bethesda, and from the Brown-Hazen Fund of the Research Corporation, New York City.

References

1. HODGE, A. J.: J. biophys. biochem. Cytol. **2**, (Suppl.), 131 (1956).
2. SPIRO, D.: J. biophys. biochem. Cytol. **2**, (Suppl.), 157 (1956).
3. HUXLEY, A. F.: Brit. med. Bull. **12**, 167 (1956).
4. — Brit. med. Bull. **12**, 171 (1956).
5. KRÜGER, P.: Tetanus und Tonus der quergestreiften Skeletmuskeln der Wirbeltiere und des Menschen. Leipzig: Akad. Verlagsges. 1952.
6. CAESAR, R., G. A. EDWARDS and H. RUSKA: J. biophys. biochem. Cytol., **3**, 867 (1957).
7. EDWARDS, G. A., H. RUSKA, P. SOUZA SANTOS and A. VALLEJO-FREIRE: J. biophys. biochem. Cytol. **2**, (Suppl.), 143 (1956).

8. EDWARDS, G. A., H. RUSKA and E. DE HARVEN: J. biophys. biochem. Cytol. 4, 251 (1958b).
9. — — — J. biophys. biochem. Cytol. 4, 107 (1958a).
10. RUSKA, H., G. A. EDWARDS and R. CAESAR: Experientia (Basel) 14, 117 (1958).
11. HESS, A.: Proc. roy. Soc. B 144, 496 (1956).
12. ROBERTSON, J. D.: J. biophys. biochem. Cytol. 2, 381 (1956).
13. ROEDER, K. D.: Insect Physiology. New York: John Wiley and Sons, Inc. 1953.
14. HOYLE, G.: Nervous control of insect muscles, in: Recent advances in invertebrate physiology, University
of Oregon Publications, Eugene, Oregon 1957.

Experimentelle Pathologie des Herzmuskels

REINHARD POCHE
Pathologisches Institut der Medizinischen Akademie Düsseldorf

Die Beurteilung und Bewertung elektronenmikroskopischer Befunde am Herzmuskel nach experimentellen Eingriffen setzt die Kenntnis der normalen submikroskopischen Morphologie des Herzmuskels voraus. Andererseits können experimentelle Untersuchungen auch unseren Einblick in die normale submikroskopische Morphologie erweitern und vertiefen. Es sei deshalb zunächst eine kurze zusammenfassende Darstellung der submikroskopischen Struktur des normalen Herzmuskels gegeben, die sich im wesentlichen auf die Arbeiten von BEAMS u. Mitarb. (1), KISCH (9, 10), LINDNER (11, 12, 13, 14, 15), VAN BREEMEN (2), HODGE u. Mitarb. (8), SJÖSTRAND u. Mitarb. (36, 38), WEINSTEIN (39), POCHE und LINDNER (16, 33), PRICE u. Mitarb. (35), EDWARDS u. Mitarb. (3), HIBBS (5), MUIR (24, 25), MOORE und RUSKA (23), POCHE (27, 28, 31), PORTER und PALADE (34), FAWCETT und SELBY (4) stützt.

Der Herzmuskel setzt sich aus individuellen Zellen zusammen, stellt also kein Syncytium dar (2, 16, 23, 28, 33, 35, 36, 38). Diese Zellen sind allseitig von einer scharf konturierten Plasmamembran, der Protomembran nach LINDNER (14), umgeben. Die Zellgrenzen zwischen den einzelnen Herzmuskelzellen verlaufen in den Glanzstreifen oder interkalaren Scheiben. Im Bereich der Zellgrenzen, an denen auch die Myofibrillen enden, sind die aneinandergrenzenden Herzmuskelzellen mehr oder weniger stark miteinander verzahnt. Dementsprechend sieht man auf elektronenmikroskopischen Bildern von Längsschnittpräparaten eine mehr oder weniger stark geschlängelte Doppelmembran, die aus den Protomembranen der beiden Herzmuskelzellen und einem wenig elektronendichten Spatium besteht. Auf Querschnitten von Glanzstreifen findet man dagegen ein aus einer derartigen Doppelmembran bestehendes Netzwerk, das sich je nach der Schnittführung in einzelne ungleichmäßige Ringe auflösen kann (23, 28, 33); [Schemata bei POCHE u. LINDNER (33) und POCHE (28)]. Eine solche Doppelmembran, die aus den Protomembranen zweier unmittelbar aneinandergrenzender Herzmuskelzellen und dem Spatium besteht, bezeichnen wir als „Mesomembran" und sprechen von einer „primären Mesomembran", wenn diese Zellgrenze keine weitere Differenzierung zeigt (14). Die auf eine Mesomembran zulaufenden Myofibrillen sind dort in einer der Innenseite der Protomembran angelagerten kontrastreichen Substanz verankert, die wir als innere Kittsubstanz bezeichnet haben (16, 33). Eine solche innere Kittsubstanz findet sich aber auch an anderen Stellen der Mesomembran, wo es auf einen besonders festen Zusammenhalt der Herzmuskelzellen ankommt (14, 31, 38). Diese besonders differenzierten Anteile der Mesomembran nennen wir mit LINDNER (14) „sekundäre Mesomembran". Im Bereich der Glanzstreifen verbinden die sekundären Mesomembranen die Myofibrillen zu einem den Zellen übergeordneten System und ermöglichen dadurch eine koordinierte Kontraktion größerer Herzmuskelabschnitte (14). Gegen die Intercellularspalten oder das interstitielle Bindegewebe ist die Herzmuskelzelle durch die „Exomembran" (14) abgegrenzt. Diese besteht aus der Protomembran, einem wenig elektronendichten Spatium und der sog. „Perimembran", die der Basalmembran der interstitiellen Blutcapillaren ähnelt und wie diese an manchen Stellen eine Verbindung mit kleinen kollagenen Fibrillen des interstitiellen Bindegewebes erkennen läßt. Die Exomembran entspricht dem Begriff des Sarkolemms von PERRY (26). Die Perimembran ist bei den einzelnen Arten verschieden deutlich ausgebildet; so ist sie z. B. bei der Ratte weniger scharf konturiert als beim Meerschweinchen, beim Hund oder beim Menschen. An der Bildung der

Glanzstreifen ist die Perimembran nicht beteiligt, sondern zieht kontinuierlich über diese hinweg. Protomembran und Perimembran bilden jedoch manchmal gemeinsame Falten oder fingerförmige Einstülpungen, die auf die Z-Streifen gerichtet sind und anscheinend als kleine Kanälchen tief in die Herzmuskelzelle eindringen können (*14, 31*). Die Protomembran der Herzmuskelzelle zeigt wie die Plasmamembran der Capillarendothelien das Phänomen der Membranvesiculation, das wir als morphologischen Ausdruck des Stoffdurchtrittes durch die Membran ansehen.

Die Kerne der Herzmuskelzellen sind meist zentral gelegen; man findet sie aber nicht selten auch dicht unter dem Sarkolemm. Das Karyoplasma enthält kleine Granula sowie feine kurze Filamente. Es ist gegen das Sarkoplasma durch eine doppelte Membran, die sog. primäre und die cytoplasmatische oder sekundäre Kernmembran abgegrenzt. Vereinzelt zeigt die Kernmembran echte Poren. Die Kernkörperchen setzen sich aus dichten Ansammlungen kleiner Granula zusammen.

Die Myofibrillen des Herzmuskels bestehen, wie die anderer quergestreifter Muskeln (*7, 8, 37*), aus kleinen Untereinheiten, den Subfibrillen oder Myofilamenten, die untereinander durch feine filamentöse Querbrücken verbunden sind. Sie zeigen die gleiche klassische Querstreifung wie die Myofibrillen der Skeletmuskelfasern. Zum Unterschied vom Skeletmuskel können sich im Herzmuskel breite Myofibrillen in schmale aufzweigen; sie bilden innerhalb der Herzmuskelzelle ein synfibrilläres System.

Mitochondrien sind im Herzmuskel wesentlich zahlreicher als im Skeletmuskel. Sie liegen in einfachen oder auch doppelten Reihen zwischen den Myofibrillen. Größere Anhäufungen von Mitochondrien finden sich perinucleär und im fibrillenfreien zentralen Sarkoplasma sowie auch unter dem Sarkolemm. Neben Mitochondrien enthält das Sarkoplasma der Herzmuskelzellen — allerdings in wesentlich geringerer Zahl — auch andere Cytosomen wie dichte Körper (dense bodies, micro bodies), sog. osmiophile Körper und granuläre Cytosomen (Pigmentkörner, Lipofuscin) sowie Fetttropfen. Ferner finden sich im Grundsarkoplasma small dense particles oder Paladesche Granula.

Die Herzmuskelzelle besitzt wie die Skeletmuskelfaser ein weitverbreitetes Sarkoplasmareticulum (*3, 14, 23, 31, 34*). In Höhe der A-Abschnitte der Sarkomere findet man meist in Längsrichtung des Myofibrillenverlaufes angeordnete kleine Bläschen oder Tubuli. Ähnliche Gebilde liegen auch dicht unter der Protomembran und in der Nähe der Kernmembran. In Höhe der I-Abschnitte der Sarkomere finden sich Tubuli, die oft quer zur Verlaufsrichtung der Myofibrillen gestellt sind, und deren Wand nach Untersuchungen am Herzmuskel des Hundes und am menschlichen Herzmuskel aus zwei Membranen besteht, von denen die äußere der Protomembran und die innere der Perimembran entspricht. Diese Tubuli haben eine solche Ähnlichkeit mit den oben erwähnten fingerförmigen Einstülpungen der Exomembran, daß sie als deren Fortsetzungen gedeutet werden können (*14, 31*). Im Herzmuskel der Ratte ist die Erkennung dieser Strukturen dadurch erschwert, daß die Perimembran hier nur mäßig ausgebildet ist. Andererseits findet man aber bei der Ratte in Höhe von i_1, d. h. der unmittelbar an die Z-Streifen angrenzenden Anteile der I-Abschnitte, Einheiten des Sarkoplasmareticulums, die aus einem rundlichen Bläschen und einem oder zwei U-förmig darum gelagerten schmalen Tubuli bestehen. Diese Einheiten werden als Diaden oder Triaden bezeichnet (*34*). Nach unserer heutigen Auffassung dient das Sarkoplasmareticulum in seiner Gesamtheit dem Stofftransport und der Erregungsleitung in die Zelle hinein. Ein Golgi-Apparat gehört nicht zu den häufigen Befunden in der Herzmuskelzelle, wird aber gelegentlich beobachtet. Ergastoplasmamembranen dagegen werden im normalen Herzmuskel gewöhnlich nicht gefunden. Wir haben sie bisher nur einmal in hypertrophierenden Herzmuskelzellen gesehen (*31*).

Morphologische Veränderungen als Folge von experimentellen Eingriffen können sich nun grundsätzlich an allen erwähnten Strukturelementen der Herzmuskelzelle manifestieren. Dabei trifft man aber auf die Schwierigkeit, daß gewisse Zellorganellen, insbesondere die Mitochondrien und das Sarkoplasmareticulum, in ihrer Struktur sehr labil sind. Sowohl das ganze Herz als auch kleine Herzmuskelstückchen oder selbst einzelne Herzmuskelfasern sind imstande, nach der Isolierung aus dem Organismus weiter zu schlagen. Die Herzmuskelzellen halten in diesem Falle ihre Funktion und damit ihren hohen Funktionsstoffwechsel unter Mangelbedingungen aufrecht. Dies führt sehr schnell zu Störungen, die in Strukturveränderungen wie Schwellung der Mitochondrien und des Sarkoplasmareticulums ihren morphologischen Ausdruck finden. Solche agonalen Veränderungen sind aber höchst unerwünscht, wenn man an diesen Zellorganellen andererseits auch experimentell-pathologische Veränderungen erwartet, wie z. B. im Sauerstoffmangelexperiment. Nach den Untersuchungen von LINDNER (*14*) kommt es nach Überdosierung von Strophanthin und Digitoxin zur Ischämie des Herzmuskels und schließlich zu einem systolischen Stillstand der Herzkammern. Elektronenmikroskopisch findet sich dabei eine Abhebung der Protomembran von der Muskelfibrille, eine Schwellung und Verklumpung der Mitochondrien, eine Vacuolisierung des Sarkoplasmas sowie eine Schwellung des Zellkernes. MÖLBERT (*21*) sah Mitochondrienschwellungen nach Sauerstoffmangel im Unterdruckversuch. Diese Beispiele zeigen, daß es schwer sein kann, agonale und experimentelle Veränderungen zu unterscheiden. Die Vermeidung agonaler Veränderungen ist eine Frage der Präparation. Bisher sind wir — wie wohl die meisten Untersucher — so vorgegangen, daß wir am narkotisierten Tier den Thorax eröffnet und dann möglichst schnell ein Stück Herzmuskel entnommen und in Osmiumtetroxydlösung gebracht haben. Die weitere Zerkleinerung des Materials zu Stückchen von etwa 1 mm Kantenlänge erfolgte dann schon in der Fixierungsflüssigkeit. Bei dieser herkömmlichen Stückfixierung lassen sich jedoch Schwellungen einzelner Mitochondrien nie ganz vermeiden. Man muß deshalb, um zuverlässige Ergebnisse zu erreichen, immer streng die gleiche Technik einhalten, viele Blöcke, und von diesen wiederum viele Schnitte untersuchen. PRICE u. Mitarb. (*35*) haben bei Hunden durch Gewebspunktion mit der Silverman-Nadel kleine Herzmuskelstückchen gewonnen, die dann sofort fixiert werden konnten. SJÖSTRAND u. Mitarb. (*38*) fixierten bei Mäusen und Fröschen das ganze Herz in situ durch Injektion des Fixierungsmittels in den Herzbeutel. LINDNER und WELLENSIEK (*17*) haben das isolierte Meerschweinchenherz in der Langendorff-Apparatur durch Perfusion mit Osmiumlösung oder Formalin fixiert. Bei allen diesen Methoden ist aber auch nicht die Gewähr gegeben, daß alle Herzmuskelzellen in jedem Falle gleichzeitig von dem Fixierungsmittel erreicht werden. Außerdem sind alle drei Methoden nur in bestimmten Spezialfällen anwendbar. Um agonale Veränderungen zu vermeiden, ist es notwendig, die vitale Struktur vor der Fixierung zu stabilisieren (*19, 20*). Eine solche Stabilisierung der vitalen Struktur läßt sich durch Kaliumcitrat erreichen. Kaliumcitrat, das in der modernen Chirurgie bei Herzoperationen mit Ausschaltung des Herzens aus dem Kreislauf angewendet wird, führt zu einem reversiblen diastolischen Herzstillstand. Dadurch werden die Kontraktionen und damit der hohe Funktionsstoffwechsel schlagartig ausgeschaltet. Wir haben in Zusammenarbeit mit dem Oberarzt der Chirurgischen Klinik der Medizinischen Akademie Düsseldorf, Herrn Doz. Dr. B. LÖHR, bei Hunden durch Injektion von etwa 30 cm³ einer 2,5 %igen Kaliumcitratlösung in die Aorta unter Abklemmung der Aorta einen diastolischen Herzstillstand erzeugt. Dabei wurde die Herzaktion im EKG verfolgt. Der Stillstand trat etwa 4 sec nach Injektionsbeginn ein. Die Hunde wurden künstlich beatmet; der CO_2-Gehalt der Ausatmungsluft betrug um 3 %. Unmittelbar vor der Injektion des Kaliumcitrats wurde das rechte Herzohr zur elektronenmikroskopischen Untersuchung exstirpiert. 2 min nach dem Stillstand wurden aus der rechten Herzkammer und 10, 20, 30 und 40 min nach dem Stillstand jeweils aus der linken Herzkammer kleine Muskelproben entnommen und sämtlich der gewöhnlichen Stückfixierung unterworfen. Die elektronenmikroskopische Untersuchung ergab, daß die Struktur der Herzmuskelzellen sehr gut erhalten war, und daß auch 40 min nach dem Stillstand noch keine Mitochondrienschwellungen aufgetreten waren (Abb. 1 und 2). Wir glauben deshalb sagen zu können, daß durch Kaliumcitrat die agonalen Veränderungen der Herzmuskelzellen weitgehend vermieden werden können, und daß sich so eine befriedigende Stabilisierung der vitalen Struktur erreichen läßt.

Bei den Befunden des Herzmuskels nach experimentellen Eingriffen kann ich mich im wesentlichen auf die am Düsseldorfer Pathologischen Institut auf Veranlassung von Herrn Prof. Dr. MEESSEN durchgeführten Untersuchungen beziehen. Bisher wurden untersucht: Die Verfettung (*29, 31*), Lipofuscinablagerungen (*31*), die Hypertrophie beim Menschen (*31*) die Hungeratrophie (*31*), die Phosphorvergiftung (*31*), Überdosierung von Strophanthin und Digitoxin (*14*), Überdosierung von Schilddrüsenhormon (*30*), der alimentäre Kaliummangel (*31*), und der Herzmuskel des Siebenschläfers während und außerhalb des Winterschlafes (*18,32*)- Untersuchungen des

Herzmuskels nach Einwirkung von Dinitrophenol, nach Überdosierung von Nebennierenrinden-
hormonen, nach Vitamin B_1-Mangelernährung sowie Untersuchungen der experimentellen Hypertrophie und des experimentellen Herzinfarktes sind im Gange. Am Freiburger Pathologischen

Abb. 1. Herzmuskel, Hund, linker Ventrikel, 40 min nach Herzstillstand durch Kaliumcitrat. Künstliche
Beatmung nach dem Herzstillstand eingestellt, Herz blieb in situ. Entnahme des Gewebes 40 min nach Stillstand,
anschließend gewöhnliche Stückfixierung in 1%iger Osmiumtetroxydlösung nach SJÖSTRAND. Mitochondrien
(*Mi*), Sarkoplasmareticulum (*SR*), Z-Streifen der Myofibrillen (*Z*). RCA-EMU-3 C-Elektronenmikroskop,
100 kV, elektronenoptisch 22500:1, Abb. 80000:1 (631 A/58)

Institut wurden der Sauerstoffmangel im Unterdruckversuch (*21*) und die experimentelle Hypertrophie (*22*) untersucht. Aus der Reihe dieser Untersuchungen möchte ich zwei wichtige Probleme

herausgreifen: Die Verfettung des Herzmuskels und die Herzmuskelveränderungen nach alimentärem Kaliummangel.

Im normalen Herzmuskel der Ratte, in dem sich lichtmikroskopisch kein Fett nachweisen läßt, findet man bei elektronenmikroskopischer Untersuchung ganz vereinzelt kleine, durchschnittlich 0,23:0,16 μ messende, rundliche, ovale oder klecksartige, homogene, dunkle Gebilde.

Abb. 2. Herzmuskel, Hund, linker Ventrikel, 40 min nach Herzstillstand durch Kaliumcitrat. Präparation wie Abb. 1. Z-Streifen (*Z*) und M-Linien (*M*) der Myofibrillen, Glanzstreifen (*Gl*), Mitochondrien (*Mi*), Blutcapillaren (*BC*), Interzellulärspalten (*ICR*). RCA-EMU-3 C-Elektronenmikroskop, 100 kV, elektronenoptisch 2150:1, Abb. 8900:1 (636 A/58)

Extraktionsversuche mit Petroläther bestätigten die allgemeine Annahme, daß es sich bei diesen Gebilden um Fett handelt (*31*). Wenn man Ratten 0,5—1,0 cm³ einer 1%igen öligen Lösung von

weißem Phosphor unter die Haut injiziert und den Herzmuskel der Tiere laufend elektronenmikroskopisch untersucht, so findet man schon nach $2^1/_2$ Std. eine leichte Zunahme sowohl der
Größe als auch der Zahl der kleinen Fetttröpfchen. Außerdem ist das Sarkoplasmareticulum bei

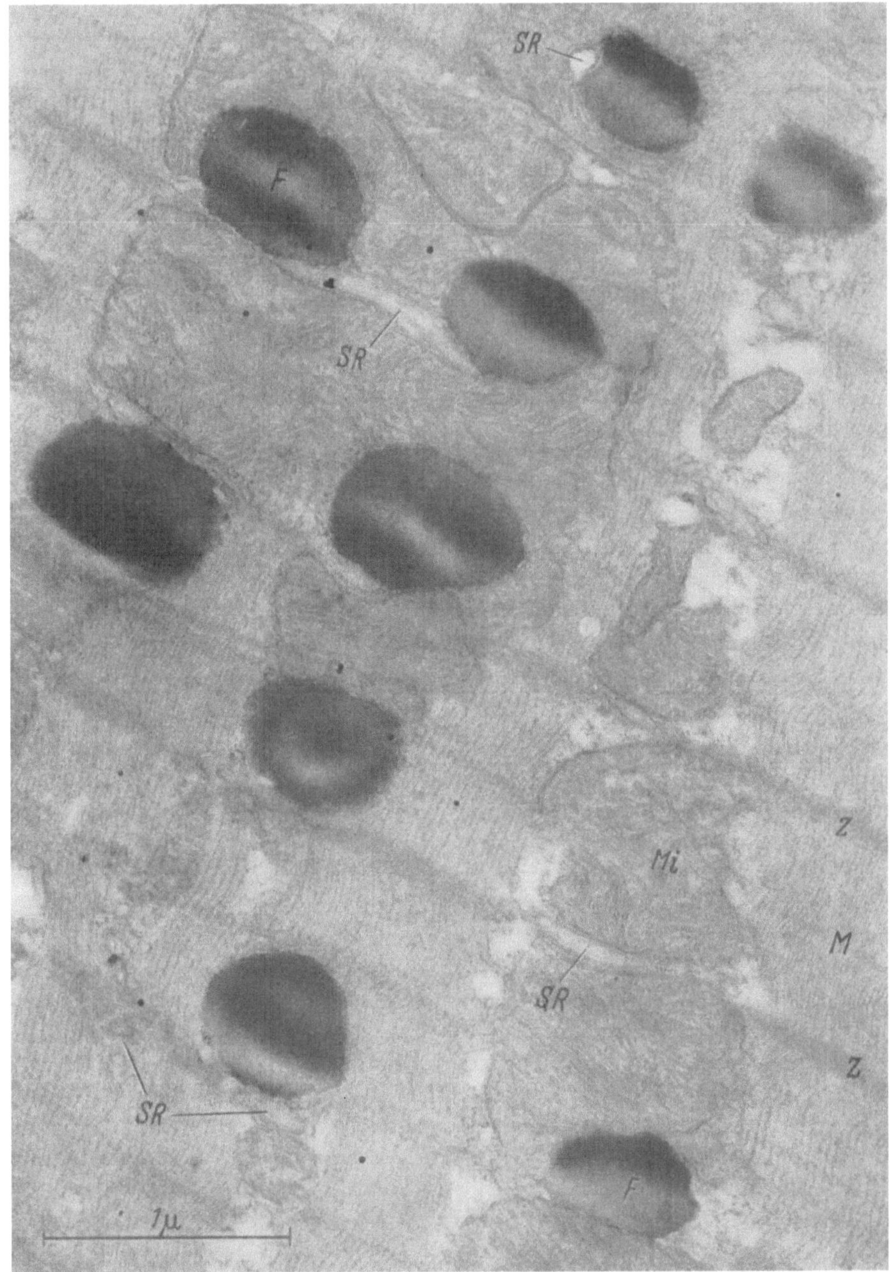

Abb. 3. Herzmuskel, Siebenschläfer, linker Ventrikel, starke Verfettung. Aus natürlichem Winterschlaf nach
5 Wochen erweckt, in Umgebungstemperatur von 18—20° C gebracht, anschließend 4 Tage lang völliger Hunger
und Durst. Entnahme des Gewebes in Äthernarkose, Stückfixierung in 1%iger Osmiumtetroxydlösung nach
SJÖSTRAND. Z-Streifen (*Z*) und M-Linien (*M*) der Myofibrillen, Mitochondrien (*Mi*), Sarkoplasmareticulum
(*SR*), Fetttropfen (*F*). Siemens-Elektronenmikroskop ÜM 100, 80 kV, elektronenoptisch 7 900:1, Abb. 33 200:1
(10 522/57)

diesen Tieren weiter als gewöhnlich und die Mitochondrien zeigen eine leichte Schwellung mit
fleckförmiger Aufhellung der Matrix. Diese Veränderungen haben nach 5 und 10 Std. weiter

zugenommen und erreichen ihren schwersten Grad etwa 20—24 Std. nach der Phosphorinjektion. Die Fetttropfen sind jetzt stark vermehrt, und ihre Größe erreicht teilweise 1 μ. Erst jetzt läßt sich auch im Lichtmikroskop eine diffuse feintropfige Verfettung des Herzmuskels nachweisen. Das Sarkoplasmareticulum ist hochgradig erweitert und die Mitochondrien konfluieren und zeigen schwerste Schwellungen. Überleben die Tiere dieses Stadium, so gehen die Veränderungen einschließlich der Verfettung im Laufe von 5—10 Tagen zum großen Teil wieder zurück (*29, 31*).

Eine physiologische Verfettung des Herzmuskels kommt beim Siebenschläfer (Myoxus glis, Glis glis) vor (*29, 32*). Da diese Tiere während des fast 7 Monate dauernden Winterschlafes weder Nahrung noch Flüssigkeit zu sich zu nehmen brauchen sondern allein von ihren Fettreserven leben, müssen sie während der Sommermonate sehr reichlich Fettgewebe ansetzen. Am Ende der Aktivitätsperiode, kurz vor Beginn des Winterschlafes, hat der Siebenschläfer seinen größten Fettreichtum erreicht. In diesem Stadium kann es auch zu einer diffusen feinsttropfigen Verfettung des Herzmuskels kommen, wie wir lichtmikroskopisch und elektronenmikroskopisch nachweisen konnten. Die stärkste Herzmuskelverfettung haben wir aber bei Siebenschläfern gefunden, die nach einem Winterschlaf von nur wenigen Wochen Dauer erweckt und anschließend noch einige Tage ohne Futter gelassen wurden, so daß sie bei dem hohen Stoffwechsel des Aktivitätszustandes weiterhin von ihren Fettreserven leben mußten (Abb. 3).

Sowohl bei der pathologischen als auch bei der physiologischen Verfettung waren die Fetttropfen niemals in irgendwelchen Zellorganellen sondern stets frei im Sarkoplasma gelegen. Sie zeigten jedoch sehr enge topographische Beziehungen zu den Mitochondrien und dem Sarkoplasmareticulum; nicht selten findet man Fetttropfen unmittelbar zwischen einem Mitochondrion und einem Tubulus des Reticulums gelegen. Unsere Untersuchungen sprechen für die schon von älteren Autoren vertretene Ansicht, daß eine Herzmuskelverfettung durch eine Fettinfiltration in die Herzmuskelzelle hinein zustande kommt.

Wenn man Ratten auf eine Kaliummangelkost setzt, die aber sonst alle notwendigen Nahrungsbestandteile einschließlich Vitaminen und Mineralien enthält, kommt es nach lichtmikroskopischen Untersuchungen zu einem Faserödem und zu kleinen Herzmuskelnekrosen. Bei der elektronenmikroskopischen Untersuchung der Herzen von Kalium-Mangel-Ratten fällt zunächst auf, daß die Myofibrillen mehr erschlafft als kontrahiert sind. Weiter findet man neben einer geringen, lichtmikroskopisch noch nicht nachweisbaren Vermehrung der Fetttropfen ein Ödem von Herzmuskelzellen sowie ein interstitielles Ödem. Auf elektronenmikroskopischen Übersichtsaufnahmen finden sich außerdem kleine Herde, die den lichtmikroskopischen Nekroseherden entsprechen. Bei stärkerer Vergrößerung erkennt man jedoch, daß es sich hier um Herzmuskelzellen handelt, deren Myofibrillen zerfallen oder in Zerfall begriffen sind, die aber noch intakte Mitochondrien enthalten. Dem Zerfall der Myofibrillen geht eine sehr starke Vermehrung von protoplasmatischen Strukturen wie Vacuolen variabler Größe und Cytomembranen vorauf. Die Veränderungen sind wahrscheinlich irreversibel, wenn der alimentäre Kaliummangel lange genug anhält. Der spätere Untergang von Herzmuskelzellen wird hier durch eine Entdifferenzierung eingeleitet, bei der zunächst die hochdifferenzierte kontraktile Substanz ihre Fibrillenstruktur einbüßt, während noch gut erhaltene Mitochondrien vorhanden sind. Diese Veränderungen lassen darauf schließen, daß der Kaliummangel primär an der kontraktilen Substanz angreift, und daß die beim alimentären Kaliummangel beobachteten Veränderungen nicht auf einen indirekten Sauerstoffmangel zurückgehen (*31*).

Wenn man die bisher bekannten Veränderungen des Herzmuskels nach den verschiedenen experimentellen Eingriffen überblickt, so zeichnet sich eine submikroskopische Cellularpathologie der Herzmuskelzelle ab, die uns das Verständnis der energetischen Herzinsuffizienz wesentlich erleichtern kann. Man darf darüber aber nicht vergessen, daß viele Herzinsuffizienzen sich aus einer ganz anderen Dimension ableiten. So können beispielsweise Störungen im Gefüge des Herzmuskels zu vektoriellen Veränderungen und damit zu einer Beeinträchtigung der mechanischen Leistungsfähigkeit führen, ohne daß die Herzmuskelzellen dabei pathologische Veränderungen ihrer submikroskopischen Struktur zu zeigen brauchen.

Literatur

1. Beams, H. W., T. C. Evans, C. D. Janney and W. W. Baker: Anat. Rec. **105**, 59 (1949).
2. Breemen, V. L. van: Anat. Rec. **117**, 49 (1953).
3. Edwards, G. A., H. Ruska, P. de Souza Santos and A. Vallejo-Freire: J. biophys. biochem. Cytol. **2**, Suppl., 143 (1956).
4. Fawcett, D. W., and C. C. Selby: J. biophys. biochem. Cytol. **4**, 63 (1958).
5. Hibbs, R. G.: Amer. J. Anat. **99**, 17 (1956).
6. Hodge, A. J.: J. biophys. biochem. Cytol. **1**, 361 (1955).
7. — J. biophys. biochem. Cytol. **2**, Suppl. 131 (1956).
8. — H. E. Huxley and D. Spiro: J. exp. Med. **99**, 201 (1954).
9. Kisch, B.: Electron microscopic histology of the heart. An application of electron-microscopic research to physiology. In collaboration with J. M. Bardet. New York: Brooklyn Med. Press 1951.
10. — Der ultramikroskopische Bau von Herz und Kapillaren. Darmstadt: D. Steinkopff 1957.
11. Lindner, E.: Zbl. Path. **89**, 41 (1952).
12. — Beitr. path. Anat. **114**, 244 (1954).
13. — Verh. dtsch. Ges. Path. **38**, 154 (1955).
14. — Z. Zellforsch. **45**, 702 (1957).
15. — Verh. anat. Ges. **54**, (1957) (im Druck).
16. — u. R. Poche: Physik. Verh. **6**, 22—23 (1955).
17. — u. H. J. Wellensiek: Verh. anat. Ges. **55** (1958) (im Druck).
18. Meessen, H.: Münch. med. Wschr. **1957**, 685 .
19. — Verh. dtsch. Ges. Path. **42**, (1958) (im Druck).
20. — Beiträge zur submikroskopischen Morphologie des Herzmuskels. Symposion aus Anlaß der 75jährigen Bestehens des Ludwig-Aschoff-Hauses, Freiburg, 1. 8. 1958.
21. Mölbert, E.: Beitr. path. Anat. **118**, 421 (1957).
22. — Verh. dtsch. Ges. Path. **42**, (1958) (im Druck).
23. Moore, D. H., and H. Ruska: J. biophys. biochem. Cytol. **3**, 261 (1957).
24. Muir, A. R.: J. Anat. **91**, 251 (1957).
25. — J. biophys. biochem. Cytol. **3**, 193 (1957).
26. Perry, S. V.: Physiol. Rev. **36**, 1 (1956).
27. Poche, R.: Vortrag auf dem Elektronenmikroskopischen Kolloquium im Anschluß an die 40. Tagung der Dtsch. Ges. Path. in Düsseldorf am 15. 4. 1956.
28. — Zbl. Path. **96**, 392 (1957).
29. — Verh. dtsch. Ges. Path. **41**, 351 (1958).
30. — Beitr. path. Anat. **118**, 407 (1957).
31. — Virchows Arch. path. Anat. **331**, 165 (1958).
32. — Über den Winterschlaf. Öffentliche Antrittsvorlesung vor der Medizinischen Akademie Düsseldorf am 21. 2. 1958.
33. — u. E. Lindner: Z. Zellforsch. **43**, 104 (1955).
34. Porter, K. R., and G. E. Palade: J. biophys. biochem. Cytol. **3**, 269 (1957).
35. Price, K. C., J. M. Weiss, D. Hata and J. R. Smith: J. exp. Med. **101**, 687 (1955).
36. Sjöstrand, F. S., and E. Andersson: Experientia (Basel) **10**, 369 (1954).
37. — — Exp. Cell. Res. **11**, 493 (1956).
38. — E. Andersson-Cedergren and M. M. Dewey: J. Ultrastructure Res. **1**, 271 (1958).
39. Weinstein, H. J.: Exp. Cell. Res. **7**, 130 (1954).

Über Beziehungen zwischen Feinbau und Funktion im glatten Muskelgewebe und im spezifischen Herzmuskelgewebe

Rudolf Caesar

Pathologisches Institut der Universität Tübingen

Im Mittelpunkt elektrophysiologischer Experimente und Hypothesen über die Erregungsbildung und Reizleitung steht die Oberflächenmembran der Zelle. Ihre Bedeutung für den Erregungsvorgang dürfte unumstritten sein. Diese Oberflächenmembran trennt Lösungen verschiedener Zusammensetzung, und zwar im Sinne eines asymmetrischen Donnanpotentials; sie schafft ein Konzentrationsgefälle für verschiedene Ionen zwischen Cytoplasma und extracellulärer Flüssigkeit. Bereits in einer unerregten Zelle herrscht infolge dieser semipermeablen Oberflächenmembran ein Grenzflächenpotential, das man als Ruhepotential bezeichnet. Bei der Erregung ändert nun diese Membran ihre Permeabilität für gewisse Ionen, es kommt zur Potentialänderung und zum Aktionspotential. Dieses Aktionspotential breitet sich als Erregungswelle weiter aus, es kann den Anstoß geben zur Erregung einer Nachbarzelle. Im Innern der gleichen Zelle folgen dem Aktionspotential sekundäre Prozesse, wie beispielsweise die Kontraktion der Myofibrillen in der Muskelfaser.

Diese polarisierte Oberflächenmembran der Zelle trug bislang vom Standpunkt des Morphologen nur hypothetischen Charakter. Mit dem verbesserten Auflösungsvermögen des Elektronenmikroskops zeigte sich jedoch, daß alle tierischen Zellen wohldefinierte Oberflächenstrukturen in der Tat besitzen. Es handelt sich dabei um etwa $10\,m\mu$ dicke Membranen, die das Cytoplasma und seine Strukturen umhüllen und gegen den extracellulären Raum abgrenzen. Diese Zellmembran, auch Plasmamembran bezeichnet, entspricht in Einzelheiten der vom physiologischen Experiment her geforderten Membran. Die Identität der beobachteten mit der kalkulierten Struktur kann kaum einen Zweifel unterliegen. Es muß hervorgehoben werden, daß wir über die chemische Komposition dieser Membran nichts Sicheres aussagen können. Wir wissen auch nicht, wie sich diese chemische Komposition von Zellart zu Zellart ändert.

Im folgenden soll ein kurzer Überblick über unser bisheriges Wissen von Arrangement und strukturellen Details von Zelloberfläche, Zellkontakt und Beziehung der Zellen zum intercellulären Raum im glatten Muskelgewebe und im Reizleitungssystem der Säuger gegeben werden. Wir beziehen uns bei den Angaben über die glatte Muskulatur, soweit nicht anders erwähnt, auf elektronenmikroskopische Untersuchungen von Bergman (1), Ganssler (2), Prosser und Sperelakis (3), Moore und Ruska (4), Caesar, Edwards und Ruska (5). Die zusammengetragenen Beobachtungen über das Reizleitungssystement stammen Untersuchungen von Poche (6), Muir (7), Caesar, Edwards und Ruska (8), Grimley und Edwards (9).

Jede einzelne Zelle des glatten Muskelgewebes und des Reizleitungssystems wird von einem wohldefinierten Oberflächenmembrankomplex gegen Nachbarzellen oder gegen das Interstitium hin abgegrenzt. Beide Gewebe sind also einwandfrei cellulärer — nicht syncytialer — Natur. Dieser Oberflächenmembrankomplex, ein Ausdruck, den wir in Anlehnung an Robertson gebrauchen, stellt sich in beiden Geweben unterschiedlich dar. Im glatten Muskelgewebe besteht er regelmäßig aus drei Anteilen:

1. einer eigentlichen Zellmembran (Plasmamembran),
2. einer Basalmembran,
3. einem schmalen Zwischenraum.

Die eigentliche Zellmembran besitzt im Gegensatz zu den anderen Schichten eine ziemlich konstante Dicke von 70—100 Å und eine recht konstante Dichte. Da sie die einzig konstante Oberflächenstruktur tierischer Zellen überhaupt ist, gehen wir sicher nicht fehl, sie für die aktive, semipermeable, polarisierte Membran der Zelloberfläche auch des glatten Muskelgewebes zu halten. Mag es auch mit verbesserter Methodik gelingen, die einzelnen Schichten dieser Membran morphologisch besser aufzulösen oder ihre chemisch-physikalische Struktur zu erfassen, so dürfte die Bezeichnung Zellmembran ihre Berechtigung dabei insgesamt nicht verlieren. Diese Zellmembran umschließt nun ein Cytoplasma mit Kern, Myofilamenten, Mitochondrien, Zentriol und spärlichem endoplasmatischem Reticulum. Sie trennt dieses Cytoplasma und sein Milieu vom extra- bzw. intercellulären Raum. An korrespondierenden Stellen erscheinen die Plasmamembranen benachbarter Muskelzellen verdickt, sie messen hier bis zu $40\,m\mu$. Die mögliche Bedeutung dieser Verdickungen soll später im Zusammenhang diskutiert werden. Zum intercellulären Raum hin liegt der Zellmembran wie im Skeletmuskel eine weitere Schicht auf, deren Realität und funktionelle Bedeutung fraglicher erscheinen muß. Einen Hinweis auf ihre Realität entnehmen wir vor allem vergleichenden Untersuchungen von Edwards, Ruska und de Harven (9) an Arthropodenskeletmuskeln. Bei diesen Muskeln ist an der Existenz einer Basalmembran bei einer Dicke von etwa $300\,m\mu$ nicht zu zweifeln. Bei Säugern handelt es sich in der glatten Muskulatur um eine unschärfer begrenzte Schicht wechselnder Dicke und Dichte (etwa $20\,m\mu$); zumeist erscheint sie deutlich blasser als die Zellmembran. Da sie sich optisch ähnlich verhält wie die Basalmembran der Capillaren und Epithelien, wurde auch ihr die Bezeichnung Basalmembran zuerkannt. Im glatten Muskelgewebe umhüllt sie jede einzelne Zelle und trägt somit zur Trennung von der Nachbarzelle bei. Während die Zellmembran sicher ein Produkt der Zelle selbst ist, wissen wir über die Herkunft der Basalmembran nichts. Sehr wahrscheinlich stellt sie eine Bildung bzw. eine zellnahe Zone von Intercellular- bzw. Grundsubstanz dar und ist für den positiven Ausfall der PAS-Schiff-Reaktion an dieser Stelle verantwortlich. Über ihren Beitrag zum Gesamtwiderstand des Oberflächenmembrankomplexes und über ihre Durchlässigkeit für Ionen können wir leider nur Vermutungen anstellen. Auf jeden Fall dürfte aber eine relativ dichte Anhäufung von Intercellularsubstanz in unmittelbarer Nähe der eigentlichen Zellmembran nicht gleichgültig sein. Zwischen Zellmembran und Basalmembran bleibt häufig ein schmaler Spalt von etwa 10 mμ Dicke frei.

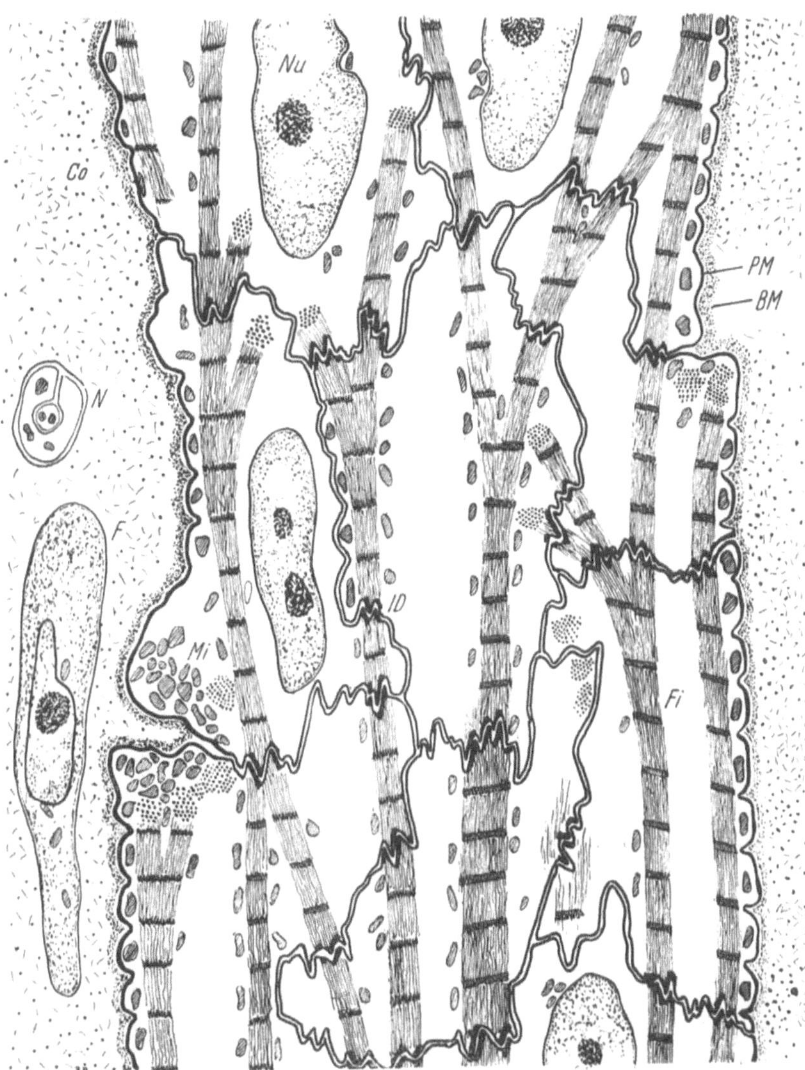

Abb. 1. Schematische Zeichnung der Anordnung der Myofibrillen und der Membranverhältnisse im Reizleitungssystem. Erklärung s. Text. [Aus CAESAR, EDWARDS und RUSKA: Z. Zellforsch. 48, 699 (1958)]

Ein anderes Anordnungsprinzip der Membranen treffen wir im Reizleitungssystem an. Es gleicht in wesentlichen Zügen dem bereits für das Herzmuskelgewebe beschriebenen. Unsere Kenntnisse über den Feinbau des Reizleitungssystems des Herzens entstammen einer Analyse von Purkinjefasern und Hisschem Bündel beim Schaf. Neuere Untersuchungen von GRIMLEY und EDWARDS (9) bestätigen eine prinzipiell gleiche Feinstruktur beim Hund und mehreren Nagern. GRIMLEY und EDWARDS (9) zeigten interessanterweise auch einen ähnlichen Aufbau des Sinus

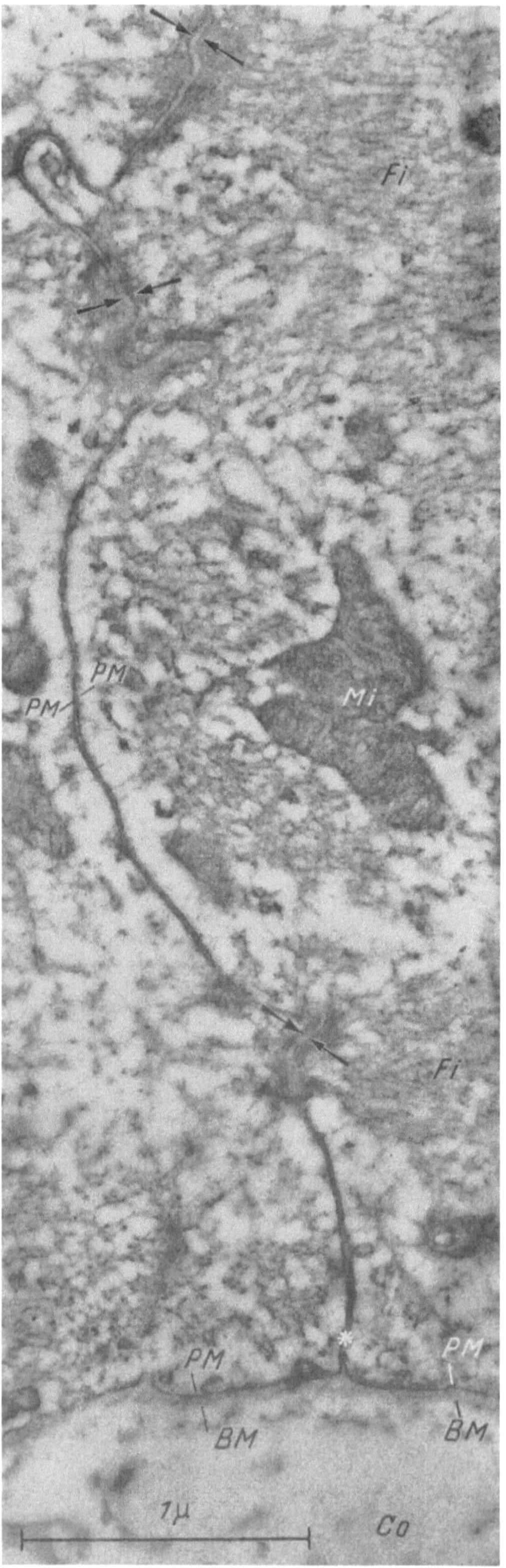

venosus der Schildkröte; entwicklungsgeschichtlich steht derselbe ja in gewissem Zusammenhang mit den Knoten des Reizleitungssystems der Säuger.

Im glatten Muskel grenzte Basalmembran an Basalmembran, zwischengeschaltet waren gelegentlich noch einige Kollagenfibrillen. Im Reizleitungssystem grenzt dagegen fast regelmäßig Zellmembran an Zellmembran. Das Reizleitungssystem wird nur in seiner Gesamtheit von einer Basalmembran überzogen, die die Zellen nach außen gegen einen gefäß- und nervenhaltigen Bindegewebsstreifen abschließt (s. Abb. 1 u. 2). Die Zellen stehen also in engerem Kontakt als im glatten Muskelgewebe.

Zwischen den einzelnen Zellen bleibt nur ein schmaler Spalt des Intercellularraumes regelmäßig sichtbar. Die parallele Lagerung zweier Zellmembranen entspricht dem lichtoptischen Begriff des Glanzstreifens. Im Cytoplasma finden sich neben dem Kern mit Nucleolus spärlich Mitochondrien und endoplasmatisches Reticulum in Nähe des Kernes und der Myofibrillen.

Die Armut an Mitochondrien und Myofibrillen sowie an Profilen des endoplasmatischen Reticulums ist recht auffällig im Gegensatz zur Herzarbeitsmuskulatur. Sie unterstreicht die geringe mechanische Bedeutung und den bereits bekannten niedrigen Sauerstoffverbrauch des spezifischen Herzmuskelgewebes. Besondere Aufmerksamkeit erfordert das Verhalten der Myofibrillen. Sie bevorzugen eine zelloberflächennahe Lage, an der Stelle eines Z-Streifens „endigen" sie an der Zellmembran wie die Myofibrillen im Herzmuskel (s. Abb. 1 u. 2). An der Stelle des Z-Streifens erscheint an der Innenseite der Zellmembran ein dichtes, teils granuläres, teils vesiculäres, teils homogenes Material, in dem die Fibrillen münden (s. Abb. 1 u. 2). Im ganzen gesehen verlaufen die Myofibrillen vorwiegend in Längsrichtung des Purkinjefadens; es kommen jedoch viele Kreuzungen und Überschneidungen der Verlaufsrichtung vor. Sehr eigenartig wirkt das Anordnungsmuster der Myofibrillen in einem größeren

Abb. 2. Zwei oberflächlich gelegene Reizleitungszellen aus einem falschen Sehnenfaden des Schafes. Beide werden gemeinsam von der Basalmembran (*BM*) überzogen und gegen Bindegewebe (*Co*) abgegrenzt. Beide Zellmembranen (*PM*) stülpen sich bei * ein und bilden den Glanzstreifen. An den mit Pfeilen markierten Stellen endigen die Fibrillen (*Fi*) an der Zellmembran. An diesen Regionen ist der Cytoplasmaseite der Zellmembranen ein dichtes, teils granuläres, teils homogenes Material angelagert. Zwischen den Zellmembranen bleibt ein schmaler Zwischenraum sichtbar. 39500 × (*Mi*) = Mitochondrien

Streifen des Reizleitungssystems: die Zellmembranen zweier Nachbarzellen laufen zumeist treppen-
und mäanderförmig „durch das Gewebe" und „schneiden" dabei immer wieder Segmente von
Myofibrillen zwischen zwei Z-Streifen förmlich heraus (s. Abb. 3). Natürlich gehören auf diese
Weise die „ausgeschnittenen" Myofibrillen verschiedenen Zellen zu. Der Gedanke drängt sich auf,

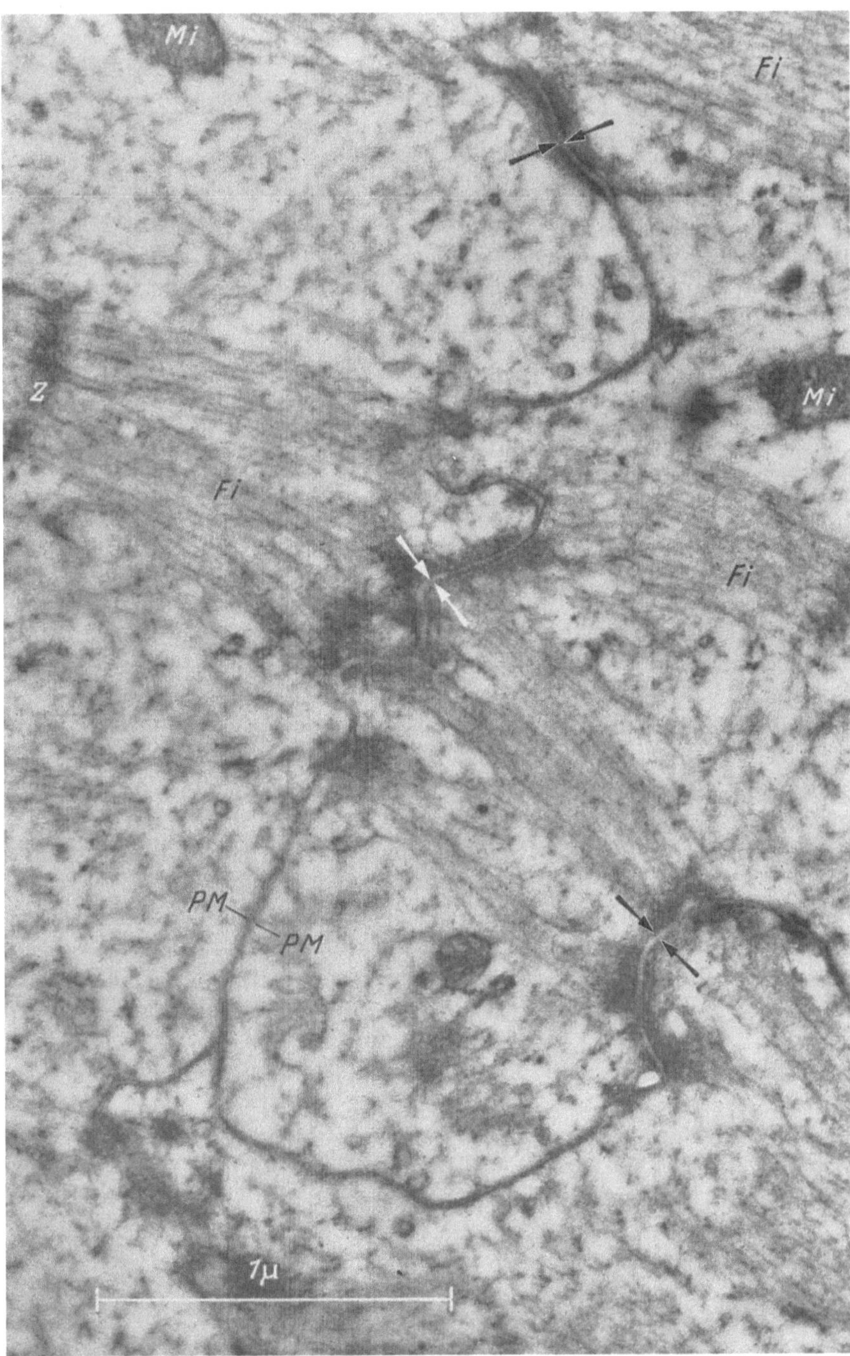

Abb. 3. 3 Reizleitungszellen aus dem Innern eines falschen Sehnenfadens. Durch die mannigfachen Windungen
der in Form des Glanzstreifens verlaufenden Zellmembranen (*PM*) werden Segmente von Myofibrillen zwischen
2 Z-Streifen herausgeschnitten; die Myofilamente endigen jeweils an der Zellmembran (mit Pfeil Mitte und
unten), die Myofibrillen gehören trotz gleicher Verlaufsrichtung verschiedenen Zellen zu. Oben rechts korrespon-
dierende Verdickungen (mit Pfeil) der Zellmembranen, ohne Myofibrillenendigungen. 48000×

daß die innigen nachbarlichen Beziehungen zwischen den Myofibrillen an diesen Regionen nicht belanglos sind.

Anlagerungen an die Zellmembranen kommen vorwiegend an den Stellen von Myofibrillenendigungen vor. Seltener sieht man im Verlauf zweier gemeinsam verlaufenden Zellmembranen an korrespondierenden Stellen Verdickungen, wie sie bereits für den glatten Muskel erwähnt wurden, ohne daß Myofibrillen endigen (s. Abb. 3). Auch von zahlreichen anderen Geweben verschiedenster Herkunft sind solche lokalen Verdickungen der Zellmembranen bekannt. Welche funktionelle Bedeutung sie haben, wissen wir nicht. Bezeichnungen, wie Intercellularbrücken oder Kittleisten, deuten eine mechanische Funktion an. Die Bezeichnung Ephapse präjudiziert eine funktionelle Beteiligung an der intercellulären Erregungsleitung.

In der Tat ist es verlockend, eine Beteiligung dieser lokalen korrespondierenden Verdickung der Zellmembranen im Glanzstreifen oder im glatten Muskelgewebe beim Übergang der Erregung von Zelle zu Zelle anzunehmen. Beiden Gewebsarten kommt eine Automatie, eine Schrittmachereigenschaft beim Erregungsvorgang zu. Nachdem ein Aktionspotential in einer Zelle abgelaufen ist, gäbe es für dessen Fortleitung nach einer Nachbarzelle hin insgesamt 3 Möglichkeiten:

1. Syncytiale Leitung.
2. Synaptische Leitung (unter Vermittlung einer chemischen Substanz).
3. Ephaptische Leitung.

An einer Theorie der unbeschränkten syncytialen Leitung kann man unseres Erachtens in beiden Geweben, da die morphologischen Befunde dem einwandfrei entgegenstehen, nicht festhalten. Längswiderstandsmessungen im Reizleitungssystem ergeben zwar einen recht geringen Widerstand für die „Quermembranen" [WEIDMANN (*11*)], andererseits erscheint es doch sehr zweifelhaft, daß morphologisch so uniforme Membranen, von denen die einzelnen Zellen umhüllt werden, in verschiedenen Richtungen verschiedene Widerstände aufweisen sollten. Es ist weiterhin kaum vorstellbar, daß die gesamte Zellmembran in Längsrichtung eine andere Permeabilität für Ionen aufweisen soll als in Querrichtung. Befunde von ENGELMANN (*12*) und von ROTHSCHUH (*13*) sprechen ebenfalls für die Existenz quer zur Richtung der Erregungsausbreitung liegender Ionenhindernisse.

Eine synaptische Übertragung mittels eines besonderen Synapsenstoffes läßt sich zwar nicht ganz sicher ausschließen, ist aber aus mehreren Gründen nicht sehr wahrscheinlich [PROSSER u. SPERELAKIS (*3*)]. Beispielsweise sind beide Gewebe relativ unempfindlich gegen eine Reihe sonst synaptisch wirksamer Substanzen.

Bei den Widersprüchen, welche sich zwischen morphologischen und physiologischen Befunden ergeben, scheint die Theorie einer ephaptischen Übertragung der Erregung dagegen beiden Seiten am gerechtesten zu werden. Die ephaptischen Regionen besäßen einen geringen Querwiderstand und die zwischenzellige Erregungsleitung liefe über diese Regionen (*3*). Es wäre möglich, daß die korrespondierenden, verdickten Regionen zweier sich gegenüberliegender Zellmembranen in diesem Sinne wirksam wären. Doch bedarf diese Theorie durchaus der experimentellen Untermauerung.

Bemerkungen über die Erregungsleitung wären unvollständig, würde man nicht auch noch der nervalen Seite einige Worte schenken. Beide Gewebe, die glatte Muskulatur und das Reizleitungssystem sind außerordentlich reich an marklosen Nervenfasern. Da die meisten Nervenfasern einen Durchmesser zwischen $0{,}1\ \mu$ und $0{,}5\ \mu$ besitzen, dürfte ein großer Teil der lichtmikroskopischen Auflösung kaum zugänglich gewesen sein. Im glatten Muskelgewebe verlaufen die marklosen Nervenfasern im Intercellularraum. Im Reizleitungssystem fanden wir sie bislang in falschen Sehnenfäden nur in der äußeren Bindegewebsschicht. Die Axone sind charakterisiert durch synaptische Bläschen, zahlreiche Mitochondrien und Neurofilamente. Zumeist sind sie von einem schmalen Cytoplasmaausläufer eines Lemnoblasten eingescheidet. Selten dagegen kommen die Axone mit den eigentlichen Erfolgszellen in unmittelbaren Kontakt. Der Kontakt der Nervenfasern mit den Zellen vollzieht sich nur «en passant», wir haben nur eine Synapse ''per distance'' vor uns.

Ein loserer Kontakt der Nervenfasern paßt gut in unsere Vorstellungen vom Autorhythmus beider Gewebe, der durch nervale Einflüsse nur modifiziert wird.

Wir sind uns dessen bewußt, daß nur erste Anfänge zu einem Verständnis des Zusammenhanges zwischen Morphe und Funktion im glatten Muskelgewebe und im Reizleitungssystem des Herzens vorhanden sind. Die elektronenmikroskopischen Befunde mögen als eine neue Ausgangsbasis für Untersuchungen der Physiologen dienen.

Literatur

1. Bergman, R. A.: I. European Regional Conference on Electron Microscopy. Stockholm, 1956. p. 43.
2. Ganssler, H.: I. European Regional Conference on Electron Microscopy. 1956, p. 43.
3. Prosser, C. L., and N. Sperelakis: Amer. J. Physiol. 187, 536 (1956).
4. Moore, D. H., and H. Ruska: J. biophys. biochem. Cytol. 3, 457 (1957).
5. Caesar, R., G. A. Edwards and H. Ruska: J. biophys. biochem. Cytol. 3, 867 (1957).
6. Poche, R.: Vortrag auf dem Elektronenmikroskopischen Kolloquium im Anschuß an die 40. Tagg. der Dtsch. Ges. Path. in Düsseldorf 1956.
7. Muir, A. R.: J. Anat. (Lond.) 91, 251 (1957).
8. Caesar, R., H. Ruska and G. A. Edwards: Z. Zellforsch. 48, 698 (1958).
9. Grimley, P., and G. A. Edwards: In Vorbereitung.
10. Edwards, G. A., H. Ruska and E. de Harven: J. biophys. biochem. Cytol. 4, 107 (1958).
11. Weidmann, S.: Elektrophysiologie der Herzmuskelfaser. Bern: Huber 1956.
12. Engelmann, T. W.: Pflügers Arch. ges. Physiol. 15, 116 (1877).
13. Rotschuh, K. E.: Elektrophysiologie des Herzens. Darmstadt: D. Steinkopff 1952.

The mechanism of contraction

H. E. Huxley

Department of Biophysics, University College, London

Recent studies, by a number of different biophysical methods, and in particular by electron-microscopy, have enabled a fairly simple hypothesis to be put forward to describe many features of the molecular basis of contraction in striated muscle. [J. biophys. biochem. Cytol. 3, 631 (1957); in "The Cell" (ed.: Brachet and Mirsky), Academic Press (in press); in "The structure and Function of Muscle (ed.: Bourne), Academic Press (in press)]. The evidence for and against this so called "sliding-filament hypothesis" was reviewed, and the relations between the physiological properties of muscles and the detailed molecular mechanisms involved in such a scheme were described. The advances in electron microscope techniques needed to take the structural studies a stage further were considered.

Die Untersuchung einiger Elemente des Muskelgewebes in seiner Histogenesis und Regeneration

V. P. Gilëv

Laboratorium für Elektronenmikroskopie, Akad. d. Wiss. der UdSSR, Moskau

Als Untersuchungsobjekte dienten Rumpfmuskulatur der Larven von Axolotl 1—10 Tage nach dem Schlüpfen aus dem Ei und Rattenmuskel, der sich aus zerkleinertem Muskelgewebe regeneriert hatte (Methode von A. N. Studitsky). Die Fixierung des Gewebes wurde in 1%igem OsO_4, durchgeführt, das mit Acetatveronal bis zu p_H 7,2 gepuffert wurde (2, 3, 4). Die Entwässerung der Objekte erfolgt in Alkoholen, die Einbettung in Methacrylaten.

Auf dem ersten, myoblastischen Stadium der Entwicklung des Regenerats (bzw. des Myotoms) waren die Muskelelemente durch spindelartige Myoblasten repräsentiert. Im Cytoplasma dieser Zellen beobachtet man ein sehr stark entwickeltes endoplasmatisches Reticulum, das durch dünnwandige Bläschen und manchmal durch Röhrchen dargestellt wird, welchen die Paladeschen Ribonucleoprotein-Granula angelagert sind. Einzelne bläschenförmige Elemente des Reticulums sind nicht selten in Kettchen angeordnet, welche lange Züge bilden, die durch die Zelle verlaufen.

Es werden keinerlei fibrilläre Strukturen im Cytoplasma der „freien" Myoblasten beobachtet, welche sich noch nicht in Stränge und Muskelröhrchen umgebildet haben. Die „Fibrillarität" des Cytoplasmas, welche im Lichtmikroskop zu sehen ist, besteht offenbar aus nicht genügend aufgelösten und miteinander verschmelzenden Kettchen von bläschenförmigen Elementen des endoplasmatischen Reticulums und seiner Röhrchen.

Das endoplasmatische Reticulum der Fibroblasten, welche in großer Menge im Regenerat vorkommen, besitzt die Form flacher aneinander gepreßter Höhlen, wodurch es sehr an die

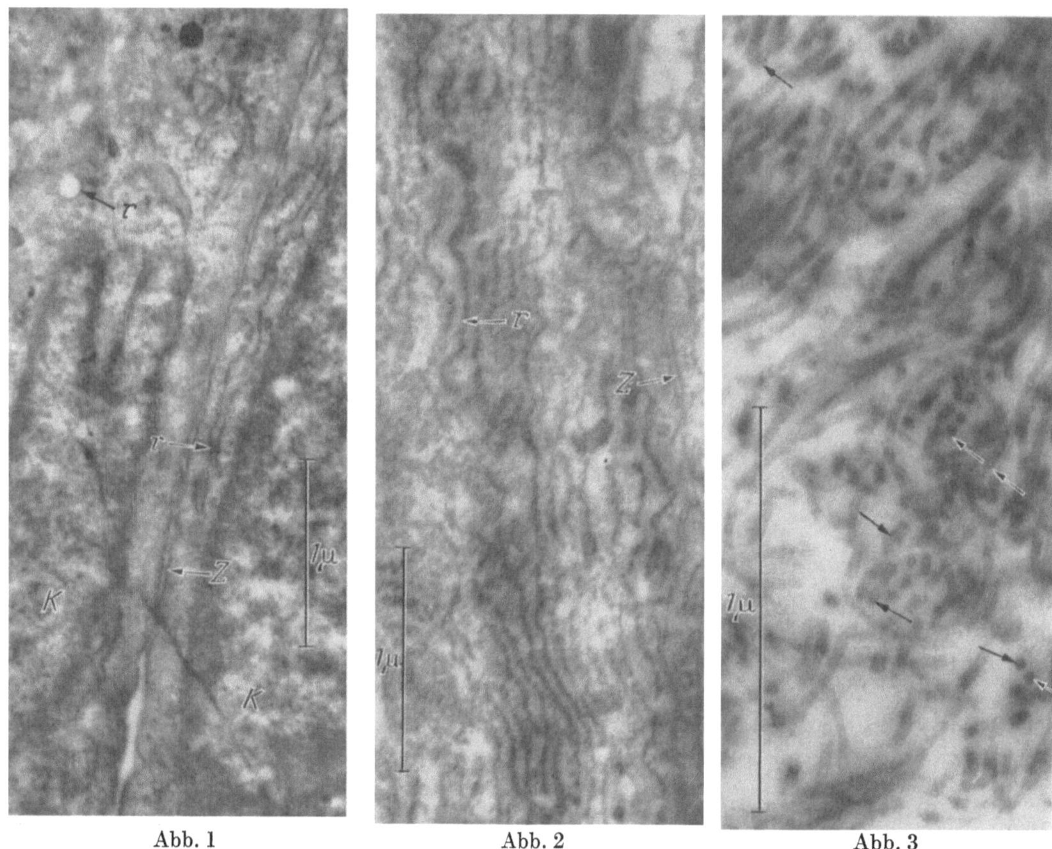

Abb. 1   Abb. 2   Abb. 3

Abb. 1. Zwei Myoblasten des Muskelregenerats der Ratte. 24000mal. In den mit Pfeilen bezeichneten Fasern ist ein helles Mittelstück zu sehen. 53000mal. K = Kern; r = endoplasmatisches Reticulum; Z = Zellmembran

Abb. 2. Ein Teil des Fibroblasten aus dem Muskelregenerat der Ratte. 29000mal

Abb. 3. Die jungen Kollagenfasern aus dem Muskelregenerat der Ratte

Struktur der intracellulären cytoplasmatischen Membranen der sekretorischen Zellen gewisser Drüsen erinnert. An der Außenseite der Grenzmembranen des Reticulums befinden sich zahlreiche RNP-Granula.

Die Mitochondrien der Fibroblasten und der Muskelelemente besitzen eine für jede Tierart typische Form: beim Axolotl sind sie mit breiten und flachen cristae mitochondriales ausgestattet, welche die Mitochondrien quer unterteilen; bei der Ratte haben die cristae eine röhrenartige Struktur und sind auf den Schnitten in Form von Kreisen und zweischichtigen Abschnitten verschiedener Länge zu sehen. Interessanterweise werden in den „freien" Myoblasten des Regenerats Mitochondrien, wenigstens in ihrer typischen Form, nicht beobachtet. Die hellen rundlichen Strukturen im Cytoplasma besitzen keine cristae und sind mehr den großen Bläschen des endoplasmatischen Reticulums ähnlich als den Mitochondrien. Das Vorhandensein einer ganzen Reihe von Übergangsformen von den großen Bläschen des endoplasmatischen Reticulums zu den Mitochondrien gestattet die Annahme, daß ein genetischer Zusammenhang zwischen diesen Strukturen besteht. Eine wahrscheinlichere Quelle der Bildung der Mitochondrien sind jedoch kleine (80—240 mμ) rundliche osmiophile Einschlußkörper. An den größten von ihnen läßt sich eine Außenmembran differenzieren, und im Inneren den Cristae ähnliche Bildungen. Im Zusammenhang mit dieser Frage sollen eigenartige große (bis zu 2μ) „Vacuolen" erwähnt werden, welche manchmal in den Muskelröhrchen von Myotomen des Axolotl anzutreffen sind. Diese rundlichen Bildungen sind mit einer dicken, läufig mehrschichtigen Membran versehen und enthalten manchmal im Inneren rundliche Körperchen, die einen hellen Inhalt aufweisen und eine zweischichtige Außen-Membran besitzen. Die Größe dieser Körperchen und die Struktur ihrer Membran erinnern an Mitochondrien, jedoch enthalten sie nicht Cristae. Ob dies ein zufälliges Zusammentreffen ist oder ob diese Strukturen irgendeine Beziehung zu den Mitochondrien besitzen, kann vorläufig nicht gesagt werden.

Der Prozeß der Neubildung der ersten Myofibrillen in den jungen Muskelelementen von Axolotl und in den Regeneraten des Rattenmuskels ist jenen Prozessen sehr ähnlich, welche bei der Entwicklung des Herzmuskels des Kükens vor sich gehen und von HIBBS (5) beschrieben werden. Im Cytoplasma der Myoblasten, welche miteinander zu verschmelzen beginnen, und in den jungen Muskelröhrchen werden dünne lockere Bündel von Fäden (der künftigen Protofibrillen) beobachtet. Diese Fäden entstehen inmitten der Elemente des endoplasmatischen Reticulums, und stehen offenbar mit ihnen in engem Zusammenhang. An diesem Prozesse haben die Mitochondrien wahrscheinlich keinen solch unmittelbaren Anteil. In späteren Stadien der Differenzierung werden die Bündel der Fäden immer dichter, und es beginnen in ihnen hellere und dunklere Anteile zu erscheinen, welche unscharf ineinander übergehen; es kommt zur Ausbildung der anisotropen und isotropen Scheiben. Zugleich mit ihnen oder etwas später entstehen die Z-Streifen. Als letzte entstehen die M-Streifen. Die Bündel der Fäden wandeln sich also allmählich in typische Myofibrillen um. Zu dieser Zeit erscheinen im Sarkoplasma, in der Nähe der Z-Streifen, die von PORTER (6) beschriebenen großen Bläschen des sarkoplasmatischen Reticulums.

In den differenzierten Muskelröhrchen, welche schon eine große Zahl von quergestreiften Myofibrillen enthalten, verschiebt sich der Bildungsort der neuen Fäden aus dem Inneren des Sarkoplasmas nach Anteilen hin, welche unmittelbar an die differenzierten Myofibrillen angrenzen. Die letzteren werden immer dicker dank des „Anschlusses" von immer neuen und neuen Protofibrillen. Diese Protofibrillen entstehen in Anteilen, welche von interfibrillären Teilen der Z-Streifen der Länge nach abgegrenzt werden. Die im Stadium der typischen Muskelröhrchen gebildeten Myofibrillen (richtiger ihre neugebildeten Teile) erfahren also ihre Ausbildung, ohne durch ein Stadium des Fehlens der Querstreifung zu gehen. Die weitere Vermehrung der Myofibrillen erfolgt auf dem Wege der Längsspaltung, deren Bilder sehr häufig zu sehen sind.

Zwischen den Zellelementen des Regenerats beobachtet man Bündel von kollagenen Fasern. Der Durchmesser der einzelnen Fasern beträgt ungefähr 280 Å. Es ist von Interesse, daß die jungen Kollagenfasern keine Gleichartigkeit ihres Querschnitts aufweisen: auf den Quer- und Längsschnitten ist deutlich eine osmiophile periphere Zone in der Breite von ungefähr 60 Å und ein helles Mittelstück mit einem Durchmesser von ungefähr 160 Å zu sehen. Ob diese Bilder die tatsächlichen Strukturen darstellen, oder ob diese Erscheinung durch ungleichmäßige Wirkung des Fixators hervorgerufen wurde, läßt sich vorläufig nicht feststellen.

Was den Ort der Entstehung der Kollagenfasern betrifft, so könnte man sich auf Grund des nicht seltenen Vorhandenseins von Strukturen vom Typus des endoplasmatischen Reticulums und sogar der Mitochondrien zwischen den Kollagenfasern der Meinung anschließen, die von den Verteidigern der exoplasmatischen Theorie der Bildung des Grundstoffes im Bindegewebe vertreten wird. Wenn man jedoch die Tatsache in Betracht zieht, daß in den Fibroblasten, welche eine deutliche Kontur besitzen, niemals eine Fibrillierung zu bemerken ist, so läßt sich vorläufig noch kein endgültiger Schluß ziehen.

Literatur

1. STUDITSKY, A. N.: Dokl. Akad. Nauk SSSR **84**, N. 2, 389 (1952).
2. PALADE, G. E.: J. exp. Med. **95**, 285 (1952).
3. SJÖSTRAND, F. S., and V. HANZON: Exp. Cell Res. **7**, 393 (1954).

4. Rhodin, J.: Diss., Stockholm, 1954.
5. Hibbs, R. G.: Amer. J. Anat. **99,** 17 (1956).
6. Porter, K. R.: J. biophys. biochem. Cytol. **4,** Suppl., 163 (1956).

On the electron microscopic structure of Z-lines

F. Guba, M. Garamvölgyi and E. Ernst

Department for Micromorphology, Research Institute for Techn. Physics, Hungarian Acad. of Sci., Budapest,
and Department of Biophysics, Med. Univ., Pécs, Hungary

Myofibrils can be regarded as the structural and physiological unit of striated muscle. The structural and biochemical investigation of myofibrils at rest and in action has been intensely

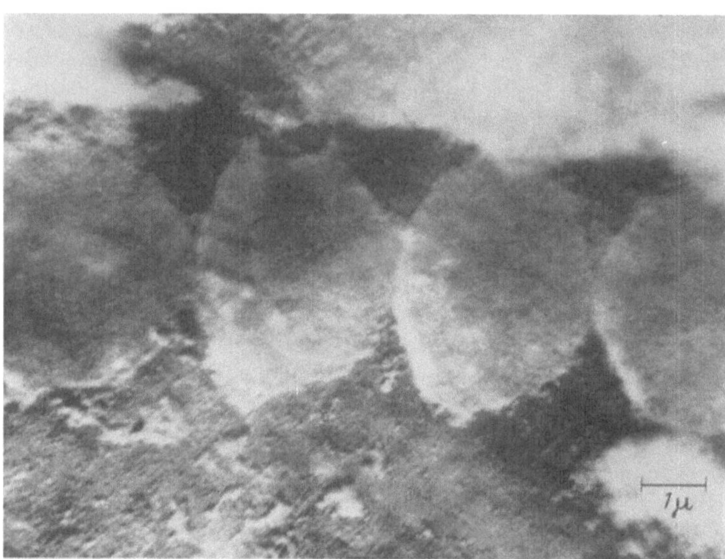

Fig. 1. The process of isolating Z-membranes. The membranes remain "spread" after the disappearance of the protofibrillar structure. 8500 ×

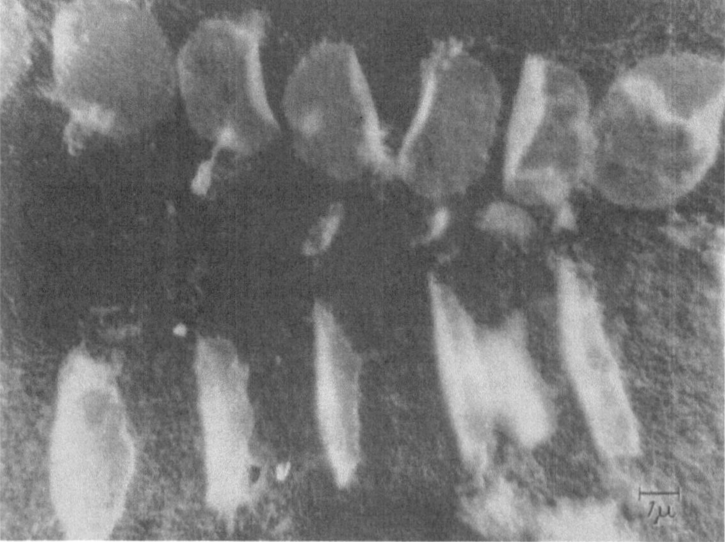

Fig. 2. The process of isolating Z-membranes. After the dissolution of the protofibrillar substance, the membranes striated in myofibril are spread in different measure. 5000 ×

carried on in the last two decades and is still being carried on, bearing remarkable results. These results cleared biological action until macromolecular dimensions.

The structure of myofibrils shows the same segmentation as the whole muscle: it includes segments A, I, etc. Physiological activity can be localized as the change of the single sarcomeres, th. i. the section between two Z-lines lying in the middle of the I segments. Little has been said in the near past about the physiological role and the morphological structure of the Z-lines bordering the single sarcomeres. Some were of the opinion that they were fixing structure, others suggested that they may have a role in communicating stimuli. As to their structure, our own investigations showed them to consist of threads; some authors describe them as discs.

The first step to any investigation of Z-lines is their isolation in measurable quantities. This aim has been reached by us in the following way:

The muscle of the honey bee's wing has been prepared, the muscle bundles minced by the WARING BLENDOR in physiological salt solution. The resulting myofibril-suspension has then been cleaned from the course debris — originating not from muscle parts — by centrifuging at low revolution. So much lactic acid has afterwards been added to the myofibril suspension as to reach an end-concentration of 0.5%. The influence of the lactic acid caused the myofibrils to solve in 30 min at 0° C, whereas the more resisting Z-lines remained. These were sedimented by centrifuging for one hour at 5000 r.p.m. and washed from the lactic acid with bidistilled water. This was the substance of our investigations.

The process of dissolution was followed in the electron microscope with the help of the dialysis-method applied in our laboratory. After dissolving the protofibrillar material from the

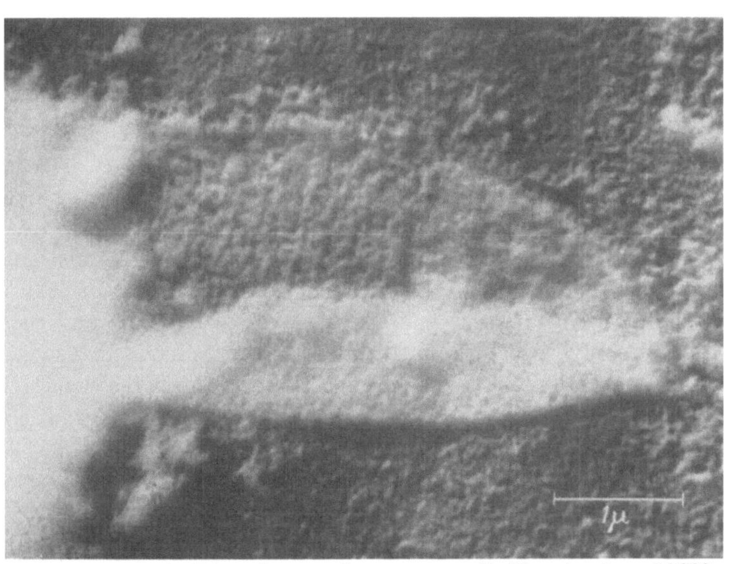

Fig. 3. The single Z-membranes show a woven fibrillar structure. 18000 ×

myofibrils, discs remain (Fig.1). Fig. 2 reveals the pattern of the differently "spreaded" membranes. It may be seen that the membranes stand at right angles to the longitudinal axis of the myofibrils. Higher magnification (Fig. 3) demonstrates that the membranes are a tissue of woven fine threads.

Our biochemical investigations were directed to the chemical structure of isolated membranes and their enzymatic behaviour. According to our former investigations the Z-discs should be proteins. The present tests confirm this statement: the structure of the membranes disrupt on the action of proteolytic enzymes (Fig. 4). Formerly, we suspected an eventual collagenic nature of the Z-membranes. Chromatography of the amino acid derived from the isolated Z-membranes did not confirm this supposition: neither of the amino acids presented itself in a distinguished quantity.

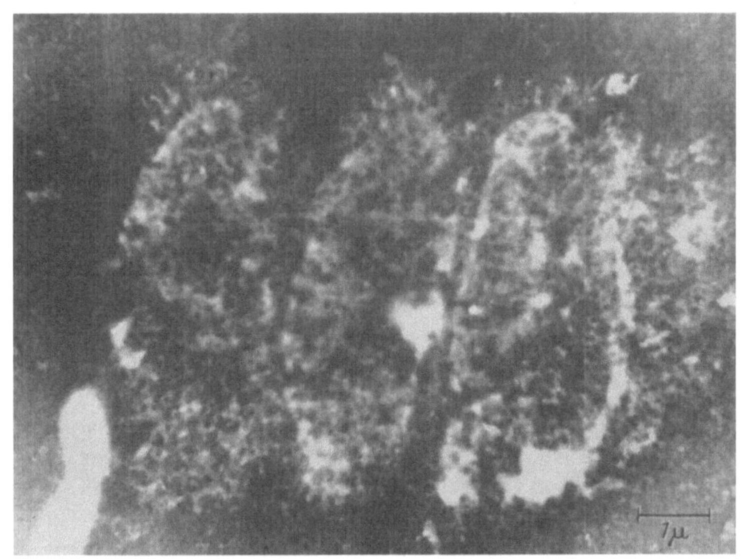

Fig. 4. The structure of Z-membranes treated with trypsin. 10000 ×

In the row of enzymatic investigations the isolated Z-membranes were faintly positive in their acetylcholinesterase activity; in fact, they showed about the same activity as the isolated myofibrils.

The investigations are still in progress relating to morphology as well as biochemically.

Elektronenmikroskopische Untersuchungen an Langendorff-Herzen unter normalen und abnormalen Bedingungen

E. Lindner und H.-J. Wellensiek

Aus dem Topographisch-Anatomischen Institut und dem Pharmakologischen Institut der
Medizinischen Akademie Düsseldorf

Nach früheren elektronenmikroskopischen Beobachtungen ist das Langendorff-Herz für morphologische Stoffwechseluntersuchungen geeignet (*1*). Männliche und weibliche Meerschweinchen (20 Tiere) von 260—425 g wurden durch Nackenschlag getötet. Über eine in die Aorta eingebundene Kanüle wurden die Herzen in der Apparatur nach Langendorff mit Tyrode-Lösung durchströmt (NaCl 0,8%, $C_6H_{12}O_6$ 0,1%, $NaHCO_3$ 0,1%, $CaCl_2$ 0,02%, KCl 0,02%; O_2-Druck von 100 cm H_2O; 37° C). Durch Schlauchklemmen konnte über ein Winkelstück auf eine zweite Perfusionsflasche umgeschaltet werden. Sie enthielt: a) mit N_2 durchperlte Tyrode-lösung unter einem Druck von 100 cm H_2O; b) Tyrodelösung mit Zusatz von 2,4-Dinitrophenol, $5 \cdot 10^{-5}$ bis 10^{-3} m; c) mit Zusatz von Natriummonojodacetat, $2 \cdot 10^{-4}$ bis 10^{-3} m. Mechanogramm und coronare Durchflußmenge wurden kymographisch registriert. Aus einem dritten Gefäß konnte das coronare Gefäßsystem mit 10 cm³ Fixierungslösung (gepufferte Osmiumsäure nach Palade (*2*), pH = 7,4 oder 10%iges neutrales Formalin) unter O_2-Druck von 100 cm H_2O durchströmt werden. Nach dem Öffnen des Hahnes am Fixierungsgefäß führten die Herzen noch 19—24 kleiner werdende Schläge aus und standen innerhalb 6—8 sec still. 10 cm³ Fixierungslösung liefen in 2—3 min durch die normalen Herzen. Herzen mit starken Kontrakturen, Nekrosen und abgesunkenem coronarem Durchfluß benötigten bis zu 16 min. Sie waren vorwiegend in der inneren Zone der linken Herzkammer nicht vom Fixierungsmittel erfaßt. Elektronenmikroskopisch wurden untersucht: rechte Herzkammer, äußere und innere Schale der linken Herzkammer, Vorhof, vereinzelt auch das Kammerseptum.

Eine Hälfte von jeder Probe wurde jeweils 2 Std. mit gepufferter Osmiumsäure nachfixiert, die anderen Stückchen wurden sofort mit Äthylalkohol entwässert. Einbettung in verpolymerisiertes Butyl-Methacrylat bei 65° C. Dünnschnitte mit Porter-Blum-Mikrotom. Aufnahmen mit Siemens-Elmiskop I bei 1 000—80 000 facher Vergrößerung. Alle Methacrylat-Blöcke und die in Paraffin eingebetteten restlichen Herzstückchen wurden lichtmikroskopisch kontrolliert.

Bei akutem, absolutem *Sauerstoffmangel* trat Herzstillstand in $4^1/_2$—5 min ein. Die coronare Durchströmung stieg steil an. Unter *Dinitrophenol* nahm die Herzamplitude bei Dosen von 10^{-5} bis 10^{-4} m um 25—55% ab und die Durchströmung stieg stark an. Bei höherer Dosis (10^{-3} m) trat Herzstillstand in 2 min ein. Abhängig von der Dauer des Herzstillstandes (4—16 min) und mit zunehmender Kontraktur sank die Durchströmung bis auf 50% des Ausgangswertes ab. Mit *Monojodacetat* schlugen die Herzen unter den genannten Versuchsbedingungen zunächst noch etwa 10 min lang mit unveränderter Hubhöhe weiter. Dann nahm die Hubhöhe rasch ab und führte in der 14.—18. min zum Herzstillstand. In 2—3 min folgte eine Kontraktur unter Abnahme des coronaren Durchflusses.

Normale Herzmuskelzellen haben vorwiegend lange oder flache, nicht geschwollene Mitochondrien. Die Myofibrillen sind kontrahiert oder erschlafft und zeigen zwei Arten von Filamenten (*3, 4*). Das Sarkolemma ist zweischichtig. Es sendet finger- oder spaltförmige Fortsätze in die Muskelzellen, die nach unserer Auffassung das transversale System in Höhe des I-Abschnittes der Fibrillen bilden (*5*). Das Sarkoplasmareticulum (*6*) steht in geordneter Beziehung zu Muskelfibrillen, Mitochondrien, Lipidgranula und Zellkern (*7*). Der Golgi-Apparat tritt deutlich hervor. Der Zellkern ist langgestreckt. Das Verhältnis der Volumina von Mitochondrien:Fibrillen:granulärem Grundcytoplasma variiert.

Bei O_2-Mangel sinkt die Anzahl der Zellen mit intaktem Gefüge parallel mit der Dauer des Herzstillstandes (1—19 min). Aber auch nach 19 min sind noch gut erhaltene Zellen besonders in den Vorhöfen, in der rechten Kammer und in der äußeren Schale der linken Kammer zu finden. Für das längere Überleben der Fasern scheint ein hoher Volumenanteil des granulären Grundcytoplasmas bedeutungsvoll zu sein. In nur teilweise veränderten Zellen liegen normale und durch Matrixschwellung stark vergrößerte Mitochondrien nebeneinander. Unter *Dinitrophenol* sind

Zellen mit normalem Gefüge bis zum Einsetzen der Kontraktur nach etwa 4 min Herzstillstand vorherrschend. Das transversale System ist weit oder erweitert, das Grundcytoplasma vielfach reichlich. In einzelnen Zellen findet man neben normalen Mitochondrien stark vergrößerte Mitochondrien mit Verlust der Matrix, Verwerfung und Schwellung der inneren Membranen und Bildung großer dunkler Granula. Nach *Monojodacetat* sieht man in den nicht in Kontraktur übergegangenen Herzabschnitten vorwiegend *diastolische* Fasern mit intaktem Zellgefüge. In manchen Zellen ist die Membranordnung in den größtenteils nicht geschwollenen Mitochondrien verändert.

Die bisher beschriebenen Veränderungen wurden in Zellen gefunden, deren intaktes Zellgefüge noch eine Lebensfähigkeit annehmen läßt. Sie dürften vielleicht in engerem Zusammenhang mit der Wirkung der angewendeten Stoffwechselgifte stehen. In allen Versuchsgruppen fanden wir darüber hinaus mit zunehmender Dosis und Versuchsdauer Zellen, die in ihrer Gesamtheit geschädigt waren. Die Fasern sind in Kontraktur mit engen Abständen der verschmierten Kontrakturstreifen. Alle Mitochondrien zeigen Matrixverlust und geringe Schwellung. Das transversale System ist eng, während das endoplasmatische Reticulum weit ist. Das Sarkolemma kann ausgeprägte Arkaden bilden. Die Blutcapillaren sind zusammengedrückt, ihre Endothelien sind geschwollen. Ein weiteres Stadium ist das allgemeine nekrotische Zellödem mit großen, geschwollenen, fast leeren Mitochondrien, passiver Verdrängung und Auflösung der Myofibrillen und völliger Zerstörung der Zuordnung von Myofibrillen, Mitochondrien und endoplasmatischem Reticulum. Schließlich verlieren die Zellkerne ihren Zusammenhang mit dem Sarkoplasmareticulum und sind von Flüssigkeit umgeben, in der feine Fäden und Körnchen erkennbar sind. Der Kerninhalt geht unter Verlust der normalen Form verloren. Der Golgi-Komplex haftet am Kern. Die primäre Kernmembran ist verdickt, die sekundäre Kernmembran liegt dicht an und ist teilweise zerstört und an manchen Stellen bläschenförmig abgehoben. Diese Veränderungen waren in den Versuchsgruppen mit Sauerstoffmangel und Dinitrophenol und gelegentlich in normalen Herzen mit thrombotischem Gefäßverschluß zu finden. Sie scheinen mit gewissen zeitlichen Variationen gleichartig zu verlaufen, soweit wir nach unseren bisherigen Erfahrungen sagen können. Ähnliche Beobachtungen machten wir früher an Froschherzen, die mit Digitalis vergiftet worden waren (*5*). In den mit Monojodacetat vergifteten Herzen, die bis zum Stadium der maximalen Kontraktur untersucht wurden, fehlten die Bilder der schweren ödematösen Zellzerstörung und die ausgeprägte Mitochondrienschwellung im Stadium der Kontraktur. Die Gruppe der geschilderten Veränderungen der gesamten Zelle möchten wir als unspezifischen Prozeß deuten, der in die Zellnekrose überleitet. Wir nehmen eine Parallele an zu der fortschreitenden Strukturzerstörung, die ROTHSCHUH (*8*) nach Querschnittsverletzungen am Skeletmuskel beschrieben hat. Wir wollten mit diesen ausgewählten Beispielen zeigen, daß die am Herzmuskel schwierige Unterscheidung zwischen den vom Stoffwechsel hervorgerufenen Zellveränderungen und den folgenden Nekroseprozessen (*9, 10, 11, 12*) durch geeignete Versuchsanordnung erleichtert wird.

Ausführliche Veröffentlichung erfolgt an anderer Stelle.

Mit Unterstützung der Deutschen Forschungsgemeinschaft und der Gesellschaft von Freunden und Förderern der Medizinischen Akademie Düsseldorf.

Die elektronenmikroskopischen Aufnahmen wurden im Institut für Elektronenmikroskopie an der Medizinischen Akademie Düsseldorf (Direktor Prof. Dr. H. RUSKA) hergestellt.

Literatur

1. LINDNER, E., u. H.-J. WELLENSIEK: Verh. anat. Ges. 55. Vers. Frankfurt 1958 (im Druck).

2. PALADE, G. E.: J. exp. Med. **95**, 285 (1952).

3. HUXLEY, H. E.: Biochim. biophys. Acta **12**, 387 (1953).

4. HANSON, J., and H. E. HUXLEY: Proc. 3. int. Conf. on Electron Microscopy. London 1954, 576 (1956).

5. LINDNER, E.: Z. Zellforsch. **45**, 702 (1957).

6. PORTER, K., R. and G. E. PALADE: J. biophys. biochem. Cytol. **3**, 269 (1957).

7. MOORE, D. H., and H. RUSKA: J. biophys. biochem. Cytol. **3**, 261 (1957).

8. ROTHSCHUH, K. E.: Pflügers Arch. ges. Physiol. **260**, 437 (1955).

9. MOORE, D. H., H. RUSKA and W. M. COPENHAVER: J. biophys. biochem. Cytol. **2**, 755 (1956).

10. POCHE, R.: Virchows Arch. path. Anat. **331**, 165 (1958).

11. — Beitr. path. Anat. **118**, 407 (1957).

12. MÖLBERT, E.: Beitr. path. Anat. **118**, 421 (1957).

The structure of certain smooth muscles which contain "paramyosin" elements

G. F. Elliott

Wheatstone Physics Laboratory, University of London King's College

The name 'paramyosin' is given to a structural element, isolated from the slow portions of the lamellibranch adductor muscles as long ribbons which, under certain conditions, show a network of scattering dots (1). The same network has been detected in the intact dried muscle by small angle X-ray diffraction (2, 3). As yet, however, only a few preliminary observations have been published on the structure of the intact muscles as seen in the electron-microscope (4, 5, 6).

Fig. 1. Longitudinal section of the white adductor muscle of the Portuguese oyster, *Gryphaea angulata*.
70,000 ×

In the phase contrast microscope the muscle cell, about 10 μ in diameter, seems to contain large numbers of fibrils parallel to the long axis of the cell, but imperfectly resolved (7). In the electron microscope, in longitudinal sections of the white adductor muscle of the Portuguese oyster, *Gryphea angulata*, the fibrils can be seen to lie parallel to the long axis in both contracted and extended muscles. The complicated bending of the fibrils mentioned by Kawaguti and Ikemoto (6) is an artifact. High resolution electron micrographs of these fibrils show a wide variety of transverse band patterns (Fig. 1). In measurements on 500 fibrils, 50% were found to have a period of between 220 and 260 Å, lying at about 70° to the fibril axis, 17% a period between 340 and 380 Å, at about 40° to the axis, 10% a transverse period of approximately 290 Å, and 5% a transverse period of approximately 145 Å. There are also "chevron" effects, combining two or more of these patterns. At first sight such a variety of band patterns is surprising in material which has a small angle X-ray pattern extending to the 50th order, showing great regularity of the underlying macro-molecular structure.

Transverse sections show no regular packing of the fibrils, which are roughly circular and of diameters between 500 and 1500 Å. In accurate cross sections the fibrils are seen to have a lamellar structure, the lamellae are spaced at about 200 Å (Fig. 2). This lamellar structure is not always seen, as a small departure from true cross sectioning causes the different levels in a section 400 Å thick to smear out the density differences on the photographic plate. Though the fibrils are approximately circular, no cylindrical symmetry has been observed in cross section. Such symmetry would be expected if the fibril were a large scale helix, which is a possible interpretation of the X-ray pattern (3). Additional evidence for the lamellar structure comes from the small angle X-ray pattern of these muscles. On the equator of the pattern a streak extends out to about 30 Å, this is presumably given by the individual macromolecules. Superimposed on this streak it is occasionally possible to detect reflections which may be interpreted as orders of a 200 Å period across the fibril (8).

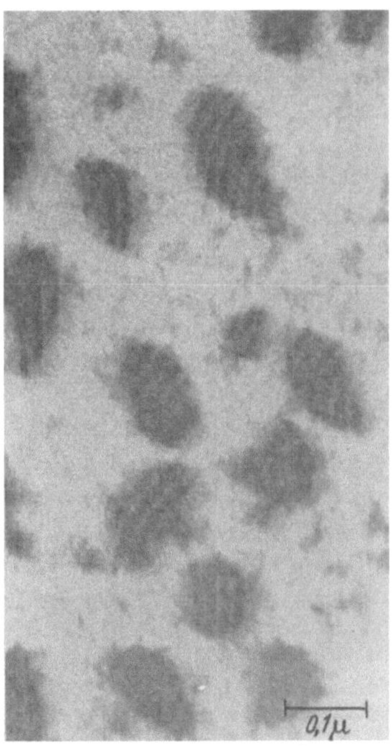

Fig. 2. Transverse section of the same muscle. 100,000 ×

It is suggested that these large fibrils in the intact muscle are made up of ribbons each having the characteristic 'paramyosin' pattern of scattering centres stacked face to face and separated by about 200 Å. The band patterns observed can then be explained in detail in terms of the ways in which this stack of ribbons can be sectioned and the scattering centres can line up from the

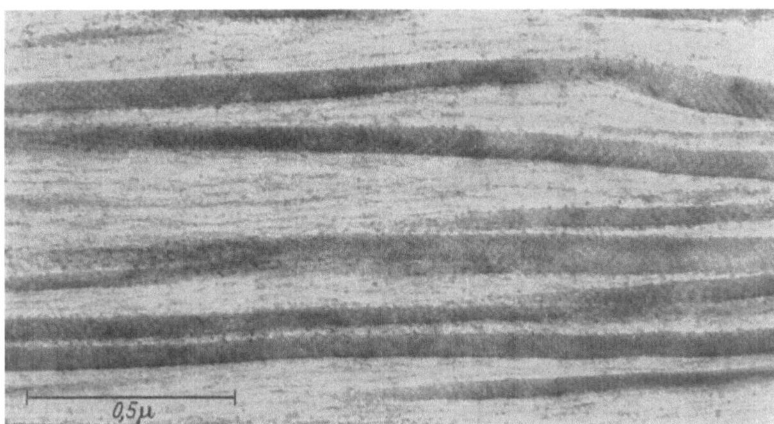

Fig. 3. Longitudinal section of the same muscle showing possible secondary filaments between the main fibrils.
57,000 ×

point of view of the observer (8). Comparison has been made between muscles fixed in neutral, 5%, formalin and stained with phosphotungstic acid which were either embedded in araldite and sectioned for the electron microscope or examined in the wet state in an optically focussing small angle X-ray camera [modified from FRANKS (9)]. Neither by X-ray diffraction nor by electron microscopy has it been possible to detect any significant change in the basic 720 Å period of the 'paramyosin' net when muscles fixed at maximal and minimal body length (a change of approximately 30%) are examined. The electron microscope statistics are not yet complete, however.

It seems probable that the structurally characterised 'paramyosin' is identical with the biochemically derived tropomyosin of BAILEY (*10, 11, 5*). It is at least certain that the tropomyosin, which is a large proportion of the structural protein of these muscles, is contained in the 'paramyosin' fibrils of the intact muscle. Some forms of the re-precipitated tropomyosin show a marked similarity to the 'paramyosin' ribbons isolated from these muscles, though the dot pattern has not been observed in re-precipitated tropomyosin (*11*). The material between the 'paramyosin' fibrils is loosely organised in the intact muscle. Sometimes it has the appearance of distinct 50—100 Å filaments nearly parallel to the fibre axis (Fig. 3.) This appearance is not always seen, however, and it is not yet clear whether these secondary filaments represent a real structure. The location of the much smaller amount of 'actomyosin'-like protein is therefore not certain. In a similar muscle, from *Pecten maximus*, RUEGG (*12*) finds that the tropomyosin is about 80% of the structural protein and has no A. T. P.-ase activity, the actomyosin is about 20% and has all the A. T. P-ase activity. The actomyosin may be in the secondary filaments, if they are real, but it may be in the 'paramyosin' fibrils themselves, or even in both these places.

A full account of this work is in preparation. The author thanks Dr.'s J. HANSON and J. LOWY for encouragement and help.

References

1. HALL, C. E., M. A. JAKUS and F. O. SCHMITT: J. appl. Phys. **16,** 459 (1945).
2. BEAR, R. S.: J. Amer. chem. Soc. **66,** 2043 (1944).
3. — and C. C. SELBY: J. biophys. biochem. Cytol. **2,** 55 (1956).
4. HODGE, A. J., H. E. HUXLEY and D. SPIRO: J. exp. Med. **99,** 201 (1954).
5. ELLIOTT, G. F., J. HANSON and J. LOWY: Nature (Lond.) **180,** 1291 (1957).
6. KAWAGUTI, S., and N. IKEMOTO: Biol. J. Okayama Univ. **3, 107,** 248 (1957).
7. HANSON, J., and J. LOWY: Nature (Lond.) **180,** 906 (1957), and Review article in Muscle (ed. G. Bourne) (Academic press, in the press).
8. ELLIOTT, G. F.: Unpublished.
9. FRANKS, A.: Proc. Phys. Soc. B **68,** 1054 (1955).
10. BAILEY, K.: Biochim. biophys. Acta **24,** 612 (1957).
11. HANSON, J., J. LOWY, H. E. HUXLEY, K. BAILEY, C. M. KAY and J. C. RUEGG: Nature (Lond.) **180,** 1134 (1957).
12. RUEGG, J. C.: Abstracts of the Vienna Congress for Biochemistry 1958.

Zur Feinstruktur der glatten Muskulatur

HEDI GANSLER

Elektronenmikroskopische Abteilung der Medizin. Fakultät Homburg/Saarland

Wie bereits früher berichtet, zeigt die glatte Muskelzelle des Rattenuterus eine unterschiedliche Feinstruktur in Abhängigkeit von den Ovarialhormonen. Wir nahmen an, daß die glatten Muskelzellen im Unterschied zur Skeletmuskulatur keine stabilen Filamentsysteme besitzen, vielmehr die kontraktile Substanz in einer labileren Form vorliegt, die eine Sol-Gel-Transformation sowie Änderungen der viscös-elastischen Eigenschaften in ähnlicher Weise ermöglichen würde, wie dies aus den in-vitro-Versuchen SZENT-GYÖRGYS und anderer bekannt ist.

Uns interessierte nun die Frage, ob die Feinstruktur der Muskelzellen anderer glattmuskulärer Organe ähnlich oder verschieden ist. Im folgenden wollen wir über Untersuchungen am Meerschweinchendarm berichten. Wir fixierten den Muskel am narkotisierten Tier in situ, um eine Kontraktion bzw. Schrumpfung des Muskels durch die Präparation weitgehend auszuschalten. — Dem Schema von GOERTTLER ist zu entnehmen, daß die Ringmuskulatur aus schuppenartig ineinandergefügten Muskelbündeln besteht, die durch Verbindung mit einem scherenartig angeordneten Netz von Gitter- und elastischen Fasern senkrecht bis tangential zur Längsachse des Darmrohres verlaufen, wodurch Enger- und Weiterstellung bewirkt wird.

Die Struktur eines solchen Muskelbündels erscheint schon an Paraffinschnitten im Lichtmikroskop recht polymorph. Die Kerne der MZ (Muskelzellen) liegen im Längsschnitt zur Zellachse langgestreckt oder korkenzieherartig gefaltet annähernd parallel, während im Querschnitt unregelmäßigere Formen beobachtet werden. Außerdem sieht man am gleichen Präparat Bereiche, in denen sich die Einzelzellen sehr gut gegeneinander abgrenzen lassen, neben Bereichen, in denen die Muskelkerne in einem syncytialen Myoplasma zu liegen scheinen.

Lassen sich schon am Paraffinschnitt im Lichtmikroskop keine exakten Aussagen machen über die *Form* der Einzelzelle, so wird dies noch schwieriger am Ultradünnschnitt bei der elektronenmikroskopischen Untersuchung. Zwar können wir in Übereinstimmung mit anderenAutoren sagen, daß im Elektronenmikroskop stets wohl definierte Einzelzellen beobachtet werden, viel schwieriger ist es dagegen, aus dem zweidimensionalen Schnittbild die dreidimensionale Struktur des Zellverbandes zu rekonstruieren. Da wir auf Grund unserer Untersuchungen glaubten annehmen zu müssen, daß die Zellen im nicht gedehnten Zustand zopfartig ineinander angeordnet seien, fixierten wir einen Teil unseres Materials unter leichter Dehnung. Wir beobachteten hierbei sehr viel längere Muskelzellen, z. T. sogar über 80 μ (!) lange Zellen am normalen Meerschweinchendarm.

Der *Zellverband* innerhalb eines Muskelbündels bildet stets eine lückenlose Einheit, wodurch der Eindruck eines Zellmosaiks entsteht, so daß zwar die Form der Einzelzellen unregelmäßig ist, stets aber die Einheit des Zellverbandes gewahrt bleibt. Zwischen den Zellen befindet sich eine mehr oder weniger breite Lamelle einer homogenen Grundsubstanz (Basalmembran) und in diese eingebettet unterschiedlich zahlreich feinste Fäserchen (Gitterfasern), die bei guter Auflösung eine Querstruktur deutlich erkennen lassen. Wir möchten annehmen, daß der Extracellularraum aus einer viscös-elastischen Grundsubstanz besteht, die in Abhängigkeit von physikalisch-chemischen Bedingungen Variationen der Zustandsform erfährt, so daß bald ein strukturelastisches Sol vorliegt, was elektronenmikroskopisch homogen erscheint, bald feinste Fäserchen „gelieren". Es ist nicht unwahrscheinlich, daß solche Änderungen des Extracellularraums den sog. „Feststellmechanismus" glattmuskulärer Systeme ermöglichen.

Die *Feinstruktur* der Einzelzellen der Darmmuskulatur ist prinzipiell die gleiche wie die der Uterusmuskulatur. Während jedoch das Myoplasma in allen Zellen eines Uterus fast gleichförmig ist und die Strukturvarianten erst in verschiedenen Uteri, die unter definierten verschiedenartigen Hormonbedingungen standen, (Kastration, Daueroestrus und Schwangerschaft) aufgefunden wurde, konnten wir alle „Strukturmuster" der Einzelzellen schon in einem einzigen Präparat der Darm-Muskulatur beobachten: Der Grund hierfür dürfte in der unterschiedlichen Nervenversorgung zu suchen sein. Während beim Uterus Nervenfasern nur im lockeren parametranen cervixnahen Bindegewebe beobachtet wurden, nie aber in der Muskulatur, lassen sich Bündelchen feinster Nervenfasern aus den großen Plexus zwischen die einzelnen Muskelbündel ziehend, auch elektronenmikroskopisch einwandfrei identifizieren. Auf Übersichtsaufnahmen lassen sich folgende „Strukturmuster" des Zellverbandes unterscheiden:

1. Der Zellverband erscheint gleichförmig, jedoch bleibt der celluläre Charakter gewahrt, da das Myoplasma kontrastärmer ist als die Zellmembran.

2. Der Zellverband erscheint gleichförmig und syncytial, da Myoplasma und Zellmembranen gleichen Kontrast zeigen.

3. Der Zellverband erscheint ungleichförmig, wodurch die celluläre Anordnung besonders hervorgehoben wird. Das Myoplasma benachbarter Zellen zeigt unterschiedliche Dichten.

Die Darstellbarkeit von Filamenten, sog. Elementarfibrillen, mit einem Durchmesser von etwa 50 Å scheint auf Grund unserer bisherigen Ergebnisse vom Dehnungszustand der Zellen abzuhängen. Wir haben den Eindruck, daß die Dehnung der Zellen die Sichtbarmachung solcher Elementarfibrillen begünstigt. Im nicht gedehnten Zustand ist das Myoplasma mehr oder weniger homogen und zeigt vor allem keine Längsorientierung.

Wir möchten annehmen, daß die besprochenen Unterschiede der Feinstruktur der Muskelzellen als unterschiedliche Aggregatzustände der kontraktilen Substanz gedeutet werden müssen. Die Leistung der Einzelzelle, für deren Verständnis man die an Modell-Lösungen erzielten Ergebnisse heranziehen kann, würde bewegungsmechanisch erst durch Koordination eines Muskelbündels zum Zellverband bedeutungsvoll. Eine solche funktionelle Einheit könnte vielleicht auch die Diskrepanz zwischen Physiologen und Morphologen erklären insofern, als das Muskelbündel zwar morphologisch cellulär ist, jedoch funktionell als Syncytium betrachtet werden muß.

Microstructure of muscles in cercariae of the digenetic trematodes Schistosoma mansoni and Tetrapapillatrema concavocorpa

F. J. Kruidenier and A. E. Vatter, Jr.*

Zoology Department and Electron Microscope Laboratory University of Illinois, Urbana, Illinois, USA

Apparent structural periodicity along the muscle fibers of the cercaria of *Schistosoma mansoni* (Digenea: Anepitheliocystidia: Schistosomatidae) stimulated the study of the ultrastructure of muscles in this stage of the life history of the Trematoda. Classic investigations (*e. g.* (*1*)] have demonstrated an outer circular, an intermediate longitudinal, and, in various species, an inner layer of diagonally oriented muscle fibers in the subcuticula of trematodes. Specialized musculature present in the suckers, and tails of the cercariae are derived from these layers. Myocytes were found to be bipolar, with myofibrillae in the polar processes, and multipolar, with simple or multibranced protoplasmic strands extending from the bodies of the myocytes to relatively distant, otherwise separate myofibrillae. More recently, striated myofibrillae "with areas comparable to the Z-line, J-disk, and Q-disk" (*2*) recognized in striated muscles of some other animals have been reported. Exellent studies of the ultrastructure of the striatet muscles of vertebrates (*4, 5, 6*) have facilitated the present interpretations.

Cercariae were collected within two hours after their emergence from snail hosts and fixed at 0—2° C in a modification of the buffered, 1% osmium solution of Palade at p_H 8.6. Dehydration and infiltration followed the procedural outline of Vatter und Wolfe (*6*). Specimens were sectioned on Sjöstrand or Porter-Blum microtomes and examined in an RCA-EMU-2E or EMU-3C electron microscope. Two species were studied, *S. mansoni* from experimentally infected snails, *Australorbis glabratus*, and *Tetrapapillatrema concavocorpa* (Epitheliocystidia: Plagiorchiidae) from naturally infected *Helisoma trivolvis*. Both species possess tailed cercariae. The tails of S. mansoni are forked (furcocercous): *T. concavocorpa* has a simple, unforked tail.

Variations in the relation of the myocytes to their myofibrillae appears much as reported by early investigators. The innermost, diagonally oriented fibers were only intermittently distinguish able in the bodies of cercariae as individual myofibrilla. The outer circular and the intermediate longitudinal musculature formed uniform layers in both the bodies and the tails. They were. therefore, studied most intensively.

The layer of circular myofibrillae is thin, approximately $^1/_3$ micra in the tails (Fig. 1, 2). It consists of a single layer of fibrillae in both species. The fibrillae are separated from one another, from the basal membrane of the overlying cuticula, and from the underlying longitudinal muscles by a distinct cell membrane. Groups of the superficial myofibrillae are surrounded by a single, continuous membrane which extends inward between units of the longitudinal muscle layer. The cell membrane involutes between adjacent fibrillae of individual myocytes. It was never observed to completely surround these individual units. No nuclei are present among the fibrillae of the encircling fibers but some sarcoplasm and occasional mitochondria were observed. The bodies of the myocytes must, therefore, be situated at some distance from these contractile elements in both the bodies and the tails of the cercariae. Preliminary observations indicate that these myocytes are mesiad to the layer of longitudinal muscles.

A very thin intracellular matrix (about 20 millimicra) separates longitudinal from circular muscles. The myofibrillae of the longitudinal layer are relatively thick (Fig. 1, 2, 4). Groups of fibrillae are contained within a myocyte but are partly separated from one another by ramifications of the cell membrane. Numerous mitochondria are present in the sarcoplasm along the inner margins of and between myofibrillae. The sarcoplasm of these myocytes within the tails of the cercariae is very abundant but that of the comparable units in the bodies of the cercariae is scanty. This appears to be associated with bipolarity of the longitudinal muscles of the tail and multipolarity of the muscles of the body.

Myofibrillae within the suckers (Fig. 5) are surrounded individually by a distinct membrane. A thin, intracellular matrix separates them from interspersed, apparently polymorphous bodies of

---

* Present address: Abbott Laboratories North Chicago, Illinois

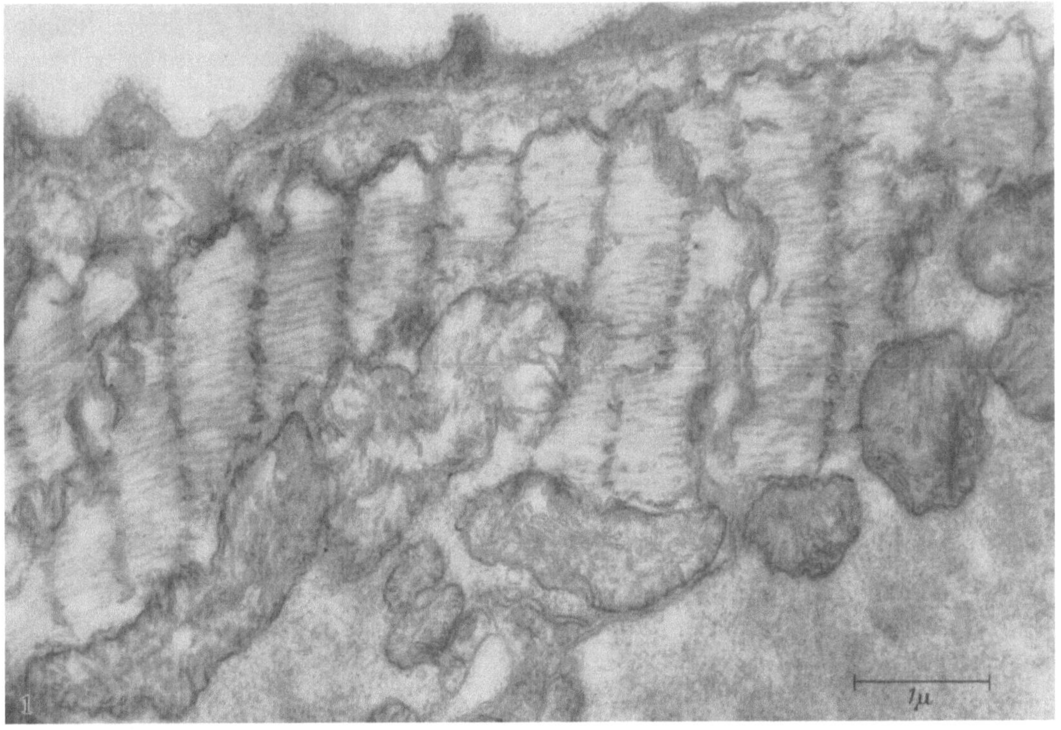

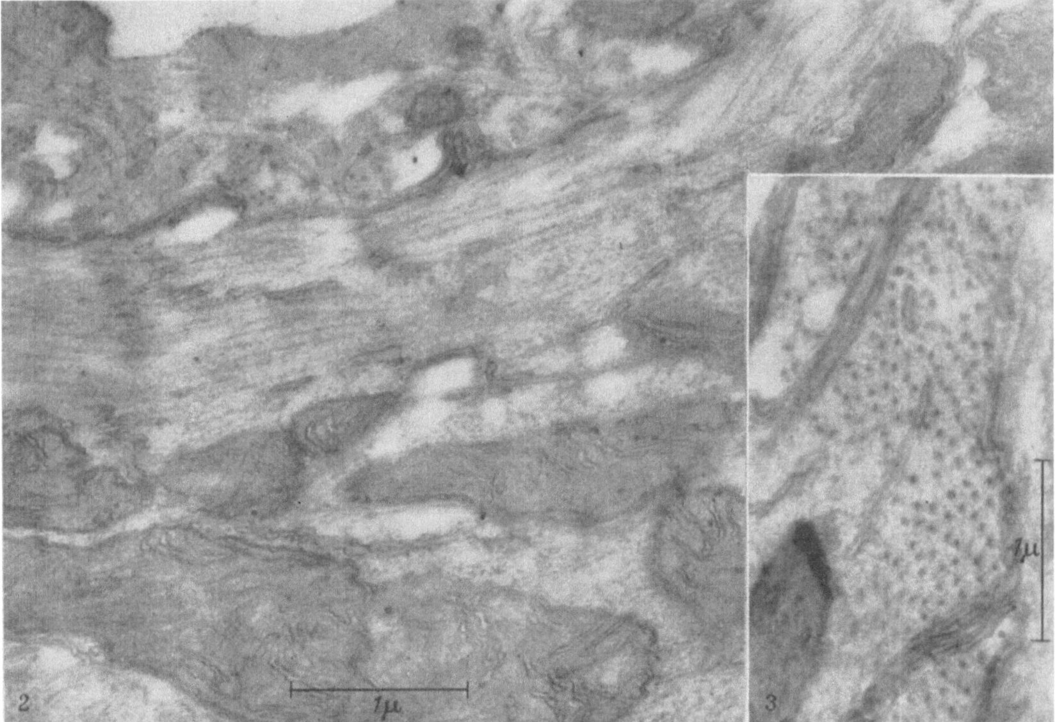

Fig. 1. Longitudinal section of the tail of *S. mansoni* demonstrating the relation of the circular and longitudinal muscles and the periodic arrangement of the sarcotubules. 19000 ×

Fig. 2. Longitudinal section of the tail of *T. concavocorpa* demonstrating circular and longitudinal myocytes. 24000 ×

Fig. 3. Cross section of the circular muscles in the body of *T. concavocorpa* demonstrating the relation of the cell membrane, sarcotubules and the myofilaments. 24000 ×

their myocytes. Individual fibrillae extend from the surface to the base of the sucker. They ar
relatively large, approximately $1^1/_2$—4 micra in diameter, and their terminations are commonly
divided into from two to several branches which insert into the basal membrane of either the
cuticula or the capsule of the sucker. Sarcoplasm along the fibrillae is sparce. Mitochondria
commonly lie immediately beneath the enclosing cell membrane but also occur in the middle of
the myofibrillae.

The sarcoplasm of the several types of myocytes is an almost uniform reticulum of minute
granules and vesicles (Fig. 1, 5). Some evidence of a Golgi apparatus (Fig. 5) needs further study.

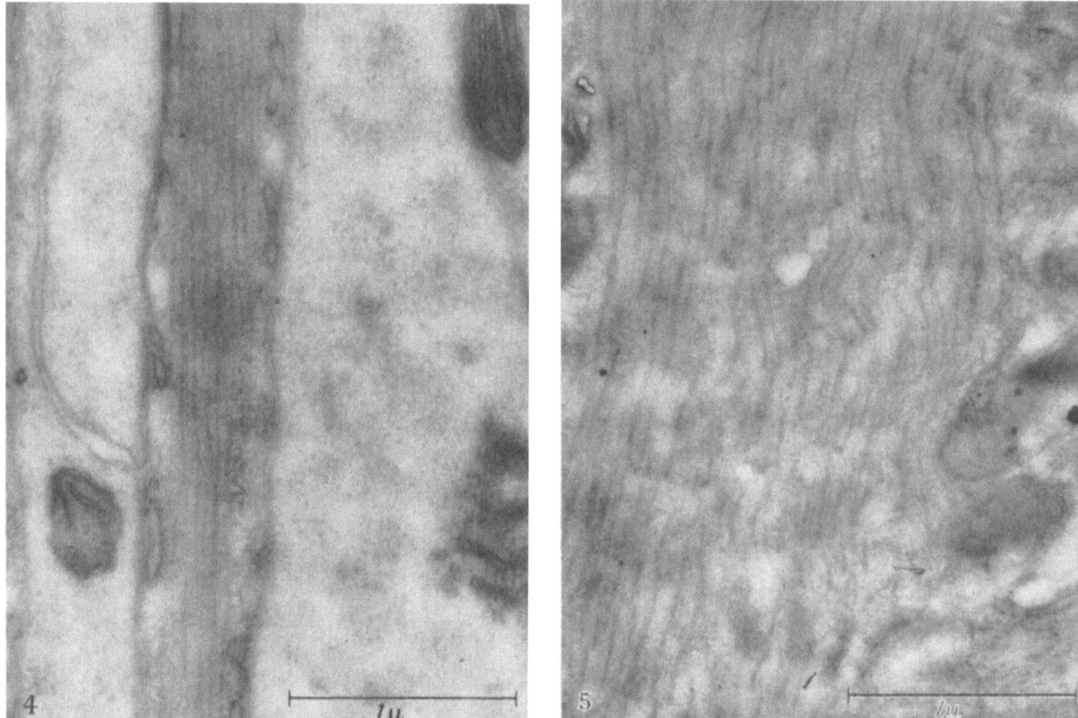

Fig. 4. Longitudinal section through the acetabulum of *T. concavocorpa* demonstrating sarcoplasm, myofilaments
sarcotubules and, possibly, a Golgi complex. A cell membrane is apparent. 31 000 ×

Fig. 5. A longitudinal section through a myofibril of the longitudinal muscle in the body of *T. concavocorpa*
demonstrating myofilaments. 31 000 ×

Each myofibril is composed of many dense primary filaments embedded in a less dense matrix.
The myofilaments appear to be in alternating rows (Fig. 2). This arrangement is frequently
distorted, apparently due to the penetration of series of micro-tubules (*v. infra*) or by the presence
of mitochondria and by contraction. No evidence of secondary myofilaments was visible in our
specimens. Indications of interconnecting strands between primary filaments was scanty and
uncertain. In no case was there an alternation of dense with less dense areas along myofibrillae or
myofilaments that could be interpreted as the familiar "striations" of voluntary or cardiac
muscles of vertebrates (and other animals).

Series of minute tubules are associated with the myofibrillae in all muscles studied. Each
tubule consists of a homogeneous, structure-free core enclosed by a membrane on the order of the
cell membranes of the myocytes. The membranes of the tubes resemble those of the tubular
constituents provisionally considered as part of the Golgi complex although the latter appear
appreciably denser in identical sections (Fig. 5). No other consistent relationship with the Golgi
apparatus was observed. The myotubules are approximately 0.1—0.2 micra in diameter, and
vary widely in lenght. The enclosing membranes are approximately 10 millimicra thick. The
distribution of this system of tubules varies in the several types of muscles studied. They appear

similar to the elements of the endoplasmic reticular system described by Porter and Palade (*6*) or the sarcotubules of Andersson (*4*) in the muscles of vertebrates. Their distribution is, of course, different from that reported by those investigators.

The tubules lie at the periphery of and parallel to the myofibrillae of the suckers. They form a complex which parallels the involuting cell membranes between the fibrillae of the outer, circular layer of muscles. Occasional tubules (Fig. 3) penetrate between and at various angles to the myofilaments of these fibrillae. Numerous complexes of the tubules are found in close association with those portions of the cell membrane which is adjacent to the myofibrillae of the sheet of longitudinal muscles. Individual tubules also penetrate these myofibrillae and occasionally run parallel to the filaments (Fig. 1, 2). A complex forms a very intimate association with the mitochondria along the myofibrillae. In no case did the tubules appear continuous with the mitochondrial membranes.

In the case of *S. mansoni*, the sarcotubules penetrate between the myofilaments of the longitudinal muscles at regular intervals (Fig. 1). Such bands of tubules extend completely across individual myofibrillae in alignment with those of the same and of adjacent cells. The interval between the bands of tubules is approximately $^1/_2$—$^3/_4$ micra. The myofilaments are not interrupted in the region of the bands. The longitudinal, subcuticular muscles of *S. mansoni* are the only elements observed in which this definite pattern was established.

It is probable that the striations reported in the muscles of the cercariae of digenetic trematodes can be attributed to sarcotubules of an endoplasmic reticulum such as described herein. Light and dark areas along the fibrillae would be caused by differential diffraction of the light through and around tubules and fibrillae. Further investigation of related families of the Anepitheliocystidia LaRue, 1957 should be made to determine whether this constitutes a constant characteristic of the super order. Further study should also reveal the constancy of the tubular systems in the Epitheliocystidia LaRue, 1957 and the value of this criterion as a differential feature.

This investigation was supported in part by a research grant (E-1319) from the National Institute of Health, Public Health Service.

References

1. Bettendorf, H.: Zool. Jahr. Abt. Anat. u. Ont. **10,** 307 (1897).
2. Pearson, J. C.: Canad. J. Zool. **34,** 295 (1956).
3. Vatter, A. E., and R. S. Wolfe: J. Bacteriol. **75,** 480 (1958)
4. Andersson, E.: Proc. Stockholm Conf. Elec. Micr. 1956. New York: Acad. Press Inc. p. 208, 1957.
5. Sjöstrand, F. S., and E. Andersson: Proc. Stockholm Conf. Elec. Micr. 1956. New York: Acad. Press Inc. p. 204, 1957.
6. Porter, K. R., and G. E. Palade: J. biophys. biochem. Cytol. **3,** 269 (1957).

3. Kollagen

On the procollagens during development of the skin

W. Th. Daems,* D. O. E. Gebhardt and G. Smits

Laboratory for Electron Microscopy and Histological Laboratory, University of Amsterdam, Holland

During the last decade a growing number of publications have appeared on the biogenesis of collagen. A Russian group of workers (*1*) were the first to isolate a protein from connective tissue by acid citrate buffer extraction. This protein was considered to be the direct precursor of mature collagen and was named "procollagen". However more recent and more differentiated work by J. Gross e. a. (*2*) and experiments with radio active glycine by Harkness and co-workers (*3*) have given evidence that the real metabolic precursor was obtained by extraction of the tissue with neutral or weakly alkaline solutions.

---

* Present adress: Laboratorium voor Electronenmicroscopie, Rijksuniversiteit, Leiden.

We have made an attempt to approach the problem of the precursors of collagen by bio-chemical and electron microscopical examination of collagenous material derived from a number of consecutive stages of development in embryonic skin.

The material was extracted with alkaline and acid citrate buffers, in order to ascertain whether, and if so to what degree, the stage of biological maturation of the fibrous material might influence the respective

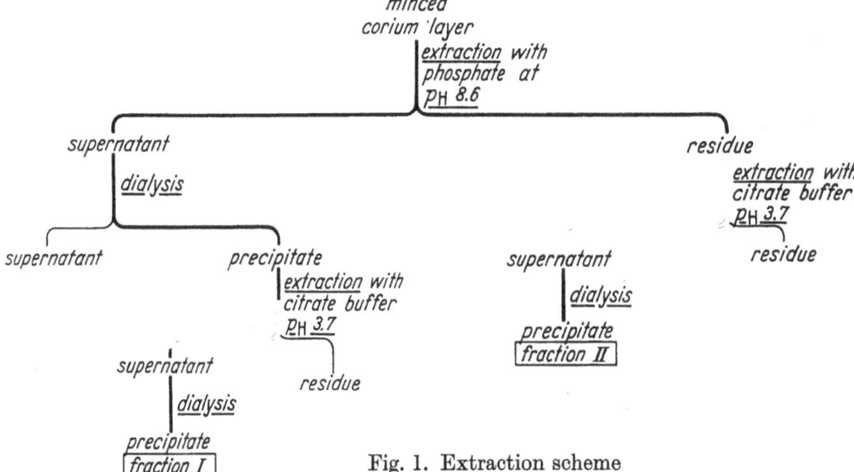

Fig. 1. Extraction scheme

yields of the alkaline and citrate soluble collagen fractions. Measurement of the width of the fibrils that precipitated from the different fractions were also made, in an attempt to establish a further characterization of these fibrils.

Skins of bovine embryos of different stages of development were obtained from the slaughter-house, where they were kept at 0°, within 24 hr after the cow had been slaughtered. The skin of a new-born calf and that of a calf aged 2 months were also used. As a rule only dorsal skin was used. In the case of the older calf however, we took skin of the forehead. The corium was prepared for extraction as follows: First the upper layer (epidermis; stratum papillare) was removed. In young embryos (crown-rump length 20—50 cm) without appreciable hair growth, this layer could easily be scraped off with a sharp scalpel. For the calf skins the upper layer was removed with a razor. The subcutaneous fatty tissue was removed in the same way. The corium layer was then cut into pieces measuring 3×3 cm and cut on a freezing microtome into 50 μ sections. In order to obtain sufficient material from

Stage of development	Fraction I (alk. sol. coll.)	Fraction II (citr. sol. coll.)
20 cm feti	2.03	0.7
20 cm feti	0.51	0.8
25 cm feti	0.46	0.8
30 cm feti	0.46	4.4
30 cm feti	0.61	5.4
40 cm feti	0.56	11.8
50 cm fetus	0.41	13.3
50 cm fetus	1.2	17.7
50 cm fetus	1.04	16.6
New born calf	0.06	16.6
New born calf	0.05	14.4
New born calf	0.02	16.3
Two months of calf	0.25	13.3
Cornea (calf)	0.1	3.0
Cornea (adult cattle)	0.04	0.2

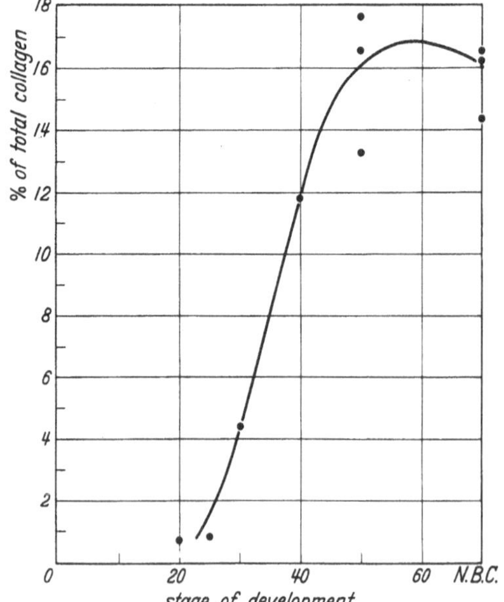

Fig. 2. Amount of citrate soluble collagen (fraction II) as percentage of total collagen

the younger embryos we pooled a number of skins of embryos of the same length in the 20, 25 and 40 cm groups. Skins of 3 embryos of 50 cm, of 3 new-born calves and of one older calf were worked up separately. For purposes of comparison, bovine corneas were also included in the investigation.

Extraction was carried out by the method of HARKNESS and co-workers (3). The alkali-soluble fraction extracted at pH 8.6 will henceforth be referred to as fraction I. This fraction is considered to be identical with the collagen soluble in neutral salt solution, as isolated by JACKSON (4, 5). The citrate-soluble fraction will be referred to as fraction II. This was obtained by extracting the phosphate-insoluble residue with acid citrate buffer.

The quantitative yields of the collagen fractions, expressed as percentages of the total collagen, were as follows (see table on left hand side).

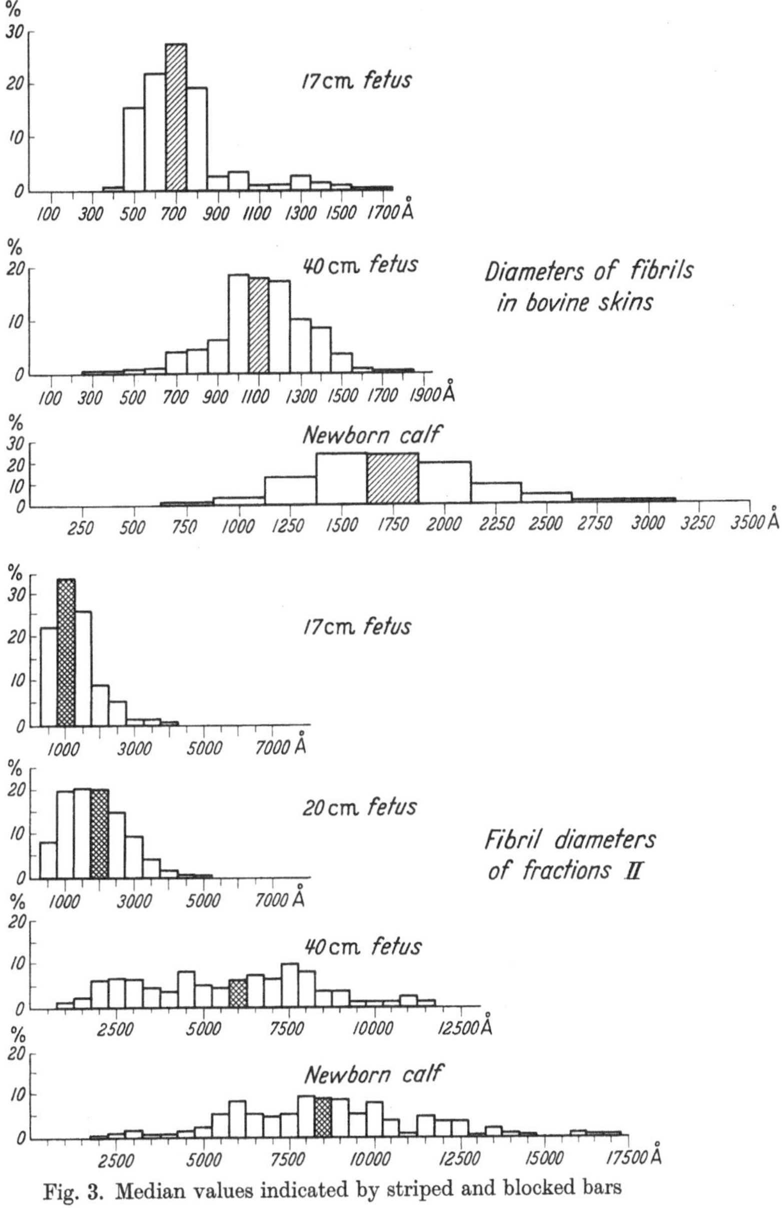

Fig. 3. Median values indicated by striped and blocked bars

Strikingly low values were found for fraction I from the newborn calf. GROSS (8) has pointed out in a recent publication that in young guinea pigs two days fasting reduced the yield of neutral salt-soluble collagen to very low values. The same phenomenon seems to have occurred in our newborn calves. The data from newborn calves are thus un-suitable for inclusion in the developmental series. It is nevertheless interesting to note that fraction II appeared to be unaffected by this starvation. This in itself might plead for the hypothesis that fraction II has already been incorporated into the for med fibrils, whereas fraction I as a true precursor consisted of free components. It will further be seen that the three determinations of fraction I on a 50 cm embryo show a very considerable fluctuation. In this connection we may point out that Gross (9) also mentioned large variations in the yield of collagen soluble in neutral salt solutions. Taking the above into consideration we are still left with the impression

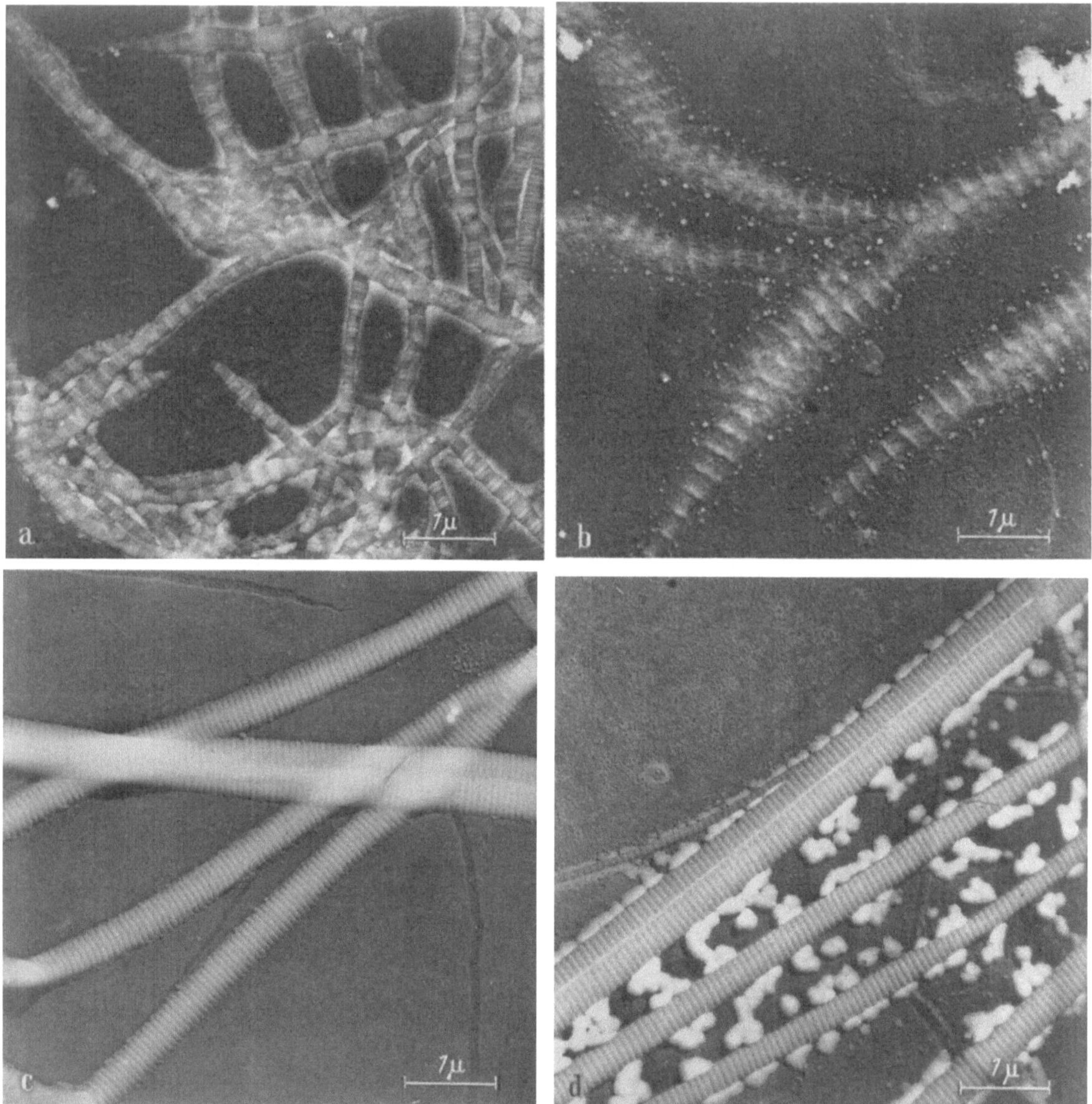

Fig. 4a—d. Fibrils from dialyzed alkaline extracts (fractions I) of skins of bovine embryos of different ages and newborn calves. a) two-band long-spacing fibrils; b) long-spacing fibrils; c) fibrils with normal 640 A periodicity; d) fibrils with normal 640 A periodicity and adherent globular material

that the proportion of fraction I tends to decrease during development. The behaviour of fraction II is quite different as is shown in the graph of Fig. 2. After an initial steep rise its percentage reaches a maximum at about the time of birth and then begins to fall. This post-natal decrease was also observed by K. D. Orekho-vitch (6) and by H. Hörmann (7), whose curve can well be connected with ours.

The fact that fibrils precipitate out of the extracts has already been mentioned in many publications, as is also the fact that the morphological appearance of the fibrils is dependent on the medium in which they are formed. As regards the influence of age on the fibrils in the body, it is known that in most kinds of connective tissue the fibril-diameter increases with age (9). This we also found in the skins examined by us (see Fig. 3). Also electron microscopy revealed characteristic differences between the fibrils precipitated out of our fractions I and II: Some of the fibrils of fraction I showed a two-band longspacing as described by RANDALL e. a. (10) and

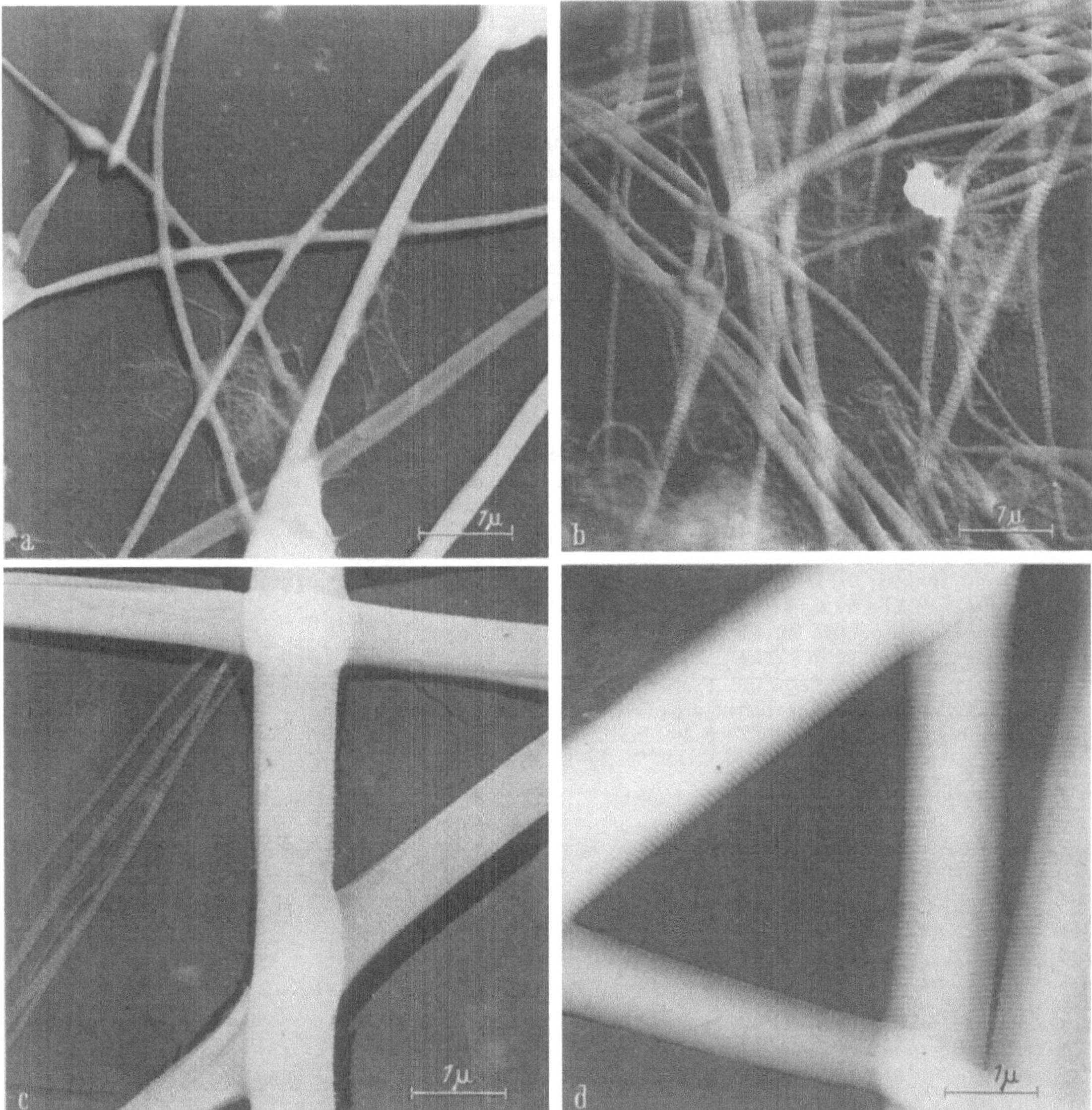

Fig. 5a—d. Fibrils from dialyzed citrate extracts (Fractions II) of: a) adult corneas; b) skins of 17 cm bovine embryos; c) skins of 40 cm bovine embryos; d) skins of newborn calves

KÜHN e. a. (11) (Fig. 4a), others showed the normal long-spacing (Fig. 4b) or the 640 Å periodicity (Fig. 4C); globular material was often seen in association with many of the fibrils (Fig. 4d). We also saw many tactiod ends. The fibrils were flat and ribbon-like and contrasted only slightly with the background. The diameter diagrams of fraction I showed no correlation between the width of the fibrils in a given skin and that of the fibrils which precipitated out of the corresponding extract.

  The fibrils which precipitated out of the citrate extracts were of a different type. There were no long-spacing fibrils; the periodicity was 640 Å as a rule, with occasionally a 210 Å cros-striation. The fibrils were more strongly contrasting and were less ribbon-like. Often they split up into very thin filaments which mostly were structureless, but sometimes showed a 210 Å periodicity. Of particular interest is the fact that the diameter diagram of the precipitated fibrils shows a notable correlation with the fibril-diameter diagram in the corresponding skins. As in the skins, we see that in the fraction II the diameter of the fibrils increases with growing age of the animal. There is also a corresponding increase in the dispersion of the diameters (Fig. 3). That this correlation is not directly related to age as such is shown by our observations on adult bovine cornea. We found that from these corneas, which, as already noted by v. D. HOOFF (12) and others, are characterized by the presence of very fine fibrils, a fraction II with again remarkably thin fibrils was obtained.

We see two plausible explanations of this remarkable parallelism in fibre thicknesses of skins and fractions II: The cause must lie either in the concentration of the collagen particles which form the precipitated fibrils, this concentration increasing with the age of the animal, or in an intrinsic property of these particles, acquired during and by the binding of the particles to the fibrils in the tissue. Preliminary experiments in which the concentration of the extract before dialysis was varied seem to indicate that the concentration is of secondary importance. The conclusion thus seems justified that some intrinsic factor is present in the constituent particles themselves, which determines the diameter of the precipitated fibrils of fractions II.

References

1. OREKHOVITCH, V. N., and V. O. SHPIKITER: Science 127, 1371 (1958).
2. GROSS, J., J. H. HIGHBERGER and F. O. SCHMITT: Proc. nat. Acad. Sci. 41, 1 (1955).
3. HARKNESS, R. D., A. M. MARKO, H. M. MUIR and A. NEUBERGER: Biochem. J. 56, 558 (1954).
4. JACKSON, D. S., and J. H. FESSLER: Nature (Lond.) 176, 69 (1955).
5. — Biochem. J. 65, 277 (1957).
6. OREKHOVITCH, K. D.: Dokl. Akad. Nauk. SSSR 71, 521 (1950)
7. HÖRMANN, H.: Beitrag zur Silikoseforschung, Sonderbd. 2, 619 (1958).
8. GROSS, J.: J. exp. Med. 107, 265 (1958).
9. — Amer. J. Path. 26, 708 (1950).
10. RANDALL, J. T., G. L. BROWN, S. FITTON JACKSON, F. C. KELLY, A. C. T. NOTHE, W. E. SEEDS and G. R. WILKINSON: In „Nature and structure of collagen" p. 213. Butterworth 1953.
11. KÜHN, K., U. HOFMANN, u. W. GRASSMANN: Z. Naturforsch. 11b, 581 (1956).
12. HOOFF, A. V. D.: Proc. Kon. Ned. Acad. Wet. C 55, 628 (1952).

Längsaufteilung kollagener Fibrillen in Elementarfibrillen

TH. NEMETSCHEK

Institut für Elektronenmikroskopie, Düsseldorf

In der elektronenmikroskopischen Literatur findet man oftmals den Hinweis, daß nach Säurequellung oder mechanischer Beeinflussung eine Längsaufspaltung des Kollagens in sogenannte Subfibrillen erfolgt (1). Diese sogenannten Subfibrillen besitzen bei einem meistens unregelmäßigen Verlauf eine Breite von 100 und mehr Å, ohne in den Aufnahmen eine hochunterteilte Querstreifung erkennen zu lassen. Wir versuchten deshalb, die fraglichen Elementareinheiten an solchen Kollagenfibrillen nachzuweisen, die alle Voraussetzungen zur Darstellbarkeit einer hochunterteilten Querstruktur besitzen und somit quer zur Längsachse bereits einen weiten Einblick in die Struktur gewähren.

Abb. 1 b—c zeigt Kollagenfibrillen, die sowohl eine hochunterteilte Querstruktur (man vergleiche hierzu Abb. 1 a) erkennen lassen, wie auch eine Untergliederung in parallel verlaufende Längseinheiten. Solche Resultate erhielten wir, wenn wir die mit Phosphorwolframsäure (PWS) vorbehandelten und auf einer Objektträgerblende befestigten, also mechanisch fixierten Kollagenfibrillen zwischen 60 und 65° C, 15—60 min der Einwirkung von 0,01 n CH_3COOH aussetzten. In Abb. 1 c besitzt diese Längsaufteilung ein büschelartiges Aussehen. In Höhe der Querstreifen, man beachte besonders die a- und b-Doublette, weisen diese Längseinheiten Verdickungen auf. Sie enden alle in einem stärker aufgeweiteten Fibrillenbereich, der Ähnlichkeit mit Ausschnitten von WOLPERS (2) gezeigten hitzegeschrumpften Kollagenfibrillen besitzt. Ähnlich wie bei der Hitzeschrumpfung (2) erfolgt auch diese Art von Aufblähung nicht gleichzeitig über den gesamten Fibrillenverlauf, sondern nur herdförmig. So findet man neben unveränderten Fibrillenbereichen auch insgesamt unbeeinflußt gebliebene Fibrillen. Ferner konnten oftmals Fibrillen angetroffen werden, die eine nicht ganz durchgehende Längsaufteilung aufweisen und einen Vergleich mit von uns beschriebenen Zersetzungsstadien an Kollagenfibrillen zulassen, die

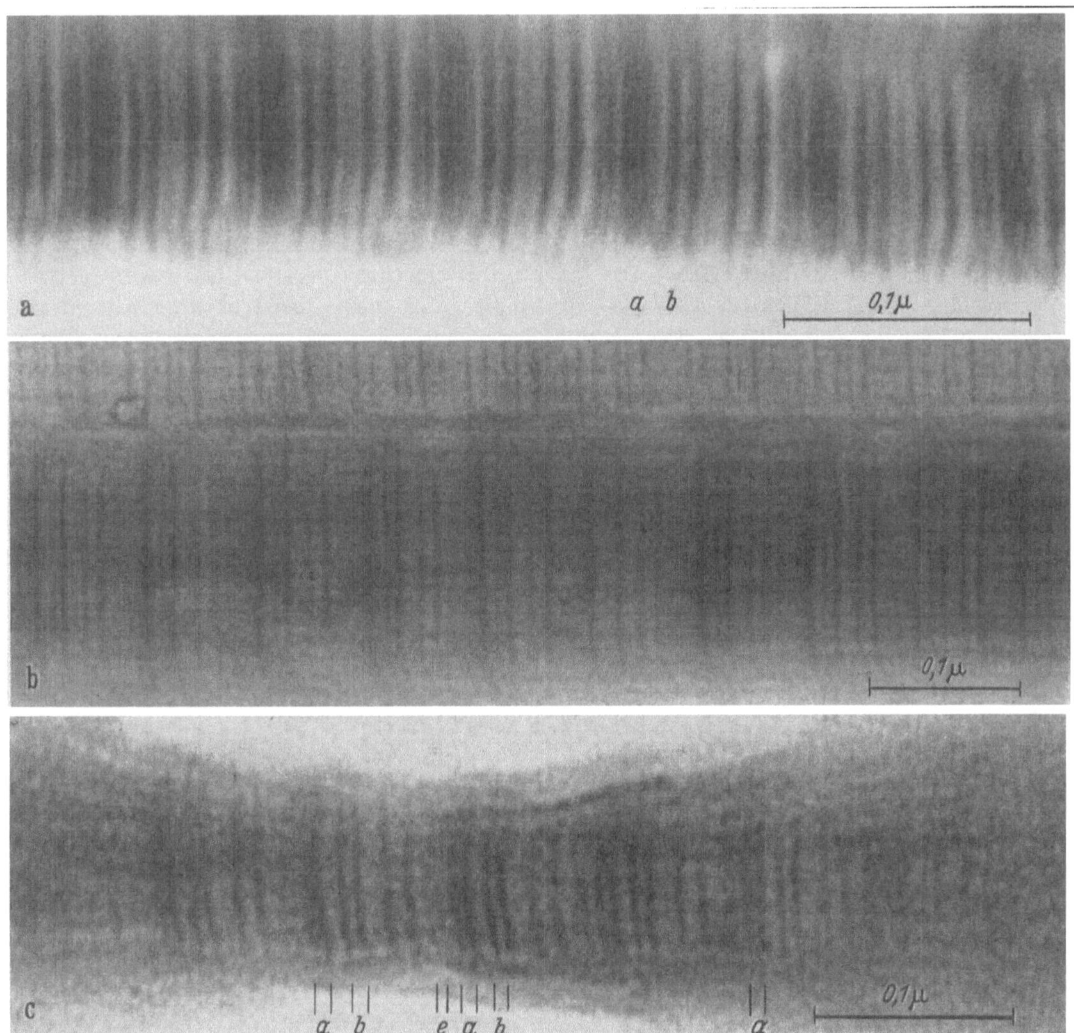

Abb. 1a—c. a) Kollagenfibrillen aus dem Nackenband des Menschen mit PWS behandelt. 11 bis 13fache Unterteilung der Querstreifung. b) und c) Kollagenfibrillen aus Rattenschwanz-Sehne mit 0,5% PWS behandelt, auf Objektträgerblende bei 60—65°C Einwirkung von 0,01-n CH_3COOH. Längsaufteilung in Elementarfibrillen

eine intensive Elektronenbestrahlung erfahren hatten (3). Abb. 1 b zeigt Fibrillen, die eine deutliche Längsaufspaltung über den gesamten Fibrillenverlauf aufweisen. Bei gut erhaltengebliebener unterteilter Querstreifung beobachtet man parallel verlaufende Längselemente von nicht genau bestimmbarer Länge, mindestens jedoch 1500 Å, die oftmals weniger als 30 Å Breite besitzen. Sowohl diese Werte als auch die Abstände voneinander unterliegen aber Schwankungen. Die Indizierung der Querstreifen gelingt mühelos an Elementar-Einheiten, die sich noch im Fibrillenverband befinden. Dieses Vorhaben versagt aber an weiter abseits liegenden ganz schmalen Elementen (etwa 30 Å Breite), da eine etwaige Folge von Bereichen unterschiedlicher Massendicke sich nur noch in Gestalt körniger Bestandteile kundtut, was in Anbetracht der Eigenkörnigkeit der Objektumgebung, vor allem bei der defokussierten Abbildung (4), keine Indizierung ermöglicht. Erst die Parallelanordnung einer ausreichenden Zahl so beschaffener Längseinheiten hat eine Vorzugsorientierung solch körniger Bestandteile in Gestalt von Querstreifen zur Folge, die dann bei guten Bildqualitäten auch gesehen werden können. Diese Längselemente müssen als übereinanderliegende *Elementarfibrillen* (*5*) aufgefaßt werden, dermaßen, daß bei der Durchstrahlung im Elektronenmikroskop eine annähernd statistische Verschmierung aus Elementarfibrillen vorliegt, die zu Längsstreifen von verschiedenen Breiten und in unterschiedlichen Abständen im Bild führt. Die in den Aufnahmen vorliegenden Größenordnungen dürften somit hinsichtlich der Breite dieser Einheiten stets ein Vielfaches der eigentlichen Elementarfibrillen darstellen, wobei das Auflösungsvermögen des Elektronenmikroskopes nicht voll ausgenutzt werden kann, da die Voraussetzung hierzu erst beim Vorliegen einer „einschichtigen" Anordnung aus solchen Elementarfibrillen gegeben wäre. Diese Aufteilung in Elementarfibrillen, die ja offenbar dadurch zustande gekommen ist, daß durch äußere Einflüsse anziehende, bzw. ausrichtende Kräfte im Fibrillenverband zum Teil überwunden wurden (*6*), läßt vermuten, daß zugleich auch Versetzungen in Längsrichtung eintreten konnten; hierbei kann angenommen werden, daß ein Ausgleich von Massendicken-Unterschieden längs dieser Elementarfibrillen und damit ein Kontrastverlust der Querstreifung erfolgt. Die zunächst etwas schwer verständlichen Kontrastverhältnisse längs dieser Einheiten sowie die Erscheinung, daß keine Querstrukturen an weit abgesprengten Elementarfibrillen zu erkennen sind, finden hierin eine Erklärung. Solche Aufnahmen führen zu dem Schluß, daß Kollagenfibrillen aus parallel zur Längsrichtung der Fibrillen verlaufenden Elementarfibrillen aufgebaut sind.

Auf einen Aufbau der Kollagenfibrillen aus Protofibrillen (Elementarfibrillen) wurde auch schon an anderen Stellen, vor allem von WASSERMANN anhand unterschiedlichen Untersuchungsmaterials hingewiesen (*7*). Auch glauben GROSS, HIGHBERGER und SCHMITT (*8*), im Tropokollagen die Bildungseinheit des Kollagens mit einer Dicke von etwa 50 Å und einer Länge von 2000 bis 3000 Å gefunden zu haben. Wir möchten annehmen, daß es sich bei diesem mit neutralen Salzlösungen, vor allem aus jugendlichem Bindegewebe extrahierbaren Baustein um solche Elementareinheiten handelt, wie wir sie mit der von uns angewandten Methode im Fibrillenverband direkt nachweisen konnten.

Diese Elementarfibrillen sind bereits Träger der Querstruktur und können durch enge Lateralanordnung die Querstreifen verstärken bzw. durch Auseinandergehen die Einzelstreifen zum Zerfall bringen. Wie wir bereits festgestellt haben, muß aus Gründen einer gegenseitigen Überlagerung damit gerechnet werden, daß die im Bild ausmeßbaren Breiten dieser Elementarfibrillen in Wirklichkeit unter 30 bzw. 25 Å liegen und somit auf Einheiten molekularer Größen bezogen werden können. So wurde nach viskosimetrischen Messungen (*9*) in Gelatinelösungen ein „Molekül" von 17 Å Durchmesser und 800 Å Länge gefunden. In Citratextrakten von Fischkollagen sollen laut BOEDTKER und DOTY (*10*) gleichfalls nach physikalisch-chemischen Untersuchungen Einheiten von 14 Å Durchmesser (runde Form) oder 17 Å Durchmesser (elliptische Form) und von 2900 Å Länge vorliegen. In jüngster Zeit gelang es HALL und DOTY (*11*), aus solchen Extrakten Fibrillen auszufällen und elektronenmikroskopisch abzubilden. Auf Grund röntgenographischer Daten hat BEAR (*12*) ein Kollagenmodell vorgeschlagen, welches Polypeptidketten von 12 bis 17 Å Durchmesser vorsieht, und TOMLIN (*13*) gibt als Fibrillenbaustein eine lange zylindrische Einheit von 12 Å Durchmesser an. Wie ersichtlich, befinden sich diese Werte in relativ guter Übereinstimmung mit den von uns angegebenen Breiten für die Elementarfibrillen, die demnach *nur* aus zwei bis drei parallel angeordneten Polypeptidketten aufgebaut sind.

Für die im Elektronenmikroskop darstellbare Querstreifung ergibt sich folgendes Bild: Elementarfibrillen, die als Strukturträger angesehen werden, besitzen Bereiche unterschiedlicher Massendicke, die günstigenfalls nur als Körnchen wahrgenommen werden könnten und erst nach lateraler Ausrichtung während der Parallelanordnung zu Bündeln solcher Elementarfibrillen den Eindruck von Querstreifen vermitteln. Bei dieser Bündelung müßten anziehende bzw. ausrichtende Kräfte die Vereinigung der Elementarfibrillen, also die Bildung kollagener Fibrillen, lenken, so daß z. B. gleichartige Strukturelemente stets zusammentreffen, etwa im Sinne der von FRIEDRICH-FREKSA (*14*) bei der Chromosomen-Konjugation angenommenen Vorgänge. Für das Zustandekommen der Bereiche unterschiedlicher Massendicke längs dieser Elementareinheiten dürfte am wahrscheinlichsten das Abwechseln geordneter und ungeordneter Bereiche in Richtung der Fibrillenachse verantwortlich sein (*15*). Entsprechend der Vorstellung von Baer (*15*) sollen in den geordneten Bereichen vor allem hydroxylhaltige Seitenketten vorliegen, während in den ungeordneten Abschnitten wahrscheinlich saure bzw. basische Säureamidseitenketten von größerer Länge lokalisiert sein sollen. Der Zusammenhalt der Elementarfibrillen müßte durch polare Kräfte und vor allem in kristallinen Bereichen (*15*) durch Wasserstoffbrücken gewährleistet sein, weshalb nur kleinste Zwischenräume, die normalerweise selbst mit Hilfe des Elektronenmikroskopes nicht erfaßt werden können, vorliegen dürften. Küntzel und Prakke (*16*) gaben für den Abstand ähnlicher Grundeinheiten, anhand röntgenographischer Untersuchungen, je nach dem Grad der Quellung des Kollagens 10 bis 16 Å an.

Die in Abhängigkeit von der Vorbehandlung im Elektronenmikroskop sichtbare einfache oder hochunterteilte Querstreifung könnte ihre Deutung darin finden, daß durch die jeweilige Vorbehandlung innerhalb dieses Elementarfibrillenverbandes eine differenziertere Lateralausrichtung begünstigt wird, etwa im Sinne des „In-Register-Bringens" von Polypeptidschrauben, wie sie Kühn, Grassmann und Hofmann für die Bedeutung der Einwirkung von PWS und Chrom-III-Verbindungen formulieren (*17*).

Literatur

1. Nutting, G. C., and R. Borasky: J. Amer. Leather. Chemists Ass. **43**, 96 (1948). — Lelli, G., e. U. Marotta: Rend. Ist. super. Sanita **13**, 518 (1950). — Randall, J. T.: Nature (Lond.) **169**, 1029 (1952); Nature and Structures of Collagen, London 1953. — Borasky, R., and J. S. Rogers: J. Amer. Leather Chemists Ass. **47**, 312 (1952). — Bahr, G. F.: Exp. Cell Res. **3**, 485 (1952). — Wassermann, F.: Ergebn. Anat. **35**, 241 (1956).
2. Wolpers, C.: Biochem. Z. **318**, 373 (1948).
3. Hofmann, U., Th. Nemetschek u. W. Grassmann: Z. Naturforsch. **7b**, 509 (1952). Nemetschek, Th., W. Grassmann u. U. Hofmann: Z. Naturforsch. **10b**, 61 (1955).
4. Gansler, H., u. Th. Nemetschek: Z. Naturforsch. **13b**, 190 (1958). Lenz, F., u. W. Scheffels: Z. Naturforsch. **13a**, 226 (1958).
5. Nemetschek, Th.: Z. Naturforsch. **13b**, 225 (1958).
6. Vergl. A. Küntzel: Kolloid-Z. **91**, 152 (1940); **96**, 273 (1941).
7. Vgl. (*1*) u. G. F. Bahr: Arch. Derm. Syph. (Berl.) **193**, 515 (1951). — Seki, M.: Arch. histol. jap. **3**, 465 (1952); **7**, 5 (1954). — Zawisch, C.: Acta anat. (Basel) **29**, 143 (1957); dort weitere Literatur.
8. Gross, J., J. H. Highberger and F. O. Schmitt: Proc. nat. Acad. Sci. USA **40**, 679 (1954); Gross, J.: J. biophys. biochem. Cytol. **2**, 261 (1956).
9. Scatchard, G., J. L. Oncley, O. W. Williams and A. Brown: J. Amer. chem. Soc. **66**, 1980 (1944).
10. Boedtker, H., and P. Doty: J. Amer. chem. Soc. **77**, 248 (1955); — Doty, P.: Proc. nat. Acad. Sci. USA **42**, 791 (1956).
11. Hall, C. E., u. P. Doty: J. Amer. chem. Soc. **80**, 1269 (1958).
12. Bear, R. S.: 120. Tag. Amer. Chem. Soc. New York 1951.
13. Tomlin, S. G.: Angew. Chem. **68**, 215 (1956).
14. Friedrich-Freksa, H.: Naturwissenschaften **28**, 376 (1940); vgl. auch W. Grassmann u. J. Trupke in: Flaschenträger-Lehnartz, Physiol. Chem. **1**, 660 (1951).
15. Bear, R. S.: Advanc. Protein Chem. **7**, 69 (1952); s. (*8*); vgl. H. Zahn: Angew. Chem. **64**, 295 (1952).
16. Küntzel, A., u. F. Prakke: Biochem. Z. **267**, 243 (1933).
17. Kühn, K., W. Grassmann u. U. Hofmann: Naturwissenschaften **44**, 538 (1958); Z. Naturforsch. **13b**, 154 (1958).

End chain and side chain interactions in the ordered aggregation of modified collagen macromolecules

A. J. Hodge and F. O. Schmitt

Department of Biology, Massachusetts Institute of Technology, Cambridge, Massachusetts, USA

Sonic irradiation of acid-soluble collagen (tropocollagen) markedly affects the aggregation properties of these macromolecules. Relatively brief irradiation prevents the formation of native-type ($\sim$ 700 Å axial period) fibrils. However, ordered aggregates of the segment long-spacing (SLS) type, in which the macromolecules are all "pointing" in the same direction and are in register with respect to their ends, are still formed on addition of ATP; these have lengths and band patterns indistinguishable from those obtained from unirradiated preparations. Dimeric and polymeric SLS-type structures are commonly found. Detailed analysis of the band patterns of these forms led to the suggestion that one or more peptide chains may extend beyond the triple helix body of the tropocollagen macromolecule at each end and that polymerization, either of the normal type or of the abnormal types observed in irradiated collagen, involves the orderly interaction of these end chains — possibly by a coiling of the chains about each other.

Longer irradiation (several hours) causes scission of the macromolecules at loci well defined by the band patterns. The resulting fragments, still possessing the native-type helical configuration of the three constituent covalent strands, are capable, upon addition of ATP, of forming well defined SLS-type aggregates having lengths of about one quarter or one half that of unirradiated macromolecules. These results are consistent with those of Nishihara and Doty that sonic irradiation has little effect on the optical rotation of collagen solutions although rapidly reducing the viscosity.

These results open new possibilities of relating detailed structure, seen in the electron microscope, with specific chemical groups in the collagen macromolecule and of discovering fundamental factors in the normal fibrogenesis of collagen.

These results are treated in more detail and with electron micrographic illustration in the Symposium discussion of one of us (A. J. HODGE, this volume p. 119).

The fine structure of certain ocular tissues*

MARIE A. JAKUS

Retina Foundation, Massachusetts Eye and Ear Infirmary and Department of Ophthalmology, Harvard Medical School, Boston, Massachusetts, USA

The vertebrate cornea consists, basically, of a multilayered epithelium with a basement membrane, a stroma of oriented collagen fibrils and interspersed stroma cells, and a single layer of mesothelial cells, commonly called the endothelium, with a thickened and modified basement membrane known as Descemet's membrane. In stained sections the stroma appears fibrous, while Descemet's membrane and often a thin layer of stroma just beneath the epithelium have no resolvable internal structure.

In view of the derivation of the corneal epithelium from the epidermis of the skin, the fact that it exhibits features in common with those observed in epidermal cells is not surprising. There are, for example, thickened, dense regions in the basal cell membranes of the innermost cells of the epithelium, upon which converge bundles of cytoplasmic filaments. Between the dense placodes and the basement membrane are filaments, very thin in some species and up to several hundred Ångstrom units thick in others. These structures are particularly clear in the cornea of the frog tadpole (strictly speaking, the "primary spectacle", since before metamorphosis the epidermal-dermal portion of the cornea is not fused with the inner, very thin, sclerotic portion) (Fig. 1), but have also been seen in corneas from many other species. Similar thickened regions, again described previously in epidermis, occur opposite each other in the membranes of adjacent epithelial cells. When the cells pull apart, as a result of shrinkage for example, the dense placodes appear to be held together by the filaments between them; these are, no doubt, the "intercellular bridges" described in stained sections of the corneal epithelium.

In most corneas so far examined, the stroma just beneath the basement membrane of the epithelium shows random orientation of its component collagen fibrils and an absence of stroma cells. When the region of disorganization reaches a thickness easily resolved with visible light, it is designated "BOWMAN's membrane"; such thick layers have been observed in human, monkey, owl and dogfish corneas.

* This investigation was supported by a research grant (B-454) from the National Institute of Neurological Diseases and Blindness, United States Public Health Service and by Atomic Energy Commission Contract AT(30—1)—1789.

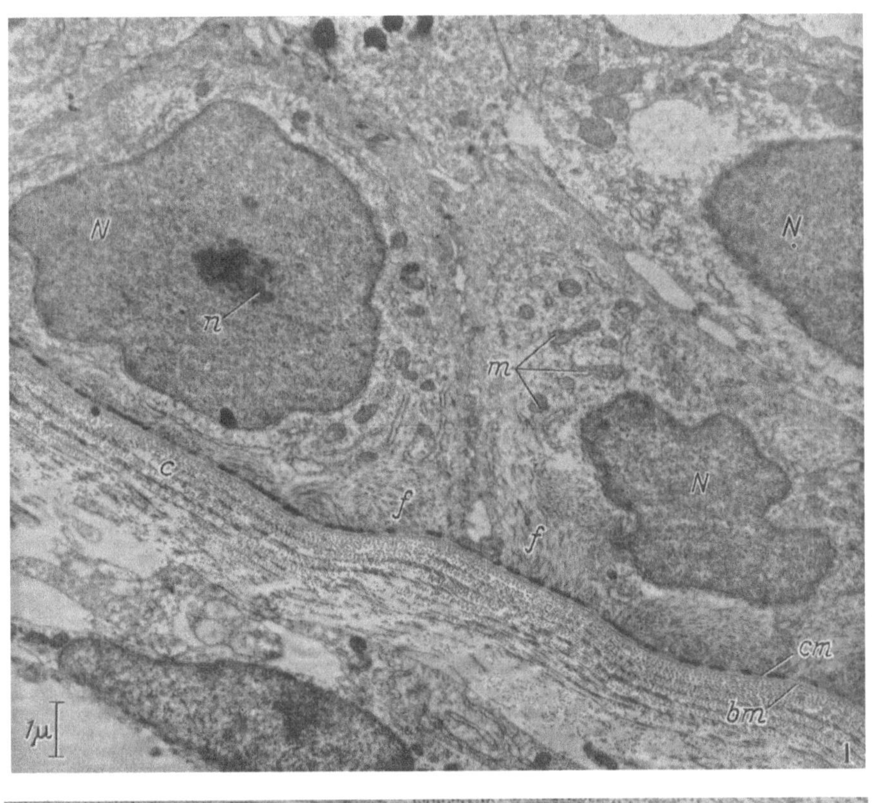

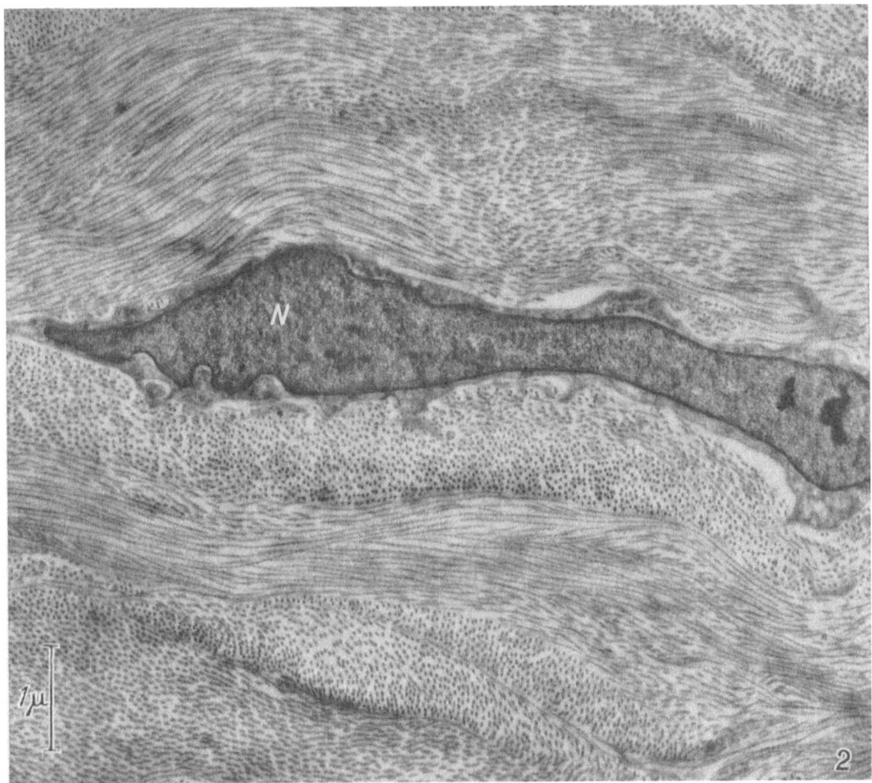

Fig. 1. Epithelial and dermal portion of frog tadpole cornea, showing nuclei (*N*), a nucleolus (*n*), mitochondria (*m*), cytoplasmic filaments (*f*) converging upon the dense, thickened regions in the basal cell membrane (*cm*), basement membrane (*bm*), and collagen fibrils (*c*) in thin layers oriented at right angles to one another. 6,525 ×

Fig. 2. Rabbit corneal stroma, showing the central portion of a stroma cell. There is only a thin layer of cytoplasm around the nucleus (*N*). 14,100 ×

In the remainder of the stroma, the collagen fibrils are organized into ribbon-like bands, or lamellae. These are the microscopically visible fibers of the stroma. In each band, the fibrils are arranged parallel to each other, and at apparent right angles to the layers above and below. The fibrils have quite uniform diameters, averaging between 200 and 300 Å in the ten species measured (measurements were of cross-sectioned fibrils in tissue which had been treated with phosphotungstic acid to increase tissue contrast). In most species the collagen macroperiod can be seen clearly in the deeper layers of the stroma; often the intraperiod banding is also visible. The periodicity is less frequently observed in the peripheral disorganized region due perhaps to a higher concentration of adsorbed material. Between the fibrils in all parts of the stroma are amorphous or finely filamentous clouds.

The stroma cells generally occur between, rather than within lamellae. They are complex in shape, with only a thin layer of cytoplasm around their nuclei (Fig. 2), but with long, thin pseudopodial extensions. Cytoplasmic inclusions appear to be sparse in mature stroma cells.

Descemet's membrane has already been described in some detail (1). It is a highly-structured layer which appears to be the hypertrophied basement membrane of the endothelium. All the evidence so far available puts Descemet's membrane into the collagen class of proteins, but as a secondarily two-dimensional member.

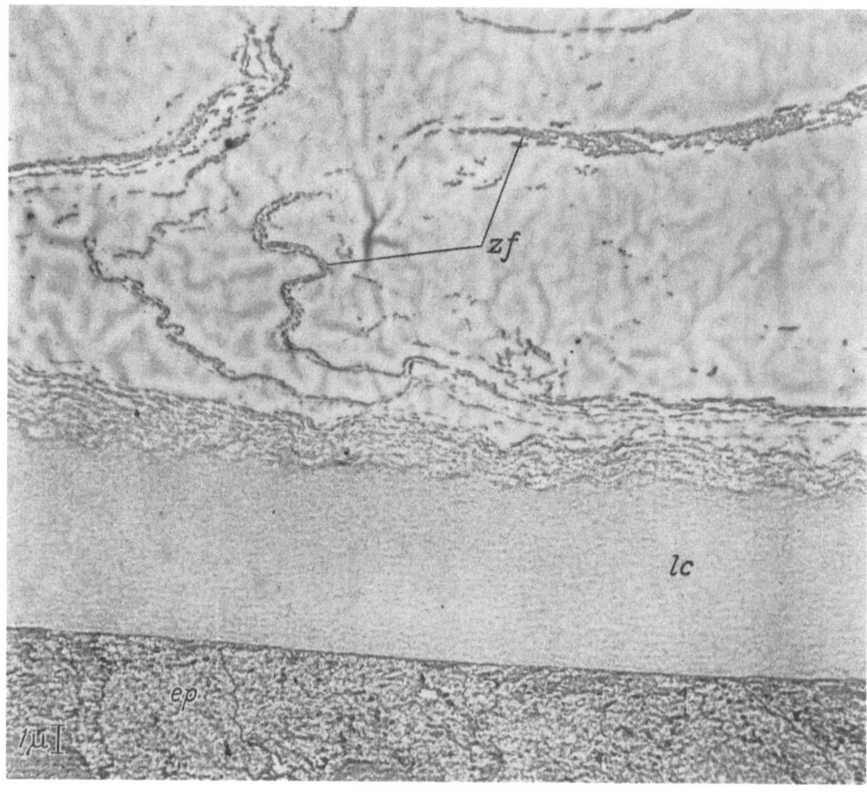

Fig. 3. Section of rabbit lens near the equator, showing some epithelial cytoplasm (ep), the lens capsule (lc) with its peripheral fibrous layer, and fibers of the zonule (zf) formed by lateral aggregation of fibrils. 3,275 ×

The cornea, then, is a predominantly collagenous structure, with the fibrous collagen of the stroma characterized by a small and quite uniform diameter. In the sclera, two other forms of collagen are found: the thin, dispersed fibrils of the cartilage, in the species which have cartilaginous scleras, and robust-looking fibrils which occur in bundles. The diameter of the latter form a wide spectrum, ranging from about 200 Å up to 1500 Å or even more. They are compact in appearance and show both the large collagen period and intraperiod banding.

Because of the similarity in staining properties and behavior in polarized light of Descemet's membrane and the capsule surrounding the lens, it was hoped that a similarity in structure might be found. So far, however, nothing comparable to the precise organization observed in beef Descemet's membrane has been seen in lens capsules. In perpendicular sections, the capsule does appear fibrous and has a suggestion of a period; this, when measurable, however, has been nearer that of the single collagen period than the 1070 Å average distance found in Descemet's membranes. In tangential sections, fine fibrils appear to be running in all directions, sometimes forming roughly stellate figures which are but poor imitations of the regular hexagonal repeating unit of Descemet's membrane.

Peripherally, both anterior and posterior capsules appear loosely fibrous, except in the region of the poles. Near the equator, the fibrils are aggregated laterally into bundles; these finally go off tangentially from the capsule to form the fibers of the zonule (Fig. 3). Zonular fibers vary considerably in diameter, the larger ones occurring at greater distances from the lens, and in degree of compactness. In longitudinal section they appear dense and straight when their component fibrils are closely packed. In these the collagen macroperiod is usually quite clear. Alternatively the fibrils may be rather loosely aggregated, due possibly to slackness in the fiber during fixation. In such fibers, the component fibrils show at best a beaded appearance. This, however, is often the case in collagen fibrils with diameters of 100 Å or less.

The fine structure, staining properties and birefringence of the lens capsule and the zonule suggest that these structures are also members of the collagen class of proteins, the lens capsule resembling Descemet's membrane in the cornea and the fibrils of the zonule similar to the filaments isolated from the vitreous body.

Reference
1. Jakus, Marie A.: J. biophys. biochem. Cytol. 2, Suppl., 241 (1956).

Mechanism of formation of polymeric compounds in tissues

K. Little
Nuffield Orthopaedic Centre, Oxford, England

Chemical analysis has shown that the important structural components of the body are high polymers, so I intend to consider their formation from the point of view of polymer chemistry. These compounds are distinguished by being of high molecular weight and formed from a limited number of structural components, which are joined together either by direct addition, or by the elimination of water, as in the case of the formation of proteins from amino-acids and polysaccharides from sugars. It is best to consider first the synthetic polymers (1), since the mechanism of their formation is known in the greatest detail, to find the type of factors which are of importance. Polyethylene is the simplest polymer, but there are variations in its physical properties according to the method of manufacture. The texture and softening point can show considerable differences with chain branching. The nylons show variations in extensibility and dye uptake, caused by small differences in physical conditions of production; but they also show variations in these and other properties, such as solubility, according to the identity and proportion of their components. These and the majority of the commercial polymers are first polymerized and then processed. The synthetic polymer, methacrylate, is polymerized in the form in which it is to be used. With a mixture containing methyl methacrylate shrinkage occurs and a badly shaped object results; if butyl or isobutyl methacrylate is chosen instead, one should get the shape of the mould reproduced, in the presence of a suitable catalyst.

The types of factor which need considering, therefore, assuming that the monomers have already been formed (the amino-acids and sugars), are possible variations in the proportions of monomeric units; the presence of catalysts (e.g. vitamins or certain amino-acids), and trace elements (e.g. oxygen or fluorine) which might alter the properties; the uniformity of conditions; whether polymerization occurs at the final site, or whether a soluble polymer is precipitated and "fixed"; and whether more than one method of deposition is used for members of the same family of polymers.

We are now in a position to consider the properties of natural polymers and see whether the electron microscope is of use in indicating the method of formation. If one looks at a cell in developing striated muscle, it seems clear from the close connection between the ridges of the nucleus and the striations, that the nucleus acts as a template to determine the positions of the striations. From photographs of fixed sections one can only guess at where the polymerization took place. A moving picture would be more helpful. Cilia are also formed within the cell. They can be seen to be both round and hollow, and to protrude through the cell walls, and have apparently originated in the outer part of the cytoplasm.

Collagen, with a less elaborate structure for the walls of the tubes, seems to be formed not within the cytoplasm, but actually on the outside of the cell walls. This is shown in such diverse localities as the loose connective tissues of invertebrates and osteocytes in mammalian long bones. Since in this case, as with the nylons, a family of polymers is involved, it is not surprising to find differing properties, such as diameter of fibrils and solubility, and consequently not surprising that in sites such as articular cartilage the collagen fibrils appear to have been deposited from solution in the tissue fluid on to the ground substance with no apparent relationship to any cells in the neighbourhood.

The other main protein of connective tissue is elastin. This is found in all sorts of irregular shapes, and although it is difficult to be dogmatic without actually watching the process, Dr. Cox (2) and I have examined in the electron microscope a large number of samples of elastica from different sites and at different stages of development, and all were consistent with the view that the elastic tissue is polymerized at the places where it is seen.

Finally I would like to mention two groups of polymers together, even though they do not seem to be related chemically, the polysaccharides of ground substance and the less dense protein of enamel matrix. The former seems to be the first component to develop faults in connective tissue conditions, as for example, osteoarthritis (3) and the latter is the faulty component in the early stages of enamel caries (4). There are a number of other similarities. Both are noncrystalline, and in the electron microscope have the appearance of dried gels, the swollen solid being the continuous phase, and capable of orientation. Both are formed in the cytoplasm, and their formation seems to be sensitive to the presence of an adequate supply of vitamins, which presumably act as catalysts. In connective tissue vitamin C is necessary if it is to be laid down correctly. For developing bones vitamin A is necessary for the matrix and vitamin D for its calcification. In dental enamel vitamin D is also necessary for calcification, and the various relationships which have been worked out between diet and dental caries are consistent with the view that a heat-sensitive vitamin is necessary for the correct formation of enamel protein in the same way that vitamin A is necessary for the bone matrix.

There is no time to develop these ideas more fully, but it seems probable that the entire formation of the structural proteins and polysaccharides can be described in terms of orthodox polymer chemistry, using concepts of differences in components, concentrations, p_H, and so on, and with well defined chemicals such as the vitamins acting as catalysts. There would then be no need to invoke the presence of hypothetical enzymes or other ill-defined entities.

References

1. LITTLE, K.: 3rd Int. Conf. Electr. Microsc. p. 165, 1954.
2. COX, R. W.: D. Phil. Thesis. Oxford 1957.
3. LITTLE, K., L. H. PIMM and J. TRUETA: J. Bone Jt. Surg. **40 B,** 123 (1958).
4. — J. dent. Res. **36,** 815 (1957).

4. Hartgewebe

The crystalline component of dental enamel

DAVID B. SCOTT

National Institute of Dental Research, National Institute of Health, Public Health Service, US Department of Health, Education, and Welfare, Bethesda 14, Maryland

It has been shown through earlier electron microscopy that the crystalline apatite which is the major component of mature dental enamel is overlaid upon a framework of very fine organic fibrils (*1, 2*). There has been difficulty, however, in observing both the inorganic and organic constituents in the same preparation. The former has been brought out best in replicas of ground sections or cleaved fragments, and the latter in thin sections of demineralized enamel. In replicas and pseudoreplicas of untreated or acid-etched specimens the crystalline elements have previously appeared flat and ribbon-like, as shown in Fig. 1, and lengths varying from 400 to 10,000 Å have been reported (*1, 2, 3, 4, 5*). Definite and uniform cross-markings at roughly 300 Å intervals have also been seen on these structures (*2, 4, 5*). This segmented appearance, illustrated in Fig. 2, has been of interest because of the possibility that it might reflect the presence of organic material within the larger crystal-like objects. Evidence that this may in fact be the case has now been derived from the examination of pseudoreplicas of enamel from which organic substance has been selectively removed. This has been accomplished through treatment of the surfaces of polished ground sections with ethylene diamine, an agent which has previously been used in the preparation of anorganic bone (*6*). In the first experiments, the samples were treated in a Soxhlet extractor, which was modified so that condensed solvent did not drip directly on to the surfaces (*7*). A complete extraction cycle took 15 min, and specimens were exposed over the time required for one to six cycles. Identical results were later obtained by simply immersing the ground sections in warm (80° C) ethylene diamine for 15 min to 2 hr. After gentle washing and drying in air, positive collodion-carbon pseudoreplicas were made by a previously described method (*8*). The ground sections treated for over 30 min were affected to such a depth that it was often necessary to remove some of the loosened crystalline material by stripping several primary films in succession, before a suitably thin pseudoreplica could be made. The finished replicas were shadowed with tungsten oxide, which is amorphous, so that the same areas that were photographed could be subjected to limited area electron diffraction.

Examination of replicas of enamel treated in the above fashion with ethylene diamine has yielded further evidence that the basic crystalline unit may be a particle on the order of 300 Å in length. In regions where disintegration can be seen at an early stage typical ribbon-like objects of varying lengths are present in large quantities, but they are observed to be made up of narrower parallel bands, which in turn are composed of small particles bound together by an amorphous material. This is illustrated in Fig. 3, particularly in the region immediately above the center and in the lower right quadrant of the micrograph. Very few of the longer structures are evident in areas of advanced disintegration, and great quantities of discrete particles are found, the smallest of which quite uniformly have lengths of abouth 300 Å, as shown at higher magnification in Fig. 4.

If these small particles are actually the crystalline units of enamel, the fibrillar organic matrix may well be involved in their orderly arrangement. Since matrix formation precedes calcification the axial fibrils within the prisms could form a mold in which crystals grow, or nucleation might occur at periodic intervals on the fibrils themselves. The present findings also suggest that the remaining intercrystalline space is taken up by amorphous organic material.

The recent conclusions (*5*) that the long ribbon-like objects, as brought out in replicas of acid-etched specimens, are crystals and that the 300 Å cross-markings are caused by periodic projections of the organic matrix fibrils are not supported by the existing information, for the latter structures have not thus far been observed, even in sections of the matrix of demineralized enamel (*1, 9*). On the other hand, the elongated objects, which have been seen in all of the various types of replicas could be accounted for in various ways other than as single crystals. They might result from the aggregation of 300 Å particles, into elements whose contours are established by the configuration of the organic matrix. It is conceivable that the structure in question might simply be fragments produced mechanically. Even if the particle alignment paralleled exactly the prism axes, and the prisms followed straight courses, a departure of only a few hundred Å from a truly longitudinal plane in grinding or cleavage would result in the exposure at the section surface of segments from more than one crystalline layer. Again, if within prisms long rows of particles were aligned along slightly undulating, angulated or twisted axes, those segments which lie in the surface plane of an exactly longitudinal ground section or cleaved fragment would similarly appear as separate objects. These latter possibilities are difficult to rule out, because of the relative

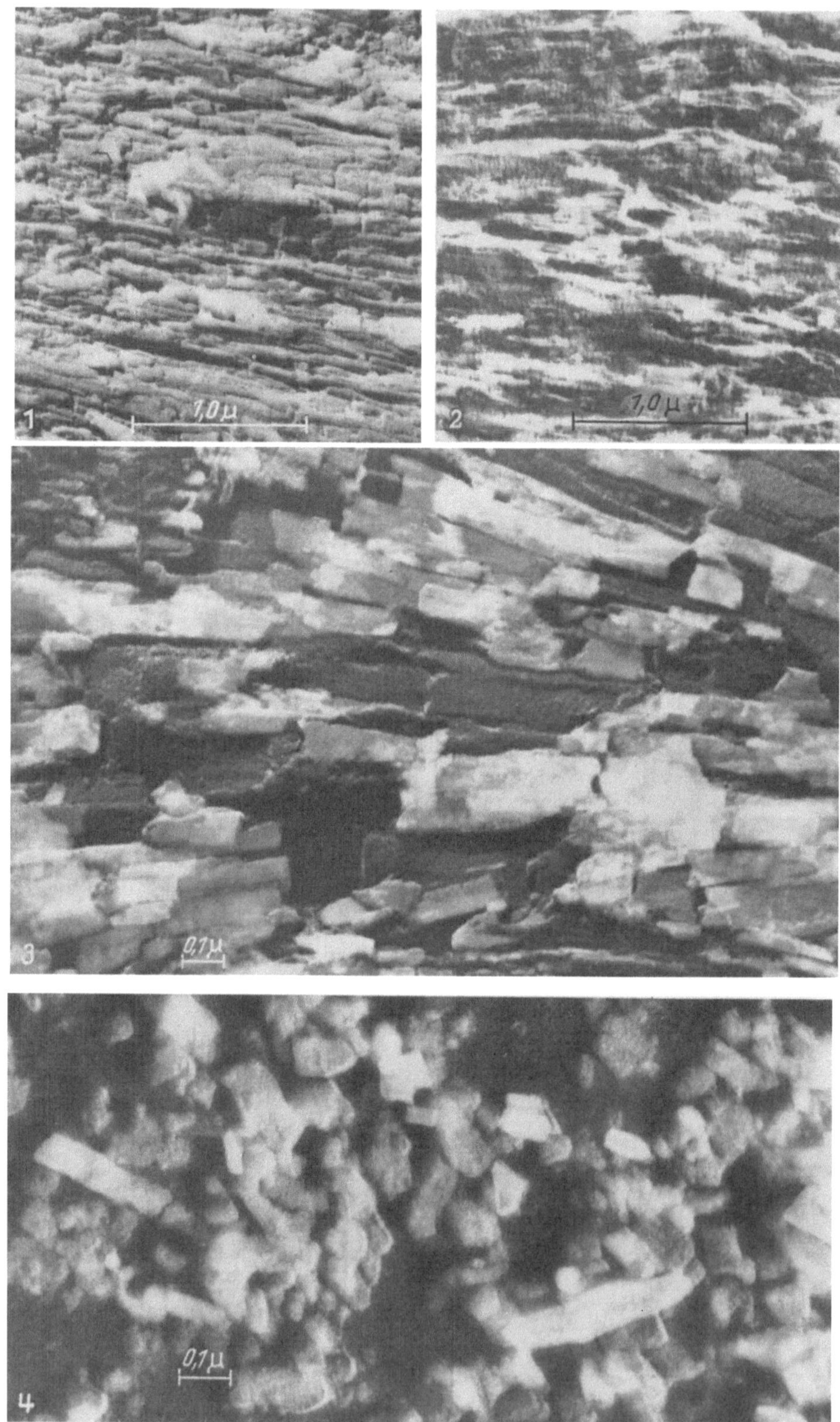

Fig. 1—4

Fig. 1. Parallel rows of crystal-like objects commonly seen in acid-etched enamel (25,000 ×). All four micrographs in this plate were made from positive carbon pseudoreplicas of ground sections, and show areas within single longitudinally sectioned prisms — Fig. 2. Regularly spaced cross-markings often evident on the crystallike objects brought out by acid-etching. 25,000 × — Fig. 3. Partial subdivision of similar crystal-like structures into narrower parallel bands and small particles as a result of a short treatment with ethylene diamine. 57,000 × — Fig. 4. Nearly complete disintegration of the enamel mineral into small particles after a longer extraction of organic matter with ethylene diamine. These particles can readily be identified as apatite by limited area electron diffraction. 72,000 ×

irregularity in prism courses and arrangement, as well as the unavoidable lack of precision in specimen preparation.

Considerable further study will be required before the dimensions of the fundamental crystalline unit of the enamel can be definitely established. Unfortunately, the only positive information derived from X-ray diffraction patterns has been that the mineral present is apatite and that the crystals are preferentially oriented in an axial direction, for the crystal sizes reported have varied from 270 Å to 10,000 Å (*1, 2, 10*). Another approach is afforded through the electron microscopic examination of sectioned developing enamel. By this means it may be possible to trace the deposition of mineral during the period of calcification. Such an investigation is now under way, but is has not yet progressed far enough to provide additional information.

References

1. Scott, D. B.: Int. dent. J. **4,** 64 (1953).

2. Frank, R. M.: Contributions à l'étude au microscope électronique des tissus calcifiés normaux et pathologiques, Strasbourg: Sutter Woerth 1957.

3. Scott, D. B.: Studies of the crystal structure of enamel by electron microscopy and electron diffraction, Proc. 25th Year Celeb. U. of Roch. Dent. Res. Fell. Program, Rochester, 1957, U. of Roch. Press, p. 175.

4. Little, K.: J. dent. Res. **34,** 778 (1955).

5. Hall, D. M.: J. dent. Res. **37,** 243 (1958).

6. Losee, F. L., and L. A. Hurley: Nav. Med. Res. Inst. Rep. **14,** 911 (1956).

7. Scott, D. B., and F. L. Losee: J. dent. Res. **37,** 23 (1958).

8. — and R. W. G. Wyckoff: J. royal Microsc. Soc. **75,** 217 (1956).

9. — M. J. Ussing, R. F. Sognnaes and R. W. G. Wyckoff: J. dent. Res. **31,** 74 (1952).

10. Posner, A. S.: Norelco Rep. **2,** 26 (1955).

Die Bildung der organischen Matrix des Schmelzes

Hans Lenz

Technische Universität Berlin und Kariesforschungsinstitut an der Johannes-Gutenberg-Universität Mainz, Außenstelle Berlin-Dahlem

Elektronenmikroskopische Untersuchungen haben gezeigt, daß der Schmelz aus anorganischen und organischen Bausteinen besteht, die in ihrer gegenseitigen Zuordnung und Orientierung den lichtmikroskopisch bekannten Summeneffekt des Schmelzbildes ergeben. Die gleichzeitige elektronenmikroskopische Darstellung dieser Bestandteile ist aber mit den bisherigen Methoden nicht möglich gewesen. Während es mit Hilfe der Abdruckmethode gelang, von den anorganischen Bausteinen, den Hydroxylapatitkristallen, eindrucksvolle Bilder zu gewinnen, konnten die Versuche, die organische Matrix, in welche die Kristalle eingelagert sind, durch Entkalkung darzustellen, nicht voll befriedigen. Denn es läßt sich schwer vorstellen, in welcher Beziehung die hierbei gefundenen Netzwerke von organischen Fibrillen ursprünglich mit den klar orientierten Kristallen gestanden haben. Es deutet alles darauf hin, daß es bei der Entkalkung des Schmelzes zu Veränderungen der organischen Substanz kommt, die es nicht erlauben, diese Netzwerke als wahres Abbild der organischen Strukturen zu betrachten. Da zu erwarten war, daß sich in der Bildungszone des Schmelzes eher Hinweise auf die ursprüngliche Struktur der organischen Matrix finden lassen, wurden Untersuchungen an Zahnkeimen vorgenommen, zumal es hier möglich ist, ohne vorherige Entkalkung Schnittpräparate herzustellen.

Untersuchungsmaterial: Zahnkeime von 4—6 Monate alten menschlichen Feten und Molarenzahnkeime von 1—3 Tage alten weißen Mäusen. Die Zahnkeime wurden unmittelbar nach Entnahme aus dem lebenden Gewebe in gepufferter OsO_4 (p_H 7,2) nach Palade fixiert und nach Behandlung in Tyrodelösung sowie Entwässerung in Alkoholreihe in Methacrylat eingebettet. Für vergleichende Zwecke wurden parallel dazu nach lichtmikroskopischen Methoden (übliche Formalinfixierung des ganzen Kiefers) hergestellte Präparate elektronenmikroskopisch nachuntersucht. Die Schnitte wurden mit einem Sjöstrand-Ultramikrotom hergestellt und in einem Siemens Elektronenmikroskop Typ 100b untersucht.

Die elektronenmikroskopische Untersuchung des inneren Schmelzepithels ergab, daß die Ameloblasten nicht, wie bisher allgemein angenommen, durch eine etwa 1 μ breite Intercellularsubstanz voneinander getrennt sind, sondern unmittelbar aneinanderliegen. Da diese Unterschiede für die Beurteilung der bekannten Theorien von der Schmelzgenese von großer Bedeutung sind, wurde nach den Ursachen dieser Differenzen gesucht. Bei vergleichenden Untersuchungen konnte festgestellt werden, daß durch die in der Literatur angegebenen lichtmikroskopischen Präparations- und Fixierungsmethoden in den Zellen Artefakte hervorgerufen werden, die bisher bei lichtmikroskopischen Beobachtungen als reale Struktur gewertet wurden.

An Schnitten von den verschiedenen Stadien der Entwicklung wurde die Ausbildung

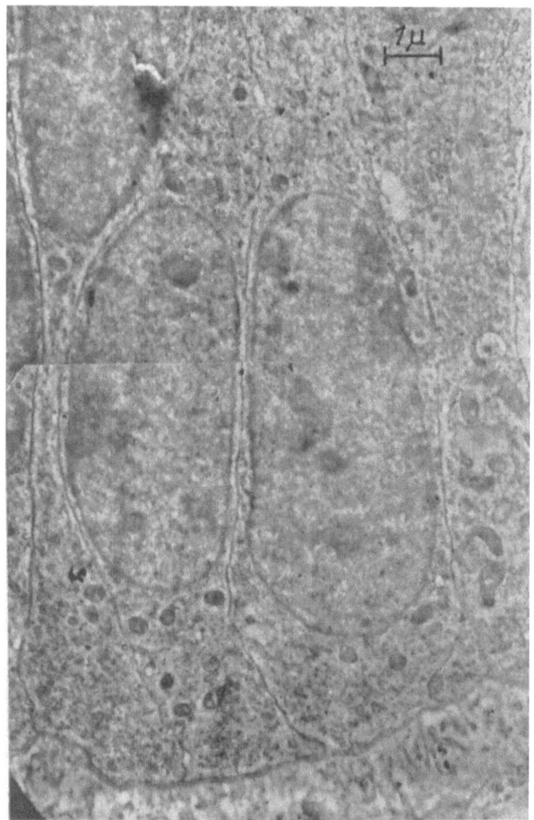

Abb. 1. Ameloblasten, Maus. 7200mal

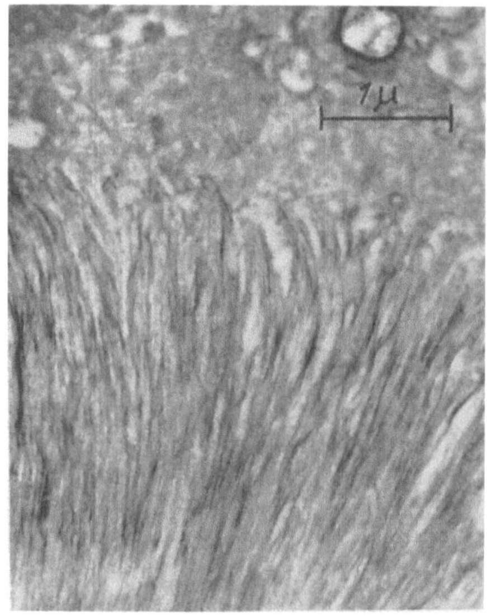

Abb. 2. Bildung der organischen Matrix des Schmelzes, Mensch. Dünnschnitt 15000mal

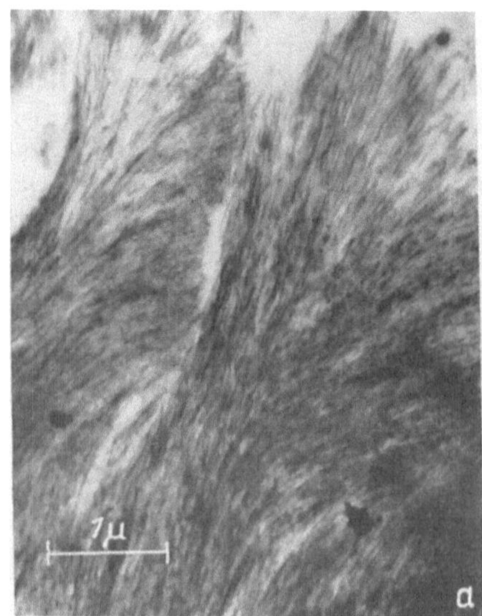

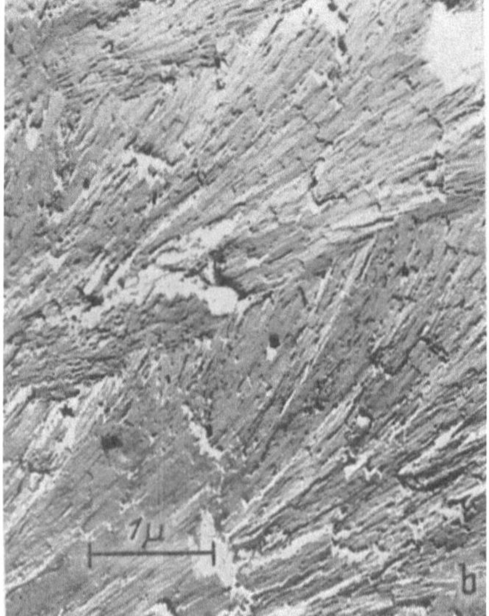

Abb. 3a u. b. Vergleich der organischen Matrix mit dem reifen Schmelz. a) Unverkalkte organische Matrix. OsO₄ fixiert. Dünnschnitt. b) Abdruck einer Schmelzbruchfläche. 15000mal

der organischen Schmelzmatrix verfolgt. Die Situation vor Beginn der Schmelzbildung gibt die Abb. 1 wieder. Die Ameloblasten liegen dicht zusammen und sind von der Pulpa durch eine durchgehende Grenzmembran getrennt. Das Protoplasma der Zellen zeigt eine granuläre Grundstruktur. Neben Mitochondrien sind nur vereinzelte Teile des endoplasmatischen Reticulums zu erkennen.

Nach Bildung der Dentinanlage und Beginn ihrer Verkalkung wird von den Ameloblasten die organische Matrix angelegt. Im Protoplasma der Zellen tritt in dieser Phase der Entwicklung das endoplasmatische Reticulum besonders stark hervor, was auf die Aktivität der Zellen schließen läßt. Die Zellgrenzen verschwinden, und in der Bildungszone werden etwa 100 Å breite Fibrillen der organischen Schmelzmatrix ausgefällt (Abb. 2). Die Fibrillen sind zu Bündeln angeordnet, die sich unmittelbar begrenzen, so daß die gesamte Schicht der organischen Matrix von ihnen ausgefüllt ist. Die Zwischenschaltung einer andersartigen 1 μ breiten interprismatischen Substanz, wie sie nach lichtmikroskopischen Befunden angenommen wird, konnte nicht beobachtet werden. Innerhalb der Bündel sind die Fibrillen fiederartig angeordnet. Ein Vergleich eines Schnittpräparates von der organischen Matrix mit Abdruckbildern vom mineralisierten Schmelz beweist eine völlige Übereinstimmung beider Muster (Abb. 3a u. b). Daraus geht hervor, daß die organische Matrix auf die sich bildenden Hydroxylapatitkristalle orientierend wirkt und die Struktur des Schmelzes von der Anordnung der organischen Matrix bestimmt wird.

Literatur: LENZ, H.: DZZ **13**, H. 17, 991 (1958).

Comparative observations on the ultra-structure of the inorganic and organic components of dental enamel

R. W. FEARNHEAD

Departments of Dental Histology and Anatomy, The London Hospital Medical College

Enamel covering the crowns of teeth consists principally of a mineral apatite which crystallises into an organic matrix of fibrous protein related to keratin. According to current opinion based on optical microscopy, fully formed enamel consists of calcified rods which are thought to be formed by columnar ectodermal cells called ameloblasts. Between the rods is a calcified inter-rod material. The origin and manner in which the ultra-structural components of the rod and inter-rod material are formed has not been satisfactorily demonstrated by optical methods.

An account is given here of some of the findings obtained from a study, of the ultra-structure of developing rat enamel, which revealed stages in the development of the organic matrix, its calcification and the relationship between the organic and inorganic components.

Developing teeth from late foetal to 10 day old rats, were fixed in PALADE's (*1*) or DALTON's (*2*) fixative, and embedded in methyl-methacrylate or epoxide resin. In order to avoid the production of artefacts as much as possible, the material was not decalcified and sections were cut using a diamond knife. In addition, fixed and unfixed fragments were dissected from the formative end of the incisors of adult rats, and mounted on perforated formvar or carbon films [SJÖSTRAND (*3*)].

Extra-cellular fibril formation. From a study of the electron micrographs of developing rat teeth it is evident that the onset of enamel matrix formation is marked by the appearance of thickened invaginated regions of the ameloblast cell membrane at its formative end. Associated with these invaginations are masses of extra-cellular granules which appear to fill the inter-cellular spaces (Fig. 1). As enamel matrix formation proceeds, the extra-cellular granular material undergoes an abrupt change into fibrils of variable length and approximately (100 Å diameter), which are embedded in a granular back-ground substance. These fibrils, which are entirely extra-cellular, surround the formative ends of the ameloblasts.

Intra-cellular fibril formation. Within the cytoplasm at the formative end of the ameloblast cells, large ovoid bodies approximately (600 Å diameter) are found packed together in a fine granular vacuolated cytoplasm. Intracellular fibrils (about 100 Å diameter) are frequently found surrounding these ovoid bodies. Later this cytoplasm becomes transformed into fibrils similar in character to the extra-cellular fibrils but having a different orientation. Both the fibrils within the rods and between the rods are embedded in a granular ground substance and in places a periodic cross linkage, similar to the rungs in a ladder, appears to be present between two and sometimes between groups of fibrils. It has not been possible to obtain electron diffraction patterns from the matrix at this stage of development.

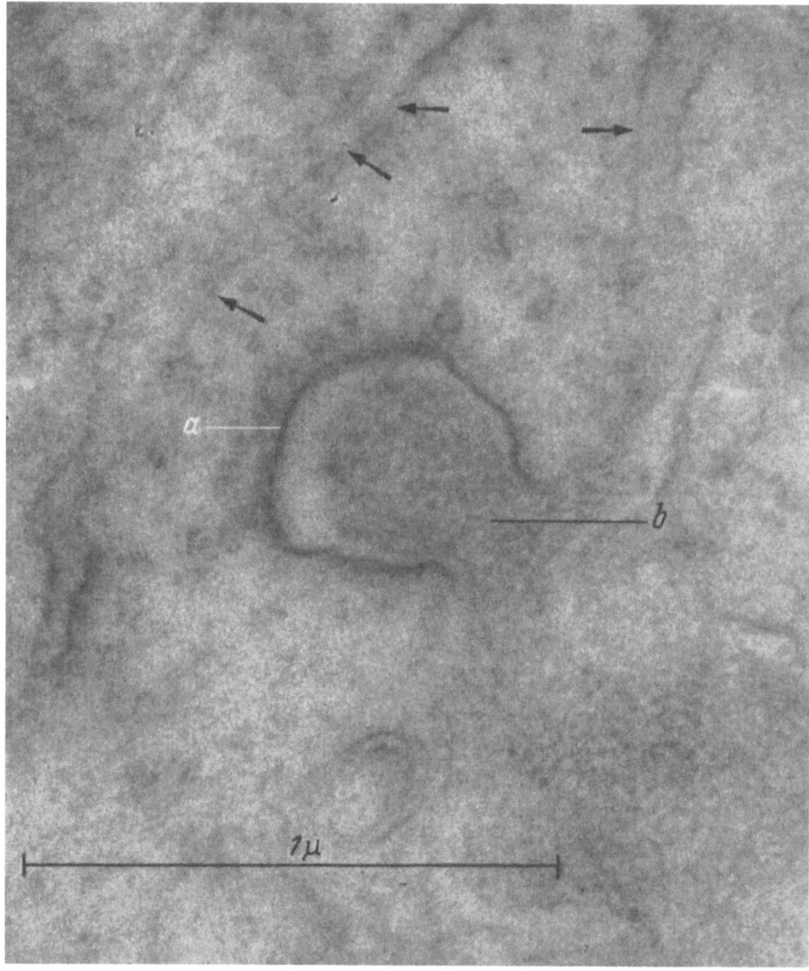

Fig. 1. Section through part of the formative region of an ameloblast. *a* thickened electron dense region of the cell wall, *b* extra cellular granules. Small arrows indicate other regions of the cellmembrane of the ameloblast

Mineralisation. In the deeper layers, further from the formative end of the ameloblast, and therefore in a more advanced state of development, the fibrous material becomes transformed into "tape-like" structures in which electron dense areas occur, and the granular ground substance present in the early stages has for the most part disappeared. It seems not unreasonable that these "tape-like" structures represent a combination of both the organic and the mineral components of the enamel and so the term "enamel crystal" is used for them in this study, in order to distinguish them from the mineral apatite crystallites. Neither fixation nor sectioning appeared to affect the "enamel crystals". A resolution better than 30 Å on the "enamel crystals" was difficult to obtain when they were supported on formvar or carbon films, the crystals being approximately 100 Å thick and the supporting films about 200 Å. High resolution was obtainable, however, on unsupported "enamel crystals", and electron micrographs from these preparations revealed structures not visible in supported crystals. Sets of electron dense lines were detected running through the crystals depending on their orientation in respect to the beam. Some of these lines lie along the axis of the crystal, (Fig. 3a) and some at about 35° to its axis (Fig. 3b and c). The lines are not perfectly regular, some are slightly curved and many vary in width and length. The spacing between the lines is larger than that to be expected from the hydroxyapatite lattice.

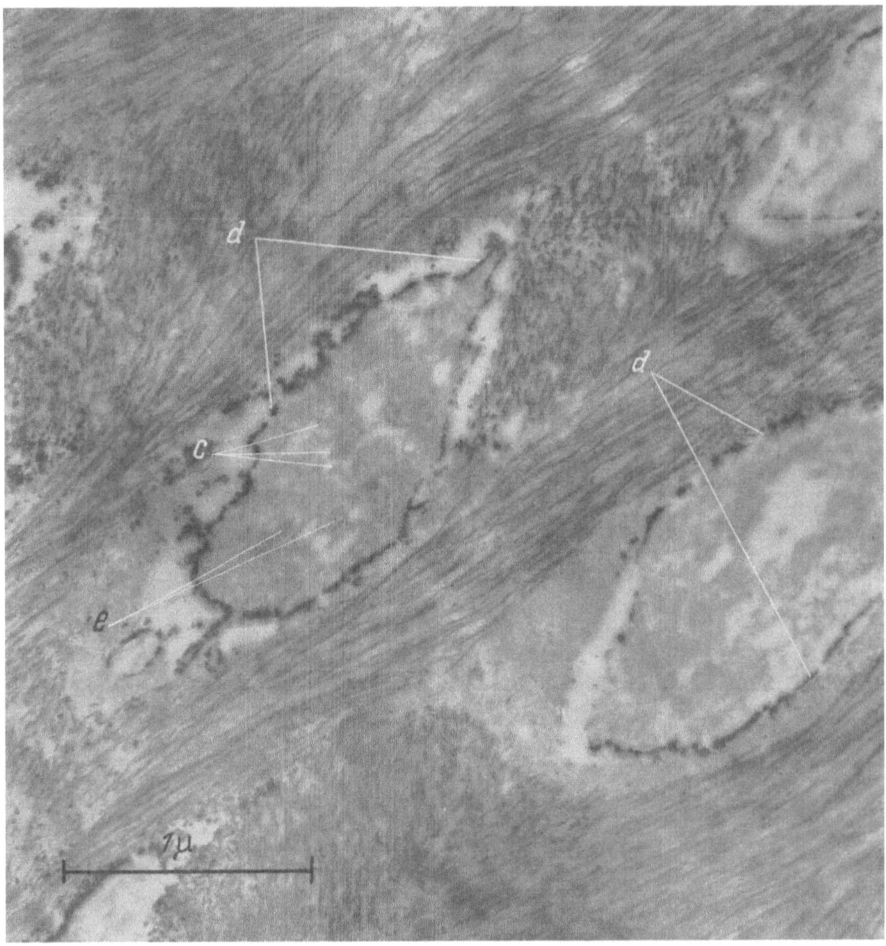

Fig. 2. Section cut obliquely through extracellular enamel matrix and the ends of the ameloblasts. *a* extra-cellular fibrils. *b* extra-cellular granules. *c* intra-cellular fibrils. *d* the formative end of the ameloblast in cross section. *e* "ovoid" bodies in the cytoplasm of the ameloblast

Discussion

BIRBECK and MERCER (*4*) have described how, in the cells of the inner root sheath of hair follicles, discrete droplets (which are regarded as trichohyaline) transform directly into fibrils. The cells of the inner root sheath and the ameloblasts are derived from the germinative layer of the epidermis. It is reasonable, therefore, to envisage a close similarity between fibril formation in the two types of cell. The evidence for intra-cellular fibril formation in ameloblasts is therefore in close agreement with BIRBECK and MERCER's observations. On the other hand, the abrupt transition of extra-cellular granules into fibrils between the cells of epithelial tissue appears not to have been recorded before. It is not possible at present to say whether the granular precursor material is formed intracellularly or extra-cellularly. Since there is, however, a precise relationship between the masses of extra-cellular granules and the invaginated, thickened, electron dense regions of the cell membrane, it seems reasonable that these special regions of the cell membrane of the ameloblast have an important role in the production of the extra-cellular granules.

In the deeper layers of the enamel matrix where mineralisation has commenced, electron diffraction patterns may be obtained from the "enamel crystals", and the 002 arc and the 211 arcs are easily identifiable and correspond in position to those of apatite. By placing the objective aperture over the diffracted 002 arc in the diffracting condition, and then focussing the crystal,

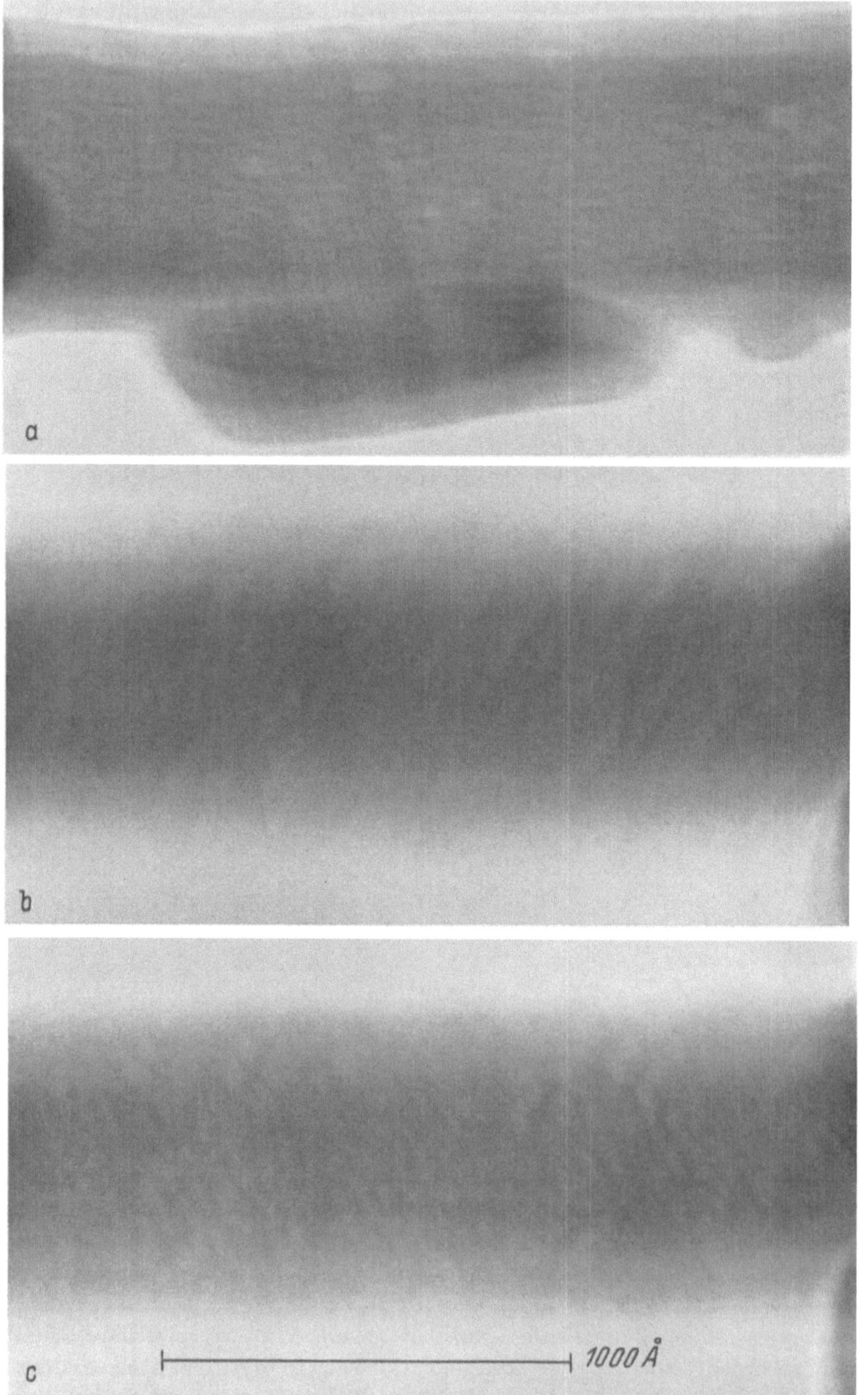

Fig. 3a—c. "Enamel crystals" without supporting film, (b and c) are taken from a through focus series of the same crystal

a dark-field image of the Bragg reflection can be obtained which is much narrower than the "enamel crystal" and which in fact corresponds in width to the crystals supported by formvar. These reflections are orientated in the same axis as the long axis of the "enamel crystal". If the selecting aperture is moved to the 211 arc the Bragg reflection comes from a series of lines at an angle to the long axis of the "enamel crystal" corresponding to the angle of the electron dense lines of the normal image (Fig. 3). Measurements from the Bragg reflections and the electron dense regions of the normal image are in close agreement. Assuming that the electron dense regions in the "enamel crystal" are caused by the apatite crystallites, a size range of about 50 Å is obtained for the crystallite size, which is in fairly close agreement with the calculations made by POSNER (5) using X-ray diffraction techniques.

THEWLIS (6) has shown that the crystallites in enamel are orientated with their C-axis at an angle to the axis of the enamel rod. Since it appears from the diffraction patterns and the Bragg reflection data that the C-axis of the apatite crystallite is co-axial with that of the "enamel crystal", the orientation of the crystallites with respect to the enamel rod presumably depends on the orientation of the fibrillar matrix of the enamel before crystallisation commences.

The electron dense regions of the "enamel crystal" are in all probability apatite crystallites, having a definite orientated relationship to an organic framework of a highly regular pattern, which presumably forms a template in which the apatite crystal growth occurs.

All the preparations were examined with a Siemens Elmiskop I electron microscope, provided by the Wellcome Trust. The work was assisted by a grant from the Medical Research Council.

References

1. PALADE, G. E.: J. exp. Med. **95**, 285 (1952).
2. DALTON, A. J.: Int. Rev. Cytol. **2**, 403 (1953).
3. SJÖSTRAND, F. S.: Electron Microscopy. Proc. Stockholm Conference, p. 120. Stockholm: Almqvist & Wiksell 1956.
4. BIRBECK, M. S. C., and E. H. MERCER: Electron Microscopy. Proc. Stockholm Conference, p. 158. Stockholm: Almqvist & Wiksell 1956.
5. POSNER, A. S.: Norelco Rep. **2**, 26. North Amer. Phillips Co. Ltd. 1955.
6. THEWLIS, J.: The structure of teeth as shown by X-ray examination. Med. Res. Council Special Rep. Ser. No. 238. H. M. Stationery Office. 1940.

Electron microscope observations on apatite crystallites in human dentine and enamel

A. MILLARD and F. G. E. PAUTARD

Department of Biomolecular Structure, University of Leeds (England)

The hardness of mineralised tissues has hindered close study of their fine structure until recent observations by FERNÁNDEZ-MORÁN and ENGSTRÖM (1) on thin sections of bone cut with a diamond knife disclosed the hydroxyapatite crystallites in the form of rod-shaped particles preferentially oriented in the direction of the collagen fibrils. The dimensions of these crystallites (about 200 Å × 30—60 Å) agree with evidence from X-ray studies, suggesting an association with one-third the collagen macro-period of about 640 Å.

We have recently succeeded in cutting thin sections of human dentine and enamel with a glass knife in a Porter-Blum ultramicrotome. The material examined was in the form of a powder ground out of first molar teeth with a corundum or steel dental burr, and consisted of particles about 1—10 μ in size which were subsequently dehydrated and sedimented and then embedded in polybutyl methacrylate. Low-magnification micrographs show clearly how the dentine particles, though often cracked or broken, are also frequently sliced into very satisfactory sections, but in the case of enamel the main result may be more of a disintegration into individual crystallites, platelets often so thin as to be easily penetrable by the electron beam. In any event, specimens prepared in this way present a panorama of the tissue cut in all directions, and fine detail can be readily distinguished in unshadowed sections viewed by transmission.

The appearance of the dentine sections is predominantly one of numerous fusiform particles arranged end-to-end, often in parallel arrays running in the same general direction, but also frequently twisted and buckled into labyrinths (Fig. 1). Where the particles are clearly separated, their tapering shape is immediately apparent, and strings of two or more, sometimes bent, sometimes straight, are common. In transverse section hexagonally-packed groups occasionally occur.

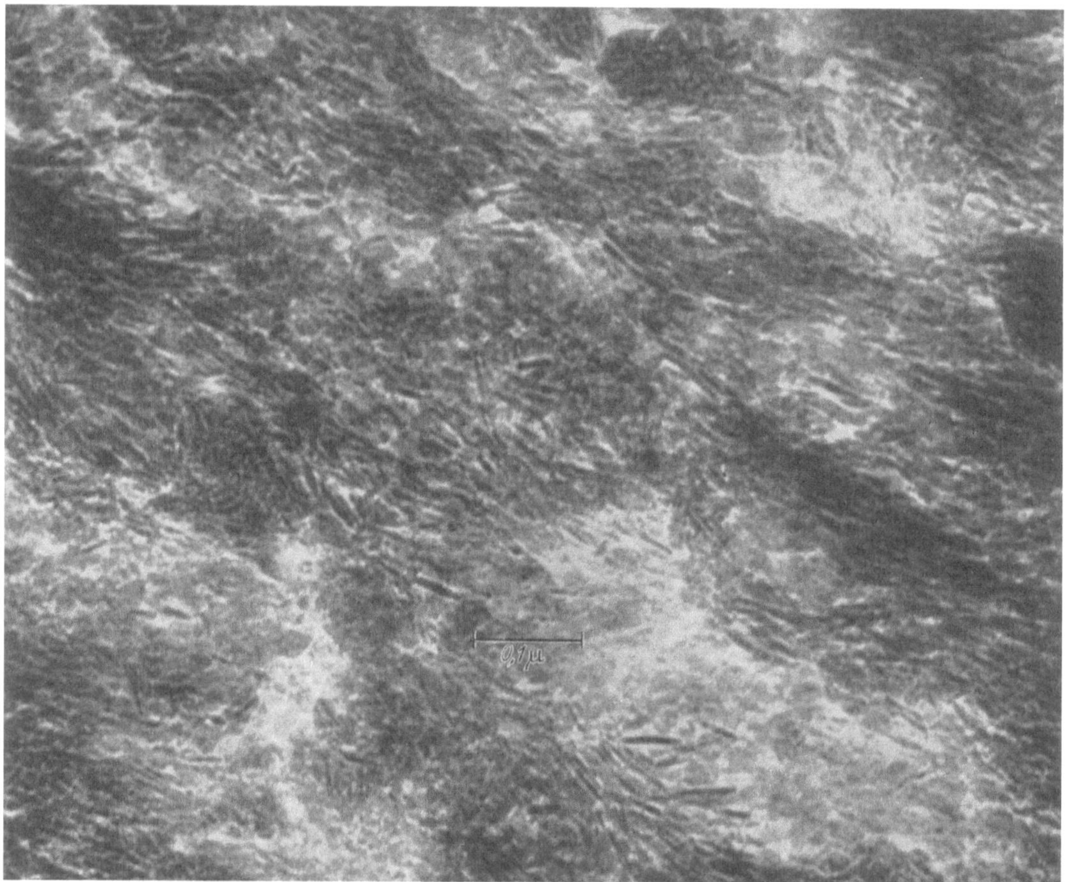

Fig. 1. Electron micrograph of a thin section of human dentine

These strings of crystallites seem to start out as rows of small rounded bodies, which eventually elongate and join up as units similar in dimensions and shape (of the order of 200 Å × 60 Å) to those from bone and from the ossicles of the protozoan *Spirostomum ambiguum* (2).

The marked fibrous layout of the chains of apatite crystallites in the dentine sections inevitably and strongly suggests that they are the mineralised expression of a corresponding arrangement of collagen filaments, and experiments are now in progress to decalcify the sections so as to try to reveal these filaments. By implication they should almost certainly be there, since it is well known that usually neither X-ray diffraction nor electron microscopy reveals the presence of collagen in untreated dentine preparations, yet it becomes quite obvious by both methods once the apatite has been removed.

The plate-like crystallites seen in the enamel sections are remarkable first on account of the extreme thinness of many of them (of the order of only 100 Å thick as indicated by shadowed preparations) but more particularly because they so frequently show rather regular arrays of circular holes, or at least thinner places, with darker peripheries. These features are also matched, in many parts of the background, by similar arrays of faint rings, like quoits, as if they were "ghosts" of such crystallites, in their thinnest possible manifestation, just beginning to appear (Fig. 2). In the background, too, there are seen many groups of small dense particles which also appear to match closely the sets of holes or thin places in the crystallites.

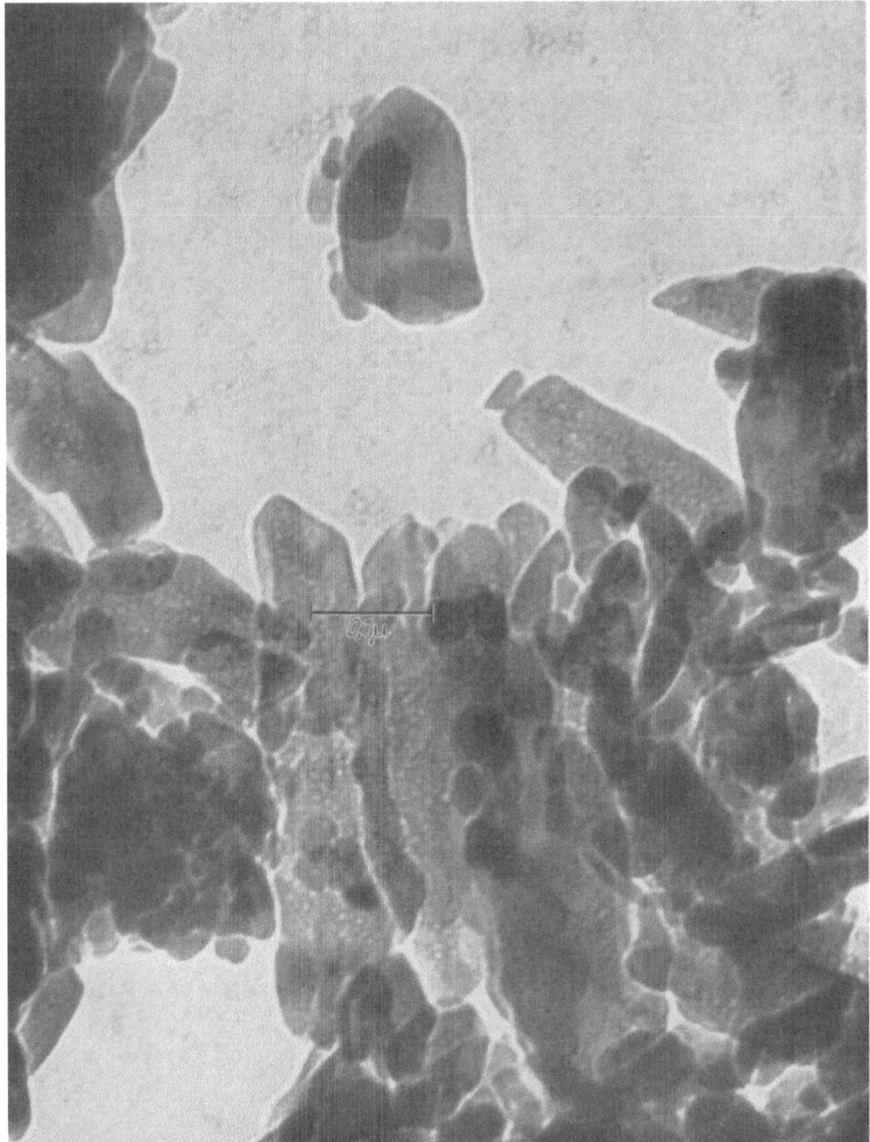

Fig. 2. Electron micrograph of the thin plate-like crystallites seen in a section of human tooth enamel

The general impression is that each thin platelet was initially "threaded on", and grew out sideways from, a group of filaments (presumably collagenous but perhaps so-called keratinous), after the manner illustrated diagrammatically in Fig. 3. Thicker platelets would be built up by the superposition of the elementary sheets formed at each level in the periodic succession of active sites along the filaments. The filaments threaded through the platelets are finally either lost or, being organic, their remnants cannot be distinguished against the surrounding apatite. It may be, though, that the small black dots also found in the background are remains of completely calcified active sites on individual filaments which did not succeed in growing a coherent platelet.

Networks of fine fibrils have already been reported by Scott and Wyckoff and others in electron microscope studies of tooth enamel, but in the light of the above observations, admittedly preliminary, we should like to suggest that here now are more precise indications of how the organic and mineral phases are actually related to each other in space. Whether or not the proposed interpretation is sound, it is striking enough how well the characteristic fine-structural

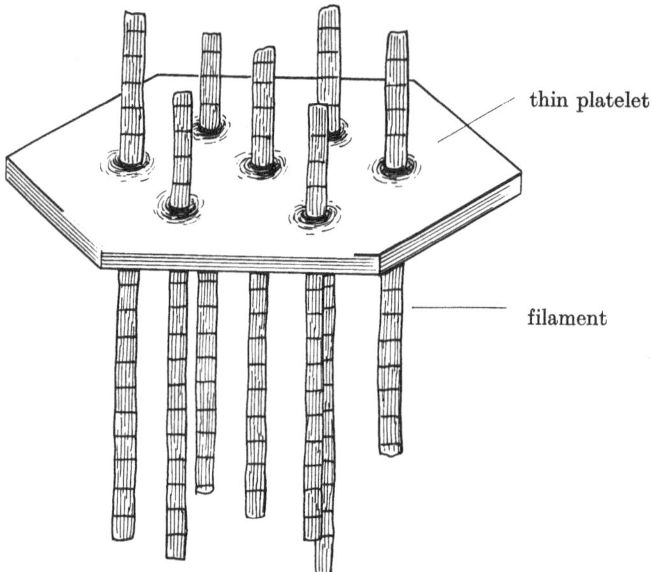

Fig. 3. Suggested scheme for the development of apatite crystallites in tooth enamel

differences between enamel and dentine are at once revealed by the technique of sectioning embedded powders, and the experiments are being extended not only in this regard but also to other materials of varying hardnesses set in correspondingly suitable resins.

References

1. Fernández-Morán, H., and A. Engström: Biochim. biophys. Acta **23,** 260 (1957).
2. Pautard, F. G. E.: In preparation.

The growth of the epiphyseal plate in mammalian long bones

J. Trueta and K. Little

Nuffield Orthopaedic Centre, Oxford/England

The purpose of this paper is partly to show how electron microscopy can be of use in filling in details in a problem where the bulk of the work is more usefully and conveniently done using other techniques; and partly to show how the injection method for tracing blood vessels is useful in electron microscopy.

For those not familiar with the growth cartilage, I will first briefly describe the normal histological appearance. It is found at the ends of the long bones, and provides the mechanism whereby the bones increase in length. At the epiphyseal side (the end of the bone) primary ossification of the original cartilage occurs, and in the rabbit and human — the two species we have examined — a bony plate lies above the growth cartilage. In places blood vessels pass through this, so as to be in immediate contact with the cartilage. This top layer of cartilage is known as the germinal layer. Its matrix is of fine texture and contains irregularly distributed cells. Next is the proliferative zone with flattened cells. Lower down they enlarge and we reach the giant cell zone, with about four to six large cuboid cells in each column. At the metaphyseal side of the growth cartilage the matrix between these cells is calcified, they die, and we find parallel trabeculae with blood vessels between.

The pattern of the blood vessels at the two sides is seen more clearly when a mixture of Berlin blue and barium sulphate is injected into the aorta or a large artery — a technique which has been described by Trueta and Harrison (1). Microradiography or, better, X-ray microscopy shows the position of calcified tissue and injected vessels. At the metaphyseal side it can be seen that before the bony trabeculae are formed the giant cells are almost completely surrounded by primary calcification.

Using these techniques much has been learnt, but there are a few problems which require the higher resolution of the electron microscope. Among these are: a) In the histological photograph of the epiphyseal side we can see osteocytes, but not their detailed relation to the canaliculi. b) There is usually no clear indication of what is vessel wall. c) In the columns there is doubt about where cell division occurs. d) At the metaphyseal side there are giant cells, primary calcification, vessels and trabeculae. There is no certainty about the precise order of events, for example, whether the cells die before or after their surroundings calcify. The electron microscope can provide further evidence on these points. In the bone plate processes from the cytoplasm of the osteocytes can be seen to pass through the canaliculi. This is best demonstrated in stereoscopic photographs.

Where the vessels penetrate the bony plate there can be: the calcified layer of bone, cells and then blood (or, more easily recognizable, the mass of fine particles of the injection medium). In places, even the single layer of cells, which is sometimes found, can be almost surrounded by calcified tissue leaving no recognizable vessel wall, or only a fine membrane. Where the vessel emerges on the growth cartilage side the injection medium can be seen apparently in contact with the matrix of the germinal layer. This matrix contains few recognizable fibrils, unlike the matrix between the columns which has well-orientated collagen fibrils along their length. The extra resolution enables us to see whether cells are separated by matrix or not, thus indicating whether there has been a recent division. An interesting point is that in a single specimen the plate-like cells and the expanding cells have not both been found to be properly fixed, suggesting that the two types of cell may have different osmotic pressures. At the metaphyseal side of the plate we find that calcification first occurs on the orientated collagen fibrils, and it appears that the order of events is first calcification and then death of the cell. We are able to make this observation because both the state of the cell and the calcification can be seen on the same section at the same time. The vessels penetrate right to the last dead giant cell. In ordinary specimens red cells, and in injected specimens the barium sulphate, can be seen in contact with the calcified matrix below the last cell. In the areas close to the bottom of the column, where there is no robust vessel wall, the injection technique is particularly useful in showing which is, in fact, the vessel. Deeper in the metaphysis vessels have definite walls.

Reference

1. Trueta, J., and W. Harrison: J. Bone Jt Surg. 35 B, 442 (1953).

5. Exokrine Drüsen

Electron microscopy of the human eccrine sweat gland with special reference to the folding of plasma membrane

Kazumasa Kurosumi, Takeshi Iijima and Tatuo Kitamura

Department of Anatomy, School of Medicine, Gunma University, Maebashi (Japan)

Infoldings of basal plasma membrane which were found in kidney tubules and other various epithelial cells (1—8) have been postulated as a special device concerning the active transport of water (6, 7). It is expected that this special structure may occur in cells of the sweat gland, which is comparable to the kidney in function that a great deal of water is excreted. We previously found basal infoldings in eccrine sweat glands of the pig (carpal organ) (9). The eccrine sweat glands in human axillary skin were observed at this work and more highly developed infoldings were found. These results may be a strong evidence to support the theory that the basal infoldings play a role in the water transport as claimed by Pease (6) and Ruska et al. (7).

Some findings on the minute structures and genesis of secretory granules of human eccrine sweat glands are also reported here.

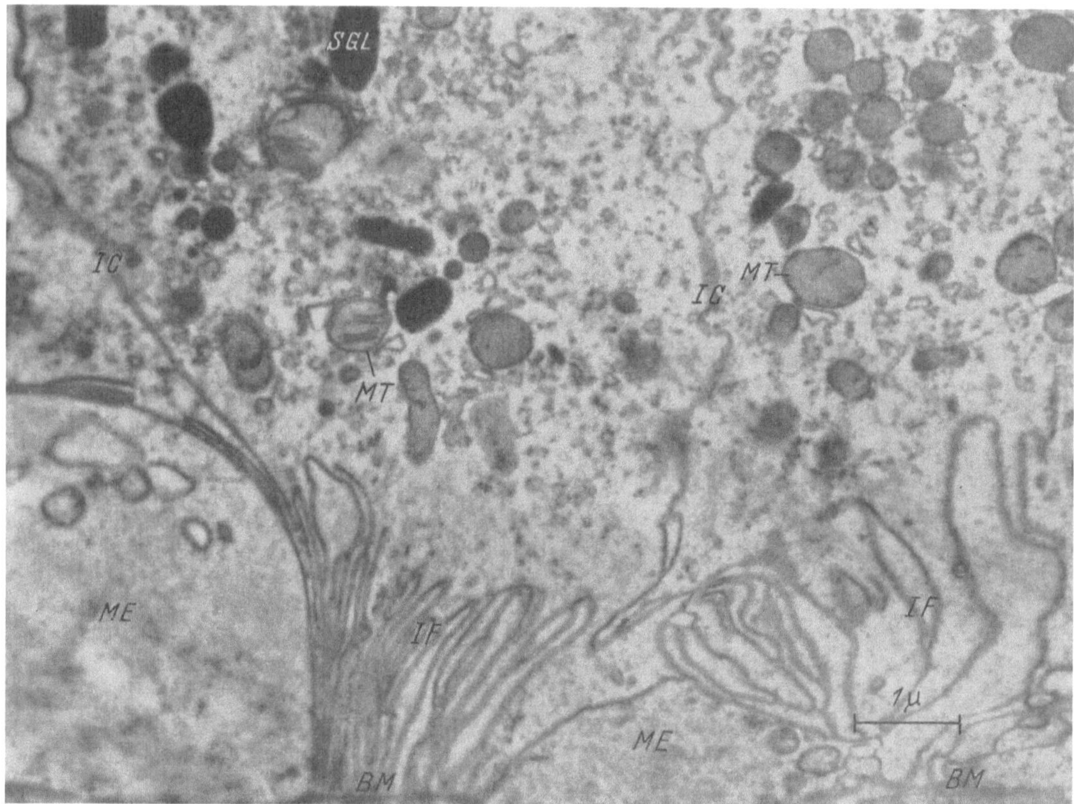

Fig. 1. Basal part of sweat gland cells abutting the myoepithelial cells (*ME*), as well as the basement membrane (*BM*). Closely packed basal infoldings (*IF*) are observed especially at the region where the glandular cell directly abuts the basement membrane. Most of the infoldings are perpendicular to the basal surface, but some are parallel to the glandular-myoepithelial cell boundaries. *IC*, intercellular boundary; *MT*, mitochondria; *SGL*, secretory granule probably of lipid nature. 14,500 ×

Small pieces of human axillary skin removed at operations for the osmidrosis were fixed immediatly after the resection in 1% osmic acid solution buffered at p_H 7.4 for about 3 hr. After washing and dehydration, the specimens were embedded in a mixture of n-butyl and methyl methacrylate (7:3) and were cut with JUM-4 ultra-microtome. Sections without extraction of embedding media were examined in an electron microscope, Type JEM-3.

The intercellular boundaries between adjacent glandular cells show a remarkable irregularity, the intercellular interdigitations (Fig. 1 and 3). Some of them resemble the microvilli in shape, i.e. finger-like processes of the cytoplasm of one cell invade into the neighbouring cell. Another form of the interdigitation looks like multiple folds, whose orientation may be either parallel or vertical to the adjoining plane of two cells. As a rule, the interdigitations are more developed at the basal than apical territories of intercellular boundaries, but few exceptions exist.

The plasma membrane of the glandular cell invaginates as double layered folds at the basal surface facing the myoepithelial cell or the basement membrane (Fig. 1). The space between paired membranes is essentially extracellular, and no part of myoepithelial cytoplasm follows the infolding, unlike the interdigitation. It is observed that the infolding is more elaborate at the region where the glandular cell directly abuts the basement membrane than at the contact surface between the glandular- and myoepithelial cells (Fig. 1). Any special accumulation of mitochondria between the infolded membranes was not observed. The orientation of infolded membranes is generally vertical to the basal surface at the place directly abutting the basement membrane, while near the boundary against myoepithelial cell, it is parallel to the contact surface (Fig. 1). This fact may be concerned with the direction of the myoepithelial contraction.

Three distinct types of granules were found in cells of the eccrine sweat gland. The first is spherical or ellipsoidal granule with relatively clear content (Fig. 2). The size of this granule is

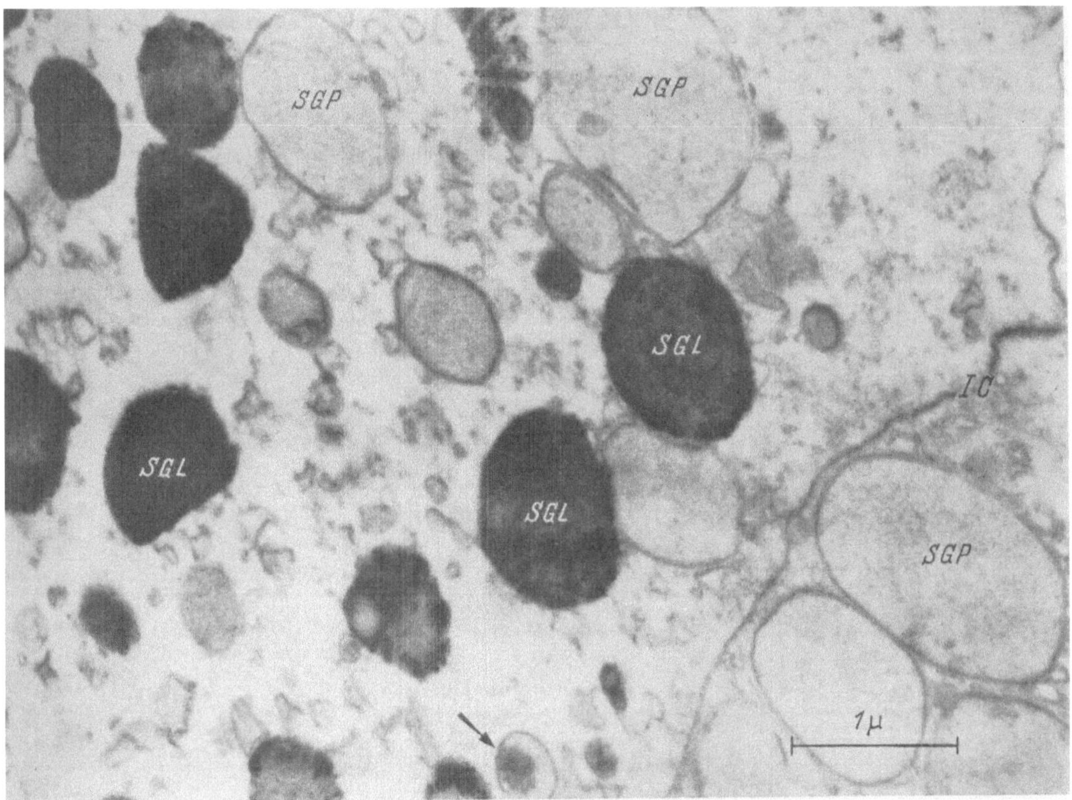

Fig. 2. Part of the cytoplasm of glandular cells showing the two distinct types of the secretory granule. Double membrane structure is obvious at the limiting membrane of less opaque granules, which is probably of protein nature (*SGP*). Another type of the granule (*SGL*) is extremely dense and considered to be of lipid nature. At the mid-lower margin, a dense mass surrounded by smooth membrane is observed as shown by an arrow, indicating the probable immature form of dense secretory granules. (*IC*) intercellular boundary. 23,000 ×

generally larger than that of mitochondria. The circumference of the granule is limited by an opaque and smooth membrane, which is depicted as a double-layered membrane in high magnification micrographs (Fig. 2). Although the content of large sized granules of this type is quite homogeneous, small or intermediate sized granules contain several straight or curved lines of double nature, whose features are similar to those of internal membranes of the mitochondria, the so-called cristae mitochondriales (Fig. 1). Therefore it may be assumed that this granule arises from the expansion of mitochondria and that this granule is probably of protein nature as an assumption from their density.

The second type of granule is highly opaque for electron beam owing partly to its strong affinity to the osmium. Additionally this granule is not homogeneous, but many particles of various sizes of extreme opacity are always found within the granule (Fig. 2). The size of this granule are widely variable, ranging from minute granules smaller than the mitochondria to enormously large ones about 2 μ in diameter, while the shape may be spherical, ovoid or rod-like and sometimes it shows an irregular contour (Fig. 1 and 2). This granule is tentatively presumed to be of lipid nature from its strong affinity to the osmium and irregular shape which is similar to lipid droplets found in other tissues.

Not infrequently, it is observed that masses of the same density as this granule are enveloped by smooth membraned vesicles or sacs of various sizes, somewhat resembling the so-called "intracisternal granules" of Palade (*10*), but in this case the enveloping sacs are always smooth surfaced (Fig. 2). Thus this type of granules may be assumed to be formed within the cisternae of smooth surfaced variety of endoplasmic reticulum.

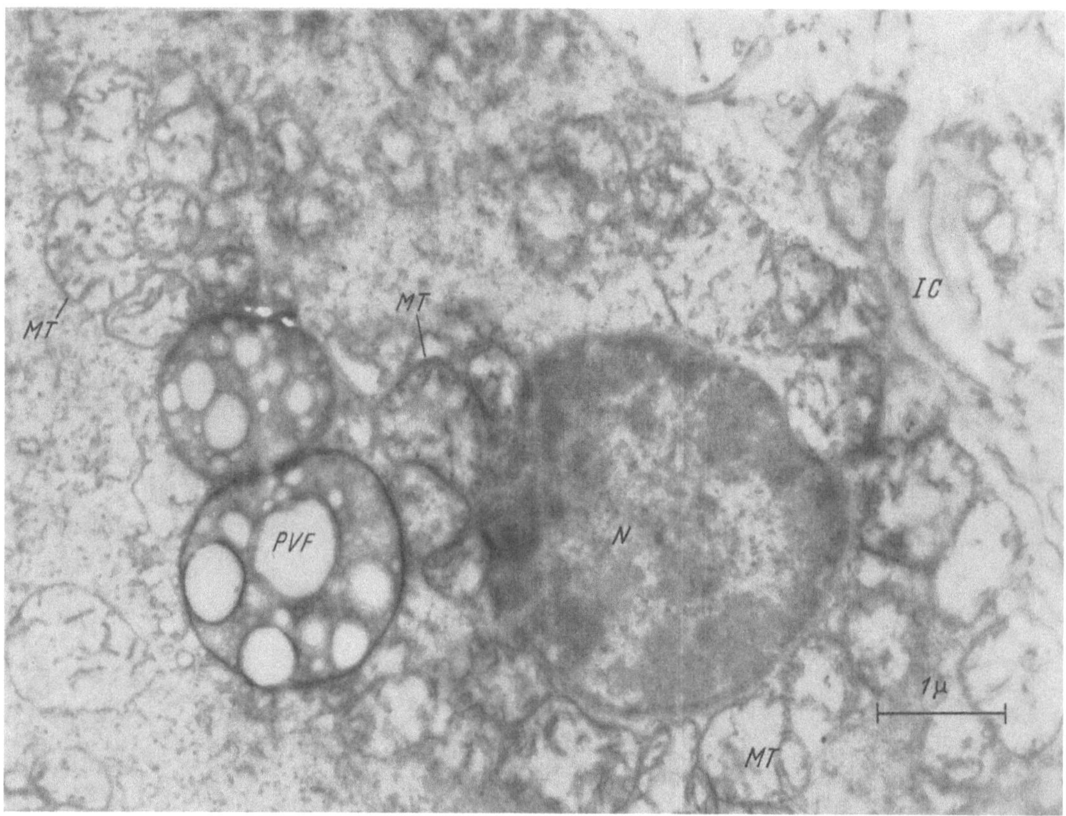

Fig. 3.   Central part of a glandular cell. The nucleus cut slightly off center (*N*) is shown at about the center of the picture. Numerous mitochondria (*MT*) are closely packed throughout the cytoplasm. Two large granules (*PVF*) containing vesicles of various sizes are seen left to the nucleus. These granules correspond to the so-called "polyvesicular fat droplets" (Ito, 1943). The cell boundary (*IC*) with interdigitation is shown at the upper right corner. 17,500 ×

Both of two types of above mentioned granules are constantly found in all of the glandular cells, so that they are thought to be secretory granules. The third type of granule, however, has been rarely observed and the relationship to the secretory activity is not determined. These are large spherical bodies (1—2 μ) found in supranuclear zone in a group of 2—3 granules. This type of granule is characterized by the facts to possess an electron dense membrane surrounding the surface and contain a great number of clear vesicles of varying sizes (40 mμ—1 μ) which shows this granule as a foam-like pattern (Fig. 3). A matrix substance among the vesicles within this granule is of high electron density. Despite the resemblance of this granule to some sort of artefact, this structure is evident to be a preexisting structure of the cell and coincides the structure reported by Ito (*11*) with the light microscope as the "polyvesicular fat droplet" of human eccrine sweat glands.

Discussion. The infolding of plasma membrane was first observed in renal tubular cells (*1, 2*) and termed "basal infolding" (*6*) or "β-cytomembrane" (*12*). This structure was then detected in various epithelia such as Malpighian tubules of insects (*3*), salivary glands (*6*), ciliary body of the eye (6), choroid plexus (5), utricle of inner ear (8) and others, all of which are known as tissues with extensive water transport either absorption or excretion. Pease (6) and Ruska et al. (7) claimed that the basal infolding may be associated with the active transport of water. Palade (13) and Bennett (14) postulated a hypothesis that the invagination and pinching off from the surface membrane are concerned with the intracellular transport of some substance. For the confirmation of these theories, the sweat gland must be examined because of its function associating with the high water excretion. Laden et al. (15) briefly reported an electron optical feature of human sweat gland, but they did not refer to this structure. In a previous report, we pointed out the basal infolding occurred in pig's sweat gland (9); and further observations on human organs revealed more elaborate feature of this structure as reported here. The present result may strongly support the theory that the basal infolding is primarily concerned with the intracellular transport of water or any other fluid material.

Two distinct secretory granules, one is probably protein nature and the other lipid, were detected in the eccrine sweat gland of man, and it is of interest that these two sorts of granules may arise from different origins.

References

1. Sjöstrand, F. S., and J. Rhodin: Exp. Cell Res. **4**, 426 (1953).
2. Pease, D. C.: Anat. Rec. **121**, 723 (1955).
3. Beams, H. W., T. N. Thamisian and R. L. Devine: J. biophys. biochem. Cytol. **1**, 197 (1955).
4. —, E. Anderson and N. Press: Cytologia **21**, 50 (1956).
5. Maxwell, D. S., and D. C. Pease: J. biophys. biochem. Cytol. **2**, 467 (1956).
6. Pease, D. C.: J. biophys. biochem. Cytol. **2**, Suppl. 203 (1956).
7. Ruska, H., D. H. Moore and J. Weinstock: J. biophys. biochem. Cytol. **3**, 249 (1957).
8. Smith, C. A.: Ann. Otol. Rhinol. Laryngol. **65**, 450 (1956).
9. Kurosumi, K., and T. Kitamura: Nature (Lond.) **181**, 489 (1958).
10. Palade, G. E.: J. biophys. biochem. Cytol. **2**, 417 (1956).
11. Ito, T.: Fol. Anat. Jap. **22**, 273 (1943).
12. Sjöstrand, F. S.: Physical Technique in Biological Research. Vol. 3, p. 241, Academic Press, N.Y. 1956.
13. Palade, G. E.: J. biophys. biochem. Cytol. **2**, Suppl. 85 (1956).
14. Bennett, H. S.: J. biophys. biochem. Cytol. **2**, Suppl. 99 (1956).
15. Laden, E. L., I. Linden and J. O. Erickson: Arch. Derm. Syph. (Chicago) **71**, 219 (1955).

Submicroscopic changes of the parotid gland caused by functional rest and secretory nerve stimulation*

E. Manni

Institute of Physiology, University of Turin

The changes produced in the salivary gland from secretion process have been studied by many. Classical are Heidenhain (1) and Langley (2) researches. With the high resolving power of the electron microscope it is possible to look for better the course of the secretory process. The researches in this branch are few: Gautier and Diomede-Fresa (3), studying the rat submaxillary found an increase of the ergastoplasm after injection of pilocarpine. I am studying with the electron microscope the changes occurring in the lamb parotid gland during the secretory activity induced by electric stimulation of the parasympathetic nerves (4), comparing with the aspect of normal or excretory-duct-ligated gland: it was observed that the excretory duct ligature of the salivary glands leads to true rest condition (5), because the glands are continuously secreting.

The pieces of lamb parotid (fasting 12 to 24 hr) were cut off in pentothal narcosis: a) from normal animals; b) from lambs whose excretory ducts had been ligated for 12 days; c) from lambs whose Moussu nerves (4) were previous stimulated during 30 min with faradic scarcely supraliminary current to produce an abundant salivary secretion.

* This work is supported with a grant of the Consiglio Nazionale delle Ricerche.

The material was fixed with buffered 2% osmium tetroxide and embedded in methyl and butyl methacrylat (3/7); the ultrathin sections were obtained with Danon and Kellenberger microtome modified by Kellenberger and observed with the electron microscope Siemens UM 100 d of the Istituto Elettrotecnico Nazionale Galileo Ferraris, Turin [1].

Normal parotid. The nuclear membrane is double with an inner and an outer layer, both osmiophilic, separated by an osmiophobic interspace (Fig. 1). The inner layer is thicker than the outer one: both possess holes of 200 Å circa. The nucleoplasm consists in a network with close stitches and with osmiophilic particles in the joint points of the network. The ground substance

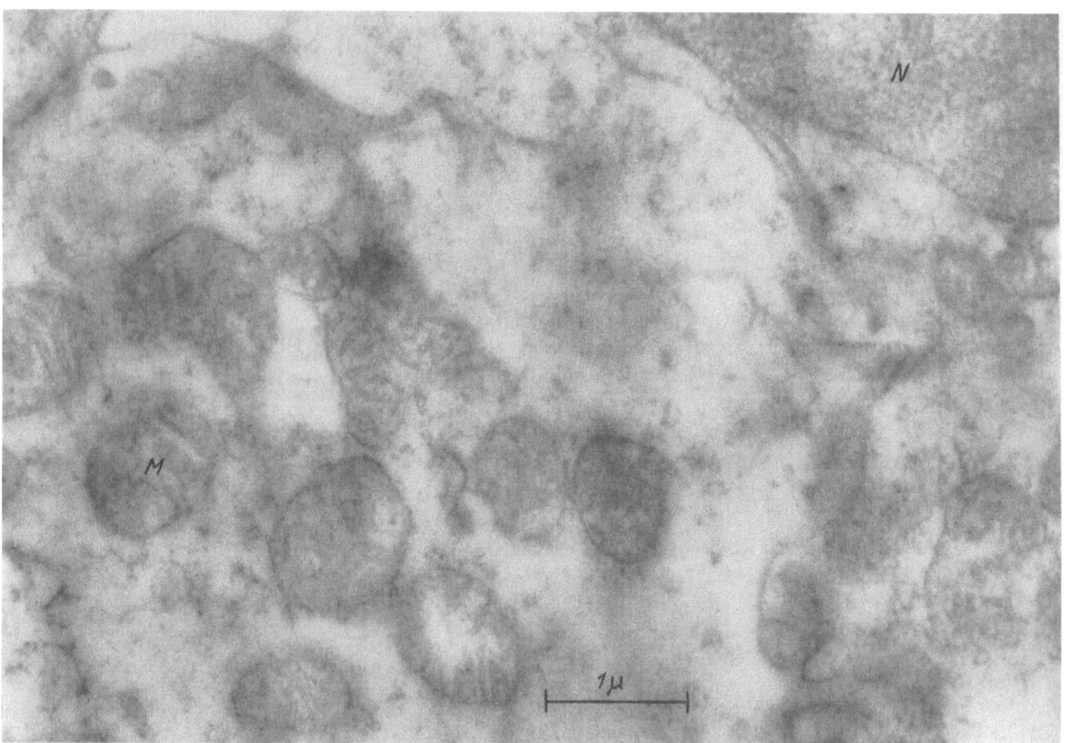

Fig. 1. Lamb parotid in normal secretory activity. N nucleus, M mitochondria. 19,000 ×

of the cytoplasm appears like a network of filaments with aspect of canaliculi, whose shape is different according to the sections and with submicroscopic particles (50—300 Å). In the cytoplasm are other filamentous formations which differ from the above described; they consist in two osmiophilic layers separated by an osmiophobic area. Some double membranes originate from the cell membrane as folds, sometimes they are S-shaped. Others have a rectolinear or curvilinear course parallely arranged: opaque particles are attached on the side of these membranes. The S-shaped double membranes bring to mind those described by Pease (6) in the submaxillary gland and could correspond to Sjöstrand β-cytomembranes (7); the parallel membranes could be homologued to the pancreas α-cytomembranes (7) and to the ergastoplasm of the previous authors. Many are the mitochondria exhibiting a typical structure, previously described by Manni (8); the zymogen granules (large opaque formations to electron ray) and the vacuoles are scarce.

Excretory duct ligature. In this experimental condition the cytoplasm shows the most striking modifications: the reticulum endoplasmaticum is enriched with particles, filaments, double

---

[1] This microtome belongs to the Istituto Chimico of Turin University: I thank Prof. Nasini, and Mrs. A. Mojoni, Ch. D., for her assistance.

membranes and zymogenic granules (Fig. 2). The mitochondria and vacuoles decrease. The duct ligature produces functional resting condition of the cell.

Faradic stimulation of the parasympathetic nerves. In this experimental condition the cell diameter increases with changes of the nucleoplasm and cytoplasm. The nuclear membrane holes appear more evident and the endonuclear network shows larger and more regular stitches while the osmiophilic particles diminish. Striking are the changes of the cytoplasm: the network of

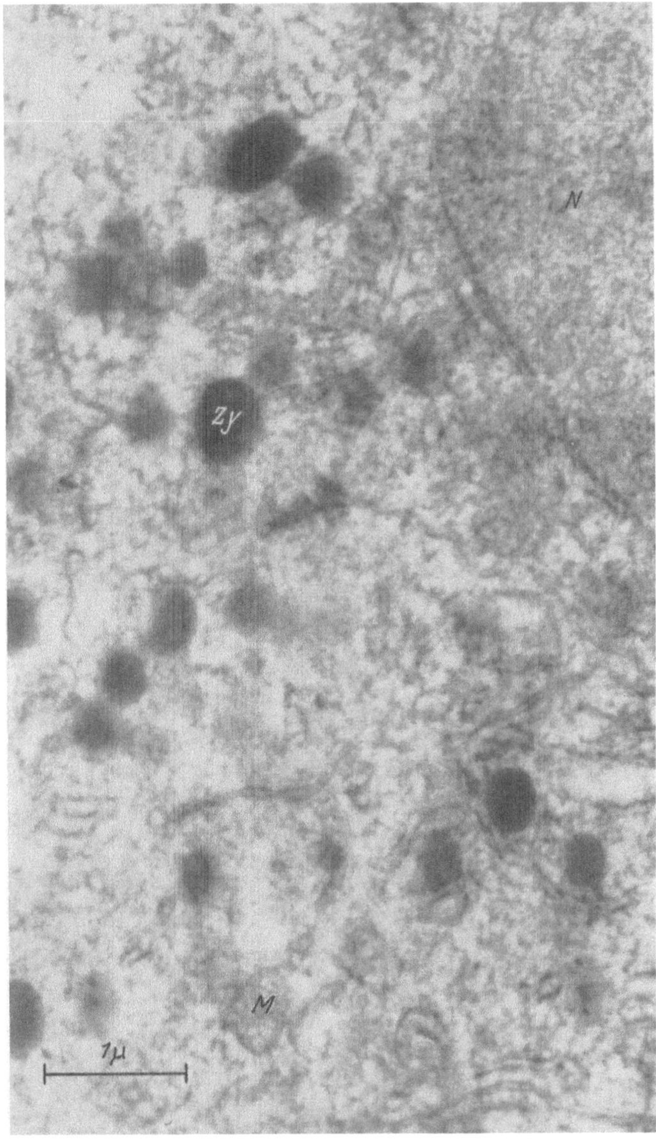

Fig. 2. The parotid gland of the same animal 12 days after the excretory duct ligature. Striking increase of cytoplasmatic formations and zymogenic granules (*Zy*); the mitochondria disappear. 19,000 ×

reticulum endoplasmaticum becomes poorer, specially roundabout the nucleus and many vacuoles appear. The S-shaped double membranes (Fig. 3) are more evident and are in touch with little and numerous vesicles or microvilli appearing in the distal part of the cell. Zymogen granules disappear. No evident changes of the mytochondria structure are observed.

To conclude in spite of the great difficulty involved in estimating the various experimental conditions, striking differences exist between the normal parotid lamb gland and the same gland whose secretory nerves have been stimulated or whose secretory duct has been ligated.

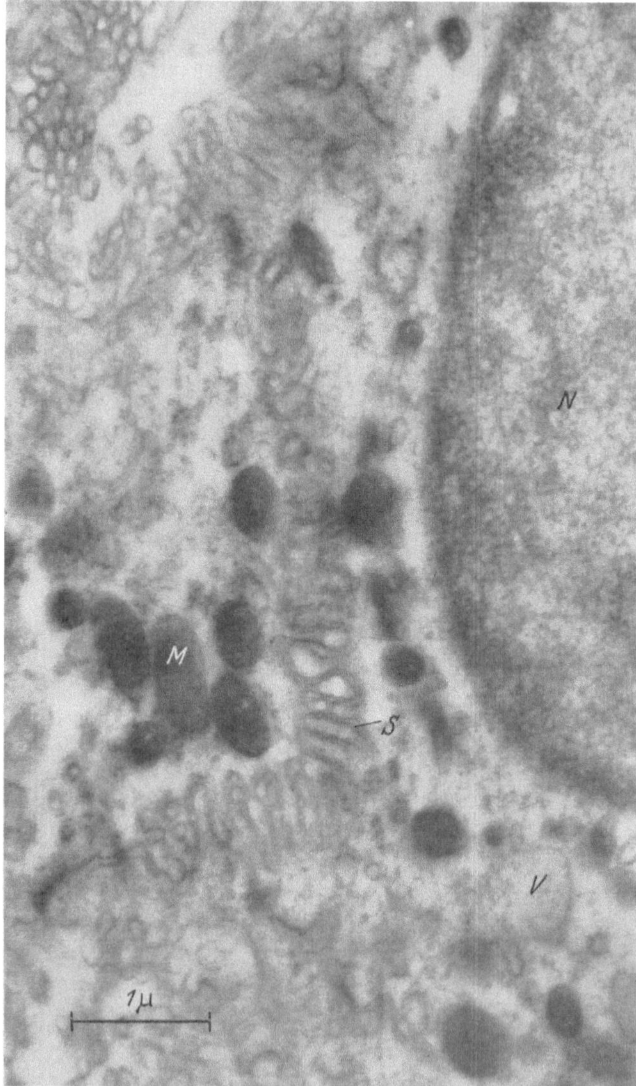

Fig. 3. Lamb parotid whose secretory nerves are stimulated during 30 min. Note the S-shaped cytomembranes (S), microvilli, vacuoles (V) and mitochondria (M). 19,000 ×

References

1. Heidenhain, R.: Pflügers Arch. ges. Physiol. 17, 1 (1878).
2. Langley, J. N.: J. Physiol. 2, 261 (1879—1880).
3. Gautier, A., et V. Diomede-Fresa: Mikroskopie 8, 23 (1953).
4. Moussu, M.: Arch. Physiol. norm. et path., S. V., 22, 68 (1890).
5. Fernandes, J. F., and L. C. U. Junqueira: Exp. Cell Res. 5, 329 (1953).
6. Pease, D. C.: J. biophys. biochem. Cytol. 2, Suppl. 203 (1956).
7. Sjöstrand, F. S.: Int. Rev. Cytol. 5, 455 (1956).
8. Manni, E.: Boll. Soc. ital. Biol. sper. 32, 890 (1956).

Das Doppellamellen-System in den Drüsenzellen der Parotis

W. Schwarz und M. Hofmann

Forschungsabteilung für Elektronenmikroskopie der Freien Universität Berlin

Die Endstückzelle der Parotis kann als Vorbild einer serösen Drüsenzelle angesehen werden, ihr Sekret besteht nämlich vorwiegend aus Eiweiß, während andere Stoffe wie Fermente und Rhodanide nur spärlich vorhanden sind. Glykoproteide befinden sich nicht im Sekret. Die beträchtliche Menge des Parotisspeichels wird mit dem Auftreten und der Verflüssigung von lichtmikroskopisch sichtbaren Sekretgranula und der Funktion des Ergastoplasmas in Zusammenhang gebracht. Gautier und Diomede Fresa (1) konnten in der Form und Verteilung des Ergastoplasmas ähnliche Verhältnisse wie in Pankreaszellen feststellen. Sie unterscheiden drei Formen im Ergastoplasma: die Wirbel-, Geflecht- und Parallelanordnung der Strukturen. Ihre ursprüngliche Deutung des Ergastoplasmas als fibrilläre Struktur ist jetzt zugunsten einer lamellären aufgegeben. Die Beziehungen der Doppelmembranen zum Kern hat Dalton (2) mehrfach untersucht und dabei auf einen innigen Kontakt zwischen Ergastoplasma und Kernmembran hingewiesen. Dieser Befund ist deshalb so wichtig, weil damit ein unmittelbares Abfließen von Ribonucleoproteiden aus dem Kern in das Ergastoplasma im Bereich des Möglichen liegt. Die Abgabe dieser Nucleoproteide an das Cytoplasma fällt zwar unter den histologischen Begriff der Kernsekretion, doch liegt die Größe der abfließenden Kernbestandteile unter dem Auflösungsvermögen

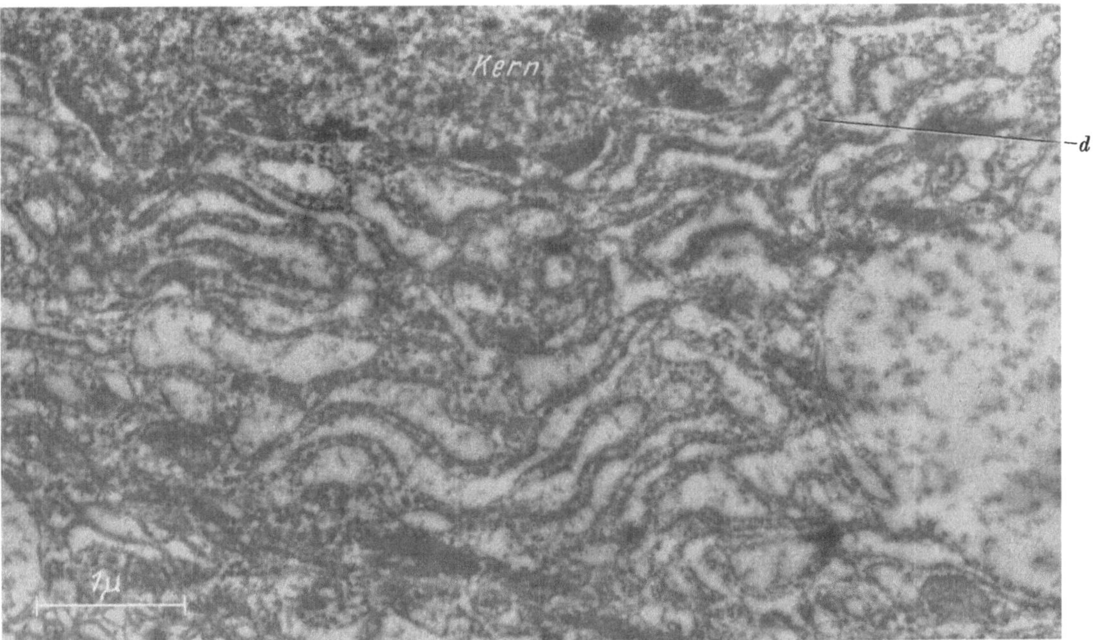

Abb. 1. Kern und benachbarter Ergastoplasmateil einer Parotiszelle. Zusammenhang zwischen Doppellamellen und Kern an den Lücken des osmiophilen Randsaumes (d). Vergrößerung 20000:1

des Lichtmikroskopes. Unter der histologisch definierten Kernsekretion versteht man deshalb meistens Veränderungen des Kernkörperchens und damit verknüpft die Abgabe lichtmikroskopisch sichtbarer basophiler Kernteile und Kernvacuolen an das Cytoplasma, wie sie besonders bei eiweißsezernierenden Zellen beobachtet wird (3, 4, 5). Bei diesen Kernbestandteilen handelt es sich um Ribonucleoproteide, die im Kernkörperchen entstehen und für die Eiweißproduktion verantwortlich sein sollen. Hiernach stammen die Ribonucleinsäuren des Ergastoplasmas aus dem Kern. Die Vacuolen des Kernes werden von Wessel (6) auf Grund elektronenmikroskopischer Untersuchungen dagegen mit pathologischen Veränderungen der Zelle in Zusammenhang gebracht. Nach Vergiftung mit Colchicin stülpen sich in Leberzellen Ergastoplasmateile und Mitochondrien in den Zellkern ein. Für die Einschleusung dieser Cytoplasmateile wird ein Kernödem verantwortlich gemacht. Ähnliche Vorgänge kommen auch bei Crocker-Sarkomzellen vor. Zur Beurteilung der Beziehungen zwischen Kernsekretion und Eiweißproduktion der Zelle sind aber

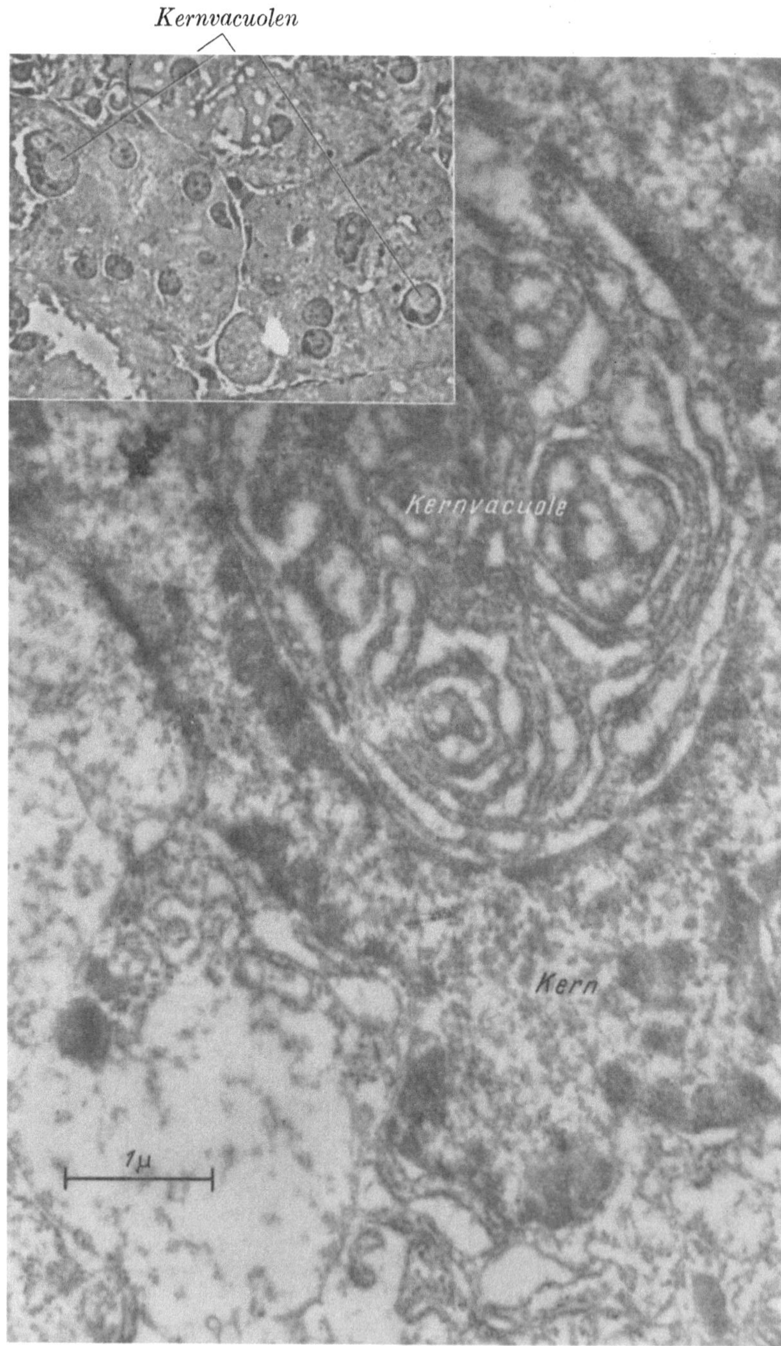

Abb. 2. Doppellamellen in der Vacuole eines Kernes einer Parotiszelle. Vergrößerung 20000:1. Oben rechts: lichtmikroskopisches Bild mehrerer Parotiszellen mit Kernvacuolen. Vergrößerung 1200:1

Untersuchungen an normalen eiweißsezernierenden Zellen notwendig. Die Parotiszellen sind dafür besonders geeignet. Wir untersuchten das Parotisgewebe der Ratte. Nach Fixierung und Einbettung in der von Palade angegebenen Weise wurden Schnitte mit dem Sjöstrand-Mikrotom hergestellt. Die Aufnahmen wurden mit einem Zeiss-Elektronenmikroskop Typ EM 8/II bei einer

Strahlspannung von 40 kV gemacht. Dickere Schnitte aus demselben Block wurden lichtmikroskopisch zur Kontrolle herangezogen.

Das *Ergastoplasma* besteht wie in den Pankreaszellen (*7, 8*) aus Systemen von Doppellamellen mit Paladegranula. Besonders auffällig ist die wirbelförmige Anlage der Lamellen. Daneben kommen vor allem in den apicalen Gebieten aber geflechtartige Anordnungen vor, die besonders viele Sekretgranula enthalten. Diese Granula werden dabei von den Lamellen umschlossen. In den basalen Zellteilen liegen dagegen die Doppelmembranen mehr parallel geordnet und umschließen hauptsächlich Mitochondrien.

Die *Kerne* zeigen von Zelle zu Zelle erhebliche Größenunterschiede und sind auch in ihrer Struktur und Dichte nicht immer gleich. Die locker gebauten Kerne haben eine geringe, aber fast gleichmäßige Dichte. In Kernnähe liegen im Cytoplasma reichlich Doppellamellen, die aber nur selten mit der Kernmembran in Beziehung treten. In anderen Zellen ist dagegen die Dichte der Kernsubstanz uneinheitlich. Sehr dichte Schollen von unterschiedlicher Größe liegen im Kern verstreut. In der Außenzone nahe der Kernmembran befindet sich ein Gürtel von sehr dichter Kernsubstanz, der an einigen Stellen unterbrochen ist. An diesen Lücken ist immer ein Zusammenhang zwischen einer Doppellamelle des Ergastoplasmas und der Kernoberfläche vorhanden (Abb. 1). Die Doppellamelle weicht hier auseinander und faßt kleine Granula von 150 Å Dicke zwischen sich, die gleich groß sind wie die Granula in den Lücken des osmiophilen Kernrandsaumes. Das Ergastoplasma in der Nachbarschaft dieser Kerne gehört zu dem geflechtartigen Typus. Im Kerninnern sieht man bereits im lichtmikroskopischen Bild der Plexiglasschnitte Vacuolen von verschiedener Größe, die lichtoptisch nicht vollkommen leer sind, sondern einige feinste osmiophile Fäden enthalten (Abb. 2). Nur die elektronenmikroskopische Untersuchung zeigt, daß der Vacuoleninhalt aus Doppellamellen besteht (Abb. 2). Mitochondrien sind nicht vorhanden, wohl aber osmiophile Kondensate, die eine gewisse Ähnlichkeit mit den Sekretgranula im Cytoplasma zeigen. Eine dichtere Kernsubstanz ist gürtelförmig auch um die Vacuole angeordnet, und einzelne Lamellen treten hier ebenfalls an einigen Stellen mit der Vacuolenwand und damit mit der Kernsubstanz in Verbindung. Das Ergastoplasma der Kernvacuole kann damit auch Nucleoproteide aus dem Kern erhalten. Die Vermutung liegt nahe, daß diese Vacuolen Einstülpungen des Cytoplasmas in den Kern darstellen. Die Kernoberfläche kann dadurch vergrößert und die Abgabe von Nucleoproteiden an das Ergastoplasma gesteigert werden. Gegen diese Auffassung sprechen sowohl histologische als auch experimentelle Befunde. Viele histologische Untersuchungen stützen vielmehr die Ansicht, daß es sich bei den Vacuolen um Veränderungen des Nucleolus handelt. Auch lassen sich die radioaktiv markierten Ribonucleinsäuren zuerst im Kern nachweisen und erst später im Cytoplasma. Beobachtungen an lebenden Fibroblasten haben gezeigt, daß die Kernvacuolen im Kern entstehen und nach außen abgegeben werden. Nach dieser Voraussetzung kommt dem ergastoplasmatischen Inhalt der Kernvacuole eine andere Bedeutung zu. Man muß vor allem daran denken, daß hier ein Bildungsort ergastoplasmatischer Lamellen sein kann. Bestimmte Befunde in anderen Kernen sprechen dafür. Mitunter sieht man Löcher in einem runden oder elliptischen Kondensat auftreten, die sich aufeinander zu vergrößern. Zwischen den offenen Stellen bleiben schließlich die Wände als Doppellamellen stehen.

Die Untersuchungen wurden mit Unterstützung der Deutschen Forschungsgemeinschaft durchgeführt.

Literatur

1. Gautier, A., et V. Diomede-Fresa: Mikroskopie 8, 23 (1953).

2. Dalton, A. J., and Marie D. Felix: Ann. N. Y. Acad. Sci. **63,** 1117 (1956).

3. Altmann, H. W.: Klin. Wschr. **1955,** 306.

4. — Handbuch der gesamten Hämatologie. 2. Aufl. Bd. 1. München, Berlin, Wien: Urban & Schwarzenberg 1957.

5. Ries, E., u. M. Gersch: Biologie der Zelle. Leipzig: B. G. Teubner 1953.

6. Wessel, W.: Virchows Arch. **331,** 314 (1958).

7. Sjöstrand, F. S., and V. Hanzon: Exp. Cell Res. **7,** 393 (1954).

8. Oberling, Ch., W. Bernhard, A. Gautier et F. Haguenau: Presse med. **35,** 719 (1953).

Electron microscope observations of degenerating and regenerating pancreas following ethionine administration*

L. Herman, P. J. Fitzgerald, M. Weiss and I. S. Polevoy

Department of Pathology, State University of New York, Downstate Medical Center, Brooklyn, N. Y.

It has been previously shown, that administration of dl-ethionine (α-amino-γ-ethylmercapto-butyric acid), an amino acid analogue of methionine, results in selective pancreatic acinar cell destruction. This is followed by a subsequent regenerative phase with almost complete restitution of the acinar cells (4, 7, 19). The progressive degenerative changes observed by light

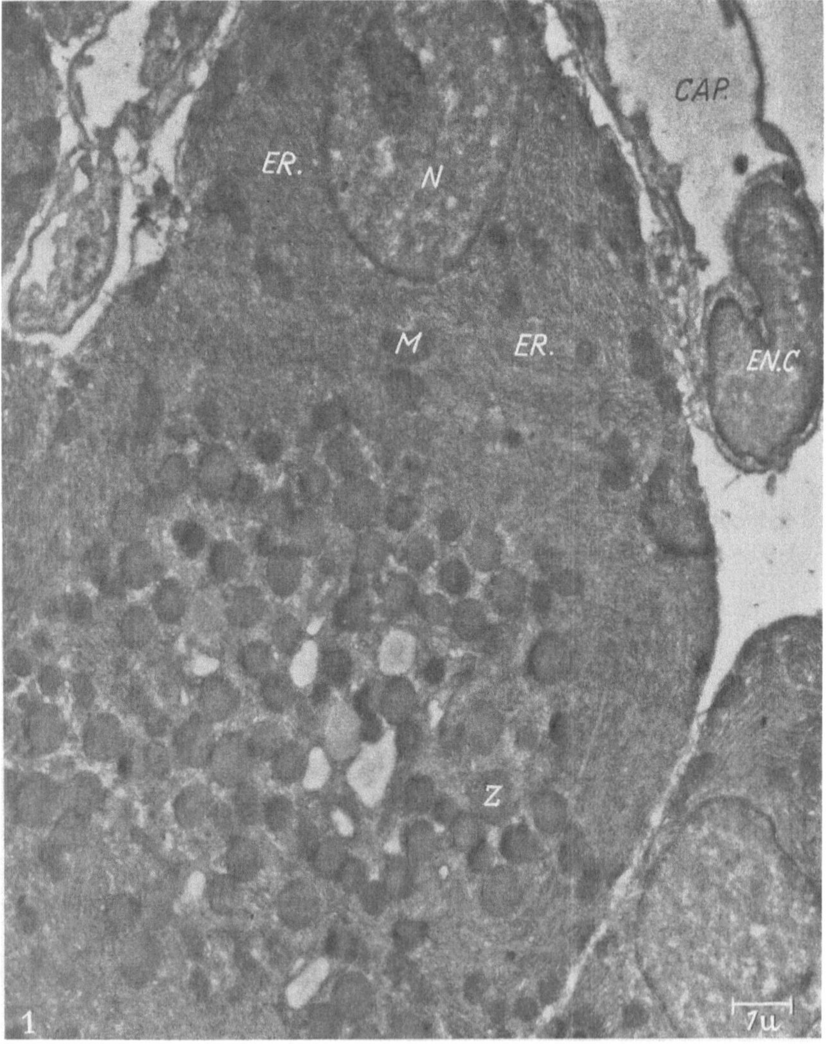

Fig. 1. Normal rat pancreas: Parts of two acinar cells are shown. Note the ergastoplasm, (*ER*); nucleus, (*N*) with a nucleolus; and mature, normal, zymogen granules, (*Z*). Part of a small capillary, (*CAP*), with an endothelial cell, (*EN. C*), appears in the upper right. 8,000 ×

microscopy have been a decreased perinuclear basophilia; eosinophilic cytoplasm, accompanied by zymogen granule destruction; a disruption of acinar cell pattern; increased vacuolization, disintegration and removal of acinar cells. Factors such as sex (*1*), protein depleted diet (*3, 18*),

* This research was supported by the US-Public Health Service, National Institute of Cancer, Grant No. C-3301.

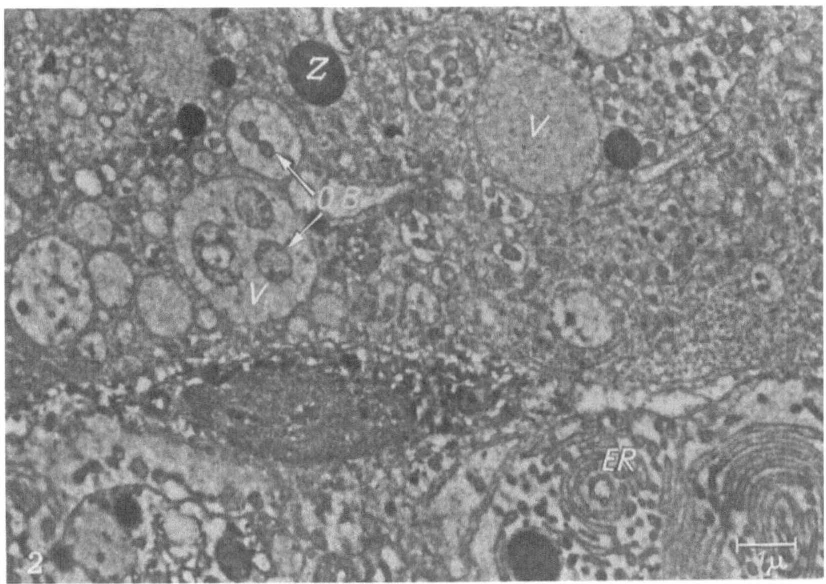

Fig. 2. Early degeneration of rat pancreas: 5 day. Note vacuoles, (*V*), with oval bodies, (*O. B*) lamellae, (*ER*), and a few zymogen granules, (*Z*). 8,000 ×

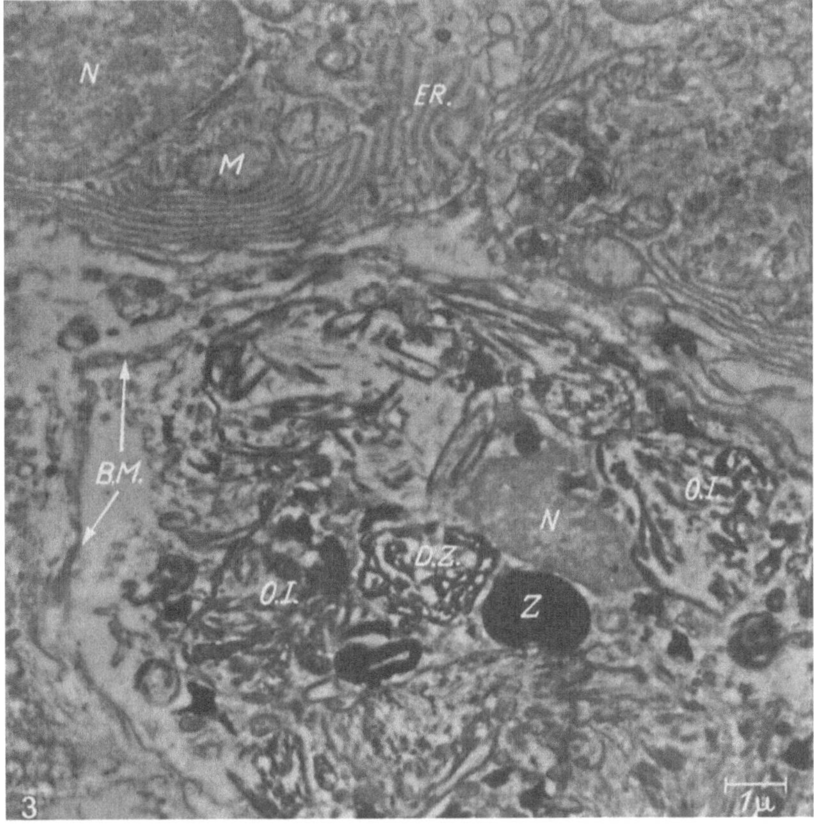

Fig. 3. Maximum degeneration of rat pancreas: 10 day. A cross sectional area of an acinus appears in the center. Note osmiophilic inclusions, (*O. I.*), surrounded by a basement membrane, (*B. M.*). Some inclusions represent fused zymogen granules, others arise from ergastoplasmic membrane fusions. Just to the right of center, a shrunken distorted nucleus, (*N*), appears. Part of the cell in the upper left appears less disturbed. A well-preserved nucleus, (*N*), some swollen mitochondria, (*M*), and "smudged" ergastoplasmic membranes, (*ER*) are present. 8,000 ×

methionine administration (2), animal weight, age, and dosage level of ethionine (3, 4) influence the ultimate course of degeneration and regeneration. A cursory review of the literature indicated that no electron microscope study of this biological model of degeneration and regeneration has been undertaken. A number of ultrastructure studies of normal, starved, refed, pilocarpine-injected and neutral red injected pancreas has been made (11, 12, 13, 14, 15, 16, 17, 20, 21).

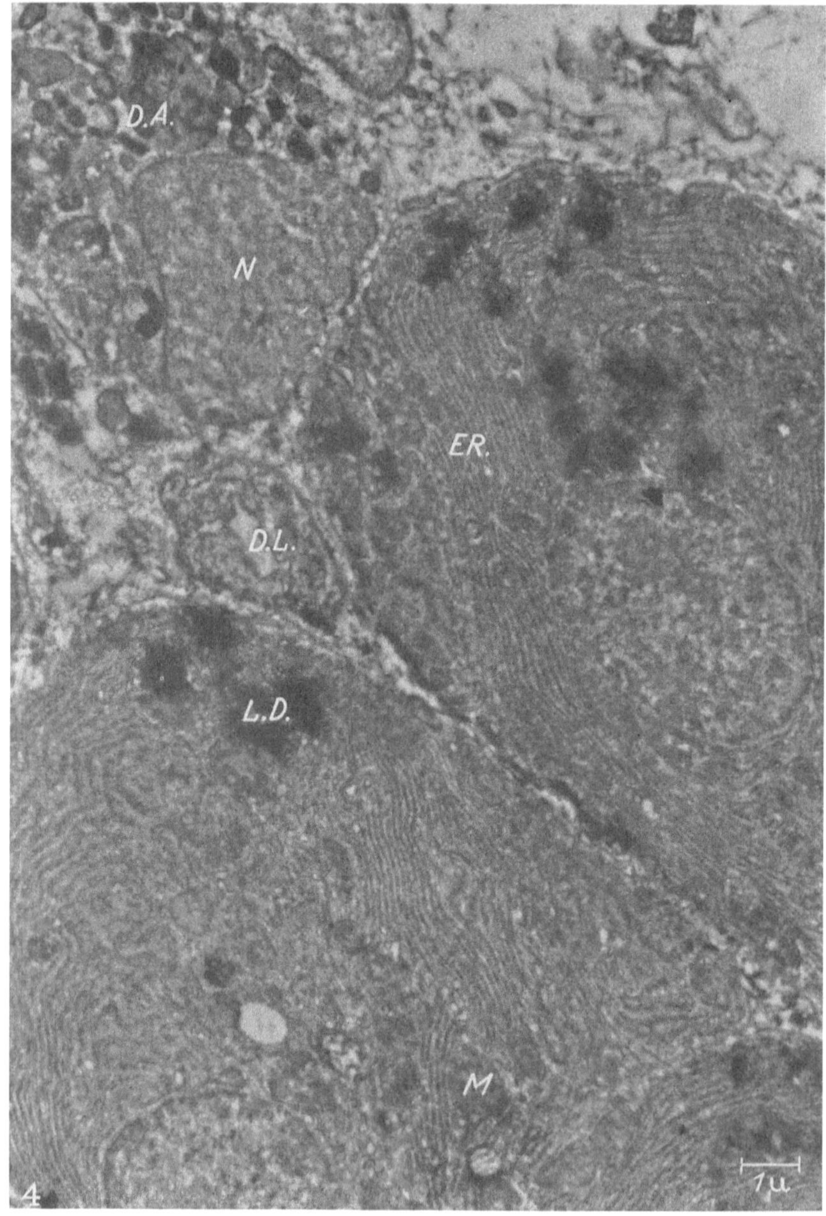

Fig. 4. Regenerative phase of rat pancreas: 8 days after the last (10th) ethionine injection. A small area of degeneration, (D. A.), persists in the upper left. Most of the micrograph is occupied by regenerating cells. Note the osmiophilic lipid droplets, (L. D.); the extensive ergastoplasm, (ER); and the lack of zymogen granules. A small duct, (D. L.), appears to the left of center. 8,000 ×

This paper presents preliminary observations on structural changes of the ethionine treated pancreas of rats and salamanders kept on a protein-free diet.

One hundred and seventy five white Wistar rats (Carworth Farms, New City, N.Y.), weighing approximately 180 g were placed on "protein-free" diet (Nutritional Biochemicals Corp.). Eighty-eight of these animals were daily injected intraperitoneally with dl-ethionine (0.7 mg/gm body weight) for a maximum period of ten days. The remaining surviving animals served as controls. Animals of each group were sacrificed at 5, 8, 10, 12, 18 and 28 days. In addition 50 salamanders (*Ambystoma tigrinum*) were divided into two groups. One group received daily 5 mg/gm body weight of dl-ethionine in amphibian Ringers solution for 10 days. The second group received 10 mg/gm body weight of dl-ethionine for the same period of time. Both groups of animals were sacrificed on the 11th day. A third group received daily equivalent volumes of amphibian Ringers solution and served as a control group. The pancreas was quickly removed and fixed for 6 hr at room temperature in phosphate buffered 10% formalin and Bouins fixative for light microscopy. Other pieces were gently diced into approximately 0.5 mm square blocks under a dissecting microscope and quickly fixed at 6° C for 1½ hr

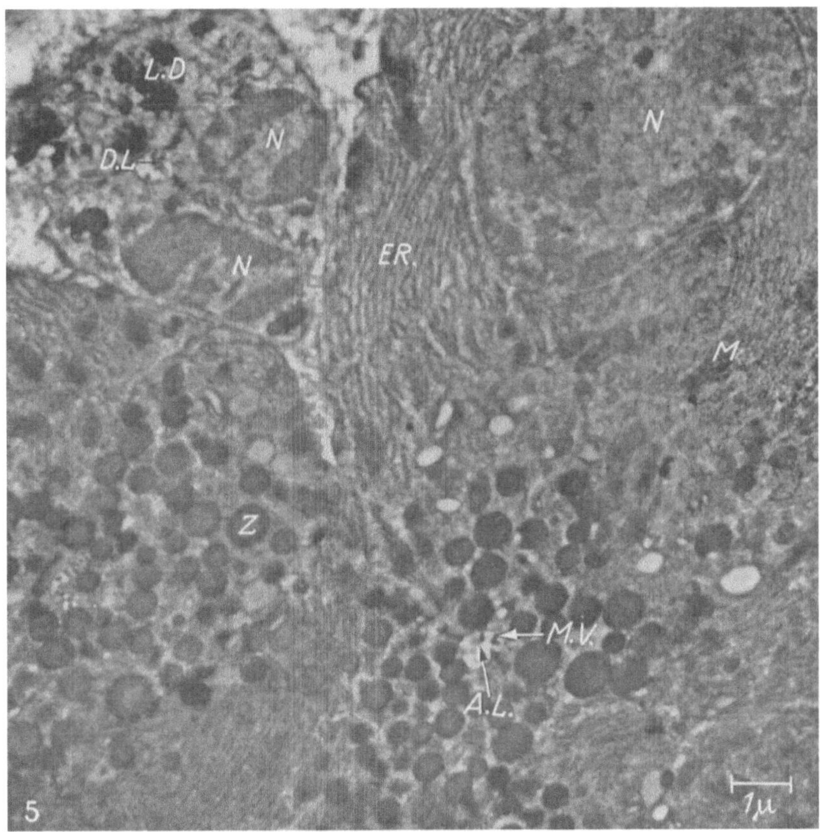

Fig. 5. Later regenerative phase of rat pancreas: 18 days after the last (10th) ethionine injection. A duct (*D. L.*) appears in the upper left containing some lipid droplets, (*L. D.*). Acini have reformed and the lumen of one at *A. L.* contains microvilli, (*M. V.*) which may be noted projecting into the lumen. Mature zymogen granules, (*Z*), have formed and appear in clusters at the apices of the cells. 8,000 ×

in a veronal-acetate buffered 1% osmium tetroxide, (*10*), to which glucose (4%) was added. These were rapidly dehydrated, infiltrated with n-butyl methacrylate, and embedded in pre-polymerized plastic. Thin sections were cut on a Servall "Porter-Blum" ultra thin microtome using a glass knife and were viewed with an RCA EMU3C electron microscope. The pancreas blocks fixed in Bouins and formalin fixatives were paraffin-embedded cut at 5 μ, and stained with hematoxylin and eosin and azure B (*5*).

Light microscope observations: the general pattern of degeneration and regeneration of acinar cells was, with a few exceptions, the same as those described by others (*4, 7, 19*). From days 5—8, there was a progressive loss of acini, cytoplasmic basophilia and zymogen granules. At day 10 almost all the acinar cells were destroyed leaving dilated ducts, islets, and supporting connective tissue. Islet cells appeared undamaged. Regenerating cells first became evident at day 12. At day 18 these cells increased in number, size and basophilia. A few cells at this stage of regeneration contained zymogen granules. At day 28 clearly formed acini were discernible. These cells contained mature zymogen granules at their apices and a heavily stained perinuclear basophilia which could be resolved into a lamellate structure with the light microscope (*6*).

Electron microscopy observations: days 5—8 of ethionine treatment (Fig. 2) showed variable stages of degeneration. Many areas still retained well-organized acini with normal ergastoplasmic lamellae, mitochondria, Golgi and zymogen granules. Microvilli were frequently seen in the lumen of acini. However, there were other regions with the beginning stages of degeneration. These were characterized by frequently occurring vacuoles which contained denser oval or round bodies within the cytoplasm; a "smudging" of ergastoplasm; and swollen mitochondria with disrupted cristae (Fig. 6). The structure of the zymogen granules was retained but they exhibited an increased osmiophilia (Fig. 2). In the treated salamander series what appeared to be zymogen granules at day 10 were more osmiophilic than normal and frequently contained electron lucent centers (Fig. 7). The degenerated granules and damaged mitochondria seen in this study bore a close resemblance to those structures described as "agglutinated granules" by Siekevitz and Palade in guinea pig pancreas (15). Zymogen granules sometimes appeared as oval aggregates in a partially fused state often surrounded by disrupted, concentrically whorled, closely packed, osmiophilic membranes. These osmiophilic membrane inclusions assumed varieties of patterns and sizes (Fig. 9). They seemed to be derived from disrupted ergastoplasm. It is conceivable that ethionine has interfered with the metabolism of the ergastoplasm with distortion of phospholipid relations and rendered the structure more osmiophilic. The osmiophilic inclusions as well as the vacuoles containing dense bodies were similar to those seen in the pancreas following neutral red injections (21). More advanced stages of degeneration of both rat and salamander pancreas resulted in patterns of clustered structures which appeared as completely or partially whorled filamentous material (Fig. 7 and 8). Maximum degeneration was observed at day 10 (Fig. 3). Connective tissue with characteristically banded collagen fibrils was frequently seen. Acinar patterns were rarely visible. Dilated ducts were frequently observed containing cuboidal cells. The duct cell was almost filled with a irregular deeply indented nucleus (Fig. 5). Many microvilli projected into the duct lumen (Fig. 5). Duct cell proliferation and suggestions of migration into neighboring tissues were observed. The mitochondria of degenerating acinar cells occasionally contained small granules (Fig. 10). Small oval bodies with whorled lamellae were also seen (Fig. 10).

Regenerating cells were first observed in the 12 day sample. This was two days after the cessation of ethionine treatment. These cells were more frequently seen at day 18 (eight days after cessation of ethionine), scattered among areas of persistant degeneration (Fig. 4). The well formed nucleus contained characteristic coarse and fine granulation and occasionally a large nucleolus. The cytoplasm exhibited clusters of highly osmiophilic lipid droplets. Almost the entire volume of the cell was packed with ergastoplasm. Interspersed between the ergastoplasmic lamellae were large mitochondria. Zymogen granules were conspicuously absent in most of these 18 day ergastoplasm-filled regenerating cells (Fig. 4). These regenerated cells were observed as single isolated units with only an occasional suggestion of acinus formation. At day 28 (18 days after cessation of ethionine), many more regenerated cells were present. The nuclei were well developed and nucleoli were more frequently encountered appearing as large dense granular aggregates.

Fig. 6. Degenerating rat pancreas: 5 day. Two swollen mitochondria with short cristae and a transverse membrane are shown. 22000 ×

Fig. 7. Degenerating salamander pancreas: *A. tigrinum* received 5 mg/d. ethionine for 10 days. Fused osmiophilic material appears at *D. A*. Whorled lamellar structures are present, (*W. L.*). Degenerating granules appear at *D. Z*. The one in the upper right contains an electron lucent center and appears to be undergoing a whorled filamentous formation similar to that seen in the degenerating rat pancreas of Fig. 8. 5500 ×

Fig. 8. Degenerating rat pancreas: 5 day. A large osmiophilic, filamentous structure with electron lucent centers, (arrows), is seen. 34500 ×

Fig. 9. Degenerating rat pancreas: 5 day. A cluster of degenerating zymogen granules, (*D. Z.*) have clumped together and the clump appears to be enveloped by ethionine disturbed ergastoplasm, (*ER*). A few zymogen granules are freely scattered about (*Z*). 11000 ×

Fig. 10. Degenerating rat pancreas: 10 day. Small dense structures are seen within mitochondria, (arrow). An occasional lamellar containing oval body is seen at *DR*. 34000 ×

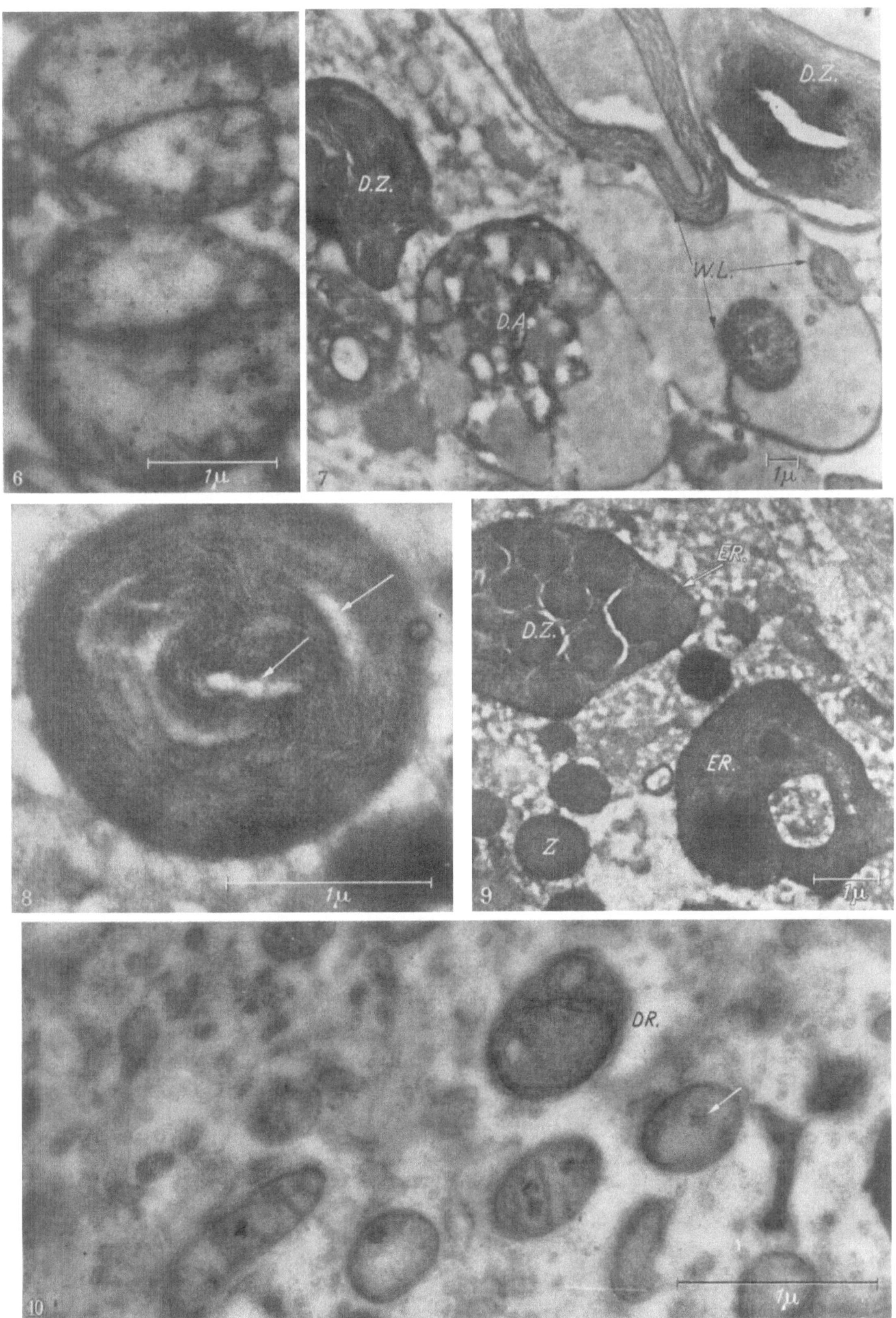

Fig. 6—10 (Legends see p. 376)

Acinar formation was common with the apex of the acinar cells exhibiting characteristic microvilli formation. These cells contained heavy perinuclear ergastoplasm and many well formed mature zymogen granules at their apices (Fig. 5).

References

1. Farber, E., D. Koch-Weser and H. Popper: Endocrinology **48**, 205 (1950).
2. — and H. Popper: Proc. Soc. exp. Biol. (N. Y.) **74**, 838 (1950).
3. Fitzgerald, P. J.: Unpublished Data.
4. — and M. Alvizouri: Nature (Lond.) **170**, 929 (1952).
5. Flax, M. H., and M. H. Himes: Physiol. Zool. **25**, 297 (1952).
6. Garnier, Ch.: J. Anat. et Physiol. **36**, 22. Paris (1900).
7. Kinney, T. D., N. Kaufman and J. V. Klavins: A. M. A. Arch. Path. **60**, 639 (1955).
8. Koch-Weser, D., E. Farber and H. Popper: A. M. A. Arch. Path. **51**, 498 (1951).
9. — and H. Popper: Proc. Soc. exp. Biol. (N. Y.) **79**, 34 (1952).
10. Palade, G. E.: J. Exp. Med. **95**, 285 (1952).
11. — J. biophys. biochem. Cytol. **1**, 59 (1955).
12. — J. biophys. biochem. Cytol. **2**, 417 (1956).
13. — and P. Siekevitz: J. biophys. biochem. Cytol. **2**, 671 (1956).
14. Robertson, J. S.: Aust. J. exp. Biol. med. Sci. **32**, 229 (1954).
15. Siekevitz, P. and G. E. Palade: J. biophys. biochem. Cytol. 4, 203 (1958).
16. Sjöstrand, F. S., and V. Hanzon: Exp. Cell. Res. **7**, 415 (1954).
17. — — Exp. Cell Res. **7**, 393 (1954).
18. Wachstein, M., and E. Meisel: Proc. Soc. exp. Biol. (N. Y.) **77**, 569 (1951).
19. — — Lab. Invest. **2**, 253 (1953).
20. Weiss, J. M.: J. exp. Med. **98**, 607 (1953).
21. — J. exp. Med. **101**, 213 (1955).

6. Endokrine Drüsen

The ultrastructure of the stimulated mouse thyroid gland

R. Ekholm

Department of Anatomy, University of Gothenburg (Sweden)

The ultrastucture of the thyroid gland in white mice maturing under normal laboratory conditions has been described previously (*1, 2, 3*). Certain cytoplasmic changes observed in cells from stimulated thyroid glands will now be presented. These findings are in many respects compatible with previous workers' (*4,5*) observations which, however, owing to recent refinements in technique now require further penetration.

Thyroid glands from three groups of adult, male white mice were studied. The animals in one group were killed after receiving 14 daily thiouracil doses of 5 mg, those in another one hour after an injection of 0.5 USP units of thyrotrophin, whilst those in the last group served as controls. All the thyroid glands were fixed in 1% osmium tetroxide solution, embedded in methacrylate, sectioned on an ultramicrotome, and examined under an RCA electron microscope, Model EMU 3 b.

The paramount effect of thiouracil stimulation is height increase of the thyroid cells. Usually having a roughly cubical shape in the controls, they resemble tall cylinders after thiouracil. The apical surface of the cell, which faces the colloid, is markedly convex (Fig. 1). Owing to the tortuous course of the plasma membrane where it borders on the pericapillary space, the basal zone of the normal thyroid cell is divided into a varying number of irregular lobes. The configuration of this basal system of membranes, Sjöstrand's β-cytomembranes, varies widely from

---

Fig. 1. Survey picture of the thyroid from a mouse treated with thiouracil for 14 days. The cells are very tall with domeshaped apical surfaces. The α-cytomembranes are richly developed and give a lace-work appearance to the cytoplasm

Fig. 2. The α-cytomembranes in a thyroid cell from a thiouracil-treated mouse showing marked distention of the spaces bounded by the particle-free sides of the membranes

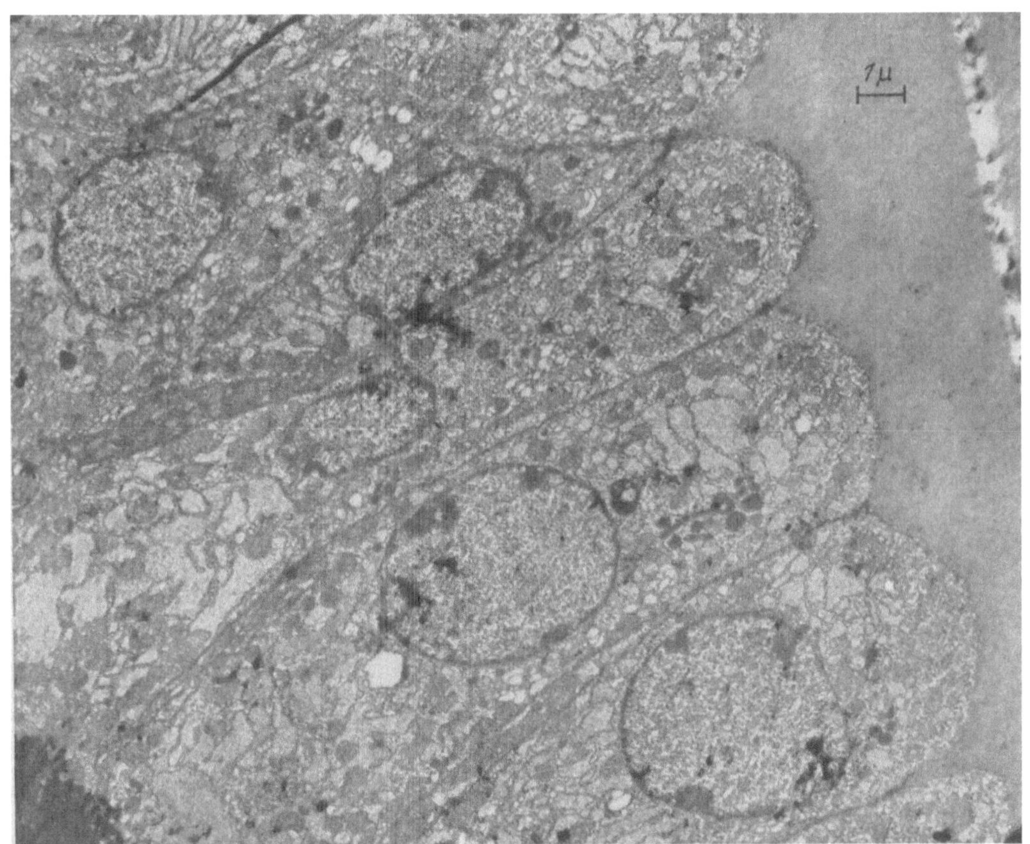

Fig. 1

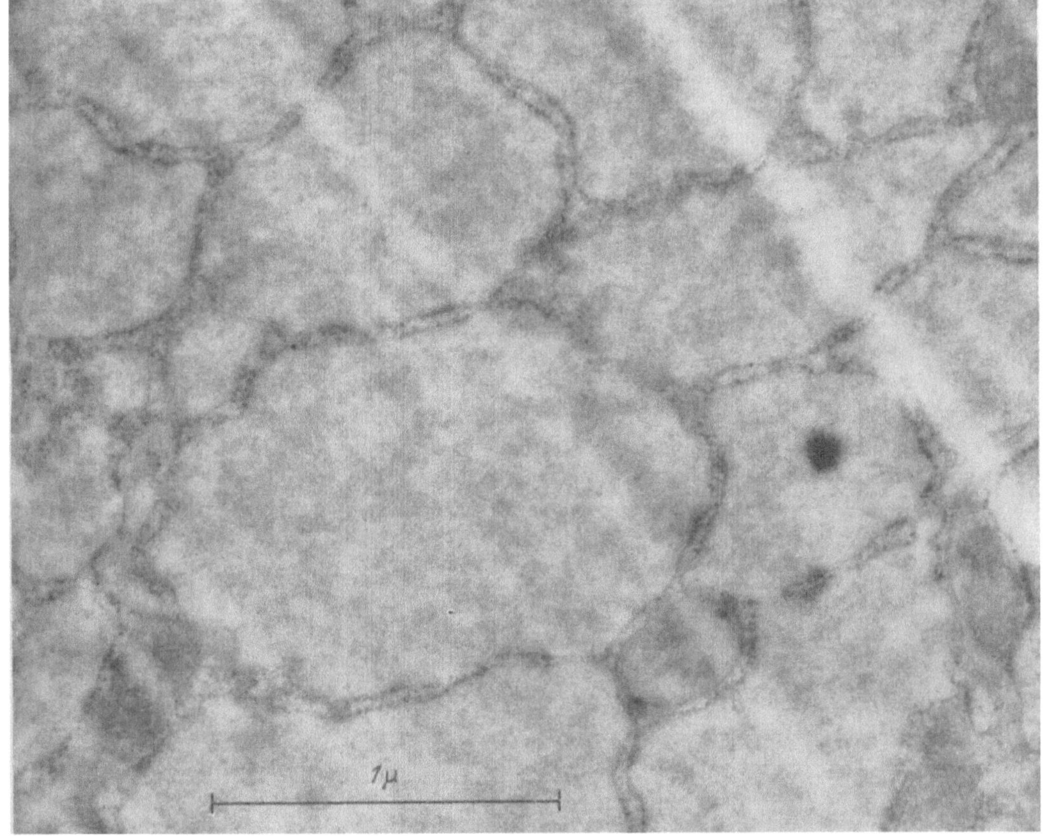

Fig. 2

(Legends Fig. 1 and 2 see p. 378)

one cell to another and also from one part to another of the same cell. After stimulation with thiouracil these membranes seem to become considerably more convoluted, implying that the area of the basal surface must increase. Thiouracil stimulation gives rise to no immediately apparent change in shape or structure of the thyroid mitochondria. But comparative measurements reveal that the mean transverse diameter of mitochondria from stimulated glands is about 15% longer than in thyroid cells from the controls. The difference is statistically significant. Although its characteristic features, viz. short membranes, vacuoles and small vesicles, regularly

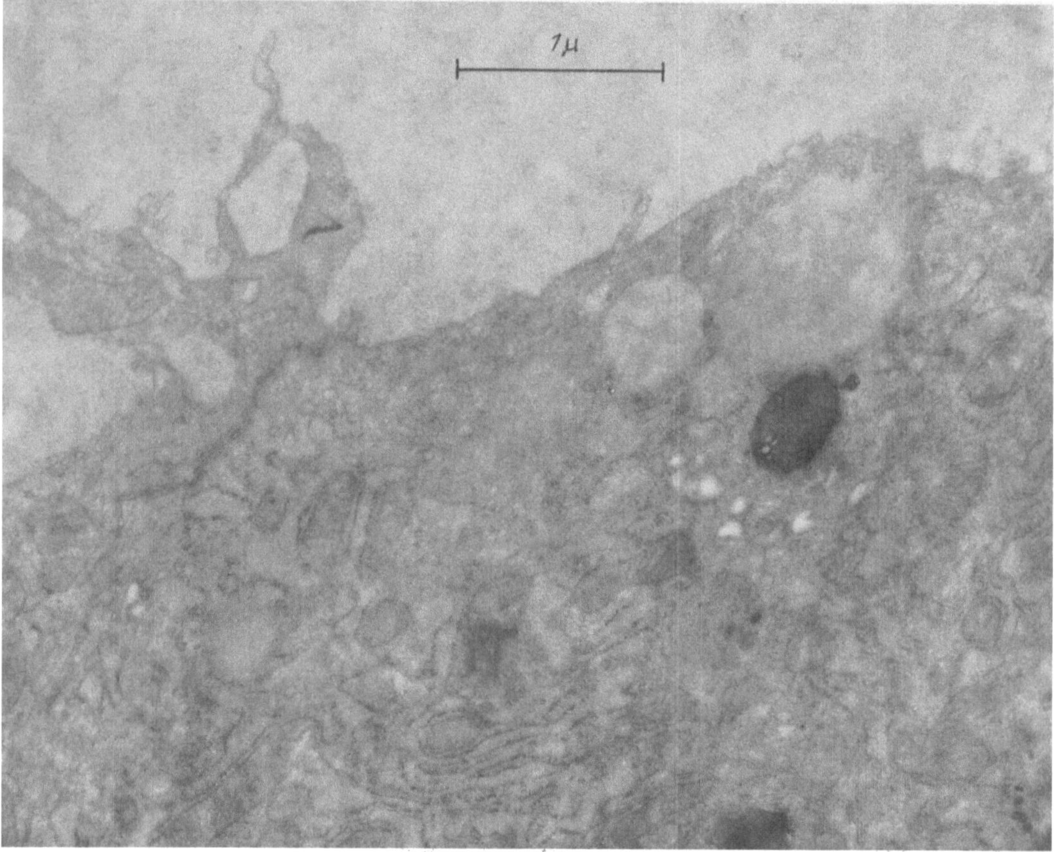

Fig. 3. Thyroid from a mouse given thyrotropin one hour previously. In the apical zone a couple of big granules, one of them enclosed in a bleb on the apical surface

can be distinguished, the Golgi apparatus in the normal thyroid cell is by no means large. After thiouracil stimulation the Golgi apparatus has become more prominent because its over-all dimensions are greater and the vacuoles are more conspicuous. Normally α-cytomembranes (the endoplasmic reticulum) occupy a major portion of the thyroid cell, predominantly the basal and middle zones. These membranes are generally arranged in pairs with the particle-bearing sides facing one another and closely spaced. The interspaces between the smooth sides have a finely and evenly granulated appearance. After thiouracil stimulation, the entire system of α-cytomembranes has expanded greatly and occupies not only the basal and middle zones but also the apical portions of the thyroid cell; in addition the spaces between the smooth sides of the membranes have become much broader (Fig. 2).

Short-term thyrotrophin stimulation predominantly affects the granules of the thyroid cell. The apical portion of the untreated cell generally contains large numbers of small, rounded granules of uniform and moderately opaque appearance. Occasional cells also contain one or more larger granules of round or slightly irregular shape. After thyrotrophin administration the thyroid cells contain several rounded granules having a diameter often exceeding one micron (Fig. 3).

These giant granules are provided with a distinct surface membrane surrounding a fine-grained area of varying density. Usually situated in the apical half of the cell, they may occasionally fill practically the entire cell. The apical surface of many cells is distended by these large granules and seems to bulge into the colloid. A varying number of microvilli take origin on the apical surface of the normal thyroid cell. Both the number and size of these microvilli are reduced by the brief thyrotrophin-stimulation.

Thus, practically all the thyroid cytoplasmic elements undergo quantitative and/or qualitative morphological changes after application of the stimuli used here.

References

1. Ekholm, R., and F. S. Sjöstrand: Proc. Stockholm Conf. Electron Microscopy, 1956, p. 171, Stockholm: Almqvist & Wicksell 1957.
2. — Z. Zellforsch. **46**, 139 (1957).
3. — and F. S. Sjöstrand: J. Ultrastructure Res. **1**, 178 (1957).
4. Braunsteiner, H., K. Fellinger and F. Pakesch: Endocrinology **53**, 123 (1953).
5. Dempsey, E., and R. Peterson: Endocrinology **56**, 46 (1956).

Observations on the fine particulate components in certain membrane-bound bodies of the rat thyroid cell

J. D. Lever

Department of Anatomy, University of Cambridge (England)

Dempsey and Peterson (*1*) and lately Ekholm and Sjöstrand (*2*) have given good preliminary accounts of the electron microscopy of the thyroid cell. However, apart from its readily identifiable cytoplasmic components this cell contains two somewhat ambiguous inclusion bodies. The first of these bodies in the rat thyroid is usually rounded membrane-bound and has a size range of 0.5 to 3.0 μ. Such inclusions contain a moderately dense material of comparable appearance to vesicular colloid (*1*) and it is noteworthy that the so-called "colloid" bodies observed by the light microscope have a comparable size range: in a current series on the rat thyroid, intracellular colloid bodies, staining positively with the periodic acid Schiff reagent, measured from $\frac{1}{2}$ to 3 μ in diameter. Work, as yet unpublished, on the normal, activated and inactive thyroid (Lever, in preparation) supports the assertion (*1*) that these 0.5 to 3.0 μ-sized bodies seen in electron micrographs of the rat thyroid are in fact the intracellular colloid bodies and that they likely originate from the endoplasmic reticulum. Space does not permit their illustration in this paper and reference should be made to the figures in earlier works (*1* and *2*).

Second Type bodies. In addition to these presumptive colloid bodies (described above) there are other somewhat ambiguous membrane-bound inclusions to be found in electron micrographs of the thyroid (Fig. 2): the term *Second Type bodies* will be used in describing these inclusions, which although they are not to be found in every cell cross-section are nevertheless widely encountered throughout any representative thyroid field. They are round to oval in shape, as are the colloid bodies, but have a different size range; from 0.4 to 1.25 μ with an average cross-sectional diameter of 0.7 μ: and are for the most part situated close by the membranes of the endoplasmic reticulum. Being moderately electron dense the gross structure of these Second Type bodies resembles that of the so-called "siderosomes" (*3*) and the "lysosomes" of the normal liver (*4*).

Ekholm and Sjöstrand (*2*) working on mouse thyroid, described inclusions similar to these Second Type bodies and observed aggregates of very dense 70 Å particles within them: a finding which is endorsed in this study of the rat thyroid (Fig. 3). These extremely dense particles in the rat gland have a size range of between 50 to 80 Å but most are about 80 Å. They occur in high concentration within the Second Type bodies but may also be found and in lower concentration lying free in the cytoplasm, and in particular near to the membranes of the endoplasmic reticulum.

There is however no possibility of confusing these comparatively regular, very dense, 80 Å particles with the less dense larger ($\pm$ 200 Å) and irregular (ribonucleic acid) granules of Palade (Fig. 2).

Within the so-called Second Type bodies the 80 Å particles may either be aggregated in large irregular collections such that their individuality is obscured, or they may occur singly, in pairs, or triplets: in this more discrete distribution individual particles may often appear to be surrounded by an electron lucid halo (Fig. 3). Particle aggregates have not been observed at large in

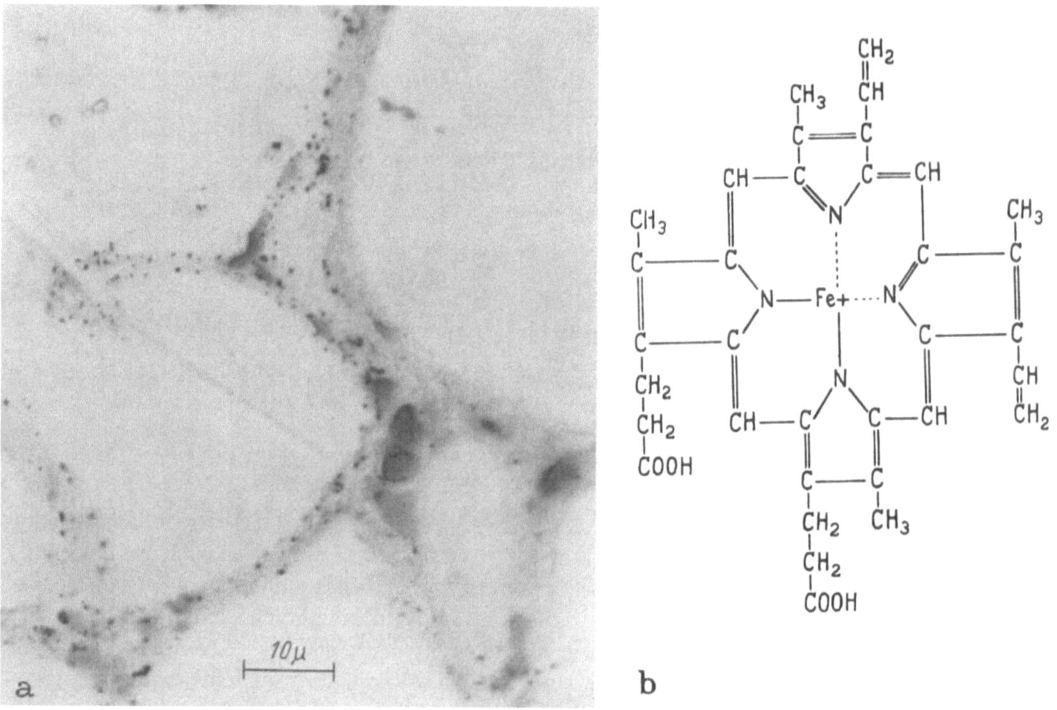

Fig. 1a. Normal rat thyroid gland stained by the α-naphthol technique to demonstrate peroxidase activity: this is indicated within the parenchymal cytoplasm by a general colouration but is most intense within rounded (0.5—1.25 μ) bodies seen in many of the thyroid cells: 725 $\times$

Fig. 1b. The structural formula of peroxidase: the protoporphyrin ring is approximately 10 Å wide

the cytoplasm where numbers of single particles are usually found. In the higher resolution micrographs a further internal form may be observed within many of the individual 80 Å particles. In some instances these may each appear to consist of a ring of smaller, 10—15 Å, sub-particles (called hereafter particulates) surrounding a central and relatively lucid area (Fig. 3), while others may contain a dense central opacity of some 20—30 Å surrounded by a less dense and poorly defined annular zone. In further instances, 10—15 Å particulates can be observed simply

Fig. 2. Electron micrograph of part of a normal rat thyroid cell. Round or oval membrane-bound bodies (A) of moderate overall electron density and ranging from 0.4—1.25 μ in size are found adjacent to the membranes of the endoplasmic reticulum. These large bodies contain numbers of very dense particles of some 80 Å in size which particles are also observed free in the cytoplasm but in lower concentration. The larger more irregular Palade granules associated with the membranes of the endoplasmic reticulum (E) cannot be confused with these particles: 35000 $\times$

Fig. 3. Part of a 0.4—1.25 μ membrane-bound body (see Fig. 2) at high magnification. The contained particles are mostly $\pm$ 80 Å in size and many are surrounded by an electron-lucid halo. A further internal structure can be resolved in many of these particles ($\rightarrow$) which are seen to consist of particulates estimated at 10—15 Å in size. This substructure is mostly without arrangement but occasionally particulates are orientated in annular fashion around a central more lucid area. 360000 $\times$

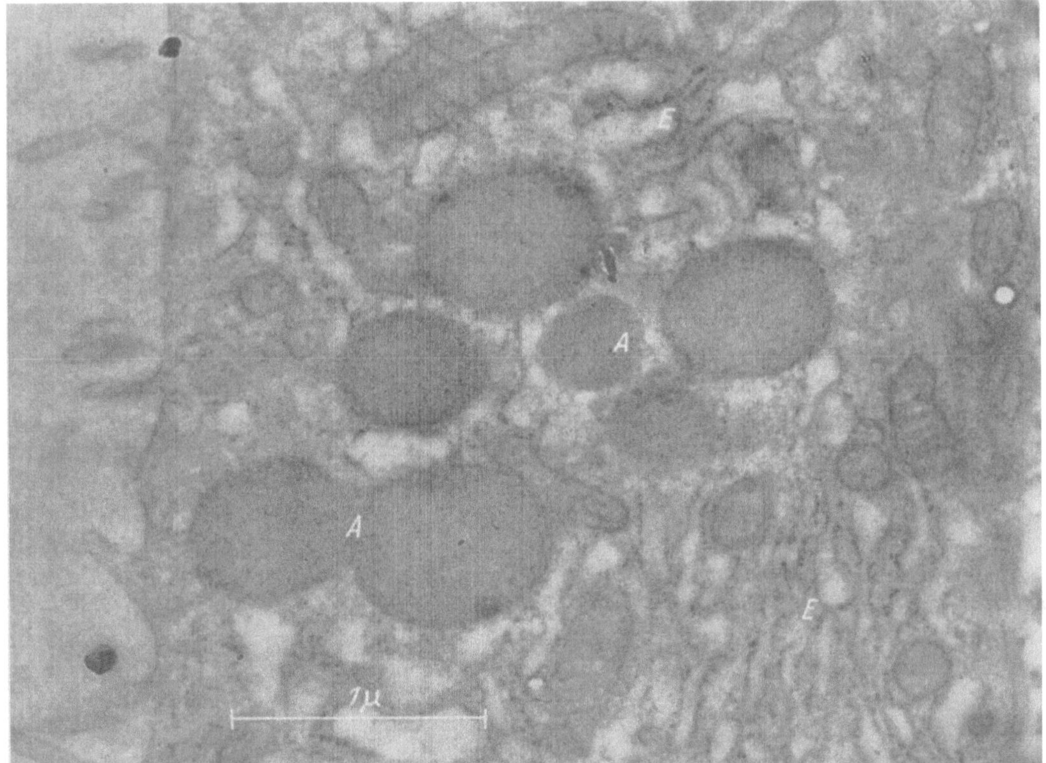

Fig. 2

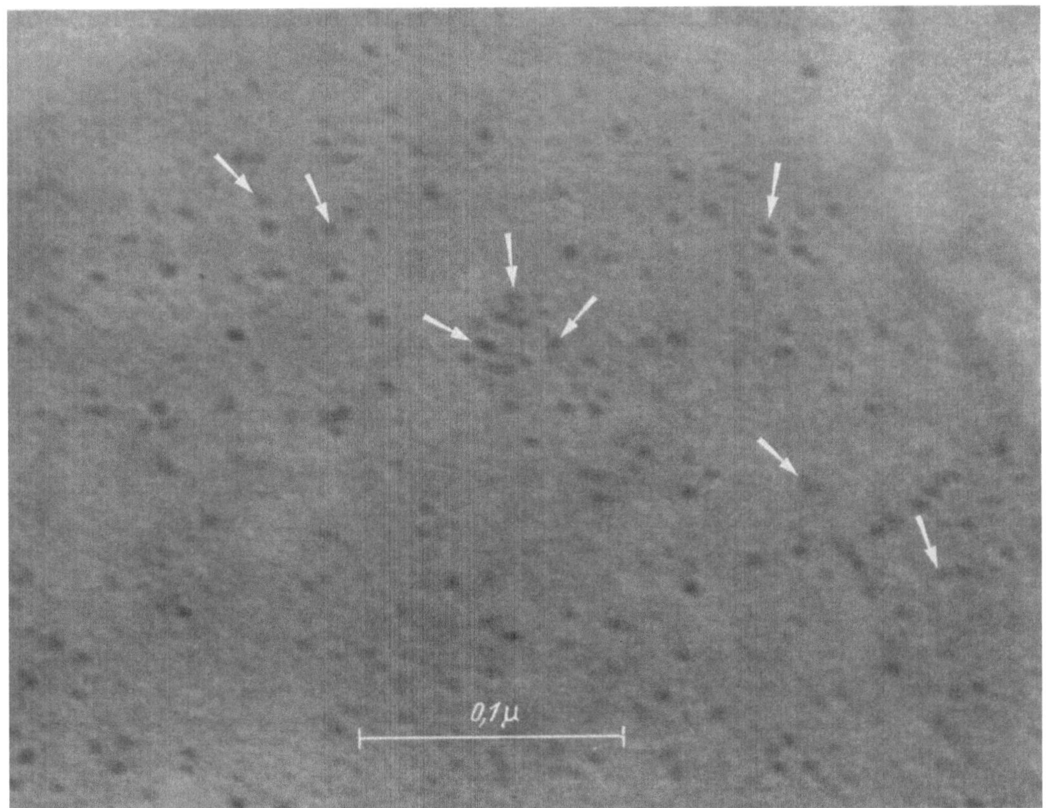

Fig. 3

(Legends Fig. 2 and 3 see p. 382)

as component structures of the larger 80 Å particles but often there is no obvious spatial arrangement of these particulates (Fig. 3). However, it must be admitted that in a large number of the 80 Å particles no obvious internal structure can be discerned. In spite of a superficial resemblance to the ferritin micelles described by Farrant (5) and Richter (3) the 80 Å particles in the thyroid cell do not contain the regular substructure of four 27 Å units which is regarded by these workers as characteristic of ferritin.

Pending further investigation, the nature of these 80 Å particles in the thyroid is at present undecided. Perhaps of relevance to the question is Dempsey's (6) finding that the thyroid contains a peroxidase in some quantity: this enzyme is involved in the processes of iodine-protein linkage within the gland. He and others have demonstrated by the benzidine and α-naphthol reactions that specific colourations for peroxidase, while widerspread but not intense in the thyroid cytoplasm at large, are very intense in certain rounded intracellular granules or bodies. In the present series of glands stained by the α-naphthol technique these bodies vary in size from 0.5 to 1.25 μ (Fig. 1 a), a range corresponding to that of the Second Type membrane-bound bodies in parallel electron micrographs. Cytochemical staining for peroxidase cannot be elicited if the tissue is preheated or if minute quantities of thiourea are included with the reagent. The enzyme peroxidase is a metallo-porphyrin containing a single iron atom in the centre of a protoporphyrin ring of some 10 Å in diameter (Fig. 1 b). It might be postulated that this molecule would produce an electron shadow slightly larger than 10 Å in size by virtue only of its high content and concentration of carbon atoms: since from their central location within the protoporphyrin ring it would seem impossible for individual iron atoms of neighbouring peroxidase molecules to be close enough together to produce electron shadows. If this postulate is correct then the dense 80 Å particles in the thyroid cell may represent concentrations of peroxidase, the molecules of which may occasionally be visualised as 10—15 Å particulates. The contention is strengthened by the fact that both the cytochemical colouration for peroxidase as seen by light microscopy, and the concentration of the 80 Å particles in electron micrographs are each maximal within intracellular bodies having an identical size range: moreover, the particle concentration and the staining for this enzyme are both minimal but unquestionably present in the cytoplasm at large.

I am indebted to Professor J. D. Boyd for his interest in this work which was in part supported by a Royal Society Grant. My thanks are also due to Dr. V. E. Cosslett and Mr. R. W. Horne for electron microscope facilities in the Cavendish laboratory. Mr. R. Parker has given valuable technical assistance.

References

1. Dempsey, E. W., and R. R. Peterson: Endocrinology **56**, 46 (1955).
2. Ekholm, R., and F. S. Sjöstrand: J. Ultrastructure Res. **1**, 178 (1957).
3. Richter, G. W.: J. exp. Med. **106**, 203 (1957).
4. Novikoff, A. B., H. Beaufay and C. deDuve: J. biophys. biochem. Cytol. **2**, No. 4, suppl., 179 (1956).
5. Farrant, J. L.: Biochim. biophys. Acta **13**, 569 (1954).
6. Dempsey, E. W.: Endocrinology **34**, 27 (1944).

Ultrastructure of the adrenal cortex in the mouse

T. Zelander

Department of Anatomy, University of Gothenburg (Sweden)

At this Department the adrenal cortex of the mouse has been studied in the electron microscope during recent years. The initial results of the investigation have been published in a preliminary report (11) dealing with the arrangement of the glandular tissue and the structure of the epithelial cells in the glomerular and fascicular zones. Our findings are on the whole similar to those published previously (1, 3, 5, 6, 7, 8, 9).

The present paper will deal with the variable appearance of the mitochondria, lipid granules and ground cytoplasm in the different adrenocortical zones and also with the distribution in these zones of Golgi apparatus, so-called globule and yellow pigment.

The adrenals were obtained from 60 male white mice between 3 and 5 months of age which had matured under normal laboratory conditions. The animals were decapitated and isotonic 1% osmium tetroxide solution buffered to p_H 7.2 was injected into the aorta. Then the adrenals were excised and put into the fixing solution where they remained until 60 min had elapsed after the decapitation. The specimens were dehydrated in ethanol, embedded in methacrylate, sectioned on the ultramicrotome, and examined in the following microscopes: RCA EMU 2 b, RCA EMU 3 b, Akashi Tronoscope TRS-50.

In the glomerular zone and outermost portion of the fascicular zone the mitochondria are oval or elongated and have internal structures of saccular or tubular type. Similar internal

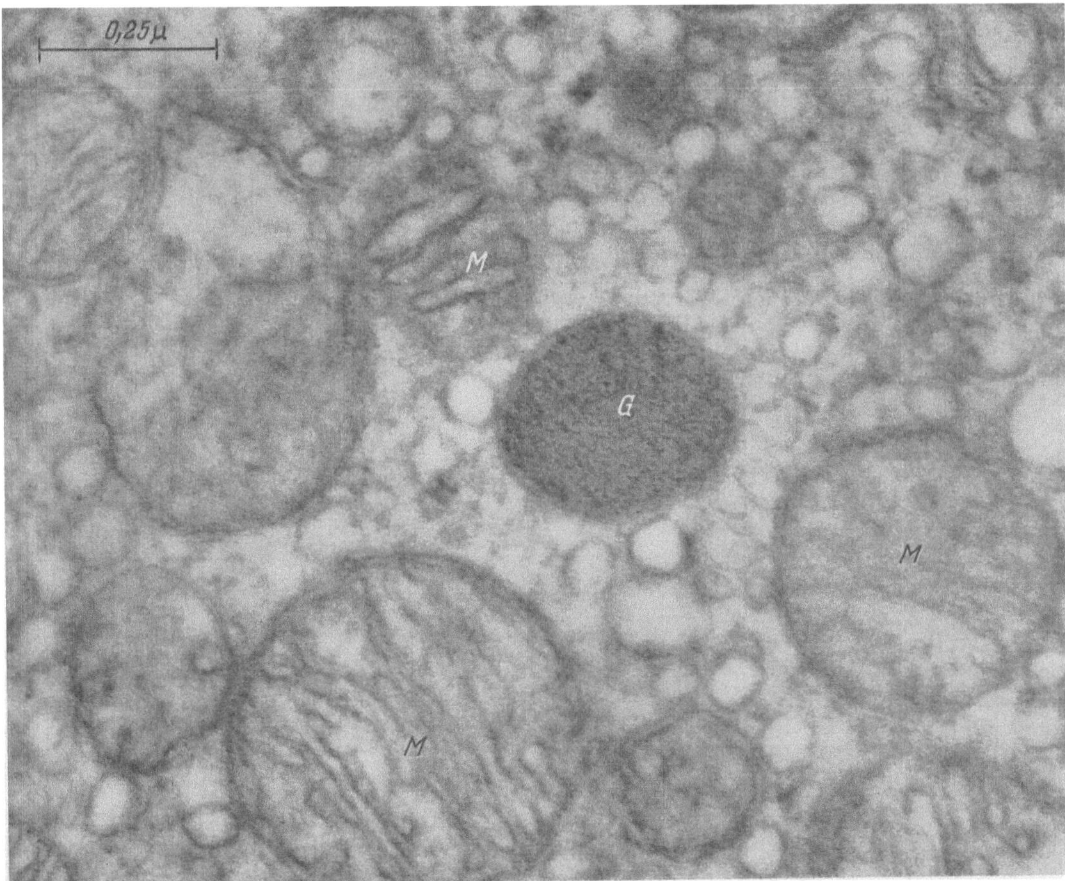

Fig. 1. Section through zona fasciculata of mouse adrenal cortex showing a globulus (G) with its double outer membrane and granular inner substance. M = mitochondria

structures are present in the more rounded mitochondria in the reticular zone. The rounded mitochondria in the innermost portion of the fascicular zone mostly exhibit a more membranous appearance, the membranes usually being disposed in a more or less parallel arrangement (Fig. 1). Occasionally the inner membranes resemble a nest of concentric circles.

Lipid granules are rather scarce in the glomerular zone; they are generally rounded and homogeneous. On the outermost portion of the fascicular zone the lipid granules begin to become increasingly irregular in shape and are often wholly or partially dissolved. The innermost portion of the fascicular zone and the reticular zone are quite poor in lipid granules, and those that do occur are often irregularly polygonal, in most cases homogeneous. Whereas the ground cytoplasm has a finely granular appearance in the glomerular zone and outermost portion of the fascicular zone, its appearance steadily becomes more vesicular in a central direction. Golgi apparatus were observed mainly in the fascicular zone, most frequently in its innermost portion. In most cases the Golgi apparatus is composed of 4 to 8 pairs of membranes, in the terminal parts of which membrane system the majority of the Golgi vacuoles are situated. Golgi vesicles are rare. Because most of the Golgi apparatus are encountered in those parts of the cortex where the ground cytoplasm is vesicular, it is no easy matter to ascertain which vesicles are associated with the ground cytoplasm and which belong to the Golgi apparatus.

In the fascicular and reticular zones so-called globules occur. These are round or oval bodies with smooth outer contours, and a double edged surface membrane surrounding a markedly osmiophilic, inhomogeneous substance. The inhomogeneous inner structure is composed partly of intensely osmiophilic and homogeneous patches and partly of granulated portions whose granules have a diameter of about 80 Å. U. and seem extremely dense against a moderately osmiophilic

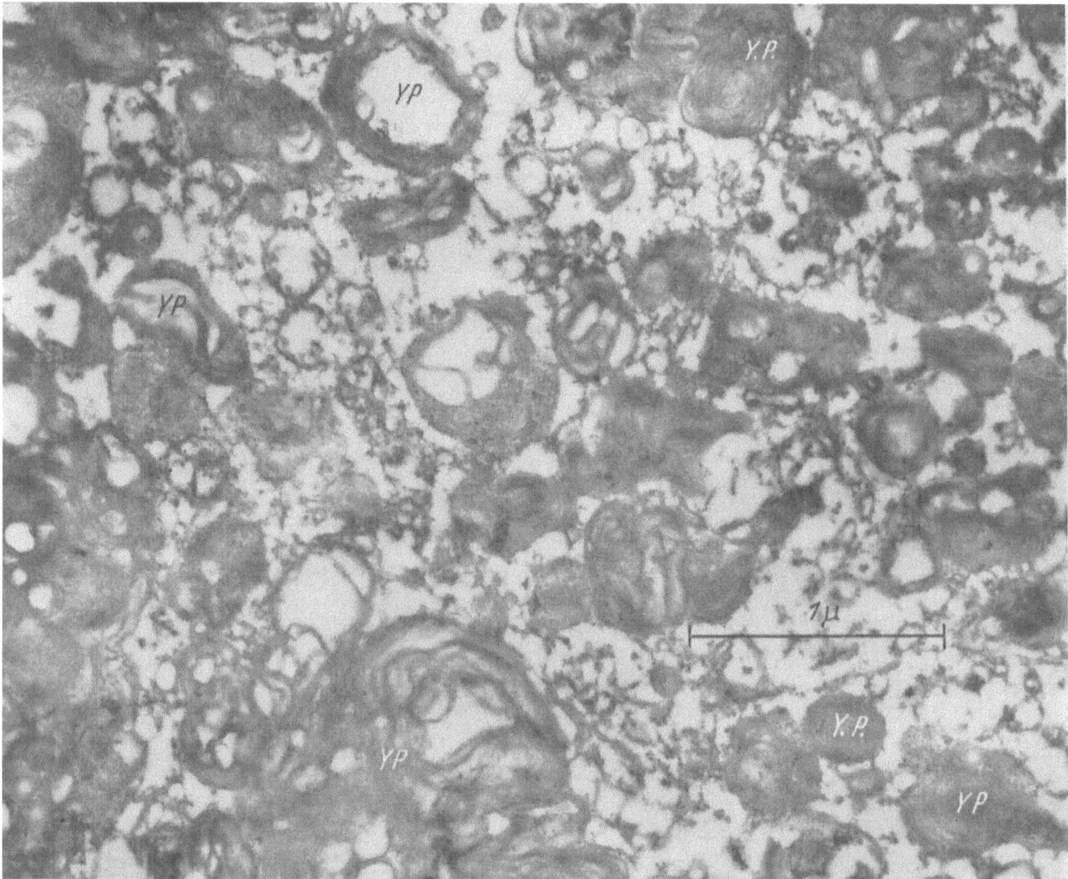

Fig. 2. Survey picture of yellow pigment bodies (*Y. P.*) in a cell of zona reticularis

background Fig. 1. The ratio between the areas occupied by homogeneous and by granular patterns varies (greatly). Occasional globules have been found to lack the homogeneous component, and a few others have been completely filled up with the homogeneous component. The relative areas occupied by the two structures is obviously a function of the plane of the section, and so the variability may be less extreme than it seems. These globules are encountered not only in the hormone-secreting cells, even though that is where they are most frequent, but also in the pericapillary cells. The globules in the latter cells tend to be somewhat larger and more elongated.

In the reticular zone and adjacent cell layers of the fascicular zone, one encounters still another cellular component, viz. yellow pigment. Investigations aided by the use of fluorescence microscopy (*4, 10*) have demonstrated that yellow pigment, or lipofucsin, normally occurs in the reticular zone of the mouse. Yellow pigment isolated from grey matter in human spinal medulla has a characteristic ultrastructure which is the subject of another address to this forum (2).

Appreciable amounts of yellow pigment are seen predominantly in those cells in the reticular zone of the adrenal cortex which seem to be imperfectly preserved by osmium tetroxid fixation and contain few and often deformed mitochondria, few if any lipid granules and a ground cytoplasm of low density and markedly vesicular structure. The size of the pigment bodies varies greatly, from a maximum of 4 microns downwards. They have a polycystic appearance with a markedly osmiophilic wall which contrasts sharply against the very osmiophobic content (Fig. 2). On the borderline between osmiophilic and osmiophobic material there is a large number of very dense granules having a diameter of about 45 Å. The cystic wall is itself a composite structure

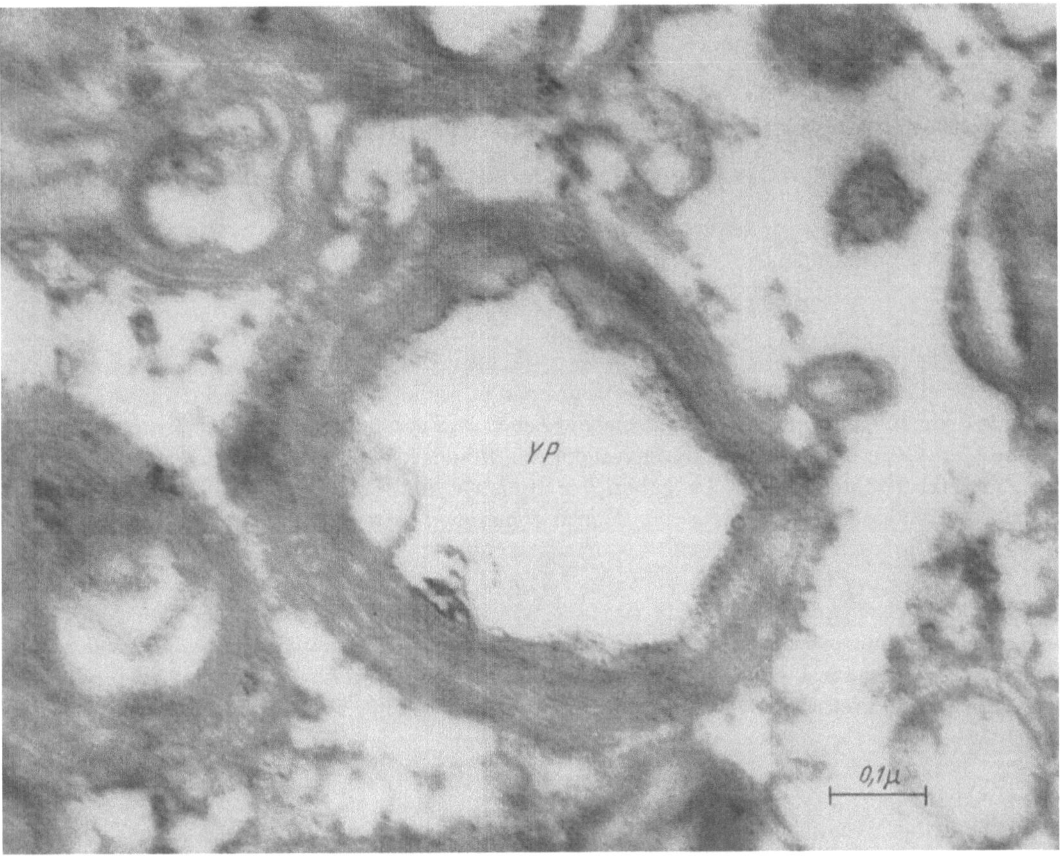

Fig. 3. Yellow pigment body ($Y\ P$) with granules on the border line of the light region and pairs of very dense layers in the osmiophilic zone

of dark and light layers resembling the appearances noted in yellow pigment from grey spinal medulla in humans, although the great regularity characterizing the striped zone in those pigment bodies was replaced by a somewhat different arrangement in the yellow pigment bodies of interest here. In the latter case the layers are usually disposed in pairs of two dark layers with an intermediate light layer. Whilest the centre-to-centre distance between two dark layers of a pair remains sensibly constant at about 60 Å, the distance between the pairs varied greatly from one pigment body to another and from one part to another of the same pigment body (Fig. 3).

As the foregoing shows, the normal morphology of the cortical cells varies markedly, implying that a large series of sections must be studied. In view of the restricted field of vision in an electron microscope, the need for a large series becomes still more pressing. In our case the sections were sufficiently numerous for an analysis which indicated that, despite their relative rarity, the two structures mentioned last are normally occurring phenomena.

25*

References

1. Belt, D. W.: Anat. Rec. **124,** 258 (1956).
2. Björkerud, S., and T. Zelander: This volume p. 437.
3. Braunsteiner, H., K. Fellinger and F. Pakesch: Wien. Z. inn. Med. **36,** 281 (1955).
4. Hamperl, H.: Virchows Arch. path. Anat. **292,** 1 (1934).
5. Lever, J. D.: Amer. J. Anat. **97,** 409 (1955).
6. — Endocrinology **58,** 163 (1956).
7. — J. biophys. biochem. Cytol. **2,** Suppl., 293 (1956).
8. — J. biophys. biochem. Cytol. **2,** Suppl., 313 (1956).
9. Luft, J., and O. Hechter: J. biophys. biochem. Cytol. **3,** 615 (1957).
10. Sjöstrand, F.: Acta anat. (Basel) Suppl. I, 1 (1944).
11. Zelander, T.: Z. Zellforsch. **46,** 710 (1957).

Elektronenmikroskopische Untersuchungen der Hoden-Zwischenzellen von normalen und hypophysektomierten Ratten

G. Wilke und E. Schuchardt

Neuropathologische Abteilung des Max-Planck-Instituts für Hirnforschung, Gießen,
und Anatomisches Institut der Universität Gießen

Eine Erweiterung unserer Kenntnisse über Struktur und Funktion der Zelle ist auch im submikroskopischen Bereich zu erwarten, wenn man es in der Hand hat, durch definierte experimentelle Bedingungen den *Funktionszustand* der Zellen *im Extrem* zu *variieren.* Als Untersuchungsobjekte bieten sich hier die hypophysengesteuerten, innersekretorischen Drüsen an.

In unserer Untersuchungsreihe haben wir zunächst die Zwischenzellen, das inkretorische Element der männlichen Keimdrüse, in Angriff genommen. Die Hoden geschlechtsreifer Ratten lieferten das Untersuchungsmaterial. — Nach Hypophysektomie kommt es zu einer Rückbildung der Zwischenzellen, die sich schon lichtmikroskopisch in Größen- und Strukturveränderungen von Kern und Plasma zeigt. In diesem Zustand ist die spezifische Funktion der Zwischenzellen, die Androgenbildung, erloschen. Durch Verabreichung von Choriongonadotropin, das vorwiegend ICSH-Aktivität besitzt, können die rückgebildeten Zellen zur Entfaltung und damit zur Wiederaufnahme ihrer inkretorischen Tätigkeit gebracht werden; unter ICSH-Einfluß stellen sich die Zwischenzellen auf ihr maximales Leistungsniveau ein.

Unser Untersuchungsmaterial wurde unter folgenden experimentellen Bedingungen gewonnen: 1. vom unbehandelten Normaltier, 2. vom hypophysektomierten Tier und 3. vom hypophysektomierten Tier nach Choriongonadotropinbehandlung, außerdem wurden die Zwischenzellen 4. nach Röntgenbestrahlung beim Normaltier und 5. nach Röntgenbestrahlung beim hypophysektomierten Tier untersucht.
Die Röntgenbestrahlung hat neben der Einführung einer weiteren Bedingung den Vorteil, daß die Auffindung der Zwischenzellen im Elektronenmikroskop erleichtert wird, weil das Samenepithel bei entsprechender Dosis degeneriert und verschwindet.

Beim *hypophysektomierten Tier* liegen die Zwischenzellen in einem lockeren *Verband,* z. T. in strangartiger Anordnung, wobei die Zellindividualität deutlich gewahrt bleibt. Die Zellen besitzen vereinzelt kurze Fortsätze, es besteht aber keine wesentliche Tendenz zur Plasmasprossung, wie das bei aktiven Zellen der Fall ist. Das *intertubuläre Bindegewebe* tritt durch feine und grobe kollagene Fibrillen hervor, offenbar durch relative Verschiebung zu ungunsten der Zwischenzellen. Die Fibrocyten sind durch schlanke, mitochondrienarme, protoplasmatische Fortsätze gekennzeichnet, die die Zwischenzellen durchsetzen und umhüllen und sich dabei flächenartig überlappen. Die *Kerne* der Zwischenzellen sind stark verkleinert und unregelmäßig gestaltet. Sogar gelappte Formen werden beobachtet. Das Nucleoplasma zeichnet sich durch besondere Dichte aus. Das Chromatin liegt in groben, fleckförmigen Schollen vor und bevorzugt eine randständige Lagerung. Die Doppelung der Kernwand ist nur stellenweise sichtbar. Der Kern ist umgeben von einem schmalen, stark kondensierten *Plasmasaum,* in dem spärliche feingranuläre und kleinvesiculäre Strukturen zu erkennen sind. Die Zellgrenzen sind sehr scharf. Die kleinen *Mitochondrien* (Abb. 2) entsprechen dem Crista-Typ und weisen scharf konturierte Außen- und Innenmembranen auf. Sie liegen dicht gepackt in dem spärlichen Cytoplasma.

Im anderen Extrem kommt es nach *Choriongonadotropinbehandlung* zu einer Wiederentfaltung der Zwischenzellen, die sich neben der Größenzunahme von Kern und Plasma in strukturellen Veränderungen äußert. Im *Verband* nehmen die Zwischenzellen eine rasenartige Ausbreitung an. Sich berührende Zellen zeigen an den Kontaktstellen scharfe Begrenzung; an der freien Oberfläche kommt es jedoch zur Plasmasprossung unter Bildung stark vacuolisierter, keulenförmiger Abschnürungen. Im *intertubulären Bindegewebe* ist ein relatives Zurücktreten der kollagenen Fibrillen im Gesamtbild zu beobachten. Die Veränderungen der Fibrocyten beruhen auf der Ausbildung breiter protoplasmatischer, mitochondrienhaltiger Fortsätze. Diese Fortsätze erscheinen auf dem Schnitt relativ breit und plump. Sie zeigen wechselnde Dicke, besitzen z. T. kleine dorn- oder kolbenartige Plasmaausstülpungen und enthalten hier und dort Vacuolen. In flächenhafter Ausbreitung durchsetzen und umhüllen sie die Zwischenzellen, wobei es zur Überlappung und Spaltbildung kommt. Die großen *Zwischenzellkerne* sind von runder bis ovaler Gestalt. Sie weisen ein lockeres, feingranuläres Nucleoplasma und kleinfleckiges, randständiges Chromatin auf. Die innere Kernmembran ist scharf, die äußere Kernmembran wird stellenweise durch perinucleäre Vacuolenbildung abgehoben. Der *Zelleib* besteht im Schnitt aus einem breiten, stark vacuolisierten Plasmagürtel, der z. T. schwammartige

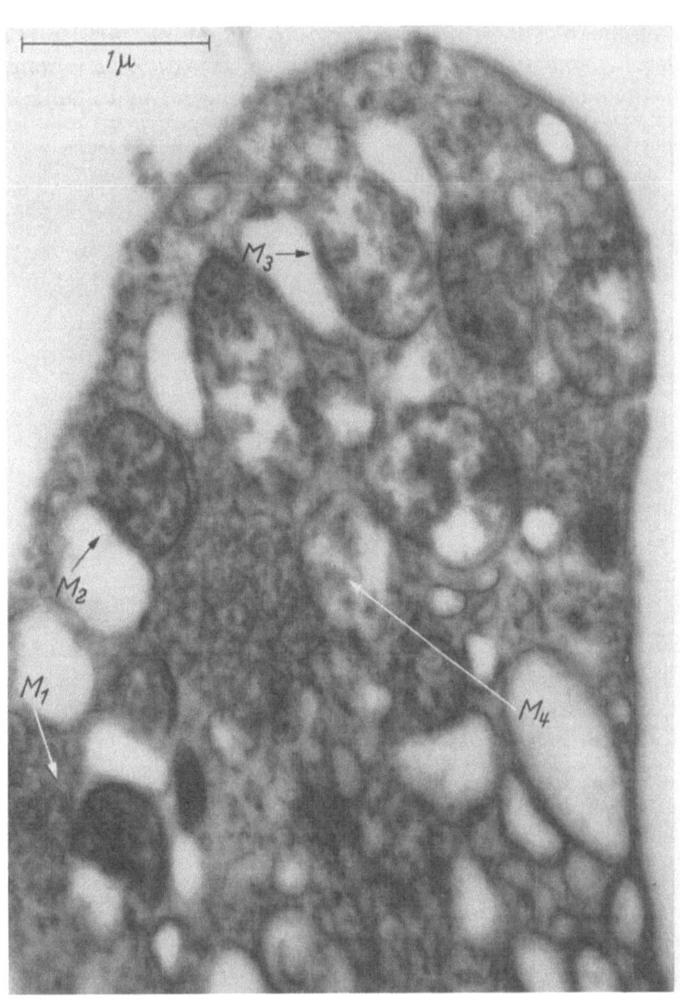

Abb. 1. *Zwischenzelle, Normaltier.* Große Mitochondrien vom Crista- und Tubulus-Typ, z. T. mit aufgelöster Außenmembran, Aufhellung der Matrix und Verbindung derselben zum umgebenden Cytoplasma, das ungleichförmig dicht ist und z. T. große cisternale Ausweitungen des Ergastoplasmas aufweist (vgl. Abb. 3), 25700:1 M_1Crista-Typ mit relativ dichter Matrix, M_2 Mitochondrien vom Tubulus-Typ, M_3 Auflösung der Außenmembran mit Aufhellung der Matrix und Verbindung zum Cytoplasma, M_4 Weitgehend leere Mitochondrien

Auflockerung zeigt und Vacuolen von verschiedener Größe und Gestalt enthält, wobei es sich offenbar um ein extrem ausgeweitetes endoplasmatisches Reticulum handelt. Das dazwischenliegende Plasma ist feingranulär und läßt kleinvesiculäre und tubuläre Strukturen erkennen. Weiterhin finden sich große elektronendichte Tropfen, wahrscheinlich Lipoideinlagerungen. Die *Mitochondrien* (Abb. 3) imponieren durch ihre Größe und besitzen eine lockere Innenstruktur mit Übergängen zu einer sehr lichten Matrix. Die äußere Wand variiert in ihrer Stärke; sie erscheint stellenweise verdünnt bis zum Schwund und mit Eröffnung zum umgebenden Cytoplasma.

Den geschilderten Verhältnissen entsprechen auch die Befunde beim *Normaltier*. Die großen und entfalteten Zwischenzellen liegen in einem lockeren, schwammartigen *Verband* ohne scharfe

Zellbegrenzung mit Neigung zur Absprengung vacuolisierter Plasmaanteile. Die umhüllenden plasmatischen Ausläufer der Fibrocyten sind mitochondrien- und vacuolenhaltig. Die Strukturbesonderheiten von *Kern und Plasma* decken sich im wesentlichen mit den Befunden beim choriongonadotropinbehandelten, hypophysektomierten Tier. Sie zeichnen sich jedoch durch eine größere Einheitlichkeit aus, die dem Hoden nach ICSH-Behandlung fehlt. Dieser läßt vielmehr noch verschiedene Phasen der Entfaltung erkennen.

Beim *röntgenbestrahlten Normaltier* vermissen wir das hochaktive Bild der Zwischenzellen. Der *Verband* stellt eine geschlossene epitheloide Zellansammlung dar, mit scharfen Zellgrenzen und nur mäßigen, finger- oder keulenförmigen Ausstülpungen. Die Zellen selbst sind mittelgroß und strukturverdichtet. Die *Kerne* sind entrundet und zeigen Neigung zur Einbuchtung. Das Nucleoplasma ist feingranulär und relativ dicht und enthält feinverteiltes, z. T. auch randständiges Chromatin. Die Kernwand ist doppelt konturiert, aber frei von perinucleärer Vacuolenbildung. Ein relativ breiter, elektronendichter, feingranulärer Plasmasaum stellt den Zelleib dar, in welchem ein gut ausgebildetes endoplasmatisches Reticulum von lamellärem, tubulärem und vesiculärem Charakter zu sehen ist. Die reichlich vorhandenen *Mitochondrien* sind von ziemlich einheitlicher Größe und typischer Struktur. Sie entsprechen dem bekannten Crista-Typ.

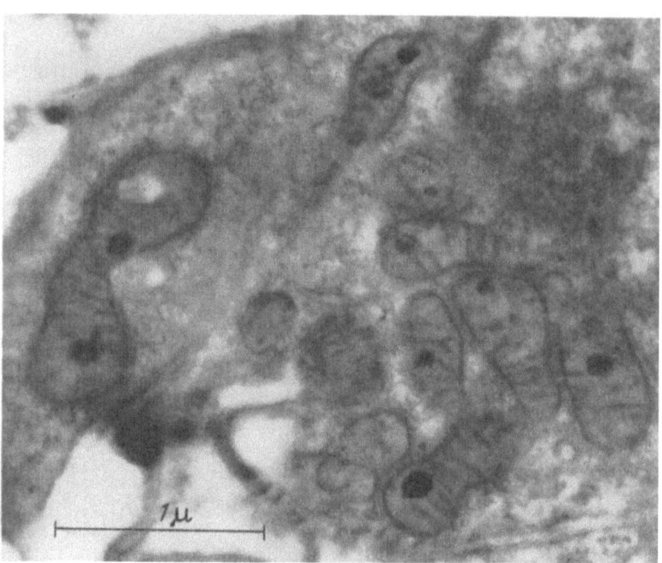

Abb. 2. *Zwischenzellen (re. Hoden), hypophysektomiertes Tier vor Choriongonadotropinbehandlung.* ♂, Hypophysektomie am 6. 11. 1956, Gew. 106,0, re. Hoden am 12. 12. 1956 operativ entfernt. Mitochondrien vom Crista-Typ mit dichter Matrix und scharf konturierten Außen- und Innenmembranen, enge Lagerung in dem spärlichen und verdichteten Cytoplasma. 28400:1

Beim hypophysektomierten und bestrahlten Tier liegt eine Atrophie der Zwischenzellen vor. Die kleinen, im Chromatin grobscholligen und strukturdichten *Kerne,* werden von schmalen, granulären, vacuolenarmen *Plasmasäumen* umgeben. Sie bilden einen lockeren, z. T. strangartigen *Verband,* der durch schlanke protoplasmatische Fortsätze der Fibrocyten umhüllt wird. Die vorhandenen *Mitochondrien* haben charakteristische Cristae mit gut kontrastierten Innen- und Außenmembranen und einer dichten Matrix. Die Mitochondrien sind aber hier beim hypophysektomiert-bestrahlten Tier *noch* kleiner als beim hypophysektomiert-unbestrahlten Tier und lassen daher auch die bei letzterem beobachtete relativ dichte Lagerung vermissen. Dieses Erscheinungsbild der Mitochondrien unter den Bedingungen der Hypophysektomie und Bestrahlung muß als Ausdruck einer stark verminderten Aktivität der Zelle angesprochen werden.

Aus der Vielzahl der bei der elektronenmikroskopischen Untersuchung der Hodenzwischenzellen gefundenen Besonderheiten, über die ausführlich an anderer Stelle berichtet wird, soll hier nur das *Verhalten der Mitochondrien* herausgestellt werden. Die funktionelle Aktivität der Zwischenzelle kommt elektronenmikroskopisch an bestimmten Veränderungen der Mitochondrien zum Ausdruck. In der ruhenden Zelle (Abb. 2) sind die Mitochondrien vom Crista-Typ und relativ klein. Bei Aufnahme der funktionellen Aktivität wird dieser Typ zunächst beibehalten, jedoch ist eine Größenzunahme zu verzeichnen, die mit einer Aufhellung der Matrix verbunden ist. Im nächsten Schritt kann es zu einer Auflösung der Matrix kommen. Damit verbunden ist das Auftreten tubulärer Innenstrukturen unter mehr oder weniger starkem Verlust der Cristae. In dieser Phase sind auch Veränderungen an den Außenmembranen festzustellen. Es treten lokale Verdünnungen auf, die zum vollständigen örtlichen Schwund führen und damit eine Kommunikationsmöglichkeit zwischen dem Inneren der Mitochondrien und dem umgebenden Cytoplasma schaffen (Abb. 1 und 3). Den Veränderungen an den Mitochondrien geht eine zunehmende Vacuolisierung

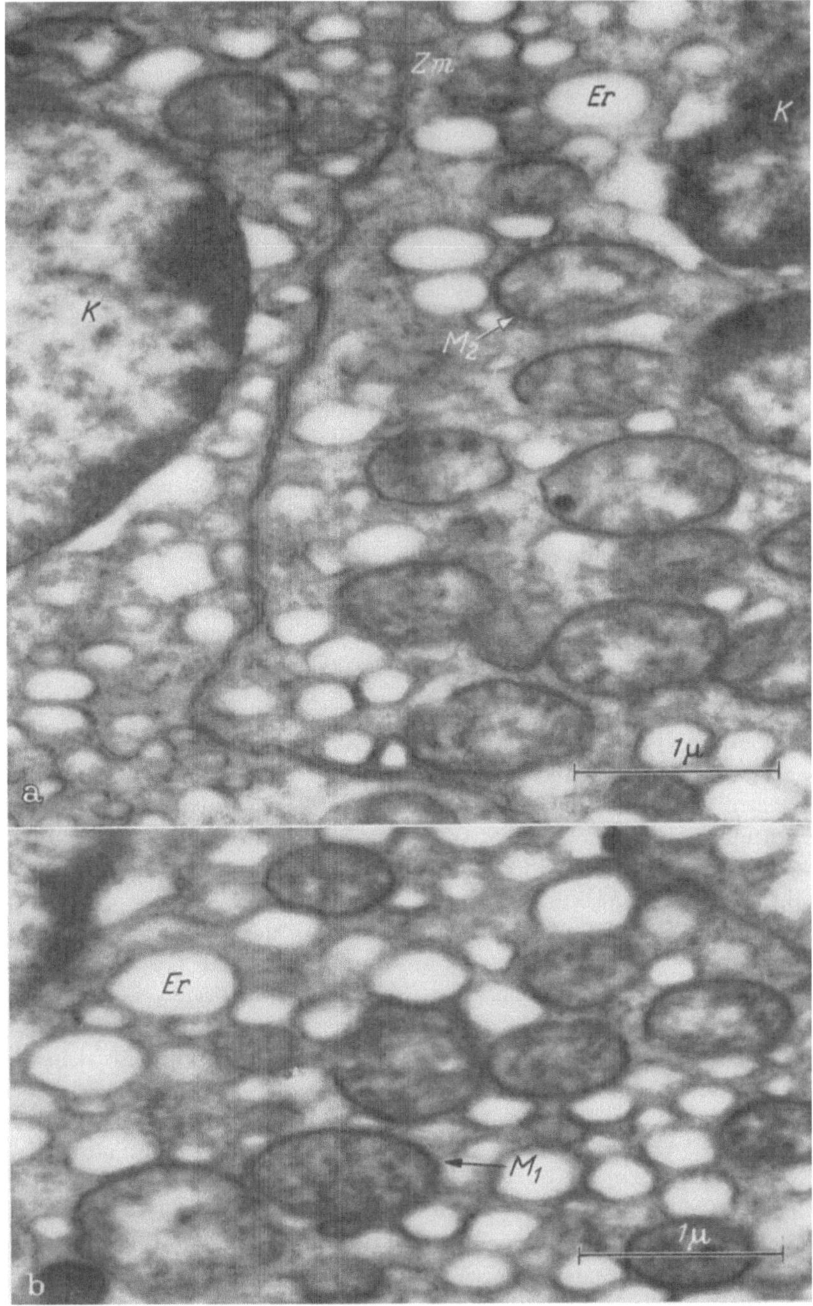

Abb. 3a u. b. *Zwischenzellen (li. Hoden), hypophysektomiertes Tier nach Choriongonadotropinbehandlung (dasselbe Tier wie Abb. 2).* R 45 39, ♂, Hypophysektomie am 6.11. 1956, Gew. 106,0, re. Hoden am 12. 12. 1956 operativ entfernt, 13.—18. 12. 1956 10 E Pregnyl pro die, † 19.12. 1956. Mitochondrien in Größe und Aussehen verändert, größer als in Abb. 2, vorwiegend vom Tubulus-Typ, Matrix aufgehellt und z. T. aufgelöst, Außenmembran der Mitochondrien stellenweise verdünnt bis zum Schwund und Verbindung der Matrix zum umgebenden Cytoplasma. Lockere Lagerung der Mitochondrien in einem großen Plasmaleib mit stark ausgeweitetem endoplasmatischem Reticulum. 28400:1. *Zm* Zellmembran, *K* Kerne der Zwischenzellen, *M*₁ Mitochondrien vom Tubulus-Typ mit quergeschnittenen Tubulusprofilen, *M*₂ Mitochondrien mit lokalem Schwund der Außenmembran, Verbindung der Matrix zum Cytoplasma. Mehr oder weniger intensive Aufhellung der mitochondrialen Matrix, *Er* Stark ausgeweitetes endoplasmatisches Reticulum des Plasmas in Gestalt einer Ausweitung der Cisternen des endoplasmatischen Reticulums parallel.

Die Untersuchungen zeigen also bezüglich der Mitochondrien, daß Cristatyp und Tubulustyp miteinander in enger funktioneller Beziehung stehen. In den inaktiven Zwischenzellen stellen die Mitochondrien vom Cristatyp zellphysiologisch das ruhende Energiepotential dar, während in den entfalteten Zellen die hochaktive Form der Mitochondrien durch den Tubulustyp vertreten wird. In dieser Struktur kommt der hohe Energiebedarf zum Ausdruck, der für die spezifische Zellleistung, die Androgensynthese, erforderlich ist.

Die Untersuchungen wurden mit Unterstützung der Deutschen Forschungsgemeinschaft ausgeführt.

7. Exkretionsorgane

L'ultrastructure des tubes de Malpighi et le problème de leur fonctionnement chez les insectes

A. Berkaloff

Laboratoire de microscopie électronique appliquée à la Biologie; Laboratoire de Botanique de l'École Normale Supérieure; Laboratoire de Physique atomique et moléculaire du Collège de France, Paris

L'ultrastructure des tubes de Malpighi a été décrite pour la première fois par Beams, Tahmisian et Devine (1) chez *Melanoplus differentialis* (Orthoptère). C'est une structure classique d'organe excréteur, très comparable à celle trouvée dans le tube contourné du rein des Vertébrés par Rhodin (2, 3). Mais l'examen des tubes de Malpighi de *Gryllus domesticus* (Orthoptère Gryllidae), ainsi d'ailleurs que d'autres insectes appartenant à des ordres très divers, nous a montré que cette structure est loin d'être stable. En effet le système de doubles-membranes périphériques ou β-cytomembranes (4) peut avoir un développement très variable. Ce système peut être représenté soit par des β-cytomembranes limitant des aires cytoplasmiques apparemment isolées du cytoplasme central soit par des lames s'enfonçant droit vers le centre cellulaire, cloisonnant alors des compartiments ouverts vers le centre. C'est alors que l'on constate le maximum de développement des β-cytomembranes. Les stades intermédiaires existent et sont caractérisés par l'existence et le développement de perforations dans les β-cytomembranes.

Par ailleurs, à l'examen in-vivo, on trouve dans les tubes peu chargés en inclusions intracellulaires (mais qui présentent alors souvent une lumière remplie de produits d'excrétion) un type cellulaire curieux, possédant un pôle chargé en inclusions, pôle dirigé vers l'extrémité aveugle du tube, tandis que le pôle dirigé vers l'extrémité ouverte est pratiquement dépourvu, de telles inclusions. A l'examen au microscope électronique ces cellules présentent un aspect caractéristique, très différent de celui des cellules que l'on pourrait appeler normales. En effet les β-cytomembranes y sont extrêmement réduites et toutes du type fermé dans la portion claire tandis que dans la portion chargée les β-cytomembranes sont semblables à celles des cellules normales quoique plus disloquées.

Pour nous, et d'après ce que nous avons pu observer dans les tubes examinés in-vivo dans le sang même de l'insecte, donc dans des conditions très proches de l'état normal, il semblerait que le développement des β-cytomembranes, donc très vraisemblablement les phénomènes d'excrétion, présentent une sorte de cycle et que l'excrétion ne soit pas continue, comme chez les Vertébrés et même chez un insecte comme *Rhodnius* (Hémiptère), mais discontinue. Ces cellules seraient des cellules qui, après avoir éliminé les produits d'excrétion figurés en commençant par l'extrémité proximale de la cellule, commenceraient à regénérer le système de β-cytomembranes qui ont plus ou moins dégénéré lors de l'élimination. L'examen in-vivo semble montrer que cette régression a lieu un peu avant l'élimination de même que le développement commence avant l'accumulation des produits figurés. L'examen des inclusions elles même semble confirmer cette hypothèse d'un fonctionnement discontinu des tubes. En effet à côté des globules de pigment intracellulaire on trouve la plupart du temps des grains tout à fait caractéristiques formés de couches concentriques épaisses. Nous avons essayé de déterminer leur nature chimique et leur évolution dans la cellule.

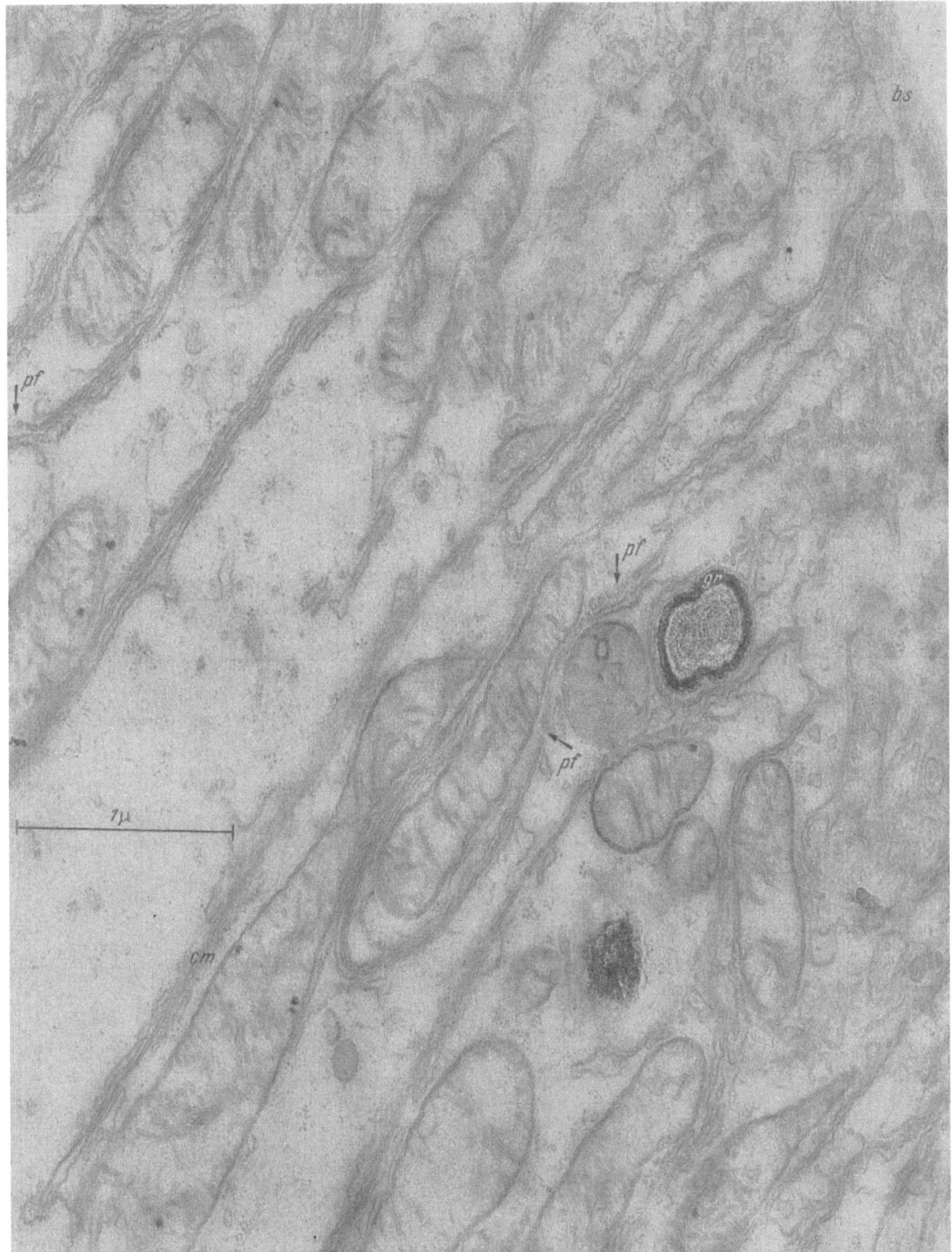

Fig. 1. Bordure externe d'un tube de Malpighi. Le système de β-cytomembranes (*cm*) est ici du type ouvert. Notez les très nombreuses perforations (*pf*) qui, dans la zone la plus externe transforment le cytoplasme en une véritable éponge. On distingue en (*gr*) un grain d'urate et en (*bs*) la basale externe du tube

Nous avons pu montrer à l'aide de la réaction de la murexide modifiée (*5*) qu'il s'agissait vraisemblablement de purines associées à d'autres substances chimiques. Du point de vue morphologique, nous avons comparé cette structure avec celle des urosphérites du rein de l'escargot

Helix pomatia. Nous avons trouvé une structure absolument équivalente qui nous indique d'ailleurs, par comparaison, la disposition des purines par rapport aux autres éléments. Les purines correspondent aux zones claires tandis que les zones sombres correspondent à des arrêts dans le dépôt. Le dépôt s'effectue sur la paroi de petites vacuoles que l'on trouve dans la région

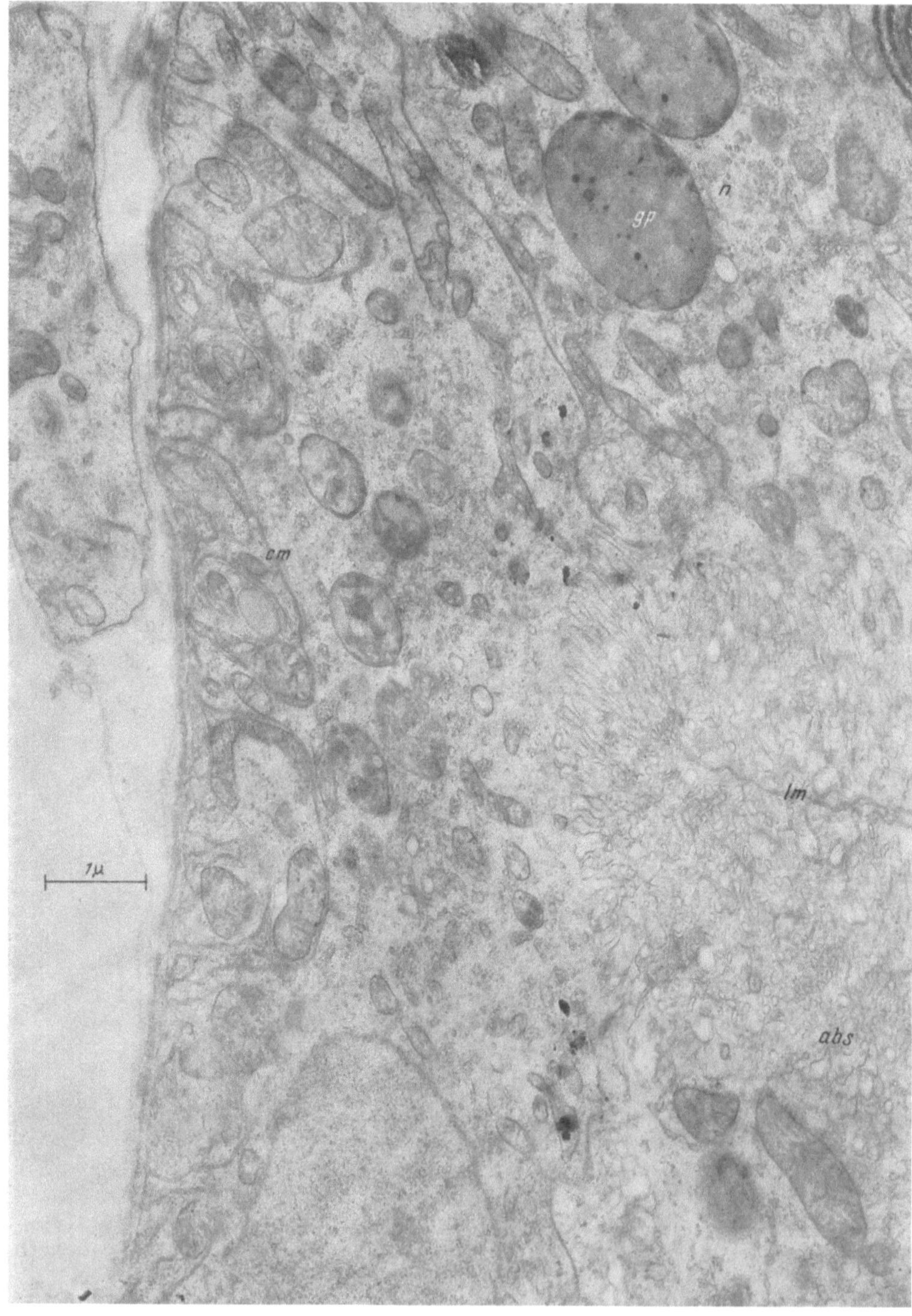

Fig. 2. Aspect général d'une cellule lors de la régénération des β-cytomembranes. Les β-cytomembranes (*cm*) sont du type fermé sans exceptions. Noter l'aspect particulier de la portion cellulaire dirigée vers la lumière (*lm*) du tube. Le cytoplasme semble y indiquer, en (*abs*), une réabsorption suivant un processus analogue à la pinocytose. En (*n*) la portion normale de la cellule voisine avec ses globules de pigments volumineux (*gp*)

périphérique des cellules. Le dépôt est simultané dans une même région de la cellule car on retrouve les même séries de couches avec la même importance dans des grains voisins. Nous avons
pu obtenir des images de l'élimination de ces grains. Celle-ci est quasiment simultanée pour les

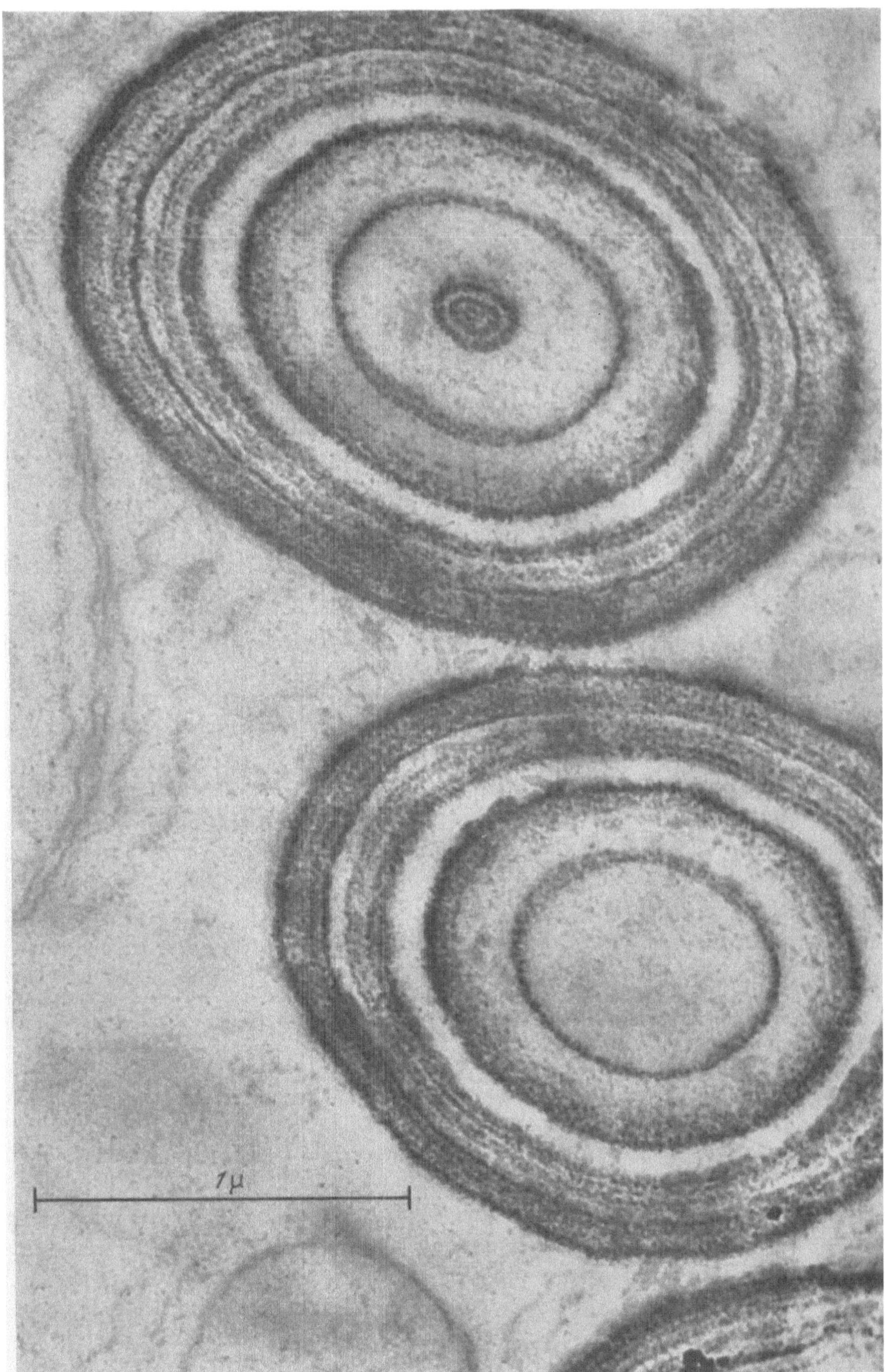

Fig. 3. Deux grains d'urate voisins. Les séries de dépôts sont identiques. On voit très nettement que les couches
se suivant dans le même ordre, deux couches équivalentes ayant la même importance

grains d'une même cellule car, malgré la faible épaisseur des coupes pratiquées on peut voir
fréquemment plusieurs grains en cours d'élimination.

Par conséquent il semble bien que, pour les produits d'excrétion figurés tout au moins, le fonctionnement soit discontinu, pour une cellule et même pour un tube donné. Ceci est important car, le cas de l'hémiptère *Rhodnius* mis à part, on ne sait pas très bien comment expliquer les phénomènes d'excrétion chez les insectes malgré les très beaux travaux de WIGGLESWORTH (*6*), de RAMSAY (*7*) et de leurs collaborateurs. En effet si chez *Rhodnius* et chez quelques autres insectes on a nettement mis en évidence des cellules excrétrices et des cellules réabsorbantes, comme dans tous les organes excréteurs, chez la plupart des autres insectes on ne trouve aucune de ces différenciations et le schéma donné pour *Rhodnius* n'est pas directement applicable. Nos observations nous inclinent à penser que le fonctionnement de ces cellules est alternatif avec une phase excrétrice et une phase réabsorbante, ces deux phases pouvant d'ailleurs plus ou moins se chevaucher.

Bibliographie

1. BEAM, H. W., T. N. TAHMISIAN et R. L. DEVINE: J. biophys. biochem. Cytol. **1**, 197 (1955).
2. RHODIN, J.: Correlation of ultrastructural organization and function in normal and experimentally changed proximal convoluted tubule cells of the mouse kidney. Aktiebolaget Godvil Stockholm 1954.
3. — Ergebnisse der elektronmikroskopischen Erforschung von Struktur und Funktion der Zelle. Verh. dtsch. Ges. Path. 41. Tagung. Stuttgart: Gust. Fischer 1958.
4. SJÖSTRAND, F. S.: Intern. Rev. Cytol. **5**, 455 (1956).
5. BERKALOFF, A.: C. R. Acad. Sci. (Paris) **246**, 2807 (1958).
6. WIGGLESWORTH, V. B.: J. exp. Biol. **8**, 411 (1931).
7. RAMSAY, J. A.: J. exp. Biol. **32**, 183 (1955).

Electron microscope studies on renal biopsies from patients with ischaemic anuria, lipoid nephrosis, multiple myelomas and diabetes mellitus

OLE Z. DALGAARD

Biophysical Institute of the University of Copenhagen and Municipal Hospital, Copenhagen (Denmark)

The introduction of the renal biopsy method by IVERSEN and BRUN in 1951 (*5*) made electron microscopy of human kidney tissue a practical possibility. RHODIN (*6, 7*) has shown that all the cellular structure undergo very rapid post-mortem changes. Renal biopsy, however, makes it possible to obtain the necessary fresh tissue. Furthermore, repeated observations can be made on the same patient.

The biopsy specimens were quickly cut into two portions. The one portion was fixed in formalin and then prepared in the usual manner for light microscopy. The histological report was prepared by Professor GORMSEN. The other portion was fixed in buffered osmium solution and embedded in plastic for electron microscopy. The electron microscope pictures were taken with an R.C.A. electron microscope at the Institute for Biophysics, under the direction of Professor KOCH and Mr. FRITZ CARLSEN, M. Sc. Photographic plates were made at magnifications from 4,000 to 13,500 times. Sections about 1 u thick were also cut from the plastic blocks for phase contrast microscopy. In this way it was possible, for example, to see the same glomerulus both by light microscopy and electron microscopy.

Our *normal material* consists of 3 biopsies. Our material of ischaemic anuria includes three cases.

The most characteristic findings in the *shock kidney*, as described for example by BRUN (*2*) in Denmark, are as follows: There is dilatation of the distal convoluted tubules, with flattening of the epithelium; cylinders are found here in particular and more distally in the nephron. Interstitial lesions are seen with small foci of cellular infiltration and oedema. The glomeruli, on the other hand, are normal.

Our first patient was a 64-year-old woman with barbituric acid poisoning, shock and ischaemic anuria. The biopsy was made on the third day of anuria. Our second patient was a 41-year-old woman who had been operated on for carcinoma of the ampulla of VATER. Shock developed following a post-operative haemorrhage. The biopsy was made immediately after death on the fourth day of anuria. Our third patient was a 58-year-old man with barbituric acid poisoning, shock and ischaemic anuria. The biopsy was taken immediately after death on the 10th day of anuria.

Light microscopy showed slight lesions, as previously described. The glomeruli were normal in all three cases. *Electron microscopy* showed no definite signs of degeneration in the glomeruli, not even in those glomeruli from the two renal biopsies taken post mortem. The capillaries were normal, with normal endothelial cells, basement membranes and epithelial cells with foot processes. It is too early to quote the results of our studies on the tubules. At the moment we are working down through the different sections of the nephron.

Our electron microscope findings of normal glomeruli in the shock kidney confirm those of previous studies, both experimental and those on patients with ischaemic anuria. Without going into details, it would appear most reasonable to imagine that the grosser changes in the tubule cells in ischaemic anuria, as shown by the light microscope, are secondary to the more delicate reversible changes in the cell structure, perhaps involving certain enzyme systems.

Our material includes two cases of so-called true (genuine) or *lipoid nephrosis*.

The most significant light microscopic findings in so-called true (genuine) or lipoid nephrosis is a thickening of the walls of the glomerular capillaries. FARQUHAR and co-workers (3), carried out electron microscope studies on kidney biopsies from sixteen children with the nephrotic syndrome, and FOLLI and co-workers (4), studied kidney biopsies from eight patients with the nephrotic syndrome.

Our first patient was a sixteen-year-old girl. She had had considerable oedema since December 1956, a low serum protein of 4.8%, excessive proteinuria of 9 $^o/_{oo}$, and elevated serum lipoid, 2200 mg-%. The kidney biopsy was taken on January 1958 after an illness lasting one year. The 24 hr creatinine clearance was normal 91 ml per minute.

Our other patient was a 17-year-old youth. Four days prior to admission he had developed considerable oedema, with low serum protein. 4.2%, excessive proteinuria, about 15 g per 24 hr, elevated serum lipoid, 1275 mg-%. The 24 hr creatinine clearance was 135 ml per minute. Kidney biopsy was carried out. The patient was then treated with Acton, 160 international units daily, for 15 days. As a result, the proteinuria fell to 0,5 g daily. A further kidney biopsy was then taken.

Light microscopy showed apparently normal glomeruli in patient no. 1, while in patient no. 2 only slight thickening of the glomerular basement membrane were seen. Our *electron microscope* findings are very similar to those of FARQUHAR (3) and FOLLI (4). The most characteristic and pronounced degeneration was found in the glomerular epithelial cells. Instead of foot processes, the capillary loops were covered by a wide layer of epithelial cytoplasm. Only occasionally, irregularly placed foot processes could be seen. The part of the epithelial cytoplasm immediately bordering on the basement membrane had a greater affinity for osmium staining than normal. The epithelial cytoplasm contained many more vacuoles than usual. There were numerous formations in the glomerular space between the epithelial cells. These had a possible resemblance to cast-off foot processes. The basement membrane was of normal thickness or thickened. In places, nodular thickenings of varying size and shape were seen, localized to the cytoplasm of the endothelial cells. Particularly in the first patient, where the condition had persisted longest, there were parts where the cytoplasm of the endothelial cells showed an accumulation of material resembling basement membrane. In places the basement membrane had a "moth-eaten" appearance, with lighter areas of low density. Following the Acton treatment of patient no. 2, whereby the proteinuria fell from 15 g to 0.5 g per day, the epithelial cells were seen to have become normal, with normal foot processes. *If* it is assumed that filtration takes place between the foot processes, then the area for the filtration must be very reduced in true or lipoid nephrosis. It could then be expected that the glomerular filtration would be correspondingly reduced in amount. However, this cannot be the case, as in both patients a normal or raised creatinine clearance was found. Of necessity, the ultrafiltrate must pass through the cytoplasm of the epithelial cells. The increased number of vacuoles could be considered as expressing an increased functional capacity, or could simply indicate an oedema of the tissue. *We have also observed* the very interesting finding of FOLLI and co-workers, namely, that the foot processes regenerate when the proteinuria is reduced by Acton treatment. This led these observers to assume that it is the changes in the epithelial cells of the glomeruli which are significant for the pathogenesis in the nephrotic syndrom, whereas previously the greatest significance was ascribed to the endothelial cells and basement membrane. We have so far not reached a conclusion on this point. It is well known, from light microscopic investigations, that in nephrosis the proximal tubule cells are filled with lipoid and hyaline drops or granules. Varying amounts of granules in the cytoplasm of the proximal tubule cells are shown by the electron microscope also. There were all variations in size, from large granules to mitochondria, and remains of the transverse bars or christae of the mitochondria were found in the granules. Some few mitochondria had become transformed only at the one end. The granules did not consist of fused mitochondria, as Rhodin (6) found was the case in mice which had received injections of protein. Among the tubule cells, we also observed cells which lacked the brush border, or had only a very few. The cytoplasm of these cells pro truded into the lumen of the tubule beyond the brush border.

We had one case of *multiple myeloma*. According to the current concept, renal insufficiency in multiple myeloma is caused by cylinders blocking the collecting tubules. The nephron therafter atrophies proximal to the cylinder. In some few cases, the renal insufficiency is caused by the accumulation of a hyaline substance in the glomerular capillaries.

Our patient was a 66-year-old man. The diagnosis of multiple myeloma had been made in the summer of 1956. The sternal marrow then showed 60% plasma cells, and serum electrophoresis pointed unmistakably to myelomatosis. The 24 hr creatinine clearance was 80 ml per minute. There was no proteinuria.

Light microscopy of two kidney biopsies showed 20 glomeruli in all with scattered, quite slight thickening of the basement membranes and only one almostly completely hyalinized glomerulus. Large, strongly stained hyaline cylinders were seen in the lumina of several distal tubules, distending the lumen. *Electron microscopy* showed varying thickening of the basement membranes in the glomeruli, the membrane being enormously thick in some places, up to 5—10 times normal. Nodular thickenings were also seen. A varying number of large granules were also seen in the cytoplasm of the proximal tubules. These were transformed mitochondria, remains of mitochondrial structure being frequently observed. It is possible that they were full of protein, just as in nephrosis. Farquhar and co-workers (3) found that in three cases of disseminated lupus erythematosus, the most characteristic feature was a pronounced thickening of the basement membrane. As we have just pointed out, a pronounced thickening of the basement membrane can also be seen in multiple myeloma. Further investigation will reveal whether this is one of the rare cases with accumulation of hyaline substance in the glomerular capillaries, or whether these lesions are general in multiple myeloma and of greater significance for the renal failure than the cylinders in the tubules.

Finally, we have one case of *diabetes mellitus*. In diabetic glomerulosclerosis, we find the well-known glomerular lesions: both Kimmelstiel-Wilson nodular masses — which must be considered microscope studies — and diffuse hyaline lesions of the glomerular basement membranes, these latter being non-specific. The exact localization of the hyaline masses was first indicated by Bergstrand and Bucht's (1) study of a single kidney biopsy. The hyaline material was shown to be deposited in the cytoplasm of the endothelial cells, and could not be distinguished from the basement membranes.

Our patient was a 63-year-old man. In the course of an admission for pneumonia, he was diagnosed as having mild diabetes mellitus, on account of an abnormal sugar tolerance curve. The duration of the diabetes was unknown. Ophthalmoscopy showed a few microaneurisms and slight double-contoured arteries. The 24 hr creatinine clearance was 71 ml per minute, a little under normal.

Light microscopy showed that one glomerulus was totally hyalinized, 11 were normal or had slight signs of diffuse glomerulosclerosis, possibly due to age. There was slight arteriosclerosis and arteriolosclerosis. *Electron microscopy* showed, as Bergstrand and Bucht had found, focal thickenings and folds in the capillary basement membrane. Localized to the cytoplasm of the endothelial cells were deposits of hyaline material. These deposits could not be distinguished from the basement membrane. The designation "intercapillary glomerulosclerosis" is therefore undoubtedly incorrect, as Bergstrand and Bucht (1) have also pointed out: In places, the hyaline substance showed fine fibrils. Proliferation of the endothelial cells was also observed.

The proximal tubule cells were normal.

References

1. Bergstrand, A., and H. Bucht: Lab. Invest. **6**, 293 (1957).
2. Brun, Claus: Acute Anuria. København: Ejnar Munksgaard 1954.
3. Farquhar, M. G., R. L. Vernier and R. A. Good: J. exp. Med. **106**, 649 (1957).
4. Folli, G., V. E. Pollak, R. T. Reid, C. L. Pirani and R. M. Kark: J. Lab. clin. Med. **50**, 813 (1957).
5. Iversen, P. and C. Brun: Amer. J. Med. **11**, 324 (1951).
6. Rhodin, J.: Diss. Stockholm 1954.
7. — Exp. Cell Res. **8**, 572 (1955).

Electron microscopic studies on human renal biopsies.
The structural basis of proteinuria *

David Spiro

Department of Pathology, Harvard Medical School and the Edward S. Webster Laboratory of the Department
of Pathology, Massachusetts General Hospital, Boston, Massachusetts

The structural alterations as observed with conventional histopathological techniques in human glomeruli obtained from patients with diseases characterized by marked proteinuria are varied. Glomeruli from such individuals may appear either within normal limits as in lipid nephrosis or else are characterized by marked thickening of the glomerular capillary walls as in chronic glomerulonephritis or amyloidosis of the kidney. These findings are rather contradictory and provide no structural basis for understanding the marked increase in permeability to plasma proteins of the glomeruli in such diseases. In order to further elucidate this problem, a study was undertaken of the fine structure of the human glomerulus obtained from biopsy material in various conditions associated with proteinuria. Nine cases of proteinuria were studied: two of juvenile lipid nephrosis, three of membranous glomerulonephritis, two of subacute and chronic glomerulonephritis, and two of amyloidosis. Marked proteinuria with a nephrotic syndrome was exhibited by one of the patients with lipid nephrosis, by one of the patients with membranous glomerulonephritis, and by both of the amyloid cases. The observations indicate that the structural basis of proteinuria is the presence of defects in the basement membrane of the glomerular capillary wall. This observation was previously reported in more detail in a study which consisted of some of the material included here (*1*).

Percutaneous renal biopsies and control material from nephrectomy specimens were divided immediately and portions were used for routine histopathological studies; the remainder was prepared for electronmicroscopy according to conventional procedures. In most cases, phosphotungstic acid staining was employed to enhance contrast particularly of the basement membrane. Sections were cut with a rotary Cantilever microtome (2) and examined in an RCA EMU 3B electron microscope. All glomeruli which were thin sectioned for electronmicroscopy were first examined by means of thicker sections under the phase microscope for supplementary diagnostic confirmation.

Normal glomerulus. The structure of the normal human glomerulus is similar to that of other mammals (see ref. 3 for review). The concensus of opinion is that of the three components of the capillary wall, the epithelial and endothelial cellular layers are interrupted, whereas the basement membrane is continuous (Fig. 1[1]). The endothelial cells present characteristic pores and the epithelial cells (podocytes) form numerous branches eventuating in discrete foot-like processes that are apposed to the external surface of the basement membrane. The basement membrane is at least 1 000 Å units in thickness and appears to have a filamentous fine structure which may be ordered. The fact that the only continuous layer making up the capillary wall is the basement membrane suggests that the latter is the ultimate diffusion barrier in the formation of the glomerular filtrate. Immature glomeruli from the nephrogenic zone of neonatal kidneys differ in structure from the mature glomeruli. In neonatal glomeruli the capillaries are invested by broader sheets of epithelial cell cytoplasm instead of by discrete foot processes (Fig. 2).

---

* Aided by a grant from the National Heart Institute of the National Institutes of Health. US Public Health Service (H 1834-C3-).

[1] *Abbreviations: Ep* epithelium, *BM* basement membrane, *En* endothelium, *BS* Bowman's space, *Vac* vacuole, *Dr* colloid droplet, *Am* amyloid

400   DAVID SPIRO:

Lipid Nephrosis. The most striking abnormality seen in lipid nephrosis relates to the basement membrane. This structure varies in thickness due to irregular peripheral strand-like foci of thickening as well as areas of marked attenuation (Fig. 3—5). The most significant lesion is the presence of actual defects in the basement membrane of the order of several hundred to a thousand Ångström units in size, where epithelial and endothelial cells are in apposition (Fig. 3, 4). In accord with previous observations (4, 5) the podocyte processes are replaced by broad sheets of epithelial cell

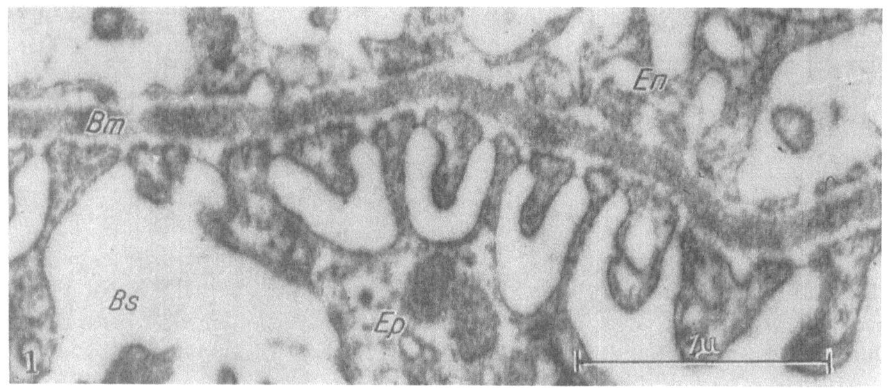

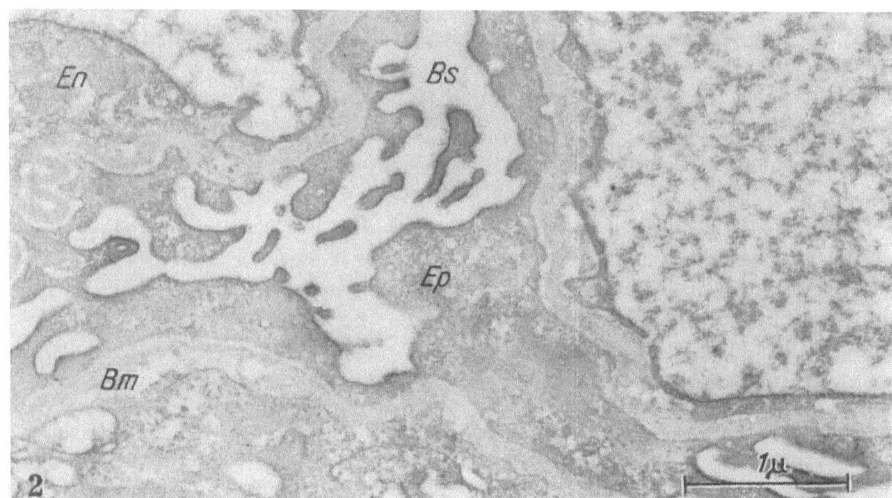

Fig. 1. Normal glomerular capillary wall disclosing discontinuous layers of endothelial cytoplasm and epithelial cell foot processes. Basement membrane appears fibrous and is the only continuous layer in the capillary wall. 34,200 ×

Fig. 2. Neonatal glomerulus which reveals epithelium that has not as yet achieved differentiation into discrete foot processes. 21,900 ×

cytoplasm which are apposed to the external aspects of the capillaries (Fig. 3—5). The epithelial cells contain numerous dense droplets often associated with vacuoles (Fig. 3). The endothelial cell layer is frequently hyperplastic. The glomeruli studied in this case were all within normal limits under the light microscope. This is not surprising because despite the changes observed under the electron microscope, the combined overall thickness of the endothelial cell basement membrane and epithelial cell foot process layers is approximately the same as in the normal. Study of a case of lipid nephrosis with minimal proteinuria, in remission presumably due to steroid therapy, disclosed glomeruli with normal structure under the electron microscope.

Glomerulonephritis. The main feature here also relates to the basement membrane which is for the most part diffusely thickened (Fig. 6). However, in many foci there is extreme attenuation of the basement membrane which is often associated with ultra-microscopic aneurysm-like out-pouchings of a capillary wall (Fig. 6—9). The most significant lesion again is considered to be

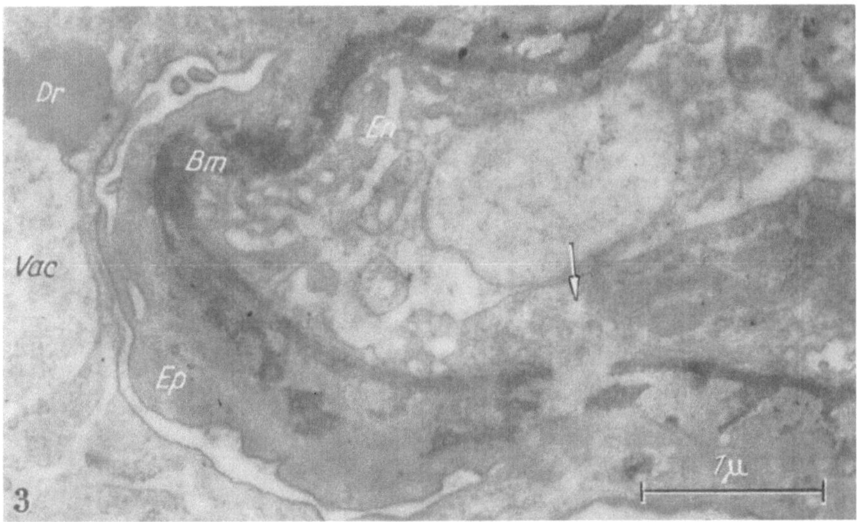

Fig. 3. Lipid nephrosis. Basement membrane is irregularly thickened. In one area (see arrow) there is a gap in the basement membrane through which endothelial cytoplasm has herniated and is in direct contact with epithelial cytoplasm. The external surface of the capillary is invested by broad confluent sheets of epithelial cytoplasm. A large vacuole and a colloid droplet are present in the epithelium. Endothelium is slightly hyperplastic. 25,400 ×

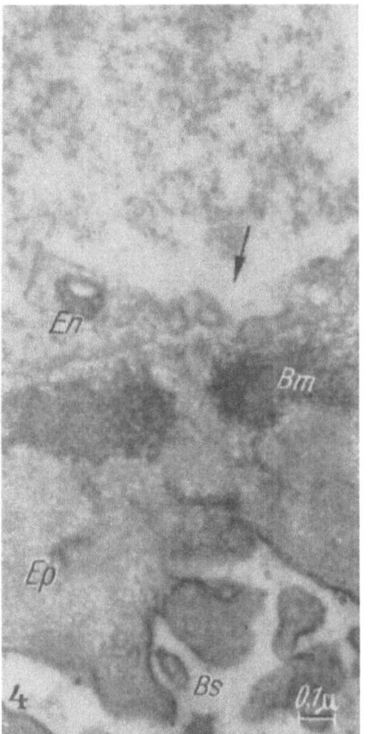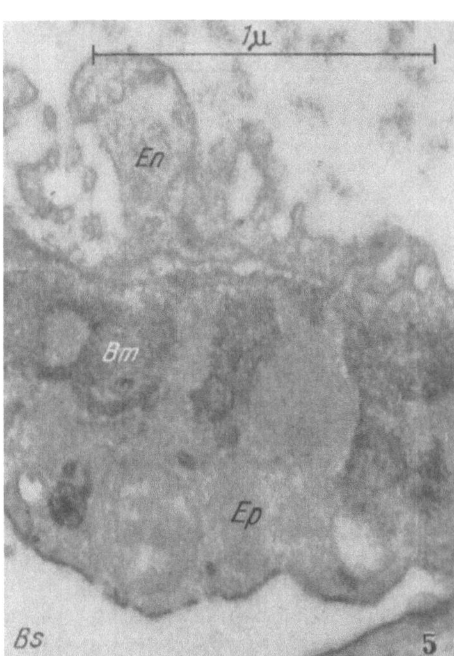

Fig. 4. Lipid nephrosis, another defect in the basement membrane is present (see arrow), permitting juxtaposition of endothelium and confluent sheets of epithelial cytoplasm. 42,300

Fig. 5. Lipid nephrosis. Basement membrane is irregularly thickened and attenuated. Epithelium is similar to Fig. 3 and 4. 46,800

the presence of gaps in the basement membrane approximating the size of those observed in lipid nephrosis (Fig. 9). Serial sections define these basement membrane defects as relatively circumscribed lesions in three dimensions. The epithelial cells present changes similar to those described

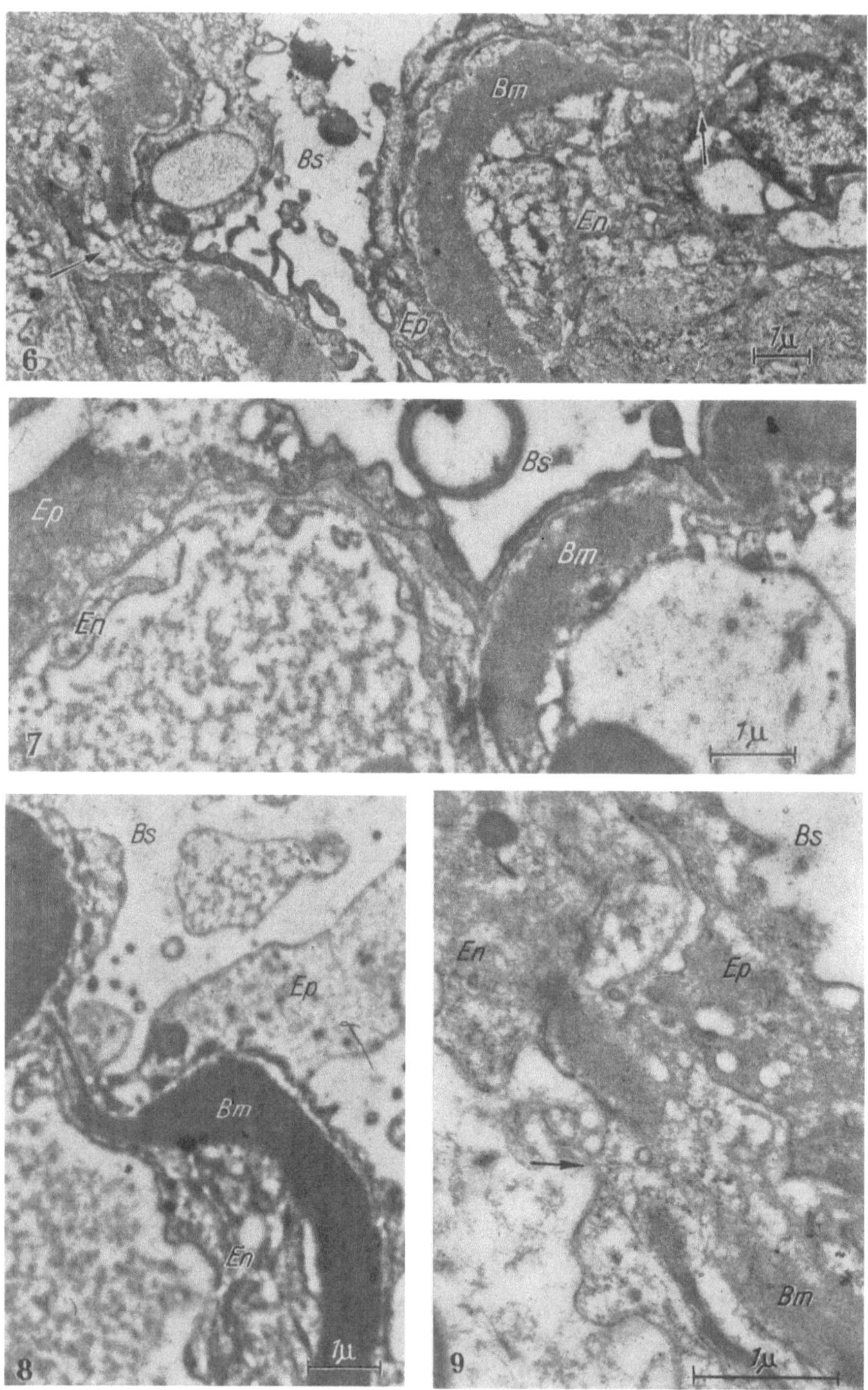

Fig. 6. Membranous glomerulonephritis. Basement membrane is generally thickened. There are, however, two areas in which the basement membrane is markedly attenuated or absent (see arrows). In these areas epithelium is markedly proliferated. An epithelial vacuole and a droplet are present. Endothelium is markedly hyperplastic. 8,120 ×

Fig. 7. Membranous glomerulonephritis. On the right the basement membrane is thickened. The segment of capillary on the left has virtually no basement membrane and is the seat of an aneurysm-like bulge. The layer of epithelial cytoplasm covering this area is markedly thickened. 12,100 ×

Fig. 8. Membranous glomerulonephritis. An area of marked attenuation of a generally thickened basement membrane is noted. 9,800 ×

Fig. 9. Membranous glomerulonephritis. A gap is present in the basement membrane (see arrow). In this region endothelial and confluent epithelial cell cytoplasm are in contact. At the top of the micrograph the basement membrane is extremely thin and protrudes outward towards Bowman's space. 20,600 ×

under lipid nephrosis which result in broad confluent sheets of cytoplasm that have replaced the foot processes. Dense droplets and vacuoles are also present in the altered podocyte layer (Fig. 6). These changes in the epithelial cells appear to be most striking in regions where the basement membrane is thinned or absent (Fig. 6, 7, 9). The endothelial cells are generally markedly proliferated. In many cases, masses of endothelial cells almost obliterate the capillary lumina. The frequency of basement membrane defects is apparently related to the degree of proteinuria

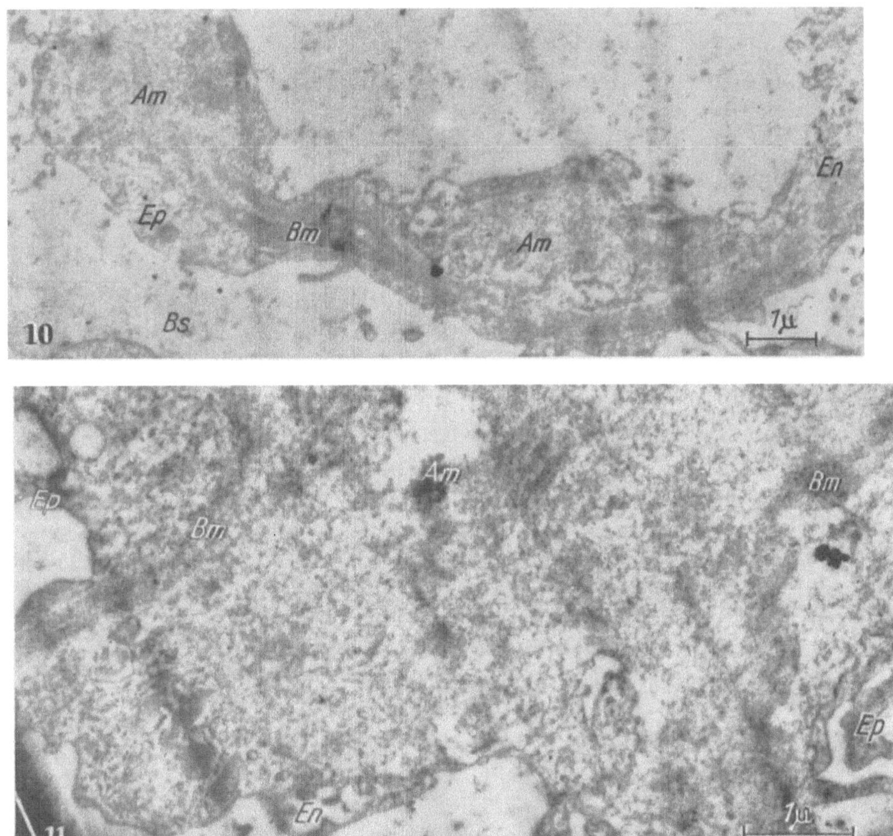

Fig. 10. Amyloidosis. A subendothelial deposit of amyloid is noted in the right-hand portion of the micrograph. The basement membrane and epithelial foot process layers are normal here. On the left a more extensive deposit of amyloid which has replaced the basement membrane is seen. Overlying this region there is confluence of epithelial cell cytoplasm. 9,500 ×

Fig. 11. Amyloid is present on both sides of a fragmentary residual basement membrane. Amyloid tissue itself consists of a loose irregular meshwork of fine filaments. 15,400 ×

existing at the time the biopsy was obtained. More advanced lesions in glomerulonephritis reveal the presence of numerous collagen fibrils in what are probably endothelial cells and result in a loss of recognizable glomerular structure.

Amyloidosis. The earliest amyloid deposits are seen subendothelially and are surrounded peripherally by a normal basement membrane and layer of podocytes (Fig. 10). In more extensive lesions, amyloid substance penetrates and replaces the basement membrane in areas and becomes contiguous to the epithelial cells (Fig. 10). The latter under these circumstances reveal the previously described change of replacement of foot processes by more extensive sheets of cytoplasm. The amyloid substance is composed of a loose irregular meshwork of thin (approximately 100 Å in diameter) filaments which are several 100 Å apart (Fig. 11).

Advanced disease results in obliteration of normal glomerular architecture by broad sheets of amyloid as well as numerous collagen fibrils.

The electronmicroscopic studies disclosed a constant glomerular structural abnormality associated with proteinuria. This abnormality consists of a loss of continuity of the normally continuous glomerular capillary basement membrane. In lipid nephrosis and in glomerulonephritis the basement membrane discontinuities consist of actual gaps or pores through which endothelial and epithelial cells are in direct contact. In amyloid disease the discontinuities are due to replacement of the basement membrane by the structurally far looser amyloid substance. These defects in the basement membrane are of appropriate size to permit diffusion of plasma proteins while excluding the cellular elements of blood and therefore may constitute the structural basis of proteinuria. The confluence of epithelial cell foot processes which has been so well described by Farquhar et al. (*4, 5*) are probably secondary to the underlying basement membrane lesions. The significance of the epithelial cell transformation is not clear but it may represent a reparative attempt to restore the filtration mechanism. The filtration pathway in the abnormal glomerulus is rather obscure, and appears to involve epithelial as well endothelial cell cytoplasm.

References

1. Spiro, D.: Amer. J. Path. **35,** 47 (1959).
2. Hodge, A. J., H. E. Huxley and D. Spiro: J. Histochem. and Cytochem. **2,** 54 (1954).
3. Rhodin, J.: Amer. J. Med. **24,** 661 (1958).
4. Farquhar, M. G., R. C. Vermies and R. A. Good: Amer. J. Path. **33,** 791 (1957).
5. — — — J. exp. Med. **106,** 649 (1957).

8. Respirationsorgane

The ultrastructure of rat lung in the pre- and postnatal period. Some remarks on the fine structure of the foetal alveolar epithelium

Janusz Groniowski and Wiktor Djaczenko

Department of Pathological Anatomy, Academy of Medicine — Poznań/Poland

Ultra-structural differences are to be expected between the cells of foetal lung epithelium and those of postnatal lung epithelium. Some definite idea upon the ultra-structure of lungs already engaged in respiration including alveolar epithelium, is given in a number of publications dealing with healthy lungs in normal animals and man (*2, 5, 7, 8, 9, 10, 15, 18*). There are also some publications describing the ultrastructure of lung tissue in certain pathological states (*11, 17, 18*). The foetal lungs have been, as yet, the object of very few electron microscope examinations. In the literature obtainable, van Breemen, Neustein and Bruns (*3*) examined the lung tissue from premature human infants and young guinea-pigs compelled to breathe oxygen in conditions of high humidity, with the object of studying the formation of hyaline membranes. The publications of Schulz (*19, 20*) deal with the pathology of mitochondria of alveolar epithelial cells in rats. This author examines here the structural differences of mitochondria to be found in foetuses and after birth.

This paper deals with lung tissue of a rat foetus 22 millimeters crown-rump length, the longest one out of a litter of five. The living foetus was obtained by laparotomy under ether anaesthesia. Clamping of the trachea prevented respiration. The thorax of the foetus was opened quickly and many sections from different parts of the lungs were taken. The material was immediately fixed in 0,5% osmium tetroxide buffered to a pH of 7.35 with veronal acetate. After dehydration the tissue material was infiltrated and embedded in n-butyl methacrylate. To polymerize, 2,4-dichlorobenzoyl peroxide up to 15% was added. The polymerization was carried out at about 40° C for about five days. To obtain ultrathin sections a Leitz rotary microtome specially adapted by us for this purpose was used. The thickness of the sections was from 150 Å to 400 Å, the latter being made thick to obtain better contrast. The electron microscope being used was an Elmi D2, Carl Zeiss Jena.

Observations. We examined, first of all, cells of the epithelium lining the alveolar ducts and the bulges being the future lung alveoli. In spite of the close relationship between these cells, they differ considerably in their cytology and morphology.

The nucleoplasm of these cells was surrounded by two membranes separated by the osmiophobic space. Not uncommonly the nuclear membranes, here and there, were considerably separated forming sacks or channels on the circumference of the nucleoplasm. Within the nucleoplasm we sometimes found two or even three focal condensations looking like nucleoli. In some

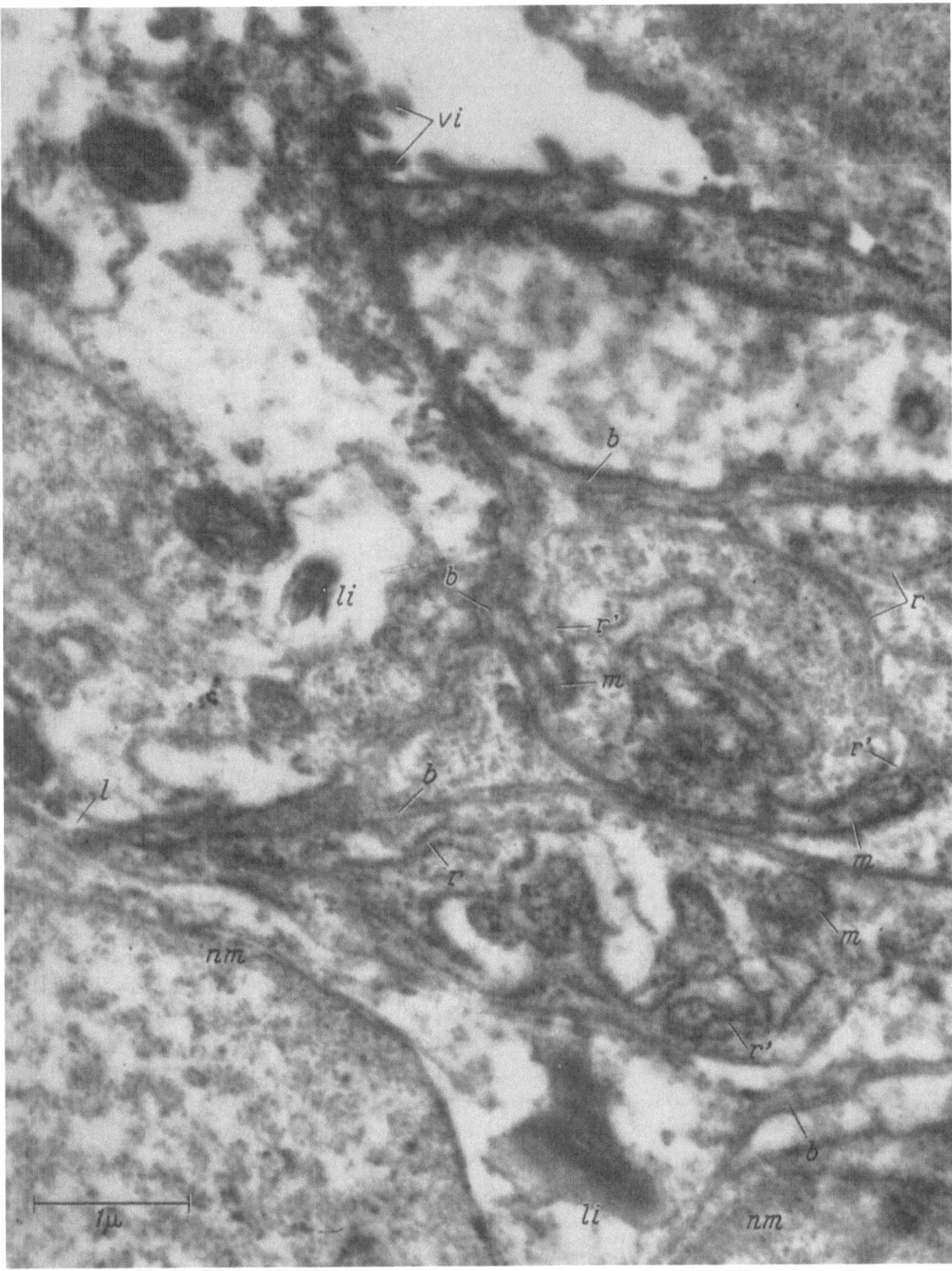

Fig. 1. The epithelial cells lining the alveolar duct or alveoli. The lumen of alveolar duct can be seen in the upper part of the figure and at *l*. There are small microvilli on the surface of the epithelial cells *vi*. Two nuclei of the cells in the lower part of the figure; the closely opposed nuclear membranes at *nm*. Intercellular boundary between epithelial cells *b*. The endoplasmic reticulum *ER* — is represented by rough surfaced elements mostly of elongated form *r*. The granular cytoplasmic matrix in their vicinity. The *ER* profiles *r'* in contact with or close to mitochondria *m*. Lipoid bodies-*li*. 23000 ×

regions of the cell membranes on the inside of the alveolar ducts, were found very fine projections of the cytoplasm which are called microvilli. In other places, the cell membrane protruded into the cytoplasm, perhaps more often on the endothelial side, forming channels. Between the membranes of the adjacent epithelial cells there was, as a rule, a narrow space, in places hardly discernible and in other places so large as to form a kind of intercellular pocket.

The above-described epithelial cells were characterized by endoplasmic reticulum having the appearance of small channels or vesicles. Their osmiophilic walls corresponded to both the profiles of a smooth surface and those of a rough one (*12*). The profiles of the smooth surface were found rather in the peripheral part of the cell and those of the rough surface in the central one rather. We could find, many a time, a morphological connexion of these structures with the outer nuclear membrane; in this way the osmiophobic space between the nuclear membranes extended into the channels or vesicles of the endoplasmic reticulum. This phenomenon is already known in other cells (*12, 14*). The endoplasmic reticulum was often found in close vicinity of mitochondria. We have not succeeded easily in defining the morphological character of their contact with the wall of the endoplasmic reticulum in the material. The above mentioned variety of appearance of the epithelial cells resulted largely from the differences in appearance of the endoplasmic reticulum. In different

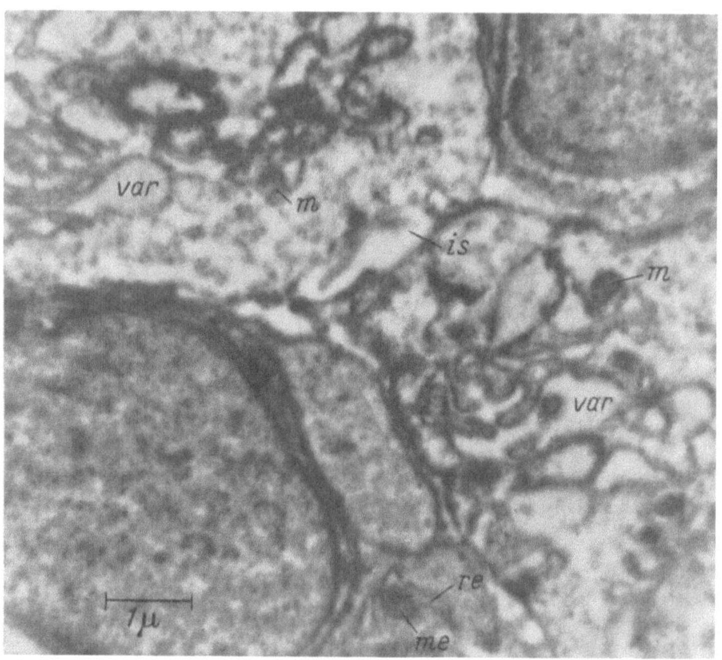

Fig. 2. The fragments of two epithelial cells; intercellular space between them *is*. At the upper right and lower left part of the figure are situated endothelial cells. In the cytoplasm of the epithelial cells-a complex of infolded vacuolar spaces and/or dilated channels of endoplasmic reticulum *var*. In the same region mitochondria *m*. Endoplasmic reticulum *re* and mitochondrium *me* in the endothelial cell. 12 000 ×

cells the endoplasmic reticulum had different shapes, was situated in different parts of the cytoplasm and appeared different in amount. There were also large vacuolar structures taking up considerable part of the cytoplasm. They appeared to be the extension of the spaces formed by the profile of the endoplasmic reticulum. Yet they did not always possess distinct boundaries, and here and there they passed gradually into the surrounding cytoplasm. In more enlarged osmiophobic spaces of the endoplasmic reticulum a very fine inner network made up of filaments of very small density could sometimes be discerned.

The appearance of the mitochondria also made the epithelial cells differ to some extent from one another. In the mitochondria we could find changes indicating different stages of lamellar transformation described by Schulz (*20*) concerning the lung epithelium. Namely we could see enlarged osmiophobic spaces between the inner membranes of the mitochondria giving rise to the vesiclelike clearings of mitochondrial matrix. Then we saw the complete disappearance of the matrix followed by breaks in the continuation of the outer membranes of the mitochondria. Finally we found disintegrated mitochondria present in the form of loosely lying strongly osmiophilic lamellae. The considerably changed mitochondria, particularly the disintegrating ones, were seen, as a rule, in the cytoplasm of diminished density, or in the above-mentioned large vacuolar structures. Unlike the mitochondria of normal appearance which were in the more central parts of the cell, the changed mitochondria, especially the disintegrating ones, were found in the peripheral parts of the cytoplasm. A rare phenomenon in the epithelial cells was the presence of single cytosomes of concentric structure, previously described in the epithelial lining of the lung alveoli in cats (2).

In many cells of the epithelium there were strongly osmiophilic lipoid bodies of different shape and size. They were usually surrounded by a delicate membrane separated from the osmiophilic mass by a osmiophobic space. Sometimes the lipoid bodies in one cell of the epithelium

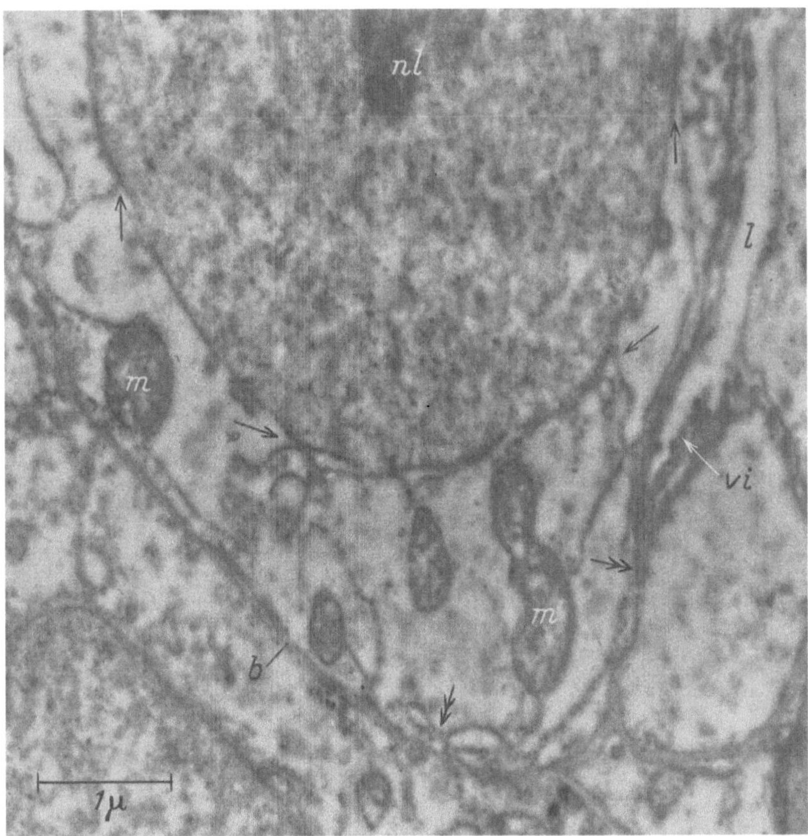

Fig. 3. The epithelial cells. At the upper right-the narrow lumen *l* of the alveolar duct or alveolus and at *vi* there are small microvilli. The nuclear membranes are mostly separated (arrows) and form very broad spaces. They reach to the cell membrane (double arrow). The same space in contact with mitochondria *m*. Intercellular boundary between epithelial cells *b*. Nucleolus *nl*. 18000 ×

were joined together by means of membranes inclosing them or projections of these membranes. We did not find any connexion of the lipoid bodies with the mitochondria or cytosomes.

Discussion. In a mature rat foetus, some rather characteristic structural properties can be found in the epithelial cells lining alveolar ducts and bulges of the future alveoli, namely: (*i*) differences in the endoplasmic reticulum, (*ii*) the presence of vacuolar structures in the cytoplasm, (*iii*) changes in the mitochondria and (*iv*) rather numerous lipoid bodies.

It is difficult to assign a function to the lungs in the foetal life. It is assumed that the future respiratory passages in a foetus are filled up to their extremes with the amniotic fluid (4). The amniotic fluid present here becomes absorbed, which was proved by many authors in experimental animals, and confirmed also in man (16). The fact that there is a developed network of lymphatic vessels in the lungs in the latter period of the foetal life (6) may be related to the absorption of amniotic fluid in the lungs, at least in the latter stages of pregnancy. Perhaps some ultra-structural features of lung epithelial lining might somehow be connected with this foetal function of the lungs. It seems that the relatively well developed endoplasmic reticulum observed by us in the

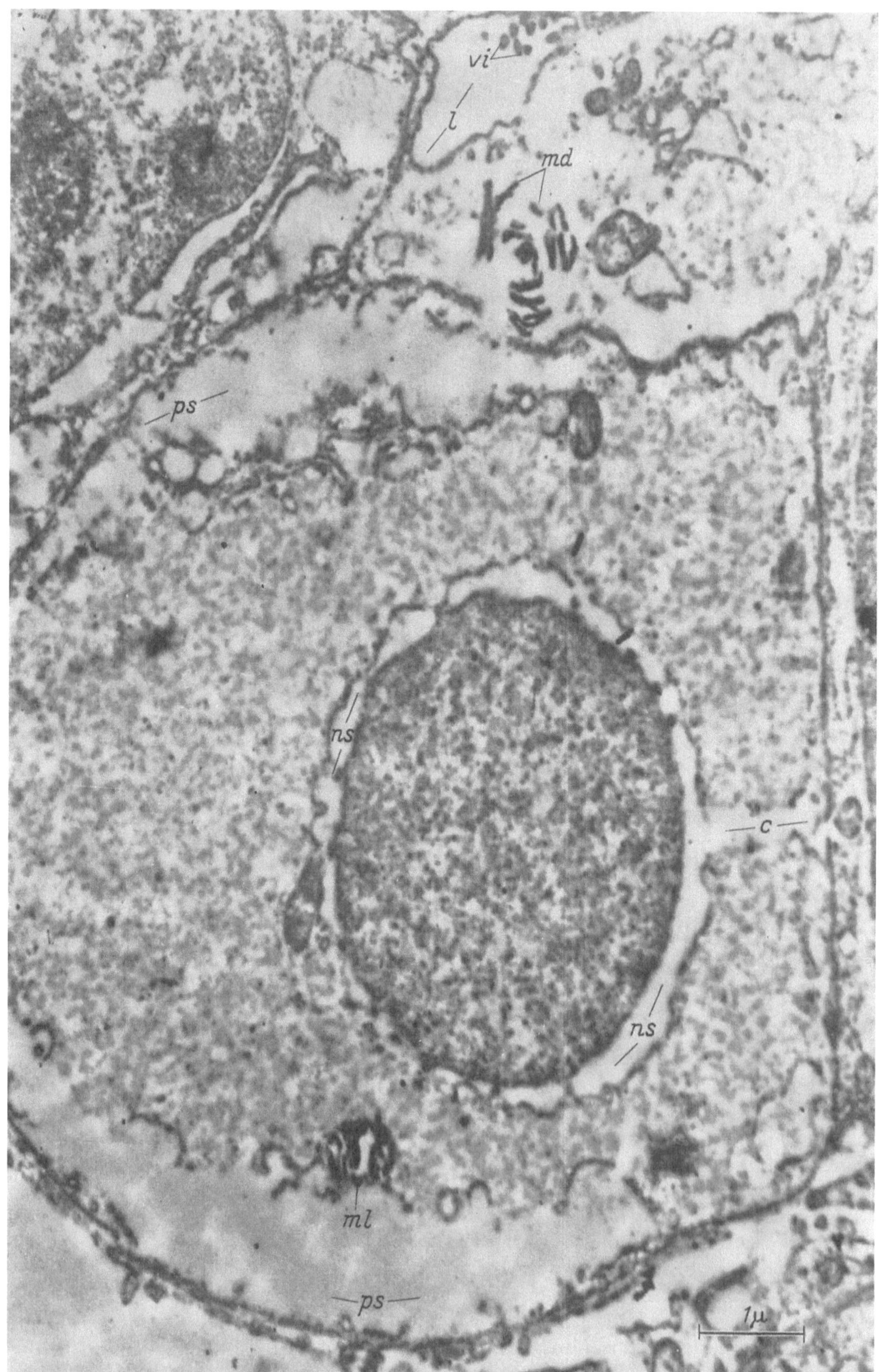

Fig. 4. The epithelial cells of the alveolar duct or alveoli. At the upper right part of the figure the narrow lumen *l* and at *vi* there are microvilli. The nuclear membranes are separated widely and form a circular space around the nucleoplasm *ns*. This space has a connexion *c* with a broad space at the periphery of the cytoplasm *ps*. Lamellar degeneration of the mitochondria *ml*. The fragments of the disintegrated mitochondrium at *md*. 17000 ×

lung epithelial lining, which is sometimes in contact with the infolding cell membrane and often is in close vicinity to the mitochondria, might have some relation to the absorption of amniotic fluid, an idea which has some analogies to the observations by Pease (*13*). The vacuolar structures as well as the lamellar transformations of the mitochondria may be the expression of the regressive changes in the lung epithelium in the examined period of the foetal life. The presence of the lipoid bodies in the lung epithelial lining in this period of the foetal life was found histochemically in different animals, and has been dealt with very fully (*1*). Our finding of numerous lipoid bodies in the cytoplasm of the epithelial cells confirms and extends the above observations.

We wish to give our best thanks to Dr. Frank N. Low for the valuable discussions we had while he was staying in Poznań.

References

1. Angeli, F., e G. de Biase: Arch. De Vecchi Anat. pat. **19**, 339 (1953).
2. Bargmann, W., u. A. Knoop: Z. Zellforsch. **44**, 263 (1956).
3. Breemen, V. L. van, H. B. Neustein and P. D. Bruns: Amer. J. Path. **33**, 769 (1957).
4. Davies, M. E., and E. L. Potter: J. Amer. med. Ass. **131**, 1149 (1946).
5. Gieseking, R.: Beitr. path. Anat. **116**, 177 (1956).
6. Hayek, H. v.: Die menschliche Lunge. Berlin: Springer 1953.
7. Karrer, H. E.: J. biophys. biochem. Cytol. **2**, 241 (1956).
8. — J. biophys. biochem. Cytol. **2**, Suppl. 115 (1956).
9. Low, Frank N.: Anat. Rec. **117**, 241 (1953).
10. — and M. M. Sampaio: Anat. Rec. **127**, 51 (1957).
11. Meessen, H., u. H. Schulz: Lungen und kleiner Kreislauf. Bad Oeynhausener Gespräche I, 19.—21. Oktober 1956. Berlin 1957.
12. Palade, G. E.: J. biophys. biochem. Cytol. **2**, Suppl. 85 (1956).
13. Pease, D. C.: J. biophys. biochem. Cytol. **2**, Suppl. 203 (1956).
14. Policard, A., et M. Bessis: Exp. Cell Res. **11**, 490 (1956).
15. — A. Collet and S. Pregermain: Electron microscopy-Proceedings of the Stockholm Conference September 1956. p. 244 Stockholm 1957.
16. Potter, E. L.: Pathology of the fetus and the newborn. Chicago 1952.
17. Schulz, H.: Naturwissenschaften **43**, 205 (1956).
18. — Virchows Arch. path. Anat. **328**, 582 (1956).
19. — Verh. dtsch. Ges. Path. 41. Tagung. Stuttgart 1958, S. 342.
20. — Beitr. path. Anat. **119**, 45 (1958).

Die Entstehung der elastischen Fasern in der embryonalen Lunge des Menschen

W. Schwarz

Forschungsabteilung für Elektronenmikroskopie der Freien Universität Berlin

Die *elastischen* Fasern treten in der Ontogenese im allgemeinen später als die *kollagenen* auf (*1*). In der Lunge dagegen entstehen die ersten elastischen Elemente bereits im 3. Schwangerschaftsmonat im Mesenchym der Lungenanlage (*2*, *3*). Reticuläre Fasern finden sich in diesem Embryonalstadium nur an wenigen hilusnahen Stellen. Wir hatten Gelegenheit, die Lungen eines menschlichen Embryos von 7,6 cm Gesamtlänge zu untersuchen. Die entodermale Lungenanlage besteht aus einem Drüsenbaum, dessen Endstücke noch ein einschichtiges Zylinderepithel besitzen. Die Zellen der *mesenchymalen* Lungenanlage sind verästelt und bilden einen schwammartigen Zellverband. Die großen Zellkerne sind rund oder oval. Der Zelleib enthält neben vielen Vacuolen nur wenige Doppellamellen und Mitochondrien. Das Cytoplasma zeigt in der Außenschicht und in der Zentralzone des Zelleibs keine Unterschiede. Man kann deshalb nicht von einem Exoplasma und Endoplasma der mesenchymalen Zellen sprechen. Alle Zellen haben Fortsätze, die in verschiedenen Ebenen des Raumes liegen. Ihr Querschnitt ist rund oder oval. Auch in den Verästelungen der Zellen sind Vacuolen und einzelne Mitochondrien zu erkennen. Eine einfache Zellmembran von 200 Å Dicke umkleidet Zelleib und Zellfortsätze. Das Mesenchym der Lungenanlage bildet keine protoplasmatische Einheit, sondern einen Verband von Zellindividuen. Mit ihren Fortsätzen stehen die einzelnen Zellen untereinander nur in Berührung. Die Kontaktstellen mit einer Dicke von 600 Å können nur mit dem EM in ihre zwei Zellmembranen und einen Zwischenraum aufgelöst werden. Die mesenchymale Anlage der menschlichen Lunge ist damit kein „echtes" Syncytium (Zellkomplex), sondern ein Zellverband (Abb. 1). Die zwischenzellige Substanz ist fast noch

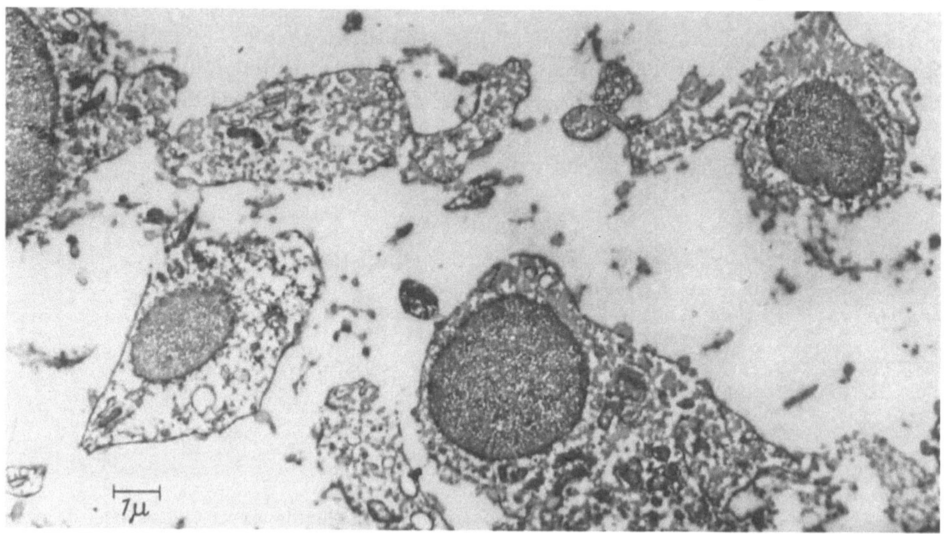

Abb. 1. Mesenchymaler Zellverband in der Lungenanlage eines 7,6 cm langen menschlichen Keimlings. 6000 mal

überall flüssig oder gelartig und füllt den Raum innerhalb des schwammartigen mesenchymalen Zellverbandes aus. Ein Teil der Zellfortsätze ist aber nicht rund, sondern breit und platt. Die Platten können dabei so dünn werden, daß sich die Zellmembranen fast berühren und kaum

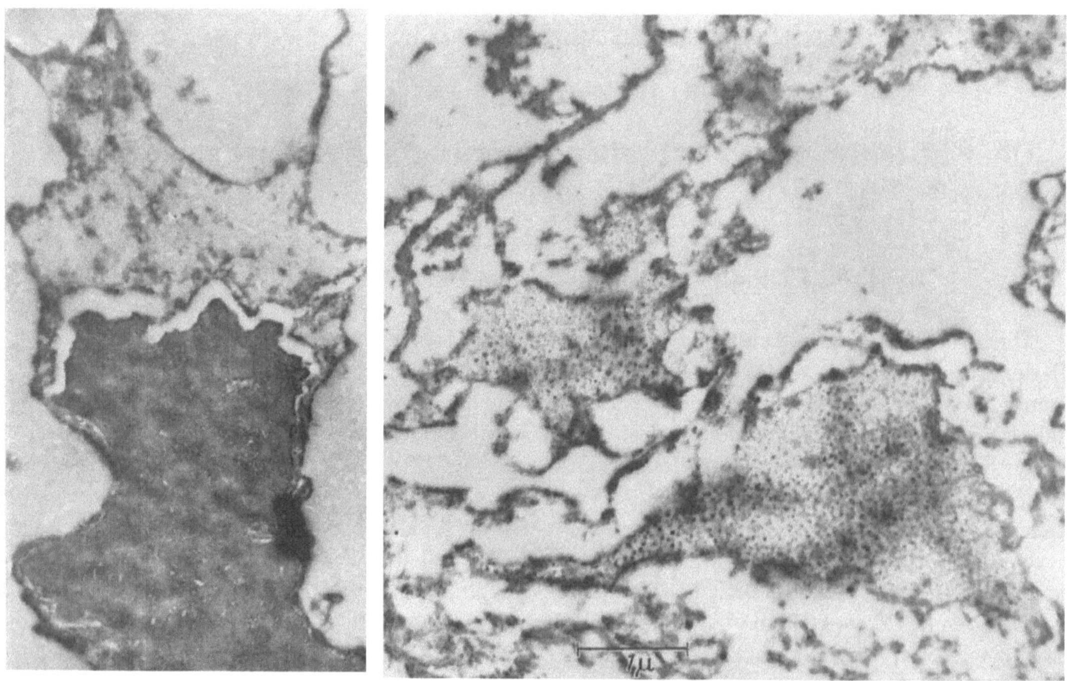

Abb. 2. Links: Elastische Faser, eingefaßt von blattförmigen Zellfortsätzen. Am oberen Pol aufgeblähter Fortsatz mit dichterem Inhalt. Rechts: Dichteres Material mit Körnchen, umgeben von Zellfortsätzen. 15000 mal

Protoplasma zwischen sich fassen. Auch an diesen Stellen kommen Vacuolen vor, jedoch keine Mitochondrien und Doppellamellen. Der Abstand zwischen Zelleib und ihren blattförmigen Fortsätzen ist an den einzelnen Stellen der Lunge verschieden groß. Mitten im Mesenchym sind diese Blätter weit von ihren zugehörigen Zellen entfernt. In der Nähe der Gefäßanlagen und der entodermalen Lungenknospen sind diese Zellausläufer kurz, und die blattförmigen Fortsätze liegen sehr zellnahe. Die Bildung der elastischen Elemente hängt eng mit den blattförmigen Zellfortsätzen zusammen. Mehrere membranartige Zellfortsätze stoßen aneinander und umspannen kanalartige Räume. Einzelne Zellfortsätze, die an der Kammerbildung beteiligt sind, blähen sich auf. Die beiden Zellmembranen werden von einer homogenen und dichten Masse auseinandergedrängt. Die Dichte dieser Flüssigkeit ist größer als die der normalen Intercellularflüssigkeit des Lungenmesenchyms (Abb. 2). Diese dichte Flüssigkeit wird in die Kammer abgegeben und damit

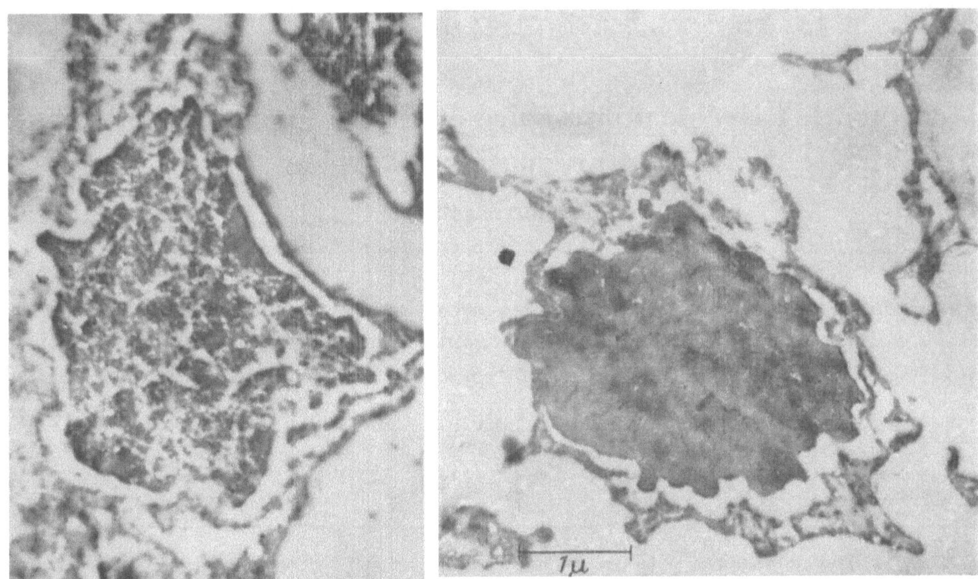

Abb. 3. Links: Anhäufung von elastischen Blöcken in einer Kammer, die sich zu Gruppen und homogenen Teilen zusammenschließen. Rechts: dichte elastische Faser, umgeben von Zellfortsätzen. 15000mal

kommt die Bildung der ersten elastischen Elemente in Gang. Später werden in den Kammern kleine runde 120—200 Å große Körnchen sichtbar, die sich zu etwa 400 Å großen viereckigen Einheiten entwickeln (Abb. 2). Diese Teile liegen zuerst regellos verstreut in der Kammer, nach und nach rücken sie näher zusammen und vereinigen sich schließlich zu Gruppen (Abb. 3). Die Zwischenräume verschwinden mehr und mehr, und schließlich sind die eckigen Einheiten vollkommen zusammengeschlossen. Der Inhalt der Kammer macht dann einen einheitlich dichten Eindruck und erinnert an die bekannten Schnittbilder der elastischen Fasern (4, 5, 6). Der Ablauf dieser Prozesse geht anscheinend nicht schlagartig vor sich. Man findet nämlich in manchen Kammern die einzelnen Abläufe der Faserbildung nebeneinander. In einigen Kammern hat man außerdem den Eindruck, daß beim Zusammenschluß der Blöckchen auch einige Filamente mit eingeschlossen werden. Das Endergebnis ist jedenfalls ein polygonales oder rundes elastisches Körnchen mit einer Größe von 1—3 μ, das noch immer von blattförmigen Zellfortsätzen umkleidet wird (Abb. 3). Die elastischen Elemente in der Lungenanlage des Menschen bilden sich damit außerhalb der Zellen in bestimmten Kammern des Intercellularraumes. Diese Räume sind vom übrigen Intercellularraum durch blattförmige Zellfortsätze getrennt und begrenzt. Die mesenchymalen Zellen können auf diese Weise die Zusammensetzung der flüssigen Intercellularsubstanz an bestimmten begrenzten Stellen verändern und so den Ort und auch die Richtung der entstehenden elastischen Fasern festlegen. Eine unmittelbare Umwandlung des Cytoplasmas in eine elastische Faser haben wir nirgends wahrnehmen können (1). Auch die Verwandlung kollagener und präkollagener Fasern zu elastischen (7) darf hier bei der Histogenese des Lungeninterstitiums verneint werden, da ein großer Teil der elastischen Elemente vor dem Auftreten der übrigen Bindegewebsfasern gebildet wird.

Die Untersuchungen wurden mit Unterstützung der Deutschen Forschungsgemeinschaft durchgeführt. Das Material für diese Untersuchung verdanke ich Herrn Prof. Dr. E. v. Schubert, Direktor der Frauenklinik der F. U. im Städt. Krankenhaus Moabit.

Literatur

1. Maximow, A.: Die embryonale Entwicklung des Blutes und Bindegewebes. In: Handbuch der mikroskopischen Anatomie des Menschen, II, 1. Berlin: J. Springer 1927.
2. Linser, P.: Anat. Hefte 13, 309 (1900).
3. Teuffel, E.: Arch. Anat. Physiol., Anat. Abt. 1902, 377.
4. Policard, A., A. Collet, L. Giltaire-Ralyte, Chr. Heuet et C. Defosset: Bull. Microsc. appl., Ser. 2, 4, 139 (1955).
5. Dettmer, N.: Z. Zellforsch. 45, 265 (1956).
6. Schulz, H.: Virchows Arch. path. Anat. 328, 582 (1956).
7. Turnbridge, R. E., and M. J. Wood: Nature (Lond.) 176, 966 (1955).

La structure de l'alvéole pulmonaire étudiée au moyen de la technique de l'imprégnation à l'argent

Vittorio Marinozzi

Institut d'Anatomie et Histologie pathologique de l'Université de Rome

Les travaux les plus récents de microscopie électronique se rapportant à la structure des alvéoles pulmonaires, sont d'accord quant à l'existence d'un revêtement cellulaire continu des parois alvéolaires. Toutefois la nature de celui-ci est interprétée de façon différente suivant les

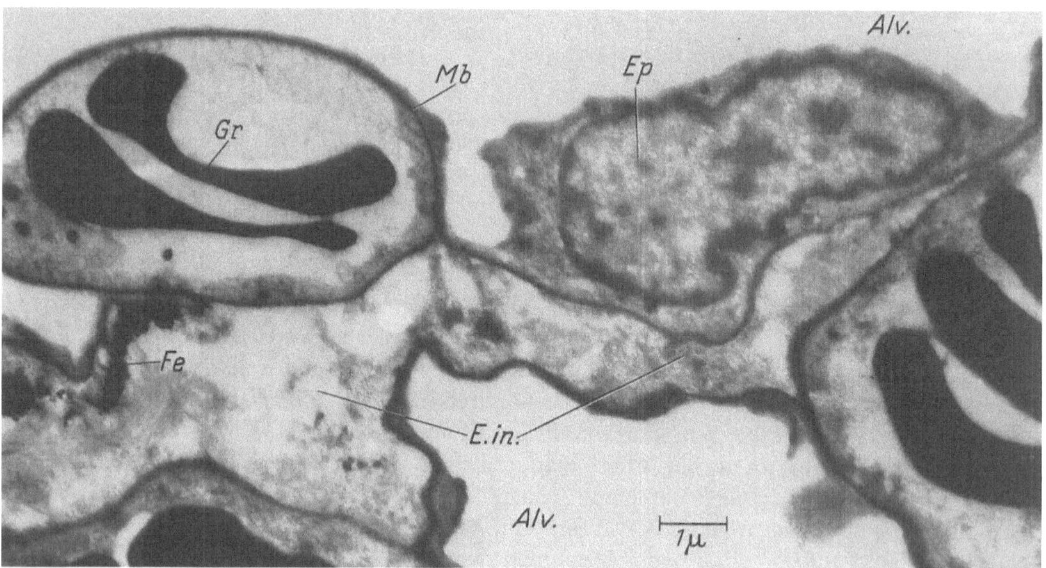

Fig. 1. *Ep* cellule épitheliale (vraie cellule de revêtement). *Alv* alvéole, *Mb* membrane basale, *Gr* globule rouge, *Fe* fibres élastiques, *E. in.* espace interstitiel[1]

auteurs: certains (*1, 2, 3, 4, 5, 6*) le considèrent d'origine épithéliale, alors que d'autres (*7, 8*) lui reconnaissent une origine mésenchymale.

Presque tous les auteurs affirment également qu'il existe une fine membrane, laquelle s'interpose entre les cellules de revêtement et l'éndothélium des capillaires. D'après d'autres recherches encore, il y aurait deux membranes basales, une épithéliale et une endothéliale (*1, 2, 3, 7*). Celles-ci sont difficilement résolubles en conditions physiologiques, alors qu' elles apparaissent bien distinctes au cours de l'édème pulmonaire expérimentale (*9*) ou dans le poumon cardiaque (*10*).

---

[1] Toutes les figures ici reproduites, se rapportent à des coupes privées du méthacrylate avant d'être soumises à la coloration à l'argent.

J'ai l'intention ici de m'entretenir brièvement sur les résultats de mes observations personnelles se rapportant à la structure du poumon de rat, étudiée au moyen d'une méthode particulière d'imprégnation à l'argent des coupes ultra-fines (*10, 11*). Cette méthode — dont j'ai eu déjà l'occasion de m'occuper au cours de ce Congrès — apparaît particulièrement intéressante du fait qu'elle permet une définition plus exacte quant à la nature des cellules alvéolaires et aux rapports histogénétiques de celles-ci avec les macrophages pulmonaires.

Au niveau de la paroi des alvéoles il existe deux types de cellules nettement distincts. Un premier type se caractérise principalement par la grande extension superficielle du cytoplasme qui

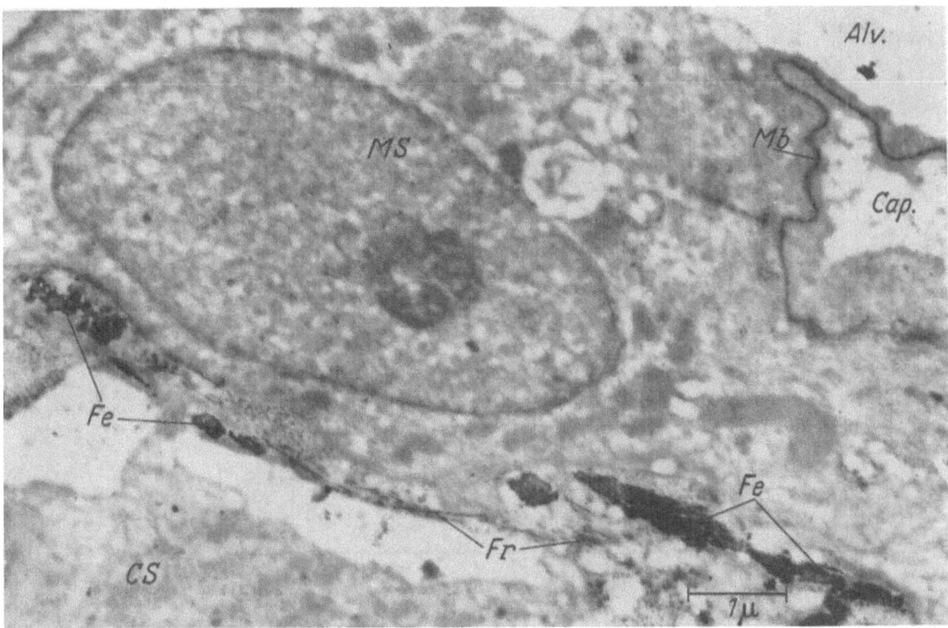

Fig. 2. *MS* macrophage septale (cellule à cytoplasme vacuolisé). *Mb* membrane basale. *Cap* capillaire. *Fe* fibres élastiques. *Fr* fibres réticulaires. *CS* cellule interstitielle (ou septale proprement dite). *Alv.* alvéole

constitue un mince halo autour du noyau et envoie des expansions laminaires qui assurent le revêtement continu de la paroi (Fig. 1). Dans le cytoplasme on reconnaît seulement de rares et petites structures sphéroïdales, qui correspondent à des mitochondries.

Les cellules du deuxième type ont, en général, une forme ovoïdale et leur noyau occupe une position centrale. Contrairement à celles du type précédent, qui développent une surface presque lisse, elles présentent de nombreux prolongements cytoplasmiques, d'épaisseur différente, sur la partie de leur surface qui s'avance librement dans l'alvéole. Dans le cytoplasme on reconnaît, à côté des mitochondries, de nombreux vacuoles de grandeur différente et, occasionnellement, des inclusions particulières, de forme arrondie et douées d'une forte argentaffinité. Celles-ci sont semblables à celles qui, en plus grand nombre, sont constamment présentes dans le cytoplasme des macrophages endoalvéolaires.

Tandis-que les cellules du premier type sont toujours en rapport avec la fine membrane basale épithéliale (très évidente à cause de son intense affinité pour l'argent), les cellules à cytoplasme vacuolisé sont constamment en rapport direct, à travers des solutions de continuité de la même membrane basale, avec les structures de l'interstice (Fig. 2) et, en particulier, avec les cellules interstitielles (ou *septales proprement dites*) (Fig. 3). La nature mésenchymale de ces dernières est démontrée par leur connection constante et intime avec les fibrilles réticulaires, dont l'imprégnation à l'argent met en évidence la structure périodique.

Bien que l'existence de deux formes de cellules alvéolaires, complètement distinctes au point de vue cyto-morphologique, aient été désormais reconnues par la plus grande partie des auteurs

(*6, 7, 8, 12, 13, 14, 15*), l'opinion toutefois dominante est qu'il s'agit d'aspects évolutifs de la même cellule; si bien, que l'on est porté à croire à l'hypothèse que les cellules à cytoplasme vacuolisé sont une forme évolutive intermédiaire entre les vraies cellules de revêtement de l'alvéole et les macrophages pulmonaires (*6, 7*).

Cependant, si d'une part mes recherches ne permettent pas de douter de la nature épithéliale des vrais cellules de revêtement de l'alvéole (c'est-à-dire des cellules que j'ai décrites comme

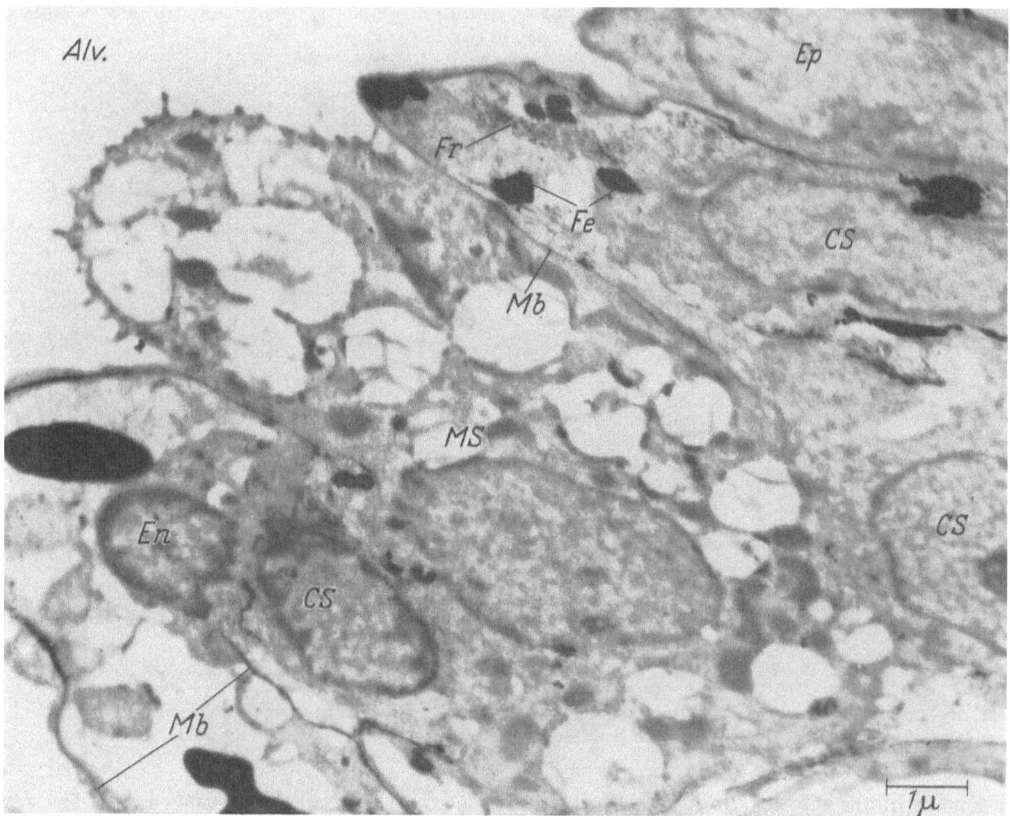

Fig. 3. *Alv* alvéole. *MS* macrophage septale (cellule à cytoplasme vacuolisé). *CS* cellules interstitielles (ou septales proprement dites). *Fr* fibres réticulaires. *Mb* membrane basale. *Ep* cellule épithéliale. *En* cellule éndothéliale

appartenant au premier type), d'autre part elles consentent de retenir comme absolument vraisemblable la nature mésenchymale des cellules à cytoplasme vacuolisé et, par conséquent, des macrophages pulmonaires qui, vraisemblablement, dérivent de ces dernières à la suite de leur détachement de la paroi alvéolaire.

Les cellules à cytoplasme vacuolisé correspondraient, ainsi, à un stade évolutif intermédiaire entre les élements mésenchymaux du stroma et les macrophages endoalvéolaires, qui partant pourraient être désignés comme *macrophages septales*.

Bibliographie

1. Low, F. N.: Anat. Rec. **113,** 437 (1952).
2. — Anat. Rec. **117,** 241 (1953).
3. — Anat. Rec. **120,** 827 (1954).
4. Bargmann, W., u. A. Knoop: Z. Zellforsch. **44,** 263 (1956).
5. Karrer, H. E.: Bull. Johns Hopk. Hosp. **98,** 65 (1956).
6. — J. biophys. biochem. Cytol. **2,** 241 (1956).
7. Policard, A., A. Collet et L. Giltaire-Ralyte: C. R. Acad. Sci. (Paris) **240,** 2363 (1955).
8. — — et S. Pregermain: Electron Microscopy. Proc. Stockholm. Conference, Sept. 1956, p. 244.
9. Schulz, H.: Electron Microscopy. Proc. Stockholm Conference. Sept. 1956, p. 240.

10. Marinozzi, V.: Rend. Acad. Naz. Lincei (Cl. Sci. fis., mat. e nat.) **24,** 600 (1958).
11. — Ces. C. R., p. 103.
12. Low, F. N., and M. M. Sampaio: Anat. Rec. **127,** 51—63 (1957).
13. Kisch, B.: Exp. Med. Surg. **13,** 101 (1955).
14. — Exp. Med. Surg. **15,** 101 (1957).
15. Swigart, R. H., and D. J. Kane: Anat. Rec. **118,** 57 (1954).

The alveolar macrophage

H. E. Karrer

Department of Pathobiology, School of Hygiene and Public Health, Johns Hopkins University, Baltimore, Maryland (USA)

The alveolar macrophages, or dust cells, of the lung are cells which lie free within the air spaces of the lung alveoles. They have been extensively studied in the light microscope [for summary see (*1*)]. Their phagocytic potentialities are likewise known from light microscopic studies, in which dyes, India ink, or oils were injected into the trachea of animals. The same cells are phagocytic also in pathologic lung conditions. The origin of these cells is not clear as yet (for reviews, see (*2, 3*)]. Most workers consider them to be modified alveolar epithelial cells, whereas others believe that they are true connective tissue macrophages.

In this study, the alveolar macrophages were examined in thin sections by means of a Siemens & Halske Elmiskop I b electron microscope. Lung tissue was obtained from normal mice as well as from mice that had previously been instilled intranasally with diluted India ink. The tissue was fixed in buffered, isotonic osmium tetroxide solution, dehydrated in acetone, and embedded in methacrylate.

Alveolar macrophages of normal lung. The alveolar macrophages of normal lung are usually in direct contact

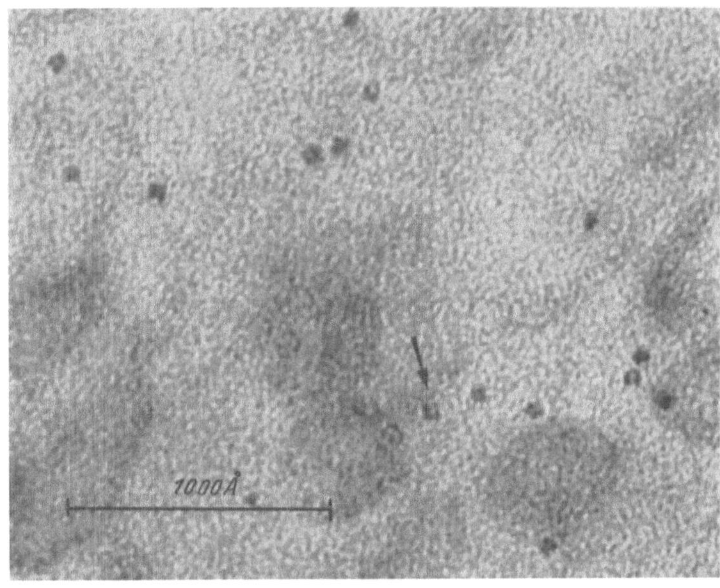

Fig. 1. Ferritin molecules in the cytoplasm of an alveolar macrophage. Only the dense iron hydroxide micelles are visible; the protein hull (apoferritin) of the molecules is not apparent in a thin tissue section. The arrow points at a molecule in which the four iron hydroxide micelles are typically arranged as corners of a square

with the epithelium of the alveolar wall, but they are independent from this epithelium. Their contact is established by means of short pseudopod-like protrusions of the cytoplasm. The cytoplasm contains small mitochondria, sparce profiles of the endoplasmic reticulum, and a variety of different inclusion bodies which are believed to be remnants of ingested materials and of phagocytized cells.

Another typical component of the cytoplasm, which is constantly observed, consists of small very dense granules distributed at random throughout the cell with the exception of the nucleus. These granules also occur in small clusters the diameter of which does not exceed 0.2 μ. The single granules, which have a diameter of about 60 Å, consist of smaller subunits approximately 10 to 25 Å in diameter which lie about 22 to 32 Å apart (Fig. 1). Rather infrequently such a granule is seen to consist of four such subunits which are arranged as corners of a square (arrow,

Fig. 1). These granules are believed to represent the iron hydroxide cores of ferritin molecules, because of their resemblance to purified ferritin (*4*). This belief is further supported by the findings of other authors describing similar granules inside a variety of cells, e.g. red blood cells (*5—10*), liver cells (*11*), and various hemosiderin-containing cells (*12—13*).

Alveolar macrophages after India ink instillation. Such alveolar macrophages are seen to engulf great masses of India ink particles. Once inside the cell, these particles are found located within membrane-bounded small vesicles (I, Fig. 2), within larger vacuoles, and within inclusion bodies (I, Fig. 3). The small vesicles are formed from invaginated portions of the plasma membrane

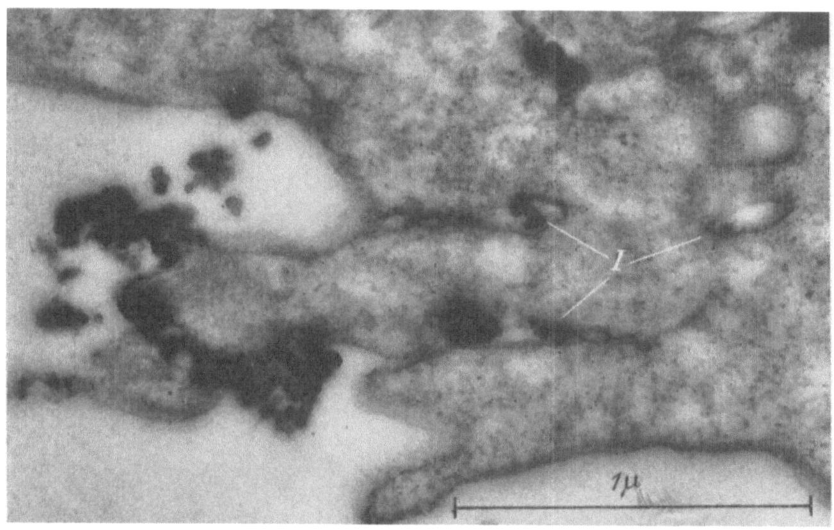

Fig. 2. Portion of an alveolar macrophage, phagocytising India ink particles. The black India ink particles are readily seen (*I*). A number of them lie outside the cell and are in contact with the plasma membrane. A few particles are in the process of being phagocytised: they lie inside membrane-bounded vesicles which are recognized as invaginations of the cell plasma membrane

(Fig. 2), the single vesicles being indistinguishable from smooth-surfaced units of the endoplasmic reticulum. In one instance a string of newly formed vesicles containing India ink was observed that seemed to be continuous with a rough-surfaced cisterna of the endoplasmic reticulum. The larger vacuoles and the inclusion bodies containing India ink also seem to form at the cell surface by a similar process, with their bounding membranes deriving from the plasma membrane. The inclusion bodies sometimes are filled with a homogeneous material of low density within which the India ink particles are embedded (Fig. 3). In other inclusions this homogeneous matrix appears vacuolated (Fig. 3) and also contains denser material which sometimes consists of 40 Å thick lamellae (arrow, Fig. 3) separated by equally wide interspaces.

The events leading to the phagocytosis of India ink particles (i.e., formation of vesicles, vacuoles and inclusions at the cell surface) are comparable to the pinocytosis process as observed in cinemicrographs of tissue culture cells (*14*). The larger vacuoles and inclusions correspond to the pinocytosis droplets that can be observed in such cinemicrographs, but the small vesicles are of a dimension that would prevent their detection in light-optical studies. The contents of vesicles, vacuoles, and inclusion bodies represent the phagocytized ("pinocytized") materials. Thus it appears that phagocytosis and pinocytosis are essentially identical processes and are based on the invagination and the detachment of bits of the plasma membrane, which is accompanied by the ingestion of particulates (phagocytosis) or of fluid (pinocytosis).

The small vesicles which arise from infoldings of the plasma membrane (Fig. 2) resemble smooth surfaced profiles of endoplasmic reticulum. One is reminded of observations made by various authors, which established that continuities may exist between the membranes of the endoplasmic reticulum and the plasma membrane. One may therefore be justified to interpret these pinocytosis vesicles (Fig. 2) as endoplasmic reticulum which is in the process of being formed from infolded bits of plasma membrane. The observed continuity of such vesicles with a rough-surfaced cisterna suggests that such cisternae are perhaps formed by this same process of infolding, a possibility which has been suggested in the past (*14*).

The vacuoles and the dense, partially lamellated masses within the matrix of inclusion bodies (Fig. 3) are considered to indicate that secondary changes attributable to the action of the cell take place within this inclusion body matrix. Such observations are perhaps in parallel with findings describing gradual changes in the phase contrast images of pinocytosis droplets (*14*).

References

1. Bargmann, W.: Handbuch der mikroskopischen Anatomie des Menschen (W. von Möllendorff, Herausgeb.), Bd. 5, Teil 3, S. 810. Berlin: Springer 1936.

2. Robertson, O. H.: Physiol. Rev. **21**, 112 (1941).

3. Rosin, A.: Beitr. path. Anat. **79**, 625 (1928).

4. Farrant, J. L.: Biochim. biophys. Acta **13**, 569 (1954).

5. Bessis, M.: Bruxelles méd. **37**, 1321 (1957).

6. Bessis, M. C., and J. Breton-Gorius: J. biophys. biochem. Cytol. **3**, 503 (1957).

7. Bessis, M., et J. Breton-Gorius: Revue d'Hématol. **12**, 43 (1957).

8. — — C. R. Acad. Sci. (Paris) **244**, 2846 (1957).

9. — — C. R. Acad. Sci. (Paris) **245**, 1271 (1957).

10. — — J. Sem. Hôp. Path. et Biol. **33**, 411 (1957).

11. Kuff, E. L., and A. J. Dalton: J. Ultrastructure Res. **1**, 62 (1957).

12. Richter, G. W.: J. exp. Med. **106**, 203 (1957).

13. — J. biophys. biochem. Cytol. **4**, 55 (1958).

14. Gey, G. O.: Harvey Lect. **50**, 154 (1954—1955).

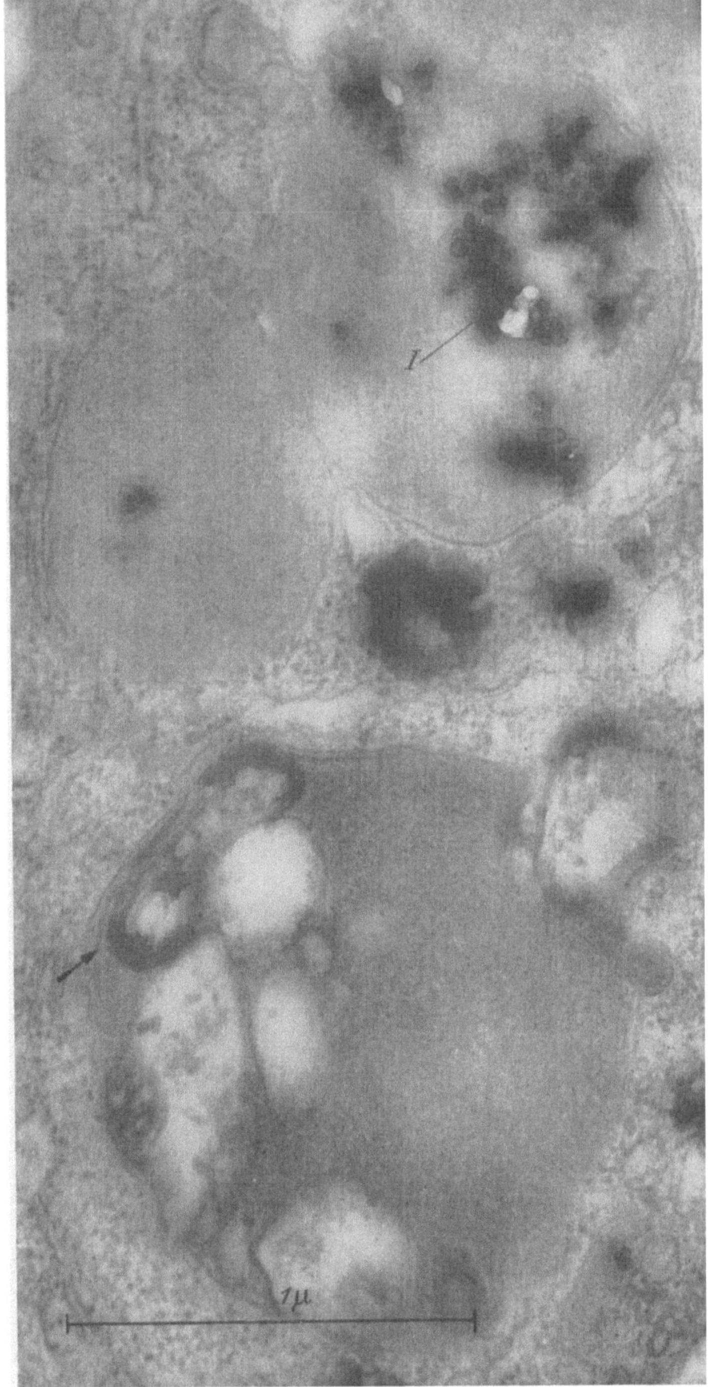

Fig. 3. Portion of an alveolar macrophage, phagocytising India ink. This portion of the cell contains large inclusion bodies, which are products of phagocytosis, and which in places contain India ink (*I*). The matrix of some of these inclusions shows vacuolization and the formation of dense material, which appears lamellated (arrow)

Elektronenmikroskopische Morphologie der Lungenalveolen des Protopterus und Amblystoma

M. de Groodt, A. Lagasse und M. Sebruyns

Institut für Histologie und Laboratorium für Elektronenmikroskopie der Universität Gent (Belgien)

Gegenstand dieser Mitteilung ist die elektronenmikroskopische Untersuchung der Lungenalveolen des *Protopterus Dolloi* und des *Axolotl*. Wir beabsichtigen durch diese Untersuchung von wenig differenzierten Lungenalveolen den komplizierten Bau der Lungenalveolen der Säuger besser zu verstehen. Vor zwei Jahren haben Bargmann und Knoop (*1*) im Anschluß an frühere lichtmikroskopische Befunde mitgeteilt, daß bei den Anuren und Säugern die Blut-Luftschranke

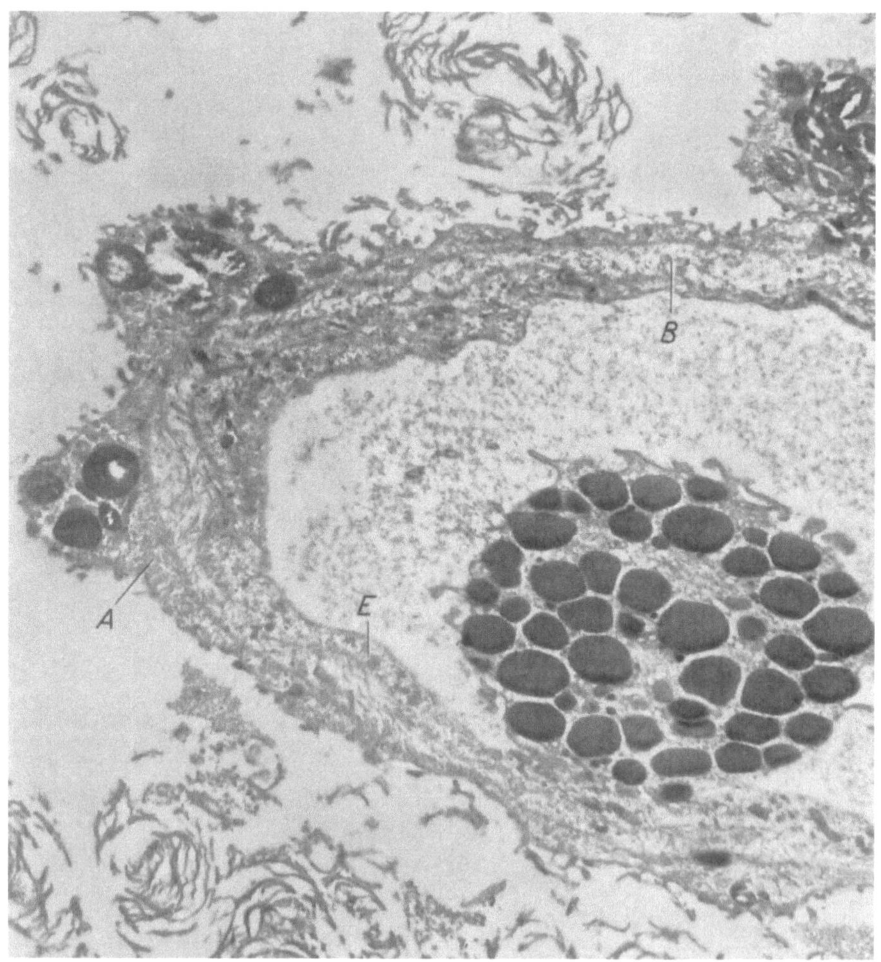

Abb. 1. Blut-Luftschranke bei *Protopterus*. Endothelschicht einer Lungencapillare (*E*), Basalmembran (*B*) und Alveolarcytoplasma (*A*). Zahlreiche Lamellen frei im Luftraum. 6000 ×

den gleichen Bauplan aufweist. Da die Auskleidung der Froschlunge zweifellos von entodermaler Herkunft ist, werden auch die Alveolardeckzellen der Säuger als Epithelzellen betrachtet. Auf Grund ihres morphologischen Verhaltens nun werden diese Alveolardeckzellen von vielen Autoren zum Bindegewebe gerechnet. Die Diskussion der Frage „Epithel oder Bindegewebe" bezüglich der Alveolarauskleidung ist Gegenstand einer ausführlichen Literatur; wir verweisen auf unseren Artikel in «Archives d'Anatomie Microscopique» (*2*).

Die Differenzierung der Alveolarwand ist bei Axolotln und Lungenfischen geringer als bei den Landtieren. Deshalb erschien es uns angezeigt, die Alveolarauskleidung bei Axolotl und Protopterus Dolloi elektronenmikroskopisch zu untersuchen.

Lungenstückchen wurden nach PALADE fixiert. Die Schnitte fertigten wir an mit dem Ultramikrotom von PORTER-BLUM. Für die Aufnahmen benutzten wir das Siemens-Elmiskop I.

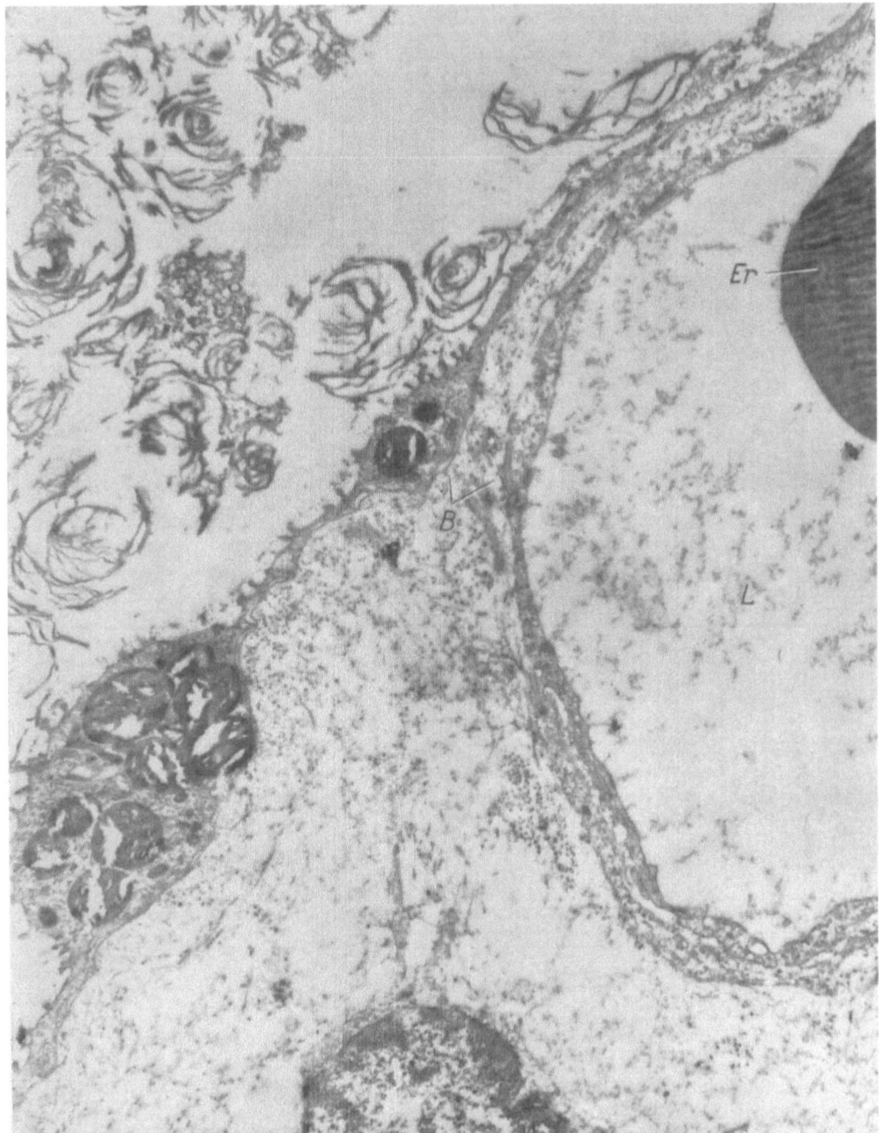

Abb. 2. *Protopterus*-Lunge. Capillarlumen (*L*) mit Erythrocyt (*Er*) die Basalmembran (*B*) geht in einen breiten Bindegewebsraum über. 6000 ×

Unsere elektronenmikroskopischen Aufnahmen zeigen bei den beiden untersuchten Tieren eine auffallende Ähnlichkeit. Beim *Axolotl* wie beim *Protopterus* buckeln sich die Capillaren nur wenig in die Alveolarlichtung vor. Die Lungenalveole ist in gleicher Weise von einem Cytoplasmabelag der Alveolardeckzellen ausgekleidet, der von einer Endothelschicht der Lungencapillaren durch eine Basalmembran getrennt ist (Abb. 1). An der Basis der Capillarvorwölbung geht diese Basalmembran in ein breites Bindegewebespatium über, in welchem Fibrillen und Bindegewebezellen zu erkennen sind (Abb. 2).

27*

Die Alveolarzellen sind an ihrer freien Oberfläche mit regelmäßigen, langen, stachelartigen Cytoplasmafortsätzen ausgestattet. Im Cytoplasma selbst beobachten wir zahlreiche runde bis ovale osmiophile Einschlüsse. Wir können zwischen kompakten Einschlußkörpern und solchen

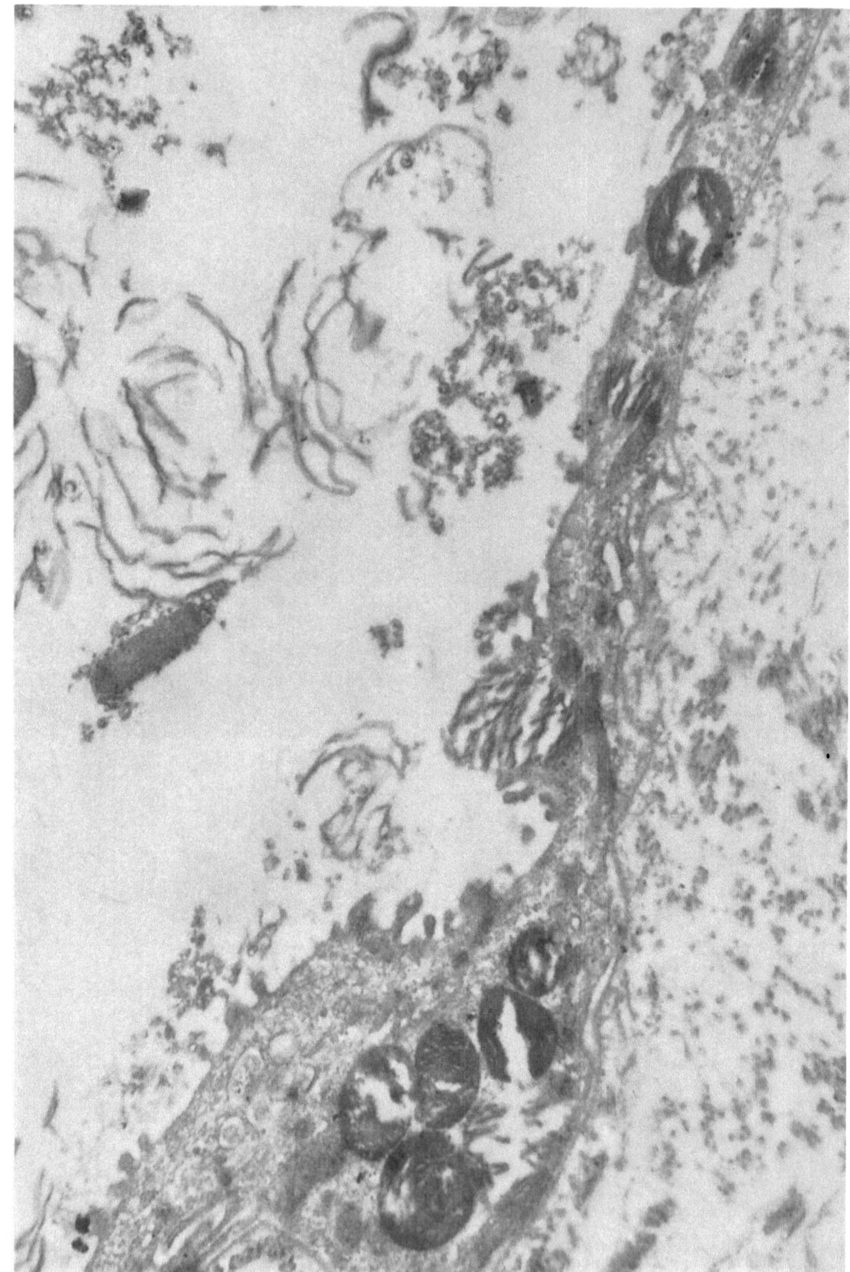

Abb. 3. Alveolarwand der Lunge des *Protopterus* mit Übergangsformen zwischen homogenen osmiophilen Gebilden und Lamellensystemen, welche in die Alveolarlichtung ausgestoßen werden. 18000 ×

unterscheiden, die sich offenbar im Zustande der Auflösung befinden. Die Auflösung tritt einmal als Spaltbildung auf, ein anderes Mal sieht man eine Aufsplitterung in feinere, konzentrisch gelagerte Lamellen. Auch sehen wir Bilder in den Alveolarzellen des *Protopterus*, welche ganz den lamellenförmig transformierten Mitochondrien gleichen, die SCHULZ (3) nach experimenteller Kohlensäurevergiftung beobachtet hat. Auf Abb. 3 sieht man bei *Protopterus* alle Übergangs-stadien zwischen den homogenen osmiophilen Einschlüssen und Lamellensystemen, welche von den Alveolarzellen ausgestoßen werden und sich in großer Anzahl frei im Lumen des Luftraumes befinden. Übergangsformen zwischen Mitochondrien einerseits und osmiophilen Einschlüssen oder lamellären Systemen andererseits wurden nicht gesehen. Zwischen den Endothelzellen der Lungencapillaren finden sich Lücken, die als vorübergehende Bildungen auch für andere Tiere beschrieben wurden (2).

Aus unseren Beobachtungen ergibt sich, daß die Blut-Luftschranke der Lunge des *Protopterus* und des *Axolotls* sich hinsichtlich des Feinbaues nicht von jenen in der Lunge der Säuger unterscheidet. Die Alveolarwand des primitivsten Lungentypus läßt denselben Strukturplan wie bei den komplizierten Lungentypen der Säuger erkennen. Dieser Befund und die Tatsache, daß die Dipnoerlunge von entodermalem Epithel ausgekleidet ist, berechtigen uns erneut, die Alveolarwand der Säugerlunge als echtes Epithel zu bezeichnen.

Das Auffinden von Übergangsformen zwischen osmiophilen Einschlüssen und lamellenförmigen Systemen im Alveolarepithel des *Protopterus*, welche im Luftraum ausgestoßen werden, läßt fragen, ob es sich hier um einen physiologischen oder pathologischen Prozeß handelt. Es könnte natürlich sein, daß die Lungen der Dipnoer eine CO_2-Retention aufweisen, Die Tatsache aber, daß Schulz (3) in der normalen embryonalen Rattenlunge gleichartige Bilder gefunden hat, wie wir bei den normalen *Protopterus*, rät uns zur Vorsicht bei der Interpretation von Strukturänderungen als pathologische Vorgänge.

Literatur

1. Bargmann, W., u. A. Knoop: Z. Zellforsch. **44**, 263 (1956).
2. Groodt, M. de, A. Lagasse et M. Sebruyns: Arch. Anat. micr. **47**, 67 (1958).
3. Schulz, H.: Beitr. path. Anat. **119**, 45 (1958).

Die submikroskopische Morphologie des Kiemenepithels*

Heribert Schulz
Pathologisches Institut der Medizinischen Akademie in Düsseldorf
(Direktor: Prof. Dr. med. H. Meessen)

Die Atmungsorgane der Fische stellen in der Regel die Kiemen dar. Der Gasaustausch spielt sich hauptsächlich in den sekundären Kiemenfältchen ab, die sich in großer Anzahl auf den Kiemenblättchen (Hemibranchien) befinden. Diese Fältchen bestehen aus zwei sehr feinen Epithellagen, die von Pilasterzellen und den Kiemencapillaren auseinandergehalten werden. In Fortführung unserer Untersuchungen an respiratorischen Epithelien (*1, 2, 3, 4, 5, 6*) interessiert uns zunächst die normale Ultrastruktur des Kiemenepithels und der Kiemencapillaren, die innerhalb des Gaswechsels liegen. Insbesondere sollen im Vergleich zur atmenden Oberfläche der Lungenalveolen der Säugetiere und des Menschen die Frage nach der Größe des Atemweges in Abhängigkeit von der Atemfunktion sowie die Frage nach der Ultrastruktur der Mitochondrien im Kiemenepithel beantwortet werden.

Kiemen von normal atmenden Fischen (*Flußbarsch, Brese, Goldfisch*) und von Labyrinthfischen (*Trichogaster trichopterus*), die unter respiratorisch ungünstigen Bedingungen leben, wurden in einer 1% igen, gepufferten Osmiumsäurelösung (pH 7,4) fixiert (*7*), in Butylmethacrylat eingebettet und mit der Ultradünnschnittmethode elektronenmikroskopisch untersucht. Außerdem wurden bei Trichogaster trichopterus die Ultrastruktur des Labyrinthorganes, das der akzessorischen Luftatmung dient, sowie von Axolotl (Amblystoma mexicanum) die Kiemenbüschel studiert. — Von etwa 300—350 Å dicken Schnitten des Kiemenepithels des Goldfisches wurden mit dem Elektronenmikroskop stereoskopische Aufnahmen angefertigt. Die Objekte wurden zur Herstellung der stereoskopischen Teilbilder im Siemens-ÜM-100 um 5° und im RCA-EMU-3C-Elektronenmikroskop um 10° in beiden Richtungen in der optischen Achse gegen die normale Mittellage gekippt. — Am Kiemenepithel des Axolotl führten wir vergleichende histochemische Untersuchungen durch. Die histochemischen Reaktionen wurden in eine Gruppe zur Prüfung auf *Mucoproteine* (PAS-Reaktion, Toluidinblau-Technik mit und ohne Einwirkung von Schwefelsäure) sowie in eine Gruppe zur Prüfung auf *Phospholipoide* (Sudan-Schwarz-B nach Formalin-Kobalt-Fixierung, Baker-Technik mit und ohne Pyridinextraktion) eingeteilt.

---

* Mit Unterstützung der Deutschen Forschungsgemeinschaft.

Die Epithelien der Kiemenfältchen von Flußbarsch, Brese und Goldfisch bestehen — im Gegensatz zur Oberfläche der Lungenalveolen der luftatmenden Tiere — aus stark ineinander

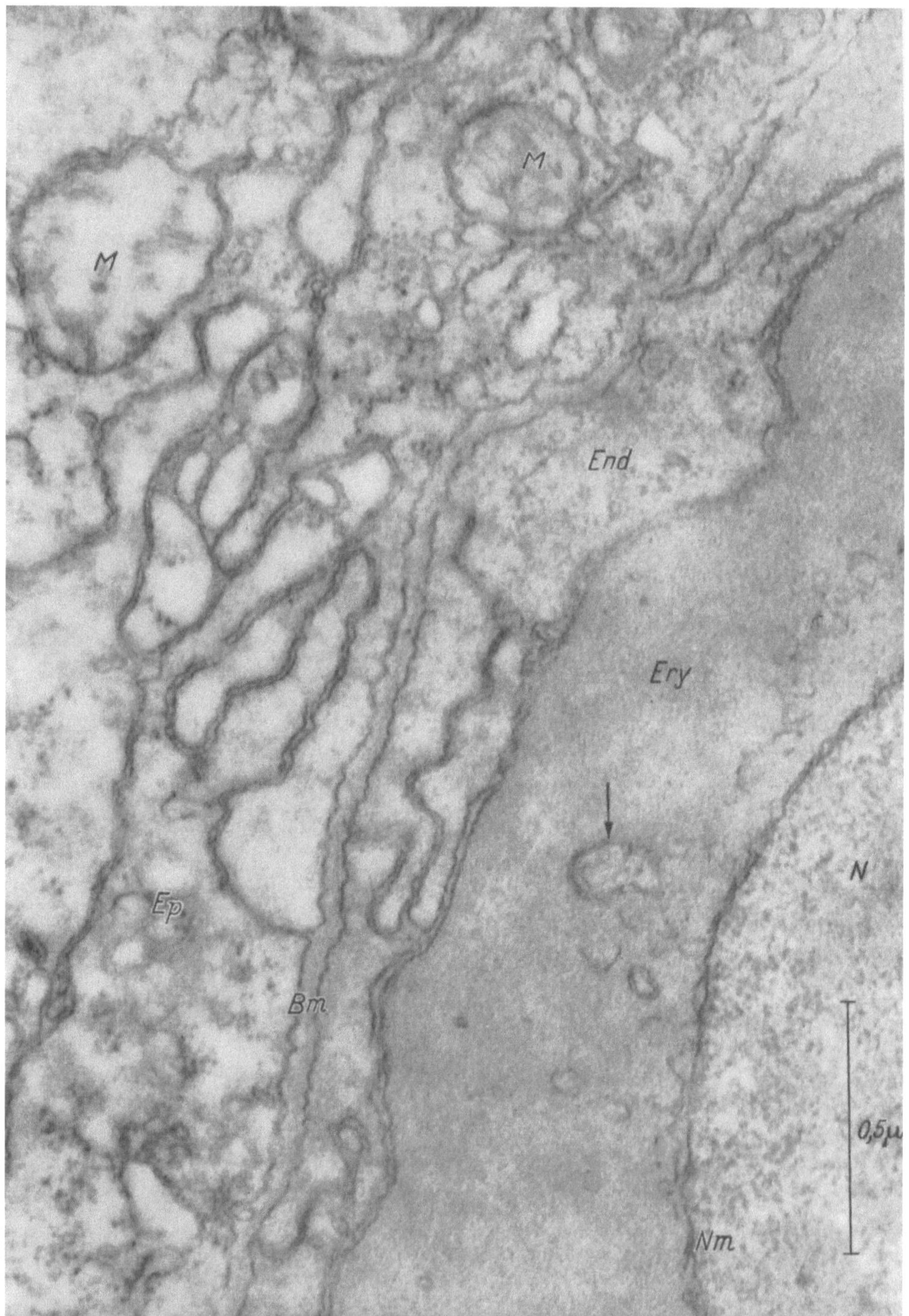

Abb. 1. Ausschnitt von Kiemenepithel und Kiemencapillare des Goldfisches. Auf der rechten Bildseite ist ein kernhaltiger Erythrocyt (*Ery*) dicht den Endothelzellen (*End*) angelagert. N = Zellkern des Erythrocyten mit grob-granulärem Nucleoplasma. Nm = primäre und sekundäre Kernmembran. Bei → Mitochondrium im Cytoplasma des Erythrocyten. Bm = homogene Basalmembran mit sich gegenüberliegenden, sehr osmiophilen gewellten Zellmembranen von Endothel und Epithel (*Ep*). Die seitlichen Zellmembranen von Epithel und Endothel sind serpentinenartig gewunden (β-Cytomembranen). M = Mitochondrien des Kiemenepithels. Archiv-Nr.: 225 E/58. RCA-EMU-3C-Elektronenmikroskop. Elektronenoptisch: 16600mal, Abb.: 68900mal

verzahnten Zellen mit zahlreichen serpentinenartig gewundenen β-Cytomembranen (Abb. 1), wie sie in den Tubulusepithelien der Niere und im Ciliarepithel beobachtet wurden. Das Vorkommen von β-Cytomembranen läßt annehmen, daß die Kiemenepithelien auch Funktionen des Ionen- und Wasserhaushaltes erfüllen. Die Basalmembran der Kiemencapillaren mißt 400—665 Å. Die dünnste Stelle des Endothels beträgt 800—1150 Å, die des Epithels 0,4 μ. Der Minimalwert für den gesamten Atemweg beträgt 0,5—0,6 μ. Das Cytoplasma der großen, ovalen kernhaltigen Erythrocyten enthält wenige Mitochondrien und agranuläre Membranen, die den Strukturen des Golgi-Apparates entsprechen. Die Kerne der Erythrocyten haben keine Nucleolen, das Nucleoplasma besteht aus 120—150 Å auffallend großen Granula und aus einer dichten homogenen Substanz. — Die Kiemenbüschel des Axolotl bestehen aus einem geschichteten Epithel und einem gefäßführenden, sehr pigmentreichen bindegewebigen Gerüst. Zwischen den großen, meist ovalen Zellen der Intermediärzone finden sich intercelluläre Straßen und Intercellularbrücken. Die superficiellen Zellen haben ein wabiges Cytoplasma und zeigen an ihrer freien Oberfläche kleine und größere, teils tentakelartige Cytoplasmafüßchen, die in einer extracellulären homogenen, filmartigen Schicht eingebettet sind. Der große Abstand von den Blutgefäßen des Kiemengerüstes bis zum Wasser, das die Kiemenepithelien umspült, läßt beim erwachsenen Axolotl auf eine geringe Atemfunktion der Kiemenbüschel schließen. — Der geringste Atemweg, den wir in unseren submikroskopischen Untersuchungen an respiratorischen Epithelien bisher beobachten konnten, findet sich im Labyrinthorgan von Trichogaster trichopterus. Der Minimalwert für den gesamten Blut-Luft-Weg beträgt hier nur 600 Å, der Mittelwert 1300—2700 Å und der Maximalwert etwa 1,2 μ. Dieser Befund deutet auf eine sehr hohe respiratorische Leistung des Labyrinthorgans. An einzelnen Stellen des Blut-Luft-Weges hat man den Eindruck, daß die Capillaren keine Basalmembran und keine Endothelien mehr aufweisen, so daß der Blutraum nur von einer dünnen Epithelschicht bedeckt ist. An anderen Stellen finden sich kernhaltige Erythrocyten mit tentakelartigen Ausläufern, die tief in das Cytoplasma der Endothelien eindringen.

Die Mitochondrien im Kiemenepithel des Goldfisches gleichen im Prinzip den Mitochondrien höherer Tiere, sie weisen jedoch in der Feinstruktur des Mitochondrienkörpers einige Unterschiede auf. In Mitochondrien von vielen Geweben der Säugetiere und des Menschen sind die Innenmembranen meist quer zur Längsachse des Mitochondriums angeordnet und bestehen aus scheibenartigen Doppellamellen, die mit ,,cristae mitochondriales" bezeichnet sind. Im Kiemenepithel finden sich im Inneren der Mitochondrien zahlreiche im Durchmesser 140—250 Å große Röhren oder Tubuli, wie sie in elektronenoptischen Untersuchungen an pflanzlichen Chondriosomen (8), an Paramecien (9, 10, 11, 12), Amöben (13) und in der Nebennierenrinde von Ratten (14) bisher beobachtet wurden. In elektronenmikroskopischen Stereoaufnahmen konnten wir ein Austreten von Tubuli in das Cytoplasma durch Sprossung der äußeren Mitochondrienhülle oder durch Ausschleusung des Mitochondrieninhaltes nachweisen (Pfeile, Abb. 2). Die abgegebenen Tubuli mitochondriales finden sich als freie Gebilde in der Grundsubstanz des Cytoplasmas. Sie sind stark geschlängelt, einige weisen eine Gabelung auf und in Querschnitten erkennt man, daß es sich um Röhren handelt, die im Innern elektronenoptisch hell sind. Die Funktion der Tubuli mitochondriales dürfte der Funktion der ,,cristae mitochondriales" identisch sein. Die Tubuli bewirken jedoch je Einheit und durch ihr vermehrtes Vorkommen eine beträchtliche Vergrößerung der Oberfläche. Da an die Innenmembranen der Mitochondrien zahlreiche oxydative Enzyme lokalisiert sind (15, 16), können die von den Mitochondrien im Kiemenepithel der Fische abgegebenen Tubuli die gleichen Enzyme besitzen. Ob die ausgeschleusten Tubuli mitochondriales als zusätzliches Membranmaterial auch mit der gleichen enzymatischen Aktivität wie innerhalb des Mitochondriums wirksam sind, ist vorerst nicht zu beurteilen. Wesentlich ist zunächst der Befund, daß ein hoher Prozentsatz der Tubuli von den Mitochondrien abgegeben wird und als submikroskopisch erkennbare Strukturen außerhalb der Mitochondrien im Cytoplasma vorkommen.

In anderen Kiemenepithelien, die durch eine starke Mikrovesikulation des Golgi-Apparates ausgezeichnet sind, finden sich Gruppen von Mitochondrien mit zahlreichen nadelförmigen, kristalloiden Innenstrukturen und aufgelockerter Mitochondrienmatrix (Abb. 3). Solche Mitochondrien betrachten wir als Modifikationsformen, die unter normalen respiratorischen Bedingungen im Kiemenepithel auftreten können. Die kristalloide Modifikation der Mitochondrien ist unseres Wissens im Schrifttum bisher noch nicht beschrieben worden. Ob es sich hierbei um

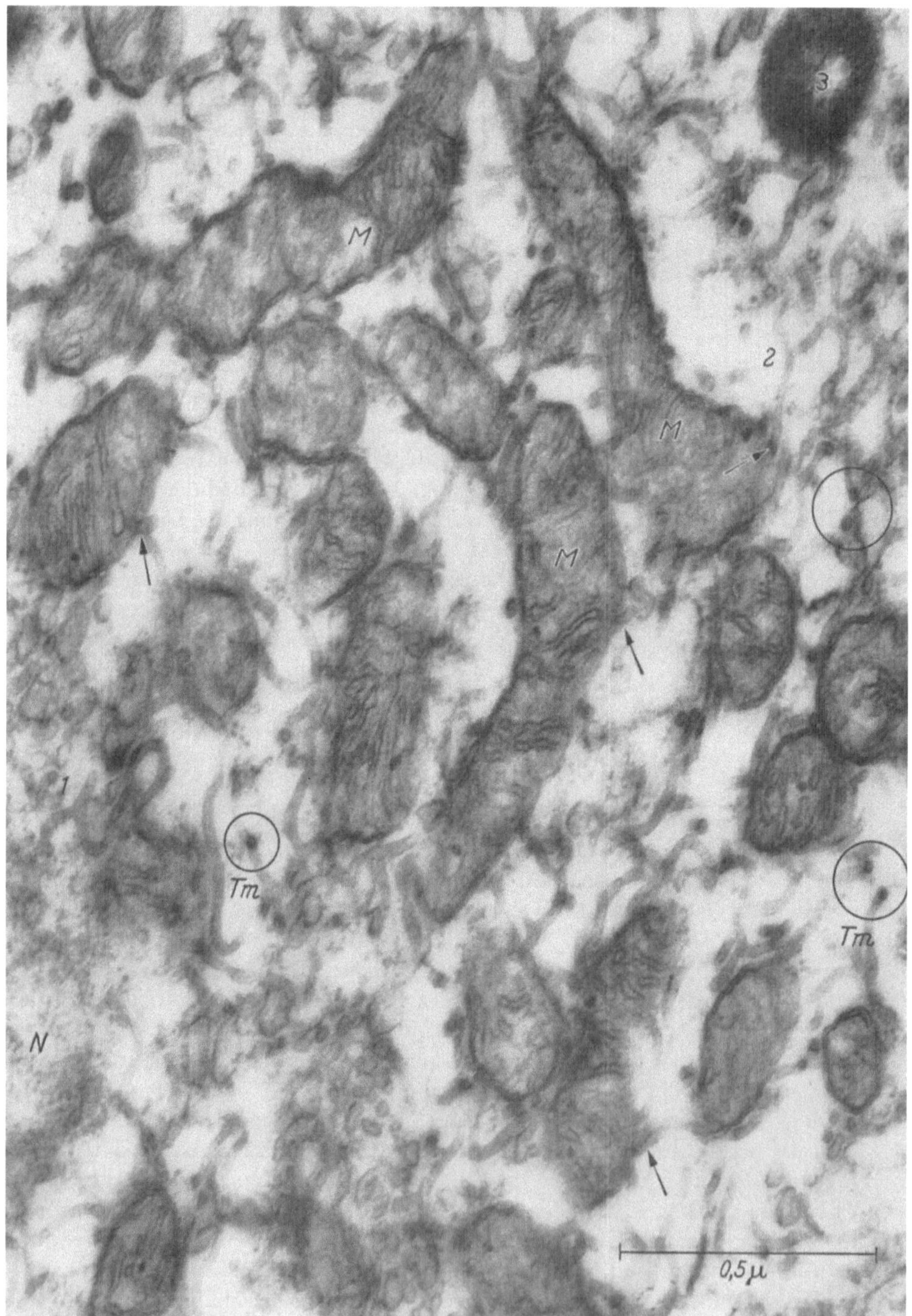

Abb. 2. Ausschnitt einer Kiemenepithelzelle des Goldfisches. Die Dicke des Schnittes beträgt etwa 300—350 Å. Im Cytoplasma längs- und quergeschnittene Mitochondrien (M) und zahlreiche in der Grundsubstanz der Zelle frei liegende stark geschlängelte Tubuli mitochondriales (Tm). In den Kreisen finden sich Querschnitte von Tubuli mitochondriales; bei den Pfeilen erkennt man an den Mitochondrien (M) eine Sprossung und Ausschleusung von Tubuli mitochondriales. $1, 2$ = gewundene Tubuli mitochondriales mit Verzweigungen. 3 = ovales osmiophiles Cytosom. N = Zellkern am linken Bildrand. Im Cytoplasma finden sich außerdem einige Mikrobläschen des Golgi-Apparates und einige Palade-Granula. Archiv-Nr.: 300 B/58, RCA-EMU-3C-Elektronenmikroskop. Elektronenoptisch: 16 600mal, Abb.: 71 000mal

Entwicklungsformen von Mitochondrien oder um degenerierende Tubuli mitochondriales handelt, kann vorerst noch nicht entschieden werden; der allgemeine Aspekt der Zellen und der Mitochondrien sprechen jedoch mehr für degenerative Veränderungen. Eine lamellenförmige Transformation der Mitochondrien, die wir in den Alveolarepithelien der menschlichen Lunge und der Rattenlunge (*1, 2, 6*) nach Kohlensäureatmung nachgewiesen haben, ist in den Kiemenepithelien

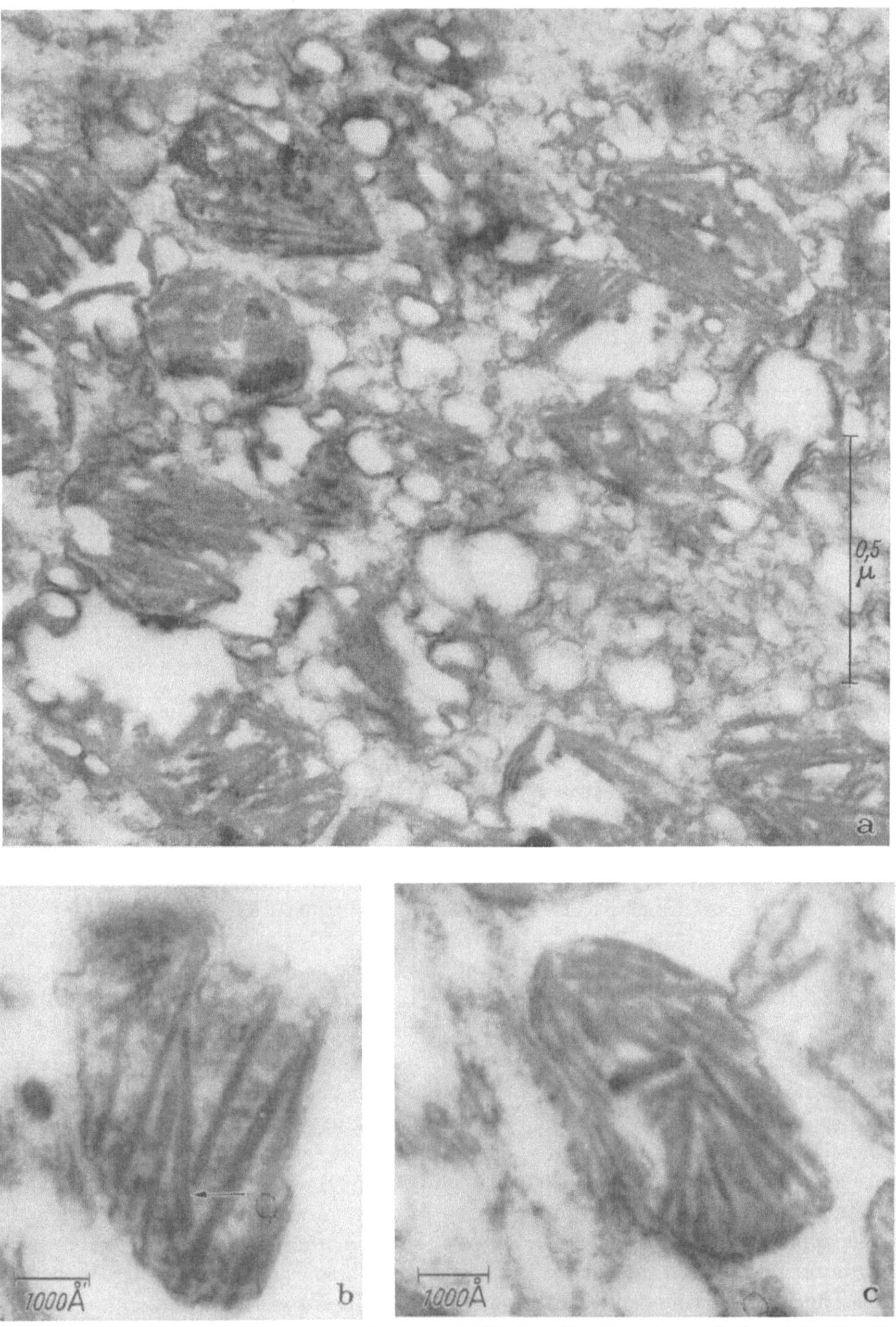

Abb. 3a—c. Ausschnitte von Kiemenepithelzellen des Goldfisches. Im Cytoplasma finden sich Mikrobläschen und veränderte Mitochondrien mit zahlreichen nadelförmigen, kristalloiden Innenstrukturen sowie mit aufgelockerter Mitochondrienmatrix. Bei → helle Zone in einem stiftförmigen Gebilde. Archiv-Nummern: a 386 A/58, b 695 E/58, c 747 C/58. RCA-EMU-3C-Elektronenmikroskop. Elektronenoptisch: 16600mal, Abb.: a 68000mal, b 107000mal, c 91300mal

der Fische nicht zu beobachten. Eine lamellenförmige Transformation von Mitochondrien ist dort auch nicht zu erwarten, da diese an das Vorhandensein von „cristae mitochondriales" gebunden ist. Es ist daher möglich, daß die kristalloide Modifikation der Mitochondrien mit tubulären Innenstrukturen ein Analogon zur lamellenförmigen Transformation darstellt.

In den Kiemenbüscheln von Axolotl und den Kiemenepithelien des Goldfisches beobachteten wir außerdem eine besondere Gruppe von submikroskopischen Strukturen, die wir an anderer Stelle gemeinsam mit DE PAOLA (17) ausführlich beschrieben haben. Diese Strukturen haben weitgehende Ähnlichkeiten zu Myelinscheiden und finden sich vorwiegend in geschichtetem Kiemenepithel, in dem eine Schleimsekretion beobachtet wird.

Literatur

1. Schulz, H.: Naturwissenschaften 43, 205 (1956).
2. — Virchows Arch. path. Anat. 328, 582 (1956).
3. Meessen, H., u. H. Schulz: Bad Oeynhausener Gespräche I vom 19.—21. Oktober 1956. S. 54. Berlin-Göttingen-Heidelberg: Springer 1957.
4. Schulz, H.: Z. Zellforsch. 46, 583 (1957).
5. — Beitr. path. Anat. 119, 71 (1958).
6. — Beitr. path. Anat. 119, 45 (1958).
7. Palade, G. E.: J. exp. Med. 95, 285 (1952).
8. Heitz, E.: Z. Naturforsch. 12b, 576 (1957).
9. Sedar, A. W., and K. R. Porter: J. biophys. biochem. Cytol. 1, 583 (1955).
10. Powers, E. L., C. F. Ehret, L. E. Roth and O. T. Minick: J. biophys. biochem. Cytol. 2, Suppl., 341 (1956).
11. Wohlfarth-Bottermann, K. E.: Z. Naturforsch. 11b, 578 (1956).
12. — Z. Naturforsch. 12b, 164 (1957).
13. Sedar, A. W., and M. A. Rudzinska: J. biophys. biochem. Cytol. 2, Suppl., 331 (1956).
14. Belt, D., and D. C. Pease: J. biophys. biochem. Cytol. 2, Suppl., 369 (1956).
15. Barnett, R. J., and G. E. Palade: J. biophys. biochem. Cytol. 3, 577 (1957).
16. Siekevitz, P., and M. L. Watson: J. biophys. biochem. Cytol. 2, 653 (1956).
17. Schulz, H., u. D. de Paola: Z. Zellforsch. 49, 125 (1958).

9. Reproduktionsorgane

The fine structure of the limiting membrane of the seminiferous tubule in the rat

Y. Clermont

Department of Anatomy, McGill University, Montreal, Canada

The limiting membrane of the seminiferous tubule in the rat testis is composed of two concentric PA-Schiff staining layers referred to as internal (epithelial side) and external (vascular side) lamellae. Enclosed between the lamellae there is a single continuous layer of flattened cells. Outside the external lamella a second continuous sheath is formed by other flattened cells. Under the electron microscope, the cytoplasm of the cells seen between the lamellae shows (a) a few organelles (mitochondria, aggregates of RNA-containing granules, and rough-surfaced vesicles) seen generally close to the nucleus; (b) smooth surfaced vesicles, which may arise from the numerous invaginations seen along the cell membrane; (c) a dense meshwork of delicate filaments. These characteristics make these elements similar to smooth muscle cells. The cytoplasm of the cells applied outside the external lamella does not contain the filamentous component and the invaginations at the cell surface are less numerous. These cells are probably connective tissue elements. The internal connective tissue lamella contain three layers; an inner and an outer layer made up of homogeneous dense substance and an intermediate layer made up of fibrils probably collagenous in nature. The external lamella is composed almost exclusively of the homogeneous substance.

An extended version of this article appeared in Exp. Cell. Res. 15, 438 (1958).

Some aspects of the oogenesis of the pondsnail Limnaea stagnalis L

A. Recourt

Zoological Laboratory University of Utrecht, and the Physical Technical X-ray Laboratorium Philips' Gloei-lampenfabrieken, Eindhoven (Holland)

The physiology and embryology of the pondsnail Limnaea stagnalis L. has formed part of the studies undertaken at the Zoological Laboratory of the State University of Utrecht. Earlier study of the oogenesis of this organism was carried out using light microscopy and this paper deals with observations made with the help of electron microscopy.

The research carried out by Bretschneider and Raven (1), was of particular interest cyto-chemically in regard to different cell inclusions, but the limitations of the light microscope resulted in many questions still remaining unresolved.

The gonad of this hermaphrodite animal consists of a great number of acini, the contents of which are extravasated in a collecting channel. The acinus wall is formed of connective tissue fibrils embedded in matter which seen through an electron microscope appears homogeneous. The innerside of the acinus membrane is covered by germ-epithelium, which is the origin of the oocytes, the spermatocytes, and the nutritive cells. The cells lining the cavity show an intense amoeboid mobility as do also the early oocytes. Afterwards, in the sessile phase, the oocytes are surrounded by nutritive cells which form a follicle around the egg. The development of the spermatides is not mentioned in this paper.

The material is fixed in Palade buffer solution with a p_H of 8 and an osmiumtetroxide concentration of 1% After dehydration and embedding in a mixture of methyl and butyl methacrylate, thin sections are made with the help of an ultramicrotome built in accordance with the principle developed by Haanstra (2). The sections are collected on the surface of carbon films which in some cases are used as nets (3). Pictures of these sections are taken with different types of Philips electron microscopes (EM 100-A, EM 100-B and EM 75-A).

It is not the intention to give a complete description of a young eggcell but a few points on the general cytology will probably be of interest. It has been found that the border between the oocyte and the follicle has a great number of membranes, and it is thought that an intensive exchange of nutritive products takes place in this area. In all probability the follicle also has a very strong influence on the morphology of the oocyte. It is clear again that membraneous structures are of great morpho-genetic importance, as has been described earlier by Birbeck and Mercer (4).

The cytoplasm of the oocyte shows various structure elements: a relatively simple endoplasmic reticulum containing vesicula and some considerably longer membranes which are covered with the so-called Palade granules, in this case about 250 Å in diameter. These grains are also found more dispersed through the cell.

The mitochondria show the well-known structure. It is perhaps possible to say that the structure of the internal membranes is in some degree more irregular than that of the mitochondria which are found in various cells of Vertebrates.

The nucleus is surrounded by a clearly-defined membrane. This membrane is seen to be double and is perforated very regularly. The nucleoplasm is much less dense than the cytoplasm outside the nucleus-membrane. In the very young stages of the oocytes two nucleoli are visible in the nucleus: the so-called eu-nucleolus which is about 4 μ large with a vacuolized structure consisting of granules of about 100—200 Å in diameter, and further the para-nucleolus which is considerably smaller than the eu-nucleolus (about 1.5 μ in diameter). Structurally the latter element is nearly the same as the former one. In the further developing stages the compact mass of the nucleoli disappears and seems to spread over the whole nucleus. Small groups of the granules move in the direction of the nucleus membrane, which shows a highly folded nature.

These particle groups possibly travel through the perforations, or are separated from the nucleus by pino-cytosis. This latter theory describes the possibility that a part of the nucleus membrane travels into the cell with the nucleolus material. In view of the fact that peforated membranes have been found in the cytoplasm this theory could be correct. However, the distances between the perforations in these membranes are not consistent with those in the structure of the nucleus-membrane. Therefore, it is also possible that these membranes are oblique sections of the endoplasmic membranes. In view of the fact that membranes of this type are generally

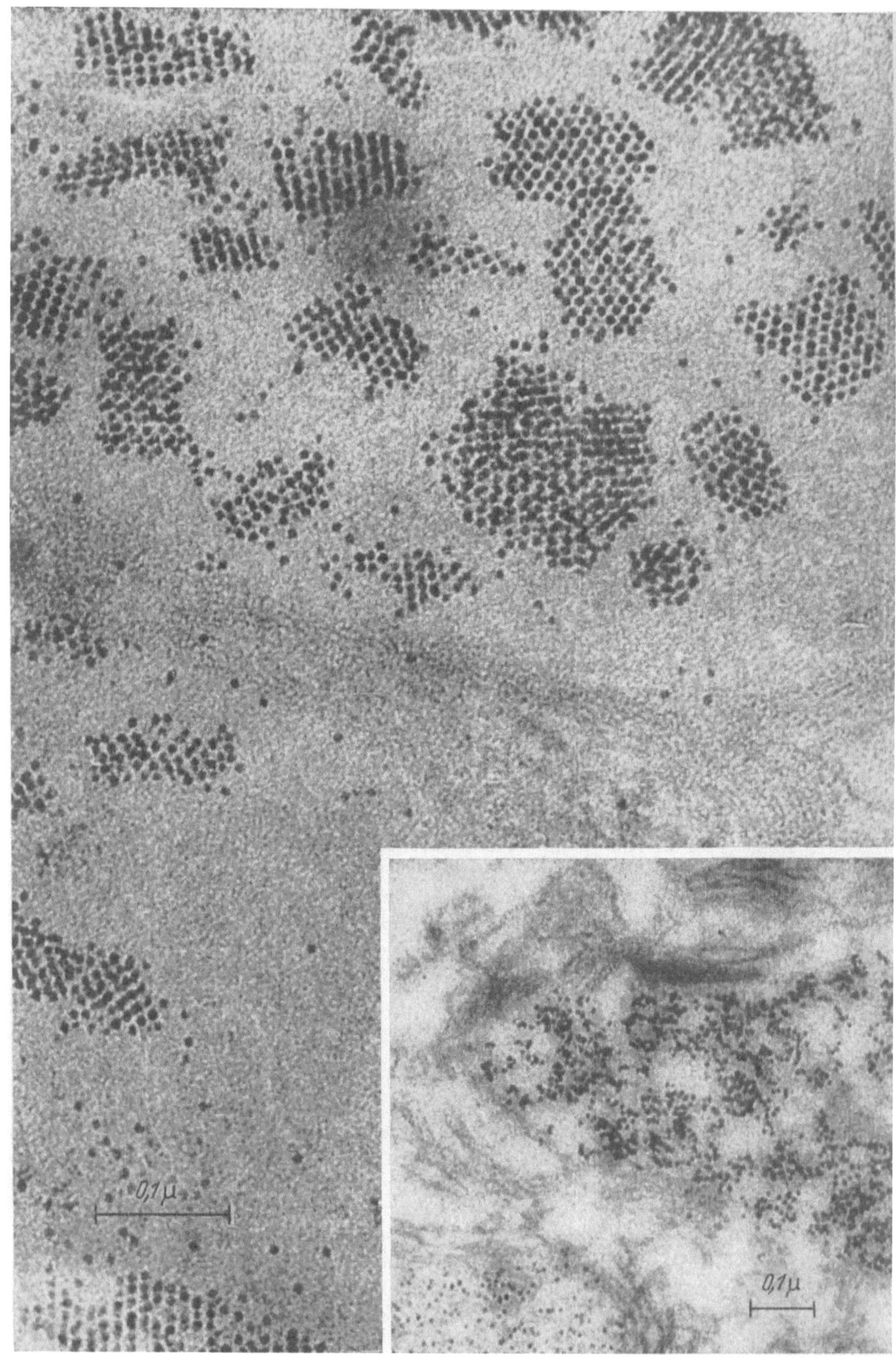

Fig. 1. Details of two yolk-granules with crystalline matter inside the membrane. Magnification 225,000 ×.
Inset: details of yolk granules showing proteinmolecules not regularly arranged in crystals. Magnification
105.000 ×

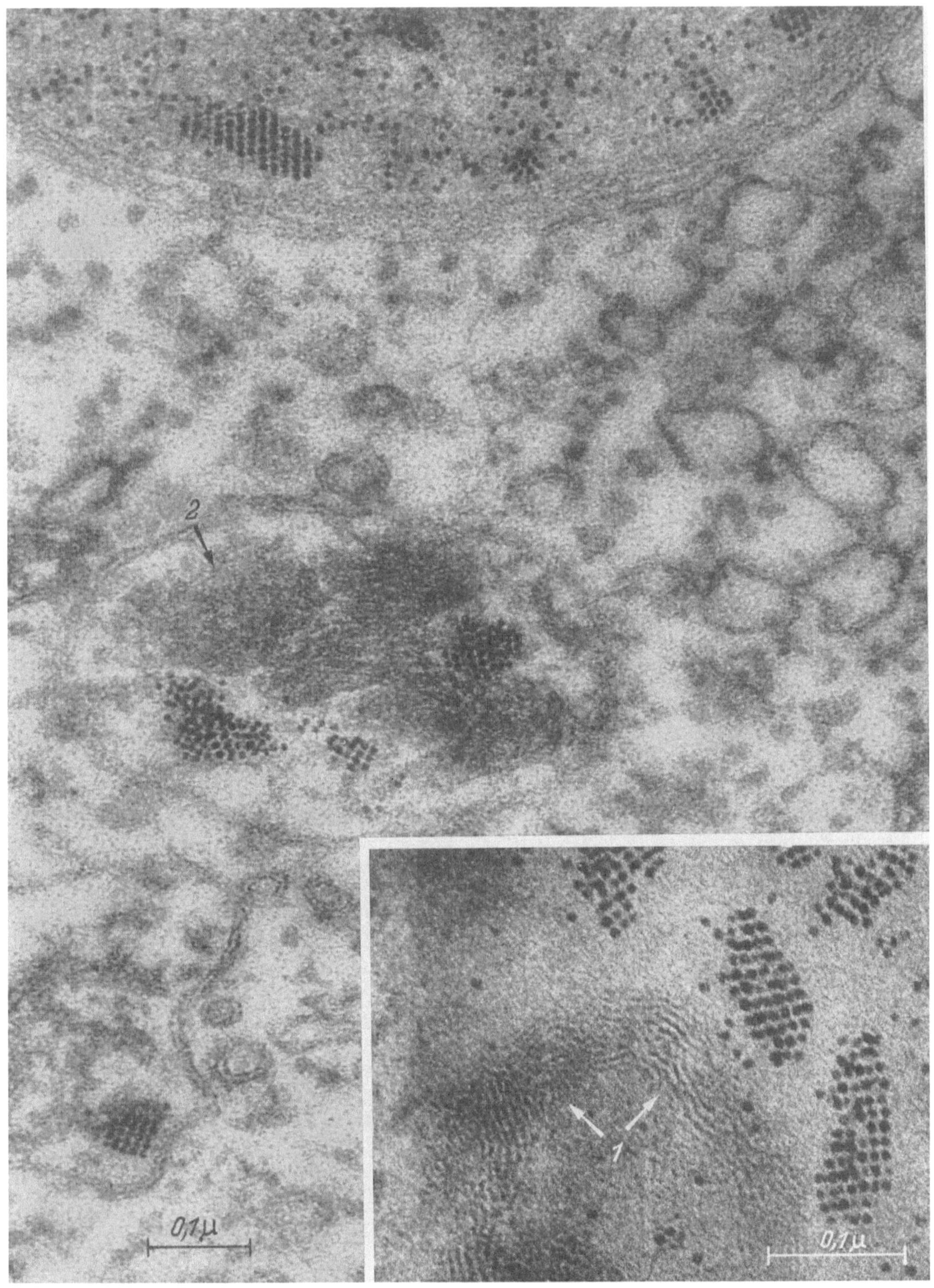

Fig. 2. Section through a young development stage of a yolk granule, showing a membraneous structure inside the granule (2). At the top of the picture an older stage. Magnification 162.000 ×. Inset: a detail of a yolk granule membrane showing numerous parallel membranes (1). Magnification 225.000 ×

not very long and are very seldom straight, it is understandable that these types of membrane sections are not found very frequently.

The following observations in connection with mitochondria are of interest when compared with those made by Bretschneider and Raven (*1*). They discovered that granular mitochondria are present in the young eggcell. At a later stage of development more and more filiform mitochondria appear; these are nearly 4 μ long. Granular swellings about 0.5 μ in size are formed at the

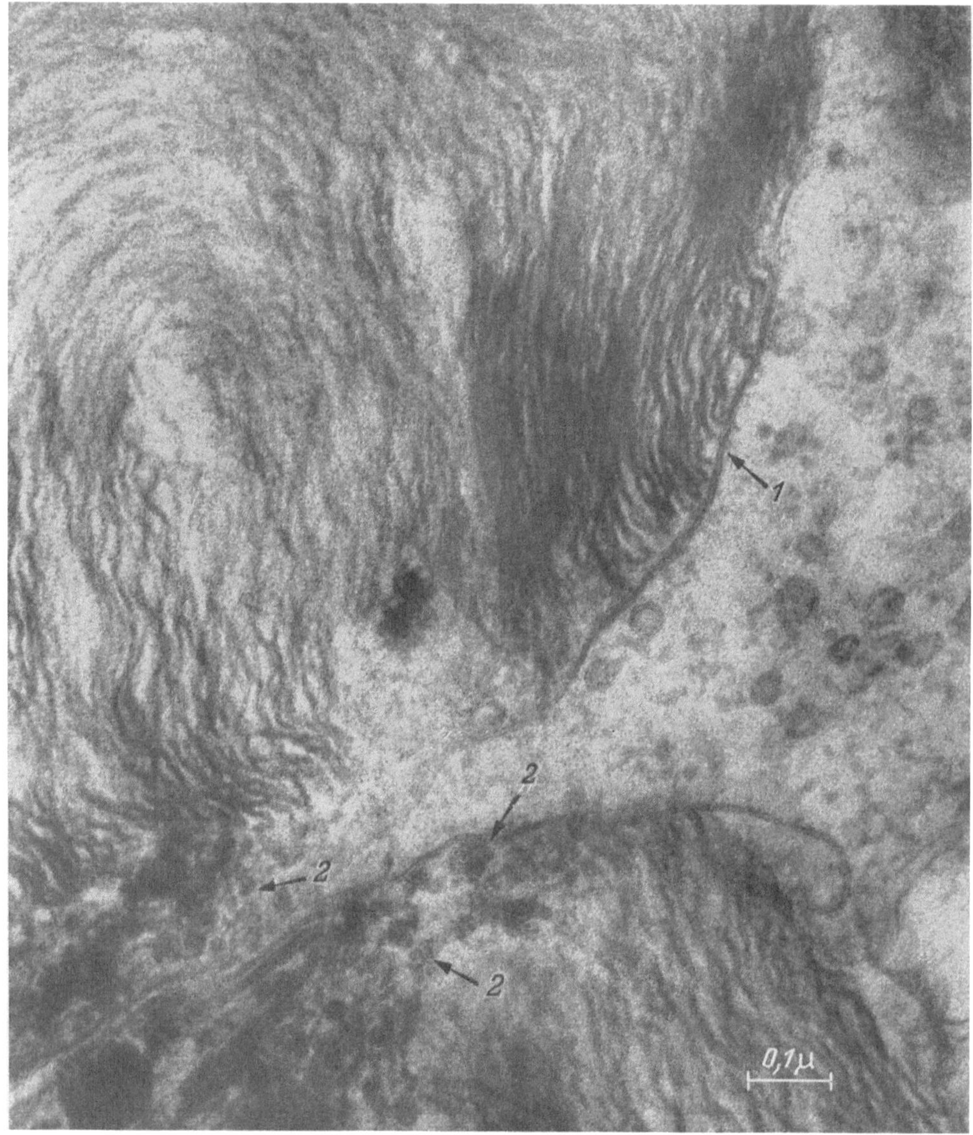

Fig. 3. Details of particles found in the follicle cells. These particles are showing many internal membranes and a double outer membrane (*1*). Note the crystalline material formed inside these particles (*2*). Magnification 105.000 ×

surface of these mitochondria. The whole mitochondrium is normally transformed into a chain of these granules which Bretschneider and Raven called α-granules. Using the electron microscope it was possible to see that these swellings are in reality particles formed by the division of the long mitochondria. It seems possible that these small mitochondria are identical with the α-granules mentioned above.

The Golgi-complex which is found in young oocytes shows the same structural elements when compared with descriptions given by a number of authors of other Golgi-complexes. In the earliest mobile oocyte the Golgi field is found next to the nucleus. Afterwards this Golgi field divides and disperses into single granules, which appear to consist of a complex of parallel double membranes. At the same time the development of the yolk-granules starts. It has never been found that the yolk-substance forms in contact with this membraneous component. In the earliest stages the yolk-granules have a membraneous periphery which observed through a light microscope, appears to be very osmiophilic. (inset Fig. 2) At this stage, a complex of very dense particles, 50—60 Å in diameter is already visible in the centre (Fig. 1). These particles are arranged in relatively big crystals, as described by Elbers (5) and Favard and Carasso (6), but they are also found in a irregular pattern (inset Fig. 1). These young yolk granule stages are developing further towards particles with a diameter of about 4 μ; the osmiophilic periphery becomes thinner, perhaps in order that this crystalline structure may form (Fig. 2.) The high contrast of these protein molecules occurs not only because of the high osmiophily: formaldehyde fixation (6,7) also gives a very high contrast. It has been thought probable that iron is absorbed into the proteinmicelles, but it has not been possible to prove this, for example, by electron diffraction or fluorescent analysis. Using light microscope techniques it has been found that in very young oocyte stages iron is dispersed through the cell. After yolk formation has begun the demonstrable quantity of iron becomes smaller and finally no iron can be detected at all. It is only after fertilization when the reserve proteins have been used, that iron is again found. Metal ions taken up in the protein molecules are not detectable in any way, and so it is possible that the higher iron concentration in the neighbourhood of stretched protein molecules play a role in changing the p_H of the medium, and this perhaps causes the conversion to the globular form. This dense globular form can be observed in the yolk-granules.

Finally it is interesting to pay attention to the presence of a great number of membranes surrounding the areas where it is obvious that a protein-synthesis or perhaps a general synthesis takes place. Granules have also been found in the follicle cells, and these granules are composed almost entirely of membranes; in between these membranes crystalline matter has also been seen (Fig. 3). These particles resemble very closely the yolk-nuclei (8). Further, descriptions, given by Afzelius (9), of yolk-nuclei as they appear in Echinondermata eggcells, and those by Palay and Palade (10) of the structure of particles found in the Nissl substance of neurons, strengthen the opinion that the synthesis of different cell products always takes place in accordance with a general pattern.

References

1. Bretschneider, L. H., and Chr. P. Raven: Arch. neerl. Zool. 10, 1 (1951).
2. Haanstra, H. B.: Philips techn. Rev. 17, 178 (1955).
3. Sjöstrand, F. S.: Proc. Stockholm Conf. on Electr. Microsc. Sept. 1956, p. 120.
4. Birbeck, M. S. C., and E. H. Mercer: Proc. Stockholm Conf. on Electr. Microsc. Sept. 1956, p. 156
5. Elbers, P. F.: Proc. Kon. Ned. Akad. Wet. Ser. C, 60, 96 (1957).
6. Favard, P., and N. Carasso: C. R. Acad. Sci. 245, 2547 (1957).
7. Recourt, A.: Will be published.
8. Rebhun, L.: J. biophys. biochem. Cytol. 2, 159 (1956).
9. Afzelius, B. A.: Exp. Cell Res. 8, 147 (1955).
10. Palay, S. L., and G. E. Palade: J. biophys. biochem. Cytol. 1, 69 (1955).

Vitellogenèse de la planorbe. Ultrastructure des plaquettes vitellines

Nina Carasso et Pierre Favard

Collège de France, Laboratoire anatomique et moléculaire et Ecole Normale Supérieure,
Laboratoire de Botanique, Paris

Au cours de la vitellogenèse de la Planorbe s'élaborent des globules vitellins dont l'ultra-structure est caractéristique. Ces globules renferment en effet des macromolécules protéiques éparses dans la substance fondamentale ou groupées en agrégats ou en cristaux.

Ces molécules présentent un fort contraste aux électrons après fixation au formol, contraste qui peut s'expliquer par la présence de fer à leur niveau comme le montrent les réactions histochimiques [cf. ferritine (*1, 2*)].

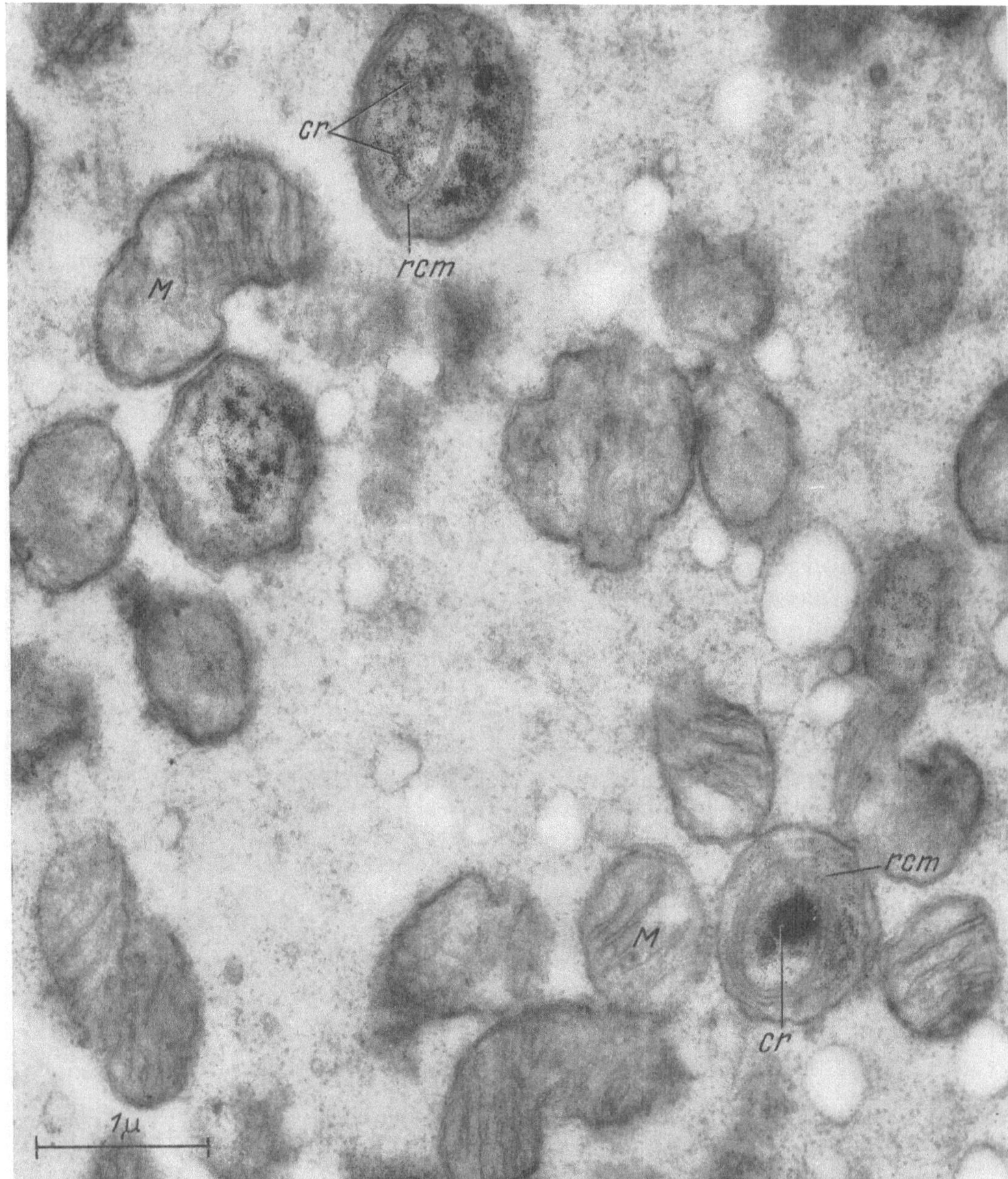

Fig. 1. Transformation de mitochondries en globules vitellins. Dans cette portion de cytoplasme d'un ovocyte, on observe des mitochondries typiques (*M*) voisinant avec des mitochondries en voie de transformation à l'intérieur desquelles on voit des macromolécules plus ou moins groupées (*cr*), et des restes de cloisons mitochondriales (*rcm*)

La présence de ces molécules (déjà signalées par ELBERS, 1957, chez la Limnée (*3*)] nous a permis de comprendre l'origine du vitellus dans le cas de la Planorbe.

Cette origine est double: d'une part, elle est purement mitochondriale; d'autre part, elle met en jeu cytoplasme, ergastoplasme, et mitochondries (*4*).

Origine mitochondriale. Dans les ovocytes jeunes, on observe, á côté de mitochondries typiques, la présence de mitochondries en train de remanier leur ultrastructure et dans la substance fondamentale desquelles apparaissent des macromolécules (Fig. 1). Ces mitochondries remaniées augmentent de volume. Leurs crêtes s'estompent ou persistent à l'état de résidus. Les macromolécules qu'elles contiennent augmentent en nombre et se groupent pour la plupart en agrégats ou en cristaux. Cette origine mitochondriale, démontrée par le microscope électronique, vient confirmer certaines hypothèses des auteurs classiques, qui, comme BRAMBELL, parlaient de «M Yolk», vitellus d'origine mitochondriale.

Origine complexe. A côté de cette origine mitochondriale, les plaquettes vitellines peuvent se former par un processus plus complexe. Des lames ergastoplasmiques sans grains de PALADE viennent ceinturer et isoler des portions de cytoplasme renfermant la plupart du temps quelques mitochondries. Des macromolécules apparaissent tant dans les portions cytoplasmiques ainsi délimitées que dans les mitochondries qui y sont incluses, et les membranes s'estompent progressivement. De la sorte se trouvent constituées des plaquettes qui augmentent encore un peu de volume cependant que de nouvelles lames ergastoplasmiques viennent s'accoler à leur périphérie. Ce sont probablement ces lames qui ont été confondues par BRETSCHNEIDER et RAVEN (*6*), chez la Limnée, avec l'appareil de GOLGI, auquel ces auteurs avaient accordé un rôle prépondérant dans la synthèse du vitellus. En fait, dans le cytoplasme des oeufs de Planorbe, l'appareil de GOLGI est représenté par des dictyosomes typiques (Fig. 3), qui ne contractent aucun rapport morphologique avec les grains vitellins.

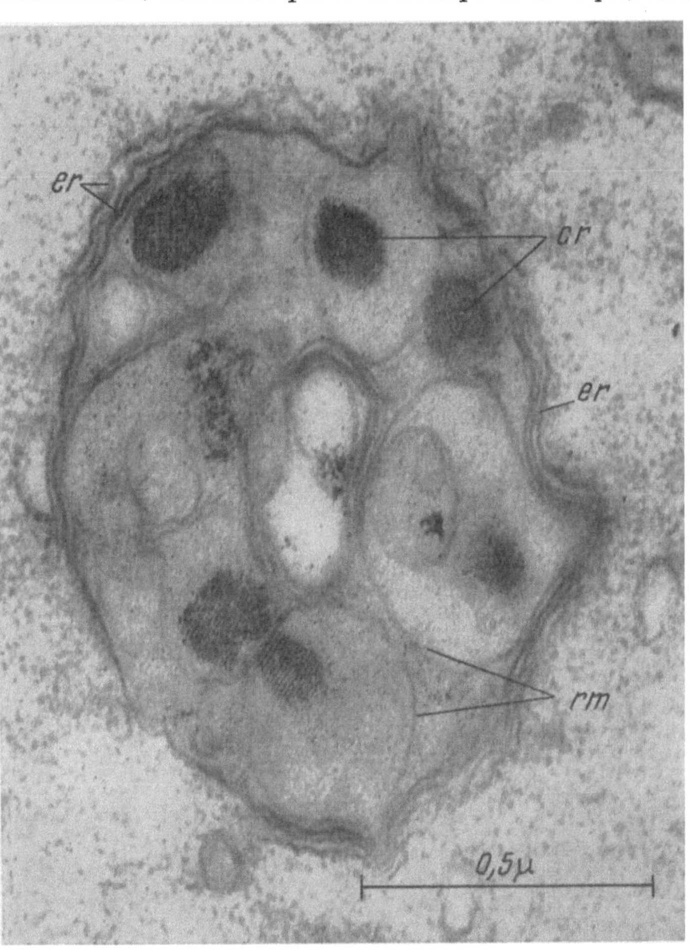

Fig. 2. Jeune globule vitellin d'origine complexe limité par des lames ergastoplasmiques (*er*), et contenant des cristaux (*cr*). De nombreux résidus de membranes (*rm*) subsistent à l'intérieur du grain. Cet ensemble évoluera pour donner un globule vitellin typique

Les plaquettes d'origine complexe ont une substance fondamentale moins osmiophile que celle des plaquettes d'origine purement mitochondriale. Plaquettes des deux origines sont par ailleurs tout à fait comparables.

Les membranes externes des globules vitellins, d'origine mitochondriale ou ergastoplasmique, délimitent à l'intérieur du cytoplasme de l'ovocyte des zones capables d'une évolution indépendante. Elles constituent de véritables barrières maintenant l'hétérogénéité de l'ovocyte. En effet, quand celles-ci disparaissent dans les cas de dégénérescence, l'ensemble du cytoplasme se transforme en un globule vitellin géant (*4*).

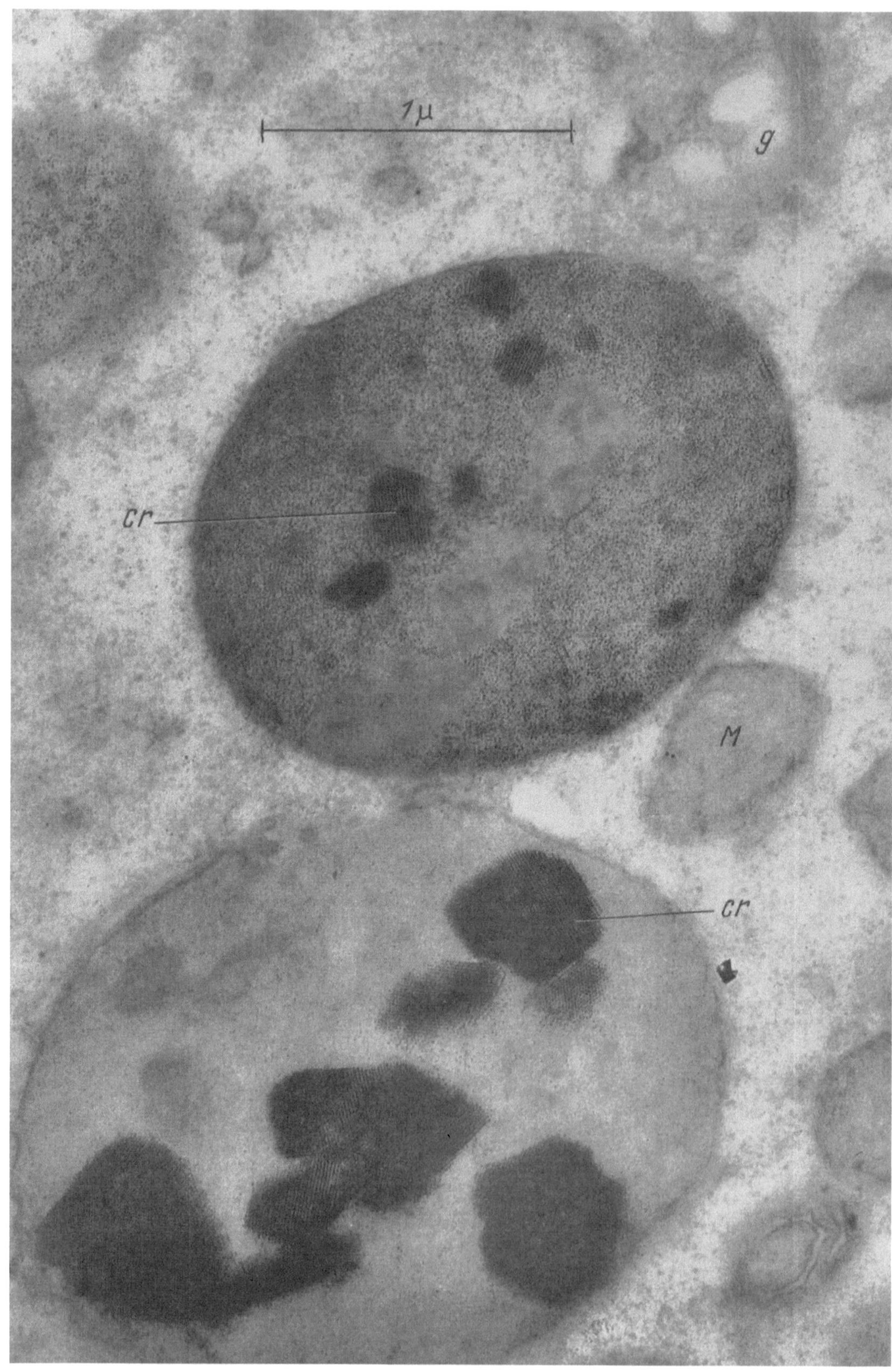

Fig. 3. Portion des cytoplasme montrant des globules vitellins typiques contenant des macromolécules éparses ou groupées en cristaux (*cr*). Mitochondries (*M*), dictyosome (*g*)

Ultrastructure des plaquettes vitellines. Les plaquettes vitellines ont une ultrastructure hétérogène:

a) la substance fondamentale, plus ou moins osmiophile selon l'origine de la plaquette, comprend des zones amorphes et parfois des zones organisées, finement striées, généralement exemptes de molécules. Les stries sont distantes les unes des autres de 80 Å. Dans certains cas, quelques molécules peuvent cependant être alignées sporadiquement le long de certaines stries. Une telle disposition peut rappeler une charge progressive en fer d'une protéine, phénomène qui s'observe pour l'apoferritine *(2)*.

L'hétérogénéité de la substance fondamentale est parfois soulignée par des arrangements en lignes ou en rosettes des macromolécules qui viennent s'y accoler, probablement par effet de charge.

b) les macromolécules sont soit dispersées dans la substance fondamentale, soit groupées en agrégats dans des zones plus osmiophiles, soit groupées en arrangements cristallins (Fig. 3, *cr*).

Chaque macromolécule semble elle-même constituée de grains plus fins situés en périphérie.

Cette étude montre donc le rôle prépondérant joué par les mitochondries dans la vitellogenèse de la Planorbe, ce dernier rôle pouvant également être joué par des ensembles plus complexes comprenant cytoplasme, ergastoplasme et mitochondries.

Le problème de la vitellogenèse a donné lieu à de nombreuses controverses. Il n'est pas possible, pour le moment, de généraliser les résultats acquis sur la Planorbe. Dans ce cas cependant, il nous faut souligner le rôle fondamental joué par les mitochondries.

Bibliographie

1. BERTHIER, J.: Bull. Biol. Fr. Belg. **82,** 61 (1948).
2. BESSIS, M., et J. BRETON-GORIUS: C. R. Acad. Sci. (Paris) **245,** 1271 (1957).
3. ELBERS, P. F.: Proc. Kon. Akad. Wet. **60,** 96 (1957).
4. FAVARD, P., et N. CARASSO: Arch. Anat. micr. Morphol. exp. **47,** 211 (1958)
5. BRAMBELL, F. W. R.: Brit. J. exp. Biol. **1,** 501 (1924).
6. BRETSCHNEIDER, L. H., et C. P. RAVEN: Arch. Neer. Zool. **10,** 1 (1951).

10. Nervengewebe

Die Feinstruktur der Kleinhirnrinde des Goldhamsters*

H. HAGER und W. HIRSCHBERGER

Hirnpathologisches Institut der Deutschen Forschungsanstalt für Psychiatrie (Max-Planck-Institut, München) und Max-Plank-Institut für Verhaltensphysiologie, Seewiesen (Obb.)

Die Rinde des Säugerkleinhirns läßt sich in die Molekularschicht, die Schicht der Purkinjezellen und die Körnerschicht gliedern. Bei schwacher Vergrößerung sind die Purkinjezellen auf Grund ihres dichteren Perikaryons gut von der Umgebung abgrenzbar. An den Ästen des geweihförmigen Dendritenbaumes finden sich enganliegende, vorwiegend longitudinal getroffene, scharf begrenzte Anschnittprofile, die Mitochondrien und wenig cytoplasmatische Strukturen enthalten. Es handelt sich um Kletterfasern, also um aus der Marklamelle stammende cerebellopetale Fasern, die im Sinne von Cajals longitudinalen axondendritischen Verbindungen mit den Baumfortsätzen in synaptischen Kontakt (glatten Streckenkontakt) treten. Das Cytoplasma der Purkinjezellen zeigt bei höherer Auflösung eine überraschend reiche Organisation. Es finden sich zahlreiche große, oval geformte Mitochondrien, ferner ein verzweigtes endoplasmatisches Reticulum. Die Golgizonen sind in atypischer Form inselförmig über das Perikaryon verteilt. In ihrem Bereich fallen in förmlichen Stapeln geschichtete Doppelmembransysteme auf. Kern und Kernmembran der Purkinjezellen zeigen keine besonderen Baumerkmale. Der zur Körnerzellschicht gerichtete Teil des Perikaryons ist von einem dichten Saum von Faseranschnittprofilen schalenförmig umgeben, die elektronenoptisch dieselben Strukturmerkmale zeigen wie die

* Mit Unterstützung der Deutschen Forschungsgemeinschaft.

28*

Kletterfasern. Es handelt sich dabei um die Korbfasern. Die Endigungsweise dieser Fasersysteme an den Purkinjeschen Zellen war lange Gegenstand heftiger Kontroversen. Held (1, 2) nahm einen kontinuierlichen Übergang der Korbfaserverästelungen in ein „pericelluläres Neuritennetz" an. Unsere elektronenoptischen Befunde geben keinen Anhaltspunkt für die Existenz des Heldschen Terminalnetzes und sprechen für synaptischen Kontakt im Sinne Cajals (3, 4). Es folgt die Betrachtung der Molekularschicht. Held (1, 2) kam seinerzeit zu der Ansicht, daß die Dendritensysteme der Purkinjeschen Zellen und der anderen Nervenzellen gemeinsam mit den Fortsätzen der Gliazellen ein allgemeines Grundnetz bilden (Neurenzytium). Elektronenoptisch zeigt die Molekularschicht äußerst verwickelte Strukturverhältnisse. Im Dünnschnittbild ist ein dichtes

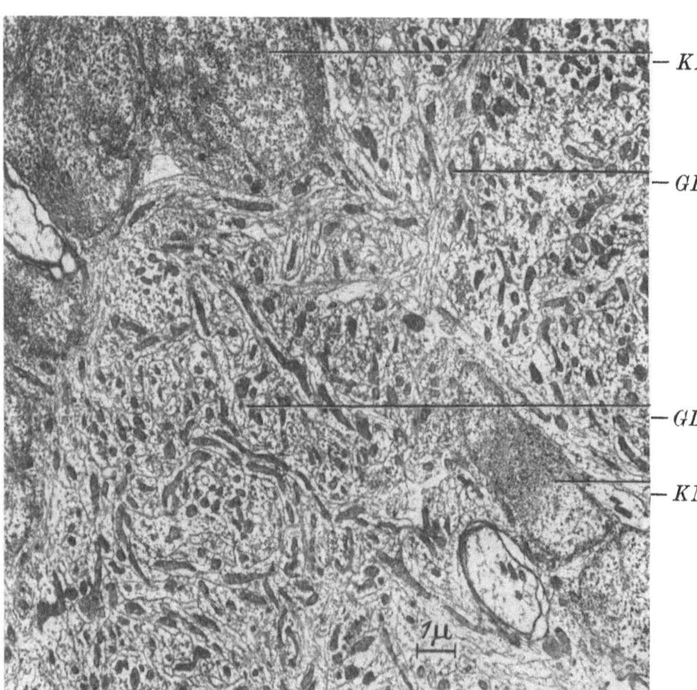

Abb. 1. Glomerulus cerebellosus (GL) aus der Körnerschicht der Kleinhirnrinde des Goldhamsters. KN Körnerzellen

Gefüge von Anschnittprofilen gröberer, feiner und feinster Zellfortsätze erkennbar, die häufig Synapsenbläschen enthalten. Die Molekularschicht der Kleinhirnrinde wird überwiegend von nervösen, z. T. auch von gliösen Zellfortsätzen gebildet, sie ist eine Region, in der die nervösen Elemente der Rinde in ungemein differenzierter Form in synaptischen Kontakt miteinander treten.

In der Körnerschicht machen die Körnerzellen den wesentlichen Anteil der Nervenzellen aus. Daneben finden sich auch noch vereinzelt liegende größere Ganglienzellen, die etwa die halbe Größe der Purkinjezellen haben und als Golgische Zellen bezeichnet wurden. Häufig weisen sie nahezu bis zum Nucleolus reichende tiefe Kernbuchten auf. Im übrigen ist ihr elektronenoptisches Bild dem der Purkinjezellen recht ähnlich. Im Nisslbild ist das Cytoplasma des Perikaryons der Körnerzellen kaum zu sehen; elektronenoptisch stellt es sich als schmaler, gut abgrenzbarer Saum dar, der nur wenige Mitochondrien und kein ausgeprägtes endoplasmatisches Reticulum enthält. Der Kern zeigt ein oder mehrere Nucleolen. Die Dendriten der Körnerzellen treten nach den klassischen Untersuchungen Cajals mit den kollateralen oder terminalen Moosfaserendigungen in gewissen inselförmigen Gebieten der Körnerschicht in Beziehung. Held (1, 2) hatte diese merkwürdigen Bildungen besonders gründlich untersucht und Glomeruli cerebellosi, d. h. Kleinhirnknäuel, genannt. Die Natur dieser Glomeruli cerebellosi war bis in jüngster Zeit noch umstritten. Elektronenoptisch zeigen die in inselförmiger Anordnung zwischen den Körnerzellen liegenden Glomeruli interessante Strukturverhältnisse (Abb. 1). Es handelt sich um Zellfortsatzknäuel, deren einzelne Komponenten, wie stärkere Vergrößerungen (Abb. 2) zeigen, von osmiophilen Membranen umgeben und durch schmale Fugen getrennt sind. Besonderheiten dieser Gewebebezirke wurden zum Teil schon lichtmikroskopisch analysiert. Held (1, 2) fand mit einer Neurosomenfärbung reichlich dicht gelagerte Körperchen. Wir beobachteten massive Ansammlungen großer Mitochondrien (Abb. 1 u. 2). Ferner fanden sich reichlich Synapsenbläschen. Daß die Dendritenbüschel der Körner und die Endigungen der Moosfasern in eine Grundsubstanz eingebettet sind, wie Held (1, 2) und sogar Cajal (3, 4) glaubten, konnten wir nicht bestätigen. Ob protoplasmatische Gliafortsätze sich an der Bildung der Glomeruli beteiligen, können wir nicht entscheiden. Kerne von Oligodendroglia-zellen finden sich entgegen den Befunden lichtmikroskopischer Untersucher (5) in den Glomeruli nicht, ebensowenig werden sie von Capillaren durchzogen. Boeke (6, 7) nahm an, daß die Grundsubstanz der Glomeruli nervös-protoplasmatischer Natur sei. In ihr sollten die Dendriten von Körnerzellen und Moosfaserendigungen aufgehen und damit in Form eines protoplasmatischen Zusammenhangs eine einzige Synapse bilden. Diese Deutungen bestehen auf Grund unserer Befunde nicht zu Recht. Es handelt sich bei den Glomeruli um besondere Synapsenformationen,

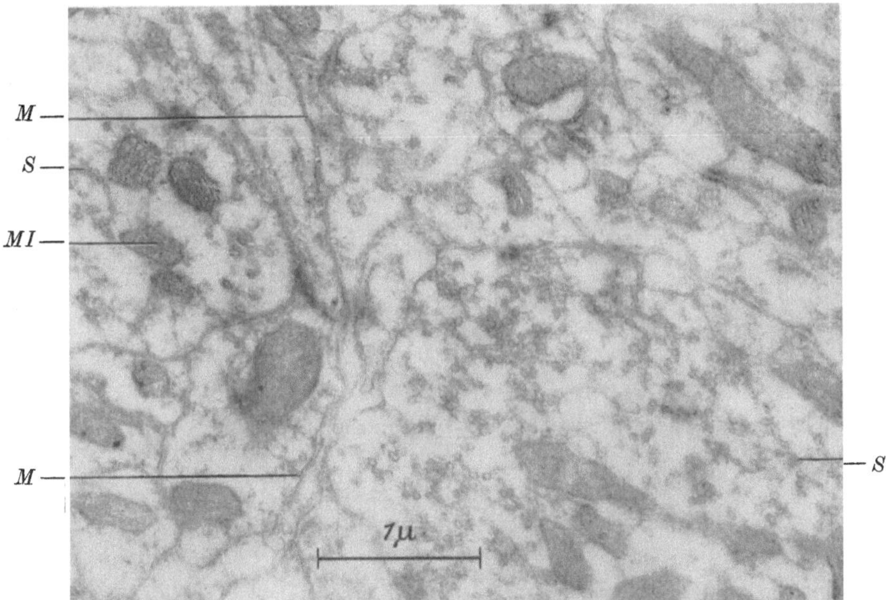

Abb. 2. Segment aus einem Glomerulus cerebellosus der Körnerschicht der Kleinhirnrinde des Goldhamsters. *S* Synapsenbläschen, *MI* Mitochondrien, *M* begrenzende Membranen

die elektronenoptisch als Anschnitte von offenbar aus Körnerzelldendriten und Moosfaserendigungen zusammengesetzten Knäueln erscheinen, in der die einzelnen Komponenten weder anastomosieren noch ein Syncytium bilden.

Die Einwände, die von den Reticularisten und Verfechtern syncytialer Bautheorien des Nervensystems auf Grund ihrer lichtmikroskopischen Befunde an den Glomeruli cerebellosi gegen die klassische Neuronentheorie erhoben wurden, bestehen demnach zu Unrecht.

Literatur

1. Held, H.: Arch. Anat. Suppl. 273 (1897).
2. — Mschr. Psychiat. **65**, 68 (1927).
3. Cajal, S. R. Y.: Trav. Lab. Rech. biol. Univ. Madrid **24**, 181 (1926).
4. — — Handbuch der Neurologie, herausgegeben von O. Bumke u. O. Förster, I. S. 920. Berlin: Springer 1935.
5. Pensa, A.: Mem. R. Accad. naz. Lincei, Cl. Sci. Fis., Mat. e Nat. Ser. VI, **5**, 25 (1931).
6. Boeke, J.: Acta neerl. Morph. **4**, 31 (1941).
7. — Schweiz. Arch. Neurol. Psychiat. **49**, 9 (1942).

Ultrastructure of the yellow pigment of human nerve cells

S. Björkerud and T. Zelander

Department of Histology and Department of Anatomy, University of Gothenburg (Sweden)

That human nerve cells contain so-called yellow pigment has been known since the end of the 19th century. It was found by light microscopical examination of human nerve tissue that the cells contained a substance deposited in the form of small granules having a pale yellow colour.

The pigment has been encountered in man, some primates, in small amounts in old guinea pigs, horses and dogs. In man it is always present after age 6—7, usually after 2—3 years and

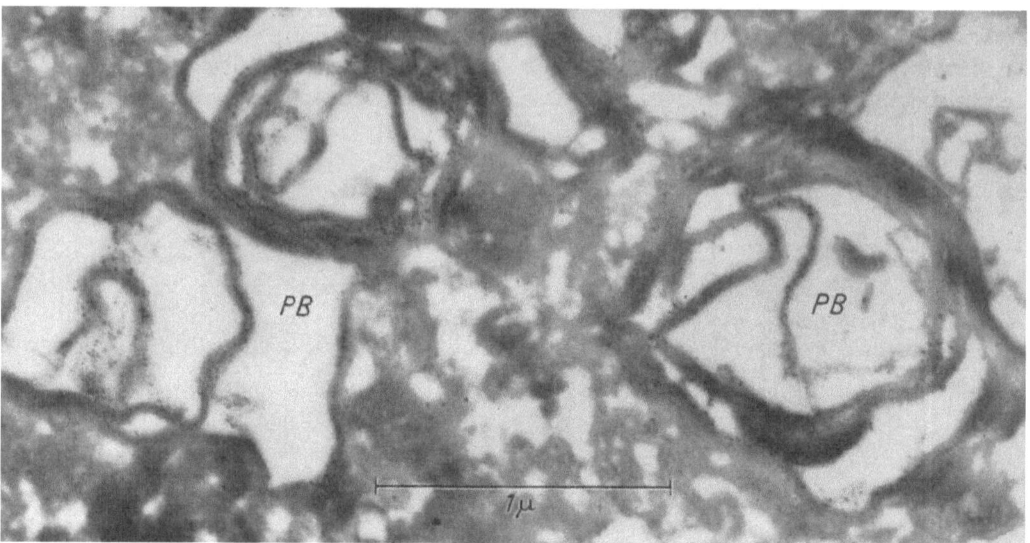

Fig. 1. A survey picture of three pigment bodies *PB* with osmiophilic zones and central regions of very low density

occasionally as soon as a few months after birth. Hence those statements are hardly relevant which were voiced by some workers during the first decades of the 20th century to the effect that the yellow pigment is a phenomen associated with old age. With advancing age it accumulates in more and more cells in increasing quantities. At age 45 or thereabouts only occasional nerve in the cell centrals nervous system are likely to be entirely free from yellow pigment.

The distribution within the nervous system of yellow pigment and its staining properties in situ have been well elucidated by previous investigations (*1, 6*).

Few investigations have been made into the physical and chemical properties of yellow pigment (*7*).

Yellow pigment has previously been isolated from myocardial tissue (*4*) but, so far as we are aware, its electron microscopical morphology has not been studied.

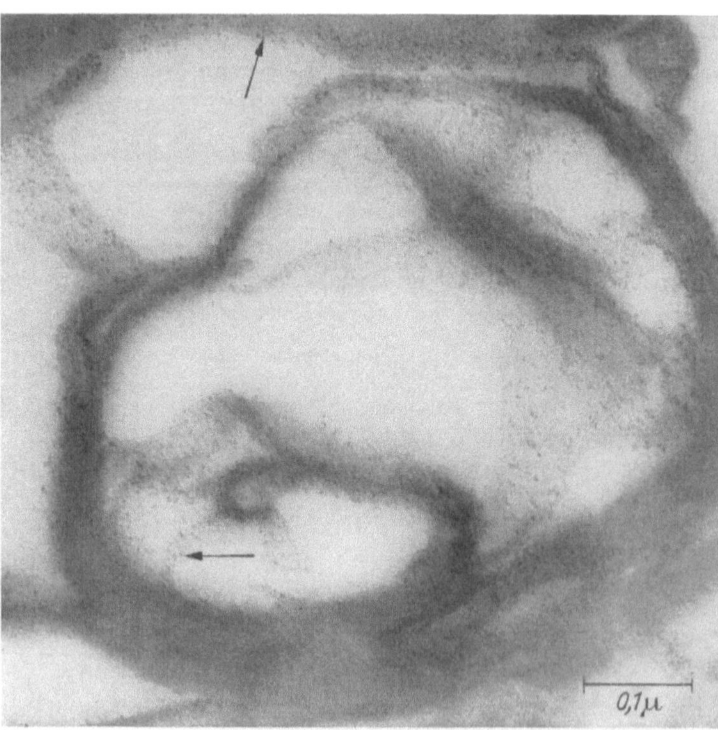

Fig. 2. Part of a pigment body with fine dense granules (arrows) at the border of the light regions

In a paper on the ultrastructure of spinal ganglion cells (*5*) it was stated that pigment granules with the same shape and distribution of osmiophilic substance as the pigment granules described here were observed. — In order to study in greater detail the morphologic and chemical structure of the pigment, one of us developed a sufficiently gentle isolation method (2). It yielded yellow pigment in amounts large enough for electron microscopy and chemical analysis.

The material was spinal medulla obtained at autopsy from corpses from 6 to 80 years of age which were known to have been free from CNS diseases and jaundice. The medulla was chilled to $+6°$ C within 2 to 4 hr and refrigerated to $-20°$ C within 12 hr after death.

Grey matter was dissected free and aspirated with physiological saline as vehicle. The saline dispersion was homogenized by bubbling gas (N_2) through it. Isolation of yellow pigment was performed by a combination of differential centrifugation and spontaneous sedimentation. It was separated with respect to its high density by centrifugation (18,000 g for 15 min at $0— +5°$ C) against a density cushion (30% sucrose solution). Consisting of whole cells, nuclei and pigment granules, the sediment was separated with respect to particle size by spontaneous sedimentation (6% sucrose solution at $+4°$ C). The pigment was present in the supernatant. After the pigment had been rinsed with physiological saline it was centrifuged until it formed a pellet at the bottom of the tube. This pellet was treated in the same way as a tissue block. Consequently it was fixed in 1% osmium tetroxide solution, dehydrated, embedded in methacrylate, sectioned on an ultramicrotome, and examined in an RCA electron microscope Model EMU 3 b.

In order to ensure that the structures in the electron microscopical picture which were interpreted as yellow pigment really were identical with the substance having the light microscopical properties of yellow pigment, the following test was made. A small piece of the isolated pellet was suspended in physiological saline and a smear of the suspension prepared. The smear was examined under a fluorescence microscope which revealed that a large number of grains of varying size had the fluorescence characterizing yellow pigment (3). The positions and shapes of the largest pigment grains were then recorded. The smear was now stained with 1% osmium tetroxide solution and re-examined under a standard light microscope. Osmiophilic grains of the same size, shape and position as the previously fluorescencing grains were then apparent. It could be seen that the osmiophilia of the largest grains was confined to their peripheral portions.

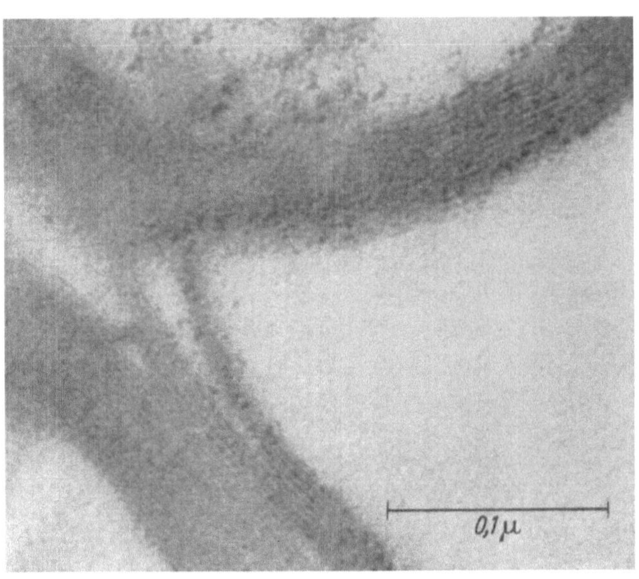

Fig. 3. The osmiophilic zone is composed of alternate very dense and less dense stripes

Observations. Sections cut from the pellet exhibit osmiophilic structures evenly distributed over a moderately dense background having a nondescript fragmented appearance. Similar pictures are observed whatever part of the pellet the sections comes from. Neither mitochondria nor mitochondrial residues were observed.

The osmiophilic structures are rounded or polygonal and exhibit a dense osmiophilic zone surrounding a small number of osmiophobic regions. Most of the pigment bodies seem to be of polycystic nature, with 4 or 5 osmiophobic patches marked off by the osmiophilic zone and spokes radiating inwards from it (Fig. 1). At the boundary between osmiophilic and osmiophobic areas, there are a good deal of small and extremely dense granules having a diameter of about 40 Å. (Fig. 2). Sections from dehydrated though not osmium-fixed pellets also displayed at the boundaries between light and dark regions numerous fine granules of the same order of magnitude as those just mentioned. The osmiophilic zone of the pigment bodies showed a regular subdivision into alternate dense and less dense stripes, the centre-to-centre distance between two adjacent dark lines being some 35 or 40 Å (Fig. 3).

The pigment from spinal medulla which fluorescence-microscopically is defined as yellow pigment possesses a characteristic ultrastructure in the form of a strongly osmiophilic zone surrounding osmiophobic regions. Furthermore, the osmiophilic zone is subdivided into alternate dark and light lines with a centre-to-centre distance of 35 to 40 Å. Lastly, at the boundary between osmiophilic and osmiophobic areas there are a large number of dense granules having a diameter of about 40 Å.

References

1. ALTSCHUL, R.: Virchows Arch. path. Anat. **301,** 273 (1938).
2. BJÖRKERUD, S.: Unpublished.
3. HAMPERL, H.: Virchows Arch. path. Anat. **292,** 1 (1934).
4. HEIDENREICH, O., and G. SIEBERT: Verh. dtsch. Ges. Path. **39,** 193 (1956).
5. HESS, A.: Anat. Rec. **123,** 399 (1955).
6. HEUCH, W.: Beitr. path. Anat. allg. Path. **54,** 68 (1912).
7. HYDÉN, H., and B. LINDSTRÖM: Faraday Soc. Disc. **9,** 436 (1950).

Recherches en vue de l'identification au microscope électronique des cellules interstitielles de CAJAL

JACQUES TAXI

Laboratoire de Biologie animale, P. C. B., Faculté des Sciences, Paris, et
Laboratoire de Microscopie électronique appliquée à la Biologie, C. N. R. S., Paris

Les «cellules interstitielles» de l'intestin des Mammifères ont été décrites par CAJAL (*1*) sous le nom de neurones interstitiels, cependant que DOGIEL (*2*) les considérait comme conjonctives. Plus tard, LAWRENTJEW (*4*) les a identifiées aux cellules de SCHWANN satellites des neurites dans

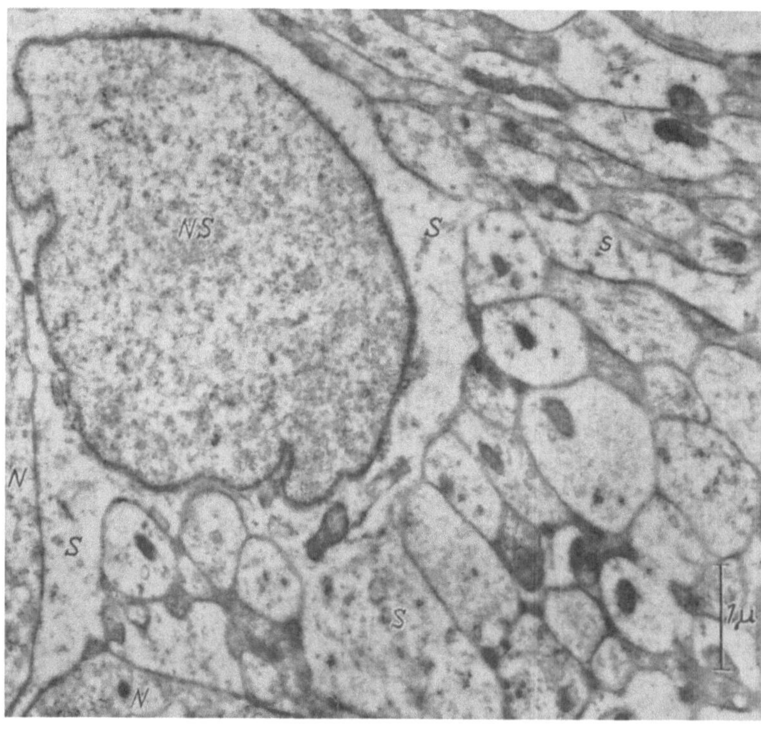

Fig. 1. Intestin de Souris. On voit un noyau Schwannien (*NS*) dans une travée du plexus d'AUERBACH. Le cytoplasme Schwannien (*S*) est moulé sur les paquets de fibres nerveuses qu'il sépare. *N*, neurone. 14 000 ×

les plexus périphériques du système nerveux autonome. Dans un travail datant des quelques années (*4*), effectué à l'aide de la méthode de GOLGI, de la coloration postvitale au bleu de méthylène et de diverses méthodes d'imprégnation argentique, nous avons pu montrer que les cellules interstitielles se distinguent nettement, au moins chez l'adulte, des cellules de SCHWANN des travées du plexus d'AUERBACH. Mais manquant d'arguments pour décider de leur nature, nous

nous contentions de leur appliquer le qualificatif de «neuronoïde», que justifient certaines analogies avec les neurones, notamment en ce qui concerne leur colorabilité.

On peut attendre d'une étude de l'ultrastructure des cellules interstitielles qu'elle aide à la solution du problème de leur nature. Pour aborder cette étude, nous avons choisi comme matériel l'intestin de Souris, déjà utilisé par Cajal et particulièrement indiqué pour la fixation osmiée à cause de sa minceur. Les pièces ont été fixées au tétroxyde d'osmium à 1 % tamponné selon Palade à pH 7,4 et enrobées au méthacrylate de n-butyle. Dans une première étape du travail, il apparaît nécessaire de rechercher des moyens de distinguer sur les micrographies électroniques les cellules interstitielles des autres cellules qui accompagnent les travées nerveuses du plexus. Les imprégnations argentiques panoptiques montrent en effet qu'il existe au niveau du plexus d'Auerbach,

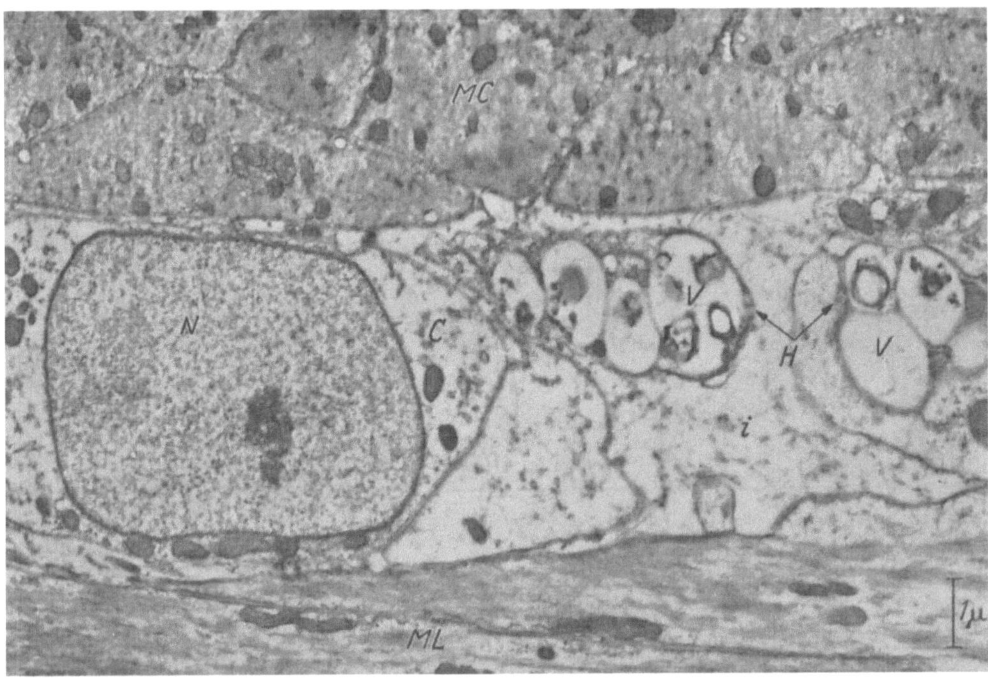

Fig. 2. Intestin d'une souris ayant subi des injections de rouge trypan pendant quatre semaines. Coupe passant par la région qui sépare les couches musculaires longitudinale (*ML*) et circulaire (*MC*). Dans la zone médiane, on observe une cellule interstitielle (*C*) avec son noyau (*N*). Deux prolongements d'histiocytes (*H*) contiennent des vésicules (*V*) dont la présence est due à l'athrocytose du rouge trypan; *i*, milieu intercellulaire. 8 900 ×

outre les neurones et les cellules interstitielles, des cellules de Schwann, des cellules endothéliales de capillaires sanguins et des histiocytes; ces derniers ont été caractérisés par Ottaviani et Cavazzana (5) grâce à leur propriété d'athrocytose vis-à-vis du bleu trypan.

Au microscope électronique, les cellules nerveuses et les cellules endothéliales de capillaires sanguins se reconnaissent en général aisément. Il en est de même pour les cellules de Schwann grâce aux connexions très étroites qui les unissent à des faisceaux plus ou moins importants de neurites (Fig. 1). L'identification des cellules interstitielles et des histiocytes pose un problème plus difficile. Deux types cellulaires observés sur les micrographies électroniques et qui diffèrent nettement des cellules nerveuses, schwanniennes et endothéliales paraissent leur correspondre. Dans l'un de ces types, le noyau est entouré d'une couche de cytoplasme généralement mince, dépassant rarement 2 microns. Ce cytoplasme présente une forte densité sur les micrographies électroniques, due en particulier à sa richesse en granulations associées ou non aux saccules ergastoplasmiques et à la présence d'assez nombreuses mitochondries. Dans l'autre type cellulaire, au contraire, la densité générale est faible, par suite de l'extrême pauvreté en granulations; les saccules couverts de granules qui caractérisent l'ergastoplasme sont très peu nombreux, et de petite taille, et l'on rencontre plus souvent une forme agranulaire du reticulum endoplasmique; entre les éléments figurés s'étendent souvent des espaces clairs d'importance variable qui suggèrent l'idée d'un gonflement oedemateux de ces cellules; cet aspect très fréquent, et qui n'est sans doute qu'un artefact de fixation, paraît caractéristique de ce type de cellules.

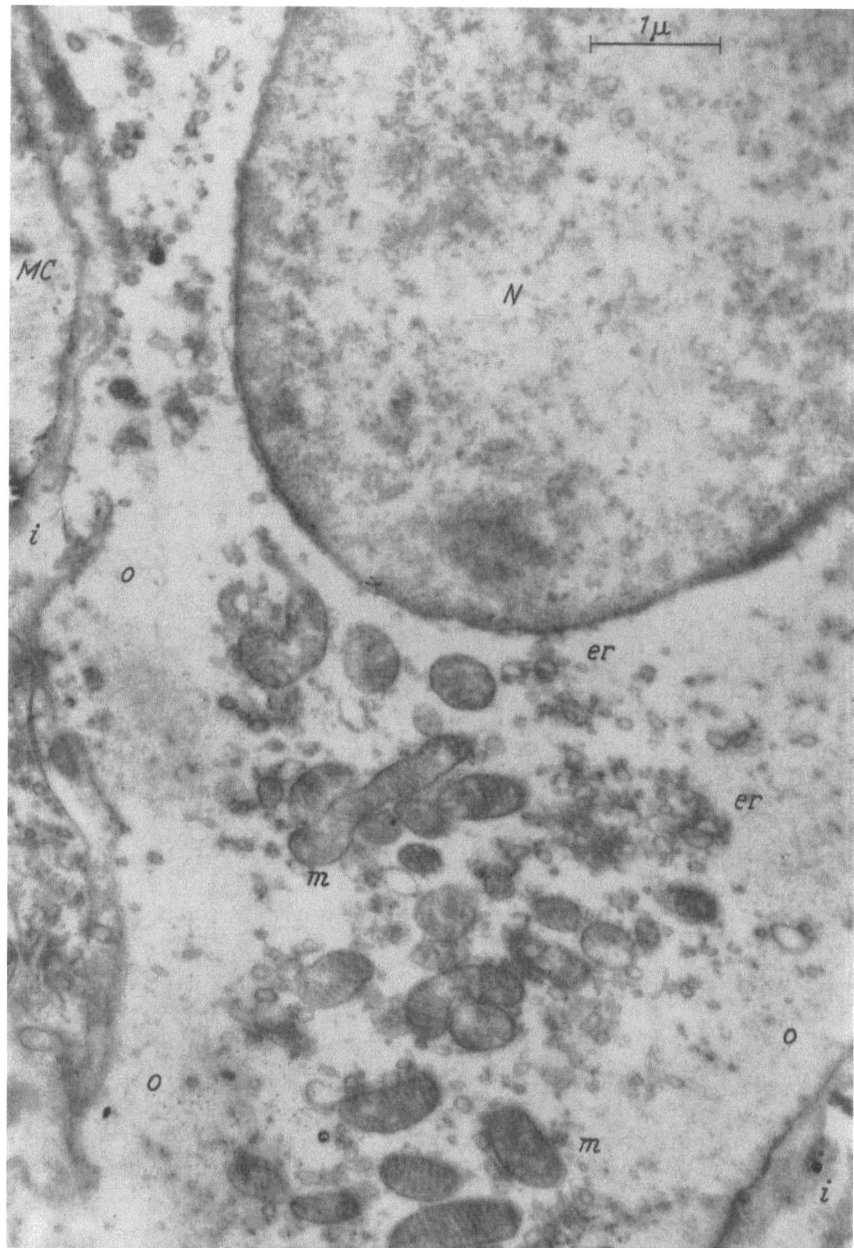

Fig. 3. Intestin de Souris. La coupe intéresse une partie d'une cellule interstitielle, avec son noyau (*N*); *MC*, musculeuse circulaire; *i*, milieu intercellulaire. Dans le cytoplasme de la cellule interstitielle, des mitochondries (*m*) et des saccules de reticulum endoplasmique agranulaire (*er*); *o*, régions du cytoplasme dépourvues d'éléments figurés et suggérant l'idée d'un certain gonflement oedemateux de la cellule. 18 000 ×

Pour établir une correspondance précise entre ces deux sortes de cellules et les cellules qui restent à identifier, à savoir les cellules interstitielles et les histiocytes, nous avons pensé pouvoir caractériser l'un deux (le type histiocytaire) par les modifications qu'il subirait chez des animaux soumis à des injections répétées de rouge trypan. Nous avions pu vérifier préalablement en microscopie ordinaire que seuls au niveau du plexus d'AUERBACH les histiocytes présentent la propriété d'athrocytose vis-à-vis de ce colorant, et que les cellules interstitielles ne sont en rien affectées, même si l'on répète les injections durant des semaines. Cette observation s'est trouvée parfaitement confirmée au microscope électronique: chez des souris de quinze grammes ayant subi pendant quatre à six semaines tous les deux jours une injection intrapéritonéale de 0,2 cm³ de rouge trypan à saturation dans du liquide de LOCKE, les cellules du premier type décrit plus haut subissent seules d'importants changements de structure, dont la phase finale se traduit par la présence au sein du cytoplasme de larges vésicules parfois confluentes contenant des formations en lamelles concentriques plus ou moins altérées (Fig. 2), qui rappellent les aspects décrits par POLICARD, COLLET et PRÉGERMAIN (6) comme «figures myéliniques», au cours des réactions inflammatoires provoquées par la silice dans le tissu pulmonaire.

Nous sommes donc en droit de considérer comme probable que le premier type décrit plus haut correspond aux histiocytes et le second aux cellules interstitielles. Les caractères cytologiques de celles-ci (Fig. 3), tels qu'ils ont été brièvement rapportès plus haut, permettent d'écarter toute assimilation des cellules interstitielles aux neurones, dont l'ultrastructure est bien connue depuis le travail de PALAY et PALADE (7). Cette interprétation étant écartée, la détermination de la nature exacte des cellules interstitielles de CAJAL nécessite d'autres recherches.

Bibliographie

1. CAJAL, S. R. y: Los ganglios y plexos nerviosos del intestino de los Mamiferos. Moya édit., Madrid 1893.
2. DOGIEL, A. S.: Anat. Anz. **10**, 517 (1895).
3. LAWRENTJEW, B. J.: Z. mikr.-anat. Forsch. **6**, 467 (1926).
4. TAXI, J.: Arch. Anat. micr. Morph. exp. **41**, 281 (1952).
5. OTTAVIANI, G., e P. VAVAZZANA: Arch. ital. Anat. Embr. **43**, 75 (1940).
6. POLICARD, A., A. COLLET et S. PRÉGERMAIN: Bull. micr. appl. **7**, 49 (1957).
7. PALAY, S. L., and G. E. PALADE: J. biophys. biochem. Cytol. **1**, 69 (1955).

Some aspects of glial function as revealed by electron microscopy*

E. D. P. DE ROBERTIS, H. M. GERSCHENFELD and FLORA WALD

Institute of General Anatomy and Embryology, Faculty of Medicine, University of Buenos Aires (Argentina)

The structural analysis of the central nervous system, with the so-called selective histological techniques, is hindered by the fact that only partial views of the total organization are obtained. Thus the methods for the demonstration of astroglia, oligodendroglia, microglia, myelin sheaths, Nissl substance and so forth, emphasize only one component at the time and do not permit a spatial integration of all the structural elements. On the contrary, with the electron microscope in thin sections fixed in osmium tetroxide, all these and other components can be visualized simultaneously and can be followed in their topographic relationships down to dimensions beyond the resolving power of the optical microscope.

With this technique the different types of glial cells present in the central nervous system may be recognized by their morphology and the relationship that these cellular elements bear with neurons, nerve fibres, nerve endings, capillaries, and so forth, can be clearly determined.

This study was carried out on a wide variety of material from different regions of the central nervous system (cerebral and cerebellar cortex, the acoustic ganglion, the spinal cord and so forth), of mammals of different age and species in normal and in several experimental conditions.

The problem of the extracellular space in the CNS. — In these electron microscope observations it was found that the glial elements fill all the intervening spaces between the neurons and the vascular elements and that no real extracellular spaces are present within the central nervous system (Fig. 1 and 3). The plasma membranes of all cellular components of the nervous tissue are in intimate contact among themselves and with the basal membranes of the capillaries (Fig. 2). A distance of only 120 to 250 Å can be observed between the adjacent membranes and this is

* Work supported by a grant of the National Multiple Sclerosis Society of New York.

filled by a material which has a definite electron density. These facts which have been observed by other investigators (*1, 2, 3, 4*) are in complete contradiction with some physiological data, which

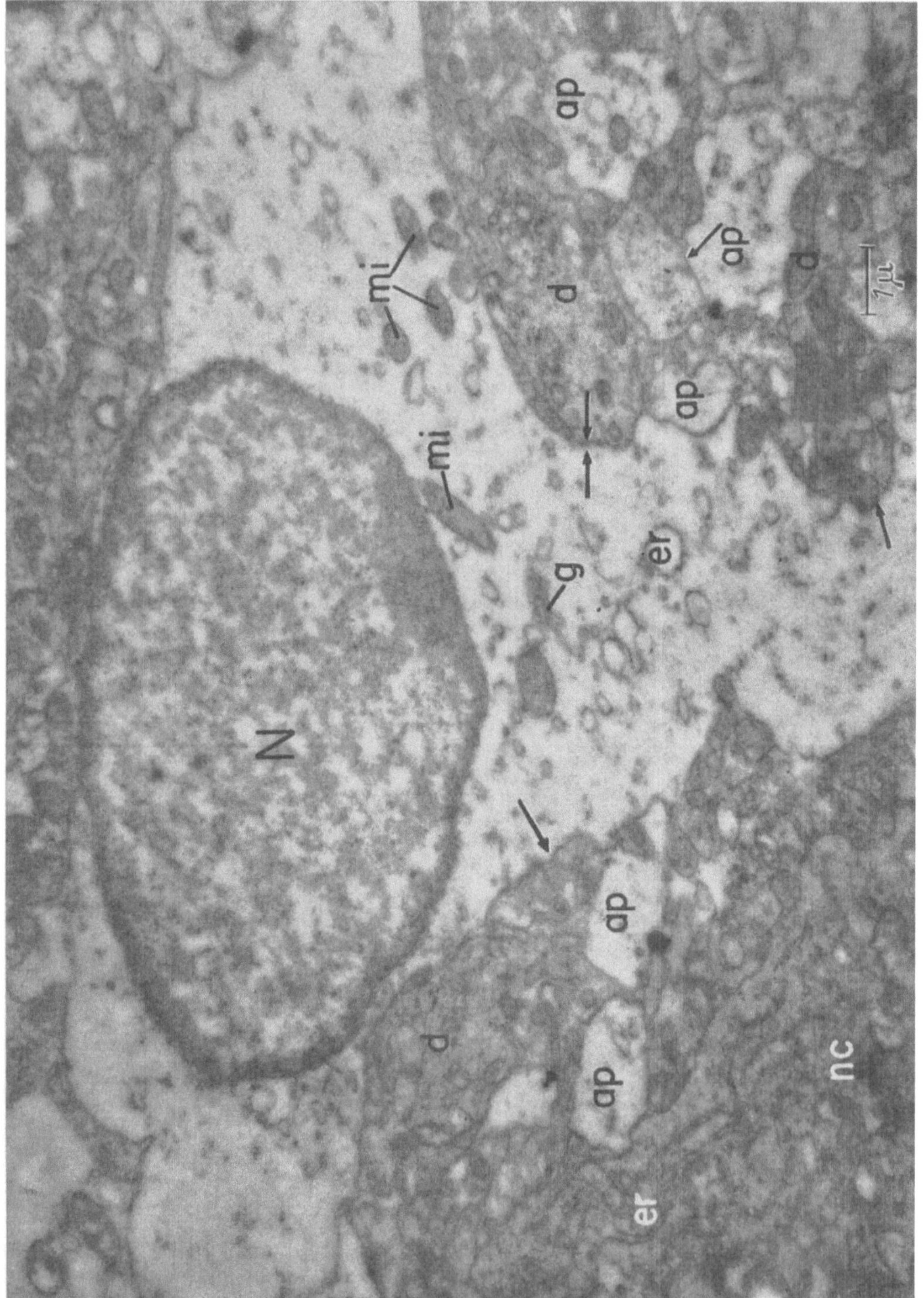

Fig. 1. Astrocyte of a rat's brain showing the typical watery cytoplasm in contrast with the dense cytoplasm of neurons. *ap*, astrocyte process; *er*, endoplasmic reticulum; *d*, dendrite; *g* Golgi substance; *mi*, mitochondria; *nc*, neurone cytoplasm; *N*, nucleus of the astrocyte. Arrows indicate some points of intimate contact between the membrane of neighbouring cellular elements. Cells are in intimate contact being the distance between the plasma membranes of 120—250 Å. 9600 ×

attribute to the CNS a definite extracellular space. This has been calculated to be of the order of 31,4% for the chloride space and 34,6% for the sodium space "in vivo" (*5*) and of 14% with inulin and 19% with ferrocyanide "in vitro" (*6*).

These apparently contradictory findings lead us to investigate the possibility of producing experimentally a true edema of the CNS by increasing its water content.

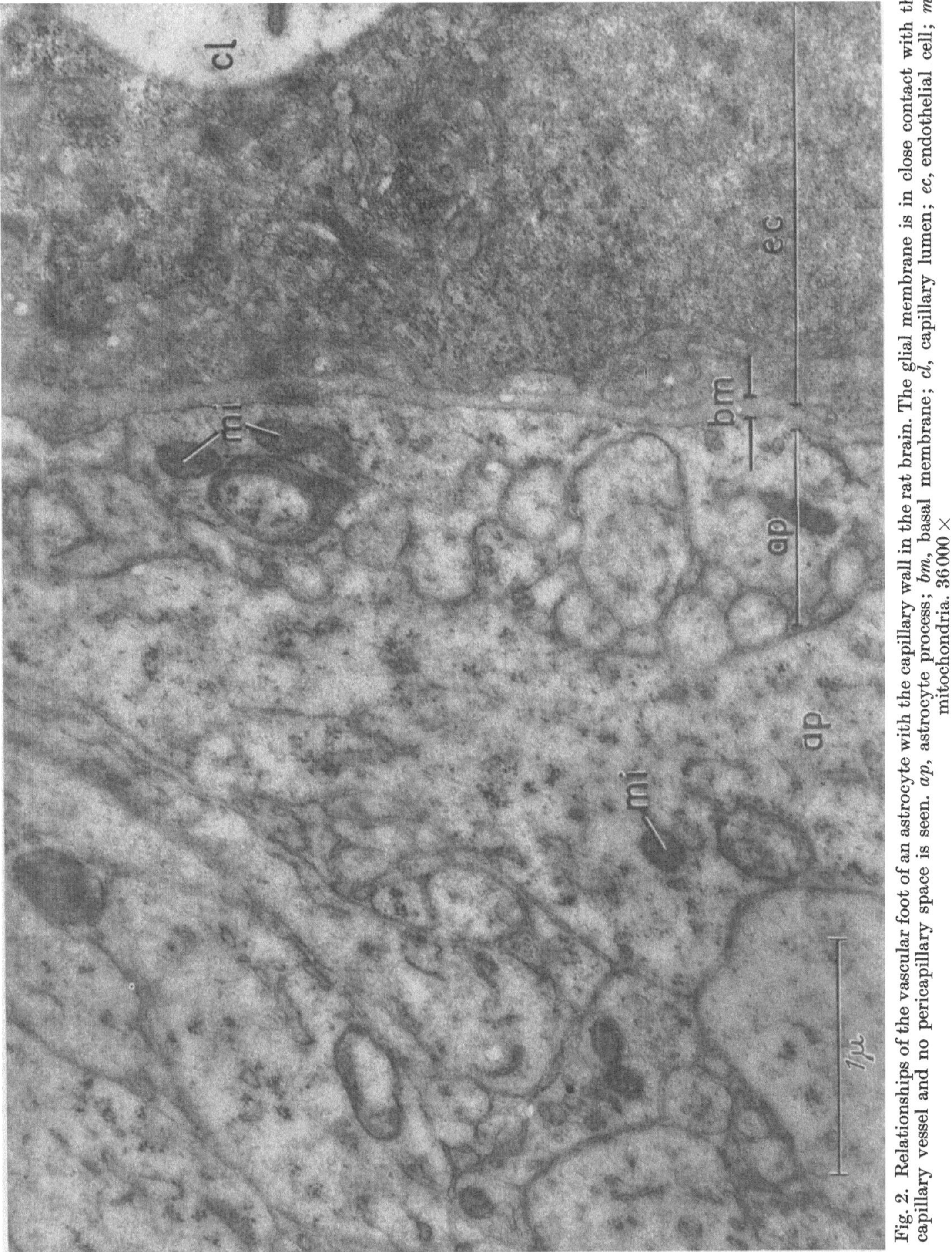

Fig. 2. Relationships of the vascular foot of an astrocyte with the capillary wall in the rat brain. The glial membrane is in close contact with the capillary vessel and no pericapillary space is seen. *ap*, astrocyte process; *bm*, basal membrane; *cl*, capillary lumen; *ec*, endothelial cell; *mi*, mitochondria. 36000 ×

The experimental procedures and the actual results will be described at length elsewhere (7). Here it will be only mentioned that no change in the water content of the brain can be produced by such drastic procedures as: a) bilateral nephrectomy followed by injection of large dosis of a saline solution enough to triplicate or quadruplicate the general extracellular space of the body; b) a similar treatment plus craniectomy; c) hydric intoxication with large injections of water and

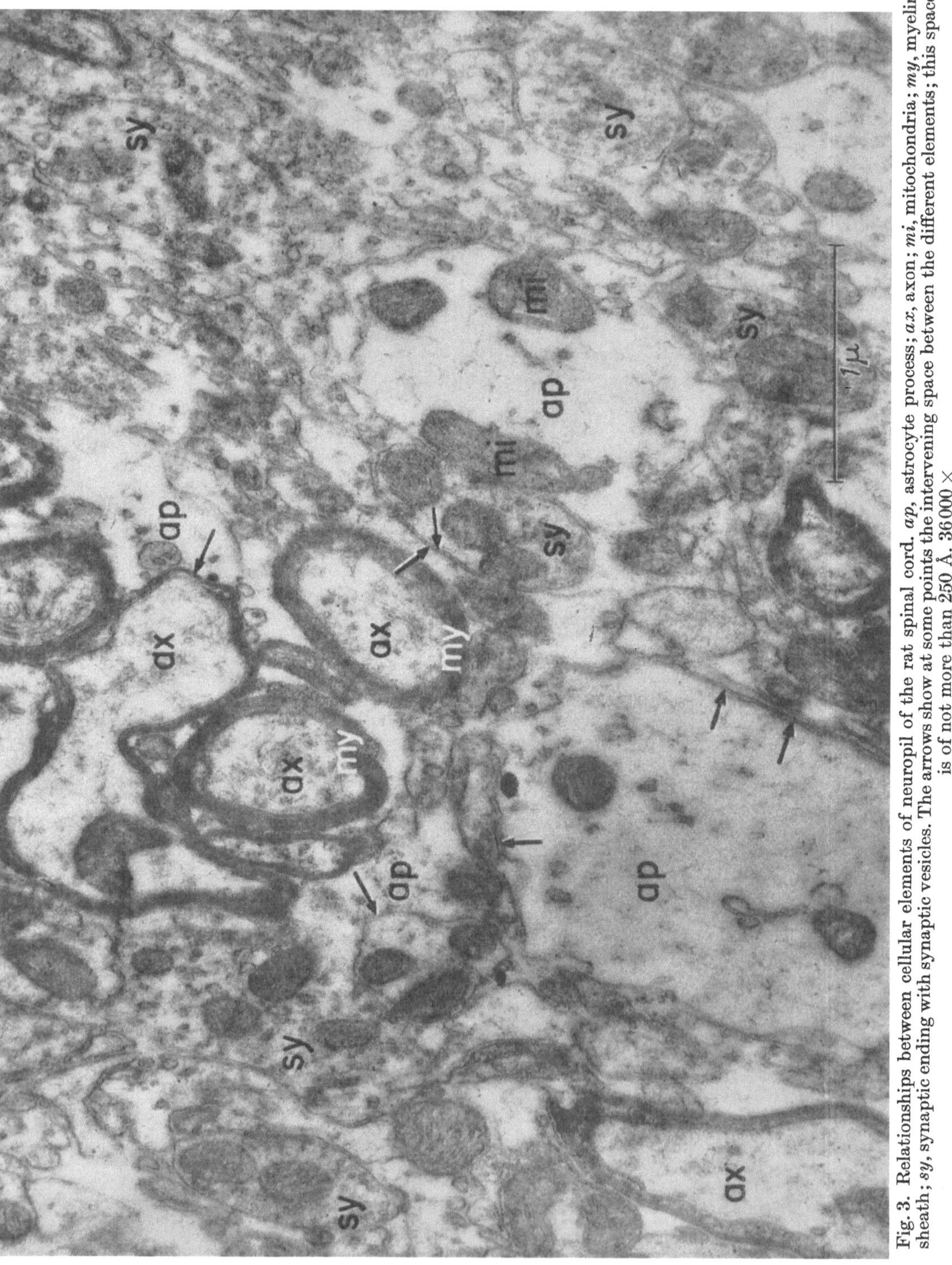

Fig. 3. Relationships between cellular elements of neuropil of the rat spinal cord. *ap*, astrocyte process; *ax*, axon; *mi*, mitochondria; *my*, myelin sheath; *sy*, synaptic ending with synaptic vesicles. The arrows show at some points the intervening space between the different elements; this space is of not more than 250 Å. 36000 ×

of pitressin as suggested by Rowntree (8) and d) the intravenous injection of hypertonic and hypotonic solutions recommended by Weed and McKibben (9).

Astroglia and the water-space of the brain. — The lack of an intercellular space leads us to think that some of the cellular elements of the glia may be involved in the transport of fluid and ions from the capillaries to the neurons. Of the different types of glial cells, that can be recognized with the electron microscope, the astrocytes seem to be particularly fitted for this function. They are, on one side, in intimate contact with the capillaries by means of their vascular feet, they cover the Virchow-Robin's spaces and the pial membrane. On the side they come into direct contact with the cell body of neurons and all their expansions. The glial cytoplasm is thus the main cellular element interposed between the neurons and the vascular system. Furthermore a high water content of the astrocytes is suggested by their extraordinary low electron density which is comparable to that observed within the capillary lumen (Fig. 1 and 3).

The probable intervention of astroglia in the water and ion metabolism of the brain is also suggested by some recent experiments which will be reported at length elsewhere. In view of the fact that true edema could not be produced in the brain by water surcharge, a reactional swelling was induced locally by a traumatic lesion (stab wound). In this case near the lesion there is a definite swelling of the tissue, which is well known to neurosurgeons and neuropathologists. However even in this case we were unable to see the formation of a real extracellular space. The swelling seems to be cellular and localized mainly in the cytoplasm of the astrocytes.

As a conclusion of all these observations and experiments we may state that in the CNS there is not a real extracellular space similar to that of other tissues. This fluid compartment is probably replaced by the astroglia cells which participate actively in the water and electrolyte transport between the vascular system and the neurons.

Oligodendroglia and myelin. — We have already demonstrated that in the CNS myelin is formed within the cytoplasm of the oligocytes by a process of membrane synthesis (*10, 11*). In this process the myelin membrane result of the coalescence and fusion of vesicular elements and they are deposited in the neighborhood of the axon-oligocytic membrane. After the formation of the first 8 to 10 myelin membranes they become aligned and the orderly structure of myelin is fixed, while the growth of new layers continues by apposition of intracellular vesicles and membranes.

These results indicate that myelin of adult nerve fibres is located intracellularly within the cytoplasm of the oligodendrocyte, where it was formed. We have now observed that during the degeneration of the nerve fibres the destruction and disposal of the myelin sheath also takes place within the cytoplasm of the oligocyte. In the stab wound experiments of the CNS it is possible to follow the different stages of degeneration of the axon and the disintegration of the sheath inside the oligocytes. The presence of migrating microglial cells has also been recognized near the wound. These cells participate in the removal of free degenerating material including myelin, but the oligocyte is the main cellular element involved in the disposal of the myelin sheath.

References

1. Wyckoff, R. G., and J. Z. Young: J. Anat. 88, 568 (1954).

2. Dempsey, E. W., and G. B. Wislocki: J. biophys. biochem. Cytol. 1, 245 (1955).

3. Farquhar, M. G., and J. F. Hartmann: J. Neuropath. exp. Neurol. 16, 18 (1956).

4. Horstmann, E.: Naturwissenschaften 44, 448 (1957).

5. Davson, H.: Physiology of the Ocular and Cerebrospinal Fluids. Boston: Little, Brown & Co. 1956.

6. Allen, J. N.: Arch. Neurol. Psychiat. (Chicago) 72, 241 (1955).

7. Gerschenfeld, H. M., F. Wald, J. A. Zadunaisky and E. D. P. de Robertis: Neurology. 1959 (in press).

8. Rowntree, L. G.: J. Pharmacol. exp. Ther. 29, 135 (1926).

9. Weed, L. H., and P. S. McKibben: Amer. J. Physiol. 49, 522 (1919).

10. Robertis, E. D. P. de, H. M. Gerschenfeld and F. Wald: Anat. Rec. 130, 292 (1958).

11. — — — J. biophys. biochem. Cytol. 4, 651 (1958).

The effect of stimulation on the axoplasm structure of a nerve fiber

V. L. Borovyagin

Institute of Biophysics, Acad. Sci., Moscow (USSR)

In studies of the physico-chemical mechanisms of the origin and propagation of stimuli across nerve fibers the main significance is usually attached to phenomena occurring in the surface membrane. The inner medium-axoplasm is considered to be of secondary importance. However, there are facts which contradict the membrane approach of the processes.

It was observed (*1*) that axoplasm exuded from the end of an axon becomes thick at the cut during stimulation. Further variations in the viscous-elastic properties of a nerve conductor produced by stimulation were found (*2*). Also changes in light scattering produced by stimulation with a series of electric pulses were observed (*3*).

The above data show that the axoplasm structure must be to some extent labile. This led us to undertake a study of the axoplasm in its various functional states by applying electron microscopy. One difficulty encountered in an attempt to solve this problem is that fixation of the substrate is such a powerful factor, that may stop the reversible changes occurring in the structure. For this reason in the initial stage of our study we used data on rapid structure variations at the moment of a pulse passage, but also those — on the residual phenomena of the afteraction. The residual phenomena gradually sum up at a stimulation frequency above 450 Hz. We used the ischiaticus of the frog and a branch of ischiaticus of the innervating sartorius muscle.

The preparations were fixed in OsO_4 and formalin. The fixator was cooled in both cases to reduce its self-stimulating action; cooling also reduced excitability of the musclenervous preparation. Similar fixation conditions were used for excitation. The electrodes were laid on the

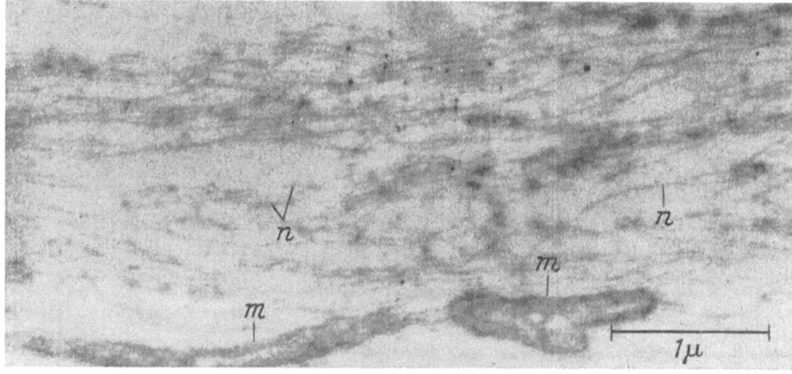

Fig. 1. Axoplasm of a non-stimulated frog nerve fiber ("rest") fixed in OsO_4; *m* mitochondria, *n* neurofibrils.
21 000 ×

nerve roots and the fixation was performed 3 min after stimulation with rectangular pulses of a frequency 570 Hz — impulse length of 45 mc/sec. We shall further define the first type of pictures as the pictures of "rest", and the second will be defined as "stimulation". The tentative nature of these should be quite clear.

The observed electron-microscope pictures show that the axoplasm structural-frame is to a certain degree labile. However, there are typical differences between "rest" (Fig. 1) and "stimulation" (Fig. 2). Keeping in mind the ways of fixation four cases must be considered:

1. The state of relative rest in case of osmium fixation. We observed mitochondria in optically rather hollow axoplasma and clot-like and deeper stained granules of the matter which tend to linear aggregation parallel to fibre axis.

Simultaneously a great deal of fine protofibrils can easily be observed, which lie along the fibre axis and form a hardly distinguishable network due to shoots and to filaments arranged at a small angle. Protofibril formation 3-30 mμ in diameter form sometimes rather long peculiar swarms.

It is difficult to judge the length of separate protofibril filaments since they leave the section plane because of their deficient parallelism.

In similar cases one can observe coarser pictures, when single protofibrils have obviously aggregated into thick plates where not only the finest web-like nets but also thick filaments 30 mμ and more in diameter can be observed. Here, in the "rest" pictures we also observed certain parallelism in the formation of the filamentous structures and in the presence of the net-like anastomosis a general trend parallel to the filament axis. Sometimes these filaments have knot-like formations indicating the initial stage of fragmentation.

2. The second case is also a "rest" state, but for formalin fixation. Here we have coarser protofibrils as in the second variant of the first case. Certain experiments show violation of more

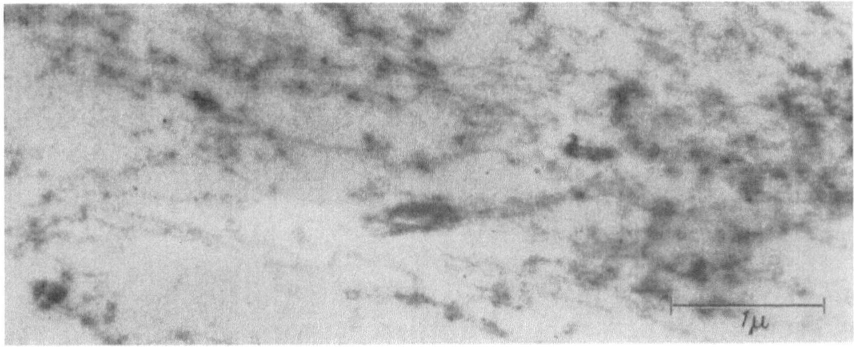

Fig. 2. Axoplasm of a stimulated frog nerve fiber ("stimulation") fixed in OsO$_4$. 21 000 $\times$

or less parallel filament netting with coagulation of the latter and transformation of the whole picture into a more confused net. One can also observe stages of more pronounced disintegration (evident fragmentation picture).

3. After osmium fixation of stimulated material one can never observe a fine and orderly protofibril picture with parallel or more or less parallel bundles of filaments. The axoplasm picture can be characterized either as protoplasmatic clots on one side, fragmentation of the finest filaments and aggregation of the disintegrated particles on the other. Sometimes association of fragments in coarser plaits stretching in the axoplasm make the structures visible that preserve parallelism of the filament axis. Certain experiments with traces of order show protofibrils consisting of knots, the parallelism of numerous filaments being sharply violated.

4. Formalin fixation after stimulation gives a picture similar to that in the three cases but with coarser structures and at first sight more disintegrated protofibril filaments.

The data obtained enable us to conclude that:

1. Axoplasm ultrastructure is labile and changes depending on the functional state.

2. Protofibril formations (in norm) grow weakly as a result of prolonged stimulation and have a stronger relative tendency to disrupt and aggregate under the action of the fixating agent than in the conditional rest state.

3. The pictures received on fixed preparations are a reflection of shifts occurring in vivo as a result of variation of the viscous-elastic properties and light scattering of a stimulated nerve fibre.

4. Simultaneously with quickly and reversibly occurring physico-chemical changes induced by stimulation residual and accumulating variations occur which can be detected by ultrastructure investigations.

These changes are in its way rudimentary "memory" of the excitable formation about a repetitive and prolonged stimulator.

References

1. FLAG, J. V.: J. Neurophysiol. **10,** 211 (1947).
2. LUDCOVSKAY, P. G., and G. M. FRANK: Dokl. Acad. Nauk USSR **87,** 389 (1952).
3. FRANK, G. M.: Izvestia Acad. Nauk USSR Biol. Serie. **1,** 26 (1958).

11. Sinnesorgane

Comparative submicroscopic morphology of rods and cones

E. de Robertis and A. Lasansky

Institute of General Anatomy and Embryology, Faculty of Medicine, University of Buenos Aires (Argentina)

A comparative study of the ultrastructure of the different segments of rods and cones in the retina of the rabbit was presented. The similarities in the general pattern of organization of both photoreceptors will be stressed. The presence of a "connecting cilium" (J. biophys. biochem. Cytol. 3, 319 (1956)] in the cone cells is indicative of a morphogenetic origin of the outer segment similar to that previously described for the rod [J. biophys. biochem. Cytol. 2, 209 (1956)]. Both the outer segments of the rods and cones are built of double membranous flattened sacs, but the dimensions of the membrane and the interspaces are definitely different. Some of these data were discussed on the basis of the concentration of solids of the outer segment and the relative position of protein and lipid, within the membrane of the sac. The comparative fine structure of the inner segment and of the synaptic region is described.

An extended version of this article appeared in J. biophys. biochem. Cytol. 4, 743 (1958).

Submicroscopic structure of photo-receptors of bird and insect eyes as revealed by electron microscopy

Gonpachiro Yasuzumi

The Electron Microscope Research Laboratory, Department of Anatomy,
Nara Medical College, Kashihara, Nara Pref. (Japan)

The submicroscopic structure of the visual cells in the retina of such vertebrates as guinea pig and perch (2, 8—12), albino rat and mouse (5—7), and frog (14) has been studied by electron microscopy and their characteristic internal structure well defined. However, the peculiarities of the inner segments of the rod and cone of the avian retina have so far attracted little attention.

Recently, the submicroscopic structure of the rhabdomere of the compound eye has been revealed by Fernández-Morán (1) in *Musca domestica*, Miller (4) in *Limulus*, Goldsmith and Philpott (3) in *Sarcophaga bullata* and *Anax junius*, Wolken et al. (13) in *Drosophila melanogaster*, and Yasuzumi and Deguchi in *Drosophila virilis* (15).

Adult *Uroloncha striata* var. *domestica Flower* and wild-type, red-eyed, adult *Drosophila virilis* were used in the present study. The specimens were fixed in 1% osmium tetroxide buffered (pH 7.3) with veronal-acetate. The tissue, embedded in a mixture of *n*-butyl- and methyl methacrylate, was cut with glass knives on a Shimadzu microtome. All sections were examined in an Akashi electron microscope at 50 kV and electron optical magnification of mostly 10,000 to 15,000 times.

In a slightly oblique cross-section through the inner segment of visual cells, a characteristic and apparently cellular apparatus is found surrounding the visual cells outside the dense, discontinuous outer limiting membrane (Fig. 1). It is a thin membrane with toothed, wheel-like projections. The projection of the membrane in the proximal portion of the inner segment measures 0.33—0.67 μ in length and 40—70 mμ in width. It consists of a double membrane; each member of the pair measures about 80—150 Å thick. The projections are spaced at intervals of 0.17—0.50 μ. In general, the projections between the adjacent membranes are observed to mesh with each other like gears, with an exception where two or three projections are wedged into a single, corresponding space. It is interesting that a part of the inner segment of the rod is surrounded by a regular or an irregular navicular profile, which is provided with long projections. The visual cells in the outer nuclear layer are connected by twisted glia fibers which occasionally cover

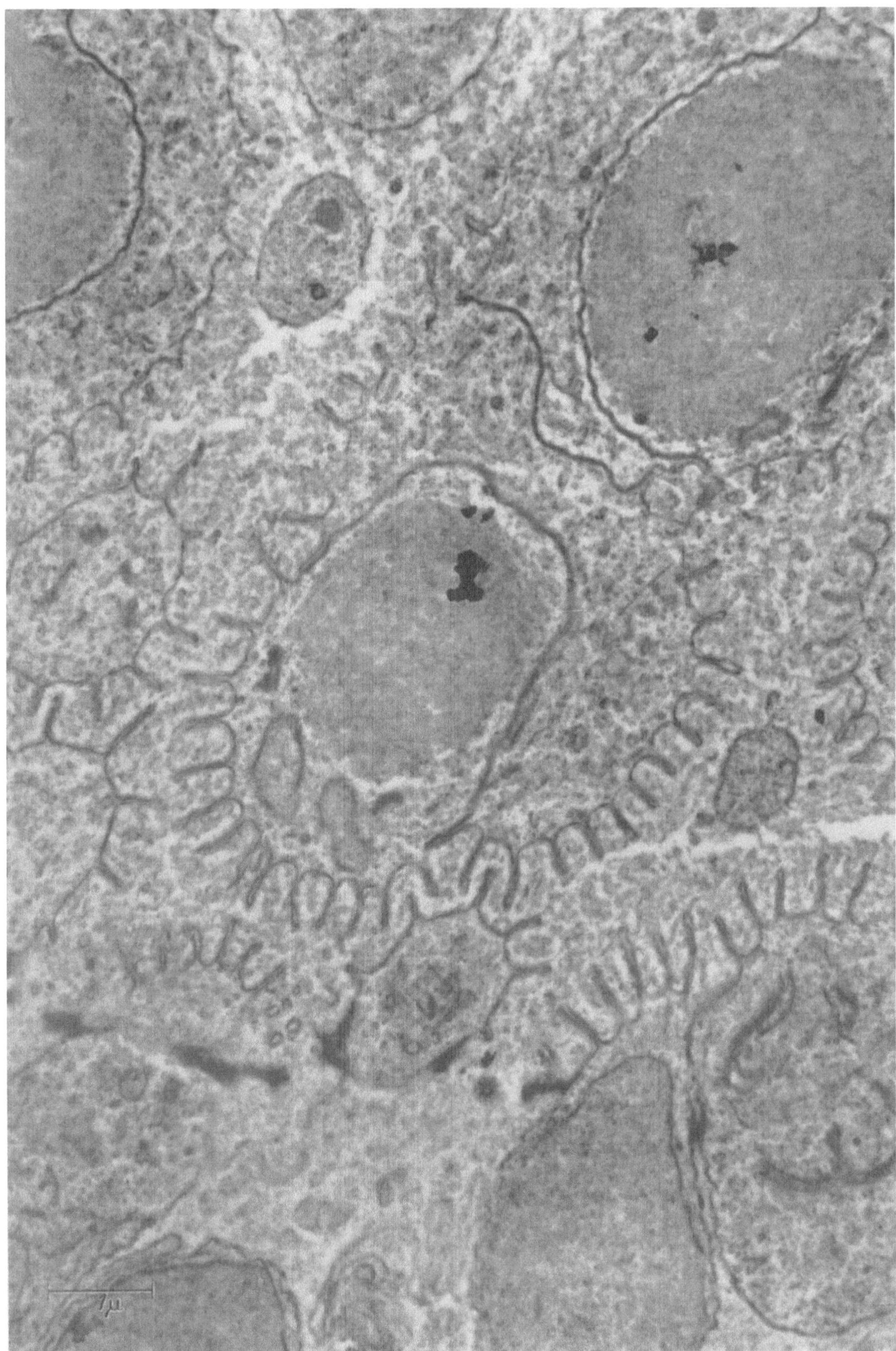

Fig. 1. A slightly oblique cross-section through the *Uroloncha* retina. 16,500 ×

the visual cells. The nucleus of the rod cell occupies most of the cell and leaves only a narrow area of the cytoplasm. The cytoplasm of the cone contains the Golgi complex and circular profiles of the endoplasmic reticulum. No membrane with pecten-like margins is observed in the outer nuclear layer. The limiting membrane with pecten-like margins, depicted here, seems to be referable to the fibrillar baskets (Faserkorb) observed previously with the light microscope. The detailed description concerning the internal fine structure of the rod will be published elsewhere in near future (*16*).

The submicroscopic structure of the ommatidium and periopticon of *Drosophila virilis* has been already published (*15*). In the present study a three dimensional drawing of the rhabdomere is presented (Fig. 2). The rhabdomere of the compound eye is enveloped by a limiting membrane and composed of two segments, a distal segment with a homogeneous structure and a proximal segment with a lamellar or polygonally reticular structure. The proximal segment of the rhabdomere consists of tightly packed tubules or rods which are arranged perpendicularly or obliquely to the long axis of the rhabdomere.

References

1. Fernández-Morán, H.: Nature (Lond.) **117**, 742 (1956).
2. Finean, J. B., F. S. Sjöstrand and E. Steinmann: Exp. Cell Res. **5**, 557 (1953).
3. Goldsmith, T. H., and D. E. Philpott: J. biophys. biochem. Cytol. **3**, 429 (1957).
4. Miller, W. H.: J. biophys. biochem. Cytol. **3**, 421 (1957).
5. Robertis, E. de: J. biophys. biochem. Cytol. **2**, 329 (1956).
6. — J. biophys. biochem. Cytol. **2**, Suppl., 209 (1956).
7. — and C. M. Franchi: J. biophys. biochem. Cytol. **2**, 307 (1956).
8. Sjöstrand, F. S.: J. cellul. comp. Physiol. **33**, 383 (1949).
9. — J. cellul. comp. Physiol. **42**, 15 (1953).
10. — J. cellul. comp. Physiol. **42**, 45 (1953).
11. — J. appl. Physics **24**, 117 (1953).
12. — Proc. Stockholm Conf. Electron Microscopy, 1956, p. 194. Stockholm: Almqvist & Wiksell, and New York: Academic Press Inc. 1957.
13. Wolken, J. J., J. Capenos and A. Turano: J. biophys. biochem. Cytol. **3**, 441 (1957).
14. Yamada, E.: Acta Anat. Nipponica (Jap.) **32**, 77 (1957).
15. Yasuzumi, G., and N. Deguchi: J. Ultrastructure Res. **1**, 259 (1958).
16. — O. Tezuka and T. Ikeda: J. Ultrastructure Res. **1**, 295 (1958)

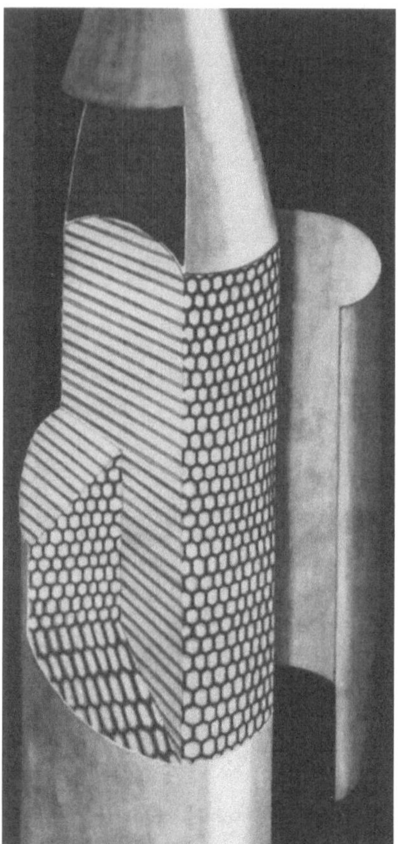

Fig. 2. Three dimensional drawings of the rhabdomere of *D. virilis*

Elektronenoptische Untersuchungen am Pigmentepithel der menschlichen Retina

H. Becher

Anatomisches Institut der Universität Münster

Zu dieser Untersuchung diente das Auge eines 84jährigen Mannes, welches wegen eines Carcinoms des unteren Augenlides entfernt werden mußte. Abgesehen von einem Altersstar war das Auge, vor allem die Retina, ganz normal. (Fixation unmittelbar nach Entfernung des Bulbus in 1%iger Osmiumsäure von p_H 7,3. Einbettung in einer Mischung von Methyl- u. Butylmethacrylat im Verhältnis 1:4. Ultramikrotom nach Sjöstrand. Mikroskop Philips 75 kV.)

Abb. 1 zeigt in vereinfachter Darstellung das elektronenoptische Bild einer Pigmentepithelzelle. Die Größe dieser Zellen beträgt an der Basisfläche 12—14 μ, ihre Höhe bis zum Beginn der Fortsätze 9—12 μ, der Durchmesser der Kerne 6—8 μ. An den Seitenflächen der Zellen bestehen von der Basis bis zum Beginn der Fortsätze deutliche Zellgrenzen. Im Bereich der Fortsätze erscheint lichtmikroskopisch ein syncytialer Zusammenhang der Zellen. Die Basalfläche der Pigmentzellen schließt mit einer gefalteten Basalmembran ab und grenzt an die Bruchsche Membran, der nach außen die Gefäße der Chorioides unmittelbar anliegen.

An der Pigmentzelle sind drei Zonen zu unterscheiden: 1. eine basale, in der längliche, geschlängelte Mitochondrien liegen, Pigmentkörnchen aber fehlen; 2. eine breite mittlere Zone, die den Zellkern enthält und in der zahlreiche Pigmentkörnchen von verschiedener Größe und Form und verschiedener Pigmentierungsintensität eingelagert sind; 3. eine obere Zone, in der ebenfalls Pigmentkörnchen liegen können und von der unregelmäßige, verschieden breite und hohe Fortsätze ausgehen, welche die Außenglieder der Stäbchen und Zapfen umfassen.

Ihr Cytoplasma ist von schwammartiger Konsistenz. Es besteht aus einem teils lockeren, teils dichteren Netzwerk von kleineren und größeren Körnchen und Bläschen, die sich perlschnurartig zu Reihen und Röhrchen ordnen. Ein eindeutiges Golgifeld konnte bisher nicht beobachtet werden. Im basalen Cytoplasma liegen schmale, lange, oft geschlängelt verlaufende Mitochondrien mit quergestellten Cristae. Bei den kleineren elliptischen oder runden Mitochondrien mag es sich um Schräg- oder Querschnitte oder um eine eigene Art kurzer Mitochondrien handeln. Einige Mitochondrien der basalen Zone erscheinen gequollen und wie flüssigkeitsgefüllte Blasen. Das Plasma der basalen Schicht zeigt eine dichte Grundstruktur, ist aber an vielen Stellen von größeren und kleineren Vacuolen durchsetzt.

In der mittleren Zellzone liegt der Kern und eine große Zahl von Pigmentgranula. Diese befinden sich in regelloser Anordnung in dem schwammartigen Plasma. Sie sind nach Größe, Form und Intensität

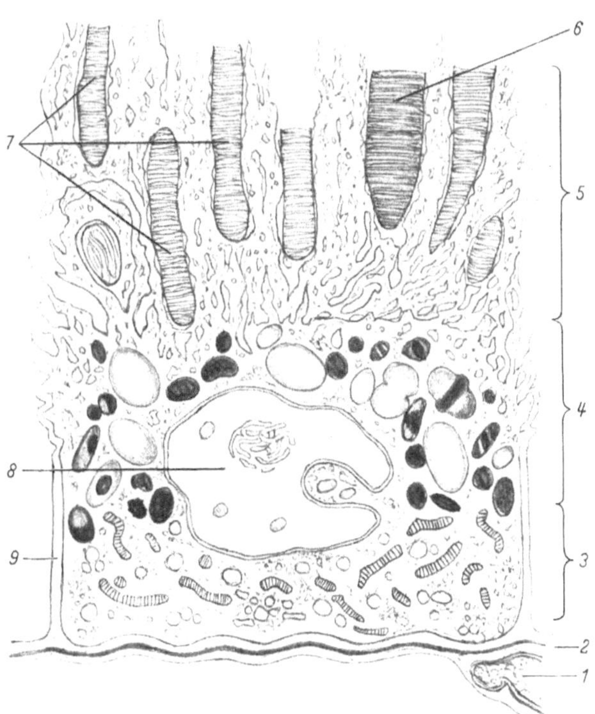

Abb. 1. Vereinfachte Darstellung einer Pigmentepithelzelle der menschlichen Retina bei Dunkeladaptation nach elektronenoptischen Ergebnissen. *1* Teil eines Endothelzellkernes der Chorioideacapillaren. *2* Bruchsche Membran. *3* Basale Zone der Pigmentzelle mit Mitochondrien im endoplasmatischen Reticulum. *4* Mittlere Zone der Pigmentzelle mit Zellkern und pigmentierten und unpigmentierten Granula verschiedener Form und Größe. *5* Obere Zone der Pigmentzelle mit Außengliedern der Stäbchen und Zapfen, die in das mit tubulären Strukturen und Gangsystemen durchsetzte Plasma hineinragen. *6* Außenglied eines Zapfens. *7* Außenglieder von Stäbchen. *8* Zellkern mit Kernkörperchen. *9* Zellgrenze.

der Pigmentierung sehr mannigfaltig. Die größten Pigmentgranula sind 1,5 μ, die kleinsten nur 0,1—0,2 μ. Die meisten Pigmentkörnchen haben elliptische Form, jedoch fehlt es nicht an kugeligen, wetzsteinförmigen oder stäbchenartigen Pigmentgranula. Bisweilen ist die Form unregelmäßig: äquatoriale Einschnürung kann Achterformen ergeben, Wellung des Außenrandes Rosetten- und Stechapfelformen entstehen lassen. Solche Formen, ebenso wie Gestaltungen mit seitlichen Ausbuckelungen oder Ringformen sind durch Schrumpfung bei der Fixation und nachfolgenden Behandlung entstanden. Neben der Verschiedenheit an Form und Größe besteht ein Unterschied der Pigmentkörnchen nach Art und Grad der Pigmentierung. Bei weitem nicht alle Pigmentkörnchen besitzen ausgefärbtes melanotisches Pigment. Manche sind nur an einem Pole dunkel oder besitzen einen runden oder unregelmäßig geformten dunkleren Anteil oder zeigen dunklere und hellere Bänderung. Die hellen Körnchen gehören durchweg zu den größeren und größten Einschlüssen und haben meist einfach elliptische oder Kugelform.

Der apikale, den Außengliedern der Stäbchen und Zapfen zugewandte Teil der Zellen ist durch Fortsätze gekennzeichnet, welche zwischen die Außenglieder der Sehzellen vorragen (Abb. 1—3). Schon im oberen Teil der Seitenränder der Pigmentzellen beginnt eine unregelmäßige Verzweigung und Fortsatzbildung, die sich über die ganze apikale Fläche erstreckt, so daß hier eine Zellbegrenzung nur teilweise deutlich ist. Die Fortsätze umfassen die Außenglieder und legen sich ihnen dicht an, mit anderen Worten, die Außenglieder ragen in das Plasma der Epithelzellen

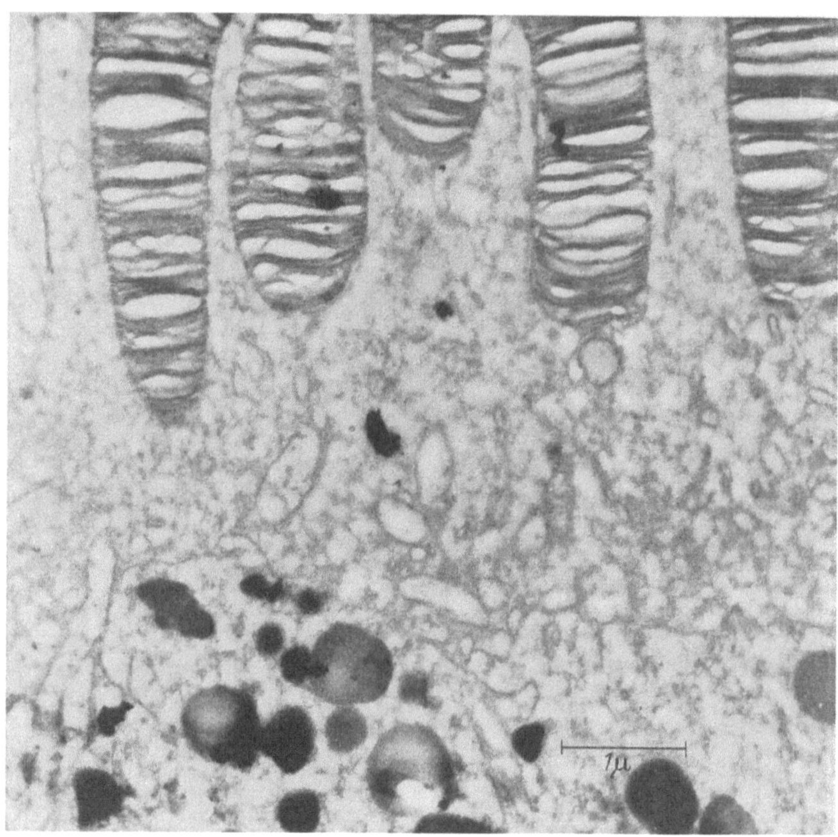

Abb. 2. Mittlere und obere Zone einer Pigmentepithelzelle. Am unteren Bildrand Pigmentgranula von verschiedener Form, Größe und Pigmentintensität. Oben 5 Stäbchenaußenglieder, die in das Plasma der Pigmentzelle hineinragen. Vergr. 16000 mal

hinein. Der Zusammenhang zwischen beiden ist so fest, daß bei einer Trennung des Pigmentepithels von der Retina Außenglieder in großer Zahl abreißen und im Pigmentepithel stecken bleiben. Die Außenglieder ragen verschieden tief zwischen die Fortsätze der Pigmentzellen (Abb. 2, 3) und die am weitesten vorragenden Außenglieder erreichen die Zone der Pigmentkörnchen. Im Bereiche der Zellfortsätze und an den Seitenrändern der Außenglieder liegen keine Pigmentkörnchen. Diese Lagerung des Pigmentes an unseren Präparaten entspricht der Dunkeladaptation (Abb. 2). Das Plasma der Fortsätze ist durch lockere Beschaffenheit und eine mehr flüssige Phase gekennzeichnet. Außer dem für den basalen Anteil der Epithelzellen geschilderten Netzwerk treffen wir in den Fortsätzen tubuläre Strukturen und Gangsysteme, die in ihrem Verlauf zu den Enden der Sehzellen ausgerichtet sind und an diese tropfenartig erweitert und sie umfließend heranreichen (Abb. 2 u. 3). Man geht wohl nicht fehl, diese röhrenförmigen und schwammartigen plasmatischen Strukturen während des Lebens als flüssigkeitsgefüllt anzunehmen und ihnen eine Aufgabe für die Stoffübertragung und im Stoffaustausch mit den Sehzellen zuzusprechen. Bisweilen umfließen die Plasmafortsätze der Pigmentepithelzellen die Außenglieder der

Sehzellen in Spiraltouren, wie besonders auf Schräg- und Querschnitten der Außenglieder zu erkennen ist (Abb. 1).

Das Pigmentepithel erfüllt mehrere Aufgaben. Allgemein wird angenommen, daß die ausgefärbten Pigmentkörnchen als Lichtschutz dienen und eine Abschirmung der Stäbchen und Zapfen bei der Helladaptation bewirken. Auch die Bedeutung des Pigmentepithels für die Entstehung des Sehpurpurs ist seit langem bekannt und man kennt die Wege der Synthese des Rhodopsins ebenso wie seinen Abbau durch die Einwirkung der Lichtquanten (*1, 2*). In der interessanten Theorie des Sehens von SEGAL (*3*) wird das Pigmentepithel als Receptor für rot angesehen. Ich möchte auf die obenerwähnte Tatsache zurückkommen, daß ein großer Teil der Pigmentkörnchen nicht melanotisch ausgefärbt ist, sondern im elektronenoptischen Bild hell erscheint, wobei gerade die größeren Körnchen kontrastarm sind. Mit-Hilfe der Fluorescenz durch Acridinorange läßt sich der Nachweis der Verschiedenartigkeit der Körnchen in den Pigmentzellen führen. Im Fluorescenzlicht erscheinen die pigmentierten Körnchen tiefbraun bis schwarz und zeigen keine Fluorescenz. Daneben aber erscheinen zahlreiche in prachtvollem Rot, Orange und Gelb fluorescierende Granula, die ein solches Bild äußerst farbenprächtig wirken lassen. Diese fluorescierenden Körnchen haben sicherlich mit einer Lichtabschirmung nichts zu tun, ihnen spreche ich vielmehr eine wichtige Aufgabe im Transport der Stoffe aus dem Blute der Gefäße der Chorioidea auf dem Wege durch das Pigmentepithel zu den Stäbchen und Zapfen zu. Ich halte diese rot fluorescierenden Granula besonders als Fermentträger für bedeutungsvoll.

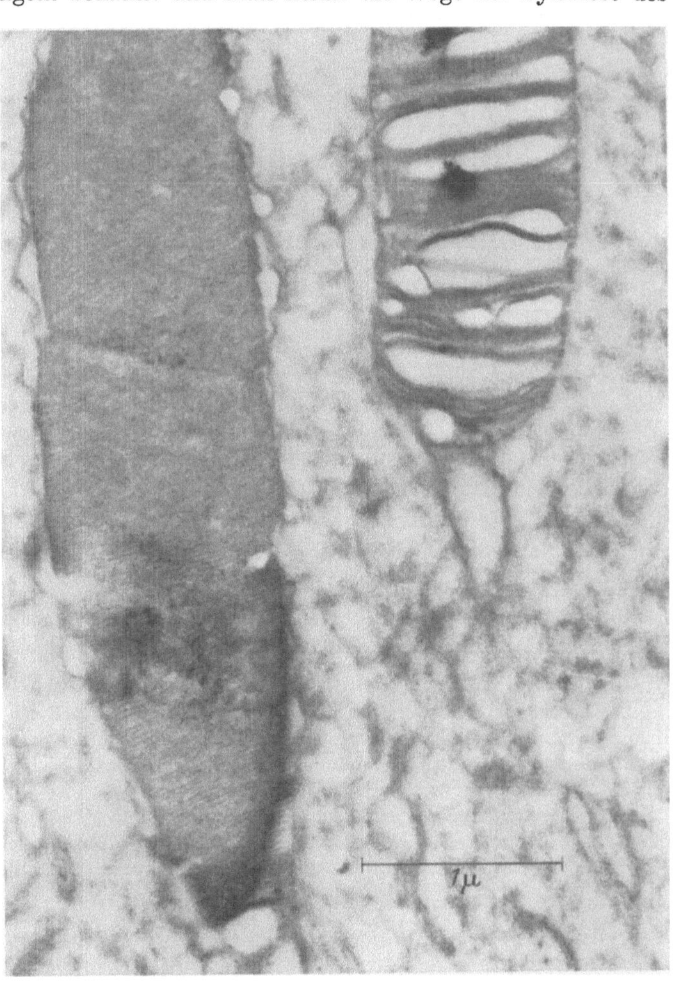

Abb. 3. Stäbchenaußenglied (rechts) und Zapfenaußenglied (links) im Plasma einer Pigmentepithelzelle. Beachte die tubulären Strukturen und Gangsysteme und deren Beziehung zu den Enden und Seitenrändern der Außenglieder, die durch deutliche Querstreifung gekennzeichnet sind. Vergr. 28000 mal

Die Rotfluorescenz dieser Körnchen ist sehr lichtempfindlich und blaßt über verschiedene Gelbintensitäten rasch ab. Diese Fluorescenzbeobachtungen am Pigmentepithel konnten sowohl bei Säugetieren wie auch bei niederen Wirbeltieren (Teleostiern und Haien) in grundsätzlich gleicher Weise erhoben werden. Man wird dem Pigmentepithel eine bedeutende Aufgabe für den Stofftransport zwischen der Chorioidea und dem Sinnesepithel schon aus dem Grunde zusprechen müssen, weil die Schicht der Stäbchen und Zapfen keine eigene Blutgefäßversorgung besitzt.

Literatur

1. BAUMGART, E.: Naturwissenschaften **39**, 388 (1952).
2. SCHENCK, G. O.: Naturwissenschaften **40**, 212 (1953).
3. SEGAL, J.: Mechanismus des Farbensehens. Jena, G. Fischer 1957.

G. Ergebnisse der Elektronenmikroskopie in der Pathologie

1. Tumorgewebe*

Elektronenmikroskopische Untersuchung von virus-ähnlichen Partikelchen in chemisch induzierten Carcinoma-Zellen**

F. DE ROM, M. SEBRUYNS, M. DE GROODT, M. THIERY und A. LAGASSE

Gynäkologisches Institut, Abt. Krebsforschung; Institut für Histologie und Laboratorium für Elektronenmikroskopie der Universität Gent (Belgien)

Im Laufe der elektronenmikroskopischen Untersuchung von experimentell erzeugtem Carcinoma cervicis uteri der C3H/A-Maus haben wir virusähnliche Partikelchen beobachtet. Da gleichartige Körperchen vorher vielfach beschrieben wurden, wurde versucht, die von uns gefundenen Partikel mit einer intra-cellulären Form eines spezifischen Virus zu identifizieren (*1*). In diesem Zusammenhang erscheint uns die Mitteilung der elektronenmikroskopischen Darstellung virusähnlicher Gebilde in Carcinomen, die mit 3,4-Benzpyren induziert wurden, von Interesse zu sein. Unseres Wissens liegt bisher keine Publikation über eine derartige Beobachtung vor.

Die Portio vaginalis uteri von weiblichen virginellen Mäusen des C3H/A-Stammes[1] wurde vom 3. Lebensmonat an mit 1%iger Lösung von 3,4-Benzpyren in Aceton, zweimal wöchentlich, gepinselt. Nach einer Behandlungsdauer von etwa 40 Pinselungen oder mehr trat eine Cervixgeschwulst auf, die histologisch als Epidermoid-Epithelioma diagnostiziert wurde.

Nach Entstehung eines palpablen Tumors wurden die Mäuse getötet und der Geschlechtsapparat in toto exstirpiert. Carcinomstückchen von 1 mm Kantenlänge wurden nach PALADE fixiert. Die Schnitte fertigten wir mit dem Ultramikrotom PORTER-BLUM. Aufnahmen: Siemens Elmiskop I.

Auf Abb. 1 kann man viele virusähnliche Partikelchen sehen. Diese haben eine annähernd kreisförmige Gestalt und eine scharfe Begrenzung. Häufig sieht man um den Rand herum noch eine etwas hellere Zone und um diese eine dunkle Linie. Diese äußere dunkle Linie, deren Breite 800—900 Å beträgt, ist besonders dort deutlich, wo viele virusähnliche Körperchen zusammenliegen. Das Partikelchen selbst kann ein kompaktes Aussehen bieten, aber auch eine helle Innenzone zeigen. In verschiedenen Zellen treten morphologische Varianten dieser Körperchen auf. Das Zentrum kann mehr oder weniger kontrastiert erscheinen, und es gibt auch Körperchen, die mit einer doppelten Membran umgeben sind. Wenn eine kompakte Innenzone vorhanden ist, liegt diese nicht immer in der Mitte eines Ringes, sondern oft an dessen Rand. Es ist also anzunehmen, daß die einzelnen Partikelchen in wahlloser Richtung durchschnitten sind und daß ein Schnitt der ober- oder unterhalb ihres „Kernes" getroffen wurde, leer oder hell erscheint (Abb. 3).

Im Grundplasma liegen (außer den bekannten Zellorganellen) kleinere oder größere Vacuolen mit einer einfachen oder doppelten membranartigen Wandstruktur (Abb. 2). In diesen Hohlräumen kann man Aggregate von virusähnlichen Partikelchen beobachten.

Für die Verteilung der virusähnlichen Körperchen im Cytoplasma ist keine bevorzugte Lokalisation zu erkennen. Sie liegen in größeren oder kleineren Gruppen beisammen; Anhäufungen

---

* Ultrastruktur der Tumorviren siehe diesen Band S. 610.

** Die Arbeit wurde unterstützt durch das „Nationaal Fonds voor Wetenschappelijk onderzoek", Brüssel.

[1] Die weiblichen C3H/A-Mäuse verdanken wir Dr. O. MÜHLBOCK, Antoni van Leeuwenhoekhuis, Amsterdam.

findet man ebensogut in der Umgebung des Zellkernes wie in der Zellperipherie oder irgendwo dazwischen.

Die Carcinomzellen zeigen eine unregelmäßige Plasmamembran mit vielen Einbuchtungen. Zwischen der welligen Oberfläche ist oft ein breiter Intercellularraum zu sehen, der viele virusähnliche Körperchen enthält. In Kernnähe findet man häufig Cytoplasmaeinschlüsse mit zahlreichen ungeordnet beisammenliegenden runden Körperchen mit einem meistens dichten, manchmal auch hellen „Kern". Dieser „Kern" wird von einer Membran konzentrisch umgeben. Bei den

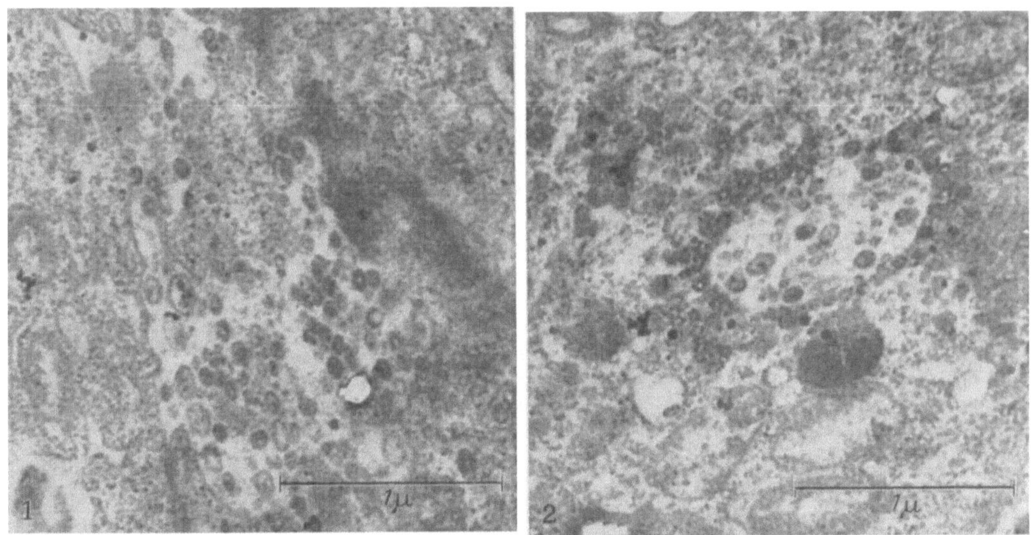

Abb. 1. Virusartige Partikelchen in chemisch induzierter Carcinomzelle. 30000 mal

Abb. 2. Hohlraum in chemisch induzierter Cervix-Carcinomzelle, viele virusartige Partikelchen enthaltend. 30000 mal

gepinselten Mäusen und bei Mäusen aus dem gleichen C3H/A-Stamm wurden auch nichtgepinselte Tiere als Kontrolle untersucht. In Dünnschnitten von Leber, Milz, Dickdarm, Mamma und Pancreas konnten keine virusähnlichen Körperchen beobachtet werden.

Die Meinungen über die Struktur und Bedeutung der virusähnlichen Partikelchen gehen auseinander [vgl. Literatur bei BERNHARD (1)].

Wir möchten zunächst darauf hinweisen, daß die von uns gesehenen Partikelchen strukturell vollkommen den sog. virusähnlichen Gebilden gleichen. Eine Einteilung in kleinere und größere

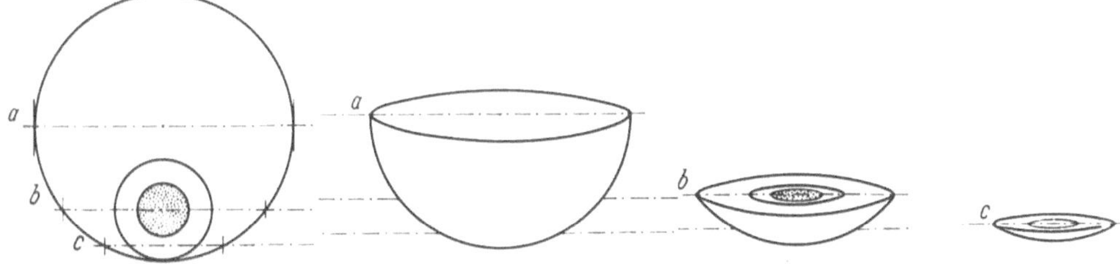

Abb. 3. Schematische Darstellung der Schnittführung durch virusartige Partikelchen

Partikelchen sowie zwischen Partikelchen mit hellem und dunklerem „Kern" erscheint uns nicht gerechtfertigt. Räumlich gesehen ist die Größe abhängig von der Schnittfläche und die beobachteten virusähnlichen Partikelchen mit hellem „Kern" können auch solche sein, deren dunkles Zentrum nicht in der Schnittebene liegt (Abb. 3).

Es ist nach unserer Meinung wenig wahrscheinlich, daß das kausale Agens, wodurch die Tumoren, deren Dünnschnitte gezeigt wurden, entstehen, ein Virus sein soll. Die carcinogenen

Eigenschaften von 3,4-Benzpyren und die strukturelle Verwandtschaft mit anderen krebserzeugenden Kohlenwasserstoffen ist genügend bekannt. Auch ein zufälliges Zusammenfallen einer Virusinfektion mit der Tumorentstehung ist wohl sehr unwahrscheinlich. Mancher Untersucher hat virusähnliche Körperchen in malignen Tumorzellen beobachtet, welche nicht chemisch induziert waren. Außerdem zeigen die Kontrollen keine virusähnlichen Gebilde.

Die Annahme, daß eine Virusinfektion ein gutes Milieu in Tumorzellen findet, wäre möglich, ist aber nicht bewiesen. Wäre dieses der Fall, dann hätte das Auffinden von Viruspartikelchen nichts mit der Tumorentstehung selbst zu tun, es wäre ein sekundärer Faktor. Zu diskutieren bleibt aber trotzdem, ob es sich bei diesen Körperchen tatsächlich um eine intracelluläre Form eines Virus oder vielleicht um das Substrat eines krebsartigen Reaktionszustandes handelt.

Literatur: *1.* Bernhard, W.: Cancer Res. 18, 491 (1958).

Electron microscope study of mouse mammary carcinoma

S. R. S. Rangan, K. J. Ranadive and Satyavati M. Sirsat

Indian Cancer Research Centre, Bombay - 12 (India)

Investigators closely connected with the field of mammary carcinoma in certain highly inbred strains of mice, are aware of the important role played by Bittner's milk agent in this type of oncogenesis. The agent has been shown to possess some, though not all, of the properties of well established viruses (*1*).

The development of electron microscope along with its allied techniques like ultra-thin sectioning of biological tissues, provided an opportunity for workers in this field to search for the morphological presence of the milk agent virus (*2—5*). This paper presents the observations of investigations carried out in our laboratories.

A total number of 35 spontaneous, transplanted and chemically induced mammary tumors from the following inbred strains of mice and one normal mid-pregnancy mammary gland obtained from a C_3H high cancer strain mouse forms the material for this investigation: — dba (old line) — 8; dba (new line) — 5; dba (—MTI) — 4; C_3H (jax) — 9; Strong A — 5; L(P) strain — 4. One tissue from the dba (old line) strain was cultivated in vitro for 7 days. The transplanted tumors examined varied in generations from the 2nd to the 12th. Observations on the biological activity of milk borne agent in these strains have been reported previously (*6—8*). It is interesting to investigate the morphological presence or absence of the agent in these tumors and interpret the findings in the light of the previous data.

Tissue bits less than 1 mm³ from the periphery of the tumors were fixed in 1% Osmium tetroxide, some as per Palade's technique (*9*) and a few others according to Rhodin's (*10*) isotonic fixation at room temperature. Tissues were processed in the usual manner in graded ethanol and embedded in methacrylate monomer mixture. Hardening of the blocks was carried out at 48° C during the early stages of the experiment and was changed to 60° C later. Thin sections as judged by the interference colours were cut either on a modified Cambridge Rocking microtome or a Porter-Blum microtome. Sections were mounted on collodion coated copper grids and viewed in an RCA EMU 2D electron microscope.

Observations. Apart from the well established cellular elements, characteristic particles of two different types namely (*a*) the intracytoplasmic particle and (*b*) the extracytoplasmic particle, have been observed in 20 out of the 35 tumors examined.

The intracytoplasmic particles are round or oval in shape measuring on an average between 500 to 700 Å in diameter. They have a typical annular appearance — a clear homogeneous space bounded by double membranes, the inner membrane clearly and thickly demarcated whilst the outer one poorly brought out.

The occurrence of the intracytoplasmic particle varied from tumor to tumor. In some tumor each cell contained the particles in greater or lesser number whereas in another tumor many cells had to be scanned before any particle was detected. For instance, only one cell of the many hundreds examined contained these particles in tumor No. T. 208 of A strain whereas in tumor No. T. 241 of the same strain nearly all the cells showed their presence.

The particles seem to have preferential locations inside the cytoplasm. Most often they are seen in small clusters next to the nucleus and near about the Golgi region. They were also seen in regular arrangement around empty vacuoles which sometimes contain small dense particulate material. Oval bodies, sometimes simulating the internal structure of mitochondria, are seen

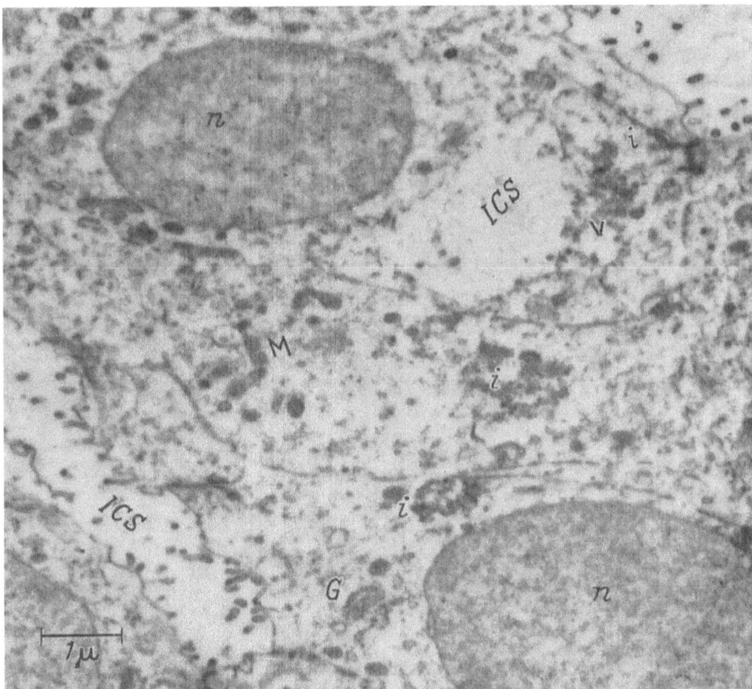

Fig. 1. Shows a group of cells from a spontaneous tumor of dba (new line) with intercellular spaces (*ICS*) with microvilli projecting into them. Intracytoplasmic particles are shown in small clusters at places marked (*i*). Empty vacuole with intracytoplasmic particles on its border is seen at (*v*). Intracytoplasmic particles are seen inside some of the microvilli also. *G* represents the Golgi region and *M* mitochondria. Note the nearness of the intracytoplasmic particles to the Golgi region and the nucleus (*n*). 11 200 ×

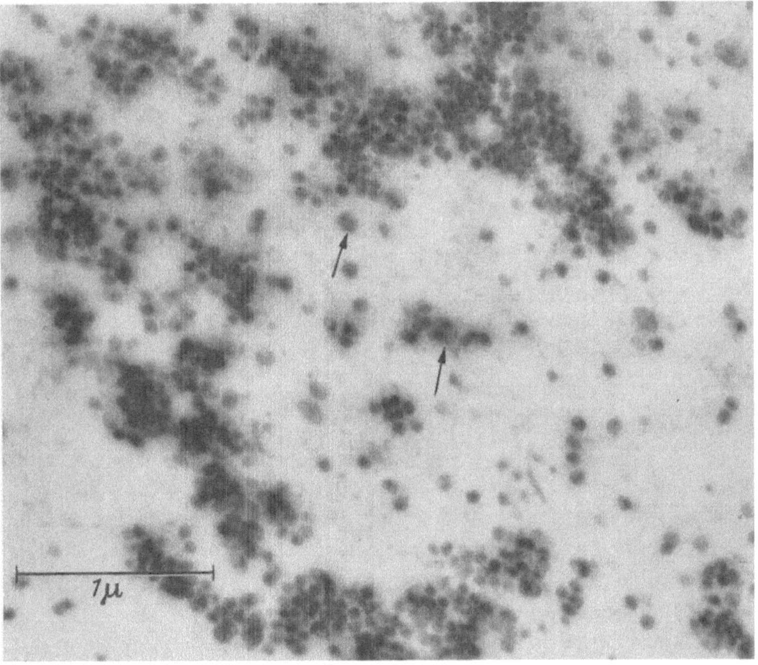

Fig. 2. Shows particles of the extracytoplasmic type, from a C_3H (jax) spontaneous tumor. At places marked by arrows more than one dense region is seen inside the particle. 26 600 ×

associated with the intracytoplasmic particles. These particles are seen in the microvilli which project into the intercellular spaces formed by a group of cells (Fig. 1).

The extracytoplasmic particles, which are quite distinct from the intracytoplasmic ones and other cellular components, are mostly seen in the intercellular spaces and in some large spaces inside the cytoplasm. Occasionally, they are observed in the cytoplasm and inside the small empty vacuoles which have the intracytoplasmic particles arranged regularly on their outside (Fig. 1).

The extracytoplasmic particles have a typical nucleoid structure. They have a central or eccentric dense region surrounded by a well defined membrane (Fig. 2). The central region is absent in some particles whereas in others more than one single dense region could be seen. The particle as a whole measures on an average about 1000 Å in diameter, the inner dense region approximating to the size of the intracytoplasmic particle. These particles are definitely distinct from the secretory granules seen in normal mammary gland (Fig. 3). Our observations regarding the particle's size, shape, location etc. agrees with those of other workers (2—5).

Table 1

Strain	Evidence of Biological Activity of Milk Factor	Tumor No.	Type of tissue	Presence or absence of particles	Strain	Evidence of Biological Activity of Milk Factor	Tumor No.	Type of tissue	Presence or absence of particles
dba (old line)	—(?)	T. 193	Spontaneous	+			T. 252	Spontaneous	+
		T. 193	Spontaneous 7 days in vitro	+			T. 255	Spontaneous	+
		T. 242	Spontaneous	+			V. 14	Spontaneous	+
		T. 180	Transplanted 2nd generation	—			V. 15	Spontaneous	+
		T. 186	Transplanted 2nd generation	—	C₃H (jax)	++	V. 18	Spontaneous	+
		T. 232	Transplanted 5th generation	—			V. 18	Transplanted 2nd generation	++
		T. 233	Transplanted 2nd generation	+			V. 18	Transplanted 6th generation	++
		T. 202	Transplanted 6th generation	—			T. 205	Transplanted 10th generation	+
							T. 229	Transplanted 12th generation	+
dba (new line)	+	V. 10	Spontaneous	+			T. 208	Spontaneous	+
		V. 10	Transplanted 2nd generation	++			T. 241	Spontaneous	+
		V. 10	Transplanted 6th generation	++	Strong A	+	T. 174	Transplanted 3rd generation	—
		T. 340	Spontaneous	+			T. 317	Transplanted 9th generation	—
		T. 340	Transplanted 2nd generation	++			P. 377	20 MCA induced	—
dba (—MTI)	—	T. 325	20 MCA induced	—					
		T. 329	20 MCA induced	—			H. 204	Spontaneous	—
		T. 378	20 MCA induced	—			H. 204	Transplanted 2nd generation	—
		T. 378	Transplanted 2nd generation	—	L (P)	—	H. 204	Transplanted 3rd generation	—
C₃H (jax)	+++	—	Mid-pregnancy mammary gland	—			P. 419	20 MCA induced	—

Table 1 gives a comprehensive picture of the results regarding the presence or absence of the particles in the material investigated. The particles have been observed consistently in all the spontaneous tumors examined except in one spontaneous tumor of strain L (P). This was the first and the only tumor of this strain developed after over 10 years of inbreeding and the strain never displayed evidence of biologically active milk agent (8).

Among the transplanted tumors, the particles are present in some and completely absent in others. In the dba (old line), except for tumor No. T. 233 which is a transplanted tumor of 2nd generation the particles have not been detected in any other transplanted tumor, whereas in the

strains dba (new line) and C₃H (jax) their presence has been observed in all the transplanted
tumors even including tumor No. T. 229 of 12th generation. This may be explained on the basis of
difference in the biological activity of the agent in the three strains (*6, 7*). In strain dba (old line) there
is no evidence of biological activity of the milk factor. The strain developed only two spontaneous
breast tumors in the year 1956 after inbreeding the large stocks for over 15 years.

Another interesting observation is the apparent increase of the particles in the transplanted
tumors Nos. V_{10} of dba (new line) and V_{18} of C₃H (jax) strains, over their corresponding spontaneous
growths. In the tissue that was grown in vitro for 7 days, only particles of the extracytoplasmic type were seen. The significance of these observations is not very clear.

Discussion. The extra chromosomal factor which plays an important role in the genesis of mammary carcinoma in certain inbred strains of mice has been found to operate through the mother's milk. A filtrable agent has been recovered from such carcinomas and has been found to induce mammary tumors on inoculation to mice without the agent. In fact the agent has been shown to be widely distributed in many of the tissues including mammary gland and blood, of mammary cancer strain mice (*1*).

The physical and chemical investigations confer the properties of a virus to the mammary tumor inciter (*1*). Electron microscope studies of spontaneous tumors

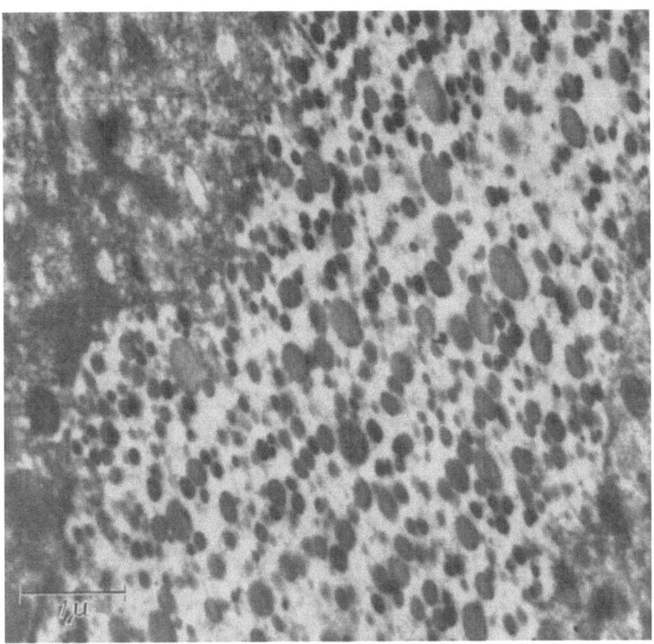

Fig. 3. Shows the secretory granules from a mid-pregnancy
mammary gland of a C₃H (jax) mouse. 13 740 ×

occurring in high cancer strain mice have revealed the virus-like particles in these tissues. A correlation between the filtrable agent and the virus-like particles seen in electron micrographs could
be arrived at easily but for two facts — the presence of such particles in tumors obtained from
mice biologically free from the agent, and their absence in tumors and mammary glands of mice
known to possess the milk agent. However such particles have been observed in tumors biologically
free of the agent(*3, 4, 11*).

Virus like particles have been observed electron microscopically in a number of neoplasms,
some of which are filtrable like the fowl tumors (*12*) mouse leukemia (*13*), frog renal adenocarcinoma (*14*), while others like Ehrlich ascites tumors (*15, 16*), hepatoma in C₃H mice (*17*) are nonfiltrable. Recently particles of the intracytoplasmic type seen in our material have been found
in a plasma cell tumor (*18*). Another contradictory phenomenon is the electron microscopical
absence of virus like particles in normal mammary gland tissue of high cancer strain mice and other
normal tissues which are biologically positive for the agent (*4*).

The questions that arise are, are we seeing the milk agent in the size and shape of the virus-like particles in the electron micrographs? If so which one of the two types, the intra- or the
extracytoplasmic particle is the infectious agent?

In view of the above facts and due to the lack of experimental data on thin section studies of
ultrafiltrate pellets, a precise answer is difficult.

Is the mammary tumor agent a virus? In certain respects like its physical and chemical
properties — yes. But in certain respects like growing in chorioallantoic membrane, where we have
contrary reports (*19, 20*), its nature is not definite.

We would like to suggest that the milk agent is an organic complex, resembling virus in structure and size, formed endogenously in the tumors and possessing carcinogenic activity. The failure to detect their presence electron microscopically in normal tissues may be due to the tissues being non-malignant. The agent becomes detectable only after initiating the disease which further goes through stages of promotion and development as a result of interaction with hormones on genetically susceptible soil.

References

1. Bittner, J. J.: Experimental aspects of Mammary Cancer in Mice. Chapter IV in Breast Cancer and its Diagnosis and Treatment. Edward F. Lewison. Baltimore: Williams and Wilkins Co. 1955.

2. Bang, F. B., I. Vellisto and R. Libert: Bull. Johns Hopk. Hosp. **4,** 255 (1956).

3. Dmochowski, L.: Acta Un. int. Cancr. **12,** 582 (1956).

4. Bernhard, W., M. Guerin et C. Oberling: Acta Un. int. Cancr. **12,** 544 (1956).

5. Suzuki, T.: Gann **48,** 139 (1957).

6. Khanolkar, V. R., and K. J. Ranadive: J. Path. Bact. **59,** 593 (1947).

7. Ranadive, K. J.: Acta Un. int. Cancr. **12,** 701 (1956).

8. — and S. A. Hakim: Brit. J. Cancer **12,** 44 (1958).

9. Palade, G. E.: J. exp. Med. **95,** 285 (1952).

10. Rhodin, J.: Correlation of ultrastructural organization and function in normal and experimentally changed proximal convoluted tubule cells of the mouse kidney. Aktiebolaget Godvil, Stockholm, Sweden 1954.

11. Bang, F. B., H. B. Andervont and I. Vellisto: Bull. Johns Hopk. Hosp. **4,** 287 (1956).

12. Epstein, M. A.: Brit. J. Cancer **12,** 248 (1958).

13. Dmochowski, L., and C. E. Grey: Ann. N. Y. Acad. Sci. **68,** 559 (1957).

14. Fawcett, D. W.: J. biophys. biochem. Cytol. **2,** 725 (1956).

15. Selby, C. C., C. E. Grey, S. Lichterberg, C. Friend, A. E. Moore and J. J. Biesele: Cancer Res. **14,** 790 (1954).

16. Adams, W. R., and A. M. Prince: J. biophys. biochem. Cytol. **3,** 161 (1957).

17. Fawcett, D. W., and J. W. Wilson: J. nat. Can. Inst. **15,** 1505 (1955).

18. Howatson, A. F., and E. H. McCulloch: Nature (Lond.) **181,** 1213 (1958).

19. Timofejevsky, A. D.: VII. int. Cancer Congress. Abstract of Proc. 1958, p. 51.

20. Dmochowski, L., L. Jordan and J. A. Sykes: VII. int. Cancer Congress. Abstract of Proc. 1958, p. 248.

Structure fine de cancers de la glande mammaire chez la femme

F. Haguenau

Institut de Recherches sur le Cancer, Villejuif/Seine (France)

Grâce au couteau de diamant qui permet malgré l'abondance du tissue collagène d'obtenir des coupes impeccables, et grâce aux progrès réalisés dans les techniques de prélèvement, *les tumeurs humaines peuvent être fixées et coupées dans des conditions aussi parfaites que le matériel expérimental.* Dans ces conditions le pathologiste moderne n'a plus aucune excuse pour négliger l'étude systématique des tumeurs humaines au microscope électronique.

A l'Institut du Cancer, au cours d'une étude que nous projetons de poursuivre en collaboration avec Mme. Vogt-Hoerner[1] et Mme. Arnoult (*4, 5*) , nous avons été amenés à examiner systématiquement 54 cas de cancers du sein et 16 glandes hyperplasiques ou supposées normales. Contrairement à une tendance trop répandue dans le monde des «électroniciens», il a été donné une grande importance à l'examen au faible grossissement, ce qui permet de prolonger en quelque sorte au microscope électronique l'examen au microscope optique. Cette *histologie électronique* permet de mieux connaître les différentes portions glandulaires et notamment les différentes variétés histologiques des cancers mammaires au microscope électronique (Fig. 1 et 2). C'est ensuite seulement que l'étude cytologique est entreprise dans le but de rechercher dans les détails non accessibles au microscope ordinaire, des structures anormales voire spécifiques (Fig. 3).

Les 54 cas déjà étudiés ne sont pas en nombre suffisant pour permettre des conclusions définitives, mais les résultats de l'analyse détaillée de chacun d'eux permettent déjà de noter les points suivants:

Il n'existe pas toujours une corrélation exacte entre les formes histologiques classiques et les types reconnaissables au microscope électronique. Autrement dit, des cancers classés au microscope

---

[1] Pathologiste du Centre Clinique de l'Institut Gustave Roussy (Prof. Denoix).

optique dans une même catégorie peuvent se présenter au microscope électronique avec des aspects très différents. La Fig. 1 et la Fig. 2 illustrent ce fait: il s'agit de deux cancers qui répondent au type classique de l'épithélioma alvéolaire. Ils sont formés au microscope électronique de

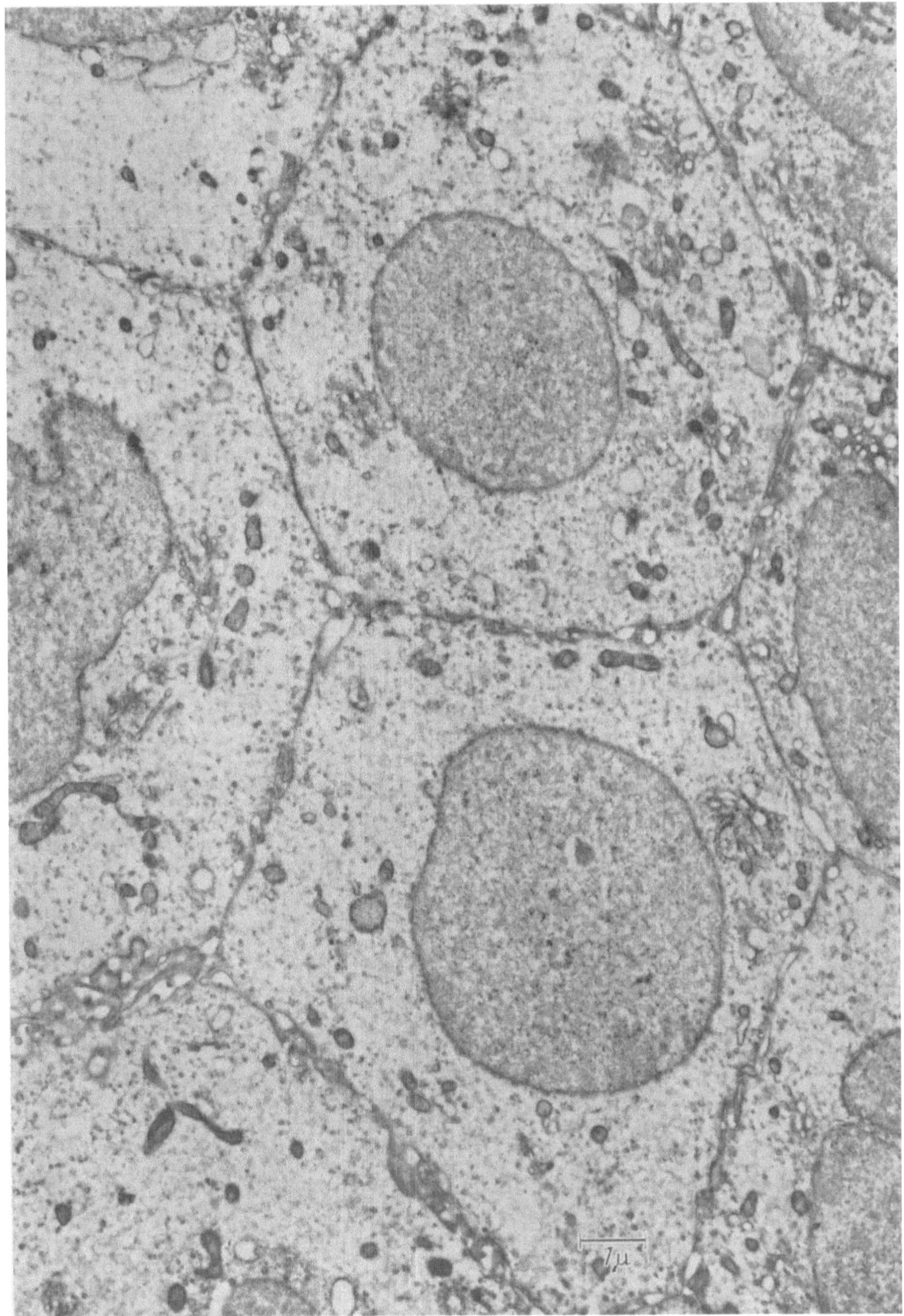

Fig. 1. Un des aspects les plus fréquents d'un cancer du sein au microscope électronique. Les cellules volumineuses, polygonales revêtent un type particulier à cytoplasme clair et organites raréfiés. Gross. 9750 fois

cellules de caractères totalement différents et qui correspondent à deux des aspects rencontrés avec le plus de fréquence au microscope électronique. Dans le premier cas (Fig. 1), il s'agit de grandes cellules polygonales claires à noyaux réguliers et membrane de surface très intriquée.

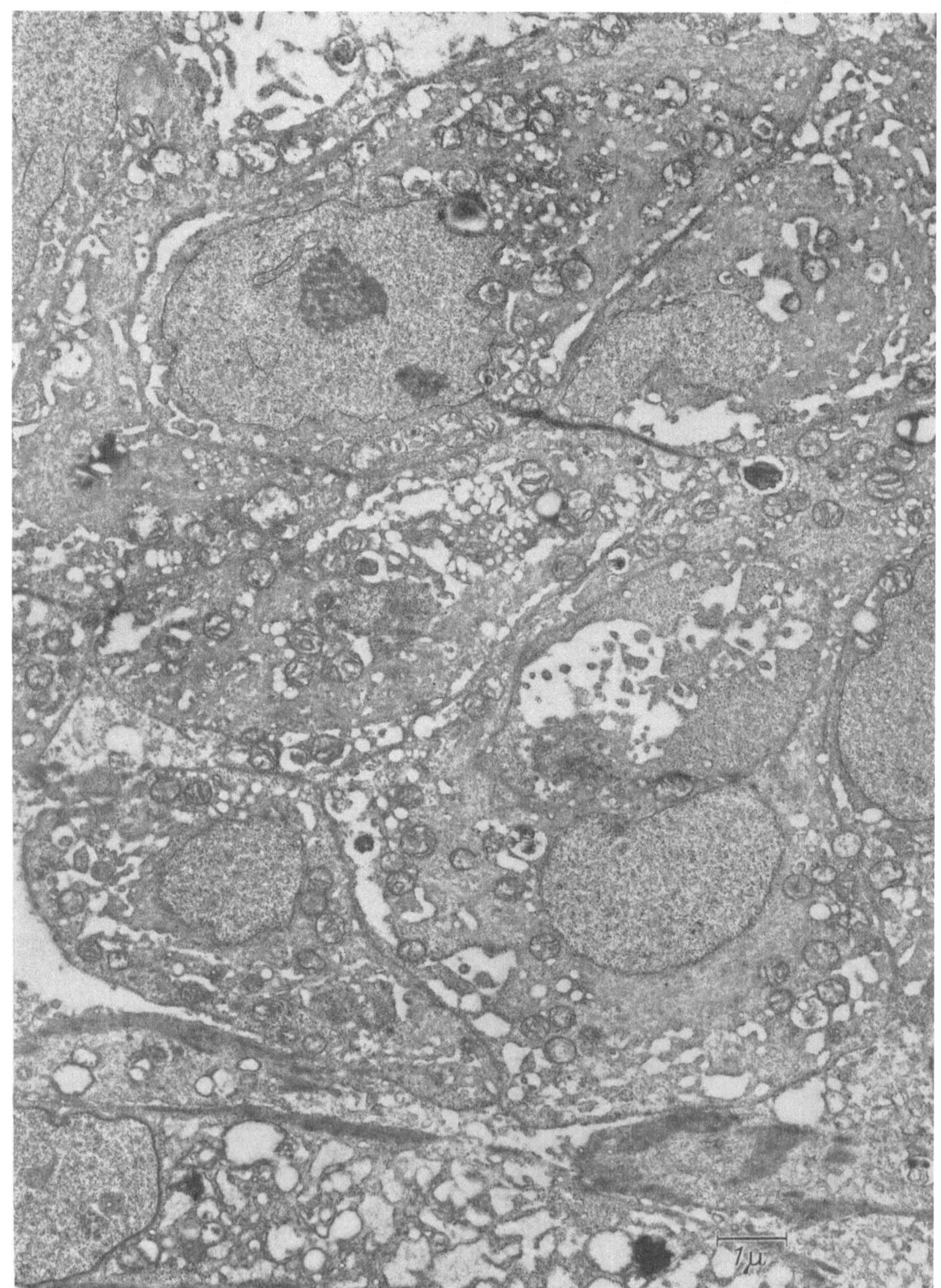

Fig. 2. Autre aspect d'un cancer du sein formé de cellules à cytoplasme dense et finement fibrillaire. Gross. 9750 fois

Leur cytoplasme très développé offre un aspect «vide». Les organites cellulaires (mitochondries, appareil de Golgi, ergastoplasme) sont en effet raréfiés. Dans le deuxième cas au contraire (Fig. 2), les cellules frappent par leur aspect dense, plein et «fonctionnel». Les noyaux sont irréguliers.

La substance fondamentale est très finement fibrillaire. Ce feutrage cytoplasmique est troué à l'
emporte-pièce par des vacuoles cernées d'une membrane très bien dessinée et qui correspondent
tant à un appareil de GOLGI hypertrophique très vacuolisé qu'à des mitochondries nombreuses et
tuméfiées. La nature de ces cellules reste à être établie. Il pourrait s'agir de cellules d'un type
proche de la cellule épidermique, de cellules myoépithéliales modifiées, ou des cellules A décrites
par CASPERSSON et SANTESSON que l'on retrouve avec une grande fréquence à la périphérie des

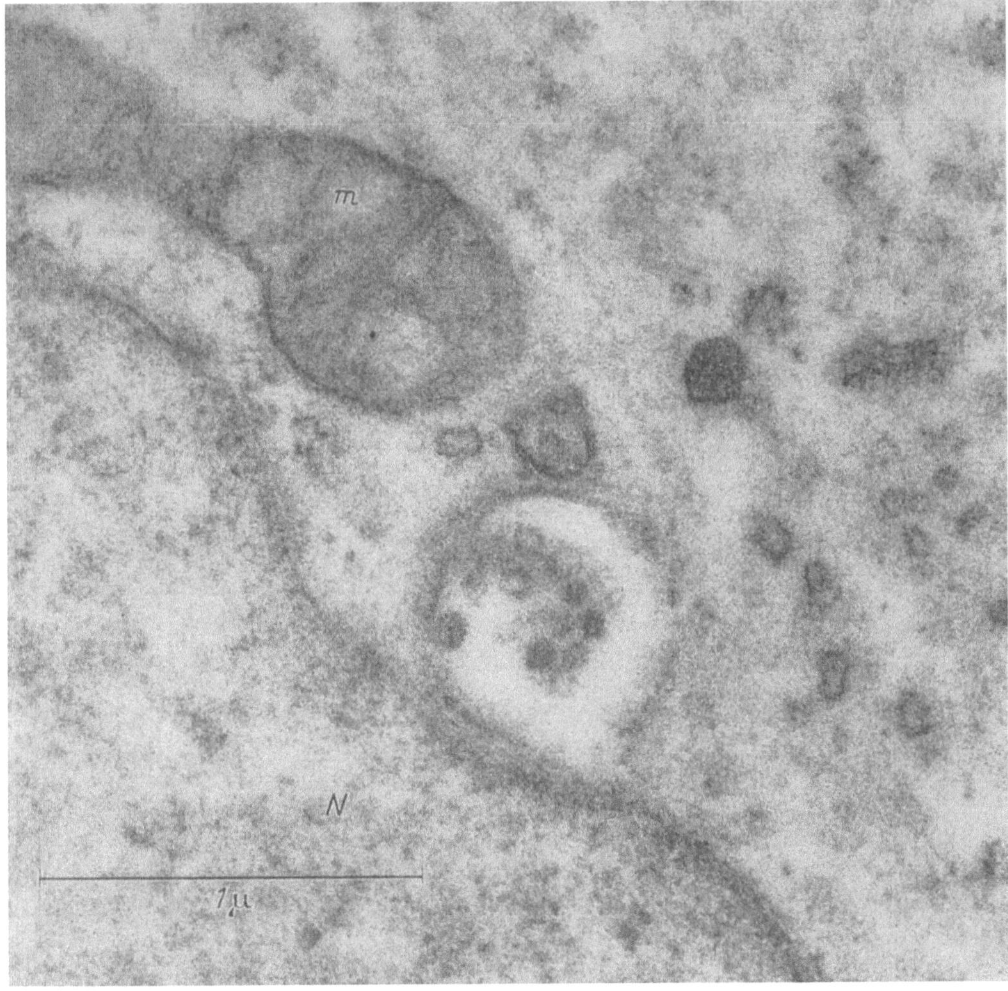

Fig. 3. Portion d'une cellule de cancer du sein. Au-dessous de la mitochondrie (*m*) et en bordure du noyau (*N*)
on voit une vacuole bordée par une membrane simple, et contenant des particules dont l'origine n'est pas
encore établie. Gross. 51 000 fois

lobules mammaires, ou encore de cellules subissant une stimulation hormonale particulière. L'
exemple des deux cas choisis ici montre ce que l'on est en droit d'espérer d'une telle histologie
électronique elle permettra peut-être de trouver la corrélation entre l'aspect morphologique des
cellules et l'évolution de la maladie que la microscopie optique s'est vainement efforcée d'établir
jusqu'ici.

A côté de l'étude *histologique*, le microscope électronique révèle des caractères *cytologiques* qui
correspondent d'abord à ceux décrits en général pour la cellule cancéreuse [voir la récente revue
de BERNHARD (*1*)]. Certains points méritent d'être soulignés pour le cancer du sein: la présence
très fréquente de canaux intracytoplasmiques, fait déjà signalé dans des cancers expérimentaux;
la grande fréquence d'inclusions de type dégénératif (figures de myéline, corps cristalloïdes) tant

dans le cytoplasme que dans le noyau; l'extraordinaire développement que peuvent prendre dans certains cas les membranes cellulaires qui tendent à devenir micropapillomateuses.

Aucune de ces particularités, cependant, n'est réellement spécifique de la cellule maligne.

Un soin particulier a naturellement été consacré à la recherche d'images rappelant les corpuscules virus présents dans les cancers mammaires de la souris (Bernhard, op. cité). Des particules ont été trouvées effectivement dans les cellules de plusieurs cas de cancers, mais leur signification à l'heure actuelle est totalement inconnue. La Fig. 3 en offre un exemple. Il s'agit de corpuscules sphériques d'une dimension de 40 mμ environ. Ils forment de petits amas en grappes dont les limites sont quelquefois indistinctes, quelquefois marquées par une membrane très nette. Ils sont toujours contenus dans des vacuoles dont la position dans la cellule est variable, soit près du noyau dans la zone de Golgi (Fig. 3), soit en plein cytoplasme ou à la périphérie. Ces vacuoles peuvent être totalement dépourvues de membrane comme si celle-ci avait été lysée.

S'agit-il de mitochondries vacuolisées et fragmentées en granules comme celles qui ont été décrites dans les cellules en culture (3) ? S'agit-il de vacuoles golgiennes et, dans ce cas, les particules pourraient éventuellement répondre à des micrograins de sécrétion ? De toutes façons, on peut éliminer les «compound vacuoles» décrites par Low (2) dans les cellules sanguines. Une dernière possibilité enfin, celle de la nature virale de ces grains, n'est nullement à rejeter, ce qui n'impliquerait pas forcément une origine virale de ces tumeurs, conclusion que seules des recherches biologiques conjuguées pourraient étayer.

Bibliographie

1. Bernhard, W.: Cancer Res. 18, 491 (1958).
2. Low, F. N., and J. A. Freeman: Electron microscopic atlas of normal and leukemic human blood. McGraw-Hill Book Company, Inc. 1958.
3. Morgan, C., H. M. Rose and D. H. Moore: Ann. N. Y. Acad. Sci. 68, 302 (1957).
4. Haguenau, F.: Pathol. Biol. 7, 989 (1959).
5. — Bull. Cancer (sous presse).

Comparison of the fine structure of two human carcinomas*

G. A. Edwards, C. Ruska, H. Ruska and J. Skiff

From the Division of Laboratories and Research, New York State Department of Health, Albany, USA; Institute of Electron Microscopy, Med.Acad. Düsseldorf; and Veterans Administration Hospital, Albany, USA

Various stages of human tumors transplanted to cortisone treated hamsters (*1*) have been examined in the electron microscope. These include a 9th generation bronchogenic carcinoma, strain A-42, and both early and late 13th generation squamous cell carcinomas of the penis, strain A-21.

The tumor nodules were dissected from the cheek pouch of the hamsters, cut into small pieces, fixed in 2% OsO_4 buffered with veronal-acetate to pH 7.4, dehydrated in graded ethanols, infiltrated with a mixture of butyl- and methyl methacrylates in a ratio of 9:1, embedded in the methacrylate mixture initiated with $1^1/_2$% Lucidol, and polymerized at 47° C for 36 hr. Sections were cut with diamond knives on Porter Blum microtomes and observed in the Siemens Elmiskop I.

Bronchogenic carcinoma (Figs. 1—4). The cells of the bronchogenic carcinoma generally are loosely applied to each other (Figs. 1 and 3), although in the light microscope the whole tumor appears as a firm, solid sheet of cells. They show little or no polarity. The cell border is extremely irregular, varying from small rugosities to long papillate projections (Fig. 1). The cell border consists of plasma membrane only and shows little pinocytotic activity. The cytoplasm of the early stage cells is filled with numerous and widely scattered tubules, vesicles and sacs of the endoplasmic reticulum, usually with RNA granules attached on the cytoplasmic side of the membrane profiles. Characteristically, the irregular, paired profiles of the reticular sacs are long and tortuous. Myriads of RNA granules occur, singly and in rosettes, throughout the cytoplasm.

* This study was aided by grants from the American Cancer Society and the Brown-Hazen Fund of the Research Corporation.

Mitochondria are relatively few in number and present a rather "empty" appearance. They contain few, peripheral, incomplete cristae and a sparse matrix of fine filaments and granules (Figs. 1 and 2). The reticular apparatus of GOLGI is large and usually situated close to the nucleus. The nuclei are large, irregular, often lobed. They contain large nucleoli, considerable chromatin aggregates, and possess numerous pores.

Cells more internal in the tumor nodule (Fig. 3) vary greatly one from another. Some cells have lobulated nuclei with a clumped, granular matrix. Others show small necrotic nuclei. The cytoplasm of the former cells possesses relatively few RNA granules, many long ergastoplasmic elements, and the mitochondria possess few or no cristae but have a finely filamentous matrix. The presumably older cells at this level of the tumor (Fig. 3) possess a degenerate nucleus, and few small cristae-less mitochondria. These cells have no Palade granules, show only few scattered vesicles of the endoplasmic reticulum, and the cytoplasmic matrix comprises scattered, extremely fine filaments; thus giving the total cell a ghost-like appearance (Fig. 3).

Anaplastic cells in the center of the tumor (Fig. 4) possess few clearly recognizable cellular components. Characteristically the sacculi and vesicles of the Golgi apparatus increase in number and breadth. The remainder of the cytoplasm is almost completely filled with fine to coarse circular filaments, among which are scattered occasional, small, dense, mitochondrial fragments. The nucleus at this stage is essentially a vestige; its small dense clump of matrix withdrawn from the irregular, thickened nuclear membranes. In the final stage of dedifferentiation the cell consists of only plasma membrane, cytoplasmic filaments and nuclear remnant.

Squamous cell carcinoma (Figs. 5—8) Cells of both early and late stages of the 13th generation squamous cell carcinoma vary within a given nodule from almost normal-appearing epidermal cells to highly keratinized, degenerating cells. The more normal cell from the surface (Fig. 5) is bordered by a plasma membrane which occasionally shows some irregularity. The cytoplasm contains numerous RNA granules, little endoplasmic reticulum, scattered tonofibrillar filaments, and a small Golgi apparatus. Masses of tonofibrils occur immediately around the nucleus. The nuclei and their constituent nucleoli are large and irregular. Characteristically, where small areas of adjacent cells become intimately apposed, desmosomes are formed. The cytoplasm near a desmosomal region contains aggregates of RNA granules, vesicles of the endoplasmic reticulum and thin, unevenly spaced tonofibrillar filaments normal to the plane of the plasma membrane. These latter filaments become continuous with the dense granular material in the cytoplasmic side of the desmosome. The desmosome (Fig. 6 shows details) appears to possess components and a structural organization similar to those of cardiac intercalated discs. There is no continuity between cells.

Legends to pages 468 and 469

Fig. 1. Cells from the surface of a human bronchogenic carcinoma. *Er* = endoplasmic reticulum, *Mi* = mitochondria, *Nu* = nucleus. 24 000 ×

Fig. 2. Detail of cytoplasm of surface bronchogenic carcinoma cell showing ergastoplasmic sacs with attached RNA granules (*Er*) and "empty" mitochondria (*Mi*). 40 000 ×

Fig. 3. Portions of three intermediate cells of bronchogenic carcinoma. Nuclei of left and upper cells have less pores than preceding stage, and mitochondria (*Mi*) possess few or no cristae. Cell at lower right is essentially filled with filaments (*F*) and only few mitochondrial and reticular fragments. 15 000 ×

Fig. 4. Detail of cell from center of bronchogenic tumor, showing degenerate nucleus (*Nu*), small mitochondria (*Mi*), enlarged Golgi apparatus (*G*) and filamentous cytoplasmic matrix (*F*). 35 000 ×

Fig. 5. Early stage squamous cell carcinoma with almost normal nucleus (*Nu*), circumnuclear masses of tonofibrils (*T*) and desmosomal (*D*) apposition of neighboring cells. 30 000 ×

Fig. 6. Detail of cytoplasm of desmosomal region. Fine tonofibrillar strands (*T*) contact inner dense granules of desmosomal disc. The desmosome (*D*) comprises apposing plasma membranes, their interspace, and parallel layers of fine granules. 60 000 ×

Fig. 7. Intermediate stage cell of squamous carcinoma. Enlarged nucleus (*Nu*) possesses numerous pores. Cytoplasm filled with increased Golgi apparatus (*G*) and ergastoplasm (*Er*). Few small mitochondria (*Mi*), lipid bodies (*L*) and concentric structures also present. 30 000 ×

Fig. 8. Late stage squamous cell carcinoma in which cytoplasm is filled with lipid, mitochondrial fragments, and numerous concentrically layered bodies. 30 000 ×

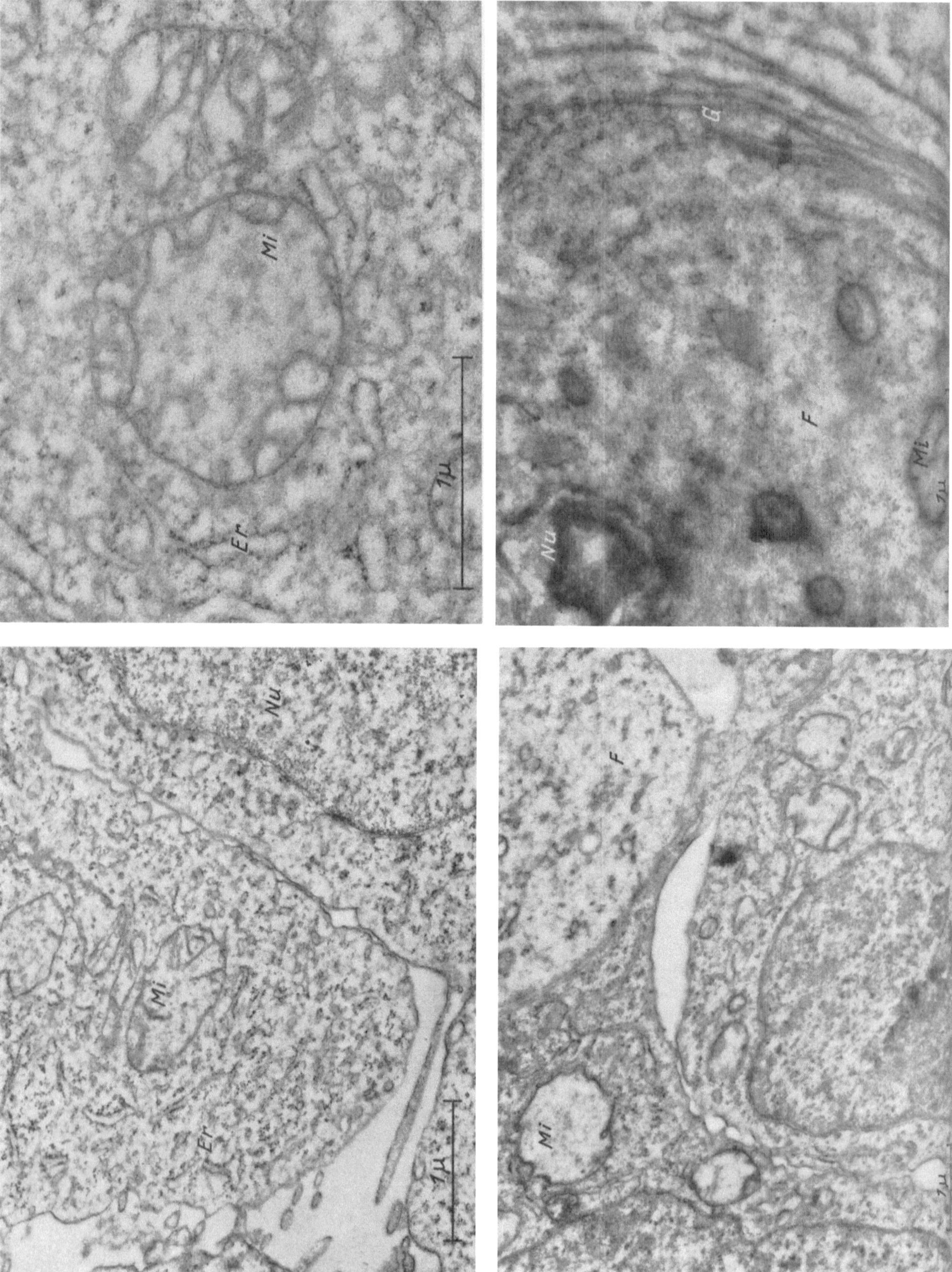

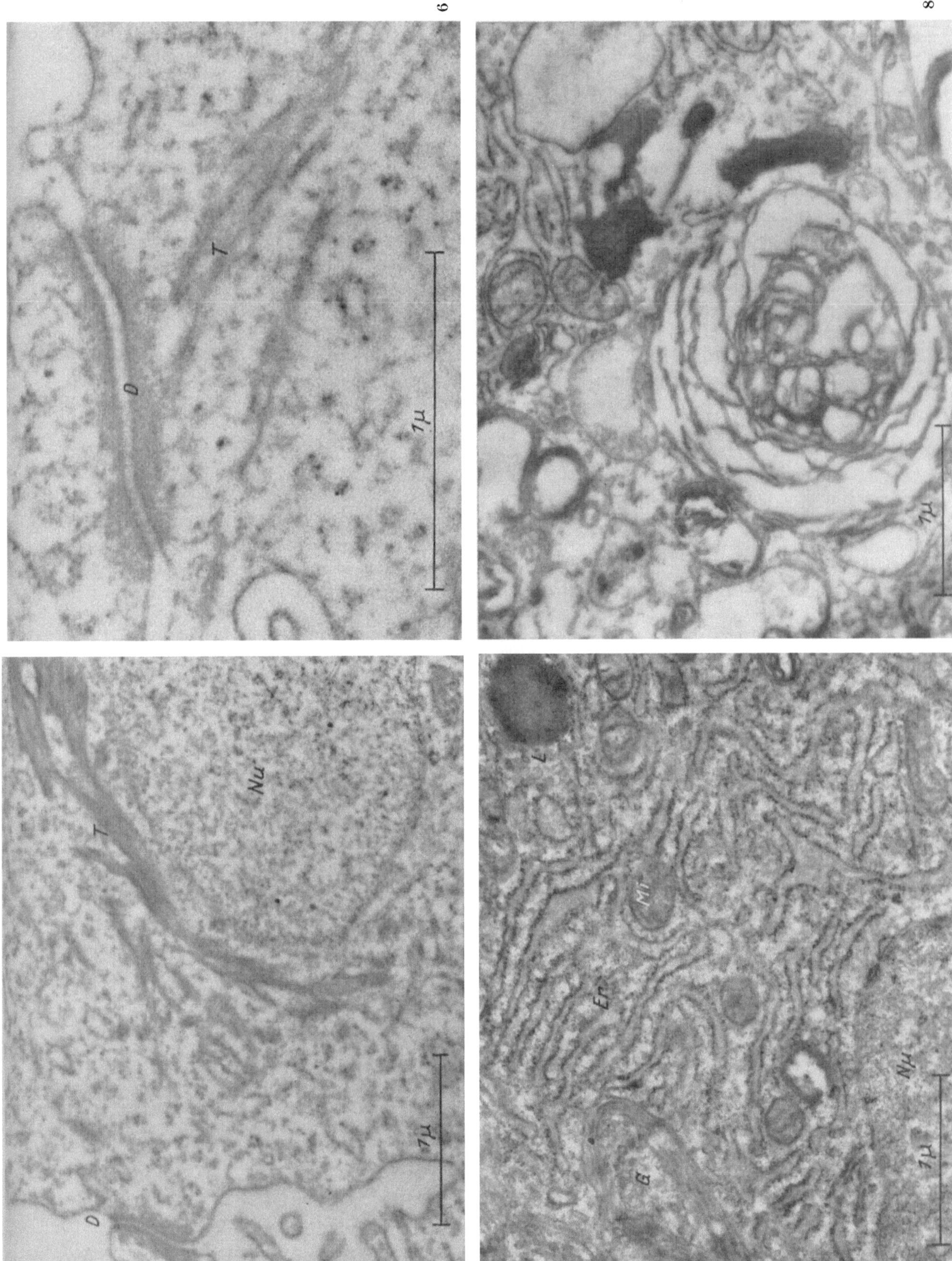

The adjoining cells are limited by their respective, thickened plasma membranes and are separated by an interspace of 30 mμ. The interspace contains some material of medium electron density with occasional short strands of denser material normal to, but not necessarily continuous with, the parallel plasma membranes. The desmosomal area of each cell comprises thickened plasma membrane, parallel layers of fine granules and an inner layer of filaments and larger, denser granules.

In the more mature tumor cells (Fig. 7) the tonofibrils become fewer in number and the cytoplasm becomes increasingly filled with long, interconnected, often parallel ergastoplasmic sacs whose internum contains a finely granular substance of medium electron density. The hypertrophied GOLGI apparatus becomes increasingly widespread, its sacculi arranged in whorls of concentric membrane profiles. In general, the invasion of the cytoplasm by ergastoplasm and GOLGI apparatus begins in a single locus of the cell and gradually spreads, rather than by multiple origins, until the entire cell is filled with membranes. Tonofibrils are not found in the membrane-filled areas, suggesting a replacement of tonofibrils by membranes. In these apparently rapidly changing cells the nuclei are large, possess numerous pores and the perinuclear cisternae vary greatly in width. The cytoplasm may also contain lipid bodies and small, empty structures of the size and shape of mitochondria but bounded by concentric membrane layers.

In anaplasia, in the center of the tumor, concentric bodies almost completely fill the cytoplasm (Fig. 8). At this stage the cytoplasm still possesses few ill-defined mitochondria, little ergastoplasm, but considerable lipid. The nucleus appears as a shriveled, dense body retracted from the nucleo-cytoplasmic bordering membranes. Further there appear small to very large spherical bodies of tightly packed, concentric membranes. These eventually become solidified, forming dense irregular structures. They are usually widely scattered but may occur in clumps. Subsequent stages involve cell breakdown and scattering of the components into the interstitium.

Bibliography

1. SKIFF, J. V., A. A. STEIN, M. MAISEL, C. HEILBRUNN and D. HERTZ: Cancer Res. **18,** 485 (1958).

Elektronenmikroskopische Untersuchungen der virusähnlichen Körperchen in bösartigen Geschwülsten des Menschen

A. SCHUBIN

Institut für experimentelle Pathologie und Therapie des Krebses, AMN USSR, Moskau

In einer Reihe von Arbeiten sind in der USSR und im Auslande virusähnliche Körperchen beschrieben worden, die in den Geweben und Gewebsextrakten verschiedener menschlicher Geschwülste gefunden wurden. Wir beschreiben die Resultate der elektronenmikroskopischen Untersuchungen an folgenden Gewebsextrakten: Magenkrebs 137 Fälle, Magenpolypen 14 Fälle, Sarkome 79 Fälle, Mammacarcinome 64 Fälle, Melanome 17 Fälle. Das Material wurde durch Zentrifugierung und Trypsinbehandlung vorbereitet. In 40—50% haben wir in den Geschwülsten virusähnliche Körperchen gefunden, in normalen Geweben aber nur in einzelnen Fällen. Weitere serologische und immunologische Untersuchungen erlauben uns, an die Virusnatur dieser Körperchen zu denken.

Dieser Beitrag erschien ausführlich in „Fragen der Ätiologie und der Pathogenese der Tumoren" (russ.) 1957, S. 49, Moskau.

Zur Feinstruktur des Mäuse-Ascites-Carcinoms nach Einwirkung von N-Lost-Benzimidazol

KURT ZAPF und KURT RINTELEN

Deutsche Akademie der Wissenschaften zu Berlin, Institut für Medizin und Biologie, Berlin-Buch

Die Feinstruktur normaler und degenerativer Zellen des Ehrlich-Mäuse-Ascites-Carcinoms (EMAC) ist nach elektronenmikroskopischen Untersuchungen von SELBY u. Mitarb. (*1, 2*), YASU-ZUMI u. Mitarb. (*3*) sowie WESSEL und BERNHARD (*4*) geklärt. Ziel unserer Untersuchung war es festzustellen, inwieweit der deutlich nachweisbare Hemmeffekt von N-Lost-Benzimidazol (β, β-Dichlor-diäthyl-aminomethyl-benzimidazol-HCl) (NLB) auf das Wachstum des EMAC (5, 6) im Strukturbild der beeinflußten Asciteszelle in Erscheinung tritt und ob daraus auf den Wirkungsmechanismus geschlossen werden kann.

Männliche Mäuse etwa gleichen Gewichts erhielten intraperitoneal 0,2 ml Ascitesflüssigkeit. Die Behandlung der Ascites-Mäuse mit N-Lost-Benzimidazolen erfolgte subcutan in Abständen von 1—3 Tagen, beginnend am Tage der Überimpfung oder später. Die Hemmwirkung wurde an über 150 Tieren geprüft. Der unterschiedliche Gewichtsverlauf behandelter und unbehandelter Ascites-Mäuse sowie die Ascites-Menge ist aus einem Kurvenbeispiel (Abb. 1) ersichtlich. Für die Feinstrukturanalyse wurden 26 Tiere verwendet. Die Ascites-Flüssigkeit wurde 6, 8, 10 und 11—14 Tage nach Überimpfung entnommen und sofort in vorgekühlter 1 % iger gepufferter Osmiumsäure bei 4° C in unterschiedlichen Zeitintervallen fixiert und in üblicher Weise in Methacrylsäureester eingebettet und geschnitten(7). Zur cytochemischen Strukturanalyse in Schnitten verwendeten wir vergleichend die Feulgen'sche Nuclealreaktion (FN) und die Desoxyribonuclease (DNAse)-Differenzierung mit und ohne Herauslösen des Einbettungsmittels.

Es war zu erwarten, daß EMAC eine Zellpopulation unterschiedlicher Empfindlichkeit (8) gegen NLB darstellt und ein Teil der Zellen zugrunde geht. Außerdem können morphologische Veränderungen variieren, so daß notwendigerweise eine große Zahl von Präparaten nnd Aufnahmen ausgewertet werden müssen, um pathologische Zellveränderungen aus dem Bauplan der Zelle zu erschließen.

Die Desintegration des Zellaufbaus behandelter Zellen zeigt sich in Veränderungen am Nucleus und an cytoplasmatischen Strukturelementen. Das elektronenmikroskopische

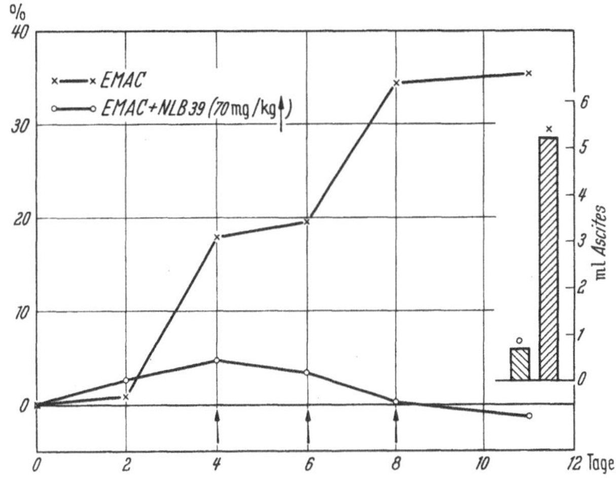

Abb. 1. Wachstum und Ascitesmenge normal und nach NLB-Behandlung

Bild einer jungen Asciteszelle zeigt nach NLB-Einwirkung oft den normalen cytologischen Aufbau (Abb. 2): Nucleus mit mehr oder weniger gleichmäßigem Verlauf und Bau der Membran (NM), eine oder mehrere Nucleolen (N) sowie Chromatinanhäufungen an der Nucleusmembran (Ch), geringe Entwicklung des endoplasmatischen Reticulum (E), unterschiedliche Verteilung der Mitochondrien (M) und Lipoide (2, 4). In NLB-behandelten Asciteszellen ist das endoplasmatische Reticulum in Kern- und Mitochondrien-Nähe nur schwach in Form von "Vesicles" und "Cisternae" ausgebildet. Die "Membranae fenestratae" (4, 9, 10, 11), die wir in geschlossener lamellärer Form in einer Anzahl von Kontrollzellen beobachteten, begegnete uns in einem typischen, parallel zur Nucleusmembran angeordneten Membransystem nur einmal. Die Ablagerung von Lipoid-Körpern erscheint häufig auch in den peripheren Lagen des Cytoplasmas vermehrt. Die Gesamtvacuolisierung des Cytoplasmas als Ausdruck degenerativer Prozesse in Abhängigkeit vom Alter und von der Stoffwechsellage der Zelle ist in NLB-behandelten Zellen gehäuft anzutreffen.

Untersuchungen an Tumormitochondrien (12) und ihre Bedeutung für den Zellstoffwechsel (13) veranlaßten uns zu einer eingehenden Untersuchung dieser Organelle. Vergleichsweise wurden isolierte normale Lebermitochondrien sowie Leberhepatom der Ratte untersucht. Mitochondrien von NLB-Tieren verschwinden sehr häufig vollständig. In Membrannähe wurden geschwollene, mit unterschiedlich reduzierten cristae mitochondriales sowie unscharf konturierten Membranen versehene Mitochondrien beobachtet, so daß sie desorganisiert und fragmentiert erschienen. Hinweise einer funktionellen Entleerung der Tumormitochondrien (11, 12) konnten wir aus unseren Befunden nicht ableiten. Das Nebeneinander verschieden stark destrukturierter Mitochondrien und ihr vollständiges Verschwinden läßt auf eine primäre Veränderung dieser Plasmaorganelle unter NLB-Einfluß schließen. Besonders auffällig sind Kernveränderungen nach NLB-Einwirkung. Das DNS-haltige Material, das wir in Schnitten mit FN und durch DNAse identifizieren konnten, liegt in der Interphase in kleinen Partikeln im Karyoplasma und in unregelmäßigen Schollen der Nucleus-Membran vermehrt gegenüber den Kontrollen an (Abb. 3). Die Chromatinsubstanz umlagert die Kernmembran nicht lückenlos, sondern es sind Zwischen-

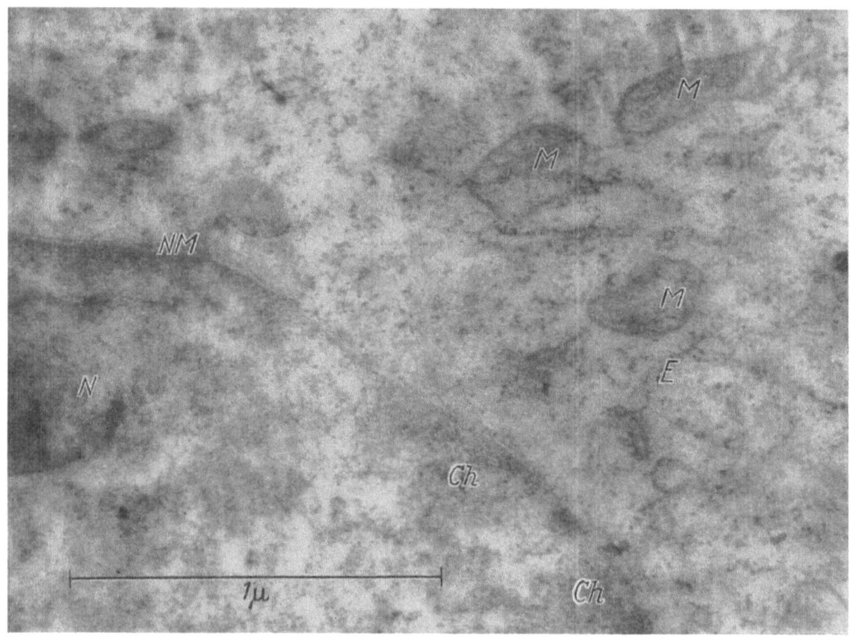

Abb. 2. Asciteszelle nach NLB-Behandlung. 6. Tag, Kernmembran NM, Nucleolus N, Chromatinanhäufung Ch, endoplasmatisches Reticulum E, Mitochondrien M

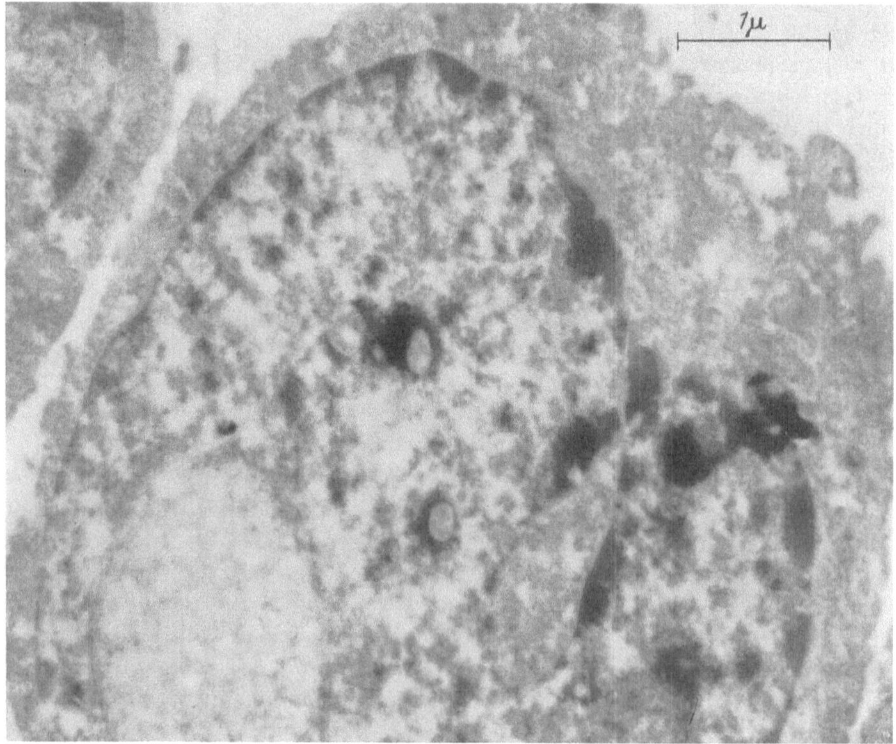

Abb. 3. Asciteszelle nach NLB-Behandlung, 10. Tag

räume von etwa 150—400 Å vorhanden, die auch im Karyoplasma bis zu den mehr oder weniger stark aufgelösten Nucleolen reichen können. Häufig ist der Nucleolus im Kern exzentrisch an die

Kernmembran angelagert, jedoch führen Kerneinstülpungen nicht immer zu einer peripheren Lage des Nucleolus. Mit dem Austritt der Nucleolarsubstanzen aus dem Kern ist oft eine starke Vacuolisierung von Kern und Cytoplasma verbunden.

In EMAC, besonders 10—14 Tage nach NLB-Einwirkung, waren Nucleolen in einigen Fällen nach Cytoplasmaeinstülpung (*4, 14, 15*) der Kernmembran angelagert (Abb. 3). Es kommt zu einer kontrollierten Abgabe von Nucleolarsubstanz durch das Porensystem der Kernmembran, wie aus der gleichen Elektronendichte des Materials und aus Anfärbungen vor und hinter den Membranporen geschlossen werden kann.

Vereinzelt wurden Nucleolarpartikel in das Cytoplasma hinein geschleust. Die Extrusion (*16*), die in verschiedenen Zellgeweben und in Ovocyten beobachtet wurde (*16, 17, 18, 19, 20, 22*), ist hier als krankhaft funktionelle Veränderung aufzufassen. Für die Annahme, daß die Ribonucleoproteide der Nucleolen über vorgebildete Leitbahnen des Interphasenchromatins zur Kernmembran gebracht werden (*16, 18*), ließ sich bei EMAC kein morphologischer Nachweis führen. Cytoplasmaeinstülpungen (*14*) traten in EMAC-Zellen 10—14 Tage nach NLB-Behandlung häufig auf.

Die morphologisch faßbaren Veränderungen an NLB-behandelten Ascitescellen lassen folgende Schlußfolgerung zu: Der Verlust cytoplasmatischer Differenzierung durch Veränderung der Ribonucleoproteide des Ergastoplasmas und durch Degeneration und Auflösung der Mitochondrien führt zu Veränderungen des Zellstoffwechsels mit einer gesteigerten Extrusion von Nucleolarsubstanz in das Cytoplasma und in einigen Fällen zur Karyorhexis und Pyknose des Kerns. Es ist wahrscheinlich, daß mit der erhöhten Abgabe von Ribonucleoproteiden aus dem Kern eine Reaktion zur Regulierung einer gestörten Kern-Plasma-Mitochondrien-Relation (*21*) eingeleitet wird.

Literatur

1. Selby, C. C.: Exp. Cell Res. **5**, 386 (1953).
2. — I. J. Biesele and C. E. Grey: Ann. N. Y. Acad. Sci. **63**, 748 (1956).
3. Yasuzumi, G., R. Sugihara, M. Kiriyama, T. Ikeda and S. Higashizawa: J. Nara Med. Ass. **7**, 135 (1956). (jap.); ref. Excerpta med. **11**, 1012 (1958).
4. Wessel, W., u. W. Bernhard: Z. Krebsforsch. **62**, 140 (1957).
5. Rintelen, K., u. W. Knobloch: Acta biol. med. germanica **1**, 109 (1958).
6. Hirschberg, E., A. Gellhorn and W. S. Gump: Cancer Res. **17**, 904 (1957).
7. Jung, F., K. Zapf, I. Quasdorf u. G. Mai: Mikroskopie (Wien) **13**, 91 (1958).
8. Lettré, H.: Z. Krebsforsch. **59**, 568 (1953).
9. Epstein, M. A.: J. biophys. biochem. Cytol. **3**, 851 (1957).
10. Palade, G. E.: J. biophys. biochem. Cytol. **1**, 59 (1955).
11. Swift, H. J.: J. biophys. biochem. Cytol. **2**, Suppl. 415 (1956).
12. Weissenfels, W.: Z. Naturforsch. **13 b**, 203 (1958).
13. Graffi, A., u. E. J. Schneider: Naturwissenschaften **43**, 376 (1956).
14. Wessel, W.: Virchows Arch. path. Anat. **331**, 314 (1958).
15. Powell, A. K.: Brit. J. Cancer **11**, 112 (1957).
16. Altmann, W.: Z. Krebsforsch. **58**, 632 (1952).
17. — Verh. Ges. Naturforsch. u. Ärzte. S. 60. Berlin-Göttingen-Heidelberg: Springer 1955.
18. Reitalu, J.: Acta path. microbiol. scand. **41**, 257 (1957).
19. Homann, W.: Z. Krebsforsch. **59**, 673 (1954).
20. Zapf, K.: Jena. Z. Naturwiss. **66**, 223 (1932).
21. Lettré, H.: Naturwissenschaften **40**, 203 (1953).
22. Rintelen, K. und K. Zapf: Naunyn-Schmiedebergs Arch. exp. Path. Pharmak. **236**, 1 (1959).

Optical and electron microscopical studies of mesenchymal tumours

Satyavati M. Sirsat

Department of Electron Microscopy, Indian Cancer Research Centre, Parel, Bombay 12 (India)

The importance of the role played by connective tissue in normal and abnormal cellular phenomena has been aptly described by Robb-Smith (*10*), who calls it the biological moderator of cellular energy. It is surprising therefore that so few reports have been published on the fine structure of collagen isolated from neoplasms, specially those of mesenchymal origin. The references to this subject available in literature (*1, 2, 6, 7*) are not agreed over the observations. This systematic study of human and animal mesenchymal tumours was carried out from two aspects : 1. observation of the fine structure of the collagen fibril and 2. elucidation of the type of alteration produced in the extracellular connective tissue, during an overall transformation of the cellular elements of the mesenchyme. The investigation was carried out at two levels. The biopsy tissue was studied optically, using differential staining techniques for connective tissue. Isolated collagen from the same biopsy was observed in the electron microscope to assess the fibrillar response to tissue alteration at the submicroscopic level.

The material consisted of seven human mesenchymal tumours of different sites — 5 fibrosarcomas, 1 fibroma, 1 myxosarcoma, and 3 mouse fibrosarcomas induced by 20-methylcholanthrene. The biopsy tissue was fixed in 10% buffered neutral formalin of Lillie and divided into two portions, one for optical and histochemical study and the other for electron microscopy.

Paraffin sections 6 μ thick were used for the optical and histochemical study. Serial sections from each biopsy were stained with Haematoxylin and Eosin, Mallory's Triple Stain and Phosphotungstic Acid Haematoxylin (PTAH), Weigert's Resorcin-fuchsin stain for elastica, Hales Colloidal Iron stain for acid mucopolysaccharides (modified by Rinehart and Abul-Haj), periodic acid Schiff, and Toluidine blue metachromasia at p_H 4.5.

For the electron microscopic study the tissue was washed for an hour in distilled water and a small bit of dermal collagen was teased out with sharp needles to form a uniform suspension of isolated collagen fibrils. Microdrops of this suspension were placed on collodion coated copper grids, air dried and shadowed ligthly with chromium. The specimens were examined in an RCA EMU-2D electron microscope using an objective aperture of internal diameter 0.001 inches.

The table shows the results of the different staining techniques on biopsies of ten mesenchymal tumours. Collagen from the human and animal undifferentiated, and one differentiated, fibrosarcomas showed normal even staining with Mallory's Triple Stain and PTAH, and no resorcin-positive areas with Weigert's stain for elastica. In two well differentiated tumours, and in the

Table 1. *Summary of results of differential staining techniques on biopsies of 10 mesenchymal tumours*

Source	No.	Case No.	Differential Staining Techniques (1) Haematoxylin and Eosin		(2) Mallory's Triple	(3) PTAH	(4) W-F-R	(5) C-I	(6) Pas	(7) Tol-Blue
			Diagnosis	Type						
Human	1	21564	Fibrosarcoma	Undifferentiated	—	—	—	+	+	+
	2	21144	Fibrosarcoma	Undifferentiated	—	—	—	+	+	+
	3	868 R	Fibrosarcoma	Well-Differentiated	—	—	—	+	+	+
	4	2050 R	Fibrosarcoma	Well-Differentiated	+	+	+	+	+	+
	5	494 R	Fibrosarcoma	Well-Differentiated	+	+	+	+	+	+
	6	18324	Fibroma	—	+	+	+	+	+	+
	7	22084	Myxosarcoma	—	—	—	—	+	+	+
Mouse (20 MCA Induced)	8	277	Fibrosarcoma	Undifferentiated	—	—	—	+	+	+
	9	283	Fibrosarcoma	Undifferentiated	—	—	—	+	+	+
	10	1255	Fibrosarcoma	Undifferentiated	—	—	—	+	+	+

MCA-Methylcholanthrene
PTAH-Mallory's Phosphotungstic-Acid Hämatoxylin
W-R-F-Weigerts Resorcin-Fuchsin stain for elastica
C-I-Hales colloidal iron technique (Modified Rinehert and Abul-Haj)
PAS-Periodic acid Schiff
Tol-blue-Toluidine blue Metachromasia

Techniques 2, 3, 4:
— Unaltered collagen, giving normal tinctoral reactions
+ Abnormal elastotically altered collagen
Techniques 5, 6, 7:
+ Increase in acid mucopolysaccharides in ground substance of mesenchyme

fibroma, the collagen showed uneven faint staining or chromophobic areas with Mallory's Triple stain and PTAH. Sections stained with PTAH also showed on occasional coarse purple or orange purple fibre (Fig. 1a). Weigert's resorcin-fuchsin stain for elastica showed areas of resorcin positive fragmented fibres and granules (Fig. 1b). A comparison of serial sections makes it evident that the areas of chromophobia with the first two techniques and chromophilia with Weigert's resorcin-fuchsin are identically located.

All the undifferentiated fibrosarcomas and one well differentiated tumour showed long strands of normal collagen associated with considerably increased amounts of amorphous material.

Collagen from two well differentiated fibrosarcomas and the fibromas showed abnormal kinking, fragmentation, partial or complete degradation of the fibril into elastin like sheets and filaments (Fig. 2). One characteristic feature of all the fibrosarcomas and the fibroma was an increase in

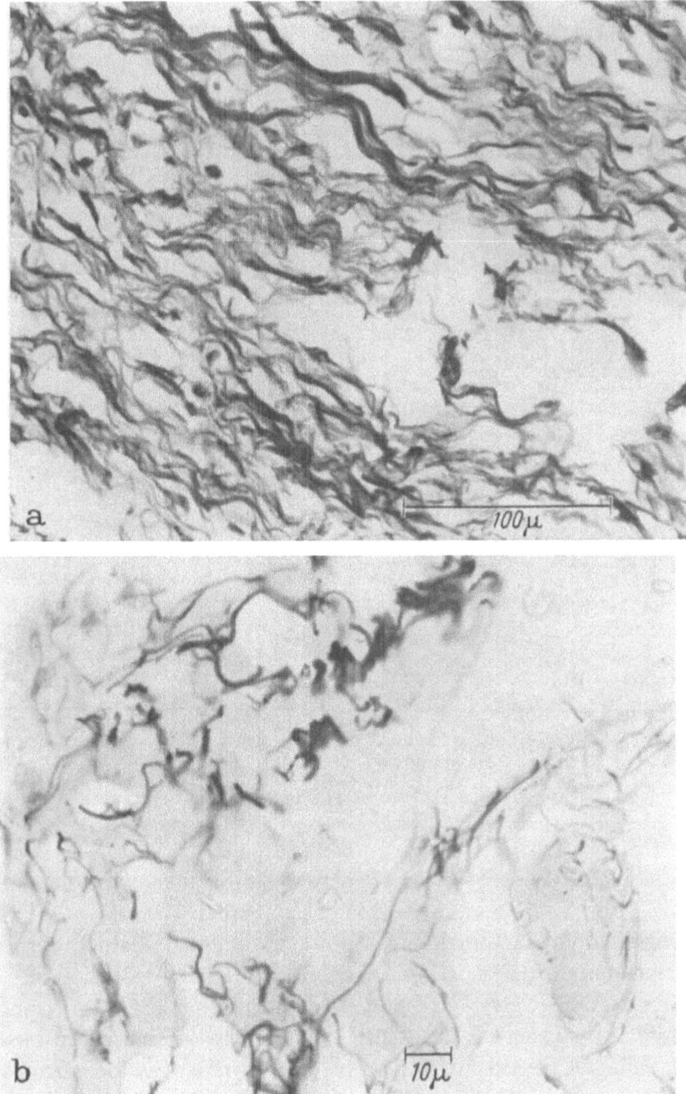

Fig. 1 a. Section of well differentiated fibrosarcoma showing unevenly stained collagen. Coarse partially or completely purple fibres are seen. Mallory's PTAH 280 ×

Fig. 1 b. Section of well differentiated fibrosarcoma showing coarse resorcin positive "pools", fragmented fibres and granules. Weigert's Resorcin fuchsin stain 560 ×

short tapered fibrils-"tactoids", in all fields. Isolated collagen from the myxosarcoma showed long strands of normal collagen. There was a large amount of amorphous material associated with the fibrils (Fig. 3).

Discussion. There is a significant increase in the number of short tapered fibrils in all biopsies of fibrosarcoma and in the fibroma. Unlike KAJIKAWA's observations on benzpyrene sarcomas (6) this result agrees with BANFIELD's findings in mesenchymal tumours (1). The frequent presence of tapered ends in actively developing tissue is understandable in view of RANDALL's suggestion that tapered ends could mean fibril growth by pyramidal accretion (9).

Isolated collagen from all the tumours studied, except the myxosarcoma showed a wide range of variation in fibril width (180 Å—1 000 Å). Occasional fibrils of 1 500 Å—2 000 Å were also seen. GALE (2) also mentions such fibrils of more than average width in his observation on fibrosarcomas. Two factors could be responsible for this great disparity in the width of collagen fibrils from the same tissue (a) active fibrogenesis, associated with the abnormal rate of cell multiplication and (b)

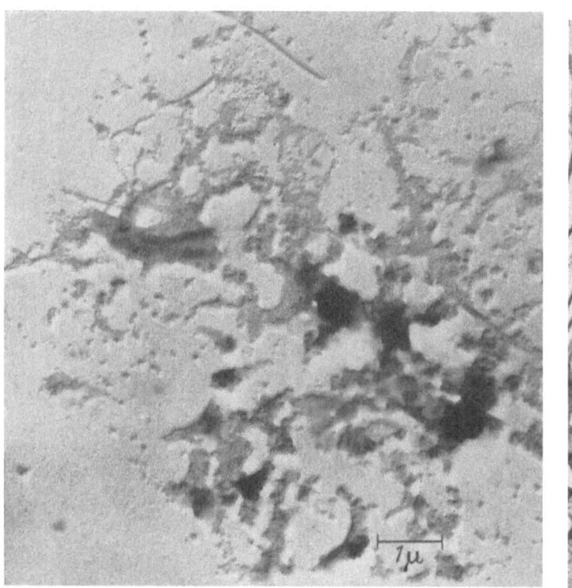

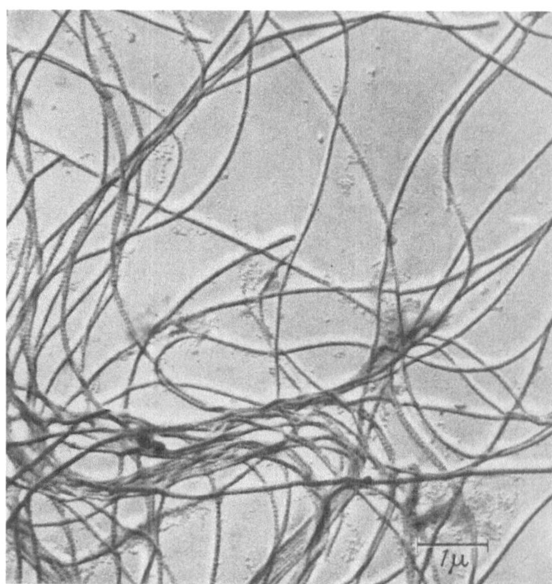

Fig. 2 Fig. 3

Fig. 2. Isolated collagen from fibroma. Picture shows short bits, partially degraded fibrils and angular bits and sheets of elastin like material. The vestigial periodicity of collagen is seen at many places in the dermis. Cr-shadowed 8,800 ×

Fig. 3. Isolated collagen from myxosarcoma. Picture shows long strands of normal collagen. Cr- shadowed 8,800 ×

swelling of the fibrils induced by a secondary degenerative process in the tumour tissue. Collagen from the myxosarcoma showed an average fibril width of 300 Å—700 Å.

In two well differentiated fibrosarcomas of low grade malignancy, and the benign fibroma the collagen showed abnormal tinctorial reactions, including a positive reaction to resorcin-fuchsin which is a staining characteristic of elastica. Isolated collagen from these same tumours showed an apparent morphological transformation to filaments and sheets similar to elastin. This non-specific alteration of collagen, occurring in trauma (4), primary connective tissue dyscrasias (12, 14, 15), and in dermal connective tissue during induced epidermal carcino-genesis (5,8) has been termed "elastotic" or "pseudoelastic" degeneration of collagen by GILLMAN and his co-workers (3, 4). In this investigation, this type of collagen alteration is not seen in any of the un-differentiated fibrosarcomas, which differ from the well differentiated type in biological behaviour as well as histology (11, 13). Wether an actual correlation does exist between the presence of abnormal exogeneous collagen and the behaviour pattern of the neoplasm is a matter of further experiment.

This work was done under the direction of Dr. V. R. Khanolkar. Thanks are due to Miss P. L. Veturkar for technical assistance.

References

1. BANFIELD, W. G.: Proc. Soc. exp. Biol. (N. Y.) 81, 658 (1952).
2. GALE, J. C.: Amer. J. Path. 27, 455 (1951).
3. GILLMAN, T., J. PENN, D. BRONKS and M. ROUX: Nature (Lond.) 174, 789 (1954).
4. — — — — A. M. A. Arch. Path. 59, 733 (1955).
5. — — — — Brit. J. Cancer 9, 272 (1955).

6. Kajikawa, K., and Y. Sumita: Acta path. jap. **3,** 75 (1953).
7. Lelli, G., e U. Maratta: R. C. Istit. sup. Sanita **13,** 503 (1950).
8. Orr, J. W.: J. Path. Bact. **46,** 495 (1938).
9. Randall, J. T., R. D. B. Fraser, A. V. W. Martin, S. F. Jackson and A. C. T. North: Nature (Lond.) **169,** 1029 (1952).
10. Robb-Smith, A. H. T.: The functional significance of connective tissue, in general pathology. Ed. Sir H. Flarey, Chapter 12, p. 255, 1958. London: Lloyd Linke Ltd.
11. Sirsat, M. V.: Ind. J. Surg. **18,** 1 (1956).
12. Sirsat, S. M., and V. R. Khanolkar: J. Path. Bact. **73,** 439 (1957).
13. Stout, A. P.: Cancer **1,** 30 (1948).
14. Tunbridge, R. E., R. N. Tattersall, D. A. Hall, W. T. Astbury and R. Reed: Clin. Sci. **11,** 315 (1952).
15. Winer, L. H.: A. M. A. Arch. Derm. **71,** 338 (1955).

2. Strahlenwirkungen

Elektronenmikroskopische Analyse von Strahlenschäden im Cytoplasma

L. Schneider

Zentrallaboratorium für angewandte Übermikroskopie der Universität Bonn

Die Wirkung von elektromagnetischen und corpusculären Strahlen auf menschliche, tierische und pflanzliche Organismen ist in heutiger Zeit von besonderem Interesse. Konnten bisher licht-mikroskopisch vor allem die Veränderungen und Schäden an Kernstrukturen weitgehend erforscht werden, so ist es nunmehr mit Hilfe des Elektronenmikroskopes möglich, auch die cytoplasmatischen Feinstrukturveränderungen nach Stahleneinwirkung zu erfassen.

Die vorliegenden Untersuchungen wurden an Parameciumzellen — *P. caudatum* und *P. aurelia* — vorgenommen. Die Normalstruktur des Cytoplasmas (Abb. 1) von *Paramecium* ist durch Untersuchungen von Wohlfarth-Bottermann (*1*) bekannt. Sie entspricht weitgehend derjenigen des Cytoplasmas von Säugerzellen. Parameciumzellen bieten jedoch gegenüber Säugerzellen den Vorteil, daß sie (neben ihrer einfachen Kultivierung) als Einzelzellen sehr gut zu fixieren sind und bei Verwendung eines Klones sowohl physiologisch als auch genetisch einheitliches Material darstellen.

Die Parameciumzellen wurden mit verschiedenen Röntgendosen, maximal bis zum Eintritt einer Strahlenlähmung, die dem Strahlentod vorausgeht, bestrahlt. Diese Dosis liegt bei 250000 r (LD 50) für P. caudatum und bei etwa 450000 r für P. aurelia. Etwa die Hälfte dieser Dosis (120000 bzw. 200000 r) bewirkt eine geringe Aktivitätssteigerung, die sich unter anderem in einer erhöhten Lokomotion bemerkbar macht. Die Strahlenintensität betrug 4000 r/min bei 100 kV, 25 mA und 0,3 mm Al-Filter. Die Fixation der Zellen erfolgte direkt nach der Bestrahlung in einer Fixierungslösung von 1% OsO_4 + 1% Kaliumbichromat. Kontrastiert wurde mit einem Gemisch von Phosphorwolframsäure (1%) und Uranylacetat (0,5%) während der Entwässerung. Als Einbettungsmittel dienten Methacrylsäureester[1] und Vestopal W nach Kellenberger[2].

Nach Applikation einer halben letalen Röntgendosis (120000 r bzw. 200000 r), die, wie gesagt, eine geringe Aktivitätssteigerung bewirkt, findet man bereits Veränderungen im Aufbau des Cytoplasmas (Abb. 2). Das Cytoplasma dieser bestrahlten Zellen weist bei etwa gleicher Strukturdichte gegenüber unbestrahlten Kontrolltieren zahlreiche kleine Vacuolen (Bläschen) und komplexe Doppelmembranstrukturen auf. Erhöht man die Röntgenbestrahlung bis zu einer letalen Dosis (Lähmungsstadium), so ist die auffallendste Strahlenwirkung eine zunehmende Strukturdichte des Cytoplasmas (Abb. 3). Die im normalen Cytoplasma sichtbaren distinkten Strukturkomponenten lassen sich bei einer letalen Strahlenschädigung nur noch schwer erkennen. Parallel hierzu verringert sich die Fähigkeit des Cytoplasmas, Kontrastmittel aufzunehmen. Dadurch ergibt sich ein fast homogener Aspekt. Lichtmikroskopisch ist dieser Zustand der Zelle als Koagulationsnekrose [Müller (*2*) u. a.] beschrieben worden.

---

[1] Fa. Dr. Theodor Schuchardt GmbH, München.
[2] Fa. M. Jaeger u. Co., Genève/Schweiz.

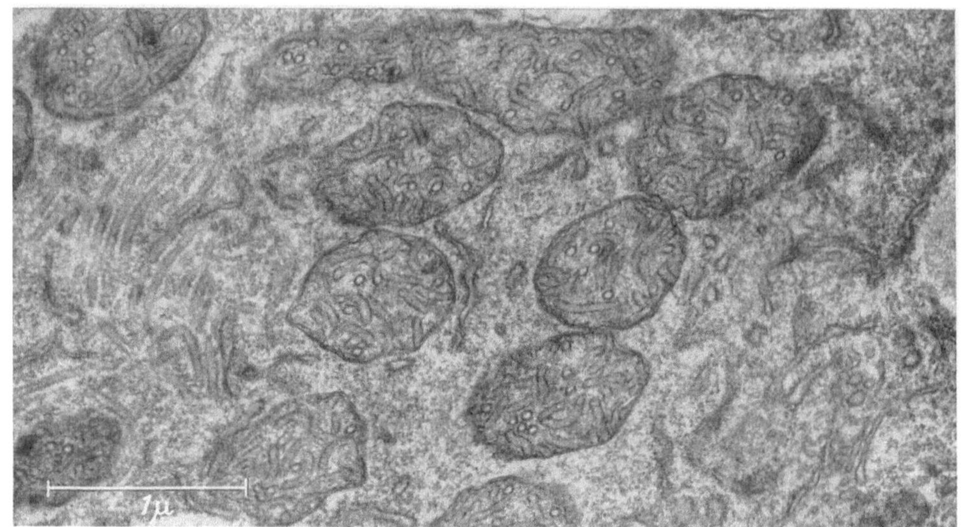

Abb. 1

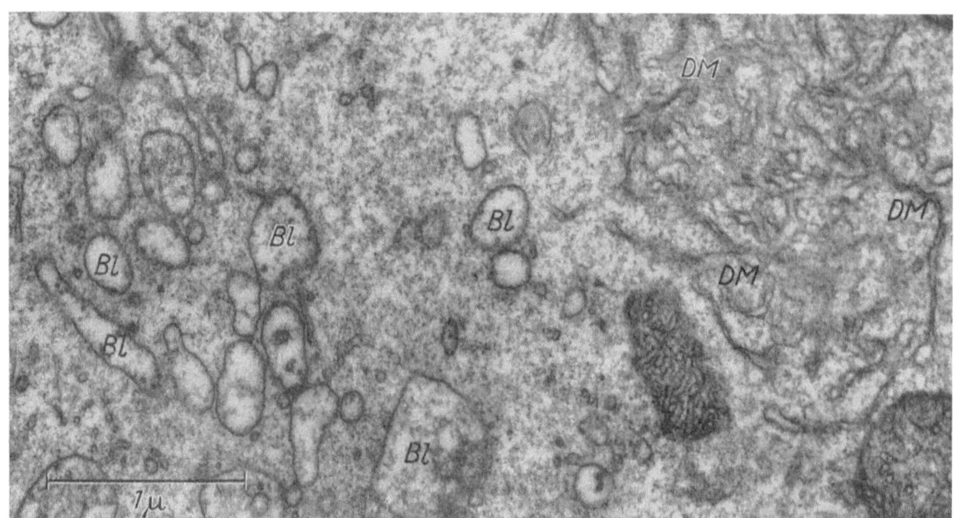

Abb. 2

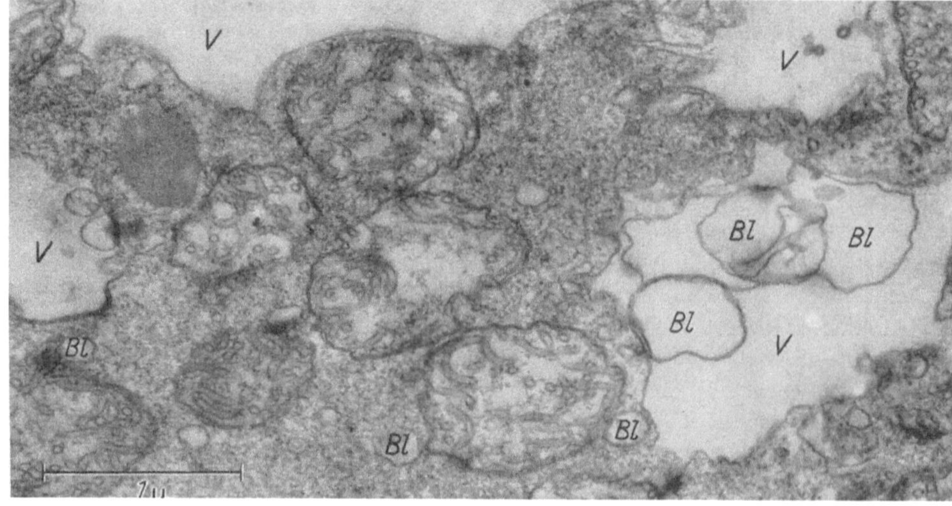

Abb. 3

(Legenden zu Abb. 1—3 siehe S. 479)

In letal bestrahltem wie in weniger stark geschädigtem Cytoplasma finden sich zahlreiche größere und kleinere Bläschen (Abb. 2 und 3). Diese sind von einer feinen Doppelmembran umgeben und elektronenoptisch entweder leer oder von einer Substanz teilweise erfüllt. Sie liegen meist regellos verstreut, in einzelnen Fällen aber, wie das Wohlfarth-Bottermann (3) erstmals beschrieben hat, in eigentümlichen Reihen an Doppelmembranen. Größere Ansammlungen von derartigen Bläschen kommen häufig in der Nähe des Cytopharynx vor. Dieser, sowie verschiedene Zellorganelle können auch von einer Bläschenreihe umsäumt werden, wie das Glauser (4) bei bestrahlten Leberzellen von Ratten am Nucleus beobachtet hat. Solche cytoplasmatischen Veränderungen nehmen mit steigender Dosis zu. Eine letale Dosis, die sich morphologisch in einem birnenförmigen Anschwellen der Paramecien äußert, bewirkt das dichteste Cytoplasma, das aber auch von den meisten und größten Vacuolen durchsetzt ist (lichtmikroskopisch als „trübe Schwellung" oder „tropfige Entmischung" des Cytoplasmas beschrieben). Es ist der Schluß naheliegend, daß bei letaler Schädigung der Inhalt vieler kleiner Bläschen zu großen Vacuolen zusammenfließt. Auch die durchschnittliche Größe der einzelnen Bläschen steigt mit zunehmender Dosis. So beträgt die durchschnittliche Größe der Bläschen bei P. caudatum nach Applikation von 120000 r 3000 Å, nach 240000 r jedoch 6000 Å, wobei die größten bis zu 2 μ messen.

Eine weitere Veränderung im cytoplasmatischen Aufbau nach Einwirkung von Röntgenstrahlen stellt offensichtlich die Bildung doppelmembranähnlicher Elemente dar (Abb. 2). Eine Neubildung von Doppelmembranstrukturen hatte sich bereits bei der eigentümlichen Reihenanordnung von Bläschen an Doppelmembranen gezeigt. Die doppelmembranähnlichen Strukturen liegen meistens regellos im Cytoplasma verstreut. Sie bestehen aus mehreren Membranpaaren, jedes Membranpaar ist an seinem Ende ösenartig geschlossen. Diese doppelmembranähnlichen Komplexe können langgestreckt oder auch aufgeknäuelt sein. Der Abstand der einzelnen Membranen eines Paares mißt etwa 150—200 Å. Diese Strukturen konnten sehr vereinzelt bei unbestrahlten Tieren, häufig dagegen bei bestrahlten Paramecien gefunden werden.

Die Trichocysten der bestrahlten Tiere waren, sofern sie fertig ausgebildet waren, größtenteils innerhalb der Zelle explodiert. Der Makronucleus wurde in einigen Fällen bei bestrahlten Tieren sehr stark gelappt und mit regelrechten Ausläufern versehen gefunden.

Zusammenfassend kann man also sagen, daß eine Strahlenschädigung im Cytoplasma elektronenmikroskopisch analysierbar ist. Sie äußert sich mit steigender Strahleneinwirkung in einer zunehmenden Strukturdichte, verbunden mit verringerter Aufnahmefähigkeit von Kontrastmitteln (Schwermetallsalzen) und im Auftreten von zahlreichen kleinen Bläschen, deren Inhalt bei starker Schädigung anscheinend teilweise zu größeren Vacuolen zusammenfließt. Außerdem scheint die Bildung von Doppelmembranstrukturen vermehrt zu sein. Die aufgezeigten cytoplasmatischen Veränderungen stellen Frühschäden dar. Sie haben sich während der Bestrahlung ausgebildet. Über Spätschäden im Cytoplasma nach Röntgeneinwirkung soll an anderer Stelle berichtet werden.

Die Untersuchungen wurden mit Unterstützung des Bundesministeriums für Atomkernenergie u. Wasserwirtschaft durchgeführt.

Dem Leiter des Zentrallaboratoriums für angewandte Übermikroskopie der Universität Bonn, Herrn Dozent Dr. K. E. Wohlfarth-Bottermann, danke ich für Förderung dieser Arbeit.

Literatur

1. Wohlfarth-Bottermann, K. E.: Protoplasma (Wien) **49,** 231 (1958).
2. Müller, E.: Der Zelltod, in Handbuch der allgemeinen Pathologie 2, Teil 1, S. 613—679. Berlin-Göttingen-Heidelberg: Springer 1955.
3. Wohlfarth-Bottermann, K. E.: Protoplasma (Wien) **50,** 82 (1958).
4. Glauser, O.: Schweiz. Z. Path. Bakt. **19,** 150 (1956).

Abb. 1. Normales unbestrahltes Cytoplasma, Paramecium aurelia, 27500:1

Abb. 2. Cytoplasma nach Strahleneinwirkung (Aktivitätsstadium). Normale Strukturdichte, Auftreten von Bläschen (*Bl*) und doppelmembranähnlichen Strukturen (*DM*). P. aurelia; Strahlendosis 200000 r, 27500:1

Abb. 3. Cytoplasma nach letaler Strahleneinwirkung (Lähmungsstadium). Auffallende Strukturdichte, geringer Kontrast, Auftreten von großen Vacuolen (*V*), die wahrscheinlich aus zahlreichen kleinen Bläschen (*Bl*) entstanden sind. P. caudatum; Strahlendosis 240000 r, 27500:1

Some observations on radiation damage in epithelial cells of the mouse intestine*, **

J. C. Hampton and H. Quastler

Department of Anatomy, University of Washington, Seattle, Washington and
Biology Department, Brookhaven National Laboratory, Upton, L. I., New York

The epithelium lining the small intestine is the seat of severe and potentially lethal radiation damage. The evolution of the pathological changes indicate that the critical primary disturbance is one of cell proliferation (*1*). The intestinal epithelium is a renewal tissue (*2*). Cells are born in the crypts and mature before emerging from the mouths of the crypts; during their functional life, they migrate from the base to the tip of the villi (*3*). The development of tritiated thymidine as a specific label for DNA, together with high resolution autoradiography, have made possible a rather detailed study of the kinetics of cell proliferation and movements in the normal animal (*4, 5*) and after irradiation (*6*). Briefly, the events in the lower ileum of the mouse following whole body irradiation with a large dose (3000 rad) are as follows: mitosis and DNA synthesis disappear rapidly. During the following $^3/_4$ days, $^1/_2$ to $^2/_3$ of the crypt cells disappear through cell death or through emigration onto the villi following precocious maturation [normally, cells mature only at a certain stage following mitosis (*5*)]. At the end of this period, the lower one-fourth of the villi is covered with cells having thus abnormally matured. During the following day, DNA synthesis is taken up again in the crypts, and no or few cells emigrate; at the same time, sloughing off of cells from the tip of the villi proceeds, and the villi shorten. Beginning about 1 $^3/_4$ days after irradiation, the remaining crypt cells begin to move out onto the villi, leaving the crypts empty except for Paneth cells. No mitosis intervenes between DNA synthesis and maturation, and the resulting cells are very large and misshapen. After they, too, have been desquamated, the villi collapse and the intestine is denuded (*1*). The Paneth cells show some early changes (*6*) but seem to recover and succumb only on the fourth day.

Turning now to electron microscopy, it will be demonstrated that submicroscopic changes in cytoplasmic components accompany the more gross alterations observed under the light microscope.

Changes observed in cells on Villus. Fig. 1 and 2 represent the distal portion of the two types of lining cells, columnar and goblet cells, fixed 24 hr after treatment with 3000 rad. These cells were mature and functional at the time of irradiation and show normal submicroscopic details. However, cells near the base of the villi show immediate response to irradiation. Fig. 3 represents such a cell fixed $^1/_2$ hr after treatment with 3000 rad. The microvilli (MV) of cells in this location are short and thick as compared to those cells in Fig. 1. This is normal. It will be noted, however, that some of the microvilli in Fig. 3 appear to have "exploded".

As stated above, cell proliferation in the generative compartment undergo a variety of alterations. The cells which cover the villi 3 days after irradiation have had a highly abnormal development, including one or two phases of DNA synthesis not followed by mitosis. Some are binucleate (Fig. 4). In addition to the two abnormal appearing nuclei (N), this cell shows various cytoplasmic abnormalities: lipid granules are large and numerous (L), endoplasmic membranes are disorganized and cells are separated at the intercellular borders (IB) except in the region of the terminal bars (TB). Fig. 5 depicts a portion of an epithelial cell fixed 3 days after treatment with 3000 rad. Many lipid granules (L). large vesicles (V), reduction in number of mitochondria (M), reduction in amount of endoplasmic reticulum (ER), and a decrease in the number of microvilli are prominent features of epithelial cells under these conditions. Fig. 6 shows several epithelial cells in the pre-denudation stage at three days post irradiation with 3000 rad. Microvilli (MV) are

* Part of this research was carried out at Brookhaven National Laboratory under the auspices of the US Atomic Energy Commission.
** This study was supported by U. S. Public Health Service research grant no. N-2698 and State of Washington Initiative 171 Fund for research in medicine and biology.

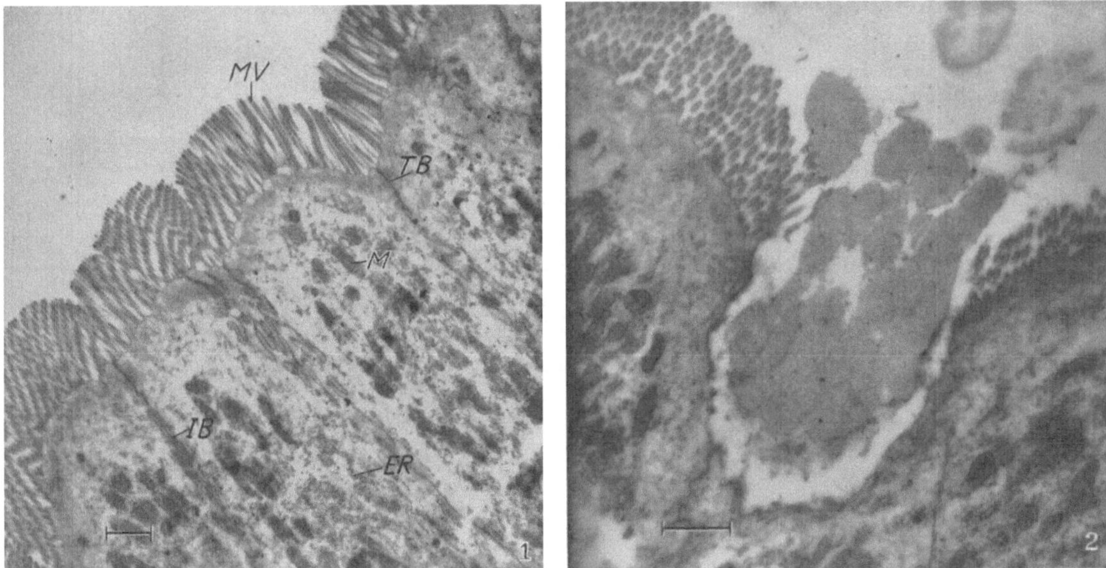

Fig. 1. Distal portions of several columnar epithelial cells of the mouse that were fixed 24 hr after treatment with 3 000 rad. These cells were mature and functional at the time of irradiation and appear to be normal

Explanation of figures. All figures presented in this paper are electron micrographs taken at initial magnifications of 2,200—5,700 and enlarged photographically as desired. The bar shown in each figure represents a distance of one micron.

Abbreviations: B Bacteria, ER Endoplasmic reticulum, G Paneth cell granule, IB Intercellular border, L Lipid granule, M Mitochondria, MV Microvilli, N Nucleus, S Space between lamina propria and epithelium, TB Terminal bar, V Vacuole

Fig. 2. Distal portion of a goblet cell from the same animal and from about the same location on the villus as that shown in Fig. 1. Mature goblet cells appear to be unaffected by radiation

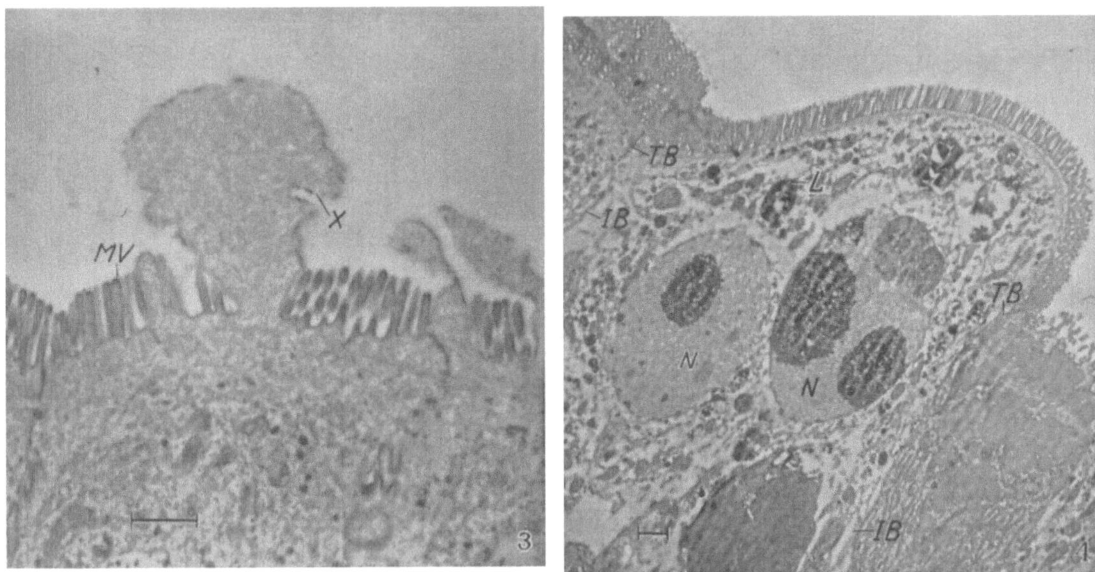

Fig. 3. The distal portion of a columnar cell situated near the base of a villus fixed one-half hour after treatment with 3 000 rad. Note enlargement of several microvilli (MV) and disruption (?) of the plasma membrane at X

Fig. 4. Shows a part of a binucleate columnar cell fixed three days after exposure to 3 000 rad. Both nuclei appear to be abnormal, the cytoplasm contains many lipid granules (L), and spaces occur between adjacent epithelial cells except at the region of terminal bars (TB)

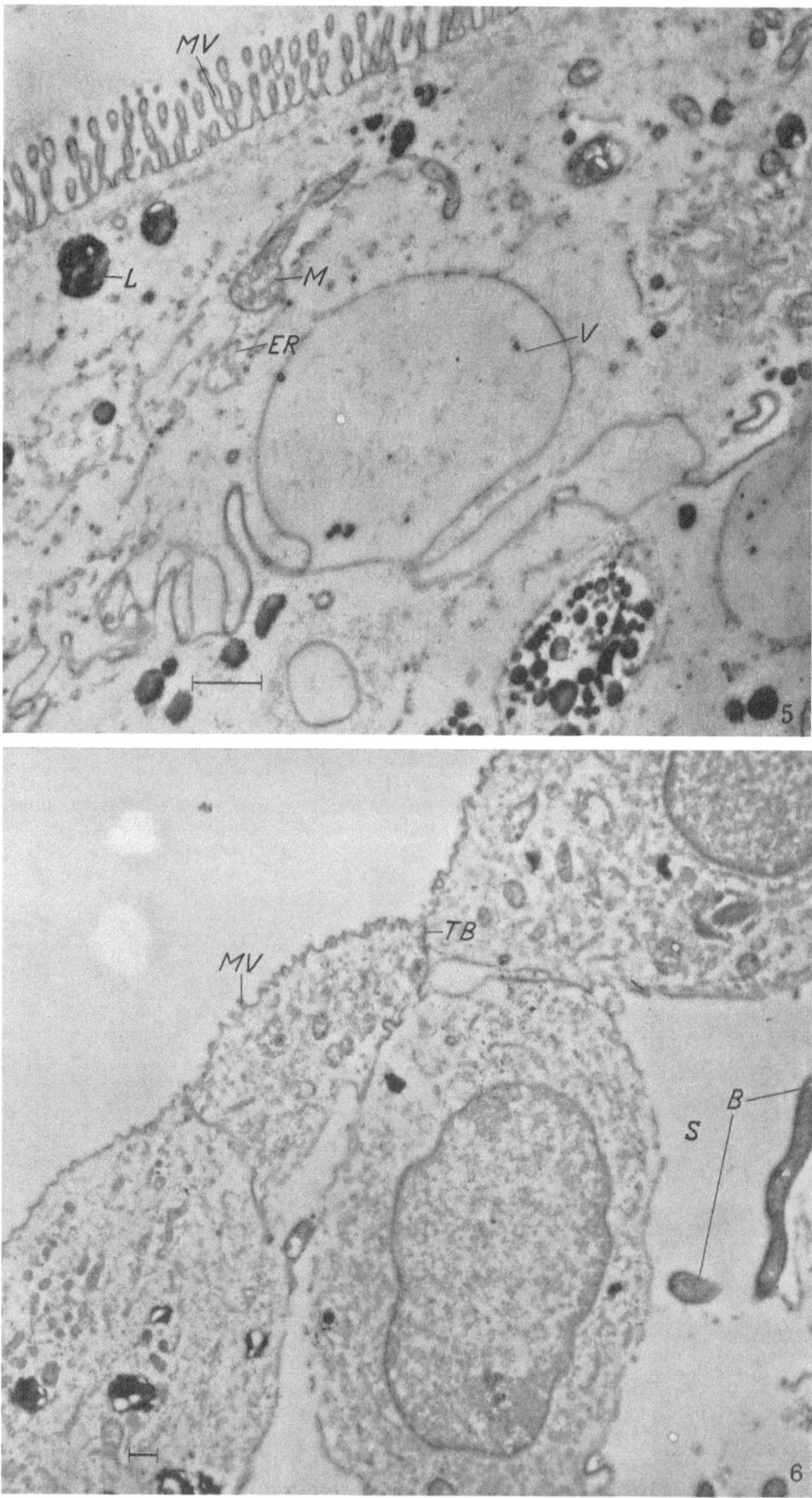

Fig. 5 and 6. These figures demonstrate the features of epithelial cells that remain at three days post-irradiation with 3000 rad. The cells become cuboidal to squamous in profile, to be bound together at the region of the terminal bars only and to separate from the lamina propria. Note the reduction in size and number of microvilli (*MV*), the large vacuoles in the cytoplasm (*V*, Fig. 5), lipid granules (*L*) and reduction in endoplasmic reticulum. Parts of several bacteria can be seen in the sub-epithelial space (*B*)

only a fraction of the normal length, the cells have separated from one another except near the terminal bars (TB) and the epithelium has become detached from the lamina propria. The resultant space (S) between epithelium and lamina propria has been invaded by bacteria (B).

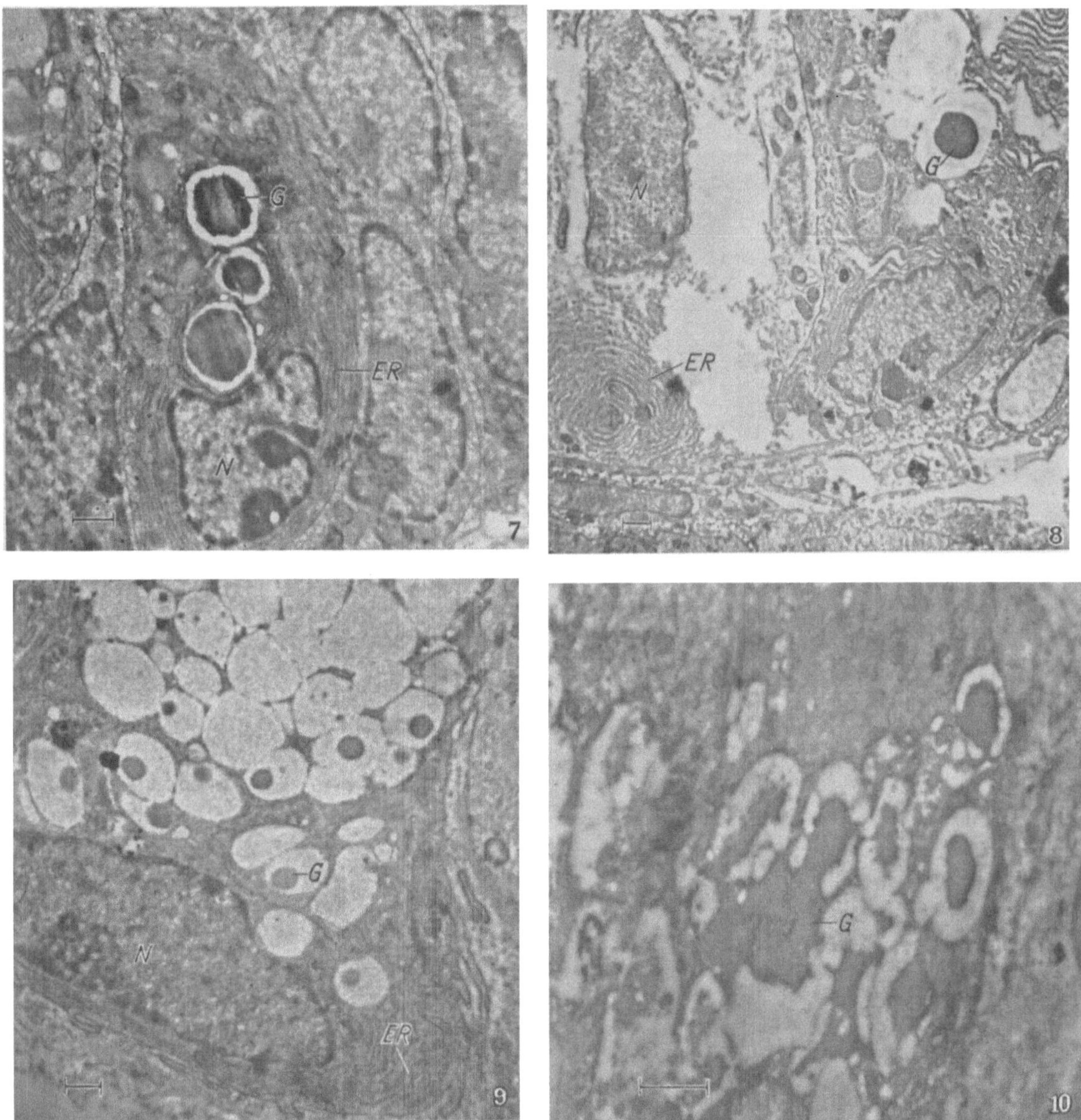

Fig. 7—10. The response of Paneth cells to 3000 rad over a three-day period is demonstrated by these figures. Fig. 7 represents a normal Paneth cell with characteristic granules (G) and arrangement of endoplasmic reticulum Fig. 8 shows disruption of endoplasmic reticulum and extrusion of granules in two Paneth cells fixed one-half hour after exposure. Paneth cells undergo partial recovery during the first 24 hr (Fig. 9) but fail to recover completely. Fig. 10 shows part of Paneth cell at three days post-irradiation in which coalescence of granules can be seen. This cell was dead or dying at the time of fixation

Observations on Paneth cells. Normal Paneth cells are characterized by their basal nucleus (N), several lamellae of endoplasmic membranes (ER), and large dense granules (G) between the nucleus and the apical end of the cell (Fig. 7). The granules usually vary in density and are contained in vacuoles surrounded by membranes which probably derive from endoplasmic reticulum. Short microvilli adorn the lumen surface of the cell. Within 12 min after irradiation disorganization of endoplasmic reticulum occurs, all granules become very dense and are extruded into the lumen

31*

of the crypt (Fig. 8). During the first 24 hr after irradiation, Paneth cells undergo a partial recovery with some reconstitution of endoplasmic reticulum and granules. However, the granules do not attain the same degree of density seen in the normal cell and an amorphous material of apparent low density surrounds the granules (Fig. 9). At three days post irradiation large vacuoles appear in the cytoplasm, endoplasmic reticulum fragments, granules coalesce and cell death occurs (Fig. 10).

References

1. QUASTLER, H.: Rad. Res. **4**, 303 (1956).
2. LEBLOND, C. P., and B. E. WALKER: Physiol. Rev. **36**, 255 (1956).
3. — C. E. STEVENS and R. BOGOROCH: Science **108**, 531 (1948).
4. HUGHES, W. L., V. P. BOND, G. BRECHER, E. P. CRONKITE, R. B. PAINTER, H. QUASTLER and F. G. SHERMAN: Proc. nat. Acad. Sci. (Wash.) **44**, 476 (1958).
5. — — J. Cell. Res. (in press).
6. LEWIS, Y. S., H. QUASTLER and G. SVIHLA: J. nat. Cancer Inst. **21**, 813 (1958).

Effects of ionising radiation on the testis of the rat with some observations on its normal morphology

DENNIS LACY and JOSEPH ROTBLAT

Departments of Zoology and Physics, St. Bartholomew's Hospital Medical College, University of London

The following is a brief account of an investigation into the effects of high doses of 15 MeV electrons on the testes of the rat. The right testes only of 20 rats were irradiated, the dose being 10,000 r. The rest of the body was screened and received only about 30 r. The animals were killed at intervals after irradiation (3 days to 3 weeks).

For study by electron microscopy small pieces of both normal and irradiated tissue were fixed for three hours in osmium tetroxide solution buffered to pH 7.2 (1), embedded in methacrylate, sectioned and examined in a Siemens Elmiskop I a. A wide variety of histochemical and histological techniques were employed for study by light microscopy.

Tissue bounding each of the seminiferous tubules. Examination of this tissue in normal animals by light microscopical techniques reveals two main non-cellular layers between which lies a cellular one. We have been able to demonstrate that the non cellular layers consist of an elaborate network of argyrophilic fibres (reticulin). Elastic fibres are not present. Examination of this region by electron microscopy reveals four main layers (Fig. 1.):

1. *Inner non cellular layer* (basement membrane), either homogeneously dense in appearance or with two dense regions applied immediately next to adjacent cell membranes and separated from each other by a clear band. Rarely we observed a faint fibrillar content. 2. *Inner cellular layer*. This consists of spindle shaped cells with various cytoplasmic components scattered along their length. Occasional bundles of very fine filaments were seen in the cytoplasm of such cells. At intervals the layer appeared two cells thick. 3. *Outer non cellular layer*. Essentially similar in composition to 1. above. 4. *Outer cellular layer*. A very thin unicellular layer. We have not identified any filaments within these cells.

Three weeks after irradiation the boundary tissue is often excessively folded due primarily to the condensation of tubular contents. In some tubules the nuclei of the inner cellular layer are crowded together and no longer elongated: not infrequently they appear triangular in shape with the apex pointing towards the tubular contents. It appears that this is a contractile layer. The filaments within the cytoplasm were seen more clearly and in greater concentration; they were approximately 65 Å in diameter and sometimes "beaded" in appearence. In some regions large numbers of fibres were seen in the originally homogeneously dense layers (Fig. 1). Such fibres were about 400 Å in diameter and had a banded pattern similar to that reported for collagen, the periodicity being 600 Å. (After 50,000 r we have observed fibres about 300 Å diameter with a banded pattern of 600 Å periodicity.) In other regions the dense material in the inner non-cellular region was reduced in amount and thrown into many small villiform folds. Irradiation also revealed the existence of occasional cells apparently lying within the innermost layer.

Effects of irradiation on the tubular contents. Even after such a high dose of irradiation the general effect produced differs for a considerable time due to variations in the conditions of individual tubules (or along the length of the same tubule) immediately prior to irradiation. To be brief we shall refer to only three main normal conditions of the tubules, illustrative of three

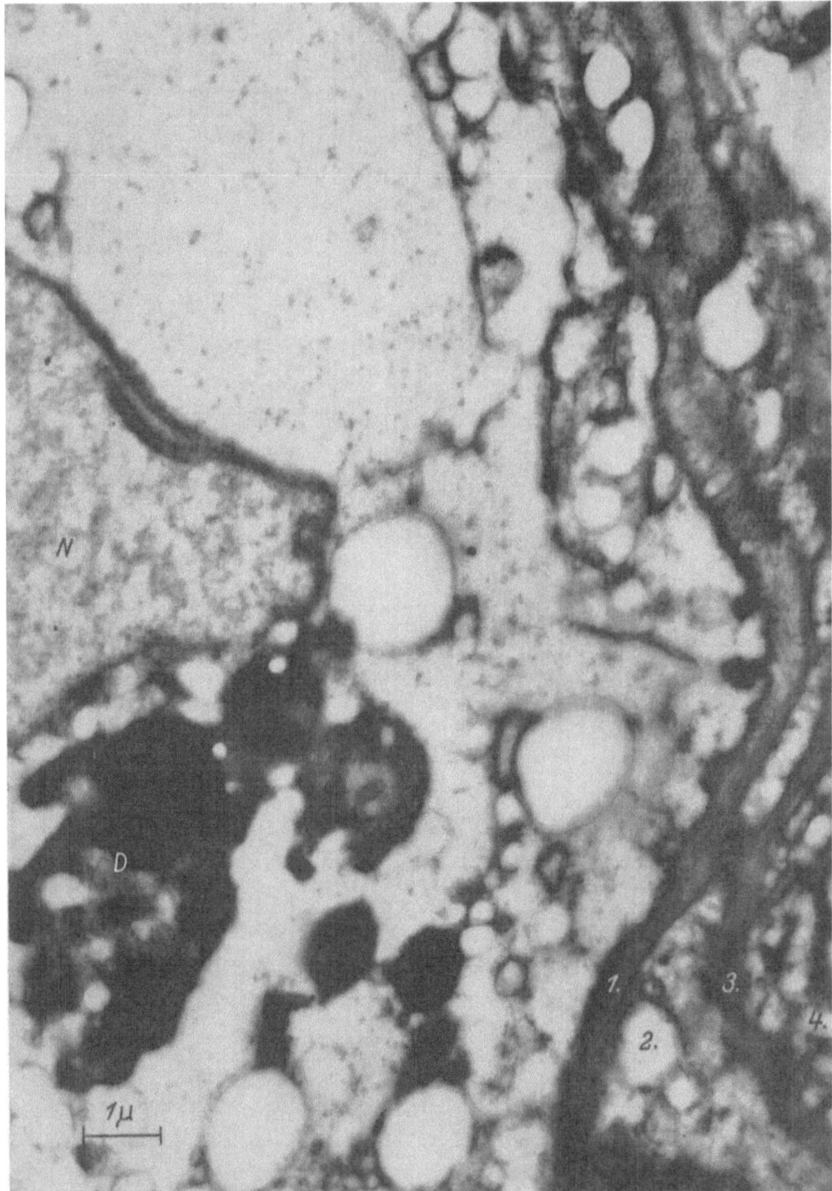

Fig. 1. Micrograph showing material bounding part of a seminiferous tubule and a Sertoli cell. 10,000 r, rat killed three weeks after irradiation. 10,500 ×, N = nucleus, D = Degenerating spermatid within which a lipoidal droplet can be seen. Nr. 1, 2, 3, 4 — see text

successive phases of spermatogenesis (2), as follows: Phase I: Large lipoidal droplets situated peripherally in the Sertoli cells: two layers of spermatocytes (juvenile and primary). Phase II: Fine lipoidal droplets radiating towards the lumen: two layers of young spermatids (Fig. 4). Phase III: lipid mainly concentrated in the residual cytoplasm of the nearly mature spermatids situated near and in the lumen (Fig. 5).

Examination of the tubules three days after irradiation revealed some with peripheral amounts of lipid in excess of that seen in normal animals. In extreme cases such tubules contained more

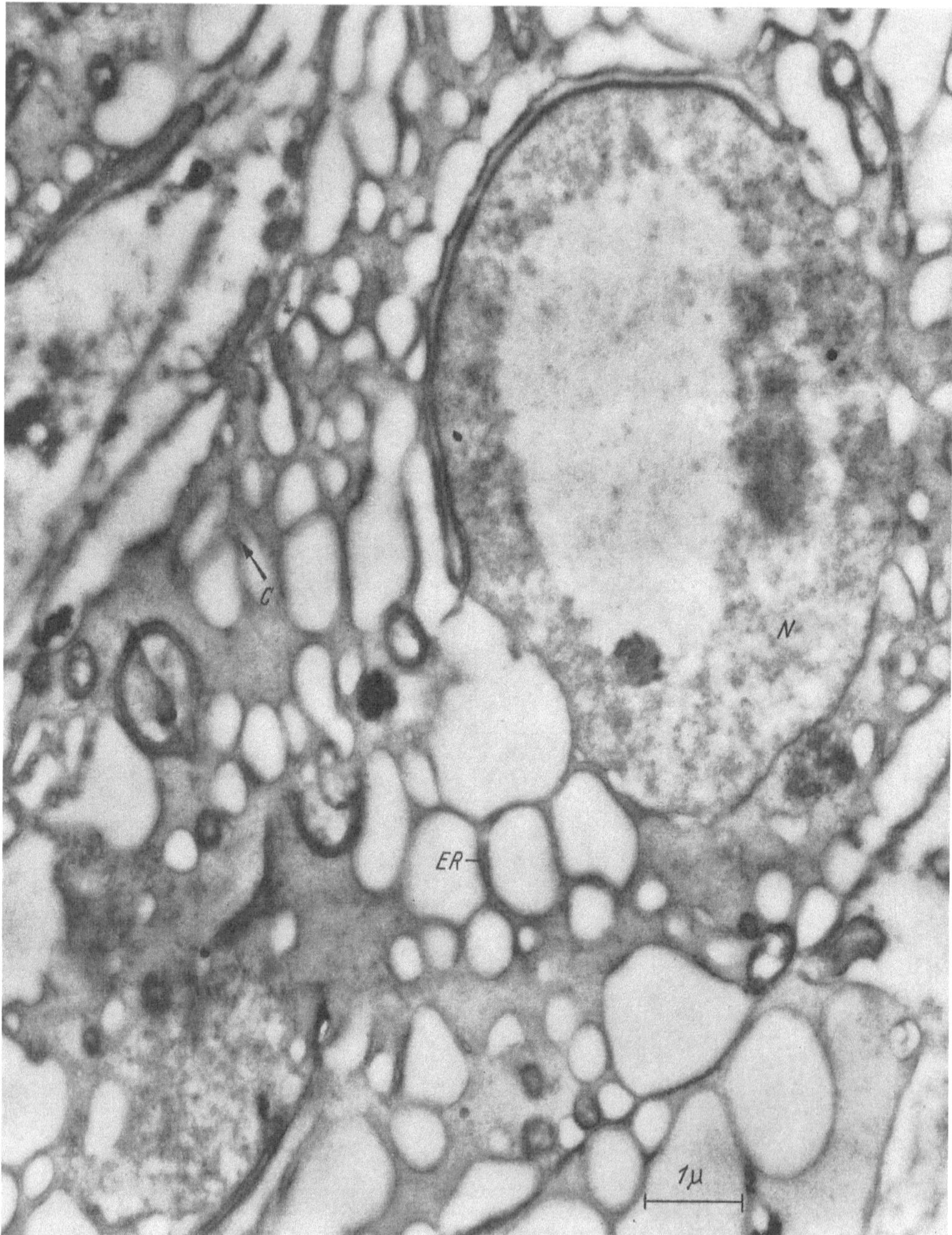

Fig. 2. Micrograph of degenerating spermatid. Serial sections have shown that the vacuoles are part of a continuous system (endoplasmic reticulum *ER*). Some continuity can be seen at *C*. The matrix is excessively granular in appearance. 10,000 r, rat killed three days after irradiation. 14.000 ×, *N* = nucleus

juvenile spermatocytes than usual, a few degenerating spermatocytes and no spermatids (Fig. 6). These tubules were probably in phase I at the moment of irradiation. Some tubules contained small amounts of peripheral lipid, one or two layers of juvenile spermatocytes and many spermatids with nuclei exhibiting karyolysis. These tubules were probably in phase II at the time of irradiation. Tubules which contained least peripheral lipid and showed least damage were those containing most mature spermatids (phase III at irradiation).

Examination of tubules at later intervals after irradiation revealed a continuation of the general pattern described above. By the end of three weeks nearly all the tubules were filled with

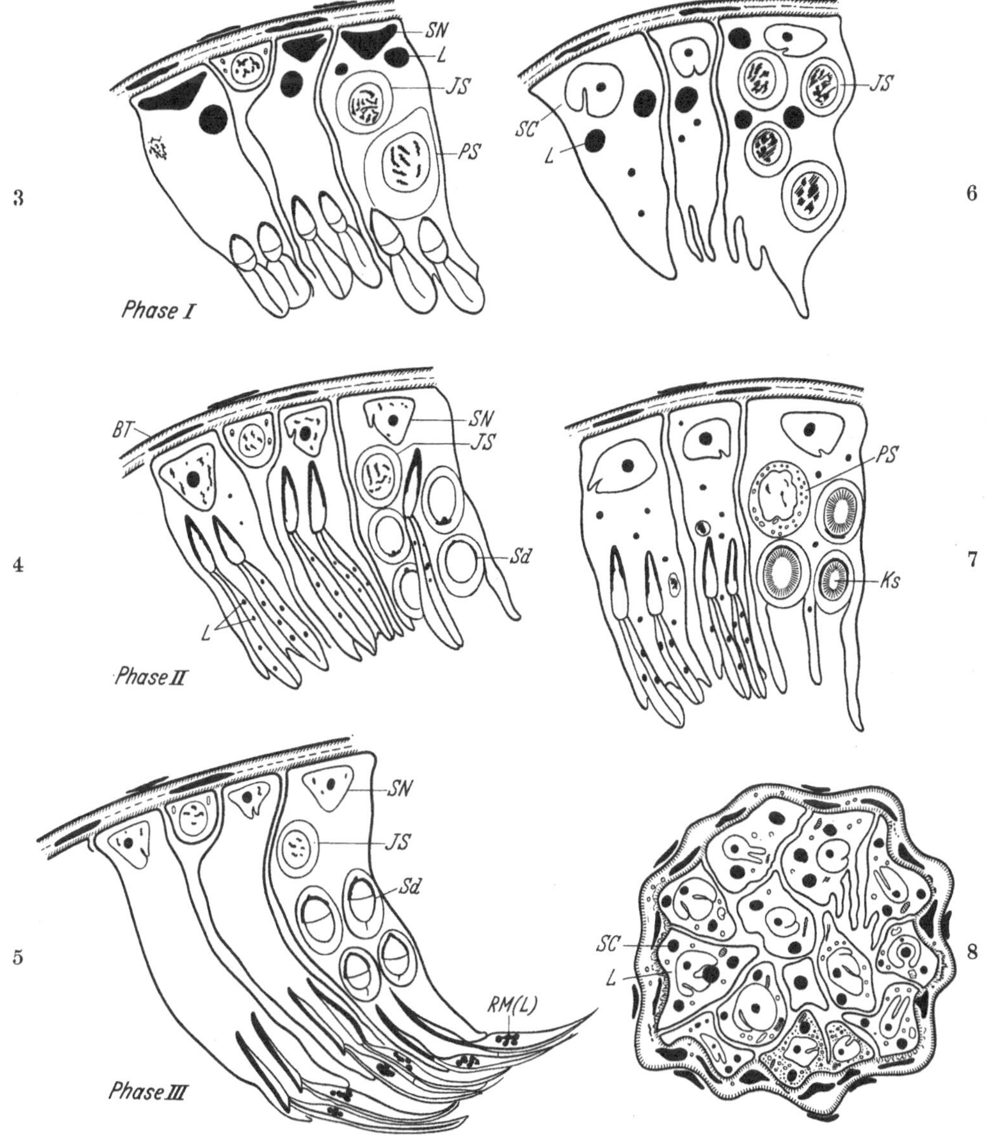

Fig. 3—8

Fig. 3—5. Normal tubules in three different phases (I, II, III) of spermatogenesis showing different locations of lipid (from SMITH and LACY, in press)

Fig. 6 and 7. Three days after irradiation. The tubule shown in Fig. 6 was probably in phase I at the moment of irradiation. The one shown in Fig. 7 illustrates the effect of radiation on spermatocytes and spermatids

Fig. 8. Three weeks after irradiation. Only Sertoli cells remain intact within the tubules

BT boundary tissue, *Ks* karyolysis, *JS* juvenile spermatocyte, *L* lipid, *PS* primary spermatocyte, *SN* Sertoli nucleus, *SC* Sertoli cell, *Sd* spermatid, *RM* residual material (lipid)

large amounts of lipid and all the germinal elements had undergone degeneration (Fig. 8). The most radio-resistant of all cells within the tubules were the Sertoli cells. Three weeks after irradiation they occurred throughout the tubules, their nuclei occurring both at the periphery and within the original lumen. Histochemical tests on the lipid revealed a gradual increase in its cholesterol content, reaching near maximum by the end of three weeks.

With particular reference to the distribution and formation of lipid the electron microscope has revealed the following main points of interest: 1. The lipid occurs within the Sertoli cytoplasm and is not extracellular. 2. Under normal conditions the germ cells (with the exception of spermatogonia) are surrounded by Sertoli cytoplasm. Upon degeneration the limiting membrane of the germ cells becomes progressively more difficult to identify. It appears therefore, that the degenerating products are passively, or otherwise, phagocytosed by the Sertoli cells. 3. Large amounts of lipid do not appear until the germinal tissue has become completely degenerate. 4. Inclusions have been observed which morphologically show transitions between finely granular bodies, those consisting of small granules and membranes arranged concentrically, concentric membranes tightly packed so as to similate a dense body and homogeneously dense inclusions (lipid droplets). This suggests that the lipid may originate from basic cytoplasmic components (endoplasmic reticulum and cytoplasmic particles).

Degenerating spermatogonia and spermatocytes appear as dense granular masses within which it is sometimes possible to identify nuclei with marked karyolysis. After about three days it is seldom possible to distinguish between nucleus and cytoplasm. In young spermatids the acrosome remains clearly evident up to about four days after irradiation. As shown in Fig. 2, karyolysis is again evident but there is no obvious increase, above the normal, of chromatin next to the nuclear membrane. In some regions the two membranes investing the nucleus remain normal in appearance but in other parts they become widely separated and form the borders of large vacuoles (nuclear blebbing) (Fig. 2). Within the cytoplasm are similar vacuoles.

Two weeks after irradiation only the Sertoli cells remain intact. In some tubules the heads of late spermatids can still be seen. Within the Sertoli cytoplasm are irregular masses of dense material within which it is sometimes possible to identify mitochondria; such masses appear to be breakdown products of the original germ cells (Fig. 1). The Sertoli nucleus contains several deep invaginations which sometimes envelop large lipoidal bodies. The cytoplasm of these cells is not consistent in appearance. In some cells it is almost devoid of content apart from the lipoid. In other cells many vacuoles (endoplasmic reticulum) and numerous fine dense particles are also present (Fig. 6.). Occasionally Sertoli syncytia are evident but for the greater part this component is essentially cellular in character.

References

1. Palade, G.: J. exp. Med. **95,** 285 (1952).
2. Smith, B. V. K., and D. Lacy: In press.

H. Ergebnisse der Elektronenmikroskopie in der Botanik

Leaf surfaces under the electron microscope

B. E. JUNIPER

Department of Botany, University of Oxford (England)

In recent years, the increasing use of selective herbicides, insecticides etc., which are applied by spraying, has stimulated interest in the nature of the leaf surface. The efficiency of these materials is affected by the characteristics of the leaf surface, i. e. the retention or shedding of spray droplets.

The wettability of leaf surfaces may depend on macro-characteristics. For example, in *Salvinia* curious protuberances, which trap air and thus prevent the leaf from being wetted, are present on the surface. In the majority of cases, however, the wettability cannot be related to structures visible in the light microscope. It was, therefore, suspected that structure beyond the resolution of optical instruments was responsible for the different characteristics. A preliminary examination

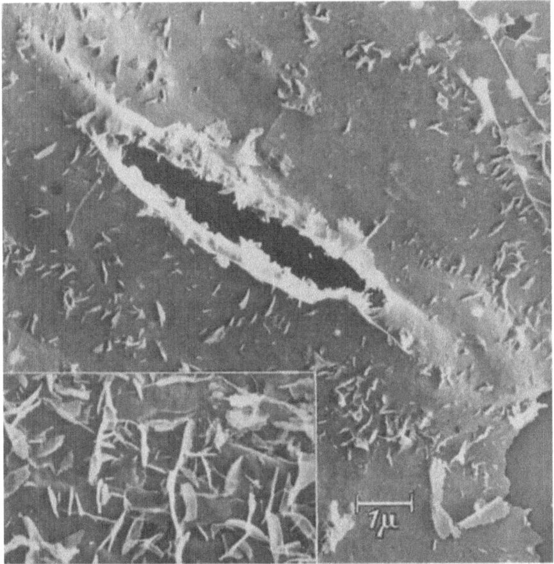

Fig. 1. *Chamaenerion angustifolium.* Shadowed carbon replica showing wax distribution around stoma

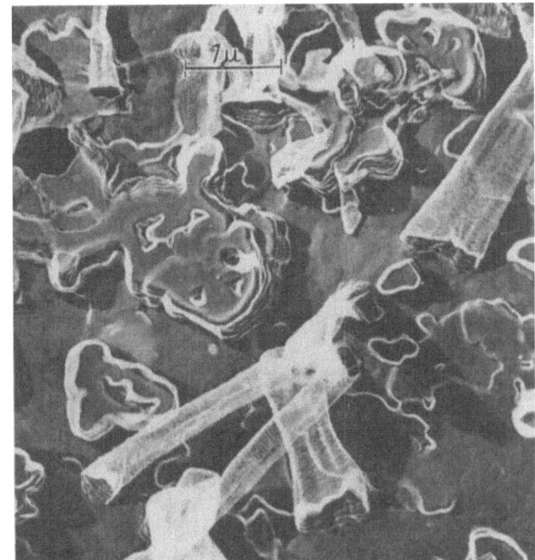

Fig. 2. *Brassica oleracea* var *capitata.* Shadowed carbon replica of adaxial surface

of leaf surfaces in the electron microscope (*1*) confirmed this. Subsequent examinations (*2, 3*) have provided information on physiological characteristics such as the effect of light intensity and soil treatment.

In the few cases when leaf surfaces have been studied in the electron microscope (*4, 5*) (other than those already cited) a double-stage plastic replica method has been adopted. This technique requires the leaf surface to be wetted at some stage with a solution of a plastic in an organic solvent. It seems likely, in the light of current work, that the waxy layer covering the leaf is destroyed or badly damaged by this treatment. The replica method used by the author avoids this difficulty and is carried out as follows.

490   B. E. Juniper: Leaf surfaces under the electron microscope

The leaf under examination is placed in a vacuum plant and coated with about 150 Å of evaporated carbon. The carbon layer is then backed with two layers of plastic. The first layer is Formvar, deposited from a solution in chloroform, and the second Bedacryl, used in a solution in benzene. When the solvents have evaporated, a strip of adhesive tape is applied to the Bedacryl surface, and the leaf removed. The carbon/plastic film remains on the tape. The carbon is separated by dissolving the Bedacryl in acetone, mounting the Formvar/carbon film on grids, and finally dissolving away the Formvar.

This process is of interest, firstly because the leaf does not gas sufficiently to affect the carbon evaporation; it is probable that the cutin/wax layer effectively seals the leaf, thus permitting the use of a vacuum deposited replica. Secondly, the backed carbon layer can be removed from the leaf very easily in most cases. This is probably caused by the action of the solvents on the wax which acts as a separating layer.

The results obtained have been consistent and in agreement with macroscopic observations on wetting.

The surface configuration of wettable and non-wettable leaves is quite different. In the first case, the surface is nearly smooth, and in the second it is covered with a great variety of fine plate-like or tubular structures usually not visible in the light microscope. Two extremes are shown in the figures, the platelets of *Chamaenerion angustifolium* (Fig. 1), and the coarse tubes of the *Brassicas* (Fig. 2).

These structures are remarkably uniform over the leaf surface. Any variation usually occurs in field specimens rather than in specimens grown in a growth cabinet. A comparison indicates that the surfaces of the field specimens may often be damaged, or their wax structures affected by microclimatic changes. An example of field variation is shown in Fig. 3, *Lupinus albus*, though the cause is not known.

The structure of a leaf surface also changes in the vicinity of stomata. This is shown in Fig. 1, the adaxial surface of a leaf of *Chamaenerion angustifolium*. The inset indicates the density of platelets further away from the stoma.

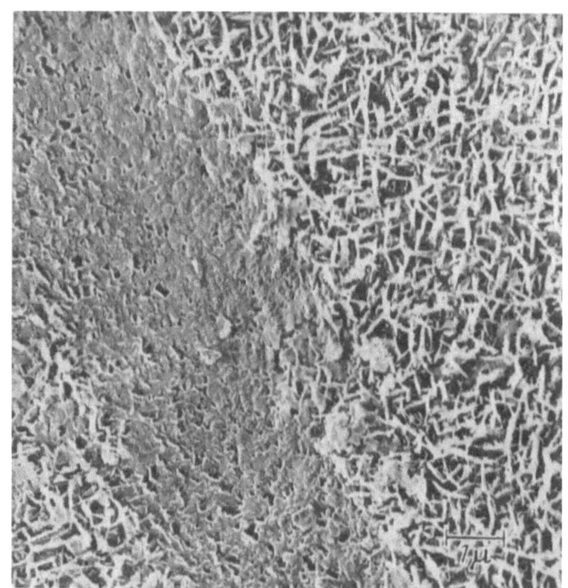

Fig. 3. *Lupinus albus*. Shadowed carbon replica of adaxial surface of leaf showing unexplained structure on field specimen

Experiments have shown that the leaf responds rapidly to changes in environmental conditions, such as light intensity. A plant grown in total darkness produces no wax structures, but the normal amount of wax is produced when the plant is returned to light. In the same way, a plant may recover from mechanical damage. In a case where a leaf was brushed with a squirrel-hair brush, removing nearly all the wax structures, recovery was complete within fourteen days.

The treatment of the soil with trichloracetic acid, a common agricultural technique used in conjunction with other herbicides, produced a reduction in the size and numbers of the wax platelets in *Pisum sativum*. There was a corresponding increase in wettability.

Little is known as yet about the method of formation and chemical composition of the wax layer. It is, however, clear that the electron microscope will be of considerable value in developing chemical sprays for plants. The investigations so far carried out have indicated the water repelling mechanism of the leaf, thus suggesting possible lines of approach in improving selective herbicides.

The author would like to thank Imperial Chemical Industries Ltd. for financial assistance in the above research, Sir Paul Fildes, F.R.S., for electron microscope facilities, Dr. F. A. L. Clowes for his advice and encouragement, and Miss J. Sampson for taking electron micrographs.

References

1. Bradley, D. E., and B. E. Juniper: Nature (Lond.) **180**, 330 (1957).

2. Juniper, B. E., and D. E. Bradley: J. Ultrastructure Res. **2**, 16 (1958).

3. — New Phytol. **58**, 1 (1959).

4. Mueller, L. E., P. H. Carr and W. E. Loomis: Amer. J. Bot. **41**, 593 (1954).

5. Schieferstein, R. H., and W. E. Loomis: Plant Physiol. **31**, 240 (1956).

Die Entstehung des Vacuolensystems in Pflanzenzellen

K. Mühlethaler

Eidgenössische Technische Hochschule Zürich/Schweiz

Von Guilliermond (*1*) und Dangeard (*2*) sind zahlreiche mikroskopische Untersuchungen über die Entstehung des Vacuolensystems in Pflanzenzellen gemacht worden. Nach diesen Autoren

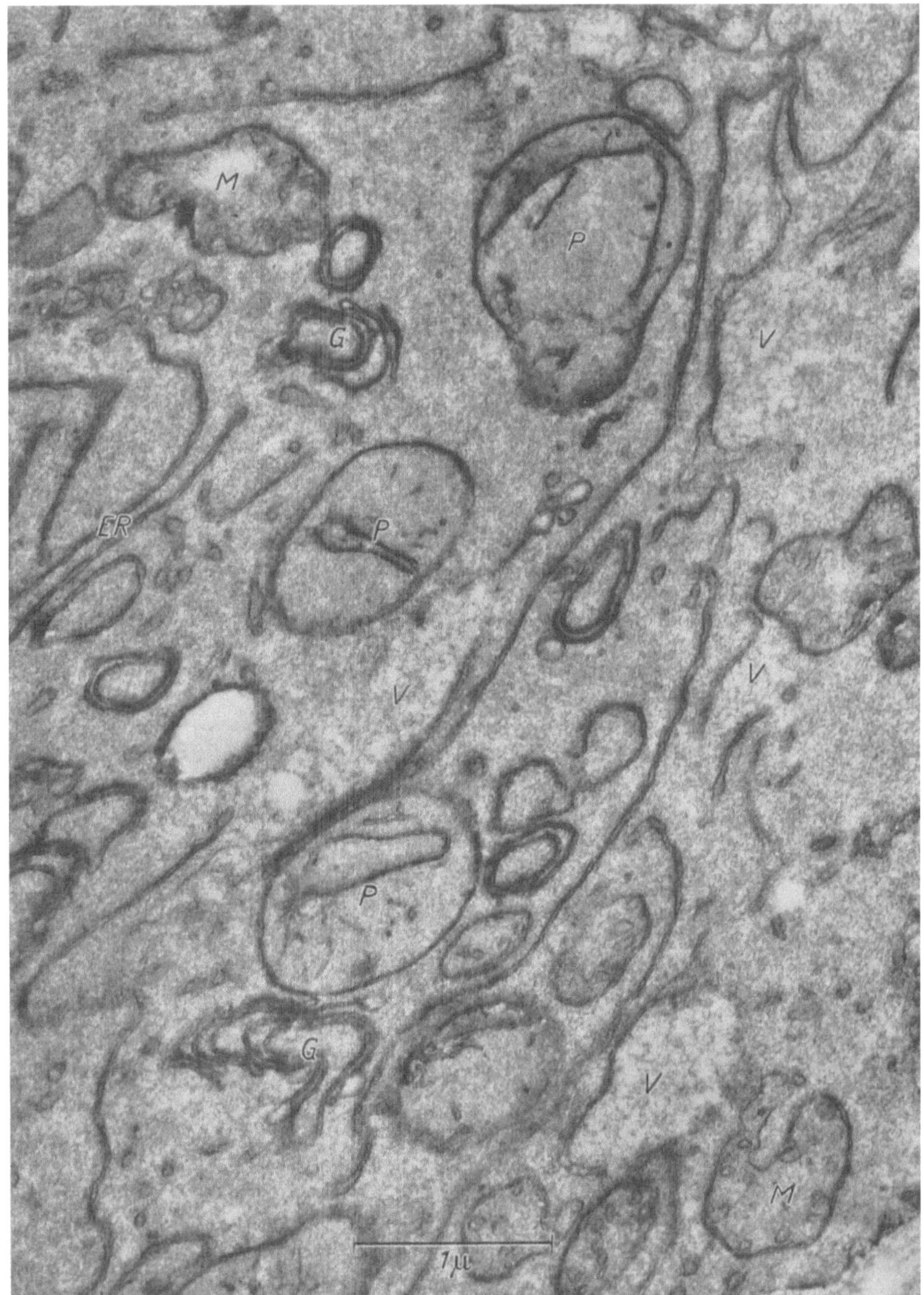

Abb. 1. Schnitt durch eine Meristemzelle aus der Zwiebelwurzel. Die Vacuolenanlagen (*V*) sind als aufgelockerte Bezirke bereits erkennbar. Als weitere Zellorganellen sind die Golgikörper (*G*), Mitochondrien (*M*), Proplastiden (*P*) und das endoplasmatische Reticulum (*ER*) zu sehen. 28 000mal

sollen im Plasma zuerst kleine Tröpfchen entstehen, die im Laufe der Entwicklung sich vergrößern und schließlich zu einem großen zentralen Saftraum zusammenfließen. Mit lichtmikroskopischen Methoden ist es aber nicht möglich, die Frühstadien der Entwicklung genau zu erfassen. Von DE VRIES (3) wurde die Theorie aufgestellt, daß die Vacuolen aus plastidenähnlichen

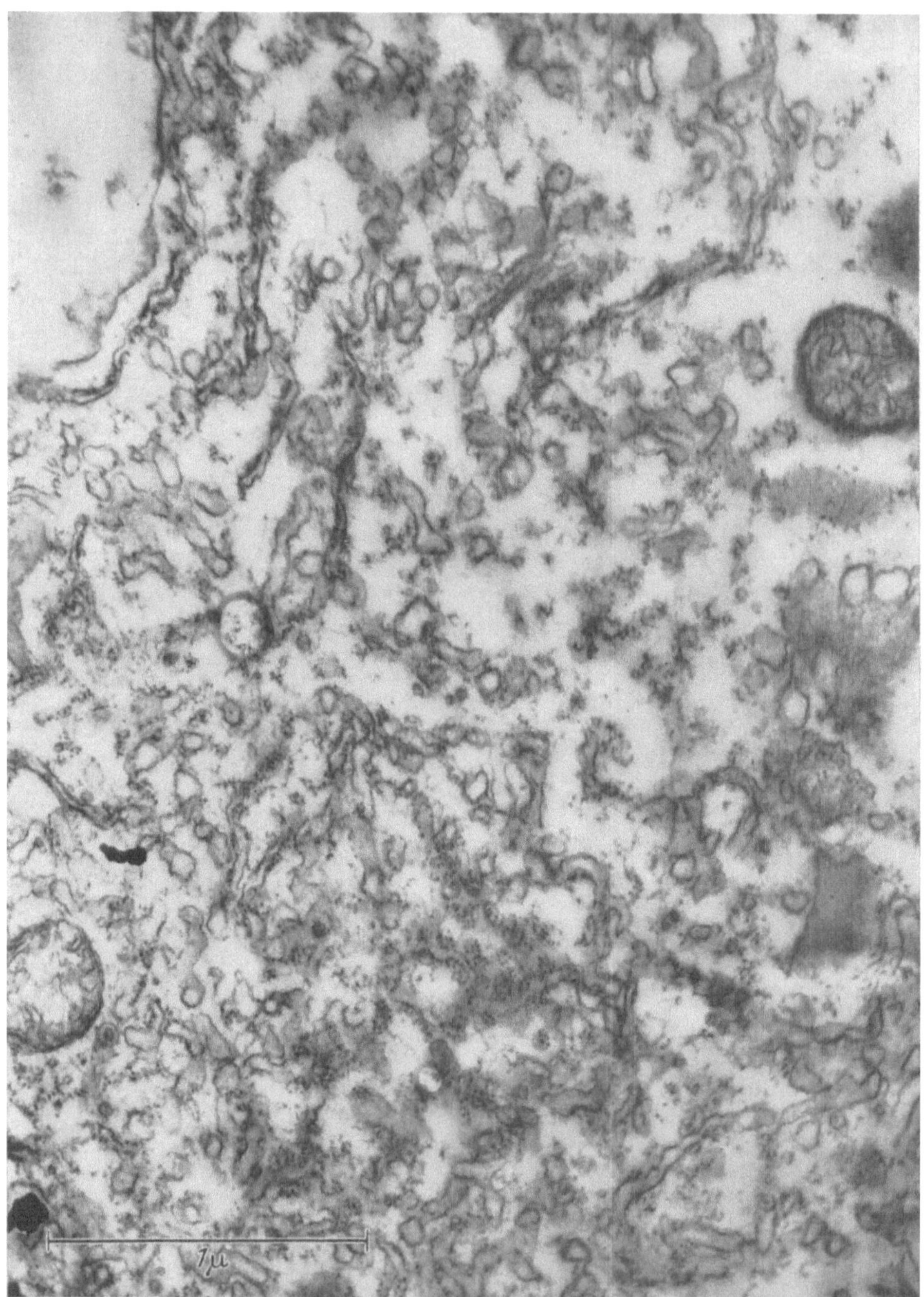

Abb. 2. Meristemzelle aus der Zwiebelwurzel. Frühes Entwicklungsstadium des endoplasmatischen Reticulums. Das Kavernensystem bildet sich durch lineare Aggregation von Bläschen. 45000 mal

Körpern, die er als Tonoplasten bezeichnete, hervorgehen. Diesen Zellelementen schrieb er, wie den Mitochondrien usw., eine genetische Kontinuität zu. PFEFFER (4) dagegen konnte an lebendem

Material nachweisen, daß die Vacuoleninitialen durch eine Entmischung des Cytoplasmas entstehen. Ihre Entstehung beruht daher auf einem kolloidchemischen Vorgang, der an keine spezifischen Partikel gebunden ist. Die vorliegende elektronenmikroskopische Untersuchung hat gezeigt, daß die von PFEFFER (4) vertretene Ansicht richtig ist.

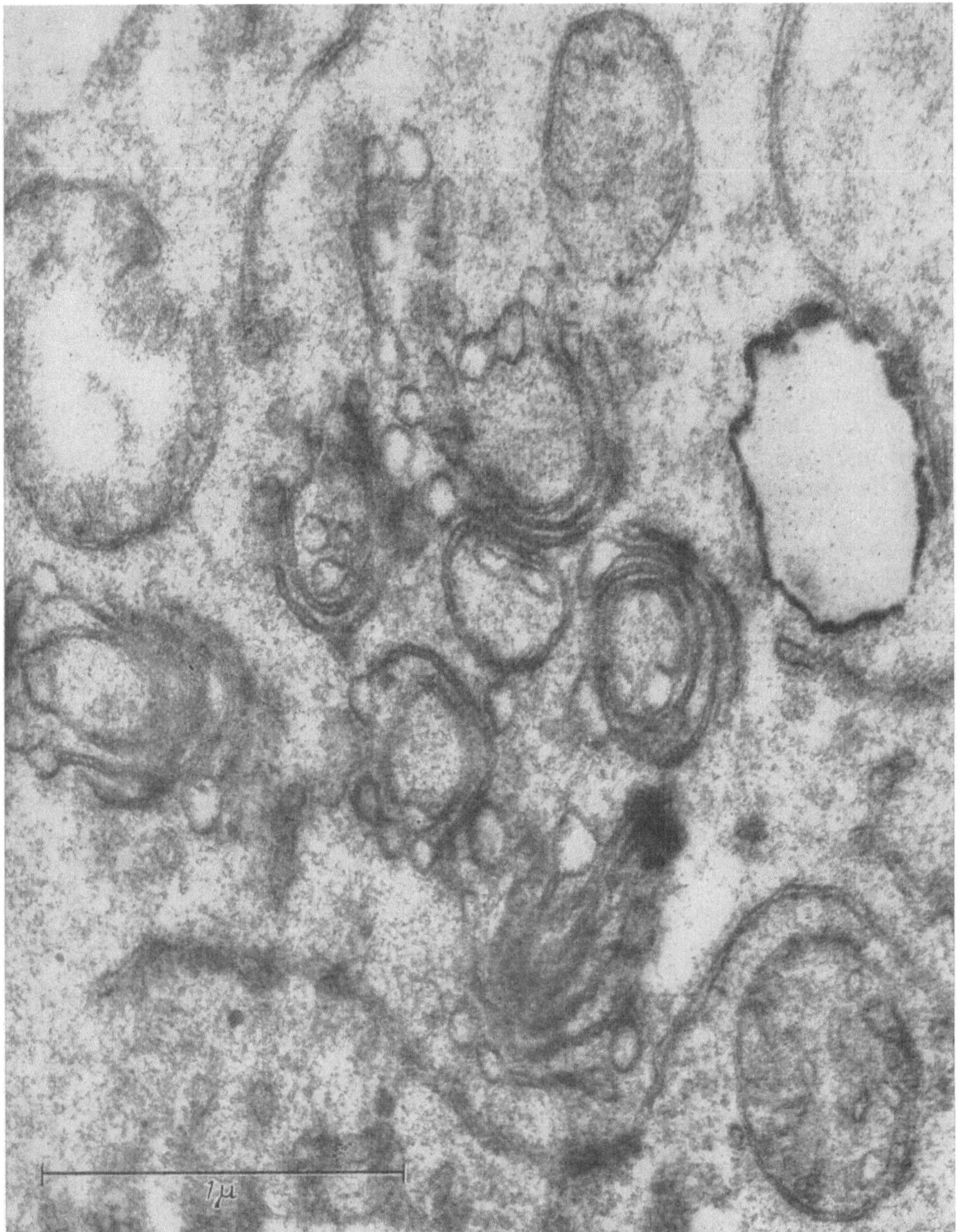

Abb. 3. Meristemzelle aus der Zwiebelwurzel. Gruppe von Golgikörpern. Vom Lamellenrande aus werden Sekrettropfen in das Plasma abgeschnürt. 50 000 mal

Als Untersuchungsmaterial verwendeten wir junge Wurzeln von *Allium cepa*. Wie Abb. 1 zeigt, sind in den Wurzelzellen nahe dem Vegetationspunkt noch keine scharf abgegrenzten Vacuolen vorhanden. Einzelne Bereiche des Grundcytoplasmas erscheinen aufgelockert und stellen, wie die spätere Entwicklung zeigt, den Beginn der Vacuolenbildung dar. Diese Ent-

mischung ist so zu denken, daß die Eiweiß-Makromoleküle das zwischen ihnen liegende Quell-
wasser nicht mehr zu binden vermögen, so daß dieses zu Tröpfchen zusammenfließt.

Die durch Fusion entstandenen Vacuolenschläuche sind von Guilliermond (1) mit den in
tierischen Zellen gefundenen Golgistrukturen in Beziehung gebracht worden. Die neueren elektronenmikroskopischen Untersuchungen haben ergeben, daß diese Zellorganellen aus geschichteten Doppellamellen bestehen, die am Rande von zahlreichen Bläschen umgeben sind. Wie Abb. 2
zeigt, finden wir auch in Pflanzenzellen solche Strukturen. Die Golgikörper bestehen meist aus
3—10 konzentrisch oder hufeisenförmig gebogenen Doppellamellen von etwa 0,5—1 μ Durchmesser. Vom Lamellenrand schnüren sich zahlreiche Sekrettröpfchen ab, deren physiologische
Bedeutung noch unbekannt ist. Aus den elektronenmikroskopischen Aufnahmen läßt sich entnehmen, daß kein Zusammenhang mit dem Vacuolensystem besteht. Die oben erwähnte An-
schauung von Guilliermond (1) ist daher abzulehnen.

Die Arbeiten von Palade und Porter (5) haben ergeben, daß die tierischen Zellen noch ein
weiteres Kavernensystem besitzen, das als endoplasmatisches Reticulum bezeichnet wurde. Es
besteht nicht aus bläschenförmigen Elementen, wie das zuerst erwähnte Vacuolensystem, sondern
aus verzweigten Doppellamellen. In einzelnen Abschnitten können diese zu größeren Kavernen
ausgeweitet sein, aber der größte Anteil besteht aus zusammengepreßten Schläuchen. Wie aus
Abb. 1 hervorgeht, ist dieses System auch in den pflanzlichen Zellen sehr gut entwickelt. Wie das
Golgi-System muß auch das endoplasmatische Reticulum bestimmte Stoffe aufbauen, die sehr
wahrscheinlich im Dienste der inneren Sekretion stehen.

Die Bildung dieser Zellorganelle geht von kleinen Bläschen aus, die von bestimmten Bildungszentren in das Grundplasma abgeschieden werden. Abb. 3 zeigt solche Bläschen in einer jungen
Zelle der Zwiebelwurzel. Die kleinen Vesikel reihen sich später linear aneinander und bilden,
nach dem Verschmelzen, die ausgeprägten Doppellamellenstrukturen, wie sie in Abb. 1 zu sehen
sind. Die Bläschen sind von Anfang an von einer Membran umgeben und können dadurch von den
Vacuoleninitialen gut unterschieden werden. Es gibt ebenfalls keine Verbindung zwischen dem
endoplasmatischen Reticulum und dem Vacuolensystem.

Zusammenfassend können wir also feststellen, daß in den pflanzlichen Zellen drei voneinander
unabhängige Kavernensysteme existieren, die wir entsprechend ihrer morphologischen Ausbildung als Vacuolen, Golgikörper und endoplasmatisches Reticulum bezeichnen. In ihrer physiologischen Funktion entspricht das Vacuolensystem einem Eliminationsorgan, während die beiden
andern im Dienste der inneren Sekretion stehen dürften. Ihre Entwicklung unterscheidet sich
vor allem darin, daß die Vacuolen durch eine Entmischung der Plasmaeiweiße entstehen, während
die Golgikörper und das endoplasmatische Reticulum aus hochorganisierten Zellorganellen hervorgehen.

Literatur

1. Guilliermond, M. A.: Arch. Anat. micr. **23**, 1 (1927).
2. Dangeard, P.: Le vacuome de la cellule végétale. Protoplasmatologia Bd. 3, D 1. Wien: Springer 1956.
3. Vries, H. de: Jb. wiss. Bot. **16**, 465 (1885).
4. Pfeffer, W.: Bot. Z. **44**, 114 (1886).
5. Palade, G. E., and K. R. Porter: J. exp. Med. **100**, 641 (1954).

L'infra-structure du cytoplasme végétal d'après les cellules des ébauches foliaires d'Elodea canadensis

R. Buvat

Laboratoire de Botanique, Ecole Normale supérieure, 24 Rue Lhomond, Paris V.

Le schéma classique de la structure des cellules végétales, tel qu'il résulta des remarquables
travaux des cytologistes du début du siècle, a été grandement complété, précisé, et modifié
depuis l'utilisation du microscope électronique. Le changement d'échelle et d'ordre de grandeur
du pouvoir séparateur ont rendu nécessaire une comparaison précise permettant de relier les

résultats de la Cytologie végétale classique, élaborée au moyen du microscope photonique, à ceux obtenus par les techniques nouvelles.

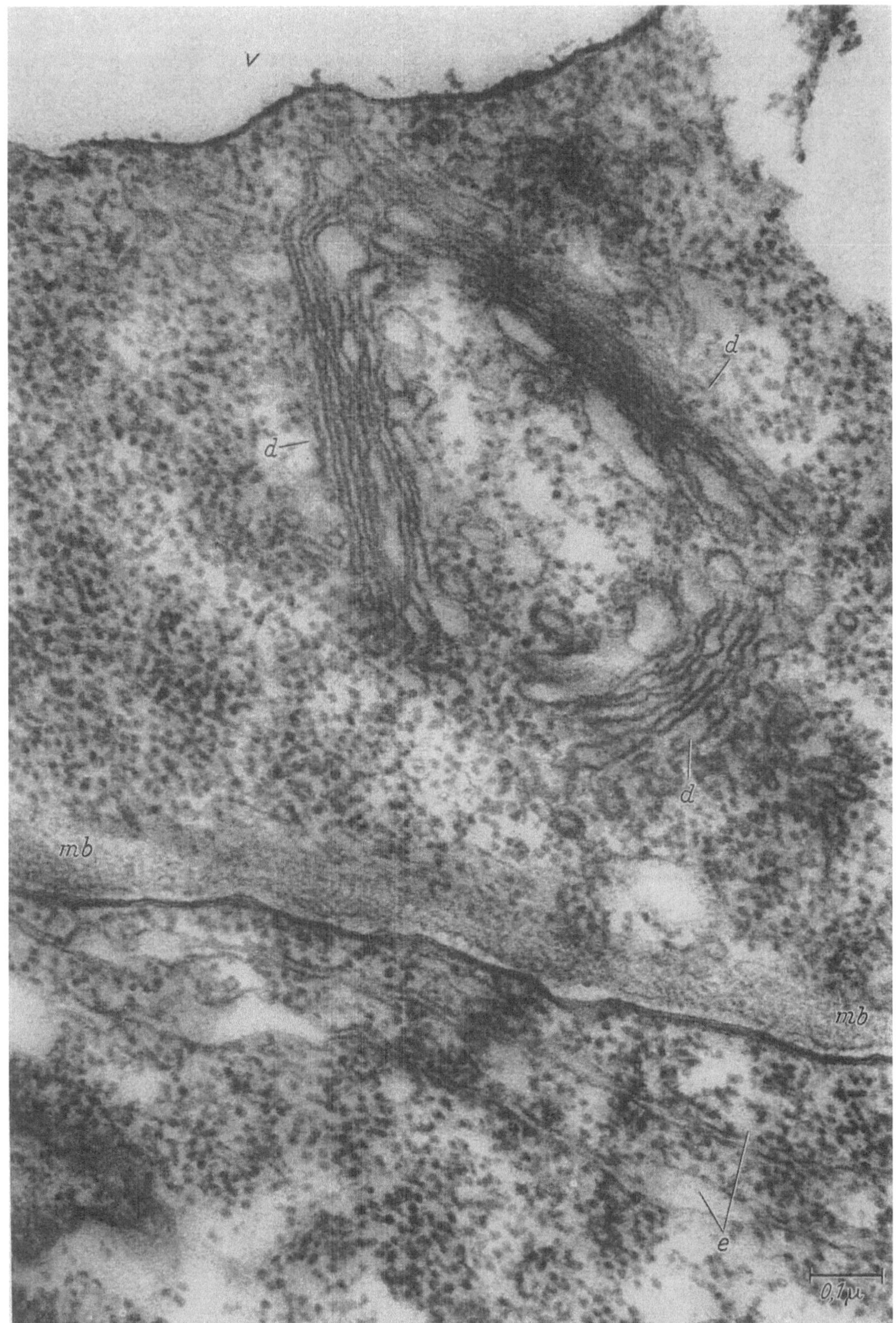

Fig. 1. *Elodea canadensis*, ébauche foliaire. Trois dictyosomes associés (*d*); *mb* membrane pectocellulosique; *v* vacuole; *e* reticulum endoplasmique. 115 000 fois

C'est pour cette raison que nous avons choisi un matériel rendu célèbre par les recherches de
Guilliermond (*1*): les cellules des ébauches foliaires d'*Elodea canadensis*. Au schéma établi par ce

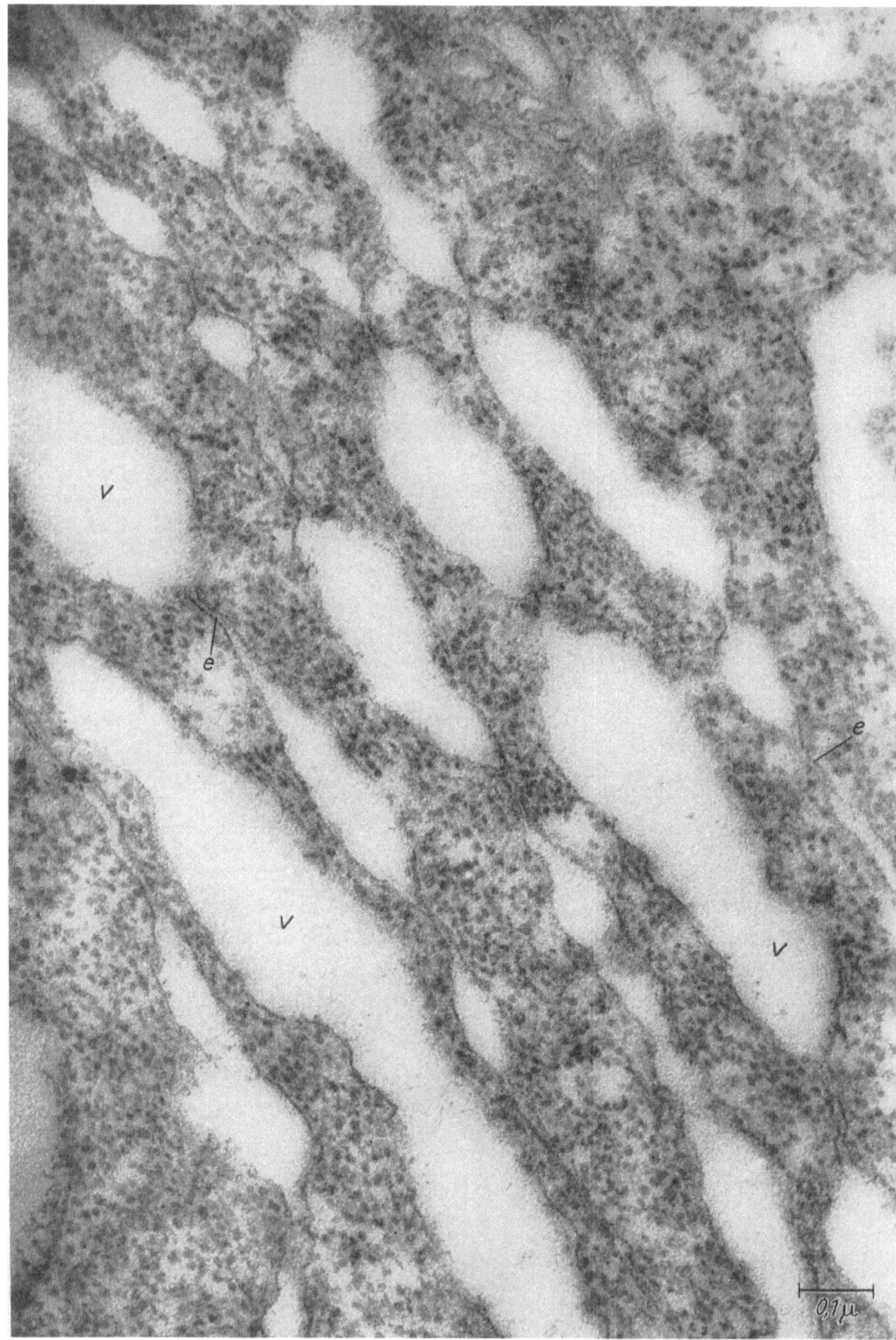

Fig. 2. id. — Formation des vacuoles méristématiques (*v*) par dilatations locales du reticulum endoplasmique *e*.
115000 fois

dernier auteur nous devons ajouter 1) des structures supplémentaires; 2) des rapports entre les structures ainsi que 3) entre les cellules elles-mêmes et 4) des indications morphologiques suggestives du rôle physiologique de certaines structures.

1°) *Structures supplémentaires.* Ce sont d'abord les dictyosomes, dont l'ensemble constitue l'appareil de GOLGI. Ils ont été vus maintenant par tous les cytologistes qui ont examiné au microscope électronique des cellules de Végétaux vasculaires et d'Algues (voir par exemple: PORTER (*2*), PERNER (*3*), SAGER et PALADE (*4*) CHARDAR et ROUILLER (*5*), BUVAT (*6, 7*) etc.). L'appareil de GOLGI est donc aussi bien un constituant du cytoplasme végétal que du cytoplasme animal (Fig. 1).

Les premières observations électroniques du cytoplasme végétal y ont également révélé l'existence du *reticulum endoplasmique* (PORTER (*2*), BUVAT et CARASSO (*8*)). Dans les cellules de l'*Elodea*, il se présente exclusivement sous la forme «smooth surfaced» (PALADE (*9*)), mais d'autres cellules végétales (*8*) le montrent sous les deux aspects, *smooth* et *rough* (ergastoplasme), comme chez les Animaux.

2°) *Rapports entre les structures intra-cellulaires.* L'analogie structurale entre la membrane nucléaire et le reticulum endoplasmique fait soupçonner l'existence de rapports analogues à ceux observés par PORTER (*2*) dans les cellules animales. Nous n'avons pas, jusqu'à ce jour, observé nettement ces rapports chez l'*Elodea*. Par contre, l'existence de pores dans la membrane nucléaire a été retrouvée, et paraît générale.

Une autre analogie structurale soulève à nouveau une question âprement discutée en Cytologie classique: celle des rapports entre les chondriosomes et les plastes. Divers travaux récents ont insisté sur les différences structurales, suggérant l'indépendance de ces deux lignées d'organites (Notamment STRUGGER (*10*), PERNER (*11*), STRUGGER et PERNER (*12*)). Nous avons recherché les proplastes dans les cellules les plus fortement méristématiques des primordiums foliaires d'*Elodea* et nous avons constaté que la distinction des plastes et des mitochondries est parfois très difficile. L'analogie structurale entre les deux catégories, *dans l'état le plus indifférencié*, est évidente au point de vue *qualitatif*. Toutefois, quelques différences *quantitatives*: dimensions légèrement supérieures, contenu plus homogène, crêtes plus réduites, permettent le plus souvent de reconnaître les proplastes. Nous pensons donc pour l'instant que, comme le soutenait GUILLIERMOND, ces deux sortes d'organites sont structuralement très voisines à l'origine, mais génétiquement distinctes,

Enfin, nos recherches nous ont permis de reconnaître des rapports entre le *reticulum endoplasmique* et les *vacuoles* (*13*). La classique évolution des vacuoles, au début de la différenciation cellulaire, permet de se convaincre que ce sont des différenciations, locales mais énormes, du reticulum endoplasmique (Fig. 2). Comme nous avons, d'autre part, retrouvé les communications entre ce dernier et la double pellicule ectoplasmique, nous arrivons à considérer les vacuoles comme d'énormes réservoirs, intercalés dans le reticulum endoplasmique, entre la pellicule ectoplasmique et la membrane nucléaire.

3°) *Rapports intercellulaires.* — La microscopie électronique n'a pas permis de retrouver de synapses (MANGENOT (*14*)), respectant l'autonomie morphologique des cytoplasmes dans les plasmodesmes (STRUGGER (*15*), BUVAT (*16*)) mais a démontré au contraire que le cytoplasme est continu le long de ces structures et donc d'une cellule à ses voisines. Les cellules d'*Elodea* nous ont permis de retrouver ces faits (*16*) et de constater notamment que la plupart des plasmodesmes, sinon tous, sont traversés par des canalicules du reticulum endoplasmique, ce qui accentue l'idée de continuité.

La membrane pecto-cellulosique ne doit donc pas être considérée comme un organe de *séparation* des protoplasmes. En outre de son rôle mécanique, elle apparaît plutôt comme un dispositif de conduction des substances utiles, assurant des surfaces de contact entre le milieu intérieur et la pellicule ectoplasmique.

4°) *Comportements suggestifs de rôles physiologiques.* Cette dernière idée est appuyée par des particularités de comportement de la pellicule ectoplasmique, que nous avons observées en étudiant l'*Elodea*, mais que nous avons retrouvées ensuite, en collaboration avec Mme. A. LANCE, dans d'autres matériels (*17*). Çà et là en effet, la pellicule ectoplasmique cesse d'être en contact avec la membrane et forme des invaginations caractéristiques (Fig. 3). Les rapports de ces invaginations avec leur voisinage excluent l'idée d'un artéfact. Par contre, les fréquences de leurs aspects, en sections ultra-fines, suggèrent fortement qu'elles s'isolent dans le cytoplasme, sous

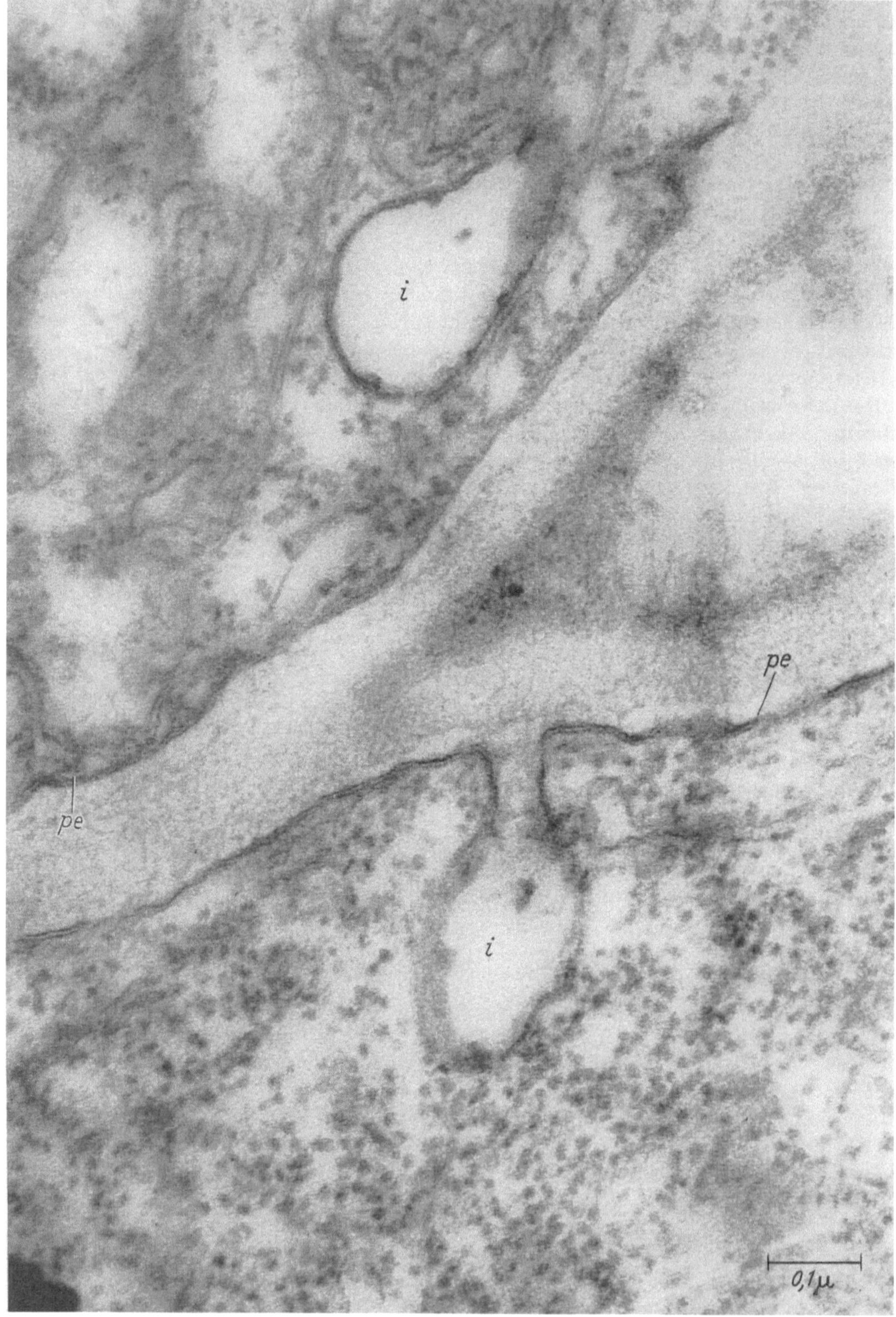

Fig. 3. id. — Invaginations (*i*) de la pellicule ectoplasmique (*pe*), suggérant des faits de pinocytose. 140 000 fois

forme de gouttelettes fluides entourées d'une double membrane. Ces inclusions ne confluent pas avec les vacuoles qui s'en distinguent par leur simple membrane. Avec des dimensions plus réduites, ces aspects rappellent les phénomènes de *pinocytose* décrits dans les cellules animales [Voir notamment Lewis (*18*)]. Des recherches ultérieures, de caractère expérimental, sont certes nécessaires pour affirmer cette analogie, mais diverses observations, relatives au contenu de ces gouttelettes, sont favorables à l'idée d'un mode d'absorption analogue à la pinocytose, dans les cellules végétales.

Ces premiers résultats laissent encore beaucoup à désirer, mais ils montrent que la microscopie électronique doit nous permettre de faire de grands progrès dans notre connaissance de la cellule végétale. L'étude, en cours, de l'origine et de l'évolution des structures, lors de la différenciation, précisera peut être des observations dont l'interprétation ne peut encore être suggérée qu'avec beaucoup de réserve.

Bibliographie

1. Guilliermond, A.: C. R. Acad. Sci. (Paris) **175,** 283 (1922).
2. Porter, K. R.: Harvey Lect. **51,** 175 (1957).
3. Perner, E. S.: Naturwiss. Rdsch. **44,** 336 (1957).
4. Sager, R., and G. E. Palade: J. biophys. biochem. Cytol. **3,** 463 (1957).
5. Chardar, R., et C. Rouiller: Rev. Cytol. et Biol. végét. **18,** 153 (1957).
6. Buvat, R.: C. R. Acad. Sci. (Paris) **244,** 1401 (1957).
7. — C. R. Acad. Sci. (Paris) **246,** 2157 (1958).
8. — et N. Carasso: C. R. Acad. Sci. (Paris) **244,** 1532 (1957).
9. Palade, G. E.: J. biophys. biochem. Cytol. **2,** suppl., 85 (1956).
10. Strugger, S.: Protoplasma **48,** 360 (1957).
11. Perner, E. S.: Proc. Stockholm Conference on Electron Microscopy. 1956, 272.
12. Strugger, S., u. E. Perner: Protoplasma **46,** 711 (1956).
13. Buvat, R.: C. R. Acad. Sci. (Paris) **245,** 350 (1957).
14. Mangenot, G.: Bull. hist. appl. **3,** 142 (1926).
15. Strugger, S.: Protoplasma **48,** 231 (1957).
16. Buvat, R.: C. R. Acad. Sci. (Paris) **245,** 198 (1957).
17. — et A. Lance: C. R. Acad. Sci. (Paris) **245,** 2083 (1957).
18. Lewis, W. H.: Amer. J. Cancer **29,** 666 (1937).

Plasmatische Lamellensysteme bei Pflanzen

E. Heitz

Max Planck-Institut für Biologie, Tübingen

Bei 12 Arten aus den verschiedensten Gruppen und Familien der Laub- und Lebermoose, sowie einem Farn wurde das regelmäßige Vorkommen von plasmatischen Lamellen nachgewiesen. Es handelt sich um Doppellamellen, die paarweise an ihrer Peripherie zu flachen Kapseln zusammengeschlossen sind (Abb. 1). Der Inhalt der Kapseln ist meistens etwas dichter als das Cytoplasma der Zelle (Abb. 1 b, d, g, h). Von beiden heben sich die dichteren Einzellamellen deutlich ab. (Kontrastierung ist überflüssig). Die Lamellensysteme haben einen Breitendurchmesser von durchschnittlich 0,5, selten bis 0,8 μ (Abb. 1 c). In den Systemen sind meistens 5—7 Kapseln vereinigt. Der Höhendurchmesser der Kapseln beträgt 0,15 bis höchstens 0,2 μ. Die paarweise Kapselbildung benachbarter Lamellen (Abb. 1 in allen Figuren), ferner das häufige Auftreten von Vacuolen in der Nähe der Doppellamellen (Abb. 1, besonders in b, c, g, h) macht diese Lamellensysteme der Pflanzen denen bei Tieren — dort als „Golgiapparat" bezeichnet, — homologisierbar.

Erstmalig haben Mercer u. Mitarb. (*1*) bei der Alge *Nitella* plasmatische Lamellen gefunden und sie mit dem endoplasmatischen Reticulum der Tiere verglichen. 1957 haben dann Buvat (*2*), Heitz (*3*), Perner (*4*) Lamellensysteme bei einigen Phanerogamen festgestellt und mit dem endoplasmatischen Reticulum, bzw. granulären Reticulum (Heitz) und Golgi-Apparat (Buvat, Perner) der Tiere homologisiert. (Neuerdings bestätigte Sitte (*5*) die Richtigkeit der Befunde und Auffassungen).

Die vorliegende Untersuchung zeigt erneut die homologe Struktur von Zellorganellen, bzw. plasmatischen Strukturen bei Pflanze und Tier.

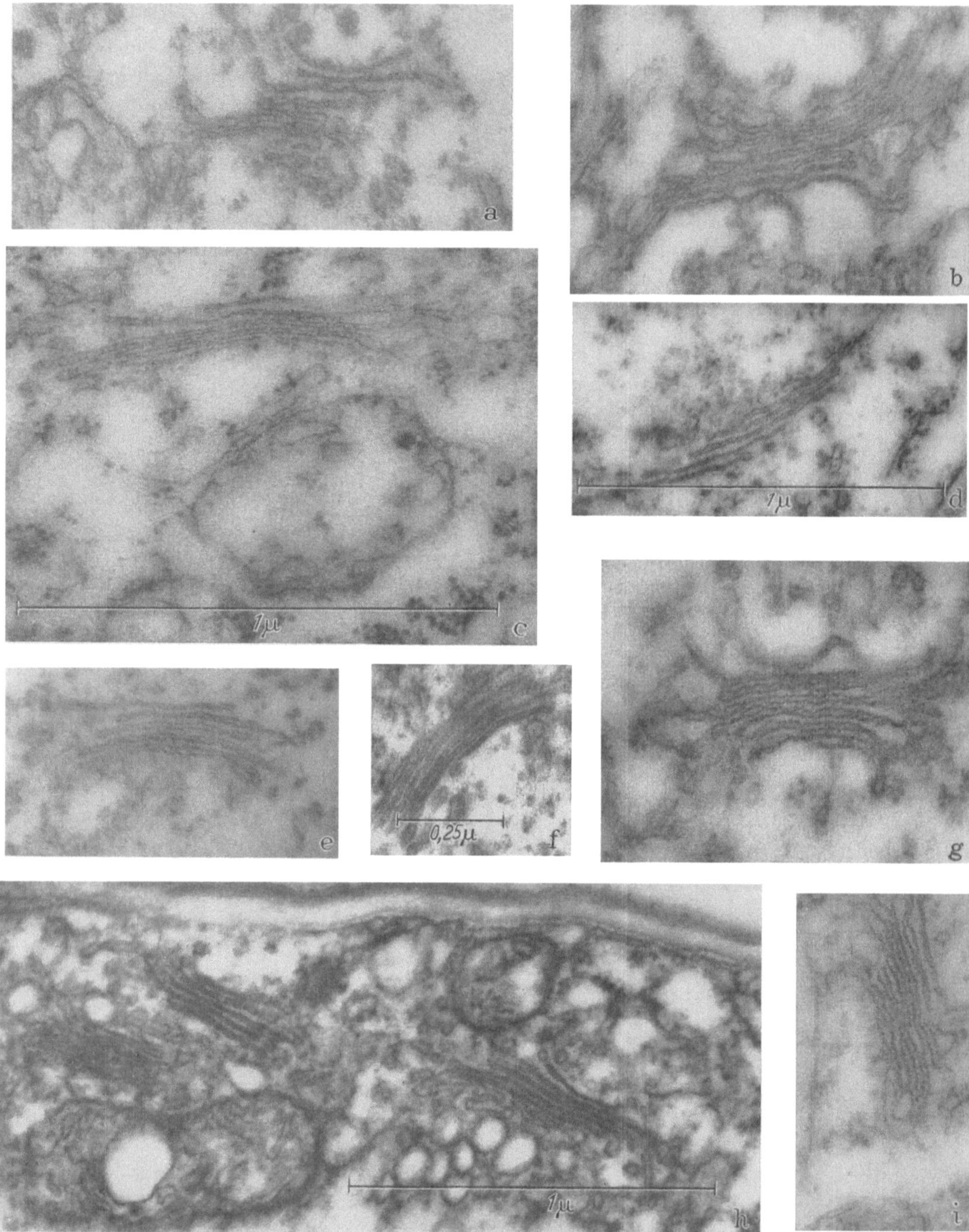

Abb. 1. Plasmatische Lamellensysteme bei 5 Laubmoosen (a—e), 3 Lebermoosen (f—h) und 1 Farn (i); überall sind 5, höchstens 7 Doppellamellen vorhanden, die durch paarweisen Zusammenschluß, am Rande des Systems flache Kapseln bilden; häufige Vacuolenbildung dicht an den Doppellamellen (vgl. b, c, e, g, h). a) *Pottia truncatula*, b) *Mnium cuspidatum*, c) *Neckera crispa*, d) *Fissidens exilis*, e) *Leucobryum glaucum*, f) *Riccardia pinguis*, g) *Plagiochila asplenioides*, h) *Conocephalus conicus*, i) *Pteridium aquilinum*. El. mikr. 20000mal; a, b, c, e, g, i auf 70000, d, h auf 54000, f auf 60000 nachvergrößert

Literatur

1. Mercer, F. V., A. J. Hodge, A. B. Hope and J. D. McLean: Aust. J. Biol. Sci. **8**, 1 (1955); zit. nach A. J. Hodge, J. D. McLean u. V. F. Mercer: J. biophys. biochem. Cytol. **2**, 597 (1956).
2. Buvat, R.: C. r. Acad. Sci. (Paris) **244**, 1401 (1957).
3. Heitz, E.: Z. Naturforsch. **12b**, 579 (1957).
4. Perner, E. S.: Naturwissenschaften **44**, 336 (1957); Protoplasma **49**, 407 (1958).
5. Sitte, P.: Protoplasma **49**, 447 (1958).

Beitrag zur Kenntnis der Chloroplastenstruktur

E. Heitz

Max Planck-Institut für Biologie, Tübingen

Durch Untersuchungen verschiedener Autoren an Chloroplasten ist der elektronenmikroskopische Nachweis des Zusammenhangs zwischen Stromalamellen und Granalamellen erbracht.

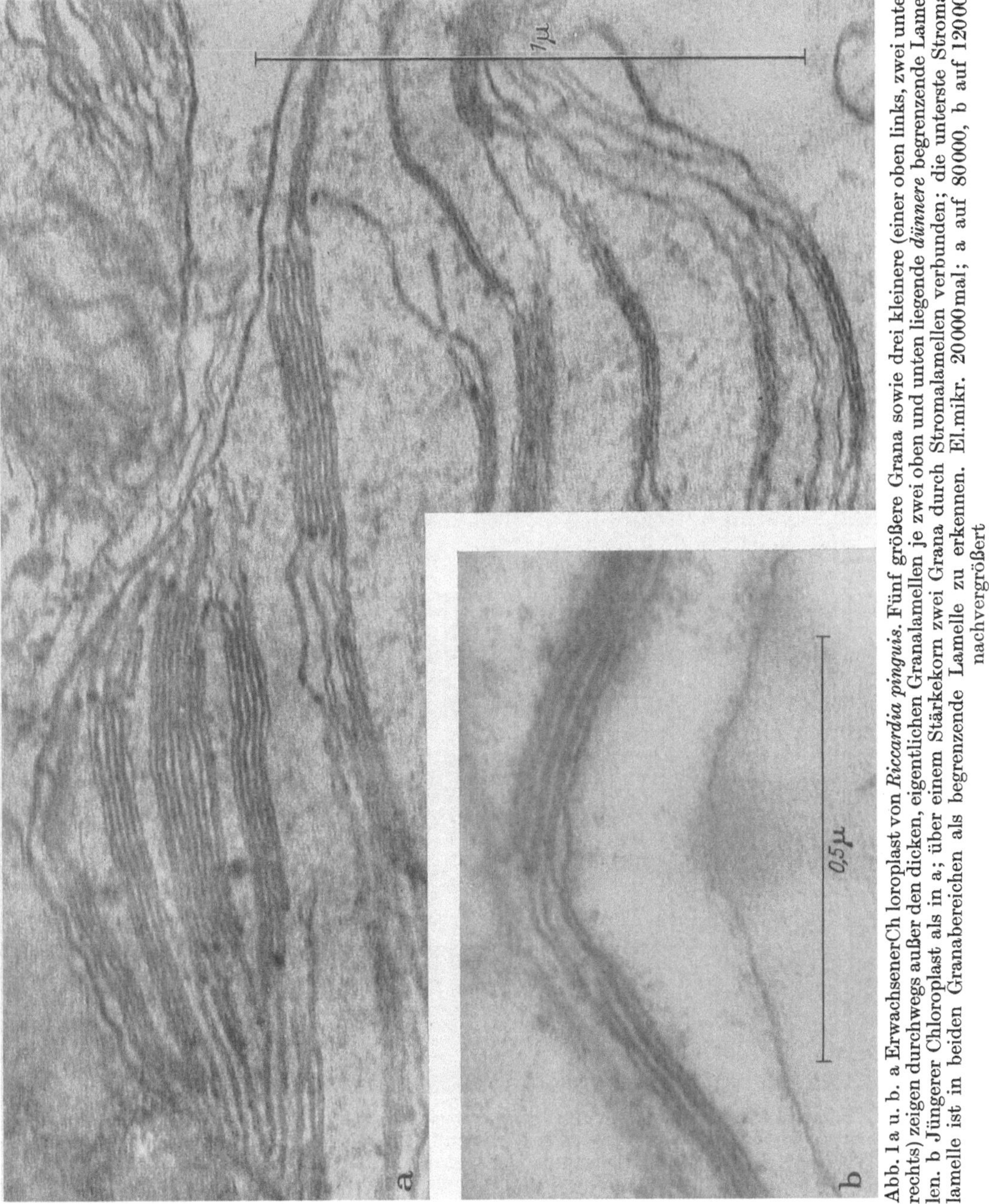

Abb. 1a u. b. a Erwachsener Chloroplast von *Riccardia pinguis*. Fünf größere Grana sowie drei kleinere (einer oben links, zwei unten rechts) zeigen durchwegs außer den dicken, eigentlichen Granalamellen je zwei oben und unten liegende *dünnere* begrenzende Lamellen. b Jüngerer Chloroplast als in a; über einem Stärkekorn zwei Grana durch Stromalamellen verbunden; die unterste Stromalamelle ist in beiden Granabereichen als begrenzende Lamelle zu erkennen. El.mikr. 20000mal; a auf 80000, b auf 120000 nachvergrößert

Die diesbezüglichen Aufnahmen zeigen die Stromalamellen als *verbindende* Lamellen *zwischen* den Granabereichen (*1, 2*).

1. Es wurde eine Pflanze gefunden (*Riccardia pinguis*, Lebermoos), bei welcher die *dünnen* Stromalamellen, als solche kenntlich, auch *innerhalb* der Granabereiche verlaufen. Sie treten hier

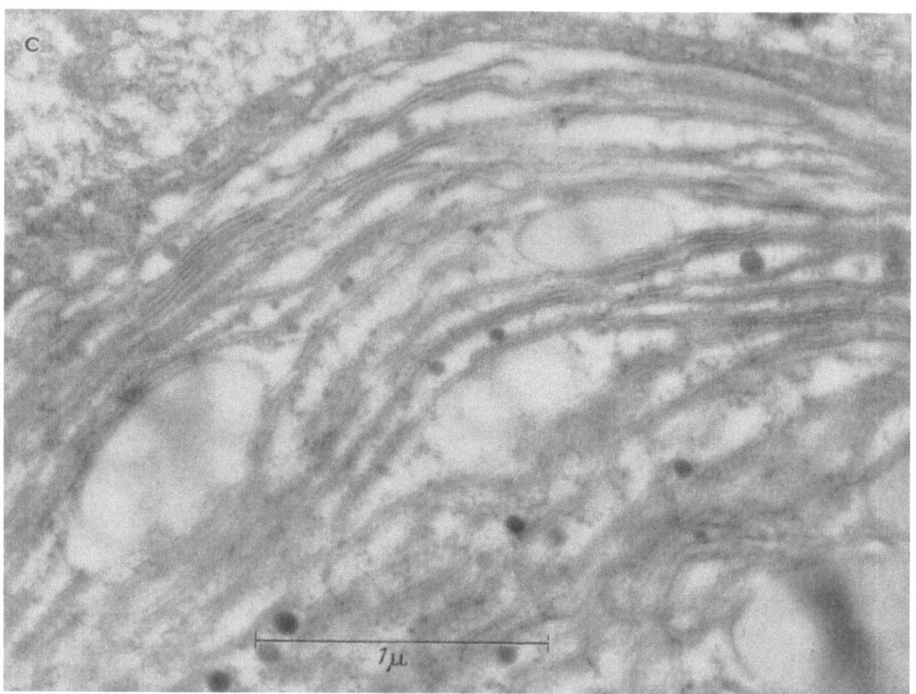

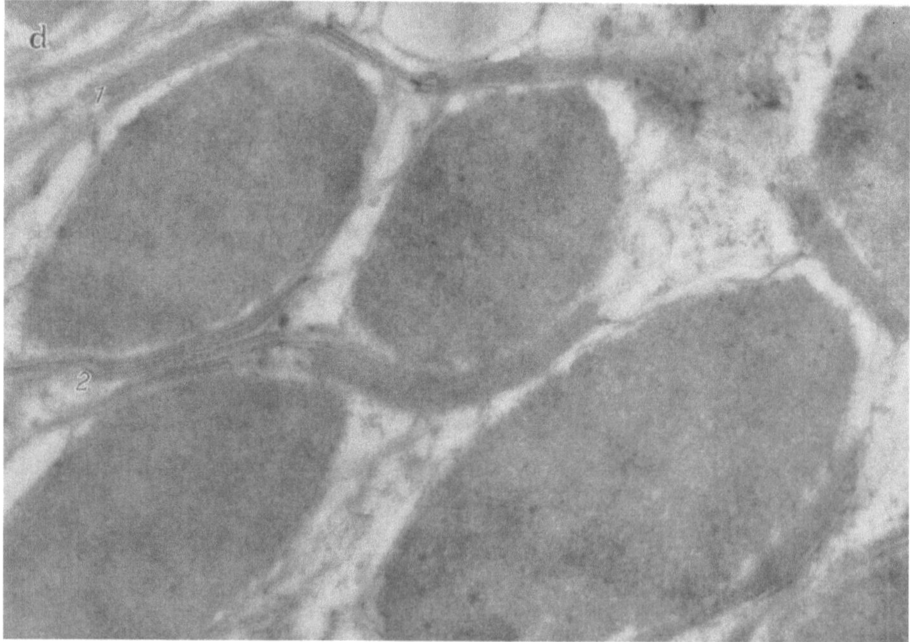

Abb. 1c u. d. Erwachsener Chloroplast von *Anthoceros crispulus*. c Stück von der äußeren Region desselben; oben *über* der doppelten Membran ein breites „Peristromium", fein granulös mit zahlreichen kleinen Vacuolen, vom Plasma durch eine dünne, einfache Membran abgegrenzt; im Stroma verlaufen lange, *dünne* und *dicke* Lamellen; zwischen diesen liegen, außer den kleineren, stark elektronenstreuenden Kugeln, große, linsenförmige Stärkekörner (hell). d Stück aus der inneren Region des Chloroplasten mit großen Pyrenoiden; dicht an diesen liegen (im ganzen 6) granaartig geschichtete Körper; bei 2 heben sich, wie bei *Riccardia pinguis*, zwei dünne begrenzende Lamellen von den dickeren inneren ab; bei 2 als Stromalamellen zu erkennen. El.mikr. 20000mal; c und d auf 40000 nachvergrößert

als obere und untere „begrenzende Lamellen" der stets dickeren Granalamellen auf (Abb. 1a und b). Dasselbe wurde (*3*) bei den Keimblattchloroplasten einer Phanerogame, *Solanum lycopersicum*, beschrieben. Im Inneren des Granabereiches treffen auf eine Granalamelle manchmal zwei Stromalamellen, wie bei *Aspidistra* (*1*), manchmal nur eine, wie bei *Zea* (*2*).

    2. Bei einem anderen Lebermoos (*Anthoceros crispulus*), bei welchem nach bisherigen Angaben (*4, 5*) ausschließlich Stromalamellen vorhanden sein sollen, konnten Anhäufungen von Lamellen zu granaartigen Gebilden in begrenzten Bezirken (direkt benachbart den Teilpyrenoiden) nachgewiesen werden (Abb. 1d bei 1 u. 2). Außerdem sind im Gegensatz zu den vorliegenden Beobachtungen (*4, 5*), die wie die betreffenden Abbildungen vermuten lassen, auf ungenügenden Schnitten beruhen, die Lamellen nicht gleich dick, sondern es wechseln dünne und dicke Lamellen miteinander ab (Abb. 1c). Ein Fall von sehr regelmäßigem Wechsel von dünnen und dicken Lamellen (zwei dickere regelmäßig begrenzt von zwei dünneren) wurde kürzlich (*6*) bei einemDinoflagellaten beschrieben. Wie bei *Riccardia pinguis* finden sich auch bei *Anthoceros crispulus*, wenigstens gelegentlich, Granabereiche mit begrenzenden, auffallend dünnen Lamellen (Abb. 1d, bei 2).

Literatur

1. Steinmann, E., u. F. S. Sjöstrand: Exp. Cell Res. 8, 15 (1955).
2. Hodge, H. J., J. B. McLean and F. V. Mercer: J. biophys. biochem. Cytol. 1, 605 (1955).
3. Lefort, Marcelle: C. R. Acad. Sci. (Paris) **244**, 2957 (1957).
4. Kaja, H.: Ber. dtsch. bot. Ges. **70**, 343 (1957).
5. Chardard, R.: Rev. Cytol. Biol. végét. 18, 360 (1957).
6. Grell, K., u. K. E. Wohlfarth-Bottermann: Z. Zellforsch. **47**, 8 (1957).

The formation of the cell plate during cytokinesis in *Allium cepa* L.

K. R. Porter and J. B. Caulfield*
The Rockefeller Institute, New York

It has long been an interest of biologists to discover the structural basis underlying the expression of cell shape or form, and the higher resolutions provided by electron microscopy have renewed hopes that the elucidation of this problem might now be possible by direct observation. In this connection the structure and behaviour of the newly-revealed systems of the cytoplasm, especially those that are membrane-limited, have been watched carefully for appearances of order which might be related to the over-all shape of the cell. Thus far, however, the correlations, if any, have not become apparent. In further pursuit of this interest we have made a few observations on a particular region of the cell, where, if at any time, the cytoplasmic systems involved in determining cell limits should show themselves, namely when the new cell wall appears during cytokinesis in plant tissues, i. e. at the cell plate.

When the chromosomes move in anaphase toward the poles of the spindle in the division of plant cells, the intervening region between the chromosome masses takes on a fibrous appearance and a line, or cell plate, appears at the equatorial plane of the structure. This formation exclusive of the plate is known as the phragmoplast. The first clear separation between the two new cells appears first at the center of the plate and progresses laterally until it reaches the side walls of the parent cell and completes the division. The phragmoplast (appearing fibrous in light microscope images) spreads out in advance of the edge of the cell plate, thus forming a ring-shaped (or barrel-shaped) structure. In so doing, it frequently loses all connection with the nucleus and the original spindle and behaves as an independent component of the dividing cell. In the light microscope, the early cell plate is said to appear as a row of granules and most observers have accepted this interpretation. Only Becker (*1*) has suggested that these granules might represent small vacuoles and further proposed that cell separation might be achieved by a simple coalescence of these vacuoles.

    * Visiting investigator from Department of Pathology and Oncology, University of Kansas Medical Center, Kansas City, Kansas, supported by Institutional Grant INSTR-60 from the American Cancer Society, Inc.

The description of cell plate development that follows is based on preliminary observations made by electron microscopy. In essence it reports the earliest phases of a study which repeats the earlier study made by Becker. We have, of course, a basic interest in tracing the cell plate back in its development to the very earliest structural manifestations of where it will appear, to see what system or systems of the cytoplasm are involved in determining its location. The observations were made, for the most part, in thin sections of root tips of the onion, *Allium cepa* L. While in a phase of rapid growth, the tips of the roots were fixed in 1% OsO_4 buffered with veronal acetate (at p_H 7.2) and rendered more or less isotonic with plant cell sap by adding sucrose (0.015 g/ml) (2). They were thereafter dehydrated and sectioned by standard methods.

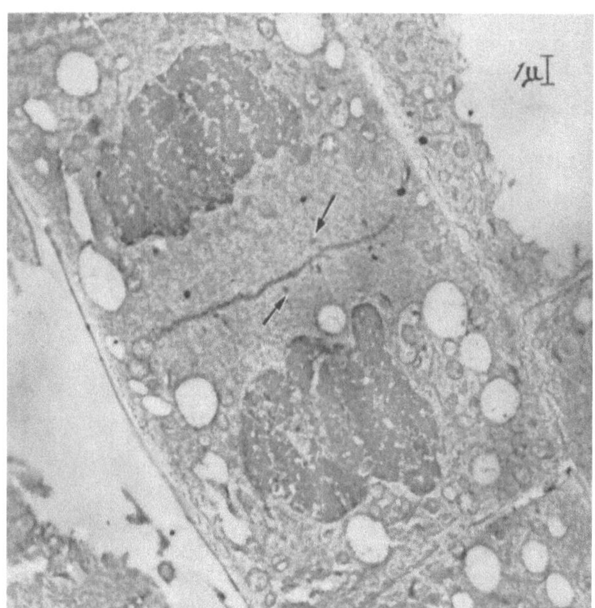

Fig. 1. Low power micrograph showing meristematic cell of onion root tip in early telophase of division. Cell plate appears as dark line across interzonal region between telophase nuclei. The plate has not, at this stage, reached the cell margins. Small dense bodies (phragmosomes), indicated by arrows, are distributed along both sides of plate. 3500 ×

In making observations on the fine structure of the cell plate, we have simply sought out cells in different stages of anaphase and early telophase and then taken micrographs of the interzonal region of the spindle. These observations are easily arranged in a natural sequence based on the appearance of the chromosomes and early telophase nuclei. We are not sure that we have observed the very earliest stages of plate formation, but this is not very important because at its margins the plate is going through stages of formation as it expands toward the lateral walls of the parent cell.

Observations on early plate formation. Images of cells in mid-anaphase, after the chromosomes of the daughter cells have completely separated, show no evidence of a cell plate. Long profiles of elements of the endoplasmic reticulum tend to be oriented parallel to the long axis of the spindle but there is no indication of a mid-line separation of the two new protoplasts. Dense bodies about $0.2\,\mu$ in diameter and without membrane or evident internal structure are scattered on opposite sides of an imaginary equator. They would probably not have been regarded as significant enough to mention if they had not appeared more prominently in the phragmoplast at later stages. It would seem most appropriate to refer to them as phragmosomes although this conflicts with the earlier use of this term (3, 4, 5). A little later, as mitosis enters early anaphase, a distinct line of greater density makes its appearance at the equator of the dividing unit (Fig. 1). Here it is possible to identify a small population of the dense bodies (about $0.2\,\mu$ in diameter) dispersed at a fairly even distance from the plate. At slightly higher magnification (Fig. 2) it becomes evident that the plate consists of a large number of small vesicles and a slight condensation (in terms of density) of the surrounding matrix material of the cytoplasm. The vesicles vary greatly in size, the smallest being about 25 mμ in diameter, the larger from 200 mμ on upward to 500 mμ. Actually there is a continuous gradient of sizes, with the smaller ones the more numerous (Fig. 3). The concentration of vesicles is greatest along the line evident in low power of the cell plate, but the limits of the planar aggregation are not sharp and grade off into the adjacent cytoplasm. The larger vesicles are located mostly along the center of the plate, where the concentration of vesicles is highest. This suggests that they have developed, in part, by a coalescence of the smaller vesicles. The sequence of events from this early stage onward is not hard to imagine and essentially confirms Becker's views. The story is told by comparing the mid portion of a plate with the structure at the margins where early stages in plate formation are represented (Fig. 3). It is evident then that the vesicles enlarge and eventually coalesce to form a continuous phase of separation between the daughter cells. Only a few strands of cytoplasm persist as a connection between the two new cells and these represent the plasmadesmata (Fig. 4). Similar observations on sequences in plate formation have just been published by BUVAT and PUISSANT (6). As the separation develops some

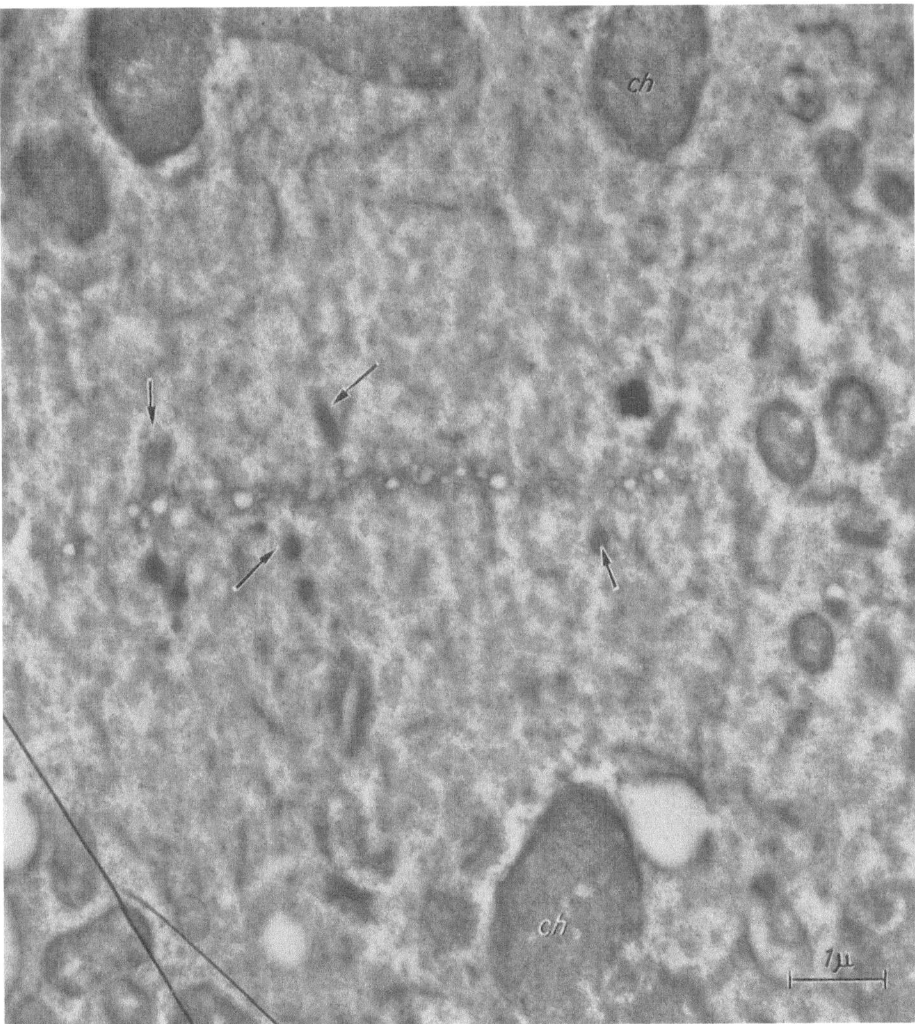

Fig. 2. Electron micrograph of early plate found in late anaphase of division. Chromosomes are included in image at *ch* along upper and lower margins of picture. The row of small vesicles across center of picture represents early cell plate. Side walls of cell are beyond limits of micrograph. Dense bodies indicated by arrows are commonly associated with plate development. 12,500 ×

dense material accumulates between the cells and represents, presumably, the pectin layer and the primary cell wall.

Cytoplasmic structures associated with the early cell plate. At this earliest stage in the development of the cell plate certain other structures of the cytoplasm are consistently associated with the plate in such a way as to suggest a correlation. The most obvious of these are dense, spheroidal to oblong bodies about $0.2\,\mu$ in diameter mentioned above. They are, in some instances, sharply defined, but usually the limits are diffuse. Internal structure is not very clear but appears to be made of small vesicles. The more obvious of these bodies (phragmosomes) tend to be a uniform distance from the cell plate but uniform spacing, with respect to one another, has not been noted.

506 K. R. PORTER and J. B. CAULFIELD: The formation of the cell plate during cytokinesis in *Allium cepa* L.

As the plate matures and grows centrifugally they disappear except at the edges of the plate, where they persist as part of the phragmoplast. How these phragmosomes contribute to plate formation is not very clear. When at some distance $(0.5\,\mu)$ from the plate they appear relatively uniform in size and density. Nearer the plate they are less so and look as though they may break up into vesicles and possibly initiate the development of the plate. Possibly they contain enzymes

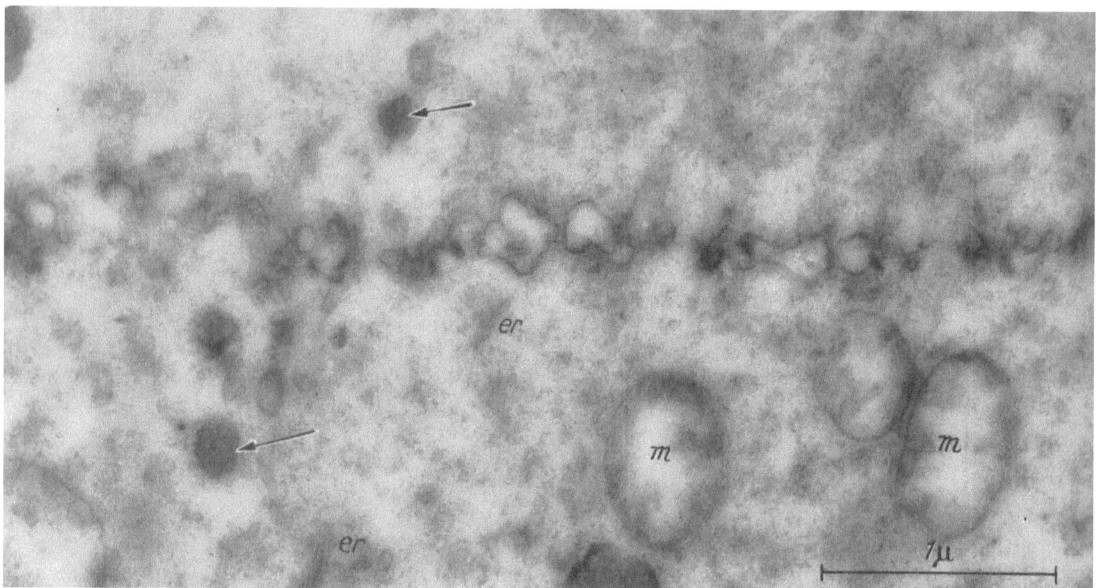

Fig. 3. Higher magnification of early cell plate. It appears as a row of vesicular elements possibly representing foci of lysis of cytoplasmic ground substance. An extreme edge of the plate is at the left. As division progresses the vesicles of the plate enlarge and fuse to give the more or less complete separation shown in Fig. 4. Arrows point to images of phragmosomes. Mitochondria are marked *m*. Profiles of small tubular elements of the endoplasmic reticulum are indicated at *er*. 32,000 ×

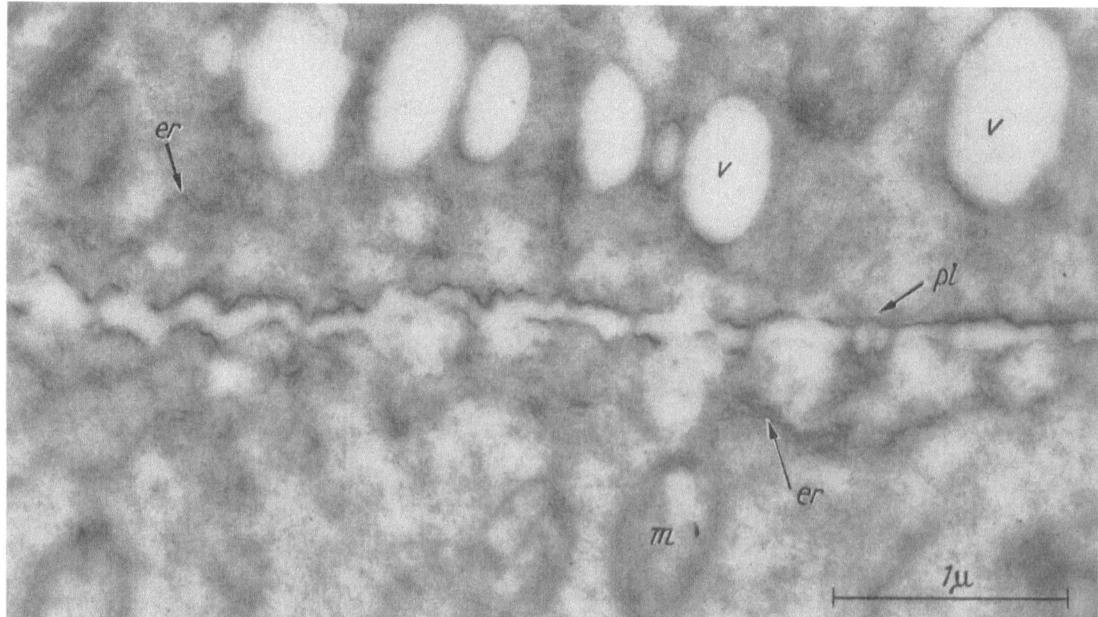

Fig. 4. Micrograph of late stage in plate development. Small connections between the cells persist as plasmadesmata (*pl*). Otherwise separation is complete. Elements of the endoplasmic reticulum are indicated by *er*, mitochondria by *m* and vacuoles by *v*. 32,000 ×

which serve to lyse the cytoplasmic matrix. Regardless of how they work, a correlation between their presence and plate development seems incontrovertible.

The only other structures showing any relation to the developing plate are certain elements of the endoplasmic reticulum. These appear in osmium-fixed preparations as slender (30 mμ in diameter) canalicular units, sometimes double, oriented normal to the plane of the plate. They seem to reach down to the plate from the interzonal region of the dividing cells. Later when the primary wall has been laid down, similar elements of the ER adopt an orientation parallel to the plate. Their intimate relation to plate formation, if any, has yet to be defined. Other evidence of the involvement of definable cytoplasmic structures or organizations in plate formation is too fragmentary to be taken seriously.

References

1. BECKER, W. A.: Acta soc. Bot. Poloniae 11, 139 (1934).
   BECKER, W. A.: Bot. Rev. 4, 446 (1938).
2. CAULFIELD, J. B.: J. biophys. biochem. Cytol. 3, 827 (1955).
3. BAILEY, I. W.: Amer. J. Bot. 7, 417 (1920).
4. SINNOT, E. W., and R. BLOCK: Amer. J. Bot. 28, 225 (1941).
5. ESAU, K.: Plant Anatomy. New York: John Wiley and Sons 1953.
6. BUVAT, R., and A. PUISSANT: C. R. Acad. Sci. (Paris) 247, 233 (1958).

Etude sur le champignon Allomyces macrogynus Em.

B. BLONDEL et G. TURIAN

Laboratoire de Biophysique et Institut de Botanique générale, Université de Genève

La basophilie du cytoplasme des cellules d'Allomyces est différemment localisée suivant le stade dans le cycle de développement de ce champignon auquel se trouvent ces cellules.

Chez les cellules mobiles (gamètes et zygotes) la totalité de la basophilie cellulaire est concentrée dans un corps paranucléaire. Celui-ci répond positivement aux tests cytochimiques pour l'acide ribonucléique (*1*). Lors de la germination du zygote le corps paranucléaire disparaît et la basophilie occupe l'ensemble du cytoplasme de la plantule (*2*). Il était donc intéressant de comparer la structure fine du cytoplasme non basophile des cellules mobiles avec celle du cytoplasme basophile de la plantule.

Méthodes. Le matériel est fixé à l'OsO$_4$, tamponné à p$_H$ 6, pendant environ 16 heures, puis généralement traité à l'acétate d'uranyle 0,5% pendant 2 heures. Il est ensuite deshydraté à l'acétone et inclus au polyester (*3*). Les coupes sont faites au couteau de verre sur un microtome selon KELLENBERGER (*4*).

Résultats. Conformément aux recherches précédentes (*5, 6*), nous trouvons dans le gamète que le corps paranucléaire est un organite distinct, dépourvu de mitochondries. Il est constitué de grains denses d'environ 170 Å et limité par une membrane double. Le cytoplasme périphérique (Fig. 1) présente une granulation fine (bien inférieure à 100 Å) dispersant faiblement les électrons. Il est parsemé de vésicules presque sphériques dont le contenu disperse moins les électrons que le cytoplasme environnant. Ce cytoplasme contient de grands granules noirs qui correspondent vraisemblablement aux granules phospholipidiques, et des mitochondries à cristae réguliers.

Le cytoplasme de la plantule en germination (Fig. 2) présente une structure fine différente de celui du gamète. Des grains relativement denses, d'environ 160 Å, sont uniformément répartis dans l'ensemble du cytoplasme basophile. Celui-ci contient de nombreuses vésicules et cisternae aux formes et dimensions variées, isolées ou en chaîne, qui sont les éléments du reticulum endoplasmique (*7*). Ces éléments n'ont pas semble-t-il, d'orientation spécifique. Le contenu des vésicules est normalement dépourvu de grains.

Nous avons donc dans le cytoplasme non basophile du gamète de ce champignon une composante «vésicules» (reticulum endoplasmique) et dans le cytoplasme basophile de la plantule: la

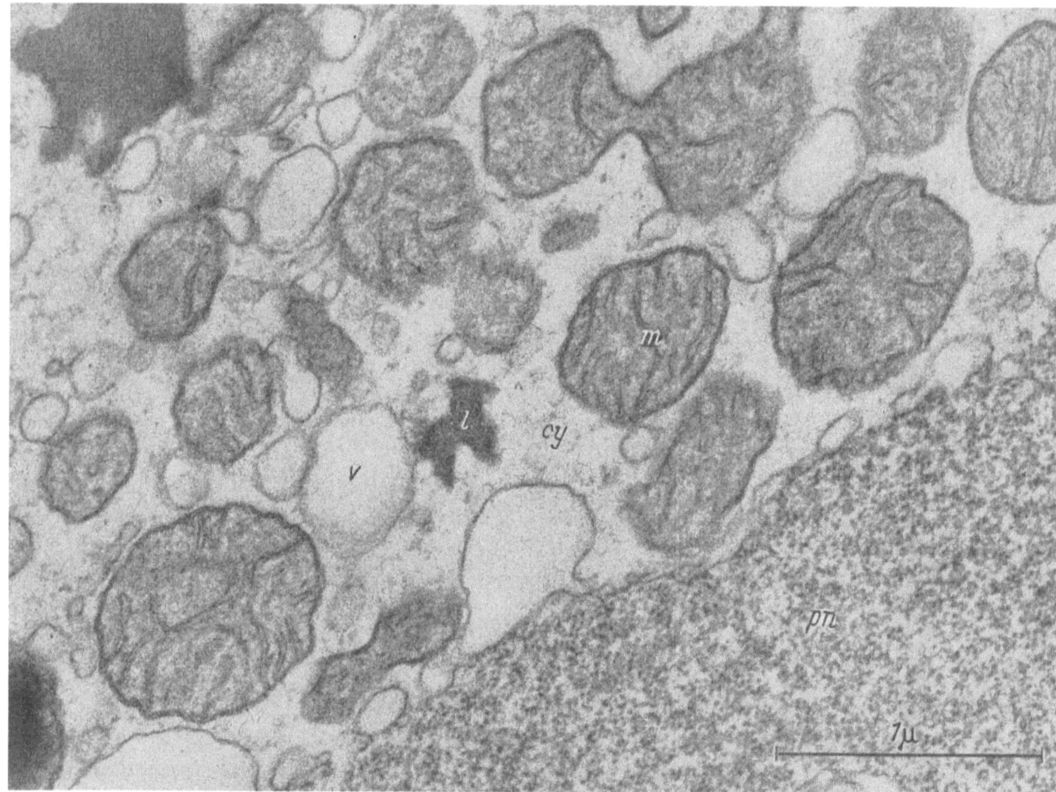

Fig. 1. Coupe d'un gamète d'Allomyces. Le corps paranucléaire basophile (*pn*) est constitué de grains denses d'environ 170 Å. Le cytoplasme non basophile (*cy*) présente une granulation fine; il est parsemé de vésicules (*v*). Les mitochondries (*m*) sont à cristae réguliers. Granules phospholipidiques (*l*)

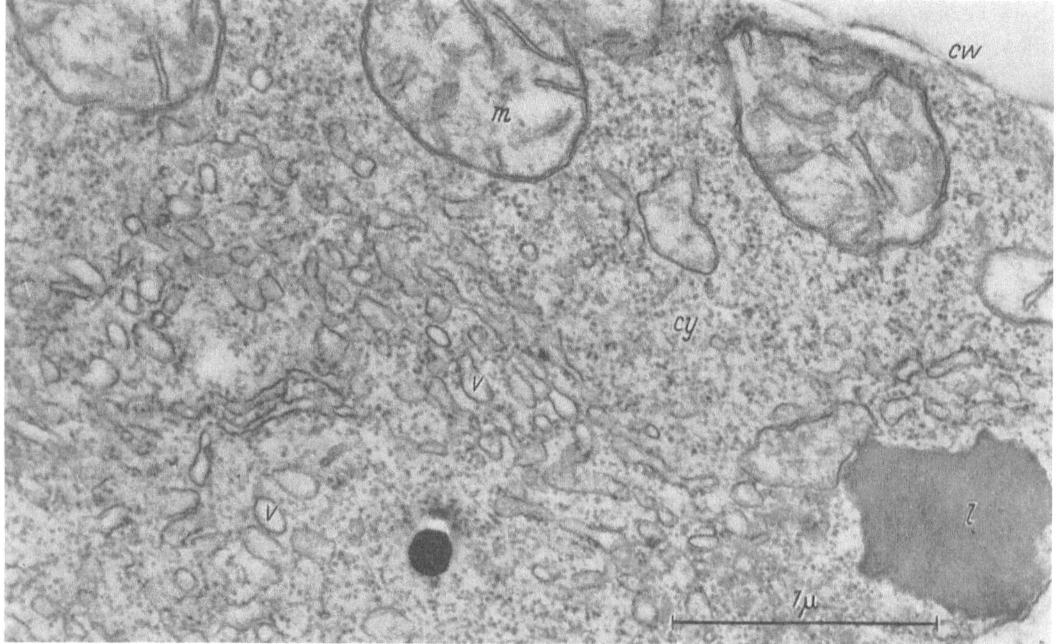

Fig. 2. Coupe d'une plantule d'Allomyces. Des grains denses d'environ 160 Å sont uniformément répartis dans tout le cytoplasme basophile (*cy*). Vésicules et cisternae (*v*) isolées ou en chaînes. Mitochondries (*m*) à cristae irréguliers et peu nombreux. — Granules phospholipidiques (*l*) — Paroi cellulaire (*cw*)

composante « vésicules » plus la composante «grains» répartie dans tout le cytoplasme. Il paraît donc évident ici que la basophilie du cytoplasme est portée par ces grains.

Nous constatons en outre que dans la plantule, les mitochondries n'ont pas le même aspect que dans le gamète. Les mitochondries de la plantule ont des cristae irréguliers, peu nombreux, généralement courts, et accompagnés de tubuli.

Cette recherche fut effectuée grâce à un subside du Fonds National Suisse de la Recherche Scientifique.

Bibliographie

1. Turian, G.: C. R. Acad. Sci. (Paris) **240,** 2343 (1955).
2. — Experientia **12,** 24 (1956).
3. Ryter, A., et E. Kellenberger: Ces C. R. p. 52
4. Kellenberger, E.: Experientia (Basel) **12,** 282 (1956).
5. Turian, G., et E. Kellenberger: Exp. Cell. Res. **11,** 417 (1956).
6. — — Proc. Stockholm Conf. on Electron Microscopy p. 276, 1956.
7. Palade, G. E.: J. biophys. biochem. Cytol. **2,** Suppl. 85 (1956).

J. Ergebnisse der Elektronenmikroskopie in der Mikrobiologie

1. Protozoologie

Ultrastructure of the pellicle and the nucleus of Leishmania donovani

J. Chakraborty and N. N. Das Gupta

Biophysics Division, Institute of Nuclear Physics, Calcutta (India)

In a number of publications (*1, 2, 3, 4, 5*) from this laboratory from 1951 to 1958, we reported on the phase, electron microscopic and cytochemical observations on *Leishmania donovani*, the Kala-azar parasite. Using water disruption, acid hydrolysis, and ultra-thin sectioning techniques, a number of protoplasmic constituents of this protozoon were studied under the electron microscope. In two recent papers, Chang (*6*) and Inoki et al. (*7*) have also described electron microscopic observations on the structural organisation of this protozoon.

The present report deals with the ultrastructure of the pellicle and nucleus of *Leishmania donovani* as observed using three different fixatives.

Cultures of *Leishmania donovani* were maintained continuously in NNN medium. 8—10 days old cultures were used in the present investigation. The parasites were concentrated by repeated centrifugation in normal saline. They were then fixed for 30 min in three alternative ways:

A. In 1% OsO_4 solution at p_H 7.4 according to the method of Palade (*8*) or Sjöstrand (*9*).
B. In 1% OsO_4 in buffered potassium dichromate at p_H 7.2.
C. In isotonic 45% acetic acid solution.

After fixation, the specimens were dehydrated in ethyl alcohol (30, 50, 70, 90 100%) and then impregnated with a mixture of 4 parts (by volume) of butyl methacrylate, 1 part (by volume) of methyl methacrylate and 2% (by weight) of Luperco CDB. After complete polymerisation (overnight 48° C), about 500 Å thick sections were cut with a Porter Blum microtome. Sections from specimens fixed according to method C above, were finally stained for 30 min in OsO_4 solution. A Siemens' Elmiskop I at 60 kV was used.

Pellicle: The pellicle or the periplast of *Leishmania donovani* is a complex structure. It consists of an outer and inner layer (*Po* and *Pi* in Fig. 1); the combined thickness of the two layers is about 400 Å with a separation of 90 Å. The outer coat is thin and continuous but the inner one is composed of a large number of fine long fibrils running obliquely over the body of the protozoon (*Pf* in Figs. 2a, 5a and 6). In transverse sections, the inner layer therefore appears to be composed of a number of well spaced dots (*Pi* in Fig. 1), or if the section is slightly more tangential, small fragments of the fibrils may also be visible (Figs. 1 and 6). These fibrils are about 140 Å thick, the separation between adjacent fibrils is also of the same order. The parallel arrangement of the pellicular fibrils can be seen more clearly in Fig. 2a where a portion of the pellicle has been cut tangentially. Fig. 2b is a contrast print of the portion within the marked rectangle in Fig. 2a. In

---

Explanation of symbols:

C Cristae within mitochondria; *Ch* Chromosomes; *f* Fibrillar connections between separating chromosomes; *Fl* Flagellum; *K* Karyosome; *Ld* Lipid droplets; *M* Mitochondria; M_1 and M_2 Elongated mitochondria; M_3 and M_4 Oval or round mitochondria; *N* Nucleus; *Nm* Nuclear membrane; *Nmo* Outer layer in the nuclear membrane; *Nmi* Inner layer in the nuclear membrane; *Nmp* Porous layer within the nuclear membrane *P* Pellicle; *Po* Outer pellicular layer; *Pi* Inner pellicular layer; *Pf* Pellicular fibrils; *Pb* Parabasal body; *Tc* Transverse connections between pellicular fibrils; *V* Vacuole

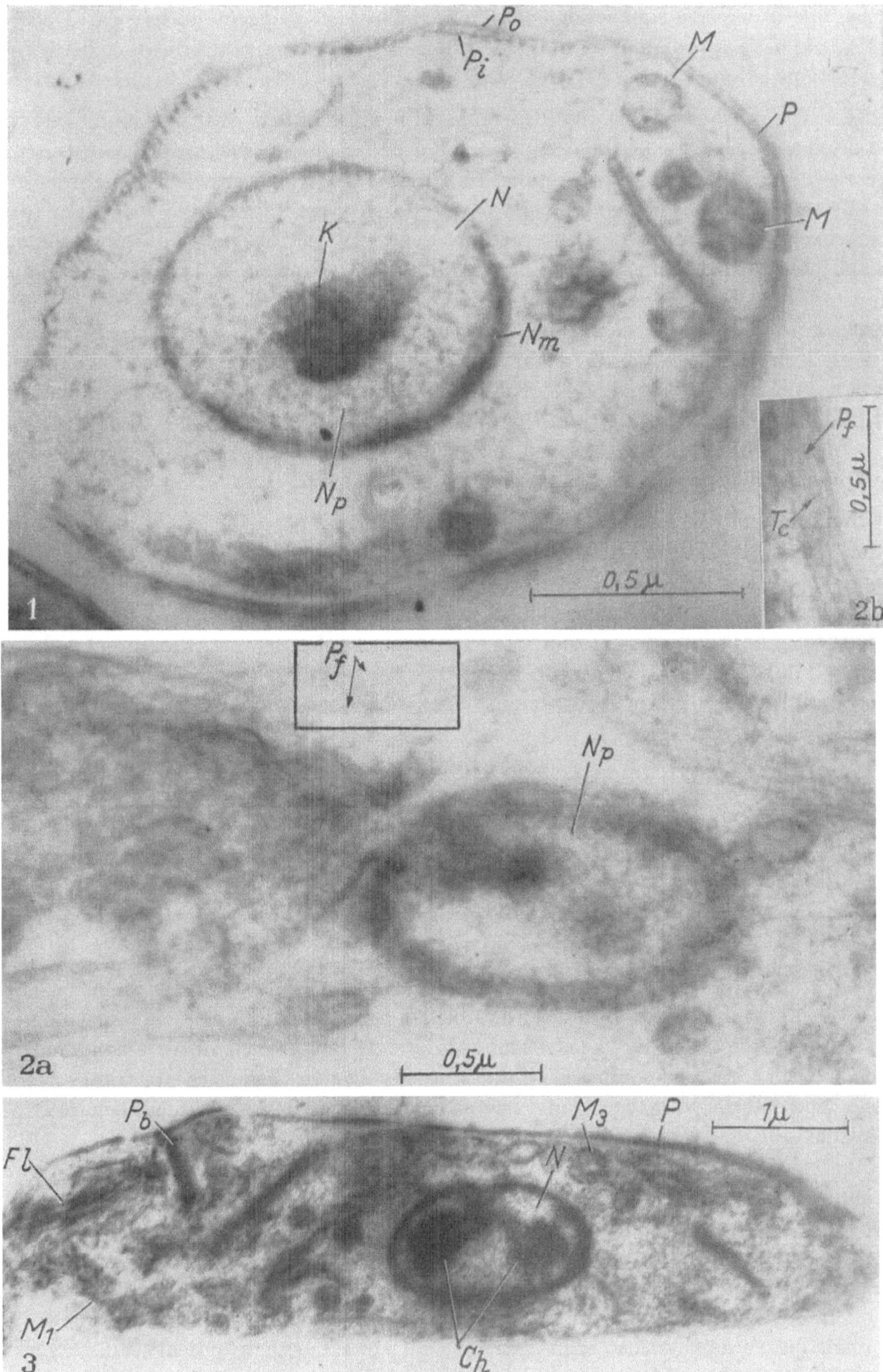

Fig. 1. Transverse section of *L. donovani*, with double pellicular layers, the outer layer *Po* is continuous and the inner *Pi* appears as a dotted line (Fixative A)

Fig. 2a. Partially tangential section through the pellicle, showing the parallel fibrils *Pf* (Fixative A)

Fig. 2b. A contrast print of the portion of Fig. 2a within the marked rectangle. The parallel fibrils *Pf* and the transverse connections *Tc* between adjacent fibrils are clearly visible

Fig. 3. Longitudinal section with two sets of chromosomes at two opposite poles and a slight kink in the nuclear membrane just before the nuclear division (Fixative A)

(Explanation of symbols see page 510)

this print, fine transverse connections can be seen between adjacent fibrils. These connections (Tc in Fig. 2 b) have approximately the same width as the fibrils and the distance between successive connections is found to vary from 500 to 2000 Å.

Nucleus. The nucleus in this species is generally a prominent oval or round body which is differentiated from the rest of the cytoplasm by a dense nuclear membrane Nm which may be seen clearly in all the micrographs reproduced in this paper. The nuclear membrane seems to consist of two separate continuous layers (Nmo and Nmi in Figs. 7 and 8). These two layers together with a fine gap between them measure about 250 Å. A honeycomb like porous layer is found associated with the inner layer (Nmp in Fig. 8). The porous layer has an extension of about 900 Å.

Compared to the nuclear membrane, the mass of nucleoplasm appears to be much lighter (Np in Figs. 1, 2 and 4). Very often a single dense body, karyosome, is found within the nucleus. It may be round, oval or elongated (K in Figs. 1, 8 and 5a). The nuclear material has also been found to be differentiated into a number of dense bodies (Ch in Figs. 3, 4, 6 and 9); four such bodies may be identified in Fig. 6, seven in Fig. 4 and eight in Fig. 9. These are the separating chromosomes, fibrillar connections between which are noticable at f in Fig. 10.

Mitochondria and other structures. The mitochondria in this species are well differentiated structures. They are scattered all over the body at random (M in Figs. 1, 3 and 4). These are sometimes long and filamentous (M_1, M_2 in Figs. 4 and 12) and sometimes oval or rounded (M_3, M_4 in Figs. 4 and 12). The mitochondria are found to be surrounded by a membrane about 120 Å thick and are usually denser than the cytoplasm. In thin sections the mitochondria appear to contain a number of tubular bodies (C in Fig. 11), the "cristae" as reported by Inoki et al. (7). A very prominent tubular mitochondrium just close to the flagellum has been found in all the three longitudinal sections reproduced in this paper (M_1 in Figs. 3, 4 and 12). In contrast to the mitochondria which seem to show some internal structures, dense round bodies are often found within the body of the parasite (Ld in Fig. 4). These are the lipid droplets.

The parabasal body (Pb in Figs. 3, 4 and 12) consists of a transversely banded structure within a vacuolated space and surrounded by a distinct membrane. It does not seem to have any apparent connection with the flagellum (Fl in Figs. 3 and 4). Apart from the flagellar and parabasal vacuole, a number of smaller vacuoles (V) with sharp membranes are also seen scattered over the body of the protozoon in Fig. 11.

Discussion. Chang (6) first reported about the double layered nature of the pellicle in *Leishman donovan bodies*, but she failed to detect the fibrillar inner membrane in the leptomonad form. Pyne (3) observed these fibrils in *Leishmania donovani*. The present observation confirms that the inner pellicular layer is composed of fine fibrils, which in their turn are transversely interconnected. These fibrils with their transverse connections possibly play an important role in the body movement of the leptomonad form.

A porous structure within the nuclear membrane has been reported by Bairati and Lehmann (10), Harris and James (11) and Greider et al. (12, 13) in *Amoeba proteus*. This is the first time that it has been observed in *L. donovani*. It should, however, be mentioned that in the majority of the micrographs, the nuclear membrane has not been so well resolved as to make visible the separate layers seen in Fig. 8 and described above.

The separated dense bodies, within the nucleus have been identified as chromosomes, between which connecting fibrils have also been noticed (Fig. 10). From their optical studies, Sen Gupta and Ray (14) detected six chromatin bodies within the nucleus of *L. donovani*. In the present study a maximum of eight such bodies has been observed.

Of the three fixatives used, C seemed to be a good fixative for revealing the chromosomes and the fibrillar connections between them (Figs. 7 and 10). Specimens in Figs. 5 and 6 were fixed in the fixative B. Although the pellicular fibrils can be seen clearly in these figures, structural details do not seem to be so well preserved. Specimens in the rest of the figures were all fixed in the fixative A. This seemed to produce uniformly good results for revealing the finer morphological details.

The authors are indebted to the Ministry of Scientific Research and Cultural Affairs, Government of India for research grants and to Shri M. L. De and Shri C. K. Pyne for help in the electron micrographs.

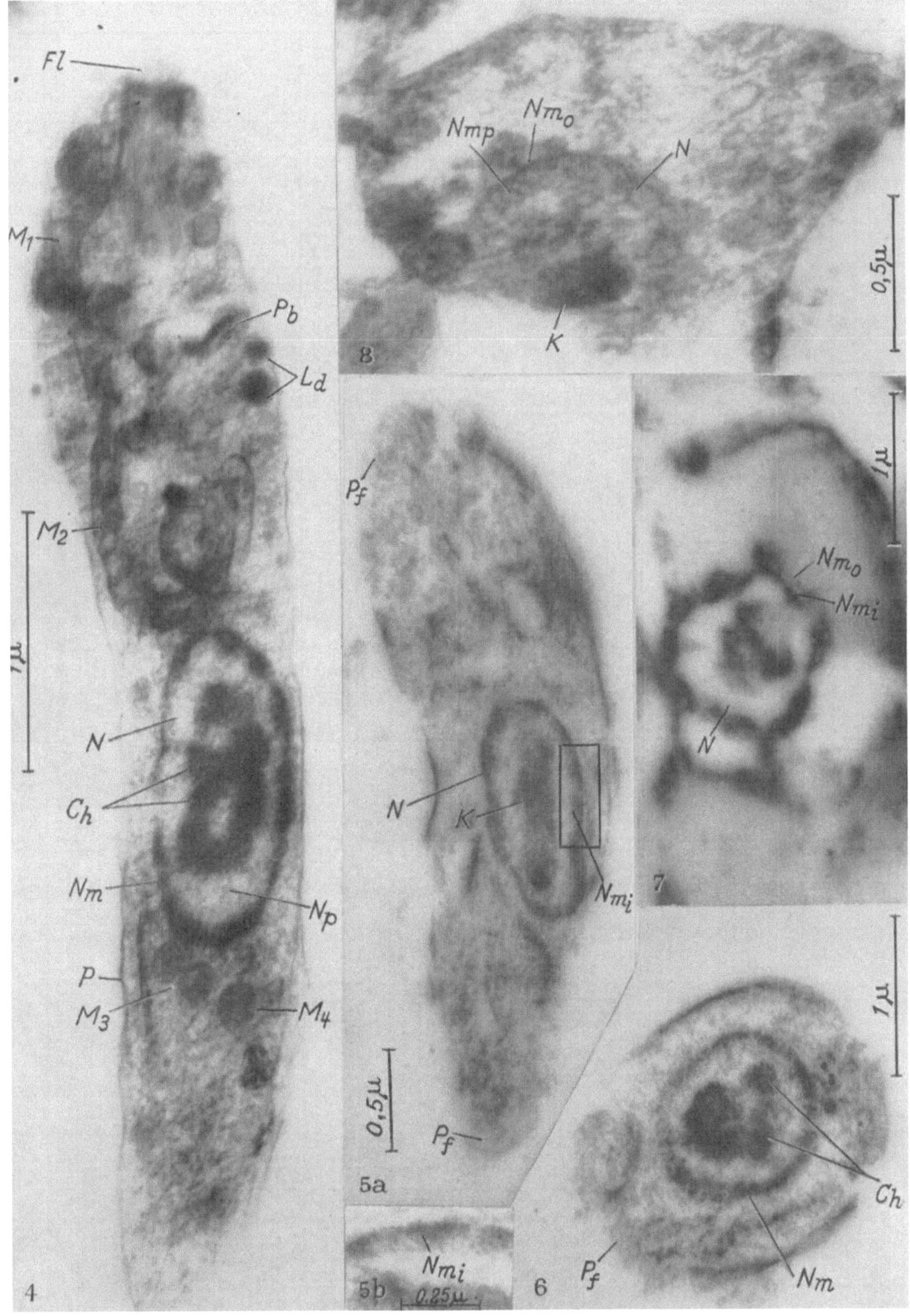

Fig. 4—8

Fig. 4. Prominent oval nucleus N with 7 chromosomes Ch. (Fixative A) — Fig. 5a. Longitudinal section showing the pellicular fibrils Pf and the nucleus with elongated karyosome K. (Fixative B) — Fig. 5b. An enlarged print of the marked portion of nuclear membrane in Fig. 5a. (Fixative B) — Fig. 6. Transverse section, four chromosomes Ch are visible within the nucleus. (Fixative B) — Fig. 7. Transverse section showing the double walled nuclear membrane with a narrow space in between them. (Fixative C) — Fig. 8. Oblique section of $L.$ $donovani.$ The outer nuclear membrane Nmo is continuous and the inner layer appears to be porous. The pores Nmp are arranged in a regular honey-comb like manner. (Fixative A). Explanation of symbols see page 510

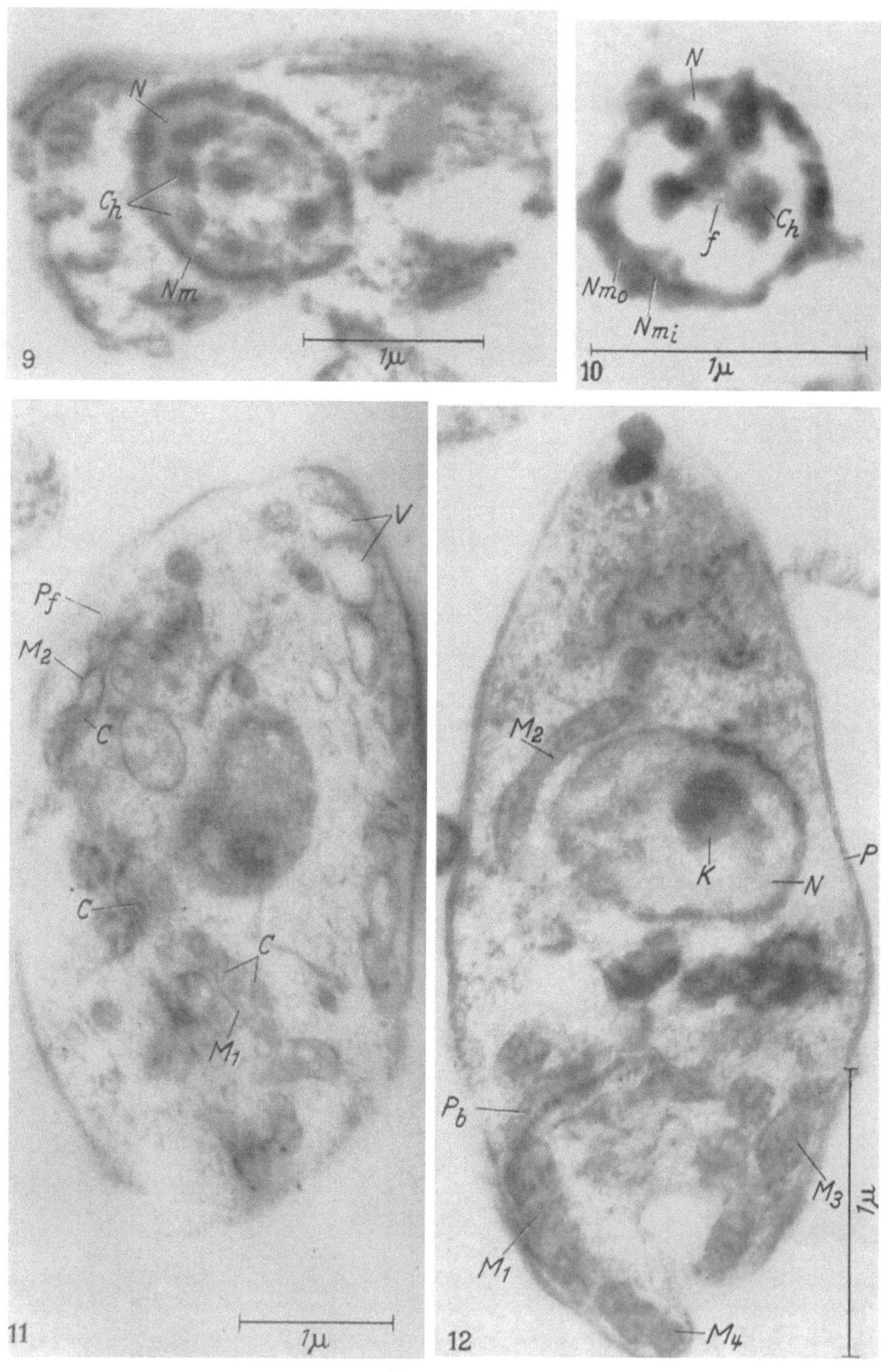

Fig. 9—12

Fig. 9. Nucleus with 8 chromosomes and a thick nuclear membrane. (Fixative B)

Fig. 10. Nucleus with inner and outer walls, five separated chromosomes and inter-connecting fibrils f between them. (Fixative C)

Fig. 11. Elongated and oval mitochondria with cristae C inside them, prominent vacuoles. (Fixative A)

Fig. 12. Longitudinal section showing the elongated, oval and round mitochondria with prominent mitochondrial membrane surrounding M_2. (Fixative A). Explanation of symbols see page 510

References

1. Das Gupta, N. N., A. Guha and M. L. De: Exp. Cell Res. **6,** 353 (1954).
2. Guha, A., C. K. Pyne and B. B. Sen: J. Histochem. and Cytochem. **2,** 212 (1955).
3. Pyne, C. K.: Exp. Cell Res. **14,** 388 (1958).
4. — and J. Chakraborty: J. Protozool. **5,** 264 (1958).
5. Sen Gupta, P. C., N. N. Das Gupta and D. L. Bhattacharya: Nature (Lond.) **167,** 1063 (1951).
6. Chang, P. C. H.: J. Parasitol. **42,** 126 (1956).
7. Inoki, S., K. Nakanishi, T. Nakabayashi and M. Ohno: Med. J. Osaka Univ. **7,** 719 (1957).
8. Palade, G. E.: J. exp. Med. **95,** 285 (1952).
9. Sjöstrand, F. S.: In: Physical Techniques in Biol. Res. **3,** 241 (1956).
10. Bairati, A., and F. E. Lehmann: Experientia (Basel) **8,** 60 (1952).
11. Harris, P., and T. W. James: Experientia (Basel) **8,** 384 (1952).
12. Greider, M. H., W. J. Kostir and W. J. Frajola: J. biophys. biochem. Cytol. **2,** Suppl. 445 (1956).
13. — — — J. Protozool. **5,** 139 (1958).
14. Sen Gupta, P. C., and H. N. Ray: Proc. Zool. Soc. **7,** 113 (1954).

Fine structure of the kinetoplast in a trypanosomid flagellate

B. A. Newton

Medical Research Council Unit for Chemical Microbiology, Biochemical Laboratory, University of Cambridge
(England)

The kinetoplast is a disc shaped organelle (ca. 1 μ in diameter) which lies adjacent to the blepharoplast or basal granule in trypanosomid flagellates. The blepharoplast is generally regarded as the starting point of the flagellum and it has long been known that the blepharoplast and the kinetoplast are intimately related (*1*). Both these organelles reproduce by bipartition at the onset of cell division. The function of the kinetoplast remains unknown. The work to be described in the present communication was undertaken in an attempt to learn something of the structural inter-relationships existing between the flagellum, blepharoplast and kinetoplast in the trypanosomid flagellate *Strigomonas oncopelti*.

Intact organisms were fixed in 1% buffered osmium tetroxide as described by Palade (*2*); 30 min fixation proved to be the most satisfactory. After dehydration and embedding in methyl-methacrylate sections were cut and examined in the Siemens 'Elmiskop' operating at 80 kV; instrumental magnifications of 5000 ×, 10,000 × and 20,000 × were used.

Examination of thin sections has revealed that the kinetoplast of *S. oncopelti* consists of two main parts; a central electron-transparent (*ET*) area and a more electron-dense (*ED*) peripheral area. The central area contains an intricate network of filaments and granules (Fig. 1—3). The diameter of the individual filaments ranged from 50—150 Å; the number of granules was found to vary considerably according to the region of the kinetoplast sectioned. Data obtained from the study of a large number of sections suggests that this network may originate from or be a continuation of the nine peripheral fibres of the flagellum (the two axial fibres terminating at a point anterior to the kinetoplast). The electron-dense material observed at the periphery of the kinetoplast resembles nuclear material in appearance and the width of this peripheral band depends upon the position of sectioning; little or no peripheral material is found in sections cut through the anterior end of the kinetoplast (Fig. 2). These observations are consistent with the electron-dense material being present as a 'cap' over the posterior end of the kinetoplast. Light microscopy of stained preparations of *S. oncopelti* and numerous other trypanosomid flagellates (*1, 3, 4, 5*) has shown that the kinetoplast has a Feulgen-positive periphery and an achromatic central region; with this in mind it is suggested that the electron-dense material seen in thin sections may represent the Feulgen-positive material of this structure. A more detailed account of this work is to be published (*6*).

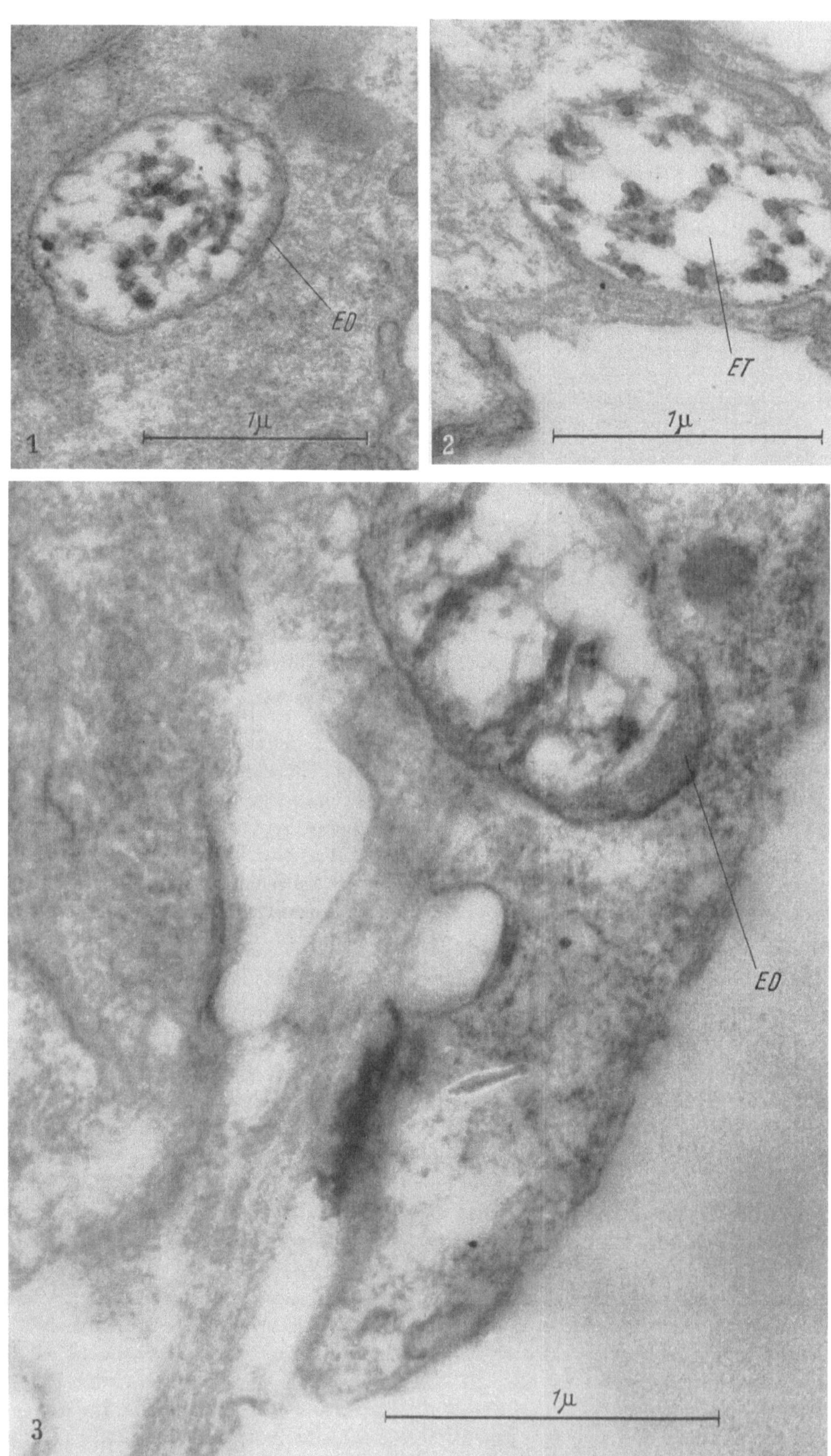

Fig. 1 and 2. Transverse sections through kinetoplasts showing the complex network of filaments and granules. Fig. 1 shows electron-dense (*ED*) peripheral material which resembles nuclear material in appearance

Fig. 3. A longitudinal section showing the flagellum within the cell and internal structure of the kinetoplast

References

1. Robertson, M.: Parasitol. **19**, 375 (1927).
2. Palade, G. E.: J. exp. Med. **95**, 285 (1952).
3. Bresslau, E., u. L. Scremin: Arch. Protistenk. **48**, 509 (1924).
4. Barrow, J. H.: Trans. Amer. Microsc. Soc. **73**, 242 (1954).
5. Lwoff, A., et M. Lwoff: Bull. biol. **65**, 170 (1931).
6. Horne, R. W., and B. A. Newton: Exp. Cell Res. **15**, 103 (1958).

Contribution à la cytologie d'Euglena viridis

Gérard de Haller

Institut de Zoologie de l'Université de Genève

Dans la série des travaux qui ont paru au sujet des Protozoaires, des Flagellés en particulier, étudiés au microscope électronique, peu traitent des Euglènes. Ceux de Wolken et Palade (*1*) et de Roth (*2*) présentent cependant des micrographies de coupes d'Euglènes, montrant les plastides, les mitochondries, le stigma, des fibrilles, le flagelle. Il s'agit dans les deux cas d'Euglena gracilis. En présentant mes observations sur Euglena viridis, je me référerai entre autres à ces publications.

Méthode: Cultivées en milieu liquide stérile, à la lumière naturelle, les Euglènes (forme active) ont été fixées au tétroxyde d'osmium et incluses au Vestopal (polyester) selon la méthode mise au point à Genève pour l'étude du noyau microbien (*3*). J'ai utilisé, au Centre de Microscopie électronique de l'Institut de Physique de l'Université de Genève, un microtome à avance mécanique (selon Danon et Kellenberger, modifié) et à couteau de verre, et un microscope RCA type EMU 2D.

Observations. Le *protoplasme* de E. viridis contient de nombreuses vésicules de toute forme et de toute dimension, parfois très allongées (100×5000 Å). L'intérieur de ces formations est toujours dépourvu de composant granuleux. Ces diverses structures constituent le *réticulum endoplasmique* (Fig. 1:*RE*), et sont parfaitement comparables à ce qui a été décrit chez d'autres Protistes (*4*).

Les micrographies électroniques montrent le *noyau* entouré d'une double membrane percée de pores. Le nucléoplasme granuleux contient des masses sombres irrégulières de $0,5 \times 1\,\mu$ en moyenne, qui représentent les *grains Feulgen-positifs* du noyau. Au centre du noyau, le *nucléole* est constitué d'un plasma clair et très finement granuleux et d'une partie réticulée plus dense rappelant un nucléolonéma.

Les *mitochondries* des Euglènes sont de forme allongée, irrégulière, parfois ramifiée. Contrairement à celles d'autres Protozoaires (*4*), elles sont du type à *cristae*. Chez E. viridis (Fig. 1 et 3:*M*) le diamètre maximum des sections de mitochondries est d'environ $1\,\mu$, leur longueur ne peut se mesurer sur des coupes minces. Les cristae sont généralement rectilignes, parfois très longues, et orientées dans tous les sens. Wolken et Palade (*1*) les avait montrées, chez E. gracilis, courtes et perpendiculaires à la paroi, alors que celles que présente Roth ont l'aspect décrit ici (*2*).

Les *corps de* Golgi (*10*) (Fig. 1:*G*) sont répartis en grand nombre dans tout le cytoplasme. Ils sont dépourvus de zone à grosses vacuoles, les vésicules périphériques ne dépassant pas $0,2\,\mu$ de diamètre. L'empilement des doubles membranes est régulier, leur section arquée. Le diamètre d'un dictyosome, c'est-à-dire la longueur maximum des lames sur les coupes, mesure $1,5\,\mu$, sa hauteur 0,5 à 1, $2\,\mu$, comprenant 18 à 50 lames, soit 9 à 25 poches.

Les *chloroplastes* (Fig. 1:*Pl*) sont entourés d'une membrane double. Les *lames* qui les composent sont constituées elles-mêmes de *4 lamelles* de 40 Å d'épaisseur, paraissant doubles par endroits, et généralement masquées par une substance granuleuse qui les recouvre. Jamais, dans les conditions de culture adoptées ici pour E. viridis, les lames ne s'épaississent ni ne se réunissent en granum, pas plus qu'il n'apparaît de pyrénoïde, alors que Wolken et Palade (*1*) montraient un granum (qu'ils appelaient pyrénoïde) dans les chloroplastes de E. gracilis. (Cependant Baker en 1933 (*5*) a obtenu à volonté des Euglènes avec ou sans pyrénoïde selon l'état des cultures.) On sait que chez l'Euglène le *paramylon* est toujours extra-plastidien; au microscope électronique il a le même aspect homogène et transparent que l'amidon. Chaque grain, de 3 μ de diamètre, se trouve dans une petite vacuole limitée par une fine membrane.

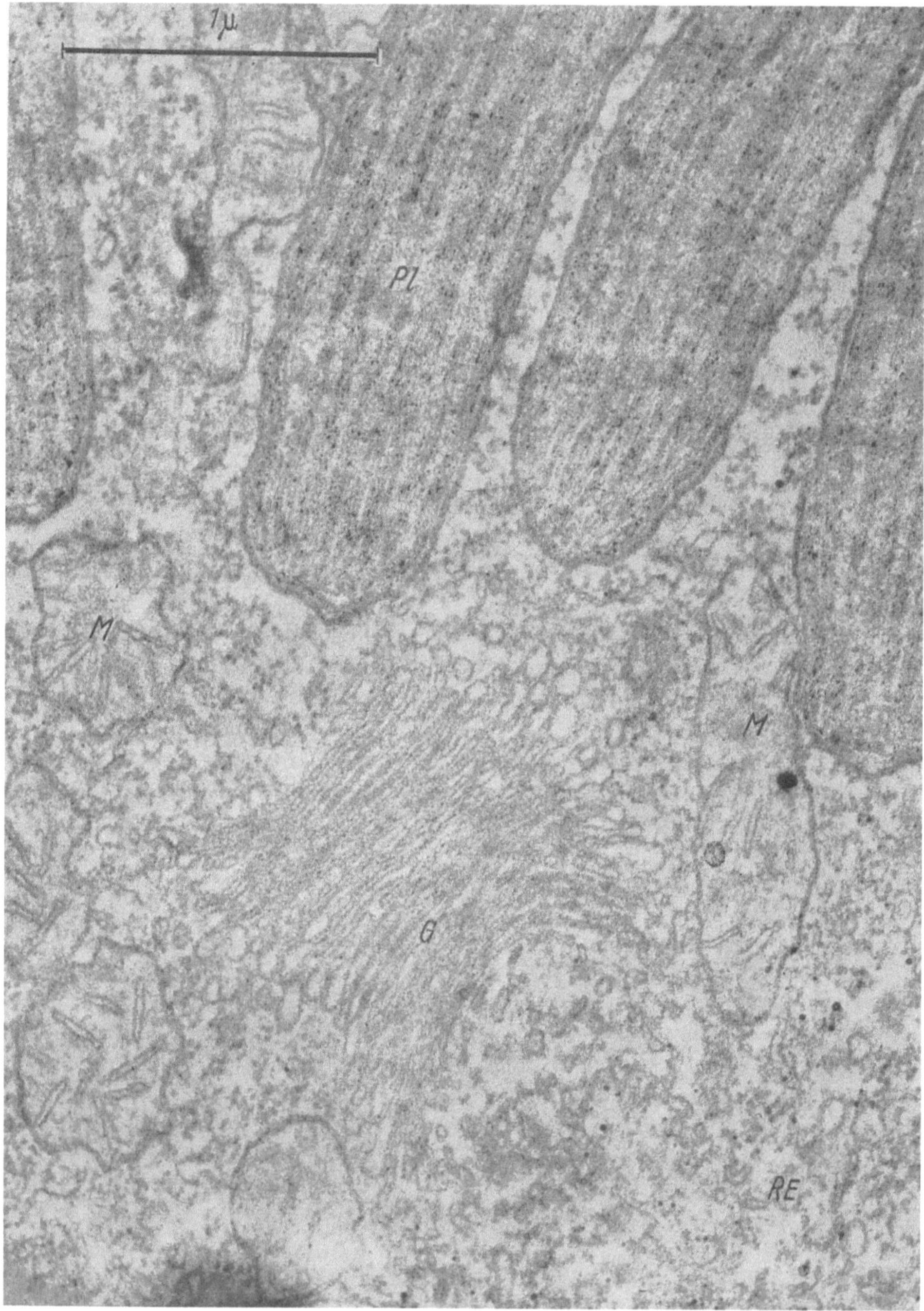

Fig. 1. Cytoplasme d'E. viridis: réticulum endoplasmique, de type vésiculaire (*RE*); corps de GOLGI, dépourvu de zône à grosses vacuoles (*G*); mitochondries très irregulières, du type à cristae (*M*); chloroplastes, sans grana (*Pl*)

Le *stigma*, situé dans le cytoplasme périvestibulaire, du côté dorsal, est composé de sphérules denses et homogènes de 0,2 à 0,3 μ de diamètre, disposées sans ordre précis en un amas en forme de cupule évasée sur ses bords, d'un diamètre total de 3,5 μ (Fig. 2: *St*). Comme cela ressort déjà des micrographies de Wolken et Palade et de Roth sur l'espèce gracilis, le stigma ne fait partie d'aucun plastide.

Ce qu'on appelle le *photorécepteur* (Fig. 2: *Ph*) se compose d'une masse protoplasmique claire, renflement de l'une des racines flagellaires, et, accollée à cette masse, d'une sorte de *lentille* à peu près plan-convexe, homogène et dense, de 1,5 μ de diamètre, désignée par Roth, comme "intra-flagellar swelling". Il est situé dans le goulot, face a la concavité du stigma. La masse claire contient

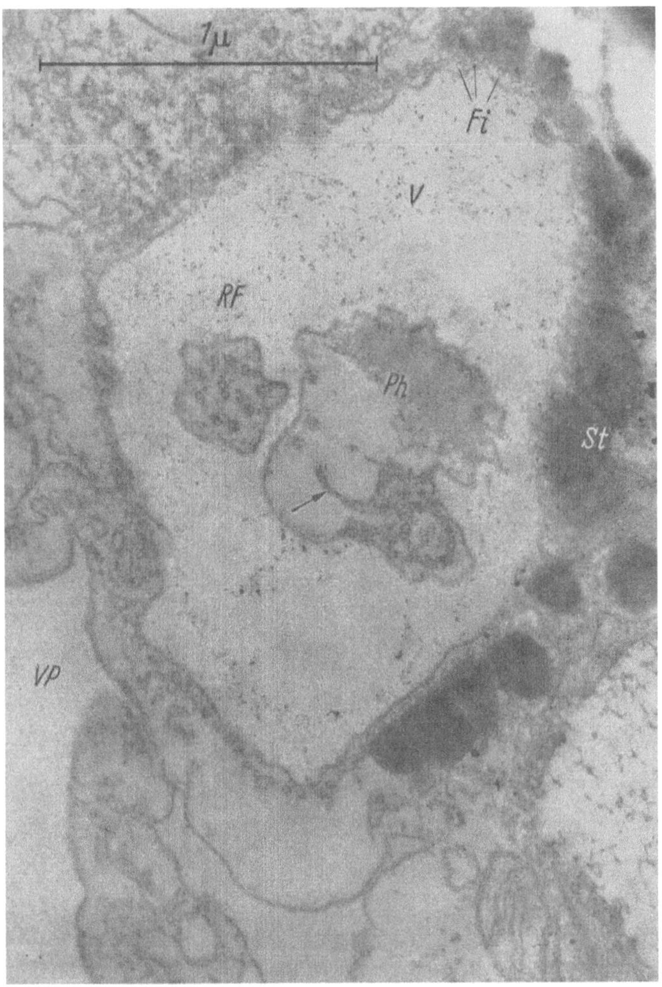

Fig. 2. Coupe d'E. viridis, perpendiculaire au goulot, au niveau du photorécepteur. Vestibule (*V*); racines flagellaires (*RF*); photorécepteur, renflement d'une des racines flagellaires (*Ph*); la flèche indique la lame courbe caractéristique; stigma (*St*); fibres sous-cuticulaires (*Fi*); vacuole pulsatile (*VP*)

des *lames contournées* (Fig. 2: flèche) d'aspect très constant et caractéristique, mais dont le rôle éventuel m'échappe.

Le *flagelle* présente la structure typique de 9 paires de fibrilles périphériques et une paire centrale. La membrane qui l'entoure est lâche (Fig. 2: *RF*, racines flagellaires). Des *mastigonèmes* latéraux très fins sont visibles sur la portion extérieure du flagelle.

Les *corps mucifères* ne sont en rien comparables à des trichocystes (*6, 7*). Ce sont des vacuoles de 2,5 × 1 μ en moyenne, à contenu grossièrement coagulé et à membrane simple, épaisse de 100 Å (Fig. 3: *CM*). La fixation et l'inclusion semblent les dilater passablement.

Roth en 1958 (*2*) décrit sous la cuticule à l'intérieur du réservoir des fibrilles parallèles de 210 Å de diamètre. Il en est de circulaires et de longitudinales. Sous la cuticule externe, cet auteur en montre des séries parallèles aux sillons cuticulaires chez Peranema trichophorum, mais pas chez E. gracilis. J'ai retrouvé chez E. viridis aussi bien les fibrilles du réservoir (Fig. 2:*Fi*) que celles qui semblent renforcer la cuticule externe (Fig. 3a:*Fi*). On en voit deux ou trois dans chaque espace compris entre deux sillons cuticulaires alors que sous la cuticule du goulot et du réservoir, placées les unes à côté des autres, elles forment une véritable palissade. L'axostyle de Pyrsonympha tel que le décrit Grasse en 1956 (*8*) est composé de fibres semblables mais disposées en couches nombreuses.

La cuticule est en outre soulignée dans le cytoplasme, à environ 2 μ de profondeur, par une couche simple de *vésicules* sphériques ou ovales, juxtaposées, visibles sur des coupes d'orientation quelconque comme une *chaînette* (Fig. 3a, b, c: *Ves*). Les éléments situés sous les sillons qui parcourent la cuticule sont plus volumineux que les autres. Ils atteignent 0,2 μ de long diamètre. J'ignore la signification de ces formations. Ont-elles un rapport avec les argyrosomes (Silberliniensystem) ? (*9*).

Une étude plus approfondie de ces structures cellulaires d' E. viridis paraîtra prochainement.

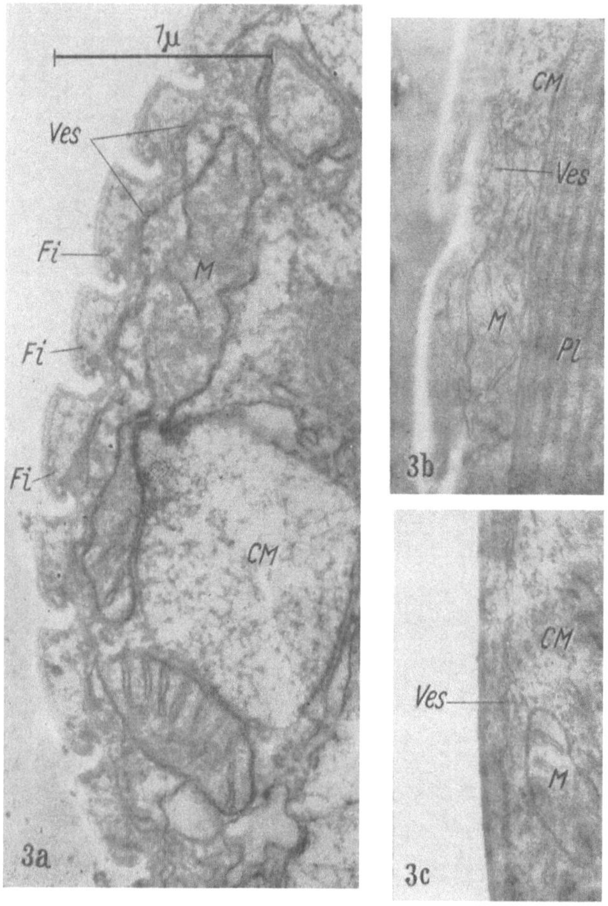

Fig. 3a—c. Coupes perpendiculaires à la cuticule: a) perpendiculaire aux sillons de la cuticule, b) oblique par rapport aux sillons, c) parallèle aux sillons. Fibrilles sous-cuticulaires (*Fi*); vésicules formant une nappe sous la cuticule, et dont les éléments situés sous les sillons sont plus importants que les autres (*Ves*); mitochondries (*M*); corps mucifères, à contenu grossièrement coagulé (*CM*); chloroplast (*Pl*)

Bibliographie

1. Wolken, J. J., and G. E. Palade: Ann. N. Y. Acad. Sci. **56**, 5, 873 (1953).
2. Roth, L. E.: J. Ultrastr. Res. **1**, 3, 223 (1958).
3. Ryter, A., et E. Kellenberger: Ces. C. R. p. 52
4. Rouiller, CH.: Scientific Instruments (RCA-News) **2**, 2, 1 (1957).
5. Baker, C. L.: Arch. Protistenk. **80**, 434 (1933).
6. Jakus, M. A., et C. E. Hall: Biol. Bull. **91**, 1, 141 (1946).
7. Dragesco, J.: Bull. Microsc. appl. IIe p. **2**, 92 (1952).
8. Grasse, P. P.: Arch. Biol. **67**, 595 (1956).
9. Chadefaud, M., et J. Arlet: C. R. Acad. Sci. (Paris) **219**, 220 (1944).
10. Grasse, P. P.: Nature (Lond.) **179**, 31 (1957).

The ultrastructure of the chromatoid bodies in Entamoeba invadens

K. Deutsch, V. Zaman and D. C. Barker

Department of Zoology, University of Edinburgh (England)

Entamoeba invadens is a parasite living in the intestine of reptiles. The cyst and the trophozoite contain rod- or bar-shaped inclusions which are up to 10 μ long. They are also found in other *Entamoebae* and are often called chromatoid bodies.

We have studied the morphology of *Entamoeba invadens* in the electron microscope (Fixative: osmium tetroxide, in one case also formalin). The micrographs show that the bodies consist of small particles which are arranged in a crystalline pattern. The diameter of the particles is about 200 Å. High magnification pictures reveal a fine structure of the particles (Fig. 1). The bodies are

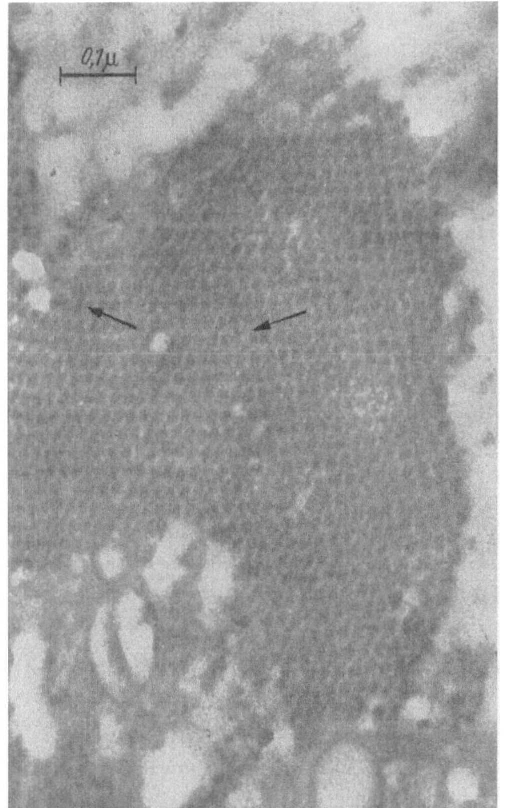

Fig. 1. Chromatoid body. Arrow indicates particles showing fine structure. Fixative: OsO_4

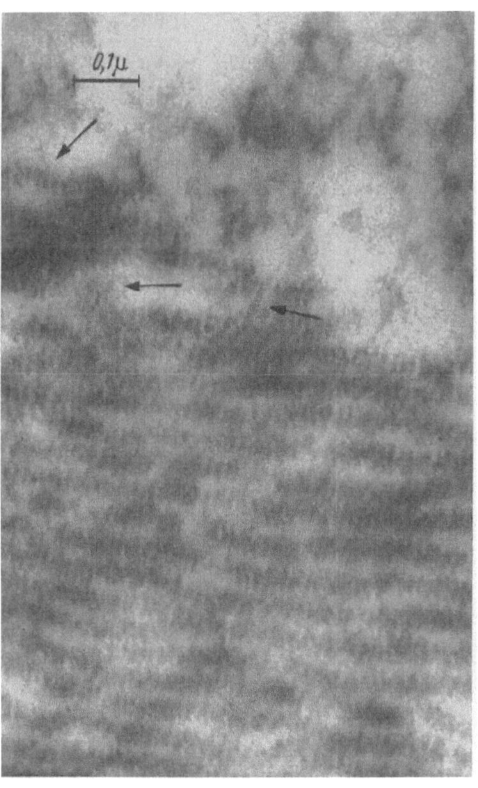

Fig. 2. Chromatoid body after treatment of the living organism with ribonuclease. Arrows indicate particles showing fine structure. Fixative: OsO_4

smaller and more numerous in the trophozoite than in the cyst, and in the trophozoite a number of particles have been found which have not aggregated yet to form bodies.

A histochemical study has shown that the bodies consist mainly of ribonucleic acid and some unspecified proteins. One of the tests which we have carried out is of particular interest as it involved the use of the electron microscope. The living organisms were treated for four hours with ribonuclease (10 mg ribonuclease in 10 cm³ water). They were then fixed in osmium tetroxide and prepared for sectioning by standard procedure. Preliminary results show that in the trophozoite originally the globular particles appear to be flattened as a result of the treatment with the enzyme (Fig. 2). The enzyme has apparently decomposed the nucleic acid contained in the particles. This tallies very well with results of histochemical tests which have demonstrated a substantial loss of ribonucleic acid in the bodies after treatment of the living organism with ribonuclease. It is interesting to note that in formalin-fixed, untreated organisms, the particles appear also to be flattened (Fig. 3). High magnification pictures of enzyme-treated organisms reveal also a fine structure of the particles: they consist

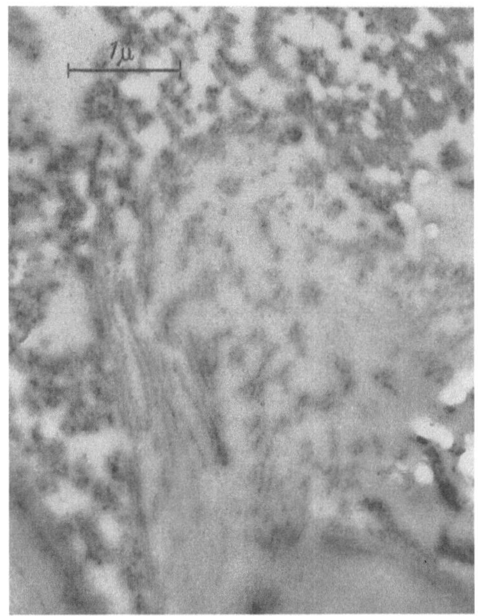

Fig. 3. Chromatoid body. Fixative: formalin

of linearly arranged subunits which in the electronmicrographs are represented as electron dense rings (diameter about 70 Å) with electrontransparent centres (diameter about 30 Å) (Fig. 2). In the cyst after treatment with ribonuclease no changes were observed, presumably the comparatively large enzyme molecules could not penetrate the cyst wall. We shall try to clarify by further investigations the significance of these peculiar inclusions.

2. Bakteriologie

Ein Beitrag zur Morphologie von Leptospiren

J. Parnas, A. Feltynowski und K. Burdzy

Lehrstuhl für Mikrobiologie der Medizinischen Akadamie, Lublin, und Staatliches Institut für Hygiene, Warszawa (Polen)

Die Morphologie von Leptospiren ist nach wie vor ein interessantes Thema für die elektronenmikroskopische Untersuchung. Es liegt eine ganze Reihe von Arbeiten bereits vor (u. a. Babudieri (*1*), Czekalowski und Eaves (*2*, *3*), Mölbert (*4*), sowie Gängel und Themann (*5*)], aber manche Fragen scheinen noch nicht geklärt zu sein. So ist z. B. die Bedeutung der von Jakob (*6*) gefundenen „Granula", der sog. Leptospirogene, immer noch Gegenstand der Diskussion. Es ist zu bemerken, daß solche Granula schon im Jahre 1952 von einem von uns ebenfalls beobachtet worden sind (*7*). Der Teilungsvorgang der Leptospiren ist auch noch nicht ganz geklärt.

Das Ziel unserer Arbeit war: 1. eine größere Anzahl von Leptospirenstämmen morphologisch zu untersuchen, um eventuelle Unterschiede festzustellen; 2. die Einzelheiten des morphologischen Aufbaus der verschiedenen Stämme auf Grund der eigenen Beobachtungen und Beobachtungen anderer Autoren zu studieren.

Untersucht wurden folgende Stämme:

1. L. andaman	15. L. icterohaemorrhagiae Wijnberg
2. L. australis A Ballico	16. L. naami Mankarso
3. L. australis B Zanoni	17. L. naami Naam
4. L. autumnalis Akiyama	18. L. poi
5. L. autumnalis Bangkinang	19. L. pomona
6. L. ballum	20. L. pyrogenes
7. L. batavia	21. L. sarmini
8. L. bovis	22. L. sax-koebing
9. L. canicola	23. L. schüffneri
10. L. grippotyphosa Andaman	24. L. sejroe
11. L. grippotyphosa Duyster	25. L. semaranga
12. L. grippotyphosa Tomaszów	26. L. sentoti
13. L. javanica	27. L. sorex
14. L. icterohaemorrhagiae Kantorowicz	28. L. mitis

Diese Stämme haben wir größtenteils von der WHO-Kollektion in Amsterdam bekommen; *L. grippotyphosa* Tomaszów ist in Polen isoliert worden; die Stämme *L. andaman* und *L. sorex* stammen aus Prag.

Die Vorbereitung der Präparate war wie folgt: Alle Leptospirenstämme waren 4 Tage alt und in Korthoff-Medium gezüchtet. Etwa 5 ml einer Leptospirenkultur wurde 15 min bei 3000 r. p. m. zentrifugiert, der Bodensatz mit gepuffertem Osmiumtetroxyd (pH 7,1—7,2) 15 min fixiert, nachher 4mal bei 8000 r. p. m. auch je 15 min zentrifugiert und mit destilliertem Wasser gewaschen. Der Bodensatz wurde mit etwas bidestilliertem Wasser versetzt und je ein Tropfen der Suspension auf elektronenmikroskopische Objektträger mit Formvarfolie aufgebracht. Die Präparate wurden mit Chrom schräg bedampft und in einem Siegbahn-Schönander-Elektronenmikroskop beobachtet.

Da die morphologischen Eigenschaften, wie Länge, Dicke, Anzahl der Windungen für einen bestimmten Stamm veränderlich sind, hat man sich bis jetzt nicht bemüht, nach Unterschieden zwischen den verschiedenen Stämmen zu suchen. Wir haben jedoch beobachtet, daß nicht bei allen Stämmen die charakteristische Hakenform (d. h. C-Form oder S-Form) (Abb. 1) des Leptospirenkörpers zu sehen ist. Die gerade Form des Körpers (Abb. 2) war charakteristisch für die Stämme *L. schüffneri*, *L. ballum*, *L. naami Naam*, *L. naami Mankarso*, *L. pyrogenes* und *L. sejroe*. Die zwei ersten Stämme wiesen die Spitzenform am Körperende auf. Ob die gerade Form des

Leptospirenkörpers eine stammspezifische Eigenschaft ist, ist schwer zu sagen. (Es ist zu bemerken, daß die Präparate von allen Stämmen in identischer Weise vorbereitet worden sind; das Alter der Kulturen war auch dasselbe.)

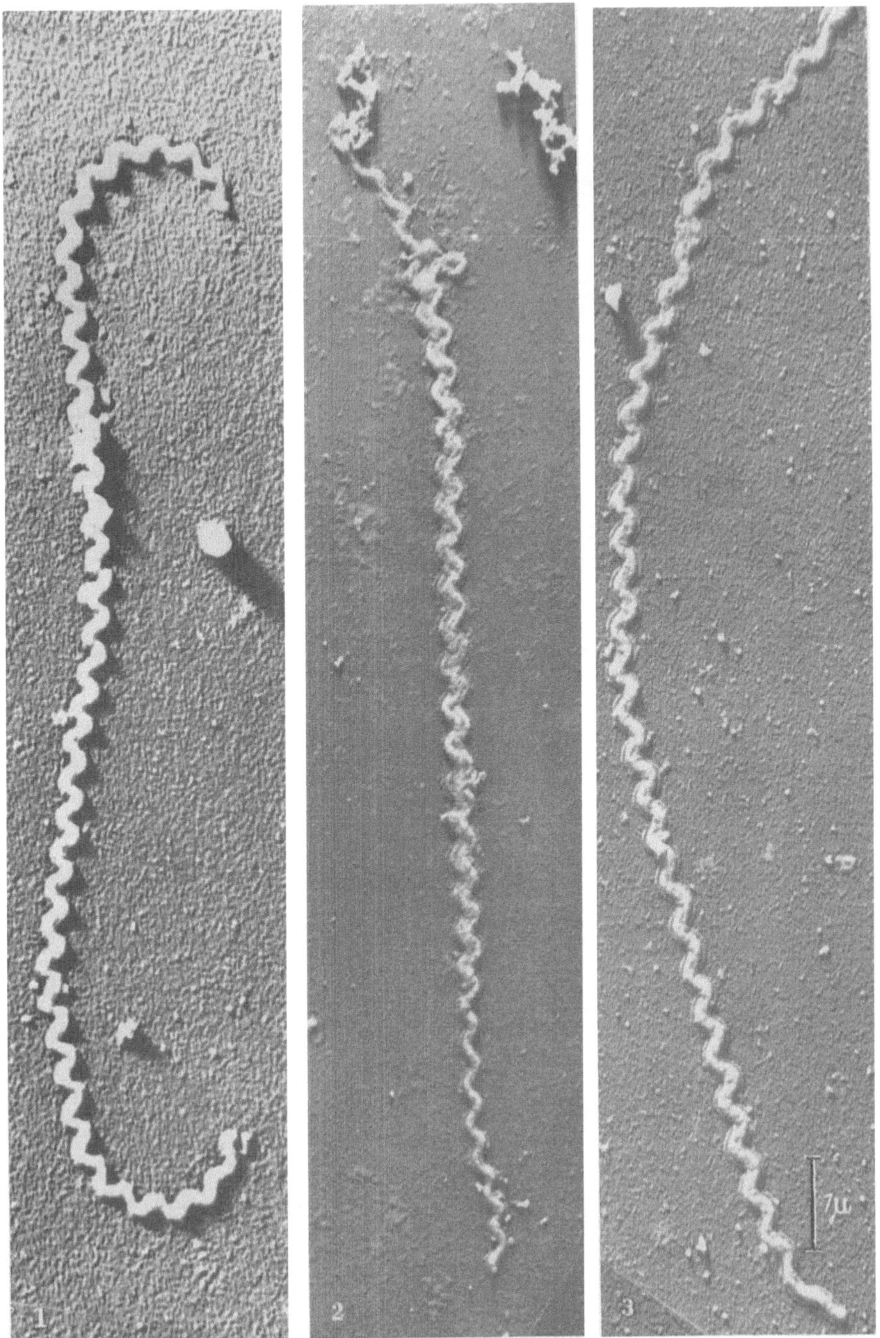

Abb. 1. L. Sax-koebing. Die Hakenform. 12000mal — Abb. 2. L. naami Naam. Die gerade Form. 12000mal
Abb. 3. L. pyrogenes. Die Zellmembran ist auf beiden Seiten des Körpers zu sehen. 12000mal

Es ist möglich, daß die Bewegung bei solchen geraden Formen z. B. durch das Drehen um die Achse zustande kommt, während sie bei S- oder C-Formen auch durch das elastische Biegen des gesamten Körpers verursacht wird.

Länge und Querdurchmesser der Leptospiren variieren auch für einen bestimmten Stamm beträchtlich. (Gemessene Werte: die Länge für *L. sorex* 10 μ, für *L. grippotyphosa* 40 μ, der Querdurchmesser 0,09 μ für *L. sentoti*, 0,14 μ für *L. pyrogenes*.)

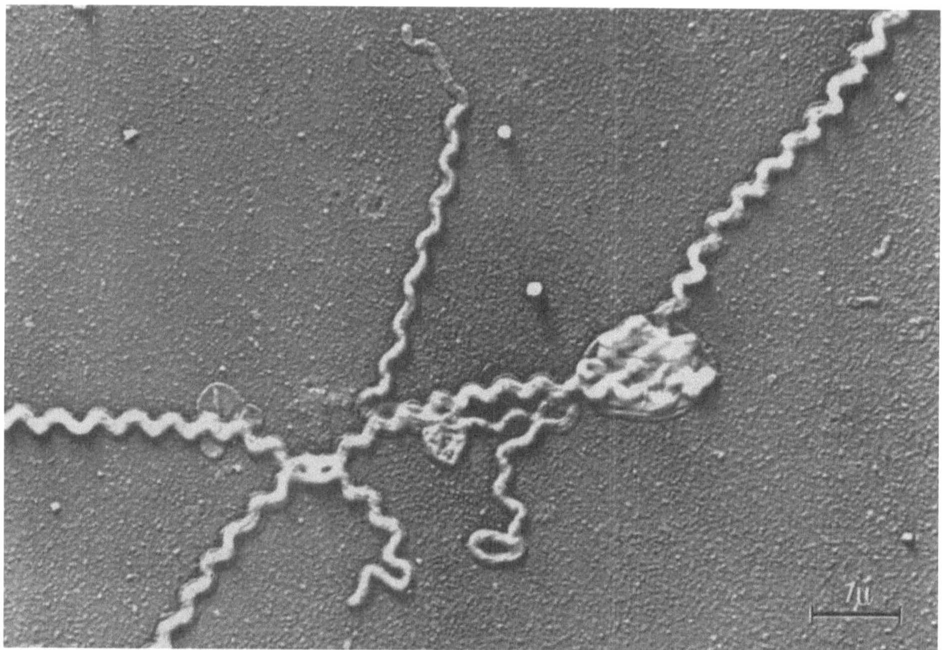

Abb. 4. *L. javanica*. Die Kugelform mit zusammengeballtem Leptospirenkörper. 12000 mal

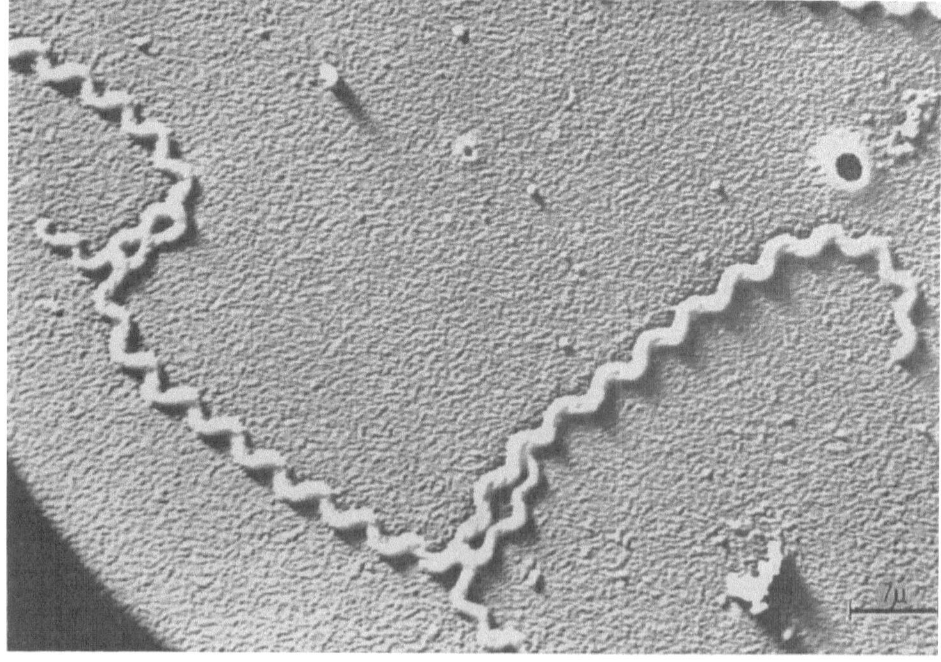

Abb. 5. *L. semaranga*. Die Schleifenbildung. 12000 mal

Die Anzahl der Windungen hängt natürlich von der Gesamtlänge des Körpers ab. Sie ändert sich bei einem bestimmten Stamme ebenfalls beträchtlich. (Gemessene Werte: für *L. grippotyphosa* 26, für *L. sax-koebing* 38, für *L. schüffneri* 26, für *L. naami Naam* 17).

Der Aufbau des Leptospirenkörpers aus Protoplasmacylinder und Achsenfaden, um den sich das periphere Plasma wendeltreppenartig schlingt, ist von uns bestätigt worden. Der Plasmacylinder ist empfindlich gegen destilliertes Wasser und unterliegt leicht der Lyse, während der Achsenfaden unbeschädigt bleibt. Gleiches kann man auch als Folge der Autolyse bei alten

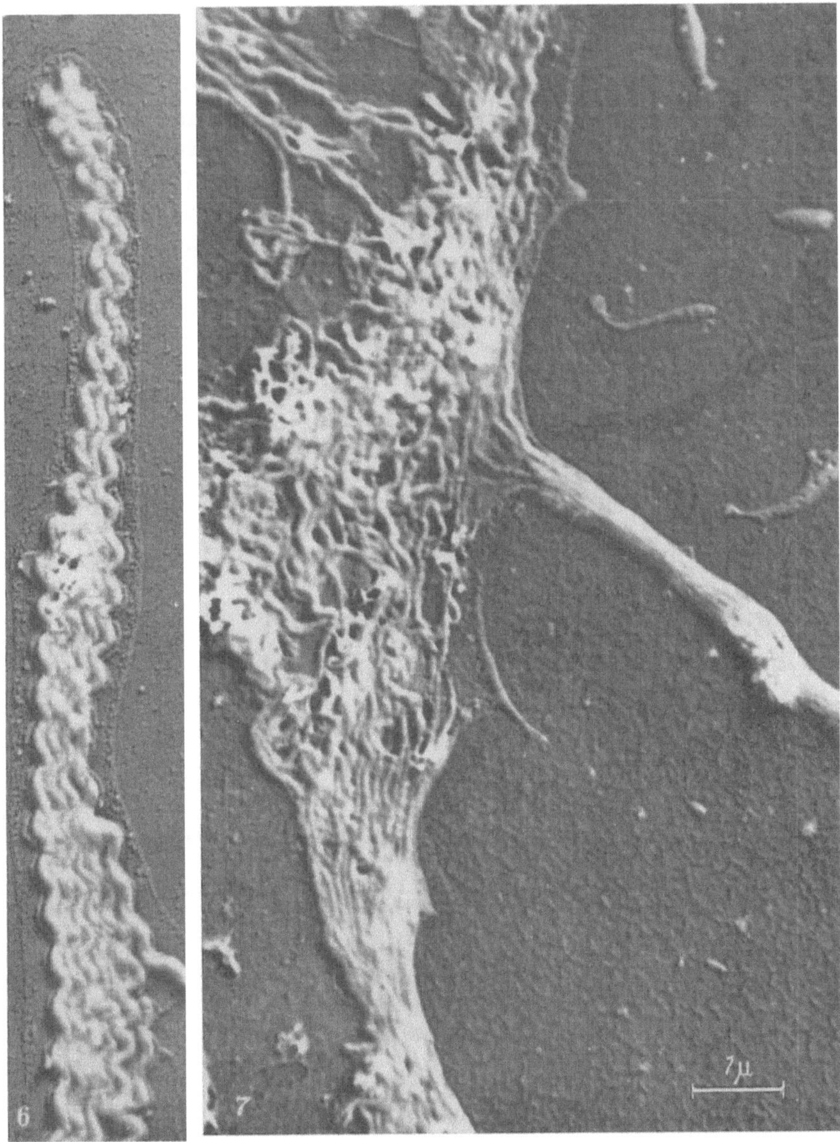

Abb. 6. L. icterohaemorrhagiae Wijnberg. Die Agglutination der Leptospiren. 12000mal

Abb. 7. Die Agglutination von L. canicola nach 48stündiger Wirkung von destilliertem Wasser. 12000mal

Kulturen beobachten. Nach der Abtrennung des Achsenfadens vom Plasmacylinder, verliert letzterer die Fähigkeit, sich zu winden und unterliegt weitgehenden Destruktionsveränderungen. Das ist ein Beweis dafür, daß die Zellmembran auch den Faden umhüllt. Der Achsenfaden hat auf beiden Enden die kugelförmigen Knöpfe ("knobs"), mit denen er in Plasma eingewachsen ist.

Die Zellmembran ist auf beiden Seiten des Leptospirenkörpers gut zu sehen (Abb. 2 u. 3). Die Sichtbarmachung der Membran gelang nur bei OsO_4-fixierten Präparaten.

Es ist interessant, daß wir auch bei den jungen 4tägigen Kulturen die Kugelformen beobachten konnten. Sie sind entweder an den Enden des Körpers oder entlang des Körpers zu sehen. Manche von diesen Formen sind groß und mit der Zellmembran umhüllt. Im Inneren ist gelegentlich ein zusammengeballter Leptospirenkörper in eine homogene Substanz eingebettet (Abb. 4). Andere dieser „Granula" sind klein und bilden eigentlich nur eine Verbreiterung des Leptospirenkörpers. Da diese kleineren „Granula" oft am Ende einer kurzen Leptospire zu finden sind, ist anzunehmen, daß sie im Rahmen eines Teilungsmechanismus eine Rolle spielen. Die Kugelformen haben wir bei allen Stämmen beobachtet.

In den jungen 4tägigen Kulturen haben wir sehr oft Verbindungen von Leptospiren in Form von Schleifen beobachtet (Abb. 5). Am Ort der Schleife ist eine feste Verbindung zwischen den Leptospiren hergestellt, wodurch evtl. ein Austausch von Substanz ermöglicht wird. Diese Schleifen können einfach oder mehrfach sein. Zusammenballungen zahlreicher Leptospiren, die auf diese Weise aneinanderhaften, fassen Gängel und Themann (5) als Brutnester auf und schreiben der Schleifenbildung eine Bedeutung im Entwicklungscyclus der Leptospiren zu.

Wir haben auch Agglutinationsformen von Leptospiren beobachtet. Diese Formen entstehen auch ohne Wirkung von Agglutinin. Das Bild (Abb. 6) zeigt die Agglutination bei *L. icterohaemorrhagiae* Wijnberg, wo die einzelnen aneinander gelagerten Leptospiren zu sehen sind. Es ist zu bemerken, daß bei geraden Formen (*L. schüffneri* und *L. ballum*) die Agglutinationsformen öfter zu finden sind und anders aussehen als bei Hakenformen. Abb. 7 zeigt die lytische Form und die Agglutinationsform von *L. canicola*, die durch 48stündige Wirkung von destilliertem Wasser verursacht wurde. Die infolge der lytischen Wirkung veränderten Leptospiren, die schon ihre Windungsform verloren haben, vereinigen sich in einem langen Streifen, ganz verschieden von der auf dem vorherigen Bild gezeigten Form.

Literatur

1. Babudieri, B.: J. Hyg. **47,** 390 (1949).
2. Czekalowski, J. W., and G. Eaves: J. Bact. **67,** 619 (1954).
3. — — J. Path. Bact. **69,** 129 (1955).
4. Mölbert, E.: Z. Hyg. **141,** 82 (1955).
5. Gängel, G., u. H. Themann: Arch. Hyg. (Berl.) **140,** 559 (1956).
6. Jakob, A.: Klin. Wschr. **27,** 364 (1949).
7. Dymowska, Z., u. A. Feltynowski: Acta microbiol. polon. **2,** 125 (1953).

Electron microscopic studies on the intracellular structures of Mycobacterium in relation to function

Tadao Toda, Kenji Takeya and Masaatsu Koike

Department of Bacteriology, School of Medicine, Kyushu University, Fukuoka (Japan)

The structure of Mycobacterium has been subjected to intensive investigation, both with light and electron microscopy. Although studies on ultrathin sections of these bacilli have been reported by the present authors (*1*), Bassermann (*2*) and Brieger and Glauert (*3, 4*), the pictures obtained were not sufficient to clarify the individual constituents of the cell. Recently, a mitochondria-like structure was demonstrated by Zapf (*5*), and Shinohara et al. (*6*).

We, in this experiment, compared various fixation methods and succeeded in finding the finer structure of each cellular constituent by use of these fixation methods. Furthermore, these structures were examined in relation to their function.

Mycobacterium avium was fixed in five different types of fixative solutions which were composed of the following: 2% OsO_4 in destilled water (*7*), 1% OsO_4 in isotonic phosphate buffer (*8*), 1% OsO_4 in isotonic veronal acetate buffer (*9*), 1% OsO_4 in isotonic dichromate pottasium buffer (*10*) and 1% $KMnO_4$ in isotonic veronal acetate buffer (*11*).

1. Nuclear apparatus. When the bacilli were fixed with OsO_4 in veronal acetate buffer, the nuclear sites were noted to be filled with networks of finely intertwisted fibrils. Dense small oval granules were often seen attached to these fibrils. When fixed with OsO_4 in destilled water, the nuclear sites were filled with finer fibrils than when fixed in veronal acetate buffer. In the cells fixed with the other three fixatives, the nuclear materials were conglomerated in the nuclear vacuoles. These facts indicate that the preservation of the nuclear apparatus is affected by the composition of the fixative as greatly as by the pH of the fixative (*12*).

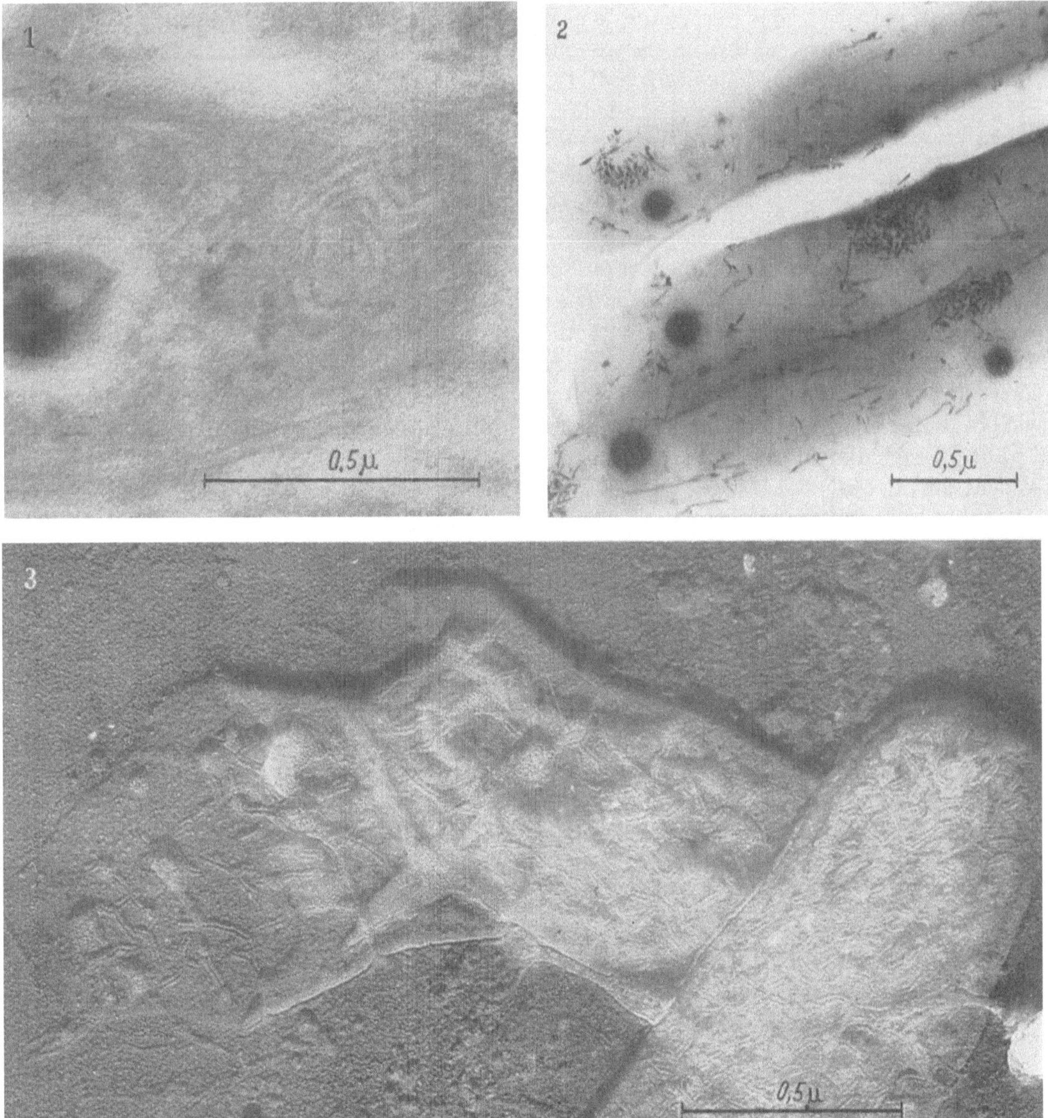

Fig. 1. A mitochondria-like structure is filled with double membranes

Fig. 2 Fine needle-like crystals of reduced tellurite are found localized in reduction sites, and large crystals are scattered throughout the cytoplasm

Fig. 3. Paired fibers are clearly seen in ghost cells

2. Cytoplasm and mitochondria-like structure. In all the above mentioned fixation methods, it was noted that the cytoplasm was composed of a sponge-like structure of uniform pattern except when fixed with $KMnO_4$ in which case coarse precipitates were observed throughout the ceii.

Using 1 % OsO_4 in dichromate potassium buffer, an organelle which appeared equivalent to the mitochondria-like structure reported by SHINOHARA et al. (*6*) was observed with much finer details. A profile of this organelle, 100 to 500 mμ in size, showed it to be filled with double membranes (Fig. 1). The two dense layers of this membrane measured 25 Å in thickness and the width of the interspace about 40 Å. These values are smaller than those of mitochondrial cristae of the higher tissue cells (*13*).

It is well known that reduced potassium tellurite is discernible in light microscopic preparations as black granules within the cell of acid-fast bacilli. In electron microscopic pictures, the same localized reduction sites were found to be centers of deposit of reduced tellurite crystals (*14*). The similar histochemical method in electron microscopy was used for demonstrating the activity of dehydrogenase system in tissue cells (*15*).

Mycobacterium avium grown on collodion film overlying 4% glycerol agar medium was replaced onto Sauton's media or 4% glycerol broth, each containing 0.05 to 0.5% potassium tellurite. Reduced tellurite localized in circumscribed reduction sites occured either as fine needle-like crystals (Fig. 2) or as small particles. Moreover, large tellurium crystals were scattered throughout the cytoplasm (Fig. 2). The reduction sites did not coincide with the electron dense granules, but the reduced tellurite was often found near the granules (Fig. 2).

3. Cell wall and cytoplasmic membrane. The cell wall was found to be 150 to 250 Å in width and to be composed of three layers. A lighter interspace was noted between the dense osmiophilic inner and outer layers. The cytoplasmic membrane enveloping the cytoplasm was also observable inside the cell wall. Pictures showing various stages of cell division were obtained. The cross septa, formed centripetally, partitioned the cell and, thereafter, thickened and split forming independent cell walls for each daughter cell. At each side of the cross septa, doubled cytoplasmic membranes were occasionally observed.

On the other hand, numerous paired fibrous structures (*16*) which have never been reported were found throughout the cell body of *Mycobacterium avium* lysed by mycobacteriophage B-1 (*17*). Further studies have revealed that this structure is also found within the "ghost cell" of healthy acid-fast bacilli, lysed either mechanically or chemically (Fig. 3). The fact that this structure has been found invariably within the cell body and never observed outside the cell suggests that this structure is related not to the cytoplasm but to the cell wall or cytoplasmic membrane.

4. Electron dense granules. Three kinds of granules, large dense, small dense in rosette-like arrangement and vacuole-like ones, have been found in acid-fast bacilli (*18*). The nature of the large dense granules was considered to be neither nucleus nor mitochondria but an accumulation of a substance, probably polyphosphate, related to bacterial metabolism as shown by the present authors (*18*). These dense granules observed in sectioned preparations were entirely electron-opaque or composed of fine dense particles. Small dense granules in rosette-like arrangement and those vacuolized by electron beam look like "Rosette and Morula" presented by Penso (*19*), as a stage of the life cycle of mycobacteriophage. The electron microscopic study of the multiplication process of mycobacteriophage B-1 failed to demonstrate the life cycle postulated by Penso.

References

1. Toda, T., M. Koike, N. Hiraki and K. Takeya: Tokyo med. J. **72,** 447 (1957); J. Bact. **73,** 442 (1956).
2. Bassermann, E. J.: Naturforsch. **11b,** 276 (1956).
3. Brieger, E. M., and A. M. Glauert: Nature (Lond.) **178,** 544 (1956).
4. Glauert, A. M., and R. H. Glauert: J. biophys. biochem. Cytol. **4,** 191 (1958).
5. Zapf, K.: Naturwissenschaften **44,** 448 (1957).
6. Shinohara, C., K. Fukushi and J. Suzuki: J. Bact. **74,** 413 (1957).
7. Chapman, G. B., and J. Hillier: J. Bact. **66,** 362 (1953).
8. Birch-Andersen, O., and A. Maaløe: Biochim. biophys. Acta **12,** 395 (1953).
9. Palade, G. E.: J. exp. Med. **95,** 285 (1952).
10. Dalton, A. J.: Anat. Rec. **121,** 281 (1955).
11. Luft, J. H.: J. biophys. biochem. Cytol. **2,** 799 (1956).
12. Kellenberger, E., and A. Ryter: Experientia (Basel) **17,** 420 (1956).
13. Sjöstrand, F. S., and V. Hanzon: Exp. Cell Res. **7,** 393 (1954).
14. Mudd, S., K. Takeya and H. J. Henderson: J. Bact. **72,** 767 (1956).
15. Barrnett, R. J., and G. E. Palade: J. biophys. biochem. Cytol. **3,** 577 (1957).
16. Takeya, K., R. Mori, M. Koike and T. Toda: To be published.
17. — and T. Yoshimura: J. Bact. **74,** 540 (1957).
18. — M. Koike, T. Uchida, S. Inoue and K. Nomiyama: J. Electronmicroscopy **2,** 29 (1954).
19. Penso, G.: Protoplasma **45,** 251 (1955).

The characteristic mitochondrial structure
of Mycobacterium tuberculosis, relating to its function

C. Shinohara, K. Fukushi, J. Suzuki and K. Sato

The Research Institute for Tuberculosis and Leprosy, Tohoku University, Sendai (Japan)

In 1957 the characteristic structures closely resembling the mitochondria of animal cells were revealed in the ultrathin sections of *Mycobacterium avium* by the authors (*1*). Subsequently, Yoshida and his co-workers (*2*) reported the similar structure in *Mycobacterium tuberculosis* var. *hominis*. The present authors also found it in *Mycobacterium tuberculosis* var. *hominis* H 37 Rv.

In the experiments presented here the particulate fractions which showed L-malate, DPNH and succinate oxidations and possessed cytochrome system were isolated from the ground homogenate of *Mycobacterium avium*. Moreover, in ultrathin sections of these particulate fractions we found the characteristic structure of mitochondria, as revealed in sections of intact mycobacterial cells.

Intact cells of *Mycobacterium avium* and *Mycobacterium tuberculosis* var. *hominis* H 37 Rv were fixed in 1 % osmium tetroxide buffered to pH 7.4 with phosphate at 16—18° C for 2 and 5 days, respectively. The fixed specimens were then dehydrated in alcohol, embedded in a mixture of 2 parts of methyl and 8 parts of butyl methacrylate and sectioned.

Electron micrographs of sections of *Mycobacterium avium* revealed mitochondria-like structures which have a limiting membrane and internal tubular structures (Fig. 1). These structures were round or elliptical in shape with diameters of 48—400 mμ. A clear limiting membrane was occasionally found between the cytoplasm and the structures. Usually one, but occasionally more,

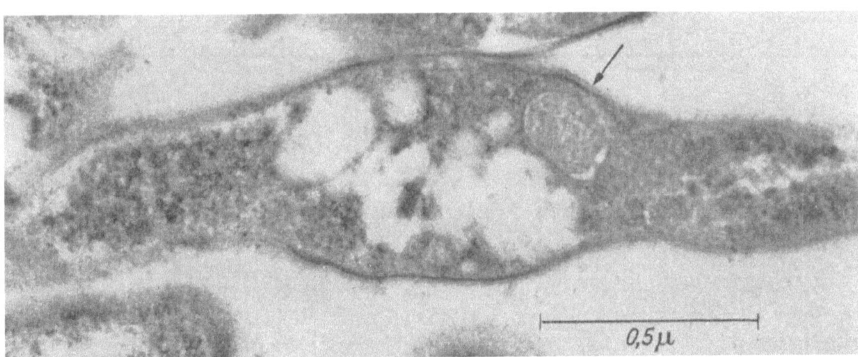

Fig. 1. An ultrathin section of *Mycobacterium avium*. The arrow shows a mitochondria-like structure.
Magnification 58 000

mitochondria-like structures were found in one section of a cell. The position in the cell was not specific. Fine tubular structures ran inside. The diameters of internal tubules ranged from 26 to 30 mμ.

The electron micrographs of sections of *Mycobacterium tuberculosis* var. *hominis* H 37 Rv showed also mitochondria-like structures (Fig. 2). The micrograph of a longitudinal section of this specimen showed the tubular feature as in the section of *Mycobacterium avium* (Fig. 2a). In the micrograph of a cross section of *Mycobacterium tuberculosis* var. *hominis* H 37 Rv, a honeycomb-like structure was revealed (Fig. 2b). The limiting membrane was not distinct in these cases.

The features of these structures resemble closely mitochondria in animal cells, particularly in the cells of *Paramecium*, *Euplotes* (*3*) and adrenal cortex (*4*).

The ultracentrifugal fractionation of *Mycobacterium avium* was carried out by the method of Yamamura and his co-workers (*5*) to obtain the fractions which have mitochondrial function.

About 50 g of cells were ground with sea sand and mixed with 0.25 M sucrose solution. Centrifuged at 3,000 r. p. m. for 30 min. Sea sand and intact cells were discarded. The supernatant was centrifuged at 10,000 g for 60 min. The supernatant thus obtained was centrifuged again at 40,000 g for 60 min. Once again the supernatant was centrifuged at 106,000 g for 60 min. The pellets were designated as R_{10}, R_{40} and R_{106}, respectively. The supernatant at 106,000 g was designated as S_{106}.

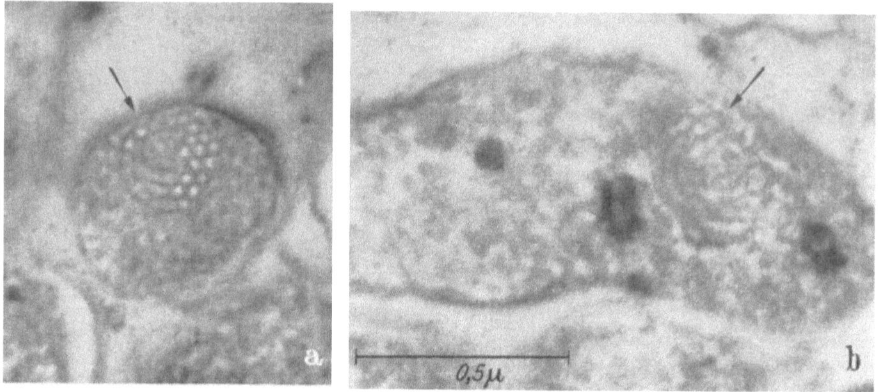

Fig. 2a and b. Ultrathin section of *Mycobacterium tuberculosis* var. *hominis* H37Rv. a) Cross section; b) Longitudinal section. The arrows show mitochondria-like structures. Magnification 58000 ×

Fraction R_{10} consists mainly of cell walls. Both R_{40} and R_{106} were beautiful reddish pellets. The enzymatic activities of each fraction were determined manometrically with the Warburg apparatus.

The rapid oxidation of L-malate and DPNH was found in fraction R_{40}. This fraction showed almost negligible oxidative activities of citrate, succinate, fumarate and lactate. However, the combination of fractions R_{40} and S_{106} fairly increased the oxidation of succinate.

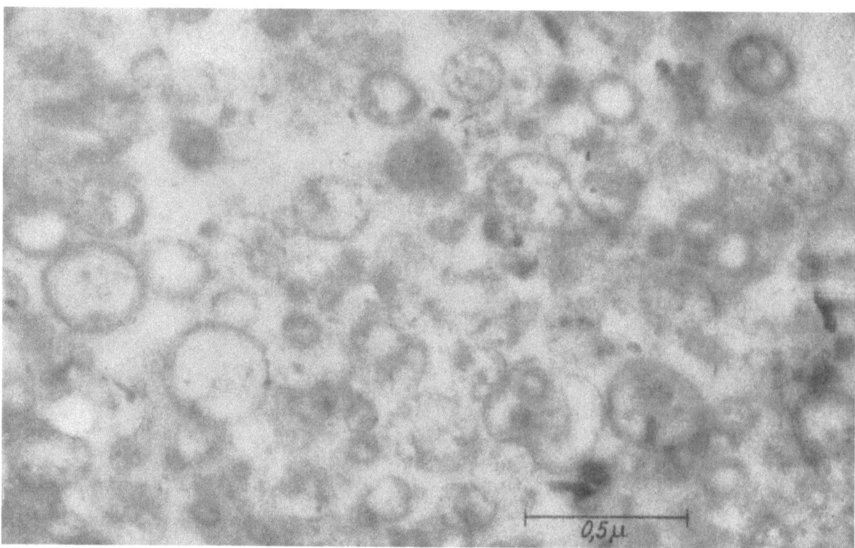

Fig. 3. An ultrathin section of particulate fraction obtained from the ground homogenate of *Mycobacterium avium*. Magnification 44000 ×

R_{106} showed similar enzymatic activities to those of R_{40}.

In the case of fraction S_{106} lactate oxidation was very rapid, that of citrate fairly rapid, but that of L-malate, succinate or DPNH was not so.

With the micro- and hand-spectroscopes the particulate fractions (R_{40} and R_{106}) in the reduced state after adding sodium hydrosulfite or malate showed distinct absorption bands around 550 and 560 mμ. Weaker absorption bands appeared around 520—530 and 590 mμ. No distinct band appeared in the case of S_{106}.

The ultrathin sections of the particulate fractions were observed simultaneously with the electron microscope. The particulate fractions were fixed in 1% osmium tetroxide buffered to pH 7.4 with phosphate at 16—18° C for 36—48 hr. The fixed specimens were then dehydrated, embedded and sectioned as in the case of intact cells. The electron micrographs of ultrathin sections of the particulate fractions revealed many ring form structures which have diameters of 30—370 mμ (Fig. 3). Some of them showed the internal tubular structures. These features closely resemble the mitochondria-like structures found in intact cells.

The particulate fractions obtained from the homogenate of *Mycobacterium avium* contained TCA cycle enzymes, DPNH oxidase and cytochromes, as mentioned above. Recently Dr. KUSU-NOSE, one of our collaborators, found that the particulate fractions contained oxidative phosphorylation properties. Moreover, the structures closely resembling the mitochondria of animal cells were revealed in the ultrathin sections of these particulate fractions. Therefore, the structures revealed in tubercle bacilli are considered to correspond to the mitochondria of plant and animal cells.

References

1. SHINOHARA, C., K. FUKUSHI and J. SUZUKI: J. Bact. **74,** 413 (1957).
2. YOSHIDA, N., K. FUKUI, T. TAMAKI, A. TANI, Y. HASHIMOTO, Y. HARA, M. TSUYOSHI and A. KAWANO: Shikoku Acta med. **11,** 628 (1957).
3. POWERS, E. L., C. F. EHRET, L. E. ROTH and O. T. MINICK: J. biophys. biochem. Cytol. **2,** No. 4, Suppl., 341 (1956).
4. LEVER, J. D.: J. biophys. biochem. Cytol. **2,** No. 4, Suppl., 313 (1956).
5. YAMAMURA, Y., M. KUSUNOSE, S. NAGAI, E. KUSUNOSE, Y. YAMAMURA jr., J. TANI, T. TERAI and T. NAGA-SUGA: Med. J. Osaka Univ. **6,** 489 (1955).

Zum Nachweis der Mitochondrienäquivalente bei Mikroorganismen

WERNER NIKLOWITZ

Deutsche Akademie der Wissenschaften zu Berlin, Institut für Mikrobiologie und experimentelle Therapie, Jena
(Direktor: Prof. Dr. med. H. KNÖLL)

Als Zentrum biochemischer Prozesse werden in den Zellen der höheren Organismen die Mitochondrien betrachtet. Obwohl man auch aus der Mikrobenzelle mit Hilfe der Homogenisierung Zellpartikel von der Größenordnung von 10—100 mμ (*1, 2*) und 30—370 mμ (*3*) erhielt, an denen biochemische Umsetzungen ablaufen, so konnten jedoch im Gegensatz dazu gleiche oder ähnliche Zellorganellen auf Grund cytomorphologischer Untersuchungen, insbesondere an ultradünnen Schnitten, in den Bakterien bisher nur bei *Mycobacterium tuberculosis* (*3*) beschrieben werden.

Im Laufe unserer elektronenmikroskopischen Substrukturforschung an Bakterien (*Escherichia coli, Sarcina ventriculi*) (*4, 5, 6*), Blaualgen (*Coelosphaerium spec., Phormidium uncinatum, Phormidium retzii, Phormidium frigidum, Oscillatoria limosa, Anabaena variabilis, Cylindrospermum licheniforme*) (*7, 8, 9, 10*) und Schleimpilzen (*Badhamia utricularis, Physarum polycephalum*) (*11, 12*) wurde neben dem Fragenkomplex zur Struktur und stofflichen Natur der Kernäquivalente bzw. Kerne dieser Organismen gleichfalls die Frage nach der Existenz von distinkten Redoxzentren einer näheren Analyse unterzogen.

Mit Hilfe cytochemischer Nachweisreaktionen (Janusgrün B, TTC, Stilben-TC) konnten lichtmikroskopisch in den Objekten Redox-Orte beobachtet werden, die eine regelmäßige Anzahl (*5, 6*) und Lokalisation (*5, 9, 10*) erkennen ließen. Ferner wurden an identischen Orten im Cytoplasma charakteristische Plasmapartien anhand von Ultradünnschnitten festgestellt, wie sie weiter unten noch näher beschrieben werden. Auf Grund oben angedeuteter lichtmikroskopischer Befunde in Verbindung mit elektronenmikroskopischen Untersuchungsergebnissen von Objekten, die vor der eigentlichen Präparation für die elektronenmikroskopische Beobachtung (Totalpräparat, Ultradünnschnitt) mit Redox-Indicatoren (TTC, Stilben-TC) behandelt wurden, konnten diese Plasmapartien als die Mitochondrienäquivalente identifiziert werden (*5*). Einerseits konnten also die Formazanablagerungen im Totalpräparat an denselben Orten erkannt werden,

34*

wie dies schon nach den lichtmikroskopischen Untersuchungen ermittelt wurde und andererseits erscheinen diese Orte im Schnitt als experimentell erzeugte „Vacuolen", da das Formazan im Alkohol (TTC) oder im Methacrylat (Stilben-TC) gelöst wird. Stets jedoch werden die Mitochondrienäquivalente, wenn sie nicht durch lange Einwirkungszeiten oder letale Konzentrationen der Indicatorlösungen verändert werden, in unmittelbarer Nähe oben erwähnter Vacuolen gefunden. Diese Ergebnisse sprechen nicht nur für die Existenz von distinkten Redox-Orten, sondern beweisen gleichzeitig, daß die Lokalisation der Reaktionsprodukte der Redox-Indicatoren den Redox-Orten und diese wiederum den Mitochondrienäquivalenten entsprechen (5).

Von allgemeinem Interesse sind die Beobachtungen zu werten, wonach die Mitochondrienäquivalente der untersuchten Bakterien und Cyanophyceen dieselben Struktureigentümlichkeiten des submikroskopischen Feinbaus aufweisen (5, 6, 9, 10). Die Mitochondrienäquivalente sind von kugeliger bis ovoider Form und erscheinen als kompakte Gebilde. Sie werden vom umgebenden Plasma durch eine osmiophile Grenzschicht abgegrenzt (Abb. 1—3). Diese Plasmapartien besitzen eine feinkörnigere Grundstruktur als das übrige Cytoplasma und werden so von diesem unterscheidbar. In den meisten Fällen konnte eine Zuordnung dieser Grundstrukturkörnchen zu „lamellenähnlichen" Strukturen beobachtet werden. Eindeutige Strukturen wie „Cristae" oder „Tubuli" sind auf Grund unserer Befunde nicht gegeben, vielleicht auch nicht zu erwarten. Vielmehr scheint es so, als ob die beschriebenen Struktureigentümlichkeiten

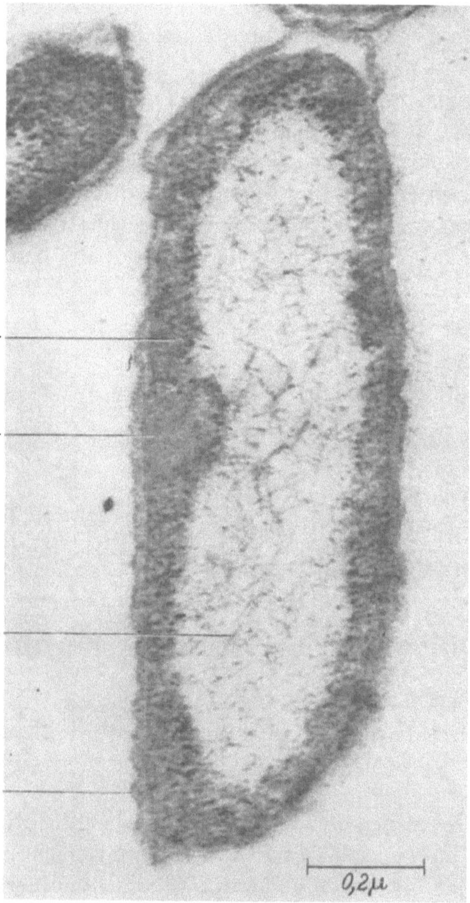

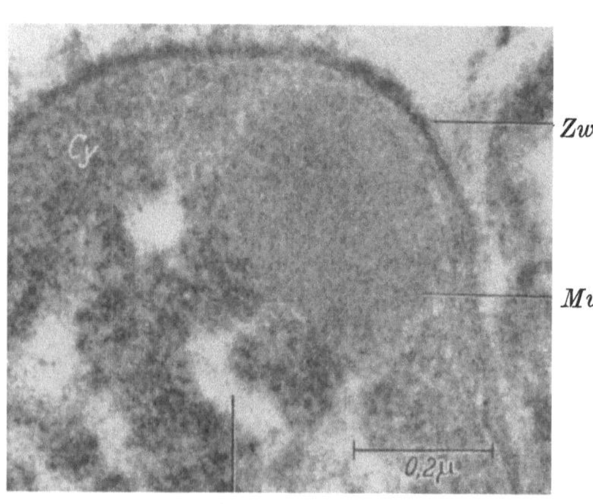

Abb. 1. *Escherichia coli.* Längsschnitt.
Mv = Mitochondrienäquivalent, *Zw* = Zellwand,
Cy = Cytoplasma, *Nv* = Kernäquivalent

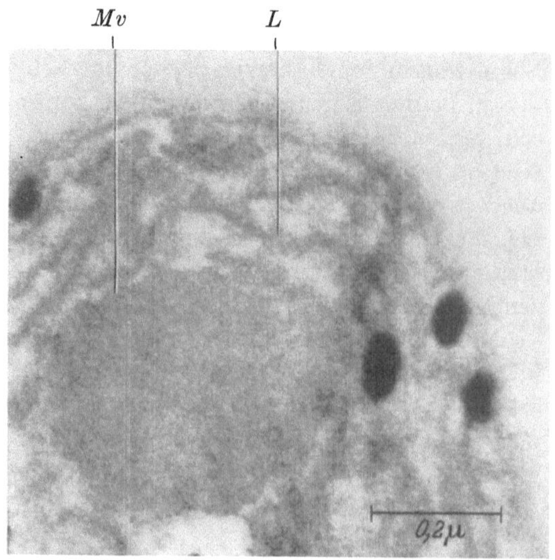

Abb. 2. *Sarcina ventriculi* (Ausschnitt)
(Erklärung s. Abb. 1)

Abb. 3. *Coelosphaerium spec.* (Blaualge) (Ausschnitt)
L = Lamellensysteme des Chromatoplasmas
(weitere Erklärung s. Abb. 1)

ein weiteres Bauprinzip der Feinstruktur von bestimmten Zellorganellen, in unserem Fall die der Mitochondrienäquivalente der Mikroorganismen (Bakterien, Cyanophyceen), darstellen. In diesem Zusammenhang soll darauf hingewiesen werden, daß die bei den Cyanophyceen zunächst als „ferment-aktive Granula" bezeichneten Zellgebilde (9, 10) demzufolge gleichfalls die Mitochondrienäquivalente dieser Organismengruppe sind und in Zukunft als solche bezeichnet werden sollen.

Die Mitochondrienäquivalente der Bakterien sind im Cytoplasma meist in der Nähe der Zellwände lokalisiert, wobei sie zum Bereich der Kernäquivalente hin von einem Cytoplasmasaum umgeben werden. Auch die Mitochondrienäquivalente der Cyanophyceen werden zellwandständig im Chromatoplasma beobachtet. Die Größenordnungen der Mitochondrienäquivalente sind in folgender Tabelle zusammengefaßt:

Tabelle 1. *Mittelwerte der Durchmesser der Mitochondrienäquivalente* (Messungen an ultradünnen Schnitten)

Objekte	E. coli	S. ventriculi Ph. frigidum	Ph. uncinatum Ph. retzii Osc. limosa	Coelosphaerium sp. A. variabilis Cyl. licheniforme
⌀ der Mv[1]	150 mμ	225 mμ	300 mμ	450 mμ

[1] Mv = Mitochondrienäquivalente

Wie aus der Tabelle ersichtlich, besitzen die Mitochondrienäquivalente der Bakterien dieselbe Größenordnung, wie sie schon für die Partikelgröße nach Homogenisierungen gefunden wurde (2, 3). Des weiteren ist die mengenmäßige Verteilung in der Zelle interessant. So besitzen die untersuchten Bakterien 1 oder 2 Redox-Orte, während sie bei den Cyanophyceen in einer größeren Anzahl vorkommen (Ausnahme: *Phormidium frigidum*). Vergleicht man die Mitochondrienäquivalente der Bakterien mit denen der Cyanophyceen und letztere untereinander, so scheint die Größe und Anzahl der Mitochondrienäquivalente in einer gewissen Korrelation zum Zellvolumen zu stehen.

Die Mitochondrien der Schleimpilze (11) weisen im Gegensatz zu den Bakterien und Cyanophyceen Feinbauverhältnisse auf, wie sie für die Protozoen beschrieben wurden (13). Von besonderem Interesse ist aber jene Beobachtung, wonach in den Mitochondrien dieser Organismen kontrastreiche Granula gefunden werden, die auf Grund verschiedener Befunde als Granula kondensierter Phosphate betrachtet werden und auf eine mögliche Entstehung dieser Granula in oder aus den Mitochondrienäquivalenten hinweisen.

Nach allem, was wir heute über die Struktur und stoffliche Natur der Mitochondrien, bzw. Mitochondrienäquivalente wissen, können wir zusammenfassend feststellen, daß auch die Mikroorganismen spezifische Zellorganellen — Mitochondrienäquivalente — besitzen, die Zentren biochemischer Prozesse auch in diesen Organismen darstellen.

Literatur

1. ALEXANDER, M.: Bact. Rev. **20,** 67 (1956).

2. STANIER, R. Y., I. C. GUNSALUS and C. F. GUNSALUS: J. Bact. **66,** 543 (1953).

3. SHINOHARA, CH., K. FUKUSHI, J. SUZUKI and K. SATO: J. Electronmicroscopy **6,** 47 (1958).

4. NIKLOWITZ, W.: Proc. Stockholm Conference Electron Microscopy, p. 115, 1956.

5. NIKLOWITZ, W.: Zbl. Bakt., I. Abt. Orig. **173,** 12 (1958).

6. KNÖLL, H., u. W. NIKLOWITZ: Arch. f. Mikrobiol. **31,** 125 (1958).

7. NIKLOWITZ, W., u. G. DREWS: Arch. f. Mikrobiol. **24,** 134 (1956).

8. — — Arch. f. Mikrobiol. **27,** 150 (1957).

9. DREWS, G., u. W. NIKLOWITZ: Arch. f. Mikrobiol. **24,** 147 (1956).

10. — — Arch. f. Mikrobiol. **25,** 333 (1957).

11. NIKLOWITZ, W.: Exp. Cell Res. **13,** 591 (1957).

12. — Physik. Verh. 8, 218 (1957).

13. WOHLFARTH-BOTTERMANN, K. E.: Z. Naturforsch. **11b,** 578 (1956).

Electron microscopy of protein crystals related to Bacillus alesti

I. M. Dawson, J. R. Norris and D. H. Watson

Department of Chemistry and Department of Bacteriology, The University, Glasgow W. 2, Scotland

The organism, *Bacillus alesti*, was originally isolated from diseased silk worms. Hannay and Fitz-James (*1*) originally noted that the cells contained spores and crystals. The crystals appeared to be protein in composition. Hannay studied some of these crystals in the electron microscope, after shadowing, and observed a regular ridging on some of them, although it was not clear whether this was a surface topographical feature or an artefact produced by the electron beam. In this work we hoped to obtain further information about this structure by use of replica techniques. The crystals proved to be regular and easily reproducible in good yield. These characteristics are

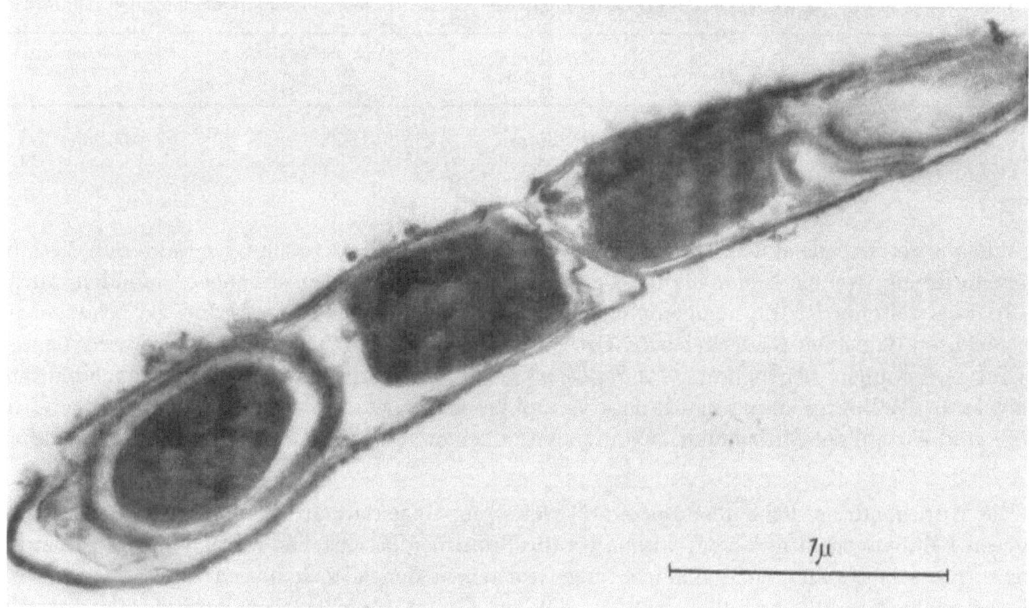

Fig. 1. Potassium permanganate fixed section of *B. alesti* cell, showing crystal and spore inside cell. The spore is surrounded by a membrane which does not surround the crystal. Other membranes are also visible

not always encountered in the electron microscopy of protein crystals and it was thought that these crystals afforded a convenient material for a bolder approach than is normally possible in this field.

The organism grows well and sporulates readily on an 'Oxoid Lab-Lemco Agar' and this medium was used throughout. Cultures were inoculated with 24 hr broth cultures consisting of young vegetative cells and incubated at 30° C. Incubation for ten days followed by harvesting into sterile distilled water gives a suspension of spores and crystals which can be used for replica work. For section cutting the incubation was continued for about thirty hours, the stage of the culture being checked by phase contrast microscopy until cells containing both spores and crystals could be seen. For section cutting, suspensions of such cells are washed, fixed in osmium tetroxide, or potassium permanganate, embedded in methacrylate or "Araldite" and cut in the normal way. (The cells are usually concentrated into a pellet by centrifuging at each stage of the preparation.)

Fig. 1 shows a micrograph of a permanganate fixed section and Fig. 2 a section which has been fixed with osmium tetroxide.

The spores of *Bacillus alesti* are surrounded by a membranous envelope, the exosporium, which Robinow (*2*) has interpreted as being the remnants of the sporangium wall. Hannay (*3*), considering the closely related *B. thuringiensis*, pointed out that if this was indeed the origin of the exosporium, the crystal would have to pass through a breach in the membrane in order to assume the free-lying position seen in old cultures. He failed, however, to observe such a rupture and we have also seen only intact exosporia in *B. alesti*. A study of ultra-thin sections of the bacterium prepared at different stages of spore and crystal synthesis is being carried out and may yield more information about the relationship of the crystal to membranes and other structures within the cell.

HANNAY (*3*) described a membrane which surrounds the spore but not the crystal and it is possible that the exosporium arises from this structure rather than from the sporangial wall. A similar membrane can be seen in Fig. 2 and both Figs. 1 and 2 show a number of membranes,

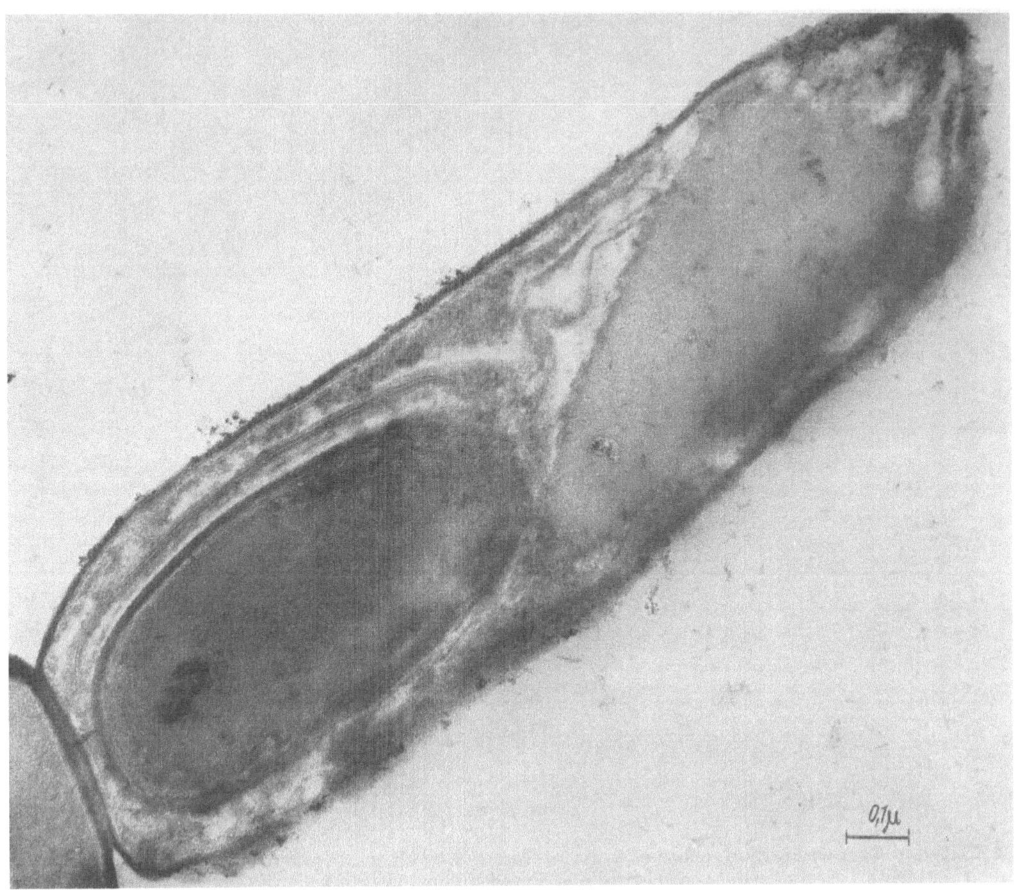

Fig. 2. Osmium tetroxide fixed section of *B. alesti* cell. Several of the membranes are seen to be double. The crystal exhibits a periodic fine banded structure at certain sites

some clearly having a double structure. The nature and origin of these membranes have not as yet been determined.

For the replica work a suspension of spores and crystals is used. A drop of this suspension is allowed to dry on a glass slide, which is then shadowed and covered with an evaporated carbon layer. The carbon-metal film can be floated off on water, mounted on grids and the replicas extracted using the potassium permanganate, potassium dichromate and sulphuric acid mixture described by BRADLEY and WILLIAMS (*4*). This, however, leaves a rather grainy shadowcast layer, presumably due to a reaction of the metal film with extraction mixture. Accordingly other extraction media were tried. Flotation of the replica film on 1% sodium hydroxide solution was found to give satisfactory extraction of the crystals in about thirty minutes, although it did not always extract the spores.

Uranium was originally used as a shadowcasting metal in this work but was unsatisfactory since it is removed by the sodium hydroxide solution used in extraction. Sodium carbonate, which had no effect on the shadowcast layer, was an unsatisfactory extraction agent. Platinium was found to be a reasonable shadowing metal from the point of view of grain and was also un-affected by the extraction medium.

Fig. 3 shows a platinum shadowed carbon replica of one of the crystals, which are tetragonal bi-pyramids. The ridging is clearly shown to be a feature of the surface topography. The periodicity is 290 Å. By analogy with the work on other protein crystals [Wyckoff (5), Dawson (6) etc.] it is conceivable that the periodic ridging is related to an axial spacing along the unique axis

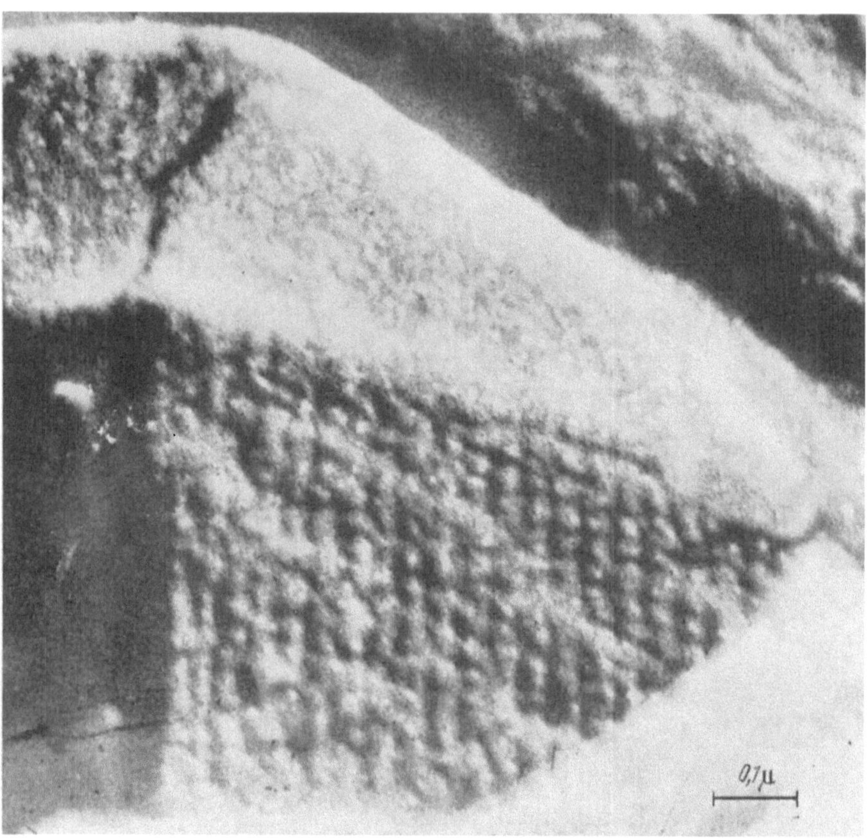

Fig. 3. Platinum shadowed carbon replica of a crystal from a *B. alesti* cell. A two dimensional array is visible on the (111) face of the tetragonal crystal

of the tetragonal crystal. On the other hand, it is also possible that it is due to some form of surface membrane, still adhering to the crystal after rupture of the cell. (The thin sections shown in Figs. 1 and 2 show a series of membrane layers round the cell.)

The shadowing direction on the crystal in Fig. 3 is such as to begin to show up detail in a direction normal to the ridging. The periodicity is 252 Å (after allowing for the tilt of the face), the ratio of the two spacings thus being 0.87. The observed morphology of several crystals suggests an axial ratio of 0.43. The value of 0.87 could, however, still be consistent with the ridges resulting from a tetragonal packing, in which there are twice as many molecules casting a shadow along the 4-fold axis of the unit cell, as there are along one of the 2-fold axes.

The topographical features observed in the replica work encouraged the hope that thin sections would exhibit corresponding structural detail on the crystals, as in the work on ferritin by Farrant and Hodge (7). No structure of this type has been observed but the crystal in the thin section of Fig. 2 does show a faint line structure whose periodicity is 40 Å.

There is some evidence from an electron microscope study of alkali degraded material to suggest that the size of fragments produced in degradation is related to this value of 40 Å. Further, some of the incompletely degraded material shows a 40 Å periodic line structure. It is of interest that structures of this dimension should appear both in crystals enclosed in the cell and in degraded material. This would suggest that the value of 40 Å refers to one dimension of one of the primary building blocks of this presumably complex protein.

We are grateful to Dr. C. L. Hannay who supplied the strain used in this work. One of us (D. H. W.) is indebted to Imperial Chemical Industries Ltd. for a research fellowship received during the course of this work.

References

1. Hannay, C. L., and P. Fitz-James: Canad. J. Microbiol. **1**, 694 (1955).
2. Robinow, C. F.: J. Bact. **66**, 300 (1953).
3. Hannay, C. L.: 'Bacterial Anatomy' 6th Symposium of Soc. for General Microbiol. p. 318, 1956.
4. Bradley, D. E., and D. J. Williams: J. gen. Microbiol. **17**, 75 (1957).
5. Wyckoff, R. W. G.: Electron Microscopy. p. 72. New York: Interscience Publ. 1949.
6. Dawson, I. M.: Nature (Lond.) **168**, 241 (1951).
7. Farrant, J. L., and A. J. Hodge: Proc. Int. Conference on E. M. London, p. 118, 1954.

A survey of the surface structure of spores of the genus Bacillus

J. G. Franklin and D. E. Bradley

National Institute for Research in Dairying, University of Reading, Berkshire, and Research Laboratory, Associated Electrical Industries Limited, Aldermaston Court, Aldermaston, Berkshire (England)

Preliminary studies (*1, 2*) of the spores of species of the genus *Bacillus* using carbon replicas revealed a diversity of surface patterns. It was suggested that these structures might be of value in the classification of members of this group and, accordingly, a full examination of the genus has been carried out.

The organisms were cultured according to the method of Williams et al. (*3*). Single stage carbon replicas were prepared by coating the spores with c. 150 Å of carbon, dissolving them away in a permanganic/chromic acid mixture, and shadowing with gold/palladium at tan $^{-1}\frac{1}{2}$.

Smith, Gordon and Clark (*4*) subdivided the genus *Bacillus* into three groups according to the morphology of the spore and sporangium and the results obtained for these groups are as follows.

Group 1. Sporangia not definitely swollen

Species examined: *B. subtilis, B. pumilus, B. firmus, B. lentus, B. licheniformis, B. megaterium, B. cereus, B. coagulans.*

In this group, spores of *B. licheniformis* and *B. cereus* are unique in their surface structure and can be identified easily. *B. licheniformis* spores are characterised by a simple longitudinal groove, the depth of which varies according to the strain, in addition to some ribbing on the surface. The spores of *B. cereus* are usually spherical and are surrounded by an 'exosporium' obscuring a surface which is probably smooth. The variety *myocides* is similar, but the variety *thuringiensis* differs in that the spores are usually cylindrical in shape.

The following pairs of species were shown to have similar spores so that any unknown

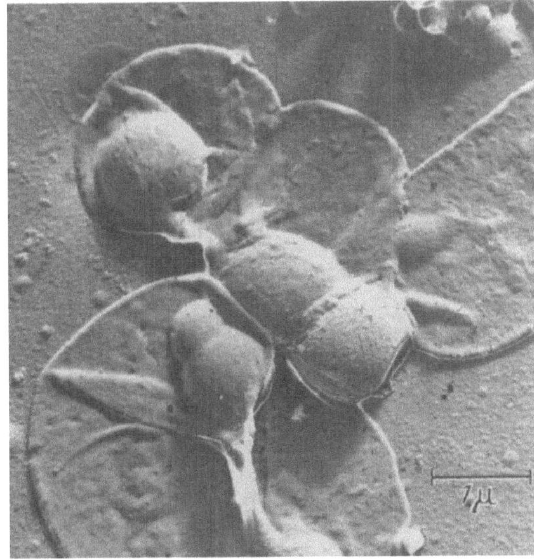

Fig. 1. Shadowed carbon replica of spores of *Bacillus cereus*, showing 'exosporium' (by permission of the J. Bact.)

organism could bei dentified as one of two species. *B. subtilis* and *B. coagulans* produce cylindrical or sometimes oval spores with the surface irregularly ribbed, the ribs often being longitudinal, but sometimes transverse at the ends of the spores. *B. coagulans* sometimes forms Group 2 type spores and these are described below. *B. pumilus* and *B. firmus* have spores with a surface configuration similar to that of *B. subtilis* and *B. coagulans*, but they are much smaller. *B. lentus* and *B. megaterium* form oval, or occasionally spherical spores with a surface which may be smooth or rough.

Group 2. Sporangia definitely swollen by oval spores.

Species examined: *B. polymyxa*, *B. macerans*, *B. alvei*, *B. circulans*, *B. brevis*, *B. laterosporus*, *B. pulvifaciens*, *B. staerothermophilus*, *B. calidolactus*, *B. coagulans*.

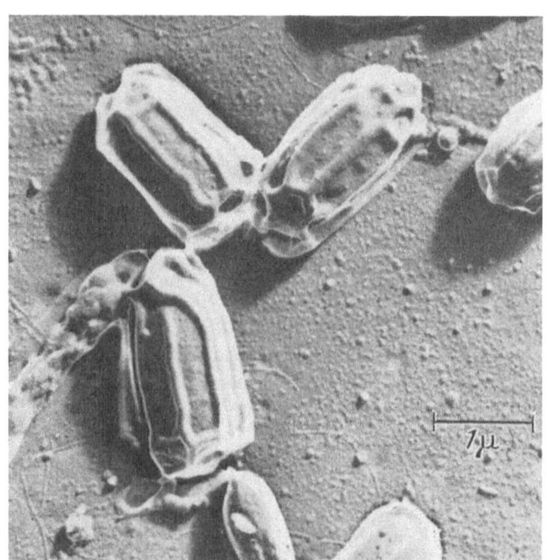

Fig. 2. Shadowed carbon replica of spores of *Bacillus polymyxa* (by permission of the J. Bact.)

The following single species of this group are unique and easily identified. *B. brevis* has oval spores with a smooth surface except for one or, occasionally, two longitudinal ribs, often terminating half-way down the spore. Spores of *B. laterosporus* are ornamented with well-defined, sparsely branching ridges; an extension of the spore coat appears as a C'-shaped protuberance on one side. The spores of *B. pulvifaciens* are characterised by four longitudinal ribs with shallow grooves on either side of the base. Spores of *B. stearothermophilus* which are comparatively large, appear smooth, but they are usually covered with a layer of contaminating material. *B. calidolactis* spores are patterned with parallel longitudinal ribs, but these are often indistinct.

The following pairs of species have similar spores. The spores of *B. polymyxa* and *B. macerans* show very pronounced ribs, usually numbering eight, but sometimes only four. These ribs are longitudinal and might be in the form of loops at the ends of the spores; alternatively the poles may be patterned with a reticulate network. It has not been possible to separate these species by electron microscopy. *B. alvei* and *B. circulans* show interesting variations. *B. alvei* spores are usually long, with indistinct parallel longitudinal ribs. The spores of *B. circulans* often exhibit similar ribs, or they may be smooth. A series of variations appear to exist, linking these species both in the surface structure of the spores, and in other characteristics (5).

Group 3. Sporangia swollen by round spores.

Species examined: *B. pantothenticus*, *B. sphaericus*, *B. pasteurii*.

The spores of each of these species are unique. *B. pantothenticus* spores are spherical to oval in shape with a slightly rough appearance. *B. sphaericus* spores are round and smooth, and difficult to obtain free of the vegetative cells. *B. pasteurii* spores are round, with a rough appearance and occasional ribbing.

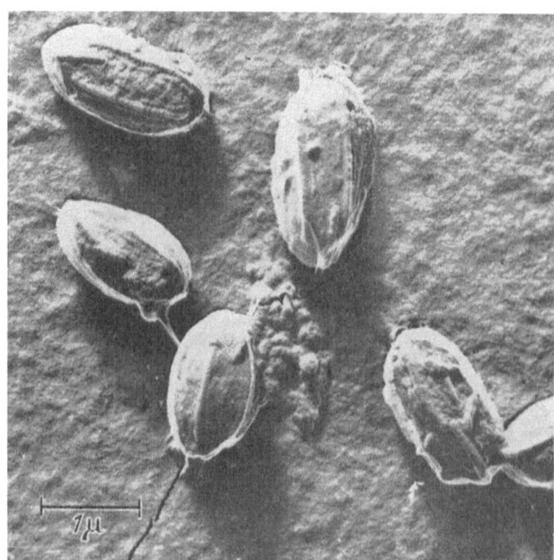

Fig. 3. Shadowed carbon replica of spores of *Bacillus pulvifaciens*

Conclusions. Only a few of the structures described can be illustrated here, but a complete series of micrographs of spores of the genus *Bacillus* has been prepared and has been published (6). The surface structure of an organism is clearly of value as an aid to its identification, and the electron microscope method has the added advantage of identifying components of a mixture, without the necessity of separation.

It might also be possible, by a more detailed examination of a large number of strains, to clarify the existing classification of a species such as *B. circulans* which is at present obscure. It may then be possible to relate the structure of the spore surface with other characteristics. The spore structures in themselves, of course, are of considerable morphological interest.

One of the authors (J. G. F) would like to thank Dr. A. T. R. Mattick and Dr. L. F. L. Clegg for the interest they have shown in this work, and the other (D. E. B.) would like to thank Dr. T. E. Allibone, F. R. S., Director of the Research Laboratory, Associated Electrical Industries, for permission to publish this paper.

References

1. Bradley, D. E., and D. J. Williams: J. gen. Microbiol. **17**, 75 (1957).
2. Franklin, J. G., and D. E. Bradley: J. appl. Bact. **20**, 467 (1957).
3. Williams, D. J., J. G. Franklin, H. R. Chapman and L. F. L. Clegg: J. appl. Bact. **20**, 43 (1957).
4. Smith, N., R. Gordon and F. Clark: U. S. D. A., Agr. Monograph No. 16, U. S. Govt. Printing Office, Washington D. C. 1952.
5. Knight, B. C. J. G., and H. Proom: J. gen. Microbiol. **4**, 508 (1950).
6. Bradley, D. E., and J. G. Franklin: J. Bact. **76**, 618 (1958).

Elektronenmikroskopische Studien über symbiontische Einrichtungen bei Insekten

Günther F. Meyer und Werner Frank

Institut für Physik und Meteorologie — Laboratorium für Elektronenmikroskopie — und Zoologisches Institut der Landwirtschaftlichen Hochschule Stuttgart-Hohenheim

Unter dem Begriff Symbiose seien in diesem Zusammenhang lediglich solche Einrichtungen verstanden, die Buchner schon seit vielen Jahren als Endosymbiosen bezeichnet hat und bei denen pflanzliche Mikroorganismen wie Hefen oder Bakterien innerhalb bestimmter, genau umrissener Zellen oder Zellbezirke des tierischen Wirtsorganismus leben. Derartige symbiontische Einrichtungen sind im Tierreich nicht besonders selten, erreichen aber ihre größte Verbreitung bei den Insekten. Die Symbionten haben eine ganz spezielle physiologische Bedeutung und dürften wohl in den allermeisten Fällen zur Produktion von Vitaminen herangezogen werden. Aus diesem Grund ist es nicht verwunderlich, daß nur diejenigen Insekten Symbionten besitzen, die eine sehr einseitige, spezialisierte Ernährung haben. Es gehören hierher alle pflanzensaft- und blutsaugenden Insekten sowie auch Horn-, Zellulosefresser usw. Symbionten finden sich aber nur dort, wo auch die Larve dieselbe Nahrung aufnimmt. So hat zwar eine Laus Symbionten, ein Floh dagegen nicht.

Eine Ausnahmestellung nimmt die Symbiose der Küchenschaben ein. Diese Tiere sind zwar Allesfresser, besitzen aber doch symbiontische Bakterien. Auf die möglichen phylogenetischen Beziehungen kann hier nicht eingegangen werden. Es sei nur kurz erwähnt, daß auch eine Termitenart (*Mastotermes darwiniensis*) aus Australien symbiontische Einrichtungen in von der Schabe kaum unterscheidbarer Anordnung besitzt. Termiten und Schaben sind aber eng verwandt und gehören wohl zu den ältesten Insekten überhaupt. Bei diesem von uns untersuchten Objekt finden sich die symbiontischen Bakterien nicht in Verbindung mit dem Darm, sondern innerhalb spezialisierter Zellen des Fettkörpers, den sogenannten Bakteriocyten. Die systematische Gruppierung aller Symbionten ist äußerst schwierig, da bis jetzt nur sehr wenige Formen kultiviert werden konnten und eine rein morphologische Zuordnung kaum möglich ist. Möglicherweise stehen die Bakterien der Schaben den Corynebakterien nahe.

Abb. 1 zeigt einen Schnitt durch eine Bakterienzelle im Fettkörper von *Blatta* und gibt einen guten Einblick in das enge Zusammenleben von tierischer Wirtszelle und pflanzlichen Mikroorganismen. Die Einlagerung von Bakterien in Einbuchtungen des Wirtszellenkernes lassen ein enges Zusammenspiel zwischen beiden Partnern vermuten. Eine Regulierung ist zu erwarten, da sich die Bakterien sonst in diesem optimalen Milieu unbegrenzt vermehren würden. Alle Bakteriocyten sind stets mehr oder weniger stark mit Symbionten angefüllt, die aber nie auf andere benachbarte Zellen übergreifen. Wie unsere Beobachtungen und auch Abb. 1 erkennen lassen, finden

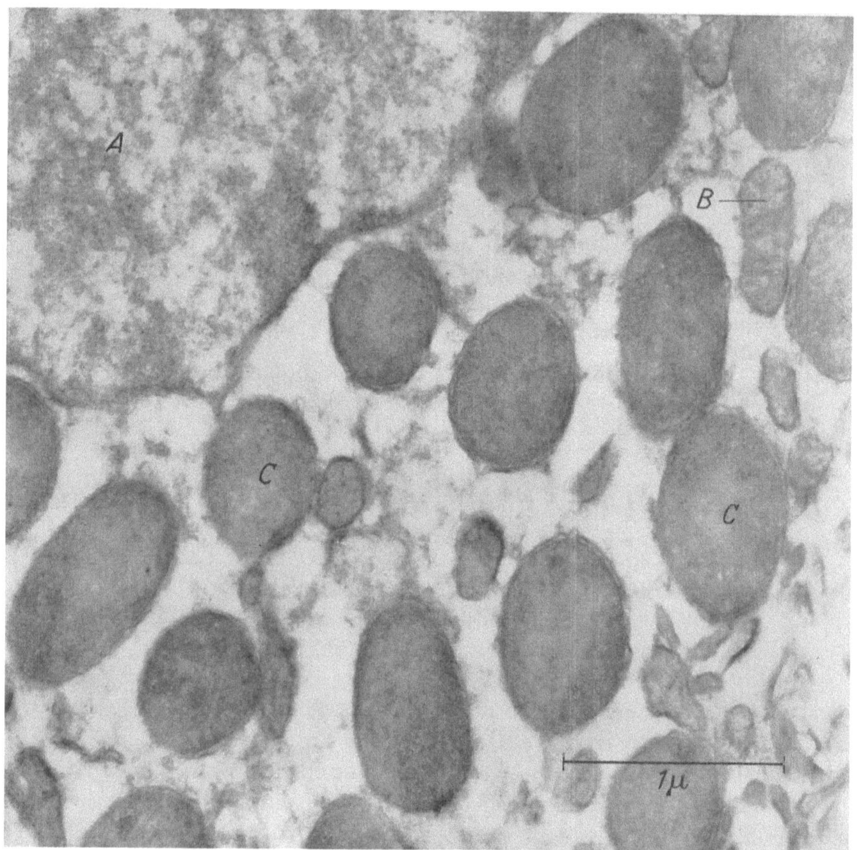

Abb. 1. Kern und Cytoplasma einer Bakteriocyte von *Blatta*. 30000mal. *A* Kern, *B* Mitochondrion, *C* Bakterien

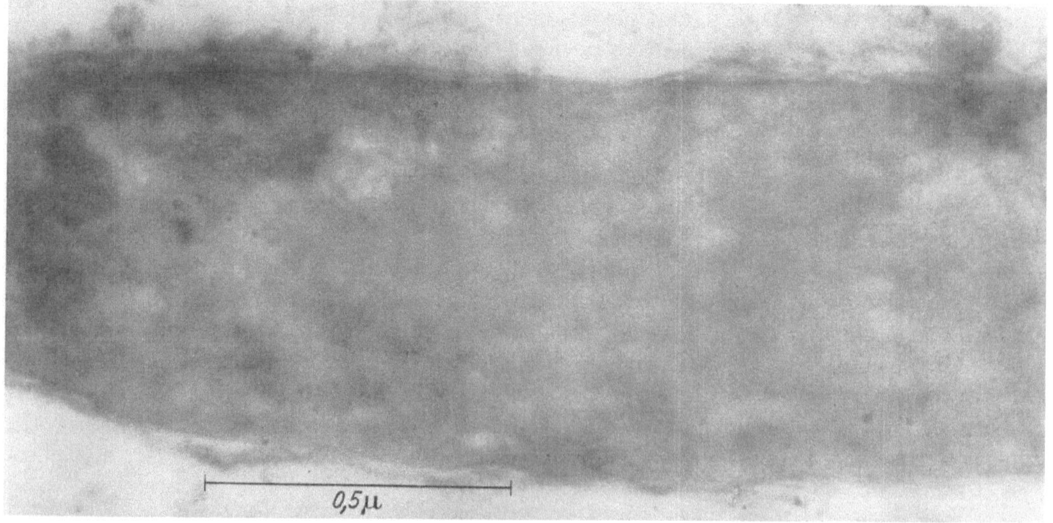

Abb. 2. Längsschnitt durch ein symbiontisches Bakterium. 85000mal. Auffallend homogenes Plasma, keine Kernäquivalente

sich auch noch in adulten Schaben zahlreiche Teilungsstadien, was darauf schließen läßt, daß ein ständiger Abbau vorhanden sein muß. Elektronenmikroskopisch lassen sich solche Abbauformen an ihrem stark granulierten Plasma erkennen. Sie weisen ähnliche Strukturen auf, wie wir sie nach Penicillineinwirkung gefunden haben.

Auffallenderweise zeigen alle symbiontischen Bakterien ein weitgehend homogenes, dichtes Plasma ohne Kernäquivalente in der bekannten Form (Abb. 2). Auch die Querschnitte durch die Bakterien in Abb. 1 sind fast völlig homogen.

Nachdem nun die normalen Verhältnisse elektronenmikroskopisch untersucht worden waren, haben uns die Veränderungen unter experimentellen Bedingungen interessiert. Wie Frank (1956)

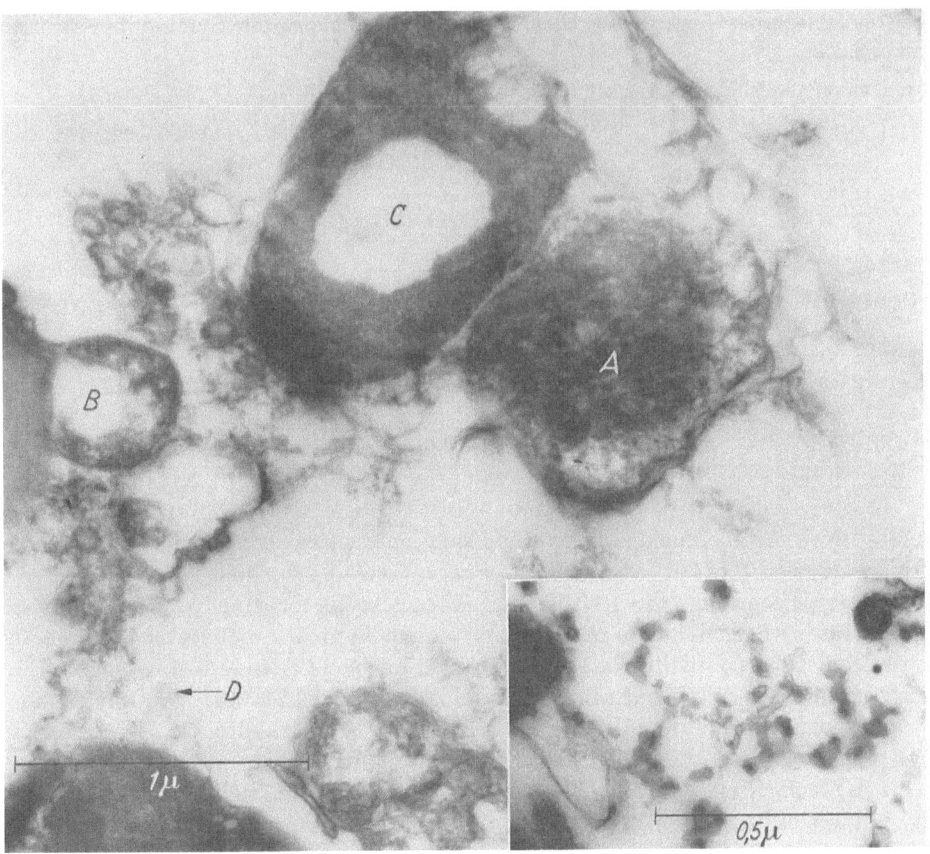

Abb. 3. Veränderte Bakterien nach Penicillineinwirkung. 40000 mal. *A* starke Granulierung, *B* Blasenbildung und *C* zentrale Lysis, *D* geschädigtes Cytoplasma. re. unten: verändertes Cytoplasma. 60000 mal

feststellen konnte, treten nach Applikation von Antibiotica bei den Schaben Wachstumsstörungen auf. Während die männlichen Tiere außer einem langsamen Wachstum keine Veränderungen erkennen lassen, ist die Schädigung der Weibchen sehr weitgehend und führt bis zur völligen Sterilität. Daneben haben die symbiontischen Bakterien auch noch Einfluß auf die Ausfärbung der Cuticula. Um eine völlige Vernichtung der Bakterien zu erreichen, sind bei Jungtieren über 3—4 Monate sehr hohe Dosen (50 i E/mg Futter) per os zu verabreichen. Jedoch kann derselbe Effekt erzielt werden, wenn man mit wenigen Injektionen (je 1000 i E) verhindert, daß die Bakterien auf das Ei weitergegeben werden. Solche Larven sind von Anbeginn an symbiontenfrei. Für die hier besprochenen Untersuchungen wurde Penicillin verfüttert, um den langsam fortschreitenden Abbau der grampositiven symbiontischen Bakterien und möglicherweise eintretende Veränderungen am Tier besser verfolgen zu können.

Abb. 3 zeigt derartige Abbauformen nach Penicillineinwirkung. Während anfangs, wie schon erwähnt, nur eine Granulierung des Plasmas eintritt, zeigen die späteren Bilder deutlich die typische formverändernde Wirkung von Penicillin mit zentraler Lysis, Blasenbildung und starker Zunahme der Osmiophilie. Die Entstehung von L-Phasen konnte aber nicht beobachtet werden.

Dies bestätigt andererseits, daß die Bakterien bis jetzt gegen Penicillin noch nicht resistent geworden sind.

Die hier besprochenen Tatsachen haben gezeigt, daß es sich bei der intracellulären Symbiose von *Blatta* um ein wirkliches harmonisches Zusammenleben von Bakterien und Wirt handelt, bei dem zwar der Nutzen für die Bakterien klar auf der Hand liegt, während eine genaue Analyse des eventuellen Nutzens für den Wirtsorganismus bisher nur auf Grund von Ausfallserscheinungen nach Ausschalten der Bakterien beurteilt werden kann. Auch elektronenmikroskopisch sind keinerlei Anzeichen vorhanden, daß von seiten des Wirtskörpers irgendwelche Abwehrreaktionen gegenüber den Bakterien — von einer Steuerung der Vermehrung der Symbionten abgesehen — vorhanden sind.

Literatur: FRANK, W.: Z. Morph. u. Ökol. Tiere **44,** 329 (1956)

3. Virologie

Sedimentation counting of particles via electron microscopy*

D. GORDON SHARP

Department of Bacteriology, School of Medicine, Univ. of North Carolina, Chapel Hill, N. C.

Electron microscope counts of virus particles have usually required highly concentrated, purified, preparations with a minimum of extraneous material (*2, 4, 6*). ISAACS has thoroughly reviewed these results recently (*5*). But, the virologist frequently does not require such purity. He generally deals with crude tissue extracts, plasmas, chorioallantoic fluids, etc., in which virus activity may be 10^6 to 10^9 ID_{50} per ml. It is rarely greater than 10^{10}. This paper will describe electron microscope counting of two viruses in just such crude form.

In certain cases where the virus particles are present in great numbers or are large and sufficiently different from contaminating material to be recognized, they can be counted without preliminary purification or concentration procedures (*7, 10*) which would certainly require more material and would probably result in loss of virus particles as well as activity. The sedimentation method of preparing suspensions of virus particles for counting in the electron microscope (*7, 8, 9, 10, 11*) is particularly effective in this work, not only because it involves concentration of the virus and is therefore quite sensitive, but because it can now be used, like the spray droplet method of BACKUS and WILLIAMS (*1*), to compare the numbers of virus particles and indicator particles in mixtures. These improvements in the technique will be discussed as they are demonstrated by counts of unpurified vaccinia virus as well as crude, and partially purified meningopneumonitis virus. Improved accuracy and precision in sedimentation counting is the result of studies to find a receiving surface for the sedimented particles which would hold not only virus but also polystyrene particles. The second objective of this paper is to describe results with protein coated collodion surfaces which gave good results with our materials and seem to have a broad field of application.

Vaccinia virus was prepared as 10% extracts of infected chorioallantoic membrane or of infected rabbit skin, by methods already described (*11*).

Meningopneumonitis virus[1] (*MV*) was counted in crude form directly from diluted chorioallantoic fluid six days after inoculation; or the fluid was partially purified. After preliminary clarification at low speed, the virus was sedimented from the chorioallantoic fluid (3,500 G for 30 min) and resuspended 25 × concentrated, in 5% sodium desoxycholate for 30 min at 27° C. It was then sedimented twice more with resuspension in distilled water.

---

* This investigation was supported in part by a research grant RGE-5177 from the Division of Research Grants, US Public Health Service and further aided by a gift of equipment from the Ivan Sorvall, Inc., Norwalk, Conn.

[1] Prepared through the kind cooperation of Dr. PHILIP MANIRE and KENDALL SMITH, M.S., of the Department of Bacteriology.

Polystyrene Latex Particles (Dow Chemical Co., Midland, Mich.) were available at 0.264 μ and 0.814 μ diameters in suspensions containing 10.2 and 11.2 grams per 100 ml respectively. The density of these particles is given as 1.06 by the manufacturer.

The Sedimentation Method for preparing a concentrated display of virus particles for counting in the electron microscope has been described elsewhere (*8, 9, 10, 11*). In this work the vaccinia virus and some *MV* counts were made from agar receiving surfaces as previously described (*10, 11*). *MV* and polystyrene particles were sedimented and more accurately counted, upon gelatin-coated collodion films. These were prepared by putting drops of Parlodion[1] (0.7% solution in amyl acetate) on the surface of dilute gelatin-water solutions. Stainless steel 200 mesh wire grids were dropped on the resulting film and removed with protein coating uppermost. After 15 min drying they were ready for use. In place of the 2 mm thick agar receiving surface which was used for vaccinia virus, a 2 mm thick piece of plastic (Lucite) with $^1/_4''$ diameter hole was substituted. Closely fitting into this hole was a $^1/_4''$ diameter, 2 mm thick disc cut from a strong bar magnet, Fig. 1. When this assembly was inserted at the periphery (bottom) of the counting rotor cell the stainless steel grids could be placed upon the magnet face where they were held tightly enough to remain in position while filling and draining the cells. Each of the eight cells of the Ivan Sorvall type SU virus counting rotor (*11*) was then filled with 1.3 ml of properly diluted particle suspensions and spun 15 min at 7,300 G. After stopping the rotor, the Lucite piece with magnet and adhering specimen screen was dried 15 min before removal of specimen.

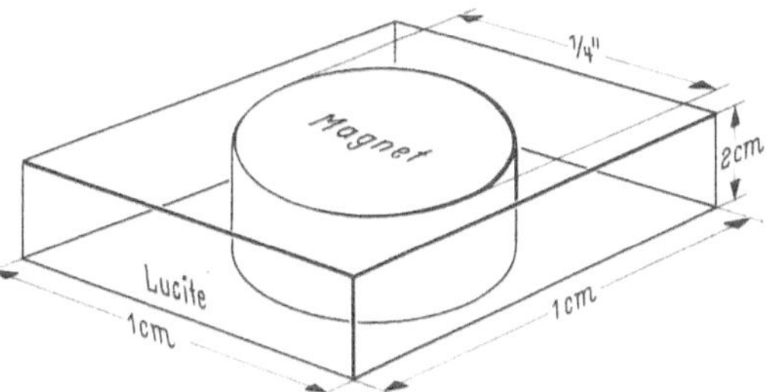

Fig. 1. Plastic (Lucite) embedded magnet for holding stainless steel grids in position at bottom of centrifuge cells

Premature removal of screens from magnet surface destroys the collodion films. All specimens were shadowcast with chromium and random fields were photographed for count in the RCA type EMU 3C electron microscope taking pictures $2'' \times 2''$, square.

The Spray Dropled Method of Backus and Williams (*1*) was used for counting some concentrated particle suspensions for comparison with results of the sedimentation method. The various mixtures of polystyrene and virus contained 0.05% bovine serum albumin to outline the drops. They were sprayed from an air gun

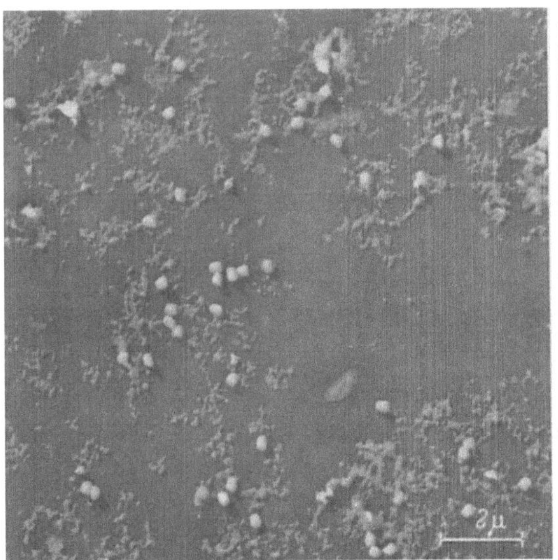
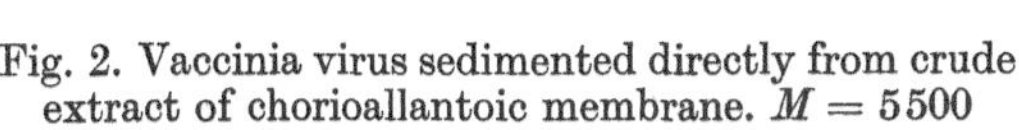

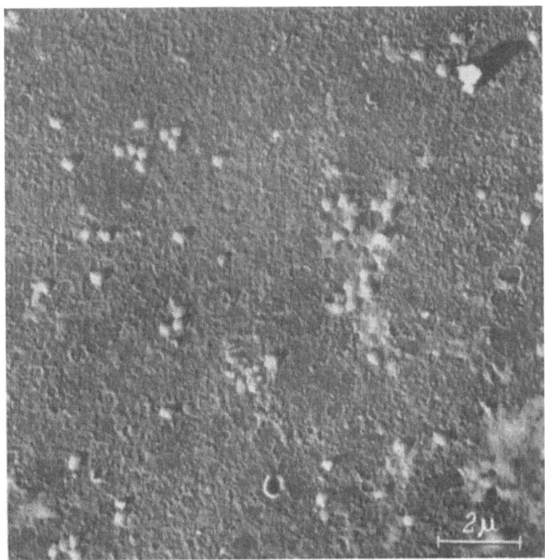

Fig. 2. Vaccinia virus sedimented directly from crude extract of chorioallantoic membrane. $M = 5500$

Fig. 3. Vaccinia virus sedimented directly from crude extract of rabbit skin. $M = 5500$

upon collodion covered grids and shadowcast. Only water dilutions of virus and polystyrene were sprayed. All measurements of volumes less than 0.5 ml, for these and the sedimentation counts, were made by weighing. Volumes 0.5 ml and larger were pipetted.

Sedimentation Counting of Crude Vaccina Virus was most effective when the 10% suspensions of infected chorioallantoic membranes were diluted 1—300 with water or 0.005 M phosphate

---

[1] Mallincrodt Chemical Works.

544   D. Gordon Sharp:

buffer, p_H 7, before sedimenting upon an agar surface. Pseudoreplicas from these surfaces regularly showed virus particles and particulate foreign matter sufficiently dispersed to allow quite accurate counting of the former. All photographs were made at magnifications less than $2000 \times$ in order to include as many virus particles as possible. A part of one of these pictures is shown in Fig. 2 enlarged sufficiently for recognition of the virus particles[1].

Ten percent extract of infected rabbit skin material were somewhat more difficult to count (*11*) because of particulate foreign matter, (Fig. 3). Dilutions of 1—500 were generally required before suitable dispersions were obtained. In cases where aggregated material was particularly troublesome, buffer dilutions (1—500) of the tissue extract were treated 5 min in the Raytheon Sonic Vibrator at 9 KC. Agitation of tissue extracts with sodium desoxycholate (*3*) has sometimes given good results and a decreased amount of sedimented debris, but often there is gel formation and the procedure has, as yet, not proven generally useful in this counting work.

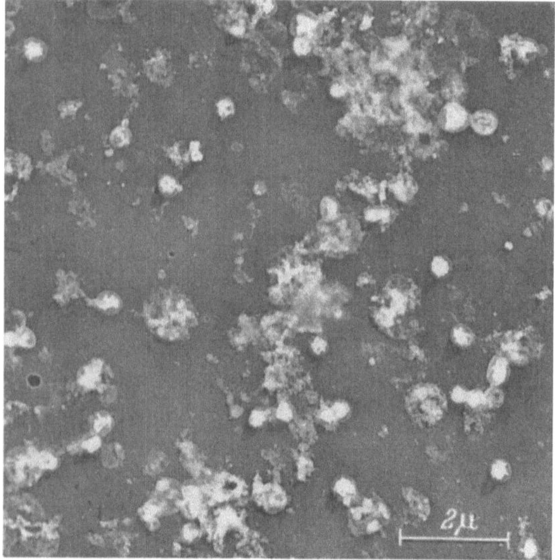

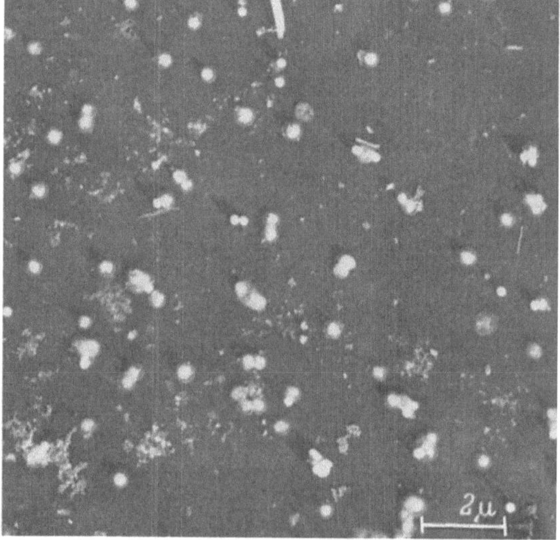

Fig. 4. Meningopneumonitis virus sedimented directly from crude chorioallantoic fluid. $M = 7350$

Fig. 5. Partially purified meningopneumonitis virus + polystyrene reference particles sedimented on gelatin-coated collodion. $M = 5500$

Crude vaccina preparations from 10% chorioallantoic membrane have yielded about 200 virus particles per photograph at 1—300 dilution. Five such pictures are sufficient to give an average count whose standard deviation is 10—15% of the mean of a series of such counts. Rabbit skin extracts of maximum potency and 1—666 dilution averaged about 50 particles per photograph. These data correspond to particle counts in the 10% extracts of 7×10^9 and 4×10^9 per ml respectively (*11*). Spray droplet particle counts made with concentrated and partially purified vaccinia virus have been in good agreement with sedimentation counts made on dilutions of the same samples.

Sedimentation Counting of Meningopneumonitis Virus from crude chorioallantoic fluid was done via collodion pseudoreplica from agar receiving surfaces. Typical pictures such as Fig. 4 from fluids diluted 1—100 have shown 100 virus particles when taken at $1700 \times$ magnification. This would correspond to a virus particle count of 1.3×10^9 per ml of crude fluid. It was soon evident, however, that this was not a reliable quantitative procedure. This virus does not replicate quantitatively from the agar surface upon collodion nor any other film tried for this purpose. In fact, when the virus is spun from 1—100 water dilutions of infected chorioallantoic fluid, upon agar, and two successive collodion films are stripped from the surface, both virus particles and foreign material appear on the first while on the second, about 10% of the virus is found

---

[1] All virus pictures are at just sufficient magnification for recognition of the virus particles. Thus, some idea of the distribution of particles and debris may be obtained from them.

with practically no foreign matter. There is, evidently, not only incomplete replication of virus but some selective action present. A wide range of diluents of various salts and various concentrations as well as diverse adjustments of p_H had no effect of improving the quantitative value of collodion pseudoreplicas beyond this point, for counting this virus.

Direct sedimentation of the virus on collodion-coated specimen grids was equally unsuccessful in producing uniformly deposited dispersions when the virus was diluted with water. When, however, 0.1 M phosphate buffer, p_H 7, was used as diluent some improvement in uniformity was observed. Areas, where the count was high, were well dispersed and yielded values in agreement with those from spray counts but overall uniformity was never obtained. Many areas were sparsely covered or almost blank.

When gelatin was added to the water upon which the collodion films were prepared a dramatic improvement was observed. Not only did MV adhere evenly to these films but polystyrene particles[1] as well (Fig. 5). The value of the latter observation was seized upon as a direct means of evaluating the overall accuracy of sedimentation counting. In the first experiments 0.5% gelatin was used in the water bath. Polystyrene (0.264 μ diameter) particles diluted 86,600; 117,000; 230,000; and 8,660,000 $\times$ respectively with 0.1 M phosphate buffer p_H 7.0 were sedimented on the gel-collodion and the number of particles counted on each of five random photographs was averaged to give a point on Fig. 6. The four points so obtained are strictly consistent with dilution and the value of latex concentration calculated from these points are 0.91, 0.89, 0.95 and 0.95 $\times$ 10^{13} particles per ml respectively. The expected number was 10^{13}. The total number counted in this experiment was about 11,000.

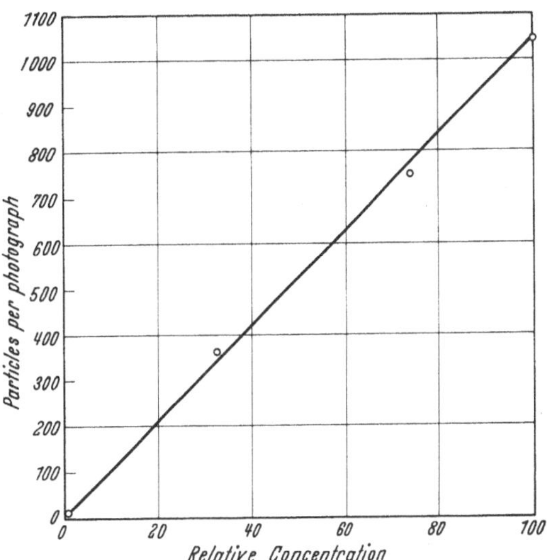

Fig. 6. Latex particles (0.264 μ diameter) per photograph, counted from 4 different dilutions when sedimented from 0.1 M phosphate buffer pH 7, upon 0.5% gelatin-coated collodion. Total particles counted 11,000

The quantity of gelatin necessary to produce satisfactory coatings on collodion was determined as follows: Collodion films were prepared on water containing 0.5, 0.05, 0.005 and 0.0005% gelatin respectively. Latex of particle size 0.264 μ was diluted with 0.1 M phosphate buffer pH 7.0

Table 1. *Counts made from one suspension of latex particles (0.264 μ diameter) sedimented on collodion coated with different concentrations of gelatin*

gelatin concentration percent	areas counted	particles per area (mean)	variance (standard deviation)	total particles counted	calculated particles per ml. of latex
0.5	40	20.8	18.2	832	0.83×10^{13}
0.05	40	22.5	23.6	901	0.91×10^{13}
0.005	40	21.5	27.8	860	0.87×10^{13}
0.0005	40	24.0	66.6	965	0.97×10^{13}

average 0.90×10^{13}
particles per ml of latex 1.0×10^{13}

to give about 165 particles per photograph at 1520 $\times$ electron magnification. Each photograph was divided into 8 equal areas and the particles on each were counted. The results, together with pertinent statistics, are presented in Table 1. Each of the 4 sets of 5 pictures was averaged to give a count for each gelatin concentration. Only the lowest gelatin concentration gave serious

---

[1] Polystyrene particles, which are so useful in comparative counting, cannot be used in agar pseudoreplica procedures because of their slight solubility in the collodion solvents.

evidence of erratic distribution of particles as shown by a value of the variance 2.8 times greater than the mean. Gelatin concentrations in the range 0.005 to 0.5% seem to be equally suitable.

A mixture of latices containing equal numbers of 0.264 μ and 0.814 μ diameter particles was made in water and sprayed as micro-droplets on collodion film. Thirty whole drops were photographed and the number of each particle's type was counted. Total particles in each drop averaged 53. Separate number ratios of small-to-large particles were calculated and are shown on Fig. 7. The mean ratio was 1.34 with a standard deviation of 0.34 for 1641 total particles counted.

This mixture was then diluted 1—377 with 0.1 M phosphate buffer pH 7, and sedimented upon collodion coated with 0.5% gelatin. Twenty pictures were taken at random. There was an average of 220 particles, total, on each picture and 20 ratios were calculated for comparison with the spray droplet data. These, also, are shown plotted on the same coordinates (Fig. 7). The mean ratio was 0.97 with standard deviation 0.14. This is in close agreement with the ratio 1.00 expected in the mixture.

Partially purified MV was sedimented from suspension in 0.05 M phosphate buffer at pH 7, upon collodion coated with 0.5% gelatin. The virus particles were well distributed, as they had never been upon other receiving surfaces. Ten photographs were made, on which the average count was 215. Similar experiments using successive 2 fold dilutions of phosphate buffer diluent gave average counts of 211 for 0.025 M phosphate, 200 for 0.12 M phosphate and 143 for 0.006 M phosphate respectively. Good distribution and uniform counts were, thus, obtained in the range of buffer concentration 0.1 M to 0.013 M but not at 0.006 M.

In one experiment a water suspension of partially purified M V and 0.264 μ polystyrene particles was photographed both in microdrops and in 1—326 dilution with 0.1 M phosphate buffer sedimented on 0.5% gelatin-coated collodion (Fig. 5). The ratios of MV to polystyrene particles were 1.47 and 2.41 for the two methods respectively. The number of polystyrene particles per photograph of the sedimented material was 95% of the number expected from dry weight. This is the first time in the author's experience that sedimentation methods have indicated a virus count substantially (64%) higher than that of the spray droplet procedure. It is difficult to account for this result without assuming MV particles were somehow lost from the spray droplets.

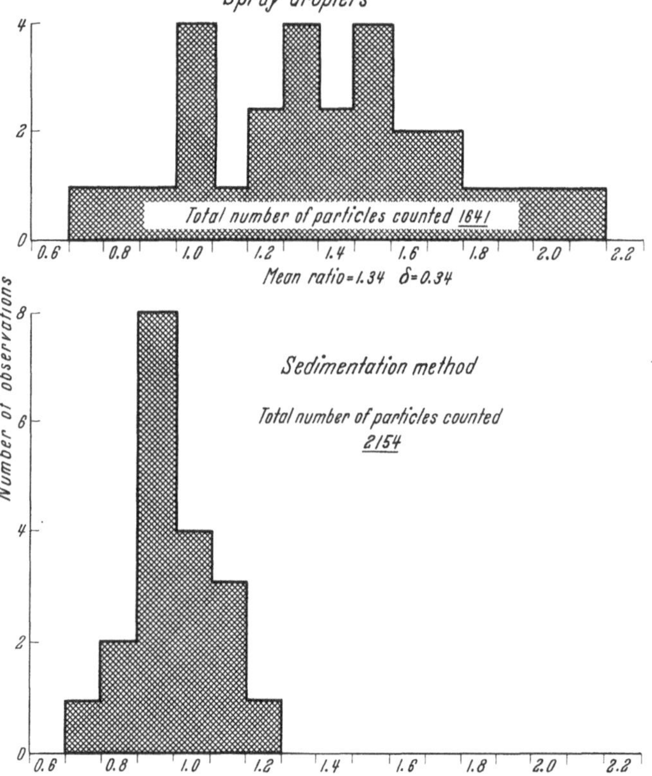

Fig. 7. Ratios of the numbers of 0.264 μ to 0.814 μ diameter latex particles counted in a mixture containing *equal* numbers of each. Top set — spray droplet method, lower set — sedimentation on gelatin-coated agar

Discussion. Agar has many advantages as a receiving surface for sedimented virus particles which are to be counted by pseudoreplica in the electron microscope, even though it is unsuited to counting of meningopneumonitis virus (MV) or polystyrene particles. Data presented here and elsewhere (*11*) show it to be excellent for counting of vaccinia virus sedimented from dilute crude suspensions of chorioallantoic membrane and rabbit skin. Treatment of tissue suspensions of vaccinia virus with sodium desoxycholate prior to sedimentation upon the agar surface sometimes reduces the amount of foreign matter sedimented and so reduces the dilution necessary to achieve a field clear enough to count. This, of course, tends to increase the sensitivity of the method for detecting virus particles in suspensions of low titer. From crude CAM extracts containing 7×10^9 vaccinia virus particles per ml., counting photographs of suitable dilutions regularly show 200 readily distinguishable particles. Rabbit skin extracts have contained more debris. One hundred particles per photograph can be distinguished in 1—600 dilutions from materials containing 8×10^9 per ml. before dilution.

In the search for a suitable receiving surface for MV, collodion coated with gelatin was found excellent not only for this virus both in crude chorioallantoic fluid (diluted 1—100) and in semi-purified form but also for polystyrene latex spheres. Best results are obtained when phosphate buffer p_H 7 is used as diluent but the concentration does not seem critical. Concentrations in the range 0.1 M to 0.012 M yield good particle distribution but pictures of particles sedimented at the higher concentration tend to show them ringed with salt. Although most of the particle count data from gelatin-coated collodion films were taken with 0.1 M phosphate buffer diluent and 0.5% gelatin treatment, survey experiments indicate lower concentrations of both may be optimum.

In several experiments, some examples of which have been described, the potential accuracy of sedimentation counting of latex and virus particles has been demonstrated. It is now reasonable to ask if the small discrepancies remaining between numbers counted and expected numbers of latex particles may be due to causes other than imperfect retention of particles by receiving surfaces. The number N, of 0.264 μ particles of density 1.06, per ml of a 10.2% latex should be 1.00×10^{13}. The corresponding number N_c calculated from sedimented films photographed at magnification M is given by,

$$N_c = \frac{n\,M^2\,D}{k\,A\,h}\,,$$

where n is the number of particles counted on picture area A, h is the height of liquid in the cell, D is the dilution factor and k is a constant cell shape factor. Clearly, N_c depends heavily upon an accurate determination of the electron magnification M[1] and N depends likewise upon the accuracy of the particle radius $(0.264\,\mu)^3$. It is pertinent, now, to reexamine the variables involved to determine methods for overall improvement. Latex particle diameter is doubtless the most sensitive variable, most of the measurement of which have been made via electron microscopy involving the factor M. Doubtless repetition of earlier estimates of particle radius from sedimentation velocity (5) should be repeated. In such experiments particle radius is proportional to the square root of the sedimentation rate which can be readily determined to $\pm 2\%$ accuracy. This could result in a more reliable value of N. Uncertainty in M has been variously estimated but frequently $\pm 5\%$ has been stated. Sedimentation counts of polystyrene spheres can now be made which approach this order of precision and they may soon be a basis for calculation of a more reliable value for the overall (average over the entire field) magnification of the electron microscope.

If an independent value for latex diameter can be established it will no longer be necessary to know M. To this end, some experiments reported here, have been directed. These include both spray droplet and sedimented samples of mixtures. Two examples of inexplicable inconsistencies have been observed. In the first, a mixture of equal numbers of 0.264 μ and 0.814 μ latex particles yielded a small-to-large particle ratio of 1.34 by spray droplet and 0.97 by sedimentation analysis. The second discrepancy arose when a mixture of MV and 0.264 μ latex particles was compared. Here, again, there was disagreement, the ratio of MV to latex was 1.47 for spray droplets and 2.41 for sedimented preparations. In this mixture there were 4.5×10^9 latex particles per ml. The number calculated from photographs of the mixture on gelatin-agar films was 4.3×10^9 per ml. There was, thus, 1.6 times more virus seen on the sedimented pictures than would have been predicted from spray droplet analysis. It is difficult to avoid the suspicion that all types of particles are not necessarily shown in their true proportions in spray droplet analysis. Further work

---

[1] Both projector and objective lenses of the electron microscope were carefully normalized before taking of each picture used for particle counting.

must certainly be done to clear up this apparent difficulty for the individual advantages of both sedimentation and spray droplet counting are valuable and neither should be discredited.

Gelatin coating on collodion films has proved exceedingly useful in improving their adhesive effect for sedimented MV particles. It is possible that other viruses may require other proteins with other isoelectric points or even oriented monolayers of proteins. The possibility that selective adhesion may occur between sedimented virus and antibody-coated collodion must not be overlooked. Isaacs has said, regarding virus counting by electron microscopy: "The difficulty is not how to count but what to count" (5). Suitable selective protein coatings may contribute to the solution of this problem.

References

1. Backus, R. C., and R. C. Williams: J. appl. Physics **21**, 11 (1950).
2. Crocker, T. T., and B. M. Bennett: J. Immunol. **69**, 183 (1952).
3. — J. Immunol. **77**, 1 (1956).
4. Dumbell, K. R., A. W. Downie and R. C. Valentine: Virology **4**, 467 (1957).
5. Isaacs, A.: Advanc. in Virus Res. **4**, 111 (1957).
6. Kellenberger, E., and W. Arber: Virology **3**, 245 (1957).
7. Overman, J. R., and I. Tamm: Proc. Soc. exp. Biol. (N. Y.) **92**, 806 (1956).
8. Sharp, D. G.: Proc. Soc. exp. Biol. (N. Y.) **70**, 54 (1949).
9. — and J. W. Beard: J. biol. Chem. **185**, 247 (1950).
10. — — Proc. Soc. exp. Biol. (N. Y.) **81**, 75 (1952).
11. — and J. R. Overman: Proc. Soc. exp. Biol. (N. Y.) **99**, 409 (1958).

Electron microscopic studies of a virus of the psittacosis-lymphogranuloma group in tissue culture cells

N. Higashi and K. Notake

Institute for Virus Research, Kyoto University (Japan)

Strain L cells were cultured at 37° C for 3 to 4 days in bottles containing a nutrient medium which was usually used for the growth of L cells. After removing the nutrient fluid and washing the cell layer surface, the bottles received inoculation of Cal 10 strain of meningopneumonitis virus. The maintenance solution did not contain antibiotics.

The bottles were reincubated at 35° C. At intervals from 10 to 48 hr after infection, the infected cells were removed, washed and centrifuged to form small pellets, which were fixed for 2 hr in OsO_4 at pH 7.4. Dehydration, embedding in methacrylate and cutting were done as usual.

The first appearance of inclusions in thin section developed at 13 hr infection. Fig. 1 illustrates very low dense cytoplasmic inclusions which are filled with varieties of developmental forms of virus. Some of them are about to divide. It is interesting to note that mitochondria lie adjacent to the border of the inclusion. This phenomenon was repeatedly recognized, but it has nothing to do with the formation of inclusions. Mitochondria may be pushed out by the process of inclusion formation. Fig. 2 shows small elementary bodies possessing an eccentric or central dense body surrounded by a limiting membrane as well as large circles or polygons showing no internal structure. As the infection proceeds the number of elementary bodies increases (Fig. 3). Therefore, elementary bodies may come to maturity from the larger polygons.

References: Gaylord, W. H.: J. exp. Med. **100**, 575 (1954).

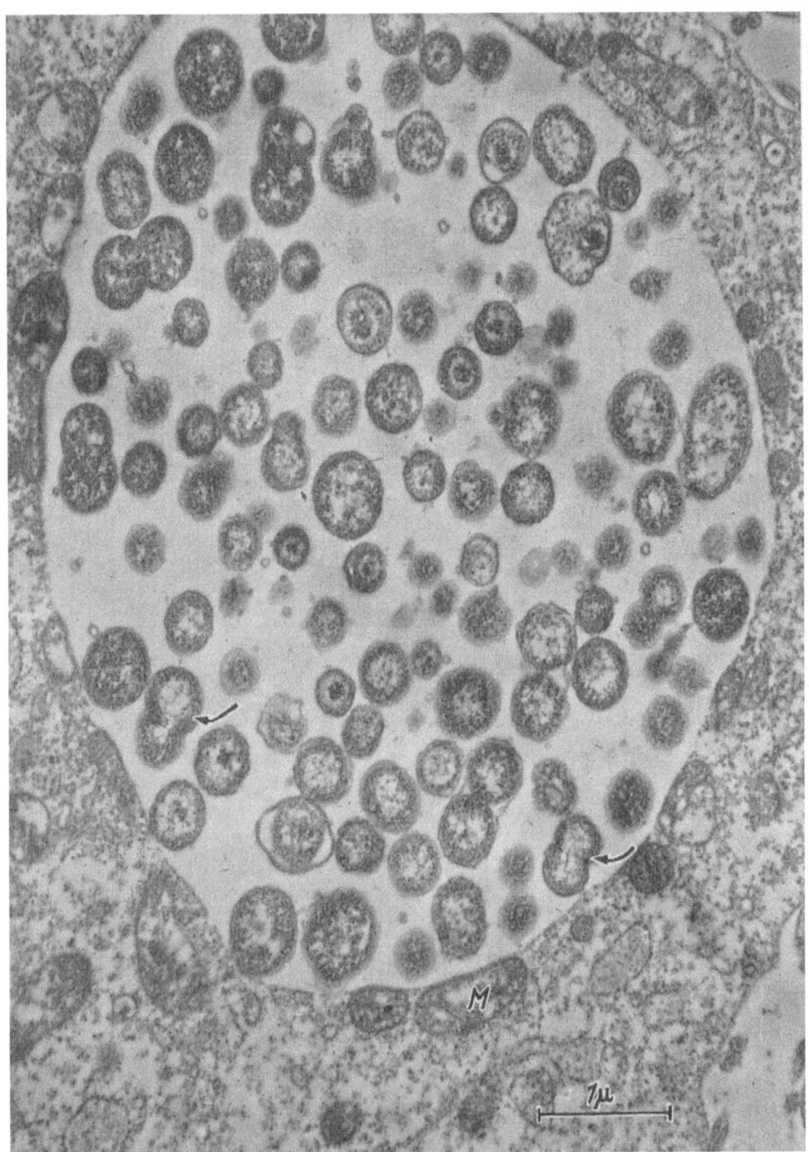

Fig. 1. Inclusion formed in meningopneumonitis virus infected strain *L* cell. The inclusion is filled with varieties of developmental forms of virus, some of which are about to divide as indicated by arrows. Mitochondria (*M*) lie adjacent to the border of the inclusion. 14 hr-infection. 18,000 ×

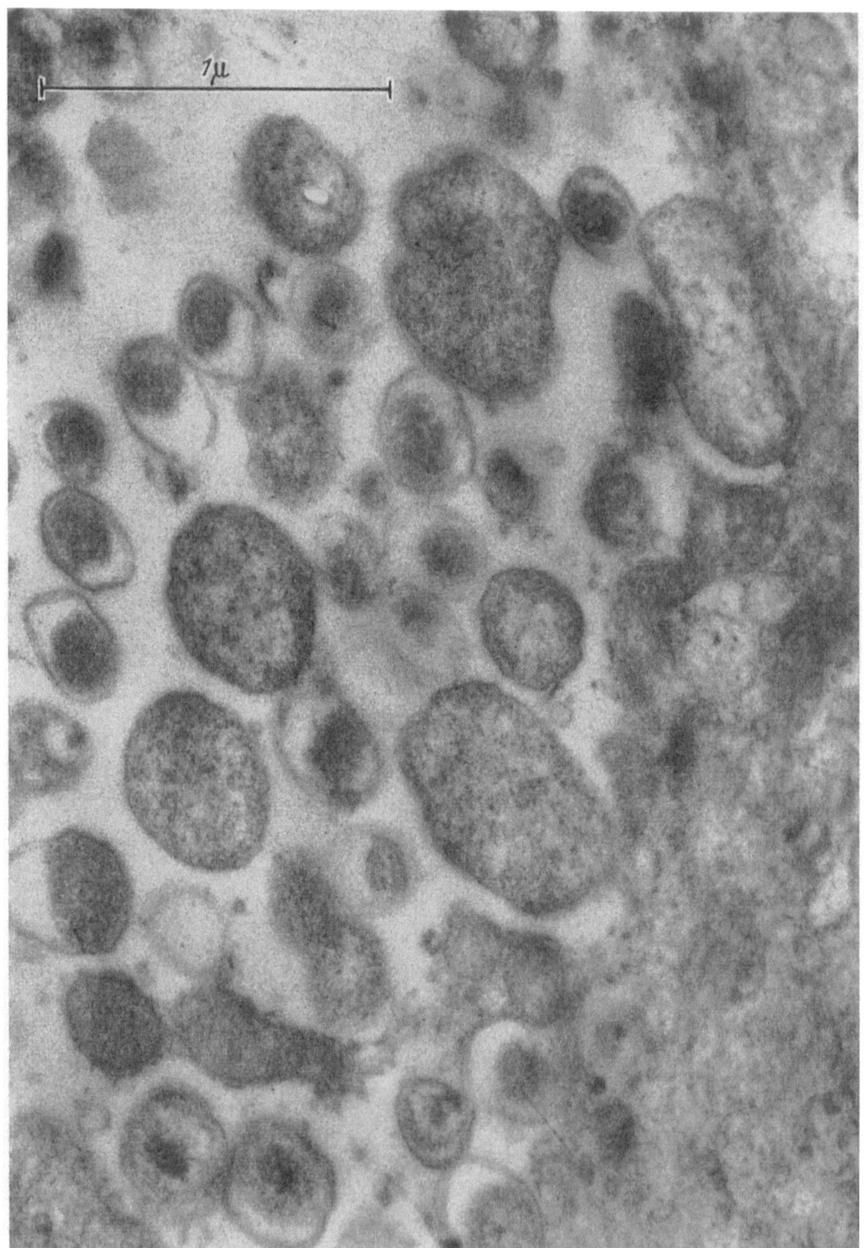

Fig. 2. Viral particles viewed at higher magnification. Small complete particles possessing an inner body and a limiting membrane, and huge polygons are shown. 35 hr-infection. 48,000×

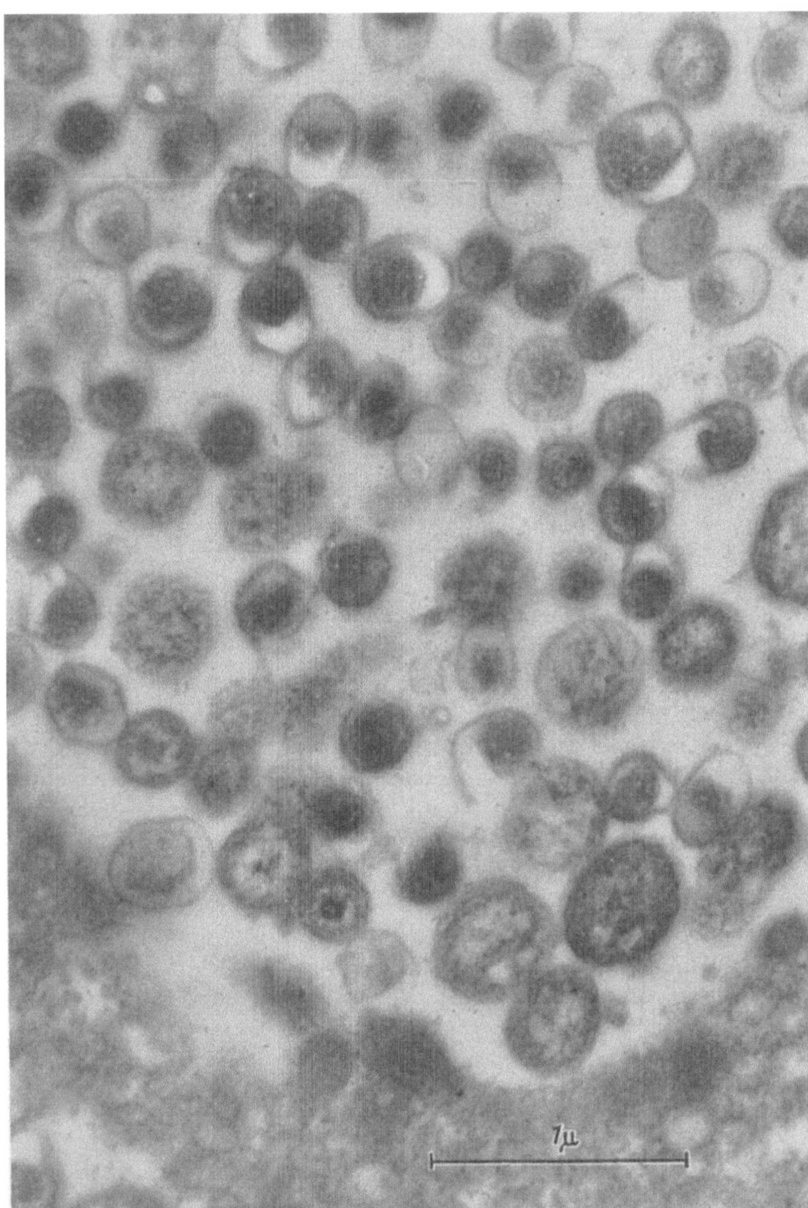

Fig. 3. Thin section through the inclusion of Cal 10 strain of meningopneumonitis virus. 48 hr-infection. 35,000 ×

Dietrich Peters:

Struktur und Entwicklung der Pockenviren

Dietrich Peters

Bernhard Nocht-Institut für Schiffs- und Tropenkrankheiten, Hamburg, Abteilung für Virusforschung

Die ersten zwanzig Jahre elektronenoptischer Strukturforschung an Viren der Pockengruppe haben mit einer Fülle von Einzelbeobachtungen, die in etwa 100 Originalarbeiten niedergelegt sind, die Morphologie der Elementarkörper (Elk.) und ihrer Entwicklung aus der Stagnation lichtoptischer Erkenntnisse herausgeführt. Noch ist das zentrale Problem, nämlich der Vermehrungsmodus der Viruspartikel, nicht in allen Stufen geklärt; insofern ist die Lage bei diesen Viren heute nicht günstiger als bei anderen. Die folgende Darstellung wird aber zeigen, daß gerade bei den Pockenviren, die sich durch typische Gestalt, Größe und Stabilität vor allen anderen auszeichnen, besonders gut fundierte Kenntnisse vorliegen. Der Gewinn dieser Entwicklung liegt jedoch nicht nur in dem vertieften Einblick in die Virusstruktur, sondern auch in der Beherrschung neuer elektronenoptischer Methoden, die möglicherweise auch auf andere Viren übertragen werden können.

Ähnlichkeiten im biologischen Verhalten der verschiedenen Pockenviren entspricht eine weitgehende Übereinstimmung ihrer Struktur. Nachdem es bereits 1938 gelungen war, typische Elk. dieser Gruppe elektronenoptisch abzubilden (*1*), beschrieb H. Ruska 1941 als strukturelles Merkmal die charakteristische Quaderform (*2*). Auf Grund des gemeinsamen Bauprinzips faßte er schließlich die Viren der Variola vera, der Vaccine, des Molluscum contagiosum, der Geflügel- und Kanarienpocken, der Ektromelie und des Kaninchenmyxoms zur Gruppe der sog. Quaderviren zusammen (*3, 4*). Spätere Arbeiten haben die nahe Verwandtschaft dieser Viren wiederholt bestätigt. Weitere Erreger, z. B. die des Kaninchenfibroms (*5, 6, 7*), der Paravaccine (*8*), der Schweine- (*9*), Schaf- (*10*) und Büffelpocken (*11*), wurden später in diese Gruppe einbezogen; vielleicht gehören auch die Viren der Stomatitis papulosa (*12*) und des Ecthyma contagiosum (Orf) (*13*) dazu.

Soweit bisher bekannt, ist die Grundstruktur der genannten Viren identisch. Unterschiede im Detail sind mangels systematischer Untersuchungen bislang nicht gesichert. Zunächst ist daher die Annahme berechtigt, daß sich Abweichungen von der Norm, über die hier und da berichtet wurde, durch Degeneration der Viruspartikel und Präparationsartefakte oder durch zu geringes Beobachtungsmaterial erklären lassen. Lediglich im weit fortgeschrittenen Stadium der Zellinfektion, im Stadium der Einschlußkörperbildung, kennen wir markante Unterschiede, auf die weiter unten eingegangen wird. Diese Phase ist aber bisher nur wenig untersucht worden, da verständlicherweise, in der Hoffnung auf Erfassung vegetativer Entwicklungszustände, die Bearbeitung früher Infektionsstadien Vorrang genoß. Im Beginn des Infektionsgeschehens verhalten sich die genannten Viren, soweit wir bisher wissen, gleich. Obwohl dies nicht in allen Teilen exakt bewiesen ist, dürften die im folgenden zu schildernden Ergebnisse, bei denen das Vaccine-Virus im Vordergrund steht, Allgemeingültigkeit besitzen.

I. Reife Elementarkörper

a) *Äußere Gestalt*

Im Höhepunkt der Infektion lassen sich die Elementarkörper ohne besondere Schwierigkeit durch Direktpräparation vom infizierten Gewebe gewinnen (z. B. *14—18*). In unseren Händen bewährte sich besonders die „indirekte" Tupfmethode (*19, 20, 21*), d. h. die Übernahme des Materials vom noch feuchten Ausstrich. Andererseits bieten angesichts der erheblichen Stabilität der Elk. hochgereinigte Suspensionen, wie man sie mit Vorteil von der Kaninchenhaut gewinnt (*22*), ein sehr geeignetes Ausgangsmaterial für systematische Studien.

Der grobe Umriß isolierter Elk. (Abb. 1a—d) wird durch einen Quader mit abgerundeten Ecken gut beschrieben. Das Achsenverhältnis Länge:Breite:Höhe beträgt im Mittel 2:1,5:1. Die Angaben über Mittelwerte der Längen (z. B. *14, 23*) liegen in der Regel zwischen 230 und 320 mμ. Virusspezifische Unterschiede in Größe und Achsenverhältnis sind zwar beschrieben

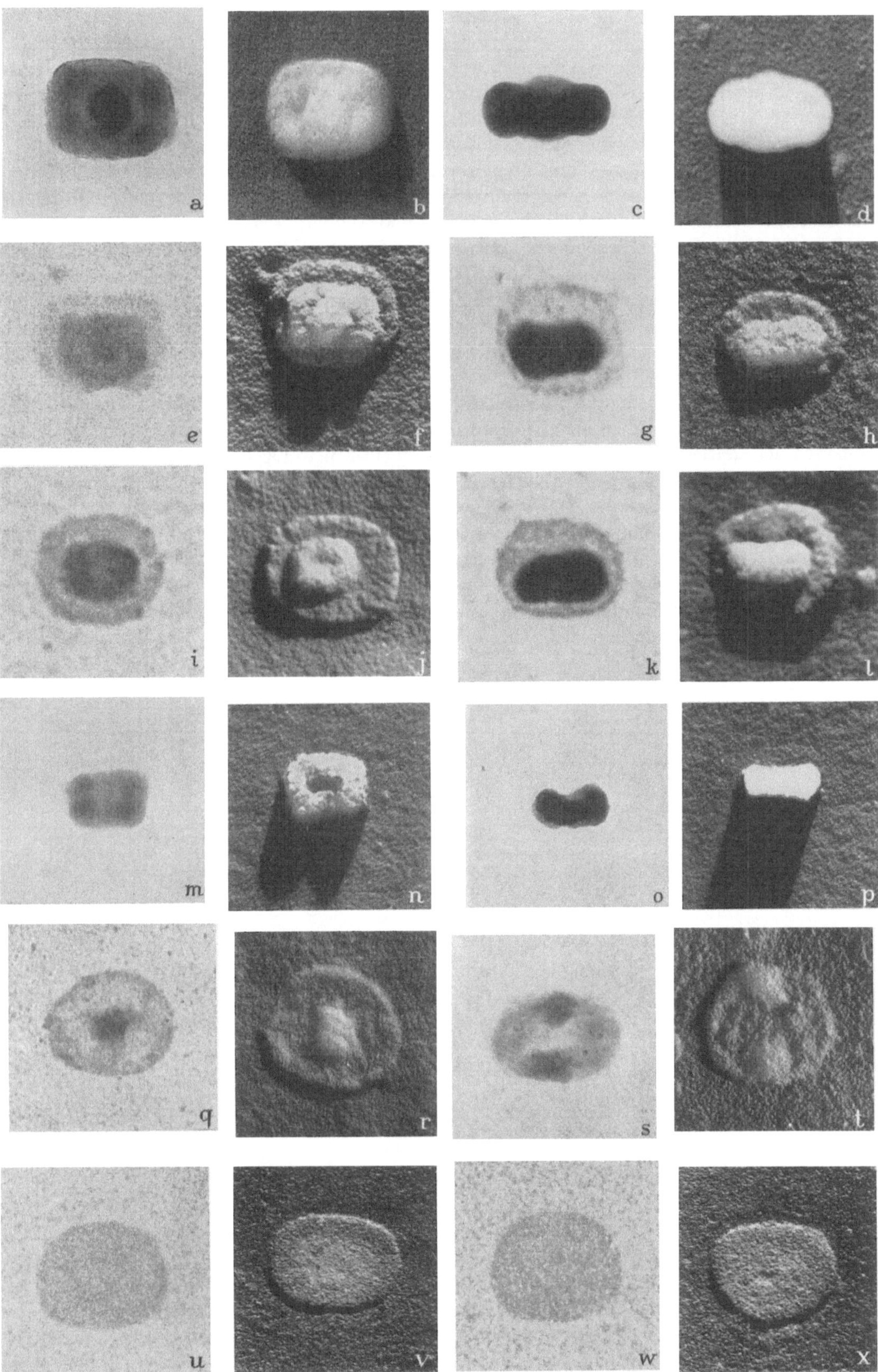

Abb. 1a—x. Enzymatischer Abbau an reifen Elementarkörpern des Vaccine-Virus in horizontaler (links) und vertikaler Lage (rechts); zum Teil Uran-bedampft; 80 000 mal.   a—d: intakte Elk.; e—h: Membran mit Innenkörper und zentraler Verdichtung; i—l: Membran mit Innenkörper; m—p: freiliegende Innenkörper; q—t: Membran mit zentraler Verdichtung; u—x: leere Membran

worden; es wurden jedoch jeweils nur wenige Elk. ausgemessen. Effekte wie belastungsabhängige Objektschrumpfung und Kontamination wurden nicht berücksichtigt. Es erscheint daher als sehr gewagt, aus Größenbestimmungen Rückschlüsse auf die Virusart ziehen zu wollen, wie es z. B. bei der Differentialdiagnose von Variola vera und Vaccine versucht worden ist.

Die Variabilität der Längen- und Breitenwerte ist — verglichen mit Kokken in der Teilungsphase — gering. Daraus wurde schon frühzeitig geschlossen, daß sich die Elk., jedenfalls im abgebildeten Stadium, nicht durch Größenzunahme und Teilung vermehren (*23*). Später an einer größeren Zahl von Elk. vorgenommene Ausmessungen ergaben allerdings auch Werte, die wegen ihrer beträchtlichen Abweichung von einer Gauß-Verteilung diese Aussage wieder einschränkten (*19*). Indessen ist der damalige Befund, daß Teilungsformen nie beobachtet wurden, wiederholt bestätigt worden. Insofern bietet sich am ehesten ein Vergleich mit ruhenden Zellen, wie Cysten oder Sporen an (*20*), deren Größenvariabilität im Vergleich mit der vegetativer Zellen bekanntlich wesentlich geringer ist.

Die übliche elektronenoptische Präparation führt fast ausschließlich zu liegenden Elk. (Abb. 1 a—b). Vermeidet man aber die Einwirkung der Grenzfläche beim Trocknen durch Anwendung der „critical point“-Methode (*24*), so beobachtet man regelmäßig in beachtlicher Zahl Viruspartikel, die ihre ursprünglich vertikale Anordnung beibehalten haben (Abb. 1 c—d) (*25*). Dieses Vorgehen ist aber an eine Hochdruckapparatur gebunden, erfordert also einigen präparativen Aufwand. Es ist daher von Interesse, daß es neuerdings gelang, die dreidimensionale Darstellung wesentlich einfacher zu erzielen, nämlich durch Trocknen aus Isopentan nach vorausgehender Dehydratisierung (*26*). Die geringe Oberflächenspannung des Isopentans (13 dyn/cm) setzt dabei die Einwirkung der Grenzfläche soweit herab, daß praktisch gleiche Ergebnisse gewonnen werden. Dabei werden an den langen Seiten der Profilansicht (Abb. 1 c—d) symmetrische

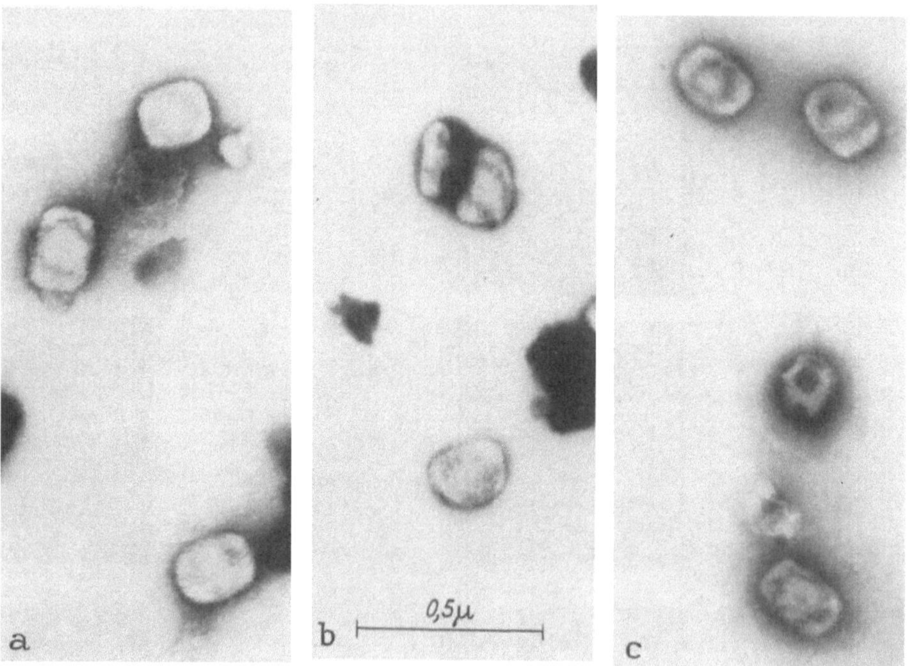

Abb. 2a—c. Abdruck reifer Vaccine-Elk., hergestellt mit Phosphormolybdänsäure; 50 000 mal

Vorwölbungen sichtbar, die einer bereits frühzeitig erkannten zentralen Verdichtung (*27*) des liegenden Elk. entsprechen (Abb. 1 a—b). Zur Erkennung dieser Verdichtung trägt wesentlich bei, daß sie von einer ringförmigen Zone geringerer Dichte umschlossen ist, die sich im bedampften Präparat als Einsenkung zu erkennen gibt (*28, 29*). Diesem Relief entspricht aber auch die Unterseite des Elk. Behandelt man nämlich Viruspartikel einer Suspension auf dem Objektträger mit Phosphormolybdänsäure (5%, p_H 1,0), so kommt es stellenweise zu einer Anhäufung

von Materie an der Peripherie des Elk. Bei der anschließenden Waschung in Wasser werden
etliche Elk. unter Hinterlassung eines Abdrucks vom Trägerfilm gelöst. Die Struktur der Matrize
zeigt deutlich, daß die untere Seite der Partikel der oberen völlig gleicht (Abb. 2) (26).

Diese Methodik zeigt Parallelen zu dem Vorgehen anderer Autoren beim Bushy stunt- (30) und beim Tabak-
mosaik-Virus (31). In diesen Fällen wurde allerdings angenommen, daß der negative Bildeindruck allein durch
die periphere Kontrastierung, also unter Verbleib des Viruspartikels auf dem Film, entstehe[1].

Der vertikal gestellte Elk. weist naturgemäß einen stärkeren Kontrast auf. Dennoch beobachtet man auch in dieser Bestrahlungsrichtung charakteristische Kontrastunterschiede. Entlang
der langen Achse hebt sich eine hantelförmige Innenstruktur durch stärkeren Kontrast deutlich

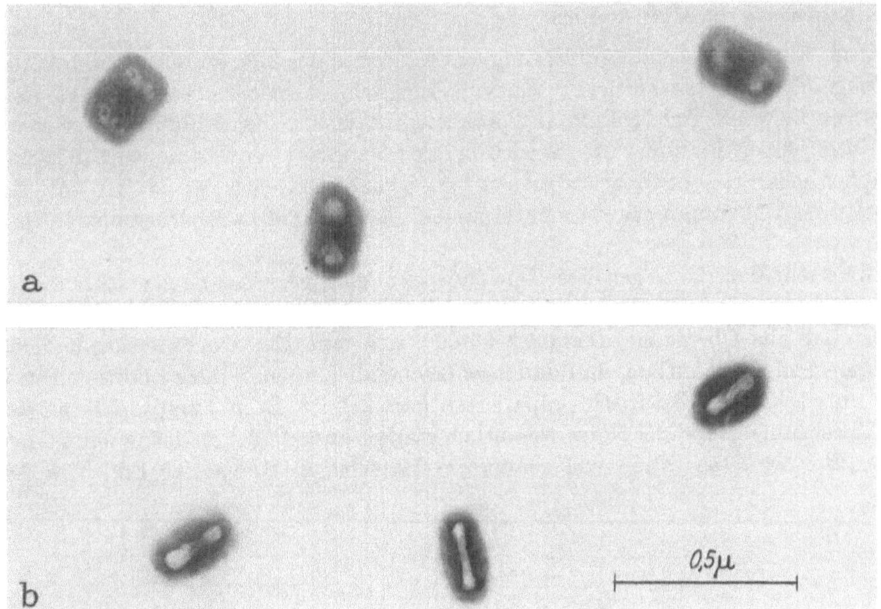

Abb. 3 a u. b. Spezifische Aufhellung des Innenkörpers reifer Elk. mit Hilfe von Tetrachlorkohlenstoff; 50 000 mal

von einer umgebenden Schicht geringerer Dichte ab (25). Bei der Betrachtung von Vertikalschnitten (Abschn. I c) werden wir dieser Innenstruktur wieder begegnen. Wesentlich markanter
kann man sie aber darstellen, wenn man sich eines elektronenoptischen Effektes bedient,
dessen Ablauf im einzelnen noch unklar ist.

Überführt man OsO_4-fixierte Elk. unter Zwischenschaltung einer Alkoholreihe in unpolare
organische Lösungsmittel wie Tetrachlorkohlenstoff oder Heptan, und bildet die Präparate
elektronenoptisch ab, ehe sie noch Feuchtigkeitsspuren aus der Luft aufnehmen konnten, so
beobachtet man charakteristische Aufhellungen, die unter der Einwirkung des Elektronenstrahls
zustande kommen (Abb. 3) (33). Im vertikal stehenden Elk. erweist sich dann die ursprünglich
kontrastreiche, hantelförmige Innenstruktur als gut durchstrahlbar, im liegenden dementsprechend eine ringförmige Zone, welche die zentrale Verdichtung umschließt. Häufig erstreckt
sich diese Veränderung nur auf kleinere Zonen in den vier Ecken; manchmal sieht man im Zentrum solcher Aufhellungen sogar noch winzige Granula liegen (Abb. 3). Sehr wahrscheinlich
sind diese besonders labilen Eckzonen identisch mit den vier Verdichtungszentren (satellite
bodies) unbehandelter Elk., über die vor Jahren schon berichtet wurde (27). In Abb. 1 a sind
diese Zonen erkennbar; nach unseren Erfahrungen treten sie in älteren Suspensionen
wesentlich deutlicher hervor. — Der Aufhellungsvorgang verdient besondere Beachtung, weil
er sich streng spezifisch nur an dem sog. Innenkörper abspielt, jener Struktur, welche die Desoxyribonucleinsäure des Elk. enthält (s. Abschn. I b 5).

---

[1] Inzwischen wurde diese Methode in abgewandelter Form mit ausgezeichnetem Erfolg auch bei Phagen
angewendet (32).

b) *Strukturanalyse durch Abbau*

1. Alkali. Schon das Studium des intakten Elk. weist demnach auf einen recht komplizierten inneren Aufbau hin. Tiefergehende Einblicke in das Bauprinzip wurden aber erst durch die Herauslösung einzelner Strukturanteile erreicht. Schon frühzeitig war erkannt worden, daß die Innenstrukturen durch Behandlung mit Alkali aus einer relativ beständigen umhüllenden Membran ("ghost") total herausgelöst werden können (*27*), wobei unter anderem ein nucleoproteidhaltiges Antigen in Lösung geht (*34*).

Nach eigenen Untersuchungen ist zwar — neben der völlig resistenten Membran — die sog. zentrale Verdichtung um einiges stabiler als die anderen Innenstrukturen. Eine saubere Abgrenzung, oder gar einen stufenweisen Abbau, der sich mit den Ergebnissen der enzymatischen Strukturanalyse vergleichen ließe, erzielten wir bei der Anwendung von Alkali aber nie.

2. Pepsin. Eine weitgehende Aufklärung der Anatomie des Elk. wurde erst durch die Anwendung hochgereinigter, kristallisierter Enzyme möglich. Wertvolle Ansatzpunkte lieferte die erstmalig von DAWSON und MCFARLANE (*35*) vorgenommene Behandlung unfixierter Elk. mit Pepsin. Unter dessen Einfluß wird nämlich bei der Mehrzahl der Elk. in sehr distinkter Weise eine peripher gelagerte, offenbar proteinhaltige Schicht gelöst (s. a. *36, 37, 38*). Neben der bereits bekannten Membran erweist sich in diesem Fall auch ein scharf begrenzter Innenkörper als stabil.

In unfixiertem Material zeigen diese Innenkörper allerdings, selbst nach erschöpfender Pepsinbehandlung, recht unterschiedliche Größen; zwischen völlig leeren Membranen und intakten Elk. finden sich alle Übergänge. Demnach können sich reife Elk. trotz gleicher äußerer Gestalt in ihrem biochemischen Aufbau, und damit wahrscheinlich auch in ihrer funktionellen Leistung, erheblich unterscheiden[1]. Bei frisch präparierten Partikeln, z. B. in Tupfpräparaten (*20, 42, 43*) sind die Abweichungen von der Norm wesentlich stärker ausgebildet als bei gereinigten Suspensionen (*29*). Bei der Präparation von gealtertem Material, z. B. Vaccine-Elk. von der Cornea

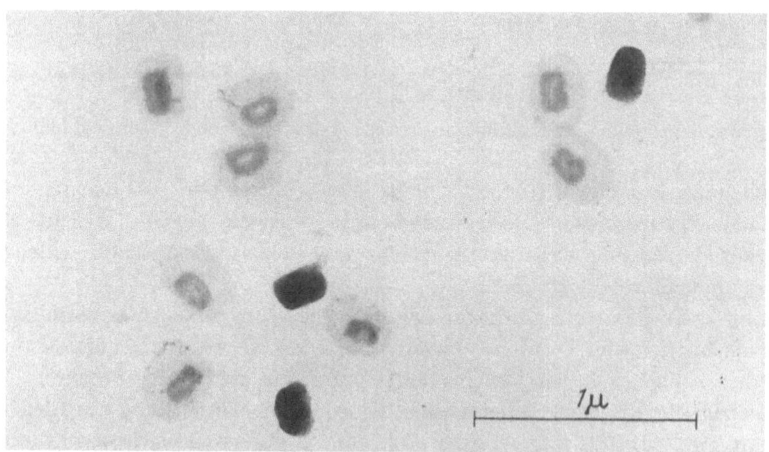

Abb. 4. Unterschiedliche Stadien reifer Elk. nach erschöpfendem Abbau mit Pepsin (AE-fixiert, Pepsin in 0,1 mol Cystein, pH 4,0); kaum angegriffene Elk. neben Membranen mit Innenkörper und solchen mit Innenkörper und zentraler Verdichtung; 30000mal

erst einige Stunden nach der Enukleation (43) oder Molluscum contagiosum-Elk. aus mehrere Wochen alten Efflorescenzen (*44*), ist erfahrungsgemäß die Zahl unabgebauter Elk. gegenüber der Norm stark vermindert. Das Auftreten unterschiedlicher Abbautypen wurde von uns als Ausdruck der Existenz von Elk.-stadien gedeutet. Beim Übergang vom resistenten zum angreifbaren Elk. könnte es sich entweder um eine Ausreifung oder um erste Anzeichen einer lytischen Degeneration handeln.

---

[1] Diese Erkenntnis dürfte einige Bedeutung haben für die Bestimmung des Verhältnisses von physikalisch faßbaren Elk. zu infektiösen Einheiten, einer Arbeitsrichtung, die in letzter Zeit mehr in den Vordergrund gerückt ist (*39, 40*, s. a. *41*).

Im Rahmen systematischer Studien (*25, 29, 45*) ergaben sich tiefere Einblicke in die Struktur
der Elk. erst nach Anwendung geeigneter Fixierungsmittel. Formaldehyd bewährte sich in Verbindung mit dem enzymatischen Abbau nicht; bei kurzen Fixierungszeiten verläuft der Abbau

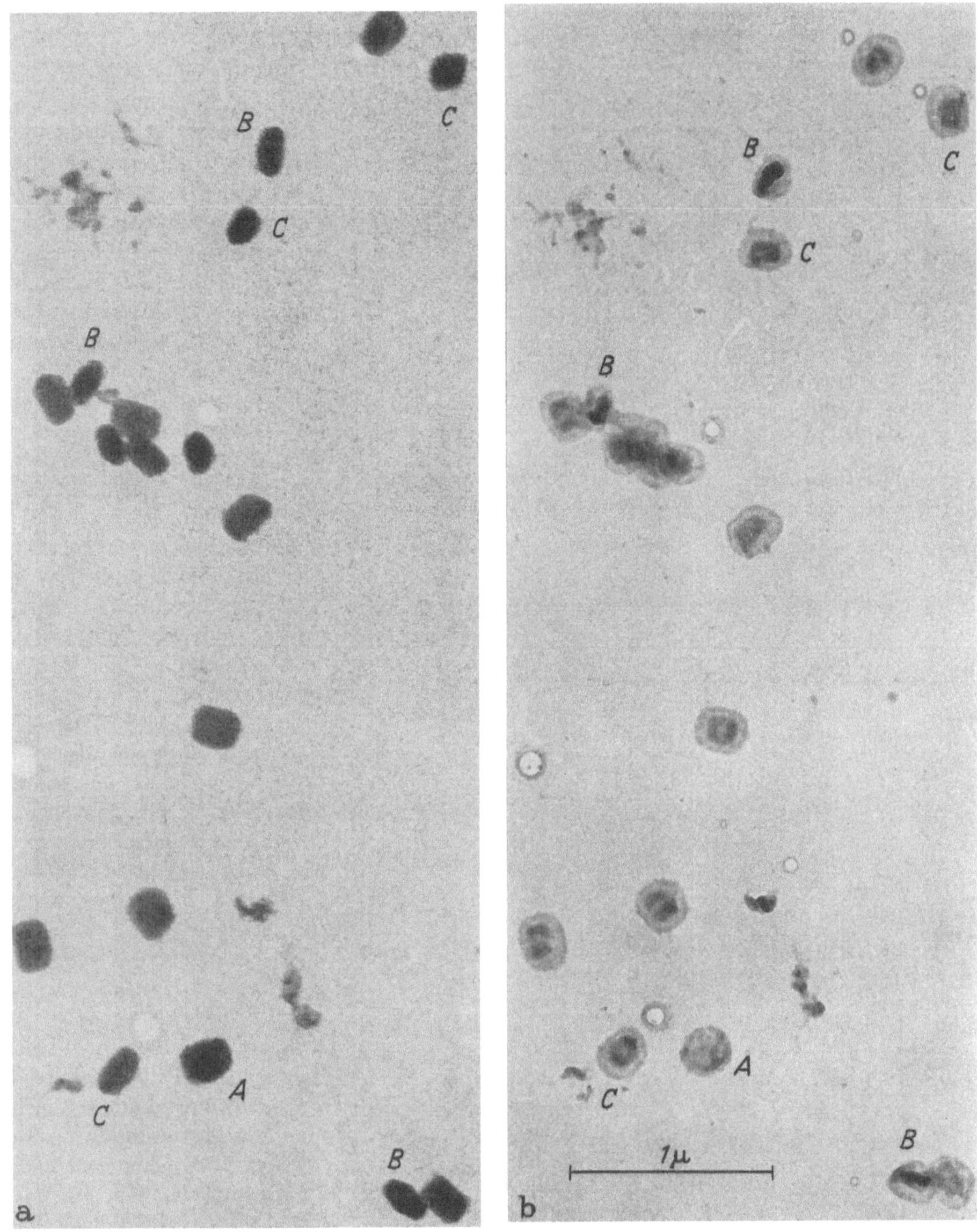

Abb. 5a u. b. Reife Elk. vor und nach Abbau mit Pepsin. Linke Aufnahme bei geringstmöglicher Objekt-
belastung aufgenommen, daher etwas unscharf. Elk., der zur leeren Membran abgebaut wurde (*A*), Elk. in
vertikaler Lage zeigen Innenkörper im Profil (*B*); ursprünglich vertikal stehende Elk., die während des Abbaus
in Horizontallage übergegangen sind (*C*); 30000mal

uncharakteristisch, bei langen ist er völlig gehemmt. Osmiumtetroxyd-Fixierung inhibiert den
Pepsinangriff ebenfalls. Allerdings läßt sich dies überwinden, indem man die Pepsinhydrolyse
in reduzierendem Milieu (Cystein oder Thioglykolsäure) ablaufen läßt. Da OsO_4 nicht zur Koagu-

lation plasmatischer Substanz führt, kann es unter der Wirkung des Pepsins leicht zu einer Schrumpfung und damit zu einer weniger differenzierten Darstellung des Innenkörpers kommen. Für die enzymatische Strukturanalyse bewährte sich in unseren Händen ein Fixierungsgemisch aus 80% Äthanol und Essigsäure (15:1) (im folgenden als AE abgekürzt) am besten.

Nach dieser Fixierung ist die Zahl pepsinresistenter Elk. — im Gegensatz zu den Verhältnissen bei unfixiertem Material — praktisch gleich Null, wenn bei optimaler Acidität (p_H 1,5—2,0) gearbeitet wird. Stellt man den p_H-Wert aber höher ein, so resultiert auch nach AE-Fixierung — wieder als Ausdruck einer Verschiedenheit der Elk. — trotz erschöpfender Behandlung ein buntes Bild (Abb. 4). Bei p_H 1,5—2,0 führt Pepsin nach AE-Fixierung, weniger deutlich auch nach OsO_4, zur Erkennung eines meist etwas länglichen, scheibenförmigen Innenkörpers, der durch eine zentrale Einsenkung ein ringartiges Aussehen erhält (Abb. 1i—j) (s. a. *17, 20*) und im Profil als kurzes oft hantelförmiges Stäbchen erscheint (Abb. 1k—l). Nicht selten sieht man Abweichungen von dieser Norm, z. B. halbmondförmige Körper und Doppelformen (Abb. 5). In Gegenwart von Cystein wird auch noch bei höheren p_H-Werten ein weitgehender Abbau erreicht; dabei beobachtet man regelmäßig mit steigendem p_H-Wert eine zunehmende Zahl von Elk., bei denen außer dem Ringkörper eine zentrale Verdichtung sichtbar bleibt (Abb. 1e—f). Diese tritt auch dann in Erscheinung, wenn bei normalem p_H-Wert nur kurz, d. h. nicht erschöpfend, mit Pepsin behandelt wird. In der Profilansicht unterscheidet sich dieses Abbaustadium vom zuvor geschilderten praktisch nicht (Abb. 1g—h). In Abb. 6a sind die Abbauverhältnisse als Funktion des p_H-Wertes graphisch

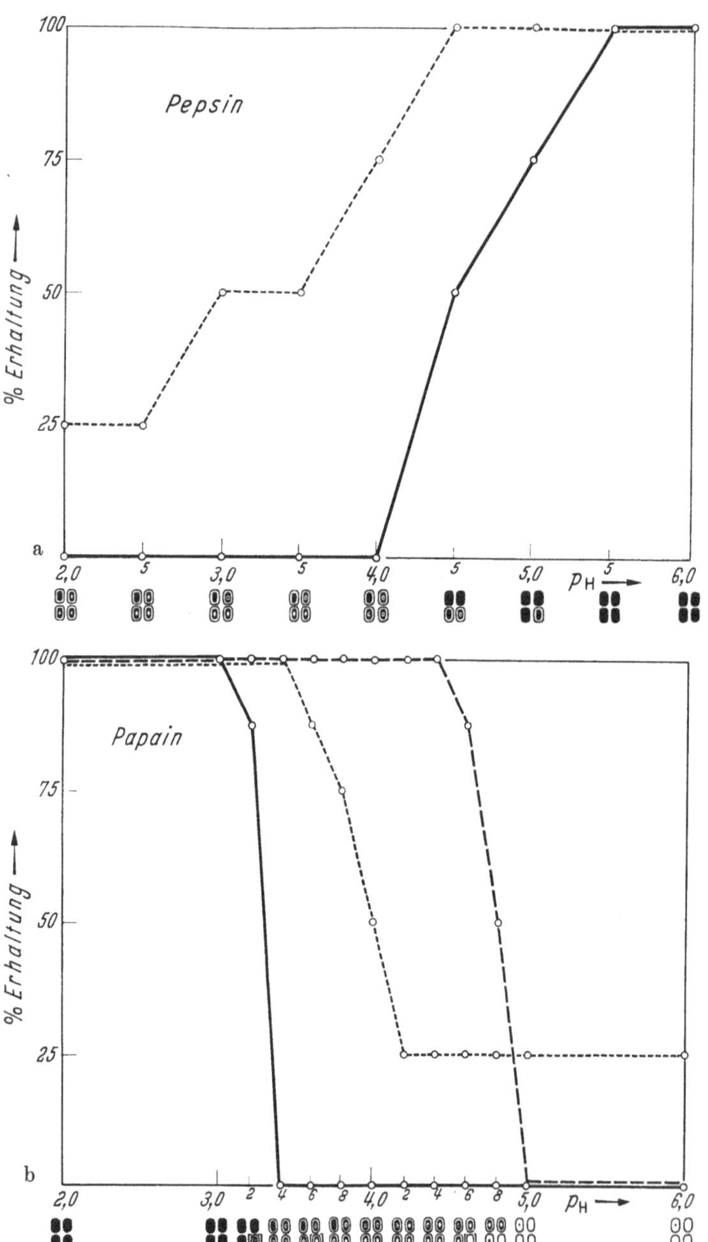

Abb. 6a u. b. Abhängigkeit des enzymatischen Abbaus AE-fixierter reifer Vaccine-Elk. vom p_H-Wert. Ordinate: Prozent Elk., bei denen der entsprechende Strukturanteil erhalten geblieben ist (Schätzwerte). Die schematische Darstellung unter der Abszisse gibt etwa die jeweils beobachtete Stadienverteilung an. ———: periphere Proteinschicht; ——○——: DNS-haltiger Innenkörper; ----○----: zentrale Verdichtung. a) krist. Pepsin in 0,1 mol Cystein, 20 Std., 37° C, b) krist. Papain in 0,1 mol Citrat und 0,1 mol Cystein, 20 Std., 37° C

dargestellt. Einen besonderen Hinweis verdient dabei die Tatsache, daß die zentrale Verdichtung unter keinen Bedingungen in sämtlichen Elk. total abgebaut wird; vielmehr erweist sie sich in etwa einem Viertel der aus Suspension präparierten Elk. als resistent. Größenunterschiede deuten Übergänge an.

Die geschilderten Vorstellungen ließen sich durch neuere Studien unterbauen (46). Im Elmiskop I läßt sich, in gewisser Analogie zu einem älteren Bericht über die Keimfähigkeit elektronenoptisch abgebildeter Sporen (47), bei Verzicht auf exakte Scharfstellung die den Elk. zugefügte Ladungsdichte so niedrig halten, daß noch nach der elektronenoptischen Aufnahme eine enzymatische Hydrolyse erfolgen kann. So ist es möglich, einzelne Präparatstellen vor und nach der Reaktion abzubilden und dadurch das Schicksal einzelner Elk. zu verfolgen. Abb. 5 zeigt außer einem Elk. (A), der in Abweichung von der Norm bis zur leeren Membran abgebaut wurde, mehrere Elk. (B), die auch nach dem Abbau ihre ursprünglich vertikale Lage beibehalten haben und daher die Profilansicht des Innenkörpers zeigen. Andere (C) haben während der weiteren Präparation die normale Horizontallage eingenommen; zwei davon standen ursprünglich auf der kleinsten Quaderfläche.

Allein die Behandlung mit Pepsin führt demnach zur Erkennung von mindestens vier strukturell unterscheidbaren Anteilen: 1. einer resistenten umhüllenden Membran, 2. einer peripheren proteinhaltigen Schicht, 3. eines resistenten scheiben- bis ringförmigen Innenkörpers und 4. einer zentralen Verdichtung, die ebenfalls Protein enthält.

3. Papain. Mit Hilfe von kristallisiertem Papain wurden diese Befunde völlig bestätigt, aber auch erweitert (25, 48). Voraussetzung für einen Angriff dieses Enzyms, das seine Wirkung im Bereich oberhalb p_H

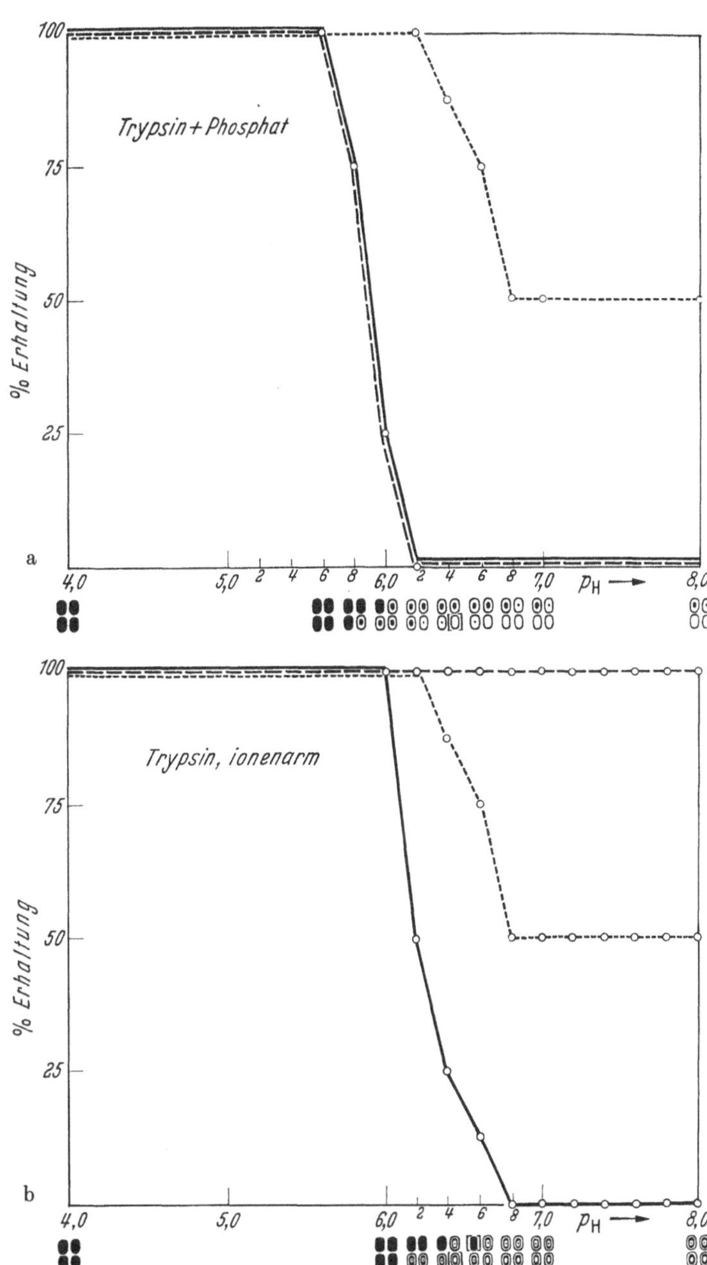

Abb. 7a u. b. Abhängigkeit des tryptischen Abbaus AE-fixierter reifer Vaccine-Elk. von p_H-Wert und Ionenkonzentration. Weitere Erklärung s. Abb. 6. a) krist. Trypsin in m/15 Phosphatpuffer, 2 Std., 37° C; b) krist. Trypsin, dialysiert, 2 Std., 37° C

3,4 entfaltet, ist — im Gegensatz zum Pepsin — eine vorausgehende Fixierung, d. h. Denaturierung. Nach Behandlung mit AE greift Papain (s. Abb. 6b) zwischen p_H 3,4 und 3,8 bei den meisten Elk. nur die periphere Schicht an (entsprechend Abb. 1e—h). Oberhalb p_H 3,6 wird in zunehmendem Maße auch die zentrale Verdichtung gelöst; auch in diesem Fall —

das sei besonders vermerkt — nie in sämtlichen Elk. Zwischen p_H 4,0 und 4,8 überwiegen Viruspartikel, die nur den ringartigen Innenkörper enthalten (entsprechend Abb. 1 i—l). Unter Verhältnissen, die wir bisher nicht exakt beherrschen, kann es bei der Papain-Reaktion auch zu einer Lyse der umhüllenden Membran kommen. Es resultieren dann entsprechend den p_H-Bedingungen freiliegende Innenkörper mit oder ohne (Abb. 1 m—p) zentrale Verdichtung. Oft zeigen die freigesetzten Innenkörper eine etwa viereckige Gestalt und ziemlich distinkte Kontrastzonen in den Ecken. Sehr wahrscheinlich sind diese identisch mit den bereits oben erwähnten Verdichtungszentren (satellite bodies) der intakten Elk.

Beim Vaccine-Virus, auch solchem, das einheitlichem Material entstammte, beobachteten wir eine Auflösung der Membran durch Pepsin allein bisher nie; wohl aber gelegentlich beim Molluscum contagiosum-Virus (20). Möglicherweise handelte es sich in diesen Fällen um Elk., deren Membranen durch das wochenlange Verweilen in den Efflorescenzen bereits vorher geschädigt waren.

Läßt man Papain bei nur wenig höheren p_H-Werten ($> 5,0$) einwirken, so beobachtet man, daß nun auch der ringartige Innenkörper gelöst wird (Abb. 6 b). Er liegt demnach im Neutralbereich als papainempfindliches Substrat vor. Infolgedessen resultieren in überwiegender Zahl leere Membranen (Abb. 1 u—x). Das unterschiedliche Verhalten des Innenkörpers ist sehr wahrscheinlich Folge unterschiedlicher Ladungsverhältnisse am Substrat; der Umschlagspunkt p_H 4,8 ist demnach ein für eine Substanz dieses Körpers charakteristischer Wert, wahrscheinlich der isoelektrische Punkt. Bei den p_H-Werten oberhalb 5,0 lassen sich unter gemilderten Reaktionsbedingungen interessante Zwischenstufen fassen. In diesem p_H-Bereich besitzt die zentrale Verdichtung nämlich eine höhere Resistenz als der Innenkörper. Bei einem solchen nicht erschöpfenden Abbau beobachtet man Membranen, hervorgegangen aus liegenden Elk., in denen einzig die zentrale Verdichtung sichtbar geblieben ist (Abb. 1 q—r). Das entspricht den geschilderten Vorstellungen. Neue Informationen vermittelt dagegen die Betrachtung der ursprünglich vertikal angeordneten Elk.; diese zeigen nämlich statt *einen* Körper deren *zwei*, und zwar in lateraler Lage. Das weist daraufhin, daß die „zentrale Verdichtung" des liegenden Elk. (Abb. 1 a) in Wirklichkeit aus zwei Gebilden besteht, deren Kontrast sich in dieser Lage addiert. Diese Gebilde sind offensichtlich Ursache der zentralen Erhebungen (Abb. 1 b, 2) bzw. der Vorwölbungen (Abb. 1 c—d) der intakten Elk.; sie sind wahrscheinlich auch verantwortlich für die bikonkave Form des Innenkörpers.

4. Trypsin. Auch mit kristallisiertem Trypsin lassen sich derartige Abbauvorgänge ausführen, falls eine Denaturierung vorausgegangen ist (49). In gepufferter Lösung ist zwar eine Differenzierung zwischen peripherer Schicht und Innenkörper nicht möglich, da beide oberhalb p_H 6,0 gleichermaßen labil sind (Abb. 7 a). Die zentrale Verdichtung erweist sich aber auch diesmal als resistenter und so resultieren neben leeren Membranen (wie Abb. 1 u—x) auch die in Abb. 1 q—t dargestellten Abbautypen. Befreit man das Enzym dagegen durch Dialyse weitgehend von Ionen, so geht der Abbau andere Wege. Für die periphere Schicht und die zentrale Verdichtung verschiebt sich die Abbauzone lediglich etwas nach dem Neutralbereich (Abb. 7 b); der Innenkörper dagegen erweist sich als stabil. Unter diesen Bedingungen führt also auch Trypsin zu den in Abb. 1 e—h und i—l abgebildeten Abbautypen.

5. Desoxyribonuclease. Die Ergebnisse der morphologischen Analyse mit proteolytisch wirksamen Fermenten werden naturgemäß durch die Eigenschaften der in die einzelnen Strukturen eingebauten Proteine bestimmt. Die geschilderten Resultate sprechen dafür, daß sämtliche Innenstrukturen der Elk. Eiweiß enthalten; wahrscheinlich ist auch die Membran nicht frei davon. Seit langem ist aber bekannt, daß auch Desoxyribonucleinsäure (DNS) zu etwa 5,6% in pepsinstabiler Bindung am Aufbau des Elk. beteiligt ist (50). Enzymatische Versuche zeigen, daß die DNS, wie zu erwarten, im sog. Innenkörper lokalisiert ist. Mit einer Methodik, die zuvor an Bakterien ausgearbeitet worden war (51, 52), gelang es nämlich, den durch Pepsin freigelegten Innenkörper mit Desoxyribonuclease und einer anschließend vorgenommenen zweiten Pepsinbehandlung spezifisch in Lösung zu bringen (20, 53, 54; s. a. 17). Wie bei Bakterien resultierten fast leere Membranen. Die Depolymerisation der DNS ist elektronenoptisch zunächst nicht zu erkennen, da die Abnahme der Massendichte zu geringfügig ist. Die Überführung des Innenkörpers aus einem DNS-geschützten pepsinstabilen in einen ungeschützten pepsinlabilen Zustand zeigt aber eindeutig, daß die Desoxyribonuclease in dieser Struktur ein Substrat, nämlich DNS, vorgefunden hat. Unter geeigneten Vorsichtsmaßnahmen läßt sich in beiden Stufen Pepsin auch durch saures Papain (p_H 4,2) ersetzen (48).

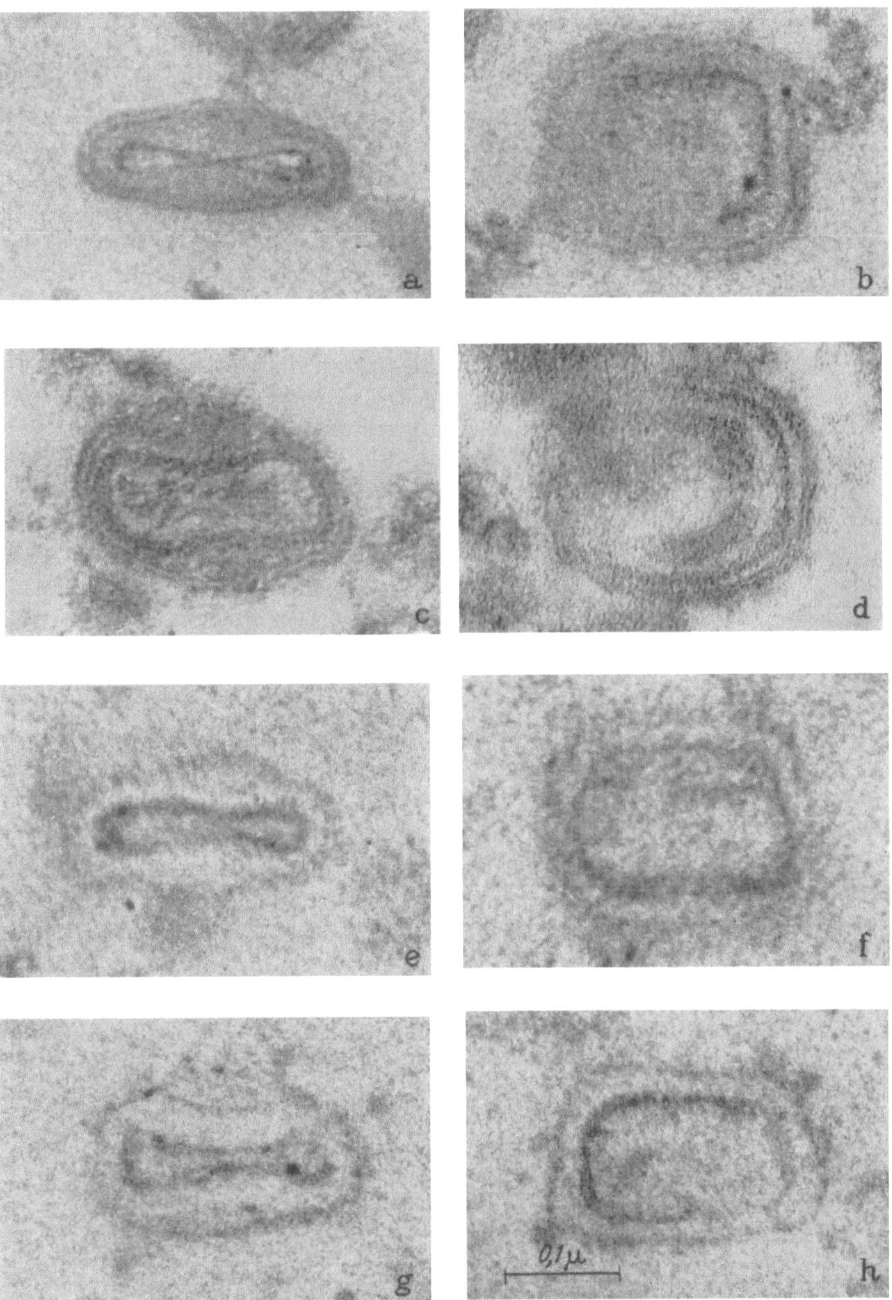

Abb. 8a—h. Vertikal- und Horizontalschnitte durch reife Vaccine Elk. nach unterschiedlicher Fixierung; a—b: OsO_4; c—d: $KMnO_4$; e—f: Formaldehyd; g—h: Alkohol-Essigsäure; 150000mal

Nach lichtoptisch-cytochemischen Erfahrungen an Protozoen und Gewebezellen sind auch die Desoxyribonucleoproteide der Zellkerne bzw. der Chromosomen gegen Pepsin und ionenarmes Trypsin resistent, gegen ionenhaltiges Trypsin und Desoxyribonuclease aber labil. In Anlehnung daran und in Analogie zu den licht- und elektronenoptischen Ergebnissen an Bakterien

(*51, 52*) liegt es nahe, den Innenkörper als Kernäquivalent aufzufassen (s. a. *55*). Eine weitere Parallele findet sich beim Vergleich der umhüllenden Membran des Elk. mit der Zellwand der Bakterien, denn beide sind gegen Pepsin und Trypsin resistent. Ein markanter Unterschied zwischen reifen Elk. und Bakterien liegt aber darin, daß Ribonucleinsäure bisher weder chemisch noch enzymatisch-elektronenoptisch mit Sicherheit im Elk. nachgewiesen werden konnte.

c) *Strukturanalyse durch Ultramikrotomie*

Zu einer ausgezeichneten Bestätigung der biochemisch-morphologischen Resultate führte die Anwendung der Ultramikrotomie. Bereits bei der Untersuchung infizierter Gewebe hatten sich charakteristische Merkmale des geschnittenen reifen Elk. ergeben (*56, 57, 58, 59*). Systematische Untersuchungen an suspendierten Elk. vervollständigten das Bild (*25, 60*). Danach ist die Grundstruktur des Körpers in ihren wesentlichen Zügen vom verwendeten Fixierungsmittel unabhängig (Abb. 8). Die Schnittbilder AE-fixierter Elk. (Abb. 8g—h) beweisen, daß auch diese Fixierung die Elk.-Struktur gut erhält; im Hinblick auf die zuvor geschilderten enzymatischen Studien, bei denen gerade diese Fixierung in weitem Maße angewendet wurde, ist das von besonderer Bedeutung. Die schematische Darstellung der Abb. 9 vermittelt die derzeitigen Kenntnisse.

Aus der Gestalt der Elk. ergibt sich erstens, daß zwei markante Schnittebenen existieren, nämlich eine horizontale und eine vertikale, und zweitens, daß diese mit unterschiedlicher Wahrscheinlichkeit getroffen werden. Tatsächlich ist die Mehrzahl der beobachteten Schnitte auf die vertikale Ebene zu beziehen und nur relativ wenige nähern sich dem idealen Horizontalschnitt. In Übereinstimmung mit den Resultaten des biochemischen Abbaus sind wiederum vier Grundeinheiten zu beobachten: 1. eine *umhüllende Membran*, nach OsO_4- bzw. $KMnO_4$ (*61*)-Fixierung deutlich, nach Formaldehyd schwach doppelt konturiert; nach AE-Fixierung nur durch eine einfache Kontur erkennbar, 2. eine homogen erscheinende *periphere Schicht*, dem pepsinlabilen Protein entsprechend, 3. ein scheibenförmiger, bikonkaver *Innenkörper* mit kontrastreicher Randzone, im Vertikalschnitt durch hantelförmige Gestalt charakterisiert, identisch mit dem DNS-haltigen Innenkörper der enzymatischen Analyse, und 4. ein mit der *zentralen Verdichtung* identisches Körperpaar, das im Vertikalschnitt zwischen Membran und Proteinschicht sichtbar ist, aber im idealen Horizontalschnitt fehlt, weil es dort nicht geschnitten wird. Diese Doppelstruktur ist zweifellos identisch mit dem durch Papain oder Trypsin freigelegten Körperpaar (s. Abb. 1s—t).

Wie die Ergebnisse nach Formaldehyd- bzw. AE-Fixierung beweisen, ist die den Innenkörper begrenzende Zone in sich sehr kontrastreich. Sie erweist sich nach OsO_4, aber auch nach Formaldehyd, gelegentlich als doppelt konturiert. Besonders deutlich und regelmäßig ist eine Doppelkontur in vergleichbarer Lage aber nach $KMnO_4$ zu beobachten. Dann erscheint die umgebende Proteinschicht jedoch merklich schmaler und der Innenkörper entsprechend voluminöser. Genauere Untersuchungen zeigten uns, daß diese Unterschiede zwischen OsO_4- und $KMnO_4$-Fixierung auch bestehen bleiben, wenn bei beiden Fixierungen die p_H-Werte zwischen 4 und 9 verschoben werden.

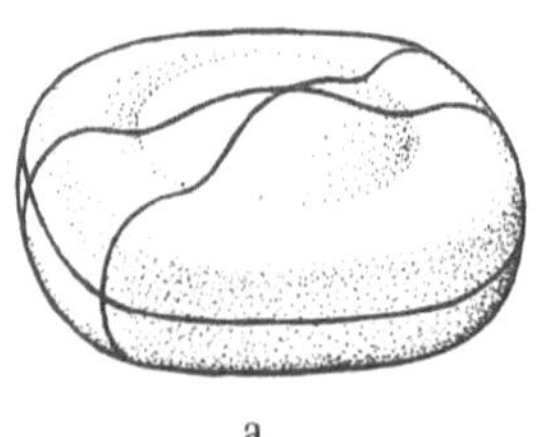

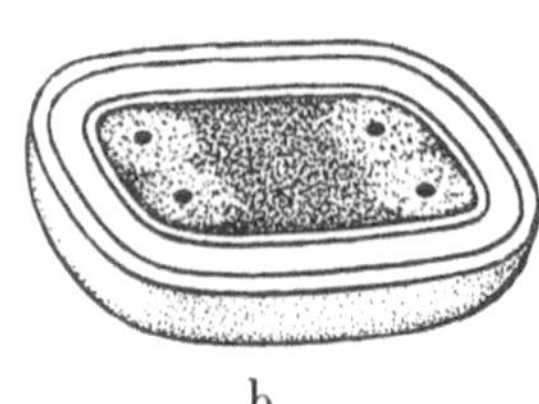

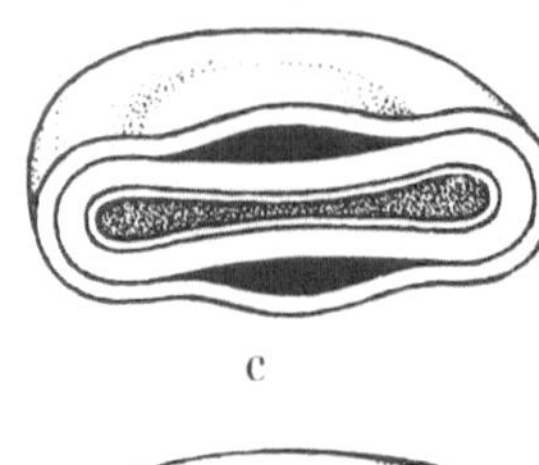

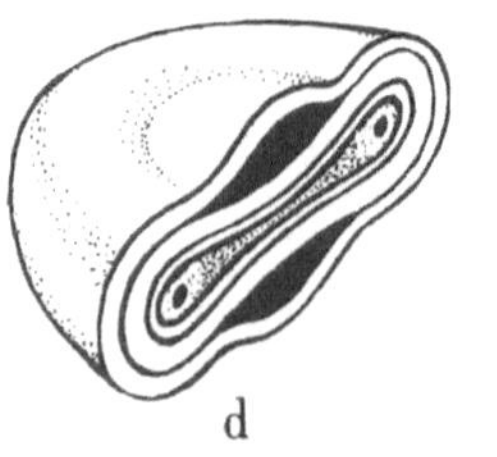

Abb. 9a—d. Schematische Darstellung des reifen Vaccine-Elk. mit Horizontal-, Vertikal- und Diagonalschnitt; OsO_4-fixiert

Demnach müssen Zweifel auftreten, ob mit beiden Fixierungsmitteln tatsächlich die gleichen Struktureinheiten kontrastiert werden. Bemerkenswert ist eine $KMnO_4$-induzierte Differenzierung innerhalb der als Innenkörper angesprochenen Zone, nämlich verstärkter Kontrast im Inneren und geringerer Kontrast an der Peripherie (Abb. 8c—d). Weitere Untersuchungen müssen entscheiden, ob $KMnO_4$ wirklich in der Lage ist, innerhalb des Innenkörpers zwischen einer zentralen Nucleinsäure-Zone und einer umgebenden Proteinzone geringerer Dichte zu differenzieren, wie kürzlich angenommen wurde (62)[1].

In Vertikalschnitten (Abb. 8g), gelegentlich auch in horizontal geführten (Abb. 8b), beobachtet man nicht selten einzeln liegende distinkte Granula von 5—12 mμ Durchmesser. Es gibt gute Gründe anzunehmen, daß diese Körper, entsprechend der schematischen Darstellung im Diagonalschnitt (Abb. 9) in den vier Ecken des Innenkörpers liegen und mit den Eckverdichtungen (satellite bodies) des intakten Elk. (s. o.) identisch sind.

Nach den geschilderten Resultaten erweist sich der reife Elk. bereits heute als ein nach strengen Bauprinzipien gestalteter, hochdifferenzierter Körper. Es ist zu erwarten, daß die Elektronenmikroskopie auch in der Zukunft noch weitere Einblicke in seine Detailstruktur vermitteln wird. Dabei wird man der Erkennung der biologischen Funktion der einzelnen Strukturanteile besondere Aufmerksamkeit widmen müssen. So liegen z. B. die durch diesen Körper ausgelösten ersten Vorgänge des Infektionsgeschehens noch völlig im Dunkeln.

II. Entwicklungsstadien

Zeigen schon die enzymatischen Studien, daß die reifen Elk. verschiedene Zustände der Enzymresistenz durchlaufen (s. Abschn. Ib), so erweitert sich das Bild ganz wesentlich durch die Beschreibung anders geformter Partikel, die heute allgemein als Vorstufen reifer Elk. angesprochen werden. Tiefere Einblicke in die Dynamik der Virusentwicklung ergaben sich durch die Einführung der Ultramikrotomie. Trotz etlicher früherer Versuche (s. Übersicht bei 64) erhielt man gesicherte Aussagen erst nach weitgehender Vervollkommnung der Methodik. Aufbauend auf den grundlegenden Arbeiten von GAYLORD und MELNICK am Vaccine-, Ektromelie- und Molluscum contagiosum-Virus (56, 57) und besonders den technisch vollendeten von MORGAN, ELLISON, ROSE und MOORE am Vaccine- und Geflügelpocken-Virus (58) haben neuere Arbeiten über die Viren der Vaccine (65, 66, 67, 68), der Variola vera (21), des Molluscum contagiosum (69, 70, 71), der Geflügelpocken (59, 72, 73), der Ektromelie (74, 75) und des Kaninchenfibroms (6, 76, 77, 78) die gewonnenen Vorstellungen bestätigt. Grundsätzlich neue Erkenntnisse über den Prozeß der Virusvermehrung haben die neueren Arbeiten aber nicht vermittelt. Alle Anzeichen sprechen dafür, daß die Vorgänge, die zur Bildung der reifen Elk. führen, bei allen Viren dieser Gruppe gleichen Gesetzen unterliegen.

Die Auswertung des bisher vorliegenden Materials und eigene Erfahrungen, die in Zusammenarbeit mit NIELSEN und ANDRES (68) gewonnen wurden, zeigen, daß sich bei der Interpretation manche Schwierigkeit ergibt. Abgesehen von den bekannten Präparationsschäden, wie Zerstörung der Ultrastruktur („explosion") und Substanzverlust, die man mehr und mehr zu beherrschen lernt, überlagern sehr wahrscheinlich auch degenerative Prozesse innerhalb der infizierten Zelle das Bild der normalen Entwicklung; und zwar sicher nicht erst im Stadium des cytopathogenen Effektes, sondern schon vom Anfang der Entwicklung an. Es ergibt sich daraus die Konsequenz, zunächst anhand der häufigsten Grundformen eine Entwicklungsreihe zu formulieren und abweichende Erscheinungsbilder als Degenerationsformen aufzufassen.

Nach den bisher vorliegenden morphologischen Fakten spielt sich der Entwicklungsprozeß dieser Viren lediglich im Cytoplasma ab. Etwa 6 Std. nach der Infektion beginnend, beobachtet man als erstes Kennzeichen, meist in Kernnähe, Verdichtungszonen recht unterschiedlicher Größe, die als Matrix oder Viroplasma bezeichnet werden. Diese stehen mit der Virusvermehrung offensichtlich in engem Zusammenhang, denn innerhalb, oft aber auch nur an der Peripherie dieser Zonen werden ziemlich regelmäßig Partikel von etwa 250 mμ Durchmesser sichtbar (Abb. 10a, 16). Diese Teilchen sind von einer etwa 9—12 mμ dicken Membran begrenzt und zeigen eigenartigerweise im Innern die gleiche granuläre Feinstruktur, wie sie auch dem umgebenden Material, der Matrix, zu eigen ist. Diese Körper werden allgemein als „unreife" Elk. interpretiert. Ihre ovale Gestalt resultiert, wie man an der Ausrichtung quer zur Schnittführung erkennt, als Folge der Schnittkompression. In Wirklichkeit sind sie wahrscheinlich rund. Bei OsO_4-fixierten Präparaten beobachtet man in vielen Partikeln einen meist exzentrisch gelegenen,

---

[1] Eine spätere Notiz (63), wonach DNS-Bereiche in Bakterien nach $KMnO_4$-Fixierung kontrastarm, in Phagen dagegen kontrastreich dargestellt werden, mahnt zur Vorsicht.

kontrastreichen Innenkörper, der des öfteren von einem hellen Hof umgeben ist (Abb. 10b—d, 14). In KMnO$_4$-fixierten Präparaten dagegen ist dieser Innenkörper wegen mangelnden Kontrastes nur schwer zu erkennen (Abb. 11, 12, 16). In gewisser Analogie zum Zellkern hat sich für

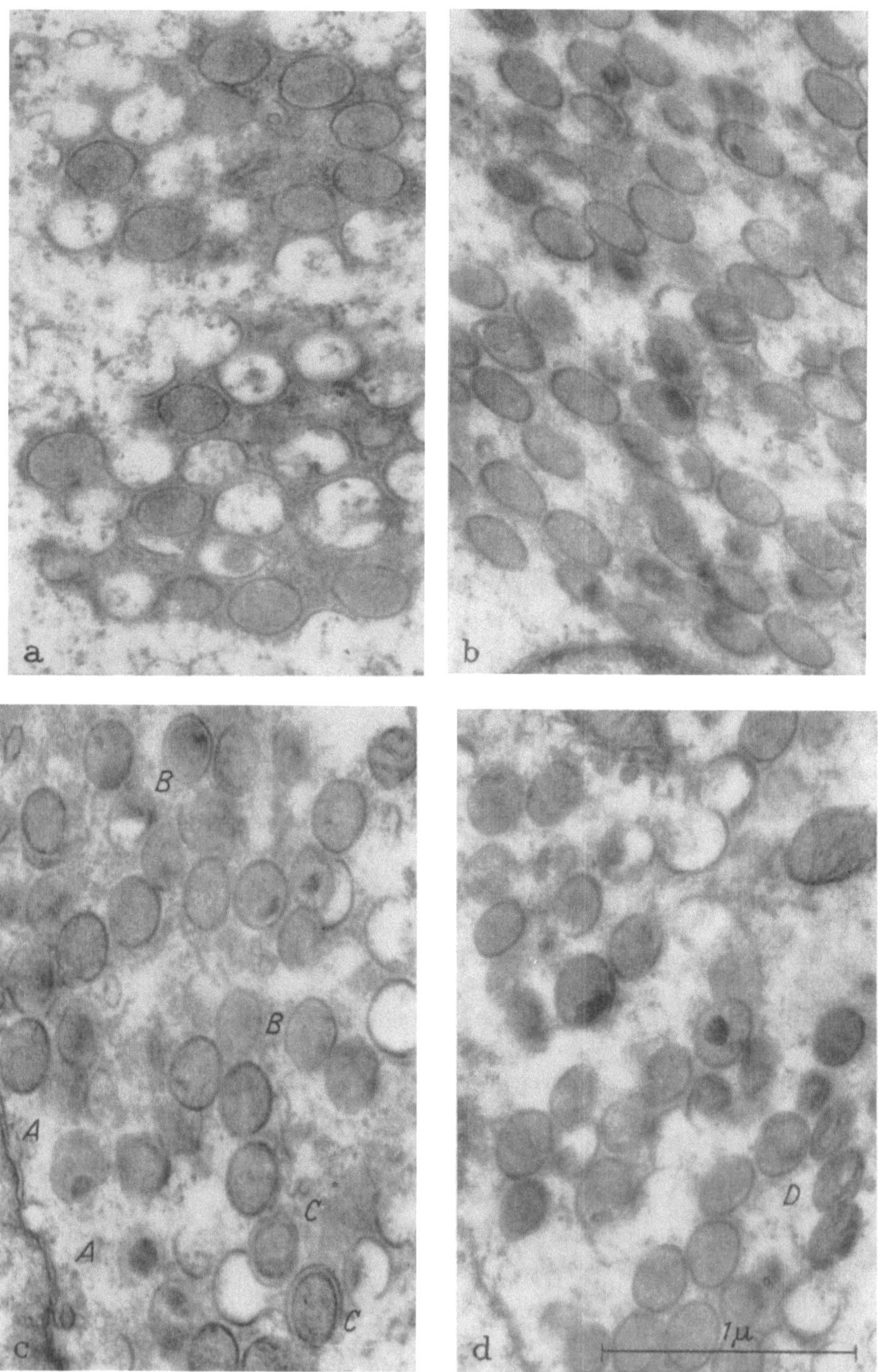

Abb. 10a—d. Schnitte durch Vaccine-infizierte HeLa-Zellen, Verdichtungszonen mit unreifen Elk. (a); unreife Elk., zum Teil mit „Nucleoiden" (b—d); von der Norm abweichende Formen (*A*, *B*, *C*); reife Elk. (*D*); OsO$_4$-fixiert; 35000mal

diese Struktur die etwas kühne Bezeichnung „Nucleoid" eingebürgert, obwohl bisher über ihre Funktion und chemische Zusammensetzung, d. h. auch über den Gehalt an Nucleinsäuren, nicht

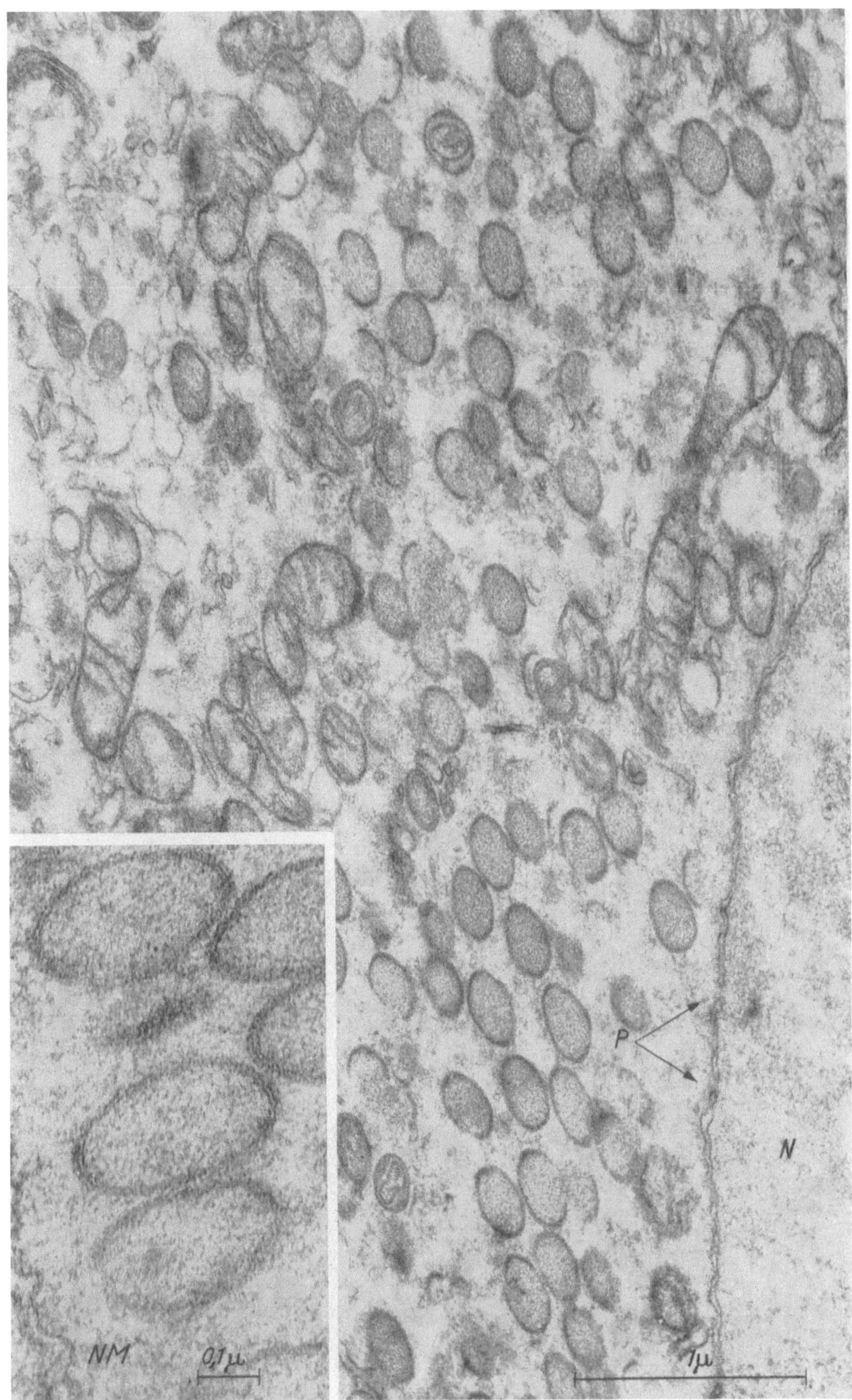

Abb. 11. Schnitt durch Vaccine-infizierte HeLa-Zelle, Bildungszentrum in Nähe des Kerns (*N*), von Mitochondrien umgeben. Zahlreiche unreife, wenige reife Elk.; umhüllende Membran unreifer Elk. doppelt konturiert (s. Einsatz); Kernmembran (*NM*) mit Poren (*P*). $KMnO_4$-fixiert; 35000 und 90000 mal

das Geringste bekannt ist. Manches spricht dafür, daß nicht alle unreifen Viruspartikel ein solches Nucleoid enthalten. Es könnte sein, daß sich dieses erst im Laufe eines Entwicklungsprozesses

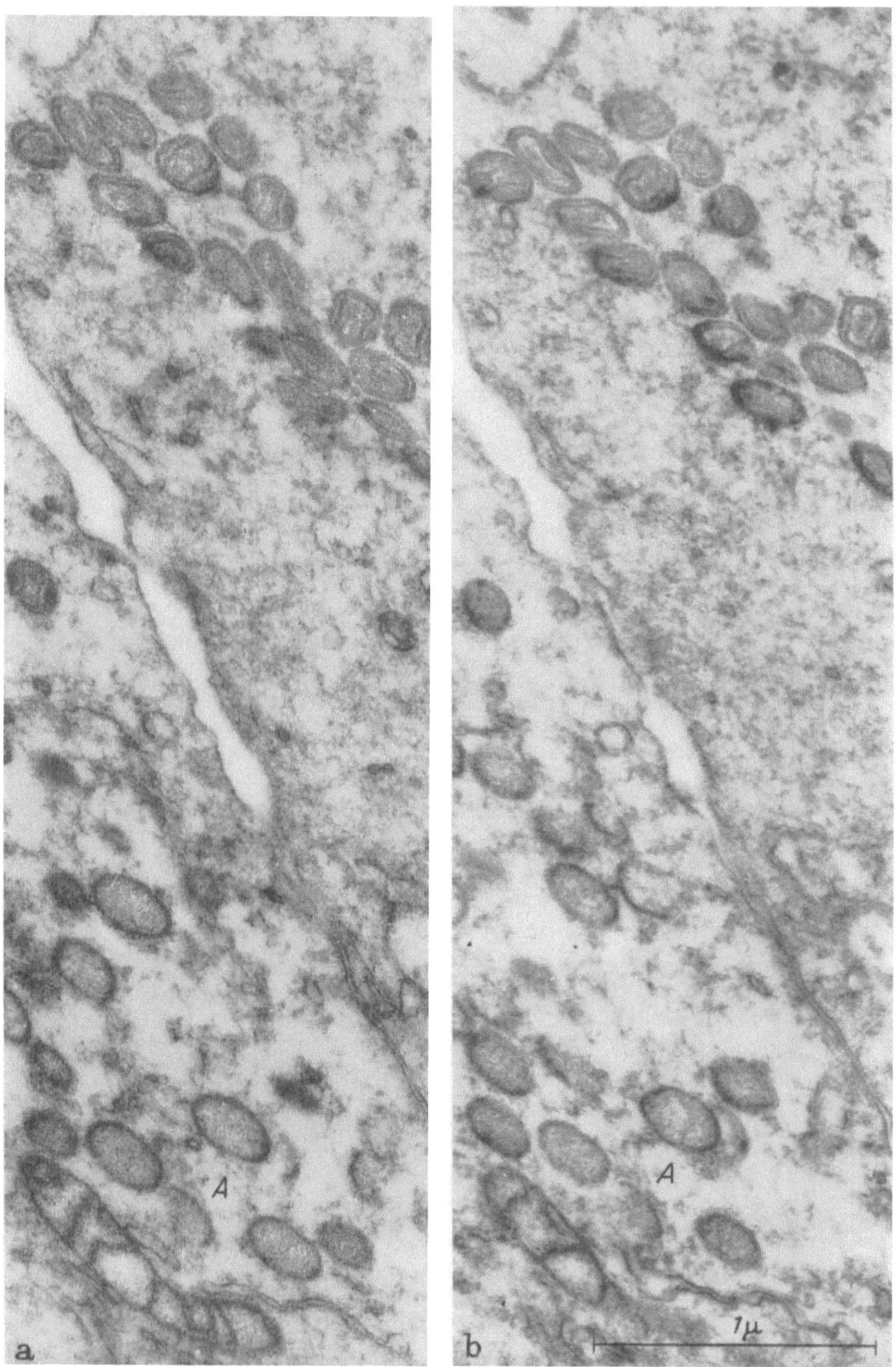

Abb. 12a u. b. Serienschnitte Vaccine-infizierter HeLa-Zellen; unterschiedliche Entwicklungsstadien der Elk. in den beiden Zellen; unvollständige Membran (*A*); KMnO₄-fixiert, 40000mal

bildet. Abb. 10c ist zu entnehmen, daß es neben den als Normalform angesprochenen Typen auch andere gibt, z. B. nucleoid-haltige Partikel ohne Membran (A), solche, bei denen Elk.-plasma und Membran durch eine helle Zone getrennt sind (B) und schließlich Teilchen, deren Plasma von zwei membranartigen Schichten umgeben ist (C).

Fast alle Autoren berichten über unvollständige Membranen, die Teile der Cytoplasma-Matrix halbmondförmig umschließen (Abb. 11, 16). Erst weitere Untersuchungen können endgültig klären, ob es sich dabei, wie von manchen vermutet wird, um Membranen im Status nascendi handelt. Anhand von Serienschnitten, bei denen unreife Elk. mindestens 5mal geschnitten wurden, machten MORGAN u. a. (79) darauf aufmerksam, daß mangelhafte Membranzeichnung jedenfalls zu einem Teil durch schnittbedingte Defekte erklärt werden kann. Abb. 12 zeigt an zwei Aufnahmen aus einer Schnittserie, daß einzelne Membranen (A) aber auch in mehreren Ebenen unvollständig sein können. Es ist sehr unwahrscheinlich, daß es sich dabei um ein Präparationsartefakt handelt. Natürlich könnten in solchen Fällen auch degenerative Prozesse verantwortlich sein.

In OsO_4-fixierten Präparaten erscheint die umhüllende Membran des unreifen Elk. in der Regel als einheitliche Struktur. Mit Hilfe der $KMnO_4$-Fixierung[1] wird aber auch bei diesem Stadium eine Doppelkontur der Membran erkennbar; wegen der Kompressionserscheinungen am deutlichsten senkrecht zur langen Achse (Abb. 11).

Mit fortschreitender Infektion finden sich neben den unreifen Partikeln in zunehmendem Maße auch reife Elk., die an der zuvor beschriebenen typischen Innenstruktur leicht zu erkennen sind (Abb. 10d [D], 11, 12, 14, 16). Von den freipräparierten reifen Elk. unterscheiden sich die zellständigen trotz grundsätzlich gleichem inneren Aufbau — geringfügig, aber doch markant — durch eine weniger ausgeprägte Quaderform. Sichere Übergangsformen zwischen unreifen und reifen Elk. sieht man nur selten. Trotzdem ist kaum daran zu zweifeln, daß die Entwicklung der reifen Formen über die unreifen verläuft; wahrscheinlich erfolgt der Übergang sehr schnell. Genauere Vorstellungen über die Stufen der Entwicklung wird man wohl erst erhalten, wenn es gelingt, den Infektionsablauf besser als bisher zu standardisieren. Abb. 12 zeigt z. B. zwei verschiedene Infektionszustände in zwei benachbarten Zellen, eine Erscheinung, wie sie für das bisher untersuchte Material typisch ist. Es liegt auf der Hand, daß es weiterer Untersuchungen bedarf, um die geschilderte Interpretation endgültig zu stützen.

Übereinstimmend ist beobachtet worden, daß die Bildungszentren im Frühstadium von angehäuften Mitochondrien umgeben sind; ein Befund, der im Hinblick auf das intensivierte Stoffwechselgeschehen verständlich ist. Für die hypothetische Annahme, daß unreife Elk. durch Kondensation des Inhalts aus kleinen Mitochondrien entstehen könnten (67), findet sich nach unseren Erfahrungen keine Stütze.

Die in Schnitten sichtbaren Verdichtungszentren mit ein- oder angelagerten unreifen Elk. entsprechen in Anordnung und Struktur den von HERZBERG und KLEINSCHMIDT (16, 17) beim Kanarienpockenvirus mittels Tupfpräparation dargestellten „Inseln" oder „Plaques". Die Frühformen erwiesen sich bei dieser Präparationsart als relativ massenarm; viele von ihnen enthielten einen Innenkörper (Nucleoid ?) (s. a. 55). Mit einem Spreitungsverfahren konnten diese Autoren auch einzeln liegende Viruspartikel aus der infizierten Zelle heraus zur Darstellung bringen. Dabei wurden neben reifen Elk. vor allem Rund- und Ovalformen freigesetzt, die ihrer flachen Auftrocknung wegen als „Scheiben" bezeichnet wurden. Diese Formen sind zwar von freipräparierten Mitochondrien kaum zu unterscheiden (s. u. a. 81); trotzdem ist es recht wahrscheinlich, daß es sich dabei um nucleoidfreie unreife Elk. gehandelt hat. — Das Nebeneinander von intakten reifen Elk., innenkörperhaltigen Frühformen und innenkörperfreien „Scheiben" legt einen Vergleich mit den oben beschriebenen Formen enzymatisch abgebauter reifer Elk. nahe und wirft damit die Frage auf, ob nicht etwa die sog. unreifen Elk. Degenerationsformen reifer Elk. seien. Dem bisherigen Anschein nach zeigen jedoch degenerierte reife Elk. im Schnitt eine andersartige Struktur als die unreifen Formen. Des weiteren ergibt sich aus zeitlich gestaffelten Untersuchungen, daß die reifen Elk. erst nach den unreifen erscheinen. Man kann die gestellte Frage demnach mit gutem Recht verneinen.

Nachdem zwei charakteristische Entwicklungsstadien der Pockenviren erkannt sind, richtet sich das Interesse in stärkerem Maße auf die Entstehungsweise und die Eigenschaften der sog. Matrix und damit auf die schwer zugänglichen frühesten Stadien der Infektion. Bisher vorliegende Berichte über einen Nucleinsäuregehalt der Matrix sind widerspruchsvoll. Auch ist es noch zweifel-

---

[1] Die übliche Methacrylat-Einbettung führt nach $KMnO_4$-Fixierung häufig zu beträchtlichen Strukturschädigungen. Gute Ergebnisse erhielten wir regelmäßig erst nach Übergang auf die Einbettung in „Plexigum" (80).

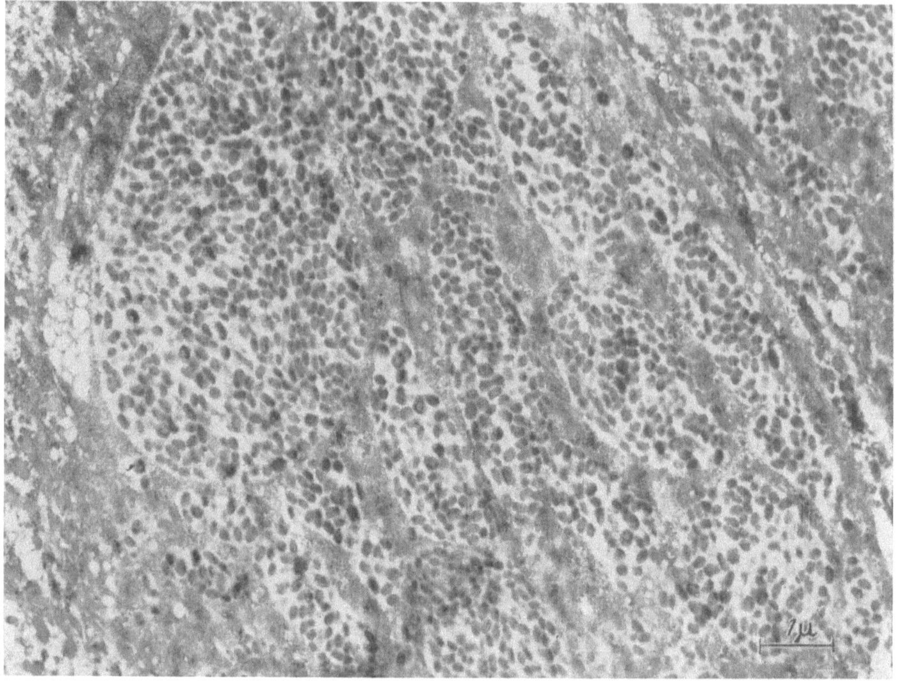

Abb. 13. Schnitt durch Molluscum contagiosum-infizierte Zelle; durch Trabekel abgetrennte Einschlußkörper; OsO_4-fixiert; 10000mal

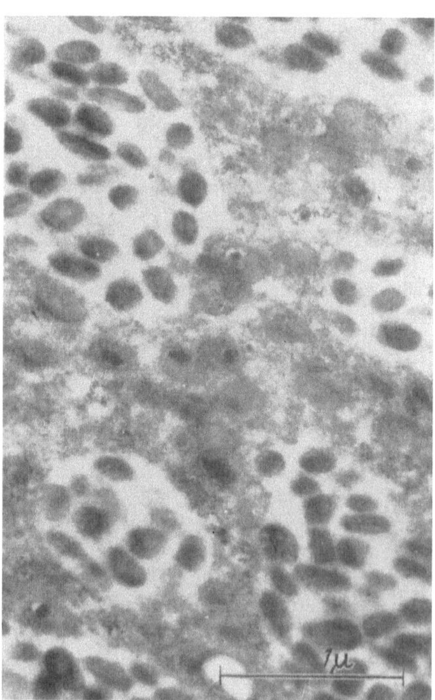

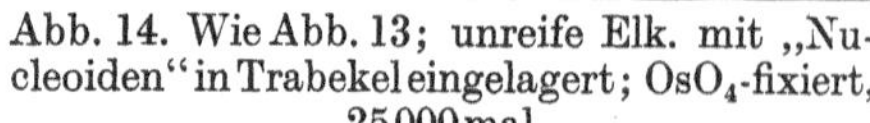

Abb. 14. Wie Abb. 13; unreife Elk. mit „Nucleoiden" in Trabekel eingelagert; OsO_4-fixiert, 25000mal

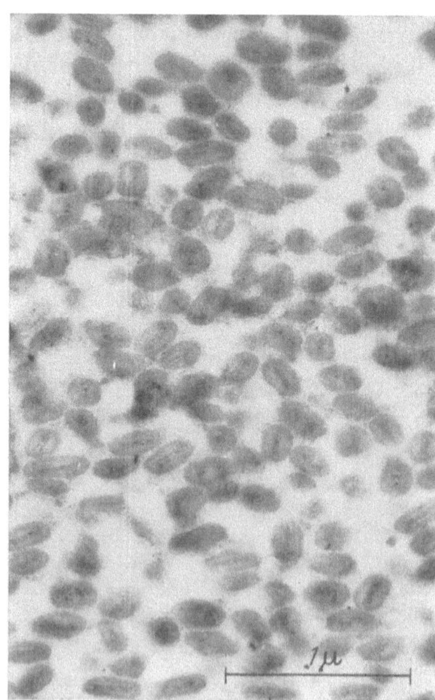

Abb. 15. Wie Abb. 13; reife Elk. typischer Gestalt im Einschlußkörper; OsO_4-fixiert; 25000mal

haft, wieweit sie mit lichtoptisch faßbaren Einschlußkörpern identifiziert werden kann. Den später auftretenden klassischen Einschlußkörpern I. Ordnung (LIPSCHÜTZ) entspricht sie sicher nicht; eher den sog. Initialkörperchen. Sehr bedeutsam ist aber, daß die Matrix vermutlich eine

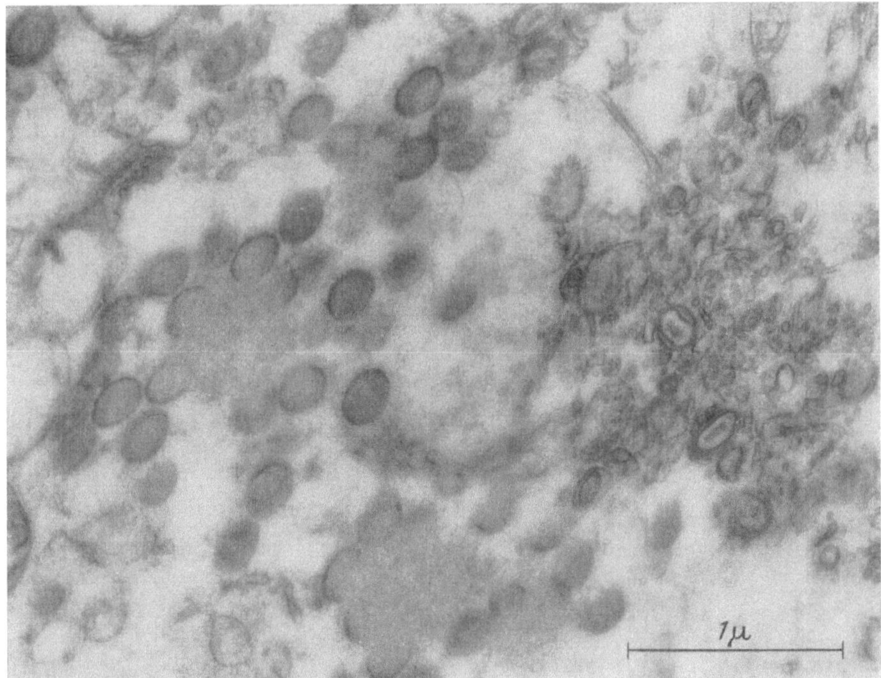

Abb. 16. Schnitt durch Geflügelpocken-infizierte Chorioallantoismembran; links zwei Verdichtungszonen (Matrix) mit unreifen Elk., darin „Nucleoide" als aufgehellte Bezirke schwach erkennbar; rechts reife Elk., umgeben von polymorphen Reaktionsprodukten; KMnO$_4$-fixiert; 30000mal

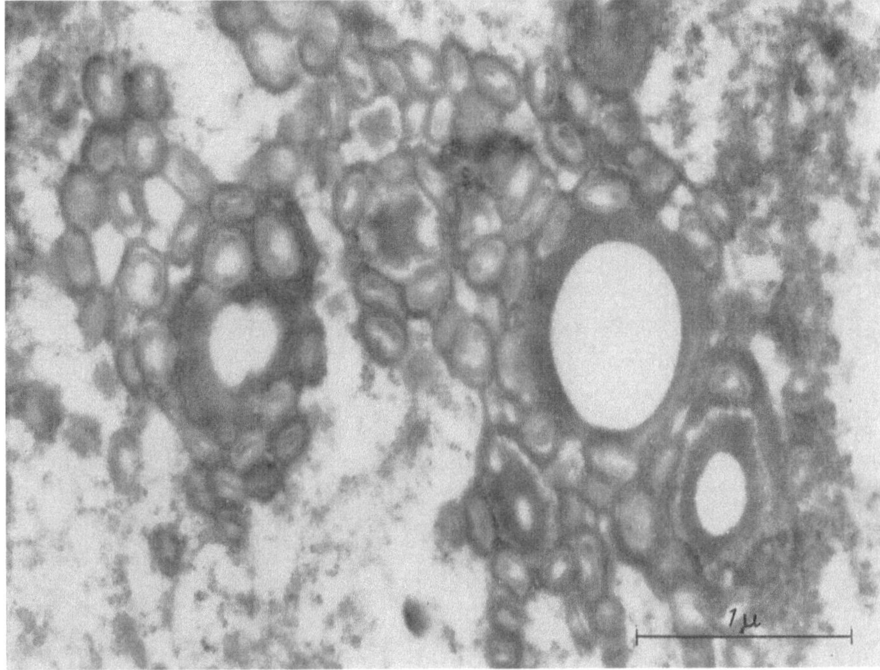

Abb. 17. Wie Abb. 16; Bollinger-Körper mit reifen Elk. und zentraler Vacuole; KMnO$_4$-fixiert. 30000mal

obligate Stufe der frühen Entwicklung von Pockenviren darstellt. Die klassischen Einschlußbildungen, wie Bollinger- und Guarnieri-Körper, repräsentieren dagegen relativ späte Stadien der Infektion, in denen die Unterschiedlichkeit der Virusstämme ebenso wie die Divergenz der

Wirtszellen ihren morphologischen Ausdruck finden. Es ist bekannt, daß die Virusvermehrung auch ohne die Bildung typischer Einschlußkörper vonstatten gehen kann. Schon die lichtoptischen

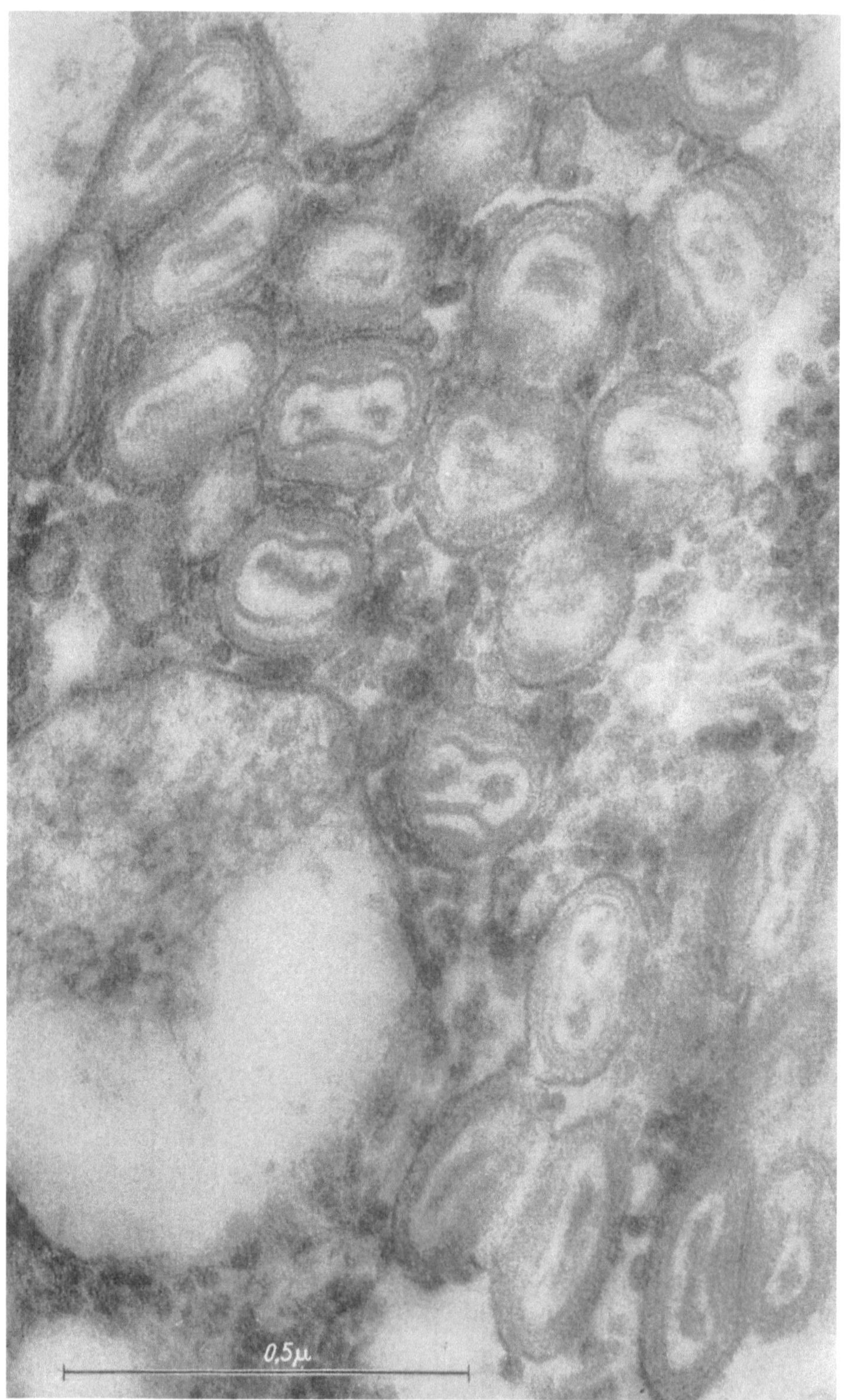

Abb. 18. Wie Abb. 16. Ausschnitt aus Bollinger-Körper; reife Elk. umsäumt von Schicht mittlerer Dichte und kontrastreicher Lamelle; $KMnO_4$-fixiert, 120000mal

Methoden haben charakteristische Unterschiede erkennen lassen, die nun durch die elektronen-optische Untersuchung bestätigt und vertieft worden sind. Das sei an zwei typischen Beispielen veranschaulicht.

Die Einschlußkörper des Molluscum contagiosum bestehen, wie auf Grund der lichtoptischen Bearbeitung zu erwarten war, ausschließlich aus Anhäufungen reifer Elk. (Abb. 13) (57, 69, 70, 71, 82), die an der Innenstruktur (Abb. 15) leicht zu erkennen sind[1]. Sie sind gegeneinander durch die lichtoptisch wohlbekannten Trabekeln abgegrenzt. Interessant ist, daß trotz der weit fortgeschrittenen Infektion in den Trabekeln, die als Restbestände der cytoplasmatischen Grund-struktur gedeutet werden, noch unreife nucleoid-haltige Elk. angetroffen werden (Abb. 14).

Ein grundsätzlich andersartiges Bild bietet sich bei der Betrachtung der Bollinger-Körper der Geflügelpocken, die bereits auf Grund lichtoptischer Studien als besonders lipoidreich be-kannt sind. Schon einzeln liegende reife Elk. sind des öfteren von andersartigen Strukturen um-geben, die vermutlich als Reaktionsprodukte der Wirtszelle aufzufassen sind. Abb. 16 zeigt das Nebeneinander von Verdichtungszonen (Matrix) und einem Focus der beschriebenen Art in einem frühen Stadium der Infektion. Später erscheinen unter Assoziation von reifen Elk. und kontrastreichen, wahrscheinlich lipoidhaltigen Reaktionsprodukten, die sowohl granulär, filamen-tös wie auch homogen strukturiert sind, typische Einschlußkörper (Abb. 17) (72, 82). Die zentrale Zone erweist sich häufig in Form einer oder mehrerer Vacuolen als aufgehellt; nicht selten sind größere Foci homogener Substanz nur an der Peripherie von reifen Elk. umsäumt. Vergleichbare Einschlußbildungen sind auch beim Virus der Ektromelie beschrieben worden (57, 74, 75). Bei höherer Vergrößerung (Abb. 18) ist zu beobachten, daß die Elk., deren äußere Begrenzung durch die doppelt konturierte umhüllende Membran deutlich gekennzeichnet ist, von einer etwa 10 bis 15 mμ dicken Schicht mittlerer Elektronendichte umsäumt und des weiteren von einer kontrastreicheren Lamelle von etwa 5—10 mμ Dicke eingeschlossen sind.

Mit derartigen Beobachtungen stehen wir erst am Anfang der morphologischen Bearbeitung aller jener Fragen, die in Beziehung zu dem Infektionsvorgang neu aufgeworfen worden sind. Wesentliche neue Erkenntnisse werden auch in Zukunft nur durch methodische Fortschritte zu erlangen sein. Es besteht gute Aussicht, daß elektronenoptisch-cytochemische Methoden weiter führen werden.

Die eigenen experimentellen Arbeiten wurden durch Sachbeihilfen der Deutschen Forschungsgemeinschaft wesentlich gefördert. Den Herren Dr. K. H. ANDRES und Dr. G. NIELSEN danke ich für ihre Mitarbeit an den in Abschnitt II dargestellten Ergebnissen.

Literatur

1. BORRIES, B. V., u. H. RUSKA: Klin. Wschr. 17, 921 (1938).
2. RUSKA, H.: Dtsch. med. Wschr. 67, 281 (1941).
3. — Arch. ges. Virusforsch. 2, 480 (1943).
4. — „Die Elektronenmikroskopie in der Virusforschung" in DOERR-HALLAUER, Handbuch der Virusforschung 2. Erg. Bd. 221. Wien 1950.
5. FENNER, F.: Nature (Lond.) 171, 562 (1953).
6. BERNHARD, W., J. HAREL et C. OBERLING: C. R. Acad. Sci. (Paris) 239, 732 (1954).
7. LLOYD, B. J., and H. KAHLER: J. nat. Cancer Inst. 15, 991 (1955).
8. NASEMANN, TH., u. E. BAUER: Klin. Wschr. 35, 62 (1957).
9. BLAKEMORE, F., and M. ABDUSSALAM: J. comp. Path. 66, 373 (1956).
10. ABDUSSALAM, M.: Amer. J. Vet. Res. 18, 614 (1957).
11. — u. D. PETERS: Unveröffentlicht.
12. RECZKO, E.: Zbl. Bakt. I. Abt. Orig. 169, 425 (1957).
13. ABDUSSALAM, M. and V. E. COSSLETT: J. comp. Path. 67, 145 (1957).
14. BOSWELL, F. W.: Brit. J. exp. Path. 28, 253 (1947).
15. ATHANASIU, P.: Ann. Inst. Pasteur 74, 418 (1948).

---

[1] Angaben, nach denen die Elk. des Molluscum contagiosum eine unklare Innenstruktur (69) oder sogar ein leeres Inneres (70, 71, s. a. 83) aufweisen, lassen sich nach unseren Erfahrungen durch Präparationsarte-fakte oder degenerative Veränderung erklären. — Sollte sich der Bericht über eine Züchtbarkeit des Molluscum contagiosum-Virus auf HeLa-Zellen (70) bestätigen lassen, dann böte sich in diesem System ein ausgezeichnetes Modell für weitere Studien. Es wurde bisher aber nur über Material berichtet, das aus der ersten und zweiten Passage stammte; die Möglichkeit, daß mitgeschlepptes Inokulat dargestellt wurde, ist daher bisher nicht aus-geschlossen.

16. HERZBERG, K., u. A. KLEINSCHMIDT: Z. Hyg. **139**, 545 (1954).
17. — — K. REQUARDT u. W. VOGELL: Arch. ges. Virusforsch. **6**, 283 (1955).
18. SKALINSKI, E. I.: Biophysik (russ.) **3**, 122 (1958).
19. PETERS, D., u. TH. NASEMANN: Z. Tropenmed. **4**, 11 (1952).
20. — u. W. STOECKENIUS: Z. Tropenmed. **5**, 329 (1954).
21. ANDRES, K. H., H. LIESKE, H. LIPPELT, E. MANNWEILER, G. NIELSEN, D. PETERS u. K. SEELEMANN: Dtsch. med. Wschr. **83**, 12 (1958).
22. HOAGLAND, C. L., J. E. SMADEL and T. M. RIVERS: J. exp. Med. **71**, 737 (1940).
23. RUSKA, H., u. G. A. KAUSCHE: Zbl. Bakt. I. Abt. Orig. **150**, 311 (1943).
24. ANDERSON, T. F.: Trans. N. Y. Acad. Sci. **13**, 130 (1951).
25. PETERS, D.: Nature (Lond.) **178**, 1453 (1956).
26. — u. R. GEISTER: In Vorbereitung.
27. GREEN, R. H., T. F. ANDERSON and J. E. SMADEL: J. exp. Med. **75**, 651 (1942).
28. SHARP, D. G., A. R. TAYLOR, A. E. HOOK and J. W. BEARD: Proc. Soc. exp. Biol. (N. Y.) **61**, 259 (1946).
29. STOECKENIUS, W., u. D. PETERS; Z. Naturforsch. **10 b**, 77 (1955).
30. HALL, C. E.: J. biophys. biochem. Cytol. **1**, 1 (1955).
31. HUXLEY, H. E.: Electron Microscopy, Proc. Stockholm Conf. 260 (1956).
32. HORNE, R. W., and S. BRENNER: Dieser Band S. 625
33. PETERS, D., R. GEISTER u. H. GIESE: In Vorbereitung.
34. SMADEL, J. E., T. M. RIVERS and C. L. HOAGLAND: Arch. Path. (Chicago) **34**, 275 (1942).
35. DAWSON, I. M., and A. S. MCFARLANE: Nature (Lond.) **161**, 464 (1948).
36. LÉPINE, P., P. ATANASIU et O. CROISSANT: C. R. Acad. Sci. (Paris) **228**, 1068 (1949).
37. BANG, F. B., E. LEVY and G. O. GEY: J. Immunol. **66**, 329 (1951).
38. FARRANT, J. L., and F. FENNER: Austral. J. exp. Biol. med. Sci. **31**, 121 (1953).
39. OVERMAN, J. R., and I. TAMM: Proc. Soc. exp. Biol. (N. Y.) **92**, 806 (1956).
40. DUMBELL, K. R., A. W. DOWNIE and R. C. VALENTINE: Virology **4**, 467 (1957).
41. SHARP, D. G.: Dieser Band S. 542.
42. PETERS, D., u. TH. NASEMANN: Naturwissenschaften **39**, 306 (1952).
43. — — Z. Naturforsch. 8 b, 547 (1953).
44. NASEMANN, TH., u. O. HUBER: Z. Tropenmed. **6**, 374 (1955).
45. PETERS, D.: Z. Naturforsch. **12 b**, 697 (1957).
46. — u. H. GIESE: In Vorbereitung.
47. ARDENNE, M. V., u. H. FRIEDRICH-FREKSA: Naturwissenschaften **29**, 523 (1941).
48. PETERS, D.: Z. Naturforsch. **12 b**, 704 (1957).
49. — In Vorbereitung.
50. HOAGLAND, C. L., G. I. LAVIN, J. E. SMADEL and T. M. RIVERS: J. exp. Med. **72**, 139 (1940).
51. PETERS, D., u. R. WIGAND: Z. Naturforsch. 8 b, 180 (1953).
52. WIGAND, R., u. D. PETERS: Z. Naturforsch. **9 b**, 586 (1954).
53. PETERS, D., and W. STOECKENIUS: Nature (Lond.) **174**, 224 (1954).
54. — — Z. Naturforsch. **9 b**, 524 (1954).
55. — Verh. dtsch. Ges. Path. **38**, 14 (1955).
56. GAYLORD, W. H., and J. L. MELNICK: Science **117**, 10 (1953).
57. — — J. exp. Med. **98**, 157 (1953).
58. MORGAN, C., S. A. ELLISON, H. M. ROSE and D. H. MOORE: J. exp. Med. **100**, 301 (1954).
59. EAVES, G., and T. H. FLEWETT: J. Hyg. **53**, 102 (1955).
60. PETERS, D.: In Vorbereitung.
61. LUFT, J. H.: J. biophys. biochem. Cytol. **2**, 799 (1956).
62. EPSTEIN, M. A.: Nature (Lond.) **181**, 784 (1958) und Brit. J. exp. Path. **39**, 436 (1958).
63. MERCER, E. H.: Nature (Lond.) **181**, 1550 (1958).
64. BANG, F. B.: Ann. Rev. Microbiol. **9**, 21 (1955).
65. EAVES, G., and T. H. FLEWETT: J. Path. Bact. **68**, 633 (1954).
66. FLEWETT, T. H.: J. Hyg. **54**, 393 (1956).
67. CROISSANT, O., P. LÉPINE et R. W. G. WYCKOFF: Ann. Inst. Pasteur **94**, 294 (1958).
68. PETERS, D., G. NIELSEN u. K. H. ANDRES: VII. Internat. Kongr. Mikrobiologie, Stockholm, Abstract 245, (1958), ausführliche Veröffentlichung in Vorbereitung.
69. TAKAKI, F., T. SUZUKI, H. YASUDA, S. TAGUCHI, P. DOHI and M. SASAO: Jikeikai med. J. **4**, 60 (1957).
70. DOURMASHKIN, R., et H. L. FEBVRE: C. R. Acad. Sci. (Paris) **246**, 2308 (1958).
71. — et B. DUPERRAT: C. R. Acad. Sci. (Paris) **246**, 3133 (1958).
72. TAJIMA, M., and Y. KUBOTA: Virology **6**, 308 (1956) und NIBS Bull. Biol. Res. **2**, 48 (1957).
73. HERZBERG, K., u. A. KLEINSCHMIDT: Dieser Band S. 573.
74. OZAKI, Y.: Acta Scholae Med. Univ. Kyoto **33**, 168 (1956).
75. DOHI, S.: Electron Microscopy Proc. I. Reg. Conf. Tokyo, 186 (1956).
76. BERNHARD, W., A. BAUER, J. HAREL et C. OBERLING: Bull. Cancer **41**, 423 (1955).
77. BAUER, A., et T. CONSTANTIN: C. R. Soc. Biol. (Paris) **150**, 246 (1956).
78. FEBVRE, H., J. HAREL et J. ARNOULT: Bull. Cancer **44**, 92 (1957).

79. Morgan, C., S. A. Ellinson, H. M. Rose and D. H. Moore: Exp. Cell Res. **9**, 572 (1955).
80. Bayer, M., u. D. Peters: J. Ultrastructure Res. **2**, 444 (1959).
81. Peters, D., u. R. Wigand: Klin. Wschr. **28**, 649 (1950).
82. — u. K. H. Andres: Unveröffentlicht.
83. Bernhard, W.: Cancer Res. **18**, 491 (1958) und dieser Band S. 610.

Dünnschnittbefunde am Kanarienpocken-Virus

K. Herzberg und A. Kleinschmidt

Hygiene-Institut der Stadt und Universität Frankfurt am Main

Über die Struktur der Elementarkörperchen des Geflügelpocken-Virus haben Arbeiten von Morgan u. Mitarb. sowie Eaves und Flewett auf Grund von Dünnschnitten an infizierten Chorioallantoismembranen des Bruteis berichtet. Die Befunde ergaben sowohl bezüglich der Struktur der Elementarkörperchen als auch der Stellen, an denen sie sich im Cytoplasma anhäuften, zwischen beiden Arbeiten Unterschiede. Wir berichten über Dünnschnittbefunde am Kanarienpocken-Virus, also einem besonderen Typ von Geflügelpocke, die an Schnitten durch das Hautödem einerseits, durch sedimentierte virushaltige Histiocyten infizierter Kanarien andererseits erhoben wurden. Unsere elektronenmikroskopischen Aufnahmen ergaben in der Anordnung von Hülle und innerem Korn z. T. Befunde wie bei Morgan u. Mitarb.; in der Beziehung der Elementarkörperchen zu den infizierten Wirtszellen (bei uns Kanarien-Histiocyten) ähnliche Bilder, wie sie Eaves und Flewett veröffentlicht haben. Wir erhielten weiterhin Aufnahmen, die zu unseren früher auf Grund von Tupfpräparaten gewonnenen Vorstellungen passen. Erkenntnisse aus der Lichtmikroskopie waren uns hierbei nützlich.

Dieser Beitrag erschien ausführlich in Zbl. Bakt. I Orig. **174**, 1 (1959).

Electron microscopic studies on the growth of pox virus in monolayer culture of strain L cells and HeLa cells

N. Higashi, Y. Ozaki and T. Fukada

Institute for Virus Research, Kyoto University (Japan)

Monolayer cell culture techniques were used for detailed examinations of the very early stage of poxvirus infection in strain L cells and HeLa cells.

L cells and HeLa cells were cultured at 37° C for 2 to 3 days in bottles containing a nutrient medium which was usually used for the growth of these cells. After microscopic examination, to insure that good cell monolayers were present, the bottles received virus inoculum. Ectromelia virus (Hampstead strain) and vaccinia virus (Handai strain) were used. Ectromelia virus was inoculated on both types of cells and vaccina virus on HeLa cells.

The bottles were reincubated and, at intervals from 3 to 24 hr following inoculation, cells which were detached from the glass surface were washed and centrifuged to form small pellets. The pellets were fixed for 1 hr at room temperature in 1% OsO_4 at p_H 7.4. Dehydration, embedding in methacrylate and thin sectioning were done as usual. The sections were examined in the JEM electron microscope. In parallel with electron microscopy of thin sections at intervals shown above, light microscopy of smeared stained preparations was done.

Experiments on the growth curve of ectromelia virus on both types of cells showed the same eclipse phase of 6 hr. Azurophilic matrix type inclusion as described below appeared to have developed 3—5 hr

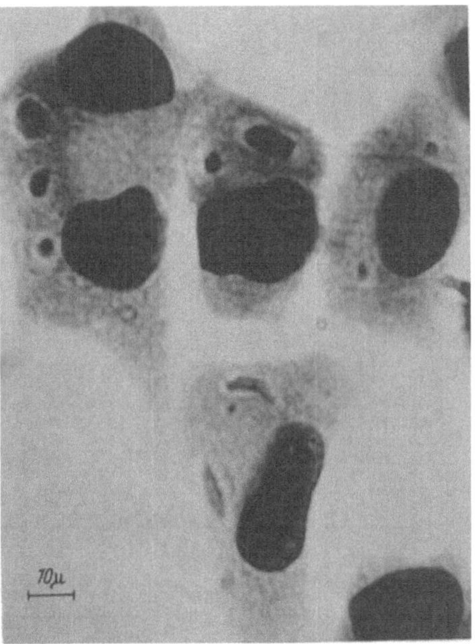

Fig. 1. Strain L cells with cytoplasmic matrix type inclusion. 8 hr-infection with ectromelia virus. (Light microscope) 600×

after infection. On the other hand, the first eosinophilic Marchal inclusions developed at 11 hr-infection only after the virus titer had almost reached maximum.

The time sequence of vaccinia virus multiplication in HeLa cells showed the eclipse phase of 8 hr. Matrix type inclusions appeared to have formed within 4 hr and virus titer reached maximum at 15 hr, but Guarnieri inclusions did not appear even after 24 hr-infection,

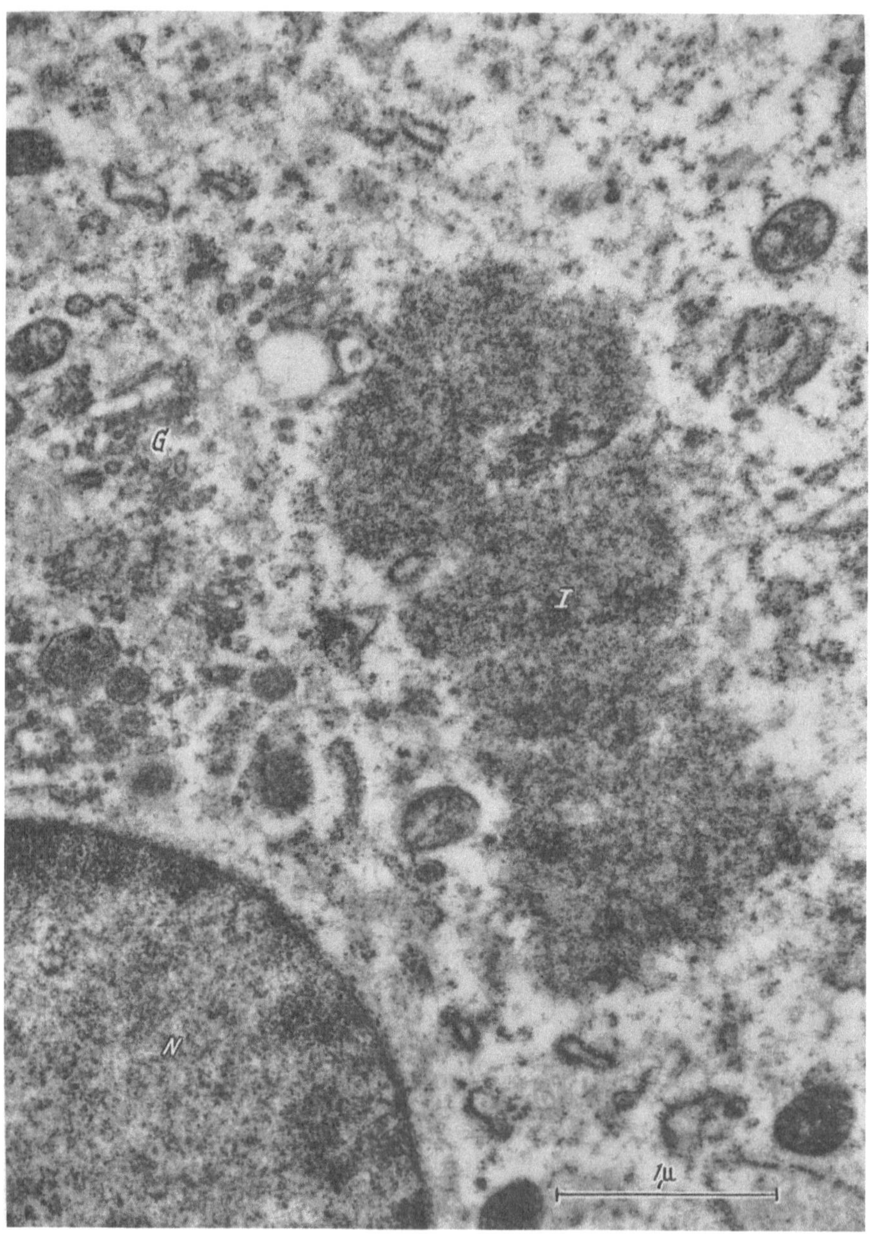

Fig. 2. Matrix type inclusion (*I*) in the L cell near nucleus (*N*) and Golgi area (*G*). 4 hr-infection with ectromelia virus. No viral components are visible within the reticulogranular matrix. 30,000×

Fig. 1 illustrates cytoplasmic inclusions of the ectromelia virus infected L cells. The inclusions were stained azurophil with Giemsa and stained blue with hematoxyline-eosine. The shape of the inclusion was not always round, sharp in its border and homogeneous in its structure. We define such an inclusion as matrix type inclusion (*I*). In as early as 6 hr-infection, nearly all the cells examined contained one or more inclusions of this type, but classical Marchal bodies quite different from those shown above never occurred. Fig. 2 shows part of an ectromelia virus infected L cell. Near nucleus and Golgi area is a characteristic region composed of dense reticulogranular material. This region is defined as matrix type inclusion, which shows strong contrast as compared with

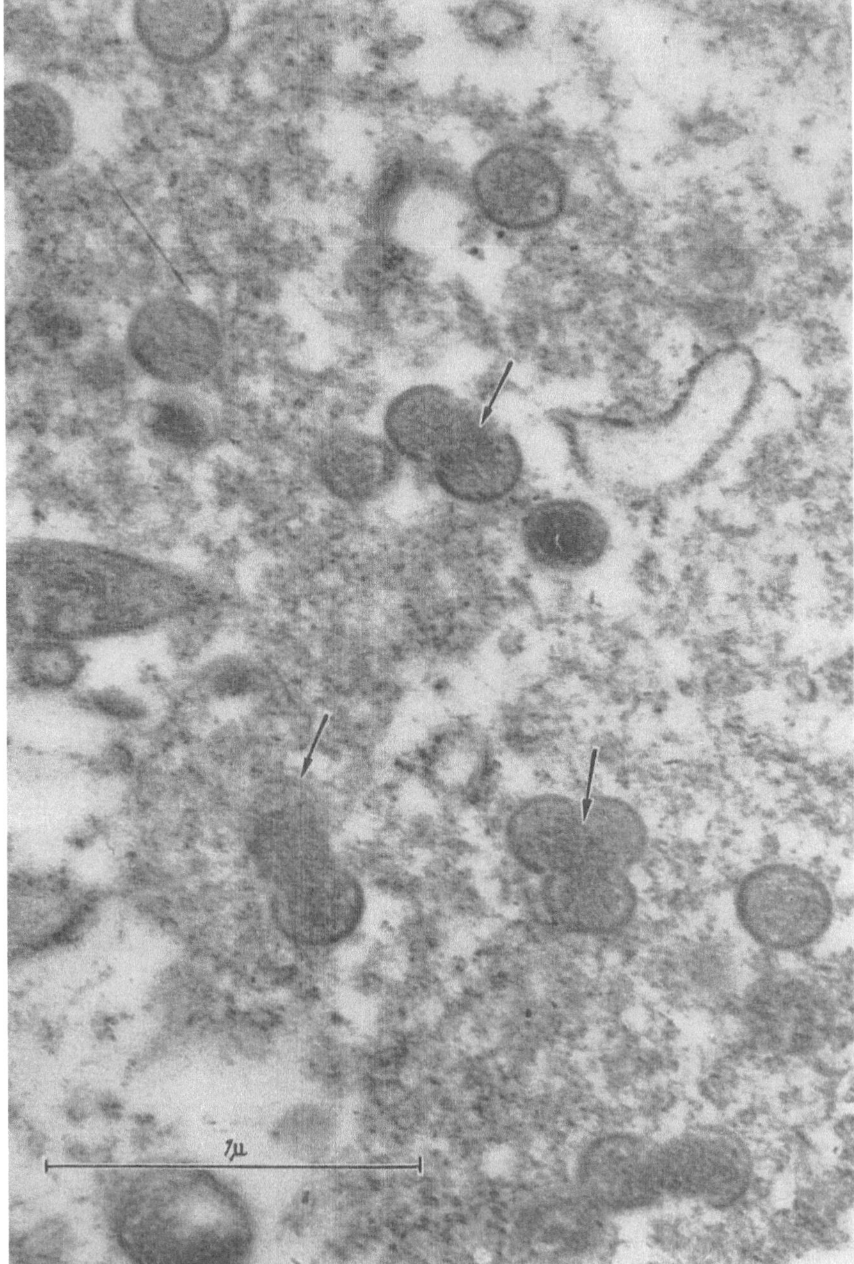

Fig. 3. Part of a matrix inclusion in the HeLa cell; 8 hr infection with vaccinia virus. Varieties of developmental forms can be seen. The viroplasmic component of some of them is not completely surrounded by the developing limiting membrane as indicated by arrows, and so viral particles appear to be in process of being surrounded by incomplete limiting membranes. 50,000 ×

Marchal inclusions possessing a dense homogeneous structure and a round or ovoid shape. Up to 4 hr-infection, no appearances considered to be the new generation of virus particles are visible in the inclusion. Vaccinia virus infected HeLa cells generally followed the same events. Fig. 3 shows part of a matrix type inclusion in the vaccinia virus infected HeLa cell. Varieties of developmental forms (2) can be seen. The viroplasmic component of some of these are poorly defined and are not completely surrounded by the developing limiting membrane as indicated by arrows. This micrograph shows that most of the particles appear to be in a progressive process of being surrounded by an incomplete limiting membrane. Some material of matrix inclusions of cells 12 hr or later after infection has almost disintegrated, dispersing viral particles throughout cytoplasm. In such a matrix regions which contained large numbers of descendant viral particles, occurrences of Marchal inclusions, were repeatedly observed. Some of them contained viral particles and some did not.

References

1. HIGASHI, N., et al.: Paper read at the 12th annual meeting of EMSJ, May 1958.
2. GAYLORD, W. H., and J. L. MELNICK: J. exp. Med. 98, 157 (1953).

Technique simple pour l'examen du virus du molluscum contagiosum au microscope électronique

ALEXANDRE KARPAROFF

Institut d'Epidémiologie et Microbiologie, Sofia (Bulgarie)

Pour l'observation du virus du molluscum contagiosum au microscope électronique on se sert de la technique suivante:

On prélève sur le malade une tumeur du molluscum contagiosum que l'on suspend dans de l'eau physiologique pour 10 min. Après on place la tumeur dans un tube contenant 0,3 ml d'eau physiologique pour 10 min encore. Pour une troisième fois on suspend la tumeur dans un tube contenant 0,2 ml d'eau physiologique pour 24 h à

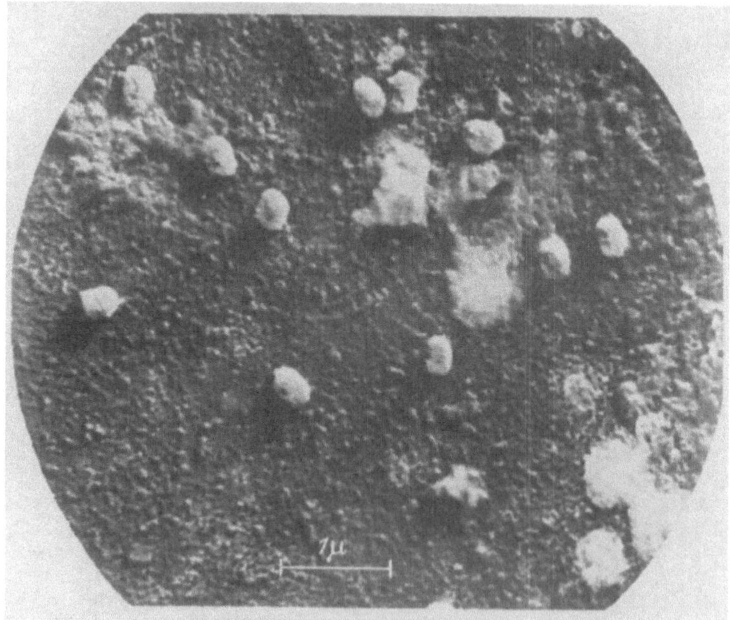

Fig. 1. Virus du molluscum contagiosum après ombrage par métallisation au chrome. 15.000:1

la température du laboratoire. On écrase la petite tumeur à l'aide d'une pipette Pasteur effilée. On centrifuge à 3.000 tours par minute pendant 20 min, puis de nouveau on écrase la tumeur et l'on ajoute 0,2 ml d'eau physiologique. On centrifuge la suspension une deuxième fois pendant 20 min à 3.000 tours par minute. On dépose une goutte du liquide surnageant sur une membrane de collodion. La préparation, séchée et ombrée au chrome est examinée au microscope électronique.

On observe les résultats obtenus sur la Fig. 1, qui a été choisie parmi beaucoup d'autres. On voit bien les corps élémentaires du virus du molluscum contagiosum avec leur forme typique, quadrangulaire avec angles arrondis et d'une taille moyenne de 250 à 300 mμ. La surface de ces corpuscules n'est pas lisse, on voit des parties saillantes et d'autres creuses ce qui montre que leur structure interne n'est pas homogène. La culture pure que l'on observe nous donne des corps élémentaires d'une ressemblance frappante.

A l'aide de cette technique simple on obtient des préparations sur lesquelles on observe des cultures pures de virus et aussi on saisit le moment òu les corpuscules du virus sortent du cytoplasme dégénéré dans lequel ils étaient enveloppés, chose qu'on ne peut observer en se servant des méthodes bien connues pour la purification du virus comme ultrafiltration, ultracentrifugation et autres.

Structure and particle counts of the influenza virus and the adenovirus

Robin C. Valentine

National Institute for Medical Research, Mill Hill, London N. W.

In recent years the concept of a virus has changed. The idea of an indivisible particle of infection has slowly given place to notions that it is an object that possesses a structure whose components differ in their nature and function. Indeed, now the complexity of virus structure is apparent, the traditional expression "a virus particle" has become inept. As an illustration of some recent ideas of virus structure it is of interest to review and contrast what is now known of the structure of two viruses both causing a respiratory infection in man and both very similar in size. One is the influenza virus, which has been the object of research over many years; the other, the more recently isolated adenovirus (originally described under the name APC virus).

General morphology

Influenza virus. The infective unit of influenza is an object whose size is about 1000 Å. This can be shown by filtration through membranes of known pore size and by centrifugation at different speeds. When suitable preparations of infective fluid are viewed with the electron microscope, large numbers of similar spherical objects with a diameter of about 1000 Å are found. Such objects are not seen in similar fluid from an uninfected source and they have been shown to possess so many of the properties of the infective unit of influenza that their identification as the virus elementary body is now beyond doubt. The influenza virus is easily distorted while being prepared for electron microscopy and in particular specimens air-dried from distilled water su pensions have a variable appearance and size and have obviously been considerably flattened in drying down on to the supporting film. For this reason, most of the early electron micrographs gave a rather unsatisfactory idea especially of particle size. Later freeze-dried preparations showed unflattened particles of a much more uniform appearance (1) as did procedures in which osmium tetroxide fixed virus was adsorbed on the supporting film (2). In these cases the mean diameter of the virus was reported to be about 800 Å.

Untreated preparations of influenza give little contrast in the electron microscope and adequate pictures require the use of the metal shadowing technique (Fig. 1) or else specimens treated with phosphotungstic acid after fixation (Fig. 2). Especially in the latter case, the surface of the influenza virus, even when it is from a highly purified suspension, appears to be rough in contrast with the smooth surface of many other viruses.

In addition to the spherical forms of the influenza virus, infective fluid often contains long thread-like objects referred to as filaments or rods (Fig. 3 and 4). Their width is about 700 Å, i.e. about the same as the diameter of the virus spheres, but in length they may measure anything up to 25 μ or even more. Since these filaments also possess most of the essential characteristics of the influenza virus, they present a number of difficult problems that will be considered in detail later. It is clear, however, that whether or not we like to consider the filament to be a form of the influenza virus, the spherical objects without a doubt must be.

Adenovirus. The adenovirus is readily cultured in HeLa cells and by filtration and centrifugation it is found that the infective particle has a size certainly not less than 500 Å in diameter.

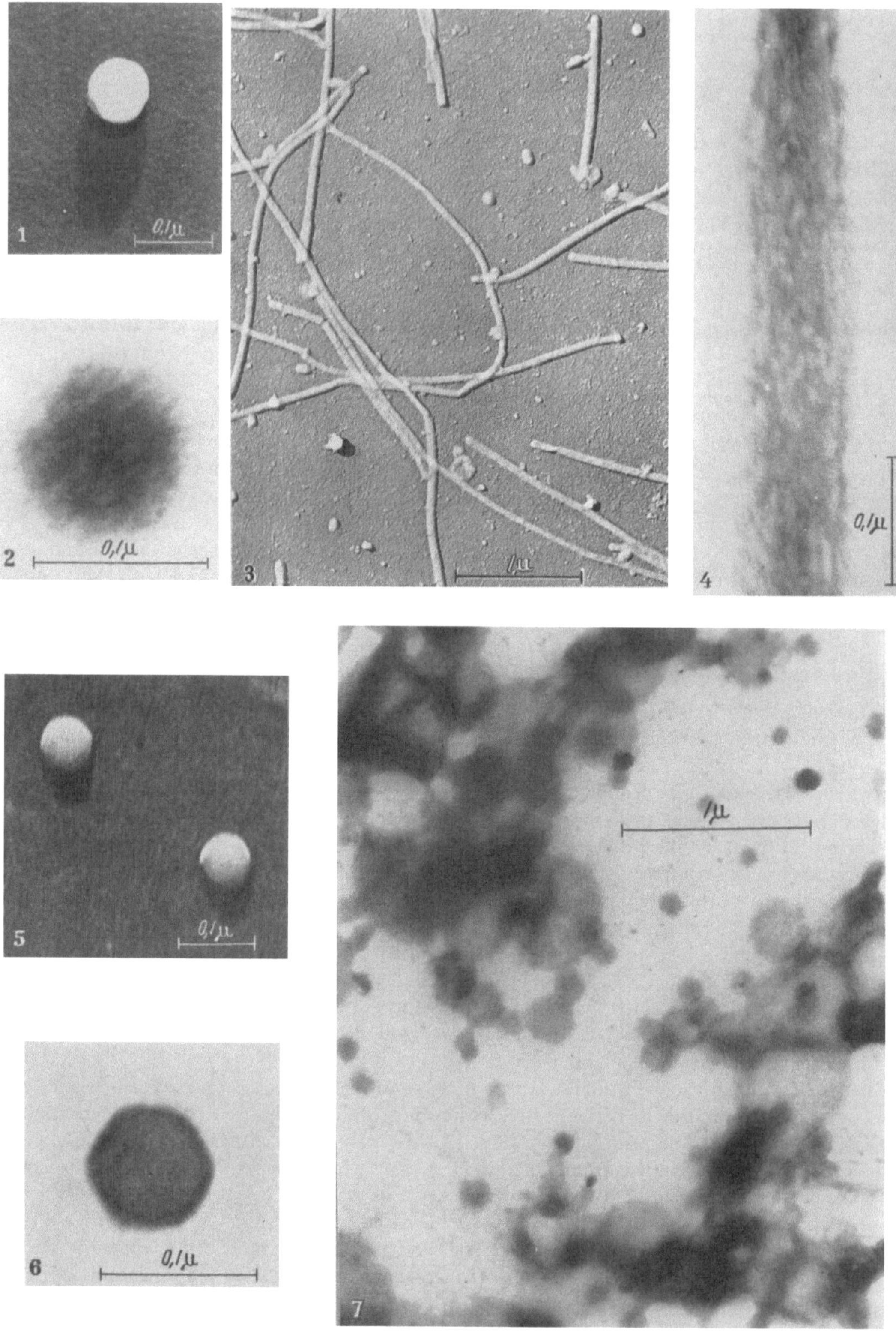

Fig. 1—7

Fig. 1. Influenza virus elementary body. 120.000 × — Fig. 2. Influenza virus elementary body. 275,000 × — Fig. 3. Influenza filaments. 20,000 × — Fig. 4. Part of an influenza filament. 200,000 × — Fig. 5. Adenovirus elementary bodies. 120,000 × — Fig. 6. Adenovirus elementary body. 250,000 × — Fig. 7. Influenza filaments breaking up in distilled water. 30,000 ×

With the electron microscope, objects with a very constant diameter of about 700 Å are found in large numbers in fluid obtained from infected cell cultures and their identification as the virus seems quite certain. When metal shadowed, the virus bodies appear to be nearly spherical (Fig. 5) and differ from the influenza virus only in having a more constant and slightly smaller diameter. However, when specimens are treated with phosphotungstic acid the surface of the virus becomes more electron dense and their smooth 6-sided outline is at once apparent (Fig. 6). This 6-sided outline is very similar to that of the rather larger insect virus, the tipula iridescent virus, which WILLIAMS has shown in a most elegant way to have the form of a regular geometrical solid, the icosahedron (3). It is thus quite possible that the adenovirus also has this same form but its smaller size and the difficulty of obtaining it highly purified have made it impossible as yet to decide between the icosahedron and the rhombic dodecahedron as the shape of the elementary bodies (4). But the regular polyhedral shape of the adenovirus is in striking contrast with the rough approximately spherical outline of the influenza virus.

Chemical Composition

Before considering the structure of the two viruses, it is useful to have some idea of their chemistry.

Influenza virus. The chemistry of the influenza virus is complex. Its nucleic acid, the essential genetic material of the virus and, in the case of some viruses at any rate, the substance that can alone carry infectivity, is of the ribonucleic acid type and forms about 1% by weight of the virus (5, 6, 7). The actual mean mass of ribonucleic acid per virus particle is about 3.3×10^{-18} g (equivalent to a molecular weight of 2×10^6) and is similar in amount to that found in all other ribonucleic acid containing viruses (8). In addition to its nucleic acid the dry weight of the influenza virus contains 61% protein, 34.5% lipid and 3.5% carbohydrate (8).

Adenovirus. A similar analysis of the adenovirus is not yet available but its nucleic acid is of the deoxyribonucleic acid type (9, 10) and, on a preliminary estimate, this forms about 40% by weight of the virus (10), giving a mass of deoxyribonucleic acid per virus particle similar in amount to that found in the vaccinia virus. In addition to the large content of deoxyribonucleic acid, the virus certainly contains protein, which may indeed prove to be its only other essential constituent.

Electron density

Photometry of electron micrographs, though a measurement seldom made, can provide useful information about the mass density of the specimen (11) especially in the case of viruses where the thickness of the object is known. With a 60 kV electron beam and using an objective aperture that rejected electrons scattered by the specimen through more than 10^{-2} radians (i.e. the majority of the elastically scattered but few of the inelastically scattered electrons), untreated freeze-dried influenza virus was found to transmit between 70% and 80% of the incident electron beam and the adenovirus 66% of the beam. A theoretical formula for the elastic scattering of electrons given by LEISEGANG (12) is in good agreement with a value we have deduced experimentally using latex particles. This formula can be used to estimate the density of the virus from the fraction of the electron beam it transmits (11,13). The mean value for the influenza virus is $0.9 \text{ g} \cdot \text{cm}^{-3}$ and for the adenovirus $1.4 \text{ g} \cdot \text{cm}^{-3}$. These figures, of course refer to the completely desiccated virus as examined in the electron microscope. The density of the hydrated particles can be estimated from their sedimentation in sucrose solutions and these suggest for the density of the wet virus; influenza $1.2 \text{ g} \cdot \text{cm}^{-3}$ and adenovirus $1.4 \text{ g} \cdot \text{cm}^{-3}$. The difference between the figures for the dry and wet densities gives the water content of the virus. We thus deduce that the influenza virus contains 25% by weight of water and that the adenovirus contains negligible water.

Biological evidence of virus structure

The essential property of a virus is that of infectivity. With both the viruses we are considering the infectivity is associated with the typical virus elementary bodies as has been already discussed. There is no evidence as yet that the infectivity of either of these viruses is ever carried by any smaller objects, though the various recent reports that infective nucleic acid has been prepared from other viruses (*14*) now makes one cautious about any statement that infectivity necessarily always requires the presence of the entire virus elementary body. It is however clear from many observations on filtered and centrifuged material that the infectivity of preparations of influenza and adenovirus is due, if not entirely, then to all but an insignificant extent to the identified elementary bodies.

The virus, however, possesses typical properties other than infectivity. Its protein has a unique antigenic structure that can be studied by various antigen-antibody reactions. The virus antigens are generally of two types: the group specific antigens which are shared by all viruses of the same group (e.g. all the influenza A viruses or all the adenoviruses) and the type specific antigens unique to the one strain or type of the virus (e.g. the Asiatic strain of influenza A or type 5 adenovirus). Preparations of influenza virus have a further important property, the ability to agglutinate a suspension of red cells. This property is removed when the type specific antibody is added to the virus. Adenovirus preparations can also agglutinate some red cells but the reaction is more variable (*15*). Although these and other additional properties of the virus are all carried by the elementary bodies, they are also to be found associated with much smaller objects. Thus it is possible to sediment the virus elementary bodies together with all the infectivity and still to detect some of the antigenic and haemagglutinating activity in the supernatant fluid. The inference drawn is that this is due to much smaller free sub-units of the virus. We thus conclude, even before examining the evidence provided by the electron microscope, that although only the entire elementary body possesses all the properties of the virus including infectivity, smaller objects that possess some of these properties may also be present in virus preparations and could be structural subunits of the elementary bodies.

Influenza virus. Confirmation that the influenza virus is composed of such subunits came when Hoyle showed that the elementary bodies could be broken down with ether (*16*). The infectivity was completely lost but the resulting fragments of the virus, estimated to have a size of about 150 Å, could be divided into two types of particle (*17*). One adsorbed reversibly on the surface of red blood cells, as does the intact virus, it contained the type specific antigen and was digested with trypsin. Chemically it was found to be a mucoprotein and did not contain nucleic acid (*6*). Particles of this type were termed haemagglutinin. The other particles did not haemagglutinate, contained the group but not the specific antigen, resisted trypsin digestion (*17*) and chemically consisted of ribonucleic acid and protein (*6*). These, or apparently essentially similar objects, have been variously referred to as the complement fixing or the soluble antigen. The rest of the influenza virus, not accounted for by these two types of particle, consists mostly of the ether soluble lipid material.

Adenovirus. Similar fractionation of the adenovirus has not yet proved possible, the elementary bodies failing to break up with ether. Virus preparations however certainly contain in addition to the elementary bodies some sort of virus subunits as well. Thus when all the infective material has been sedimented in a centrifuge the supernatant contains particles carrying the group antigen. As with the similar subunits of influenza, this antigen is not destroyed by trypsin. The supernatant fluid has another interesting property (*18*). Although not infectious in the sense that no virus is produced when it is added to a cell culture, the cells so treated nevertheless do show a typical cytopathic effect. The "cytopathic factor" in the supernatant fluid responsible for this is inhibited with specific antiserum and can be destroyed with trypsin. It is a particle whose size is much smaller than the virus and it could thus well be a protein subunit of the adenovirus. It is interesting to speculate whether it might be related to the protein inclusions that have been demonstrated in infected cells (*19*).

Electron microscope evidence of virus substructure

We have surveyed a little of the evidence obtained by virologists and chemists on the subunits of the virus of influenza and the adenovirus. It must now be seen how far the electron microscope can be used to identify these various objects and show how they are built into the elementary virus bodies.

Electron micrographs of the intact elementary bodies certainly give little hint of internal structure. There is some suggestion, especially from the filaments (Fig. 4), that the influenza virus contains thread-like material; the adenovirus particles however appear to be almost entirely homogeneous.

Three lines of attack are possible if one is to probe the structure of the virus. 1. To cut thin sections of the elementary bodies. 2. To break them down into components, e.g. with enzymes or ether. 3. To combine specific "electron stains" with interesting components of the virus such as its nucleic acid.

The first approach has been adequately covered especially by MORGAN and his colleagues who have published exceptionally fine pictures of both influenza virus (*20*) and adenovirus particles (*9*) as seen in thin sections of infected tissue. Their papers should be referred to for a full discussion of the findings some of which will be considered later. No clear picture of the structural components of the virus has yet come from this approach however.

The second of the suggested lines of investigation, the breakdown of the particles with chemicals, has yielded some results of interest in the case of the influenza virus (*2, 21*). The elementary bodies were first adsorbed on to the films on the specimen supports and then subjected to various treatments. The main findings will be outlined.

Particle breakdown

Influenza virus. 1. *Effect of distilled water.* The elementary bodies of influenza are not stable in distilled water and the infectivity and haemagglutination titres eventually fall to zero. When distilled water suspensions are examined with the electron microscope the first change in appearance is shown by the filaments. Large swellings appear along their length (Fig. 7) and the filaments then break down entirely leaving general debris. Later the more stable spherical virus forms also start to break down and clearly show a thread-like component (Fig. 8).

2. *Effect of dilute acid.* A few minutes treatment with 0.1 N hydrochloric acid does not alter the appearance of the spherical elementary bodies but it produces an interesting change in the filaments. These break up to form objects which, in some cases, bear a marked resemblance to the virus spheres (Fig. 9).

3. *Effect of ether.* The disintegration of the influenza virus with ether into two distinct components has been described above. Treatment of the elementary bodies on the specimen supports has not given clear results since the two components are not separated. Examination of the fractions produced by disintegrating the virus in suspensions has however been described (*22*). The haemagglutinin fraction consisted of particles most of which had a diameter of about 120 Å. The second fraction, the so-called "soluble antigen" which contains the nucleic acid, was also reported to consist of 120 Å diameter particles (*22*). The published picture, however, suggests that the fraction is more likely to be similar to that obtained in the same way from the related fowl plague virus, of which SCHÄFER and his colleagues have produced excellent electron micrographs (*23*). These show the nucleic acid containing component as rods of material about 150 Å wide.

4. *Effect of trypsin and nucleases.* Trypsin has little effect on the appearance of the influenza virus unless the virus is first denatured by a brief treatment with 0.1 N hydrochloric acid. The elementary bodies are then easily digested to leave an irregular but often typically polygonal ring of resistent thread-like material (Fig. 10). This frequently expands during the digestion to be slightly larger than the intact virus particle. The influenza filaments seem to be entirely digested by this treatment. If they leave any trace, it can only be an occasional ring indistinguishable from those left by the spheres. The trypsin-resistent rings when further treated with ribonuclease show no obvious change but they then become trypsin sensitive; a further treatment with dilute acid and trypsin will almost completely remove them. We conclude from this behaviour that the rings consist of protein threads stabilized by the virus ribonucleic acid.

Though ribonuclease has little effect on the appearance of the rings from influenza virus, despite the fact that it must have stripped the nucleic acid off to make them sensitive to trypsin, it has an interesting effect on similar rings that can be obtained from Newcastle disease virus. This is a rather larger virus than influenza and the rings obtained from it are usually circular

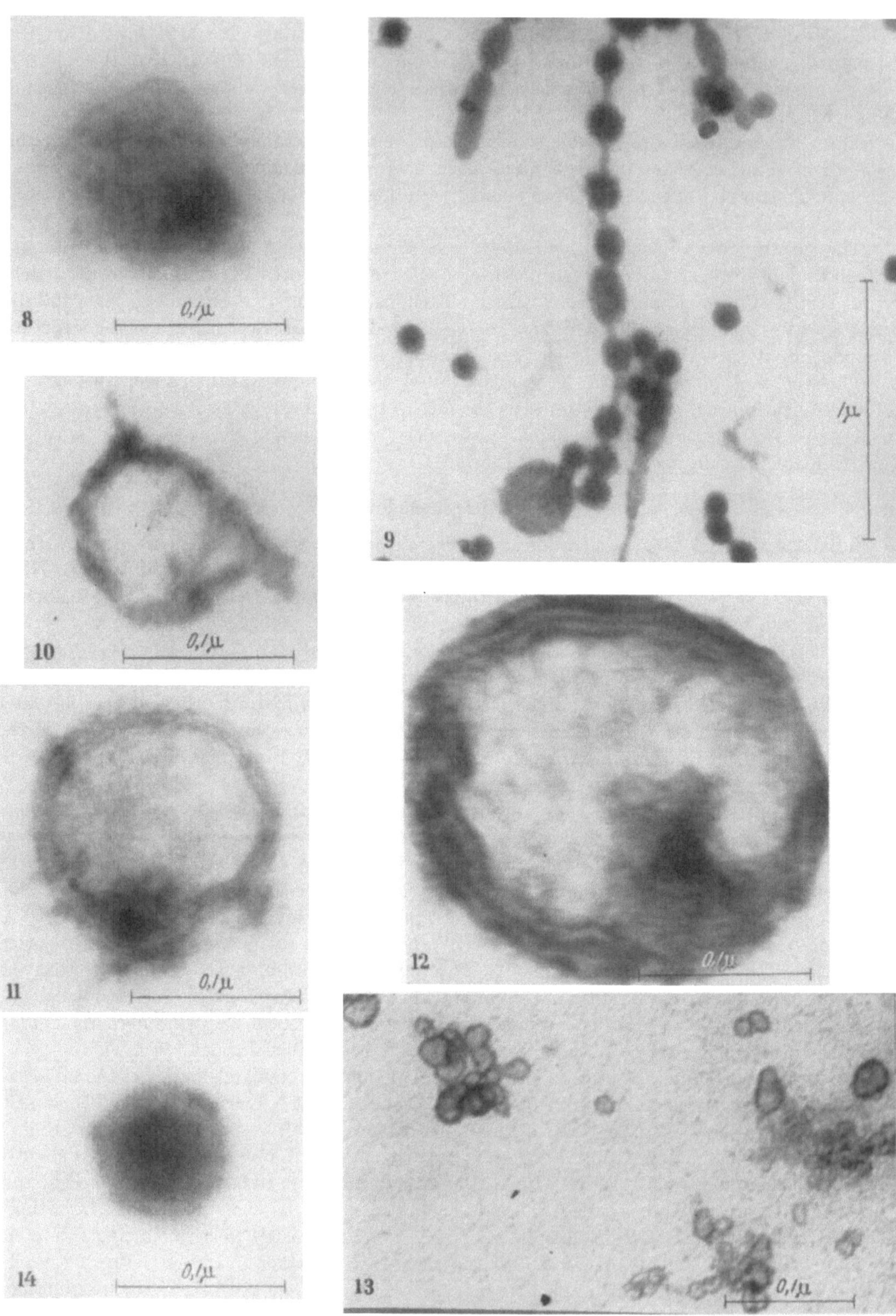

Fig. 8—14

Fig. 8. Influenza sphere breaking up in distilled water. 275,000 × — Fig. 9. Influenza filaments treated with 0.1 N hydrochloric acid. 40,000 × — Fig. 10. Influenza sphere after digestion with trypsin. 275,000 × — Fig. 11 and 12. Newcastle disease virus after digestion with trypsin and ribonuclease. 275,000 × — Fig. 13. Particles in supernatant of an adenovirus preparation. 200,000 × — Fig. 14. Adenovirus elementary body. Treated with 5% uranyl acetate. 275,00 ×

(Fig. 11) or oval rather than polygonal. Following ribonuclease treatment, but not before, these rings "stain" when treated with phosphotungstic acid to show distinct dense lines running round the ring (Fig. 12). It is well known that ribonuclease has the effect of increasing the uptake of acidic material, such as phosphotungstic acid, by regions of a cell that contains ribonucleic acid, indeed this is one of the best histological tests for ribonucleic acid. The explanation given for the effect is that the nuclease uncovers acidophilic groups in the underlying protein to which the nucleic acid had been bound. We thus interpret the lines of phosphotungstic acid that are seen on the nucleoprotein rings after nuclease treatment as in fact marking the position of the nucleic acid on the ring.

Adenovirus. None of the treatments mentioned above, that have shown something of the components of the influenza virus, has any effect on the adenovirus. The elementary bodies are remarkably stable against attack by enzymes and dilute acid and alkali. They do slowly break down in 50% alcohol but without showing any obvious structure. It has already been mentioned that when a suspension of the virus is centrifuged, the supernatant contains particles smaller than the elementary bodies which carry some of the properties of the virus and which might be subunits. It is therefore of interest to examine such supernatants with the electron microscope. Unfortunately, even if they do show a characteristic particle it is very difficult to establish that this is the object being sought. A characteristic type of particle can indeed frequently be found in the form of a ring between 70 and 150 Å in diameter (Fig. 13) but its significance has yet to be established.

"Electron stains"

In order to increase the electron density of an object as small as a virus to any appreciable extent it is necessary nearly to double its weight density (*24*). The only hope of a useful "electron stain" therefore is a substance of very high atomic weight. The most interesting constituent of a virus is probably its nucleic acid and since this has an affinity for uranium salts it was worth testing the effect of uranyl acetate on the virus. Uranium salts cannot be regarded as a very specific stain for nucleic acid (*25*) but any marked increase of electron density not also given by phosphotungstic acid, which has no affinity for nucleic acid, would most likely be locating a region with a high nucleic acid content.

Adenovirus. A central region of the adenovirus about 40 Å across was found in fact to darken appreciably after treatment with uranium salts (*13*) (Fig. 14) and no similar increase of electron density is found after phosphotungstic acid treatment (Fig. 6). It thus seems very likely that the deoxyribonucleic acid of the adenovirus is mostly or entirely contained in the centre of the virus and is surrounded by the virus protein. It is interesting that thin sections of the adenovirus frequently show a denser central region.

Influenza virus. Only the surface of the influenza virus has any affinity for uranium and this is almost certainly not a specific effect. The nucleic acid content of the virus is in fact far too low (1%) to be detectable by direct combination with an electron stain (*24*). The demonstration of the position of the nucleic acid in Newcastle disease virus by "staining" the underlying basic protein after nuclease treatment was discussed above.

Particle counts with the electron microscope

There are various ways to determine the count per unit volume of any type of particle identified with the electron microscope in a virus preparation. These methods have been reviewed elsewhere (*26, 27*). It is also possible to determine by a biological titration the number of infective units in the same preparation. In the case of bacteriophage particles these two counts prove to be identical to within a factor of two and this provided additional support for identifying the bodies seen with the electron microscope as the infective units. If any type of object has a particle count significantly less than the infective titre then clearly this would rule out the object as being the virus particle. The usual situation, however, with animal viruses is to find the reverse with considerably more particles than infective units. In such cases particle counts can provide little evidence either for or against the particles being the elementary virus bodies. If the particles in fact are the virus the difference between particle count and infective units can be due either to some of the particles being non-infectious for some reason or to the biological titration failing to be 100% efficient. Nor is the finding of a constant ratio of particles to infective units of much value in identifying the particles as the virus. For it is possible that the ratio might be variable even if the particles were the virus owing to a variation in the efficiency of measuring infectivity. Again a non-virus particle associated with the infection might well have a count bearing a constant ratio to the true virus count. The value of making particle counts with the electron microscope, therefore, is not to confirm that any given type of particle is the virus but rather to enable any determination, biological or chemical, made on a virus suspension to be related to a basic measurement, that of the number of elementary virus bodies. A ratio of particular importance is that of particle count to infectivity.

The most satisfactory method of estimating the particle count, when it is applicable, is the spray droplet method of WILLIAMS and BACKUS (28) using standard polystyrene latex indicator particles. But it is always desirable that the count found by any method should be confirmed by some other entirely independent technique even if this is a much less accurate one. It may be unwise, in view of many possible sources of error, to place too much reliance on any count when this is not done.

Influenza virus. There are two good methods for counting influenza virus particles. In addition to the spray drop method, the particles can be adsorbed on to the membranes of a calibrated suspension of lysed red cells and the mean count per red cell found with the electron microscope. These two methods agree very well (29, 30).

Particle counts made in this way have an error of probably 10% to 20%. Unfortunately, due to the scatter in the end points, the measures of infectivity made on the same suspensions never achieve anything like the same precision. They may be uncertain by as much as a factor of two. However, a considerable number of titrations made quite independently (29, 30) have agreed in finding that, as normally titrated, each infective unit of influenza contains about 10 elementary bodies as seen in the electron microscope.

Adenovirus. Particle counts on adenovirus preparations have also been made by the spray drop technique (31). The infectivity titrations were made in HeLa cell cultures (4 weeks incubation) and these showed between 10 and 100 particles per infective dose. The variation probably arose from varying efficiencies in measuring the infectivity.

With both influenza virus and the adenovirus, the ratios of particle count to infectivity are thus very similar. Ratios of the same order have in fact now been found with a number of different animal viruses.

Conclusions

Although our knowledge of the structure of both these viruses is obviously still rudimentary, from the observations reviewed above it is possible to form some sort of general, if tentative, picture.

Influenza virus structure. Since the influenza virus is capable of agglutinating red cells and can then be obtained from them still intact, and since the intact virus itself can be agglutinated with specific antiserum, it is clear that the haemagglutinating subunits that carry the specific antigens must be at the surface of the virus. These objects when obtained by disrupting the virus with ether are irregular round bodies about 120 Å across. No similar particle, however, can be seen on the surface of the intact virus and what can be seen appears rather to have the form of threads some 30 Å wide. If these threads are the haemagglutinin of the virus, which chemically is reported to be a mucoprotein, then such mucoprotein threads might well form a random coil with a roughly spherical shape when liberated from the virus.

The question of wheather the influenza virus has any limiting outer membrane raises a problem for if so the agglutinin must be arranged on or in this membrane. HOYLE (17) suggested that the influenza virus is surrounded by a lipid-containing membrane obtained from the host cell and that ether disrupted the virus by dissolving this membrane. However, the lipid content of the virus (35%) is far larger than could be accounted for by any membrane and it may well act rather as a general "cement" than as an enclosing bag — certainly neither the particles digested with enzymes (Fig. 10) nor those disintegrating in distilled water (Fig. 8) show any evidence of a membrane, — certainly nothing in the nature of that which so obviously surrounds the vaccinia virus. MORGAN, on the other hand, reports (20) that thin sections of influenza virus do show a definite membrane some 30 Å thick surrounded by a diffuse outer coat. So thin a membrane might well pass unseen in our pictures of intact particles. The various views could be reconciled by supposing that threads of mucoprotein carrying the specific antigens, the ability to adhere to red cells and the enzyme necessary for the virus to penetrate into a susceptible cell lie on a very thin supporting membrane.

Somewhere inside must be the nucleoprotein complement fixing structure (the so-called soluble antigen) that carries the group antigens. If contempory views are right, this structure alone may carry the power to infect, the rest of the virus acting only to protect it and effect its entry into a cell. Our enzyme studies have suggested that this nucleoprotein structure has the form of a ring of thread-like material. Since the virus and its nucleoprotein obviously expand readily during digestion, it is not quite apparent exactly how this ring lies within the intact particle. From its behaviour with enzymes, however, it seems clear that the ribonucleic acid is firmly bound to the protein structure. This contrasts with the phages where the nucleic acid is thought to lie free within an enclosing coat of proteins. In the case of a virus related to influenza, that of Newcastle disease, we have an indication of the path of the nucleic acid around the protein ring.

The problem still remains as to the nature of the filaments. At one time, many were tempted to believe that they were the precursors of the spheres and certainly, as we have shown, it is easy to convert filaments into rows of objects very like the spheres by the action of dilute acid. However, the arguments against this view now seem overwhelming. In thin sections of infected tissue the two forms, spheres and filaments, can be seen developing quite independently at the cell surface (20). The infectivity of filaments when broken into short lengths by ultrasonics is certainly not that of the equivalent number of spheres. In fact, we have no evidence that the filaments contain any nucleoprotein structure, though enzyme studies do not rule out the possibility that one might lie, say, at each tip. This provokes the fundamental question, that has been debated over many years, of whether the filaments are infectious. It is certainly well-established that they haemagglutinate and carry the specific influenza antigens but, in the writer's opinion, their infectivity has never been demonstrated. The difficulty lies in the apparent impossibility of obtaining a preparation of the filaments apart from the spheres; even attempts to increase the percentage of filaments appreciably above the usual 20% (or less) have not met with success. An apparent increase is often only the result of breaking them. Now the precision with which infectivity can be titrated is comparitively poor — certainly less than ± 20%. All that the results to date show is that filaments are certainly not much more infectious than spheres. Arguments (32) claiming that they prove them to have some infectivity, however, do not bear close statistical examination. On present evidence it seems safest to regard the filament as essentially non-infectious but carrying many of the other typical properties of the virus.

Adenovirus structure. The adenovirus has a regular crystal-like shape and this strongly argues that it is built up of a regular array of subunits. This regular structure, combined with a small or zero hydration, probably accounts for the resistance of the virus to enzyme attack. The picture obtained after treatment with uranyl acetate suggests that the deoxyribonucleic acid of the virus (which accounts for about 40% of its mass) is concentrated in the centre of the particle and not uniformly distributed through it. Around this central core, which may well be pure nucleic acid, will be arranged the protein "bricks" in some sort of regular array. It is tempting to think that these protein subunits can also exist free in suspensions of the virus and are responsible for the activities that remain in the supernatant when all the elementary bodies have been sedimented. But whether this is so and whether the rings found in such supernatants (Fig. 13) or similar objects seen in arrays in thin sections of infected cells (19) are in fact the virus subunits can at this moment only be speculation.

References

1. Williams, R. C.: Advanc. in Virus Res. **2**, 183 (1954).
2. Valentina, R. C., and A. Isaacs: J. gen. Microbiol. **16**, 195 (1957).
3. Williams, R. C., and K. M. Smith: Biochim. biophys. Acta **28**, 464 (1958).
4. Valentine, R. C., and P. K. Hopper: Nature (Lond.) **180**, 928 (1957).
5. Ada, G. L., and B. T. Perry: J. gen. Microbiol. **14**, 623 (1956).
6. Frisch-Niggenmeyer, W., and L. Hoyle: J. Hyg. **54**, 201 (1956).
7. Burke, D. C., A. Isaacs and J. Walker: Biochim. biophys. Acta **26**, 576 (1957).
8. Frisch-Niggemeyer, W.: Nature (Lond.) **178**, 307 (1956).
9. Morgan, C., C. Howe, H. M. Rose and D. H. Moore: J. biophys. biochem. Cytol. **2**, 351 (1956).
10. Burke, D. C.: (1958). (Private communication).
11. Valentine, R. C.: J. Photographic Sci. (**1959**). (In the press).
12. Zeitler, E., and G. F. Bahr: Exp. Cell Res. **12**, 44 (1957).
13. Valentine, R. C.: J. roy. micr. Soc. (1959). (In the press).
14. Huppert, J., and F. K. Sanders: Nature (Lond.) **182,** 515 (1958).
15. Rosen, L.: Virology **5**, 574 (1958).
16. Hoyle, L.: J. Hyg. **48**, 277 (1950).
17. — J. Hyg. **50,** 229 (1952).
18. Pereira, H. G.: Virology **6**, 601 (1958).
19. Morgan, C., G. C. Godman, H. M. Rose, C. Howe and J. S. Huang: J. biophys. biochem. Cytol. **3**, 505 (1957).
20. — H. M. Rose and D. H. Moore: J. exp. Med. **104,** 171 (1956).
21. Valentine, R. C., and A. Isaacs: J. gen. Microbiol. **16**, 680 (1957).
22. Hoyle, L., R. Reed and W. T. Astbury: Nature (Lond.) **171,** 256 (1953).
23. Schäfer, W., K. Munk and M. Mussgay: Z. Naturforsch. **11 b**, 330 (1956).
24. Valentine, R. C.: Nature (Lond.) **181,** 832 (1958).
25. Watson, M. L.: J. biophys. biochem. Cytol. **4**, 475 (1958).
26. Isaacs, A.: Advanc. in Virus Res. **4**, 111 (1957).
27. Sharp, D. G.: This Vol. p. 542.
28. Williams, R. C., and R. C. Backus: J. Amer. chem. Soc. **71**, 4052 (1949).
29. Donald, H. B., and A. Isaacs: J. gen. Microbiol. **10**, 457 (1954).
30. Tyrrell, D. A. J., and R. C. Valentine: J. gen. Microbiol. **16**, 668 (1957).
31. Pereira, H. G., and R. C. Valentine: J. gen. Microbiol. **19**, 178 (1958).
32. Ada, G. L., B. T. Perry and M. Edney: Nature (Lond.) **180**, 1134 (1957).

Neue morphologische Elemente in den Kulturen des Grippe-Virus

S. B. Stephanow

Laboratorium für Elektronenmikroskopie der Abteilung für biologische Wissenschaften der Akademie der Wissenschaften der UdSSR, Moskau

In den Kulturen des Grippe-Virus gibt es sphärische Körper und Filamente. In den letzten Jahren wurde die Aufmerksamkeit auf Schwellungen an den Filamenten und große sphärische Bildungen gelenkt. Es wurden von uns große kugelförmige Körper, welche in dünne Fäden zerfallen, und dicke fadenförmige Körper in den Kulturen des Grippe-Virus vom Typus D beschrieben (*1—5*). Jetzt haben wir in Kulturen von Grippe-Virus D eine neue Form gefunden: stäbchenförmige Teilchen, die ungefähr 100—150 mμ dick und bis 500 mμ lang sind. Die Grundmasse des Stäbchens färbt sich intensiv mit OsO_4. An einem Ende ist ein durchsichtiges, ungefärbtes Bläschen zu sehen (Durchmesser 200—400 mμ). Der gefärbte Teil des Stäbchens bildet bei Beschattung einen langen Schatten, während das Bläschen fast keinen Schatten gibt. In den Kulturen des Grippe-Virus begegnet man also mehreren Typen morphologischer Elemente. Die morphologischen Beziehungen zwischen diesen äußerlich nicht ähnlichen Elementen wurden von uns untersucht.

Zusammenhänge zwischen Filamenten und sphärischen Teilchen wurden von vielen Autoren erwähnt. Das Filament kann in eine Kette sphärischer Teilchen zerfallen, welche den typischen infektiösen Teilchen des Virus ähnlich sind. Es kommen jedoch auch Ketten vor, welche aus größeren ovalen oder polymorphen Elementen bestehen. Die Elemente der Kette können sich in verschiedenem Grade mit OsO_4 färben und besitzen verschiedene Größe. Die Ketten liegen

gewöhnlich an den Enden, manchmal jedoch auch im mittleren Teil des Filaments. Die Elemente
in den Ketten des Virus vom Typus D zeichnen sich durch besonderen Polymorphismus aus. Sie

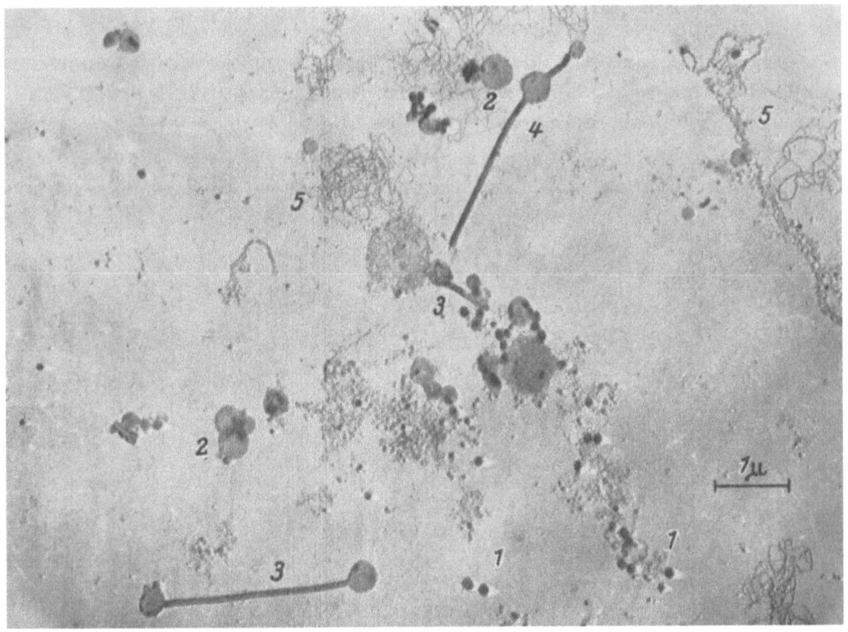

Abb. 1. Sphärische Teilchen des Grippe-Virus vom Typus A_1 (*1*), große Kugeln (*2*), Filamente mit Schwellungen
am Ende (*3*), und im Mittelteil (*4*), dünne Fäden (*5*), vergr. 10000mal

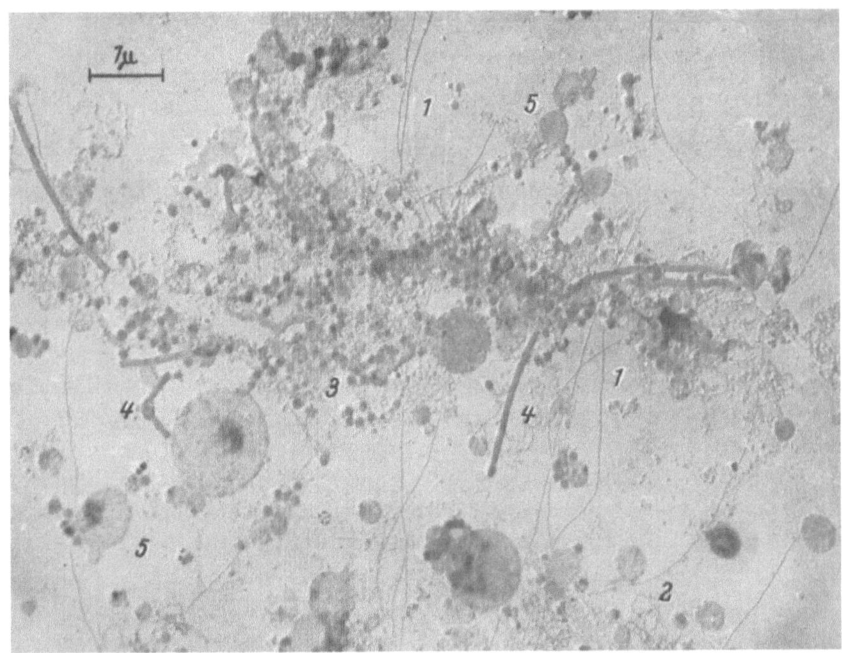

Abb. 2. Gerade (*1*) und geschlängelte (*2*) dünne Fäden, sphärische Teilchen (*3*), Filamente mit Schwellungen (*4*),
große Kugeln (*5*), Grippe-Virus vom Typus A, vergr. 10000mal

bestehen aus einer dichten zentralen Masse und einer hellen peripheren Zone. Zwischen den
Elementen sind deutlich Verbindungsbrücken zu sehen. Die Elemente der Ketten erinnern
häufig an die oben beschriebenen Stäbchen mit Bläschen an den Enden.

In den Kulturen des Virus vom Typus D besitzt die Mehrzahl der sphärischen Teilchen einen Durchmesser von ungefähr 200 mμ und dementsprechend besitzen die Filamente ebenfalls eine Dicke von ungefähr 200 mμ (5).

Es wurde von uns eine einfache Methode ausgearbeitet, welche gestattet, elektronenmikroskopische Präparate ohne Waschung oder andere physiko-chemische Einwirkungen herzustellen. In den nach dieser Methode hergestellten Präparaten werden Schwellungen auf den Filamenten nur sehr selten angetroffen (6). Stellt man jedoch aus diesen Kulturen Präparate unter Anwendung von destilliertem Wasser her (Abb.1—3), so erhält man bis 90% Schwellung der Filamente. Ein Teil der Filamente weist keine Schwellungen auf, auch nicht nach längerer Maceration in Wasser.

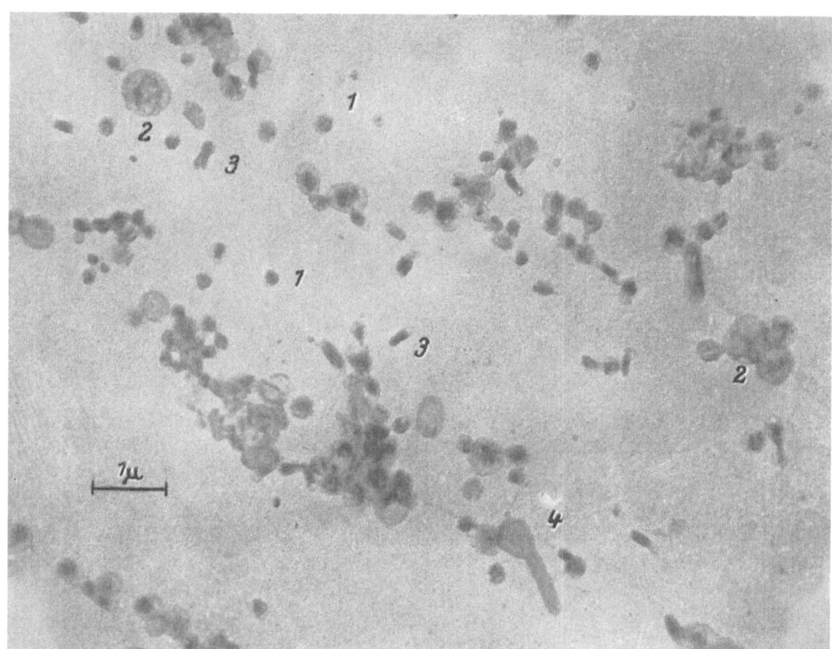

Abb. 3. Sphärische Teilchen (1), große Kugeln (2), Stäbchen mit Bläschen am Ende (3), Filament mit Schwellung (4). Grippe-Virus vom Typus D, vergr. 10000mal

Interessant ist, daß in 90% der Fälle die Schwellung an einem Ende des Filaments gelegen ist, seltener an seinem Mittelstück. Sehr selten begegnet man Filamenten mit zwei Schwellungen. Letztere verdienen besondere Beachtung, weil sie in ihrer Form, Struktur und Dichte großen Kugeln ähnlich sind, worauf wir früher hingewiesen haben (1, 5). Der Durchmesser einer großen Kugel kann die Größe von 1—2 μ erreichen, am häufigsten trifft man jedoch Kugeln mit e'nem Durchmesser von ungefähr 500—800 mμ, das heißt, wie bei den Schwellungen an den Filamenten. Häufig begegnet man Kugeln mit einem sehr kurzen, knospenförmigen Fortsatz, der in seiner Breite der des Filaments ähnelt. Solche Tatsachen sprechen ebenfalls für die Möglichkeit eines Zusammenhangs zwischen Kugeln und Schwellung.

Andererseits ist ein Zusammenhang zwischen den großen Kugeln und den typischen sphärischen Teilen des Virus eindeutig festzustellen. Es gibt eine Menge von Teilchen, welche ihrer Form, Größe und Struktur nach Zwischenglieder zwischen den einen und den anderen Elementen darstellen. Große Kugeln können, wie früher gezeigt worden war (4), in dünne Fäden und kleine Körnchen zerfallen. Die dünnen Fäden besitzen eine Dicke von 10—50 mμ und eine Länge bis zu einigen μ. Sie sind entweder geradlinig oder besitzen starke Biegungen. Man beobachtet regelmäßig ein Zerfallen des dünnen Fadens in Körnchen. Folglich besteht ein morphologischer Zusammenhang zwischen den großen Kugeln und den dünnen Fäden, wie auch zwischen den dünnen Fäden und den kleinen Körnchen.

Weiterhin kann man auch Zwischenformen feststellen, welche zwischen den großen Kugeln und den Stäbchen mit Bläschen einzuordnen sind. Häufig begegnet man Kugeln mit sehr dichten und sehr hellen Polen. Im gleichen Präparat sind ferner Körper zu beobachten, die durch eine leicht gestreckte oder durch eine stark gestreckte, stäbchenartige Form charakterisiert sind. Bei vielen Stäbchen bildet sich in der Mitte eine Einschnürung, welche an eine Teilung des Stäbchens denken läßt. Man trifft aber auch Körper, die aus einer Kugel und einem Stäbchen bestehen und mit ihren verdichteten Teilen aneinander grenzen. Die dichte Masse innerhalb eines solchen Körpers scheint gelegentlich nicht unterbrochen zu sein, in anderen Fällen sind sie durch eine Einschnürung getrennt. Das Stäbchen kann zusammen mit den sphärischen Teilchen zu einer Kette angeordnet sein.

Zwischen den verschiedenartigen Formen in den Kulturen des Grippe-Virus gibt es also Zwischenformen, welche eine fast ununterbrochene Kette von Übergängen bilden. Die Gesamtheit all dieser Tatsachen bestätigt die Ansicht, daß in der Kultur des Grippe-Virus komplizierte Prozesse der Entwicklung vor sich gehen.

Literatur

1. DRAGANOW (STEPHANOW S. B.): Bull. exp. biol. Med. (russ.) **10**, 53 (1953).
2. — J. allg. Biol. (russ.) **17**, 13 (1956).
3. — Biophysika (russ.) **1**, 370 (1956).
4. — Biophysika (russ.) **3**, 348 (1957).
5. — Proc. I. Reg. Conf. Asia a. Oceania, Tokyo, 1956, 190.
6. STEPHANOW, S. B.: Im Druck.

Studies on the structure of infectious and non-infectious influenza virus

A. BIRCH-ANDERSEN and K. PAUCKER

Statens Serum-Institut, Copenhagen (Denmark)

Non-infectious forms of influenza virus can be obtained in the laboratory under a variety of conditions and from various sources. They have been shown to differ in a number of biological and chemical properties from fully infectious particles. In the present study an attempt was made to differentiate between incomplete and infectious virus units on the basis of their appearance in the electron microscope. The PR 8 strain of influenza A was used throughout this investigation. Virus pellets were obtained on centrifugation of various materials which had been subjected to extensive purification and concentration procedures. These pellets were embedded in methacrylate and ultrathin sections were prepared for microscopic study. The examination of sections of infectious elementary bodies derived from standard seeds reveals a majority of particles with marked electron-dense centers which are surrounded by two well-defined concentrically arranged limiting layers. The latter are separated by a zone of lesser density. A second component which exhibits centers of low electron density can also be discerned although it appears to be present in minor proportion only. However, incomplete virus obtained on serial undiluted passage of allantoic fluid or in single egg passage of mildly heat-inactivated standard seed, as well as in HeLa cells, represents a heterogeneous aggregation of particles which differ considerably in shape and size. While their outer coats resemble those of the fully infectious units, little or no internal structural detail can be observed in most of them and the overall impression is that of hollow virus bodies. These differences are not seen when the infectivity of the virus is largely destroyed by prolonged exposure to 37° C in vitro. Sections of such particles cannot be distinguished from the originally infectious units. This lends further support to the hypothesis that incomplete virus arises by an altered replicating process rather than as the result of rapid inactivation in vivo.

An extended version of this paper is published in Virology 8, 21 (1959).

Adenoviruses and herpes simplex virus, with particular reference to intracellular crystals *

Councilman Morgan and Harry M. Rose

Departments of Microbiology and Medicine, College of Physicians und Surgeons, Columbia University, New York City, New York

The initial observation of crystals associated with a viral infection was made by Bunting (*1*), in 1953, during electron microscopic studies of human skin papillomata. Inspection of the published micrographs leads one to agree with the author that the intranuclear arrays of particles probably represent viral crystals. Unfortunately, the thickness of the sections precluded sufficient resolution to determine whether the particles posses internal structure. Although further investigation should be carried out with techniques currently available, the acquisition of material suitable for examination of developmental stages, which contribute important information to our understanding of any virus, is rendered difficult by virtue of the fact that no experimental animal or tissue culture system has been found which is capable of supporting growth of the papilloma virus.

The opportunity to study viral crystals in tissue cultures was provided in 1955 by Kjellén, Lagermalm, Svedmyr and Thorsson (*2*), when they encountered particles arranged in "crystalline-like patterns" within nuclei of HeLa cells infected with *adenovirus*. This observation was soon confirmed by Harford, Hamlin and Parker (*3*). Subsequently detailed examination of viral structure and types of host cell reactions were published (*4, 5, 6*).

The adenoviruses possess, in common, a characteristic antigen which can be demonstrated by the complement fixation reaction; they also contain type-specific antigens which may be distinguished by both complement fixation and neutralization tests. Of the more than 16 serotypes (*7*) which have been identified, 8 are under study in our laboratory. Certain results of these investigations, which are still in progress, will be reported and discussed in the following remarks.

By correlation of the cellular morphology observed in the electron micrographs with the sequential changes identified by light microscopy, it has been possible to recognize consecutive stages of viral development[1]. The virus differentiates within the nucleus, where characteristic changes occur. Initially, aggregates of granular and reticular material form, adjacent to which clusters of viral particles appear. Fig. 1 illustrates part of four lobes of a nucleus. Near the upper right corner is a bundle of double membranes, the significance of which remains unclear. Scattered through the nuclear matrix are remnants of aggregated reticulum with clusters of virus. As infection proceeds the number of viral particles increases, whereas the amount of reticulum diminishes, suggesting that the latter contributes to propagation of the virus. Close inspection of Fig. 1 reveals that the viral particles have formed crystalline arrays. The patterns of distinct and indistinct particles reflect orientation of the crystalline lattice with respect to the angle of cutting, particles central to the plane of section appearing sharply demarcated, those which are eccentric being ill-defined. Although impact of the microtome knife occasionally distorts the crystals, it has been possible, by constructing models, to determine that a cubic body-centered lattice best fits the patterns which have been encountered (*9*). In addition to variations in appearance of the virus introduced by the relative level of sectioning, the actual structure of the particles themselves differs, as is well illustrated by the crystal just below the center of the figure. Although gradations exist, two major morphologic types, which can be distinguished by their relative density, characterize all the adenoviruses so far studied. At higher magnification (Fig. 2) it becomes clear that after fixation in osmium tetroxide many of the particles possess an internal body of low density to the electron beam, whereas others appear dense with ill-defined central

---

* These studies were conducted under the auspices of the Commission on Influenza, Armed Forces Epidemiological Board, and were supported in part by the Office of the Surgeon General, Washington, D.C.

[1] The methods and preliminary results of the study of infected cells in contiguous thick and thin sections in the light and electron microscope have been published elsewhere (*8*). The present data, gathered by extension of this work in collaboration with Dr. Gabriel C. Godman, have confirmed the developmental sequence tentatively proposed in a previous paper (*4*).

components. On the other hand, formalin fixation (Fig. 3) brings out a relatively opaque central core enclosed by a sharply defined, frequently incomplete, peripheral membrane. The problem

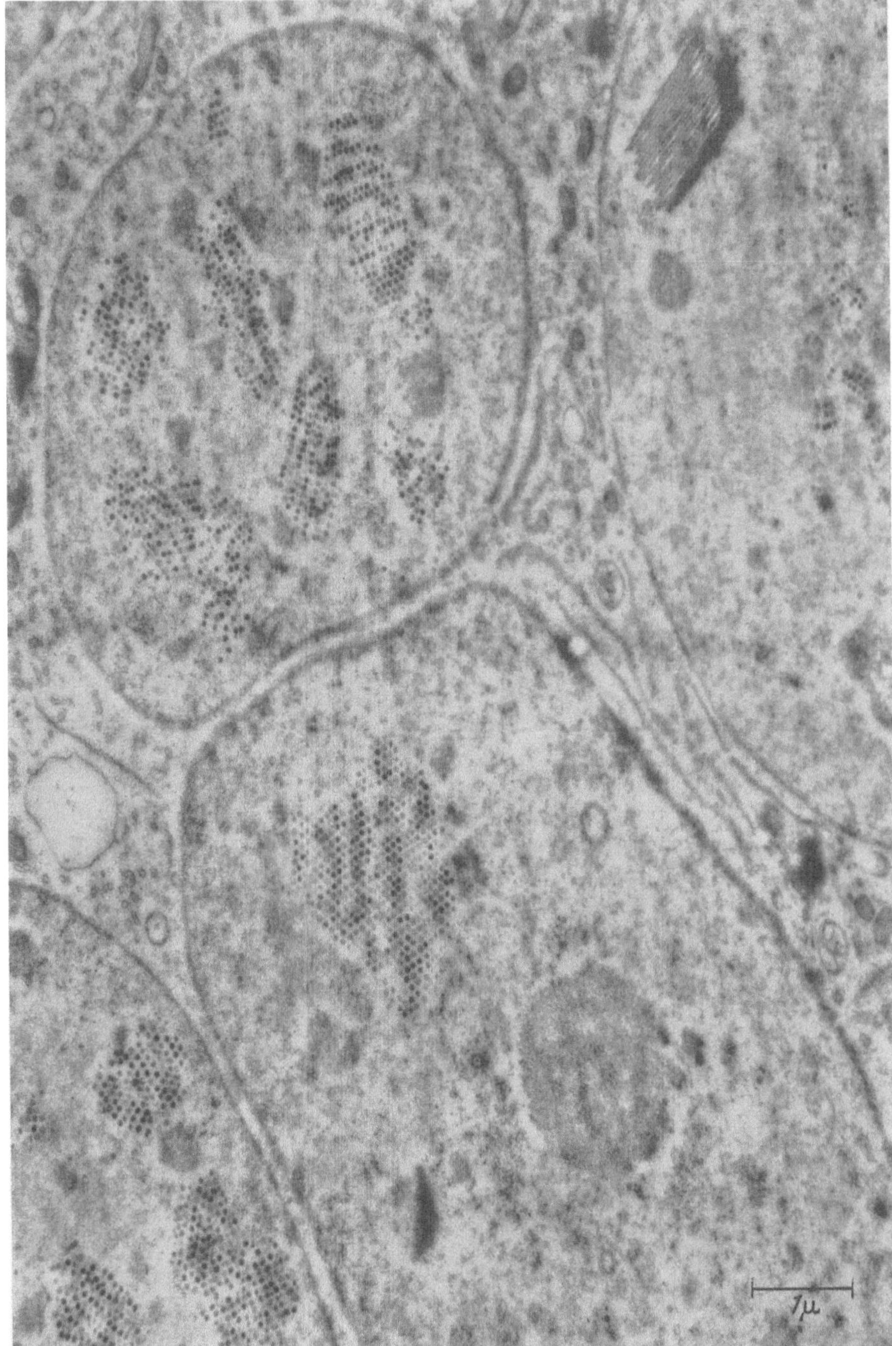

Fig. 1. Four lobes of a nucleus containing reticular aggregates and small, adenoviral crystals. 14,000 ×

of defining structure is further complicated by the appearance of crystals sectioned within cells after quick-freezing[1] (Fig. 4). Crystals composed of virus can be readily identified, but the striking

---

[1] The cells were frozen by quenching in iso-pentane at —180° C and the water substituted by ethyl alcohol at —50° C for 7 days (*10*). Subsequently the tissue was embedded in methacrylate and sectioned in the usual manner.

difference in particle density is not apparent. Turning to the shape of the virus, there is additional confusion, since it has been reported (*11, 12*) that osmium, as well as formalin, fixed particles dried from suspension may exhibit a polyhedral shape. In sections, on the other hand, the virus

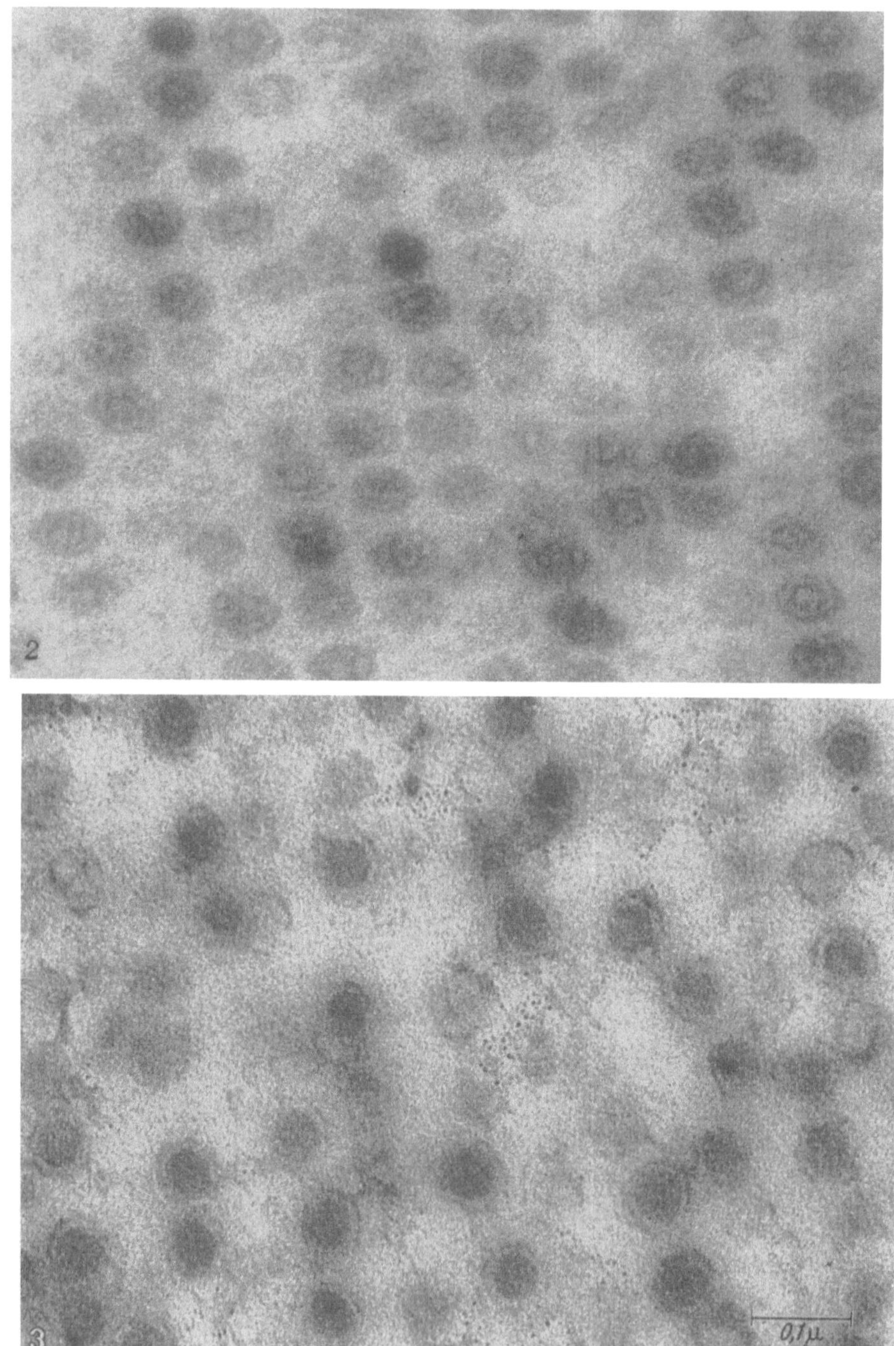

Fig. 2. An adenoviral crystal fixed in osmium tetroxide. A clearly defined, internal body of low density is visible in many of the particles. The ellipsoidal shape reflects distortion by impact of the microtome knife. 135,000 ×

Fig. 3. A similar crystal after formalin fixation. Many of the particles are poorly preserved; others exhibit a dense limiting membrane. 135,000 ×

generally appears spherical or ellipsoidal, depending upon the degree of distortion produced by impact of the microtome knife. One is led, by observations such as those presented above, to the conclusion that detailed discussion of adenoviral structure will continue to be a purely academic

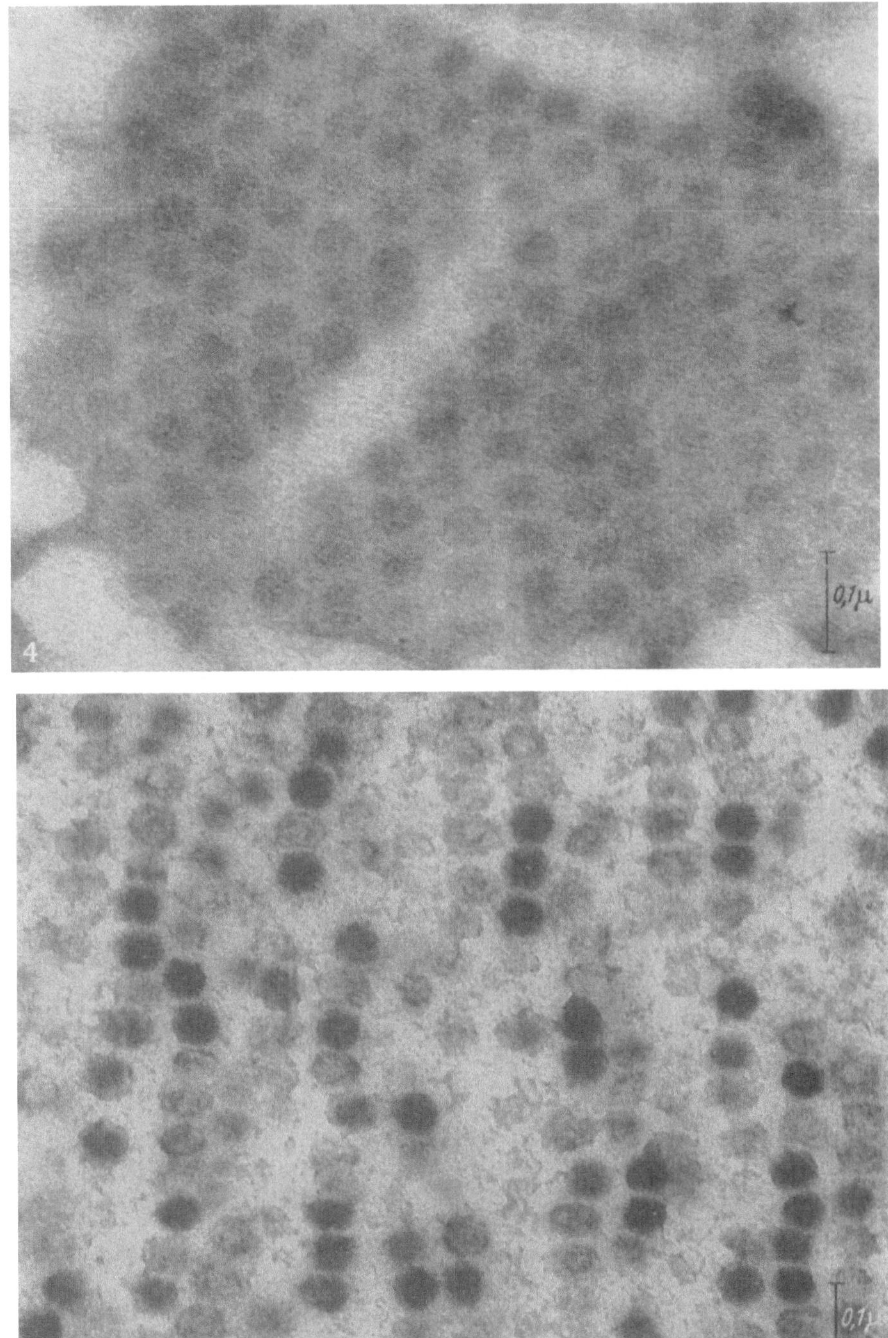

Fig. 4. An adenoviral crystal after quick-freezing. 133,000 × — Fig. 5. Portion of a crystal encountered in a cell showing advanced necrosis. 90,000 ×

exercise until a clearer understanding is achieved of the physico-chemical effects of different preparatory methods or until specific function can be related to reproducible, morphologic components.

The precise size of the virus is unknown, since electron microscopic examination necessitates dessication. If the hydrated viral crystals can be studied by X-ray diffraction, accurate measurement of unit cell dimensions should be possible. It is important, therefore, to ascertain whether

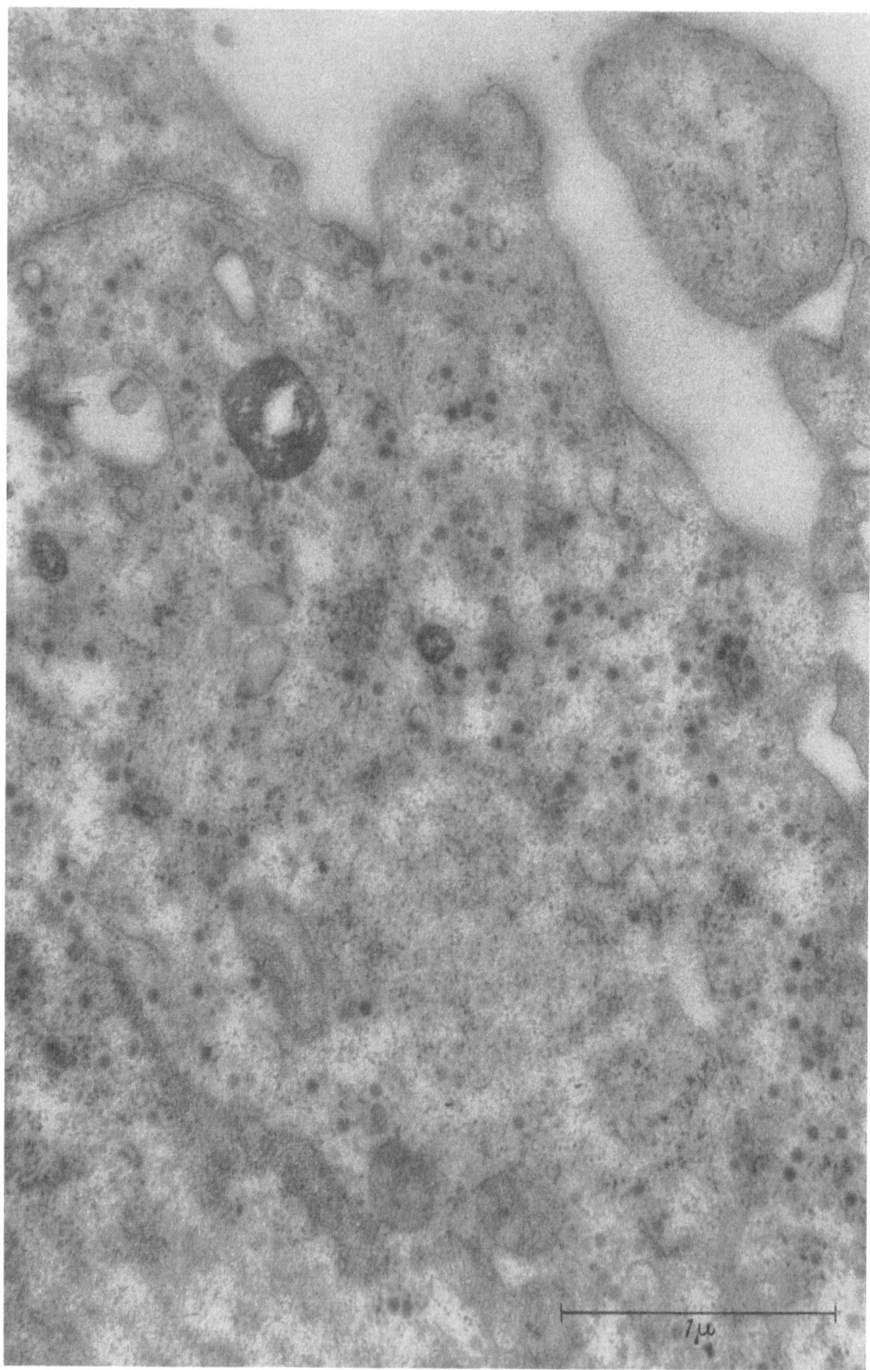

Fig. 6. Virus dispersed within the cytoplasm. The cell border traverses the upper portion of the field. 37,000 ×

or not the particles are contiguous. A suggestion that contiguity does exist is provided by the observation that the diameter (650 Å) of the peripheral viral membrane rendered visible by formalin fixation[1] approximates the most commonly encountered center to center spacing of the particles after osmium fixation. In other words, it is possible that the particles are closely packed but that osmium fails to outline the periphery of the virus.

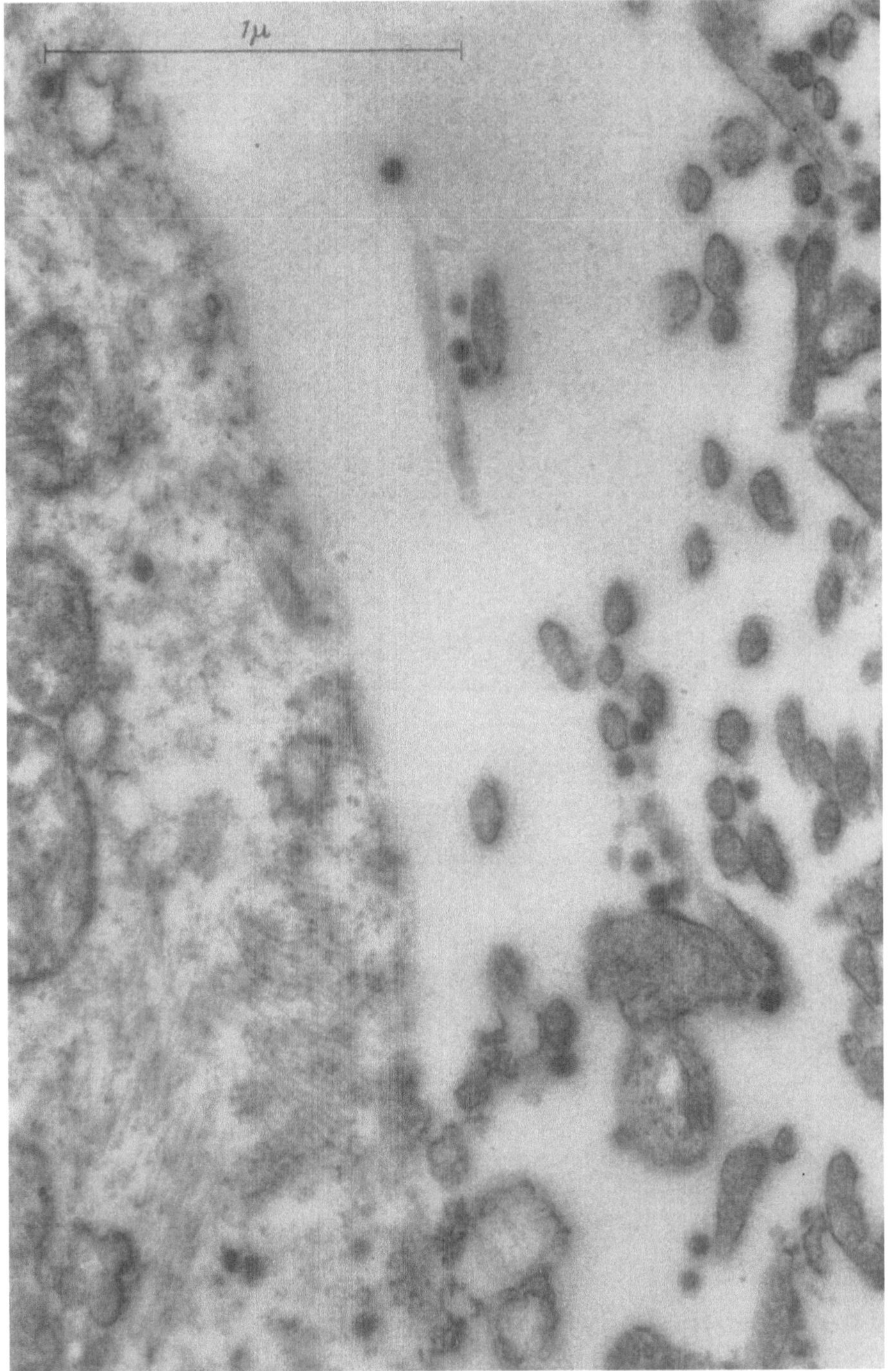

Fig. 7. Virus both within the cytoplasm (at left) and extracellular space. Cytoplasmic extensions from a neighboring cell have been cut at different angles. 56,000 ×

[1] Unfortunately, formalin fixation results in such distortion of the crystals that viral spacin cannot beg determined with accuracy.

38*

The differing forms of virus revealed by osmium fixation appear to vary with respect to their stability *in situ*. Fig. 5 shows part of a crystal in a cell which exhibited advanced pathologic

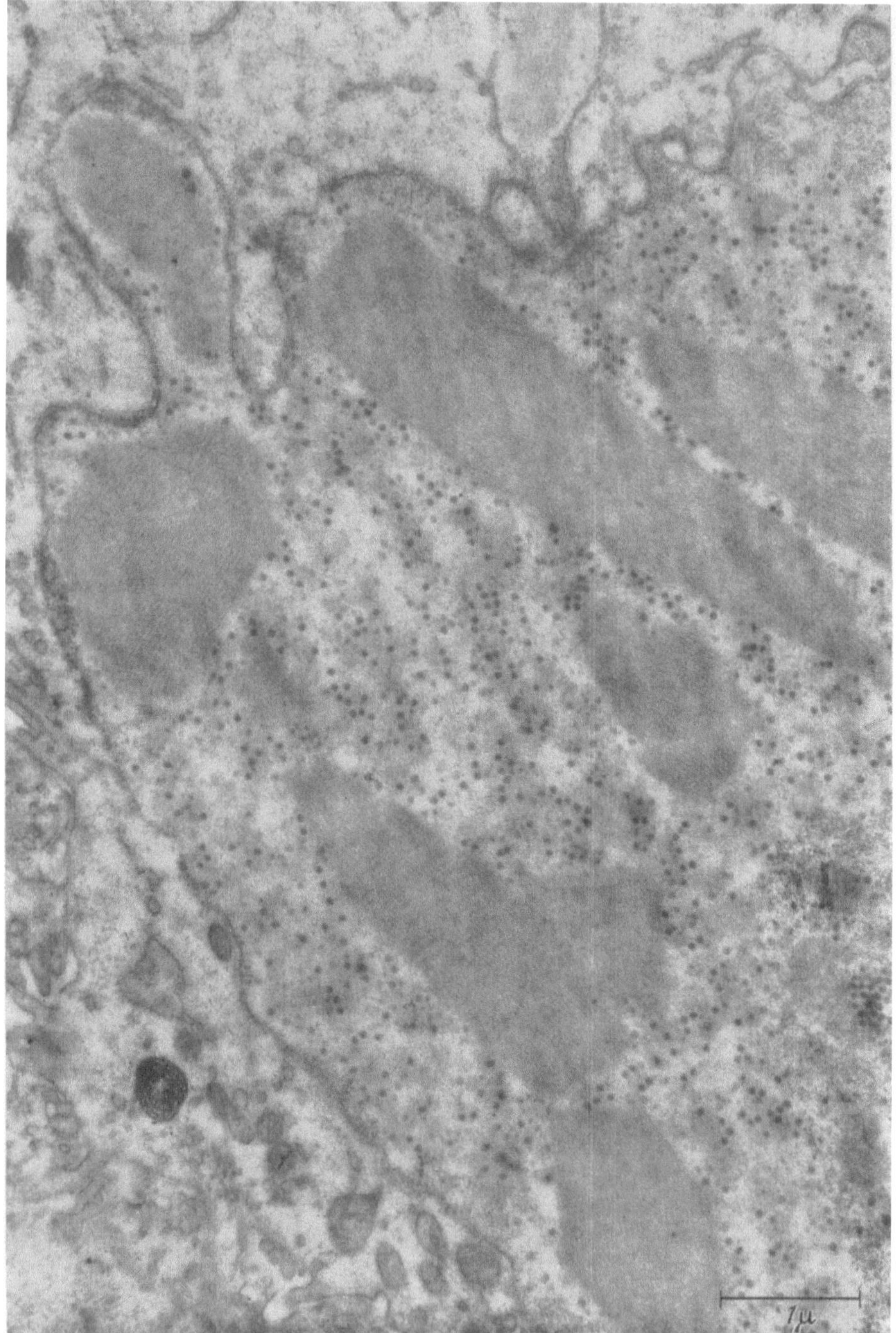

Fig. 8. Part of a nucleus containing type 5 adenovirus as well as non-viral, protein crystals. 19,000 ×

changes. The dense particles are sharply defined, whereas those of lesser density are frequently distorted with irregular, fragmented margins. The crystalline lattice has gaps containing granular debris, presumably remnants of disintegrated virus.

Following disruption of the host cell nuclei, the crystals are released into the cytoplasm where they disintegrate with dispersal of the virus. Fig. 6 illustrates the cytoplasm of a cell containing

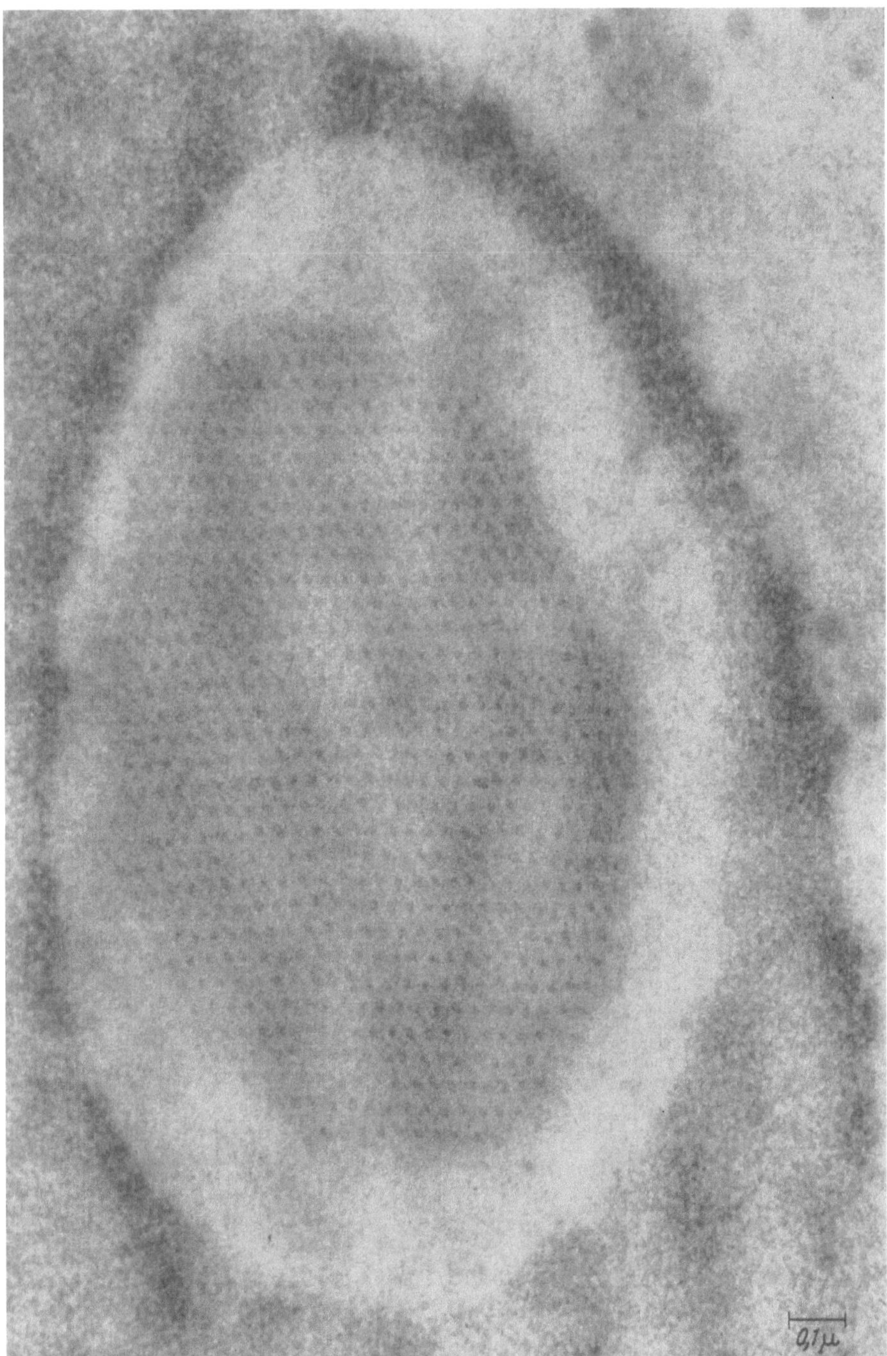

Fig. 9. A non-viral crystal in cross-section. The dense points probably reflect the molecular lattice. 83,000 ×

scattered viral particles. The cell surface is visible near the top. At the upper left the cytoplasm of a contiguous cell appears devoid of virus. In Fig. 7, cytoplasm with a few viral particles and clusters of filaments, whose nature is unknown, occupies the left portion of the field. To the right, in the extracellular space, virus is visibly lodged among cytoplasmic extensions, which have been cut at differing angles. Virus located in the cytoplasm and extracellularly is morphologically indistinguishable from forms encountered within the nucleus, suggesting that further viral development does not occur outside the nucleus of the host cell.

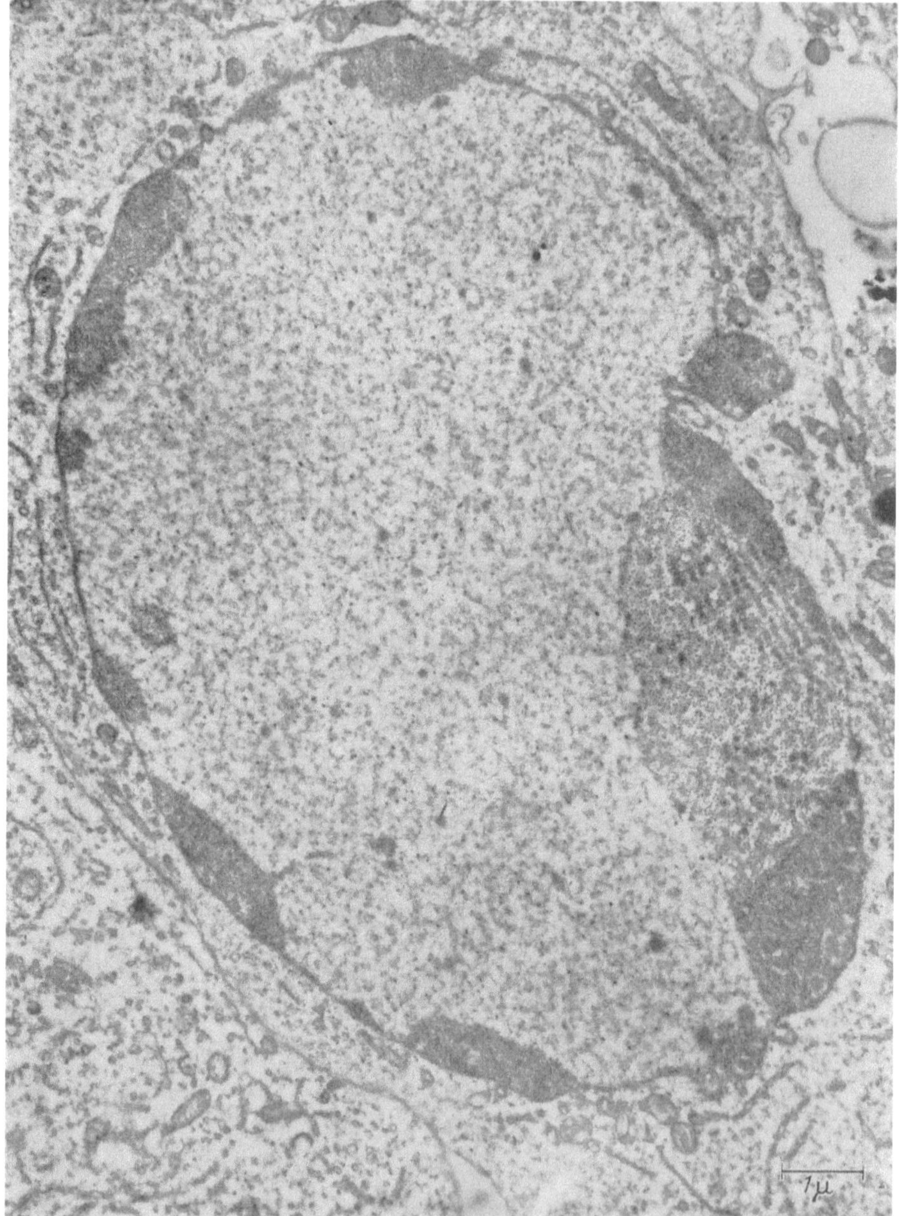

Fig. 10. A nucleus infected with the J. M. strain of herpes simplex virus. At the right is an aggregate of viral particles. 11,000 ×

Surprisingly enough, crystals have also been found which, though associated with type 5 adenovirus, are themselves devoid of recognizable viral particles (*13*). Fig. 8 illustrates a portion of a nucleus containing such crystals. The linear patterns are believed to reflect superimposition of the molecules composing the crystalline lattice. In the vicinity of the crystals, but rarely contained within them, are scattered viral particles and at the right margin a small viral crystal may be seen. Considering the thinness of the section, the nucleus obviously holds a large number of crystals as well as viral particles. When viewed in cross-section (Fig. 9), the non-viral crystals are frequently hexagonal in shape and exhibit a regular array of dense points, which presumably represent either parts of the molecules or the interstices between molecules. The spacing of c. 400 Å exhibited by these crystals is much greater than the 65 Å spacing observed in thin sections of insect polyhedral crystals (*14*). In both instances, however, the crystals develop in

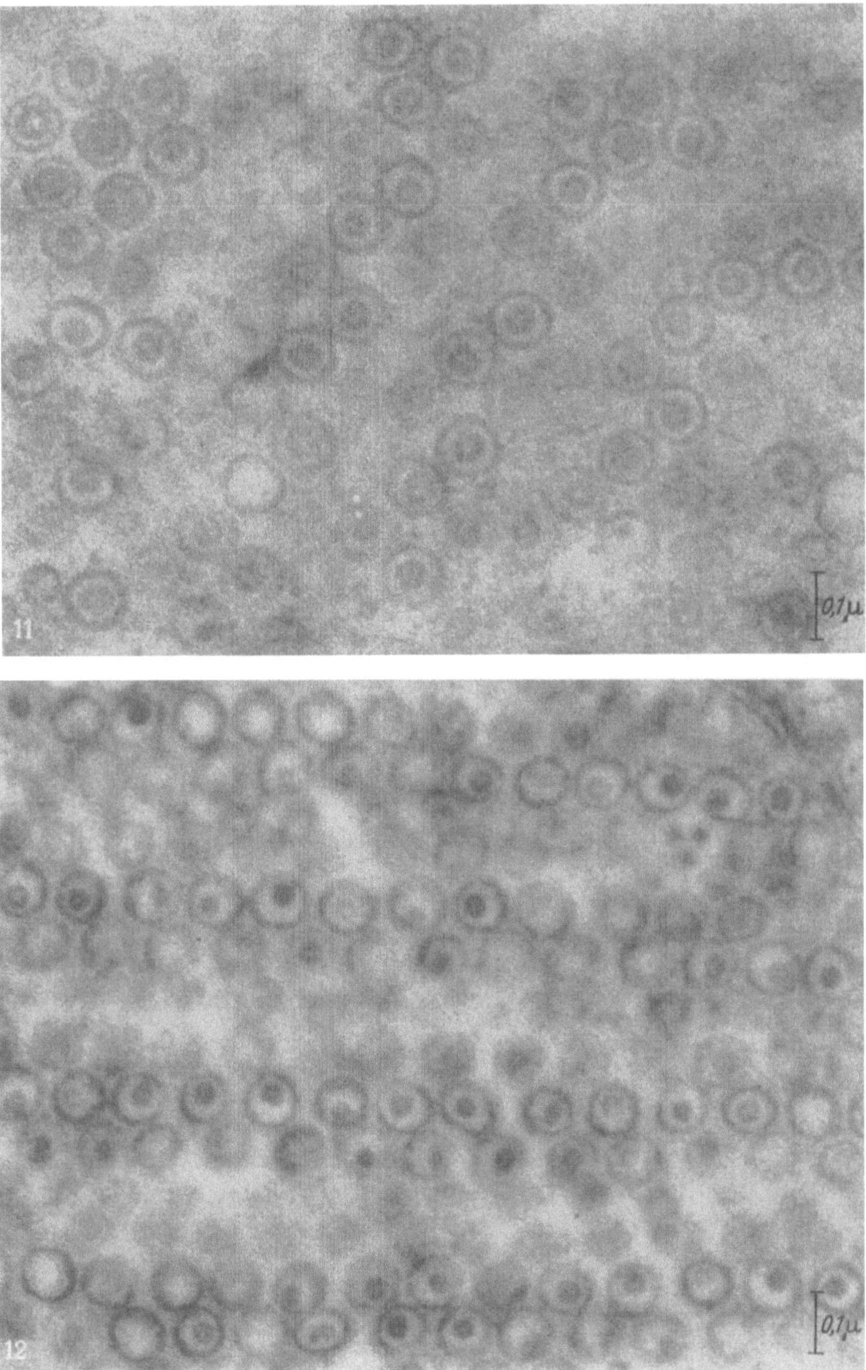

Fig. 11. Herpes simplex virus at that stage of development which is characterized by a dense core enclosed by a single limiting membrane. 87,000 ×

Fig. 12. Part of a large crystal. The pattern of distinct particles reflects orientation of the crystalline lattice with respect to the plane of section. 77,000 ×

nuclei, and are composed of protein which is Feulgen negative (*13, 15*). Why the crystals associated with type 5 adenovirus contain few viral particles and disintegrate upon release from the nucleus, whereas the insect polyhedra enclose numerous viral particles and are remarkably stable, remains to be determined.

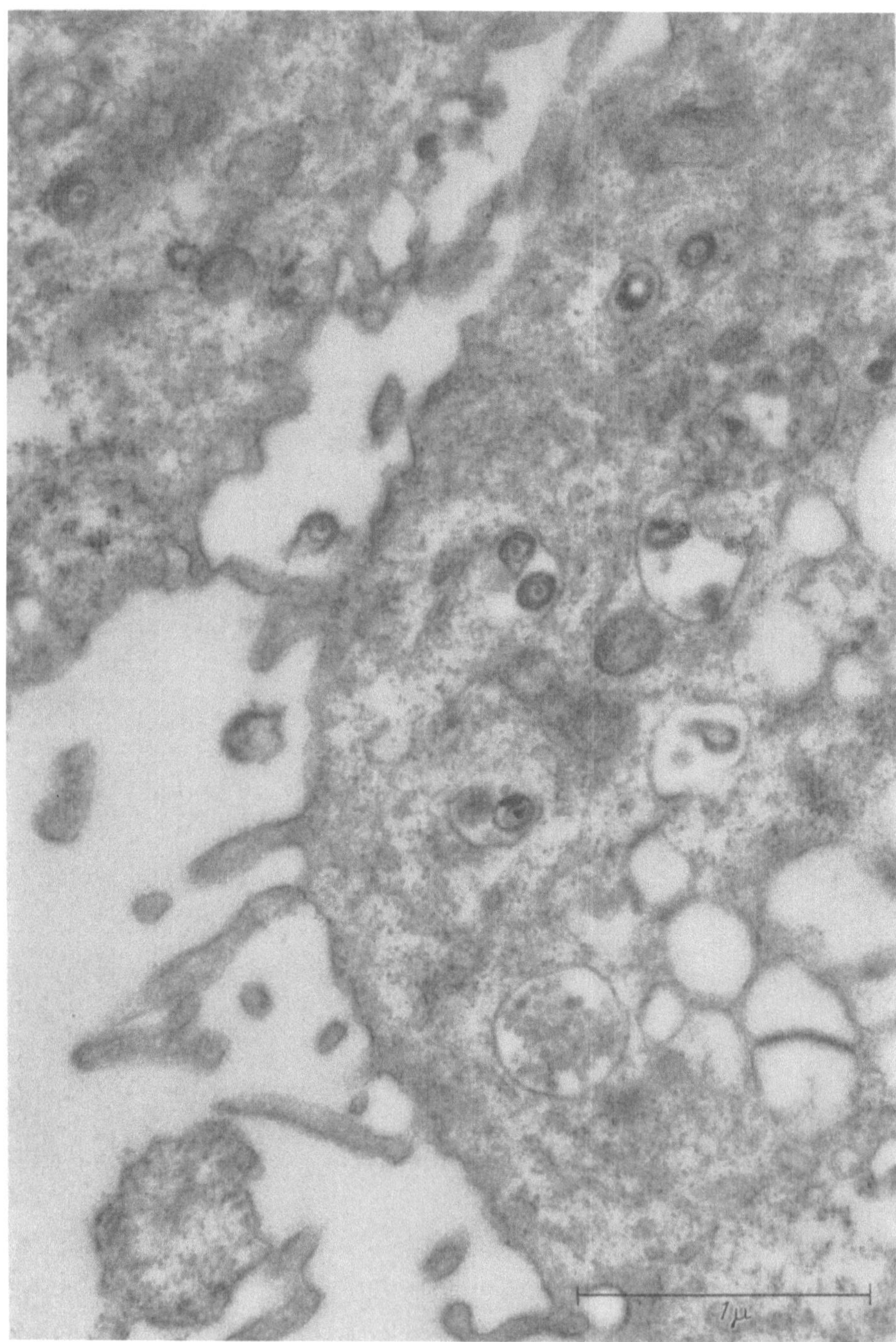

Fig. 13. Viral particles with double limiting membranes. Several of these particles lie in cytoplasmic vacuoles and one is visible in the extracellular space between two cells. 40,000 ×

Turning to *herpes simplex virus*, one is again confronted with a virus which may crystallize in the nucleus (*16*). Morphologically, however, herpes simplex virus (17) differs from the adenoviruses.

Fig. 10 illustrates a nucleus with irregular clumps of marginated chromatin. On the right is an aggregate of the small, dense "primary" bodies which have been previously described (*17*). Virus characterized by a central core and a single, limiting membrane appears to be arranged in irregular rows. At higher magnification (Fig. 11) the viral structure is clearly visible and it becomes evident that groups of viral particles are oriented in crystalline arrays. The crystals may attain considerable size. Fig. 12 shows part of a large crystal wherein rows of sharply defined particles alternate with zones of indistinct particles. Serial sections reveal that, as in the case of adenoviruses, the position of the virus in the plane of section determines the definition, and hence the patterns which the crystals exhibit largely reflect the angle at which the lattice has been cut. The crystals subsequently disintegrate and the virus acquires a second peripheral membrane. Fig. 13 illustrates virus in the cytoplasm after release from the nucleus. The particles, generally consisting of a central core and double limiting membranes, lie within walled vacuoles. One extracellular particle is visible in the left third of the field.

The existence of intracellular crystals raises many questions which cannot be answered in the present state of our knowledge. Nevertheless, certain tentative conclusions and hypotheses may be advanced. The crystals form within the nucleus and disintegrate, probably rather quickly, upon entering the cytoplasm. None have yet been encountered in the extracellular space. The nucleus, therefore, must provide an environment which is propitious for the operation of crystallizing forces. The degree of crystallization varies only slightly in the cell lines (HeLa, Hep-2 and human amnion) which have been examined[1]. It is important to realize, however, that so far the crystals of herpes simplex virus and the adenoviruses have been studied solely in tissue cultures.

Viral types differ with respect to their ability to crystallize. In the case of adenoviruses, for example, types 3, 4, 7, and 8 form numerous large crystals, whereas types 1, 2, 5, and 6 exhibit less tendency to crystallize. Viral strains also behave differently. In contrast to the J. M. strain of herpes simplex virus, which was illustrated above, the H. R. strain does not appear to form crystals, although the morphologic response of the host cells to both strains exhibits basic similarities.

Why is it, then, that one virus may crystallize whereas another, structurally indistinguishable, does not ? Perhaps the difference lies in the rapidity with which virus differentiates at the template sites. Close inspection of type 5 adenovirus reveals that the particles adjacent to the reticular masses are generally few in number and separated by fragments of nuclear matrix. Type 3 adenovirus, on the other hand, is encountered in large aggregates with little or no material interspersed between the particles. These observations hold for each of the viruses studied and suggest that when many viral particles form quickly within a relatively small locus the adjacent nuclear matrix is displaced. The close proximity of one particle to another would then permit the electrostatic forces carried by the virus to operate before dispersal could occur. Even momentary purity should result in crystallization. There may well be other factors which influence the process, but if so they are not reflected in the morphology demonstrable by electron microscopy.

In conclusion, a few remarks seem warranted regarding the significance of crystals with respect to the investigation of viruses. It is obvious that if particles are scattered through the cell or clump in small, amorphous groups, they cannot be recognized in the light microscope. Crystals, on the other hand, may be readily identified by virtue of their flat, angulated faces. If such crystals are found by electron microscopic examination to be composed of virus, it is likely that the reactions they exhibit to histochemical stains, as well as to other procedures available in light microscopy, will depend to a large extent on the chemical composition of the virus. For example, study of contiguous thick and thin sections in the light and electron microscopes reveals that specific crystals made up of adenoviral particles stain strongly with the Feulgen reaction (*8*). Presumably, then, the virus contains deoxyribose nucleic acid. Using similar techniques, the non-viral crystals associated with type 5 adenovirus were observed to contain protein devoid of DNA (*13*).

---

[1] Preliminary investigations indicate that antibodies added to the culture medium after the initiation of infection exert no demonstrable effect, at least on the crystallization of types 2 and 7 adenoviruses.

In the future, perhaps a means will be devised whereby the crystals can be isolated from their host cells and examined directly by quantitative chemical and immunologic methods. The information thus obtained, when integrated with the results of morphologic studies, should provide a significant contribution to our knowledge of viral structure and development.

References

1. Bunting, H.: Proc. Soc. exp. Biol. (N. Y.) 84, 327 (1953).
2. Kjellén, L., G. Lagermalm, A. Svedmyr and K.-G. Thorsson: Nature (Lond.) 175, 505 (1955).
3. Harford, C. C., A. Hamlin and E. Parker: Trans. Ass. Amer. Phycns 68, 82 (1955).
4. Morgan, C., C. Howe, H. M. Rose and D. H. Moore: J. biophys. biochem. Cytol. 2, 351 (1956).
5. Harford, C. G., A. Hamlin, E. Parker and T. Ravenswaay: J. exp. Med. 104, 443 (1956).
6. Lagermalm, G., L. Kjellén, K.-G. Thorsson and A. Svedmyr: Arch. ges. Virusforsch. 7, 221 (1957).
7. Rowe, W. P., R. J. Huebner and J. A. Bell; Ann. N. Y. Acad. Sci. 67, 255 (1957).
8. Bloch, D. P., C. Morgan, G. C. Godman, C. Howe and H. M. Rose: J. biophys. biochem. Cytol. 3, 1 (1957).
9. Low, B. W., and P. R. Pinnock: J. biophys. biochem. Cytol. 2, 483 (1956).
10. Deitch, A. D., and G. C. Godman: Anat. Rec. 123, 1 (1955).
11. Tousimis, A. J., and M. R. Hilleman: Virology 1, 499 (1957).
12. Valentine, R. C., and P. K. Hopper: Nature (Lond.) 180, 928 (1957).
13. Morgan, C., G. C. Godman, H. M. Rose, C. Howe and J. S. Huang: J. biophys. biochem. Cytol. 3, 505 (1957).
14. — G. H. Bergold, D. H. Moore and H. M. Rose: J. biophys. biochem. Cytol. 1, 187 (1955).
15. Bergold, G. H.: Viruses of insects. Handbuch Virusforschung. Bd. 4: (Suppl. No. 3) 60 (1958).
16. Morgan, C., E. P. Jones, M. Holden and H. M. Rose: Virology 5, 568 (1958).
17. — S. A. Ellison, H. M. Rose and D. H. Moore: J. exp. Med. 100, 195 (1954).

Beobachtungen an Adenovirus (Typ 3) -infizierten HeLa-Zellkulturen nach Fixierung mit Kaliumpermanganat

K. H. Andres und G. Nielsen

Bernhard Nocht-Institut für Schiffs- und Tropenkrankheiten, Hamburg, Abteilung für Virusforschung

In den vergangenen Jahren waren die Adeno-Viren wiederholt Gegenstand elektronenoptischer Studien (*1—10*). Im Gegensatz zu den Voruntersuchern, die mit OsO_4-fixiertem Material arbeiteten, benutzten wir das von Luft (*11*) in die Elektronenmikroskopie eingeführte Kaliumpermanganat.

Nach üblicher Weise gezogene HeLa-Zellkulturen wurden mit Adenovirus (Typ 3) beimpft und nach Intervallen von 24 Std. bis zu 7 Tagen nach Infektionsbeginn elektronenoptisch untersucht. Die Gewebeproben wurden mit $KMnO_4$-Lösung (0,6% ig in Michaelispuffer; p_H 7,2; 4° C) fixiert und in Plexigum (*12*) eingebettet. Die Blöcke wurden mit dem Porter-Blum Mikrotom geschnitten; Auswertung der Schnitte im Elmiskop I (Siemens).

Die im Verlauf einer Adeno-Infektion auftretenden morphologischen Veränderungen der HeLa-Zellen sind von Lagermalm u. Mitarb. (*8*) eingehend beschrieben worden. Etwa 20 Std. nach der Infektion erkennt man eine schollige Auflockerung des Kernplasmas, der bald eine fleckweise Verdichtung zentraler und membrannaher Kernplasmaanteile folgt. Die Kernoberfläche wird eingebuchtet und in einigen Zellkernen werden erste, noch solitär gelegene, typische polyedrische Elementarkörper gefunden. Mit zunehmender Infektionsdauer bilden sich kleine Virus-Kristalloide, die an Größe zunehmen und stellenweise zu größeren Aggregaten zusammenwachsen. Jetzt hebt sich die Kernplasmaverdichtung in Membrannähe besonders deutlich ab, wobei häufig kontrastreiche, ovale, glattrandige Körper zu sehen sind. Im weiteren Verlauf unterliegt die Zelle zunehmenden degenerativen Vorgängen, wie Kernfragmentierung und -auflösung. Dabei gelangen die Kristalloide teils in das umgebende Cytoplasma, teils lösen sie sich zu solitären Viren auf. Im Plasma liegende Kristalloide werden oftmals von myelinähnlich geschichteten Membranen umgeben (Abb. 1 und 2). Am 6.—7. Tag nach Infektionsbeginn ist die Zelle zugrunde gegangen. Viruselementarkörper werden dann zwischen den sich auflösenden Zellstrukturen gefunden.

An günstig getroffenen Virusquerschnitten beobachtet man eine regelmäßige, hexagonale Konfiguration mit einer kontrastreichen Zone im Zentrum. Ein Formwandel der Adenoviren (Typ 3) wurde von uns nicht festgestellt.

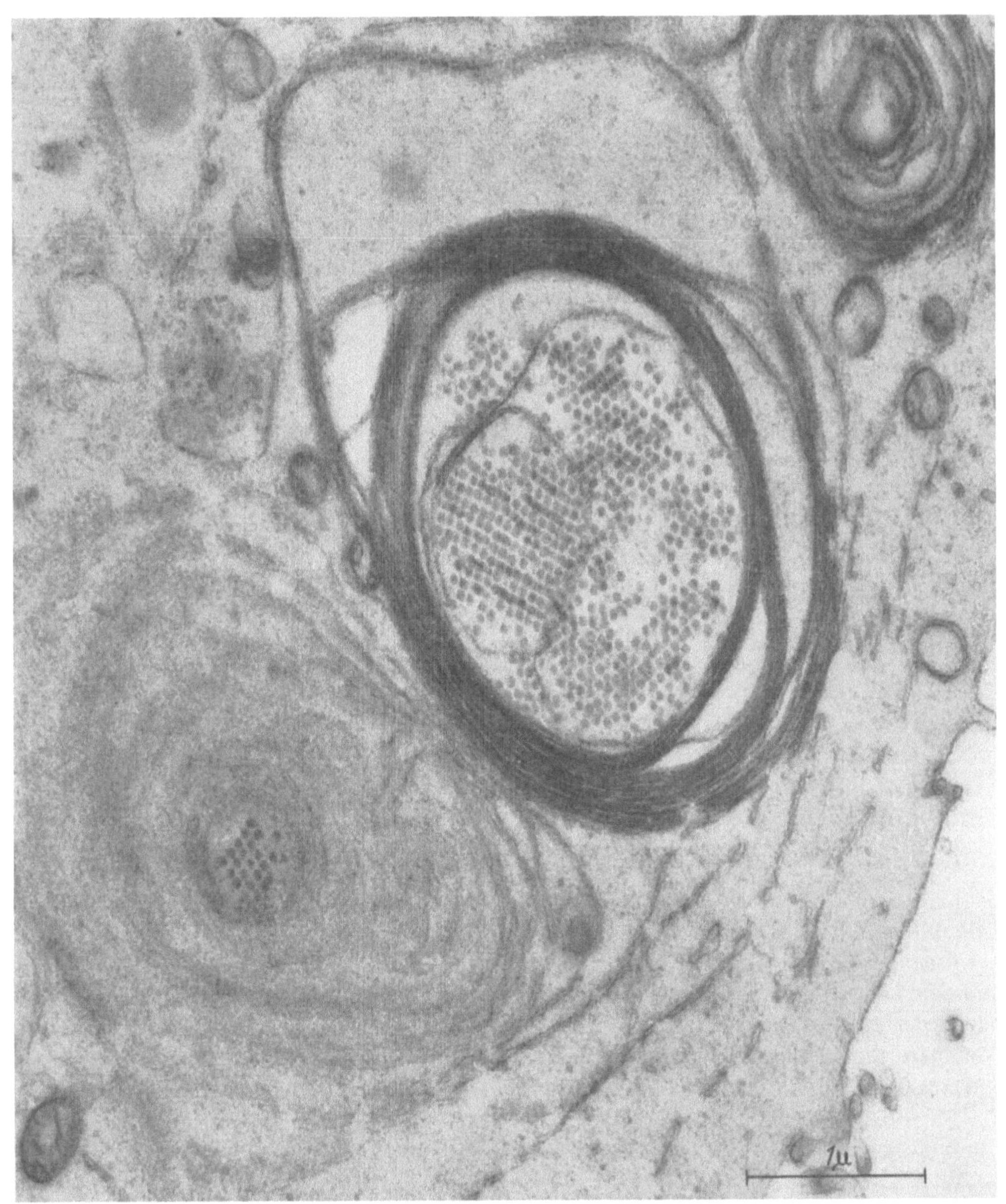

Abb. 1. Adeno-Virus (Typ 3) infizierte HeLa-Zelle, 120 Std. nach Infektion. Im Cytoplasma von Membranen einges chlossene Elementarkörper. 25000mal

Über die Gestalt des Adeno-Elementarkörpers herrscht noch keine Klarheit. Während TOUSIMIS und HILLEMAN (9) eine Ikosaederform annehmen, diskutierten VALENTINE und HOPPER (10) auch die Möglichkeit der Rhombendodekaederform, ohne sich zu entscheiden. Der Arbeitskreis um MORGAN (4, 5, 6, 7) hält eine gobuläre bis ellipsoide Form für wahrscheinlich. Bei einer Schnittdicke von etwa 300 Å wird ein Adeno-Elementarkörper von 600—700 Å Durchmesser im günstigsten Fall von 2—3 Schnitten erfaßt. Hierdurch wird die Auflösung von Einzelstrukturen des Elementarkörpers begrenzt. Aus Modellbetrachtungen folgt, daß sowohl Kubus, wie Ikosaeder und Rhombendodekaeder im Schnitt gleichseitige Sechsecke zu liefern vermögen. Während dieser Befund für Kubus und Ikosaeder nur bei einer definierten Schnittführung zu erzielen ist, liefert beim Rhombendodekaeder jeder Schnitt zwischen zwei Grenzen senkrecht zur dreizähligen Achse ein gleichseitiges Sechseck. Nach kristallographischen Gesichtspunkten ist ein dem Adeno-Kristalloid ähnliches Gitter mit einem Ikosaeder-Elementarkörper nicht zu erreichen. Ein Rhombendodekaeder dagegen erlaubt eine solche Aggregation. Die Abweichung der Schnittebene von der oben erwähnten Bedingung ergibt elongierte Sechsecke (Abb. 3 a

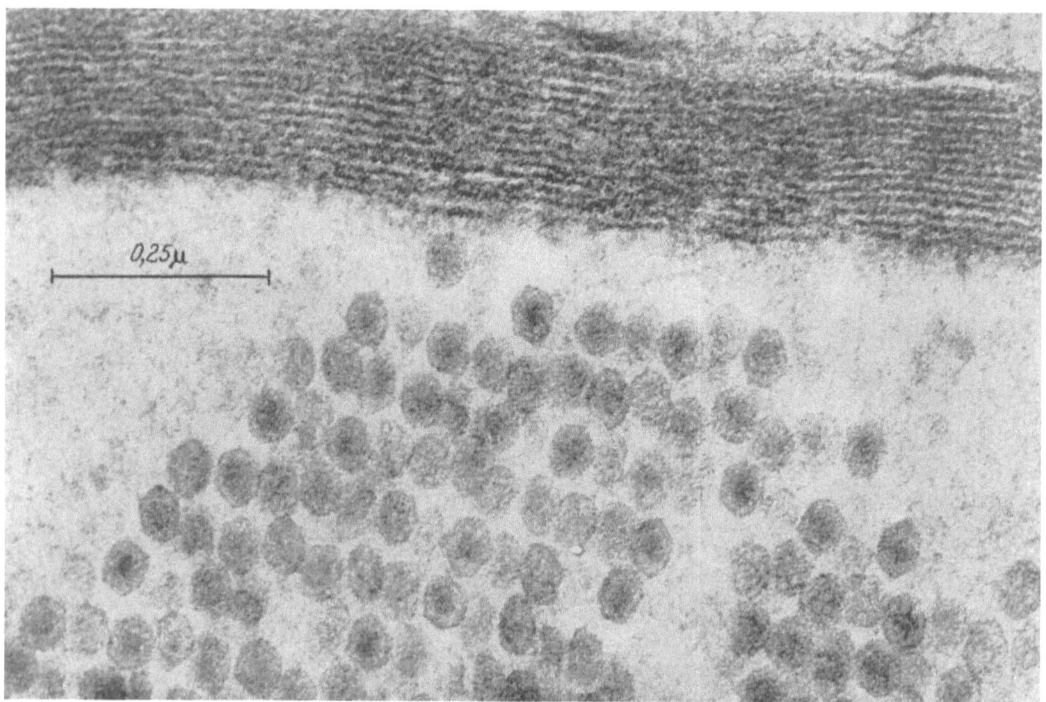

Abb. 2. Myelinfigurenähnliche Zusammenlagerung von Membranen. Hexagonale Schnittbilder von Viruselementarkörpern. 120000 mal

und b) bis zur Viereckform (Abb. 3 c). Unter Berücksichtigung der Schnittdickenverhältnisse ist dann die im Schnitt erkennbare sterische Anordnung gegeben.

Unser unter diesen Gesichtspunkten ausgewertetes Material läßt eine weitgehende Übereinstimmung mit den theoretischen Vorstellungen erkennen (Abb. 3). Wir halten daher die Rhombendodekaederform des Adeno-Elementarkörpers (Typ 3) für die wahrscheinlichste. Der im Schnitt erkennbare rhombische Aufbau der Kristalloide unterstützt diese Vorstellung.

Wir danken Herrn Dr. D. PETERS für das fördernde Interesse, das er unseren Arbeiten entgegenbrachte, und der Deutschen Forschungsgemeinschaft für die materielle Unterstützung.

Literatur

1. KJELLÉN, L., G. LAGERMALM, A. SVÉDMYR and K.-G. THORSSON: Nature (Lond.) 175, 505 (1955).
2. HARFORD, C. C., A. HAMLIN and E. PARKER: Trans. Ass. Amer. Physicns 68, 82 (1955).
3. — — — and T. v. RAVENSWAAY: J. exp. Med. 104, 443 (1956).
4. MORGAN, C., C. HOWE, H. M. ROSE and D. H. MOORE: J. biophys. biochem. Cytol. 2, 351 (1956).
5. LOW, B. W., and P. R. PINNOCK: J. biophys. biochem. Cytol. 2, 483 (1956).
6. BLOCH, D. P., C. MORGAN, G. C. GODMAN, C. HOWE and H. M. ROSE: J. biophys. biochem. Cytol. 3, 1 (1957).
7. MORGAN, C., G. C. GODMAN, H. M. ROSE, C. HOWE and J. S. HUANG: J. biophys. biochem. Cytol. 3, 505 (1957).
8. LAGERMALM, G., L. KJELLÉN, K.-G. THORSSON and A. SVEDMYR: Arch. Virusforsch. 7, 221 (1957).
9. TOUSIMIS, A. J., and M. R. HILLEMAN: Virology 4, 499 (1957).
10. VALENTINE, R. C., and P. K. HOPPER: Nature (Lond.) 180, 928 (1957).
11. LUFT, J. H.: J. biophys. biochem. Cytol. 2, 799 (1956).
12. BAYER, M., u. D. PETERS: J. Ultrastructure Res. 2, 444 (1959).

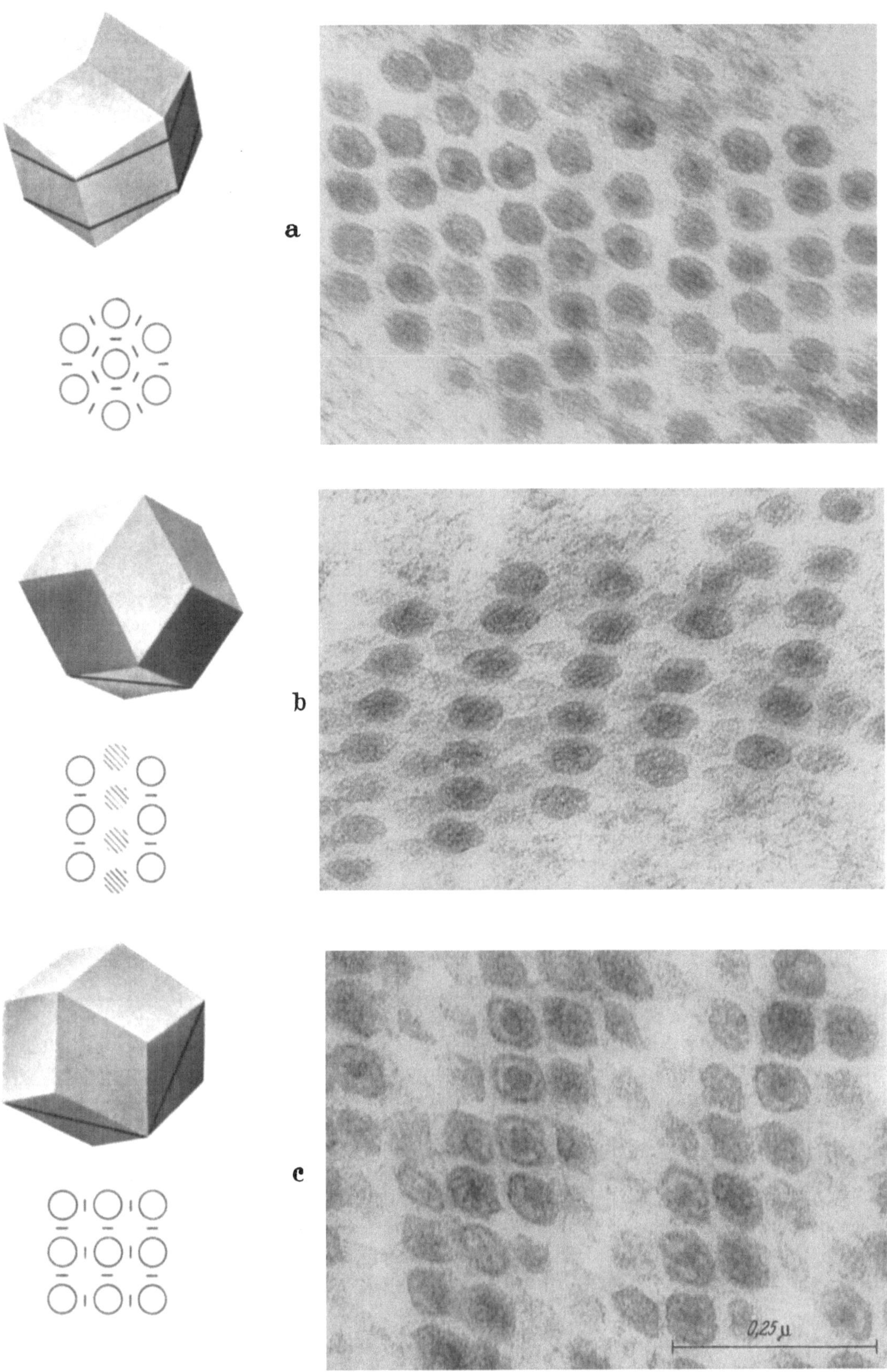

Abb. 3a—c. Modellschnitte durch Rhombendodekaeder in verschiedenen Ebenen mit zugeordneten Schnitten durch Viruskristalloide 135000mal. (Im unteren Beispiel weicht die dargestellte Schnittebene etwas von der Idealebene ab.)

Über die cytologischen Veränderungen von Herpes-B-Virus infizierten Affennieren-Gewebekulturen *

R. Mauler und V. Dostal**

Virus-Abteilung der Behringwerke, Marburg a. d. Lahn, und Hygiene-Institut der Albert Ludwigs-Universität, Freiburg i. Br.

Das Herpes B-Virus wurde erstmals von Sabin und Wright (1) 1933 von einem Patienten isoliert, der an einer tödlich verlaufenden, ascendierenden Myelitis erkrankte, nachdem er von einem klinisch gesunden Affen gebissen worden war. Andere Stämme dieses Virus konnten Black und Melnick (2) aus dem Zentralnervensystem eines Rhesus-Affen und Krech und Lewis (3) aus der Kulturflüssigkeit von Affennierenzellkulturen isolieren.

Über elektronenoptische Studien des Multiplikationsvorganges in Affennieren-Zellkulturen wurde erstmals von Reissig und Meinick (4) berichtet. Untersuchungen am Herpes-Simplex-Virus wurden von C. Morgan u. Mitarb. (5, 6) ausgeführt. Reissig und Melnick haben die morphologischen Veränderungen bei der Virusmultiplikation in Korrelation gebracht mit dem Ansteigen des Virustiters in der flüssigen Phase der infizierten Gewebekulturen und festgestellt, daß das erste Auftreten von virusähnlichen Partikeln (virus like bodies) in der Zelle mit dem Anstieg der Viruskonzentration zusammenfällt. Die frühen morphologischen Veränderungen bestanden in einem Anschwellen und einer Vacuolisierung der Mitochondrien, einem Verschwinden des Nucleolus, gefolgt von einem Randständigwerden des Kernchromatins. Als Virusvorstufen wurden Partikel angesehen, die einen dichteren Zentralkörper aufwiesen, der von einer einfachen Membran umgeben ist, und diffus im Zellkern verteilt waren. Eine „kristall"- oder paketähnliche Anordnung der Viruspartikel im Kern haben Reissig und Melnick nicht gesehen.

Die methodischen Einzelheiten über die Anlage von Gewebekulturen und der elektronenoptischen Präparation sind dem Beitrag „Über die cytologischen Veränderungen von ECHO-Virus Typ-9 infizierten Affennierengewebekulturen" S. 615 zu entnehmen. Die Infektion der Gewebekulturen erfolgte mit einem Stamm des B-Virus, der von uns in den Gewebezuchtlaboratorien der Behringwerke aus der flüssigen Phase normaler Nierengewebekulturen von Rhesus-Affen isoliert werden konnte[1]. Die Viruskonzentration bei der Infektion betrug etwa 10^5 ID 50/ml .

Die ersten Veränderungen an der Gewebekulturzelle sind bereits wenige Stunden nach der Infektion mit dem Herpes B-Virus festzustellen, zu einem Zeitpunkt, an dem ein meßbarer Virustiter in der Kulturflüssigkeit noch nicht vorhanden ist. Die Mitochondrien (Abb. 1) zeigen eine deutliche Schwellung und bei vielen von ihnen fällt eine osmiophile homogene Struktur auf. Im Kern ist der Nucleolus nicht mehr nachweisbar und das Kernchromatin zeigt eine Verlagerung zur Kernmembran hin. Zu einem späteren Zeitpunkt nach der Infektion sind besonders die Veränderungen am Kern deutlich zu sehen (Abb. 2). Das Chromatin ist zu Haufen angeordnet, die z. T. randständig sind und alle Übergänge von feinkörniger bis zu grobscholliger Struktur zeigen. An einer Stelle kann man ein Paket von ovalen bis runden Partikeln erkennen, die in ihrer Lage zueinander eine „kristallähnliche" Anordnung zeigen. Sie haben eine osmiophile Außenmembran und einen mehr oder weniger dichten Innenkörper. Einzelne solcher Partikel von einer mittleren Größe von etwa 80 mμ, die wir als Virusvorstufen ansehen, sind über den Kern verteilt und auch im Cytoplasma zu finden. Beim Auftreten der ersten Partikel im Kern sind auch Veränderungen an der Kernmembran festzustellen, an der wir jetzt bis zu fünf osmiophile Lamellen beobachten. Ein noch späteres Stadium nach der Infektion zeigt Abb. 3. Die Kernmembran zeigt wiederum bis zu fünf osmiophile Lamellen, die an einer Stelle aufgelöst erscheinen. In diesem Bezirk erkennt man aber eine Anhäufung von Partikeln, die anscheinend eine Ausschleusung in das Cytoplasma erfahren. Einzelne Partikel sind bereits mit einer Doppelmembran versehen

* Die Untersuchungen im Hygiene-Institut Freiburg i. Br. wurden mit Unterstützung der Deutschen Forschungsgemeinschaft durchgeführt.

** Wir danken Fräulein M. Hahn aus den Behringwerken Marburg a. d. Lahn für die Mithilfe bei der Präparation und den elektronenoptischen Untersuchungen.

[1] Wir danken Herrn Dr. U. Krech und Fräulein Dr. H. Wulff vom Schweizerischen Serum- und Impfinstitut Bern für die Typisierung des Herpes B-Virus.

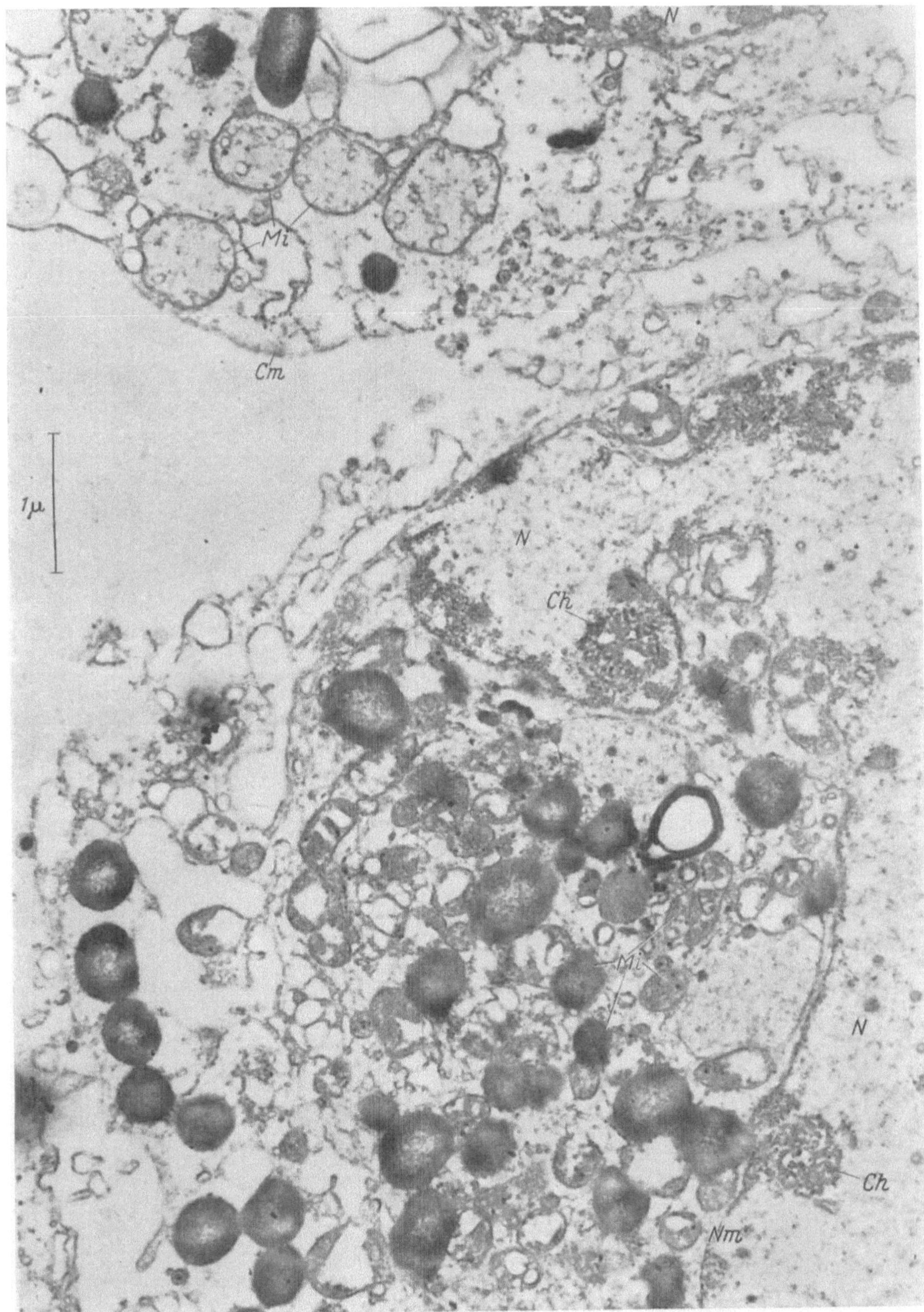

Abb. 1

Abb. 1—3. *N* Nucleus, *C* Cytoplasma, *Cm* Zellmembran, *Nm* Kernmembran, *Mi* Mitochondrium, *Ch* Chromatin, *Pn* Partikel im Kern, *Pc* Partikel im Cytoplasma. Elektronenoptische Vergrößerung bei allen Abbildungen 7200:1, kV 60, Aperturblende 30 μ

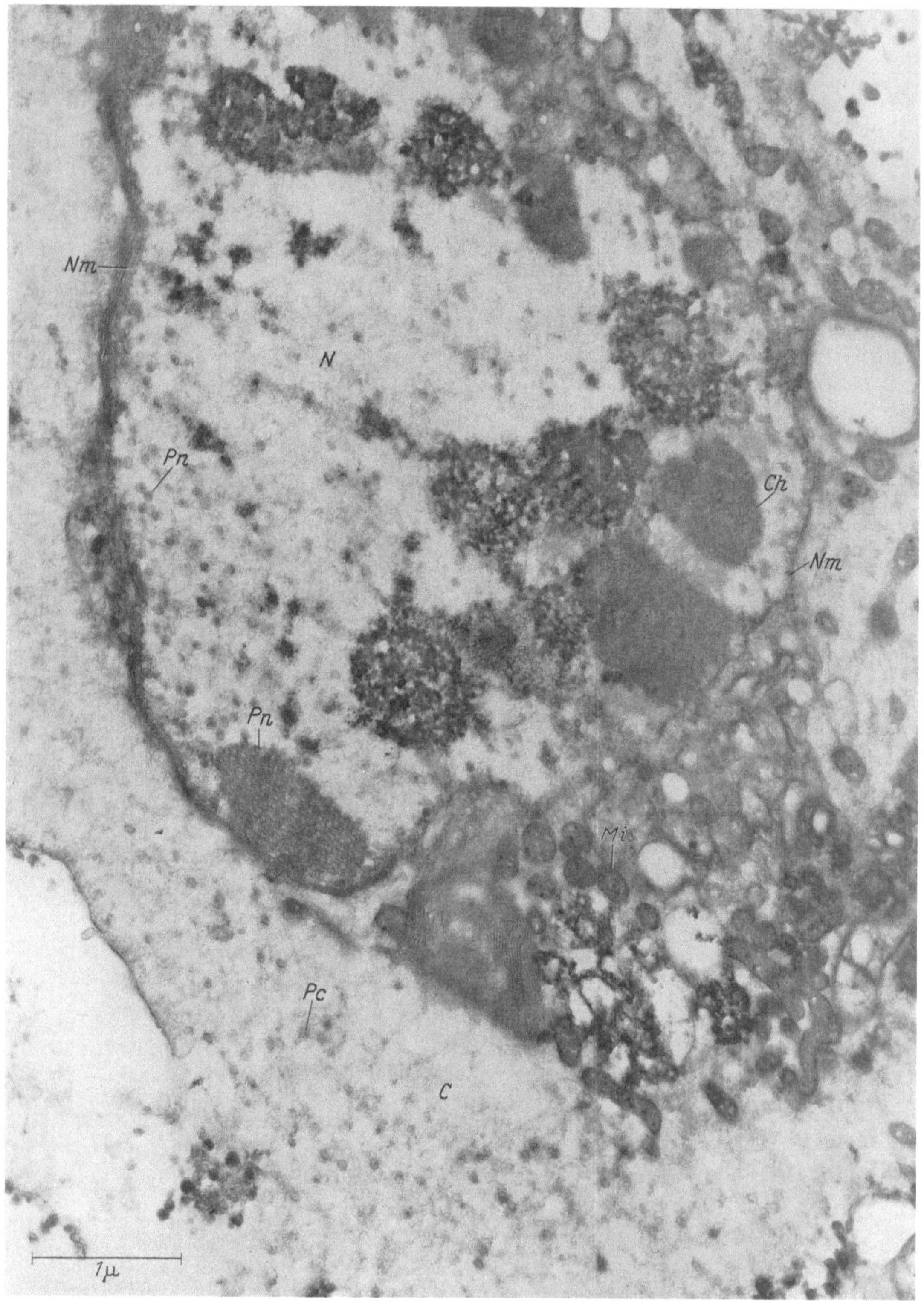

Abb. 2

und haben eine durchschnittliche Größe von 100 mμ. Im Cytoplasma sehen wir außer Vacuolen auch mehrere Partikel, die von einer gemeinsamen „Membran" (α-Cytomembran?) umgeben sind.

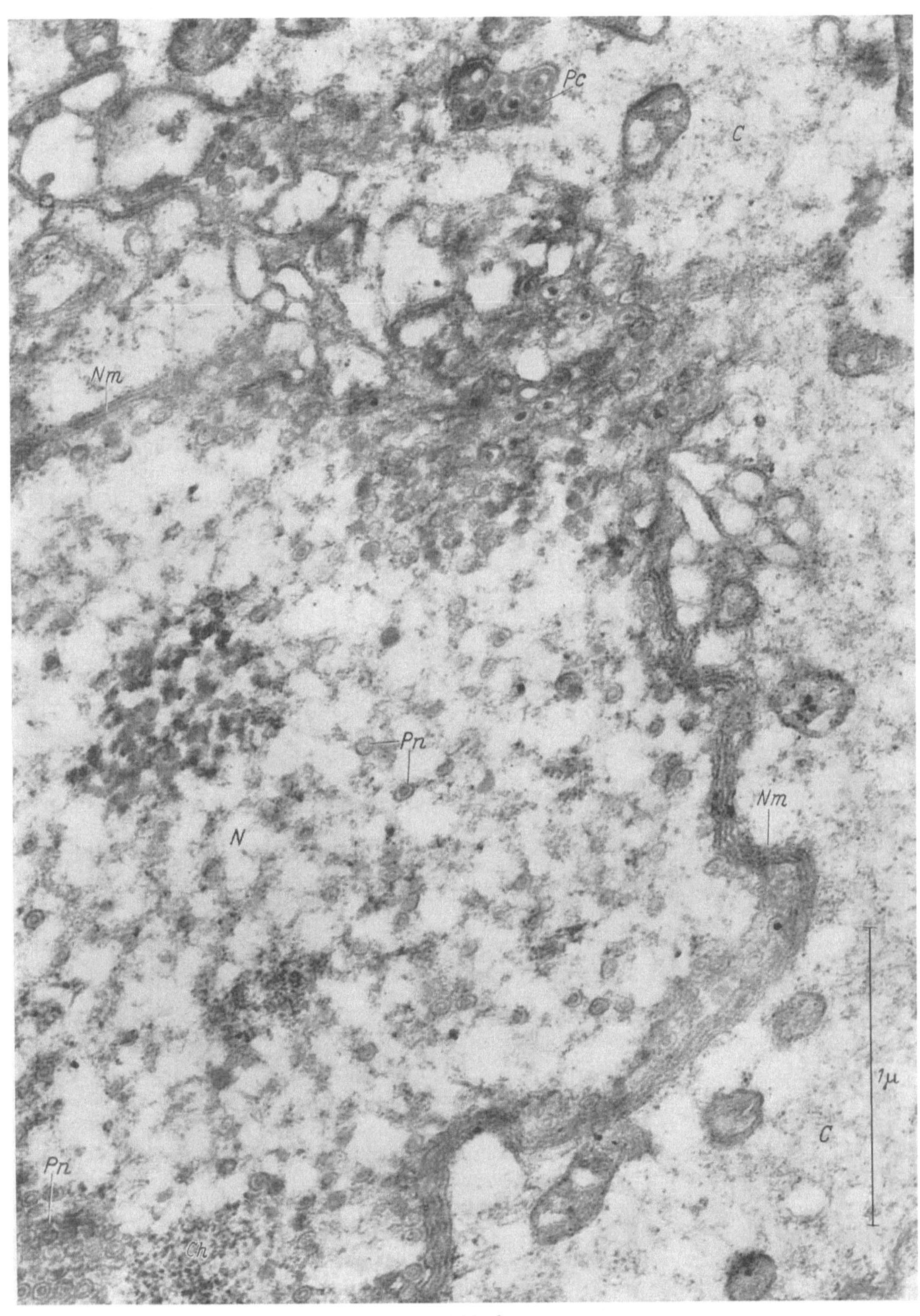

Abb. 3

Wir sehen die Partikel, die eine osmiophile Innenstruktur und eine Doppelmembran aufweisen, als komplettes (reifes) Herpes B-Virus an.

39

Literatur

1. Sabin, A. S., and A. M. Wright: J. exp. Med. **59,** 115 (1934).
2. Black, F. L., and J. L. Melnick: Fed. Proc. **13,** 487 (1954).
3. Krech, U., and L. J. Lewis: Proc. Soc. exp. Biol. (N. Y.) **87,** 174 (1954).
4. Reissig, M., and J. L. Melnick: J. exp. Med. **101,** 341 (1955).
5. Morgan, C., S. A. Ellison, H. M. Rose and D. H. Moore: J. exp. Med. **100,** 195 (1954).
6. — E. P. Jones, M. Holden and H. M. Rose: Virology **5,** 568 (1958).

L'ultrastructure des virus oncogènes*

(Résumé de la Conférence)

W. Bernhard

Institut de Recherches sur le Cancer, Villejuif/Seine (France)

Grâce aux études antérieures qui ont permis de caractériser morphologiquement plusieurs groupes de virus classiques, la mise en évidence de l'agent causal de certaines tumeurs expérimentales est devenue relativement aisée. On sait que le nombre de tumeurs à virus découvertes, surtout chez les oiseaux, mais également chez les mammifères, augmente presque chaque année.

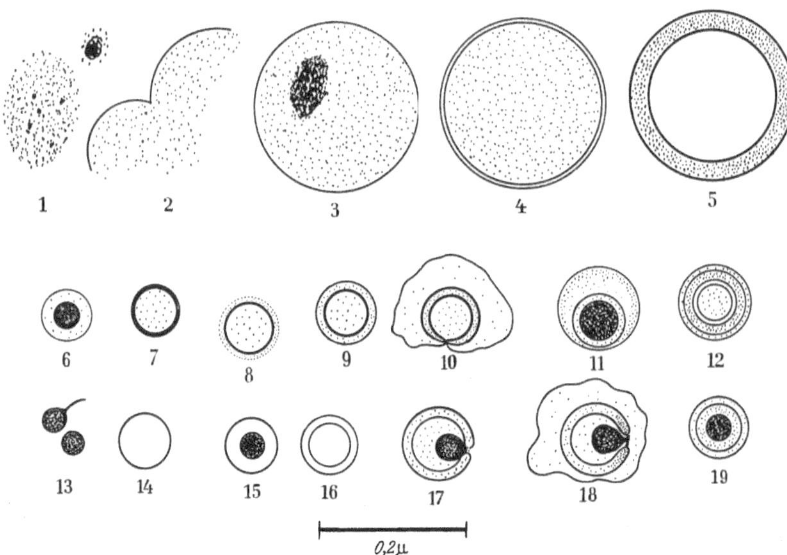

Fig. 1. Dessin destiné à montrer d'une manière schématique les principales formes de virus ou particules d'aspect viral mis en évidence dans des tumeurs bénignes ou malignes. Stades évolutifs variés. 1—5. Fibrome de Shope et Molluscum contagiosum. 6—12. Tumeur mammaire de la souris (facteur de Bittner ?). 13—18. Virus de la tumeur rénale de grenouille (Lucké) d'après Fawcett. 19. Virus trouvé dans tous les tissus tumoraux et les leucémies du poulet

Du point de vue théorique, ces cancers sont du plus grand intérêt, puisqu'il n'existe aucun autre agent cancérigène qui puisse transformer en quelques heures, comme certains de ces virus, une cellule normale en cellule tumorale (Oberling). Il est permis de penser que lorsque l'on connaîtra à fond le mécanisme de cette transformation, nos connaissances des secrets de la cancérisation en général seront considérablement enrichies. Déjà les trois dernières années ont permis la mise en évidence sur des micrographies électroniques du virus du fibrome ou fibrosarcome de Shope chez le lapin, du molluscum contagiosum de l'homme, du facteur lacté de Bittner dans la tumeur mammaire de la souris, de particules-virus associés avec le sarcome de Rous, la tumeur de Murray-Begg, la tumeur de Fujinami, la lymphomatose et des leucoses erythro-myéloblastiques chez le poulet. De plus, le virus associé avec l'adénocarcinome de la

---

* Weitere Ergebnisse über Tumorviren s. Abschnitt G 1, ds. Bd. S. 456.

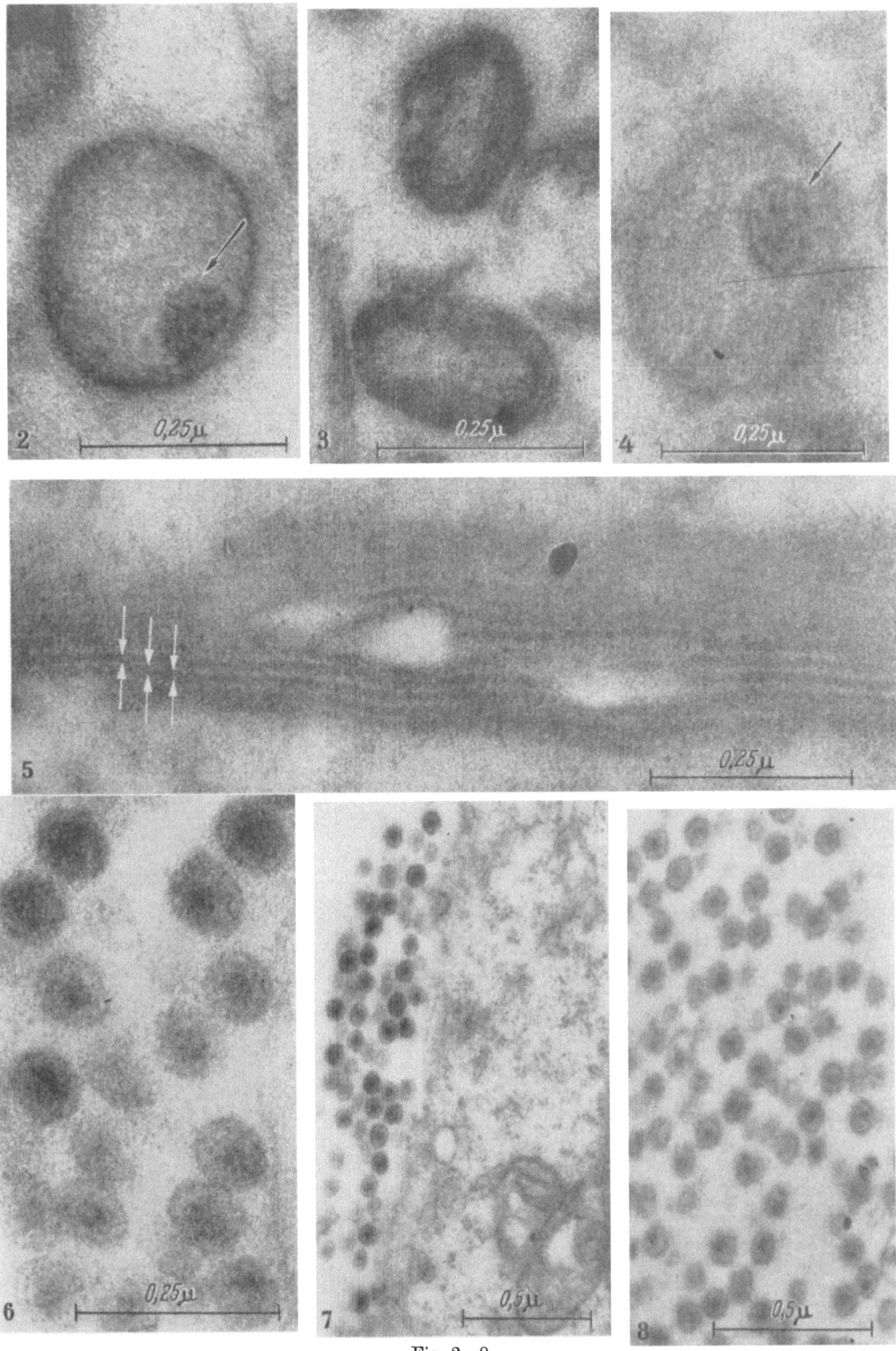

Fig. 2—8

Fig. 2. *Fibrome de Shope du lapin*. Virus à membrane unique avec nucléoïde (———➤). Gross. 130.000 × —
Fig. 3. *Fibrome de Shope du lapin*. 2 virus plus évolués à membranes multiples. Gross. 130.000 × — Fig. 4. *Molluscum contagiosum d'une lésion cutanée de l'homme*. Virus á membrane unique avec nucléoïde (——➤). Gross. 130.000 ×
Fig. 5. *Fibrome de Shope du lapin*. Formation lamellaire fréquemment rencontrée dans les corps d'inclusion à virus. L'épaisseur d'une lamelle est de l'ordre de 70 Å. Gross. 130.000 × — Fig. 6. *Sarcome de Rous de la poule*. Groupe de virus extracellulaires avec nucléoïdes bien visibles. Gross. 130.000 — Fig. 7. *Leucémie érythroblastique de la poule*. Groupe de virus extracellulaires. Moëlle osseuse. Gross. 40.000 × — Fig. 8. *Leucémie myéloblastique de la poule (Beard)*. Plasma leucémique purifié contenant presque exclusivement des particules-virus. Gross. 51.000 ×

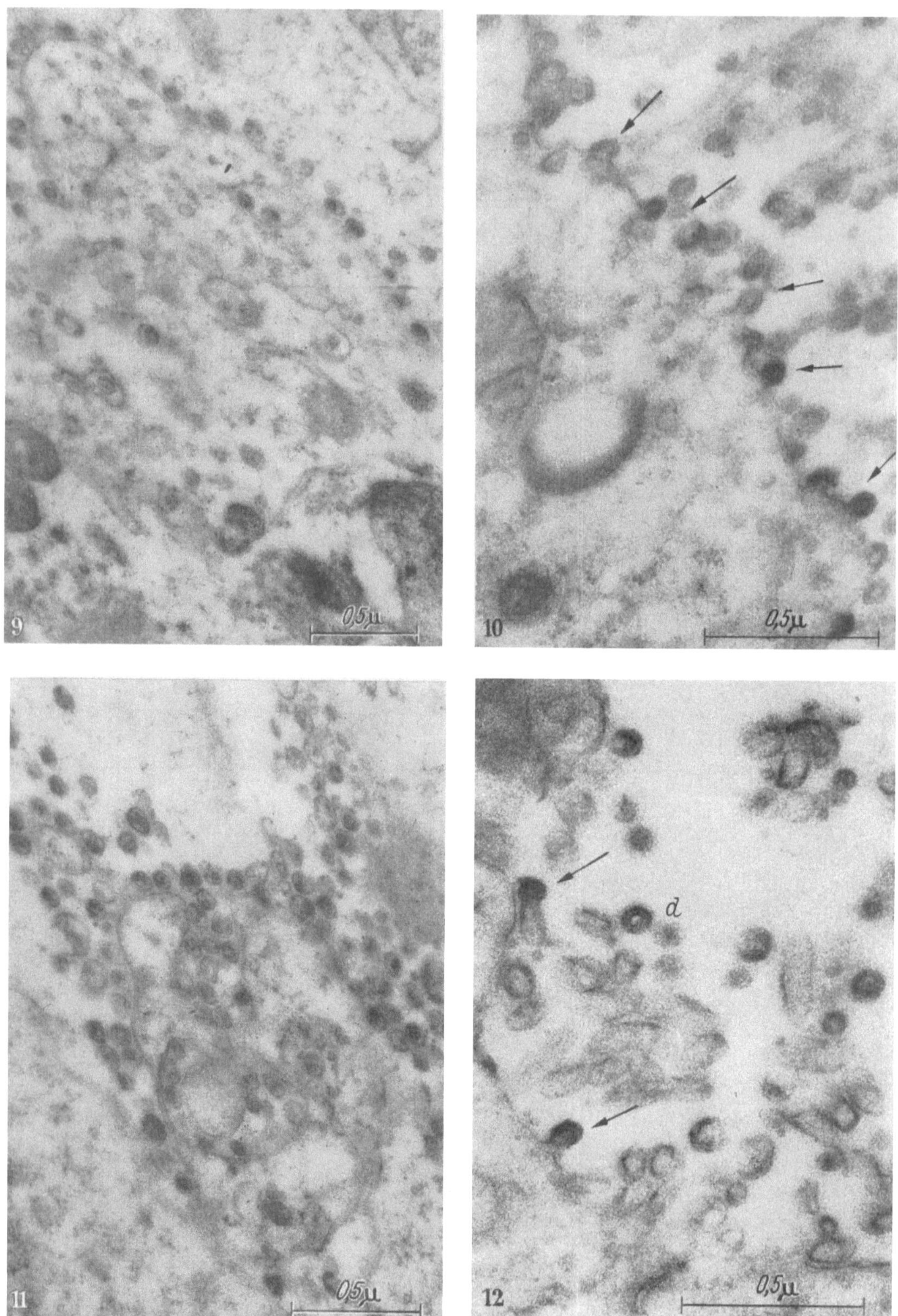

Fig. 9—12. *Leucémie spontanée de souris* (souche AK). Tissus divers contenant des particules d'aspect variable.

Fig. 9. *Ganglion lymphatique*. Particules régulièrement rangées à la surface cellulaire. Gross. 35.000 ×

Fig. 10. *Thymus*. Processus de bourgeonnement à la surface d'un thymocyte. Gross. 55.000 ×

Fig. 11. *Rate*. De nombreuses particules assez polymorphes couvrant la surface cellulaire. Gross. 41.000 ×

Fig. 12. *Tumeur de la parotide d'une souris leucémique*. (———⟶) indique un processus de bourgeonnement. *d* = forme du type 'doughnut' frequemment rencontrée dans les leucémies de souris. Gross. 60.000 ×

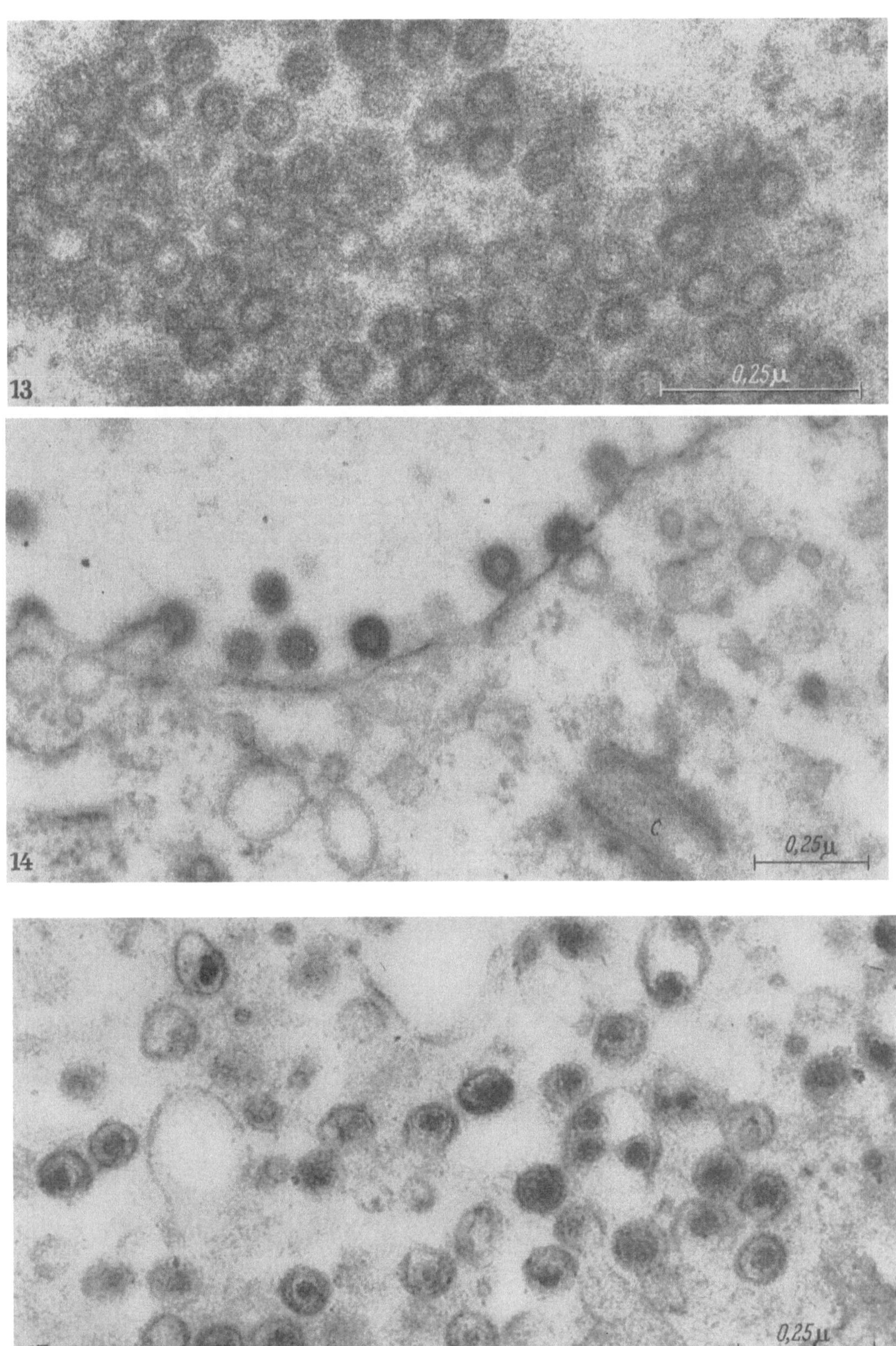

Fig. 13—15. *Tumeur mammaire spontanee de la souris*. Particules d'aspect viral formant un corps d'inclusion intracellulaire (Fig. 13), traversant la membrane cellulaire (Fig. 14) et particules extracellulaires dans un canal galactophore (Fig. 15). La forme du type 'doughnut' se transforme en particules pourvues d'un nucléoïde excentrique

Fig. 13. Gross. 130.000 ×   —   Fig. 14. Gross. 74.000 × (c = centriole)   —   Fig. 15. Gross. 88.000 ×

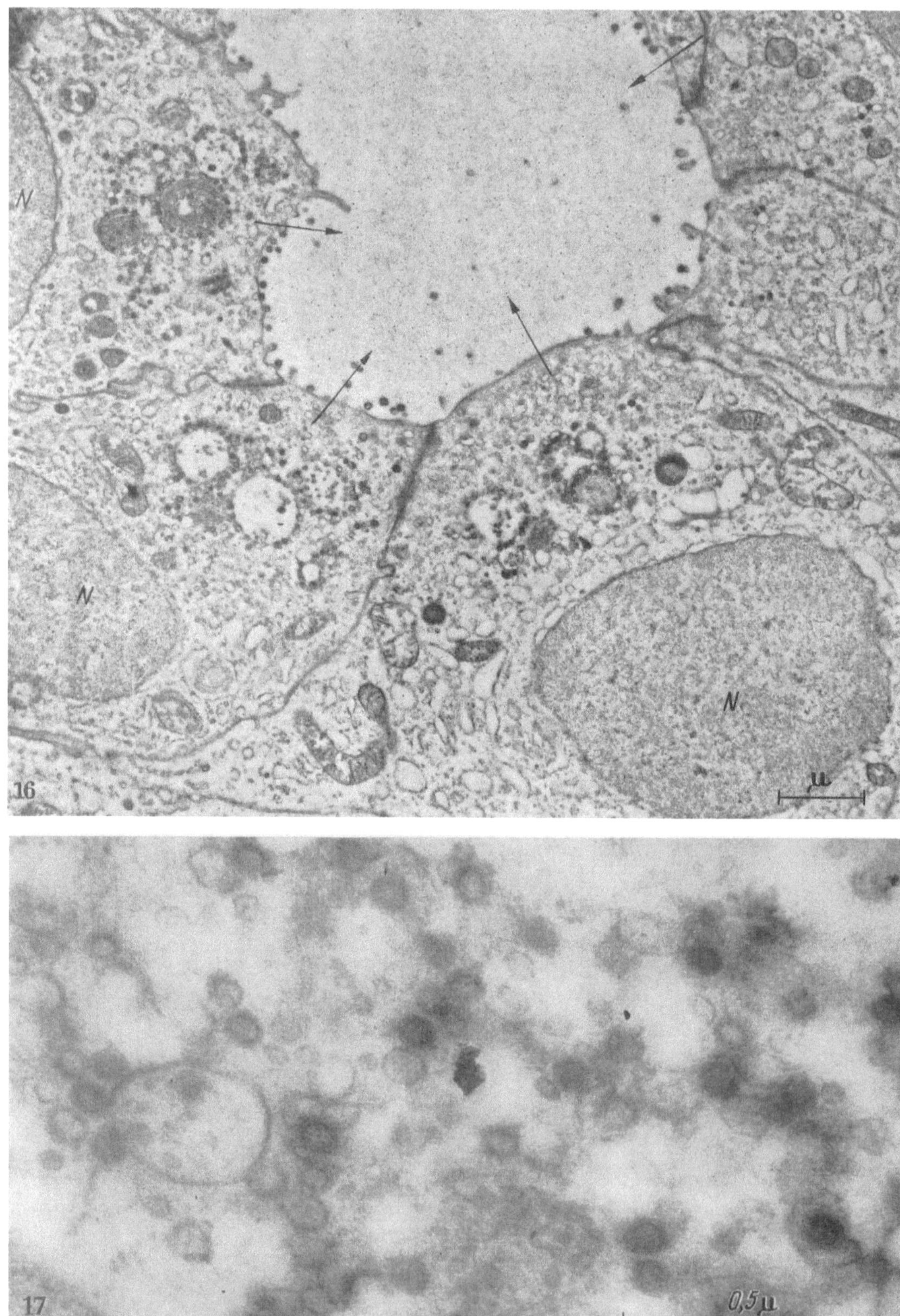

Fig. 16. *Tumeur mammaire de la souris*. Groupe de cellules tumorales ayant conservé leur polarité et formant un pseudo-canalicule. Les corps d'inclusion se trouvent à l'endroit normalement occupé par l'appareil de Golgi. Gross. 13.000 ×

Fig. 17. Zone golgienne d'une cellule tumorale infestée par des particules virus. Gross. 81.600 ×

grenouille (tumeur de Lucké) a également été découvert. Enfin, des images révélant des particules d'aspect viral dans des cellules provenant de différentes leucémies de la souris, suggèrent l'existence, chez cet animal, d'une vaste famille de particules morphologiquement semblables, mais ayant des propriétés biologiques distinctes. Il existe des cycles évolutifs complexes pour certains de ces virus. La présence chez certains animaux normaux des mêmes particules pose le difficile problème de leur signification pathologique.

Ce travail a été publié, in extenso, dans Cancer Res. 18, 491—509 (1958) (W. Bernhard, Electron Microscopy of Tumor Cells and Tumors Viruses). On y trouvera, également, la bibliographie.

Über die cytologischen Veränderungen von ECHO-Virus Typ-9 infizierten Affennieren-Gewebekulturen *

V. Dostal und R. Mauler**

Hygiene-Institut der Albert Ludwigs-Universität Freiburg i. Br. und Virus-Abteilung der Behringwerke Marburg a. d. Lahn

Wir haben bereits vor längerer Zeit zusammen mit M. Aust (1) begonnen, den Virus-Multiplikationszyklus des ECHO-Virus Typ-9 in der infizierten Affennieren-Gewebekultur elektronenoptisch zu verfolgen. Über die ersten Untersuchungsergebnisse haben wir vor einem Jahr auf der 7. Tagung der Deutschen Gesellschaft für Elektronenmikroskopie in Darmstadt (2) berichtet. Bei diesen Untersuchungen war uns aufgefallen, daß man nach der Infektion der Zellen mit dem ECHO-Virus Typ-9 im Zellgewebe kompakte, ovale Bezirke vorfindet, die von einer „Membran" umgeben sind. Diese Stellen zeigten disseminiert eine Vacuolisierung. In fortgeschrittenen Stadien war nur noch ein Reststroma zu finden, das keine weitere Differenzierung erlaubte. Die fortgeführten Arbeiten haben eindeutig gezeigt, daß im Cytoplasma der Zellen Veränderungen zu finden sind, die auf die Multiplikation des ECHO-Virus Typ-9 schließen lassen.

Material und Methoden. Für die Anlage der Gewebekulturen verwendeten wir Nieren von Rhesus-Affen, die nach der Methode von Youngner (3) mit Trypsin aufgeschlossen wurden. Die erhaltene Zellsuspension wurde unter möglichst optimalen Bedingungen (4,5) in Kulturgefäßen (Fernbachkolben) zum Auswachsen gebracht und anschließend infiziert. Wir verwendeten einen nicht für Baby-Mäuse pathogenen ECHO-Virus Typ-9 Stamm[1]. Das ECHO-Virus lag in einer Konzentration von etwa 10^8 ID 50/ml vor. Nach zweistündiger Inkubation im Brutschrank bei 37° C wurde die überstehende flüssige Phase entfernt und der Geweberasen mit einer Salzlösung mehrmals gewaschen. Anschließend wurden die Kulturen mit etwa dem gleichen Volumen eines eiweißfreien Nährmediums beschickt. Während der Multiplikationsphase des Virus entnahmen wir für die Bestimmung des Anstieges der Viruskonzentration Proben. Von gleichen Zeitpunkten wurde jeweils das Gewebe eines Fernbachkolbens nach der von uns angegebenen Methode präpariert (6), d. h. nach der Osmiumfixierung wurde der Zelldetritus durch Waschen aus dem Kolben entfernt und danach das Gewebe mit einem Gummischaber vom Glasboden des Kulturgefäßes abgekratzt. Das erhaltene Sediment ist nach den üblichen Methoden (7) weiterbehandelt worden.

Ergebnisse. Im Cytoplasma der Zelle kommt es an einer Stelle zu einer Verdichtung der Grundsubstanz (Abb. 1) und zur Bildung von kleinen, ovalen bis runden Vacuolen. Die Mitochondrien werden strukturlos. Ferner sieht man osmiophile Rundkörper (Lipoidkörper?). Es können verschiedene Grade der vacuoligen Degeneration festgestellt werden. Zu einem späteren Zeitpunkt zeigen die beobachteten infizierten Gewebekulturzellen lokal begrenzte Bezirke, die noch deutlicher für den Multiplikationsort sprechen. Es fallen relativ viele Mitochondrien auf, die in dem veränderten Zellbereich vorgefunden werden. Teilweise ist die Innenstruktur der Mitochondrien noch gut erkennbar (Abb. 2); daneben findet man Mitochondrien, die homogen erscheinen. Diese

* Die Untersuchungen im Hygiene-Institut Freiburg i. Br. wurden mit Unterstützung der Deutschen Forschungsgemeinschaft durchgeführt.
** Wir danken Fräulein M. Hahn aus den Behringwerken Marburg a. d. Lahn für die Mithilfe bei der Präparation und den elektronenoptischen Untersuchungen.
[1] Der ECHO-Virus Typ-9 Stamm wurde uns freundlicherweise von Herrn Dr. U. Krech vom Schweizerischen Serum- und Impfinstitut, Bern, zur Verfügung gestellt.

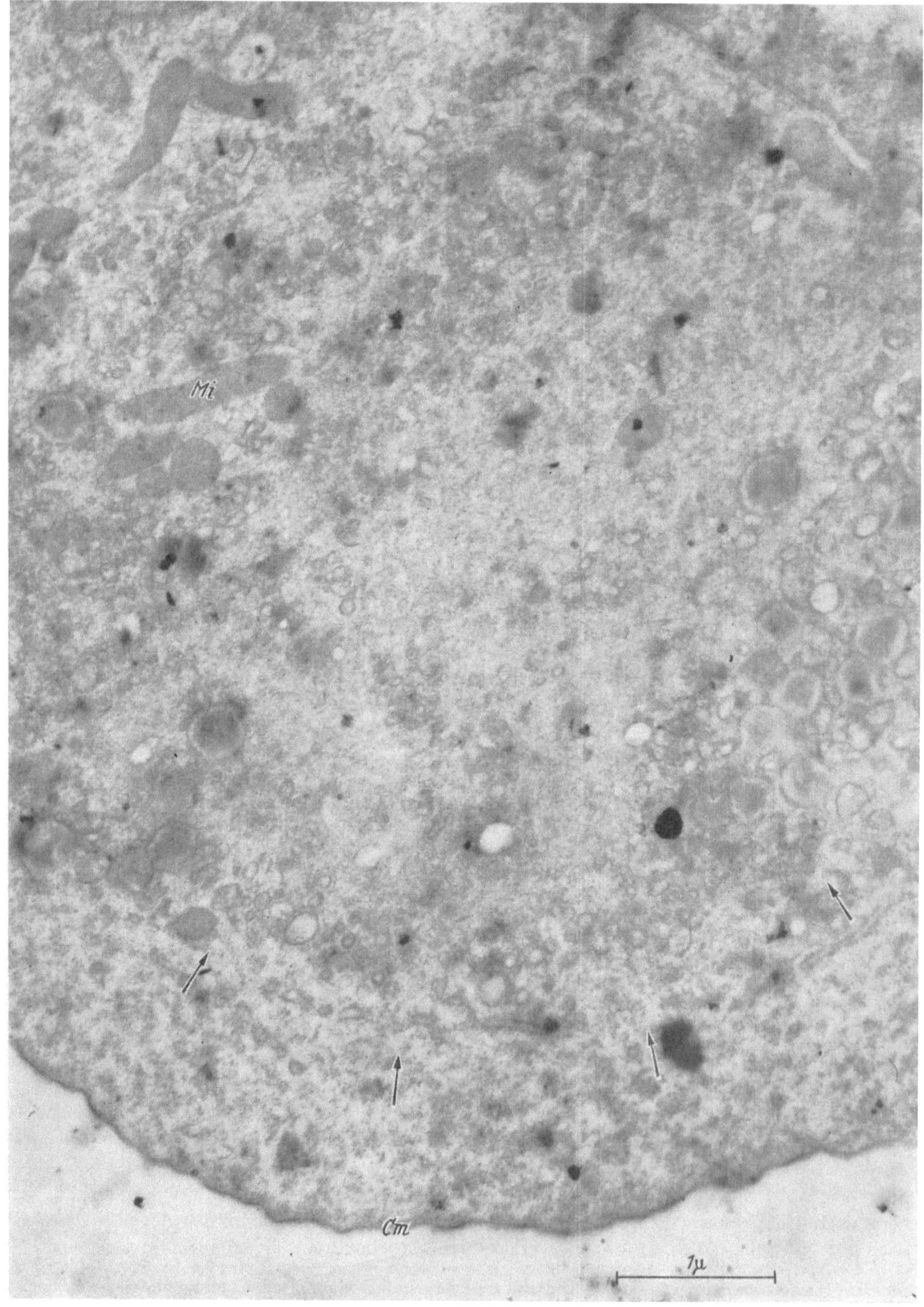

Abb. 1

Abb. 1—3. *C* Cytoplasma, *Cm* Zellmembran, *Mi* Mitochondrium, *Bl* Blase mit Viruspartikeln. Elektronen-optische Vergrößerung bei allen Abb. 7200 mal, 60 kV, Aperturblende 30 μ

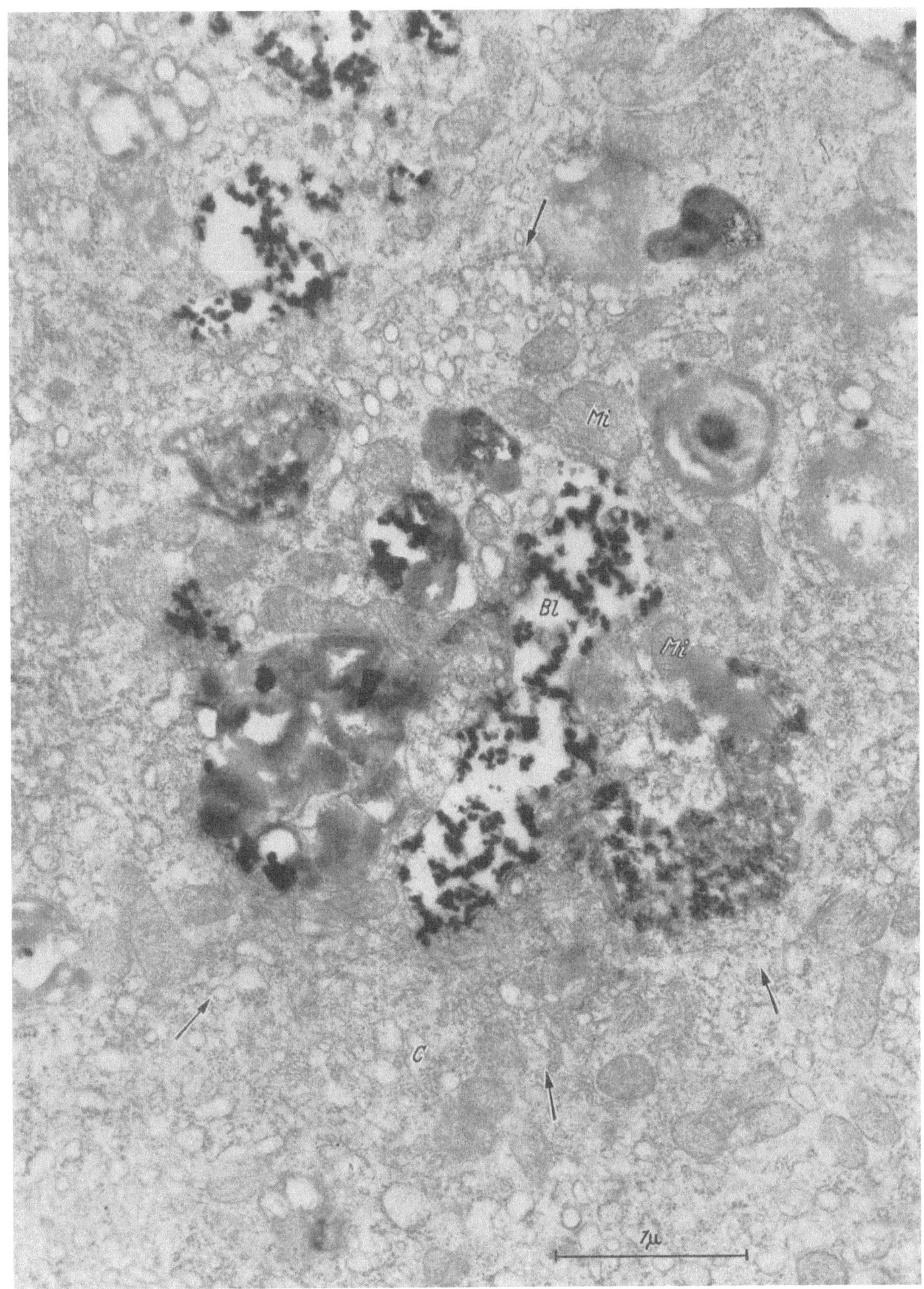

Abb. 2

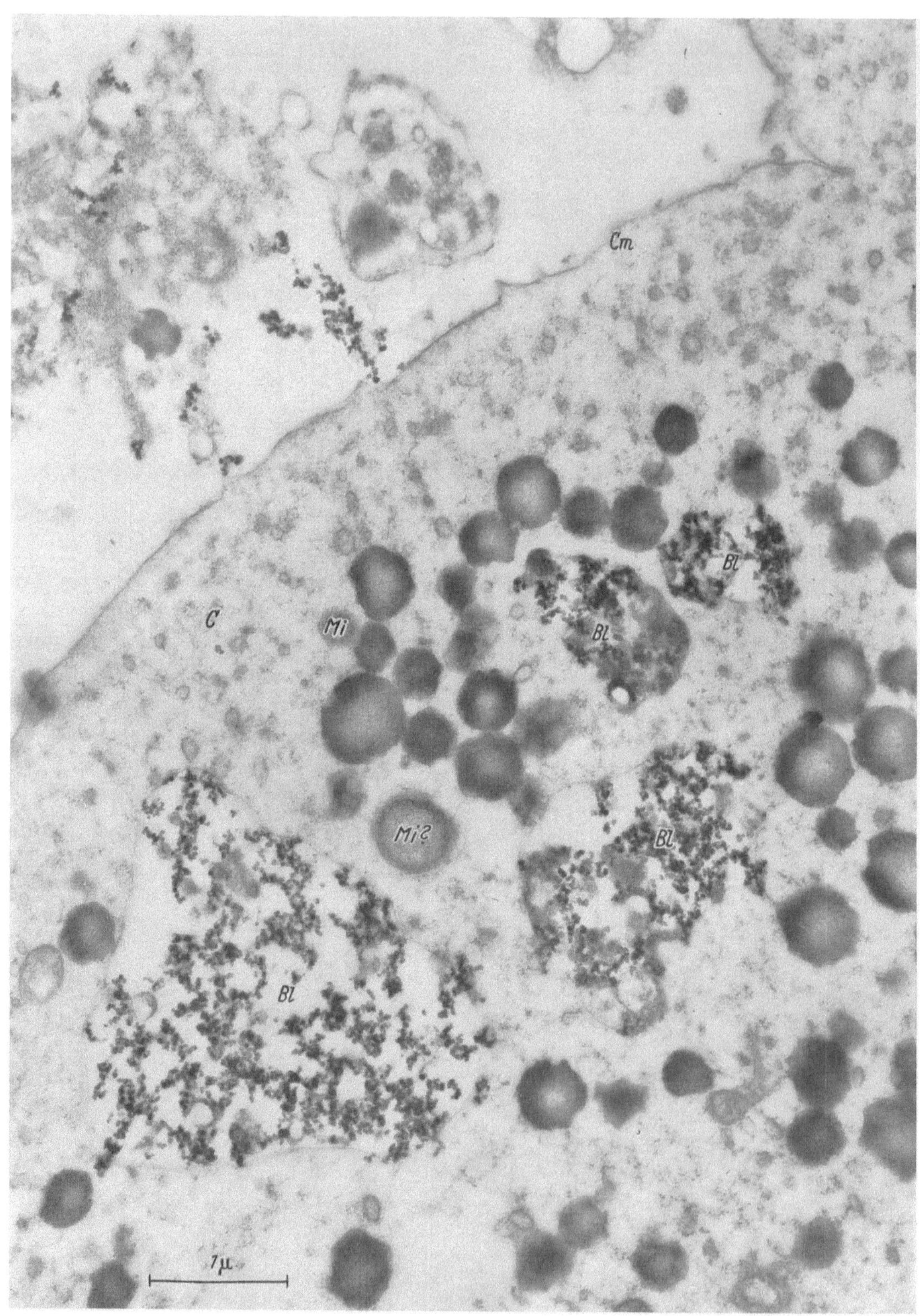

Abb. 3

Teile stehen in Verbindung mit einer stark osmiophil strukturierten Partie. Es entstehen blasige Gebilde, die von einer „Membran" umgeben sind (α-Cytomembran?). Etwa 30 Std. nach der Infektion sieht man sehr viele Zellen, die im Plasma die oben beschriebenen blasigen Veränderungen (Abb. 3) zeigen. In den Blasen findet man einzelne und auch Haufen von etwa 30 mμ großen, schwächer und stärker osmiophilen Partikeln. Sie werden als unreife und reife ECHO-Virus Typ-9 Partikel angesehen. In diesen Gebilden sind außer den Viruspartikeln Reste einer körnigen Grundsubstanz zu finden.

Die ersten morphologischen Veränderungen, die für eine Multiplikation des Virus sprechen, wurden bereits 4 Std. nach der Infektion gefunden. Zu diesem Zeitpunkt war in der Kulturflüssigkeit bereits eine meßbare Viruskonzentration nachweisbar. Diese Beobachtungen sprechen für einen relativ schnellen Ablauf der ECHO-Virus Typ-9 Vermehrung in der Affennierenzelle.

Literatur

1. Aust, M.: Inaug.-Diss. Freiburg 1958 (im Druck).
2. Dostal, V., u. M. Aust: 7. Tagung der Dtsch. Ges. für Elektronenmikroskopie in Darmstadt, 1957.
3. Younger, J. S.: Proc. Soc. exp. Biol. (N. Y.) **85**, 202 (1954).
4. Robbins, F. C., Th. Weller and J. F. Enders: J. Immunol. **69**, 673 (1952).
5. Dostal, V.: Laboratoriumsbl. **7**, 1, 1 (1957).
6. — Proc. Stockholm Conference Electron Microscopy, p. 117 (1957).
7. Newmann, S. B., E. Borysko and M. Swerdlow: Science **110**, 66 (1949).

Identification and subsequent studies of foot-and-mouth disease virus

S. S. Breese jr. and H. L. Bachrach

Plum Island Animal Disease Laboratory, Animal Disease and Parasite Research Division, Agricultural Research Service, US-Department of Agriculture, Greenport, New York

Foot-and-mouth disease virus, type A, strain 119, (FMDV-A 119) was purified from infectious tissue culture fluid, bovine vesicular fluid, and tongue epithelium and identified by analytical electron microscopy as a particle approximately 23 mμ in diameter. In addition, a smaller particle (8 $\pm$ 2 mμ) closely associated with the virus particle, particularly in animal-derived material, was found in centrifugally-separated fractions which possessed complement-fixing activity.

FMDV-A 119 adapted to bovine kidney monolayer cultures was purified and concentrated therefrom by successive application of chemical and physical procedures (1). Virus was precipitated at —3° C from tissue culture fluid with 20% methyl alcohol. After overnight storage the precipitate was collected by low-speed centrifugation and resuspended at pH 7.5 in a lesser volume of isotonic buffer or ammonium acetate. This was extracted with an n-butanol-chloroform mixture yielding a virus-containing aqueous phase. The virus was pelleted from this phase by a 135- to 150-min run at

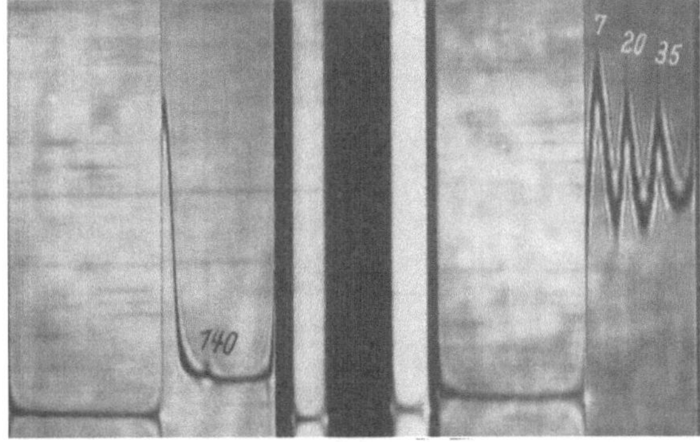

Fig. 1. Moving boundary centrifugation of purified FMDV concentrates prepared from infectious tissue cultures. Analytical centrifuge cell schlieren photograph at left reveals small virus peak with sedimentation rate, $s_{20,w}$ of 140 S with large peak of impurities still at meniscus, which in later photograph on the right has been resolved into 7, 20 and 35 S components

78400 g and resuspended in a small volume of buffer for nitrogen determinations or in ammonium acetate for electron microscopy. Further purification was achieved by moving-zone centrifugation in a D_2O gradient (1.01 to 1.07 gr/ml) contained in a Spinco SW 39 swinging-bucket rotor. Degrees of purification for fractions derived from 78400 g runs (purified virus concentrates) were established by infectivity and protein-nitrogen assays. The zone centrifugation was designed to separate the virus (140 S component) from slower moving components (7 S, 20 S and 35 S) which were shown by analytical centrifugation (Fig. 1) to represent more than 95% of the protein of purified virus concentrates. This relationship of virus to impurities was confirmed independently by comparing protein-nitrogen assay values with those calculated from electron microscope data (*1*). Recoveries of infectivity through the purified concentrate stage ranged from 22 to 82% with concomitant 68-fold increase in infectivity to protein-nitrogen ratios.

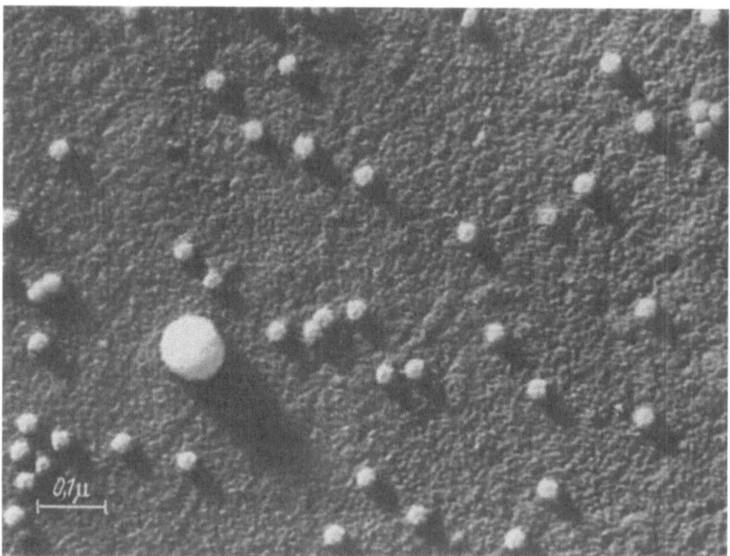

Fig. 2. Section of spray-droplet pattern containing 23 ± 2 mμ foot-and-mouth disease virus particles and one 88 mμ polystyrene latex sphere. 90000 ×, uranium shadowed

Purified virus fractions were appropriately mixed with bovine serum albumin and 88 mμ polystyrene latex spheres and sprayed onto collodion coated grids, shadowed with uranium, and examined in the electron microscope. The spray-droplet patterns contained large numbers of discrete 23 ± 2 mμ particles (Fig. 2). These were identified as FMDV particles on the basis that 1) none were found in control concentrates prepared from fluids and extracts of uninfected tissue cultures; 2) the ratios of 23 mμ particles to plaque-forming infective units averaged 690:1 for 11 virus concentrates, being as low as 33:1 in one instance; 3) the 23 mμ particles, as revealed by electron microscopy, were specifically aggregated by antiviral bovine serum but not by normal serum; and 4) identical particles were found in concentrates derived from infectious bovine vesicular fluids and tongue epithelium extracts.

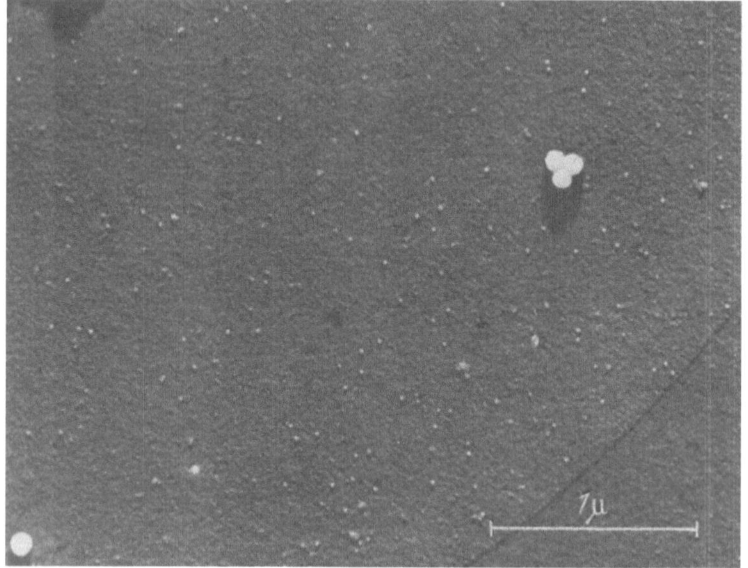

Fig. 3. Section of spray-droplet pattern containing many of the smaller 8 ± 2 mμ non-infective particles associated with foot-and-mouth disease. 28400 ×, uranium shadowed, 88 mμ polystyrene latex spheres

Fig. 2 reveals that the surface of the viral particles has knobs and dimples to varying degress much like turnip yellow mosaic virus (*2*).

The smaller $8 \pm 2\ m\mu$ particles (Fig. 3) predominated in supernatant fluids recovered from infectious bovine vesicular fluids after removal of virus particles by high-speed centrifugation. They were not always observed in the purified virus concentrates prepared from infectious tissue cultures. When bovine vesicular fluid was subjected to zone centrifugation, the small particles were recovered in a zone 1—2 cm above the virus layer. Tentative results indicate small particles may arise by disintegration of virus particles by heat or other means. It may be that the apparent obular structure of the virus is due to an assemblage of small particles.

References

1. Bachrach, H. L., and S. S. Breese jr.: Proc. Soc. exp. Biol. (N. Y.) **97,** 659 (1958).
2. Steere, R. L.: J. biophys. biochem. Cytol. **3,** 45 (1957).

Structure of bacteriophage

S. Brenner

Medical Res. Council, Unit for Molecular Biology, Cavendish Laboratory, Cambridge (England)

Structural components of bacteriophages T 2, T 4 and T 6 have been isolated and purified. The components have been characterized by their physical, chemical and serological properties. Each phage consists of a hexagonal head, which encloses the DNA, and a tail. The tails are constructed of three components: tail fibres, sheaths and cores. The sheaths are capable of "contraction" and contracted sheaths show a helical structure in the electron microscope. The cores are hollow; the hole down the middle is about 30 Å in diameter. A new technique of preparation for electron microscopy, which will be briefly described, permits a study of the fine structure of intact and disintegrated phage and also a study of the assembly of the particle, by interfering with assembly by means of proflavin and mutational changes. Previous work was discussed in the light of the new findings.

An extended version of this article will appear in J. Molecular Biology, 1 (1959).

Polymerity in the structural organization of bacteriophages

A. E. Kriss

Laboratory of Electron Microscopy of the Academy of Sciences of the USSR, Moscow

It has been shown previously (*2*), that in the phage preparations of Bac. mycoides, we succeeded in observing phage particles of spermatozoon-like form the caudal portion of which was clearly seen as a chain of closely adjoined rounded elements. According to the data available in the literature, corpuscles with the clearly exhibited granular or bead-shaped structure of the "tail" are not peculiar for the phage of Bac. mycoides; they were observed in other phages as well, in particular in the phages of the T group of E. coli. Irregularity of these pictures was explained by the author by the absence of an equivalent shadowing effect for phage particles not only in different preparations but in one field of vision as well.

Fig. 1 shows clearly that, depending on the arrangement character of phage particles with respect to the source of shadowing, their caudal portion may look either structureless or like a chain of rounded elements. In phage corpuscles which were arranged with their caudal portion perpendicularly to the stream of metal particles (designated by *A*) the shadowing did not clear up, but disguised the bead-like shape of the tail. This metal excess was absent when the tail of phage corpuscles was turned along the stream of metal particles (designated by *B*). Such an arrangement clearly contributes to a softer shadowing as a result of which the structural peculiarities of the phage tail could be found out.

Under the resolution power at which the work was carried out the rounded elements forming the caudal portion of phage particles were approximately of the same shape and size. It is of

interest that after the exposure to hydrostatic pressure of some thousands of atmospheres, in
phagolysates of Bac. mycoides filaments of different length with clearly exhibited periodicity
along their full length were found (*3*). These linear chains of particles of rounded form are of phage
origin being connected by intermediate forms with typical spermatozoon-like bacteriophage
corpuscles.

It is supposed by the authors that the polymerity of structural organization is typical not
only of the caudal portion of the phage corpuscle but of its head as well. Photograph of granular
structure of the head of a phage corpuscle was presented in a previous paper (*2*). In this respect
the photograph of phage particles after the treatment of phagolysates of Bac. mycoides with 4%

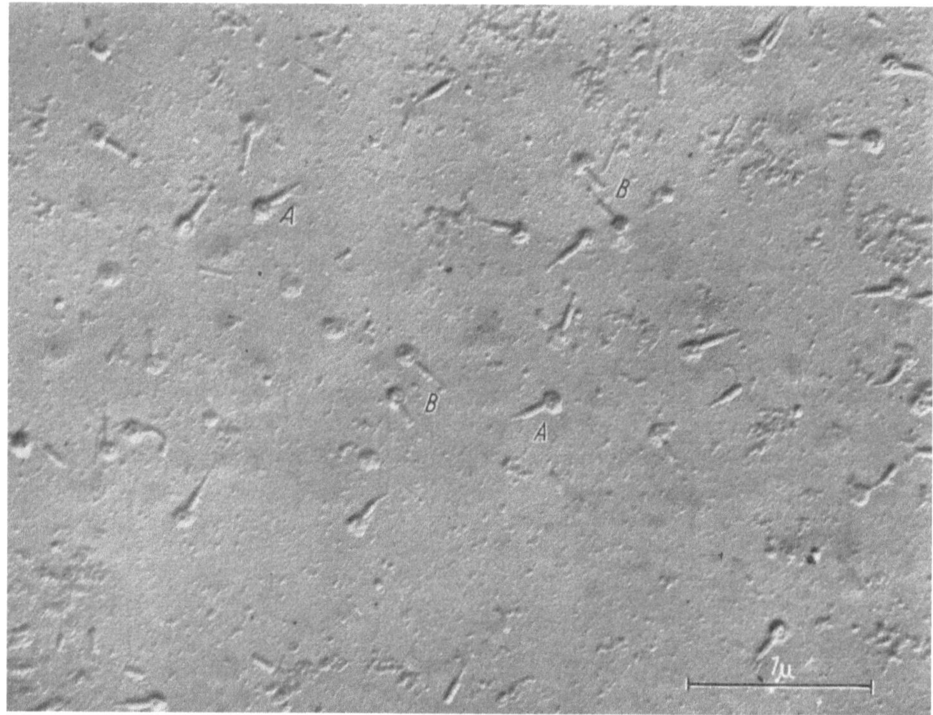

Fig. 1. Structural differences of the tail phage corpuscles of Bac. mycoides, depending on the arrangement
character with respect to the source of shadowing. 25,000 ×

trimethylphenylammonium is of particular interest (Fig. 2). The structure of intact heads,
transitions to a complete decomposition of heads into rounded particles, found in the same field
of vision undoubtedly prove, that the head, as well as the tail, may represent an aggregation
of individual spherical elements. Assuming that the phage head is formed by a spirally rolled up
chain of globular bodies which are similar in their appearance we regard this type of structural
organization as a type of spiral-shaped aggregation of "monomers", in contrast to the linear type
characteristic of some plant viruses. The cases, when rod-like figures of these viruses have a bead-
shaped configuration are not so seldom. However, the pictures making the impression that a phage
corpuscle is formed by a rolled up chain of spherical elements of a similar appearance are not so
often, this circumstance being, apparently, due not only to the difficulties in obtaining optimal
shadowing for the elucidation of fine structural peculiarities of the object or to the insufficient
resolution power of an electron microscope.

When comparing pictures observed on a large number of phage electronmicrographs obtained
by A. S. TIKHONENKO one can not help drawing the conclusion that the "polymorphism" of phage
particles, often observed in one field of vision, may be determined also by the lability of structural
elements forming phage corpuscles. This may be illustrated by some examples pertaining to the
phage of Bac. mycoides.

In the preparations from phagolysates unexposed to any treatment, in preparations purified by the method of "drop dyalisis" (*1*) by means of which the phage particles are less affected compared with other methods — a significant diversity of head structure is observed. Compact homogenous heads may be seen next to those of a clearly seen ring-shaped structure. When judging by their "shade", both are significantly raised above the film. The central portion of ring-shaped heads may be raised or look as deeply sagged. In the last case the head seems to

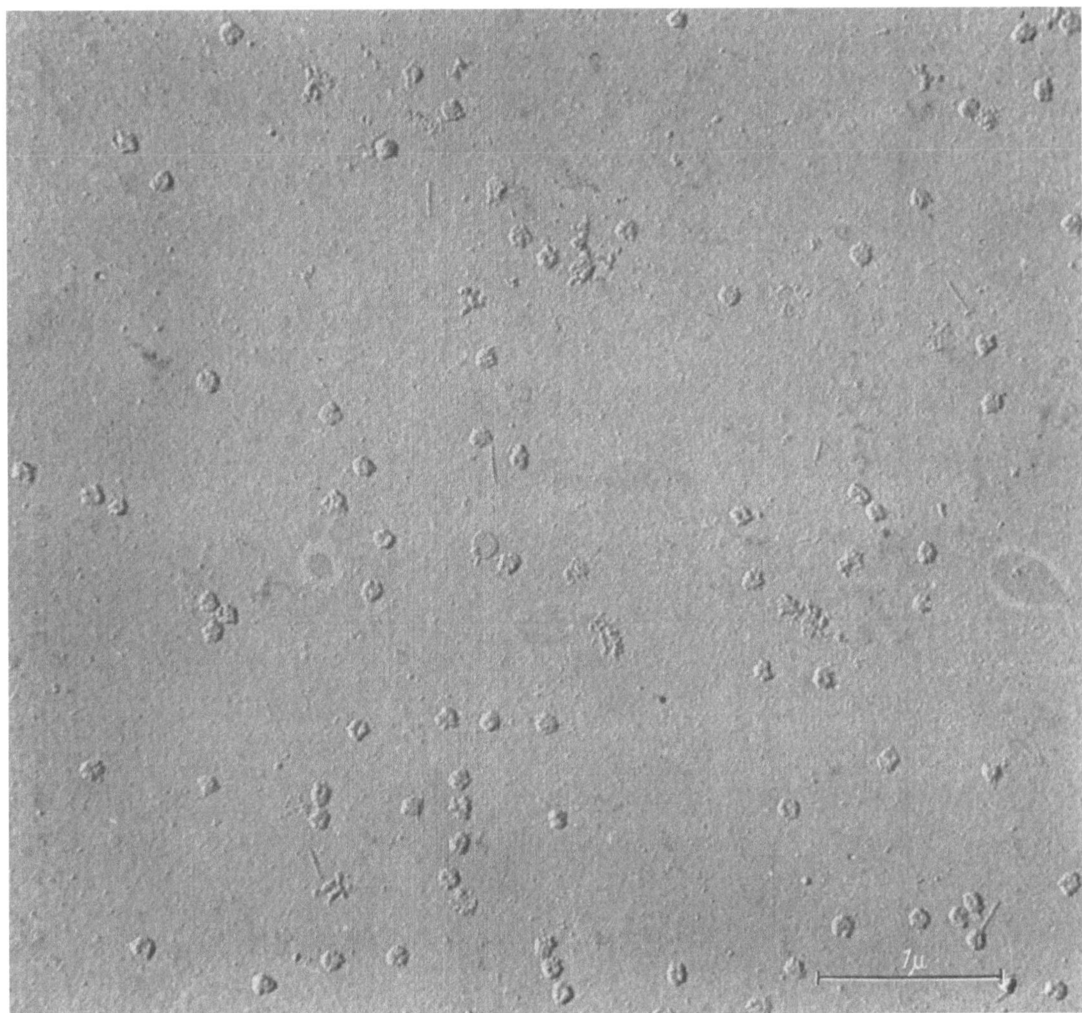

Fig. 2. Granular structure of phage heads and their granular decomposition after treatment of Bac. mycoides phagolysate with trimethylphenylammonium. 25,000 ×

have a stiff frame on the periphery which does not let it fall down completely. Fibrillar structure of the head may be not only of a ring-shape, but of that of "eight" or of other figures.

The next photograph (Fig. 3) shows the tubercular type of heads lacking a stiff frame. This relief may be retained in the case of flattening of the heads, though in a significantly smoothed form. The process of heads flattening goes on, apparently, with the density change of head forming substances. Fibrillar and tubercular structures are replaced by the gel-like ones (Fig. 4). This photograph shows the flattening of the head to be accompanied by the increase of its dimensions and by the loss of its contours, as if it runs on the films, and in this state even the fusion of gel-like masses into which the heads are transformed, is possible.

The pictures presented do not prove the widespread idea that the flattening of heads is a result of their drying out or emptying. In view of the phenomena of running and loosing contours

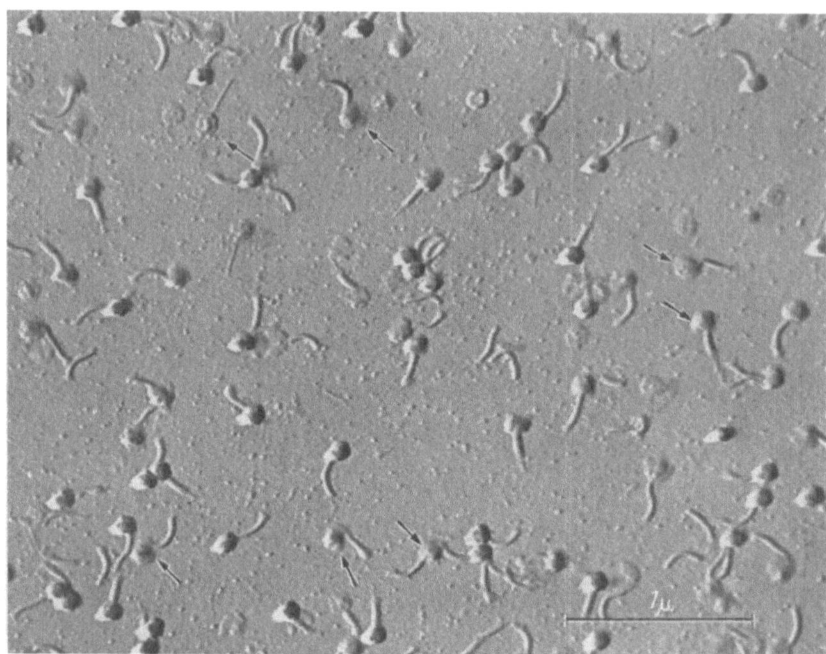

Fig. 3. Diversity of structures of phage head. Tubercular head type of Bac. mycoides phage is shown by arrows.
25,000 ×

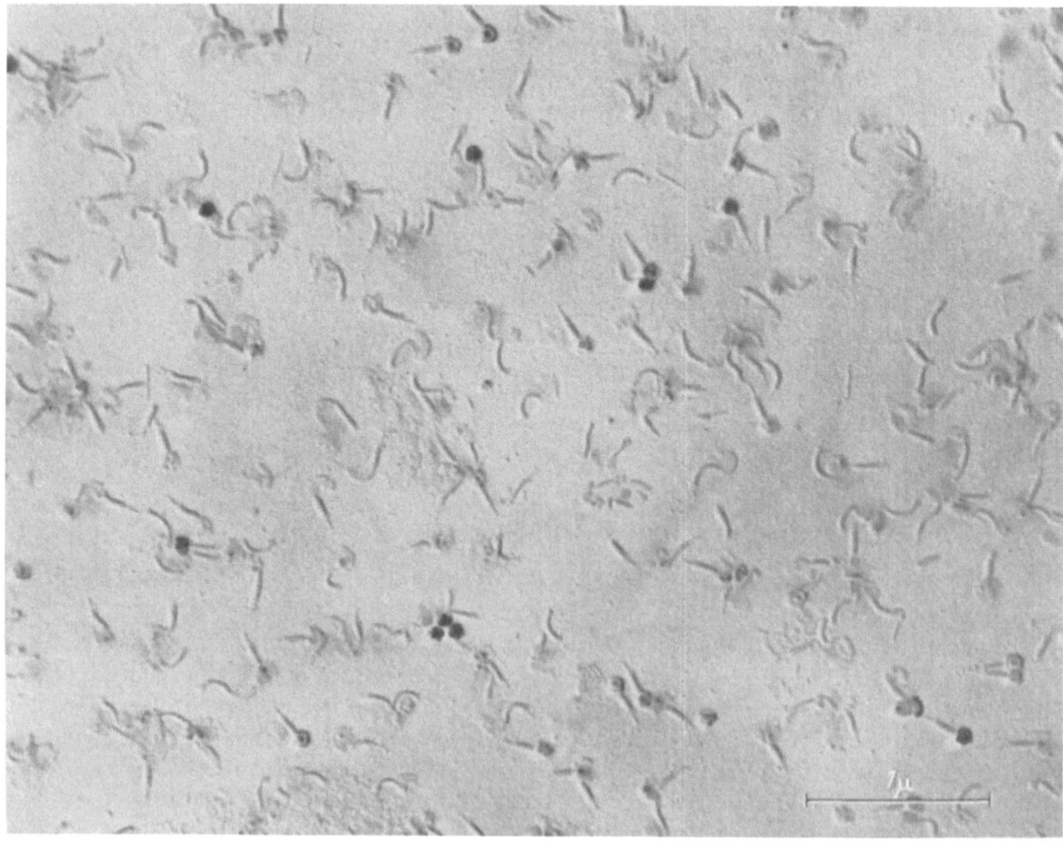

Fig. 4. Diversity of structure of phage heads. Pictures of running and loss of contours of the heads of Bac.
mycoides phage corpuscles. 25,000 ×

of heads of phage particles one is not sure that the phage head possesses a membrane. Attention is drawn to the circumstances that the number of phage corpuscles in which the head is not dense and convex but looks as if deformed and denatured is a very large one in phage preparations which were not exposed to any treatment. It is very likely that this phenomenon is due to the effect of drop dyalisis, since analogous pictures of structural changes of the head are observed under the effect of detergents and other denaturing agents.

Up to the present we can say nothing about the character of physico-chemical transformations taking place in the spermatozoon-like particle of the phage which determine the transition from its polymerous type of structure to those extreme steps when the density of structure forming the phage head is almost completely lost, and the latter gets an amorphous, running appearance. This is the task of our further investigations.

References

1. Kriss, A. E., V. I. Biryuzova and A. M. Zolkover: Microbiologia **17**, 484 (1948).
2. — and A. S. Tikhonenko: Dokl. Akad. Nauk SSSR **86**, 421 (1952).
3. — — Dokl. Akad. Nauk SSSR **93**, 353 (1953).

A negative staining technique for high resolution of viruses

R. W. Horne and S. Brenner

Cavendish Laboratory, University of Cambridge (England)

The aim of this study is to discuss the possibility of using negative staining techniques on unsectioned material. During investigations concerned with the detailed structure of the protein

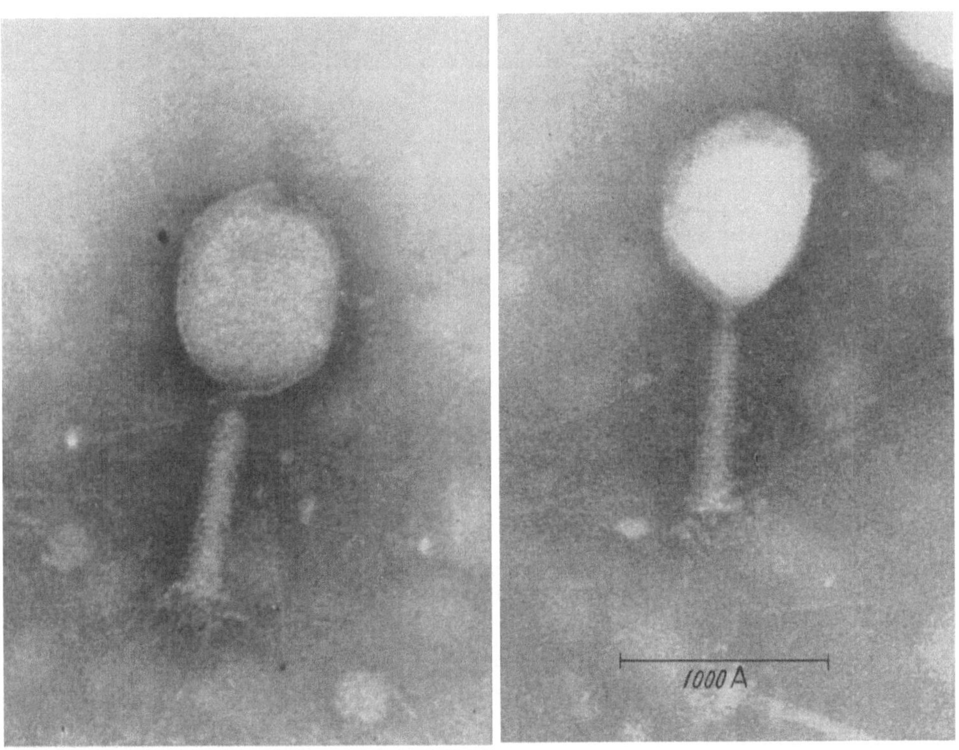

Fig. 1. Intact T2 phage particles in PTA droplet. Striations along the tail are clearly visible. No staining of the protein is evident

components of bacteriophage, serious limitations were encountered using existing shadowcasting and sectioning techniques.

At final magnification approaching 500,000 x—700,000 x much of the protein sub-unit detail is obscured by background granulation, even if evaporation is carried out under carefully controlled conditions (*1*). With sectioning methods the random orientation of the phages together with poor staining made interpretation difficult, furthermore the problem of the fixation of phage, dehydration embedding and cutting is a relatively long process. Hall (*2*) pointed out that virus structure could be resolved by surrounding the preparation with a dense material of suitable scattering power. Tobacco mosaic virus rods were examined by Huxley (*3*) after staining with phosphotungstic acid also potassium chloride. The rods were observed to have the hollow centre predicted by X-ray studies.

A negative staining technique has been developed primarily for the study of bacteriophages. In view of the large number of preparations to be examined after various stages of treatment, the technique must satisfy two requirements:

a) simple preparation involving the minimum amount of time and apparatus.

b) good preservation combined with high contrast.

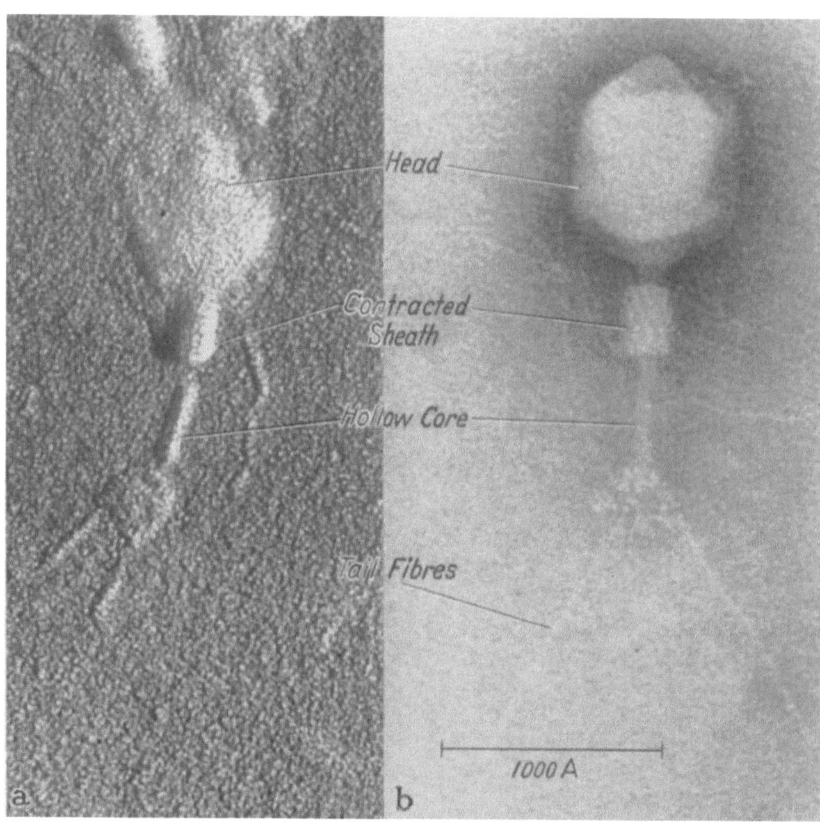

Fig. 2a and b. a) "Triggered"T 2 phage particle shadowed after air drying. b) The same preparation after spraying in phosphotungstate showing preservation of the head and tail components

A 1% solution of phosphotungstic acid (PTA) is made up and adjusted to p_H 7.2 using n KOH. 1 ml of virus suspension in water or 1% ammonium acetate is added to 0.5 ml of PTA. The final solution was poured into an atomizer spray gun and droplets sprayed directly on to prepared specimen grids. Carbon substrates prepared in evaporators using oil diffusion pumps appeared to produce a greater background structure compared with films obtained from well trapped Hg units. Specimens were examined in the Siemens Elmiskop I using double condenser illumination with an illuminating area approximately 15 μ in diameter. The electron micrographs were taken at instrumental magnifications of 40,000 x—80,000 x at 80 kV.

The detail visible with the negative PTA preparation is shown in Fig. 1 suggesting a regular periodicity along the phage tail of approximately 35—40 Å. It will be shown that the tail periodicity undergoes considerable change during the various stages of contraction and will be discussed in detail by BRENNER (4).

Shadowed preparations of phage after treatment with $Cd(CN)_3$ (Fig. 2a) show the contraction of the sheath component, but considerable distortion has occurred on drying down etc. The same preparation is shown in Fig. 2b after mixing and spraying with PTA. It is possible to resolve

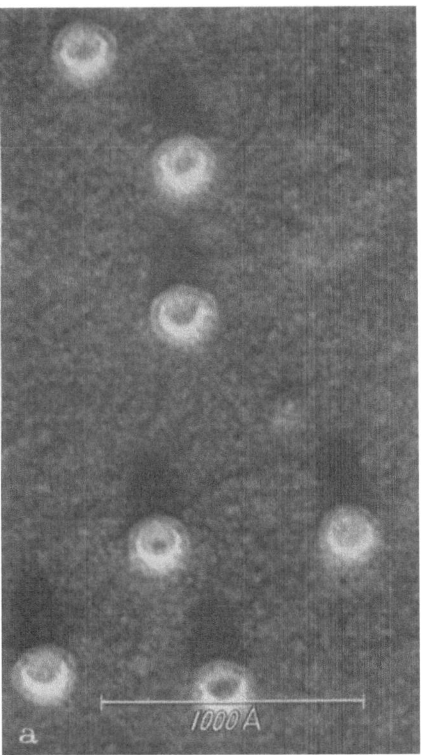
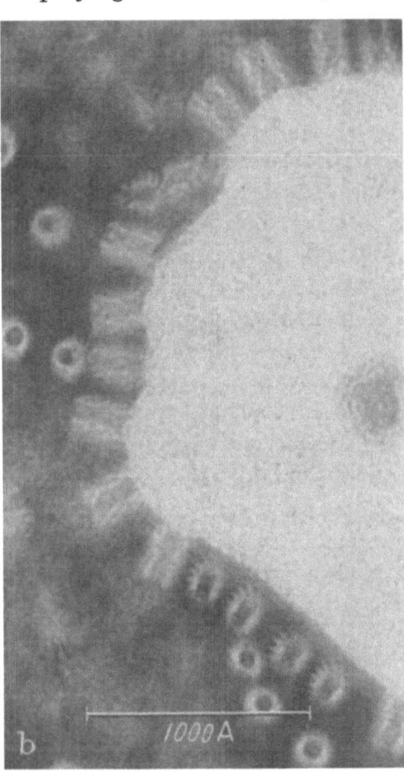

Fig. 3a and b. a) Purified sheath components sprayed and shadowed. b) The same preparation in phosphotungstate. Evidence of the sub-unit structure is visible in a number of sheaths viewed along the axis

far more detail at higher magnification using this method. Good preservation with little distortion is suggested by the angular structure of the head. Tail fibres have remained attached to the core component. Microelectron diffraction pictures taken from droplet patterns show that the areas are amorphous. Much of the background structure is due to phase contrast effects, carbon substrates and carbon deposits during examination.

Examination of isolated head and tail components using shadow casting methods revealed their presence, but added little to the sub-unit structure. Components photographed at the edge of a PTA droplet show tail components and head membranes from a similar preparation in much greater detail.

Isolated sheath components were examined by using both shadowing (Fig. 3a) and PTA methods (Fig. 3b). The "cog-wheel" sub-unit arrangement suggested in Fig. 3a is clearly visible in Fig. 3b.

Suspensions of tobacco mosaic virus, turnip yellow mosaic virus and turnip crinkle virus have been examined using the same negative staining technique. The hollow centre in TMV shown by HUXLEY has been confirmed.

References

1. COSSLETT, V. E., and R. W. HORNE: Vacuum 5, 109 (1955).
2. HALL, C. E.: J. biophys. biochem. Cytol. 1, 1 (1955).
3. HUXLEY, H. E.: Electron Microscopy Proc. Stockholm Conference, September 1956, p. 260.
4. BRENNER, S.: This volume p. 621.

Considérations quantitatives sur des coupes ultraminces de bactéries infectées par du bactériophage

Janine Séchaud, Antoinette Ryter et Edouard Kellenberger

Laboratoire de Biophysique, Université de Genève

Au cours de recherches sur le développement intracellulaire du bactériophage T2 de l'Escherichia coli B, nous avons été amenés au problème de l'utilisation quantitative des coupes ultraminces. Une relation numérique a alors été établie entre le nombre moyen de phages apparaissant sur une bactérie coupée et le nombre moyen de phages contenus par bactérie entière.

Pour ce faire, l'épaisseur des coupes a été déterminée par un calcul statistique faisant entrer en jeu les nombres relatifs de phages apparaissant sur 2 ou sur 3 coupes consécutives dans une série, et les dimensions de ces phages. Cette épaisseur est de 400 à 500 Å et semble varier peu d'une bonne coupe à l'autre, malgré le choix subjectif.

La détermination du facteur numérique a été ensuite faite par 2 voies différentes. D'une part nous avons procédé à un étalonnage expérimental en comparant le nombre de phages sur les coupes avec le nombre de phages dans les bactéries entières mesuré au microscope électronique dans le lysat de ces bactéries. D'autre part nous avons calculé ce facteur à partir de la relation entre le volume moyen d'une coupe de bactérie et le volume moyen d'une bactérie entière. Les 2 résultats concordent à environ 10% près. Les erreurs statistiques de la méthode sont d'environ 15%, tandis que les erreurs systématiques sont plus élevées. Une appréciation prudente de ces erreurs permet cependant d'affirmer que les résultats sont valables à l'intérieur d'un facteur de 2.

L'application de cette méthode nous a permis, à certains stades de développement du bactériophage, de constater la présence dans les bactéries de particules immatures ressemblant aux phages quant à la concentration d'ADN et à la forme, mais qui ne comportent pas encore d'enveloppe protéique visible dans le lysat.

Les détails de la détermination quantitative sont exposés ailleurs (2) ainsi que les résultats de l'application de cette méthode à l'étude de la multiplication des bactériophages (1).

Bibliographie

1. Kellenberger, E., J. Séchaud et A. Ryter: Virology 8, 478 (1959).
2. Séchaud, J., A. Ryter et E. Kellenberger: J. biophys. biochem. Cytol. 5, 469 (1959).

Plant viruses: Quantitative assay methods and fine structure of the characteristic particles*

Russell L. Steere **

Virus-Laboratory, University of California, Berkeley

In 1956, Gierer and Schramm (1) and Fraenkel-Conrat (2) independantly and by different procedures isolated from purified tobacco mosaic virus a ribonucleic acid fraction capable of initiating the characteristic disease in susceptible hosts. This was a fundamental advancement in our knowledge of viruses, and has resulted in a greatly increased interest in the fine structure of these pathogens.

It is my intention to review here a few facts about several of the plant viruses and to include a few new observations from our recent studies. In order to prevent it from becoming

---

* Aided by grant number C-2245 (C4). National Cancer Institutes, National Institute of Health, US Public Health Service.

** New address: Plant Virology Laboratory, U.S. Dept. of Agriculture ,Crops Research Division, Beltsville, Md.

too lengthy, this discussion will be limited intentionally to four plant viruses: tobacco mosaic virus (TMV), tobacco ringspot virus (RSV), turnip yellow mosaic virus (TYMV) and tomato bushy stunt virus (BSV). Throughout the discussion, the term "virus particles" will be used to denote those characteristic particles, which can be isolated from infected tissues, and with which the ability to initiate new infections has been associated.

In order to avoid misunderstanding or confusion, I shall review briefly the procedure for assay of these mechanically transmitted plant viruses. In 1929, HOLMES (3) discovered that localized chlorotic or necrotic infections (local lesions) develop on the inoculated leaves of certain species of tobacco a few days after they are inoculated with a suspension of TMV. Other reports have appeared, describing the use of additional local lesion hosts for TMV, and of local lesion hosts for many other viruses including each of those to be discussed here. One of the most important things to keep in mind with respect to local lesion studies with plant viruses is that inoculation is accomplished by rubbing the inoculum over the surface of the test leaves of a suitable host. Excess inoculum is generally washed from the surface of the leaves, and the local lesions, which develop in a few days, are counted. By use of such a procedure, most of the particles in an inoculum fail to enter cells and are, therefore, unable to initiate an infection. Nevertheless, one can conduct very satisfactory assays by comparing test suspensions with standard suspensions. This is accomplished by applying test inoculum and standard to opposite halves of the same leaves or, in some instances, to opposite leaves. Although KUNKEL (4) has demonstrated that a local lesion is often initiated by a single infectious unit, there is good reason to believe that very few of the infectious units present in an inoculum are capable of initiating local lesions, because of the inefficient testing procedure. It is fortunate indeed that the inaccuracies in the procedure do not prevent reliable assays altogether. It has been shown that one normally gets no more than one lesion for 50,000 characteristic particles (5). However, this is not presumed to be due to a high percentage of inactive particles in a purified suspension, but to the unsatisfactory assay methods.

Another method for the biological assay of plant viruses involves the use of hosts which develop systemic infections. The inoculum is rubbed on the leaf surfaces as in local lesion studies. One then observes the number of plants which become systemically infected. The sensitivity of the systemic hosts is often greater than that of the local lesion host. For assays, however one must use dilution end point techniques, in which adsorption of virus to the walls of containers may result in inaccuracies. Also, accidental contamination may be very difficult to avoid, particularly in studies with TMV. SCHRAMM and ENGLER (6) reported that infections have been obtained with as few as ten characteristic particles of TMV in the inoculum. One is hesitant to accept such a single isolated report because of the likelihood of errors due to contamination. Furthermore, the report shows that only 16% of the plants inoculated with a concentration of 10^{-10} g/ml became infected, whereas at greater dilutions to as low a concentration as 10^{-16} g/ml higher percentages of the inoculated plants became infected. This certainly suggests that inaccuracies were present in the procedure used by these workers.

If one is to relate the number of infectious units to the number of characteristic particles present in an inoculum, he must have a procedure, whereby the total number of characteristic particles can be determined. In plant virus studies, two primary counting methods have been employed: The spray-drop method of BACKUS and WILLIAMS (7) and the monofilm procedure of HARTMAN (8). The sedimentation procedure of SHARP (9) and the procedure of ARBER and KELLENBERGER (10) should prove applicable also. Both of the methods customarily employed appear to give reliable and reproducible results (7, 10, 11, 12).

Since we are unable to approach a $^1/_1$ ratio between the number of characteristic particles and the number of infectious units, we must rely on the homogeneity of our purified virus suspensions, and our inability to separate infectivity from these characteristic particles (except by degrading them to get infectious nucleic acid) as our means of identifying the characteristic particles as infectious units. It is with this background of the difficulty of relating infectivity directly with each characteristic particle that we turn now to a discussion of the structure of the particles of four plant viruses.

Tobacco mosaic virus

Because of its shape, stability and ease of purification, and because of the great amount of effort, which has been expended in the study of its properties, more is known about the fine structure of tobacco mosaic than about any other plant virus. Our current knowledge of the fine structure of TMV comes from four main sources of information: 1. Biochemical studies, 2. X-ray diffraction studies, 3. ultracentrifugal studies, and 4. electron microscopy.

The alkaline degradation of TMV to obtain a soluble protein fraction, the "A" protein, (14) which was capable of reaggregation upon lowering of the p_H into small platelets or into rods (15) was one of the first steps to reveal the fine structure. A similar protein, the "X" protein (16) has been isolated from infected plants. The rods, which from in vitro from either of these proteins are similar to those of TMV except in average length, but they lack nucleic acid, and fail to initiate infection when inoculated to host plants. As already mentioned, the isolation of an active ribonucleic acid fraction from purified TMV suspensions was reported in 1956 (1, 2). The observation that ribonuclease was capable of destroying the activity of partially degraded TMV (17) or infectious nucleic acid (37) was further evidence for the assumption that the nucleic acid is indeed that portion of the virus particle which is responsible for its ability to initiate infection. From the evidence which is now available, one can assume, with a fair degree of certainty, that the infective TMV particle is rod-shaped with a diameter of 150 Å, and a length of approximately 3000 Å. One can further assume with good reason that this rod is a hollow cylinder with a single thread of ribonucleic acid coiled in helical fashion within the protein cylinder with its phosphate-sugar backbone at a distance of 40 Å from the particle axis. There is no question about the rod-like nature of the particle, and its diameter has been accepted as close to 150 Å, ever since Bernal and Fankuchen (18) reported this value for the inter-particle spacing in dried oriented gels. Kahler and Lloyd (19) obtained this same value by electron microscopy. Recently, however, X-ray measurements (20) have been obtained, which suggest that the individual rods have a diameter of at least 180 Å. As far as I know, however, this diameter has not been confirmed by electron microscopy.

The length of the TMV rod has been a matter of considerable discussion, but the bulk of evidence to date suggests that the infective unit is approximately 3,000 Å long. Whereas some preparations have fairly broad length distributions, there is generally a pronounced peak near 3,000 Å. Two published reports (21, 22) have appeared, in which TMV prepared by two quite different procedures was found to have highly uniform particle lengths of approximately 3,000 Å. It is possible that both preparation methods were selective for the rods of length 3,000 Å, but this is not very likely. Furthermore, it has been shown (23) that the TMV rods, as packed in the crystalline inclusions, have a length of 3,000 Å, and that the infectivity of TMV in such crystalline inclusions is of the same order as that of clarified infectious juice containing an equal quantity of characteristic particles (24). Visual evidence, in the form of electron micrographs for the presence of a hollow core in TMV was obtained by Huxley (25) who treated a virus suspension with a solution containing phosphotungstic ions of high electron scattering power. Mild washing faieled to remove the heavy metal from the core, and permitted him to obtain electron micrographs, which revealed a dark central line. Evidence of this hollow core was also obtained by studies of X-ray fiber diagrams (26) and its radius was reported to be 20 Å. Evidence that the ribonucleic acid of TMV exists as a single unit is to be found in the work of Gierer (27). Hart (28) presented additional evidence and suggests that it exists as a single strand. This latter work was based upon electron microscopy of partially degraded TMV, which was dried from an ammonium acetate suspension. From his observations and calculations, Hart (28) suggests that the single strand forms a helix within the virus particle, which has the same pitch as the protein shell, and which is bound to the inside of each protein subunit. He assumes the 40 Å radius from the axis of the rod reported by Franklin and Holme (26) as the position occupied by the ribonucleic acid strand.

As far as surface structure is concerned, conclusions have been drawn from X-ray analysis (29) which indicate that there is a helical structure with external grooves every 23 Å. Reports of periodic surface structure, as revealed by electron microscopy, have also appeared (*30, 31*) but it is questionable whether they show real periodicity or not. One always finds a pebbly surface structure of TMV particles unless a very carefully purified preparation is used. Most likely, this pebbliness is due to the presence of small contaminating protein particles of uniform size. Shadowing of such preparations gives one the impression that there is a detectable periodicity, but careful observation shows that the periodicity depends upon the shadow direction. When clean TMV preparations are used and carefully shadowed to avoid granularity in the shadowing film, we fail to see what could be interpreted as true periodicity. Furthermore, we fail to detect even the point of contact of two particles when they combine to form one double length particle. In cross section, the TMV rod may be hexagonal as suggested by WILLIAMS (*32*). The small size, however, makes it difficult to be certain of its cross sectional contour by electron microscopy and X-ray evidence does not suggest such a structure. Further improvements in technique may clarify this question.

Comparison between tobacco ringspot virus (RSV) and turnip yellow moasic virus (TYMV)

Two plant viruses, which have been obtained in a relatively pure suspension, and which are almost identical in size, but possess quite different properties, are tobacco ringspot and turnip yellow-mosaic. During the past year, infective nucleic acid has been obtained from both these viruses (*33, 34*) and a soluble protein has been obtained from one, TYMV (*34*). Although fewer studies have been conducted with the nucleic acid fractions of these viruses than with those of TMV, sufficient evidence has been obtained to ascertain that the activity of the preparations is not due to residual whole virus. Incubation with ribonuclease destroys the activity of the nucleic acid fractions of both viruses, whereas the whole virus of either is relatively unaffected when incubated under the same conditions. Electron microscopy of the active nucleic acid from tobacco ringspot virus failed to reveal the presence of any particles characteristic for this virus. However, micrographs of the same nucleic acid fraction showed the presence of many characteristic particles when whole virus was added in an amount sufficient to give one tenth the infectivity level of the original nucleic acid fraction. This showed that the ribonucleic acid was not covering up and preventing the observation of whole virus.

Fig. 1a and b. Frozen replicas of a) Turnip yellow mosaic virus from distilled water suspension; b) Tobacco ringspot virus from distilled water suspension. Both micrographs 200,000 ×

In order to find out whether or not the bumps observed on TYMV particles in frozen replicas (*23*), were an artifact due to the presence of salt in the TYMV crystal, replicas were made of preparations in distilled water. The bumps were there as before (Fig. 1a). Tobacco ringspot virus suspended in distilled water was also prepared in the same manner for comparison. Although the RSV showed signs of irregular surface contour (Fig. 1b) it differed considerably in appearance from the TYMV.

In order to increase the sharpness of shadows, a double aperture system was used (STEERE, unpublished). Chromium was evaporated from the open end of a coil filament through two thin aluminium plates, each with an aperture of $^1/_2$ mm diameter. The first aperture was 1 cm from the filament and the second 1 cm from the first. The two apertures were lined up through the filament, then the prepared specimen grid was placed in line and at the right angle for the desired shadows, 2—4 cm from the second aperture. With such a system, sharp shadows as seen in Fig. 2 were obtained. Recent studies of frozen dried preparations of these two viruses revealed that both are unquestionably polyhedral in surface contour. The RSV particles present a hexagonal silhouette when shadowed vertically, then removed from the supporting film (Fig. 3b), whereas

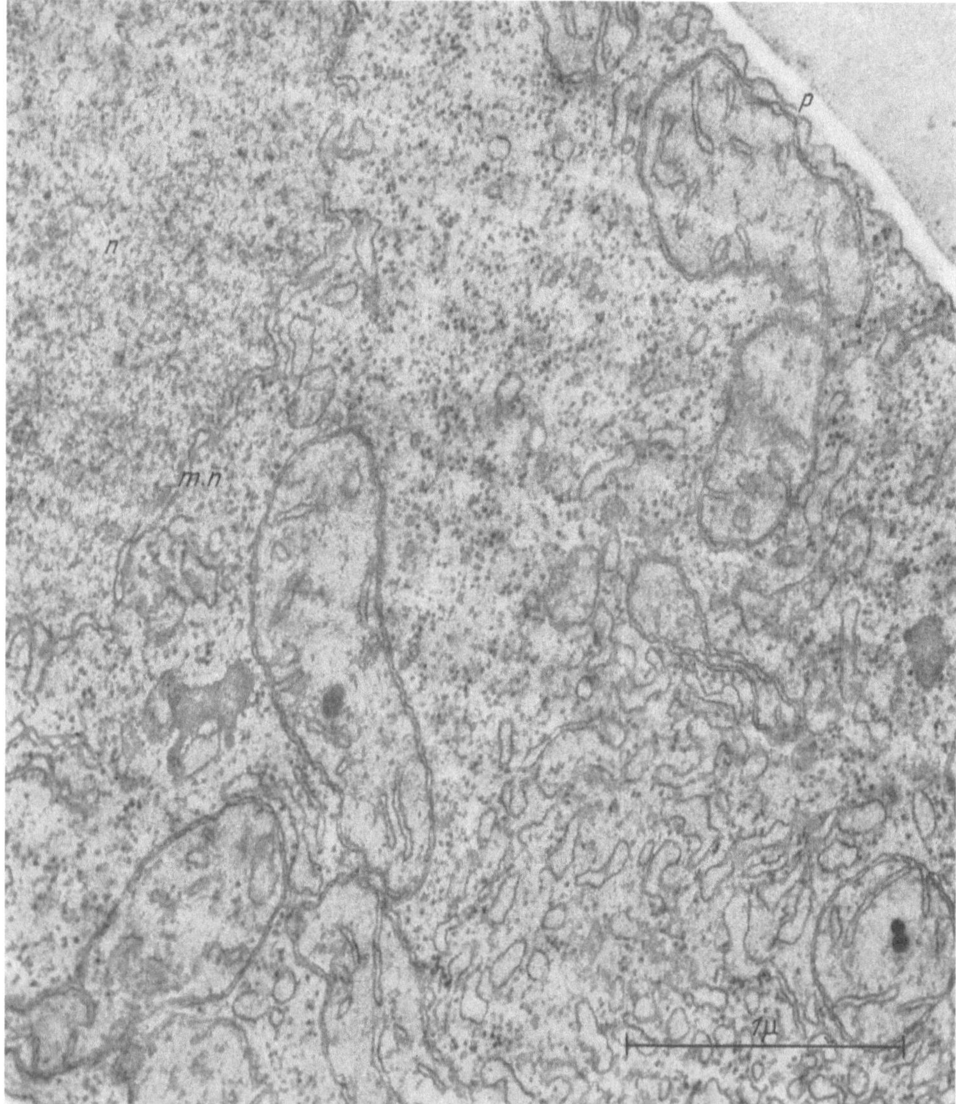

Fig. 2. Air dried tobacco ringspot virus shadowed through double aperture. Note particularly sharp shadows on the polystyrene indicator particles. Chromium shadows 90,000 ×

the TYMV particles present silhouettes which are nearly circular (Fig. 3c). Regardless of the shadow angle, the particles of RSV always cast sharply angular shadows (Fig. 4b) whereas those of TYMV approach those of a sphere, unless the shadows are greater in length than the diameter of the particles (Fig. 4c). In addition to the highly angular particles of RSV in frozen dried preparations, it has recently been possible to demonstrate that the angularity is observable in air dried preparations (Fig. 2). This confirms the belief that the polyhedral nature of this virus is not an artefact of freeze drying.

Because of the distinct differences in the apparent angularity of these two viruses, which have particles of the same size, it was considered likely that two different crystalline structures are represented. Careful study of electron micrographs, suggested the possibility that the particles of RSV have the surface contour of either a rhombic dodecahedron or an icosahedron and that the characteristic particles of TYMV may possibly have the shape of a cube octahedron. Because

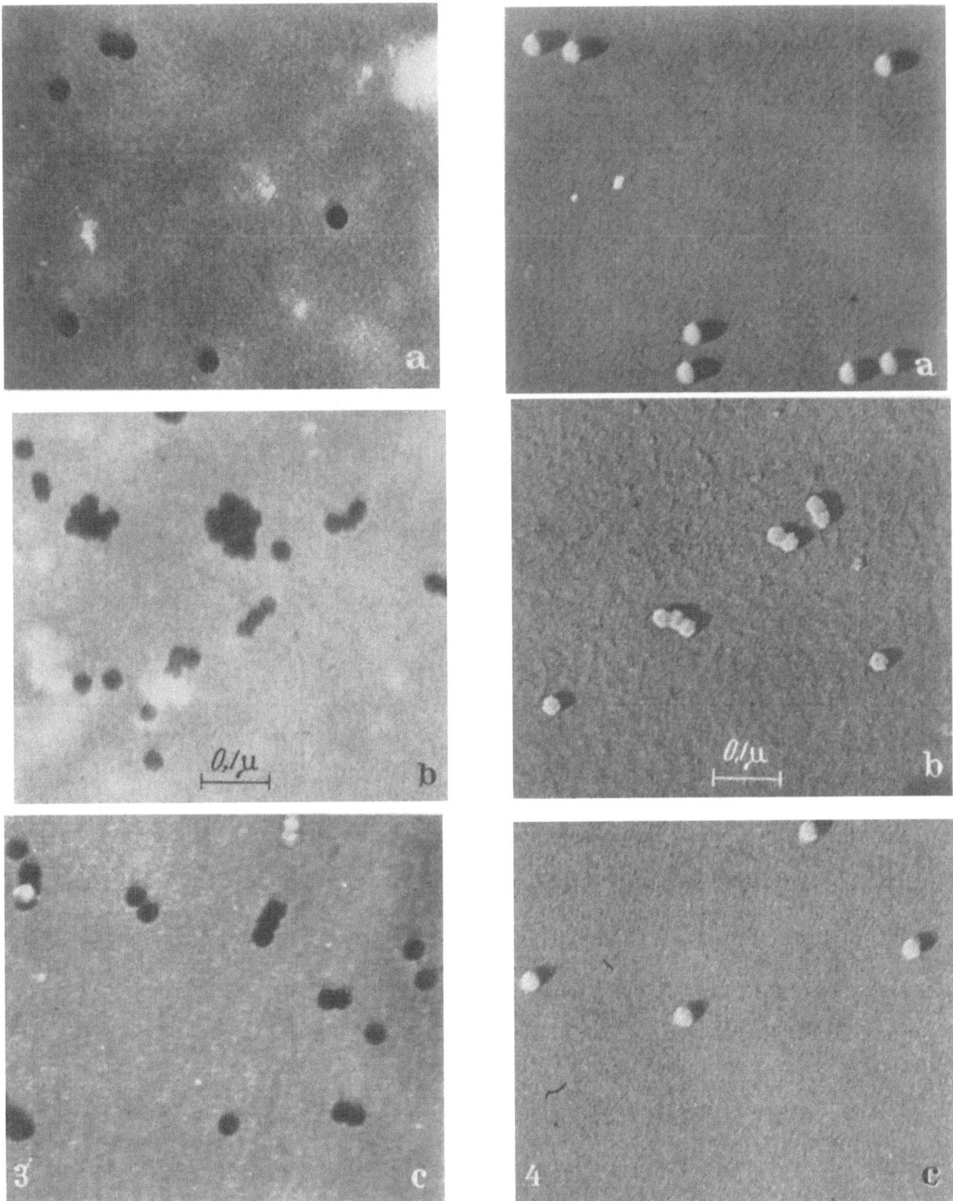

Fig. 3a—c. Silhouette micrographs of the characteristic particles of three plant viruses obtained by shadowing frozen dried preparations vertically with chromium followed by removal of the virus particles. a) Tomato bushy stunt virus; b) Tobacco ringspot virus; c) Turnip yellow mosaic virus. The more pronounced angularity of the tobacco ringspot virus is noticeable. 90,000 ×

Fig. 4a—c. Frozen-dried, chromium-shadowed particles of three plant viruses demonstrating the polyhedral nature of the particles as revealed by angular shadows. a) Tomato bushy stunt virus; b) Tobacco ringspot virus; c) Turnip yellow mosaic virus. 90,000 ×

of the small size of these particles, it is very difficult to obtain shadows sufficiently distinct to allow definitive determination of their structure. It is possible that RSV is an icosahedron, the form suggested by KAESBERG (35) for squash mosaic, wild cucumber mosaic, and bromgrass mosaic particles, and the form demonstrated so beautifully by Williams and Smith (*36*) as the shape of the iridescent virus of *Tipula paludosa*.

The appearance of bumps on the TYMV particles when prepared by the frozen replica method remains unexplained. If the particles are truly polyhedral in nature as suggested here, then two explanations are possible. The chromium shadow film may pile up preferentially on the points of the particles where facets meet, or it may be that the facets are not smooth but built up of smaller units in such a manner that the center is slightly raised and acts as a point for the piling up of the shadowing metal. Additional preparations obtained with the use of thin shadow films deposited through a collumating aperture may shed additional light on this problem.

Tomato bushy stunt virus (BSV)

Because of their larger size, 300 Å in diameter as compared with the 260 Å for RSV and TYMV, a few preparations of the particles of BSV were examined in an effort to study their structure. A silhouette micrograph of BSV particles (Fig. 3a) and micrographs with shadows two times the diameter of the particle (Fig. 4a) revealed their angularity. As was true with TYMV, however, the angularity is less pronounced than that of RSV. This and the lack of a definite hexagonal outline leads one to suggest that the particle of BSV has some form other than that of a rhombic dodecahedron or an icosahedron. There is no longer any doubt about the polyhedral nature of many of the so-called "spherical" plant viruses. The current question is not — are they polyhedral structures ? but — what is the precise shape of these particles ? It is proposed that at least two distinct forms exist, but positive identification of these shapes awaits more definitive studies.

To Professor Robley C. Williams I wish to express my appreciation for helpful suggestions and encouragement. I also wish to thank Mr. J. M. Parsons for the excellent assistance in the preparation of specimens.

References

1. Gierer, A., and G. Schramm: Z. Naturforsch. **11b**, 138 (1956).
2. Fraenkel-Conrat, W.: J. Amer. chem. Soc. **78**, 882 (1956).
3. Holmes, F. O.: Bot. Gaz. **87**, 39 (1929).
4. Kunkel, L. O.: Phytopathology **24**, 13 (1934).
5. Steere, R. L.: Phytopathology **45**, 196 (1955).
6. Schramm, G., and R. Engler: Nature (Lond.) **181**, 916 (1958).
7. Backus, R. C., and R. C. Williams: J. appl. Physics **21**, 11 (1950).
8. Hartman, R. E., T. D. Green, J. B. Bateman, C. A. Senseney and G. E. Hess: Jour. appl. Physics **24**, 90 (1953).
9. Sharp, D. G.: Proc. Soc. exp. Biol. (N. Y.) **70**, 54 (1949).
10. Kellenberger, E., and W. Arber: Virology **3**, 245 (1957).
11. Williams, R. C., and R. C. Backus: J. Amer. chem. Soc. **71**, 4052 (1949).
12. Steere, R. L.: Amer. J. Bot. **39**, 211 (1952).
13. Desjardins, P. R., C. A. Senseney and G. E. Hess: Phytopathology **43**, 687 (1953).
14. Schramm, G.: Z. Naturforsch. **2b**, 249 (1947).
15. — and W. Zillig: Nature (Lond.) **175**, 549 (1955).
16. Takahashi, W. N., and M. F. Ishii: Phytopathology **42**, 690 (1952).
17. Hart, R. G.: Virology **1**, 402 (1955).
18. Bernal, J. D., and J. Fankuchen: J. gen. Physiol. **25**, 111 (1941).
19. Kahler, W., and B. J. Lloyd, Jr.: J. appl. Physics **21**, 699 (1950).
20. Caspar, D. L. D.: Nature (Lond.) **177**, 928 (1956).
21. Williams, R. C., and R. L. Steere: J. Amer. chem. Soc. **73**, 2057 (1950).
22. Hall, C. E.: J. Amer. chem. Soc. **80**, 2556 (1958).
23. Steere, R. L.: J. biophys. biochem. Cytol. **3**, 45 (1957).
24. — and R. C. Williams: Amer. J. Bot. **40**, 81 (1953).
25. Huxley, H. C.: Stockholm Conference on Electron Microscopy. p. 260. New York: Acad. Press. Inc. 1956.
26. Frakklin, R. E., and K. C. Holmes: Acta crystallogr. **2**, 213 (1958).
27. Gierer, A.: Nature (Lond.) **179**, 1297 (1957).
28. Hart, R. G.; Biochem. biophys. Acta **28**, 457 (1958).
29. Franklin, R. E., and A. Klug: Biochem. biophys. Acta **19**, 403 (1956).
30. Baker, R. F.: Nature (Lond.), **178**, 636 (1956).
31. Matthews, R. E. F.: Nature (Lond.) **178**, 635 (1956).
32. Williams, R. C.: Biochem. biophys. Acta **8**, 227 (1952).

33. Kaper, J. M., and R. L. Steere: In press.
34. — — unpublished.
35. Kaesbegg, P.: Science **124**, 626 (1956).
36. Williams, R. C., and K. M. Smith: Biochem. biophys. Acta **28**, 464 (1958).
37. Fraenkel-Conrat, H., B. Singer and R. C. Williams: Biochem. Biophys. Acta **25**, 87 (1957).

Une nouvelle technique pour l'examen du virus de la mosaique du tabac

Alexandre Karparoff

Institut d'Epidémiologie et Microbiologie, Sofia (Bulgarie)

Pour pouvoir observer le virus de la mosaique du tabac plus facilement au microscope électronique on se sert de la technique suivante:

On triture dans un mortier des feuilles de plantes malades (Nicotiana tabacum) avec de l'eau distillée dans la proportion de la 1 à 5. On filtre ensuite la suspension sur des filtres Seitz, puis on centrifuge 3 fois le surnageant à 3000 t/min pendant 30 min. On conserve la préparation pendant 10 jours à 4° C, dans une éprouvette fermée par un bouchon de coton lâche. On centrifuge ensuite de nouveau pendant 30 min à 3000 t/min. On dépose une goutte du liquide surnageant sur une membrane de formvar (0.2%). La préparation séchée et ombrée au palladium est examinée au microscope électronique.

Sur la Fig. 1, on voit les résultats obtenus, le virus de la mosaique du tabac sous la forme classique de bâtonnets de dimension 300×15 mμ. On rencontre aussi habituellement les formes morphologiques suivantes: 1. des particules sphériques de 30—40 mμ; 2. ces mêmes particules groupées par paires; 3. de courts bâtonnets d'une longueur de 70—80 mμ, disposés parallèlement; 4. des bâtonnets dont la longueur dépasse 1000 mμ; 5. de longs filaments très fins qui se groupent par une de leurs extrémités en un ou deux granules prolongés par un bâtonnet court.

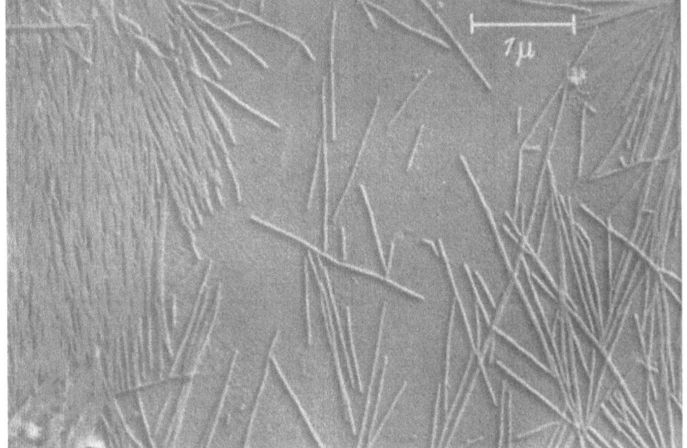

Fig. 1. Virus de la mosaique du tabac; ombrage au palladium; $12{,}000 \times 1$

On peut penser que ces formes différentes réprésentent des stades différents du développement du virus de la mosaique du tabac.

Les résultats obtenus montrent qu'avec une technique simple développée pour ce virus particulier, on obtient des virus purifiés et en concentration suffisante pour être étudiés au microscope électronique.

The formation of tobacco mosaic virus in an infected cell

V. A. Smirnova

Laboratory of Electron Microscopy, Acad. of Sciences of the USSR, Moscow

Until recently the rod 280—300 mμ in length and 15 mμ in diameter has been considered an elementary particle of the tobacco mosaic virus (TMV). The composition of the rod is fairly complex. It has a two-component structure consisting of a protein casing and a ribonucleic acid axis. It is quite obvious that the rod-like particle with this complex structure must be formed from units of lesser size and must have its own history of development. However, there have been no morphological investigations undertaken and published, which would reveal the formation process of the TMV particle.

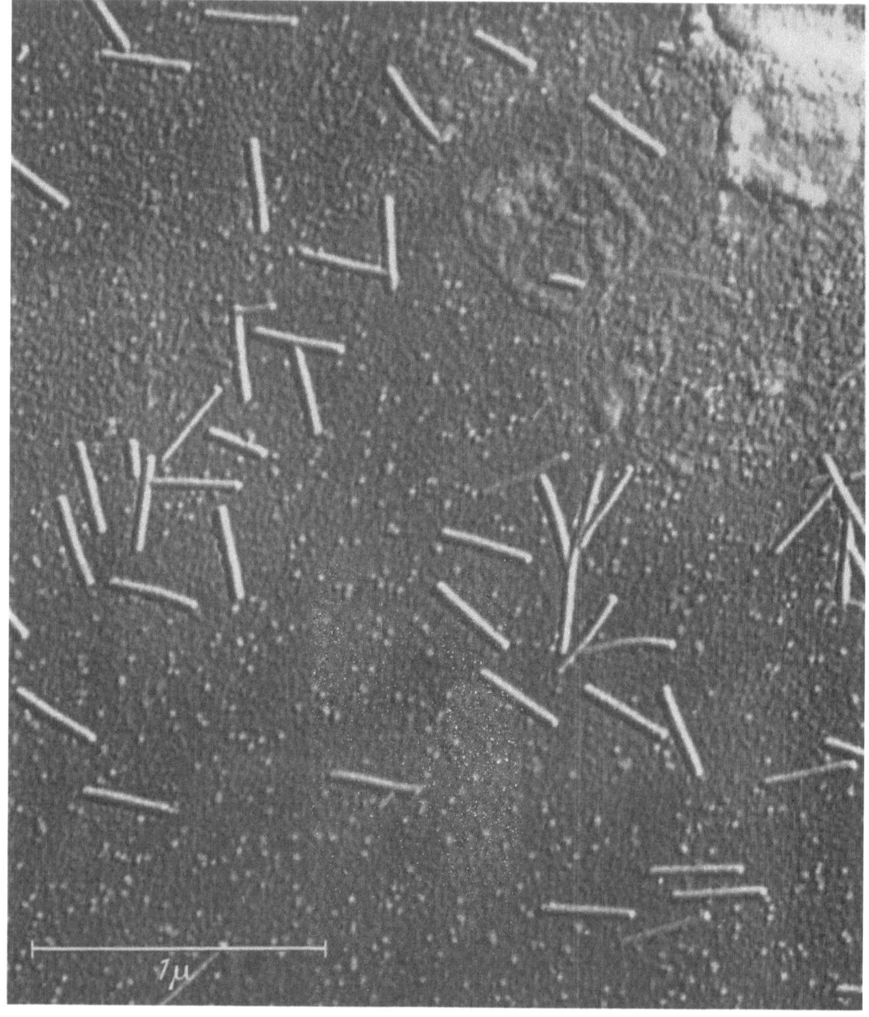

Fig. 1. Suspension of the leaves of a recently infected tomato. One can see rod-like particles and many rounded bodies

A few years ago we noted that during the intensive virus reproduction at early stages of infection, in the diseased plant extracts there was, in addition to rod-like particles, a great number of tiny rounded bodies. Their diameter closely approached the cross-section of the virus particle (Fig. 1). At later stages of infection the number of rounded bodies decreased, while that of the rod-like

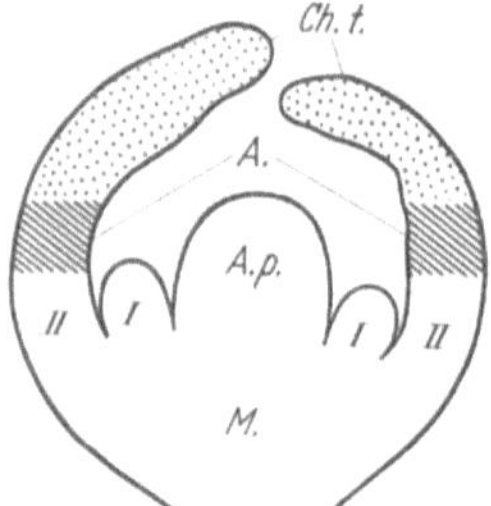

Fig. 2. The structure of a tomato vegetative bud (a scheme). *M* Meristematic tissue ≪ − ≫. *A.p.* Apical point ≪ − ≫. *I* 1-st leaf from the apical point (embrionic knot) ≪ − ≫. *II* 2-nd leaf from the apical point ≪ + − ≫. *Ch.t.* Chlorophyll-bearing tissue ≪ + ≫. *A* Area with early forms of TMV. Sign. ≪ + ≫ indicates the presence of rod-like TMV particles. Sign ≪ − ≫ indicates the absence of rod-like TMV particles

particles increased. Numerous observations showed that this phenomenon had its regularity. An assumption suggested itself that these rounded bodies were precursors of the TMV rod.

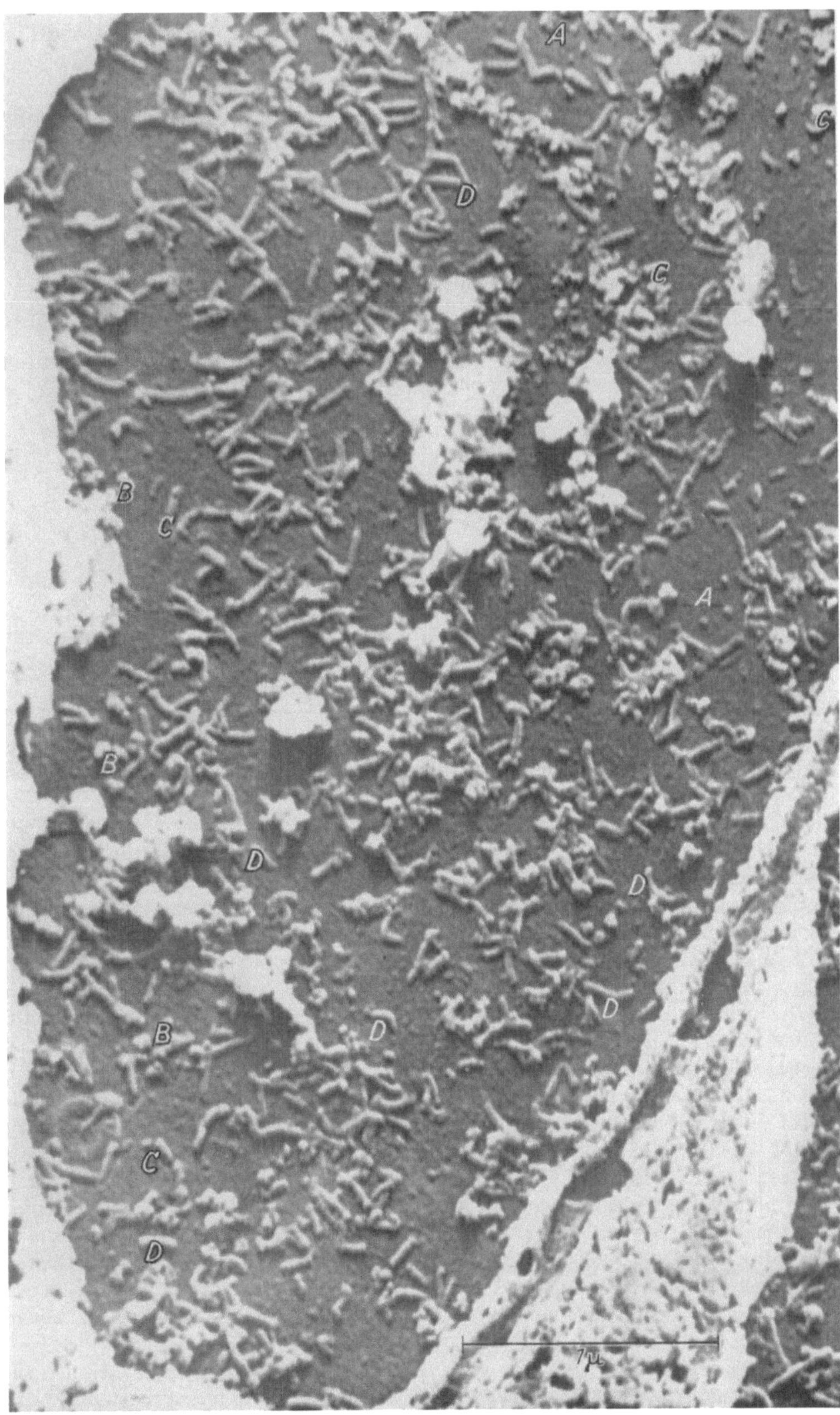

Fig. 3. The slice of a diseased tomato cell with virus particles at different stages of development

However, all efforts to discover the transitional forms from grain to rod in preparations from extracts were without result. These forms seemed to be very labile and were destroyed in the course of extracting. In rare cases the rods could be seen totally or partly composed of rounded elements.

The absence of transitional forms of the virus in extracts called for a study of the initial stages of the TMV within the cell by means of ultra-thin cuts. To do this, a tissue with a great number of recently infected cells had to be provided. In studying the vegetative buds of diseased tomatoes (*1, 2, 3*) it was found that there were no rod-like TMV particles in the apical meristem. These were found in great quantities in the upper part of the second leaf after the apical point. The lower part consisted of the meristematic tissue containing no virus rods (Fig. 2). Investigating the ultra-thin slices of the leaf an area was discovered, where the young cells appeared to be infected with the virus carried by the current of nutrients from the lower-lying diseased tissues of the plant.

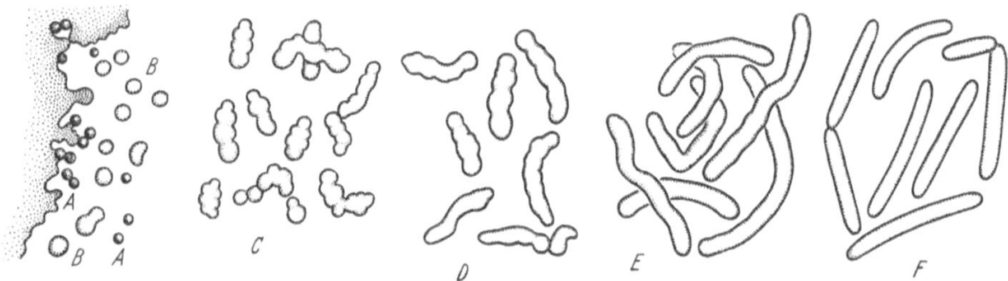

Fig. 4. The scheme of process of forming a virus particle

The cells of this area adjacent to the meristematic one disclosed the presence of differently shaped particles (Fig. 3). They must be regarded as early stages in the virus development, since they are absent from the similar areas of healthy leaves. In this area as well as in tissues of diseased adult leaves (*4, 5*) the adjacent cells differed both in the quantity and the condition of the virus. But, by systematising the obtained pictures one can assume that the process of forming a virus particle unfolds according to the scheme demonstrated, as a working hypothesis (Fig. 4). The letters on the scheme and the photos are used to indicate similar forms of virus particles.

Initially, the visible virus particle represents an oval dense body with a clear-cut outline (*A*). These bodies can be observed more frequently in the immediate proximity to the cell plasma as if gemmating from it. Then the oval particle grows larger, its outline becomes less clearly defined, it seems to be enveloped by some substance like by a casing (*B*). Further changes of the virus element proceed within the confines of such a "casing". One can see there double figures, short chains of 3 to 4 and more separate elements (*C*). The elements going to compose the short chains can be seen clearly enough. As the particle increases in size, its bead-like structure becomes smooth (*D*). Rods were frequently observed which partly had lost the bead-like shape of their one end, but which displayed well the rounded components of their structure at the opposite end. Especially clearly visible was the distal segment of the particle. These forms allow an assumption to be made to the effect that the growth of the virus particle may go forth one end. The linear extension of the particle may proceed rather far, building up structures (*E*) considerably exceeding in length the normal rod of the TMV (Fig. 5). The long, vermicular structures appear to be very elastic, since they are easily bent a any angle and interweave among themselves into a great variety of whimsical patterns. Such interweaving thick filaments forming a net can be observed where great masses of the virus occurr. As the virus "matures", the long formations seem to be broken into fragments 280—300 mμ long, because we observed none of the longer-than-usual particles in older cells where the formation of the virus is completed. It is doubtless that a certain proportion of rods arises omitting the longer-form stage. Their growth ends, as soon as they reach their normal size.

Later on, separate particles seem to become less loosely-formed and acquire a regular shape. However, they remain elastic and their ends are always rounded off. According to our observations, the rounded ends are a characteristic feature of particles that find themselves within the cells.

Particles extracted from the sap have abruptly cut ends irrespective of their length, therefore it is possible to assume that this particular trait may be acquired during the process of drying the preparations, as well as the monolythic straight-line form of the "conventional" TMV particle.

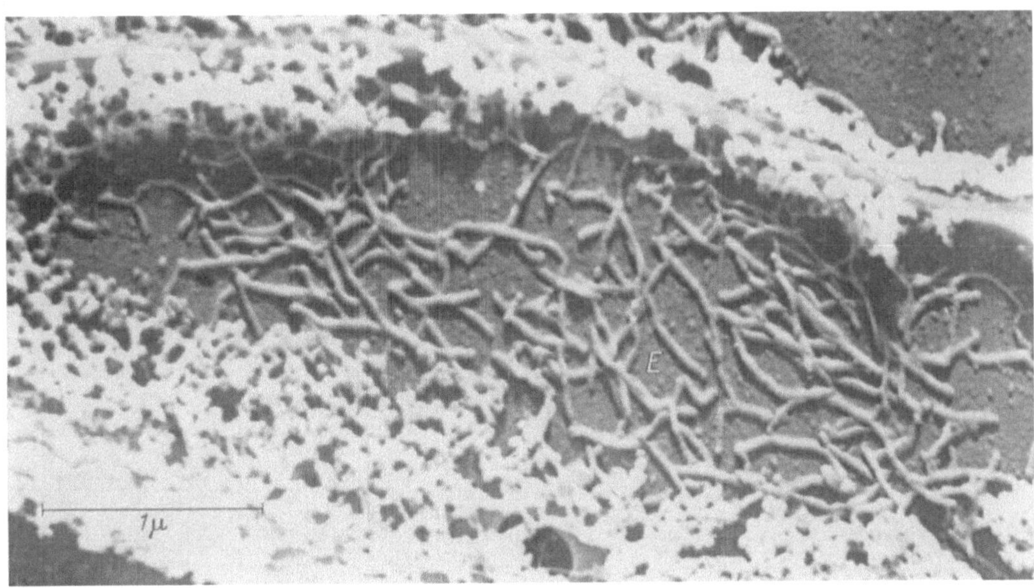

Fig. 5. The slice of a mosaic tomato cell with vermicular-shaped TMV

Taking into account that the TMV rod is a protein formation, we believe that the mellow outline and elasticity observed in the particles are in good agreement with their native condition, rather than treatment artefacts.

It is possible that the discovered spherical bodies initiating a rod are not the smallest units to build up a virus particle; it is quite likely that these elementary units were beyond the resolving capacity of our microscope.

References

1. SMIRNOVA, V. A.: Proceedings of the First Regional Conference on Electron-microscopy in Asia and Oceania, Tokyo, 1956, 195.
2. — Dokl. Akad. Nauk USSR **118,** 407 (1958).
3. — Biophysika (russ.) **3,** 252 (1958).
4. BRANDES, I.: Naturwissenschaften, **42,** 101 (1955).
5. SMIRNOVA, V. A.: Mikrobiologia (russ.) **15,** 718 (1956).

SONDERDRUCK AUS
VIERTER INTERNATIONALER KONGRESS
FÜR ELEKTRONENMIKROSKOPIE
FOURTH INTERNATIONAL CONFERENCE
ON ELECTRON MICROSCOPY
QUATRIÈME CONGRÈS INTERNATIONAL
DE MICROSCOPIE ÉLECTRONIQUE
BERLIN 10.-17. SEPTEMBER 1958
VERHANDLUNGEN
HERAUSGEGEBEN VON
W. BARGMANN · G. MÖLLENSTEDT · H. NIEHRS · D. PETERS · E. RUSKA · C. WOLPERS
BAND II BIOLOGISCH-MEDIZINISCHER TEIL
SPRINGER-VERLAG / BERLIN · GÖTTINGEN · HEIDELBERG 1960
PRINTED IN GERMANY

Problems in methacrylate embedding

DAN H. MOORE

The Rockefeller Institute New York 21, N. Y.

Inhaltsübersicht Band I

Physikalisch-technischer Teil

Eröffnungs-Ansprache. Von Ernst Ruska

Opening remarks. By V. E. Cosslett

Festvortrag: Geschichte des Elektrons. Von M. von Laue

---

* Die speziell für biologische Präparation bestimmte Technik, insbesondere Mikrotomie, siehe in Band II.
** Feldemissionsmessungen siehe unter A./1. Kathoden.

Inhaltsübersicht Band II

Biologisch-medizinischer Teil

---

* Übrige Präparationstechnik siehe Band I.

In zweisprachiger Ausgabe erschien im Dezember 1959

Die submikroskopische Anatomie und Pathologie der Lunge
The Submicroscopic Anatomy and Pathology of the Lung

Von Dr. med. HERIBERT SCHULZ, Assistent am Pathologischen Institut der Medizinischen Akademie in Düsseldorf. Englische Übersetzung von Dr. F. DALLENBACH

Mit 95 Abbildungen in 205 Einzeldarstellungen. IX, 199 Seiten 4°. 1959. Ganzleinen DM 178,—

Inhaltsübersicht

A. Material und Methode

B. Die submikroskopische Anatomie der Lunge
Die Lungenalveole. Die vergleichende submikroskopische Anatomie der Lungen und des Blut-Luft-Weges. Die Lungengefäße. Das Lungengerüst. Die Tracheal- und Bronchialschleimhaut. Die embryonale Lunge.

C. Die submikroskopische Pathologie der Lunge
Die Pathologie der Mitochondrien. Die Pathologie der Cytosomen in den Alveolarmakrophagen der Lunge. Das experimentelle Lungenödem. Die pulmonalen hyalinen Membranen. Die plasmacelluläre und eosinophile Entzündung des Alveolarseptums. Die leukocytäre Alveolitis. Die Lunge bei extrakorporalem Kreislauf. Thrombocyten und Thrombose. Die Lunge bei Mitralstenose. Die Ultrastruktur der nichtdurchströmten Lungencapillare. Die Lunge im Winterschlaf. Die Atelektase der Lunge. Die experimentelle akute Überblähung der Lunge. Transport und Ablagerung von Schwermetallen sowie von Tusche in der Lunge. Die Ablagerung von Ferritin und Hämosiderin in Alveolarmakrophagen. Die Pneumokoniosen. Die Ultrastruktur der Pneumocystis Carinii. Die Lungentuberkulose. Die Ultrastruktur und Entwicklung von lungenpathogenen Viren. — Literaturverzeichnis. Namenverzeichnis. Sachverzeichnis.

Table of Contents

A. Material and Methods

B. The Submicroscopic Anatomy of the Lung
The Pulmonary Alveolus. The Comparative Submicroscopic Anatomy of the Lung and of the Blood-Air-Pathway. The Pulmonary Vessels. The Pulmonary Stroma. The Tracheal and Bronchial Mucous Epithelium. The Embryonal Lung.

C. The Submicroscopic Pathology of the Lung
The Pathology of the Mitochondria. The Pathology of the Cytosomes in the Alveolar Macrophages of the Lung. The Experimental Lung Edema. The Pulmonary Hyaline Membranes. Inflammatory Infiltrates of the Alveolar Septum by Plasma Cells and Eosinophilic Leucocytes. The Polymorphonuclear Leucocytic Alveolitis. The Lung during Extracorporeal Circulation. Thrombocytes and Thrombosis. The Lung in Mitral Stenosis. The Ultrastructure of the Non-perfused Resting Pulmonary Capillary. The Lung during Hibernation. Atelectasis of the Lung. Experimental Acute Hyperinflation of the Lung. Transport and Deposition of Heavy Metals and India Ink in the Lung. The Deposition of Ferritin and Hemosiderin in Alveolar Macrophages. The Pneumoconioses. The Ultrastructure of the Pneumocystis Carinii. Pulmonary Tuberculosis. The Ultrastructure and Development of Pathogenic Viruses of the Lung. — Bibliography. Author Index. Subject Index.

Elektronenmikroskopische Untersuchungs- und Präparationsmethoden

Von Dr. LUDWIG REIMER, Dozent für Physik an der Universität Münster i. Westf.

Mit 135 Abbildungen und 20 Bildtafeln. VIII, 300 Seiten Gr.-8°. 1959.   Ganzleinen DM 58,—

Inhaltsübersicht

A. Untersuchungsmethoden
Elektronenoptische Grundlagen des Durchstrahlungsmikroskopes. Andere Abbildungsverfahren. Messung wichtiger optischer Konstanten. Stereoabbildungen. Entstehung des Bildkontrastes. Elektronenbeugung. Scheinstrukturen durch Interferenzeffekte. Präparatveränderungen unter Elektronenbeschuß. Bildaufzeichnung und Intensitätsmessungen.

B. Präparationsmethoden
Objektblenden und Trägernetze. Herstellung und Eigenschaften von Trägerfolien. Grundlagen der Hochvakuum- und Aufdampftechnik. Oberflächenabdrücke. Schrägbeschattung. Zielpräparation. Herstellung durchstrahlbarer Metallfolien. Anorganische disperse Systeme. Organische disperse Systeme. Fixierung und Kontrastierung. Gefriertrocknung. Entwässerung und Einbettung, Ultramikrotomie. — Bezugsquellen für apparative und präparative Hilfsmittel. Bildanhang. Sachverzeichnis.

SONDERDRUCK AUS
VIERTER INTERNATIONALER KONGRESS FÜR ELEKTRONENMIKROSKOPIE
FOURTH INTERNATIONAL CONFERENCE ON ELECTRON MICROSCOPY
QUATRIÈME CONGRÈS INTERNATIONAL DE MICROSCOPIE ÉLECTRONIQUE
BERLIN 10.-17. SEPTEMBER 1958
VERHANDLUNGEN
HERAUSGEGEBEN VON
W. BARGMANN · G. MÖLLENSTEDT · H. NIEHRS · D. PETERS · E. RUSKA · C. WOLPERS
BAND II BIOLOGISCH-MEDIZINISCHER TEIL
SPRINGER-VERLAG / BERLIN · GÖTTINGEN · HEIDELBERG 1960
PRINTED IN GERMANY

Open face flat embedding technique

E. Borysko[*, **]

New York University College of Dentistry. The Murry and Leonie Guggenheim Foundation. Institute for Dental Research, New York

* Present address: Ethicon, Inc., Somerville, New Jersey.

** This work was supported by Research Grants D-368 and D-448 from the US Public Health Service National Institutes of Health.

Inhaltsübersicht Band I

Physikalisch-technischer Teil

Eröffnungs-Ansprache. Von Ernst Ruska

Opening remarks. By V. E. Cosslett

Festvortrag: Geschichte des Elektrons. Von M. von Laue

---

* Die speziell für biologische Präparation bestimmte Technik, insbesondere Mikrotomie, siehe in Band II.
** Feldemissionsmessungen siehe unter A./1. Kathoden.

SONDERDRUCK AUS
VIERTER INTERNATIONALER KONGRESS FÜR ELEKTRONENMIKROSKOPIE
FOURTH INTERNATIONAL CONFERENCE ON ELECTRON MICROSCOPY
QUATRIÈME CONGRÈS INTERNATIONAL DE MICROSCOPIE ÉLECTRONIQUE
BERLIN 10.-17. SEPTEMBER 1958
VERHANDLUNGEN
HERAUSGEGEBEN VON
W. BARGMANN · G. MÖLLENSTEDT · H. NIEHRS · D. PETERS · E. RUSKA · C. WOLPERS
BAND II BIOLOGISCH-MEDIZINISCHER TEIL
SPRINGER-VERLAG / BERLIN · GÖTTINGEN · HEIDELBERG 1960
PRINTED IN GERMANY

Inclusion au polyester

Antoinette Ryter et Edouard Kellenberger

Laboratoire de Biophysique, Université de Genève

Inhaltsübersicht Band I

Physikalisch-technischer Teil

Eröffnungs-Ansprache. Von Ernst Ruska

Opening remarks. By V. E. Cosslett

Festvortrag: Geschichte des Elektrons. Von M. von Laue

A. Elektronen- und ionenoptische Elemente, Geräte und Verfahren:
1. Kathoden — 2. Linsen und Ablenksysteme — 3. Objekteinrichtungen — 4. Bildaufzeichnungsverfahren — 5. Photographische Emulsionen (und Elektronenwirkung auf Silbersalze) — 6. Stereoaufnahme — 7. Vakuum, Strahlspannung, Linsendurchflutung — 8. Durchstrahlungsmikroskope — 9. Reflexions- und Emissionsmikroskopie — 10. Interferenzmikroskopie und Interferometrie — 11. Röntgen-Projektionsmikroskopie — 12. Elektronen- und Röntgen-Rastermikroskopie — 13. Materialbearbeitung mit Elektronenstrahlen

B. Einwirkung des Objekts auf Strahl und Bild:
1. Streuung am Objekt und Bildkontrast — 2. Abbildung von Kristallgitter-Perioden — 3. Mehrfachbeugung am Objekt und Entstehung von Moirés

C. Elektronenmikroskopische Präparationstechnik*:
1. Trägerfolien — 2. Dünne Objektschichten — 3. Oberflächen — 4. Aufdampf- und Abdruck-Verfahren

D. Ergebnisse der Elektronenmikroskopie in der Technologie (Kristallographie, Metallographie, Chemie):
1. Kristallgitter-Strukturen — 2. Kristallwachstum — 3. Kristalloberflächen — 4. Kondensierte Schichten — 5. Kristallbau-Fehler und Versetzungen — 6. Umwandlungs- und Ausscheidungsvorgänge in Metallen — 7. Natürliche und künstliche technologische Fasern — 8. Verschiedene Produkte der chemischen Technik — 9. Staube und Rauche — 10. Spuren-Nachweis

E. Feldemissionsmikroskopie:**
1. Feldelektronen-Mikroskopie von Metalloberflächen — 2. Adsorptionsuntersuchungen an Feldkathoden — 3. Feldionen-Mikroskopie

---

* Die speziell für biologische Präparation bestimmte Technik, insbesondere Mikrotomie, siehe in Band II.
** Feldemissionsmessungen siehe unter A./1. Kathoden.